# EPILEPSY

## Mechanisms, Models, and Translational Perspectives

# EPILEPSY

## Mechanisms, Models, and Translational Perspectives

### Edited by
### Jong M. Rho • Raman Sankar • Carl E. Stafstrom

CRC Press
Taylor & Francis Group
Boca Raton London New York

CRC Press is an imprint of the
Taylor & Francis Group, an **informa** business

Cover illustration courtesy of the Barrow Neurological Institute (Michael Hickman and Mark Schornak) (Copyright, Barrow Neurological Institute 2009)

CRC Press
Taylor & Francis Group
6000 Broken Sound Parkway NW, Suite 300
Boca Raton, FL 33487-2742

---

**Library of Congress Cataloging-in-Publication Data**

---

Epilepsy : mechanisms, models, and translational perspectives / editors, Jong M. Rho, Raman Sankar, and
    Carl E. Stafstrom.
        p. ; cm.
    Includes bibliographical references and index.
    ISBN 978-1-4200-8559-4 (alk. paper)
        1. Epilepsy. 2. Anticonvulsants--Therapeutic use. 3. Gene therapy. I. Rho, Jong M. II. Sankar, Raman. III. Stafstrom, Carl Ernest. IV. Title.
        [DNLM: 1. Epilepsy--physiopathology. 2. Epilepsy--therapy. 3. Anticonvulsants--therapeutic use. 4. Epilepsy--genetics. 5. Gene Therapy. WL 385 E6432 2010]

RC372.E668 2010
616.8'53--dc22                                                                                    2010000991

**Visit the Taylor & Francis Web site at**
**http://www.taylorandfrancis.com**

**and the CRC Press Web site at**
**http://www.crcpress.com**

*Dedicated to all who have devoted their lives
to furthering our understanding of epilepsy and
advancing the care of those afflicted by this disorder*

# Contents

## SECTION I    Scientific Foundations

## SECTION II    Antiepileptic Drugs

# SECTION IV  Alternative Therapies

# SECTION V  Other Modulators of the Epileptic State

# SECTION VI  The Future of Epilepsy Therapy

# Preface

Epilepsy is an episodic neurological disorder that has afflicted humankind throughout recorded history; yet, throughout the millennia, it has never been properly acknowledged as a disease with a biological basis. In ancient times, epilepsy was referred to, somewhat ironically, as the "Sacred Disease," as it was imbued with negative references to the supernatural. Epilepsy was later believed to represent a form of demonic possession and thus resulted in social stigmatization and persecution. It was only late in the 19th century that epilepsy began its long and arduous journey to being justly recognized as a physical illness with complex pathophysiological substrates. Even today, the public is not fully apprised of the true nature of the epilepsies (as they are now considered), and efforts to expand awareness of this condition have been thwarted in large measure by deeply rooted preconceptions promulgated through the ages.

Within the last half-century, significant progress has been made in our basic understanding of the epileptic brain. Pivotal advances in drug development and surgical techniques, as well as the emergence of innovative approaches such as electrical stimulation of the nervous system, have led to a substantial reduction in the morbidity and mortality of patients with epilepsy (both children and adults). At the same time, remarkable developments in the basic neurosciences have enhanced our understanding of brain structure and function at ever finer levels of molecular, cellular, and genetic detail.

The intrinsic complexities associated with attempts at understanding normal brain structure and function lie at the heart of the challenges investigators face in deciphering the epileptic brain. The development of universally effective therapeutic approaches for epilepsy patients has been the elusive goal of clinicians and researchers since the early twentieth century. Yet, despite the availability of many new pharmacological agents within the last generation, at least one third of the people with epilepsy remain refractory to medical therapy, and an even smaller number of these individuals are potential candidates for epilepsy surgery. It is this last frustrating reality that has been the focus of many professionals in the epilepsy field.

Within the research arena, increasing focus has been placed on "translational" research (i.e., that which bridges the gap between the laboratory and patient bedside); however, effective communication and interchange between clinicians and basic researchers have been difficult to achieve on a widespread basis. It is clear that such interaction is paramount in the development of novel treatments based on a detailed knowledge of fundamental mechanisms. This volume incorporates new translational advances in bringing epilepsy therapies from the laboratory bench to the bedside and back again.

We wish to collectively thank our mentors, colleagues, students, and, most of all, our patients and their families for providing the inspiration and encouragement to help facilitate this "translational" dialog. Additionally, we thank the publisher and our families for the support they have given us throughout this project. Finally, we acknowledge the expert editorial and administrative assistance provided by Pat Roberson and Heather Milligan.

**Jong M. Rho, MD**
**Raman Sankar, MD, PhD**
**Carl E. Stafstrom, MD, PhD**

# Editors

**Jong M. Rho, MD,** is a senior staff scientist at the Barrow Neurological Institute and St. Joseph's Hospital & Medical Center in Phoenix, Arizona. After obtaining an undergraduate degree in molecular biophysics and biochemistry at Yale University, Dr. Rho received his medical degree from the University of Cincinnati. Following a pediatric residency at the University of Southern California Children's Hospital of Los Angeles and a neurology residency at the University of California, Los Angeles (UCLA), School of Medicine, he completed fellowships in pediatric neurology at the UCLA School of Medicine and in neuropharmacology at the National Institutes of Health (NIH). Prior to his current position, he held faculty appointments at the University of Washington and the Children's Hospital & Regional Medical Center in Seattle and the University of California at Irvine. Dr. Rho's main research interests are the mechanisms underlying the anticonvulsant and neuroprotective effects of the ketogenic diet, neuropharmacology of anticonvulsant compounds, and the study of surgically resected human epileptic tissue. His research activities have been sponsored by several NIH research grants, as well as a variety of intramural and extramural public and private sector sources. Dr. Rho has served on the editorial boards of *Epilepsia* and *Epilepsy Currents* and has been a regular reviewer for research grants submitted to the NIH. In addition to an extensive list of publications in basic science and pediatric neurology peer-reviewed journals, Dr. Rho has written numerous book chapters and edited several books and is a popular national and international guest lecturer.

**Raman Sankar, MD, PhD,** is a professor of neurology and pediatrics and chief of pediatric neurology at the David Geffen School of Medicine at UCLA. He holds the Rubin Brown Distinguished Chair in Pediatric Neurology. Dr. Sankar obtained his doctorate from the University of Washington in medicinal chemistry and was involved in teaching and research for several years prior to entering Tulane Medical School, where he obtained his medical degree. He trained in pediatrics at the Children's Hospital of Los Angeles and completed his training in neurology and pediatric neurology at UCLA. His laboratory research pertains to mechanisms of seizure-induced injury and epileptogenicity in the developing brain and is funded by the National Institute of Neurological Disorders and Stroke (NINDS), National Institutes of Health (NIH). Current research includes investigations on improving the throughput for screening compounds for antiepileptogenic action on the developing brain and modeling the comorbidity of depression that accompanies epileptogenesis. He is one of the investigators in a NINDS-sponsored multicenter study on childhood absence epilepsy. Dr. Sankar is a reviewer for research grants submitted to the NIH, Epilepsy Foundation, CURE, American Epilepsy Society, and other organizations. He is a member of an active pediatric epilepsy program at UCLA that is well known internationally for many advances in pediatric epilepsy surgery. Dr. Sankar has authored more than 160 research articles, reviews, and book chapters and has served on the editorial boards of *Epilepsia* and *Epilepsy Currents*. Dr. Sankar serves on the professional advisory board of the Epilepsy Foundation. He is a member of the Commission on Neurobiology of the International League Against Epilepsy

**Carl E. Stafstrom, MD, PhD,** is a professor of neurology and pediatrics and chief of the division of pediatric neurology at the University of Wisconsin Medical School, Madison. He received his AB from the University of Pennsylvania, followed by his MD and PhD degrees (the latter in physiology and biophysics) from the University of Washington, Seattle. After completing a residency in pediatrics at the University of Washington Children's Hospital, Dr. Stafstrom trained in adult and pediatric neurology at Tufts–New England Medical Center in Boston, followed by fellowships in neurology research, epilepsy, and clinical neurophysiology at Children's Hospital, Harvard Medical

School, Boston. Prior to his current position, he held faculty appointments at Duke University Medical Center and Tufts University School of Medicine. Dr. Stafstrom's main research interests are the pathophysiological mechanisms of epilepsy in the developing brain, the consequences of seizures on cognition and behavior, and alternative epilepsy therapies such as the ketogenic diet. He actively pursues these interests in both the clinic and the laboratory. Dr. Stafstrom is the author of more than 150 publications on epilepsy and its mechanisms. He served as chair of the scientific review committee of Partnership for Epilepsy Research and is on the scientific advisory boards of the Epilepsy Foundation and the Charlie Foundation. He is the former chair of the Investigators' Workshop Committee of the American Epilepsy Society. Dr. Stafstrom is associate editor of the journals *Epilepsy Currents* and *Epilepsia*. He places a high priority on the education of medical students and residents and is the recipient of numerous teaching awards.

# Contributors

**Marina V. Abramova**
Department of Neurosurgery
LSU Health Sciences Center
Louisiana State University
New Orleans, Louisiana

**Gail D. Anderson**
Department of Pharmacy
University of Washington
Seattle, Washington

**Eleonora Aronica**
Department of (Neuro)Pathology
Academic Medical Centre, University
    of Amsterdam
Amsterdam, the Netherlands

**Stéphane Auvin**
Pediatric Neurology Department
Hôpital Robert Debré
Institut National de la Santé et
    de la Recherche Médicale U676
Paris, France

**Silvia Balosso**
Department of Neuroscience, Laboratory
    of Experimental Neurology
Mario Negri Institute for
    Pharmacological Research
Milano, Italy

**Gregory K. Bergey**
Department of Neurology
Epilepsy Research Laboratory
The Johns Hopkins Epilepsy Center
The Johns Hopkins University School
    of Medicine
Baltimore, Maryland

**Edward H. Bertram**
Department of Neurology
University of Virginia
Charlottesville, Virginia

**Kristopher J. Bough**
National Institute on Drug Abuse
National Institutes of Health
Bethesda, Maryland

**Mark Bower**
Mayo Systems Electrophysiology Laboratory
Department of Neurology
Mayo Clinic
Rochester, Minnesota

**Russell J. Buono**
Coatesville Veteran's Affairs Medical Center
Coatesville, Pennsylvania

**Rochelle Caplan**
Department of Psychiatry and
    Behavioral Sciences
Semel Institute for Neuroscience and
    Human Behavior
University of California
Los Angeles, California

**Jose D. Carrillo-Ruiz**
Unit for Stereotaxic, Functional Neurosurgery,
    and Radiosurgery
Hospital General de México
Mexico City, Mexico

**Guillermo Castro**
Unit for Stereotaxic, Functional Neurosurgery,
    and Radiosurgery
Hospital General de México
Mexico City, Mexico

**Carlos Cepeda**
Mental Retardation Research Center
David Geffen School of Medicine at UCLA
University of California
Los Angeles, California

**Kevin Chapman**
Pediatric Neurology
Barrow Neurological Institute
St. Joseph's Hospital and Medical Center
Phoenix, Arizona

**James Cloyd**
Center for Orphan Drug Development
College of Pharmacy
University of Minnesota
Minneapolis–St. Paul, Minnesota

**Luca Cucullo**
Departments of Cell Biology and
    Cerebrovascular Research
Cleveland Clinic Foundation
Cleveland, Ohio

**Manoela Cuellar**
Department of Physiopharmacology
Center for Research and Advanced Studies
Instituto Politécnico Nacional
Mexico City, Mexico

**Dennis J. Dlugos**
Neurology and Pediatrics
University of Pennsylvania School of Medicine
Pediatric Regional Epilepsy Program
The Children's Hospital of Philadelphia
Philadelphia, Pennsylvania

**F. Edward Dudek**
Department of Physiology
University of Utah School of Medicine
Salt Lake City, Utah

**Vincent Fazio**
Departments of Cell Biology
    and Cerebrovascular Research
Cleveland Clinic Foundation
Cleveland, Ohio

**Thomas N. Ferraro**
Department of Psychiatry
Center for Neurobiology and Behavior
University of Pennsylvania School of Medicine
Philadelphia, Pennsylvania

**Piotr J. Franaszczuk**
Department of Neurology
Epilepsy Research Laboratory
The Johns Hopkins Epilepsy Center
The Johns Hopkins University School
    of Medicine
Baltimore, Maryland

**William A. Friedman**
Department of Neurosurgery
McKnight Brain Institute
University of Florida
Gainesville, Florida

**Christopher C. Giza**
Department of Neurosurgery
UCLA Brain Injury Research Center
David Geffen School of Medicine at UCLA
University of California
Los Angeles, California

**Michael M. Haglund**
Departments of Surgery (Neurosurgery)
    and Neurobiology
Duke University Medical Center
Durham, North Carolina

**Daryl W. Hochman**
Departments of Surgery and Pharmacology
    and Cancer Biology
Duke University Medical Center
Durham, North Carolina

**Vijay Ivaturi**
Center for Orphan Drug Development
College of Pharmacy
University of Minnesota
Minneapolis–St. Paul, Minnesota

**Damir Janigro**
Departments of Neurological Surgery,
    Molecular Medicine, Cell Biology,
    and Cerebrovascular Research
Cleveland Clinic Foundation
Cleveland, Ohio

**Jaromir Janousek**
Research Associate
Mid-Atlantic Epilepsy and Sleep Center
Bethesda, Maryland

**Fiacro Jimenez**
Unit for Stereotaxic, Functional Neurosurgery,
    and Radiosurgery
Hospital General de México
Mexico City, Mexico

**Christophe C. Jouny**
Department of Neurology
Epilepsy Research Laboratory
The Johns Hopkins Epilepsy Center
The Johns Hopkins University School
    of Medicine
Baltimore, Maryland

**Jaideep Kapur**
Department of Neurology
University of Virginia
Charlottesville, Virginia

**Hyunmi Kim**
Division of Child Neurology
University of Alabama–Birmingham
    School of Medicine
Birmingham, Alabama

**Pavel Klein**
Epilepsy Center
Mid-Atlantic Epilepsy and Sleep Center
Bethesda, Maryland
and
Department of Neurology
The George Washington University
Washington, D.C.

**Robert C. Knowlton**
Director, Division of Epilepsy
UAB Epilepsy Center
University of Alabama–Birmingham
    School of Medicine
Birmingham, Alabama

**Xu Maisano**
Department of Biology
Program in Neuroscience and Behavior
Wesleyan University
Middletown, Connecticut

**Nicola Marchi**
Departments of Cell Biology and
    Cerebrovascular Research
Cleveland Clinic Foundation
Cleveland, Ohio

**Gary W. Mathern**
Mental Retardation Research Center
Department of Neurosurgery
David Geffen School of Medicine at UCLA
University of California
Los Angeles, California

**Janice R. Naegele**
Department of Biology
Wesleyan University
Middletown, Connecticut

**Dean K. Naritoku**
Departments of Neurology and Pharmacology
University of South Alabama College
    of Medicine
Mobile, Alabama

**Minh-Tri Nguyen**
Departments of Cell Biology and
    Cerebrovascular Research
Cleveland Clinic Foundation
Cleveland, Ohio

**A. LeBron Paige**
UAB Epilepsy Center
University of Alabama–Birmingham
    School of Medicine
Birmingham, Alabama

**Manisha Patel**
Department of Pharmaceutical Sciences
University of Colorado
Aurora, Colorado

**Asuri N. Prasad**
Pediatrics and Clinical Neurosciences
Children's Hospital of Western Ontario
University of Western Ontario
London, Ontario, Canada

**Chitra Prasad**
Medical Genetics Program of
    South Western Ontario
Children's Hospital of Western Ontario
University of Western Ontario
London, Ontario, Canada

**Mark Quigg**
F.E. Dreifuss Comprehensive
    Epilepsy Program
Department of Neurology
National Science Foundation Center
    for Biological Timing
University of Virginia
Charlottesville, Virginia

**John E. Rash**
Department of Biomedical Sciences
Colorado State University
Fort Collins, Colorado

**Teresa Ravizza**
Department of Neuroscience, Laboratory
  of Experimental Neurology
Mario Negri Institute for Pharmacological
  Research
Milano, Italy

**Ryan D. Readnower**
Department of Anatomy and Neurobiology
Spinal Cord and Brain Injury Research Center
University of Kentucky
Lexington, Kentucky

**Doodipala S. Reddy**
Department of Neuroscience and
  Experimental Therapeutics
College of Medicine
Texas A&M Health Science Center
College Station, Texas

**Erich O. Richter**
Department of Neurosurgery
LSU Health Sciences Center
Louisiana State University
New Orleans, Louisiana

**Luisa Rocha**
Department of Physiopharmacology
Center for Research and Advanced Studies
Instituto Politécnico Nacional
Mexico City, Mexico

**Michael A. Rogawski**
Department of Neurology
School of Medicine
University of California, Davis
Sacramento, California

**Steven N. Roper**
Department of Neurosurgery
McKnight Brain Institute
University of Florida
Gainesville, Florida

**Raman Sankar**
Pediatric Neurology
Mattel Children's Hospital UCLA
David Geffen School of Medicine at UCLA
University of California
Los Angeles, California

**Steven C. Schachter**
Department of Neurology
Harvard Medical School Osher Research Center
and
Beth Israel Deaconess Medical Center
Boston, Massachusetts

**Philip A. Schwartzkroin**
Department of Neurological Surgery
University of California
Davis, California

**Li-Rong Shao**
Department of Physiology
University of Utah School of Medicine
Salt Lake City, Utah

**Jerry J. Shih**
Comprehensive Epilepsy Program
Department of Neurology
Mayo Clinic
Jacksonville, Florida

**Timothy A. Simeone**
Department of Pharmacology
Creighton University
Omaha, Nebraska

**Carl E. Stafstrom**
Department of Neurology and Pediatrics
University of Wisconsin
Madison, Wisconsin

**S. Matt Stead**
Division of Epilepsy and
  Electroencephalography
Division of Pediatric Neurology
Mayo Systems Electrophysiology Laboratory
Department of Neurology
Mayo Clinic
Rochester, Minnesota

**Patrick G. Sullivan**
Department of Anatomy and Neurobiology
Spinal Cord and Brain Injury Research Center
University of Kentucky
Lexington, Kentucky

**David Trejo**
Unit for Stereotaxic, Functional Neurosurgery,
    and Radiosurgery
Hospital General de México
Mexico City, Mexico

**Matthew M. Troester**
Barrow Neurological Institute
Children's Health Center
St. Joseph's Hospital and Medical Center
Phoenix, Arizona

**Ana Luisa Velasco**
Unit for Stereotaxic, Functional Neurosurgery,
    and Radiosurgery
Hospital General de México
Mexico City, Mexico

**Francisco Velasco**
Unit for Stereotaxic, Functional Neurosurgery,
    and Radiosurgery
Hospital General de México
Mexico City, Mexico

**Marcos Velasco**
Unit for Stereotaxic, Functional Neurosurgery,
    and Radiosurgery
Hospital General de México
Mexico City, Mexico

**Annamaria Vezzani**
Department of Neuroscience, Laboratory
    of Experimental Neurology
Mario Negri Institute for Pharmacological
    Research
Milano, Italy

**Judith L.Z. Weisenberg**
Department of Neurology
Washington University School of Medicine
St. Louis, Missouri

**Michael P. Weisend**
The Mind Research Network
Albuquerque, New Mexico

**James W. Wheless**
Department of Pediatric Neurology
The University of Tennessee Health Science
    Center
and
Division of Pediatric Neurology
LeBonheur Comprehensive Epilepsy Program
    and Neuroscience Center
LeBonheur Children's Medical Center
and
Department of Pediatric Neurology
St. Jude Children's Research Hospital
Memphis, Tennessee

**H. Steve White**
Anticonvulsant Screening Project
Department of Pharmacology and Toxicology
University of Utah
Salt Lake City, Utah

**Karen S. Wilcox**
Department of Pharmacology and Toxicology
University of Utah
Salt Lake City, Utah

**Michael Wong**
Department of Neurology
Hope Center for Neurological Disorders
Washington University School of Medicine
St. Louis, Missouri

**Greg Worrell**
Division of Epilepsy and
    Electroencephalography
Mayo Systems Electrophysiology Laboratory
Department of Neurology
Mayo Clinic
Rochester, Minnesota
Phoenix, Arizona

# Introduction

Epilepsy is a common episodic neurological condition that is heterogeneous in its clinical presentation, yet characterized at a more fundamental level by the common denominators of neuronal hyperexcitability and hypersynchrony. Our understanding of epilepsy has advanced significantly over the past several decades, and the treatment options (both medical and surgical) have expanded greatly as well. Progress in the basic neurosciences has translated to ever-growing observations of molecular and cellular changes that are associated with the epileptic condition, some of which may be critical to the processes of epileptogenesis, such as the pathological changes that ensue over a latent period, ultimately resulting in spontaneous recurrent seizures and their negative consequences. Further, within the past decade, advances in molecular genetics have defined not only more clearly the role of seizure susceptibility genes and multigene influences, but also genetic mutations that are specifically linked to certain, albeit rare, epilepsy syndromes.

Nevertheless, despite such exciting developments, the clinical practice of epilepsy remains largely empiric, and few insights from the research bench have had meaningful clinical impact. The relative dearth of true "translational" (i.e., clinic to bench and back to clinic) research has hampered our ability to move beyond the limited trial-and-error approach of antiepileptic drug (AED) therapy to one that is based on a detailed understanding of how specific molecular changes might dictate truly rational and targeted pharmacotherapy. Yet, despite this, and even as the mainstay of epilepsy therapy continues to be represented by AEDs, clinicians have brought forth other novel drug approaches and nonpharmacological considerations to the treatment armamentarium, including innovative surgical interventions (e.g., deep brain stimulation).

Many books deal with the subject of epilepsy, with some focusing on clinical diagnosis and treatments and others exploring the pathological substrates of the various epilepsies. Also, some noteworthy volumes provide comprehensive overviews and discussions about both basic science and clinical topics in the field of epilepsy. However, there remains a need for additional references that integrate the most relevant research developments with clinical issues that impact directly on therapeutics. Such books would define the scientific basis of clinical practice and pose a set of challenging questions and considerations that could help shape not only the future of clinical research but also provide novel insights and avenues into more fundamental investigations that would yet again make us go "back to the bench and return to the clinic." Thoughtful clinicians, who can appreciate insights drawn from the fundamental neurosciences, can and should take a more rational approach toward the treatment of patients with epilepsy. Incorporating exciting research developments into their knowledge base will empower clinicians to "think outside the box" and to test clinical hypotheses derived from implications of basic research findings.

This volume is divided into six sections. The first section begins with a broad overview of the basic anatomic and functional substrates of seizure genesis. This is followed by half a dozen chapters highlighting novel pathogenic concepts that have both emerged and have been validated experimentally. These include (1) the role of the blood–brain barrier, (2) central nervous system inflammation, (3) the critical role of metabolism in seizure genesis, (4) the mechanistic basis of drug resistance in epilepsy, (5) complex genetics underlying epileptic conditions, and (6) the unique pathophysiological basis of certain developmental epilepsies.

Chapters in the second section are related to antiepileptic drug therapy and include a current discussion on the molecular targets of AED action and the possibility that certain AEDs may exert protective effects on the disease process itself (rather than simply suppress recurrent seizures). Other considerations in the use of AEDs, both clinically available and investigational, include an appreciation for nonsynaptic mechanisms yielding potent anticonvulsant effects, pharmacokinetic and pharmacodynamic effects, and the genetic underpinnings of AED treatment and development. Finally, with a better understanding of drug interactions and attendant toxicities, beyond what can be established as mechanisms explaining clinical efficacy, the clinician can undoubtedly optimize the long-term care of patients suffering from epilepsy.

The third section focuses on surgical treatments for epilepsy (resective or otherwise), beginning with advances in the fields of structural and functional neuroimaging which have helped enormously in the selection of epilepsy surgery candidates and improving postsurgical outcomes. At the same time, there are emerging approaches for nonsurgical ablation of epileptic tissue and a greater understanding of the molecular and cellular bases of seizure genesis based on studies of such tissues. The fourth section reviews the variety of nontraditional therapeutic options, many of which have established efficacy in the treatment of medically refractory epilepsies (such as the ketogenic diet and the vagus nerve stimulator), but the particular clinical niches of others remain to be defined (e.g., immunomodulators, neurosteroids, herbs, botanicals).

The fifth section deals with neuroendocrine, hormonal, and biobehavioral factors that influence seizure susceptibility—information that should be incorporated into the design of treatment algorithms on an individualized basis. Finally, the last section of this book provides a glimpse of what future epilepsy therapies might look like, from novel mechanisms of drug delivery to gene and stem-cell therapies for epilepsy to seizure detection methods, which provide the pretext for highly targeted and early preventative intervention. Along these lines, the final chapter provides an overview of what has become the holy grail of epilepsy therapeutics over the past decade: the goal of preventing epilepsy itself by first identifying populations at risk and, perhaps more importantly, the critical mediators and influences that in a causal manner produce an enduring epileptic condition. Such knowledge would then be employed to intervene during critical windows of disease ontogeny, as well as during brain development.

The idea for this book was inspired by our collective desire to promote bridging of the so-called "translational divide"—that is, covering innovative treatment strategies based on scientific principles that have yet to be tested rigorously in the clinical setting but yet may provide practitioners with new approaches toward epilepsy therapeutics. It is in this spirit that we earnestly hope that the reader will benefit from this volume.

# Section I

## Scientific Foundations

# 1 Pathophysiological Mechanisms of Seizures and Epilepsy: A Primer

*Carl E. Stafstrom*

## CONTENTS

## INTRODUCTION

This chapter reviews the cellular and synaptic basis for focal and generalized seizure generation with an emphasis on ion channels and synaptic physiology. This background is useful for understanding the scientific basis of epilepsy and its treatment, as discussed in greater detail in subsequent chapters of this book.

A *seizure*, or *epileptic seizure*, is a temporary disruption of brain function due to the excessive, abnormal discharge of cortical neurons. The clinical manifestations of a seizure depend on the specific region and extent of brain involvement and may include an alteration in alertness, motor function, sensory perception, or autonomic function, or all of these. Any person can experience a seizure in the appropriate clinical setting (e.g., meningitis, hypoglycemia), attesting to the innate capacity of even a normal brain to support epileptic discharges, at least temporarily. *Epilepsy* is the condition of recurrent (two or more), unprovoked seizures, usually due to a genetic predisposition or chronic acquired pathologic state (e.g., cerebral dysgenesis, brain trauma). *Epilepsy syndrome* refers to a constellation of clinical characteristics that consistently occur together, with seizures as a primary manifestation. Features of an epilepsy syndrome might include similar age of onset, electroencephalogram (EEG) findings, etiology, inheritance pattern, natural history of the symptoms, and response to particular antiepileptic drugs. Mechanisms leading to the generation of a seizure (*ictogenesis*) may differ from those predisposing to epilepsy, the condition of recurrent, unprovoked seizures (i.e., *epileptogenesis*) (Dichter, 2009).

A seizure is characterized by aberrant electrical activity within the brain. Such electrical activity is the net product of biochemical processes at the cellular and subcellular levels occurring in the context of large neuronal networks. Seizures often involve interplay between cortical and subcortical structures (Blumenfeld, 2003). The surface EEG is the primary clinical tool with which normal and abnormal electrical activity in the brain is measured.

At the cellular level, the two hallmark features of epileptic activity are neuronal hyperexcitability and neuronal hypersynchrony. *Hyperexcitability* is the abnormal responsiveness (e.g., lower threshold) of a neuron to excitatory input; a hyperexcitable neuron tends to fire bursts of multiple action potentials instead of just one or two. *Hypersynchrony* refers to the recruitment of large numbers of neighboring neurons into an abnormal firing mode. Ultimately, a seizure is a network phenomenon that requires participation of many neurons firing synchronously. Conventional EEG techniques can detect cortical areas exhibiting hypersynchronous discharges in the form of interictal sharp waves or spikes. Using specialized EEG recording techniques in humans and animals with epilepsy, bursts of very localized discharges have been detected that are not detected by usual EEG methods (Engel et al., 2009). These so-called "fast ripples" (250 to 600 Hz) reflect abnormal interictal discharges in restricted cortical areas which could synchronize and lead to a seizure (see Chapter 21, this volume).

## CLASSIFICATION OF SEIZURE TYPES AND EPILEPSY SYNDROMES

Epileptic seizures are broadly divided into two groups, depending on their site of origin and pattern of spread (Figure 1.1). Partial seizures arise from a localized region of the brain, and the associated clinical manifestations relate to the function ordinarily subserved by that area. Focal discharges can spread locally through synaptic and nonsynaptic mechanisms or propagate distally to subcortical structures or through commissural pathways to involve the entire cortex. A seizure arising from the left motor cortex, for example, may cause jerking movements of the right upper extremity. If epileptic discharges subsequently spread to adjacent areas and eventually encompass the entire brain, a secondarily generalized seizure may ensue.

In contrast, generalized seizures begin with abnormal electrical discharges in both hemispheres simultaneously. Thus, the EEG signature of a primary generalized seizure is bilateral synchronous spike–wave discharges seen across all scalp electrodes. Primary generalized seizures critically involve reciprocal thalamocortical connections. The manifestations of such generalized epileptic activity can range from brief impairment of consciousness (as in an absence seizure) to rhythmic jerking movements of all extremities accompanied by loss of posture and consciousness (as in a generalized tonic–clonic seizure).

Although different mechanisms underlie partial vs. generalized seizures, it is useful to view any seizure activity as a perturbation in the normal balance between neuronal inhibition and neuronal excitation. Such an excitation/inhibition imbalance may occur in a localized region of brain, in multiple brain areas (which might be linked into a multinodal network), or simultaneously throughout the whole brain (McCormick and Contreras, 2001; Faingold, 2004). This imbalance is likely the consequence of a combination of increased excitation and decreased inhibition. It is useful to conceptualize excitation/inhibition imbalance as critical for seizures and epilepsy, but this notion may be overly simplistic when brain microcircuitry is analyzed in detail. In some circumstances, for example, increased inhibition can lead to enhanced hyperexcitability (see Inhibitory Synaptic Transmission section, later in this chapter) (Mann and Mody, 2008; Yu et al., 2006).

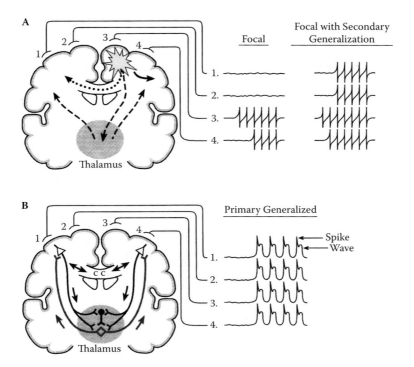

**FIGURE 1.1** Coronal brain sections depicting seizure types and potential routes of seizure spread. (A) Focal area of hyperexcitability (star under electrode 3) and spread to adjacent neocortex (solid arrow under electrode 4), via corpus callosum (dotted arrow) or other commissural pathways to the contralateral cerebral hemisphere or via subcortical pathways (e.g., thalamus, upward dashed arrows). Accompanying EEG patterns show brain electrical activity under electrodes 1 to 4. Focal epileptiform activity is maximal at electrode 3 and is also seen at electrode 4 (left traces). If a seizure secondarily generalizes, activity may be seen synchronously at all electrodes, after a delay (right traces). (B) Primary generalized seizure begins simultaneously in both hemispheres. The characteristic bilateral synchronous spike–wave pattern on EEG is generated by reciprocal interactions between the cortex and thalamus, with rapid spread via corpus callosum (CC) contributing to the rapid bilateral synchrony. One type of thalamic neuron (dark neuron) is a GABAergic inhibitory cell that displays intrinsic pacemaker activity. Cortical neurons (open triangles) send impulses to both thalamic relay neurons (open diamond) and to inhibitory neurons, setting up oscillations of excitatory and inhibitory activity and giving rise to the rhythmic spike waves on EEG. (From Stafstrom, C.E., in *Epilepsy and the Ketogenic Diet*, Stafstrom, C.E. and Rho, J.M., Eds., Humana Press, Totowa, NJ, 2004, p. 3. With permission.)

It is important to recognize that epilepsy is not a singular disease but rather a heterogeneous spectrum in terms of clinical expression, underlying etiologies, and pathophysiology. As such, specific mechanisms and pathways underlying specific seizure phenotypes may vary when perturbations at a given hierarchical level lead to structural and functional changes at either higher (e.g., network) or lower (e.g., molecular) levels of analysis.

## CELLULAR ELECTROPHYSIOLOGY

### REGIONAL DIFFERENCES IN EXCITABILITY

Brain regions differ in their intrinsic propensity to generate and propagate seizure activity, based on factors such as cell density, intrinsic membrane properties, laminar arrangement of neurons, and pattern of cellular interconnectivity. Even within the same brain region, physiological differences among various neuron types endow the region with variable excitability (Steriade, 2004).

The neocortex and hippocampus are especially prone to generating seizures. The hippocampal formation has been investigated extensively with regard to basic and epilepsy-related electrophysiological studies, and hippocampal pyramidal cells are among the most intensively studied cell types in the central nervous system. The orderly and relatively simple organization of hippocampal circuits makes them amenable for studying synaptic and nonsynaptic mechanisms relevant to seizure genesis. Furthermore, the intrinsic ability of neurons in the CA3 to fire action potentials in bursts augments the hyperexcitability of this circuit (Traub et al., 1991). Details regarding the electrophysiology of the hippocampal formation are found in classic reviews (Schwartzkroin and Mueller, 1987).

The hippocampal formation consists of the dentate gyrus, the hippocampus proper (Ammon's horn, with subregions CA1, CA2, and CA3), the subiculum, and the entorhinal cortex (Figure 1.2). These four regions are linked by excitatory, unidirectional feedforward connections. There are also some reverse projections from the entorhinal cortex to Ammon's horn and from CA3 to the dentate gyrus. The predominant forward-projecting trisynaptic circuit begins with neurons in layer II of the entorhinal cortex which project axons to the dentate gyrus along the perforant pathway, where they synapse on granule cell (and interneuron) dendrites. Granule cells, the principal cell type of the dentate gyrus, send their axons, called *mossy fibers*, to synapse on cells in the hilus and in the CA3 field of Ammon's horn. Several classes of inhibitory interneurons within the dentate hilus modulate ongoing excitatory neural activity (Lawrence and McBain, 2003). CA3 pyramidal cells project to other CA3 pyramidal cells via local collaterals, to the CA1 field of Ammon's horn via Schaffer collaterals, and to the contralateral hippocampus. CA1 pyramidal cells send their axons into the subiculum, and neurons of the subiculum project to the entorhinal cortex (as well as to other cortical and subcortical targets), thus completing the circuit. For this reason, limbic system structures such as hippocampus, subiculum, and entorhinal cortex are endowed with structural and functional features that predispose them to seizures and epilepsy (Jutila et al., 2002; Sloviter, 2008; Stafstrom, 2005).

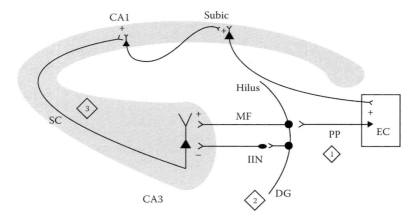

**FIGURE 1.2** Schematic of hippocampal circuitry. Major pathways of excitatory transmission in the hippocampal trisynaptic pathway begin in neurons of entorhinal cortex (EC). EC neurons send axons to the dentate gyrus (DG) via the perforant path (PP) (1), where they synapse on granule cell dendrites. Dentate granule cells project their axons (mossy fibers [MF]) (2) to synapse on cells of the hilus (particularly inhibitory interneurons [IIN]) and in the CA3 field of Ammon's horn. CA3 pyramidal neurons then project to neurons of the CA1 field of Ammon's horn via Schaffer collaterals (SC) (3). Finally, CA1 neurons send projections outward through the fornix to other brain regions, including the subiculum (Subic), which then completes the circuit by exciting EC neurons. For simplicity, only feedforward excitatory projections of the classic trisynaptic pathway are depicted. Omitted are backward projections and local circuit interactions. *Note:* +, excitatory projection; −, inhibitory projection.

## Ion Channels

Neuronal excitation depends on the number, type, and distribution of ion channels within the neuronal membrane. Two major types of ion channels are responsible for the inhibitory and excitatory activity comprising normal neuronal function: voltage-gated channels and ligand-gated channels. Voltage-gated sodium and calcium channels depolarize the cell membrane toward the action potential threshold. Voltage-gated potassium channels largely dampen neuronal excitation by repolarizing the membrane potential after an action potential or by opposing depolarizing conductances to keep the membrane potential below threshold. A variety of ion conductances operative in the subthreshold voltage range also sculpt neuronal activity. Voltage-gated channels are activated by membrane potential changes, which subsequently alter the conformational state of the channel and allow selective passage of charged ions through a pore.

Ligand-gated receptors include those mediating excitation (glutamate receptors) and inhibition (γ-aminobutyric acid, or GABA, receptors). A neurotransmitter (prepackaged in vesicles) is released from a presynaptic terminal (following presynaptic depolarization and calcium influx) into the synaptic cleft; the neurotransmitter then binds with selective affinity to a membrane-bound receptor on the postsynaptic membrane. Binding of a neurotransmitter to its receptor activates a cascade of events, including a conformational shift to reveal an ion-permeant pore. Passage of ions across these channels results in either depolarization (e.g., inward flux of cations) or hyperpolarization (e.g., inward flux of anions or outward flux of cations). Excitability can also be modified posttranslationally, for example, by receptor phosphorylation or by second-messenger pathways and modified gene expression.

## Voltage-Dependent Membrane Conductances

### Depolarizing (Excitatory) Conductances

A rapidly inactivating inward sodium conductance underlies the depolarizing phase of the action potential, and a non-inactivating, persistent sodium current can augment cell depolarization (e.g., produced by excitatory synaptic input) in the voltage range immediately subthreshold for spike initiation (Crill, 1996; Hille, 2001). Dysfunction of either the rapidly inactivating sodium current or persistent sodium current can alter neuronal excitability and enhance the propensity for epileptic firing (George, 2005; Stafstrom, 2007b).

Sodium channels consist of a complex of three polypeptide subunits; a major α-subunit forms the channel pore, and two smaller β-subunits influence the assembly and kinetic properties of the α-subunit. The shape of the action potentials is determined by the types of α- and β-subunits present in an individual neuron (Catterall et al., 2005). Many anticonvulsants act in part through interactions with voltage-dependent sodium channels, including phenytoin, carbamazepine, oxcarbazepine, felbamate, and lamotrigine (Rogawski and Löscher, 2004).

Neurons also display voltage-gated inward calcium conductances. In the hippocampus, prominent calcium currents occur in CA3 pyramidal cells, especially in dendrites, and underlie burst discharges in these cells (Wong and Prince, 1978). Activation of voltage-dependent calcium channels contributes to the depolarizing phase of the action potential and can affect neurotransmitter release, gene expression, and neuronal firing patterns. There are several distinct subtypes of calcium channels, distinguished on the basis of electrophysiological properties, pharmacological profile, molecular structure, and cellular localization (Catterall et al., 2003). The molecular structure of voltage-gated calcium channels is similar to that of sodium channels. Voltage-dependent calcium channels are hetero–oligomeric complexes comprised of a principal pore-forming $\alpha_1$-subunit and one or more smaller subunits ($\alpha_2$, β, γ, and δ) that are not obligatory for normal activity but can modulate the kinetic properties of the channel.

## Hyperpolarizing (Inhibitory) Conductances

Depolarizing sodium and calcium currents are counterbalanced by an array of voltage-dependent hyperpolarizing currents, primarily via potassium channels. Potassium channels represent the largest and most diverse family of voltage-gated ionic channels and function to inhibit or decrease excitation in the nervous system (Hille, 2001). The prototypic voltage-gated potassium channel is composed of four membrane-spanning $\alpha$-subunits and four regulatory $\beta$-subunits, which are assembled in an octameric complex to form an ion-selective pore (Gutman et al., 2005). In hippocampal neurons, potassium conductances include: (1) a leak conductance, which is a major determinant of the resting membrane potential; (2) an inward rectifier (involving the flux of other ions), which is activated by hyperpolarization; (3) a large set of delayed rectifiers, which are involved in the termination of action potentials and repolarization of the neuron's membrane potential; (4) an A-current, which helps determine inter-spike intervals and thus affects the rate of cell firing; (5) an M-current, which is sensitive to cholinergic muscarinic agonists and affects the resting membrane potential and rate of cell firing; and (6) a family of calcium-activated potassium conductances that are sensitive to intracellular calcium concentration and affect the cell firing rate and interburst interval.

Although facilitation of these hyperpolarizing conductances could be viewed as potentially anticonvulsant, none of the traditional anticonvulsants is thought to act directly and principally on voltage-gated potassium channels. By contrast, newer anticonvulsants appear to act in part by affecting potassium channel function (Rogawski, 2000; Wickenden, 2002); for example, topiramate induces a steady membrane hyperpolarization mediated by a potassium conductance (Herrero et al., 2002), and levetiracetam blocks sustained repetitive firing by paradoxically decreasing voltage-gated potassium currents (Madeja et al., 2003). Retigabine, an investigational compound with broad efficacy in animal seizure models, appears to enhance activation of KCNQ2 and KCNQ3 potassium channels (members of the so-called Kv7 family), thus increasing the effectiveness of the M-type potassium current, which acts as a brake on repetitive neuronal firing (Miceli et al., 2008; Rogawski and Bazil, 2008). This is a particularly intriguing finding given that mutations in genes encoding these proteins have been linked to a rare form of inherited epilepsy, benign familial neonatal convulsions (Biervert et al., 1998).

## SYNAPTIC PHYSIOLOGY

### Inhibitory Synaptic Transmission

Synaptic inhibition in the hippocampus is mediated by two basic circuit configurations: (1) Feedback or recurrent inhibition occurs when excitatory principal neurons synapse with and excite inhibitory interneurons, which, in turn, project back to the principal neurons and inhibit them (i.e., a negative-feedback loop). (2) Feedforward inhibition occurs when axons projecting into the region synapse with and directly activate inhibitory interneurons, which then inhibit principal neurons. Both feedforward and feedback inhibitory circuits abound in the hippocampal formation and utilize GABA as the neurotransmitter.

GABA, the principal inhibitory neurotransmitter in the mammalian central nervous system, is a neutral amino acid synthesized from glutamic acid by the rate-limiting enzyme glutamic acid decarboxylase (GAD). GAD requires pyridoxine (vitamin $B_6$) as a cofactor; inherited deficiency of pyridoxine or diminished responsiveness to pyridoxine is a cause of intractable neonatal seizures (Gospe, 2006). GABA, released from axon terminals, binds to at least two classes of receptors—GABA$_A$ and GABA$_B$—which are found on almost all cortical neurons (Martin and Olsen, 2000). In addition, GABA$_A$ receptors are found on glia, where they may exert a role in the regulation of excitability (Sierra-Paredes and Sierra-Marcuño, 2007).

The GABA$_A$ receptor is a macromolecular receptor complex consisting of an ion pore as well as binding sites for agonists and a variety of allosteric modulators, such as benzodiazepines and barbiturates, each differentially affecting the kinetic properties of the receptor (Olsen and Sieghart,

2008). The $GABA_A$ receptor is a heteropentameric complex composed of combinations of several polypeptide subunits arranged in topographical fashion to form an ion channel. This channel is selectively permeable to chloride (and bicarbonate) ions. To date, seven different subunits ($\alpha$, $\beta$, $\gamma$, $\delta$, $\epsilon$, $\pi$, $\rho$) have been described, each with one or more subtypes. Although several thousand receptor isoforms are possible from differential expression and assembly of these various subtypes, there is likely to be only a limited number of functional combinations, but the precise subunit composition of native $GABA_A$ receptors has yet to be identified. Most functional $GABA_A$ receptors follow the general motif of containing either $\alpha$ and $\beta$ or $\alpha$, $\beta$, and $\gamma$ subunits with uncertain stoichiometry. Because individual subunits might be differentially sensitive to pharmacological agents, GABA receptor subunits represent potentially useful molecular targets for new anticonvulsants.

Activation of $GABA_A$ receptors on the somata of mature cortical neurons generally results in the influx of $Cl^-$ and consequent membrane hyperpolarization, thus inhibiting cell discharge. In hippocampal cell dendrites and in the immature brain, however, $GABA_A$ receptor activation causes *depolarization* of the postsynaptic membrane. This reversal of the conventional $GABA_A$ effect is thought to reflect a reversed $Cl^-$ electrochemical gradient, a consequence of the immature expression of the $K^+$–$Cl^-$ cotransporter KCC2, which ordinarily renders GABA hyperpolarizing (Rivera et al., 1999; Staley, 2006).

In addition to $GABA_A$ receptors, "metabotropic" $GABA_B$ receptors are located on both postsynaptic membrane and on presynaptic terminals. Metabotropic receptors do not form an ion channel pore; instead, $GABA_B$ receptors act through GTP-binding proteins to control calcium or potassium conductances. Whereas $GABA_A$ receptors generate fast high-conductance inhibitory postsynaptic potentials (IPSPs) close to the cell body, $GABA_B$ receptors on the postsynaptic membrane mediate slow, long-lasting, low-conductance IPSPs, primarily in hippocampal pyramidal cell dendrites. Perhaps of greater functional significance, activation of $GABA_B$ receptors on axon terminals blocks synaptic release of neurotransmitter. It is thought that $GABA_B$ receptors are associated with terminals that release GABA onto postsynaptic $GABA_A$ receptors. In such cases, activation of $GABA_B$ receptors reduces the amount of GABA released, resulting in disinhibition (Simeone et al., 2003).

## Excitatory Synaptic Transmission

Glutamate, an excitatory amino acid, is the principal excitatory neurotransmitter of the mammalian central nervous system. Glutamatergic pathways are widespread throughout the brain, and excitatory amino acid activity is critical to normal brain development and activity-dependent synaptic plasticity (Simeone et al., 2004). There are two classes of glutamate receptors: ionotropic and metabotropic. Ionotropic glutamate receptors are broadly divided into $N$-methyl-D-aspartate (NMDA) and non-NMDA receptors, based on biophysical properties and pharmacological profiles (Dingledine et al., 1999). Each subtype of glutamate receptor consists of a multimeric assembly of subunits that determine its distinct functional properties. Glutamate receptor channel subunits are currently classified into six subfamilies based on amino acid sequence homology.

The NMDA receptor contains a binding site for glutamate (or NMDA) and a recognition site for a variety of modulators (e.g., glycine, polyamines, MK-801, zinc). NMDA receptors also display voltage-dependent block by magnesium ions (Collingridge et al., 1988). When the membrane is depolarized and the magnesium block of the NMDA receptor is alleviated, activation of the NMDA receptor results in an influx of calcium and sodium ions. Calcium entry is central to the initiation of a number of second-messenger pathways—for example, stimulation of a variety of kinases that subsequently activate signal transduction cascades leading to changes in transcriptional regulation. Activation of the NMDA receptor leads to generation of relatively slow and long-lasting excitatory postsynaptic potentials (EPSPs). These synaptic events contribute to epileptiform burst discharges, and NMDA receptor blockade results in the attenuation of bursting activity in many models of epileptiform activity (Gean, 1990; Kalia et al., 2008).

Non-NMDA ionotropic receptors are divided into α-amino-3-hydroxy-5-methyl-4-isoxazoleproprionic acid (AMPA) and kainate receptors (Dingledine et al., 1999). The AMPA receptor is responsible for the major part of the EPSP—fast-rising and brief in duration—generated by the release of glutamate onto postsynaptic neurons. In addition, the depolarization generated via AMPA receptors is necessary for effective activation of NMDA receptors. Consequently, AMPA receptor antagonists block most excitatory synaptic activity in pyramidal neurons.

Metabotropic glutamate receptors represent a large, heterogeneous family of G-protein-coupled receptors, which subsequently activate various transduction pathways, such as phosphoinositide hydrolysis and activation of adenylate cyclase and phospholipases C and D (Conn, 2003). These metabotropic receptors are important modulators of voltage-dependent potassium and calcium channels, nonselective cation currents, and ligand-gated receptors (i.e., GABA and glutamate receptors) and can regulate glutamate release. Hence, it is not surprising that they have been invoked in a wide variety of neurological processes (e.g., long-term potentiation, or LTP) and disease states (including epilepsy) (Ure et al., 2006; Wong et al., 2004). Different metabotropic glutamate receptor subtypes are specific for different intracellular processes. Although ubiquitous within the central nervous system, subtypes of metabotropic receptors appear to be differentially localized.

## PATHOPHYSIOLOGY OF EPILEPTIC FIRING

### ABNORMAL NEURONAL FIRING

At the neuronal cellular and network levels, there has been a concerted effort, extending over several decades, to understand the mechanisms governing the transition from normal firing to interictal epileptiform bursts to an ictal state, as well as the evolution of electrophysiological changes that terminate a seizure and underlie postictal changes (Lado and Moshe, 2008; Stafstrom, 2004). Likewise, mechanisms underlying epileptogenesis, the transition from normal brain to epileptic brain, represents both a critical knowledge gap and an opportunity for selective therapeutics (Clark and Wilson, 1999; Dichter, 2009; Dudek and Sutula, 2007). Understanding the scientific basis of both seizure generation and epileptogenesis requires correlated laboratory and clinical investigations. Much of our understanding of epilepsy mechanisms comes from *in vivo* animal models and *in vitro* electrophysiological studies.

As discussed earlier, consideration of the imbalance between excitatory and inhibitory factors helps to guide the approach to the mechanisms involved. Mutations have been identified in genes coding for ion channel proteins in both humans and animal models, many of which express seizures as a phenotype. These so-called epilepsy "channelopathies" represent a window for dissecting mechanisms of seizure genesis (Helbig et al., 2008; Reid et al., 2009). In addition, mutations in genes coding for proteins responsible for neurotransmitter transport and biogenesis, as well as receptor trafficking to the correct membrane location, are expanding the repertoire of epilepsy mechanisms (Hirose, 2006; Macdonald and Kang, 2009).

Figure 1.3 depicts EEG and intracellular changes that can be seen in normal, interictal, and ictal states. In the normal situation, action potentials (which represent "all-or-none" events) are generated when the neuronal membrane potential reaches the threshold for firing (approximately −40 mV). These discharges may influence the activity of adjacent neurons through electrical field (i.e., ephaptic) or synaptic mechanisms, resulting in an EPSP. Nearby inhibitory interneurons may also be activated, after a brief delay, giving rise to an IPSP. The activity recorded in the target neuron (neuron 2 of Figure 1.3B) will reflect the temporal and spatial summation of both EPSP and IPSP inputs. When extrapolated to multiple synaptic contacts, the sculpting of individual cellular responses modulated by various degrees of inhibition can be envisioned. Further, when considering that a single inhibitory interneuron can connect with hundreds or thousands of pyramidal

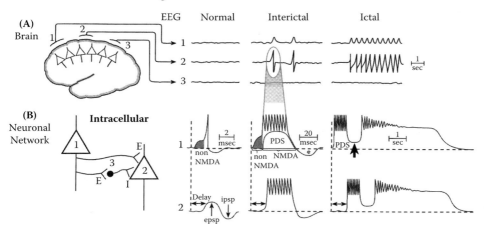

**FIGURE 1.3** Abnormal neuronal firing at the levels of (A) the brain and (B) a simplified neuronal network, consisting of two excitatory neurons (1 and 2) and an inhibitory interneuron (filled black circle, 3). EEG (top set of traces) and intracellular recordings (bottom set of traces) are shown for the normal (left column), interictal (middle column), and ictal (right column) conditions. Numbered traces refer to like-numbered recording sites. Note time-scale differences in different traces. (A) Three EEG electrodes record activity from superficial neocortical neurons. In the normal case, activity is low voltage and desynchronized (neurons are not firing together in synchrony). In the interictal condition, large spikes are seen focally at electrode 2 (and to a lesser extent at electrode 1, where they might be termed "sharp waves"), representing synchronized firing of a large population of hyperexcitable neurons (expanded in time below). The ictal state is characterized by a long run of spikes. (B) At the neuronal network level, the intracellular correlate of the interictal EEG spike is the paroxysmal depolarization shift (PDS). The PDS is initiated by a non-NMDA-mediated fast EPSP (shaded area) but is maintained by a longer, larger NMDA-mediated EPSP. The post-PDS hyperpolarization (asterisk) temporarily stabilizes the neuron. If this post-PDS hyperpolarization fails (right column, thick arrow), ictal discharge can occur. The lowermost traces, recordings from neuron 2, show activity similar to that recorded in neuron 1, with some delay (double-headed horizontal arrow). Activation of inhibitory neuron 3 by the firing of neuron 1 prevents neuron 2 from generating an action potential (the IPSP counters the depolarization caused by the EPSP). If neuron 2 does reach the firing threshold, additional neurons will be recruited, leading to an entire network firing in synchrony (seizure). (From Stafstrom, C.E., in *Epilepsy and the Ketogenic Diet*, Stafstrom, C.E. and Rho, J.M., Eds., Humana Press, Totowa, NJ, 2004, p. 3. With permission.)

neurons, it is straightforward to see how hypersynchronous behavior can be influenced by even a single cell. For localized hyperexcitability to spread to adjacent areas, the epileptic firing must overcome the powerful inhibitory influences that normally keep aberrant excitability in check (i.e., the "inhibitory surround").

## PAROXYSMAL DEPOLARIZATION SHIFT

The intracellular correlate of the focal interictal epileptiform discharge on EEG is known as the paroxysmal depolarization shift (PDS) (Ayala, 1983). The PDS is seen when recording changes in the membrane potential of a single neuron with a microelectrode while simultaneously recording a focal spike on EEG (Figure 1.3B). Initially, there is a rapid shift in the membrane potential in a depolarizing direction initiated by synaptic forces, followed by a burst of repetitive action potentials on a depolarizing plateau potential lasting several hundred milliseconds (Johnston and Brown, 1984). The initial depolarization is mediated by non-NMDA glutamate receptors (i.e., AMPA receptors), while the sustained depolarization is a consequence of NMDA receptor activation. Afterward, the PDS terminates with a repolarization phase, primarily as a consequence of inhibitory potassium and chloride conductances, carried by voltage-gated potassium channels and GABA receptors, respectively. Of note, there is a prolonged period of hyperpolarization following the PDS, again mediated by inhibitory conductances, constituting a refractory period. Failure of mechanisms to terminate a PDS could lead to massive prolongation of the abnormal neuronal discharge—that is, a seizure.

## SYNCHRONIZING MECHANISMS

The hippocampus normally displays robust neuronal synchronization. Sharp waves, dentate spikes, theta activity, 40-Hz oscillations, and 200-Hz oscillations are all forms of neuronal synchronization that can be recorded in various regions of the hippocampal formation (Buzsáki and Draguhn, 2004). Thus, synchronization of neuronal activity appears to be how the brain performs many of its normal functions; however, exaggerated neuronal synchronization is a hallmark of epilepsy. In addition, normal forms of synchronized activity that do not trigger seizures in normal tissue might trigger epileptiform discharges in a brain region that has undergone selective neuronal loss, synaptic reorganization, or changes in receptor expression.

In the hippocampus, synchronizing mechanisms include inputs from subcortical nuclei as well as intrinsic interneuron-mediated synchronization. For example, high-amplitude theta activity (4 to 7 Hz) is a salient feature of the hippocampus. The theta rhythm represents synchronized activity of hippocampal neurons and is largely dependent on input from the septum (Buzsáki and Draguhn, 2004). Subcortical nuclei such as the septum have divergent inputs that target hippocampal interneurons. In turn, the divergent axon projections of interneurons, and the powerful effect of the $GABA_A$-receptor-mediated conductances that they produce, enable interneurons to entrain the activity of large populations of principal cells. These characteristics make interneurons a very effective target for subcortical modulation of hippocampal principal cell activity. In addition, mutual inhibitory interactions among hippocampal interneurons can produce synchronized discharges (Jefferys et al., 1996).

Recurrent excitatory circuits are another basis for neuronal synchronization in the hippocampus. Recurrent excitatory collaterals are a normal feature of the CA3 region. CA3 pyramidal cells form direct, monosynaptic connections with other CA3 pyramidal cells. These recurrent excitatory interactions contribute to the synchronized burst discharges that characterize this region of Ammon's horn (Traub and Miles, 1991). In the epileptic temporal lobe, synaptic reorganization and axonal sprouting lead to aberrant recurrent excitation, providing a synchronizing mechanism in other parts of the hippocampal formation, including the CA1 region, subiculum, entorhinal cortex, and dentate gyrus. In the dentate gyrus of normal hippocampus, for example, granule cells form few, if any, monosynaptic contacts with neighboring granule cells. In epileptic hippocampus, mossy fiber sprouting results in direct excitatory interactions among granule cells (Figure 1.4) (Nadler, 2003; Sutula, 2002).

Finally, mechanisms independent of chemical synaptic transmission might synchronize hippocampal neuronal firing under some circumstances (Dudek et al., 1998); for example, gap junctions allow electrical signals to pass directly between cells (Traub et al., 2004). Gap junctions are upregulated in epileptic brain, and blockade of gap junctions significantly affects the duration of seizure activity (Nemani and Binder, 2005). Another example is illustrated by electrical field effects generated through current flow within the extracellular space. The potential synchronizing effect of such "ephaptic" interactions suggests that manipulations altering the extracellular space volume (thus affecting current flow through this compartment) can impact epileptogenicity (Schwartzkroin et al., 1998). In addition, changes in extracellular ion concentrations can affect excitability. Increases in extracellular potassium concentration have long been known to affect epileptogenic excitability and synchronization (Fröhlich et al., 2008; Traynelis and Dingledine, 1988).

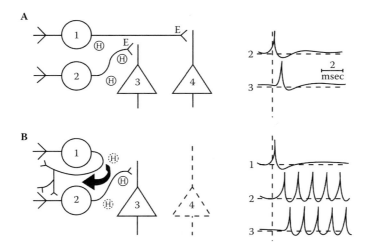

**FIGURE 1.4**  Simplified depiction of axonal sprouting in the hippocampal dentate gyrus. (A) Normal situation. Left: Dentate granule neurons (1, 2) make excitatory synapses (E) onto dendrites of hippocampal pyramidal neurons (3, 4). Right: Activation of dentate neuron 2 causes single action potential in pyramidal neuron 3 after a synaptic delay. (B) As a consequence of status epilepticus, many pyramidal neurons die (4, dashed outline), leaving axons of dentate neuron 1 without a postsynaptic target; those axons then "sprout" and innervate the dendrites of other granule neurons (thick curved arrow), creating the substrate for a hyperexcitable circuit. Now, when neuron 1 is activated, multiple action potentials are fired in neuron 2 and, therefore, in neuron 3 (right traces). As described in the text, this diagram is simplified and, in fact, numerous types of interneurons in the dentate hilus (labeled H) are also involved in the outcome of seizure-induced synaptic plasticity. The resultant circuit function will depend on the character (excitatory or inhibitory) and connectivity of these interneurons. (From Stafstrom, C.E., in *Epilepsy and the Ketogenic Diet*, Stafstrom, C.E. and Rho, J.M., Eds., Humana Press, Totowa, NJ, 2004, p. 3. With permission.)

## GLIAL MECHANISMS FOR MODULATING EPILEPTOGENICITY

The critical role of glia in epilepsy is now receiving due attention (Wetherington et al., 2008). Because the ionic balance between intracellular and extracellular compartments is altered after neuronal activity (especially after sustained repetitive discharges seen with seizures), there must exist mechanisms to restore ionic homeostasis. Otherwise, even normal neuronal activity would cease.

Astrocytes are perhaps most closely associated with regulation of extracellular potassium levels (potassium buffering), as it is clear that glia membranes are preferentially permeable to potassium. A variety of inwardly rectifying $K^+$ channels provide an appropriate means for potassium uptake, and the association of glial endfeet with brain microvasculature provides a convenient "sink" for potassium release. Glial membrane potential changes are directly correlated with changes in extracellular potassium, and blockade of potassium channels selective to glia results in neuronal hyperexcitability. Thus, it seems certain that glia help to modulate neuronal discharges through their regulation of extracellular potassium.

Despite the evidence that glia participate in potassium regulation, glia play an even more important role in the maintenance of neuronal excitability by transporting glutamate (released from neuronal terminals to excite other neurons) out of the extracellular space. Glia are uniquely equipped for this role, having at least two powerful glutamate transport molecules in their membranes (Rothstein et al., 1994). Rapid and efficient removal of extracellular glutamate characterizes normal healthy brain tissue and is essential because residual glutamate would continue to excite surrounding neurons. Indeed, blockade of glutamate transporters (or "knockout" of the genes for these transport proteins) results in epilepsy or excitotoxicity (Tanaka et al., 1997).

Glia contribute to modulation of excitability in a number of other ways. First, they play a critical role of regulating extracellular pH via a proton exchanger and bicarbonate transporter mechanisms. Even low levels of neuronal activity create significant pH transients. Further, pH modulates receptor function, particularly that of the NMDA receptor, which appears to play such an important role in epileptic discharge. Second, glia are also now thought to release powerful neuroactive agents into the extracellular space. Studies have shown that glial glutamate release can excite neighboring neurons (Tian et al., 2005), and other glia-related factors, such as the cytokine interleukin-1 (IL-1), can have profound anticonvulsant efficacy (Vezzani et al., 2008).

## PHYSIOLOGY OF ABSENCE EPILEPSY

Absence seizures represent a subtype of generalized-onset seizures with a distinct pathophysiological substrate. An absence seizure is characterized by a temporary loss of awareness and usually a sudden cessation of motor activity. These seizures are generally short lived (lasting less than 20 seconds in most cases), begin without an aura, and end abruptly without postictal changes.

The 3-Hz generalized spike–wave discharges seen with an absence seizure reflect a widespread, phase-locked oscillation between excitation (i.e., spike) and inhibition (i.e., slow wave) in mutually connected thalamocortical networks (Huguenard and McCormick, 2007; McCormick and Contreras, 2001). As depicted in Figure 1.5, neurons of layer VI of the neocortex have excitatory projections to thalamic relay (TR) neurons as well as to inhibitory GABAergic neurons comprising the nucleus reticularis thalami (NRT). In turn, excitatory outputs of TR neurons impinge upon the apical dendrites of layer VI pyramidal neurons in neocortex. This so-called thalamocortical relay is a critical substrate for the generation of cortical rhythms and is influenced by sensory inputs (e.g., from the retina). In addition, ascending projections from several brainstem nuclei, including cholinergic, noradrenergic, and serotonergic inputs, modulate thalamocortical activity (Chang and Lowenstein, 2003). This reciprocal circuitry, which is responsible in large part for normal EEG oscillations during wake and sleep states, could become overactive to generate generalized spike–wave discharges or could be dampened to reduce or eliminate spontaneous cortical rhythms. Further, the anatomy implies that spike–wave discharges can be interrupted at either cortical or thalamic levels.

Although multiple ionic conductances are involved in rhythmic pacemaking activity, two specific channels are believed to play a key role in regulating thalamocortical activity. The first is a subtype of voltage-gated calcium channel known as the "low-threshold" or T-type calcium channel (Perez-Reyes, 2003). The channel is so named because it can be activated by small membrane depolarizations. In many neurons, calcium influx through these channels triggers low-threshold spikes which in turn activate a burst of action potentials (McCormick and Contreras, 2001). Such an excitatory burst is believed to underlie the "spike" portion of a generalized spike–wave oscillation.

The second important ion channel involved in the regulation of thalamocortical rhythmicity is the hyperpolarization-activated cation channels (HCN channels), responsible for the so-called H-current. HCN channels, densely expressed in the thalamus and hippocampus, are activated by hyperpolarization and produce a depolarizing current carried by an inward flux of $Na^+$ and $K^+$ ions (Robinson and Siegelbaum, 2003; Wahl-Schott and Biel, 2009). This depolarization helps to bring the resting membrane potential toward threshold for activation of T-type calcium channels which in turn produces a calcium spike and a burst of action potentials. HCN channels are highly expressed in dendrites and to a lesser extent in the soma.

Unlike other voltage-gated conductances that can be labeled either inhibitory or excitatory, H-currents are both inhibitory and excitatory (Poolos, 2004). HCN channels possess an inherent negative-feedback property; hyperpolarization produces an activation of HCN channels, which then leads to depolarization which then deactivates these channels. The net effect of HCN channel activation is a decrease in the input resistance of the membrane, which is the voltage change produced by a given synaptic current. The H-current tends to stabilize the neuronal membrane potential toward the resting potential against both hyperpolarizing and depolarizing inputs. HCN channels

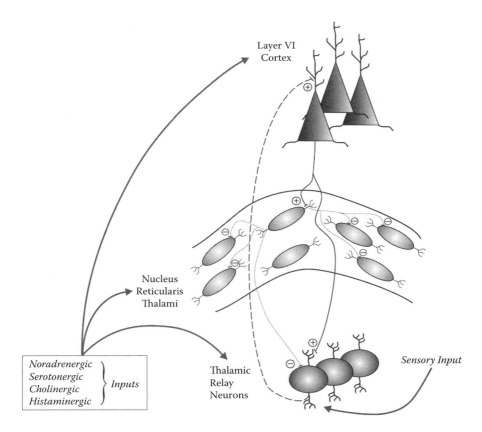

**FIGURE 1.5** Thalamocortical circuitry believed to form the basis of normal oscillatory rhythms that, when perturbed, can produce generalized spike–wave discharges seen with absence seizures. This circuitry involves excitatory projections from layer VI neocortical pyramidal neurons onto both thalamic relay (TR) neurons and inhibitory neurons comprising the nucleus reticularis thalami (NRT). TR neurons, in turn, send excitatory axons back to the neocortex. Activation of NRT neurons results in recurrent inhibition to both adjacent neurons as well as TR neurons. The neurotransmitters at the excitatory (++) and inhibitory (–) synapses are thought to be glutamate and γ-aminobutyric acid (GABA), respectively. Low-threshold (T-type) calcium channels and hyperpolarization-activated cation (HCN) channels in both TR and NRT neurons help regulate intrinsic rhythmicity. Extrinsic modulatory influences to this circuitry include inputs from both forebrain and brainstem nuclei, as well as sensory inputs from other thalamic nuclei. (From Rho, J.M. and Stafstrom, C.E., in *Pediatric Neurology: Principles and Practice*, 4th ed., Swaiman, K.F. et al., Eds., Mosby, Philadelphia, PA, 2006, p. 991. With permission.)

also appear to be involved in other types of epilepsy such as febrile seizures and temporal lobe epilepsy (Chen et al., 2001; Shin et al., 2008).

Anticonvulsants known to be clinically effective against absence seizures (e.g., ethosuximide and valproic acid) can block T-type calcium currents (Coulter et al., 1989). The relevance of HCN channels in the pathogenesis of absence seizures was underscored by the demonstration that lamotrigine, an anti-absence agent, enhanced activation of dendritic H-currents in hippocampal pyramidal neurons (Poolos et al., 2002). Furthermore, antagonists of GABA$_B$ receptors and dopaminergic agonists can also interrupt abnormal thalamocortical discharges in experimental absence epilepsy models (Snead, 1995). Finally, GABA$_B$ receptors are involved in mediating long-lasting thalamic IPSPs involved in the generation of normal thalamocortical rhythms, whereas brainstem monoaminergic projections disrupt these rhythms. Thus, absence seizures appear to result from the combined dysfunction of T-type calcium channels, HCN channels, and GABA$_B$ receptors, as well as other pathophysiological mechanisms.

## Increased Seizure Susceptibility of the Immature Brain

Seizure incidence is highest during the first decade and especially during the first year of life (Hauser, 1994). Multiple physiological factors contribute to the increased susceptibility of the developing brain to seizures (Ben-Ari and Holmes, 2006; Sanchez and Jensen, 2001; Silverstein and Jensen, 2007; Stafstrom, 2007a). Each factor alters the brain excitatory/inhibitory balance in favor of enhanced excitation and involves a multiplicity of substrates and mediators, including ion channels, neurotransmitters and their receptors, structural changes in the maturing brain, and ionic gradients. Seizure propensity in the very young involves a complex interplay between the timing of these cellular and molecular changes. In addition, excitability in the developing brain varies by brain region and cell type. Much of this information is derived from experimental epilepsy models in rodents, in which a window of heightened excitability in the second postnatal week gives rise to a lowered seizure threshold. This period of hyperexcitability in the rodent is approximately analogous to the first year of life in the human infant (Avishai-Eliner et al., 2002).

For many reasons, the excitatory/inhibitory balance in the brain changes dramatically over the course of early development. The disadvantage of these physiological adaptations is that the brain is especially vulnerable to hyperexcitability and seizure generation during a critical window of development. Nevertheless, these neurophysiological idiosyncrasies of early brain development also provide the opportunity for producing novel, age-specific therapies, some of which are undergoing development (Dzhala et al., 2008; Mazarati et al., 2008). These topics are explored in further depth by Weisenberg and Wong in Chapter 7 in this volume.

## SUMMARY

Epileptic seizures arise from a multiplicity of factors that regulate neuronal excitability and synchrony. Although much has been learned about the molecular and cellular alterations that produce or accompany seizure activity, we have yet to fully integrate these findings with our (largely phenomenological) electroclinical observations and treatment responses in experimental animal models of epilepsy, as well as in humans. In dissecting the basic neurophysiology of epilepsy, it is tempting to invoke causality on any identifiable alteration that could theoretically enhance neuronal excitation; however, investigators frequently discover that it is not easy to identify the critical mediators and pathways in the processes of ictogenesis and epileptogenesis. Ultimately, seizures are a reflection of a complex array of perturbations occurring at multiple hierarchical levels of cell structure and function and, perhaps more importantly, as yet unpredictable products of large neuronal network activity. The subsequent chapters of this volume explore many of these topics in greater detail.

## REFERENCES

Avishai-Eliner, S., K.L. Brunson, C.A. Sandman, T.Z. Baram. (2002). Stressed-out, or in (utero)? *Trends Neurosci.*, 25:518–524.

Ayala, G.F. (1983). The paroxysmal depolarizing shift. *Prog. Clin. Biol. Res.*, 124:15–21.

Ben-Ari, Y., G.L. Holmes. (2006). Effects of seizures on developmental processes in the immature brain. *Lancet Neurol.*, 5:1055–1063.

Biervert, C., B.C. Schroeder, C. Kubisch, S.F. Berkovic, P. Propping, T.J. Jentsch, O.K. Steinlein. (1998). A potassium channel mutation in neonatal human epilepsy. *Science*, 279:403–406.

Blumenfeld, H. (2003). From molecules to networks: cortical/subcortical interactions in the pathophysiology of idiopathic generalized epilepsy. *Epilepsia*, 44(Suppl. 2):7–15.

Buzsáki, G., A. Draguhn. (2004). Neuronal oscillations in cortical networks. *Science*, 304:1926–1929.

Catterall, W.A., J. Striessnig, T.P. Snutch, E. Perez-Reyes. (2003). International Union of Pharmacology. Xl. Compendium of voltage-gated ion channels: calcium channels. *Pharmacol. Rev.*, 55:579–581.

Catterall, W.A., A.L. Goldin, S.G. Waxman. (2005). International Union of Pharmacology. XLVII. Nomenclature and structure–function relationships of voltage-gated sodium channels. *Pharmacol. Rev.* 57:397-409.

Chang, B.S., D.H. Lowenstein. (2003). Epilepsy. *N. Engl. J. Med.*, 349:1257–1266.

Chen, K., I. Aradi, N. Thon, M. Eghbal-Ahmadi, T.Z. Baram, I. Soltesz. (2001). Persistently modified H-channels after complex febrile seizures convert the seizure-induced enhancement of inhibition to hyperexcitability. *Nat. Med.*, 7:331–337.

Clark, S., W.A. Wilson. (1999). Mechanisms of epileptogenesis. *Adv. Neurol.*, 79:607–630.

Collingridge, G.L., C.E. Herron, R.A. Lester. (1988). Synaptic activation of N-methyl-D-aspartate receptors in the Schaffer collateral–commissural pathway of rat hippocampus. *J. Physiol.*, 399:283–300.

Conn, P.J. (2003). Physiological roles and therapeutic potential of metabotropic glutamate receptors. *Ann. N.Y. Acad. Sci.*, 1003:12–21.

Coulter, D.A., J.R. Huguenard, D.A. Prince. (1989). Specific petit mal anticonvulsants reduce calcium currents in thalamic neurons. *Neurosci. Lett.*, 98:74–78.

Crill, W.E. (1996). Persistent sodium current in mammalian central neurons. *Annu. Rev. Physiol.*, 58:349–362.

Dichter, M.A. (2009). Emerging concepts in the pathogenesis of epilepsy and epileptogenesis. *Arch. Neurol.*, 66:443–447.

Dingledine, R., K. Borges, D. Bowie, S.F. Traynelis. (1999). The glutamate receptor ion channels. *Pharmacol. Rev.*, 51:7–61.

Dudek, F.E., T.P. Sutula. (2007). Epileptogenesis in the dentate gyrus: a critical perspective. *Prog. Brain Res.*, 163:755–773.

Dudek, F.E., T. Yasumura, J.E. Rash. (1998). 'Non-synaptic' mechanisms in seizures and epileptogenesis. *Cell Biol. Int.*, 22:793–805.

Dzhala, V.I., A.C. Brumback, K.J. Staley. (2008). Bumetanide enhances phenobarbital efficacy in a neonatal seizure model. *Ann. Neurol.*, 63:222–235.

Engel, J., Jr., A. Bragin, R. Staba, I. Mody. (2009). High-frequency oscillations: what is normal and what is not? *Epilepsia*, 50:598–604.

Faingold, C.L. (2004). Emergent properties of CNS neuronal networks as targets for pharmacology: application to anticonvulsant drug action. *Prog. Neurobiol.*, 72:55–85.

Fröhlich, F., M. Bazhenov, V. Iragui-Madoz, T.J. Sejnowski. (2008). Potassium dynamics in the epileptic cortex: new insights on an old topic. *Neuroscientist*, 14:422–433.

Gean, P.W. (1990). The epileptiform activity induced by 4-aminopyridine in rat amygdala slices: antagonism by non-N-methyl-D-aspartate receptor antagonists. *Brain Res.*, 530:251–256.

George, A.L., Jr. (2005). Inherited disorders of voltage-gated sodium channels. *J. Clin. Invest.*, 115:1990–1999.

Gospe, S.M., Jr. (2006). Pyridoxine-dependent seizures: New genetic and biochemical clues to help with diagnosis and treatment. *Curr. Opin. Neurol.*, 19:148–153.

Gutman, G.A., K.G. Chandy, S. Grissmer, M. Lazdunski, D. McKinnon, L.A. Pardo, G.A. Robertson, B. Rudy, M.C. Sanguinetti, W. Stuhmer, X. Wang. (2005). International Union of Pharmacology. LIII. Nomenclature and molecular relationships of voltage-gated potassium channels. *Pharmacol. Rev.*, 57:473–508.

Hauser, W.A. (1994). The prevalence and incidence of convulsive disorders in children. *Epilepsia*, 35(Suppl. 2):S1–S6.

Helbig, I., I.E. Scheffer, J.C. Mulley, S.F. Berkovic. (2008). Navigating the channels and beyond: unravelling the genetics of the epilepsies. *Lancet Neurol.*, 7:231-245.

Herrero, A.I., N. Del Olmo, J.R. Gonzalez-Escalada, J.M. Solis. (2002). Two new actions of topiramate: inhibition of depolarizing GABA(A)-mediated responses and activation of a potassium conductance. *Neuropharmacology*, 42:210–220.

Hille, B. (2001). *Ion Channels of Excitable Membranes*, 3rd ed. Sunderland, MA: Sinauer Associates.

Hirose, S. (2006). A new paradigm of channelopathy in epilepsy syndromes: intracellular trafficking abnormality of channel molecules. *Epilepsy Res.*, 70(Suppl. 1):S206–S217.

Huguenard, J.R., D.A. McCormick. (2007). Thalamic synchrony and dynamic regulation of global forebrain oscillations. *Trends Neurosci.*, 30:350–356.

Jefferys, J.G., R.D. Traub, M.A. Whittington. (1996). Neuronal networks for induced '40 Hz' rhythms. *Trends Neurosci.*, 19:202–208.

Johnston, D., T.H. Brown. (1984). The synaptic nature of the paroxysmal depolarizing shift in hippocampal neurons. *Ann. Neurol.*, 16(Suppl.):S65–S71.

Jutila, L., A. Immonen, K. Partanen, J. Partanen, E. Mervaala, A. Ylinen, I. Alafuzoff, L. Paljarvi, K. Karkola, M. Vapalahti, A. Pitkanen. (2002). Neurobiology of epileptogenesis in the temporal lobe. *Adv. Tech. Stand. Neurosurg.*, 27:5–22.

Kalia, L.V., S.K. Kalia, M.W. Salter. (2008). NMDA receptors in clinical neurology: excitatory times ahead. *Lancet Neurol.*, 7:742–755.

Lado, F.A., S.L. Moshe. (2008). How do seizures stop? *Epilepsia*, 49:1651–1664.

Lawrence, J.J., C.J. McBain. (2003). Interneuron diversity series: containing the detonation–feedforward inhibition in the CA3 hippocampus. *Trends Neurosci.*, 26:631–640.

Macdonald, R.L., J.Q. Kang. (2009). Molecular pathology of genetic epilepsies associated with GABA(A) receptor subunit mutations. *Epilepsy Curr.*, 9:18–23.

Madeja, M., D.G. Margineanu, A. Gorji, E. Siep, P. Boerrigter, H. Klitgaard, E.J. Speckmann. (2003). Reduction of voltage-operated potassium currents by levetiracetam: a novel antiepileptic mechanism of action? *Neuropharmacology*, 45:661–671.

Mann, E.O., I. Mody. (2008). The multifaceted role of inhibition in epilepsy: seizure-genesis through excessive GABAergic inhibition in autosomal dominant nocturnal frontal lobe epilepsy. *Curr. Opin. Neurol.*, 21:155–160.

Martin, D.L., R.W. Olsen. (2000). *GABA in the Nervous System: The View at Fifty Years.* Philadelphia, PA: Lippincott Williams & Wilkins.

Mazarati, A., J. Wu, D. Shin, Y.S. Kwon, R. Sankar. (2008). Antiepileptogenic and antiictogenic effects of retigabine under conditions of rapid kindling: an ontogenic study. *Epilepsia*, 49:1777–1786.

McCormick, D.A., D. Contreras. (2001). On the cellular and network bases of epileptic seizures. *Annu. Rev. Physiol.*, 63:815–846.

Miceli, F., M.V. Soldovieri, M. Martire, M. Taglialatela. (2008). Molecular pharmacology and therapeutic potential of neuronal Kv7-modulating drugs. *Curr. Opin. Pharmacol.*, 8:65–74.

Nadler, J.V. (2003). The recurrent mossy fiber pathway of the epileptic brain. *Neurochem. Res.*, 28:1649–1658.

Nemani, V.M., D.K. Binder. (2005). Emerging role of gap junctions in epilepsy. *Histol. Histopathol.*, 20:253–259.

Olsen, R.W., W. Sieghart. (2008). International Union of Pharmacology. LXX. Subtypes of gamma-aminobutyric acid(A) receptors: classification on the basis of subunit composition, pharmacology, and function. Update. *Pharmacol. Rev.*, 60:243–260.

Perez-Reyes, E. (2003). Molecular physiology of low-voltage-activated T-type calcium channels. *Physiol. Rev.*, 83:117–161.

Poolos, N.P. (2004). The yin and yang of the H-channel and its role in epilepsy. *Epilepsy Curr.*, 4:3–6.

Poolos, N.P., M. Migliore, D. Johnston. (2002). Pharmacological upregulation of H-channels reduces the excitability of pyramidal neuron dendrites. *Nat. Neurosci.*, 5:767–774.

Reid, C.A., S.F. Berkovic, S. Petrou. (2009). Mechanisms of human inherited epilepsies. *Prog. Neurobiol.*, 87:41–57.

Rho, J.M., C.E. Stafstrom. (2006). Neurophysiology of epilepsy. In *Pediatric Neurology: Principles and Practice*, 4th ed. (pp. 991–1007), K.F. Swaiman, S. Ashwal, and D.M. Ferriero, Eds. Philadelphia, PA: Mosby.

Rivera, C., J. Voipio, J.A. Payne, E. Ruusuvuori, H. Lahtinen, K. Lamsa, U. Pirvola, M. Saarma, K. Kaila. (1999). The K$^+$/Cl$^-$ co-transporter KCC2 renders GABA hyperpolarizing during neuronal maturation. *Nature*, 397:251–255.

Robinson, R.B., S.A. Siegelbaum. (2003). Hyperpolarization-activated cation currents: from molecules to physiological function. *Annu. Rev. Physiol.*, 65:453–480.

Rogawski, M.A. (2000). KCNQ2/KCNQ3 K$^+$ channels and the molecular pathogenesis of epilepsy: implications for therapy. *Trends Neurosci.*, 23:393–398.

Rogawski, M.A., C.W. Bazil. (2008). New molecular targets for antiepileptic drugs: alpha(2)delta, SV2A, and K(v)7/KCNQ/M potassium channels. *Curr. Neurol. Neurosci. Rep.*, 8:345–352.

Rogawski, M.A., W. Löscher. (2004). The neurobiology of antiepileptic drugs. *Nat. Rev. Neurosci.*, 5:553–564.

Rothstein, J.D., L. Martin, A.I. Levey, M. Dykes-Hoberg, L. Jin, D. Wu, N. Nash, R.W. Kuncl. (1994). Localization of neuronal and glial glutamate transporters. *Neuron*, 13:713–725.

Sanchez, R.M., F.E. Jensen. (2001). Maturational aspects of epilepsy mechanisms and consequences for the immature brain. *Epilepsia*, 42:577–585.

Schwartzkroin, P.A., A.L. Mueller. (1987). Electrophysiology of hippocampal neurons. In *Cerebral Cortex: Further Aspects of Cortical Function, Including Hippocampus* (Vol. 6, pp. 295–343), E.G. Jones and A. Peters, Eds. New York: Plenum Press.

Schwartzkroin, P.A., S.C. Baraban, D.W. Hochman. (1998). Osmolarity, ionic flux, and changes in brain excitability. *Epilepsy Res.*, 32:275–285.

Shin, M., D. Brager, T.C. Jaramillo, D. Johnston, D.M. Chetkovich. (2008). Mislocalization of H-channel subunits underlies H channelopathy in temporal lobe epilepsy. *Neurobiol. Dis.*, 32:26–36.

Sierra-Paredes, G., G. Sierra-Marcuño. (2007). Extrasynaptic GABA and glutamate receptors in epilepsy. *CNS Neurol. Disord. Drug Targets*, 6:288–300.

Silverstein, F.S., F.E. Jensen. (2007). Neonatal seizures. *Ann. Neurol.*, 62:112–120.

Simeone, T.A., S.D. Donevan, J.M. Rho. (2003). Molecular biology and ontogeny of gamma-aminobutyric acid (GABA) receptors in the mammalian central nervous system. *J. Child Neurol.*, 18:39–49.

Simeone, T.A., R.M. Sanchez, J.M. Rho. (2004). Molecular biology and ontogeny of glutamate receptors in the mammalian central nervous system. *J. Child Neurol.*, 19:343–361.

Sloviter, R.S. (2008). Hippocampal epileptogenesis in animal models of mesial temporal lobe epilepsy with hippocampal sclerosis: the importance of the "latent period" and other concepts. *Epilepsia*, 49(Suppl. 9):85–92.

Snead, O.C., 3rd. (1995). Basic mechanisms of generalized absence seizures. *Ann. Neurol.*, 37:146–157.

Stafstrom, C.E. (2004). An introduction to seizures and epilepsy: cellular mechanisms underlying classification and treatment. In *Epilepsy and the Ketogenic Diet* (pp. 3–30), C.E. Stafstrom and J.M. Rho, Eds. Totowa, NJ: Humana Press.

Stafstrom, C.E. (2005). The role of the subiculum in epilepsy and epileptogenesis. *Epilepsy Curr.*, 5:121–129.

Stafstrom, C.E. (2007a). Neurobiological mechanisms of developmental epilepsy: translating experimental findings into clinical application. *Semin. Pediatr. Neurol.*, 14:164–172.

Stafstrom, C.E. (2007b). Persistent sodium current and its role in epilepsy. *Epilepsy Curr.*, 7:15–22.

Staley, K.J. (2006). Wrong-way chloride transport: is it a treatable cause of some intractable seizures? *Epilepsy Curr.*, 6:124–127.

Steriade, M. (2004). Neocortical cell classes are flexible entities. *Nat. Rev. Neurosci.*, 5:121–134.

Sutula, T. (2002). Seizure-induced axonal sprouting: assessing connections between injury, local circuits, and epileptogenesis. *Epilepsy Curr.*, 2:86–91.

Tanaka, K., K. Watase, T. Manabe, K. Yamada, M. Watanabe, K. Takahashi, H. Iwama, T. Nishikawa, N. Ichihara, T. Kikuchi, S. Okuyama, N. Kawashima, S. Hori, M. Takimoto, K. Wada. (1997). Epilepsy and exacerbation of brain injury in mice lacking the glutamate transporter GLT-1. *Science*, 276:1699–1702.

Tian, G.F., H. Azmi, T. Takano, Q. Xu, W. Peng, J. Lin, N. Oberheim, N. Lou, X. Wang, H.R. Zielke, J. Kang, M. Nedergaard. (2005). An astrocytic basis of epilepsy. *Nat. Med.*, 11:973–981.

Traub, R.D., R. Miles. (1991). *Neuronal Networks of the Hippocampus*. New York: Cambridge University Press.

Traub, R.D., R.K. Wong, R. Miles, H. Michelson. (1991). A model of a CA3 hippocampal pyramidal neuron incorporating voltage-clamp data on intrinsic conductances. *J. Neurophysiol.*, 66:635–650.

Traub, R.D., H. Michelson-Law, A.E. Bibbig, E.H. Buhl, M.A. Whittington. (2004). Gap junctions, fast oscillations and the initiation of seizures. *Adv. Exp. Med. Biol.*, 548:110–122.

Traynelis, S.F., R. Dingledine. (1988). Potassium-induced spontaneous electrographic seizures in the rat hippocampal slice. *J. Neurophysiol.*, 59:259–276.

Ure, J., M. Baudry, M. Perassolo. (2006). Metabotropic glutamate receptors and epilepsy. *J. Neurol. Sci.*, 247:1–9.

Vezzani, A., S. Balosso, T. Ravizza. (2008). The role of cytokines in the pathophysiology of epilepsy. *Brain Behav. Immun.*, 22:797–803.

Wahl-Schott, C., M. Biel. (2009). HCN channels: structure, cellular regulation and physiological function. *Cell Mol. Life Sci.*, 66:470–494.

Wetherington, J., G. Serrano, R. Dingledine. (2008). Astrocytes in the epileptic brain. *Neuron*, 58:168–178.

Wickenden, A.D. (2002). Potassium channels as anti-epileptic drug targets. *Neuropharmacology*, 43:1055–1060.

Wong, R.K., D.A. Prince. (1978). Participation of calcium spikes during intrinsic burst firing in hippocampal neurons. *Brain Res.*, 159:385–390.

Wong, R.K., S.C. Chuang, R. Bianchi. (2004). Plasticity mechanisms underlying mGluR-induced epileptogenesis. *Adv. Exp. Med. Biol.*, 548:69–75.

Yu, F.H., M. Mantegazza, R.E. Westenbroek, C.A. Robbins, F. Kalume, K.A. Burton, W.J. Spain, G.S. McKnight, T. Scheuer, W.A. Catterall. (2006). Reduced sodium current in GABAergic interneurons in a mouse model of severe myoclonic epilepsy in infancy. *Nat. Neurosci.*, 9:1142–1149.

# 2 Blood–Brain Barrier, Blood Flow, Neoplasms, and Epilepsy: The Role of Astrocytes

*Luca Cucullo, Nicola Marchi, Vincent Fazio, Minh-Tri Nguyen, and Damir Janigro*

## CONTENTS

## INTRODUCTION

Glial cells are non-neuronal cells that play a major supportive and modulatory role in a variety of brain functions, spanning a multiplicity of functions from physical support for neurons to regulation of the brain environment. In general, glial cells help maintain the homeostasis necessary for neuronal function. Glial cells also play an important developmental role by guiding neuronal migration during early stages of maturation and by modulating the growth of axons and dendrites. Although different glial cells perform different functions, the astrocyte family is the main regulator of neurovascular interactions, the primary focus of this review.

At the brain microvascular level, astrocytes play a primary role in the differentiation of cerebrovascular endothelial cells into blood–brain barrier endothelium. Astrocytes also contribute to the modulation of mature blood–brain barrier (BBB) function and activity. At the level of arterioles and venules, astrocytes regulate vascular tone by the release of vasoactive substances, thus contributing to cerebral blood flow regulation. In general, it is reasonable to assume that a healthy astrocyte is a reflection of a properly functioning brain. Conversely, the phenomenon of *reactive gliosis* is a sign of pathophysiological derangements that may revert or progress further into neurological disorders.

Consistent with their role as regulators of brain homeostasis, glial cells also play a major modulatory role in inflammation by releasing proinflammatory cytokines that act as chemotactic agents for peripheral immune cells. Glia also facilitate leukocyte extravasation into the perivascular space and even infiltrate into the brain parenchyma. Whereas mature glia undergo mitosis reluctantly, misguided proliferation of glia can lead to the development of devastating brain tumors such as high-grade gliomas.

In this chapter, we cover various aspects of glial cell functions, starting from their role in the development and regulation of the BBB. We also describe their involvement in the regulation and maintenance of brain homeostasis, with particular focus on drug resistance, drug metabolism, and inflammation. Finally, we discuss the pathogenesis of brain tumors originating from glial cells and some recent views on the correlations among electrical excitability, cell proliferation, and drug resistance that can be exploited for the development of novel therapeutic strategies.

## ASTROCYTES AND THE BLOOD–BRAIN BARRIER

Glial cells are numerically the predominant cell type in the brain, and the glial/neuron ratio increases dramatically with brain complexity and size (Oberheim et al., 2006). Astrocytes, a specific subtype of glial cell, play an important role in regulating cerebral ionic homeostasis (Allen and Barres, 2009) and transmitter regulation (Newman, 2003). They also contribute to the maintenance of the BBB (Ballabh et al., 2004) and provide structural as well as metabolic support of neuronal cells—for example, by providing the glucose–lactate shuttle (Magistretti, 2006). At the vascular level, astrocytes extend larger processes, known as *endfeet*, whose terminations cover 99% of the abluminal vascular surface of capillaries, arterioles, and venules present in the cerebrovascular network (Simard et al., 2003). At the brain microcapillary level, these cells become one of the main building blocks of the BBB, a highly specialized dynamic and functional interface between the blood and the brain that plays a critical role in controlling and modulating the homeostasis of the central nervous system (CNS). The BBB greatly restricts the permeability of the paracellular pathways and regulates the passage of bloodborne substrates (e.g., ions, nutrients, amino acids) between the blood and the brain. The BBB also provides a defensive line against the passage of potentially harmful xenobiotic substances and modulates the immune response both at vascular and perivascular levels (see Figure 2.1).

The differentiation of vascular endothelial cells into a BBB phenotype and the induction of BBB properties are heavily dependent upon their association with perivascular glial cells (Prat et al., 2001; Utsumi et al., 2000) or astrocytes. These glia regulate protein expression, modulate endothelium differentiation, and are critical for the induction and maintenance of tight junctions and BBB properties (Grant and Janigro, 2004; Zlokovic, 2008). In fact, when BBB endothelial cells are isolated from their original environment and cultured *in vitro*, they start dedifferentiating into a phenotype similar to that of peripheral vascular endothelial cells. This occurs despite the fact that the genetic characteristics of the original phenotype are maintained (Reinhardt and Gloor, 1997). Conversely, when non-BBB endothelial cells are cocultured under exposure to intraluminal flow and in the presence of abluminal glia, they start developing BBB properties. This is clearly demonstrated by the development of higher transendothelial electrical resistance (Cucullo et al., 2002), lower permeability to paracellular markers (Stanness et al., 1997), the polarized expression of transporters such as glucose transporter type 1 (GLUT-1) (McAllister et al., 2001), amino acid transporters (Grant et al., 2003; Parkinson et al., 1998), and functional tight junctions (Stanness et al., 1997), which form a diffusion barrier that selectively excludes most bloodborne and xenobiotic substances from entering the brain. Other markers typically related to the formation of a functional BBB, including gamma-glutamyl transpeptidase ($\gamma$GTP), transferrin receptor, and P-glycoprotein, are also upregulated in endothelial cells when cocultured with astrocytes (Virgintino et al., 1998).

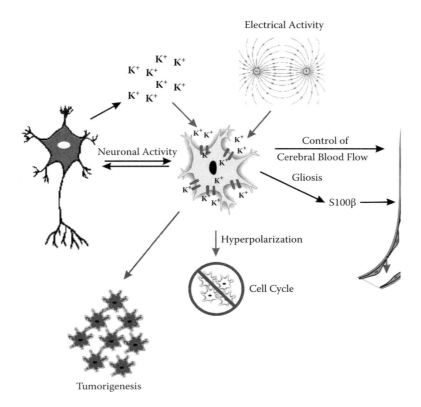

**FIGURE 2.1**    Astrocytes are essential players in normal and abnormal brain. On the one hand, they protect against insults, control synaptic transmission and perhaps blood flow, and also contribute to BBB maintenance. On the other hand, astrocytes also give rise, albeit reluctantly, to potentially deadly neoplasms.

Another important element of the BBB is the basal lamina, or extracellular matrix (ECM), whose functions range from mechanical support for cell adhesion and transmembrane migration of leukocytes (Zlokovic, 2008) to the regulation of cellular modulation across the BBB (Wolburg et al., 2009). The basal lamina is generated and maintained by perivascular astrocytes in cooperation with the BBB endothelium (del Zoppo and Milner, 2006; Wolberg et al., 2009).

## DO ASTROCYTES REGULATE CEREBRAL BLOOD FLOW?

Functional hyperemia, which is the active process of vasodilation in response to the rising metabolic demands of increased neuronal activity, is a critically important element that is fundamental to normal brain function (Kuschinsky, 1997). Recent studies have proposed that astrocytes play a dynamic role at the neurovascular level (Volterra and Meldolesi, 2005); the anatomical correlates of this important function are the strategically located contact with the neuronal synapses on one end and a link to the vascular network by means of the astrocytic endfeet (Ransom et al., 2003; Volman et al., 2007).

The primary role of astrocytes as neurovascular elements is critical to coupling the metabolic demands of neurons (Magistretti, 2006) to cerebral blood vessel responses where variations in neuronal activity trigger localized changes in blood flow (Haydon and Carmignoto, 2006; Takano et al., 2006; Zonta et al., 2003). In this process, the release of glutamate through the stimulation of metabotropic glutamate receptors (mGluRs) raises intracellular $Ca^{2+}$ levels in astrocytic endfeet and leads to changes in the vascular tone of neighboring intracerebral arterioles. For example, the increase in $Ca^{2+}$ in astrocyte endfeet promotes the opening of astrocytic $K_{Ca}$ channels and the release of $K^+$, which could then act on vascular inward rectifier potassium ($K_{ir}$) channels to produce vasodilation; however, the vascular response observed in conjunction with an elevation of intracellular $Ca^{2+}$ levels in astrocytic endfeet is equivocal and can lead to vasodilation or vasoconstriction (Gordon et al., 2007). A similar bell-shaped dose response is seen when potassium is directly applied to the vessel (Faraci and Heistad, 1993; Janigro et al., 1996; Nguyen et al., 2000). In the case of $K_{Ca}$ channels, the elevation of intracellular levels of $Ca^{2+}$ leads to the production of arachidonic acid (AA) by $Ca^{2+}$-sensitive phospholipase $A_2$ ($PLA_2$). The conversion of AA to 20-hydroxyeicosatetraenoic acid (20-HETE) leads to vasoconstriction, and the conversion to epoxyeicosatrienoic acid (EET) or prostaglandin $E_2$ ($PGE_2$) leads to vasorelaxation. Other factors that play an important role in determining the propensity toward one or the other pathway are nitric oxide (NO) and the release of inhibitory enzymes of the arachidonic acid pathway such as cytochrome P450 (CYP) ω-hydroxylase, CYP epoxygenase, and cyclooxygenase-1 (Blanco et al., 2008; Koehler et al., 2009). However, additional astrocytic mechanisms that are not dependent on astrocytic $Ca^{2+}$ signaling such as those involved in the energy-metabolism-dependent signals (e.g., glutamate transport) are also likely involved (Haydon and Carmignoto, 2006). All of these data strongly suggest a primary role for astrocytes in the regulation of cerebral blood flow during neuronal activation; however, additional studies are required to fully elucidate the number of complex mechanisms that regulate this critical process.

A word of caution must be made regarding recent studies utilizing *in vivo* and *in vitro* preparations to study the regulation of cerebral blood flow by glia (recently reviewed elsewhere, but see Haydon and Carmignoto, 2006; Iadecola and Nedergaard, 2007). Some of the studies have shown changes in vascular diameter that are consistent with post-arterial (perhaps capillary?) structures. Although these changes in diameter (and in intraluminal flow) may account for astrocyte-mediated regulation of flow, it is also possible that under conditions of electrical stimulation such as those used to evoke the release of vasoactive mediators by glia, ionic shifts cause osmotic changes that translate to changes in vascular shape. This phenomenon, if reproduced, may hold significant relevance in pathologies where intracranial pressure or altered extracellular space ion composition causes reduced perfusion and venous return, contributing to ischemic-like pathology.

## REGULATION OF CEREBRAL BLOOD FLOW AND BLOOD–BRAIN BARRIER INTEGRITY

From the studies noted above, it is clear that both astrocytes and neurons may release vasoactive substances in response to electrical activity. The mediators involved are all products of synaptic transmission and include ions (protons and potassium; see Table 2.1), metabolic intermediates (adenosine), and neurotransmitters (glutamate). It is important to remember that these signaling molecules are functional only under optimal conditions (i.e., when their levels are regulated to permit subsequent signaling). In many ways, these conditions resemble the "refractory period" that follows the generation of an action potential. If, as in many neurological disorders, the BBB is breached (Grant and Janigro, 2004; Krizanac-Bengez et al., 2004; Oby and Janigro, 2006; Zlokovic, 2008), the reestablishment of a baseline level becomes problematic. For example, the concentration of potassium will increase and the levels of adenosine will decrease after blood–brain barrier disruption. Thus, further release of potassium by neurons may result in a vasoconstriction, which is the opposite of what the system is supposed to do under physiological conditions.

These considerations lead to the important and recently therapeutically exploited strategy of BBB repair to treat neurological diseases (Granata et al., 2008). The rationale for these studies, and similar rodent model experiments (Fabene et al., 2007, 2008), was amelioration of BBB function, but the fact that endothelial repair will also improve brain perfusion must be taken into account. By contrast, procedures and therapies used to improve cerebral blood flow may positively impact BBB function by restoring metabolic support to glia and endothelial cells.

**TABLE 2.1**
**Ionic Concentrations and Osmolarity in Serum and Brain**

|  | Brain | Serum | Ratio |
|---|---|---|---|
| Total osmolarity (mOsm/L) | 295 | 295 | 1 |
| Water (%) | 99 | 93 | 1.06 |
| Glucose (mg/dL) | 60 | 90 | 0.66 |
| Na (mEq/L) | 138 | 138 | 1 |
| K (mEq/L) | 2.8 | 4.5 | 0.62 |
| Ca (mEq/L) | 2.1 | 4.5 | 0.4 |
| Mg (mEq/L) | 0.3 | 1.7 | 0.17 |
| Cl (mEq/L) | 119 | 102 | 1.16 |
| pH | 7.33 | 7.41 | 0.98 |

*Note:* See text for details.

## GLIAL CELLS, S100β, AND PERIPHERAL MARKERS OF BLOOD–BRAIN BARRIER DISRUPTION

Although it is uncertain whether an intact BBB is important for the development of neurological diseases, expanding evidence implicates a modulatory role in seizure genesis, multiple sclerosis, and Alzheimer's disease (Engelhardt and Ransohoff, 2005; Man et al., 2007; Marchi et al., 2009; Oby and Janigro, 2006; Vezzani and Granata, 2005; Zlokovic, 2008). One of the problems in BBB research has been the lack of reliable methods to measure BBB integrity (Marchi et al., 2003a, 2004a). Opening of the BBB allows molecules normally present in blood open passage into the CNS; however, this opening, unless it involves specific transporters, works bidirectionally. Proteins normally present in blood are free to diffuse into the CNS; in turn, proteins ordinarily present in high concentrations in the CNS are free to diffuse down concentration gradients into the blood. These peripheral substrates can be detected in the blood to evaluate the permeability characteristics of the BBB at any given time. In recently published review articles, Marchi et al. (2003b, 2004a) discussed the ideal properties of a peripheral marker of BBB disruption. Such proteins should have low or undetectable plasma levels in normal subjects, be normally present in cerebrospinal fluid (CSF), and have a higher normal concentration in the CSF than in plasma. Additionally, the CSF concentration of the protein should increase in response to insult or injury. The protein should be normally blocked by the BBB and exhibit flux across the BBB during barrier disruption. Several proteins, including S100β, neuron-specific enolase (NSE), and glial fibrillary acidic protein (GFAP), have been evaluated for this purpose, but only S100β meets the characteristics of having very low plasma levels with a concentration less than that found in the CSF of normal subjects (see Figure 2.2).

Glial cells secrete a number of important factors for brain development (e.g., migration of neurons), establishing the BBB, and, in the mature brain, ensuring the correct function of the brain cells. Among these, the calcium-binding protein S100β plays an important role. S100β is primarily synthesized by the endfeet process of the astrocytes and plays a fundamental role in the trophism of neurons (Nishiyama et al., 2002; Selinfreund et al., 1991). Historically, increases in S100β levels in serum and CSF have been associated with neuronal damage, thus conferring to S100β the clinical relevance of a marker of neuronal dysfunction. However, because neurological diseases are often accompanied by increased BBB permeability, the markers thought to indicate neuronal damage may in fact indicate BBB defects (Marchi et al., 2003a,b; 2004a). Markers of brain damage include NSE and GFAP. In normal subjects, NSE is more concentrated in brain, and S100β is primarily present in CNS fluids. BBB damage in the absence of neuronal damage would be expected to markedly

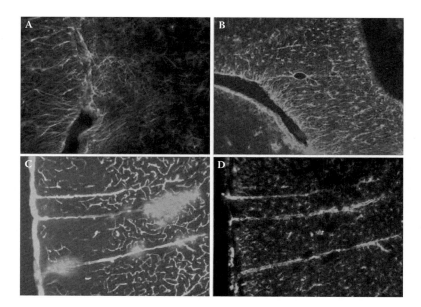

**FIGURE 2.2    (See color insert following page 458.)** Reactive astrocytosis in an animal model of inflammation and seizures: (A, B) treatment with lithium chloride; (C, D) treatment with pilocarpine. Note the enhanced GFAP staining (A, B, D) in specific regions and the leakage of a protein marker (C) in the proximity of gliosis. (From Marchi, N. et al., *Epilepsia*, 48(10), 1934–1946, 2007; Marchi, N. et al., *Neurobiol. Dis.*, 33, 171–271, 2009. With permission.)

increase serum S100β levels while leaving NSE levels unchanged. The fact that serum S100β can be used as a marker of BBB integrity is not necessarily at odds with the notion that S100β is also a marker of brain damage, as both phenomena (BBB failure and brain damage) are temporally and topographically associated (Marchi et al., 2004a).

The plasma levels of S100β are normally a third of those found in the CSF and are nearly undetectable (Janigro et al., 2002). Several diseases cause an elevation in plasma levels of S100β, which can be detected and used for both diagnostic and prognostic purposes, as well as for evaluation of disease progression. Plasma S100β levels increase in cerebral ischemia, with peak levels occurring approximately 3 days after infarction (Buttner et al., 1997; Jonsson et al., 2001). These levels have served as a useful marker of both infarct size and long-term clinical outcome (Dassan et al., 2009; Petzold et al., 2008). Traumatic brain injury has also been shown to increase S100β levels in plasma (Biberthaler et al., 2001a,b), with a positive correlation between the extent of damage after head injury and elevation in plasma S100β.

## ROLE OF GLIAL CELLS IN THE DISEASED BRAIN

Over the years, the involvement of glial cells in the pathogenesis and sustenance of CNS diseases has been increasingly demonstrated (De Simoni and Imeri, 1998; De Simoni et al., 2002; Takano et al., 2009; Vezzani et al., 2008a,b). In particular, stroke and epilepsy researchers have focused on the significance of proinflammatory mediators released by glial cells and the possibility of targeting these molecules for therapeutic purposes. In addition to the production of active mediators, glial cells are also involved in chronic changes, either congenital or acquired, including alteration of drug transporters, cytochrome P450 protein expression, and alteration of cell cycle function (Marchi et al., 2004b, 2006; Marroni et al., 2003; Oby et al., 2006).

## GLIAL CELLS AND CYTOKINES

In healthy brain, astrocytes control glutamate levels and ion and water homeostasis, release neu-rotrophic factors, and are a fundamental anatomical–physiological part of the blood–brain barrier (Iadecola and Nedergaard, 2007; Nedergaard et al., 2002). During pathological events, not only are these physiological functions perturbed but reactive phenomena can also take place (Takano et al., 2009). The release of proinflammatory molecules and overexpression of specific inflamma-tion-related receptors are hallmarks of the diseased brain. Whether this reaction is beneficial or detrimental is still a matter of debate. Controversial results have been obtained when modulating proinflammatory responses during CNS pathological events such as ischemic stroke or epileptic seizures (De Simoni and Imeri, 1998; De Simoni et al., 2002; Deng and Poretz, 2003). In addition, our knowledge of how ischemic episodes or seizures affect physiological astrocytic functions (e.g., astrocytic glutamate uptake) is incomplete (Nedergaard et al., 2002).

Glial cells are exquisitely sensitive to neurotransmitters. Astrocytes express neurotransmitter receptors and respond to neuronal activity by increasing cytosolic $Ca^{2+}$ levels according to two distinct modalities—oscillations and propagation (Nedergaard et al., 2002; Newman, 2003). Glial cells are also capable of releasing adenosine triphosphate (ATP) (Rossi et al., 2007). Several modali-ties of ATP release have been proposed, including channel-mediated release, exocytosis, and P2X7-mediated receptors (Bernardino et al., 2008). Glutamate and ATP/adenosine were demonstrated to be key mediators of astrocytic–neuronal communication. Under normal conditions, ATP released by the astrocytes is converted in the extracellular space by specific nucleotidases into adenosine. Adenosine acts as a neurotransmitter with inhibitory effects on neuronal activity. During a local injury, excessive astrocytic release of ATP activates microglial cells through P2Y12 and P2Y6 recep-tors. This event leads to the enhancement of phagocytic activity and cytokine release, setting the initial stage of the inflammatory response (Bernardino et al., 2008; Rossi et al., 2007). Adenosine is also a powerful vasodilator (Morii et al., 1986, 1987; Ngai and Winn, 1993).

Recent evidence obtained in animal models of epilepsy and by analysis of surgically resected human epileptic specimens from the temporal lobe has shown that glial cells can reactively produce proinflammatory cytokines in association with ongoing seizure activity. Patients with hippocampal sclerosis display astrocytic activation of the IL-1β system (Vezzani et al., 2008a). Even though the production of proinflammatory molecules has been demonstrated, it is unclear whether blockade of this mechanism could be beneficial in reducing seizure burden, especially considering the etiologi-cal and pathophysiological variability inherent among the various epilepsies and even seizure types. Interestingly, acute seizures are prevented by IL-1β antagonists (Marchi et al., 2009), but this effect may be due to intravascular events without involvement of glia. It seems possible, therefore, that acute events are due to the release of IL-1β by white blood cells (or the BBB itself), and chronic conditions characterized by reactive gliosis may involve astrocytes.

## GLIAL CELLS, MULTIDRUG TRANSPORTER PROTEINS, AND METABOLIC ENZYMES

Blood–brain barrier expression of multidrug transporter proteins has been extensively studied. In particular, the BBBs of brain tumors and pharmacoresistant epileptic tissues display elevated expres-sion of multidrug-resistant protein 1 (MDR1) (Löscher, 2007; Marchi et al., 2004b, 2006; Marroni et al., 2003). MDR1 was demonstrated to be expressed in astrocytes in brain slices obtained from medically refractory epileptic brain tissues. The question remains whether seizure activity induces MDR expression or MDR is an etiologic factor in epileptogenesis (Marroni et al., 2003). It appears that markers previously associated with chemoresistance of tumor cells are present in epileptic brain, raising the interesting possibility that some overlap exists among tumorigenesis, MDR, and epilepsy. Histopathologically, the most frequent lesions found in drug-resistant patients with epilepsy include gangliogliomas, and glioneuronal malformations (e.g., hamartias or hamartomas), with a significant overlap between markers of tumorigenesis and epileptogenesis being observed (Marchi et al., 2004b; Marroni et al., 2003).

It was recently proposed that in drug-resistant epilepsy, cellular alterations associated with neoplasms may be present. These include loss of functional expression of p53 in glial cells; thus, in epileptic brain astrocytes, loss of p53 occurs in those cells overexpressing MDR1 proteins (Marchi et al., 2004b; Marroni et al., 2003). In addition to MDR1, epileptic glia express other MDR proteins, including leucine-responsive regulatory protein (Lrp) and multidrug resistance-associated protein 1 (MRP1); these are normally found in tumor cells, suggesting a possible link between drug-resistant epilepsy and low-grade tumors. This evidence suggests that "epileptic" glial cells may have gained a distinct survival advantage. This concept is also supported by evidence showing that blockade of MDR1 function was associated with enhanced drug-induced cytotoxicity. A positive correlation between neuronal and astrocytic expression of MDR1 and a lack of nuclear condensation, a marker of apoptosis and irreversible cell damage, was also observed (Marchi et al., 2004b; Marroni et al., 2003). These studies support the hypothesis that expression of MDR1 in glial cells may protect against toxic xenobiotics or against endogenous compounds that enter the brain under pathological conditions.

Of interest is the relationship existing between MDR1 and P450 metabolic enzymes (CYPs) (Pal and Mitra, 2006). MDR1 and CYPs are under the control of the pregnane X receptor (PXR), a nuclear receptor family regulating a number of enzymes and transporters in mammals. Generally, MDR1, MRPs, and CYPs together constitute a highly efficient barrier for many drugs (Meyer et al., 2007; Pal and Mitra, 2006). CYP glial expression could reinforce the cellular protective effect of MDR1 by metabolizing possible toxins. CYP could also regulate the bioavailability of drugs reaching the brain in a fashion similar to organs such as the gut and liver. CNS expression of CYP1A1, CYP1B1, epoxide hydrolase, and UDP–glucuronosyltransferase was confirmed using rodents models and in human brain tissue (Ghersi-Egea and Strazielle, 2001; Ghersi-Egea et al., 1995, 2001); however, the native pattern of CYP expression in the diseased brain remains elusive.

## ELECTRICAL EXCITABILITY HAMPERS GLIOMA GROWTH: SEIZURE AND TUMOR PROLIFERATION

As discussed throughout this chapter, electrical activity exerts profound effects on astrocytic function. In fact, at this stage of modern neuroscience research, it is fair to say that astrocytes are influenced by neuronal activity as much as neighboring neurons. The question that arises is whether effects of electrical field potentials beyond those listed above exist. We noted that cells that sporadically give rise to neoplasms are frequently electrically excitable (cardiac or skeletal muscle, neurons) or surrounded by active cells (pericardium, glia) (Jemal et al., 2004). This has led to investigations into the broad changes in glial cell physiology during or after electrical stimulation. Although we are still far from obtaining a clear picture of the complex mechanisms regulating tumor proliferation, several lines of evidence suggest that ion channels are involved in cancer progression and pathology, including migration and survival (Abdul and Hoosein, 2002; Arcangeli et al., 2009; Fiske et al., 2006; Huang et al., 2009; Le Guennec et al., 2007). Environmental or physiological stimuli that may alter the resting potential of these cells and affect the electrochemical gradient of ions across the cellular membrane are also likely to affect their proliferative state. Correlative *in vivo* data, for example, suggest that conditions of altered electrical activity of the brain, such as during epileptic seizures, may be sufficient to affect the proliferative status of tumor glial cells (Luyken et al., 2003; Majores et al., 2008; Schramm et al., 2004).

The most common electrical event indicating exaggerated brain electrical activity is a seizure. Epileptic seizures are characterized by an abnormal excessive or synchronous neuronal activity in the brain (Fisher et al., 2005). If the hypothesis linking electrical stimulation to proliferation is correct, one may expect that epileptic seizures counter tumor growth. Clinical evidence has shown that long-term chronic epilepsy that precedes the formation of gliomas significantly decreases mortality (Luyken et al., 2003), suggesting that synchronous, oscillatory, and periodic abnormal electrical activity is not permissive for cell proliferation (Schramm et al., 2004). In addition, long-term epilepsy-associated tumor (LEAT) patients bearing a grade II astrocytoma have a better survival rate than non-LEAT patients; a lower rate of tumor recurrence has also been shown in these patients (Schramm et al., 2004). These findings strongly suggest that the low incidence of tumors or other neoplastic disorders in excitable tissue can be associated with exposure to synchronous, oscillatory, and periodic abnormal electrical activity that may not allow cell proliferation.

A recent report has shown that, at least *in vitro*, low-frequency (comparable to that of seizure activity *in vivo*) and very low-intensity electrical stimulation decreases the proliferative activity of rat glioma and human prostate cancer cells (Cucullo et al., 2005); a mechanism involving potassium channels was shown to be involved. This study provided additional evidence that electrical field potentials and potassium fluxes such as those originating from the neuronal tissue can affect the proliferation of cells that dwell in the proximity of neurons (e.g., glial cells). Although we are not in a position to catalog tumors as channelopathies arising directly from $Na^+$, $K^+$, and $Ca^{2+}$ ion channels, several studies have shown that tumors may indeed modulate their functional expression and activity (Bubien et al., 1999, 2004; Olsen et al., 2005; Ross et al., 2007).

Ion channel functions are not limited to excitability and ion homeostasis. In fact, many studies suggest a clear association between the expression of specific classes of ion channels and modulation of different aspects of cellular activity. In *weaver* mice, for example, the mutation in the gene coding for G-protein-activated inwardly rectifying potassium channel subunit 2 (GIRK2) causes the loss of ion selectivity. This leads to the induction of cell death in the external germinal layer of the cerebellum and alterations in neurite extension and cell migration (Migheli et al., 1999; Patil et al., 1995). Ion channels also have a major role in the cellular regulation of proliferative activity comparable to glial cells.

Recent studies have shown that the inhibition of $K_{ir}$ channels reactivates the proliferative activity in quiescent glia (MacFarlane and Sontheimer, 2000). This suggests that the downregulation of $K_{ir}$ may promote cell cycle progression, while its premature expression or overexpression is associated with cell cycle arrest in G1/G0. This regulatory activity works both ways, however, such that the expression level of a specific ion channel is also regulated by the proliferative state in which the cell resides at a particular time; for example, the arrest of spinal-cord astrocytic growth at defined stages of the cell cycle causes significant changes in the expression of voltage-activated $Na^+$ and $K^+$ currents (MacFarlane and Sontheimer, 2000). Different types of ion channels, such as $K^+$, $Cl^-$, and $Ca^{2+}$, appear involved in normal cell cycling in different cell types. It is commonly held that the activation of $Ca^{2+}$ channels is often related to increases in intracellular $Ca^{2+}$ ($[Ca^{2+}]_i$) during cell cycle progression, whereas the inhibition of $Cl^-$ or $K^+$ channels often blocks the cell cycle without affecting cell viability.

$K^+$ channels represent a large and heterogeneous family with over 80 genes encoding for various proteins that control membrane potential. Their role in the regulation of cell cycle progression has been widely studied, especially for their involvement in tumor cell proliferation (Brevet et al., 2008). Thus, $K^+$ channels have become potential targets for the development of novel therapeutic treatments (Villalonga et al., 2007); for example, with respect to tumor growth, voltage-gated and ATP-sensitive $K^+$ channels have been involved with the cell proliferation of gliomas (Lee et al., 1994; Preussat et al., 2003). More recently, experiments performed *in vitro* have highlighted a previously unrecognized role for GIRK2 (or $K_{ir}3.2$) in human glia (Cucullo et al., 2005). In an experiment involving electrical stimulation, the investigators found a direct causal role between the expression of this potassium channel and the antiproliferative effect of low-frequency, ultra-low-intensity alternating electric current (Cucullo et al., 2005). Large-conductance $Ca^{2+}$-activated $K^+$ currents have also been characterized in glioblastoma cell lines (Olsen et al., 2005), and a mislocalization of $K_{ir}$ channels primarily into the nucleus has been described in malignant glioma cell lines (Lee et al., 1994).

Taken together, these findings strengthen the importance of potassium channels in tumor differentiation and proliferation and explain their current status as major pharmaceutical targets for the development of novel antineoplastic and antiepileptic agents (Arcangeli et al., 2009; Felipe et al., 2006; Le Guennec et al., 2007).

## CONCLUSIONS

After decades of neglect, astrocytes have become one the hottest items in neurological research. The major pitfall at this juncture appears to be over-interpretation of results. The data cited above and related to cerebral blood flow regulation by astrocytes, for example, are in partial disagreement with the neuroanatomy that shows a separation of glia and vascular smooth muscle at the Virchow–Robin triad (del Zoppo, 2008). In addition, the caliber of vessels involved suggests recruitment of capillary cells, which by definition are devoid of vascular smooth muscle. Are pericytes involved, or is the observed change due to shifting osmotic gradients? Further studies will clarify this. The results linking cell cycle to electrical stimulation may similarly propose mechanisms that bear little significance to tumor treatment. It is hoped that *in vivo* experiments will shed light on this. It is also possible that the link among cell cycle, glia, and ion channels goes beyond cancer; for example, a loss of $K_{ir}$ has been reported in a variety of neurological disorders, such as epilepsy (astrocytes, but perhaps other cell types) (Bordey and Sontheimer, 1998; Bordey et al., 2001) and ischemia (vascular smooth muscle cells) (Marrelli et al., 1998). Finally, the discovery of additional noninvasive peripheral markers—including serum protein of astrocytic origin other than S100β—would greatly enhance the clinical significance of these laboratory studies.

## ACKNOWLEDGMENTS

This work was supported by NIH-2RO1 HL51614, NIH-RO1 NS43284, and NIH-RO1 NS38195 [DJ].

## REFERENCES

Abdul, M., N. Hoosein. (2002). Voltage-gated sodium ion channels in prostate cancer: expression and activity. *Anticancer Res.*, 22:1727–1730.

Allen, N.J., B.A. Barres. (2009). Neuroscience: glia—more than just brain glue. *Nature*, 457:675–677.

Arcangeli, A., O. Crociani, E. Lastraioli, A. Masi, S. Pillozzi, A. Becchetti. (2009). Targeting ion channels in cancer: a novel frontier in antineoplastic therapy. *Curr. Med. Chem.*, 16:66–93.

Ballabh, P., A. Braun, M. Nedergaard. (2004). The blood–brain barrier: an overview—structure, regulation, and clinical implications. *Neurobiol. Dis.*, 16:1–13.

Bernardino, L., S. Balosso, T. Ravizza, N. Marchi, G. Ku, J.C. Randle, J.O. Malva, A. Vezzani. (2008). Inflammatory events in hippocampal slice cultures prime neuronal susceptibility to excitotoxic injury: a crucial role of P2X7 receptor-mediated IL-1beta release. *J. Neurochem.*, 106:271–280.

Biberthaler, P., T. Mussack, E. Wiedemann, T. Gilg, M. Soyka, G. Koller, K.J. Pfeifer, U. Linsenmaier, W. Mutschler, C. Gippner-Steppert, M. Jochum. (2001a). Elevated serum levels of S-100β reflect the extent of brain injury in alcohol intoxicated patients after mild head trauma. *Shock*, 16:97–101.

Biberthaler, P., T. Mussack, E. Wiedemann, K.G. Kanz, M. Koelsch, C. Gippner-Steppert, M. Jochum. (2001b). Evaluation of S-100β as a specific marker for neuronal damage due to minor head trauma. *World J. Surg.*, 25:93–97.

Blanco, V.M., J.E. Stern, J.A. Filosa. (2008). Tone-dependent vascular responses to astrocyte-derived signals. *Am. J. Physiol. Heart Circ. Physiol.*, 294(6):H2855–H2863.

Bordey, A., H. Sontheimer. (1998). Properties of human glial cells associated with epileptic seizure foci. *Epilepsy Res.*, 32:286–303.

Bordey, A., S.A. Lyons, J.J. Hablitz, H. Sontheimer. (2001). Electrophysiological characteristics of reactive astrocytes in experimental cortical dysplasia. *J. Neurophysiol.*, 85:1719–1731.

Brevet, M., A. Ahidouch, H. Sevestre, P. Merviel, Y. El Hiani, M. Robbe, H. Ouadid-Ahidouch. (2008). Expression of $K^+$ channels in normal and cancerous human breast. *Histol. Histopathol.*, 23:965–972.

Bubien, J.K., D.A. Keeton, C.M. Fuller, G.Y. Gillespie, A.T. Reddy, T.B. Mapstone, D.J. Benos. (1999). Malignant human gliomas express an amiloride-sensitive Na$^+$ conductance. *Am. J. Physiol.*, 276:C1405–C1410.

Bubien, J.K., H.L. Ji, G.Y. Gillespie, C.M. Fuller, J.M. Markert, T.B. Mapstone, D.J. Benos. (2004). Cation selectivity and inhibition of malignant glioma Na$^+$ channels by psalmotoxin 1. *Am. J. Physiol. Cell Physiol.*, 287(5):C1282–C1291.

Buttner, T., S. Weyers, T. Postert, R. Sprengelmeyer, W. Kuhn. (1997). S-100 protein: serum marker of focal brain damage after ischemic territorial MCA infarction. *Stroke*, 28:1961–1965.

Cucullo, L., M. McAllister, K. Kight, L. Krizanac-Bengez, M. Marroni, M. Mayberg, K. Stanness, D. Janigro. (2002). A new dynamic *in vitro* model for the multidimensional study of astrocyte–endothelial cell interactions at the blood–brain barrier. *Brain Res.*, 951:243.

Cucullo, L., G. Dini, K.L. Hallene, V. Fazio, E.V. Ilkanich, C. Igboechi, K.M. Kight, M.K. Agarwal, M. Garrity-Moses, D. Janigro. (2005). Very low intensity alternating current decreases cell proliferation. *Glia*, 51:65–72.

Dassan, P., G. Keir, M.M. Brown. (2009). Criteria for a clinically informative serum biomarker in acute ischaemic stroke: a review of S100β. *Cerebrovasc. Dis.*, 27:295–302.

De Simoni, M.G., L. Imeri. (1998). Cytokine–neurotransmitter interactions in the brain. *Biol. Signals*, 7:33–44.

De Simoni, M.G., P. Milia, M. Barba, A. De Luigi, L. Parnetti, V. Gallai. (2002). The inflammatory response in cerebral ischemia: focus on cytokines in stroke patients. *Clin. Exp. Hypertens.*, 24:535–542.

del Zoppo, G.J. (2008). Virchow's triad: the vascular basis of cerebral injury. *Rev. Neurol. Dis.*, 5(Suppl. 1):S12–S21.

del Zoppo, G.J., R. Milner. (2006). Integrin–matrix interactions in the cerebral microvasculature. *Arterioscler. Thromb. Vasc. Biol.*, 26:1966–1975.

Deng, W., R.D. Poretz. (2003). Oligodendroglia in developmental neurotoxicity. *Neurotoxicology*, 24:161–178.

Engelhardt, B., R.M. Ransohoff. (2005). The ins and outs of T-lymphocyte trafficking to the CNS: anatomical sites and molecular mechanisms. *Trends Immunol.*, 26:485–495.

Fabene, P.F., F. Merigo, M. Galie, D. Benati, P. Bernardi, P. Farace, E. Nicolato, P. Marzola, A. Sbarbati. (2007). Pilocarpine-induced status epilepticus in rats involves ischemic and excitotoxic mechanisms. *PLoS ONE*, 2(10):e1105.

Fabene, P.F., M.G. Navarro, M. Martinello, B. Rossi, F. Merigo et al. (2008). A role for leukocyte–endothelial adhesion mechanisms in epilepsy. *Nat. Med.*, 14:1377–1383.

Faraci, F.M., D.D. Heistad. (1993). Role of ATP-sensitive potassium channels in the basilar artery. *Am. J. Physiol.*, 264:H8–H13.

Felipe, A., R. Vicente, N. Villalonga, M. Roura-Ferrer, R. Martinez-Marmol, L. Sole, J.C. Ferreres, E. Condom. (2006). Potassium channels: new targets in cancer therapy. *Cancer Detect. Prev.*, 30:375–385.

Fisher, R.S., B.W. van Emde, W. Blume, C. Elger, P. Genton, P. Lee, J. Engel, Jr. (2005). Epileptic seizures and epilepsy: definitions proposed by the International League Against Epilepsy (ILAE) and the International Bureau for Epilepsy (IBE). *Epilepsia*, 46:470–472.

Fiske, J.L., V.P. Fomin, M.L. Brown, R.L. Duncan, R.A. Sikes. (2006). Voltage-sensitive ion channels and cancer. *Cancer Metastasis Rev.*, 25:493–500.

Ghersi-Egea, J.F., N. Strazielle. (2001). Brain drug delivery, drug metabolism, and multidrug resistance at the choroid plexus. *Microsc. Res. Tech.*, 52:83–88.

Ghersi-Egea, J.F., B. Leininger-Muller, R. Cecchelli, J.D. Fenstermacher. (1995). Blood–brain interfaces: relevance to cerebral drug metabolism. *Toxicol. Lett.*, 82–83:645–653.

Ghersi-Egea, J.F., N. Strazielle, A. Murat, J. Edwards, M.F. Belin. (2001). Are blood–brain interfaces efficient in protecting the brain from reactive molecules? *Adv. Exp. Med. Biol.*, 500:359–364.

Gordon, G.R., S.J. Mulligan, B.A. MacVicar. (2007). Astrocyte control of the cerebrovasculature. *Glia*, 55:1214–1221.

Granata, T., L. Obino, F. Ragona, I. Degiorgi, N. Marchi, S. Binelli, D. Janigro. (2008). Steroid treatment is effective in the treatment of status epilepticus in children. *Epilepsia*, 49(Suppl. 7):121.

Grant, G.A., D. Janigro. (2004). The blood–brain barrier. In *Youmans Neurological Surgery* (pp. 153–174), H.R. Winn, Ed. Philadelphia, PA: W.B. Saunders.

Grant, G.A., J.R. Meno, T.S. Nguyen, K.A. Stanness, D. Janigro, R.H. Winn. (2003). Adenosine-induced modulation of excitatory amino acid transport across isolated brain arterioles. *J. Neurosurg.*, 98:554–560.

Haydon, P.G., G. Carmignoto. (2006). Astrocyte control of synaptic transmission and neurovascular coupling. *Physiol. Rev.*, 86:1009–1031.

Huang, L., B. Li, W. Li, H. Guo, F. Zou. (2009). ATP-sensitive potassium channels control glioma cells proliferation by regulating ERK activity. *Carcinogenesis*, 30:737–744.

Iadecola, C., M. Nedergaard. (2007). Glial regulation of the cerebral microvasculature. *Nat. Neurosci.*, 10:1369–1376.

Janigro, D., T.S. Nguyen, E.L. Gordon, H.R. Winn. (1996). Physiological properties of ATP-activated cation channels in rat brain microvascular endothelial cells. *Am. J. Physiol.*, 270:H1423–H1434.

Janigro, D., V. Fazio, L. Cucullo, N. Marchi. (2002). Markers of BBB Damage and Inflammatory Processes. Paper presented at the 32nd Annual Meeting of the Society for Neuroscience, Orlando, FL, November 2–7.

Jemal, A., R.C. Tiwari, T. Murray, A. Ghafoor, A. Samuels, E. Ward, E.J. Feuer, M.J. Thun. (2004). Cancer statistics 2004. *CA Cancer J. Clin.*, 54(1):8–29.

Jonsson, H., P. Johnsson, M. Birch-Iensen, C. Alling, S. Westaby, S. Blomquist. (2001). S100β as a predictor of size and outcome of stroke after cardiac surgery. *Ann. Thorac. Surg.*, 71:1433–1437.

Koehler, R.C., R.J. Roman, D.R. Harder. (2009). Astrocytes and the regulation of cerebral blood flow. *Trends Neurosci.*, 32:160–169.

Krizanac-Bengez, L., M.R. Mayberg, D. Janigro. (2004). The cerebral vasculature as a therapeutic target for neurological disorders and the role of shear stress in vascular homeostasis and pathophysiology. *Neurol. Res.*, 26:846–853.

Kuschinsky, W. (1997). Neuronal-vascular coupling. In *Optical Imaging of Brain Function* (pp. 167–176), A. Villringer and U. Dirnagl, Eds. New York: Plenum Press.

Le Guennec, J.Y., H. Ouadid-Ahidouch, O. Soriani, P. Besson, A. Ahidouch, C. Vandier. (2007). Voltage-gated ion channels, new targets in anti-cancer research. *Recent Pat. Anticancer Drug Discov.*, 2:189–202.

Lee, Y.S., M.M. Sayeed, R.D. Wurster. (1994). *In vitro* antitumor activity of cromakalim in human brain tumor cells. *Pharmacology*, 49:69–74.

Löscher, W. (2007). Drug transporters in the epileptic brain. *Epilepsia*, 48(Suppl. 1):8–13.

Luyken, C., I. Blumcke, R. Fimmers, H. Urbach, C.E. Elger, O.D. Wiestler, J. Schramm. (2003). The spectrum of long-term epilepsy-associated tumors: long-term seizure and tumor outcome and neurosurgical aspects. *Epilepsia*, 44:822–830.

MacFarlane, S.N., H. Sontheimer. (2000). Changes in ion channel expression accompany cell cycle progression of spinal cord astrocytes. *Glia*, 30:39–48.

Magistretti, P.J. (2006). Neuron–glia metabolic coupling and plasticity. *J. Exp. Biol.*, 209:2304–2311.

Majores, M., M. von Lehe, J. Fassunke, J. Schramm, A.J. Becker, M. Simon. (2008). Tumor recurrence and malignant progression of gangliogliomas. *Cancer*, 113:3355–3363.

Man, S., E.E. Ubogu, R.M. Ransohoff. (2007). Inflammatory cell migration into the central nervous system: a few new twists on an old tale. *Brain Pathol.*, 17:243–250.

Marchi, N., P.A. Rasmussen, M. Kapural, V. Fazio, K. Kight et al. (2003a). Peripheral markers of brain damage and blood–brain barrier dysfunction. *Restor. Neurol. Neurosci.*, 21:109–121.

Marchi, N., V. Fazio, L. Cucullo, K. Kight, T. Masaryk et al. (2003b). Serum transthyretin monomer as a possible marker of blood-to-CSF barrier disruption. *J. Neurosci.*, 23:1949–1955.

Marchi, N., M. Cavaglia, V. Fazio, S. Bhudia, K. Hallene, D. Janigro. (2004a). Peripheral markers of blood–brain barrier damage. *Clin. Chim. Acta*, 342:1–12.

Marchi, N., K.L. Hallene, K.M. Kight, L. Cucullo, G. Moddel, W. Bingaman, G. Dini, A. Vezzani, D. Janigro. (2004b). Significance of MDR1 and multiple drug resistance in refractory human epileptic brain. *BMC Med.*, 2:37.

Marchi, N., G. Guiso, S. Caccia, M. Rizzi, B. Gagliardi et al. (2006). Determinants of drug brain uptake in a rat model of seizure-associated malformations of cortical development. *Neurobiol. Dis.*, 24:429–442.

Marchi, N., E. Oby, N. Fernandez, L. Uva, M. de Curtis et al. (2007). *In vivo* and *in vitro* effects of pilocarpine: relevance to epileptogenesis. *Epilepsia*, 48(10):1934–1946.

Marchi, N., A. Batra, Q. Fan, E. Carlton, I. Caponi, I. Najm, T. Granata, D. Janigro. (2009). Antagonism of peripheral inflammation prevents status epilepticus. *Neurobiol. Dis.*, 33:171–271.

Marrelli, S.P., T.D. Johnson, A. Khorovets, W.F. Childres, R.M. Bryan. (1998). Altered function of inward rectifier potassium channels in cerebrovascular smooth muscle after ischemia/reperfusion. *Stroke*, 29:1469–1474.

Marroni, M., M.L. Agrawal, K. Kight, K.L. Hallene, M. Hossain, L. Cucullo, K. Signorelli, S. Namura, W. Bingaman, D. Janigro. (2003). Relationship between expression of multiple drug resistance proteins and p53 tumor suppressor gene proteins in human brain astrocytes. *Neuroscience*, 121:605–617.

McAllister, M.S., L. Krizanac-Bengez, F. Macchia, R.J. Naftalin, K.C. Pedley, M.R. Mayberg, M. Marroni, S. Leaman, K.A. Stanness, D. Janigro. (2001). Mechanisms of glucose transport at the blood–brain barrier: an *in vitro* study. *Brain Res.*, 904:20–30.

Meyer, R.P., M. Gehlhaus, R. Knoth, B. Volk. (2007). Expression and function of cytochrome p450 in brain drug metabolism. *Curr. Drug Metab.*, 8:297–306.

Migheli, A., R. Piva, S. Casolino, C. Atzori, S.R. Dlouhy, B. Ghetti. (1999). A cell cycle alteration precedes apoptosis of granule cell precursors in the weaver mouse cerebellum. *Am. J. Pathol.*, 155:365–373.

Morii, S., A.C. Ngai, H.R. Winn (1986). Reactivity of rat pial arterioles and venules to adenosine and carbon dioxide: with detailed description of the closed cranial window technique in rats. *J. Cerebr. Blood Flow Metab.*, 6:34–41.

Morii, S., A.C. Ngai, K.R. Ko, H.R. Winn. (1987). Role of adenosine in regulation of cerebral blood flow: effects of theophylline during normoxia and hypoxia. *Am. J. Physiol.*, 253:H165–H175.

Nedergaard, M., T. Takano, A.J. Hansen. (2002). Beyond the role of glutamate as a neurotransmitter. *Nat. Rev. Neurosci.*, 3:748–755.

Newman, E.A. (2003). New roles for astrocytes: regulation of synaptic transmission. *Trends Neurosci.*, 26:536–542.

Ngai, A.C., H.R. Winn. (1993). Effects of adenosine and its analogues on isolated intracerebral arterioles: extraluminal and intraluminal applications. *Circ. Res.*, 73:448–457.

Nguyen, T.S., H.R. Winn, D. Janigro. (2000). ATP-sensitive potassium channels may participate in the coupling of neuronal activity and cerebrovascular tone. *Am. J. Physiol. Heart Circ. Physiol.*, 278:H878–H885.

Nishiyama, H., T. Knopfel, S. Endo, S. Itohara. (2002). Glial protein S100$\beta$ modulates long-term neuronal synaptic plasticity. *Proc. Natl. Acad. Sci. U.S.A.*, 99:4037–4042.

Oberheim, N.A., X. Wang, S. Goldman, M. Nedergaard. (2006). Astrocytic complexity distinguishes the human brain. *Trends Neurosci.*, 29:547–553.

Oby, E., D. Janigro. (2006). The blood–brain barrier and epilepsy. *Epilepsia*, 47:1761–1774.

Oby, E., S. Caccia, A. Vezzani, G. Moeddel, K. Hallene et al. (2006). *In vitro* responsiveness of human drug-resistant tissue to antiepileptic drugs: insights into the mechanisms of pharmacoresistance. *Brain Res.*, 1086:201–213.

Olsen, M.L., A.K. Weaver, P.S. Ritch, H. Sontheimer. (2005). Modulation of glioma BK channels via ErbB-2. *J. Neurosci. Res.*, 81:179–189.

Pal, D., A.K. Mitra. (2006). MDR- and CYP3A4-mediated drug–drug interactions. *J. Neuroimmune Pharmacol.*, 1:323–339.

Parkinson, F.E., K.A. Stanness, C.M. Anderson, D. Janigro. (1998). Nucleoside transporter subtypes in rat brain endothelial cells and astrocytes [abstract]. *Society for Neuroscience Abstracts*.

Patil, N., D.R. Cox, D. Bhat, M. Faham, R.M. Myers, A.S. Peterson. (1995). A potassium channel mutation in weaver mice implicates membrane excitability in granule cell differentiation. *Nat. Genet.*, 11:126–129.

Petzold, A., P. Michel, M. Stock, M. Schluep. (2008). Glial and axonal body fluid biomarkers are related to infarct volume, severity, and outcome. *J. Stroke Cerebrovasc. Dis.*, 17:196–203.

Prat, A., K. Biernacki, K. Wosik, J.P. Antel. (2001). Glial cell influence on the human blood–brain barrier. *Glia*, 36:145–155.

Preussat, K., C. Beetz, M. Schrey, R. Kraft, S. Wolfl, R. Kalff, S. Patt. (2003). Expression of voltage-gated potassium channels Kv1.3 and Kv1.5 in human gliomas. *Neurosci. Lett.*, 346:33–36.

Ransom, B., T. Behar, M. Nedergaard. (2003). New roles for astrocytes (stars at last). *Trends Neurosci.*, 26:520–522.

Reinhardt, C.A., S.M. Gloor. (1997). Co-culture blood–brain barrier models and their use for pharmatoxicological screening. *Toxicol. In Vitro*, 11:513–518.

Ross, S.B., C.M. Fuller, J.K. Bubien, D.J. Benos. (2007). Amiloride-sensitive Na$^+$ channels contribute to regulatory volume increases in human glioma cells. *Am. J. Physiol. Cell Physiol.*, 293:C1181–C1185.

Rossi, D.J., J.D. Brady, C. Mohr. (2007). Astrocyte metabolism and signaling during brain ischemia. *Nat. Neurosci.*, 10:1377–1386.

Schramm, J., C. Luyken, H. Urbach, R. Fimmers, I. Blumcke. (2004). Evidence for a clinically distinct new subtype of grade II astrocytomas in patients with long-term epilepsy. *Neurosurgery*, 55:340–347.

Selinfreund, R.H., S.W. Barger, W.J. Pledger, L.J. Van Eldik. (1991). Neurotrophic protein S100 beta stimulates glial cell proliferation. *Proc. Natl. Acad. Sci. U.S.A.*, 88:3554–3558.

Simard, M., G. Arcuino, T. Takano, Q.S. Liu, M. Nedergaard. (2003). Signaling at the gliovascular interface. *J. Neurosci.*, 23:9254–9262.

Stanness, K.A., L.E. Westrum, E. Fornaciari, P. Mascagni, J.A. Nelson, S.G. Stenglein, T. Myers, D. Janigro. (1997). Morphological and functional characterization of an *in vitro* blood–brain barrier model. *Brain Res.*, 771:329–342.

Takano, T., G.F. Tian, W. Peng, N. Lou, W. Libionka, X. Han, M. Nedergaard. (2006). Astrocyte-mediated control of cerebral blood flow. *Nat. Neurosci.*, 9:260–267.

Takano, T., N. Oberheim, M.L. Cotrina, M. Nedergaard. (2009). Astrocytes and ischemic injury. *Stroke*, 40:S8–12.

Utsumi, H., H. Chiba, Y. Kamimura, M. Osanai, Y. Igarashi, H. Tobioka, M. Mori, N. Sawada. (2000). Expression of GFRalpha-1, receptor for GDNF, in rat brain capillary during postnatal development of the BBB. *Am. J. Physiol. Cell Physiol.*, 279:C361–C368.

Vezzani, A., T. Granata. (2005). Brain inflammation in epilepsy: experimental and clinical evidence. *Epilepsia*, 46:1724–1743.

Vezzani, A., S. Balosso, T. Ravizza. (2008a). The role of cytokines in the pathophysiology of epilepsy. *Brain Behav. Immun.*, 22:797–803.

Vezzani, A., T. Ravizza, S. Balosso, E. Aronica. (2008b). Glia as a source of cytokines: implications for neuronal excitability and survival. *Epilepsia*, 49(Suppl. 2):24–32.

Villalonga, N., J.C. Ferreres, J.M. Argiles, E. Condom, A. Felipe. (2007). Potassium channels are a new target field in anticancer drug design. *Recent Pat. Anticancer Drug Discov.*, 2:212–223.

Virgintino, D., D. Robertson, P. Monaghan, M. Errede, G. Ambrosi, L. Roncali, M. Bertossi. (1998). Glucose transporter GLUT1 localization in human foetus telencephalon. *Neurosci. Lett.*, 256:147–150.

Volman, V., E. Ben-Jacob, H. Levine. (2007). The astrocyte as a gatekeeper of synaptic information transfer. *Neural Comput.*, 19:303–326.

Volterra, A., J. Meldolesi. (2005). Astrocytes, from brain glue to communication elements: the revolution continues. *Nat. Rev. Neurosci.*, 6:626–640.

Wolburg, H., S. Noell, A. Mack, K. Wolburg-Buchholz, P. Fallier-Becker. (2009). Brain endothelial cells and the glio–vascular complex. *Cell Tissue Res.*, 335:75–96.

Zlokovic, B.V. (2008). The blood–brain barrier in health and chronic neurodegenerative disorders. *Neuron*, 57:178–201.

Zonta, M., M.C. Angulo, S. Gobbo, B. Rosengarten, K.A. Hossmann, T. Pozzan, G. Carmignoto. (2003). Neuron-to-astrocyte signaling is central to the dynamic control of brain microcirculation. *Nat. Neurosci.*, 6:43–50.

# 3 Metabolic Regulation of Seizures and Epileptogenesis

*Manisha Patel*

## CONTENTS

## INTRODUCTION

*Metabolism* is defined as an organized set of chemical reactions comprising metabolic pathways that are used by living organisms to transduce and store energy and thereby maintain life. Metabolic processes are tightly regulated by enzyme levels, their catalytic activities, and the availability of substrates. Metabolic processes are critical in maintaining neuronal excitability; therefore, synchronized neuronal firing associated with seizure activity *per se* represents a "metabolic challenge." Acute seizure activity associated with status epilepticus (SE) and chronic seizures constitute metabolic irregularities. Metabolism operates in discrete yet interrelated biochemical circuits with the central goal of transducing energy as well as information via signal transduction. Metabolic pathways have remarkable adaptability, capacity, and flexibility that allow maintenance of homeostasis and therefore make metabolic disorders often difficult to discern.

The role of metabolism is recognized in many human conditions such as diabetes, cancer, and various neurodegenerative diseases such as Parkinson's disease, Huntington's disease, Alzheimer's disease, and amyotrophic lateral sclerosis (ALS) (Beal, 2003; Hall et al., 1998; Halliwell, 1992; Schapira et al., 1993). Common mechanisms that form the metabolic hypothesis in these neurodegenerative diseases include aging, metabolic impairment, mitochondrial dysfunction, oxidative damage, excitotoxicity, and selective vulnerability (Beal, 1998, 2000, 2003). Each of these mechanisms is thought to contribute to the pathogenesis of temporal lobe epilepsy (TLE). In fact, the importance of metabolism in the epilepsies has been appreciated for over a century and is based on strong neurochemical, imaging, and physiological data.

## METABOLIC REGULATION OF EPILEPTIC BRAIN DAMAGE

The role of cell loss in epilepsy has been recognized for over a century. Important confirmation of the occurrence of neuronal loss in epilepsy came from work by Meldrum and colleagues (1983, 2002), who showed that seizures *per se* but not associated systemic complications resulted in hippocampal lesions. Several key studies have confirmed this (Olney, 1986; Slovitor, 1983) and further described lesions in multiple brain areas by seizure activity (Ben-Ari et al., 1981; Lothman, 1990; Nadler, 1979; Olney et al., 1983). The recognition that seizure-related brain damage was pathologically similar to excitotoxicity (Olney, 1986) is in itself evidence for the metabolic hypothesis, given the early observation that pathological and mechanistic features of glutamate-induced excitotoxic cell death involve energy-dependent processes (Henneberry et al., 1989a,b; Olney et al., 1983). The role of mitochondrial bioenergetics is now widely recognized in excitotoxic cell death (Nicholls and Budd, 2000).

In addition to excitotoxic mechanisms, apoptotic pathways are also activated by prolonged seizures in human and experimental epilepsy (Henshall and Murphy, 2008). The inherent energy dependence and mitochondrial involvement of apoptotic pathways—particularly the intrinsic pathway—provide yet another mechanism through which metabolic factors influence epileptic brain damage. Additionally, the intricacy of apoptotic signaling provides numerous therapeutic targets for controlling seizure-induced brain damage.

Although cell loss can occur following prolonged seizure activity, recent studies suggest that it is not necessary for metabolic dysfunction (Cohen-Gadol et al., 2004; Hugg et al., 1996; Vielhaber et al., 2008). In fact, metabolic impairment is prevalent in surviving neuronal subfields as well as non-neuronal cells in human TLE (Vielhaber et al., 2008). This suggests that neuronal loss *per se* is not the underlying cause of metabolic dysfunction in the epileptic brain. Because metabolic impairment can contribute to seizure-induced neuronal death and can occur independent of neuronal death, therapeutic targeting of metabolism may provide dual benefits.

## ROLE OF GLYCOLYSIS IN SEIZURE CONTROL

Oxidation of glucose is the major source of cellular energy in the brain, although the brain can efficiently utilize alternative fuels such as ketones and fatty acids. Seizure activity produces dramatic increases in glucose uptake and metabolism. This increase is unparalleled by most other conditions. Cerebral blood flow increases to match this hypermetabolism. The increased rate of glycolysis exceeds pyruvate utilization by pyruvate dehydrogenase, resulting in an increased lactate buildup. Several intriguing studies have shown a close link between seizure activity and high glucose concentrations, as well as its utilization via glycolysis. First, high glucose concentrations in the blood exacerbate seizures (Schwechter et al., 2003), whereas moderately low glucose levels have the opposite effect (Greene et al., 2001). Second, the collective information obtained from human imaging studies suggests that *hypermetabolism* occurs in human epileptic foci during ictal phases (seizure episodes) and *hypometabolism* during interictal phases. Third, 2-deoxyglucose (2-DG), a nonmetabolizable sugar, or bypassing glycolysis via fructose-1,6-bisphosphate has been shown to exert anticonvulsant effects *in vivo* (Garriga-Canut et al., 2006; Lian et al., 2007). Finally, the switching of fuels from glucose to ketones (thus bypassing glycolysis) with the ketogenic diet also results in an anticonvulsant effect (Melo et al., 2006).

Collectively, these studies link seizure activity with high glucose concentrations. Although recent progress has been made regarding the potential mechanisms by which 2-DG and the ketogenic diet exert an anticonvulsant effect, the precise mechanism by which glycolysis exacerbates seizures remains to be determined. Because glycolysis provides carbon sources for mitochondrial energy production, mitochondrial mechanisms may ultimately underlie the damaging effect of glycolysis on seizure activity.

## ROLE OF MITOCHONDRIA IN EPILEPSY

The involvement of mitochondria in normal and excessive neuronal excitability is obvious given the bioenergetic requirements of the process. Mitochondria burn dietary calories with oxygen to produce work and heat. A byproduct of this process is the production of reactive oxygen species (ROS),

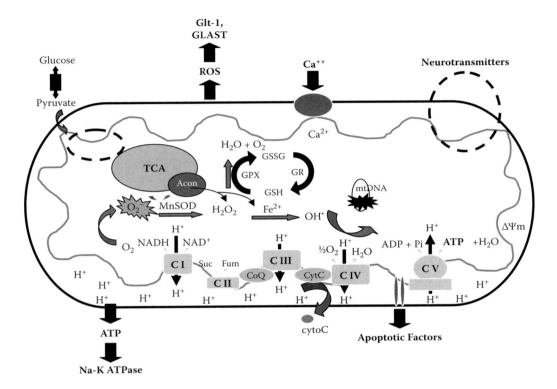

**FIGURE 3.1**   (**See color insert following page 458.**) Overview of important mitochondrial functions relevant to neuronal excitability.

which can damage cellular DNA, proteins, and lipids. In addition to adenosine triphosphate (ATP) and ROS production, several other mitochondrial functions, such as fatty acid biosynthesis, amino acid cycling, neurotransmitter biosynthesis, and ionic homeostasis (particularly calcium buffering), could also impact neuronal excitability (Figure 3.1). Mitochondrial decline and mitochondrial DNA (mtDNA) damage are thought to play a central role in the etiology of age-related metabolic and degenerative diseases (Wallace, 2005). ROS are thought to play a key role in this process by influencing the extent of mitochondrial decline. The increased incidence of epilepsy with advancing age as well as the progressive nature of some acquired epilepsies (e.g., TLE) strongly suggest that mitochondrial decline may be a critical mechanism in its etiology, much like more commonly understood neurodegenerative diseases.

## IMAGING STUDIES OF HUMAN AND ANIMAL TISSUE

Human imaging studies have by far provided the most compelling evidence for mitochondrial dysfunction in epilepsy. A detailed review of human imaging studies and neurotransmitter changes can be found in Pan et al. (2008). Collectively, these imaging studies demonstrate dramatic metabolic and bioenergetic changes based on measurement of glucose, oxygen, and mitochondrial *N*-acetylaspartate (NAA) levels. The strongest evidence for mitochondrial involvement in mesial temporal lobe epilepsy comes from measurement of NAA using magnetic resonance spectroscopy (Cendes et al., 1994; Hetherington et al., 1995; Petroff et al., 2002). The loss of NAA, which is known to be specifically synthesized by neuronal mitochondria, has been shown by various groups to be subregion specific, reversible by surgical or pharmacological intervention, and unrelated to neuronal death (Clark, 1998; Vielhaber et al., 2003). The collective interpretation of imaging studies demonstrating the decrease in NAA in human TLE points toward mitochondrial dysfunction as the cause rather than neuronal loss (Vielhaber et al., 2008). Severe metabolic dysfunction characterized by biphasic abnormal nicotinamide adenine dinucleotide phosphate (NADPH) fluorescence transients and changes in mitochondrial membrane potential ($\Delta\Psi$) have been observed in *ex vivo* preparations from both chronically epileptic rats and human subjects (Kovacs et al., 2002).

## Mitochondrial Dysfunction: A Consequence of Acute and Chronic Seizures

An important combustion byproduct of mitochondrial metabolism is the production of ROS. While abundant and overlapping endogenous antioxidants exist to overcome normal cellular ROS production, excessive production of ROS can overwhelm antioxidant defenses, resulting in oxidation of vulnerable cellular targets. Using surrogate markers of target oxidation in the kainic acid model, work from this laboratory has demonstrated that SE can oxidatively damage mtDNA and susceptible mitochondrial proteins and increase cellular lipid peroxidation (Jarrett et al., 2008a; Liang and Patel, 2006; Liang et al., 2000; Patel et al., 2001, 2008). The increased vulnerability of mtDNA to seizure-induced damage in comparison to nuclear DNA is consistent with the appearance of 8-hydroxy-2-deoxyguanosine (8-OHdG), the oxidative base lesion in mitochondrial (but not nuclear) fractions, and a greater frequency of mtDNA lesions. Expression of mitochondrial base excision repair enzymes (8-oxoguanine glycosylase and DNA polymerase gamma), reflected by increased mtDNA repair capacity, occurs shortly following an episode of SE; however, mitochondrial ROS production and mtDNA damage emerge again during recurrent seizures associated with the chronic phase of epilepsy concomitant with the failure of repair processes. In addition to being an acute consequence of SE, mitochondrial ROS production resurfaces during chronic epilepsy, suggesting that ROS formation and mtDNA damage could contribute to epileptogenesis (Jarrett et al., 2008a).

Whether and how mitochondrial ROS and the resulting dysfunction lower seizure threshold and contribute to epileptogenesis remain unknown. Assessment of mitochondrial dysfunction several weeks after SE was also addressed in pilocarpine-induced SE (Kann et al., 2005; Kudin et al., 2002). A prominent finding from these studies is the decrease in complex I activity of the electron transport chain (ETC), accompanied by lowered mitochondrial membrane potential measured by rhodamine-123 fluorescence in the CA1 and CA3 subfields of the hippocampus, perhaps due to a decreased mitochondrial DNA copy number that results in downregulation of oxidative phosphorylation (OXPHOS) enzymes encoded by mtDNA.

Together, these independent studies in two distinct animal models suggest that the acute effects of SE (i.e., increased mitochondrial oxidative stress) may result over time in oxidative damage to mtDNA (as suggested by increased levels of 8-OHdG) and decreased expression of mitochondrially encoded proteins required for the functioning of the ETC. Seizure-induced accumulation of oxidative mitochondrial DNA lesions and resultant somatic mtDNA mutations could, over a period of time, render the brain more susceptible to subsequent epileptic seizures, particularly in the context of advancing age. Human studies have confirmed at least some aspects of mitochondrial dysfunction seen in animal models. Complex I deficiency, which is a leading cause of increased superoxide production, and inhibition of aconitase, a marker of ROS, have been observed in the seizure foci and surviving CA3 cell layer, respectively, in tissue from TLE patients (Vielhaber et al., 2008). The increase in nitric oxide (NO) after seizure activity can either directly or indirectly (via reaction with superoxide and peroxynitrite formation) modulate the activity of the ETC complexes and thereby contribute to the metabolic changes associated with epilepsy.

Mechanistic studies assessing the role of mitochondrial functions on neuronal excitability have been conducted by measuring seizure-like events (SLEs) in hippocampal explant cultures. The SLEs are usually generated by a combination of electrical stimulation of the axons of the granule cells (i.e., the mossy fibers) and lowering the $Mg^{2+}$ concentration in the bathing media (0 $Mg^{2+}$) to relieve the voltage-dependent $Mg^{2+}$ blockade of the *N*-methyl-D-aspartic acid (NMDA) receptors. The SLEs are associated with increased mitochondrial calcium accumulation, mitochondrial depolarization, decreased NADPH autofluorescence, and superoxide generation. Superoxide production measured by dihydroethidium progressively increases during and with each consecutive SLE, along with increases of both cytosolic and mitochondrial calcium, thus providing a link between mitochondrial calcium, free-radical production, and neuronal death.

The mechanism by which seizure activity increases mitochondrial ROS remains incompletely understood. Although changes in mitochondrial calcium levels may explain the increased ROS formation, a discrepancy exists between ROS production due to seizure-induced changes in the mitochondrial membrane potential and calcium elevation. Seizure-induced mitochondrial ROS production occurs in the face of a loss of mitochondrial membrane potential, which is usually associated with decreased ROS production via the ETC complexes. This suggests that alternative mechanisms of ROS production may occur following seizures, such as inactivation of the tricarboxylic acid (TCA)-cycle enzyme aconitase which can release hydrogen peroxide.

## MITOCHONDRIAL DYSFUNCTION AND NEURONAL EXCITABILITY

Although mitochondrial dysfunction is an acute and chronic consequence of epileptic seizures, whether mitochondrial dysfunction can be a causative factor in epileptogenesis is unknown. The precedence for mitochondrial dysfunction initiating epilepsy has already been established from our knowledge regarding the molecular basis of certain inherited epilepsies (i.e., the mitochondrial encephalopathies). The most prominent example of mitochondrial dysfunction causing epilepsy is the occurrence of epilepsy in mitochondrial disorders due to mutations in mitochondrial DNA or nuclear DNA (DiMauro et al., 1999). Myoclonic epilepsy with ragged red fibers (MERRF) is a syndrome wherein a single mutation of the mitochondrially encoded tRNALys results in a disorder consisting of myoclonic epilepsy and a characteristic myopathy with ragged red fibers (Shoffner et al., 1990; Wallace et al., 1988). Defects in complex I and complex IV of mitochondrial OXPHOS are the leading mechanisms by which this mitochondrial gene mutation produces MERRF (Wallace et al., 1988).

Several normal functions of mitochondria, ranging from its bioenergetic functions to metabolic functions, can impact neuronal excitability. These include cellular ATP production, ROS formation, synthesis and metabolism of neurotransmitters, fatty acid oxidation, calcium homeostasis, and control of apoptotic/necrotic cell death. These vital functions are closely interrelated, but which of these factors contributes to the seizures associated with mitochondrial dysfunction remains unclear. Although mitochondrial encephalopathies due to genetic causes are rare, they may provide important lessons regarding the mechanisms underlying acquired epilepsy, such as temporal lobe epilepsy. By contrast with the gene defects of mitochondrial proteins, acquired epilepsy may arise due to chronic mitochondrial dysfunction that results in damage to the mitochondrial genome or mitochondrial bioenergetics. Factors that can promote acute and chronic mitochondrial dysfunction, which may in turn damage the mitochondrial genome, include hypoxia, trauma, and aging itself, as well as comorbid neuronal diseases such as stroke and Alzheimer's disease. ROS may be an important weapon triggered by each of these conditions that is capable of inducing direct damage to the mitochondrial genome. This can result in decreased expression of mtDNA-encoded OXPHOS subunits, known to result in epileptic seizures.

Another prominent target that links mitochondrial dysfunction and resultant ATP depletion with increased neuronal excitability is the plasma-membrane-bound sodium–potassium ATPase. Inhibition of the sodium–potassium ATPase has been shown to result in seizure activity (Grisar et al., 1992), and a defect in the $\alpha 3$ isoform of this enzyme renders the sodium pump dysfunctional and contributes to hyperexcitability and seizures (Clapcote et al., 2009). Other targets of ROS that may increase excitability are glutamate transporter 1 (GLT-1) and glutamate aspartate transporter (GLAST), which are known to be redox sensitive (Trotti et al., 1998) and play a crucial role in maintaining low levels of synaptic glutamate. Age-dependent seizures occurring in the context of chronic mitochondrial oxidative stress in Sod2(–/+) mice are accompanied by decreases in the hippocampal expression of the glial glutamate transporters GLT-1 and GLAST, which may explain their increased vulnerability to epileptic seizures (Liang and Patel, 2004). This is consistent with increased hippocampal extracellular glutamate levels that have been found in epileptic patients (During and Spencer, 1993).

## METABOLIC INTERVENTIONS

Anticonvulsant drugs remain frontline therapies for controlling epilepsies and serve to decrease neuronal excitability. Metabolic approaches may provide anticonvulsant effects in humans as well, but this has yet to be fully validated. A recent investigational drug screening approach is to search for antiepileptogenic, rather than antiepileptic, drugs that would target underlying processes that lead to the development of epilepsy. Antiepileptogenic therapies aimed at mitochondrial bioenergetics and oxidative stress pathways have been largely limited to animal studies. Clinical trials of vitamin E as an add-on therapy for refractory epilepsy have yielded conflicting results, with largely failed attempts to reduce epileptic seizures in pediatric patients. Creatine, an endogenous guanidine, functions with phosphocreatine and a mitochondrial form of creatine kinase as a spatial energy buffer between the cytosolic and mitochondrial compartments. Creatine has been shown to be protective in animal models of central nervous system (CNS) injury, including trauma, ischemia, the 3-nitropropionic acid model, 1-methyl-4-phenyl-1,2,3,6-tetrahydropyridine (MPTP)-induced parkinsonism, and ALS (Klivenyi et al., 2003; Matthews et al., 1998, 1999). Creatine supplementation has been effective in reducing hypoxia-induced seizures in both rats and rabbits (Holtzman et al., 1998, 1999).

The clinical management of intractable epilepsies with strategies such as caloric restriction or modification (e.g., through the use of a ketogenic diet) supports a role for mitochondrial bioenergetics in the protective effects. The ketogenic diet has been shown to produce mitochondrial biogenesis and bioenergetically efficient mitochondria, to increase mitochondrial glutathione levels, and to lower levels of ROS (Bough et al., 2006; Jarrett et al., 2008b; Sullivan et al., 2004). The lower levels of ROS have been shown to occur due to upregulation of the mitochondrial uncoupling protein 2 (UCP2) in mice fed a ketogenic diet (Sullivan et al., 2004). In these experimental studies, seizure-induced brain damage and subsequent epilepsy could be inhibited by caloric restriction or a ketogenic diet (Greene et al., 2001; Todorova et al., 2000).

Diet modification by ketogenic diet or caloric restriction represents a nonpharmacologic strategy that decreases seizure frequency in the epileptic EL mouse (Todorova et al., 2000). Interestingly, these paradigms are also known to limit free-radical formation. Consistent with this effect, recent studies in our laboratory have shown an upregulation of glutathione biosynthesis in animals fed a ketogenic diet (Jarrett et al., 2008b). In this study, it was demonstrated that the ketogenic diet specifically increases mitochondrial glutathione levels, stimulates de novo glutathione (GSH) biosynthesis, and improves mitochondrial redox status in the hippocampus, resulting in decreased mitochondrial ROS production and protection of mtDNA (Jarrett et al., 2008b). Thus, increased mitochondrial GSH may represent a possible candidate mechanism underlying the protection afforded by the ketogenic diet. In summary, the recent resurgence of research on the mechanisms of the ketogenic diet may identify novel neuroprotective and anticonvulsant therapeutic targets.

Several newer antiepileptic drugs such as zonisamide, topiramate, and levetiracetam inhibit mitochondrial dysfunction and possess antioxidant properties, raising the possibility that inhibition of mitochondrial dysfunction may in part underlie their actions. Oxidative stress and neuronal damage induced by kainate can be ameliorated by at least two types of superoxide dismutase (SOD) mimetics: the manganese porphyrin MnTBAP and the salen–manganese compound EUK-134. However, acute administration of catalytic antioxidants does not alter chemoconvulsant-induced behavioral seizure severity. Previous generations of manganese porphyrins (e.g., MnTBAP) showed limited ability to cross the blood–brain barrier (BBB), which necessitated intracerebral administration, and it remains to be determined if newer classes that are BBB permeable, orally active, and validated in animal models of neurodegeneration modulate seizure-induced injury or epileptogenesis. Other compounds with antioxidant properties that inhibit seizure-induced brain injury include the hormone melatonin, a nitrone spin trap, and vitamin C.

It should be noted that because ROS play an important physiological role in cell signaling, their removal with antioxidants may have deleterious consequences on cellular functions during chronic administration. Although iron chelators are poorly tolerated *in vivo*, chelation of the mitochondrial pool may be another avenue for therapeutic intervention. One promising example is the use of *N,N'-bis*(2-hydroxybenzyl) ethylenediamine-*N,N'*-diacetic acid (HBED), a lipophilic iron chelator that penetrates mitochondria following *in vivo* administration and inhibits seizure-induced hippocampal injury (Liang et al., 2008).

# REFERENCES

Beal, M.F. (1998). Excitotoxicity and nitric oxide in Parkinson's disease pathogenesis. *Ann. Neurol.*, 44:S110–S114.

Beal, M.F. (2000). Energetics in the pathogenesis of neurodegenerative diseases. *Trends Neurosci.*, 23:298–304.

Beal, M.F. (2003). Mitochondria, oxidative damage, and inflammation in Parkinson's disease. *Ann. N.Y. Acad. Sci.*, 991:120–131.

Ben-Ari, Y., E. Tremblay, D. Riche, G. Ghilini, R. Naquet. (1981). Electrographic, clinical and pathological alterations following systemic administration of kainic acid, bicuculline or pentetrazole: metabolic mapping using the deoxyglucose method with special reference to the pathology of epilepsy. *Neuroscience*, 6:1361–1391.

Bough, K.J., J. Wetherington, B. Hassel, J.F. Pare, J.W. Gawryluk, J.G. Greene, R. Shaw, Y. Smith, J.D. Geiger, R.J. Dingledine. (2006). Mitochondrial biogenesis in the anticonvulsant mechanism of the ketogenic diet. *Ann. Neurol.*, 60:223–235.

Cendes, F., F. Andermann, M.C. Preul, D.L. Arnold. (1994). Lateralization of temporal lobe epilepsy based on regional metabolic abnormalities in proton magnetic resonance spectroscopic images. *Ann. Neurol.*, 35:211–216.

Clapcote, S.J., S. Duffy, G. Xie, G. Kirshenbaum, A.R. Bechard, V.R. Schack, J. Petersen, L. Sinai, B.J. Saab, J.P. Lerch, B.A. Minassian, C.A. Ackerley, J.G. Sled, M.A. Cortez, J.T. Henderson, B. Vilsen, J.C. Roder. (2009). Mutation I810N in the alpha3 isoform of Na$^+$,K$^+$-ATPase causes impairments in the sodium pump and hyperexcitability in the CNS. *Proc. Natl. Acad. Sci. U.S.A.*, 106(33):14085–14090.

Clark, J.B. (1998). *N*-acetyl aspartate: a marker for neuronal loss or mitochondrial dysfunction. *Dev. Neurosci.*, 20:271–276.

Cohen-Gadol, A.A., J.W. Pan, J.H. Kim, D.D. Spencer, H.H. Hetherington. (2004). Mesial temporal lobe epilepsy: a proton magnetic resonance spectroscopy study and a histopathological analysis. *J. Neurosurg.*, 101:613–620.

DiMauro, S., R. Kulikova, K. Tanji, E. Bonilla, M. Hirano. (1999). Mitochondrial genes for generalized epilepsies. *Adv. Neurol.*, 79:411–419.

During, M.J., D.D. Spencer. (1993). Extracellular hippocampal glutamate and spontaneous seizure in the conscious human brain. *Lancet*, 341:1607–1610.

Garriga-Canut, M., B. Schoenike, R. Qazi, K. Bergendahl, T.J. Daley, R.M. Pfender, J.F. Morrison, J. Ockuly, C. Stafstrom, T. Sutula, A. Roopra. (2006). 2-Deoxy-D-glucose reduces epilepsy progression by NRSF-CtBP-dependent metabolic regulation of chromatin structure. *Nat. Neurosci.*, 9:1382–1387.

Greene, A.E., M.T. Todorova, R. McGowan, T.N. Seyfried. (2001). Caloric restriction inhibits seizure susceptibility in epileptic EL mice by reducing blood glucose. *Epilepsia*, 42:1371–1378.

Grisar, T., D. Guillaume, A.V. Delgado-Escueta. (1992). Contribution of Na$^+$,K(+)-ATPase to focal epilepsy: a brief review. *Epilepsy Res.*, 12:141–149.

Hall, E.D., P.K. Andrus, J.A. Oostveen, T.J. Fleck, M.E. Gurney. (1998). Relationship of oxygen-radical-induced lipid peroxidative damage to disease onset and progression in a transgenic model of familial ALS. *J. Neurosci. Res.*, 53:66–77.

Halliwell, B. (1992). Reactive oxygen species and the central nervous system. *J. Neurochem.*, 59:1609–1623.

Henneberry, R.C., A. Novelli, J.A. Cox, P.G. Lysko. (1989a). Neurotoxicity at the *N*-methyl-D-aspartate receptor in energy-compromised neurons: an hypothesis for cell death in aging and disease. *Ann. N.Y. Acad. Sci.*, 568:225–233.

Henneberry, R.C., A. Novelli, M.A. Vigano, J.A. Reilly, J.A. Cox, P.G. Lysko. (1989b). Energy-related neuro-toxicity at the NMDA receptor: a possible role in Alzheimer's disease and related disorders. *Prog. Clin. Biol. Res.*, 317:143–156.

Henshall, D.C., B.M. Murphy. (2008). Modulators of neuronal cell death in epilepsy. *Curr. Opin. Pharmacol.*, 8:75–81.

Hetherington, H., R. Kuzniecky, J. Pan, G. Mason, R. Morawetz, C. Harris, E. Faught, T. Vaughan, G. Pohost. (1995). Proton nuclear magnetic resonance spectroscopic imaging of human temporal lobe epilepsy at 4.1 T. *Ann. Neurol.*, 38:396–404.

Holtzman, D., A. Togliatti, I. Khait, F. Jensen. (1998). Creatine increases survival and suppresses seizures in the hypoxic immature rat. *Pediatr. Res.*, 44:410–414.

Holtzman, D., I. Khait, R. Mulkern, E. Allred, T. Rand, F. Jensen, R. Kraft. (1999). *In vivo* development of brain phosphocreatine in normal and creatine-treated rabbit pups. *J. Neurochem.*, 73:2477–2484.

Hugg, J.W., R.I. Kuzniecky, F.G. Gilliam, R.B. Morawetz, R.E. Fraught, H.P. Hetherington. (1996). Normalization of contralateral metabolic function following temporal lobectomy demonstrated by 1H magnetic resonance spectroscopic imaging. *Ann. Neurol.*, 40:236–239.

Jarrett, S.G., L.P. Liang, J.L. Hellier, K.J. Staley, M. Patel. (2008a). Mitochondrial DNA damage and impaired base excision repair during epileptogenesis. *Neurobiol. Dis.*, 30:130–138.

Jarrett, S.G., J.B. Milder, L.P. Liang, M. Patel. (2008b). The ketogenic diet increases mitochondrial glutathione levels. *J. Neurochem.*, 106:1044–1051.

Kann, O., R. Kovacs, M. Njunting, C.J. Behrens, J. Otahal, T.N. Lehmann, S. Gabriel, U. Heinemann. (2005). Metabolic dysfunction during neuronal activation in the *ex vivo* hippocampus from chronic epileptic rats and humans. *Brain*, 128:2396–2407.

Klivenyi, P., G. Gardian, N.Y. Calingasan, L. Yang, M.F. Beal. (2003). Additive neuroprotective effects of creatine and a cyclooxygenase 2 inhibitor against dopamine depletion in the 1-methyl-4-phenyl-1,2,3,6-tetrahydropyridine (MPTP) mouse model of Parkinson's disease. *J. Mol. Neurosci.*, 21:191–198.

Kovacs, R., S. Schuchmann, S. Gabriel, O. Kann, J. Kardos, U. Heinemann. (2002). Free-radical-mediated cell damage after experimental status epilepticus in hippocampal slice cultures. *J. Neurophysiol.*, 88:2909–2918.

Kudin, A.P., T.A. Kudina, J. Seyfried, S. Vielhaber, H. Beck, C.E. Elger, W.S. Kunz. (2002). Seizure-dependent modulation of mitochondrial oxidative phosphorylation in rat hippocampus. *Eur. J. Neurosci.*, 15:1105–1114.

Lian, X.Y., F.A. Khan, J.L. Stringer. (2007). Fructose-1,6-bisphosphate has anticonvulsant activity in models of acute seizures in adult rats. *J. Neurosci.*, 27:12007–12011.

Liang, L.P., M. Patel. (2004). Mitochondrial oxidative stress and increased seizure susceptibility in Sod2(–/+) mice. *Free Radic. Biol. Med.*, 36:542–554.

Liang, L.P., M. Patel. (2006). Seizure-induced changes in mitochondrial redox status. *Free Radic. Biol. Med.*, 40:316–322.

Liang, L.P., Y.S. Ho, M. Patel. (2000). Mitochondrial superoxide production in kainate-induced hippocampal damage. *Neuroscience*, 101:563–570.

Liang, L.P., S.G. Jarrett, M. Patel. (2008). Chelation of mitochondrial iron prevents seizure-induced mitochondrial dysfunction and neuronal injury. *J. Neurosci.*, 28:11550–11556.

Lothman, E. (1990). The biochemical basis and pathophysiology of status epilepticus. *Neurology*, 40:13–23.

Matthews, R.T., L. Yang, B.G. Jenkins, R.J. Ferrante, B.R. Rosen, R. Kaddurah-Daouk, M.F. Beal. (1998). Neuroprotective effects of creatine and cyclocreatine in animal models of Huntington's disease. *J. Neurosci.*, 18:156–163.

Matthews, R.T., R.J. Ferrante, P. Klivenyi, L. Yang, A.M. Klein, G. Mueller, R. Kaddurah-Daouk, M.F. Beal. (1999). Creatine and cyclocreatine attenuate MPTP neurotoxicity. *Exp. Neurol.*, 157:142–149.

Meldrum, B.S. (1983). Metabolic factors during prolonged seizures and their relation to nerve cell death. *Adv. Neurol.*, 34:261–275.

Meldrum, B.S. (2002). Concept of activity-induced cell death in epilepsy: historical and contemporary perspectives. *Prog. Brain Res.*, 135:3–11.

Melo, T.M., A. Nehlig, U. Sonnewald. (2006) Neuronal–glial interactions in rats fed a ketogenic diet. *Neurochem. Int.*, 48:498–507.

Nadler, J.V. (1979). Kainic acid: neurophysiological and neurotoxic actions. *Life Sci.*, 24:289–299.

Nicholls, D.G., S.L. Budd. (2000). Mitochondria and neuronal survival. *Physiol. Rev.*, 80:315–360.

Olney, J.W. (1986). Inciting excitotoxic cytocide among central neurons. *Adv. Exp. Med. Biol.*, 203:631–645.

Olney, J.W., T. deGubareff, R.S. Sloviter. (1983). "Epileptic" brain damage in rats induced by sustained electrical stimulation of the perforant path. II. Ultrastructural analysis of acute hippocampal pathology. *Brain Res. Bull.*, 10:699–712.

Pan, J.W., A. Williamson, I. Cavus, H.P. Hetherington, H. Zaveri, O.A. Petroff, D.D. Spencer. (2008). Neurometabolism in human epilepsy. *Epilepsia*, 49(Suppl. 3):31–41.

Patel, M., L.P. Liang, L.J. Roberts, 2nd. (2001). Enhanced hippocampal F2-isoprostane formation following kainate-induced seizures. *J. Neurochem.*, 79:1065–1069.

Patel, M., L.P. Liang, H. Hou, B.B. Williams, M. Kmiec, H.M. Swartz, J.P. Fessel, L.J. Roberts, 2nd. (2008). Seizure-induced formation of isofurans: novel products of lipid peroxidation whose formation is positively modulated by oxygen tension. *J. Neurochem.*, 104:264–270.

Petroff, O.A., L.D. Errante, D.L. Rothman, J.H. Kim, D.D. Spencer. (2002). Neuronal and glial metabolite content of the epileptogenic human hippocampus. *Ann. Neurol.*, 52:635–642.

Schapira, A.H., A. Hartley, M.W. Cleeter, J.M. Cooper. (1993). Free radicals and mitochondrial dysfunction in Parkinson's disease. *Biochem. Soc. Trans.*, 21:367–370.

Schulz, J.B., R.T. Matthews, T. Klockgether, J. Dichgans, M.F. Beal. (1997). The role of mitochondrial dysfunction and neuronal nitric oxide in animal models of neurodegenerative diseases. *Mol. Cell Biochem.*, 174:193–197.

Schwechter, E.M., J. Veliskova, L. Velisek. (2003). Correlation between extracellular glucose and seizure susceptibility in adult rats. *Ann. Neurol.*, 53:91–101.

Shoffner, J.M., M.T. Lott, A.M. Lezza, P. Seibel, S.W. Ballinger, D.C. Wallace. (1990). Myoclonic epilepsy and ragged-red fiber disease (MERRF) is associated with a mitochondrial DNA tRNA(Lys) mutation. *Cell*, 61:931–937.

Sloviter, R.S. (1983). "Epileptic" brain damage in rats induced by sustained electrical stimulation of the perforant path. I. Acute electrophysiological and light microscopic studies. *Brain Res. Bull.*, 10:675–697.

Sullivan, P.G., N.A. Rippy, K. Dorenbos, R.C. Concepcion, A.K. Agarwal, J.M. Rho. (2004). The ketogenic diet increases mitochondrial uncoupling protein levels and activity. *Ann. Neurol.*, 55:576–80.

Todorova, M.T., P. Tandon, R.A. Madore, C.E. Stafstrom, T.N. Seyfried. (2000). The ketogenic diet inhibits epileptogenesis in EL mice: a genetic model for idiopathic epilepsy. *Epilepsia*, 41:933–940.

Trotti, D., N.C. Danbolt, A. Volterra. (1998). Glutamate transporters are oxidant-vulnerable: a molecular link between oxidative and excitotoxic neurodegeneration? *Trends Pharmacol. Sci.*, 19:328–334.

Vielhaber, S., A.P. Kudin, T.A. Kudina, D. Stiller, H. Scheich, A. Schoenfeld, H. Feistner, H.J. Heinze, C.E. Elger, W.S. Kunz. (2003). Hippocampal *N*-acetyl aspartate levels do not mirror neuronal cell densities in creatine-supplemented epileptic rats. *Eur. J. Neurosci.*, 18:2292–2300.

Vielhaber, S., H.G. Niessen, G. Debska-Vielhaber, A.P. Kudin, J. Wellmer, J. Kaufmann, M.A. Schonfeld, R. Fendrich, W. Willker, D. Leibfritz, J. Schramm, C.E. Elger, H.J. Heinze, W.S. Kunz. (2008). Subfield-specific loss of hippocampal *N*-acetyl aspartate in temporal lobe epilepsy. *Epilepsia*, 49:40–50.

Wallace, D.C. (2005). A mitochondrial paradigm of metabolic and degenerative diseases, aging, and cancer: a dawn for evolutionary medicine. *Annu. Rev. Genet.*, 39:359–407.

Wallace, D.C., X.X. Zheng, M.T. Lott, J.M. Shoffner, J.A. Hodge, R.I. Kelley, C.M. Epstein, L.C. Hopkins. (1988). Familial mitochondrial encephalomyopathy (MERRF): genetic, pathophysiological, and biochemical characterization of a mitochondrial DNA disease. *Cell*, 55:601–610.

# 4 Brain Inflammation and Epilepsy

*Teresa Ravizza, Silvia Balosso,*
*Eleonora Aronica, and Annamaria Vezzani*

## CONTENTS

## INTRODUCTION

Inflammation includes a variety of protective processes that have evolved to activate a defensive attack against noxious stimuli, thus representing an important endogenous homeostatic mechanism of the organism. Usually, the outcome of the inflammatory program is a rapid repair of tissue damage; however, if these processes are not properly controlled in timing and extent, then inflammation becomes deleterious, thus leading to permanent tissue damage and cellular dysfunction, as suggested to occur in chronic neurodegenerative disorders and in epilepsy.

This chapter provides a brief overview of the experimental and clinical evidence linking brain inflammation to epilepsy. We describe three groups of inflammatory mediators—namely, cytokines, the complement system, and cyclooxygenase-2 (COX-2). Their expression is increased in rodent and human epileptogenic tissue, such as in temporal lobe epilepsy and malformations of cortical development that do not feature a typical inflammatory pathophysiology. Pharmacological attempts have been made in experimental models to understand their functional role in seizure activity, epileptogenesis, and seizure-induced neuronal cell death. Current knowledge suggests an involvement of these specific inflammatory pathways in the pathogenesis of seizures, thus highlighting new potential therapeutic strategies. In this chapter, we first describe changes in the expression of cytokines, the complement system, and COX-2 in epileptogenic tissue of experimental models and humans. We next discuss the pharmacological data addressing the functional consequences of brain expression of these inflammatory mediators on seizures, epileptogenesis, and cell loss. Finally, recent evidence describing the mechanisms by which brain inflammation can alter neuronal circuit excitability and seizure activity is reported.

## SEIZURE-INDUCED EXPRESSION OF INFLAMMATORY MEDIATORS

### EXPERIMENTAL MODELS

#### Proinflammatory Cytokines

A novel concept emerging from the literature is that seizure activity induced in rodents increases the production of various inflammatory molecules in the brain. In particular, a rapid inflammatory response has been detected in response to seizures induced by chemoconvulsant application, electrical stimulation, or kindling in brain areas recruited in the generation and spread of seizures (De Simoni et al., 2000; Gorter et al., 2006; Plata-Salaman et al., 2000; Shinoda et al., 2003; Vezzani et al., 1999). This response includes a rapid increase of proinflammatory cytokines, such as interleukin-1β (IL-1β), IL-6, and tumor necrosis factor-α (TNF-α) prominently in glia, and also involves endothelial cells and neurons (De Simoni et al., 2000; Ravizza et al., 2008a; Vezzani et al., 1999). This phenomenon is followed by a cascade of downstream inflammatory events, such as upregulation of Toll-like receptors (TLRs), activation of NFκB, chemokine production, complement system activation, and increased expression of adhesion molecules (Aronica et al., 2007; Fabene et al., 2008; Librizzi et al., 2007; Turrin and Rivest, 2004; Vezzani and Granata, 2005; Vezzani et al., 2008). Using experimental models of status epilepticus (SE) in which epileptogenesis is triggered and spontaneous seizures occur, we analyzed the temporal evolution of inflammatory changes in the brain. We studied the immunohistochemical pattern of IL-1β and its signaling receptor IL-1 receptor type 1 (IL-1R1) as prototypical markers of inflammation (Figure 4.1).

IL-1β is barely detectable in healthy brain, but during the acute phases of SE it is highly expressed by microglia (Figure 4.1A) and reactive parenchymal (Figure 4.1B) and perivascular astrocytes (Ravizza et al., 2008a). IL-1β upregulation persists during epileptogenesis (Figure 4.1C), in the absence of ongoing seizure activity, and in chronic epileptic tissue characterized by spontaneous and recurrent seizures (Figure 4.1D) (Ravizza et al., 2008a). IL-1R1 upregulation involves both neurons (Figure 4.1E–G) and astrocytes (Figure 4.1H) (Ravizza and Vezzani, 2006; Ravizza et al., 2008a), suggesting that this cytokine released by glia exerts both autocrine and paracrine actions. IL-1β and IL-1R1 are highly expressed during epileptogenesis, both in the astrocytic endfeet impinging on brain microvasculature and in endothelial cells of the blood–brain barrier (BBB). These events occur in areas of BBB leakage and neuronal damage (Ravizza et al., 2008a), thus suggesting that inflammation may be responsible for these pathological changes.

The production of proinflammatory molecules is usually accompanied by the concomitant synthesis of antiinflammatory mediators to modulate the inflammatory response and avoid the occurrence of deleterious effects. In this respect, IL-1 receptor antagonist (IL-1Ra), a naturally occurring IL-1β antagonist, also increases during seizures (Figure 4.1I). IL-1Ra is induced in the brain several

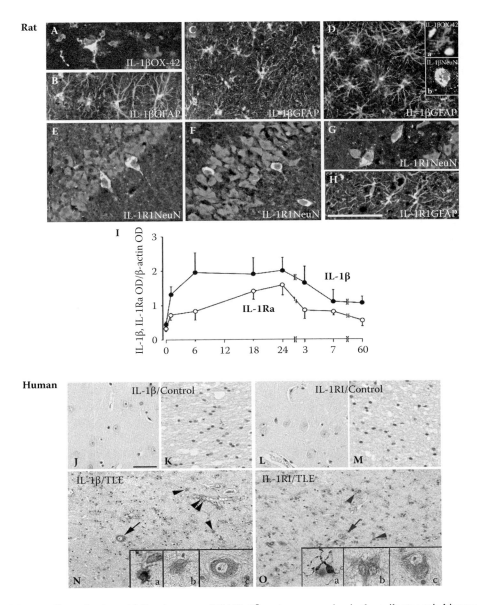

**FIGURE 4.1** **(See color insert following page 458.)** IL-1β system expression in the epileptogenic hippocampus of rats and temporal lobe epilepsy (TLE) patients. Panels A to H depict double–immunofluorescence micrographs showing the cellular localization of IL-1β (A–D) and IL-1R1 (E–H) in the rat hippocampus after status epilepticus (SE). During the acute phases of SE, IL-1β is expressed by reactive microglia cells (A) and astrocytes (B). Astrocytic upregulation persists during epileptogenesis (C) and in chronically epileptic tissue (D), where microglia (D, inset a) and neurons (D, inset b) also express the cytokine. IL-1R1 upregulation occurs both in neurons (E–G) and astrocytes (H). Panel I shows IL-1β and IL-1ra mRNA expression at different time points after electrically induced SE. Note the rapid increase in IL-1β levels, which are still higher than the control level 60 days after SE in rats with spontaneous seizures. IL-1ra, the endogenous IL-1R1 antagonist, is induced with a delayed time course and to a lesser extent as compared to IL-1β. Panels J to O present photomicrographs of immunohistochemical staining of IL-1β (J, K, N) and IL-1R1 (L, M, O) in the hippocampus of TLE patients. In epileptic specimens (N, O), note the strong activation of IL-1β (N) and IL-1R1 (O) in glial cells (N and O, insets a and b) and in neurons (N and O, inset c). In control tissue obtained at autopsy from patients without a history of seizures or other neurological diseases, no IL-1β (J, K) or IL-1R1 (L, M) staining was observed. (Adapted from De Simoni, M.G. et al., *Eur. J. Neurosci.*, 12(7), 2623–2633, 2000; Ravizza, T. et al., *Neurobiol. Dis.*, 29(1), 142–160, 2008.)

hours after IL-1β and never in excess of IL-1β (Figure 4.1I) (De Simoni et al., 2000), whereas it should be produced concomitantly and ≥100-fold in excess of IL-1β to block the effects of the inflammatory cytokine (Dinarello, 1996). Thus, the effects of IL-1β upon its production in brain cannot be rapidly and effectively controlled by IL-1ra.

TNF-α and IL-6 also increase in glial cells similarly to IL-1β, but their expression declines to basal levels within 48 to 72 hours from the onset of SE (De Simoni et al., 2000). The IL-6 signaling transducer protein Gp130 is increased after seizures in granule cells and in astrocytes of the hippocampus (Choi et al., 2003; Lehtimaki et al., 2003), but no information is available on the cellular expression of TNF-α receptors.

The cascade of inflammatory mediators intrinsic to brain parenchyma can recruit adaptive immunity components, thus leading to infiltration into the brain parenchyma of leukocytes. T cells and B cells, however, are not prominent in brain tissue of experimental models of temporal lobe epilepsy (TLE) (Ravizza et al., 2008a; Turrin and Rivest, 2004) or in human epileptogenic tissue from TLE patients (Ravizza et al., 2008a), in contrast to what is observed in Rasmussen's encephalitis (Bien et al., 2002) or tuberous sclerosis (Boer et al., 2008).

Importantly, seizures can induce cytokine expression independently of ongoing degenerative processes. Thus, cytokine induction by experimental seizures precedes, by several hours, the onset of detectable neurodegeneration (Ravizza and Vezzani, 2006), and cytokines are induced when epileptic activity is not associated with neurodegenerative effects (Dubé et al., 2005; Vezzani et al., 1999). On the other hand, the available evidence suggests that persistent inflammation can contribute to seizure-induced neuronal cell loss (Bernardino et al., 2005; Rizzi et al., 2003; Viviani et al., 2003; for review see Allan et al., 2005).

## Complement System

Although the synthesis of complement system components occurs predominantly in the liver, both glia and neurons can express these inflammatory mediators in pathological conditions (Barnum, 1995; Morgan and Gasque, 1996). Rozovsky et al. (1994) reported that ClqB and C4 messenger RNAs increased within 48 hours after kainate injection, particularly in the CA3 pyramidal layer, and C1q immunoreactivity was detected in CA1 pyramidal neurons. These changes were blocked by barbiturates that prevented seizures and neurodegeneration. Using quantitative polymerase chain reaction (PCR), an increase in the expression of the complement-related factors C1q, C3, and C4 has been reported in the rat hippocampus during the acute phases of SE, in epileptogenesis, and in chronic epileptic tissue (Aronica et al., 2007). Immunohistochemical analysis showed strong C1q (Figure 4.2B) and C3d (Figure 4.2B inset) upregulation, with these proteins being expressed mainly by microglia and neurons as well as by astrocytic endfeet surrounding blood vessels (Aronica et al., 2007). Weak complement immunostaining was observed in control tissue (Figure 4.2A). Inhibitory proteins of the complement cascade (Serping1, CD59, and Crry) were also induced during epileptogenesis in rat hippocampus, although to a lesser extent (Aronica et al., 2007).

## COX-2

COX-2 is expressed in discrete populations of neurons under basal conditions and is enriched in the cortex and hippocampus (Hurley et al., 2002; Yamagata et al., 1993). COX-2 and its diffusible prostanoid products participate in the inflammatory response (Seibert and Masferrer, 1994), neuronal death (Nakayama et al., 1998), neuronal hyperexcitability (Willingale et al., 1997), and astrocytic activation (Voutsinos-Porche et al., 2004); also, COX-2 is dramatically increased in the brain in various neurological disorders (for review, see Minghetti, 2007). It has also been suggested that COX-2 and prostaglandins play a role in epilepsy based on their brain expression and on pharmacological studies (see later). Indeed, COX-2 expression is induced in the hippocampus after kindling (Tu and Bazan, 2003), in the genetically susceptible EL mice (Okada et al., 2006), and in kainate-induced seizures (Chen et al., 1995; Hirst et al., 1999; Sandhya et al., 1998). Rapid cell-specific upregulation

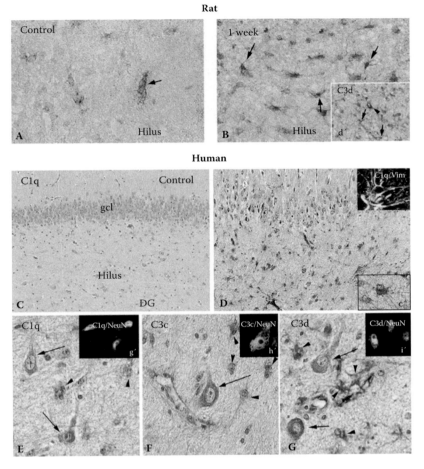

**FIGURE 4.2** **(See color insert following page 458.)** Complement factor expression in epileptogenic rat hippocampus and in temporal lobe epilepsy (TLE) patients. Panels A and B depict C1q and C3d (inset in B) immunostaining in the rat hippocampus after status epilepticus (SE). Control hippocampus shows weak C1q immunoreactivity in the hilar region (A), whereas 1 week after SE onset a strong induction of C1q (B) and C3d (B, inset) was observed in glial cells. Panels C to G represent photomicrographs of immunohistochemical staining for C1q (C–E), C3c (F), and C3d (G) in the hippocampus of TLE patients. In control tissue, C1q immunostaining is not detectable (C), whereas in epileptic specimens a strong C1q signal was observed in reactive astrocytes (D and insets) and in neurons (E, inset). Panels F and G are photomicrographs showing C3c (F) and C3d (G) in epileptic tissue. C3c and C3d immunostaining was found in neurons (F and G, arrow and inset) and astrocytes (F and G, arrowheads). (Adapted from Aronica, E. et al., *Neurobiol. Dis.*, 26(3), 497–511, 2007.)

of COX-2 expression was observed in the mouse hippocampus after SE (Lee et al., 2007); early COX-2 induction was found in neurons, whereas delayed COX-2 expression was detected in astrocytes and microglia (Figure 4.3A).

## HUMAN EPILEPTIC TISSUE

### Proinflammatory Cytokines

Clinical insights into a possible role of inflammation in epilepsy are provided by the evidence that antiinflammatory drug therapies, including steroids, result in anticonvulsant effects (Vezzani and Granata, 2005). Moreover, several studies showed increased levels of inflammatory cytokines in

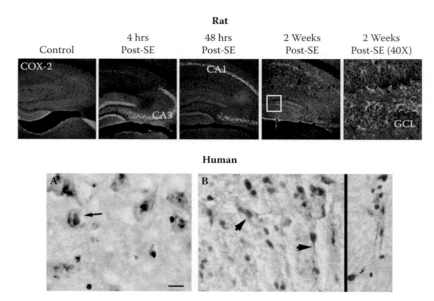

**FIGURE 4.3   (See color insert following page 458.)** COX-2 expression in the epileptogenic rat hippocampus and in TLE patients. Immunofluorescence micrographs depict COX-2 immunoreactivity at various time points after status epilepticus (SE) onset (4 hours to 2 weeks). Rapid neuronal upregulation of COX-2 expression was observed starting from 4 hours to 48 hours post-SE (second and third panels). In contrast, from 2 to 14 days after SE, COX-2 was observed in both astrocytes and microglia (fourth and fifth panels). Panels A and B are photomicrographs of immunohistochemical staining for COX-2 in the hippocampus of TLE patients. COX-2 immunostaining was found in neurons (A, arrow) and in astrocytes in TLE patients with hippocampal sclerosis (B, arrowheads). (Adapted from Desjardins, P. et al., *Neurochem. Int.*, 42(4), 299–303, 2003; Lee, B. et al., *Neurobiol. Dis.*, 25(1), 80–91, 2007.)

cerebral spinal fluid (CSF) and plasma of epileptic patients. Among the various cytokines measured, IL-6 is the only one consistently and significantly upregulated after different types of seizures in humans. Because the concentration of IL-6 is higher in the CSF than in plasma, the most likely origin of this cytokine appears to be the brain (Lehtimaki et al., 2004; Peltola et al., 2002). Elevation in cytokine levels has also been described in West syndrome patients and in children with febrile seizures (Vezzani and Granata, 2005).

In TLE hippocampal specimens, strong activation of the IL-1β system (Figure 4.1N,O) (Ravizza et al., 2008a) and NFκB (Crespel et al., 2002) was observed in reactive glia and surviving neurons, whereas these inflammatory molecules were not detected in postmortem tissue from neurologically unaffected subjects (Figure 4.1J–M) (Crespel et al., 2002; Ravizza et al., 2008a) or in surgically removed tissue from epileptic patients with an extrahippocampal lesion (Ravizza et al., 2008a).

Because inflammatory processes occur in human chronic epileptic tissue and during epileptogenesis in experimental models, it is possible that their activation precedes the onset of epilepsy, possibly playing an etiopathogenic role in the occurrence of spontaneous seizures. Evidence of inflammatory processes in lesional tissue is also reported in epilepsy associated with malformation of cortical development (Boer et al., 2006; Maldonado et al., 2003; Ravizza et al., 2006a). Moreover, a positive correlation was found between the density of IL-1β-positive cells and the frequency of seizures before tissue resection in focal cortical dysplasia and glioneuronal tumor specimens (Ravizza et al., 2006a). The presence of cells of adaptive immunity (B, T, and NK cells) was observed depending on the etiology of epilepsy. A notable absence of T and B cells is reported in TLE specimens (Ravizza et al., 2008a), while in tuberous sclerosis (Boer et al., 2008) and Rasmussen's encephalitis (Bien et al., 2002) these cells have been detected in brain parenchyma and may play a prominent role in the neuropathology.

## Complement System

In TLE patients with hippocampal sclerosis, strong immunoreactivity of various complement factors including C1q (Figure 4.2D,E), C3c (Figure 4.2F), C3d (Figure 4.2G), and C5b–C9 has been described in glial cells and, in some instances, in neurons in the same areas showing cell loss and prominent gliosis (Aronica et al., 2007). The expression of CD59, an inhibitor of the complement cascade, was increased in microglia, but the CD59 signal in neurons was modest or not detectable (Aronica et al., 2007).

## COX-2

In hippocampal biopsies from patients with refractory TLE, immunocytochemical analysis showed increased COX-2 immunoreactivity in neurons (arrow in Figure 4.3B). Interestingly, additional astrocytic expressions was observed only in patients with hippocampal sclerosis (arrowheads in Figure 4.3B), confirming the evidence obtained in experimental models (see above). Increased cerebrospinal fluid levels of prostaglandins have been also measured in patients with epilepsy (Wolfe and Mamer, 1975), supporting an increased activation of COX-2 in human epilepsy.

# FUNCTIONAL CONSEQUENCES OF UPREGULATION OF INFLAMMATORY MOLECULES IN BRAIN

## PROINFLAMMATORY CYTOKINES

### IL-1β

The application of IL-1β in rodent brain shortly before the convulsive stimulus, using concentrations within the range of those endogenously produced by seizures, results in proconvulsant effects mediated by IL-1R1 (Dubé et al., 2005; Vezzani et al., 1999, 2000). In contrast, the intracerebral injection of IL-1ra mediates powerful anticonvulsant activity (Vezzani et al., 1999, 2000), and mice overexpressing IL-1ra in astrocytes show a reduced susceptibility to seizures (Vezzani et al., 2000). Because IL-1ra inhibits IL-1β effects, these data highlight the important contribution of endogenous brain IL-1β to seizures. These data are reinforced by the evidence that inhibition of IL-1β synthesis in rodents, using selective blockers of interleukin-converting enzyme (ICE, or caspase-1), significantly reduces seizure frequency and their duration and arrests kindling development (Ravizza et al., 2008b). Recently, systemic injection of IL-1Ra decreased the likelihood of SE precipitation in one rat model of seizures associated with peripheral inflammatory reactions (Marchi et al., 2009). Intracerebral application of IL-1β in immature rodents decreased the threshold of seizure induction in two models of febrile convulsions induced by hyperthermia (Dubé et al., 2005) or by systemic injection of lipopolysaccharide (LPS) (Heida and Pittman, 2005). Moreover, the seizure threshold for febrile convulsions was increased in IL-1R1-deficent mice or in IL-1ra-injected rats (Dubé et al., 2005). These data suggest that IL-1β signaling is significantly involved in fever-associated hyperexcitability underlying the occurrence of febrile seizures.

### TNF-α

The role of TNF-α in seizures is still controversial. Nanomolar amounts of mouse recombinant TNF-α injected into the mouse hippocampus significantly reduced seizures by interacting with TNF-α type II receptors (Balosso et al., 2005). Accordingly, transgenic mice with low to moderate overexpression of TNF-α in astrocytes showed decreased susceptibility to seizure (Balosso et al., 2005). Mice with high overexpression of TNF-α develop signs of neurological dysfunction, including seizures (Akassoglou et al., 1997). Recently, a protective role of TNF-α on seizures was detected using mice with a genetic deletion of the TNF-α type I receptors (Lu et al., 2008). These findings highlight the importance of the brain concentration of this cytokine and the receptor subtype predominantly involved in its action with regard to the final outcome of seizure activity.

## IL-6

A set of experimental data supports an anticonvulsant role of this cytokine. IL-6 attenuates hyperthermia-induced seizures in developing rats, and adult mice lacking the IL-6 gene showed increased susceptibility to various chemoconvulsants and an enhanced propensity to develop audiogenic seizures (De Luca et al., 2004); however, a proconvulsant activity of IL-6 has been reported after pentylenetetrazol-induced seizures (Fukuda et al., 2007). Also, transgenic mice overexpressing IL-6 in astrocytes showed an increased susceptibility to seizures induced by glutamatergic agonists (Samland et al., 2003) and sporadic spontaneous seizures (Campbell et al., 1993), indicating a dual role for this cytokine in neuronal excitability.

## COMPLEMENT SYSTEM

The role of activation of the complement system in seizures is suggested by a study showing that sequential intrahippocampal injection of the complement factors C5b6, C7, C8, and C9, to induce the formation of the membrane attack complex (MAC), promotes behavioral and electrographic seizures in rats and pronounced hippocampal cell loss (Xiong et al., 2003).

## COX-2

The use of transgenic mice with increased neuronal expression of COX-2 has highlighted a potential proconvulsant role for this enzyme, as these transgenic mice were more susceptible to kainic-acid-induced SE (Kelley et al., 1999). Accordingly, COX-2-deficient mice, or mice treated with a COX-2 inhibitor, have a reduced susceptibility to kindling (Takemiya et al., 2003). Tu and Bazan (2003) showed that kindling induces COX-2 expression as well as the expression of cytosolic phospholipase A(2), an enzyme that catalyzes the synthesis of arachidonic acid, the substrate for COX-2 in prostaglandin synthesis. They also found that a COX-2 selective inhibitor attenuated kindling development. Detailed pharmacological studies using selective COX-2 inhibitors clearly indicate that induction of COX-2 can mediate either proictogenic or anticonvulsant effects, likely depending on the type of prostaglandins produced in the various models of seizures. This concept is exemplified by the finding that pretreatment of rats with COX-2 inhibitors exacerbates pilocarpine- and kainate-induced SE (Baik et al., 1999; Kim et al., 2008; Naffah-Mazzacoratti et al., 1995). Meanwhile, post-SE treatment with COX-2 inhibitors was found to mediate neuroprotective (Kunz and Oliw, 2001) and antiepileptogenic (Jung et al., 2006) effects by decreasing the frequency of spontaneous seizures. However, a lack of effect of COX-2 inhibition on epileptogenesis and spontaneous seizures have been recently reported (Holtman et al., 2009). The profile of prostaglandin induction in the early phase of SE (when COX-2 is predominantly expressed by neurons) shows a preponderance of prostaglandin $F_2$ ($PGF_2$) vs. prostaglandin $E_2$ ($PGE_2$), but the opposite is true during the epileptogenesis phase, when COX-2 is also expressed by astrocytes (Naffah-Mazzacoratti et al., 1995). Therefore, the dual effect of COX-2 inhibition in seizures can be explained if one considers that $PGF_2$ has anticonvulsant properties (Kim et al., 2008) and $PGE_2$ is a proneurotoxic and proconvulsant prostaglandin (Oliveira et al., 2008; Takemiya et al., 2006). Induction of COX-2 is also known to contribute to $N$-methyl-D-aspartate (NMDA)-mediated neuronal cell death in cultured neurons (Hewett et al., 2000), and transgenic mice overexpressing neuronal COX-2 are more susceptible to kainic acid excitotoxicity (Kelley et al., 1999).

## SEIZURE SUSCEPTIBILITY AND SYSTEMIC INFLAMMATION/INFECTION

Several lines of evidence have shown that a preexisting inflammatory state in the periphery or in the brain can increase seizure susceptibility and have an impact on brain excitability in rodents. In particular, systemic LPS, a bacterial toxin mimicking Gram-negative bacteria infection, facilitates spike–wave discharge activity in genetically seizure-prone rats (Kovacs et al., 2006) and decreases the threshold to pentylenetetrazole (PTZ)-induced seizures in mice (Sayyah et al., 2003). Moreover, adult rats injected intracolonicly with 2,4,6-trinitrobenzene sulfonic acid to mimic inflammatory bowel disease showed increased susceptibility to PTZ-induced seizures involving brain TNF-α induction (Riazi et al., 2008). IL-1β and TNF-α have also been involved in the increased seizure susceptibility occurring in rodents after the exposure to specific pathogens mimicking systemic infection, such as *Shigella dysenteriae*, or in the rat neonatal model of pneumococcal meningitis (Vezzani and Granata, 2005). Lipopolysaccharide injection during a critical postnatal period (between postnatal days 7 and 14) in rats induces a long-lasting increase in their seizure susceptibility to various convulsive challenges in adulthood, associated with evidence of enhanced field excitatory postsynaptic potential (EPSP) slopes after Schaffer collateral stimulation and increased epileptiform burst-firing activity in CA1 after 4-aminopyridine application to *ex vivo* hippocampal slices (Galic et al., 2008). This effect appears to involve predominantly LPS-induced production of TNF-α in brain. In adult rats exposed to systemic inflammation during their infancy, seizures induce a higher brain inflammatory reaction and a more pronounced cell loss (Galic et al., 2008). Finally, systemic LPS administration in 7- or 14-day-old rats increases SE-induced neuronal cell loss, showing that inflammation exacerbates seizure-induced injury in immature brain (Lee et al., 2000; Sankar et al., 2007). A recent report showed that LPS may cause epileptiform activity after rat cortical application with a rapid onset time-frame (Rodgers et al., 2009).

## INFLAMMATION-RELATED MECHANISMS UNDERLYING HYPEREXCITABILITY AND NEUROTOXICITY

### PROINFLAMMATORY CYTOKINES

The deleterious effects of cytokines on neuronal excitability and cell survival are likely to be mediated by their ability to induce the production of toxic mediators via autocrine or paracrine mechanisms (Allan et al., 2005). A novel view is that cytokines can alter the function of classical neurotransmitters by modulating their receptor assembly at neuronal membranes (Viviani et al., 2007).

### IL-1β

IL-1R1 colocalizes on hippocampal neurons with NMDA receptors, a subtype of glutamate receptors involved in the onset and spread of seizures. IL-1β, via the activation of neuronal IL-1R1, promotes Src kinase-mediated tyrosine phosphorylation of the NR2B subunit of the NMDA receptors. This action leads to an increase of NMDA-mediated $Ca^{2+}$ influx into neurons, resulting in an enhanced excitotoxicity (Viviani et al., 2003) and contributing to seizure generation (Balosso et al., 2008). IL-1β may also increase extracellular glutamate levels via at least two mechanisms: (1) the inhibition of glutamate reuptake by astrocytes (Casamenti et al., 1999; Hu et al., 2000), and (2) the increase of astrocytic glutamate release either directly (to be demonstrated) or via induction of TNF-α release by microglia (Bezzi et al., 2001). In this regard, it has been recently reported that astrocytic glutamate release contributes significantly to seizure-like events (Fellin et al., 2006; Tian et al., 2005). IL-1β also inhibits γ-aminobutyric acid (GABA)-mediated Cl⁻ fluxes, thus possibly contributing to hyperexcitability by reducing inhibitory transmission (Wang et al., 2000; Zeise et al., 1997).

The presence of IL-1β and IL-1R1 on endothelial cells and perivascular astrocytes strongly suggests that inflammation can impair the permeability function of the BBB. In this respect, IL-1β can promote the disassembling of the tight junctions, the production of nitric oxide (NO), and the activation of matrix metalloproteinases in endothelial cells (Allan et al., 2005). These changes may have significant consequences on brain excitability, as the extravasation of serum albumin or immunoglobulin G (IgG) in brain parenchyma following BBB damage results in chronic neuronal excitability and neuronal injury (Rigau et al., 2007; Seiffert et al., 2004). Of note, the BBB damage can promote seizures (Fabene et al., 2008; Marchi et al., 2007), and a positive correlation has been reported between the extent of BBB damage and the frequency of spontaneous seizures in rats (van Vliet et al., 2007).

## TNF-α

A functional interaction between TNF-α and α-amino-3-hydroxy-5-methyl-4-isoxazoleproprionic acid (AMPA) receptors has been recently described that involves TNF-α type I receptor (Beattie et al., 2002; Stellwagen et al., 2005). Thus, TNF-α increases the mean frequency of AMPA-dependent miniature excitatory postsynaptic currents in hippocampal neurons and decreases GABA$_A$-mediated inhibitory synaptic strength. These effects are mediated by the ability of TNF-α to activate the recruitment of AMPA receptors lacking the GluR2 subunit at neuronal membranes. This produces a molecular conformation that favors Ca$^{2+}$ influx into neurons, resulting in an amplification of the glutamate response. Conversely, TNF-α promotes the endocytosis of GABA$_A$ receptors, thus decreasing inhibition.

These fast posttranslational effects of inflammatory cytokines represent novel and nonconventional pathways by which inflammatory molecules produced in diseased tissue can affect neurotransmission and possibly contribute to hyperexcitability and the associated neuropathology.

### COMPLEMENT SYSTEM

*In vitro* studies suggest a mechanism by which MAC (C5–C9) can trigger seizures and cytotoxicity. MAC assembly in the erythrocyte membrane leads to the formation of channel conductances, resulting in Ca$^{2+}$ and Na$^+$ influx and K$^+$ efflux, with the net effect of depolarizing the membrane potential (Wiedmer and Sims, 1985). Based on these data, we can speculate that the formation of MAC on neuronal membranes may trigger similar ion fluxes, thus resulting in depolarization and increased neuronal excitability. The same ion fluxes can also induce cell swelling and subsequent osmotic lysis (Wiedmer and Sims, 1985). In addition, complement proteins can also increase vascular permeability (for review, see Lucas et al., 2006); thus, their expression in perivascular astrocytes can be involved in the increased BBB permeability properties observed during seizures and epileptogenesis.

## COX-2

The available data mostly relate to the effects of PGE$_2$ on synaptic transmission and neuronal excitability (Cole-Edwards and Bazan, 2005). In particular, there is evidence that COX-2, which is expressed in postsynaptic dendritic spines, regulates via PGE$_2$ synaptic signaling. Somatic and dendritic membrane excitability was significantly reduced in CA1 pyramidal neurons in hippocampal slices when endogenous PGE$_2$ was eliminated with a selective COX-2 inhibitor. Accordingly, the exogenous application of PGE$_2$ produced significant increases in the frequency of firing and EPSP amplitudes, most likely by reducing potassium currents in CA1 neurons (Chen and Bazan, 2005). In addition, COX-2 inhibition decreases basal excitatory transmission in CA1 hippocampal slices, an effect prevented by CB1 blockade and independent of inhibitory transmission (Slanina and Schweitzer, 2005), thus suggesting an additional mechanism by which COX-2 may modulate hippocampal excitatory transmission. COX-2-mediated prostaglandin synthesis leads to the production of free radicals as intermediate products; these, in turn, can potentiate glutamate-mediated effects (Dawson et al., 1991). The production of PGE$_2$ from TNF-α-activated astrocytes can mediate astrocytic Ca$^{2+}$-dependent glutamate release (Bezzi et al., 2001), thus contributing to ictal activity and excitotoxicity (Ding et al., 2007; Fellin et al., 2006).

## CONCLUSIONS

The experimental and clinical data indicate that specific inflammatory pathways are chronically activated in epileptogenic brain tissue. Prolonged stimulation of proinflammatory signals by recurrent seizures and the contribution of injured cells to perpetuate inflammation may play significant roles in the establishment of a pathological substrate, including neuronal hyperexcitability, cell loss, and blood–brain barrier damage. The initiation of inflammatory responses in the brain can be a consequence of intrinsic injurious events or may first originate in the periphery. Both scenarios can be envisaged, and they are not mutually exclusive (Vezzani and Baram, 2007; Vezzani and Granata, 2005). The identification of proinflammatory pathways activated by seizures and during the epileptogenic process following an initial precipitating event, as well as the elucidation of the role of inflammatory mediators in brain, are instrumental to understanding the consequences of inflammation on brain function. Additionally, pathway identification is necessary to envisage the development of novel antiepileptic pharmacological approaches targeting specific detrimental processes without compromising the homeostatic role of inflammation.

## ACKNOWLEDGMENTS

The authors acknowledge the contribution of Drs. M. Rizzi, F. Noé, and D. Zardoni (Mario Negri Institute, Milano) to part of these studies. We also acknowledge EPICURE (LSH-CT-2006-037315 [AV]), Dana Foundation [AV], and NeuroGlia (EU-FP7-202167 [EA]).

## REFERENCES

Akassoglou, K., L. Probert, G. Kontogeorgos, G. Kollias. (1997). Astrocyte-specific but not neuron-specific transmembrane TNF triggers inflammation and degeneration in the central nervous system of transgenic mice. *J. Immunol.*, 158(1):438–445.

Allan, S.M., P.J. Tyrrell, N.J. Rothwell. (2005). Interleukin-1 and neuronal injury. *Nat. Rev. Immunol.*, 5(8):629–640.

Aronica, E., K. Boer, E.A. van Vliet, S. Redeker, J.C. Baayen, W.G. Spliet, P.C. van Rijen, D. Troost, F.H. da Silva, W.J. Wadman, J.A. Gorter. (2007). Complement activation in experimental and human temporal lobe epilepsy. *Neurobiol. Dis.*, 26(3):497–511.

Baik, E.J., E.J. Kim, S.H. Lee, C. Moon. (1999). Cyclooxygenase-2 selective inhibitors aggravate kainic acid induced seizure and neuronal cell death in the hippocampus. *Brain Res.*, 843(1–2):118–129.

Balosso, S., T. Ravizza, C. Perego, J. Peschon, I.L. Campbell, M.G. De Simoni, A. Vezzani. (2005). Tumor necrosis factor-alpha inhibits seizures in mice via p75 receptors. *Ann. Neurol.*, 57(6):804–812.

Balosso, S., M. Maroso, M. Sanchez-Alavez, T. Ravizza, A. Frasca, T. Bartfai, A. Vezzani. (2008). A novel non-transcriptional pathway mediates the proconvulsive effects of interleukin-1beta. *Brain*, 131(Pt. 12):3256–3265.

Barnum, S.R. (1995). Complement biosynthesis in the central nervous system. *Crit. Rev. Oral Biol. Med.*, 6(2):132–146.

Beattie, E.C., D. Stellwagen, W. Morishita, J.C. Bresnahan, B.K. Ha, M. Von Zastrow, M.S. Beattie, R.C. Malenka. (2002). Control of synaptic strength by glial TNFalpha. *Science*, 295(5563):2282–2285.

Bernardino, L., S. Xapelli, A.P. Silva, B. Jakobsen, F.R. Poulsen, C.R. Oliveira, A. Vezzani, J.O. Malva, J. Zimmer. (2005). Modulator effects of interleukin-1beta and tumor necrosis factor-alpha on AMPA-induced excitotoxicity in mouse organotypic hippocampal slice cultures. *J. Neurosci.*, 25(29):6734–6744.

Bezzi, P., M. Domercq, L. Brambilla, R. Galli, D. Schols, E. De Clercq, A. Vescovi, G. Bagetta, G. Kollias, J. Meldolesi, A. Volterra. (2001). CXCR4-activated astrocyte glutamate release via TNFalpha: amplification by microglia triggers neurotoxicity. *Nat. Neurosci.*, 4(7):702–710.

Bien, C.G., J. Bauer, T.L. Deckwerth, H. Wiendl, M. Deckert, O.D. Wiestler, J. Schramm, C.E. Elger, H. Lassmann. (2002). Destruction of neurons by cytotoxic T cells: a new pathogenic mechanism in Rasmussen's encephalitis. *Ann. Neurol.*, 51(3):311–318.

Boer, K., W.G. Spliet, P.C. van Rijen, S. Redeker, D. Troost, E. Aronica. (2006). Evidence of activated microglia in focal cortical dysplasia. *J. Neuroimmunol.*, 173(1–2):188–195.

Boer, K., F. Jansen, M. Nellist, S. Redeker, A.M. van den Ouweland, W.G. Spliet, O. van Nieuwenhuizen, D. Troost, P.B. Crino, E. Aronica. (2008). Inflammatory processes in cortical tubers and subependymal giant cell tumors of tuberous sclerosis complex. *Epilepsy Res.*, 78(1):7–21.

Campbell, I.L., C.R. Abraham, E. Masliah, P. Kemper, J.D. Inglis, M.B. Oldstone, L. Mucke. (1993). Neurologic disease induced in transgenic mice by cerebral overexpression of interleukin 6. *Proc. Natl. Acad. Sci. U.S.A.*, 90(21):10061–10065.

Casamenti, F., C. Prosperi, C. Scali, L. Giovannelli, M.A. Colivicchi, M.S. Faussone-Pellegrini, G. Pepeu. (1999). Interleukin-1β activates forebrain glial cells and increases nitric oxide production and cortical glutamate and GABA release *in vivo*: implications for Alzheimer's disease. *Neuroscience*, 91(3):831–842.

Chen, C., N.G. Bazan. (2005). Lipid signaling: sleep, synaptic plasticity, and neuroprotection. *Prostaglandins Other Lipid Mediat.*, 77(1–4):65–76.

Chen, J., T. Marsh, J.S. Zhang, S.H. Graham. (1995). Expression of cyclo-oxygenase 2 in rat brain following kainate treatment. *NeuroReport*, 6(2):245–248.

Choi, J.S., S.Y. Kim, H.J. Park, J.H. Cha, Y.S. Choi, J.E. Kang, J.W. Chung, M.H. Chun, M.Y. Lee. (2003). Upregulation of gp130 and differential activation of STAT and p42/44 MAPK in the rat hippocampus following kainic acid-induced seizures. *Brain Res. Mol. Brain Res.*, 119(1):10–18.

Cole-Edwards, K.K., N.G. Bazan. (2005). Lipid signaling in experimental epilepsy. *Neurochem. Res.*, 30(6–7):847–853.

Crespel, A., P. Coubes, M.C. Rousset, C. Brana, A. Rougier, G. Rondouin, J. Bockaert, M. Baldy-Moulinier, M. Lerner-Natoli. (2002). Inflammatory reactions in human medial temporal lobe epilepsy with hippocampal sclerosis. *Brain Res.*, 952(2):159–169.

Dawson, V.L., T.M. Dawson, E.D. London, D.S. Bredt, S.H. Snyder. (1991). Nitric oxide mediates glutamate neurotoxicity in primary cortical cultures. *Proc. Natl. Acad. Sci. U.S.A.*, 88(14):6368–6371.

De Luca, G., R.M. Di Giorgio, S. Macaione, P.R. Calpona, S. Costantino, E.D. Di Paola, A. De Sarro, G. Ciliberto, G. De Sarro. (2004). Susceptibility to audiogenic seizure and neurotransmitter amino acid levels in different brain areas of IL-6-deficient mice. *Pharmacol. Biochem. Behav.*, 78(1):75–81.

De Sarro, G., E. Russo, G. Ferreri, B. Giuseppe, M.A. Flocco, E.D. Di Paola, A. De Sarro. (2004). Seizure susceptibility to various convulsant stimuli of knockout interleukin-6 mice. *Pharmacol. Biochem. Behav.*, 77(4):761–766.

De Simoni, M.G., C. Perego, T. Ravizza, D. Moneta, M. Conti, F. Marchesi, A. De Luigi, S. Garattini, A. Vezzani. (2000). Inflammatory cytokines and related genes are induced in the rat hippocampus by limbic status epilepticus. *Eur. J. Neurosci.*, 12(7):2623–2633.

Desjardins, P., A. Sauvageau, A. Bouthillier, D. Navarro, A.S. Hazell, C. Rose, R.F. Butterworth. (2003). Induction of astrocytic cyclooxygenase-2 in epileptic patients with hippocampal sclerosis. *Neurochem. Int.*, 42(4):299–303.

Dinarello, C.A. (1996). Biologic basis for interleukin-1 in disease. *Blood*, 87(6):2095–2147.

Ding, S., T. Fellin, Y. Zhu, S.Y. Lee, Y.P. Auberson, D.F. Meaney, D.A. Coulter, G. Carmignoto, P.G. Haydon. (2007). Enhanced astrocytic Ca$^{2+}$ signals contribute to neuronal excitotoxicity after status epilepticus. *J. Neurosci.*, 27(40):10674–10684.

Dubé, C., A. Vezzani, M. Behrens, T. Bartfai, T.Z. Baram. (2005). Interleukin-1beta contributes to the generation of experimental febrile seizures. *Ann. Neurol.*, 57(1):152–155.

Fabene, P.F., G. Navarro Mora, M. Martinello, B. Rossi, F. Merigo et al. (2008). A role for leukocyte-endothelial adhesion mechanisms in epilepsy. *Nat. Med.*, 14(12):1377–1383.

Fellin, T., M. Gomez-Gonzalo, S. Gobbo, G. Carmignoto, P.G. Haydon. (2006). Astrocytic glutamate is not necessary for the generation of epileptiform neuronal activity in hippocampal slices. *J. Neurosci.*, 26(36):9312–9322.

Fukuda, M., T. Morimoto, Y. Suzuki, C. Shinonaga, Y. Ishida. (2007). Interleukin-6 attenuates hyperthermia-induced seizures in developing rats. *Brain Dev.*, 29(10):644–648.

Galic, M.A., K. Riazi, J.G. Heida, A. Mouihate, N.M. Fournier, S.J. Spencer, L.E. Kalynchuk, G.C. Teskey, Q.J. Pittman. (2008). Postnatal inflammation increases seizure susceptibility in adult rats. *J. Neurosci.*, 28(27):6904–6913.

Gorter, J.A., E.A. van Vliet, E. Aronica, T. Breit, H. Rauwerda, F.H. Lopes da Silva, W.J. Wadman. (2006). Potential new antiepileptogenic targets indicated by microarray analysis in a rat model for temporal lobe epilepsy. *J. Neurosci.*, 26(43):11083–11110.

Heida, J.G., Q.J. Pittman. (2005). Causal links between brain cytokines and experimental febrile convulsions in the rat. *Epilepsia*, 46(12):1906–1913.

Hewett, S.J., T.F. Uliasz, A.S. Vidwans, J.A. Hewett. (2000). Cyclooxygenase-2 contributes to *N*-methyl-D-aspartate-mediated neuronal cell death in primary cortical cell culture. *J. Pharmacol. Exp. Ther.*, 293(2):417–425.

Hirst, W.D., K.A. Young, R. Newton, V.C. Allport, D.R. Marriott, G.P. Wilkin. (1999). Expression of COX-2 by normal and reactive astrocytes in the adult rat central nervous system. *Mol. Cell Neurosci.*, 13(1):57–68.

Holtman L., E.A. van Vliet, R. van Schaik, C.M. Queiroz, E. Aronica, J.A. Gorter. (2009). Effects of SC58236, a selective COX-2 inhibitor, on epileptogenesis and spontaneous seizures in a rat model for temporal lobe epilepsy. *Epilepsy Res.*, 84(1):56–66.

Hu, S., W.S. Sheng, L.C. Ehrlich, P.K. Peterson, C.C. Chao. (2000). Cytokine effects on glutamate uptake by human astrocytes. *Neuroimmunomodulation*, 7(3):153–159.

Hurley, S.D., J.A. Olschowka, M.K. O'Banion. (2002). Cyclooxygenase inhibition as a strategy to ameliorate brain injury. *J. Neurotrauma*, 19(1):1–15.

Jung, K.H., K. Chu, S.T. Lee, J. Kim, D.I. Sinn, J.M. Kim, D.K. Park, J.J. Lee, S.U. Kim, M. Kim, S.K. Lee, J.K. Roh. (2006). Cyclooxygenase-2 inhibitor, celecoxib, inhibits the altered hippocampal neurogenesis with attenuation of spontaneous recurrent seizures following pilocarpine-induced status epilepticus. *Neurobiol. Dis.*, 23(2):237–246.

Kelley, K.A., L. Ho, D. Winger, J. Freire-Moar, C.B. Borelli, P.S. Aisen, G.M. Pasinetti. (1999). Potentiation of excitotoxicity in transgenic mice overexpressing neuronal cyclooxygenase-2. *Am. J. Pathol.*, 155(3):995–1004.

Kim, H.J., J.I. Chung, S.H. Lee, Y.S. Jung, C.H. Moon, E.J. Baik. (2008). Involvement of endogenous prostaglandin F2alpha on kainic acid-induced seizure activity through FP receptor: the mechanism of proconvulsant effects of COX-2 inhibitors. *Brain Res.*, 1193:153–161.

Kovacs, Z., K.A. Kekesi, N. Szilagyi, I. Abraham, D. Szekacs, N. Kiraly, E. Papp, I. Csaszar, E. Szego, K. Barabas, H. Peterfy, A. Erdei, T. Bartfai, G. Juhasz. (2006). Facilitation of spike–wave discharge activity by lipopolysaccharides in Wistar Albino Glaxo/Rijswijk rats. *Neuroscience*, 140(2):731–742.

Kunz, T., E.H. Oliw. (2001). The selective cyclooxygenase-2 inhibitor rofecoxib reduces kainate-induced cell death in the rat hippocampus. *Eur. J. Neurosci.*, 13(3):569–575.

Lee, B., H. Dziema, K.H. Lee, Y.S. Choi, K. Obrietan. (2007). CRE-mediated transcription and COX-2 expression in the pilocarpine model of status epilepticus. *Neurobiol. Dis.*, 25(1):80–91.

Lee, S.H., S.H. Han, K.W. Lee. (2000). Kainic acid-induced seizures cause neuronal death in infant rats pretreated with lipopolysaccharide. *NeuroReport*, 11(3):507–510.

Lehtimaki, K.A., J. Peltola, E. Koskikallio, T. Keranen, J. Honkaniemi. (2003). Expression of cytokines and cytokine receptors in the rat brain after kainic acid-induced seizures. *Brain Res. Mol. Brain Res.*, 110(2):253–260.

Lehtimaki, K.A., T. Keranen, H. Huhtala, M. Hurme, J. Ollikainen, J. Honkaniemi, J. Palmio, J. Peltola. (2004). Regulation of IL-6 system in cerebrospinal fluid and serum compartments by seizures: the effect of seizure type and duration. *J. Neuroimmunol.*, 152(1–2):121–125.

Librizzi, L., M.C. Regondi, C. Pastori, S. Frigerio, C. Frassoni, M. de Curtis. (2007). Expression of adhesion factors induced by epileptiform activity in the endothelium of the isolated guinea pig brain *in vitro*. *Epilepsia*, 48(4):743–751.

Lu, M.O., X.M. Zhang, E. Mix, H.C. Quezada, T. Jin, J. Zhu, A. Adem. (2008). TNF-alpha receptor 1 deficiency enhances kainic acid-induced hippocampal injury in mice. *J. Neurosci. Res.*, 86(7):1608–1614.

Lucas, S.M., N.J. Rothwell, R.M. Gibson. (2006). The role of inflammation in CNS injury and disease. *Br. J. Pharmacol.*, 147(Suppl. 1):S232–S240.

Maldonado, M., M. Baybis, D. Newman, D.L. Kolson, W. Chen, G. McKhann, 2nd, D.H. Gutmann, P.B. Crino. (2003). Expression of ICAM-1, TNF-alpha, NF kappa B, and MAP kinase in tubers of the tuberous sclerosis complex. *Neurobiol. Dis.*, 14(2):279–290.

Marchi, N., L. Angelov, T. Masaryk, V. Fazio, T. Granata, N. Hernandez, K. Hallene, T. Diglaw, L. Franic, I. Najm, D. Janigro. (2007). Seizure-promoting effect of blood–brain barrier disruption. *Epilepsia*, 48(4):732–742.

Marchi, N., Q. Fan, C. Ghosh, V. Fazio, F. Bertolini, G. Betto, A. Batra, E. Carlton, I. Najm, T. Granata, D. Janigro. (2009). Antagonism of peripheral inflammation reduces the severity of status epilepticus. *Neurobiol. Dis.*, 33(2):171–181.

Minghetti, L. (2007). Role of COX-2 in inflammatory and degenerative brain diseases. *Subcell. Biochem.*, 42:127–141.

Morgan, B.P., P. Gasque. (1996). Expression of complement in the brain: role in health and disease. *Immunol. Today*, 17(10):461–466.

Naffah-Mazzacoratti, M.G., M.I. Bellissimo, E.A. Cavalheiro. (1995). Profile of prostaglandin levels in the rat hippocampus in pilocarpine model of epilepsy. *Neurochem. Int.*, 27(6):461–466.

Nakayama, M., K. Uchimura, R.L. Zhu, T. Nagayama, M.E. Rose, R.A. Stetler, P.C. Isakson, J. Chen, S.H. Graham. (1998). Cyclooxygenase-2 inhibition prevents delayed death of CA1 hippocampal neurons following global ischemia. *Proc. Natl. Acad. Sci. U.S.A.*, 95(18):10954–10959.

Okada, K., U. Yamashita, S. Tsuji. (2006). Cyclooxygenase system contributes to the maintenance of post convulsive period of epileptic phenomena in the genetically epileptic EL mice. *J. Uoeh.*, 28(3):265–275.

Oliveira, M.S., A.F. Furian, L.F. Royes, M.R. Fighera, N.G. Fiorenza, M. Castelli, P. Machado, D. Bohrer, M. Veiga, J. Ferreira, E.A. Cavalheiro, C.F. Mello. (2008). Cyclooxygenase-2/PGE2 pathway facilitates pentylenetetrazol-induced seizures. *Epilepsy Res.*, 79(1):14–21.

Peltola, J., J. Laaksonen, A.M. Haapala, M. Hurme, S. Rainesalo, T. Keranen. (2002). Indicators of inflammation after recent tonic–clonic epileptic seizures correlate with plasma interleukin-6 levels. *Seizure*, 11(1):44–46.

Plata-Salaman, C.R., S.E. Ilyin, N.P. Turrin, D. Gayle, M.C. Flynn, A.E. Romanovitch, M.E. Kelly, Y. Bureau, H. Anisman, D.C. McIntyre. (2000). Kindling modulates the IL-1beta system, TNF-alpha, TGF-beta1, and neuropeptide mRNAs in specific brain regions. *Brain Res. Mol. Brain Res.*, 75(2):248–258.

Ravizza, T., A. Vezzani. (2006). Status epilepticus induces time-dependent neuronal and astrocytic expression of interleukin-1 receptor type I in the rat limbic system. *Neuroscience*, 137(1):301–308.

Ravizza, T., K. Boer, S. Redeker, W.G. Spliet, P.C. van Rijen, D. Troost, A. Vezzani, E. Aronica. (2006a). The IL-1beta system in epilepsy-associated malformations of cortical development. *Neurobiol. Dis.*, 24(1):128–143.

Ravizza, T., S.M. Lucas, S. Balosso, L. Bernardino, G. Ku, F. Noe, J. Malva, J.C. Randle, S. Allan, A. Vezzani. (2006b). Inactivation of caspase-1 in rodent brain: a novel anticonvulsive strategy. *Epilepsia*, 47(7):1160–1168.

Ravizza, T., B. Gagliardi, F. Noe, K. Boer, E. Aronica, A. Vezzani. (2008a). Innate and adaptive immunity during epileptogenesis and spontaneous seizures: evidence from experimental models and human temporal lobe epilepsy. *Neurobiol. Dis.*, 29(1):142–160.

Ravizza, T., F. Noe, D. Zardoni, V. Vaghi, M. Sifringer, A. Vezzani. (2008b). Interleukin converting enzyme inhibition impairs kindling epileptogenesis in rats by blocking astrocytic IL-1beta production. *Neurobiol. Dis.*, 31(3):327–333.

Riazi, K., M.A. Galic, J.B. Kuzmiski, W. Ho, K.A. Sharkey, Q.J. Pittman. (2008). Microglial activation and TNFalpha production mediate altered CNS excitability following peripheral inflammation. *Proc. Natl. Acad. Sci. U.S.A.*, 105(44):17151–17156.

Rigau, V., M. Morin, M.C. Rousset, F. de Bock, A. Lebrun, P. Coubes, M.C. Picot, M. Baldy-Moulinier, J. Bockaert, A. Crespel, M. Lerner-Natoli. (2007). Angiogenesis is associated with blood–brain barrier permeability in temporal lobe epilepsy. *Brain*, 130(Pt. 7):1942–1956.

Rizzi, M., C. Perego, M. Aliprandi, C. Richichi, T. Ravizza, D. Colella, J. Veliskova, S.L. Moshe, M.G. De Simoni, A. Vezzani. (2003). Glia activation and cytokine increase in rat hippocampus by kainic acid-induced status epilepticus during postnatal development. *Neurobiol. Dis.*, 14(3):494–503.

Rodgers, K.M., M.R. Hutchinson, A. Northcutt, S.F. Maier, L.R. Watkins, D.S. Barth. (2009). The cortical innate immune response increases local neuronal excitability leading to seizures. *Brain*, 132(9):2478–2486.

Rozovsky, I., T.E. Morgan, D.A. Willoughby, M.M. Dugichi-Djordjevich, G.M. Pasinetti, S.A. Johnson, C.E. Finch. (1994). Selective expression of clusterin (SGP-2) and complement C1qB and C4 during responses to neurotoxins *in vivo* and *in vitro*. *Neuroscience*, 62(3):741–758.

Samland, H., S. Huitron-Resendiz, E. Masliah, J. Criado, S.J. Henriksen, I.L. Campbell. (2003). Profound increase in sensitivity to glutamatergic- but not cholinergic agonist-induced seizures in transgenic mice with astrocyte production of IL-6. *J. Neurosci. Res.*, 73(2):176–187.

Sandhya, T.L., W.Y. Ong, L.A. Horrocks, A.A. Farooqui. (1998). A light and electron microscopic study of cytoplasmic phospholipase A2 and cyclooxygenase-2 in the hippocampus after kainate lesions. *Brain Res.*, 788(1–2):223–231.

Sankar, R., S. Auvin, A. Mazarati, D. Shin. (2007). Inflammation contributes to seizure-induced hippocampal injury in the neonatal rat brain. *Acta Neurol. Scand.*, 186(Suppl. 4):16–20.

Sayyah, M., M. Javad-Pour, M. Ghazi-Khansari. (2003). The bacterial endotoxin lipopolysaccharide enhances seizure susceptibility in mice: involvement of proinflammatory factors: nitric oxide and prostaglandins. *Neuroscience*, 122(4):1073–1080.

Seibert, K., J.L. Masferrer. (1994). Role of inducible cyclooxygenase (COX-2) in inflammation. *Receptor*, 4(1):17–23.

Seiffert, E., J.P. Dreier, S. Ivens, I. Bechmann, O. Tomkins, U. Heinemann, A. Friedman. (2004). Lasting blood–brain barrier disruption induces epileptic focus in the rat somatosensory cortex. *J. Neurosci.*, 24(36):7829–7836.

Shinoda, S., S.L. Skradski, T. Araki, C.K. Schindler, R. Meller, J.Q. Lan, W. Taki, R.P. Simon, D.C. Henshall. (2003). Formation of a tumour necrosis factor receptor 1 molecular scaffolding complex and activation of apoptosis signal-regulating kinase 1 during seizure-induced neuronal death. *Eur. J. Neurosci.*, 17(10):2065–2076.

Slanina, K.A., P. Schweitzer. (2005). Inhibition of cyclooxygenase-2 elicits a CB1-mediated decrease of excitatory transmission in rat CA1 hippocampus. *Neuropharmacology*, 49(5):653–659.

Stellwagen, D., E.C. Beattie, J.Y. Seo, R.C. Malenka. (2005). Differential regulation of AMPA receptor and GABA receptor trafficking by tumor necrosis factor-alpha. *J. Neurosci.*, 25(12):3219–3228.

Takemiya, T., K. Suzuki, H. Sugiura, S. Yasuda, K. Yamagata, Y. Kawakami, E. Maru. (2003). Inducible brain COX-2 facilitates the recurrence of hippocampal seizures in mouse rapid kindling. *Prostaglandins Other Lipid Mediat.*, 71(3–4):205–216.

Takemiya, T., M. Maehara, K. Matsumura, S. Yasuda, H. Sugiura, K. Yamagata. (2006). Prostaglandin E$_2$ produced by late induced COX-2 stimulates hippocampal neuron loss after seizure in the CA3 region. *Neurosci. Res.*, 56(1):103–110.

Tian, G.F., H. Azmi, T. Takano, Q. Xu, W. Peng, J. Lin, N. Oberheim, N. Lou, X. Wang, H.R. Zielke, J. Kang, M. Nedergaard. (2005). An astrocytic basis of epilepsy. *Nat. Med.*, 11(9):973–981.

Tu, B., N.G. Bazan. (2003). Hippocampal kindling epileptogenesis upregulates neuronal cyclooxygenase-2 expression in neocortex. *Exp. Neurol.*, 179(2):167–175.

Turrin, N.P., S. Rivest. (2004). Innate immune reaction in response to seizures: implications for the neuropathology associated with epilepsy. *Neurobiol. Dis.*, 16(2):321–334.

van Vliet, E.A., S. da Costa Araujo, S. Redeker, R. van Schaik, E. Aronica, J.A. Gorter. (2007). Blood–brain barrier leakage may lead to progression of temporal lobe epilepsy. *Brain*, 130(Pt. 2):521–534.

Vezzani, A., T.Z. Baram. (2007). New roles for interleukin-1beta in the mechanisms of epilepsy. *Epilepsy Curr.*, 7(2):45–50.

Vezzani, A., T. Granata. (2005). Brain inflammation in epilepsy: experimental and clinical evidence. *Epilepsia*, 46(11):1724–1743.

Vezzani, A., M. Conti, A. De Luigi, T. Ravizza, D. Moneta, F. Marchesi, M.G. De Simoni. (1999). Interleukin-1beta immunoreactivity and microglia are enhanced in the rat hippocampus by focal kainate application: functional evidence for enhancement of electrographic seizures. *J. Neurosci.*, 19(12):5054–5065.

Vezzani, A., D. Moneta, M. Conti, C. Richichi, T. Ravizza, A. De Luigi, M.G. De Simoni, G. Sperk, S. Andell-Jonsson, J. Lundkvist, K. Iverfeldt, T. Bartfai. (2000). Powerful anticonvulsant action of IL-1 receptor antagonist on intracerebral injection and astrocytic overexpression in mice. *Proc. Natl. Acad. Sci. U.S.A.*, 97(21):11534–11539.

Vezzani, A., S. Balosso, T. Ravizza. (2008). The role of cytokines in the pathophysiology of epilepsy. *Brain Behav. Immun.*, 22(6):797–803.

Viviani, B., S. Bartesaghi, F. Gardoni, A. Vezzani, M.M. Behrens, T. Bartfai, M. Binaglia, E. Corsini, M. Di Luca, C.L. Galli, M. Marinovich. (2003). Interleukin-1beta enhances NMDA receptor-mediated intracellular calcium increase through activation of the Src family of kinases. *J. Neurosci.*, 23(25):8692–8700.

Viviani, B., F. Gardoni, M. Marinovich. (2007). Cytokines and neuronal ion channels in health and disease. *Int. Rev. Neurobiol.*, 82:247–263.

Voutsinos-Porche, B., E. Koning, H. Kaplan, A. Ferrandon, M. Guenounou, A. Nehlig, J. Motte. (2004). Temporal patterns of the cerebral inflammatory response in the rat lithium–pilocarpine model of temporal lobe epilepsy. *Neurobiol. Dis.*, 17(3):385–402.

Wang, S., Q. Cheng, S. Malik, J. Yang. (2000). Interleukin-1beta inhibits gamma-aminobutyric acid type A (GABA(A)) receptor current in cultured hippocampal neurons. *J. Pharmacol. Exp. Ther.*, 292(2):497–504.

Wiedmer, T., P.J. Sims. (1985). Effect of complement proteins C5b-9 on blood platelets: evidence for reversible depolarization of membrane potential. *J. Biol. Chem.*, 260(13):8014–8019.

Willingale, H.L., N.J. Gardiner, N. McLymont, S. Giblett, B.D. Grubb. (1997). Prostanoids synthesized by cyclo-oxygenase isoforms in rat spinal cord and their contribution to the development of neuronal hyperexcitability. *Br. J. Pharmacol.*, 122(8):1593–1604.

Wolfe, L.S., O.A. Mamer. (1975). Measurement of prostaglandin F2alpha levels in human cerebrospinal fluid in normal and pathological conditions. *Prostaglandins*, 9(2):183–192.

Xiong, Z.Q., W. Qian, K. Suzuki, J.O. McNamara. (2003). Formation of complement membrane attack complex in mammalian cerebral cortex evokes seizures and neurodegeneration. *J. Neurosci.*, 23(3):955–960.

Yamagata, K., K.I. Andreasson, W.E. Kaufmann, C.A. Barnes, P.F. Worley. (1993). Expression of a mitogen-inducible cyclooxygenase in brain neurons: regulation by synaptic activity and glucocorticoids. *Neuron*, 11(2):371–386.

Zeise, M.L., J. Espinoza, P. Morales, A. Nalli. (1997). Interleukin-1beta does not increase synaptic inhibition in hippocampal CA3 pyramidal and dentate gyrus granule cells of the rat *in vitro*. *Brain Res.*, 768(1–2):341–344.

# 5 Drug Resistance in Epilepsy and Status Epilepticus

*Jaideep Kapur and Edward H. Bertram*

## CONTENTS

## INTRODUCTION

Resistance to pharmacotherapy is a common occurrence among patients with chronic epilepsy and during prolonged status epilepticus (SE). It is estimated that about one third of patients with epilepsy will be resistant to pharmacotherapy (Jacobs et al., 2001). For patients with the prolonged seizures of SE, the numbers are more difficult to determine, as resistance is tied to the duration of the seizures as well as the etiology, but about one third to one half of patients in SE will not respond to the first lines of treatment (Alldredge et al., 2001; Treiman et al., 1998). Why do some patients respond readily, while others resist all attempts to control the seizures? It is a question that has plagued patients and their physicians since the first effective therapies appeared in the latter part of the nineteenth century. In this chapter, we will discuss what is meant by resistance and what some of the mechanisms might be. Because the mechanisms behind resistance are likely different in chronic epilepsy and in SE, we will discuss the two conditions separately. One should, however, bear in mind that many of the mechanisms discussed are hypothetically valid, but most, especially in chronic epilepsy, have not been proven as contributing to therapy resistance. In the following paragraphs, we will define resistance and identify some broad possible causes for the condition, and then we will discuss what is known for each of these possible causes.

**TABLE 5.1**
**Levels of Drug Resistance**

| Resistance Level | Effect |
|---|---|
| Complete drug resistance | No effect of any drug on any aspect of the seizures |
| Partial drug resistance/incomplete responsiveness | Some aspect of seizure affected |
| Frequency reduction | Fewer seizures, seizure features unaffected |
| Severity reduction | Seizures less intense or shorter, frequency unaffected |

## RESISTANCE IN CHRONIC EPILEPSY: THE CLINICAL SPECTRUM

Although physicians treating patients with seizures and epilepsy have a general concept of what comprises drug resistance, it is our common experience that there are many gradations of this condition and different factors that contribute to resistance. The simplest definition for *resistance* could be "failure of the drug to control seizures completely." This definition implies resistance to a single compound, which does occur, but a common experience is that a patient that is resistant to a single compound may be resistant to multiple drugs. Some patients, however, are resistant to one compound but have seizures brought under control by other drugs. Although the mechanisms underlying resistance to a single compound and multiple-drug resistance may be similar, we will focus on multidrug resistance, as single-drug resistance is not a significant clinical problem over the long term.

Under the definition of resistance are several subcategories of this condition that are worth considering (Table 5.1). The first is *complete drug resistance*, in which patients do not respond to any pharmacological therapy. Their seizure frequencies, clinical severity, and duration are unaffected by the available drugs. This condition is likely quite rare, but no good data are available. The second and probably more common category is *partial drug resistance*, which implies that some features of the seizures respond to therapy but not completely. Under this category are several additional divisions. The first is *frequency reduction*, in which some but not all seizures are prevented. The second is *severity reduction*. In this situation, the seizure frequency is unchanged, but the duration or the clinical severity is reduced. A common example would be the patient with secondarily generalized seizures that are reduced to partial complex or partial simple seizures. For many reasons, we should not consider these patients to be drug resistant, as they do respond; rather, we should define them as *incompletely responsive*, as the drugs work only up to a point. Recognizing the phenomenon of incomplete responsiveness raises another series of questions about the possible mechanisms of resistance and the pathophysiology of epilepsy. It is very likely that the mechanisms behind these different types of treatment failure are quite different, and we will speculate about these possible causes in subsequent sections.

Two additional categories are *resistance secondary to intolerable side effects* and *resistance secondary to choosing the wrong medication*. In the former situation, the patient has seizures that were or would be suppressed, but side effects were sufficiently severe that the drug dose had to be reduced into a subtherapeutic range or it was necessary to stop the one therapy that controlled the seizures. An example of the second type is when patients with generalized epilepsy are prescribed medications that are better suited for focal seizure disorders. Their seizures are readily controlled when the appropriate medication is prescribed. Although these patients may be considered therapy resistant, the real issue is one of correct diagnosis and treatment.

Why are patients resistant to therapy, specifically to multiple drugs? At the moment, we have no good answers, especially for patients with chronic epilepsy. In SE, there is more known about how alterations in receptor function may alter the responsiveness of the condition to specific treatments, but chronic epilepsy has so far not had similar insights, although there is growing evidence that

**TABLE 5.2**
**Potential Mechanisms of Resistance**

Wrong molecular target
    Overall
    Particular isoform
Exclusion of therapeutic concentrations from brain
Maldistribution of drug in brain: drug goes to wrong regions
Induced tolerance to drug

**TABLE 5.3**
**Reasons for Toxicity Preceding Efficacy**

Regional variability in blood–brain barrier, with higher concentrations in "normal" regions
Greater affinity for drug of receptor or channel isoform in "normal" regions
Regional variations in distribution of target receptor
Sequestration of drug away from target sites

resistance may be multifactorial. Possible mechanisms for resistance in epilepsy can be broadly divided into four categories: (1) wrong molecular target (i.e., channel or receptor, either in general or a particular isoform), (2) exclusion of therapeutic concentrations from the brain, (3) maldistribution of the drug within the brain so that it does not reach the key targets, and (4) induced tolerance so that a specific compound has reduced efficacy over time (Table 5.2). There is some overlap of the middle two in concept and mechanism, but they are sufficiently different to warrant separate discussion. In the following sections, we will discuss the possible basis for each of these mechanisms, the evidence that exists at the moment, and, at the end of this chapter, possible research directions that will allow for a better understanding of drug resistance and ways that we might overcome it.

## BALANCING EFFICACY WITH SIDE EFFECTS

One concept that we will emphasize is the balance between side effects and therapeutic efficacy, not so much from the mechanisms of side effects but rather from what this relationship might tell us about some of the causes of resistance. This relationship is well known to any physician treating patients with epilepsy: toxic side effects such as dizziness, sedation, psychiatric problems, or double vision appear before the seizures are controlled. This common difficulty in the therapy of epilepsy should be expected, because we prescribe drugs that are designed to affect the central nervous system but which (we hope) will affect seizures before they have a significant effect on normal function. For the majority of patients successfully treated with the current antiepileptics, this relationship is true. For the rest, the remaining third, the relationship is reversed. The reasons are unclear, but the hypothetical causes can give us some directions for investigating the mechanisms that underlie therapeutic resistance in general. The basis for the breakdown in the therapeutic/toxic balance could lie in several areas: the blood–brain barrier and selective permeability, variable regional affinity of receptor and channel subtypes for a particular drug, differing potency of the compound at different receptor and channel subtypes, and, finally, regional variation in the sequestration of a compound that prevents therapeutic levels at the desired target, whereas toxic levels are reached at other regions at the same time (Table 5.3). Resolving this issue will depend, as we will discuss later, on a better understanding of the function of the blood–brain barrier and the pharmacology of the relevant channels. Some of these problems in brain pharmacokinetics and dynamics may be altered in the epileptic condition, which is associated with many structural abnormalities.

## RECEPTORS, CHANNELS, AND RESISTANCE

One of the problems in understanding drug resistance is that we do not have a good understanding of how drugs work to prevent seizures. Although many possible mechanisms are ascribed to anti-epileptics, the actions most commonly attributed to these drugs are the blockade of voltage-gated sodium channels or the enhancement of GABAergic function. Other purported mechanisms include the inhibition of the low-threshold (T-type) calcium currents (for absence therapy) the reduced release of excitatory neurotransmitters (Macdonald and Kelly, 1995) and binding to synaptic vesicle proteins. For a few drugs, a mechanism of action has not been determined. All of these mechanisms could be an important means for suppressing seizure activity, but it is also possible that the real basis for seizure suppression is completely unrelated to any of these demonstrated drug actions.

Part of the difficulty in determining the mechanisms for these compounds is that the true mechanisms underlying epilepsy, especially underlying the initiation of a seizure, are unknown. A number of studies using animal models of epilepsy, as well as some postsurgical and postmortem tissue from in or near the seizure focus, have shown multiple changes in the physiology and pharmacology of specific channels and receptors, often due to alterations in the expression of particular isoforms of the channels or receptors. These observations are important for several reasons that affect pharmacotherapy. First, as described later, the studies clearly indicate that there is more than one change in the brain that can predispose one to epilepsy. Second, the pharmacology of these altered receptors and channels is likely to be or has already been shown to be quite different from the native state. These changes could have significant implications for the relative efficacy and potency of the current group of antiepileptic drugs, most of which were developed on normal nervous systems using acute seizure models. Finally, the changes are not uniform throughout the presumed seizure circuit, an observation that raises the possibility that the antiepileptic drugs will have differing potency at different points in the circuit, depending on which channel or receptor is expressed.

The changes that occur in the seizure circuit have been described best in limbic epilepsy (also known as mesial temporal lobe epilepsy), for which there are a number of animal models. Most of the work has focused on the hippocampus, and changes have been described in the $\gamma$-aminobutyric acid (GABA) receptor, the $N$-methyl-$D$-aspartate (NMDA) receptor, the $\alpha$-amino-3-hydroxy-5-methyl-4-isoxazoleproprionic acid (AMPA) receptor, and voltage-gated sodium and calcium channels. The GABAergic system in the hippocampus has been studied most extensively, and within this region alone a number of changes have occurred in the subunit composition, as have been determined by anatomical, physiological, and pharmacological studies. Although we will not describe the changes in detail, there are several key findings across the studies. First, as determined by mRNA expression and immunocytochemical labeling, there are significant changes in the subunit composition of the GABA receptor (Brooks-Kayal et al., 1998; Mathern et al., 1997; Schwarzer et al., 1997; Tsunashima et al., 1997). Second, the GABA-induced inhibitory potentials are altered, with many having significantly reduced duration, resulting in epileptiform evoked responses (Gibbs et al., 1997; Mangan and Lothman, 1996; Mangan et al., 1995; Rajasekaran et al., 2007). Third, the response of the GABA currents to a variety of neuromodulators is significantly altered, so compounds that are quite effective in particular neuronal types from normal animals have greatly reduced potency in neurons from epileptic animals (Gibbs et al., 1997; Mtchedlishvili et al., 2001; Rajesekaran et al., 2009; Shumate et al., 1998). Finally, the changes are not uniform throughout the hippocampus, an observation that complicates the development of pharmacotherapy that is selective and effective for the epileptic condition (Rempe et al., 1995). There are some preliminary reports that changes in GABA-induced responses are also found in other limbic sites involved in seizures such as the amygdala (Mangan et al., 2000). Recent studies have shown that there are also changes in GABA physiology and pharmacology in rats with limbic epilepsy in several of the thalamic nuclei that are key components of the limbic seizure circuits. Of note, as in the hippocampus, the changes in these

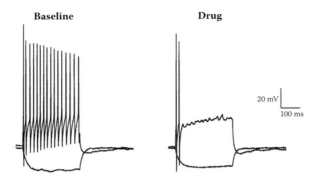

**FIGURE 5.1** Current clamp recording from hippocampal pyramidal cells from a rat with limbic epilepsy demonstrating selective blockade of repetitive firing in epileptic animals. Application of YW-I92 blocked repetitive firing in neurons from epileptic animals but had no effect on neuronal firing in normal animals. (From Jones, P.J. et al., *J. Pharmacol. Exp. Ther.*, 328, 201–212, 2009. With permission.)

nuclei are different from one another. These multiple changes in multiple regions complicate the issue of choosing a target for a drug, as the key change in the critical region has yet to be identified (Rajasekaran et al., 2007, 2009).

There is now evidence for multiple changes in voltage-gated sodium channels in limbic epilepsy. In one study, long-lasting (i.e., over 3 months), increased expression of type II and III alpha isoforms of the sodium channel was found in the hippocampus following an episode of SE in rats. This change persisted after the animals developed spontaneous seizures (Aronica et al., 2001). A follow-up study examining the sodium currents in these neurons demonstrated a significant change associated with chronic epilepsy with lower thresholds for activation and an increase in the overall sodium currents (Ketelaars et al., 2001). These changes were, as has been seen in the GABA receptor, region specific, with clear changes being observed in CA1 but no such changes in the dentate granule cells. Sodium channel mutations have also been described in several of the familial epilepsy syndromes (Escayg et al., 2001; Lerche et al., 2001; Steinlein, 2001; Wallace et al., 2001). At least one of these mutations has been created in a transgenic mouse, and these animals have had documented spontaneous seizures (Kearney et al., 2001). Although these findings are important in our quest to understand the pathophysiology of epilepsy, they may also be important in our attempts to determine the basis for drug resistance. There is now good experimental evidence for the altered pharmacology of the voltage-gated sodium channel in epilepsy. Ketelaars and colleagues (2001), as well as Remy and Beck (2006), have shown that standard sodium-channel antagonists such as phenytoin, carbamazepine, and lamotrigine were much less effective in blocking the sodium currents. Jones and colleagues (2009) showed that drugs can be designed that are specific for the sodium channels expressed in at least some forms of epilepsy. In these experiments, one of the compounds blocked depolarization-induced burst firing of hippocampal pyramidal cells from epileptic animals, but the compound had no effect on the same type of neuron from normal rats. Lamotrigine, by contrast, was effective in the neurons from normal animals but much less effective in the neurons from epileptic rats (Figure 5.1) (Jones et al., 2009). These observations of altered physiology and pharmacology in the epilepsy-associated sodium channels suggest that future approaches to overcoming pharmacoresistance may lie in designing drugs that are specific for those channels and receptors that are expressed in epilepsy. This concept of using drugs that are specific for epilepsy-associated changes, as logical and attractive as it may appear, will ultimately require validation in clinical trials.

Changes have also been identified for the glutamate receptor family, AMPA, and NMDA, either by anatomic demonstration of altered subunit expression or physiologically by showing that prolonged depolarizations can be significantly reduced through channel blockade (Aronica et al., 2000; Lothman et al., 1995; Mathern et al., 1997). At this time, much less is known about the specifics for modulating the activity of the receptors. Although some attempts have been made to use the glycine modulatory site on the NMDA receptor as a target for seizure control, in general, attempts to affect seizures through direct inhibition of these receptors have been plagued by significant neuropsychological side effects that have limited the potential usefulness of this approach (Zhang et al., 2005). Until more selective approaches to modulation of excitatory neurotransmission are developed, this path to treatment may not be as fruitful as others, but it may be possible, as with sodium channels and GABA receptors, to identify the specific changes associated with epilepsy and target them pharmacologically. By this approach, we may be able to avoid resistance that results from side effects appearing before clinical efficacy.

## MULTIPLE DRUG-RESISTANCE PROTEINS

Multiple drug resistance (MDR) proteins were first described in the cancer literature; they received their name because their presence was associated with resistance to chemotherapeutics, presumably by keeping the drug from reaching the desired targets. There have been numerous reports that these proteins are found in tissue from patients with intractable epilepsy as well as in animal models of drug-resistant epilepsy. These observations have raised the question about the potential role of these proteins in drug resistance in epilepsy, and in the following paragraphs we will review what is currently known. We will not explore MDR proteins in any depth, but for the interested reader a number of reviews on the subject can greatly expand on the points raised here (Elsinga et al., 2004; Hermann et al., 2006; Litman et al., 2001; Matheny et al., 2001; Miller et al., 2008; Tanigawara, 2000).

The MDR proteins transport compounds across membranes against concentration gradients using adenosine triphosphate (ATP) in the process, hence the name of the ABC (ATP-binding cassette) family of membrane transporters. The first one described in association with drug resistance, and the one of possibly greatest relevance for epilepsy, was P-glycoprotein (P-gp), which is also known as multidrug-resistant protein 1 (MDR1). Other proteins found in normal and neoplastic tissues include the multidrug-resistance-associated family of proteins (MRP1–7), the mitoxantrone-resistance protein (MXR), the breast cancer resistance protein (BCRP) (other nomenclature systems are used, but these appear to be most commonly accepted) (Litman et al., 2001). Another recent discovery includes the major vault protein (MVP) (Steiner et al., 2006). Each has specific substrates, activators, and inhibitors, and each is most commonly associated with specific cell lines and tissue types.

P-glycoprotein, or MDR1, is the protein that is most commonly associated with the brain and one that has been associated with intractable epilepsy, with increased expression in neurons and astrocytes. In normal brain, MDR1 is typically seen at the blood–brain barrier, where presumably it will keep compounds out of the extracellular space of the brain (Litman et al., 2001; Sisodiya et al., 1999; 2002; Tishler et al., 1995). The expression of a particular MDR protein may depend on the pathology associated with the epilepsy; for example, MRP1, not normally expressed in brain, has been found in the neurons and astrocytes of dysplastic tissue in patients with epilepsy (Sisodiya et al., 1999). BCRP has been associated with tumor-related epilepsies but not hippocampal sclerosis or cortical dysplasias (Aronica et al., 2005). MVP has been found in an animal model of limbic epilepsy; the distribution was not uniform, as it was seen in some regions such as the piriform cortex that are part of the seizure circuitry but not others (van Vliet et al., 2004). It is quite likely that there are other proteins, described and not described, that may play roles in this process as well.

There has been a steady evaluation of drugs that may be subject to transport and those compounds that are not. Within the family of MDRs, multiple compounds are transported, but the proteins are selective in what each will transport (Litman et al., 2001; Matheny et al., 2001; Tanigawara, 2000). Although there is some overlap in substrate compounds, some substances can only be transported by a specific protein, a situation in which some cells may be resistant to the effects of a particular drug and others may not because they lack the protein that is specific for that drug. This selectivity is of some importance in epilepsy because not all antiepileptic drugs are substrates for P-gp/MDR1. Phenytoin has been well described as one compound that is a substrate for P-gp/MDR1 (Marchi et

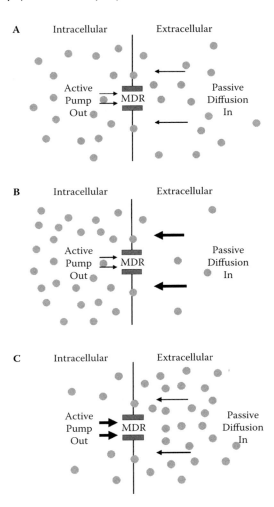

**FIGURE 5.2**  Multidrug resistance proteins and drug distribution. Drug concentrations on either side of the membrane will depend on relative rates of passive diffusion into the cell and active pumping out. (A) Relative rates are equal, and there is an equal concentration across the membrane. (B) Inward passive diffusion is much greater, and the intracellular concentration is higher. (C) Active pumping out is faster, and the extracellular concentration is greater.

al., 2006; Potschka and Löscher, 2001a,b). There are conflicting reports about carbamazepine, but little information about the other antiepileptic drugs is available (Luna-Tortos et al., 2008; Owen et al., 2001; Potschka et al., 2001). Inhibitors of these transporter proteins have been shown to cause an increase in the concentrations of phenytoin and carbamazepine in the brain (Potschka and Löscher, 2001a; Potschka et al., 2001; van Vliet et al., 2007).

Although it is clear that MDR proteins can affect drug distribution and move drugs against concentration gradients, the actual effect on drug efficacy in epilepsy remains uncertain. One uncertainty is the rate of passive diffusion in one direction as opposed to the rate of active protein pumping in the other. If a compound does not readily diffuse, the proteins will have little effect. If the drug rapidly diffuses into a cell, its inward diffusion rate may far exceed the maximum rate that can be achieved through active transport through the protein, and, in this situation, the presence of the protein will also have little effect. Only if the protein pumping rate exceeds the diffusion rate will the proteins clearly alter drug distribution (Figure 5.2). In the brain, the issue is further complicated by the role of the blood–brain barrier in allowing access to the brain extracellular space. The role that the drug-resistant proteins play at the blood–brain barrier in keeping compounds out of the brain in epilepsy is unclear. There are several reports that inhibition of these proteins results in increases in the amount of recoverable phenytoin as measured by microdialysis (Potschka and Löscher, 2001a; Potschka et al., 2001; van Vliet et al., 2007).

The primary question remains unanswered: "Does the presence of multiple drug-resistant proteins cause drug resistance in chronic epilepsy?" Although there is no answer from clinical trials, several animal studies have raised the possibility that the presence of these proteins can alter the response of seizures to drugs. Van Vliet and colleagues (2006) studied rats with limbic epilepsy that had increased expression of P-gp/MDR1 to determine whether the response of their seizures was improved by the addition of the specific P-gp inhibitor tariquidar. They found that the addition of tariquidar resulted in an improved seizure control that also showed a dose–response relationship. A concomitant increase in the brain/plasma ratio of phenytoin occurred following the addition of tariquidar, suggesting that P-gp has a role in reducing brain phenytoin concentrations and that blocking P-gp could be a new therapeutic target. In a follow-up study, van Vliet et al. (2007) demonstrated that the effect of tariquidar was specific to the regions with P-gp overexpression, providing further evidence that this protein does play a role in keeping the drug out of the brain. Additional work is needed to examine this important question, but the possibility that this family of proteins reduces the efficacy of antiepileptic drugs remains open.

## DRUG MALDISTRIBUTION

Getting a drug to the correct target is the key to successful therapy; however, doing so in the central nervous system (CNS) has always been complicated by the presence of the blood–brain barrier, which is well known for its ability to exclude a variety of potentially useful compounds. The general assumption has been that once a compound crosses the blood–brain barrier, it is evenly distributed throughout the brain. Is this assumption correct? Is the blood–brain barrier regionally selective; that is, does it allow drugs through more effectively in some areas than others? These questions are important because, if there is nonuniform distribution, a potentially effective compound may go to an area outside of the therapeutic target in far greater concentrations. Such a scenario could be responsible for the observation of significant neurological side effects in the absence of clinical efficacy. What is the evidence that such a maldistribution occurs? The few reports concerning this issue have not really been linked to any clinical phenomena, but the findings are important for their illustrative value.

In a series of experiments that used microdialysis to determine the relative concentrations of antiepileptic drugs in the frontal cortex and the hippocampus, investigators found that some of the drugs (levetiracetam, lamotrigine) were relatively evenly distributed (Tong and Patsalos, 2001; Walker et al., 2000), whereas others (phenytoin, vigabatrin) were not (Tong et al., 2000, 2009; Walker et al., 1996). In the latter group, phenytoin had an increased area under the concentration curve (AUC) compared to the frontal lobe, whereas vigabatrin had significantly higher concentrations (as well as concomitant GABA concentrations) in the frontal lobe compared to the hippocampus. These findings emphasize the potential for regional variation in the distribution of a drug in the brain and that this variable distribution may be drug dependent.

Another study examined the distribution of phenytoin in the whole rat brain, with the intent of defining specific phenytoin receptors (Geary et al., 1987). Radiolabeled phenytoin was given intravenously to rats, and it was also exposed to *ex vivo* brain slices. The slice-binding study showed a fairly even distribution of binding that was nonspecific throughout the gray matter; however, when given intravenously to rats, the phenytoin at steady state was found overwhelmingly in the white matter. This study revealed several key points about the distribution of phenytoin in the brain. First, the distribution of a drug *in vivo* may not be in any way similar to the binding studies that are performed on slices, emphasizing that active processes in the living brain may have a great influence over drug distribution. Second, the drug may have a preferential distribution (in this case, white matter) that is presumably quite different from the target (in this case, the gray matter, or, better, a specific region in the gray matter, such as the hippocampus) (van Vliet et al., 2007).

Taken together, the microdialysis studies and the phenytoin distribution study emphasize that the drug concentrations can vary considerably across the brain in a drug-specific manner. Although we have no supportive evidence at this time, there is a very real possibility that resistance is, at least in part, related to the failure of a drug to reach the appropriate target. The issue of drug distribution in the epileptic brain is further complicated by the overexpression of multidrug-resistant proteins that may alter the relative distribution of the drug in the central nervous system. Proof that the maldistribution of a drug contributes to pharmacoresistance is much further away, because one must first define the regional localization of a drug, determine the existence of drug resistance, and then show that the resistance is significantly ameliorated by providing the drug greater access to the epileptogenic sites.

## TOLERANCE

Another possible contributor to the refractory nature of seizures is the development of tolerance to the anticonvulsant effect of the drugs. Some patients experience "honeymoon" periods during which they will respond to a new drug well, only to have the seizures reappear. After the seizures break through, the patients typically fall into one of the categories of incompletely responsive described earlier. How resistance might develop after a period of successful treatment for most of the anti-epileptic drugs is a matter of speculation, as is the issue of how many patients actually have this pattern of transient response. While tolerance toward many classes of anticonvulsants may develop, it is most consistently observed with benzodiazepines, such as clonazepam, diazepam, and nitrazepam. Some clinical studies suggest that diazepam loses its antiepileptic effect in as many as 40% of patients within 4 to 6 months of initiating treatment. Tolerance to the anticonvulsant action of benzodiazepines appears to develop in partial epilepsies and in generalized epilepsies. The development of tolerance to benzodiazepines is functional and not metabolic, as the serum benzodiazepine levels do not decrease during long-term treatment (Schmidt, 1989).

Tolerance to the anticonvulsant effect of benzodiazepines is not only observed in patients but can also be reproduced in laboratory animals (Rosenberg et al., 1985, 1989). Mechanisms of benzodiazepine tolerance have been investigated intensively in experimental animals. Benzodiazepines potentiate fast, GABA-mediated inhibitory neurotransmission (Macdonald and Olsen, 1994) by binding to an allosteric site on $GABA_A$ receptors and increasing their affinity for GABA. A dysfunction of the $GABA_A$ receptor appears to underlie benzodiazepine anticonvulsant tolerance (Poisbeau et al., 1997). $GABA_A$ receptors are composed of various combinations of five subunit proteins, derived from multiple families with multiple variants. The nature of the $GABA_A$ receptor dysfunction that is central to benzodiazepine tolerance remains uncertain. Several studies suggest that changes in subunit gene expression underlie a fundamental change in $GABA_A$ receptor subunit composition, whereas others indicate that a posttranslational mechanism mediates a change in $GABA_A$ receptor function. For example, changes in specific $GABA_A$ receptor subunit mRNA and protein expression localized to specific regions occur after chronic benzodiazepine treatments that result in anticonvulsant tolerance (Pesold et al., 1997; Tietz et al., 1999). In contrast, other laboratories demonstrate that exposure to the benzodiazepine antagonist flumazenil can rapidly reverse some of the functional and biochemical changes associated with chronic diazepam treatment (Primus et al., 1996). These findings suggest that a conformational or other posttranslational change may play a key role in mediating tolerance.

At present it is uncertain whether tolerance to the effects of other antiepileptic drugs contributes to therapy resistance. The phenomenon of drug honeymoons, during which a patient has a transient benefit from a particular compound after which the seizures return to pretreatment frequency, is a common experience for physicians who treat epilepsy. A recent report suggests that at least one animal model of limbic epilepsy can develop tolerance to the seizure-suppressant effect of levetirac-etam (van Vliet et al., 2008). Whether this finding is limited to this model or to this seizure type or could be a phenomenon that occurs more commonly across seizure types and species is unknown. Clearly, the drug does exert efficacy in people, but tolerance could explain why it fails in some individuals. In this scenario, the development of tolerance may be something that happens in some individuals, but not others.

It is unknown whether this phenomenon represents the development of tolerance or the drug induces another mechanism (such as a drug-resistant protein) that reduces its efficacy, or if it is a progression of the underlying disorder. It is unclear how many patients experience this phenomenon, as the drug is effective in controlling seizures over the long term in many patients.

## PROBLEM OF DRUG REFRACTORINESS IN STATUS EPILEPTICUS

Prolonged, self-sustaining seizures are commonly referred to as *status epilepticus*. SE is often refractory to currently used first- and second-line agents in humans and experimental animals. Studies on experimental animals have provided many insights into the mechanisms of refractori-ness and suggest novel treatments for refractory SE.

There are at least 126,000 to 195,000 episodes of SE associated with 22,000 to 42,000 deaths each year in the United States (DeLorenzo et al., 1996). Results of two prospective, randomized, double-blind clinical trials suggest that as many as 35 to 50% of patients with generalized convul-sive SE are refractory to benzodiazepines, the first-line agents for treatment of SE (Alldredge et al., 2001; Treiman et al., 1998). When benzodiazepines fail to control SE, it is commonly treated with one of the second-line agents: phenytoin, fosphenytoin, or phenobarbital. Analysis of data from the Veterans Affairs cooperative study suggests that a small fraction of patients refractory to the first agent are likely to respond to the second agent. These patients are comatose and have subtle facial or eye movements and ictal discharges on electroencephalogram (EEG). The Veterans Affairs cooperative study and others suggest that 80 to 90% of the patients with subtle SE are refractory to first- and second-line agents. A retrospective analysis of patients in convulsive and subtle SE in a large academic hospital found that 31% of seizures are refractory to a combination of a benzodiaz-epine and a second-line agent (phenytoin, fosphenytoin, or phenobarbital) (Mayer et al., 2002). In addition to convulsive SE, there exists a subtle or nonconvulsive form of SE. If the rate of refractory SE is conservatively estimated at 30% of all episodes, then 38,000 to 54,500 episodes of refractory SE occur in the United States each year.

## STATUS EPILEPTICUS MAY BE REFRACTORY DUE TO UNDERLYING CAUSE OR PROLONGED SEIZURES

Refractoriness of SE may relate to the underlying cause. There has been no systematic study of causes of refractory SE; however, causes of SE in the general population can be compared with those in patients who require continuous intravenous infusion with propofol, midazolam, or pen-tobarbital for refractory SE. Anticonvulsant withdrawal, remote neurological insult, and stroke are common causes of SE overall (DeLorenzo et al., 1995). In contrast, most common causes of refrac-tory SE appear to be encephalitis, toxic-metabolic encephalopathy, and intracerebral hemorrhage (Prasad et al., 2001).

In addition to the underlying cause, it is a common clinical observation that SE appears to become refractory as it lasts longer. Studies show that the likelihood of a patient responding to first-line therapy declines as the time between the onset of seizures and treatment initiation increases (Lowenstein and Alldredge, 1993). In the Veterans Affairs cooperative study, patients in subtle SE were enrolled 5.2 hours after onset and were refractory to treatment, while patients in overt SE were enrolled 2.8 hours after onset and responded more often to therapy.

## PHARMACOLOGICAL BASIS OF TREATMENT AND REFRACTORINESS

In order to understand the mechanism of drug refractoriness in SE, it is important to recognize that current treatment of SE largely focuses on a single mechanism: enhancement of GABA-mediated inhibition. Benzodiazepines, lorazepam, and diazepam are commonly used first-line agents for the treatment of SE. The therapeutic doses of these agents enhance $GABA_A$ receptor-mediated inhibition (Treiman, 1990). Barbiturates used to treat SE (phenobarbital, pentobarbital, and other agents such as propofol) primarily exert their effects by enhancing $GABA_A$ receptor-mediated inhibition (Macdonald and Olsen, 1994). Only two agents used to treat status epilepticus—phenytoin and fosphenytoin—have a distinct mechanism of action on the voltage-gated sodium channel. Clearly, the current treatment of SE strongly emphasizes enhancement of $GABA_A$ receptor-mediated inhibition.

## MECHANISMS OF REFRACTORY STATUS EPILEPTICUS: ANIMAL MODELS

It is not possible to study the mechanisms of refractory SE in humans; however, key features of SE can be replicated in experimental models. The salient features of animal models of SE have been discussed in detail in other reviews (Dudek et al., 2006; Fountain and Lothman, 1995; Mazarati et al., 2006). Walton and Treiman (1988) first demonstrated drug refractoriness in an experimental model of SE. In lithium/pilocarpine-induced SE, the pattern of electrographic seizures changes progressively over time. Initially, EEG recordings show discrete electrographic seizures, followed by waxing and waning seizures. Later electrographic seizures become continuous. Discrete seizures occurred in the first 5 minutes of SE; 10 and 20 minutes after the onset, waxing and waning seizures were present; and continuous seizures were present in all rats 45 minutes after onset. Rats in each of the three groups were treated with diazepam. Diazepam stopped all seizures in all rats that showed a discrete seizure pattern on EEG; in contrast, seizures were controlled in only 17% of rats that showed continuous seizure pattern on EEG. This study clearly demonstrated that SE rapidly becomes refractory to benzodiazepines. Several groups have confirmed the findings of this study in lithium/pilocarpine and other models. One study demonstrated a substantial reduction of diazepam potency for termination of the seizures with the passage of time (Kapur and Macdonald, 1997); however, time-dependent resistance to anticonvulsants is not restricted to benzodiazepines. In experimental animals, seizures become progressively resistant to treatment with phenobarbital (Borris et al., 2000).

In addition to GABAergic agents, resistance to phenytoin also develops during the course of SE. In an electrical stimulation model of SE, phenytoin became less effective with the passage of time (Mazarati et al., 1998a). Phenytoin was effective when administered 10 minutes after the onset of SE, but not so when administered 40 minutes after the onset of SE. Clearly, first-line agents, such as benzodiazepine, diazepam, phenytoin, and lorazepam, are quite effective in terminating early SE; however, when seizures have lasted 45 to 60 minutes, first- and second-line agents fail. More recent studies indicate that an EEG pattern showing continuous ictal discharge (Class 3) is the most reliable indicator of benzodiazepine resistance (Wang et al., 2009).

The cellular and molecular mechanisms of refractory SE are explored in the following sections. In theory, SE could become refractory to current treatments if GABA-mediated inhibition collapses or if seizures are sustained by other mechanisms such as excitatory neurotransmission. There is evidence from studies in experimental models of SE for both loss of GABA-mediated inhibition and seizures sustained by excitatory transmission.

## THE GABA HYPOTHESIS

The structures involved in the genesis of SE have been delineated by use of the $^{14}C$-2-deoxyglucose mapping technique combined with EEG recordings in experimental animals. In the late stages of SE, the hippocampal formation, subiculum, entorhinal cortex, amygdala, pyriform cortex, substantia nigra, and midline thalamic nuclei show increased glucose utilization in a bilateral symmetrical fashion (Lothman et al., 1991; VanLandingham and Lothman, 1991). Other studies indicate that sustained seizures of SE can be maintained in a combined entorhinal cortex–hippocampal slice preparation (Rafiq et al., 1993; Stanton et al., 1987). These studies suggested that the hippocampus is an attractive site to study the mechanisms of refractory SE.

GABA-mediated inhibition in the CA1 region and dentate gyrus of the hippocampus plays a critical role in the pathogenesis of refractoriness of SE. In many experimental models of SE, a loss of GABAergic inhibition has been documented. Early studies used the paired-pulse inhibition technique to study the relationship of GABA-mediated paired-pulse inhibition in the CA1 region of the hippocampus and evolution of SE (Kapur and Lothman, 1989; Michelson et al., 1989). After measurement of GABA-mediated inhibition in the CA1 region, SE was induced by electrical stimulation of the hippocampus. The stimulated animals that developed hippocampal SE showed loss of paired-pulse inhibition. The remaining animals showed a mild reduction in paired-pulse inhibition after stimulation and did not develop self-sustaining SE. Paired-pulse inhibition in the CA1 region of the hippocampus could be reduced by diminished release of GABA or diminished postsynaptic $GABA_A$ receptor (GABAR) responsiveness. While the release of GABA from neurons during SE has not been studied, there is growing evidence that responsiveness of GABARs to GABA is diminished during SE.

## $GABA_A$ RECEPTOR PLASTICITY DURING STATUS EPILEPTICUS

The reduction in the potency of diazepam during SE occurs at the level of a single principal neuron. Whole-cell GABAR currents were obtained from dentate granule cells isolated acutely from the hippocampus of control rats and rats undergoing 45 minutes of SE. In dentate granule cells from control animals, when 10-$\mu M$ GABA was co-applied with 300-n$M$ diazepam, GABAR currents were markedly enhanced in all neurons. In contrast, in dentate granule cells from animals undergoing SE, 300-n$M$ diazepam inconsistently enhanced 6- or 10-$\mu M$ GABA-evoked GABAR currents. Diazepam concentration–response curves were obtained for enhancement of GABAR currents from neurons from both the naïve and SE-treated animals. In neurons from control animals, 1-$\mu M$ diazepam enhanced GABAR currents by 92%, but in neurons from animals undergoing SE, 3-$\mu M$ diazepam only enhanced currents by 52%. The median effective concentration ($EC_{50}$) for diazepam enhancement of GABAR currents in neurons from control animals was 195 n$M$, and those in neurons from animals undergoing SE was 4.4 $\mu M$. Thus, the prolonged seizures of SE reduced the potency and efficacy of diazepam for enhancement of dentate granule cell GABAR currents. This study suggests that a rapid alteration in functional properties of the GABARs expressed on the surface of the dentate granule cells had occurred during SE (Kapur and Macdonald, 1996).

The genes for the GABAR subunits are divided into families based on sequence homology and have been named $\alpha$, $\beta$, $\gamma$, $\delta$, $\epsilon$, $\pi$, $\theta$, and $\rho$ subunits. Each subunit has a similar transmembrane topology, and there is a large (approximately 200 amino acid) extracellular N-terminal domain that forms the GABA binding site. Functional receptors are formed by the co-assembly of five subunits around the central channel axis. It is currently believed that the majority of GABARs are composed of two $\alpha$ subunits, two $\beta$ subunits, and one $\gamma_2$ or $\delta$ subunit.

The benzodiazepine sensitivity of GABARs is dependent on their subunit composition. The presence of the $\gamma_2$ subunit is a requirement for GABAR benzodiazepine sensitivity. The $\gamma_2$ subunit can be substituted by the $\delta$ subunit, and receptors containing this subunit are benzodiazepine insensitive (Saxena and Macdonald, 1994; 1996). The $\delta$ subunit commonly assembles with the $\alpha_4$ or the $\alpha_6$ subunit, and both of these also render $GABA_A$ receptors benzodiazepine insensitive, even when combined with the $\gamma_2$ subunit (Knoflach et al., 1996; Wafford et al., 1996).

Multiple specific $GABA_A$ receptor isoforms are expressed on the surface of hippocampal dentate granule cells. *In situ* hybridization and immunohistochemical studies have demonstrated that dentate granule cells express mRNAs for $\alpha_1$, $\alpha_2$, $\alpha_4$, $\beta_1$, $\beta_3$, $\gamma_1$, $\gamma_2$, and $\delta$ GABAR subunit subtypes (Sperk et al., 1997; Wisden et al., 1992). Other studies investigating the subcellular distribution of $\alpha_1$, $\alpha$, $\gamma_2$, and $\delta$ subunit protein in hippocampal dentate granule cells demonstrated that diazepam-sensitive ($\alpha_1$- and $\gamma_2$-subunit-containing) receptors are present at the synapse, whereas diazepam-insensitive ($\alpha_4$- and $\delta$-subunit-containing) receptors are present in the extrasynaptic locations (Sun et al., 2004, 2007; Wei et al., 2003; Zhang et al., 2007). The functional significance of these extrasynaptic receptors is discussed below.

The synaptically located diazepam-sensitive GABARs mediate fast synaptic inhibition. Synaptic GABARs containing the $\gamma_2$ subunit have a low affinity for GABA and thus require high concentrations of GABA for activation. Analysis of the kinetics of synaptic currents and rapid-application experiments suggest that presynaptic terminals release vesicles containing high concentrations of GABA. Synaptic GABARs open rapidly after GABA binding, are followed by desensitization over a period of time, and end by closure and the dissociation of GABA. The entire sequence of events from activation to deactivation lasts less than 300 milliseconds, resulting in periodic or "phasic" inhibition of the postsynaptic neuron. A detailed analysis of the effect of SE on drug modulation of synaptic GABARs on dentate granule cells was recently reported. Animals were studied immediately after a Class 3 seizure on the Racine scale or 30 minutes after the first Class 3 seizure. These time points were chosen because a previous study demonstrated that resistance to diazepam termination of SE occurs within minutes of the first Class 3 seizure (Jones et al., 2002). The amplitude and charge transfer of action-potential-independent miniature inhibitory postsynaptic currents (mIPSCs) were diminished compared to controls in animals studied immediately after the first Class 3 seizure; however, these reductions in synaptic currents were restored when animals were studied 30 minutes after the first Class 3 seizure. The mIPSCs recorded from animals immediately after the first Class 3 seizure were less sensitive to diazepam and zinc modulation than those from sham controls and animals studied 30 minutes after the first Class 3 seizures. The authors concluded that there were substantial plastic changes in synaptic GABAR function and their allosteric modulation during SE.

Other studies investigating the impact of SE on GABAergic synaptic transmission reached a similar conclusion that the population of synaptic GABARs is modified during SE and extended the findings reviewed above. These investigations demonstrated that the rapid modification in synaptic GABARs is partially the result of activity-dependent, subunit-specific trafficking of GABARs during SE (Goodkin et al., 2005). The effects of prolonged epileptiform bursting on GABA-mediated synaptic transmission were studied in an *in vitro* model of SE; the period of prolonged epileptiform bursting resulted in an approximately 35% reduction in the mIPSC amplitude. Using an antibody-feeding technique, it was demonstrated that constitutive internalization of $\beta_{2/3}$-subunit-containing GABARs was rapid, with 50% of the receptors internalized in less than approximately 11 minutes. Under conditions of prolonged epileptiform bursting, the percentage of receptors accumulated increased to approximately 60%, and 50% of the receptors were internalized within approximately 7 minutes. This study suggested that activity-dependent trafficking was a potential mechanism to explain the reduction in GABA-mediated inhibition and the development of benzodiazepine pharmacoresistance that occurs during SE.

These findings were confirmed in an *in vivo* model of SE (Naylor et al., 2005). In lithium/pilocarpine-induced SE of 1-hour duration, mIPSCs were recorded from the dentate granule cells of animals. The amplitudes of mIPSCs recorded from dentate granule cells of animals in SE were approximately 30% smaller than those recorded from controls. Diazepam enhances mIPSCs in dentate granule cells by prolonging their decays. In this study, diazepam enhancement of mIPSCs was evident in dentate granule cells from SE-treated and control animals; however, diazepam did not restore total charge transferred during mIPSCs (a measure of inhibition) in SE animals to the level observed in controls. The mIPSC data were fitted to a multistate model of synaptic GABARs to derive the number of GABARs expressed per dentate granule cell inhibitory synapse in the dentate gyrus of control and SE-treated animals. The authors suggested that, during SE, the number of receptors expressed at the synapse had declined to 18 receptors per synapse compared to the 36 receptors per synapse in control animals. Using immunocytochemical methods, they were also able to demonstrate a redistribution of GABARs that contained either a $\beta_{2/3}$ or $\gamma_2$ subunit to the intracellular compartment.

In other studies, a biotinylation pull-down assay was used to demonstrate that the increase in intracellular accumulation of these GABARs resulted in a net reduction of surface-expressing, benzodiazepine-sensitive GABARs containing the $\gamma_2$ subunit (Goodkin et al., 2008; Terunuma et al., 2008). In their study, Terunuma et al. (2008) were able to demonstrate that the reduction in the surface expression of GABARs containing the $\beta_3$ subunits that occurs during SE was due to a deficit in phosphorylation of these GABARs at serine residues 408/9, which resulted in an unmasking of a patch-binding motif for the clathrin adaptor AP2 and the endocytosis of these receptors.

During SE, the signaling cascade that results in a reduction in the surface expression of the benzodiazepine-sensitive $\gamma_2$-subunit-containing GABARs could be initiated by a ligand-independent process such as an increase in neuronal excitability as the result of stimulation of excitatory amino acid receptors (Kapur and Lothman, 1989) or initiated by excessive extracellular GABA (ligand-dependent). In one study (Goodkin et al., 2008), a combination of high external potassium and NMDA receptor activation was sufficient to reduce the surface expression of $\gamma_2$-subunit-containing GABARs in an organotypical hippocampal culture model, consistent with a ligand-independent process. Although NMDA receptor activation was necessary to induce the reduction, NMDA receptor activation was not sufficient to induce a statistically significant decrease in the surface expression of the $\gamma_2$ subunits. These findings suggest that other changes in neuronal excitability induced by the high external concentration of potassium were required for the reduction in the surface expression of this subunit.

Taken together these studies performed in different laboratories using multiple models of SE reached remarkably similar conclusions: GABAergic synaptic inhibition is diminished during SE, and this reduction relates to a reduction in the surface expression of benzodiazepine-sensitive $\gamma_2$ subunit-containing GABARs.

## EFFECT OF STATUS EPILEPTICUS ON DIAZEPAM-INSENSITIVE GABARS

In addition to the $\gamma_2$ subunit, dentate granule cells express the $\alpha_4$ and $\delta$ subunit, and, as explained above, these subunits are exclusively expressed in the extrasynaptic space. Studies of recombinant GABARs consisting of these subunits demonstrated that receptors composed of these subunits have a high affinity for GABA, desensitize slowly, and are benzodiazepine insensitive (Mtchedlishvili and Kapur, 2006; Sun et al., 2004). This unique set of properties enables $\alpha_4$- and $\delta$-subunit-containing receptors to mediate a novel nonsynaptic form of benzodiazepine-insensitive GABAergic inhibition, commonly referred to as *tonic inhibition*. The impact of SE on the tonic inhibition of dentate granule cells has also been investigated. In one study (Naylor et al., 2005), the amplitude of tonic inhibition recorded from dentate granule cells of animals that had undergone SE was larger than that recorded from controls. The authors modeled their results and suggested that this increase in tonic inhibition was due to increased GABA in the extracellular space. Another study confirmed that the tonic inhibition of dentate granule cells measured in the presence of GABA uptake blockers is enhanced during SE (Goodkin et al., 2008). The surface expression of the GABAR $\delta$ subunit was not decreased in the hippocampal slices acutely obtained from animals in SE compared to the hippocampal slices obtained from naïve animals (Goodkin et al., 2008; Terunuma et al., 2008).

## ROLE OF EXCITATORY TRANSMISSION

In addition to loss of inhibition, there is growing evidence that excitatory neurotransmission is enhanced in the self-sustaining phase of SE. Several recent studies suggest that prolonged SE, refractory to benzodiazepines and phenobarbital, is effectively controlled by NMDA receptor antagonists. NMDA receptor antagonists can prevent the neuropathology and delayed development of epilepsy-associated SE. Finally, neuropeptides such as galanin and substance P that modulate prolonged, refractory SE are known to exert their cellular effects on excitatory transmission.

A growing number of studies demonstrate the effectiveness of NMDA receptor antagonists in the treatment of refractory SE. In the perforant path stimulation model of SE, seizures that rapidly become refractory to benzodiazepines were controlled by NMDA receptor antagonists (Mazarati and Wasterlain, 1999). Other studies indicate that NMDA receptor antagonist MK-801 prevents development of refractoriness to diazepam in the lithium/pilocarpine model of SE (Rice and DeLorenzo, 1999). Another study found that the NMDA receptor antagonist ketamine did not control seizures when administered in the early phase of SE but was effective after prolonged seizures (Bertram and Lothman, 1990; Borris et al., 2000). Noncompetitive antagonists of NMDA receptors appear to be more effective than competitive and allosteric site inhibitors (Yen et al., 2004). Ketamine can work synergistically with diazepam to terminate SE (Martin and Kapur, 2008).

## NEUROPEPTIDES

Neuropeptides have emerged as candidates for termination of refractory SE. Neuropeptide Y suppresses epileptiform activity in rat hippocampal slices (Klapstein and Colmers, 1997). Mice lacking neuropeptide Y have uncontrollable seizures in response to the chemoconvulsant kainic acid, and 93% progressed to death; however, death was rare in wild-type littermates. Intracerebroventricular neuropeptide Y prevented death due to seizures induced by kainic acid (Baraban et al., 1997). These studies taken together suggest a critical role for neuropeptide Y in seizure termination.

Galanin is a 29- or 30-amino-acid residue neuropeptide present in the septal projections to the hippocampus and has an inhibitory effect on the hippocampus by inhibiting excitatory neurotransmission. Galanin immunoreactivity in the hippocampus is diminished following limbic SE (Mazarati et al., 1998b). Injection of galanin into the hippocampal dentate hilus prevents the onset of limbic SE and can stop established limbic SE (Mazarati et al., 2000). It appears that galanin can act as an endogenous anticonvulsant, stopping SE.

In SE, it is clear that a number of acute changes occur that limit the response to standard therapy. Experimental models can help discover new therapies for some forms of SE, but they cannot help (for the moment) with some forms that are associated with specific conditions such as encephalitis. For now, the studies emphasize the need to treat early and aggressively so the patients do not have a chance to develop a refractory condition. It is unclear whether certain causes of SE are resistant because of the duration of the seizure activity or because of the underlying cause. This is an area that requires more information. Perhaps the best approach would be to ensure that all patients are treated aggressively at the first onset of prolonged seizure activity.

## MOVING TOWARD A SOLUTION

How do we get beyond the problem of pharmacotherapy resistance in epilepsy and SE? The obvious answer is to understand the pathophysiology associated with the many causes of the two conditions. Because the basis for these conditions is multifactorial in complex neural circuits and because (in SE, at least) the pathophysiology is constantly evolving, it is unlikely that we will arrive at a very detailed understanding in the near future. For this reason, we have to be much more empiric in our approach while recognizing the many caveats that may adversely affect the performance of a promising drug *in vivo*. The issues that need to be addressed in this process are what we described above.

*Are the compounds appropriate for the receptors and channels that we are targeting?* The changes in these membrane proteins are well described, with many more yet to be elucidated. It is essential that drugs are designed for the specific channels that play important roles in seizure initiation and evolution, so the drugs are maximally effective at these channels but much less potent at related channels. Solving this problem may also resolve the problem of clinical side-effects before clinical efficacy.

*Are the drugs reaching the appropriate targets?* This issue may be influenced by the drug-resistance proteins and the blood–brain barrier, which may, in some cases, have resulted in regional variation in the distribution of the drugs. Compounds should be designed with these issues in mind so they actually reach the intended targets, preferably with reduced distribution to other sites.

*Do effective drugs lose their efficacy over time?* Tolerance is the most difficult issue to identify and to design for, especially in the preclinical phase. The benzodiazepines are known for the rapid development of tolerance, an issue that limits the utility of an otherwise very effective drug. Whether loss of efficacy over time for the other drugs is the result of tolerance is unknown, but there are anecdotes to suggest that this process may occur in some patients, and there is evidence from some animal studies that tolerance to standard antiepileptic drugs can develop. Designing studies to determine whether tolerance develops in the course of treatment is an important step. This information will let us know to avoid certain mechanisms of drug action that may be prone to the induction of tolerance or to develop a method for preventing tolerance.

How do we resolve these seemingly insurmountable problems? The first step is, as it is for so many challenges, being aware of what the problems are. One problem lies in the methods that have been used to identify drugs. In general, the process has depended on normal animals that experience acute seizures induced by some exogenous stimulus (as opposed to the natural state of spontaneous seizures that occur in an abnormal bran region). It is unknown whether the acute induced seizure models used at present will identify compounds that will be effective in preventing spontaneous seizures, especially the critical seizure initiation phase that likely has a unique set of mechanisms. It is also unclear whether the distribution of the drugs in these normal brains approximates the distribution in an epileptic brain that has alterations in glia, upregulation of drug-resistant proteins, and possible changes in blood–brain barrier function. Because of the multiple changes occurring in the receptors and channels in many of the epileptic models that have been examined, it is not known whether the mechanisms identified in the acute induced models are even relevant for epilepsy.

Perhaps the first step in resolving the issue of model relevance is to begin using models that are as realistic as possible at some point in the drug discovery process. Such models may simultaneously test for efficacy of the mechanism and appropriate targeting. The downside to this approach is that, for most of the realistic models, the spontaneity of seizures makes timing of the drug application in relation to seizure onset problematic, especially in animals that have rapid metabolic turnover of the compounds. The beauty of the acute models is that they are predictable in response and the timing of the seizure, and drug administration can be precisely and consistently coordinated. The practicality of screening large numbers of compounds will dictate the maintenance of some of the older models, but at some point the more realistic models must be incorporated.

In summary, there are a number of ways a drug can fail in controlling the seizures of epilepsy and SE, and the real underlying reasons continue to elude us. There are now some definable and testable hypotheses about mechanisms that may have considerable impact on drug resistance. The first step in resolving the issue is to start investigating drug resistance in SE and chronic epilepsy as a specific phenomenon using appropriate models. Only then can we begin to make headway in this area.

## REFERENCES

Alldredge, B.K., A.M. Gelb, S.M. Isaacs, M.D. Corry, F. Allen, S. Ulrich, M.D. Gottwald, N. O'Neil, J.M. Neuhaus, M.R. Segal, D.H. Lowenstein. (2001). A comparison of lorazepam, diazepam, and placebo for the treatment of out-of-hospital status epilepticus. *N. Engl. J. Med.*, 345:631–637.

Aronica, E., E.A. van Vliet, O.A. Mayboroda, D. Troost, F.H. da Silva, J.A. Gorter. (2000). Upregulation of metabotropic glutamate receptor subtype mGluR3 and mGluR5 in reactive astrocytes in a rat model of mesial temporal lobe epilepsy. *Eur. J. Neurosci.*, 12:2333–2344.

Aronica, E., B. Yankaya, D. Troost, E.A. van Vliet, F.H. Lopes da Silva, J.A. Gorter. (2001). Induction of neonatal sodium channel II and III alpha-isoform mRNAs in neurons and microglia after status epilepticus in the rat hippocampus. *Eur. J. Neurosci.*, 13:1261–1266.

Aronica, E., J.A. Gorter, S. Redeker, E.A. van Vliet, M. Ramkema, G.L. Scheffer, R.J. Scheper, P. van der Valk, S. Leenstra, J.C. Baayen, W.G. Spliet, D. Troost. (2005). Localization of breast cancer resistance protein (BCRP) in microvessel endothelium of human control and epileptic brain. *Epilepsia*, 46:849–857.

Baraban, S.C., G. Hollopeter, J.C. Erickson, P.A. Schwartzkroin, R.D. Palmiter. (1997). Knock-out mice reveal a critical antiepileptic role for neuropeptide Y. *J. Neurosci.*, 17:8927–8936.

Bertram, E.H., E.W. Lothman. (1990). NMDA receptor antagonists and limbic status epilepticus: a comparison with standard anticonvulsants. *Epilepsy Res.*, 5:177–184.

Borris, D.J., E.H. Bertram, J. Kapur. (2000). Ketamine controls prolonged status epilepticus. *Epilepsy Res.*, 42:117–122.

Brooks-Kayal, A.R., M.D. Shumate, H. Jin, T.Y. Rikhter, D.A. Coulter. (1998). Selective changes in single cell GABA(A) receptor subunit expression and function in temporal lobe epilepsy. *Nat. Med.*, 4:1166–172.

DeLorenzo, R.J., J.M. Pellock, A.R. Towne, J.G. Boggs. (1995). Epidemiology of status epilepticus. *J. Clin. Neurophysiol.*, 12:316–325.

DeLorenzo, R.J., W.A. Hauser, A.R. Towne, J.G. Boggs, J.M. Pellock, L. Penberthy, L. Garnett, C.A. Fortner, D. Ko. (1996). A prospective, population-based epidemiologic study of status epilepticus in Richmond, Virginia. *Neurology*, 46:1029–1035.

Dudek, F.E., S. Clark, P.A. Williams, H.L. Grabenstatter. (2006). Kainate-induced status epilepticus: a chronic model of acquired epilepsy. In *Models of Seizures and Epilepsy* (pp. 415–432), A. Pitkanen, P.A. Schwartzkroin, and S.L. Moshe, Eds. San Diego, CA: Academic Press.

Elsinga, P.H., N.H. Hendrikse, J. Bart, W. Vaalburg, A. van Waarde. (2004). PET studies on P-glycoprotein function in the blood–brain barrier: how it affects uptake and binding of drugs within the CNS. *Curr. Pharm. Des.*, 10:1493–1503.

Escayg, A., A. Heils, B.T. MacDonald, K. Haug, T. Sander, M.H. Meisler. (2001). A novel SCN1A mutation associated with generalized epilepsy with febrile seizures plus—and prevalence of variants in patients with epilepsy. *Am. J. Hum. Genet.*, 68:866–873.

Fountain, N.B., E.W. Lothman. (1995). Pathophysiology of status epilepticus. *J. Clin. Neurophysiol.*, 12:326–342.

Geary, W.A., 2nd, G.F. Wooten, J.B. Perlin, E.W. Lothman. (1987). *In vitro* and *in vivo* distribution and binding of phenytoin to rat brain. *J. Pharmacol. Exp. Ther.*, 241:704–713.

Gibbs, J.W., 3rd, M.D. Shumate, D.A. Coulter. (1997). Differential epilepsy-associated alterations in postsynaptic GABA(A) receptor function in dentate granule and CA1 neurons. *J. Neurophysiol.*, 77:1924–1938.

Goodkin, H.P., J.L. Yeh, J. Kapur. (2005). Status epilepticus increases the intracellular accumulation of GABA$_A$ receptors. *J. Neurosci.*, 25:5511–5520.

Goodkin, H.P., S. Joshi, Z. Mtchedlishvili, J. Brar, J. Kapur. (2008). Subunit-specific trafficking of GABA(A) receptors during status epilepticus. *J. Neurosci.*, 28:2527–2538.

Hermann, D.M., E. Kilic, A. Spudich, S.D. Kramer, H. Wunderli-Allenspach, C.L. Bassetti. (2006). Role of drug efflux carriers in the healthy and diseased brain. *Ann. Neurol.*, 60:489–498.

Jacobs, M.P., G.D. Fischbach, M.R. Davis, M.A. Dichter, R. Dingledine, D.H. Lowenstein, M.J. Morrell, J.L. Noebels, M.A. Rogawski, S.S. Spencer, W.H. Theodore. (2001). Future directions for epilepsy research. *Neurology*, 57:1536–1542.

Jones, D.M., N. Esmaeil, S. Maren, R.L. Macdonald. (2002). Characterization of pharmacoresistance to benzodiazepines in the rat Li–pilocarpine model of status epilepticus. *Epilepsy Res.*, 50:301–312.

Jones, P.J., E.C. Merrick, T.W. Batts, N.J. Hargus, Y. Wang, J.P. Stables, E.H. Bertram, M.L. Brown, M.K. Patel. (2009). Modulation of sodium channel inactivation gating by a novel lactam: implications for seizure suppression in chronic limbic epilepsy. *J. Pharmacol. Exp. Ther.*, 328:201–212.

Kapur, J., E.W. Lothman. (1989). Loss of inhibition precedes delayed spontaneous seizures in the hippocampus after tetanic electrical stimulation. *J. Neurophysiol.*, 61:427–434.

Kapur, J., R.L. Macdonald. (1996). Status epilepticus: a proposed pathophysiology. In *The Treatment of Epilepsy* (pp. 258–268), S. Shorvon, Ed. Oxford: Blackwell Science.

Kapur, J., R.L. Macdonald. (1997). Rapid seizure-induced reduction of benzodiazepine and Zn$^{2+}$ sensitivity of hippocampal dentate granule cell GABA$_A$ receptors. *J. Neurosci.*, 17:7532–7540.

Kearney, J.A., N.W. Plummer, M.R. Smith, J. Kapur, T.R. Cummins, S.G. Waxman, A.L. Goldin, M.H. Meisler. (2001). A gain-of-function mutation in the sodium channel gene *Scn2a* results in seizures and behavioral abnormalities. *Neuroscience*, 102:307–317.

Ketelaars, S.O., J.A. Gorter, E.A. van Vliet, F.H. Lopes da Silva, W.J. Wadman. (2001). Sodium currents in isolated rat CA1 pyramidal and dentate granule neurones in the post-status epilepticus model of epilepsy. *Neuroscience*, 105:109–120.

Klapstein, G.J., W.F. Colmers. (1997). Neuropeptide Y suppresses epileptiform activity in rat hippocampus *in vitro*. *J. Neurophysiol.*, 78:1651–1661.

Knoflach, F., D. Benke, Y. Wang, L. Scheurer, H. Luddens, B.J. Hamilton, D.B. Carter, H. Mohler, J.A. Benson. (1996). Pharmacological modulation of the diazepam-insensitive recombinant γ-aminobutyric acid$_A$ receptors $\alpha_4\beta_2\gamma_2$ and $\alpha_6\beta_2\gamma_2$. *Mol. Pharmacol.*, 50:1253–1261.

Lerche, H., K. Jurkat-Rott, F. Lehmann-Horn. (2001). Ion channels and epilepsy. *Am. J. Med. Genet.*, 106:146–159.

Litman, T., T.E. Druley, W.D. Stein, S.E. Bates. (2001). From MDR to MXR: new understanding of multidrug resistance systems, their properties and clinical significance. *Cell. Mol. Life Sci.*, 58:931–959.

Lothman, E.W., E.H. Bertram, 3rd, J.L. Stringer. (1991). Functional anatomy of hippocampal seizures. *Prog. Neurobiol.*, 37:1–82.

Lothman, E.W., D.A. Rempe, P.S. Mangan. (1995). Changes in excitatory neurotransmission in the CA1 region and dentate gyrus in a chronic model of temporal lobe epilepsy. *J. Neurophysiol.*, 74:841–848.

Lowenstein, D.H., B.K. Alldredge. (1993). Status epilepticus at an urban public hospital in the 1980s. *Neurology*, 43:483–488.

Luna-Tortos, C., M. Fedrowitz, W. Löscher. (2008). Several major antiepileptic drugs are substrates for human P-glycoprotein. *Neuropharmacology*, 55:1364–1375.

Macdonald, R.L., K.M. Kelly. (1995). Antiepileptic drug mechanisms of action. *Epilepsia*, 36(Suppl. 2):S2–S12.

Macdonald, R.L., R.W. Olsen. (1994). GABA$_A$ receptor channels. *Annu. Rev. Neurosci.*, 17:569–602.

Mangan, P.S., E.W. Lothman. (1996). Profound disturbances of pre- and postsynaptic GABA$_B$-receptor-mediated processes in region CA1 in a chronic model of temporal lobe epilepsy. *J. Neurophysiol.*, 76:1282–1296.

Mangan, P.S., D.A. Rempe, E.W. Lothman. (1995). Changes in inhibitory neurotransmission in the CA1 region and dentate gyrus in a chronic model of temporal lobe epilepsy. *J. Neurophysiol.*, 74:829–840.

Mangan, P.S., C.A. Scott, J.M. Williamson, E.H. Bertram. (2000). Aberrant neuronal physiology in the basal nucleus of the amygdala in a model of chronic limbic epilepsy. *Neuroscience*, 101:377–391.

Marchi, N., G. Guiso, S. Caccia, M. Rizzi, B. Gagliardi, F. Noe, T. Ravizza, S. Bassanini, S. Chimenti, G. Battaglia, A. Vezzani. (2006). Determinants of drug brain uptake in a rat model of seizure-associated malformations of cortical development. *Neurobiol. Dis.*, 24:429–442.

Martin, B.S., J. Kapur. (2008). A combination of ketamine and diazepam synergistically controls refractory status epilepticus induced by cholinergic stimulation. *Epilepsia*, 49:248–255.

Matheny, C.J., M.W. Lamb, K.R. Brouwer, G.M. Pollack. (2001). Pharmacokinetic and pharmacodynamic implications of P-glycoprotein modulation. *Pharmacotherapy*, 21:778–796.

Mathern, G.W., E.H. Bertram, 3rd, T.L. Babb, J.K. Pretorius, P.A. Kuhlman, S. Spradlin, D. Mendoza. (1997). In contrast to kindled seizures, the frequency of spontaneous epilepsy in the limbic status model correlates with greater aberrant fascia dentata excitatory and inhibitory axon sprouting, and increased staining for *N*-methyl-D-aspartate, AMPA, and GABA(A) receptors. *Neuroscience*, 77:1003–1019.

Mayer, S.A., J. Claassen, J. Lokin, F. Mendelsohn, L.J. Dennis, B.F. Fitzsimmons. (2002). Refractory status epilepticus: frequency, risk factors, and impact on outcome. *Arch. Neurol.*, 59:205–210.

Mazarati, A.M., C.G. Wasterlain. (1999). *N*-methyl-D-aspartate receptor antagonists abolish the maintenance phase of self-sustaining status epilepticus in rat. *Neurosci. Lett.*, 265:187–190.

Mazarati, A.M., R.A. Baldwin, R. Sankar, C.G. Wasterlain. (1998a). Time-dependent decrease in the effectiveness of antiepileptic drugs during the course of self-sustaining status epilepticus. *Brain Res.*, 814:179–185.

Mazarati, A.M., H. Liu, U. Soomets, R. Sankar, D. Shin, H. Katsumori, U. Langel, C.G. Wasterlain. (1998b). Galanin modulation of seizures and seizure modulation of hippocampal galanin in animal models of status epilepticus. *J. Neurosci.*, 18:10070–10077.

Mazarati, A.M., J.G. Hohmann, A. Bacon, H. Liu, R. Sankar, R.A. Steiner, D. Wynick, C.G. Wasterlain. (2000). Modulation of hippocampal excitability and seizures by galanin. *J. Neurosci.*, 20:6276–6281.

Mazarati A.M., K.W. Thompson, L. Suchomelova, R. Sankar, Y. Shirasaka, J. Nissinen, A. Pitkanen, E.H. Bertram, C. Wasterlain. (2006). Status epilepticus: electrical stimulation models. In *Models of Seizures and Epilepsy* (pp. 449–464), A. Pitkanen, P.A. Schwartzkroin, and S.L. Moshe, Eds. San Diego, CA: Academic Press.

Michelson, H.B., J. Kapur, E.W. Lothman. (1989). Reduction of paired pulse inhibition in the CA1 region of the hippocampus by pilocarpine in naive and in amygdala-kindled rats. *Exp. Neurol.*, 104:264–271.

Miller, D.S., B. Bauer, A.M. Hartz. (2008). Modulation of P-glycoprotein at the blood–brain barrier: opportunities to improve central nervous system pharmacotherapy. *Pharmacol. Rev.*, 60:196–209.

Mtchedlishvili, Z., J. Kapur. (2006). High-affinity, slowly desensitizing GABA$_A$ receptors mediate tonic inhibition in hippocampal dentate granule cells. *Mol. Pharmacol.*, 69:564–575.

Mtchedlishvili, Z., E.H. Bertram, J. Kapur. (2001). Diminished allopregnanolone enhancement of GABA(A) receptor currents in a rat model of chronic temporal lobe epilepsy. *J. Physiol.*, 537:453–465.

Naylor, D.E., H. Liu, C.G. Wasterlain. (2005). Trafficking of GABA(A) receptors, loss of inhibition, and a mechanism for pharmacoresistance in status epilepticus. *J. Neurosci.*, 25:7724–7733.

Owen, A., M. Pirmohamed, J.N. Tettey, P. Morgan, D. Chadwick, B.K. Park. (2001). Carbamazepine is not a substrate for P-glycoprotein. *Br. J. Clin. Pharmacol.*, 51:345–349.

Pesold, C., H.J. Caruncho, F. Impagnatiello, M.J. Berg, J.M. Fritschy, A. Guidotti, E. Costa. (1997). Tolerance to diazepam and changes in GABA(A) receptor subunit expression in rat neocortical areas. *Neuroscience*, 79:477–487.

Poisbeau, P., S.R. Williams, I. Mody. (1997). Silent GABA$_A$ synapses during flurazepam withdrawal are region specific in the hippocampal formation. *J. Neurosci.*, 17:3467–475.

Potschka, H., W. Löscher. (2001a). *In vivo* evidence for P-glycoprotein-mediated transport of phenytoin at the blood–brain barrier of rats. *Epilepsia*, 42:1231–1240.

Potschka, H., W. Löscher. (2001b). Multidrug resistance-associated protein is involved in the regulation of extracellular levels of phenytoin in the brain. *NeuroReport*, 12:2387–1389.

Potschka, H., M. Fedrowitz, W. Löscher. (2001). P-glycoprotein and multidrug resistance-associated protein are involved in the regulation of extracellular levels of the major antiepileptic drug carbamazepine in the brain. *NeuroReport*, 12:3557–560.

Prasad, A., B.B. Worrall, E.H. Bertram, T.P. Bleck. (2001). Propofol and midazolam in the treatment of refractory status epilepticus. *Epilepsia*, 42:380–386.

Primus, R.J., J. Yu, J. Xu, C. Hartnett, M. Meyyappan, C. Kostas, T.V. Ramabhadran, D.W. Gallager. (1996). Allosteric uncoupling after chronic benzodiazepine exposure of recombinant gamma-aminobutyric acid(A) receptors expressed in Sf9 cells: ligand efficacy and subtype selectivity. *J. Pharmacol. Exp. Ther.*, 276:882–890.

Rafiq, A., R.J. DeLorenzo, D.A. Coulter. (1993). Generation and propagation of epileptiform discharges in a combined entorhinal cortex/hippocampal slice. *J. Neurophysiol.*, 70:1962–1974.

Rajasekaran, K., J. Kapur, E.H. Bertram. (2007). Alterations in GABA(A) receptor mediated inhibition in adjacent dorsal midline thalamic nuclei in a rat model of chronic limbic epilepsy. *J. Neurophysiol.*, 98:2501–2508.

Rajasekaran, K., C. Sun, E.H. Bertram. (2009). Altered pharmacology and GABA-A receptor subunit expression in dorsal midline thalamic neurons in limbic epilepsy. *Neurobiol. Dis.*, 33(1):119–132.

Rempe, D.A., P.S. Mangan, E.W. Lothman. (1995). Regional heterogeneity of pathophysiological alterations in CA1 and dentate gyrus in a chronic model of temporal lobe epilepsy. *J. Neurophysiol.*, 74:816–828.

Remy, S., H. Beck. (2006). Molecular and cellular mechanisms of pharmacoresistance in epilepsy. *Brain*, 129:18–35.

Rice, A.C., R.J. DeLorenzo. (1999). *N*-methyl-D-aspartate receptor activation regulates refractoriness of status epilepticus to diazepam. *Neuroscience*, 93:117–123.

Rosenberg, H.C., E.I. Tietz, T.H. Chiu. (1985). Tolerance to the anticonvulsant action of benzodiazepines: relationship to decreased receptor density. *Neuropharmacology*, 24:639–644.

Rosenberg, H.C., E.I. Tietz, T.H. Chiu. (1989). Tolerance to anticonvulsant effects of diazepam, clonazepam, and clobazam in amygdala-kindled rats. *Epilepsia*, 30:276–285.

Saxena, N.C., R.L. Macdonald. (1994). Assembly of GABA$_A$ receptor subunits: role of the delta subunit. *J. Neurosci.*, 14:7077–7086.

Saxena, N.C., R.L. Macdonald. (1996). Properties of putative cerebellar gamma-aminobutyric acid A receptor isoforms. *Mol. Pharmacol.*, 49:567–579.

Schmidt, D. (1989). Benzodiazepines: diazepam. In *Antiepileptic Drugs* (pp. 735–764), R.H. Levy, Ed. New York: Raven Press.

Schwarzer, C., K. Tsunashima, C. Wanzenbock, K. Fuchs, W. Sieghart, G. Sperk. (1997). GABA(A) receptor subunits in the rat hippocampus. II. Altered distribution in kainic acid-induced temporal lobe epilepsy. *Neuroscience*, 80:1001–1017.

Shumate, M.D., D.D. Lin, J.W. Gibbs, 3rd, K.L. Holloway, D.A. Coulter. (1998). GABA(A) receptor function in epileptic human dentate granule cells: comparison to epileptic and control rat. *Epilepsy Res.*, 32:114–128.

Sisodiya, S.M., J. Heffernan, M.V. Squier. (1999). Over-expression of P-glycoprotein in malformations of cortical development. *NeuroReport*, 10:3437–3441.

Sisodiya, S.M., W.R. Lin, B.N. Harding, M.V. Squier, M. Thom. (2002). Drug resistance in epilepsy: expression of drug resistance proteins in common causes of refractory epilepsy. *Brain*, 125:22–31.

Sperk, G., C. Schwarzer, K. Tsunashima, K. Fuchs, W. Sieghart. (1997). GABA(A) receptor subunits in the rat hippocampus. I. Immunocytochemical distribution of 13 subunits. *Neuroscience*, 80:987–1000.

Stanton, P.K., R.S. Jones, I. Mody, U. Heinemann. (1987). Epileptiform activity induced by lowering extracellular [Mg$^{2+}$] in combined hippocampal–entorhinal cortex slices: modulation by receptors for norepinephrine and *N*-methyl-D-aspartate. *Epilepsy Res.*, 1:53–62.

Steiner, E., K. Holzmann, L. Elbling, M. Micksche, W. Berger. (2006). Cellular functions of vaults and their involvement in multidrug resistance. *Curr. Drug Targets*, 7:923–934.

Steinlein, O.K. (2001). Genes and mutations in idiopathic epilepsy. *Am. J. Med. Genet.*, 106:139–145.

Sun, C., W. Sieghart, J. Kapur. (2004). Distribution of alpha1, alpha4, gamma2, and delta subunits of GABA$_A$ receptors in hippocampal granule cells. *Brain Res.*, 1029:207–216.

Sun, C., Z. Mtchedlishvili, A. Erisir, J. Kapur. (2007). Diminished neurosteroid sensitivity of synaptic inhibition and altered location of the alpha4 subunit of GABA(A) receptors in an animal model of epilepsy. *J. Neurosci.*, 27:12641–12650.

Tanigawara, Y. (2000). Role of P-glycoprotein in drug disposition. *Ther. Drug Monit.*, 22:137–140.

Terunuma, M., J. Xu, M. Vithlani, W. Sieghart, J. Kittler, M. Pangalos, P.G. Haydon, D.A. Coulter, S.J. Moss. (2008). Deficits in phosphorylation of GABA(A) receptors by intimately associated protein kinase C activity underlie compromised synaptic inhibition during status epilepticus. *J. Neurosci.*, 28:376–384.

Tietz, E.I., X.J. Zeng, S. Chen, S.M. Lilly, H.C. Rosenberg, P. Kometiani. (1999). Antagonist-induced reversal of functional and structural measures of hippocampal benzodiazepine tolerance. *J. Pharmacol. Exp. Ther.*, 291:932–942.

Tishler, D.M., K.I. Weinberg, D.R. Hinton, N. Barbaro, G.M. Annett, C. Raffel. (1995). MDR1 gene expression in brain of patients with medically intractable epilepsy. *Epilepsia*, 36:1–6.

Tong, X., P.N. Patsalos. (2001). A microdialysis study of the novel antiepileptic drug levetiracetam: extracellular pharmacokinetics and effect on taurine in rat brain. *Br. J. Pharmacol.*, 133:867–874.

Tong, X., M.T. O'Connell, N. Ratnaraj, P.N. Patsalos. (2000). Vigabatrin and GABA concentration inter-relationship in rat extracellular fluid (ECF) from frontal cortex and hippocampus. *Epilepsia*, 41(Suppl.):95.

Tong, X., N. Ratnaraj, P.N. Patsalos. (2009). Vigabatrin extracellular pharmacokinetics and concurrent gamma-aminobutyric acid neurotransmitter effects in rat frontal cortex and hippocampus using microdialysis. *Epilepsia*, 50:174–183.

Treiman, D.M. (1990). The role of benzodiazepines in the management of status epilepticus. *Neurology*, 40:32–42.

Treiman, D.M., P.D. Meyers, N.Y. Walton, J.F. Collins, C. Colling, A.J. Rowan, A. Handforth, E. Faught, V.P. Calabrese, B.M. Uthman, R.E. Ramsay, M.B. Mamdani. (1998). A comparison of four treatments for generalized convulsive status epilepticus. Veterans Affairs Status Epilepticus Cooperative Study Group. *N. Engl. J. Med.*, 339:792–798.

Tsunashima, K., C. Schwarzer, E. Kirchmair, W. Sieghart, G. Sperk. (1997). GABA(A) receptor subunits in the rat hippocampus. III. Altered messenger RNA expression in kainic acid-induced epilepsy. *Neuroscience*, 80:1019–1032.

van Vliet, E.A., E. Aronica, S. Redeker, J.A. Gorter. (2004). Expression and cellular distribution of major vault protein: a putative marker for pharmacoresistance in a rat model for temporal lobe epilepsy. *Epilepsia*, 45:1506–1516.

van Vliet, E.A., R. van Schaik, P.M. Edelbroek, S. Redeker, E. Aronica, W.J. Wadman, N. Marchi, A. Vezzani, J.A. Gorter. (2006). Inhibition of the multidrug transporter P-glycoprotein improves seizure control in phenytoin-treated chronic epileptic rats. *Epilepsia*, 47:672–680.

van Vliet, E.A., R. van Schaik, P.M. Edelbroek, R.A. Voskuyl, S. Redeker, E. Aronica, W.J. Wadman, J.A. Gorter. (2007). Region-specific overexpression of P-glycoprotein at the blood–brain barrier affects brain uptake of phenytoin in epileptic rats. *J. Pharmacol. Exp. Ther.*, 322:141–147.

van Vliet, E.A., R. van Schaik, P.M. Edelbroek, F.H. da Silva, W.J. Wadman, J.A. Gorter. (2008). Development of tolerance to levetiracetam in rats with chronic epilepsy. *Epilepsia*, 49:1151–1159.

VanLandingham, K.E., E.W. Lothman. (1991). Self-sustaining limbic status epilepticus. I. Acute and chronic cerebral metabolic studies: limbic hypermetabolism and neocortical hypometabolism. *Neurology*, 41:1942–1949.

Wafford, K.A., S.A. Thompson, D. Thomas, J. Sikela, A.S. Wilcox, P.J. Whiting. (1996). Functional characterization of human gamma-aminobutyric acidA receptors containing the alpha 4 subunit. *Mol. Pharmacol.*, 50:670–678.

Walker, M.C., M.S. Alavijeh, S.D. Shorvon, P.N. Patsalos. (1996). Microdialysis study of the neuropharmacokinetics of phenytoin in rat hippocampus and frontal cortex. *Epilepsia*, 37:421–427.

Walker, M.C., X. Tong, H. Perry, M.S. Alavijeh, P.N. Patsalos. (2000). Comparison of serum, cerebrospinal fluid and brain extracellular fluid pharmacokinetics of lamotrigine. *Br. J. Pharmacol.*, 130:242–248.

Wallace, R.H., I.E. Scheffer, S. Barnett, M. Richards, L. Dibbens, R.R. Desai, T. Lerman-Sagie, D. Lev, A. Mazarib, N. Brand, B. Ben-Zeev, I. Goikhman, R. Singh, G. Kremmidiotis, A. Gardner, G.R. Sutherland, A.L. George, Jr., J.C. Mulley, S.F. Berkovic. (2001). Neuronal sodium-channel alpha1-subunit mutations in generalized epilepsy with febrile seizures plus. *Am. J. Hum. Genet.*, 68:859–865.

Walton, N.Y., D.M. Treiman. (1988). Response of status epilepticus induced by lithium and pilocarpine to treatment with diazepam. *Exp. Neurol.*, 101:267–275.

Wang, N.C., L.B. Good, S.T. Marsh, D.M. Treiman. (2009). EEG stages predict treatment response in experimental status epilepticus. *Epilepsia*, 50:949–952.

Wei, W., N. Zhang, Z. Peng, C.R. Houser, I. Mody. (2003). Perisynaptic localization of delta subunit-containing GABA(A) receptors and their activation by GABA spillover in the mouse dentate gyrus. *J. Neurosci.*, 23:10650–10661.

Wisden, W., D.J. Laurie, H. Monyer, P.H. Seeburg. (1992). The distribution of 13 GABA$_A$ receptor subunit mRNAs in the rat brain. I. Telencephalon, diencephalon, mesencephalon. *J. Neurosci.*, 12:1040–1062.

Yen, W., J. Williamson, E.H. Bertram, J. Kapur. (2004). A comparison of three NMDA receptor antagonists in the treatment of prolonged status epilepticus. *Epilepsy Res.*, 59:43–50.

Zhang, D.X., J.M. Williamson, H.Q. Wu, R. Schwarcz, E.H. Bertram. (2005). *In situ*-produced 7-chlorokynurenate has different effects on evoked responses in rats with limbic epilepsy in comparison to naive controls. *Epilepsia*, 46:1708–1715.

Zhang, N., W. Wei, I. Mody, C.R. Houser. (2007). Altered localization of GABA$_A$ receptor subunits on dentate granule cell dendrites influences tonic and phasic inhibition in a mouse model of epilepsy. *J. Neurosci.*, 27:7520–7531.

# 6 Epilepsy with Complex Genetics: Looking Beyond the Channelopathies

*Asuri N. Prasad and Chitra Prasad*

## CONTENTS

## BACKGROUND

Epilepsy is a chronic neurological disorder affecting nearly 1% of the human population worldwide. With an incidence of approximately 50 per 100,000 population and the prevalence of active epilepsy being around 8.2 per 1000, it is estimated that at least 50 million people worldwide suffer from this condition at any one time; the lifetime prevalence is estimated to be at 100 million people (WHO, 2001). A large proportion of individuals with epilepsy have no clear underlying cause that can be identified on their clinical assessment. In some, the etiology may be hidden (cryptogenic), but a substantial proportion are considered to have "idiopathic" epilepsy. In the Rochester epidemiology project, the cause of epilepsy was presumed to be idiopathic in 68% of the incidence cases (1935 to 1984), signifying that familial/genetic factors were considered as being contributory (Annegers et al., 1996). The genetics of epilepsy as well as gene hunting in epilepsy have seen considerable enthusiasm and effort since the first epilepsy gene (mutation involving a nicotinic cholinergic receptor subunit) underlying an idiopathic partial epilepsy (autosomal dominant nocturnal frontal lobe epilepsy) was identified in 1995 (Steinlein et al., 1995). In the following discussion, we examine the genetic basis of the more common epilepsies (i.e., idiopathic generalized and partial). Despite the application of much effort and technical expertise, the yield in terms of "genes" that explain the basis of common epilepsy has been disappointingly small. The field is replete with a large volume of

**TABLE 6.1**
**Concordance Rates for Generalized and Localization-Related Epilepsy in Selected Twin Studies**

| Study Authors | Nature of Epilepsy in Proband | Population Cohorts | Proband-Wise Concordance Rates in Twins | |
| --- | --- | --- | --- | --- |
| | | | Monozygotic | Dizygotic |
| Berkovic et al. (1998) | Generalized epilepsy | Australian | 0.82 | 0.26 |
| | Localization-related epilepsy | | 0.36 | 0.05 |
| Kjeldsen et al. (2003) | Generalized epilepsy | Danish | 0.65 | 0.12 |
| | Localization-related epilepsy | | 0.30 | 0.10 |

literature, but we have taken a more selective approach for the clinician with an interest in this field, noting key insights into new developments, the progress made in our understanding of the genetics of human epilepsy, and the challenges ahead.

## GENETIC SUSCEPTIBILITY IN HUMAN EPILEPSY

There is little doubt at this point in time of the heritable nature of human epilepsy. A considerable volume of evidence from family- and population-based studies has been extensively reviewed (Ottman et al., 1996; Winawer and Shinnar, 2005). Epilepsy has been shown to aggregate in families, and family studies have traditionally provide a starting point for the investigation of the genetic basis of epilepsies. There is a two- to fourfold increase in risk for epilepsy in patients whose families have first-degree relatives affected with epilepsy (Winawer, 2002; Winawer and Shinnar, 2005). Twin studies have been popular, as twins tend to share similar environmental influences superimposed on either a monozygotic or dizygotic twin background. In childhood absence epilepsy (CAE), the high concordance rates in population-based studies of twins with epilepsy—monozygotic (MZ) vs. dizygotic (DZ)—provide strong supportive evidence for a genetic contribution (Table 6.1) (Berkovic et al., 1998; Kjeldsen et al., 2003). Application of the contemporary International League Against Epilepsy (ILAE) classification schema to historical seizure data on twins collected by Lennox also provides similar evidence for a high concordance rate for idiopathic generalized epilepsies (MZ, 0.80; DZ, 0.00) (Vadlamudi et al., 2004). Twin studies allow us to differentiate between the effects of shared genetic background and shared environmental influences. Population-based twin studies offer investigators the ability to carefully phenotype individuals, although twins may not necessarily be representative of the entire population. Such limitations may involve problems of bias due, for example, to an excess of female monozygotic twins within volunteer registries.

Family-based studies have also included the study of "multiplex families," where several individuals within a single pedigree appear to be affected by the same epilepsy syndrome. Advances in molecular genetics have permitted the identification of mutations that account for specific epilepsy syndromes using carefully selected families. In the last 15 years, investigator groups have been successful in identifying several "epilepsy genes." Using large affected families and the strategies of linkage analysis and positional cloning, gene loci have been mapped and mutations affecting genes coding for different channel proteins identified (Table 6.2). These epilepsy syndromes are relatively rare but usually demonstrate a phenotype whose parameters can be defined consistently in each affected individual. A closer look reveals that these are mostly "monogenic" epilepsies; that is, they involve families showing a clear pattern of segregation consistent with Mendelian laws of inheritance. The concept of inherited epilepsy as a channelopathy took shape on the basis of these studies (Mulley et al., 2003; Ptacek, 1997). The genes identified through such studies have significantly influenced our understanding of the molecular mechanisms leading to epilepsy.

**TABLE 6.2**
**Genes Causing Monogenic Epilepsy Syndromes Identified by Linkage Analysis and Positional Cloning**

| Gene | Gene Locus | Epilepsy Syndrome | Seizure Types | Functional Effects |
|---|---|---|---|---|
| SCN1A | 2q24 | GEFS + SMEI | Febrile, myoclonic, absence, tonic clonic, partial | Sodium influx in the neuronal soma and dendrites |
| SCN2A | 2q24 | GEFS + BFNIC | Febrile, generalized tonic, tonic–clonic | Fast sodium influx effects on action potential initiation and propagation |
| SCN1B | 19q13 | GEFS+ | Febrile, absence, tonic–clonic, myoclonic | Modulate α-subunit |
| GABRA1 | 5q34 | ADJME | Tonic–clonic, myoclonic, absence | Inhibition of GABA-activated currents |
| GABARG2 | 5q31 | FS, CAE, GEFS+ | Febrile, absence, tonic–clonic, myoclonic, clonic, partial | Rapid GABAergic inhibition |
| GABRD | 1p36 | GEFS+ | Mixed febrile and afebrile seizures | Decreased amplitudes of GABA$_A$ receptor currents |
| BRD2 | 6p21 | JME | Generalized tonic–clonic, myoclonic | Nuclear transcriptional regulator unknown mechanisms |
| EFHC1 | 6p12-p11 | JME | Generalized tonic–clonic, myoclonic | Protein with an EF hand motif |
| CLCN2 | 3q26 | IGEs | Tonic–clonic, myoclonic, absence | Voltage-gated chloride channel regulating neuronal chloride efflux |
| KCNQ2 KCNQ3 | 20q13 8q24 | BFNC | Neonatal seizures | Modulates M current |
| CHRNA4 CHRNA2 CHRNB2 | 20q13 8q 1p21 | ADNFLE | Nocturnal partial seizures | Nicotinic current modulation and interaction among different subunits |
| LGI | 10q24 | ADPEAF | Partial seizures with auditory features and hallucinations | Influences glial–neuronal interactions? |

Even within families considered to display Mendelian inheritance, significant phenotypic variability and lack of consistency in genotype–phenotype correlations continue to pose challenges to our complete understanding of their genetics. The majority of the causative mutations identified seem to affect ion- or ligand-gated channels, with the exception of the *LGI1* gene (Steinlein, 2004). These families for the most part demonstrate a dominant inheritance pattern but may differ in expression of the phenotype among the affected individuals in the type and severity of seizures and their response to treatment. Furthermore, penetrance of the condition in these families is incomplete; in other words, there may be members who carry the electroencephalogram (EEG) trait but never experience seizures, thus suggesting a role for other genetic or environmental modifying influences.

Mutations involving different gene loci may be expressed through a single phenotype (locus heterogeneity), whereas mutations involving the same gene may be expressed through completely different phenotypes at an individual level (variable expressivity). For example, the syndrome of generalized epilepsy with febrile seizures plus (GEFS+) is known to be extremely variable in its clinical expression with multiple seizure-type combinations associated with and without fever described in individuals within the same family. Mutations in three different genes are associated with the disorder: two sodium channel genes, *SCN1B* (Wallace et al., 1998) and *SCN1A* (Escavg et al., 2000), and a γ-aminobutyric acid (GABA) receptor gene, *GABRG2* (Baulac et al., 2001a). In addition, at least three additional gene loci (2p24, 21q22, 8p23-p21) (Harkin et al., 2007; Kanai et al., 2004; Zucca et al., 2008) linked to this epilepsy phenotype have also been identified (Audenaert et al., 2005; Baulac et al., 2008; Hedera et al., 2006). Mutations of *SCN1A* are also causally associated with severe myoclonic epilepsy of infancy (SMEI); these are mostly *de novo* and truncating mutations (Claes et al., 2001, 2003), while the mutations leading to GEFS+ tend to be missense in type (Kanai et al., 2004). Mutations of the *SCN1A* gene have also been identified in an expanding spectrum of epilepsy, including infantile epileptic encephalopathies (Harkin et al., 2007) as well as cryptogenic generalized and focal epilepsy (Zucca et al., 2008).

Finally, establishing definitive phenotype–genotype correlations even within specific epilepsy syndromes with single gene defects can be challenging. Mutations identified in the *LGI1* have been found in 50% of families with two or more affected individuals with autosomal dominant partial epilepsy with auditory features (ADPEAF); however, in sporadic patients with the same clinical features, none was shown to carry an inherited mutation (Berkovic et al., 2004). Thus, the syndrome is dominant in some families but in others a complex inheritance is likely. These issues highlight the difficulties in fitting genetics of epilepsy into simpler schemes of Mendelian inheritance.

In addition to the monogenic epilepsies described above, nontraditional inheritance models also account for disorders with an inherited basis where epilepsy is a prominent feature of the overall phenotype. These include imprinting disorders (differential expression of genes based on a parent of origin effect), such as Angelman syndrome; mitochondrial inheritance, such as myoclonic epilepsy with ragged red fibers (MERRF); triplet repeat expansion, such as dentatorubropallidoluysian atrophy (DRPLA); and dodecamer repeats, such as progressive myoclonic epilepsy. These models have been extensively reviewed and discussed elsewhere (Prasad and Prasad, 2008a; Prasad et al., 1999).

## EPILEPSY, GENETICS, MENDELISM, AND BEYOND

The laws of inheritance as propounded by Gregor Mendel in 1886 laid the basis of our understanding of inheritance patterns according to broadly applicable rules of transmission. The idea of a single gene–single trait was not found to be universally applicable to all human traits, particularly those of a quantitative nature (these traits follow a Gaussian frequency distribution curve), as pointed out by Francis Galton and others. In 1918, the statistician Sir Ronald Fisher suggested that the observed discrepancies in quantitative traits could be explained by invoking smaller additive effects of multiple genes (Fisher, 1918). In addition, environmental effects were also considered to influence the phenotype, giving rise to the concept of a multifactorial–polygenic model of complex inheritance. This paradigm includes not only the qualitative and quantitative effects of all genetic and environmental variables that affect the trait but also the effects of their mutual interactions. This model poses significant methodological challenges to geneticists. Even though, in principle, individual genetic components may be identified and localized on a map, environmental influences are difficult to analyze and quantify, and their interactive effects are even more difficult to predict. Further, the genetic architecture is likely to be dependent on the distribution of gene and genotype frequencies, age, and gender in a particular population. Also, identical complex traits may have different biological causes in different populations (NIGMS, 1998). Tools to dissect and analyze the genetic contribution to human epilepsy and other chronic disorders at a population level are still being developed (Rao, 2008).

In the case of epilepsy, the traits are not quantitative but are dichotomous (i.e., well vs. affected). The more common idiopathic epilepsies show unpredictable and nonlinear patterns of transmission, even when there are multiple affected individuals in the same family. The risk varies among relatives of idiopathic generalized epilepsy (IGE) probands; siblings tend to carry a higher risk in comparison to second-degree relatives (Marini et al., 2004). There appears to be a greater decline in the risk based on distance from the proband. Affected individuals differ in terms of their age of onset and offset, seizure type and progression, EEG findings, and response to treatment. The problem of multiple phenocopies (individuals with acquired epilepsy) within the same family is another confounder. Finally, the occurrence of *de novo* mutations leading to sporadic forms of epilepsy merits consideration of a different inheritance model in patients with idiopathic epilepsy.

## MODELING EPILEPSY WITH COMPLEX GENETICS (COMPLEX EPILEPSY)

The recognition of an epilepsy susceptibility gene is based on identification of a mutation or an allelic variant that is associated with an elevation in seizure propensity. The evidence for this association may come from any one of several approaches involving linkage, association studies, and mutation analysis. In addition, the allelic variant or mutation should meet the requirement of biological plausibility. The supportive evidence usually comes via functional studies. Functional studies involve demonstrating the effects of defective gene mutants or variants in different expression systems (*in vitro* or through knockout models). These systems of models should convincingly demonstrate that the gene defect or variant under consideration results in a functional change in neuronal network behavior that favors seizure generation and propagation.

The above approach has worked for single gene defects in rare families with Mendelian inheritance patterns, but multiple susceptibility genes and a triggering effect from the environment may be involved in complex epilepsy. There are no clear segregation patterns, and the phenomenon of epistasis (intragenic interactions) is possible. Pleiotropic effects and epigenetic modification of expression also influence the final phenotype. As pointed out earlier, the phenotypic variability is such that stratification may be a problem. An oligogenic or digenic model of inheritance has been proposed for IGE (Durner et al., 2001). Although one affected locus common to all IGE reduces seizure threshold, involvement of a second genetic locus at a different site is necessary for the final expression of the seizure phenotype. This model appears to be uncommon and applicable only to rare families (Baulac et al., 2001b).

If the monogenic and oligogenic models have not been found to be broadly applicable to the common idiopathic epilepsy syndromes, then what are the alternative possibilities? A net functional effect of contributions from variants or polymorphisms at multiple alleles (loci) is presently considered likely. The crux of the problem is to determine the extent to which each of these variables makes a contribution to epileptogenesis (Pritchard, 2001).

This brings the discussion back to the nature of genetic variation and the role of natural selection in the occurrence of complex traits. The underlying genetic variation in the causation of epilepsy is unknown. Genetic variants linked to epilepsy could be common (allele frequencies > 1%) or rare (allele frequencies < 1%) in the population. The number and frequency of the susceptibility alleles are influenced by the occurrence of mutation, genetic drift, and natural selection (Di Rienzo, 2006).

The common disease–common variant (CDCV) model assumes that the variants have resisted natural selection to achieve a significant threshold within the population. These variants should then be detectable through association studies (Reich and Lander, 2001). Both theoretical and empirical arguments support this notion. First, epilepsy as a chronic condition is not evolutionarily disadvantageous; second, if multiple susceptibility alleles are involved, the pressure from natural selection may be diluted; finally, as polygenic disorders are more common, population genetics predicts the occurrence of a higher frequency for at least some common causal variants in the population because of demographics (Gabriel et al., 2002). In practice, however, this model has not found much support in terms of the identification of many common variants underlying human epilepsy (Dibbens et al., 2007).

The candidate genes for several single gene defects underlying monogenic epilepsy seem to code for proteins that make up ion- or ligand-gated channels. It would be natural to assume that variants in the genes coding for ion channels and their receptors that have been associated with Mendelian forms of inherited epilepsy would also be involved in complex epilepsy.

Initial studies of variant polymorphisms with functional consequences—calcium-channel subunit CACNA1H, CAE; GABA receptor subunit GABARD, GEFS+; GABA receptor subunit GABARB3, IGE; and potassium channel subunit KCND2, temporal lobe epilepsy (TLE)—show these variants as carrying weak effects, as far as their contribution to susceptibility to epilepsy is concerned. None of these variants has been shown individually to have a causal association with idiopathic epilepsy. These variants are linked to seizure susceptibility on the basis of functional studies and are suggested to be consistent with the multiple rare variant complex epilepsy (MRVCE) model (Dibbens et al., 2007; Mulley et al., 2005). This model for complex epilepsy involves causative variants in susceptibility genes drawn from a population of genes (with rare and low-frequency polymorphic alleles) for which none of the variants is sufficient by itself to give rise to an epilepsy phenotype independently. The susceptibility alleles may give rise to the phenotype in different combinations, with each combination leading to a different endophenotypes. These low-frequency variants would not be detectable through allelic association studies.

From the foregoing discussion, both rare variant and common variant models for epilepsy individually seem not to be broadly applicable to common epilepsy. To get around this problem, a blended approach has been suggested involving both the MRVCE (rare functional variants that are too weak to be individually an effector of epilepsy) and the ancestral common variant complex epilepsy (ACVCE) model. The polygenic heterogeneity model has been proposed to reconcile the differences in various models. This new paradigm suggests that the final expression is determined by permutations and combinations of both common and low-frequency polymorphic variants (Di Rienzo, 2006; Dibbens et al., 2007).

With this background it can be stated (at the risk of some oversimplification) that current thinking regarding genetic influences on the risk of epilepsy conceptually consists of the net effects resulting from a mix of susceptibility genes with large effects and those with smaller effects, the former producing Mendelian inheritance patterns and the latter complex inheritance patterns. The detection and the construction of a composite of all genes with different contributions to determining susceptibility to epilepsy are the challenge for investigators.

With the sequencing of the human genome, additional findings regarding the contribution of genomic elements to human diversity have come to light. A significant part of the human genome is conserved through evolution; even within these highly conserved regions there are variations involving insertions and deletions. The discovery that nonsynonymous single-nucleotide polymorphisms (SNPs) (i.e., those that are capable of altering or leading to changes in amino acid sequences) can produce subtle effects on the translated protein structure and function has spurred an additional body of research. SNPs affecting splicing sites or regulatory regions within introns may change the final gene expression. Changes within exons could result in an amino acid substitution that may or may not prove to be deleterious in determining the effect on the protein transcribed. Nearly a third (29 to 30%) of the SNP mutations are predicted to be neutral (i.e., not of pathological significance), 30 to 42% are considered to be moderately deleterious, and the remainder are considered to be lethal (Boyko et al., 2008). There is growing evidence that SNPs relevant to epilepsy are interspersed throughout the genome (Crino, 2007; Venter et al., 2001). This has led to the speculation that an additive dose effect of several SNPs throughout the genome (with or without a mutation in a gene carrying a major effect) may play a key role in the final determination of seizure susceptibility. Again, this approach may face further challenges, particularly if variants of other common gene families expressed in the neuron and synapse, in addition to susceptibility alleles involving ion channels, are also involved in determining susceptibility to epilepsy. Progress in identifying susceptibility alleles to common/idiopathic epilepsy has been frustratingly slow utilizing the current approaches. These approaches, the successes, and challenges are discussed next.

## APPROACHES TO IDENTIFYING EPILEPSY GENES: SUCCESSES AND CHALLENGES

### LINKAGE MAPPING AND POSITIONAL CLONING

The success in identifying human epilepsy genes in the last two decades can be attributed largely to linkage mapping and positional cloning strategies in single gene disorders. These strategies have not been successful in the identification of genes underlying complex epilepsy. Linkage mapping relies on the cosegregation of the disease and marker alleles within a family; however, the mere finding of linkage to a particular locus does not imply a causal association. Most of the gene loci have been identified using a haplotype segregation analysis, a technique that is prone to mistakes introduced by low penetrant genes and the presence of phenocopies. With this approach, erroneous conclusions regarding the precise extent of the candidate regions can occur. As a result, the correct gene may defy identification if it lies beyond the borders identified through such linkage studies. If the same loci are identified as being strongly linked to a particular phenotype in different families, then the search for a mutation can be carried out with greater level of certainty. The next steps involve the use of molecular genetics to identify mutations in both coding as well as in the noncoding regions of genes.

Present-day techniques involve the use of multiplex amplification and probe hybridization (MAPH) and multiplex ligation-dependent probe amplification (MLPA), which are more sensitive to smaller and medium-sized intragenic deletions and duplications that might otherwise be missed using earlier methods (Sellner and Taylor, 2004). Complementary methods using comparative genomic hybridization (CGH) as well as high-resolution bacterial artificial chromosome (BAC) arrays may also be helpful in picking up small-sized deletions and duplications (Cowell, 2004). If these methods lead to identification of a gene or genetic variant allele on the haplotype, it remains to be shown whether the variant in question is indeed the pathogenic mutation. One way of addressing this issue is to examine if the variant in question is absent in sufficient numbers of matched controls (Collins and Schwartz, 2002). On the other hand, if the variant is found only in a rare family, then functional studies will be necessary to confirm a substantive biological effect. Stronger evidence for confirming causality involves the discovery of a number of mutations being identified in the same gene in multiple affected individuals in many families sharing an identical phenotype and the absence of the mutation or variant and phenotype in controls.

## ASSOCIATION STUDIES

Association analysis aims to correlate traits or disorders with allele frequencies between cases and controls drawn from families or between families in a population using a variety of sophisticated multivariate statistical methods. These analytical techniques are continually evolving, and a discussion of these methods is beyond the scope of this discussion.

The chief advantage in association analysis lies in the fact that cases do not have to be necessarily related to one another, and the disorder does not have to be caused by a single gene defect. Association analysis tends to be highly sensitive and will identify alleles with very minor or subtle effects, while genes with major effects may be missed. The existence of linkage disequilibrium is critical for association analysis to be successful. Association analysis works on the assumption that the marker allele is closely linked to the disease gene and the mutation affecting the disease gene occurred only once through natural history. If the same mutation occurred multiple times in the population or the same gene was influenced by different mutations producing the same phenotype (a situation that may well occur when different populations are mixed), association analysis may lead to false-negative results. Thus, stratification by ethnicity and or genetic ancestry becomes important (Rao, 2001, 2008).

Tan et al. (2004) reviewed the results of more than 50 association studies and found that they led to conflicting and nonreplicable findings or failed to meet the requirement of biologic plausibility. The authors identified the variables that require close attention when proceeding with such association studies to avoid the pitfalls mentioned above. In the study design, it is particularly important to give careful consideration to defining the phenotype, source for controls, and sample size. Molecular testing should be performed with blinding. Batch testing should include both cases and controls. During statistical analyses, attention should be given to clearly defining an *a priori* hypothesis. If multiple hypotheses are involved, these should be stated, and multiple testing, if done, should be specified and corrections applied. Furthermore, details of population stratification and haplotype construction, if used, should be described. In discussing results, attention needs to be paid to the strength of association, positive and negative results, and evidence for biologic plausibility. Excellent reviews are available to help readers learn how to interpret results of association studies; for example, see Pearson and Manolio et al. (2008).

Association studies have used the availability of an ever-increasing number of SNPs and haplotype maps. Both methods are affected by confounding variables such as heterogeneity and incorrect assignment of phenotype. Errors in ascertainment of pedigree relationships, genotyping errors are additional contributors to the lack of success in identifying a disease gene or allele. Methods to address these problems are often overlooked, leading to false negative or positive results. Careful attention to stratifying population samples on the basis of clinical characteristics can help address the problem of heterogeneity (Tan et al., 2006).

Although it is conceivable for association studies to pick up significant subsequent associations in specific genes or polymorphic variants, failure to replicate these results in other studies has led to skepticism regarding results of genetic-association studies for the reasons discussed above. It is now increasingly recognized that replicability of results in different studies, although highly desirable, may be impossible to accomplish. There are several examples where evidence of susceptibility genes may arise from one or two of the different approaches adopted but the findings may fail to be biologically plausible. Likewise, the reverse may be true in that there may be evidence of biological plausibility but no evidence for association or linkage across different patient populations. Conflicting results are not uncommon when dealing with results of association studies. *BRD2* is considered to be a susceptibility gene for juvenile myoclonic epilepsy (JME) at the *EJM1* locus (Pal et al., 2003). This finding, however, could only be replicated in English and Irish samples but not in families with JME from South India or Australia in another multicenter association study (Cavalleri et al., 2007a). Furthermore, data from mutational and functional analysis are still lacking in supporting a biological effect of this gene on seizure susceptibility. It has been argued that evidence using complementary approaches to increase coherence may permit one to strengthen the implications of findings from association studies or to move on to reject associations with weak supportive evidence (Pal et al., 2008).

In an attempt to overcome one of the key limitations to association studies, multicenter efforts to apply current strategies to a large population sample of cases are currently being carried out (e.g., EPIGEN, EPICURE). The results of one such study are of particular interest (Cavalleri et al., 2007b). Investigators pooled patients from various centers (United Kingdom, Australia, Ireland, and Finland) in the EPIGEN consortium. SNPs and common set haplotypes were used to probe potential associations between 279 prime candidate genes (receptors, metabolizers, and transporters of currently known neurotransmitters and for antiepileptic drugs) and selected endophenotypes in 2717 cases with epilepsy and 1118 controls.

The endophenotypes included five epilepsy syndromes: (1) epilepsy as defined by the ILAE; (2) juvenile myoclonic epilepsy (JME); (3) idiopathic generalized epilepsy excluding JME (IGE non-JME); (4) mesial temporal lobe epilepsy associated with hippocampal sclerosis (MTLE-HS); and (5) all other focal neocortical epilepsies (other). Five seizure classifications were used: (1) generalized tonic–clonic seizures (GTCSs), occurring only in the context of a syndromic diagnosis of an IGE; (2) myoclonic seizures; (3) absence seizures; (4) secondarily GTCS (SGTCSs); and (5) partial seizures (either simple or complex).

Two significant observations resulted. First, the yield for specific indisputable common genetic risk factors to common epilepsy is low, supporting the idea that low-frequency, rare variant polymorphisms likely underlie susceptibility alleles for all epilepsy. Second, the story is different when one looks at epilepsy subphenotypes. Polymorphisms in five genes (*KCNAB1*, *GABRR2*, *KCNMB4*, *SYN2*, and *ALDH5A1*) accounted for 23 of 50 significant *p* values. SNPs within these five genes bear the most significant associations to epilepsy subphenotypes in a population-specific manner; for example two intronic tagging SNP variants associated with *KCNAB1* were associated significantly with a wide variety of epilepsy manifesting with generalized as well as partial seizures in the set-association analysis in the Finnish cohort. Also, two intronic SNPs involving the *GABRR2* receptor gene are involved as risk factors for JME and other myoclonic-epilepsy-related phenotypes in the Irish cohort.

This study highlights the challenges currently experienced by investigators and suggests ways to improve upon current studies. Newer methods of modeling and analysis are continually evolving, such as the use of constrained Bayesian networks for pathway-based SNPs and composite likelihood to model complex systems (Rao, 2008). Such techniques can be accompanied by multicenter cooperation and pooling of sample sizes, careful application of standardized and rigorous endophenotyping of cases (not currently practiced in clinics), and population stratification strategies that could further lead to identification of specific susceptibility alleles in at least a proportion of cases with idiopathic epilepsy.

A search of the online database Online Mendelian Inheritance in Man (OMIM) using the term "idiopathic generalized epilepsy" disclosed linkages to multiple loci for the different epilepsy syndromes entered in the database. These are listed in Table 6.3 and provide readers with an idea of the extent of genetic heterogeneity that exists within idiopathic epilepsy. Unfortunately, the use of different terminology in the database as opposed to that of the ILAE classification raises additional problems for non-epileptologists using OMIM.

## GENETIC DIVERSITY AND PATHWAYS TO EPILEPSY

There may be other inherited mechanisms that could potentially account for the phenotypic variability of common epilepsy and lead to the identification of gene targets as well as their interactions within entire molecular signaling pathways. These include somatic mosaicism, a role for copy number variants (CNVs), modifier genes, epigenetic factors, and links to the metabolome.

### SOMATIC MUTATIONS AS A CAUSE FOR SPORADIC EPILEPSY

Mutations occurring in somatic cell DNA acquired during brain development, or even in germline DNA, could potentially provide a pathogenic mechanism for certain kinds of sporadic epilepsies. An influence on tissue patterning may lead to the occurrence of two populations of cells that either possess or lack the mutation. Ultimately, these mutational alterations may lead to structural changes in cellular cytoarchitecture or in the absence of such recognizable structural changes that may only be detectable in their functional effects on specific brain regions where such mutations reside (e.g., at a neurotransmitter receptor subunit, or uptake site), ultimately leading to epilepsy (Lindhout, 2008).

Epilepsy associated with mental retardation in females is X linked (EFMR; Xq22 locus) and provides a source for interesting observations, as an example (Tan et al., 2006). Here, the pedigrees and linkage analysis show an inverse pattern of inheritance; hemizygous females are affected while heterozygous males are unaffected. In the family described by Scheffer et al. (2008), the patterns of X inactivation were normal. The variable clinical phenotype of EFMR is not explicable on the grounds of random X inactivation patterns alone. Other possible hypotheses include: (1) the presence of a putative protective functional homolog or gene on the Y chromosome (Page et al., 1984), (2) a functional disomy of X-linked genes resulting from regional interference with the X-inactivating process and leading to the variant phenotypes of Rett syndrome (Rosenberg et al., 2001), or (3) metabolic interference between two allelic variants due to different cell populations expressing two different proteins (Johnson, 1980). The finding of microscopic cortical dysplasia in a frontal lobe surgical specimen (as compared to other normal areas) from a patient with EFMR suggests abnormal tissue patterning in this condition.

The condition of craniofrontal nasal dysplasia caused by mutations in the ephrin-B1 gene shares similar X-linked inheritance patterns (Twigg et al., 2006). The ephrin-B1 gene plays an important role in cell-to-cell interactions through regulation of gap junction communication involving connexin 43. The heterozygous mouse model shows abnormal patterning in brain development. Thus,

**TABLE 6.3**
**Idiopathic Generalized Epilepsy Syndromes and Gene Loci**

| OMIM Number | Epilepsy Syndrome | Gene Loci Based on Linkage/Association Studies |
|---|---|---|
| 606904 | Epilepsy, Juvenile Myoclonic (JME)[a] | 5q34-q35<br>3q26-qter<br>2q22-q23<br>1p36<br>6p12-p11 |
| 608816 | Myoclonic Epilepsy, Juvenile 3 (EJM3)[b] | 6p21 |
| 604827 | Myoclonic Epilepsy, Juvenile 2 (EJM2)[b] | 15q14 |
| 254770 | Myoclonic Epilepsy, Juvenile 1 (EJM1)[a] | 6p12-p11 |
| 611364 | Myoclonic Epilepsy, Juvenile 4 (EJM4)[b] | 5q12-q14 |
| 604233 | Generalized Epilepsy with Febrile Seizures Plus (GEFS+) | 19q13 |
| | Generalized Epilepsy with Febrile Seizures Plus Type 1 (GEFS+1)[a] | 5q31.1-q33.1<br>2q24<br>1p36.3<br>2q23-q24.3<br>2p24<br>8p23 |
| 600669 | Epilepsy, Idiopathic Generalized (EIG)[a] | 18q21<br>14q23<br>9q32-q33<br>8q24<br>6q24-q25<br>2q22-q23 |
| 606970 | Epilepsy, Idiopathic Generalized Susceptibility to 1 (EIG1)[b] | 8q24 |
| 606972 | Epilepsy, Idiopathic Generalized Susceptibility to 2 (EIG2)[b] | 14q23 |
| 608762 | Epilepsy, Idiopathic Generalized Susceptibility to 3 (EIG3)[b] | 9q32-q33 |
| 609750 | Epilepsy, Idiopathic Generalized Susceptibility to 4 (EIG4)[b] | 10q25-q26 |

| | | |
|---|---|---|
| 611934 | Epilepsy, Idiopathic Generalized Susceptibility to 5 (EIG5)[b] | 10p11.22 |
| 607208 | Severe Myoclonic Epilepsy of Infancy (SMEI) | 5q31.1-q33.1 2q24 |
| | Epilepsy, Intractable Childhood with Generalized Tonic–Clonic Seizures (ICEGTC)[a] | |
| 609446 | Generalized Epilepsy and Paroxysmal Dyskinesia (GEPD)[a] | 10q22.3 |
| 611942 | Epilepsy, Idiopathic Generalized Susceptibility to 6 (EIG6)[a] | 16p13.3 |
| 609572 | Photoparoxysmal Response 2 (PPR2)[b] | 13q31.3 |
| 605021 | Myoclonic Epilepsy Infantile[b] | 16p13 |
| 607628 | Epilepsy with Grand Mal Seizures on Awakening[a] | 3q26-qter |
| 607745 | Seizures, Benign Familial Neonatal-Infantile[a] | 2q23-q24.3 |
| 607682 | Epilepsy, Childhood Absence 3 (ECA3)[a] | 3q26-qter |
| 121201 | Epilepsy, Benign Neonatal 2 (EBN2)[a] | 8q24 |
| 607681 | Epilepsy, Childhood Absence 2 (ECA2)[a] | 5q31.1-q33.1 |
| 600131 | Epilepsy, Childhood Absence 1 (ECA1)[b] | 8q24, 16p12-p13.1 |
| 612269 | Epilepsy, Childhood Absence 5 (ECA5)[a] | 15q11.2-q12 |
| 607682 | Epilepsy, Childhood Absence 3 (ECA3)[a] | 3q26-qter |
| 611136 | Epilepsy, Childhood Absence 4 (ECA4)[a] | 5q34-q35 |
| 607631 | Epilepsy, Juvenile Absence (JAE)[a] | 6p12-p11, 3q26-qter |

*Note:* The list was generated through a search of the online database OMIM using the terms "idiopathic generalized epilepsy." The data represent a catalog of reported gene loci on the basis of linkage or association studies. OMIM terminology styles differ from the ILAE classification for epilepsy syndromes which may cause confusion for non-neurologists.

[a] A descriptive entry, usually of a phenotype, and does not represent a unique locus.

[b] A confirmed Mendelian phenotype or phenotypic locus for which the underlying molecular basis is not known.

while EFMR is rare, the occurrence and adverse interactions between (epi)genetically different cell populations leading to an epileptogenic focus may not be uncommon after all. In a recent report of two Italian families with two siblings affected with SMEI, the parents were unaffected or had only experienced febrile seizures during childhood (Gennaro et al., 2006). In one family, DNA assays on parental lymphocytes demonstrated evidence of somatic mosaicism, and in the other family there was evidence for germline mosaicism in the mother.

The presence of long interspersed nuclear element 1 (LINE-1, or L1) in the mammalian genome is also of interest in understanding the mechanisms of emergence of somatic mosaicism and variation in phenotypic expression. These retrotransposon elements are mobile segments of DNA that migrate around in the genome of developing neurons. These have long been thought to contribute to the diversity of phenotypes and individuality (Ostertag and Kazazian, 2005). In an experiment where L1 retrotransposons introduced into neuronal precursor cells (NPCs) of rats and a transgenic mouse model, neurons undergoing retrotransposon events were altered in the expression of their genes through chromatin modification via epigenetic mechanisms (also discussed later), in particular involving reduction in SOX2 expression (Muotri et al., 2005). The findings that neuronal genomes were not static and that their differentiation process can lead to somatic mosaicism through such occurrences provide yet another window into nature's experiments in producing diversity. Although doubts existed initially as to whether the process could be maintained evolutionarily in humans, a recent report of an X-linked progressive eye disorder (choroideremia) mediated through such retrotransposon events occurring early in human embryonic development suggests that the process can occur in humans (van den Hurk et al., 2007).

## A ROLE FOR MODIFIER GENES?

The observation of variable expressivity in idiopathic epilepsies within members of the same family raises the issue of modifier genes that influence the expression of the clinical phenotype. Such genes might serve to act synergistically or antagonistically with other susceptibility alleles, while not carrying a major causative effect on phenotypic expression. Support for this notion comes from animal models where strain backgrounds can show differences in seizure onset and expression even though the same mutation has been engineered (Bergren et al., 2005; Nabbout et al., 2007). Human studies in families with febrile seizures and later development of absence seizure phenotype also suggest the linkage to a gene locus on 3p and a modifier gene effect through an 18p locus in genome-wide expression studies (Nabbout et al., 2007). This bears similarity to the concept of *synergistic heterozygosity*, wherein the pathogenic effects of partial defects of a multiple nature or at multiple points in the same biochemical pathway can lead to significant deficits in enzymatic function leading to pathogenic effects (Vockley et al., 2000).

## EPIGENETIC MODIFICATION

The biological explanation for the diversity of cellular proteins and their function lies in understanding the process of modulation and regulation of gene expression during cell development and proliferation. The rapidly expanding field of *epigenetics*, a term originally introduced by C.H. Waddington (Van Speybroeck, 2002), sheds new light on our understanding of phenotypic variability. Epigenetic processes in the form of DNA methylation events and modifications to the histone proteins do not affect the genetic code; however, they modulate gene expression through changes in chromatin accessibility. These processes are stable and are also passed on during mitotic cell division. DNA methylation occurs at cytosines that reside as dinucleotide CpG-rich islands at the 5′ end of regulatory regions.

Through the addition of a methyl group to these cytosines, the gene is effectively silenced; thus, methylation and demethylation processes effectively function as off/on switches for gene expression. Rett syndrome, a neurodegenerative disorder with significant epilepsy, results from heterozygous mutations in the X-linked methyl CpG binding protein 2 (*MECP2*) gene. The *MECP2* gene product functions as a transcriptional repressor by binding to methylated DNA. The genetic deregulation that follows the occurrence of mutations in the X-linked *MECP2* genes has been shown to affect the expression of genes (UBE3 ligase and *GABRB3*) at other locations (e.g., 15q11–q13; Angelman syndrome) through possible histone modifications (Makedonski et al., 2005; Samaco et al., 2005). Both Angelman syndrome and Rett syndrome share significant features of their clinical phenotype, particularly the occurrence of seizures.

In addition to methylation, reversible posttranslational modifications of histone proteins are also involved in regulating gene expression. Histone-modifying enzymes—histone deacetylases (HDACs) or histone acetylases, as well as histone methyltransferases—are involved in dynamic interactions with chromatin in determining whether the genes remain accessible for transcription (activation) or in a closed inaccessible state (silenced). Cellular methylation patterns are affected by changes in diet (availability of folate), hormone levels, and inherited genetic polymorphisms (Rodenhiser and Mann, 2006). These aspects of epigenetics open up a Pandora's box of potential issues pertaining to the effects of dietary modifications.

Inherited genetic polymorphisms (such as the C677T polymorphism) have been linked to elevated homocysteine levels following antiepileptic drug (AED) therapy as well as an increased risk of epilepsy in offspring born to Scottish women homozygous to the C677T susceptibility allele (Dean et al., 2008). The direct or indirect effects of the antiepileptic drugs on methylation patterns and epigenetic regulation are not fully understood. Valproic acid (VPA), used commonly to treat epilepsy, has been of particular interest, as it actively enhances replication-independent demethylation of DNA. VPA has been known to block seizure-induced neurogenesis through its effect on HDACs (Jessberger et al., 2007). Its effectiveness thus may be linked to the prevention of aberrant neurogenesis that has been associated with the pathological consequences of status epilepticus. VPA and several other antiepileptic drugs are also folate antagonists, and the effects of folate deficiency on methylation patterns require fuller elucidation. Nevertheless, one could speculate that these effects may be related to the teratogenic effects on fetal development as a consequence of pregnancy-related exposure.

## COPY NUMBER VARIANTS

Although the entire genome has been decoded, the sequence information by itself is of limited use without an understanding of the cellular processes involved in the timing and translation of this code. Recent research using advances in genome scanning technology has shed light on characterizing structural variations in human DNA that are ~1 kb to ~3 Mb in size throughout the human genome. These copy number variations (CNVs) or internal rearrangements involving duplications or deletions occur at high frequencies when compared to other classes of cytogenetically detected variations or rearrangements. These variants, existing in a heterozygous state, lie between neutral polymorphisms and lethal mutations. Their position may directly interrupt genes or may influence neighboring gene function by virtue of a position effect. As such, CNVs or copy number polymorphisms (CNPs) are capable of influencing biochemical, morphological, physiological, and pathological processes, thus yielding potentially new mechanisms to explain diversity in human genetic diseases (Estivill and Armengol, 2007; Gurnett and Hedera, 2007; Kehrer-Sawatzki, 2007). Both CNVs and CNPs have been used to identify genetic susceptibility loci and potential candidate genes successfully for complex genetic disorders such as schizophrenia, autism, and severe speech and language disorder (Redon et al., 2006). Using array comparative genomic hybridization, researchers were able to identify significant gains in copy number (expressed as percentage gained or loss in following parentheses) at several gene loci involving IGE—1p (60%), 5p (55%), and 10q (55%); partial epilepsy—11p (45%) and 21q (45%); and febrile seizures—8q (55%). Losses in copy number at 7q (55%) were also noted (Kim et al., 2007). The precise mechanisms through which such changes in copy number act to change gene function and increase seizure susceptibility remain unclear.

## LINKING METABOLOME TO EPILEPSY

Several inborn errors of metabolism (e.g., nonketotic hyperglycinemia, pyridoxine dependency states) present with seizures as a prominent part of the phenotype. A meta-analysis of published studies indicates numerous points of intersection between heritable enzyme deficiencies and their consequences resulting in seizure susceptibility (Prasad and Prasad, 2008). There are several examples where direct metabolic disruptions due to mutations in mitochondrial proteins involved in the tricarboxylic acid cycle are linked to a reduced seizure threshold and an epileptic phenotype. Studies in the *Drosophila* bang-sensitive seizure mutant are of particular interest. One of the mutations, *knockdown (kdn)*, targets the enzyme citrate synthase, a key enzyme in the Krebs cycle. Quantitative assays show reduced adenosine triphosphate (ATP) levels in these mutants. What is most interesting is that the mutant animals show differential sensitivity to seizure induction, with the neuronal giant fiber pathway demonstrating a direct link between metabolic disruptions and lowered seizure threshold (Fergestad et al., 2006).

Mutations in the phosphoglycerate kinase (PGK) enzyme also confer seizure susceptibility at elevated temperatures in the *Drosophila* mutant *nubian* (Wang et al., 2004). PGK is required for ATP generation in the terminal stage of the glycolytic pathway. Brain extracts from *nubian* mutant strains show a threefold reduction in resting ATP levels compared with controls. Microarray analysis reveals transcriptional changes at genes implicated in glucose and lipid metabolism. The loss of ATP generation in *nubian* mutants leads to a temperature-dependent defect in neuronal activity, with initial seizure activity followed by an activity-dependent loss of synaptic transmission.

Whereas the typical inborn error of metabolism is recessively inherited and rare, the effects and consequences of allelic variants/nonsynonymous polymorphisms and their effects in terms of metabolic alterations have not yet been explored in the clinic. For the most part, investigations into biochemical markers for such perturbations in the epilepsy clinic occur in limited and selected cases in the pediatric and neonatal population. Biochemical pathways, such as transcriptional and signaling pathways, are complex and interlinked. Disturbances at one point or at multiple places in the same or different pathways can result in functional abnormalities significant enough to cause disease.

In the recessively inherited methylenetetrahydrofolate reductase (MTHFR) deficiency, the neurotoxicity is linked not only to hyperhomocysteinemia but also to the widespread effects that the associated methionine deficiency has on DNA methylation within the cell. In the case of the genetic polymorphism (C677T) in the gene coding for MHTHFR, the effect on the variant enzyme has been investigated (Yamada et al., 2001). The variant enzyme has the same level of catalytic activity as the wild type *in vitro* but loses its activity in a state of folate depletion. A low folate status in individuals carrying the polymorphic variant homozygous allele increases the risk for hyperhomocysteinemia and its attendant risks of neural tube defects or cardiovascular risks, and through hitherto uncharacterized epigenetic interactions may also influence seizure susceptibility (Dean et al., 2008). Thus, the study of genotypic variants and their influence on phenotypic diversity in health and disease through effects on biochemical pathways offers another window into the numerous avenues leading to seizure susceptibility.

## THE FUTURE: EVOLVING STRATEGIES AND CHALLENGES

When the first gene involved in inherited human epilepsy was identified in 1995, researchers were optimistic that the holy grail of identifying the few genes involved in epilepsy susceptibility seemed within reach. The succeeding years have involved much work, with a few successes and significant challenges remaining. Human epilepsy involves a huge diversity of causes, with numerous pathways leading to phenotypic variability (Figure 6.1). Epilepsy is now viewed in terms of an additive effect of nature (polygenic low-frequency susceptibility variants) and nurture (environmental/dietary effects). It is evident from the above discussion that there are several pathways to seizure susceptibility and epilepsy beyond the channelopathies. What does this imply for the future of epilepsy genetics?

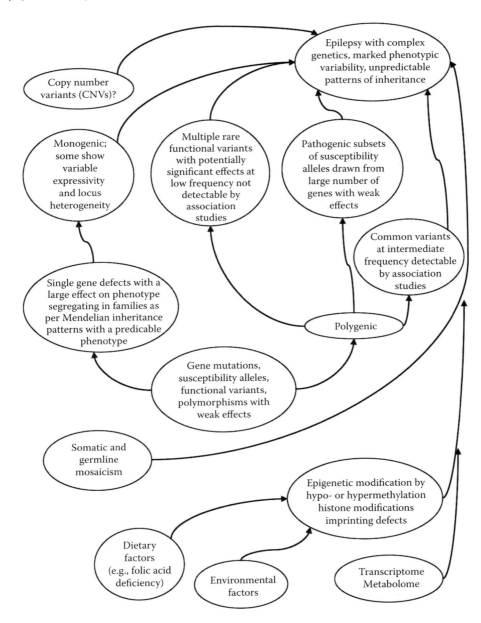

**FIGURE 6.1** Potential mechanisms leading to epilepsy with complex genetics.

Some investigators question whether association studies are the optimal way forward for chronic disorders such as epilepsy; however, it is likely that association studies will continue to be refined. Genome-wide association studies will likely be carried out (assuming sufficient funding) using haplotype mapping and the pooling of patients through multicenter collaborations. The use of bioinformatics and functional genomics will improve our understanding of the effects of these variant alleles within various biological systems involving the entire genome (e.g., transcriptome, proteome, metabolome). In this context, the widespread use of RNA interference technologies with their ability to artificially silence genes will also accelerate our understanding of the function of genes within the context of the interrelated cellular processes. It is hoped that their use will lead to the characterization of potential therapeutic targets, in addition to unraveling the molecular signaling pathways leading to epilepsy.

A study using an RNA interference (RNAi) screen in *Caenorhabditis elegans* disclosed 90 genes involved in GABA synapses; 21 of these genes have been previously identified or implicated in epilepsy (Vashlishan et al., 2008). These technologies may be more effective in identifying a wider range of potential candidate genes than was hitherto possible; however, there remain several challenges in terms of delivery of these exogenous molecules into the nervous system and addressing potential side effects. What is needed for the future is a systems approach that integrates our understanding of the behavior of biological networks in health and disease, such as are currently being developed for other complex disorders (Baranzini, 2004; Tegner et al., 2007; Xiong et al., 2005). Future developments and complementary approaches using epidemiology, population genetics, different mapping strategies, and newer technologies (DNA- and RNA-based) should lead to better understanding and less confusion in the field of genetics of complex epilepsy.

## ACKNOWLEDGMENTS

The authors are indebted to D. Rodenhiser, Ph.D., Associate Professor, Department of Pediatrics, Oncology and Biochemistry, University of Western Ontario, and Scientist, Child Health Research Institute, London, Ontario, for reviewing the manuscript and for his insightful comments and suggestions.

## REFERENCES

Annegers, J.F., W.A. Rocca, W.A. Hauser. (1996). Causes of epilepsy: contributions of the Rochester epidemiology project. *Mayo Clin. Proc.*, 71:570–575.

Audenaert, D., L. Claes, K.G. Claeys, L. Deprez, T. Van Dyck, D. Goossens, J. Del-Favero, W. Van Paesschen, C. Van Broeckhoven, P. De Jonghe. (2005). A novel susceptibility locus at 2p24 for generalised epilepsy with febrile seizures plus. *J. Med. Genet.*, 42:947–952.

Baranzini, S.E. (2004). Gene expression profiling in neurological disorders: toward a systems-level understanding of the brain. *Neuromolecular Med.*, 6:31–51.

Baulac, S., G. Huberfeld, I. Gourfinkel-An, G. Mitropoulou, A. Beranger, J.F. Prud'homme, M. Baulac, A. Brice, R. Bruzzone, E. LeGuern. (2001a). First genetic evidence of GABA(a) receptor dysfunction in epilepsy: a mutation in the gamma2-subunit gene. *Nat. Genet.*, 28:46–48.

Baulac, S., F. Picard, A. Herman, J. Feingold, E. Genin, E. Hirsch, J.F. Prud'homme, M. Baulac, A. Brice, E. LeGuern. (2001b). Evidence for digenic inheritance in a family with both febrile convulsions and temporal lobe epilepsy implicating chromosomes 18qter and 1q25–q31. *Ann. Neurol.*, 49:786–792.

Baulac, S., I. Gourfinkel-An, P. Couarch, C. Depienne, A. Kaminska, O. Dulac, M. Baulac, E. LeGuern, R. Nabbout. (2008). A novel locus for generalized epilepsy with febrile seizures plus in French families. *Arch. Neurol.*, 65:943–951.

Bergren, S.K., S. Chen, A. Galecki, J.A. Kearney. (2005). Genetic modifiers affecting severity of epilepsy caused by mutation of sodium channel SCN2A. *Mamm. Genome*, 16:683–690.

Berkovic, S.F., R.A. Howell, D.A. Hay, J.L. Hopper. (1998). Epilepsies in twins: genetics of the major epilepsy syndromes. *Ann. Neurol.*, 43:435–445.

Berkovic, S.F., P. Izzillo, J.M. McMahon, L.A. Harkin, A.M. McIntosh, H.A. Phillips, R.S. Briellmann, R.H. Wallace, A. Mazarib, M.Y. Neufeld, A.D. Korczyn, I.E. Scheffer, J.C. Mulley. (2004). *LGI1* mutations in temporal lobe epilepsies. *Neurology*, 62:1115–1119.

Boyko, A.R., S.H. Williamson, A.R. Indap, J.D. Degenhardt, R.D. Hernandez, K.E. Lohmueller, M.D. Adams, S. Schmidt, J.J. Sninsky, S.R. Sunyaev, T.J. White, R. Nielsen, A.G. Clark, C.D. Bustamante. (2008). Assessing the evolutionary impact of amino acid mutations in the human genome. *PLoS Genet.*, 4:e1000083.

Cavalleri, G.L., N.M. Walley, N. Soranzo, J. Mulley, C.P. Doherty et al. (2007a). A multicenter study of *BRD2* as a risk factor for juvenile myoclonic epilepsy. *Epilepsia*, 48:706–712.

Cavalleri, G.L., M.E. Weale, K.V. Shianna, R. Singh, J.M. Lynch et al. (2007b). Multicentre search for genetic susceptibility loci in sporadic epilepsy syndrome and seizure types: a case-control study. *Lancet Neurol.*, 6:970–980.

Claes, L., J. Del-Favero, B. Ceulemans, L. Lagae, C. Van Broeckhoven, P. De Jonghe. (2001). *De novo* mutations in the sodium-channel gene *SCN1A* cause severe myoclonic epilepsy of infancy. *Am. J. Hum. Genet.*, 68:1327–1332.

Claes, L., B. Ceulemans, D. Audenaert, K. Smets, A. Lofgren et al. (2003). *De novo* SCN1A mutations are a major cause of severe myoclonic epilepsy of infancy. *Hum. Mutat.*, 21:615–621.

Collins, J.S., C.E. Schwartz. (2002). Detecting polymorphisms and mutations in candidate genes. *Am. J. Hum. Genet.*, 71:1251–1252.

Cowell, J.K. (2004). High throughput determination of gains and losses of genetic material using high-resolution BAC arrays and comparative genomic hybridization. *Comb. Chem. High Throughput Screen*, 7:587–596.

Crino, P.B. (2007). Gene expression, genetics, and genomics in epilepsy: some answers, more questions. *Epilepsia*, 48(Suppl. 2):42–50.

Dean, J.C., Z. Robertson, V. Reid, Q. Wang, H. Hailey, S. Moore, A.D. Rasalam, P. Turnpenny, D. Lloyd, D. Shaw, J. Little. (2008). A high frequency of the MTHFR 677C>T polymorphism in Scottish women with epilepsy: possible role in pathogenesis. *Seizure*, 17:269–275.

Di Rienzo, A. (2006). Population genetics models of common diseases. *Curr. Opin. Genet Dev.*, 16:630–636.

Dibbens, L.M., S.E. Heron, J.C. Mulley. (2007). A polygenic heterogeneity model for common epilepsies with complex genetics. *Genes Brain Behav.*, 6:593–597.

Durner, M., M.A. Keddache, L. Tomasini, S. Shinnar, S.R. Resor et al. (2001). Genome scan of idiopathic generalized epilepsy: evidence for major susceptibility gene and modifying genes influencing the seizure type. *Ann. Neurol.*, 49:328–335.

Escayg, A., B.T. MacDonald, M.H. Meisler, S. Baulac, G. Huberfeld, I. An-Gourfinkel, A. Brice, E. LeGuern, B. Moulard, D. Chaigne, C. Buresi, A. Malafosse. (2000). Mutations of *SCN1A*, encoding a neuronal sodium channel, in two families with GEFS+2. *Nat. Genet.*, 24:343–345.

Estivill, X., L. Armengol. (2007). Copy number variants and common disorders: filling the gaps and exploring complexity in genome-wide association studies. *PLoS Genet.*, 3:1787–1799.

Fergestad, T., B. Bostwick, B. Ganetzky. (2006). Metabolic disruption in *Drosophila* bang-sensitive seizure mutants. *Genetics*, 173:1357–1364.

Fisher, R. (1918). The correlation between relatives on the supposition of Mendelian inheritance. *Trans. R. Soc. Edinburgh*, 52:399–433.

Gabriel, S.B., S.F. Schaffner, H. Nguyen, J.M. Moore, J. Roy et al. (2002). The structure of haplotype blocks in the human genome. *Science*, 296:2225–2229.

Gennaro, E., F.M. Santorelli, E. Bertini, D. Buti, R. Gaggero et al. (2006). Somatic and germline mosaicisms in severe myoclonic epilepsy of infancy. *Biochem. Biophys. Res. Commun.*, 341:489–493.

Gurnett, C.A., P. Hedera. (2007). New ideas in epilepsy genetics: novel epilepsy genes, copy number alterations, and gene regulation. *Arch. Neurol.*, 64:324–328.

Harkin, L.A., J.M. McMahon, X. Iona, L. Dibbens, J.T. Pelekanos et al. (2007). The spectrum of *SCN1A*-related infantile epileptic encephalopathies. *Brain*, 130:843–852.

Hedera, P., S. Ma, M.A. Blair, K.A. Taylor, A. Hamati, Y. Bradford, B. Abou-Khalil, J.L. Haines. (2006). Identification of a novel locus for febrile seizures and epilepsy on chromosome 21q22. *Epilepsia*, 47:1622–1628.

Jessberger, S., K. Nakashima, G.D. Clemenson, Jr., E. Mejia, E. Mathews, K. Ure, S. Ogawa, C.M. Sinton, F.H. Gage, J. Hsieh. (2007). Epigenetic modulation of seizure-induced neurogenesis and cognitive decline. *J. Neurosci.*, 27:5967–5975.

Johnson, W.G. (1980). Metabolic interference and the +/− heterozygote: a hypothetical form of simple inheritance which is neither dominant nor recessive. *Am. J. Hum. Genet.*, 32:374–386.

Kanai, K., S. Hirose, H. Oguni, G. Fukuma, Y. Shirasaka, T. Miyajima, K. Wada, H. Iwasa, S. Yasumoto, M. Matsuo, M. Ito, A. Mitsudome, S. Kaneko. (2004). Effect of localization of missense mutations in *SCN1A* on epilepsy phenotype severity. *Neurology*, 63:329–334.

Kehrer-Sawatzki, H. (2007). What a difference copy number variation makes. *Bioessays*, 29:311–313.

Kim, H.S., S.V. Yim, K.H. Jung, L.T. Zheng, Y.H. Kim, K.H. Lee, S.Y. Chung, H.K. Rha. (2007). Altered DNA copy number in patients with different seizure disorder type: by array-CGH. *Brain Dev.*, 29:639–643.

Kjeldsen, M.J., L.A. Corey, K. Christensen, M.L. Friis. (2003). Epileptic seizures and syndromes in twins: the importance of genetic factors. *Epilepsy Res.*, 55:137–146.

Lindhout, D. (2008). Somatic mosaicism as a basic epileptogenic mechanism? *Brain*, 131:900–901.

Makedonski, K., L. Abuhatzira, Y. Kaufman, A. Razin, R. Shemer. (2005). MECP2 deficiency in Rett syndrome causes epigenetic aberrations at the PWS/AS imprinting center that affects UBE3A expression. *Hum. Mol. Genet.*, 14:1049–1058.

Marini, C., I.E. Scheffer, K.M. Crossland, B.E. Grinton, F.L. Phillips et al. (2004). Genetic architecture of idiopathic generalized epilepsy: clinical genetic analysis of 55 multiplex families. *Epilepsia*, 45:467–478.

Mulley, J.C., I.E. Scheffer, S. Petrou, S.F. Berkovic. (2003). Channelopathies as a genetic cause of epilepsy. *Curr. Opin. Neurol.*, 16:171–176.

Mulley, J.C., I.E. Scheffer, L.A. Harkin, S.F. Berkovic, L.M. Dibbens. (2005). Susceptibility genes for complex epilepsy. *Hum. Mol. Genet.*, 14(2):R243–R249.

Muotri, A.R., V.T. Chu, M.C. Marchetto, W. Deng, J.V. Moran, F.H. Gage. (2005). Somatic mosaicism in neuronal precursor cells mediated by l1 retrotransposition. *Nature*, 435:903–910.

Nabbout, R., S. Baulac, I. Desguerre, N. Bahi-Buisson, C. Chiron, M. Ruberg, O. Dulac, E. LeGuern. (2007). New locus for febrile seizures with absence epilepsy on 3p and a possible modifier gene on 18p. *Neurology*, 68:1374–1381.

NIGMS. (1998). *The Genetic Architecture of Complex Traits Workshop*. Bethesda, MD: National Institute of General Medical Sciences (http://www.nigms.nih.gov/news/reports/genetic_arch.html).

Ostertag, E.M., H.H. Kazazian. (2005). Genetics: lines in mind. *Nature*, 435:890–891.

Ottman, R., J.F. Annegers, N. Risch, W.A. Hauser, M. Susser. (1996). Relations of genetic and environmental factors in the etiology of epilepsy. *Ann. Neurol.*, 39:442–449.

Page, D.C., M.E. Harper, J. Love, D. Botstein. (1984). Occurrence of a transposition from the X-chromosome long arm to the Y-chromosome short arm during human evolution. *Nature*, 311:119–123.

Pal, D.K., O.V. Evgrafov, P. Tabares, F. Zhang, M. Durner, D.A. Greenberg. (2003). *BRD2* (*RING3*) is a probable major susceptibility gene for common juvenile myoclonic epilepsy. *Am. J. Hum. Genet.*, 73:261–270.

Pal, D.K., L.J. Strug, D.A. Greenberg. (2008). Evaluating candidate genes in common epilepsies and the nature of evidence. *Epilepsia*, 49:386–392.

Pearson, T.A., T.A. Manolio. (2008). How to interpret a genome-wide association study. *JAMA*, 299:1335–1344.

Prasad, A.N., C. Prasad. (2008a). Genetic influences on the risk for epilepsy. In *Pediatric Epilepsy: Diagnosis and Therapy* (pp. 117–134), 3rd ed., J.M. Pellock, W.E. Dodson, and B.F.D. Bourgeois, Eds. New York: Demos Medical Publishing.

Prasad, A.N., C. Prasad. (2008b). Linking biochemical pathways to seizure susceptibility in early life; lessons from inborn errors of metabolism. In *Biology of Seizure Susceptibility in Developing Brain* (p. 232), T. Takahashi and Y. Fukuyama, Eds. Paris: John Libbey Eurotext.

Prasad, A.N., C. Prasad, C.E. Stafstrom. (1999). Recent advances in the genetics of epilepsy: insights from human and animal studies. *Epilepsia*, 40:1329–1352.

Pritchard, J.K. (2001). Are rare variants responsible for susceptibility to complex diseases? *Am. J. Hum. Genet.*, 69:124–137.

Ptacek, L.J. (1997). Channelopathies: ion channel disorders of muscle as a paradigm for paroxysmal disorders of the nervous system. *Neuromusc. Disord.*, 7:250–255.

Rao, D.C. (2001). Genetic dissection of complex traits: an overview. *Adv. Genet.*, 42:13–34.

Rao, D.C. (2008). An overview of the genetic dissection of complex traits. *Adv. Genet.*, 60:3–34.

Redon, R., S. Ishikawa, K.R. Fitch, L. Feuk, G.H. Perry et al. (2006). Global variation in copy number in the human genome. *Nature*, 444:444–454.

Reich, D.E., E.S. Lander. (2001). On the allelic spectrum of human disease. *Trends Genet.*, 17:502–510.

Rodenhiser, D., M. Mann. (2006). Epigenetics and human disease: translating basic biology into clinical applications. *CMAJ.* 174:341–348.

Rosenberg, C., C.H. Wouters, K. Szuhai, R. Dorland, P. Pearson, B.T. Poll-The, R.M. Colombijn, M. Breuning, D. Lindhout. (2001). A Rett syndrome patient with a ring X chromosome: further evidence for skewing of X inactivation and heterogeneity in the aetiology of the disease. *Eur. J. Hum. Genet.*, 9:171–177.

Samaco, R.C., A. Hogart, J.M. LaSalle. (2005). Epigenetic overlap in autism-spectrum neurodevelopmental disorders: *MECP2* deficiency causes reduced expression of *UBE3A* and *GABRB3*. *Hum. Mol. Genet.*, 14:483–492.

Scheffer, I.E., S.J. Turner, L.M. Dibbens, M.A. Bayly, K. Friend et al. (2008). Epilepsy and mental retardation limited to females: an under-recognized disorder. *Brain*, 131:918–927.

Sellner, L.N., G.R. Taylor. (2004). MLPA and MAPH: new techniques for detection of gene deletions. *Hum. Mutat.*, 23:413–419.

Steinlein, O.K. (2004). Genes and mutations in human idiopathic epilepsy. *Brain Dev.*, 26:213–218.

Steinlein, O.K., J.C. Mulley, P. Propping, R.H. Wallace, H.A. Phillips, G.R. Sutherland, I.E. Scheffer, S.F. Berkovic. (1995). A missense mutation in the neuronal nicotinic acetylcholine receptor alpha 4 subunit is associated with autosomal dominant nocturnal frontal lobe epilepsy. *Nat. Genet.*, 11:201–203.

Tan, N.C., J.C. Mulley, S.F. Berkovic. (2004). Genetic association studies in epilepsy: "the truth is out there." *Epilepsia*, 45:1429–1442.

Tan, N.C., J.C. Mulley, I.E. Scheffer. (2006). Genetic dissection of the common epilepsies. *Curr. Opin. Neurol.*, 19:157–163.

Tegner, J., J. Skogsberg, J. Bjorkegren. (2007). Thematic review series: systems biology approaches to metabolic and cardiovascular disorders. Multi-organ whole-genome measurements and reverse engineering to uncover gene networks underlying complex traits. *J. Lipid Res.*, 48:267–277.

Twigg, S.R., K. Matsumoto, A.M. Kidd, A. Goriely, I.B. Taylor et al. (2006). The origin of *EFNB1* mutations in craniofrontonasal syndrome: frequent somatic mosaicism and explanation of the paucity of carrier males. *Am. J. Hum. Genet.*, 78:999–1010.

Vadlamudi, L., E. Andermann, C.T. Lombroso, S.C. Schachter, R.L. Milne, J.L. Hopper, F. Andermann, S.F. Berkovic. (2004). Epilepsy in twins: insights from unique historical data of William Lennox. *Neurology*, 62:1127–1133.

van den Hurk, J.A., I.C. Meij, M.C. Seleme, H. Kano, K. Nikopoulos et al. (2007). L1 retrotransposition can occur early in human embryonic development. *Hum. Mol. Genet.*, 16:1587–1592.

Van Speybroeck, L. (2002). From epigenesis to epigenetics: the case of C.H. Waddington. *Ann. N.Y. Acad. Sci.*, 981:61–81.

Vashlishan, A.B., J.M. Madison, M. Dybbs, J. Bai, D. Sieburth, Q. Ch'ng, M. Tavazoie, J.M. Kaplan. (2008). An RNAi screen identifies genes that regulate GABA synapses. *Neuron*, 58:346–361.

Venter, J.C., M.D. Adams, E.W. Myers, P.W. Li, R.J. Mural et al. (2001). The sequence of the human genome. *Science*, 291:1304–1351.

Vockley, J., P. Rinaldo, M.J. Bennett, D. Matern, G.D. Vladutiu. (2000). Synergistic heterozygosity: disease resulting from multiple partial defects in one or more metabolic pathways. *Mol. Genet. Metab.*, 71:10–18.

Wallace, R.H., D.W. Wang, R. Singh, I.E. Scheffer, A.L. George, Jr. et al. (1998). Febrile seizures and generalized epilepsy associated with a mutation in the Na+-channel beta1 subunit gene *SCN1B*. *Nat. Genet.* 19:366–70.

Wang, P., S. Saraswati, Z. Guan, C.J. Watkins, R.J. Wurtman, J.T. Littleton. (2004). A *Drosophila* temperature-sensitive seizure mutant in phosphoglycerate kinase disrupts ATP generation and alters synaptic function. *J. Neurosci.*, 24:4518–4529.

WHO. (2001). *Epilepsy: Aetiology, Epidemiology and Prognosis*, Fact Sheet No. 165. Geneva: World Health Organization (https://apps.who.int/inf-fs/en/fact165.html).

Winawer, M.R. (2002). Epilepsy genetics. *Neurologist*, 8:133–151.

Winawer, M.R., S. Shinnar. (2005). Genetic epidemiology of epilepsy or what do we tell families? *Epilepsia*, 46(Suppl. 10):24–30.

Xiong, M., C.A. Feghali-Bostwick, F.C. Arnett, X. Zhou. (2005). A systems biology approach to genetic studies of complex diseases. *FEBS Lett.*, 579:5325–5332.

Yamada, K., Z. Chen, R. Rozen, R.G. Matthews. (2001). Effects of common polymorphisms on the properties of recombinant human methylenetetrahydrofolate reductase. *Proc. Natl. Acad. Sci. U.S.A.*, 98:14853–14858.

Zucca, C., F. Redaelli, R. Epifanio, N. Zanotta, A. Romeo et al. (2008). Cryptogenic epileptic syndromes related to *SCN1A*: twelve novel mutations identified. *Arch. Neurol.*, 65:489–494.

# 7 Pathophysiology of Developmental Epilepsies

*Judith L.Z. Weisenberg and Michael Wong*

## CONTENTS

## INTRODUCTION

Epilepsy in children often has unique characteristics that differentiate it from adult-onset epilepsy. The immature brain exhibits an increased susceptibility to seizures, resulting in a peak incidence of epilepsy in the first year of life. Most of the common idiopathic epilepsy syndromes typically have their onset in childhood. Other prominent seizure disorders, such as febrile seizures and infantile spasms, also exhibit a distinct predilection for the pediatric years. Furthermore, many causes of symptomatic epilepsies are relatively specific to the pediatric population, such as certain types of malformations of cortical development.

Significant advances have recently been made in understanding the neurobiological mechanisms that cause developmental differences in seizures and account for the unique properties of many childhood epilepsies. Developmental processes that guide the growth and maturation of the nervous system under normal conditions may contribute to the increased seizure susceptibility of the immature brain. Genetic, biochemical, and structural abnormalities may initially become symptomatic or have distinctive effects in the developing brain, manifesting as specific pediatric epilepsy syndromes. Both clinical studies and animal models of epilepsy have helped contribute to our understanding of the developmental characteristics of seizures in the immature brain. Most importantly, many of these findings are beginning to be applied to the development of novel, more effective therapies for pediatric epilepsies.

This chapter summarizes our understanding of some of these concepts by examining various epilepsy syndromes in a chronological fashion across the pediatric years. Particular attention is focused on how our understanding of these syndromes is translating into potential therapies.

**TABLE 7.1**

**Developmental Phenomena and Their Proposed Mode of Action Resulting in a Hyperexcitable State in the Neonatal Brain**

| Developmental Phenomenon | Proposed Mechanism of Action |
|---|---|
| Increased number of synapses | Excessive synaptic and electrical activity |
| Increased AMPA and NMDA receptor expression | Increased glutamate-mediated excitation |
| Increased NMDA NR2A subunit expression | Decreased magnesium sensitivity, resulting in increased NMDA receptor activity |
| Absent AMPA GluR2(B) subunit expression | Increased calcium permeability of AMPA receptors |
| Increased NKCC1 expression | Increased intracellular chloride and GABA excitatory effect |
| Decreased KCC2 expression | Increased intracellular chloride and GABA excitatory effect |

## NEONATAL SEIZURES

The high incidence of neonatal seizures provides clear proof of the role of development in the pathophysiology of epilepsy. Neonatal seizures, occurring up to 44 weeks of gestational age, are quite common, with a reported frequency between 1.8 and 3.3 per 1000 (Lanska et al., 1995; Ronen et al., 1999; Saliba et al., 1999). This translates into a peak in new-onset seizures during the first year of life, with a rapid decline during subsequent years (Hauser et al., 1993). The incidence also appears to be higher in infants who have lower birth weights or are preterm. The etiology, which is usually symptomatic, and clinical manifestations of neonatal seizures can be quite varied and subtle. While treatment with anticonvulsants is standard, there is limited evidence to support its efficacy or benefits for long-term neurological outcome. In fact, a recent concern has been that some standard treatments such as barbiturates and benzodiazepines may have detrimental effects on brain development. Furthermore, it has become clear that, although treatment can be effective at eliminating the clinical manifestations of the seizures, abnormal electrographic activity may persist—a phenomenon known as *electroclinical dissociation*. Fortunately, our understanding of the mechanisms underlying the unique properties of seizures in this age group has greatly advanced in recent years and is starting to translate into more targeted therapies.

A number of key biological processes are inherent to the tendency for seizures in the neonatal period (Table 7.1). One of these is synaptogenesis, which is necessary for normal brain development. Multiple developmental histological analyses of neocortex in primate and human brain have demonstrated that the number of synapses formed in the postnatal period rapidly exceeds the number seen in the adult human brain (Huttenlocher et al., 1982; Rakic et al., 1986). Subsequently, a complex process of synapse formation and elimination occurs, with ultimate reduction in synapse density in a process known as *synaptic pruning*. There is evidence from *in vitro* and *in vivo* models that the process of synaptic competition depends at least in part on relative electrical activity of competing synapses (Szabo and Zoran, 2007). Synapse maintenance may be strengthened by bursts of patterned neuronal signaling between pre- and postsynaptic elements similar to the putative memory mechanism of long-term potentiation in the mature brain (Hua and Smith, 2004). It is likely that this dependence on synchronous electrical activity for synapse formation would also increase the susceptibility of the immature brain to seizures.

Another important process in early brain development is the evolution of expression of various excitatory and inhibitory neurotransmitter systems. The major excitatory neurotransmitter is glutamate, which works through three classes of ionotropic glutamate receptor channels: α-amino-3-hydroxy-5-methyl-4-isoxazoleproprionic acid (AMPA), kainate, and N-methyl-D-aspartate (NMDA) receptors, as well as metabotropic glutamate receptors coupled to second-messenger systems. During early postnatal development, the levels of expression of these receptor types differ. In rats, expression of several AMPA and NMDA protein subunits is low at birth, rises to a peak around the second postnatal week (corresponding to the neonatal period in humans), and then declines back to an intermediate level for adulthood (Insel et al., 1990; McDonald et al., 1990; Monyer et al.; 1994; Pellegrini-Giampietro et al., 1991; Zhong et al., 1995). This transient increase in excitatory glutamate receptors could contribute to a decreased seizure threshold in the developing brain.

In addition to changes in total receptor expression, there is also developmental regulation of the specific subunit composition of receptors. All neurotransmitter receptors are composed of a heteromeric combination of different protein subunits. For example, in the adult brain, NMDA receptors most commonly have the receptor subunit NR2A, and the immature brain tends to have increased expression of the NR2B, NR2C, NR2D, and NR3A subunits. The NMDA subunits expressed predominantly during early development possess properties that enhance excitability of the receptor in the immature brain, such as reduced sensitivity to magnesium block (Flint et al., 1997; Hollmann and Heinemann, 1994; Monyer et al, 1994; Wong et al., 2002; Zhong et al., 1995). In contrast, subsequent upregulation of the NR2A subunit imparts physiological characteristics to NMDA receptors that reduce excitability in the adult brain. Similarly, in the adult brain AMPA receptors usually contain a GluR2(B) subunit, whereas in the immature brain the GluR2(B) subunit is typically absent (Hollmann and Heinemann, 1994; Kumar et al., 2002; Pellegrini-Giampietro et al., 1991, 1992; Sanchez et al., 2001; Talos et al., 2006a,b). A major function of the GluR2(B) subunit is to impart calcium impermeability to the AMPA receptor. It is likely that increased calcium permeability through AMPA receptors lacking GluR2(B) in the immature brain results in activation of multiple calcium-dependent cellular processes that are important for normal development but increases the risk of neuronal hyperexcitability.

Complementing the developmental changes in excitatory glutamate systems, a major area of interest has been the $\gamma$-aminobutyric acid (GABA)$_A$ receptor. GABA is the major inhibitory neurotransmitter in the brain and acts primarily by binding to postsynaptic ionotropic GABA$_A$ receptors. In the adult brain, these receptors typically hyperpolarize the membrane through inward flow of chloride ions due to higher extracellular vs. intracellular chloride concentrations. In contrast, in the developing brain, activation of GABA$_A$ receptors results in depolarization of the membrane due to a reversed chloride gradient (Cherubini et al., 1990; Khazipov et al., 2004; Loturco et al., 1995; Owens et al., 1996). This excitatory activity of GABA in the neonatal brain is necessary for synaptogenesis during normal brain development (Ben-Ari, 2002; Pfeffer et al., 2009) and gradually converts to the inhibitory action of the adult brain with advancing age (Khazipov et al., 2004; Owens et al., 1996). However, this paradoxical excitation of GABA in the neonatal brain increases susceptibility to seizures (Dzhala and Staley, 2003; Khazipov et al., 2004).

The fact that GABA$_A$ receptors produce excitation rather than inhibition early in development gives rise to an interesting clinical controversy, as the most common medications used in this age group—phenobarbital and benzodiazepines—are GABA agonists. Phenobarbital exhibits limited efficacy in controlling neonatal seizures (Painter et al., 1999). Furthermore, concern has been raised that these medications may induce neuronal death in the developing brain. In neonatal rats, multiple medications, including phenobarbital, diazepam, and clonazepam, were associated with widespread apoptotic neurodegeneration (Bittigau et al., 2002, 2003). This response is dose dependent and seen at levels equivalent to those used at therapeutic doses in humans (Bittigau et al., 2003). Thus, based on both human and animal model data, it is questionable whether the common clinical use of GABA agonists represents optimal treatment for neonatal seizures.

A key development that may result in more effective therapies for neonatal seizures has been an understanding of the mechanisms mediating the developmental changes in GABA actions. The basic principle underlying the depolarizing or excitatory effect of the GABA$_A$ receptor in the neonatal period is that the intracellular neuronal chloride concentration is relatively elevated. Thus, GABA$_A$ receptor activation causes chloride efflux and depolarization in immature neurons rather than the influx and hyperpolarization that occur in the adult brain. Over the past decade, the primary mechanism by which this inversion in the chloride gradient occurs has been elucidated. This mechanism is based on the evolution of the developmental expression of two chloride transporters.

A series of *in vitro* and *in vivo* experiments in rat neocortex has clearly demonstrated that the expression of the chloride transporter $Na^+–K^+–Cl^-$ cotransporter isoform 1 (NKCC1), which transports chloride intracellularly, is high at birth and decreases over time (Clayton et al., 1998; Plotkin et al., 1997; Yamada et al., 2004). In contrast, $K^+–Cl^-$ cotransporter isoform 2 (KCC2), a chloride transporter that transports chloride extracellularly, is more highly expressed in mature neurons of rat cortex (Wang et al., 2002). KCC2 expression increases during development; this correlates with the expected decrease in intracellular chloride concentration (Yamada et al., 2004). In a functional study, Rivera et al. (1999) demonstrated that blocking KCC2 expression resulted in a depolarizing effect of the $GABA_A$ receptor. Conversely, it has been demonstrated that increased KCC2 expression causes chloride efflux in human dorsal root ganglion cells (Staley et al., 1996). Finally, a similar developmental pattern of expression of the NKCC1 and KCC2 transporters occurs within human cortex, involving a higher expression of NKCC1 than KCC2 at birth, a gradual decrease in NKCC1, and corresponding increase in KCC2 to adult levels over the first year (Dzhala et al., 2005).

Interestingly, there is some evidence for a gender difference in $GABA_A$ effects in the neonatal period. Multiple studies in rat models have demonstrated that the depolarizing effect of the $GABA_A$ receptor persists during development longer in males than females (Galanopoulou, 2005, 2008; Nunez and McCarthy, 2007). Furthermore, this change has been correlated with sex differences in KCC1 and NKCC2 expression (Galanopoulou, 2008; Nunez and McCarthy, 2007). This could play a role in gender-specific differences in the long-term effects of neonatal seizures and in susceptibility to injury secondary to neonatal insults (Galanopoulou, 2008). In addition, maturation of KCC1 and NKCC2 expression occurs in a caudal to rostral direction (Volpe, 2008). This would result in GABA-mediated inhibition occurring in the spinal cord and brainstem prior to the cerebral cortex and might explain the tendency for electroclinical dissociation to occur after treatment for neonatal seizures, with the behavioral output of seizures inhibited at the brainstem/spinal cord level but the cortical seizure activity persisting.

What is probably most exciting about this recent understanding of the developmental changes in chloride transporter expression is that there may be a ready clinically translatable therapy. NKCC1 is extremely sensitive to blockade by the diuretic bumetanide (Hannaert et al., 2002; Issenting et al., 1998). By inhibiting chloride transport by NKCC1, bumetanide prevents increased chloride concentration and decreases the depolarizing action of GABA in immature neurons (Dzhala et al., 2005; Sipila et al., 2006; Yamada et al., 2004). In a key study, it has been demonstrated that the addition of bumetanide to phenobarbital markedly reduced seizure activity in rat hippocampal preparations (Dzhala et al., 2008). Bumetanide is already used clinically as a diuretic for neonates, and a phase II pilot study of bumetanide therapy for neonatal seizures is currently underway (http://clinicaltrials.gov/archive/NCT00830531/2009_01_29). Of course, caution must be exercised in assuming that this will be a successful therapy. The safety profile of bumetanide in the central nervous system is not completely established, and one paper has demonstrated toxic effects of high concentrations of bumetanide in adult human hippocampal neurons (Pond et al., 2004). Nevertheless, the possibility of a readily translatable therapy for neonatal seizures derived from basic insights into brain development remains very exciting.

## INFANTILE SPASMS

Shifting just beyond the neonatal period, infantile spasms, also known as West syndrome, is a developmental epilepsy syndrome of infancy characterized by the classic triad of the seizure type of epileptic spasms, developmental delay, and hypsarrhythmia on electroencephalography (EEG). Although the seizures themselves are usually not severe from a semiological standpoint, infantile spasms can be a devastating disorder that is usually considered a catastrophic epilepsy due to the typically poor prognosis in terms of neurodevelopmental outcome and chronic intractable epilepsy. Selective treatments, such as adrenocorticotropic hormone (ACTH) and vigabatrin, can sometimes reduce or eliminate spasms, but there is little evidence that such treatments improve the long-term neurological prognosis. Thus, there is a strong interest in obtaining a better understanding of the pathophysiology of infantile spasms, with the hope of developing more effective therapies. Unfortunately, despite decades of research, the pathogenesis of infantile spasms mostly remains a mystery.

Efforts to gain insights into the pathophysiology of infantile spasms must consider a few of its unique clinical features. First, infantile spasms constitute an epilepsy syndrome that is truly developmentally specific. The spasms typically begin around 4 months of age and usually resolve by 1 to 2 years of age, even without treatment, although patients still maintain a high risk for long-term developmental deficits and chronic epilepsy involving other seizure types. Second, in addition to cryptogenic cases of unknown cause, a large variety of different symptomatic etiologies can cause infantile spasms, such as hypoxic ischemia, central nervous system infections, malformations of cortical development, and numerous genetic and metabolic disorders. Thus, a popular concept is that infantile spasms represent a "final common pathway" in response to any significant insult or injury to the brain occurring at this specific stage of infant brain development.

One of the still unresolved fundamental issues about infantile spasms is the underlying neuronal circuits and anatomy that generate spasms and hypsarrhythmia. EEG recordings typically show bilateral or diffuse ictal electrographic changes during spasms, consistent with generalized seizures. However, as with other generalized seizure types, it is not clear whether cerebral cortex or subcortical structures, such as the brainstem, are primarily responsible for generating spasms. The stereotypical electrodecremental ictal EEG pattern of spasms suggests that subcortical structures may be the primary generator, which then secondarily suppresses cortical activity. An additional level of complexity is that some etiologies, such as tuberous sclerosis complex (TSC), have discrete focal cortical lesions that appear to cause spasms. Thus, it is likely that there is a complex interplay of cortical and subcortical circuits that are prone to generating spasms due to various perturbations of the network during a selective period of brain development.

The other major consideration in formulating a hypothesis about the pathophysiology of infantile spasms is their unique responsiveness to ACTH. This is a relatively specific association, as most conventional seizure medications are marginally effective for spasms and ACTH is rarely used for any other type of epilepsy. A popular unifying hypothesis that coordinates the involvement of diffuse neuroanatomical circuits and explains why multiple different etiologies may result in spasms relates to the role of ACTH and associated neurochemicals in modulating intrinsic brain responses to stress (Brunson et al., 2001a). In particular, corticotrophin-releasing hormone (CRH) is a stress-activated neurotransmitter/hormone that is released in multiple brain regions in response to various pathophysiological stressors and brain injuries. CRH is known to have excitatory physiological actions on neurons and can induce severe seizures in infant rats (Baram et al., 1992). CRH receptors are found in a number of brain regions that are particularly prone to generating seizures, such as the hippocampus and amygdala. Most intriguingly, the expression of CRH receptors is developmentally regulated, with high levels of expression during infancy, which could explain the temporal selectivity of spasms to occur in this developmental period (Avishai-Eliner et al., 1996). Finally, ACTH can suppress CRH production in the brain, which could be the basis for the efficacy of ACTH as a treatment for spasms (Brunson et al., 2001b). Thus, activation and release of CRH within the brain in response to pathophysiological stress may account for many of the unique clinical features of infantile spasms, including the developmental specificity, the propensity for diverse etiologies to cause spasms, and the responsiveness to ACTH.

Despite the attractiveness of the CRH hypothesis, definitive evidence for a critical role of CRH in spasms is lacking, and the involvement of other mechanisms has not been ruled out. For a long time, one problem with performing mechanistic studies on infantile spasms has been a lack of suitable animal models for spasms, despite extensive discussion and recognition of the need for such models (Stafstrom et al., 2006). For example, although CRH induces robust seizures in infant rats (Baram et al., 1992), the behavioral and electrographic features of these seizures do not resemble spasms in people. Fortunately, significant progress has been made in recent years in developing and characterizing animal models that more closely mimic the semiologic properties of infantile spasms. Several rodent models of spasms have emerged almost simultaneously that involve different modes of "injury" to induce the spasms (Cortez et al., 2009; Lee et al., 2008; Marsh et al., 2009; Price et al., 2009; Velisek et al., 2007). These newer animal models are encouraging in that they not only evoke stereotypical behavioral seizures that resemble flexor or extensor spasms but also reproduce characteristic EEG features, including ictal electrodecremental patterns and, in some cases, an interictal hypsarrhythmia-like background. Furthermore, the multiplicity of inciting stimuli that results in a similar spasm-like phenotype, including various pharmacological and genetic manipulations, parallels the situation in people and supports the concept that spasms may represent a final common pathway in response to brain injury during early brain development. Finally, although evidence for unifying pathophysiological mechanisms for spasms has yet to be derived from these models, one model does lend some support to the CRH hypothesis. Administering the glutamate agonist N-methyl-D-aspartate by itself induces spasms in infant/juvenile rats, but prenatal exposure to the glucocorticoid betamethasone to mimic pathophysiological stress increases the sensitivity to subsequent NMDA exposure and, more interestingly, renders the spasms sensitive to ACTH therapy (Velisek et al., 2007). Future work with these and other animal models will undoubtedly provide more definitive evidence for underlying pathophysiological mechanisms of spasms and potentially lead to more effective therapeutic approaches to spasms.

## IDIOPATHIC EPILEPSY SYNDROMES OF CHILDHOOD

The idiopathic epilepsy syndromes of childhood include an assorted array of disorders that have characteristic ages of onset ranging from infancy (benign familial neonatal convulsions, or BFNCs) to adolescence (juvenile myoclonic epilepsy, or JME), particular seizure types, and specific EEG abnormalities, and they are not associated with other symptomatic neurologic etiologies. The etiology of idiopathic epilepsy syndromes has long been presumed to be genetic based on familial cases. In most cases, it appears that the inheritance pattern is not a simple Mendelian but rather a sporadic or complex inheritance pattern.

Over the past decade and a half, specific genes causing familial forms of different idiopathic epilepsy syndromes have been identified. While at this point these genes have only clearly been demonstrated to be the etiology in these rare familial cases, it is reasonable to assume that they will also contribute to our understanding of the pathophysiology of other, nonfamilial cases of idiopathic epilepsy. Interestingly, the vast majority of genetic mutations that have been identified involve genes for ion channels. In addition to epilepsy, channelopathies include a variety of disorders such as congenital myasthenia gravis, hyperekplexia, paramyotonia congenita, periodic paralysis, episodic ataxia, and familial hemiplegic migraine. Notably, all of these disorders are paroxysmal neurologic conditions.

Given the role of ion channels in controlling neuronal excitability, it is easy to hypothesize putative pathophysiological mechanisms for many of the ion channel mutations identified in idiopathic epilepsy. For example, mutations in potassium channels or ionotropic GABA receptors, which mediate membrane hyperpolarization and inhibitory postsynaptic potentials, could directly translate into a hyperexcitable state. Mutations in channels such as voltage-gated sodium channels, which are traditionally viewed as excitatory on the cellular level, might cause increased excitability by decreasing inactivation of the channel or predominantly affecting channels on inhibitory neurons.

On the other hand, although at first glance there is a mechanistic simplicity and logic to channelopathies causing epilepsy, further examination of this concept raises many questions. How does a genetic channel defect that is chronically present lead to brief, intermittent clinical symptoms? How do mutations in the same gene result in different clinical syndromes or different severities of the same syndrome? Conversely, how do mutations in different genes result in the same clinical syndrome? From a developmental perspective, how do these mutations tend to result in a characteristic age of onset and, in some cases, age of offset?

## TABLE 7.2
## Genes Identified in Pediatric Epilepsy Syndromes

| Gene | Protein | Function | Epilepsy Syndrome |
|---|---|---|---|
| **Neonatal/infantile seizures** | | | |
| KCNQ2/KCNQ3 | M-type potassium channel | Membrane hyperpolarization | Benign familial neonatal convulsions |
| SCN2A | $\alpha_2$ subunit of sodium channel | Action potential, depolarization | Benign familial neonatal/infantile convulsions |
| ATP1A2 | $\alpha_2$ subunit of Na$^+$–K$^+$–ATPase | Maintains electrochemical gradient | Benign familial infantile convulsions; familial hemiplegic migraine |
| **Febrile seizures and associated epilepsy syndromes** | | | |
| SCN1A | $\alpha$ subunit of sodium channel | Action potential, depolarization | GEFS+, SMEI/Dravet |
| SCN1B | $\beta_1$ ancillary subunit of sodium channel | Modulates Na channel gating, regulates level of channel expression | GEFS+ |
| GABRG2 | $\gamma_2$ subunit of GABA$_A$ receptor | Inhibitory postsynaptic potential | Febrile seizures/GEFS+, SMEI/Dravet |
| GABRD | $\delta$ subunit of GABA$_A$ receptor | Inhibitory postsynaptic potential | GEFS+ |
| **Idiopathic generalized epilepsies of childhood and adolescence** | | | |
| CACNA1H | Low threshold, T-type calcium channel | Calcium influx, depolarization | CAE; other IGEs |
| CACNB4 | $\beta_4$ subunit of voltage-gated calcium channels | Calcium influx, depolarization | JME |
| GABRA1 | $\alpha_1$ subunit of GABA$_A$ receptor | Inhibitory postsynaptic potential | JME, CAE |
| GABRG2 | $\gamma_2$ subunit of GABA$_A$ receptor | Inhibitory postsynaptic potential | CAE |
| GABRD | $\delta$ subunit of GABA$_A$ receptor | Inhibitory postsynaptic potential | JME |
| CLCN2 | Voltage-gated chloride channel | Membrane hyperpolarization | Multiple IGEs |
| HCN1 and HCN2 | Hyperpolarization-activated cyclic nucleotide-gated channels | Contribute to membrane resting potential | IGE |
| EFHC1 | EFHC1 protein | Apoptosis, calcium channel modulator | JME |
| **Idiopathic partial epilepsies** | | | |
| ELP4 | Elongator protein complex | Transcription, cell migration | Benign Rolandic epilepsy |

*Abbreviations:* CAE, childhood absence epilepsy; GEFS+, generalized epilepsy with febrile seizures plus; IGE, idiopathic generalized epilepsy; JME, juvenile myoclonic epilepsy; SMEI, severe myoclonic epilepsy of infancy. See text for specific references.

Although these various questions about channelopathies in familial Mendelian cases of idiopathic epilepsy remain a challenge, it should also be emphasized that the etiology of the vast majority of idiopathic epilepsy syndromes is likely much more complex than a single gene defect. Furthermore, a number of mutations in idiopathic epilepsies have also been identified that are not associated with ion channels. Most recently, a mutation in elongator protein complex 4 (ELP4), which regulates gene transcription and cell migration, has been associated with Rolandic epilepsy (Strug et al., 2009). Table 7.2 summarizes identified genes and associated protein functions in cases of idiopathic epilepsies of childhood, including both channel- and non-channel-related genes. A number of comprehensive reviews (Reid et al., 2009; Weber and Lerche, 2008) and other chapters in this book describe the genetics and pathophysiological mechanisms of channelopathies and other idiopathic epilepsies in more detail. The remainder of this section focuses on pediatric idiopathic epilepsy syndromes and emphasizes developmental issues contributing to their pathophysiology.

## BENIGN FAMILIAL NEONATAL CONVULSIONS

Benign familial neonatal convulsions (BFNCs) is a rare autosomal dominant idiopathic epilepsy that presents in the neonatal period. The typical presentation is the onset of brief multifocal clonic or tonic seizures in the first week of life in the absence of any other significant neurological deficits or underlying etiologies. Seizures usually spontaneously resolve by 2 months of age, and neurocognitive outcome is typically normal, although there is a slightly increased risk for developing epilepsy later in life.

To date, mutations in two genes encoding the voltage-gated potassium channel subunits KCNQ2 and KCNQ3 have been identified as being causative in BFNCs (Reid et al., 2009). Heteromeric assembly of KCNQ2 and KCNQ3 gene products leads to the formation of a particular type of potassium channel known as the M-channel (Wang et al., 1998). The resultant M-current is a slow activating and noninactivating potassium current that tends to limit repetitive firing of action potentials. Multiple types of mutations in the *KCNQ2* and *KCNQ3* genes have been reported that either decrease total current amplitude or alter gating properties (Reid et al., 2009).

Interestingly, a number of potential explanations as to why these mutations result in a transient epilepsy syndrome early in development have been proposed. First, there is developmental variation in *KCNQ2* and *KCNQ3* expression, with a peak in the neonatal period that then declines into adulthood (Kanaumi et al., 2008). An even more intriguing explanation, however, is that seizure remission coincides with the switch from GABA being primarily excitatory in the neonatal brain to inhibitory, as discussed above. This would suggest that early in development inhibition based on $K^+$ channels may be more significant than GABA-mediated inhibition; therefore, loss-of-function potassium channel mutations would be more likely to result in a hyperexcitable phase during the neonatal period. This theory has been demonstrated in at least one rodent model (Okada et al., 2003).

Excitingly, a translatable therapy is already under development related to this specific potassium channel. Retigabine is a novel anticonvulsant drug that acts by activating potassium currents mediated by *KCNQ2* and *KCNQ3*. This medication might be expected to be effective for BFNC and has been demonstrated in a number of Phase 2 trials to be efficacious in partial-onset seizures in adults (Rogawski and Bazil, 2008).

## GENERALIZED EPILEPSY WITH FEBRILE SEIZURES PLUS

No discussion of developmentally related epilepsies is complete without including the topics of febrile seizures and sodium channel mutations—the first because febrile seizures are so common and the second because there has been an explosion of knowledge about sodium channel mutations in the past few years. Febrile seizures affect approximately 3% of all children. Traditionally, they are not viewed as an epilepsy syndrome, as the majority of children outgrow them by age 5. Further, they are typically not associated with significant morbidity, acutely or chronically. Although febrile seizures most commonly occur in isolation, recently a condition known as generalized epilepsy with febrile seizures plus (GEFS+) has been described (Scheffer and Berkovic, 1997). GEFS+ is an idiopathic epilepsy syndrome that is inherited in an autosomal dominant fashion but has incomplete penetrance. Characteristically, patients with GEFS+ present with febrile seizures in infancy or early childhood but later develop epilepsy. They usually have generalized epilepsy with various seizure types, such as generalized tonic–clonic, absence, or myoclonic seizures. In addition, in large families with GEFS+, different family members may have a mixture of febrile seizures and other seizure types. Interestingly, it is now apparent, based on genetic studies identifying overlapping sodium channel mutations, that Dravet syndrome, also known as severe myoclonic epilepsy of infancy (SMEI), likely represents an extreme form along the spectrum of GEFS+. Furthermore, of considerable clinical and medicolegal interest, genetic evidence suggests that many children with alleged vaccine-induced encephalopathy and seizures actually have underlying SMEI due to sodium channel mutations (Berkovic et al., 2006).

A unifying feature of the spectrum from febrile seizure to GEFS+ to SMEI is the role of hyperthermia in triggering seizures. The fact that febrile seizures are so common and that the vast majority are sporadic would suggest that there are both environmental and genetic components to the pathophysiology. Multiple putative environmental triggers for febrile seizures have been reported, including direct effects of hyperthermia on the brain, increased interleukin-1$\beta$ in the setting of fever, and hyperthermia-induced hyperventilation resulting in alkalosis (Dube et al., 2009).

From a genetic perspective, a number of genes have been identified and associated with GEFS+. The first was in an accessory $\beta$ subunit of the sodium channel (*SCN1B*) (Wallace et al., 1998). In addition, a number of mutations have been identified in GABA$_A$ receptor subunits $\gamma_2$ and $\delta$ (Reid et al., 2009). However, probably the biggest focus of attention has been on the *SCN1A* gene, which encodes the $\alpha_1$ subunit of fast-inactivating voltage-dependent Na$^+$ channels. Interestingly, one of the mutations in GABA$_A$ receptor subunit $\gamma_2$ and one in *SCN1A* have been demonstrated to result in trafficking deficits that are temperature sensitive, suggesting a potential mechanism by which a genetic defect could result in sensitivity to hyperthermia (Kang et al., 2006; Rusconi et al., 2007).

As remarked above, there has been a tremendous focus on the *SCN1A* gene recently. Both loss- and gain-of-function mutations have been identified in the *SCN1A* gene associated with the full spectrum of GEFS+ to Dravet syndrome. Loss-of-function mutations have more consistently been associated with the severe end of the spectrum. It is not difficult to hypothesize as to how gain-of-function mutations in excitatory sodium channels might result in hyperexcitability, and this has been confirmed in functional studies of several *SCN1A* mutations demonstrating reduced inactivation of sodium currents (Lossin et al., 2002). Less obvious are the underlying mechanisms by which a loss-of-function mutation in *SCN1A* might result in seizures.

Animal models have contributed to our understanding of putative mechanisms for a loss-of-function sodium channel mutation resulting in hyperexcitability. One model, a heterozygous knock-in mouse with a mutation also found in humans, has been studied at an ultrastructural level (Ogiwara et al., 2007). In this study, it was demonstrated that the SCN1A protein deficit was isolated to a subpopulation of inhibitory interneurons in the developing neocortex. A loss of inhibitory neuron activity within a network could certainly result in hyperexcitability and may also explain the role of mutations in GABA receptors in GEFS+. Further, the fact that this deficit was expressed in developing neocortex may explain the early age of onset of this syndrome; however, this has yet to be specifically demonstrated. It should be noted that it was recently demonstrated that the presence of an *SCN1A* gene mutation in humans is associated with an earlier age of onset of febrile seizures in GEFS+ (Sijben et al., 2009). This reinforces the suspicion that a developmental regulation of *SCN1A* expression likely contributes to the age of onset of clinical symptoms in GEFS+ and other related seizure syndromes. Much work remains to be done to demonstrate the mechanism for this and to ultimately translate it into new therapies.

## CHILDHOOD-ONSET IDIOPATHIC GENERALIZED EPILEPSIES

Until this point, this chapter has primarily focused on epilepsies that present in the neonatal or infant period; however, a group of idiopathic generalized epilepsies (IGEs) presenting in childhood or adolescence, including childhood absence epilepsy (CAE), juvenile absence epilepsy (JAE), and juvenile myoclonic epilepsy (JME), are among the most characteristic pediatric epilepsy syndromes. CAE and JAE are characterized primarily by the onset of absence seizures in childhood, although generalized tonic–clonic seizures can also develop. The classic 3-Hz generalized spike-and-wave pattern is typically seen on EEG. The response to treatment, prognosis for remission of seizures, and neurocognitive outcome are generally good. JME is characterized by myoclonic seizures, but generalized tonic–clonic and absence seizures are also common. EEG typically shows 4- to 6-Hz generalized polyspike and wave. Spontaneous remission of JME is rare, but seizures typically respond well to anticonvulsant drugs.

As outlined in Table 7.2, multiple genes have been associated with many types of IGE, including the CAE and JME subtypes. These genes express voltage-gated channels, ligand-gated ion channels, and nonchannel proteins, demonstrating that there is no single mechanism resulting in the expression of the IGE phenotype. One gene of particular interest is *CACNA1H*, which encodes a low-threshold, T-type $Ca^{2+}$ channel. An extensive number of mutations in this gene have been associated with CAE (Chen et al., 2003).

There are strong arguments for the role of T-type $Ca^{2+}$ channels in CAE; for example, 3-Hz spike-and-wave discharges involve inappropriate oscillations of thalamocortical networks (Crunelli and Lersche, 2002). The exact site of origin and networks by which absence seizures are generated remains quite controversial. A longstanding theory has implicated a subcortical focus in the thalamus, potentially the reticular nucleus, acting as a pacemaker and resulting in a thalamo–cortico–thalamic network (van Luitjelaar and Stinikova, 2006). Alternatively, a "cortical focus" theory has also been proposed. In 2002, a study in epileptic rats demonstrated that an intact thalamocortical network is essential for the generation of spike–wave discharges (Mereen et al., 2002). Moreover, in this model, there was a consistent focus in the perioral region of the somatosensory cortex that generalized rapidly over the cortex and thereafter drove the thalamus in the thalamo–cortico–thalamic circuit. Multiple studies using EEG–functional magnetic resonance imaging (fMRI) are now underway in an attempt to further clarify the potential mechanism by which generalized seizures, and specifically absence seizures, are generated (Moeller et al., 2008).

Regardless of the primary site of origin triggering absence seizures, T-type $Ca^{2+}$ channels have been clearly demonstrated to mediate low-threshold spikes, which trigger bursts and are involved in regulating neuronal excitability in the thalamus, where they are highly expressed (Perez-Reyes, 2003). It has been shown in multiple rodent models of absence epilepsy that there is an increase in expression and function of T-type $Ca^{2+}$ channels (Reid et al., 2009). One animal model of particular interest is the knockout mouse for *CACNA1G*, another T-type $Ca^{2+}$ channel isoform. This mouse model is resistant to absence seizures (Kim et al., 2001). Conversely, another model involving a transgenic mouse with enhanced expression of *CACNA1G* exhibited frequent spontaneous absence seizures (Ernst et al., 2009).

Returning to the *CACNA1H* gene specifically, it is clear that the pathophysiology of mutations is not as simple as loss or gain of function. It encodes multiple alternative splice variants of the calcium channel. There are 12 to 14 alternate splice sites, with the resultant potential to create over 4000 alternative mRNA sequences (Zhong et al., 2006). A number of these variants were confirmed to encode functional channels with variable gating properties. A number of mutations linked to CAE and IGE are near regions of the gene associated with variable splicing. Quite recently, a point mutation has been identified and confirmed to have a splice-variant effect in a rat model (Powell et al., 2009).

Certainly, many questions remain to be answered regarding the pathophysiology of the IGEs and even regarding the *CACNA1H* gene specifically. For example, does some age-dependent factor affect the likelihood of a specific variant being spliced or expressed? Or, is there an age-dependent variation in the voltage threshold for activation of the channel? Similarly, what molecular changes account for the typical remission of CAE by adolescence? Most importantly, how can this mechanistic information be used to develop new treatments? In any case, it happens that one targeted therapy already exists. Ethosuximide, a standard treatment for absence seizures, blocks T-type $Ca^{2+}$ channels (Coulter et al., 1989). Similar to the case of retigabine potentiating potassium channels involved in BFNC, it is likely that future pharmacological and molecular genetic research will identify additional therapies that may specifically target or correct the underlying pathophysiological mechanisms causing IGEs of childhood.

## MALFORMATIONS OF CORTICAL DEVELOPMENT

Impressive progress has been made in determining the molecular basis of idiopathic epilepsy syndromes of childhood, and equally exciting advances have also occurred in elucidating the pathophysiological basis of symptomatic epilepsies typically presenting in the pediatric years. Although symptomatic etiologies causing seizures in childhood are as diverse as the types of epilepsies and epilepsy syndromes themselves, recently malformations of cortical development (MCDs) have emerged as a major category of disorders associated with pediatric epilepsy. With advances in molecular, genetic, pathological, and especially brain imaging methods, MCDs have been increasingly recognized as a cause for epilepsy presenting in infancy and childhood and can be especially problematic due to the frequency of medical intractability of the epilepsy. In this section, recent insights into mechanisms of epileptogenesis in some MCDs and corresponding clinical implications for new therapeutic strategies for epilepsy are reviewed.

From a pathological and radiographic standpoint, MCDs constitute a diverse group of developmental disorders that can result from a defect or insult occurring at any stage of normal cortical development. Although the terminology and classification of MCDs have often been confusing, a major classification scheme has recently been proposed based on the stage of cortical development that is primarily disrupted, including cellular proliferation and differentiation, neuronal migration, and cortical organization (Barkovich et al., 2005). During the proliferative stage, immature cortical neurons and glia are primarily generated from embryonic progenitor cells in the germinal ventricular zone or subcortical ganglionic eminences. During the neuronal migration stage, neurons produced during the proliferative stage migrate either radially along radial glia or tangentially to their final location in the cortex. During the cortical organization stage, cortical neurons develop mature dendrites and axons and form synaptic connections. Disruptions of cortical development at discrete stages are believed to result in specific types of MCDs, such as focal cortical dysplasia at the proliferative stage, heterotopias at the migration stage, and lissencephaly at the cortical organization stage.

Most types of MCDs can cause epilepsy presenting in the pediatric years but likely involve different, diverse mechanisms of epileptogenesis. Most notably, the subgroup of MCDs attributed to defects in the proliferative stage of cortical development, including focal cortical dysplasia with balloon cells, tuberous sclerosis complex, hemimegencephaly, and ganglioglioma, has a particularly strong association with epilepsy, especially medically intractable epilepsy. These MCDs exhibit abnormalities that are consistent with a primary defect in proliferation of neuroglial progenitor cells and may also suggest an overlapping pathogenesis for epilepsy (Wong, 2008). In particular, substantial progress has recently been made in tuberous sclerosis complex (TSC) and focal cortical dysplasia (FCD) in identifying potential mechanisms of epileptogenesis and corresponding novel therapeutic approaches for epilepsy.

Both TSC and FCD with balloon cells (FCD type IIB, per the pathological classification of Palimini et al., 2004) and, to a lesser degree, hemimegencephaly and ganglioglioma share a number of histopathological, cellular, and molecular features that indicate primary defects in neuroglial proliferation and differentiation and may also promote epileptogenesis (Crino, 2007; Wong, 2008). Pathological specimens of FCD and tubers of TSC demonstrate a loss of normal cortical lamination, astrogliosis, and an array of abnormal, dysmorphic cell types that have predominantly neuronal or glial features (Mizuguchi and Takashima, 2001; Palmini et al., 2004; Talos et al., 2008). In addition, distinctive cytomegalic cells, including balloon cells of FCD and giant cells of TSC, appear to be poorly differentiated cells with immature neuroglial markers. Most of these cellular abnormalities can likely be attributed to defects in specific molecular signaling pathways that control cell growth, proliferation, and differentiation. In particular, the mammalian target of rapamycin (mTOR) pathway represents a major signaling pathway that regulates protein synthesis related to cell growth and proliferation (Sandsmark et al., 2007). It was recently discovered that the two genes that cause TSC normally inhibit mTOR; thus, a mutation of one of the *TSC* genes leads to hyperactivation

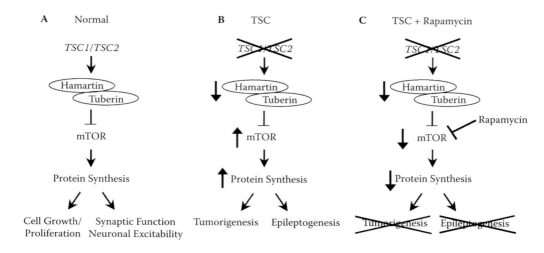

**FIGURE 7.1**    The role of mTOR signaling in tumorigenesis and epileptogenesis in tuberous sclerosis complex (TSC). (A) The *TSC1* and *TSC2* genes encode the proteins hamartin and tuberin, respectively, which form a complex that normally inhibits the mammalian target of rapamycin (mTOR) protein. mTOR functions to activate a cascade of other kinases and proteins (not shown) that regulates protein synthesis, which is ultimately involved in a variety of functions, such as cell growth and proliferation, synaptic transmission, and neuronal excitability. Thus, the *TSC* genes normally limit excessive cell growth, proliferation, and excitability. (B) In the disease of TSC, mutation of one of the *TSC* genes results in a deficiency of the hamartin–tuberin complex, causing disinhibition or hyperactivation of the mTOR pathway and resultant tumorigenesis and epileptogenesis. (C) Rapamycin, by inhibiting mTOR, may be able to counteract the effects of *TSC* gene mutations, thus potentially preventing tumor formation and epileptogenesis.

of the mTOR pathway and a resulting increase in cell size and proliferation (Figure 7.1A,B). This molecular signaling defect could account for many of the neuropathological findings in tubers of TSC, as well as explain the tendency to form tumors in various organs in this disease. Interestingly, patients with FCD, as well as ganglioglioma, can also exhibit polymorphisms in the *TSC* genes and abnormalities in the mTOR pathway, suggesting that mechanisms similar to those in TSC may be operational in these other focal cortical malformations, at least in a more limited fashion in the brain (Becker et al., 2006).

Whereas aberrant mTOR signaling represents a rational mechanism for causing the cellular and pathological abnormalities of TSC and possibly other focal cortical malformations, abnormal mTOR activation might also directly contribute to mechanisms of epileptogenesis in these MCDs. Besides regulating cell growth and proliferation, mTOR has also been implicated in controlling neuronal excitability and synaptic function. Furthermore, epileptogenesis often involves activation of a number of cellular and molecular processes that depend on protein synthesis, such as the production of neurotransmitter receptors and ion channels. Many of these protein-synthesis-dependent processes have been implicated in contributing to epileptogenesis in TSC and FCD, such as alterations in glutamate or GABA receptors and other neurotransmitter-related proteins in both neurons and astrocytes (Wong, 2008). Thus, in addition to effects on cell growth and proliferation that may indirectly influence neuronal excitability and seizures, the mTOR pathway could mediate other epileptogenic mechanisms due to alterations in the synthesis of relevant proteins (Figure 7.1A,B).

The potential clinical and translational applications of these basic mechanistic findings may turn out to be very significant. mTOR is a protein kinase that is specifically inhibited by the drug rapamycin, approved and used clinically as an immunosuppressant. The availability of rapamycin provides the opportunity to test the involvement of mTOR in epileptogenesis and to consider it as a potential novel therapy for epilepsy in TSC and the related focal MCDs (Figure 7.1C). In various animal models of TSC, rapamycin has been shown to prevent or reverse histological, cellular, and molecular brain abnormalities, such as neuronal cytomegaly, astrocyte proliferation, and abnormal glutamate transporter expression (Meikle et al., 2008; Zeng et al., 2008). Furthermore, rapamycin improves a number of behavioral and functional measures in these rodent TSC models, including learning deficits (Ehninger et al., 2008). Finally, rapamycin prevents the development of epilepsy in young mice that have not yet experienced seizures and substantially decreases seizure frequency in older mice that already have epilepsy (Zeng et al., 2008). Overall, these novel findings support an antiepileptogenic action of rapamycin in TSC. By comparison, currently available seizure medications are believed to suppress seizures symptomatically but do not seem to correct the underlying brain abnormalities causing epileptogenesis or to alter the natural history of epilepsy. In contrast, rapamycin appears to prevent epileptogenesis by reversing the underlying neuropathological, cellular, and molecular abnormalities, at least in rodent models. mTOR inhibitors already have been shown to have efficacy for tumor growth in TSC patients (Bissler et al., 2008) and will likely also be tested for epilepsy and other neurological measures in future clinical trials. If rapamycin is shown to be effective for epilepsy in TSC, the utility of mTOR inhibitors for epilepsy due to other MCDs with aberrant mTOR signaling could also be considered. If validated, this type of therapeutic approach to developmental epilepsies would represent a significant advance in not just simply treating the symptoms (seizures) but potentially correcting the underlying developmental brain defects.

## CONCLUSIONS

In this chapter, we have summarized the current understanding of the complex pathophysiology of epilepsies presenting throughout infancy and childhood. In recent years, there has been an explosion of knowledge and insights into maturational changes and developmental processes that influence the expression of epilepsy in the pediatric age group. This knowledge has revealed that the underlying mechanisms involve all aspects and levels of central nervous system functions from genetic mutations, neurotransmitter receptors, and ion transporters/channels to cellular trafficking and regulation of protein synthesis, ultimately affecting the balance between inhibitory and excitatory signals at a neuronal and network level. Excitingly, although there is much to be learned, translatable therapies are beginning to emerge. With perhaps just one exception, however, these treatments are still symptomatic. Ultimately, a greater understanding of the pathophysiology of developmental epilepsies may result in neuroprotective and antiepileptogenic strategies that prevent or cure epilepsy.

## REFERENCES

Avishai-Eliner, S., S.J. Yi, T.Z. Baram. (1996). Developmental profile of messenger RNA for the corticotrophin-releasing hormone receptor in the rat limbic system. *Dev. Brain Res.*, 91:159–163.

Baram, T.Z., E. Hirsch, O.C. Snead III, L. Schultz. (1992). Corticotropin-releasing hormone-induced seizures in infant rats originate in the amygdala. *Ann. Neurol.*, 31:231–236.

Barkovich, A.J., R.I. Kuzniecky, G.D. Jackson, R. Guerrini, W.B. Dobyns. (2005). A developmental and genetic classification for malformations of cortical development. *Neurology*, 65:1873–1887.

Becker, A.J., I. Blumcke, H. Urbach, V. Hans, M. Majores. (2006). Molecular neuropathology of epilepsy-associated glioneuronal malformations. *J. Neuropathol. Exp. Neurol.*, 65:99–108.

Ben-Ari, Y. (2002). Excitatory actions of GABA during development: the nature of the nurture. *Nat. Rev. Neurosci.*, 3:728–739.

Berkovic, S.F., L. Harkin, J.M. McMahon, J.T. Pelekanos, S.M. Zuberi, E.C. Wirrell, D.S. Gill, X. Iona, J.C. Mulley, I.E. Scheffer. (2006). *De novo* mutations of the sodium channel gene SCNA1A in alleged vaccine encephalopathy: a retrospective study. *Lancet Neurol.*, 5:488–492.

Bissler, J.J., F.X. McCormack, L.R. Young, J.M. Elwing, G. Chuck, J.M. Leonard, V.J. Schmithorst, T. Laor, A.S. Brody, J. Bean, S. Salisbury, D.N. Franz. (2008). Sirolimus for angiomyolipoma in tuberous sclerosis complex or lymphangioleiomyomatosis. *N. Engl. J. Med.*, 358:140–141.

Bittigau, P., M. Sifringer, K. Genz, E. Reith, D. Pospischil, S. Govindarajalu, M. Dzietko, S. Pesditschek, I. Mai, K. Dikranian, J.W. Olney, C. Ikonomidou. (2002). Antiepileptic drugs and apoptotic neurodegeneration in the developing brain. *Proc. Natl. Acad. Sci. U.S.A.*, 99:15089–15094.

Bittigau, P., M. Sifringer, C. Ikonomidou. (2003). Antiepileptic drugs and apoptosis in the developing brain. *Ann. N.Y. Acad. Sci.*, 993:103–114.

Brunson, K.L., M. Eghbal-Ahmadi, T.Z. Baram. (2001a). How do the many etiologies of West syndrome lead to excitability and seizures? The corticotrophin releasing hormone excess hypothesis. *Brain Dev.*, 23:533–538.

Brunson, K.L., N. Khan, M. Eghbal-Ahmadi, T.Z. Baram. (2001b). ACTH acts directly on amygdala neurons to down-regulate corticotrophin releasing hormone gene expression. *Ann. Neurol.*, 29:304–312.

Chen, Y., J. Lu, H. Pan, Y. Zhang, H. Wu, K. Xu, X. Liu, Y. Jiang, X. Bao, Z. Yao, K. Ding, W.H. Lo, B. Qiang, P. Chan, Y. Shen, X. Wu. (2003). Association between genetic variation of *CACNA1H* and childhood absence epilepsy. *Ann. Neurol.*, 54:239–243.

Cherubini, E, C. Rovira, J.L. Gairasa, R. Corradetti, Y. Ben Ari. (1990). GABA mediated excitation in immature rat CA3 hippocampal neurons. *Int. J. Dev. Neurosci.*, 8:481–490.

Clayton, G.H., G.C. Owens, J.S. Wolff, R.L. Smith. (1998). Ontogeny of cation-Cl- cotransporter expression in rat neocortex. *Dev. Brain Res.*, 109:281–292.

Cortez, M.A., L. Shen, Y. Wu, I.S. Aleem, C.H. Trepanier, H.R. Sadeghnia, A. Ashraf, A. Kanawaty, C.C. Liu, L. Stewart, O.C. Snead 3rd. (2009). Infantile spasms and Down syndrome: a new animal model. *Pediatr. Res.*, 65: 499–503.

Coulter, D.A., J.R. Hueguenard, D.A. Prince. (1989). Characterization of ethosuximide reduction of low-threshold calcium current in thalamic neurons. *Ann. Neurol.*, 25: 582–593.

Crino, P.B. (2007). Focal brain malformations: a spectrum of disorders along the mTOR cascade. *Novartis Found. Symp.*, 288:260–272.

Crunelli, V., N. Leresche. (2002). Childhood absence epilepsy: genes, channels, neurons and networks. *Nat. Rev. Neurosci.*, 3:371–382.

Dube, C.M., A.L. Brewster, T.Z. Baram. (2009). Febrile seizures: mechanisms and relationship to epilepsy. *Brain Dev.*, 31:366–371.

Dzhala, V.I., K.J. Staley. (2003). Excitatory actions of endogenous released GABA contribute to initiation of ictal epileptiform activity in the developing hippocampus. *J. Neurosci.*, 23:1840–1846.

Dzhala V.I., D.M. Talos, D.A. Sdrulaa, A.C. Brumback, G.C. Mathews, T.A. Benke, E. Delpire, F.E. Jensen, K.J. Staley. (2005). NKCC1 transporter facilitates seizures in the developing brain. *Nat. Med.*, 11:1205–1213.

Dzhala, V., A.C. Brumback, K.J. Staley. (2008). Bumetanide enhances phenobarbital efficacy in a neonatal seizure model. *Ann. Neurol.*, 63:222–235.

Ehninger, D., S. Han, C. Shilyansky, Y. Zhou, W. Li, D.J. Kwiatkowski, V. Ramesh, A.J. Silva. (2008). Reversal of learning deficits in a Tsc2+/− mouse model of tuberous sclerosis. *Nat. Med.*, 14:843–848.

Ernst, W.L., Y. Zhang, J.W. Yoo, S.J. Ernst, J.L. Noebels. (2009). Genetic enhancement of thalamocortical network activity by elevating $\alpha_1$G-mediated low-voltage-activated calcium current induces pure absence epilepsy. *J. Neurosci.*, 29:1615–1625.

Flint, A.C., U.S. Maisch, J.H. Weishaupt, A.R. Kriegstein, H. Monyer. (1997). NR2A subunit expression shortens NMDA receptor synaptic currents in developing neocortex. *J. Neurosci.*, 17:2469–2476.

Galanopoulou, A.S. (2005). GABA-A receptors as broadcasters of sexually differentiating signals in the brain. *Epilepsia*, 46(Suppl. 5):107–112.

Galanopoulou, A.S. (2008). Dissociated gender-specific effects of recurrent seizures on GABA signaling in CA1 pyramidal neurons: role of GABA$_A$ receptors. *J. Neurosci.*, 28:1557–1567.

Hannaert, P., M. Alvarez-Guerra, D. Pirot, C. Nazaret, R.P. Garay. (2002). Rat NKCC2/NKCC1 cotransporter selectivity for loop diuretic drugs. *NaunynSchmiedbergs Arch. Pharmacol.*, 365:193–199.

Hauser, W.A., J.R. Annegers, L.T. Kurland. (1993). Incidence of epilepsy and unprovoked seizures in Rochester, Minnesota, 1935–1984. *Epilepsia*, 34:453–468

Hollmann, M., S. Heinemann. (1994). Cloned glutamate receptors. *Annu. Rev. Neurosci.*, 17:31–108.

Hua, J.Y., S.J. Smith. (2004). Neural activity and the dynamics of central nervous system development. *Nat. Neurosci.*, 7:327–332.

Huttenlocher, P.R., C. deCourten, L.J. Garey, H. Van der Loos. (1982). Synaptogenesis in human visual cortex: evidence of synapse elimination during normal development. *Neurosci. Lett.*, 33:247–252.

Insel, T.R., L.P. Miller, R.E. Gelhard. (1990). The ontogeny of excitatory amino acid receptors in rat forebrain 1,*N*-methyl-D-aspartate and quosqualate receptors. *Neuroscience*, 35:31–43.

Issenting P., S.C. Jacoby, J.A. Payne, B. Forbush 3rd. (1998). Comparison of Na–K–Cl cotransporters NKCC1, NKCC2, and the HEK cell Na–L–Cl cotransporter. *J. Biol. Chem.*, 273:11295–11301.

Kanaumi, T., S. Takashima, H. Iwasaki, M. Itoh, A. Mitsudome, S. Hirose. (2008). Developmental changes in *KCNQ2* and *KCNQ3* expression in human brain: possible contribution to the age-dependent etiology of benign familial neonatal convulsions. *Brain Dev.*, 30:362–369.

Kang, J.Q., W. Shen, R.L. Macdonald. (2006). Why does fever trigger febrile seizures? GABA$_A$ receptor gamma2 subunit mutations associated with idiopathic generalized epilepsies have temperature-dependent trafficking deficiencies. *J. Neurosci.*, 26:2590–2597.

Khazipov R., I. Khalilov, R. Tyzio, E. Morozova, Y. Ben-Ari, G.L. Holmes. (2004). Developmental changes in GABAergic actions and seizure susceptibility in the rat hippocampus. *Eur. J. Neurosci.*, 19:590–600.

Kim, D., I. Song, S. Keum, T. Lee, M.J. Jeong, S.S. Kim, M.W. McEnery, H.S. Shin. (2001). Lack of burst firing of thalamocortical relay neurons and resistance to absence seizures in mice lacking alpha (1G) T-type Ca(2+) channels. *Neuron*, 31:35–45.

Kumar, S.S., A. Bacci, V. Kharazia, J.R. Huguenard. (2002). A developmental switch of AMPA receptor subunits in neocortical pyramidal neurons. *J. Neurosci.*, 22:3005–3015.

Lanska, M.J., D.J. Lanska, R.J. Baumann, R.J. Kryscio. (1995). A population-based study of neonatal seizures in Fayette County, Kentucky. *Neurology*, 45:724–732.

Lee, C.L., J.D. Frost, J.W. Swann, R.A. Hrachovy. (2008). A new animal model of infantile spasms with unprovoked persistent seizures. *Epilepsia*, 49:298–307.

Lossin, C., D.W. Wang, T.H. Rhodes, C.G. Vanoye, A.L. George, Jr. (2002). Molecular basis of an inherited epilepsy. *Neuron*, 34:877–884.

Loturco, J.J., D.F. Owens, M.J. Heath, M.B. Davis, A.R. Kriegstein. (1995). GABA and glutamate depolarize cortical progenitor cells and inhibit DNA synthesis. *Neuron*, 15:1287–1298.

Marsh, E., C. Fulp, E. Gomez, I. Nasrallah, J. Minarcik, J. Sudi, S.L. Christina, G. Marcini, P. Labosky,W. Dobyns, A. Brooks-Kayal, J.A. Golden (2009). Targeted loss of Arx results in a developmental epilepsy mouse model and recapitulates the human phenotype in heterozygote females. *Brain*, 132:1563–1576.

McDonald, J.W., M.V. Johnston, A.B. Young. (1990). Differential ontogenic development of three receptors comprising the NMDA receptor/channel complex in the rat hippocampus. *Exp. Neurol.*, 110:237–247.

Meikle, L., K. Pollizzi, A. Egnor, I. Kramvis, H. Lane, M. Sahin, D.J. Kwiatkowski. (2008). Response of a neuronal model of tuberous sclerosis to mammalian target of rapamycin (mTOR) inhibitors: effects on mTORC1 and Akt signaling lead to improved survival and function. *J. Neurosci.*, 28:5422–5432.

Mereen, H.K., J.P. Pijn, E.L. van Luijtelaar, A.M. Coenen, F.H. Lopes da Silva. (2002). Cortical focus drives widespread corticothalamic networks during spontaneous absence seizures in rats. *J. Neurosci.*, 22:1480–1495.

Mizuguchi, M., Takashima S. (2001). Neuropathology of tuberous sclerosis. *Brain Dev.*, 23:508–515.

Moeller, F., H.R. Siebner, S. Wolff, H. Muhle, O. Granert, O. Jansen, U. Stephani, M. Siniatchkin. (2008). Simultaneous EEG–fMRI in drug-naïve children with newly diagnosed absence epilepsy. *Epilepsia*, 49:1510–1519.

Monyer, H., N. Burnashev, D.J. Laurie, B. Sakmann, P.H. Seeburg. (1994). Development and regional expression in the rat brain and functional properties of four NMDA receptors. *Neuron*, 12:529–540.

Nunez, J., M.M. McCarthy. (2007). Evidence for an extended duration of GABA-mediated excitation in the developing male versus female hippocampus. *Dev. Neurobiol.*, 67:1879–1890.

Ogiwara, I., H. Miyamoto, N. Morita, N. Atapour, E. Mazaki, I. Inoue, T. Takeuchi, S. Itohara, Y. Yanagawa, K. Obata, T. Furuichi, T.K. Hensch, K. Yamakawa. (2007). Na(v)1.1 localizes to axons of parvalbumin-positive inhibitory interneurons: a circuit basis for epileptic seizures in mice carrying an *Scn1a* gene mutation. *J. Neurosci.*, 27:5903–5914.

Okada, M., G. Zhu, S. Hirose, K.I. Ito, T. Murakami, M. Wakui, S. Kaneko. (2003). Age-dependent modulation of hippocampal excitability of KCNQ channels. *Epilepsy Res.*, 53:81–94.

Owens, D.F., L.H. Boyce, M.B. Davis, A.R. Kriegstein. (1996). Excitatory GABA responses in embryonic and neonatal cortical slices demonstrated by gramicidin perforated-patch recordings and calcium imaging. *J. Neurosci.*, 16:6414–6423.

Painter, M.J., M.S. Scher, A.D. Stein, S. Armatti, Z. Wang, J.C. Gardiner, N. Paneth, B. Minnigh, J. Alvin. (1999). Phenobarbital compared with phenytoin for the treatment of neonatal seizures. *N. Engl. J. Med.*, 341:485–489.

Palmini, A., I. Najm, G. Avanzini, T. Babb, R. Guerrini, N. Foldvary-Schaefer, G. Jackson, H.O. Lüders, R. Prayson, R. Spreafico, H.V. Vinters. (2004). Terminology and classification of the cortical dysplasias. *Neurology*, 62(Suppl. 3):S2–S8.

Pellegrini-Giampietro, D.E., M.V. Bennett, R.S. Zukin. (1991). Differential expression of three glutamate receptor genes in developing rat brain: an *in situ* hybridization study. *Proc. Natl. Acad. Sci. U.S.A.*, 88:4157–4161.

Pellegrini-Giampietro, D.E., M.V. Bennett, R.S. Zukin. (1992). Are Ca(2+)-permeable kainate/AMPA receptors more abundant in immature brain? *Neurosci. Lett.*, 144:65–69.

Perez-Reyes, E. (2003). Molecular physiology of low-voltage-activated t-type calcium channels. *Physiol. Rev.*, 83:117–161.

Pfeffer, C.K., V. Stein, D.J. Keating, H. Maier, I. Rinke, Y. Rudhard, M. Hentschke, G.M. Rune, T.J. Jentsch, C.A. Hübner. (2009). NKCC1-dependent GABAergic excitation drives synaptic network maturation during early hippocampal development. *J. Neurosci.*, 29:3419–3430.

Plotkin, M.D., E.Y. Synder, S.C. Herbert, E. Delpire. (1997). Expression of the Na–K–2Cl cotransporter is developmentally regulated in postnatal rat brains: a possible mechanism underlying GABA's excitatory role in immature brain. *J. Neurobiol.*, 33:781–795.

Pond, B.B., F.F. Galeffi, R. Ahrens, R.D. Schwartz-Bloom. (2004). Chloride transport inhibitors influence recovery from oxygen–glucose deprivation-induced cellular injury in adult hippocampus. *Neuropharmacology*, 47:253–262.

Powell, K.L., S.M. Cain, C. Ng, S. Sirdesai, L.S. David, M. Kyi, E. Garcia, J.R. Tyson, C.A. Reid, M. Bahlo, S.J. Foote, T.P. Snutch, T.J. O'Brien. (2009). A Cav3.2 T-type calcium channel point mutation has splice-variant-specific effects on function and segregates with seizure expression in a polygenic rat model of absence epilepsy. *J. Neurosci.*, 29:371–380.

Price, M.G., J.W. Yoo, D.L. Burgess, F. Deng, R.A. Hrachovy, J.D. Frost, Jr., J.L. Noebels. (2009). A triplet repeat expansion genetic mouse model of infantile spasms syndrome, Arx(GCG)10+7, with interneuronopathy, spasms in infancy, persistent seizures, and adult cognitive and behavioral impairment. *J. Neurosci.*, 29:8752–8763.

Rakic, P., J.P. Bourgeois, M.F. Eckenhoff, N. Zecevic, P.S. Goldman-Rakic. (1986). Concurrent overproduction of synapses in diverse regions of primate cortex. *Science*, 232:232–235.

Reid, C.A., S.A. Berkovic, S. Petrou. (2009). Mechanisms of human inherited epilepsies. *Prog. Neurobiol.*, 87:41–57.

Rivera, C., J. Viopio, J.A. Payne, E. Ruusuvuori, H. Lahtinen, K. Lamsa, U. Pirvola, M. Saarma, K. Kaila. (1999). The K+/Cl− co-transporter KCC2 renders GABA hyperpolarizing during neuronal maturation. *Nature*, 397:251–255.

Rogawski, M.A., C.W. Bazil. (2008). New molecular targets for antiepileptic drugs: $\alpha_2\delta$, SV2A, and $K_v7$/KCNQ/M potassium channels. *Curr. Neurol. Neurosci. Rep.*, 8:345–352.

Ronen, G.M., S. Penney, W. Andrews. (1999). The epidemiology of clinical neonatal seizures in Newfoundland: a population-based study. *J. Pediatr.*, 134:71–75.

Rusconi, R., P. Scalmani, R.R. Cassulini, G. Giunti, A. Gambardella, S. Franceschetti, G. Annesi, E. Wanke, M. Mantegazza. (2007). Modulatory proteins can rescue a trafficking defective epileptogenic Nav1.1 Na+ channel mutant. *J. Neurosci.*, 27:11037–11046.

Saliba, R.M., F.J. Annegers, D.K. Waller, J.E. Tyson, E.M. Mizrahi. (1999). Incidence of neonatal seizures in Harris County, Texas, 1992–1994. *Am. J. Epidemiol.*, 150:763–769.

Sanchez, R.M., S. Koh, C. Rio, C. Wang, E.D. Lamperti, D. Sharma, G. Corfas, F.E. Jensen. (2001). Decreased glutamate receptor 2 expression and enhanced epileptogenesis in immature rat hippocampus after perinatal hypoxia-induced seizures. *J. Neurosci.*, 21:8154–8163.

Sandsmark, D.K., C. Pelletier, J.D. Weber, D.H. Gutmann. (2007). Mammalian target of rapamycin: master regulator of cell growth in the nervous system. *Histol. Histopathol.*, 22:895–903.

Scheffer, I.E., S.F. Berkovic. (1997). Generalized epilepsy with febrile seizures plus: a genetic disorder with heterogeneous clinical phenotypes. *Brain*, 120:479–490.

Sijben, A.E., P. Sithinamsuwan, A. Radhakrishnan, R.A. Badawy, L. Dibbens, A. Mazarib, D. Lev, T. Lerman-Sagie, R. Straussberg, S.F. Berkovic, I.E. Scheffer. (2009). Does a *SCN1A* gene mutation confer earlier age of onset in febrile seizures in GEFS+? *Epilepsia*, 50:953–956.

Sipila, S.T., S. Schuchmann, J. Voipo, J. Yamada, K. Kaila. (2006). The cation–chloride cotransporter NKCC1: NKCC1 promotes sharp waves in the neonatal rat hippocampus. *J. Physiol.*, 573:765–73.

Stafstrom, C.E., S.L. Moshe, J.W. Swann, A. Nehlig, M.P. Jacobs, P.A. Schwartzkroin. (2006). Models of pediatric epilepsies: strategies and opportunities. *Epilepsia*, 47:1407–1414.

Staley, K., R. Smith, J. Schaak, C. Wilcox, T.J. Jentsch. (1996). Alteration of GABA$_A$ receptor function following gene transfer of the CLC-2 chloride channel. *Neuron*, 17:543–551.

Strug, L.J., T. Clarke, T. Chiang, M. Chien, Z. Baskurt et al. (2009). Centrotemporal sharp wave EEG trait in Rolandic epilepsy maps to elongator protein complex 4 (ELP4). *Eur. J. Hum. Genet.*, 17(9):1171–1181.

Szabo, T.M., M.J. Zoran. (2007). Transient electrical coupling regulates formation of neuronal networks. *Brain Res.*, 1129:63–71.

Talos, D.M., R.E. Fishman, H. Park, R.D. Folkerth, P.L. Follett, J.J. Volpe, F.E. Jensen. (2006a). Developmental regulation of alpha-amino-3-hydroxy-5-methyl-4-isoxazole-propionic acid receptor subunit expression in forebrain and relationship to regional susceptibility to hypoxic/ischemic injury. I. Rodent cerebral white matter and cortex. *J. Comp. Neurol.*, 497:42–60.

Talos, D.M., P.L. Follett, R.D. Folkerth, R.E. Fishman, F.L. Trachtenberg, J.J. Volpe, F.E. Jensen. (2006b). Developmental regulation of alpha-amino-3-hydroxy-5-methyl-4-isoxazole-propionic acid receptor subunit expression in forebrain and relationship to regional susceptibility to hypoxic/ischemic injury. II. Human cerebral white matter and cortex. *J. Comp. Neurol.*, 497:61–77.

Talos, D.M., D.J. Kwiatkowski, K. Cordero, P.M. Black, F.E. Jensen. (2008). Cell-specific alterations of glutamate receptor expression in tuberous sclerosis complex cortical tubers. *Ann. Neurol.*, 63:454–465.

van Luijtelaar, G., E. Sitnikova. (2006). Global and focal aspects of absence epilepsy: the contribution of genetic models. *Neurosci. Behav. Rev.*, 30:983–1003.

Velisek, L., K. Jehle, S. Asche, J. Veliskova. (2007). Model of infantile spasms induced by $N$-methyl-$\delta$-aspartic acid in prenatally impaired brain. *Ann. Neurol.*, 61:109–119.

Volpe J.J. (2008). *Neurology of the Newborn*, 5th ed. Philadelphia, PA: Elsevier.

Wallace, R.H., D.W. Wang, R. Singh, K. Yamakawa, T. Sugawara, E. Mazaki-Miyazaki, S. Hirose, G. Fukuma, A. Mitsudome, K. Wada, S. Kaneko. (1998). Febrile seizures and generalized epilepsy associated with a mutation in the Na$^+$-channel beta 1 subunit gene *SCN1B*. *Nat. Genet.*, 19:366–370.

Wang, C., C. Shimuizu-Okabe, K. Watanabe, A. Okabe, H. Matsuzaki, T. Ogawa, N. Mori, A. Fukuda, K. Sato. (2002). Developmental changes in KCC1, KCC2, and NKCC1 mRNA expression in the rat brain. *Brain. Res. Dev. Brain Res.*, 139:59–66.

Wang, H.S., Pan, Z., Shi, W., B.S. Brown, R.S. Wymore, I.S. Cohen, J.E. Dixon, D. McKinnon. (1998). KCNQ2 and KCNQ3 potassium channel subunits: molecular correlates of the M-channel. *Science*, 282:1890–1893.

Weber, Y.G., H. Lerche. (2008). Genetic mechanisms in developmental epilepsies. *Dev. Med. Child Neurol.*, 50:648–654.

Wong, H.K., X.B. Liu, M.F. Matos, S.F. Chan, I. Pérez-Otaño, M. Boysen, J. Cui, N. Nakanishi, J.S. Trimmer, E.G. Jones, S.A. Lipton, N.J. Sucher. (2002). Temporal and regional expression of NMDA receptor subunit NR3A in the mammalian brain. *J. Comp. Neurol.*, 450:303–317.

Wong, M. (2008). Mechanisms of epileptogenesis in tuberous sclerosis complex and related malformations of cortical development. *Epilepsia*, 49:8–21.

Yamada, J., A. Oakabe, H. Toyoda, W. Kilb, H.J. Luhmann, A. Fukuda. (2004). Cl⁻ uptake promoting depolarizing GABA actions in immature rat neocortical neurons is mediated by NKCC1. *J. Physiol.*, 557:829–841.

Zeng, L.H., L. Xu, D.H. Gutmann, M. Wong. (2008). Rapamycin prevents epilepsy in a mouse model of tuberous sclerosis complex. *Ann. Neurol.*, 63:444–453.

Zhong, J., D.P. Carrozza, K. Williams, D.B. Pritchett, P.B. Molinoff. (1995). Expression of mRNAs encoding subunits of the NMDA receptor in developing rat brain. *J. Neurochem.*, 64:531–539.

Zhong, X., J.R. Liu, J.W. Kyle, D.A. Hanck, W.S. Agnew. (2006). A profile of alternative RNA splicing and transcript variation of *CACNA1H*, a human T-channel gene candidate for idiopathic generalized epilepsies. *Hum. Mol. Genet.*, 15:1497–1512.

# Section II

**Antiepileptic Drugs**

# 8 Mechanisms of Antiepileptic Drug Action

*Timothy A. Simeone*

## CONTENTS

## INTRODUCTION

Epileptic seizures reflect enduring alterations in structure and function of the brain that support recurrent, spontaneous abnormal neuronal discharges and can lead to clinically overt changes in motor control, sensory perception, behavior, or autonomic function. At a cellular level, two hallmark features of epileptiform activity are neuronal hyperexcitability and neuronal hypersynchrony. It is important to understand that a single abnormally discharging neuron is insufficient to produce a clinical seizure. For a seizure to manifest electroclinically, there must be synchronous activity of large populations or networks of neurons (McCormick and Contreras, 2001). In this context, seizures can be considered emergent properties of neuronal networks, similar to the emergence of cognition (Faingold, 2004). Notwithstanding this important concept, determination of the mechanisms of action of antiepileptic drugs (AEDs) usually begins with, or is reduced to, modulation of membrane-bound ion channels.

The neuronal cell membrane is populated by a variety of ion channels that convert chemical signals into electrical activity. As such, ion channels occupy an important position in neuronal excitability. Incoming signals act through ligand-gated and voltage-gated ion channels to generate excitatory or inhibitory responses, which may be subthreshold or suprathreshold for subsequent action potential initiation and which, in turn, give rise downstream to action-potential-propagated neurochemical signals. Traditionally, three major classes of molecular targets are believed to be relevant for limiting epileptic activity: (1) voltage-gated sodium and calcium channels, (2) γ-aminobutyric acid type A (GABA$_A$) receptors, and (3) ionotropic glutamate receptors. Most clinically useful

AEDs do modulate one or more of these targets, but, as will be presented later, many of the AEDs also affect additional molecular targets (Bialer et al., 2007; Meldrum and Rogawski, 2007; Rho and Sankar, 1999; Rogawski, 2006; White et al., 2007).

## METHODS USED TO IDENTIFY AED MECHANISMS

Once a new AED is identified, a wide array of techniques are available to determine its possible mechanisms of action, including (but not limited to) binding studies, imaging, electrophysiology, molecular biology, and molecular genetics. Competitive binding studies determine the specificity of the interaction between a radiolabeled AED and its binding sites by examining the ability of various neurotransmitters, hormones, or even other drugs to compete for the binding sites. If no radiolabeled AED is available, then the inverse experiment can be performed using radiolabeled neurotransmitters, hormones, or drugs with known binding sites. Importantly, this technique has limited or no sensitivity if the AED binding site is completely novel. Also, binding studies require the AED to exhibit high-affinity binding (<1 $\mu M$) to ensure a discernable signal from background binding. This naturally precludes identification of lower affinity binding sites that are most likely within the AED therapeutic concentration range.

In recent years, high-throughput fluorescent imaging techniques, such as the Fluorometric Imaging Plate Reader (FLIPR; Molecular Devices, Union City, CA), have been used to ascertain test drug effects on cellular calcium concentration or membrane potential. Typically, these experiments are performed by adding a receptor agonist, such as the neurotransmitter glutamate, or a depolarizing agent, such as potassium, to induce a change in fluorescence in cells. The effectiveness of sequential concentrations of a drug of interest, in either reducing or preventing the change in fluorescence, is then quantified. Understandably, these experiments, when performed using isolated primary neuronal cultures, have low specificity and are unable to reveal a putative binding site for the AED, thus providing limited information about the mechanism of action. Specificity can be increased if the study uses a cell line expressing individual recombinant ion channels or receptors.

Electrophysiological evidence remains the gold standard for discerning AED mechanisms of action, even though this approach is time consuming and labor intensive. Fundamentally, electrophysiological techniques record changes in voltage or current in a group of neurons, within individual cells, or associated with single ion channels. These experiments are generally performed *in vitro* with acutely dissected brain slices or cultured neurons. Extracellular recordings provide information about AED effects on population events and provoked epileptiform activity. Whole-cell patch-clamp techniques allow the investigator to determine AED effects on the single-cell resting membrane potential, input resistance, action potential characteristics (i.e., threshold, duration, and peak), and action potential firing characteristics (i.e., frequency, adaptation, and repetitive firing susceptibility). Furthermore, AED effects on specific voltage- and ligand-gated ion channels are resolved by isolating ionic currents, changing the membrane potential, or applying a receptor/channel agonist. When a particular channel has been identified as an AED target, single-channel recording techniques establish AED effects on channel opening and closing kinetics and detail the exact manner in which the AED modulates ion channel function.

The disciplines of molecular biology and molecular genetics have become essential in establishing and determining AED binding sites on specific molecular targets. Molecular biology techniques have provided the technology to overexpress a known ion channel in a cell line that is essentially devoid of endogenous voltage- and ligand-gated channels. Combined with electrophysiology, such molecular approaches provide unequivocal evidence of AED action. Site-directed mutagenesis of ion-channel proteins permits a detailed map of the molecular binding site for the AED. Molecular genetics has allowed investigators to identify specific mutations in particular ion channels that are associated with various epilepsies, such as severe myoclonic epilepsy of infancy (SMEI) and autosomal dominant nocturnal frontal lobe epilepsy (ADNFLE), among many others (see Chapter 6, this volume). Studies of the functional consequences of such natural mutations have been possible using site-directed mutagenesis and gene knockout and knockin technology. AED modulation of these mutated ion channels can be assessed and possibly correlated to refractoriness to individual AEDs. The growing structural and functional characterization of AED molecular targets and increasing understanding of how AEDs modulate these targets provide a framework for creating improved epilepsy therapies and have encouraged many investigators in their attempts to develop drugs for specific molecular targets.

## TRADITIONAL TARGETS

### SODIUM CHANNELS

Voltage-gated sodium channels are heteromultimeric proteins consisting of a major $\alpha$ subunit and one or more $\beta$ subunits that influence the kinetic properties of the $\alpha$ subunit. Sodium channels activate upon membrane depolarization and underlie the fast upstroke of the action potential. The sodium current is a self-limiting process in that the channel inactivates within milliseconds, thus terminating the action potential. Repolarization of the neuronal membrane primes the sodium channel for subsequent activation. The kinetics of the onset and recovery from inactivation allow sustained, high-frequency action potential firing during normal and epileptic activity. Under normal conditions, there is an additional mechanism in the form of slow inactivation, which prevents excessive action potential firing. Slow inactivation occurs during prolonged depolarizations in which full onset and recovery occur over several seconds. Consequently, these lengthy depolarizations can participate in the termination of repetitive action potential bursts.

A number of AEDs effective in the treatment of partial-onset epilepsy inhibit voltage-gated sodium channels and block sustained, high-frequency, repetitive firing of neurons (Figure 8.1). Examples include:

- *Phenytoin* (Errington et al., 2008; Kuo, 1998; Kuo and Bean, 1994; Kuo and Lu, 1997; Kuo et al., 1997; Lang et al., 1993; Remy et al., 2003; Segal and Douglas, 1997; Song et al., 1996; Spadoni et al., 2002; Xie et al., 2001)
- *Carbamazepine* (Errington et al., 2008; Kuo, 1998; Kuo et al., 1997; Lang et al., 1993; Song et al., 1996; Sun et al., 2007)
- *Oxcarbazepine* (Huang et al., 2008; Schmutz et al., 1994)
- *Lamotrigine* (Kuo, 1998; Kuo and Lu, 1997; Lang et al., 1993; Remy et al., 2003; Spadoni et al., 2002; Xie et al., 2001)
- *Zonisamide* (Schauf, 1987b)
- *Felbamate* (Pisani et al., 1995; Taglialatela et al., 1996)
- *Topiramate* (DeLorenzo et al., 2000; McLean et al., 2000; Taverna et al., 1999; Sun et al., 2007; Zona et al., 1997)
- *Valproate* (Fohlmeister et al., 1984; Remy et al., 2003; Schauf, 1987a; Spadoni et al., 2002; Van dan Berg et al., 1993)

One of the newer AEDs, lacosamide, acts in a more distinct manner, enhancing slow inactivation (Errington et al., 2008). Most of these AEDs appear to inhibit sodium channels in a voltage- and use-dependent manner (i.e., the block is more effective at depolarized rather than hyperpolarized potentials, and the potency of the block increases with increasing action potential frequency or prolonged depolarizations). These observations suggest that these AEDs act in a manner similar to, or modulate the kinetics of, either fast or slow inactivation, thus accounting for the selective block of high-frequency firing without affecting normal neuronal action potential discharge.

Upon the decay of the transient sodium current, a remaining, or persistent, sodium current ($I_{NaP}$) is observable that constitutes less than 4% of the total sodium current (Brumberg et al., 2000; reviewed in Stafstrom, 2007). It has been proposed that the non-inactivating $I_{NaP}$ is either produced

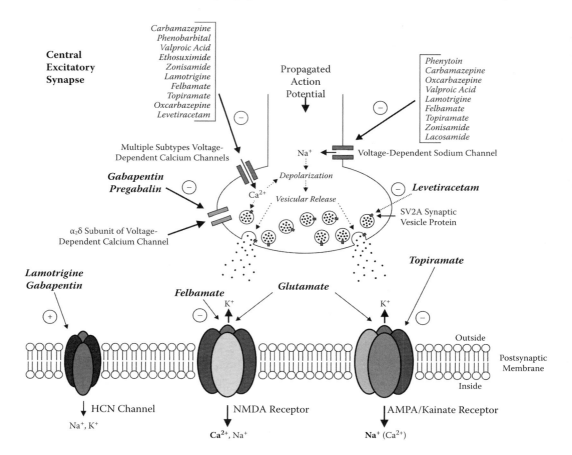

**FIGURE 8.1** Schematic representation of an excitatory synapse in the central nervous system, and the putative major sites of action of various antiepileptic drugs (AEDs). Plus or minus signs denote activation/potentiation or inhibition, respectively. AMPA, α-amino-3-hydroxy-5-methyl-4-isoxazoleproprionic acid; HCN, hyperpolarization-activated cyclic nucleotide-regulated cation channels; NMDA, *N*-methyl-D-aspartate; SV2A, synaptic vesicle protein 2A. (Adapted from Rho, J.M. and Sankar, R., *Epilepsia* 40:11, 1471–1483, 1999. With permission.)

by the overlap between activation and inactivation voltage ranges of sodium channels (resulting in a "window current") or generated by sodium channels entering an infrequently visited open state that lacks fast inactivation. In any case, the voltage range of $I_{NaP}$ activation is immediately subthreshold for action potential initiation and is thus repeatedly crossed during an action potential train. Accordingly, $I_{NaP}$ occupies a unique position, and its presence or absence can regulate repetitive firing and the spread of hyperexcitability. Experimental results suggest that the following AEDs inhibit this non-inactivating, persistent sodium current, thus providing an additional mechanism of preventing epileptiform activity:

- *Phenytoin* (Berger and Lüscher, 2004; Chao and Alzheimer, 1995; Lampl et al., 1998; Niespodziany et al., 2004; Segal and Douglas, 1997)
- *Carbamazepine* (Sun et al., 2007)
- *Topiramate* (Sun et al., 2007; Taverna et al., 1999)
- *Lamotrigine* (Berger and Lüscher, 2004; Spadoni et al., 2002)
- *Ethosuximide* (Leresche et al., 1998; Niespodziany et al., 2004)
- *Valproate* (Taverna et al., 1998)

## Calcium Channels

Like sodium channels, multiple subunits combine to form voltage-gated calcium channels. Membrane depolarization opens the calcium-permeable pore, allowing calcium to enter the neuron. There are two types of calcium channels: high-voltage-activated (HVA) and low-voltage-activated (LVA). As the name suggests, HVA calcium channels (L-, R-, P/Q-, and N-types) require strong membrane depolarization to induce opening and are predominantly responsible for the regulation of calcium entry and neurotransmitter release from presynaptic axon terminals in response to the arrival of an action potential. LVA calcium channels (T-type) are named as such because they can be activated by small depolarizations of the plasma membrane. T-type calcium channels regulate neuronal firing by participating in bursting and intrinsic oscillations. Thus, HVA and LVA calcium channels span many aspects of electrochemical signaling and represent important targets for exogenous modulation of neuronal excitability.

Each HVA calcium channel consists of one $\alpha_1$ subunit (with seven known subtypes), which forms the channel pore, and various auxiliary subunits (including the $\alpha_2\delta$ subunit). Two structural analogs of GABA, gabapentin and pregabalin, are effective in the treatment of partial seizures, and have been shown to selectively interact with $\alpha_2\delta$-1 and $\alpha_2\delta$-2 auxiliary subunits (reviewed in Dooley et al., 2007; Taylor et al., 2007). Mutant mice expressing a point mutation of arginine to alanine at the 217th amino acid in the $\alpha_2\delta$-1 subunit have significantly reduced radioligand binding of gabapentin and pregabalin in all relevant brain regions (Bian et al., 2006). Despite strong evidence for this unique site of action, neither gabapentin nor pregabalin has been consistently shown to exert an effect on calcium currents. This result is perhaps not surprising, as $\alpha_2\delta$-1 and $\alpha_2\delta$-2 auxiliary subunits can combine with many of the seven subtypes of HVA $\alpha_1$ subunits, yielding tremendous heterogeneity in molecular composition—and, hence, pharmacosensitivity. Even so, gabapentin and pregabalin can decrease presynaptic neurotransmitter release, at both inhibitory GABAergic and excitatory glutamatergic synapses. In contrast, various subtypes of HVA channels are consistently inhibited by the following, reducing neurotransmitter release in most cases:

- *Phenobarbital*, L-type (ffrench-Mullen et al., 1993; Gross and Macdonald, 1988)
- *Levetiracetam*, N-type (Lukyanetz et al., 2002; Martella et al., 2008; Niespodziany et al., 2001)
- *Lamotrigine*, N- and P/Q-types (Hainsworth et al., 2003; Martella et al., 2008; Stefani et al., 1996b; Wang et al., 1996, 1998)
- *Felbamate*, L-type (Stefani et al., 1996a)
- *Oxcarbazepine*, L- and N-types (Schmutz et al., 1994)
- *Carbamazepine*, L-type (Ambrósio et al., 1999)
- *Topiramate*, N-, P-, L-, and R-types (Kuzmiski et al., 2005; Martella et al., 2008; McNaughton et al., 2004; Zhang et al., 2000)

Absence seizures are a unique seizure type associated with brief lapses of consciousness that begin and terminate abruptly and are a result of generalized spike–wave discharges seen on the scalp electroencephalogram (EEG). Absence seizures reflect widespread and heightened phase-locked oscillations between excitation (i.e., spike) and inhibition (i.e., slow wave) in mutually connected thalamocortical networks. T-type calcium channels generate low-threshold calcium currents that depolarize the membrane and trigger bursts of action potentials (reviewed in Crunelli and Leresche, 2002; Huguenard, 2002). Such excitatory bursts are believed to underlie the spike portion of a generalized spike–wave oscillation. Inhibition of T-type calcium channels by the AEDs ethosuximide (Broicher et al., 2007; Coulter et al., 1989; Gomora et al., 2001; Lacinová, 2004; Todorovic and Lingle, 1998), valproate (Broicher et al., 2007; Lacinová, 2004; Todorovic and Lingle, 1998), and zonisamide (Matar et al., 2009) terminates the 3-Hz spike–wave discharges and correlates with efficacy in the treatment of generalized absence seizures. The relevance of the low-threshold (T-type) calcium channel as a therapeutic target is reinforced by the demonstration that mice lacking the *CACNA1G* gene product lack burst firing in thalamocortical relay neurons and are resistant to absence seizures (Kim et al., 2001). However, studies with the Han Chinese populations suggested that *CACNA1H* gene mutations may be involved (Chen et al., 2003) but similar mutations were not encountered in European patients (Chioza et al., 2006). Therefore, the role of T-type calcium channels in the pathogenesis of human childhood absence epilepsy remains unsettled.

Recent attention has involved the potential role of hyperpolarization-activated cyclic nucleotide-gated (HCN) cation channels (mediating $I_h$) in modulating the excitability of the thalamocortical network. This concept is detailed in a later section on novel targets.

## Synaptic Excitation

Neuronal excitation in the central nervous system is mediated by ionotropic glutamate receptors—α-amino-3-hydroxy-5-methyl-4-isoxazoleproprionic acid (AMPA), kainate, and *N*-methyl-D-aspartate (NMDA)—and metabotropic glutamate receptors (Figure 8.1) (Simeone et al., 2004). Inhibition of excitatory glutamatergic neurotransmission would appear to be the most obvious way to limit excessive excitatory activity, and this conceptual notion has been supported by many *in vivo* and *in vitro* preclinical studies. However, in clinical trials, selective glutamate receptor antagonists, while efficacious, have been hampered by unacceptable side effects. Of the currently available AEDs, seven appear to exert activity, in part, on ionotropic glutamate receptors—and none interacts principally with metabotropic glutamate receptors.

*N*-methyl-D-aspartate receptors are macromolecules formed by coassembly of an obligatory NR1 subunit and at least one type of NR2 subunit. Felbamate, at therapeutically relevant concentrations, inhibits NMDA-evoked currents and appears to be a low-affinity, open-channel blocker (Chang and Kuo, 2007a,b; Kuo et al., 2004; Pisani et al., 1995; Rho et al., 1994, 1997; Subramaniam et al., 1995). Studies conducted in *in vitro* expression systems have demonstrated that felbamate is a more potent antagonist of NMDA receptors containing NR2B subunits (Chang and Kuo, 2008; Harty and Rogawski, 2000; Kleckner et al., 1999). The NR2B subunit is abundantly expressed in the immature brain and largely restricted to the forebrain in the adult. These expression patterns may contribute to the effectiveness of felbamate in pediatric epilepsies and low neurobehavioral side effects in adults. Additionally, phenytoin (Wamil and McLean, 1993), carbamazepine (Hough et al., 1996; Lampe and Bigalke, 1990), and gabapentin (Hara and Sata, 2007) have been reported to inhibit NMDA receptor function, but subunit composition specificity has not yet been ascertained.

As the primary mediators of fast excitatory neurotransmission, AMPA and kainate receptors have differential but important roles in seizure genesis and spread. Phenytoin has been found to competitively block non-NMDA receptors expressed in *Xenopus* oocytes (Kawano et al., 1994), and lamotrigine inhibits native AMPA-mediated currents in dentate granule cells (Lee et al., 2008); however, phenytoin, but not lamotrigine, inhibited AMPA-mediated currents in cortical neurons (Phillips et al., 1997). Also, AMPA- and kainate-mediated currents are inhibited by the AEDs levetiracetam (Carunchio et al., 2007) and topiramate (Angehagen et al., 2004, 2005; Gibbs et al., 2000; Gryder and Rogawski, 2003; Kaminski et al., 2004; Poulsen et al., 2004; Skradski and White, 2000). Kainate receptors are particularly interesting as AED targets because they are not restricted to postsynaptic membranes; they can also be found on presynaptic axon terminals. Presynaptically, activation of kainate receptors modulates glutamate release and inhibits GABA release. As such, kainate-receptor-mediated inhibition represents a very potent mechanism of seizure termination. Topiramate selectively inhibits GluR5-containing kainate receptors (Gryder and Rogawski, 2003), whereas the effects of levetiracetam are attributed to inhibition of AMPA receptors (Carunchio et al., 2007).

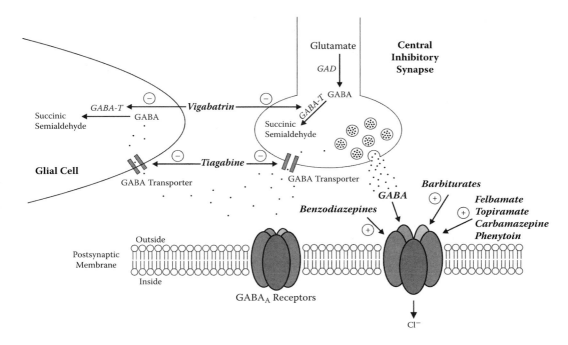

**FIGURE 8.2**   Schematic representation of an inhibitory synapse in the central nervous system, and the putative major sites of action of various antiepileptic drugs (AEDs). Plus or minus signs denote activation/potentiation or inhibition, respectively. GABA, γ-aminobutyric acid; GABA-T, GABA transaminase; GAD, glutamic acid decarboxylase. (Adapted from Rho, J.M. and Sankar, R., *Epilepsia* 40:11, 1471–1483, 1999. With permission.)

## Synaptic Inhibition

Neuronal inhibition in the central nervous system is primarily achieved by the presynaptic release of the neurotransmitter GABA, which binds to ionotropic $GABA_A$ and metabotropic $GABA_B$ receptors (Figure 8.2). $GABA_A$ and $GABA_B$ receptors mediate fast and slow inhibition in the brain, respectively. During normal neuronal activity, communication between inhibitory GABAergic interneurons and excitatory glutamatergic neurons provides the critical balance that is necessary to prevent the development of synchronized epileptiform discharges—discharges that would naturally arise in an unchecked, recurrently connected excitatory network. However, the role of inhibition with respect to epileptiform activity is not simple (McCormick and Contreras, 2001); for example, it appears that inhibitory neurotransmission may actually participate in the synchronization of large populations of neurons and the subsequent propagation of seizures to adjacent brain regions. Thus, AEDs that act principally to augment GABAergic neurotransmission may prevent the initiation of epileptiform activity. In some cases, however, such agents may also exacerbate the synchronization and spread of seizure activity, notably in thalamocortical circuits (Sohal et al., 2003).

The paradoxes in the role played by GABAergic inhibition are resolved to a large extent when one differentiates between types of inhibition. *Phasic inhibition* is mediated by synaptic receptors of distinct subunit stoichiometry that confer low GABA sensitivity (hence requiring an acute rise in local GABA concentrations from synaptic release) and rapid inactivation (thus suitable for receiving further signals). *Tonic inhibition* is mediated by extrasynaptic receptors that display high GABA sensitivity to small changes in ambient GABA levels and very slow inactivation (ideal to control overall level of inhibition) (Mody and Pearce, 2004; Nusser and Mody, 2002; Stell and Mody, 2002). The synaptic receptors tend to be those containing γ subunits and display benzodiazepine sensitivity (see following paragraphs), while extrasynaptic receptors do not. Extrasynaptic receptors typically contain the stoichiometry of $\alpha_{4/6}\beta_{2/3}\delta$. It appears that augmenting tonic inhibition enhances burst firing of thalamocortical neurons (Cope et al., 2005) and may promote absence epilepsy, whereas an increase in phasic inhibition, such as is possible by benzodiazepines acting on synaptic (e.g., $\alpha_1\beta_{2/3}\gamma$) receptors, tends to limit spike–waves.

The various distinct ways in which GABAergic inhibition can be augmented include: (1) increased GABA$_A$ and GABA$_B$ receptor function, (2) enhanced GABA synthesis, (3) decreased GABA degradation, and (4) inhibition of GABA reuptake into neuronal and glial cells (Figure 8.2). GABA$_A$ receptors are oligopentameric, macromolecular protein complexes consisting of many different subunits ($\alpha_{1-6}$, $\beta_{1-3}$, $\gamma_{1-3}$, $\delta$, $\epsilon$, $\pi$, $\theta$, $\rho$), resulting in a dizzying array of possible combinations; most GABA$_A$ receptors consist of two $\alpha$ subunits, two $\beta$ subunits, and one $\gamma$ subunit that combine to form a pore selectively permeable to chloride and bicarbonate ions. Biophysical and pharmacological properties of GABA$_A$ receptors are determined by their inherent subunit composition (Simeone et al., 2003). In general, GABA$_A$ receptor inhibitors are convulsants and positive modulators are anticonvulsants. The following AEDs have been experimentally shown to modulate GABA$_A$ receptors in differential ways:

- *Phenytoin* (Granger et al., 1995)
- *Carbamazepine* (Granger et al., 1995)
- *Phenobarbital* (Macdonald et al., 1989; Rho et al., 1996; Twyman et al., 1989)
- *Diazepam* (Kemp et al., 1987; Pritchett et al., 1989; Sigel and Baur, 1988; Sigel et al., 1990)
- *Felbamate* (Rho et al., 1994, 1997; Simeone et al., 2006a)
- *Topiramate* (Gordey et al., 2000; Simeone et al., 2006b; White et al., 1997, 2000)

GABA$_A$ receptor function can be enhanced by increasing the channel conductance or open state of the channel (either channel burst frequency or open-state duration). Accordingly, the barbiturate phenobarbital increases mean open duration (Macdonald et al., 1989; Twyman et al., 1989), the benzodiazepine diazepam increases open-state frequency without affecting duration (Twyman et al., 1989), and topiramate increases both open-state and burst frequency and duration (White et al., 1997, 2000). Felbamate enhances GABA-mediated currents in hippocampal and cortical neurons, and, although the precise effects on single-channel kinetics remain unclear, evidence suggests a barbiturate-like action (Rho et al., 1996).

Studies of GABA$_A$ receptors expressed in oocytes or mammalian cells, and comprising different subunit combinations, have revealed subunit-specific modulation of GABA$_A$ receptors by AEDs. Diazepam binds to a recognition site juxtaposed between the $\alpha$ and $\gamma$ subunits and enhances GABA activation of most receptors comprised of an $\alpha\beta\gamma$ subunit combination with varying affinities (Sigel et al., 1990); however, it is ineffective on $\alpha_{4/6}\beta\gamma$ combinations (Derry et al., 2004). Therapeutic concentrations of topiramate enhance $\alpha\beta_{2/3}\gamma_2$ combinations and directly activate $\alpha_4\beta_3\gamma_2$, whereas higher concentrations of this AED directly activate $\alpha\beta_{2/3}\gamma_2$ combinations and inhibit $\alpha\beta_1\gamma_2$ combinations (Simeone et al., 2006b). Studies utilizing the *Xenopus* oocyte expression system indicate that a therapeutic concentration of felbamate (300 $\mu M$) exclusively enhances GABA$_A$ receptors comprised of $\alpha_{1/2}\beta_{2/3}\gamma_2$ and is either ineffective or inhibits every other subunit combination tested (Simeone et al., 2006a). Also, phenytoin and carbamazepine specifically potentiate $\alpha_1\beta_2\gamma_2$ receptors (Granger et al., 1995). In contrast, barbiturates such as pentobarbital positively modulate every GABA$_A$ receptor subunit combination, although potency does depend on $\alpha$ and $\beta$ subunit subtypes (Thompson et al., 1996). It is safe to speculate that such subunit specificity, along with possible variations in local AED concentrations, may contribute to the unique anticonvulsant profile of each AED and its effectiveness against particular seizure types.

GABA$_B$ receptors are coupled via a guanosine 5′-triphosphate-binding protein to calcium or potassium channels (Simeone et al., 2003). The GABA structural analog, gabapentin, has been reported to modulate inwardly rectifying potassium channels and N-type calcium channels via preferential interaction with GABA$_B$ receptors containing GABA$_{B1a}$ subunits. The subunit specificity of gabapentin was initially thought to contribute to its anticonvulsant activity by selectively reducing excitatory neurotransmitter release with a simultaneous sparing of inhibitory neurotransmission. Subsequent studies have been unable to replicate these findings, having failed to identify binding at radioligand sites related to $GABA_B$ receptors, to demonstrate an inhibitory effect on gabapentin actions by $GABA_B$ receptor antagonists, and to prevent gabapentin action by co-application of the $GABA_B$ agonist baclofen (reviewed in Taylor et al., 2007).

Ambient levels of extracellular GABA are controlled by enzymes involved in the GABA life cycle and GABA uptake by transporters. Two pyridoxal-5'-phosphate-dependent enzymes, glutamate decarboxylase (GAD) and GABA transaminase (GABA-T), are involved in GABA synthesis and degradation, respectively. The AED vigabatrin (γ-vinyl GABA) is a GABA analog that has no activity on GABA receptors but acts by inhibiting GABA-T, which results in significant elevations in brain GABA levels (Figure 8.2). Vigabatrin increases ambient extracellular GABA either by reversal of GABA transporters or by shifting the concentration equilibrium (reviewed in Rogawski and Löscher, 2004). Other GABA analogs, such as gabapentin and pregabalin, were once thought to enhance GAD and inhibit GABA aminotransferase, resulting in increased GABA concentrations, but subsequent studies have disproved these interactions at therapeutically relevant drug concentrations (reviewed in Taylor et al., 2007).

The AED tiagabine increases GABA concentrations through a different mechanism. Tiagabine inhibits the GABA transporter GAT1 (Schousboe et al., 2004), resulting in prolonged postsynaptic inhibitory currents that are enhanced during repetitive interneuron activation—activity that is expected to occur during epileptiform discharge (Thompson and Gähwiler, 1992). The pharmacologic consequence of an increase in ambient GABA may depend on the generator (seizure network); AEDs that function in this manner tend to be anticonvulsants in models of localization-related epilepsies but are proconvulsant in generalized spike–wave models (Bouwman et al., 2007; Coenen et al., 1995). Such AEDs likely increase tonic inhibition, enhance synchronization, and thus contribute to an increase in the frequency and duration of spike–wave paroxysms (Bouwman et al., 2007; Coenen et al., 1995; Vergnes et al., 1984), making them ill suited for use in generalized epilepsies.

## NEW MOLECULAR TARGETS

### HCN CHANNELS

The hyperpolarization-activated cation (HCN) channel consists of four subunits that form a pore permeable to $Na^+$ and $K^+$ ions. HCN channels are highly expressed in neuronal dendrites (and to a lesser extent in the soma) in both thalamus and hippocampus (Figure 8.1) (reviewed in Poolos, 2005). Unlike other voltage-gated conductances that are classified as either inhibitory or excitatory, h-currents ($I_h$) are both inhibitory and excitatory. HCN channels possess an inherent negative-feedback property; that is, hyperpolarization produces an activation of HCN channels, which then leads to depolarization and immediate deactivation of these channels. The net effect of HCN channel activation is a decrease in the input resistance of the membrane, which is a measure of the voltage change produced by a given synaptic current. The h-current tends to stabilize the neuronal membrane potential toward the resting potential against both hyperpolarizing and depolarizing inputs.

Lamotrigine, an AED that is highly effective against absence seizures, enhances activation of dendritic h-currents in hippocampal pyramidal neurons by shifting the voltage-dependent activation toward depolarized potentials. Increasing $I_h$ reduces input resistance, length constant, and temporal summation and suppresses action potential initiation by dendritic inputs (Berger and Lüscher, 2004; Poolos et al., 2002; Ying et al., 2007). Further supportive evidence that $I_h$ participates in absence seizures is that deletion of the HCN subunit type 2 results in absence epilepsy and sinus dysrhythmia in mice (Ludwig et al., 2003). Gabapentin has also been shown to enhance $I_h$ amplitude by increasing the channel conductance, which results in a slight membrane depolarization and reduced input resistance (Surges et al., 2003, 2004).

## POTASSIUM CHANNELS

Potassium channels represent a large and heterogeneous family of ion channels and generally function to dampen neuronal excitation by inducing membrane hyperpolarization; hence, they represent a natural target for AED development. Previously, it was thought that none of the AEDs available for clinical use today acts directly on voltage-gated potassium channels; however, recently published studies implicate potassium channels as part of the mechanism of action of several AEDs. Topiramate, for example, induces a steady membrane hyperpolarization mediated by a potassium conductance (Herrero et al., 2002), whereas levetiracetam (Madeja et al., 2003) and oxcarbazepine (Huang et al., 2008) block sustained repetitive firing by paradoxically decreasing delayed rectifier voltage-gated potassium currents. Phenytoin inhibits A-type potassium channels by increasing the probability of the inactivated state (Pisciotta and Prestipino, 2002) and also inhibits the open-channel probability of delayed rectifier potassium channels (Nobile and Lagostena, 1998; Nobile and Vercellino, 1997). Furthermore, lamotrigine blocks hippocampal neuronal A-type potassium channels by stabilizing the inactivated state (Grunze et al., 1998a,b; Huang et al., 2004) but potentiates 4-aminopyridine-sensitive, $K^+$-mediated hyperpolarizing conductances in cultured cortical neurons (Zona et al., 2002). Delayed rectifier and A-type potassium channels participate in membrane repolarization after an action potential and play a role in determining the frequency of repetitive firing. Presumably, inhibition of either potassium channel slows membrane repolarization, thereby decreasing sodium channel de-inactivation and effectively preventing responses to fast depolarization by limiting the frequency of repetitive firing.

Carbamazepine, at therapeutically relevant concentrations, blocks recombinant rat brain small-conductance, calcium-activated potassium channels, which underlie the slow after-hyperpolarization of action potentials that mediate the phenomenon of spike frequency adaptation (Dreixler et al., 2000). In contrast, large-conductance, calcium-activated potassium channels participate in spike repolarization and fast hyperpolarization and are activated by zonisamide and acetozolamide in a manner that shifts voltage sensitivity to less positive potentials and decreases mean closed durations, thereby increasing open probability (Huang et al., 2007; Tricarico et al., 2004). Interestingly, mutations in large-conductance, calcium-activated potassium channels are associated with generalized epilepsies (Díez-Sampedro et al., 2006; Du et al., 2005; Lorenz et al., 2007).

An especially important novel target pertaining to potassium conductance appears to be the Kv7.2/7.3 channels that mediate the so-called M-current (Figure 8.3) (Cooper and Jan, 2003). These channels, encoded by *KCNQ2* and *KCNQ3* genes, have been linked to benign neonatal convulsions (Biervert et al., 1998; Singh et al., 1998). These channels are highly localized to the axon initial segment (and in the nodes of Ranvier in the peripheral nervous system) (Devaux et al., 2004), and their conductance serves to powerfully modulate the firing properties of neurons (Shah et al., 2008). The anticonvulsant effect of pharmacologic enhancement of the M-current has been shown *in vitro* (Yue and Yaari, 2004) as well as *in vivo* (Raol et al., 2009). The description of mouse models of human *KCNQ2* and *KCNQ3* mutations can be found in Chapter 10 of this volume. The Kv7.2/7.3 channel seems to show promise as an antiepileptogenic target as evidenced by a study describing the effect of retigabine, an investigational AED that enhances M-currents (Figure 8.3), on rapid kindling in developing rats (Mazarati et al., 2008).

## CARBONIC ANHYDRASE INHIBITION

Several AEDs, including acetazolamide, topiramate, and zonisamide, act either principally or in part by inhibiting certain carbonic anhydrase isoforms (De Simone et al., 2005; Dodgson et al., 2000; Shank et al., 2008). The clinical use of carbonic anhydrase inhibitors covers the last half century, but it is unclear how anticonvulsant effects are achieved with these agents. Interestingly,

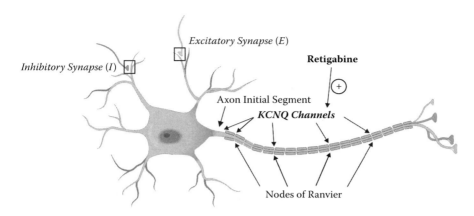

**FIGURE 8.3** Illustration of a central neuron depicting the axon initial segment, nodes of Ranvier, and localization of KCNQ channels, which represent the putative molecular target of the novel anticonvulsant compound retigabine. The unique subcellular localization of KCNQ channels accounts for their potent modulatory role in action potential propagation. The plus sign denotes activation of KCNQ channels. Figures 8.1 and 8.2 represent more macroscopic views of central excitatory (E) and inhibitory (I) synapses in the dendritic tree of the neuron.

the demonstration that carbonic anhydrase II knockout mice have decreased susceptibility to pentylenetetrazole (PTZ)- and fluorothyl-induced seizures emphasizes the relevance of this family of enzymes in modulation of neuronal excitability (Velísek et al., 1993a,b). Inhibition of brain carbonic anhydrase results in an increase in total $CO_2$ concentrations in neurons, and increased pH and a decreased $HCO_3^-$ concentration in neuroglia. One downstream effect of intracellular alkalinization by acetazolamide is an upregulation of $I_h$ in thalamocortical neurons due to a shift in voltage-dependent activation of HCN channels to more positive potentials (Munsch and Pape, 1999). This pH-mediated upregulation of $I_h$ results in decreased membrane responses to hyperpolarizing stimuli and a reduced ability to generate rebound calcium-mediated spikes. These changes also lower interstitial $K^+$ concentrations, as well as pH. It is well known that a reduction in extracellular $K^+$ leads to diminished neuronal excitability, but the impact of decreased extracellular pH is less clear, despite the fact that, depending on subunit composition, extracellular protons can block NMDA receptor-mediated excitation and enhance $GABA_A$ receptor function.

## SYNAPTIC VESICLE PROTEIN

One of the newest AEDs has challenged our conceptual understanding of relevant mechanisms of action. Unlike traditional agents, levetiracetam failed two standard screening tests in animals—the maximal electroshock (MES) and PTZ seizure threshold tests—yet had a profound effect in retarding amygdala kindling in rats. Levetiracetam's principal molecular target, originally identified as a specific high-affinity neuronal binding site, was recently demonstrated to be a specific synaptic vesicle protein, SV2A, which is involved in neurotransmitter release (Gillard et al., 2006; Lynch et al., 2004). The functional consequences of an interaction between SV2A and levetiracetam remain to be elucidated; however, a putative augmentation by levetiracetam can be tentatively hypothesized based on the observation that mice lacking SV2A have severe seizures and demonstrate a profound diminution in GABAergic transmission (Crowder et al., 1999; Custer et al., 2006; Xu and Bajjalieh, 2001). Furthermore, SV2A heterozygous mice have lower thresholds in several seizure models along with reduced anticonvulsant efficacy of levetiracetam (Kaminski et al., 2009).

## CONCLUSIONS AND FUTURE DIRECTIONS

Although we have elucidated many individual molecular targets of AEDs, there is much that remains unknown. There is an emerging consensus that the clinical activity of many AEDs across a broad spectrum of seizure types may be a consequence of the modulatory promiscuity for any given AED. For example, topiramate's actions on voltage-gated sodium and calcium channels, combined with its effects on both $GABA_A$ and glutamate receptors, are consistent with efficacy against multiple seizure types. Conversely, AED promiscuity has also been implicated in the associated cognitive side effects that accompany most drugs (Sankar and Holmes, 2004). This has led to the concept that future AEDs rationally designed to interact with a specific ion channel or protein will have not only greater efficacy but also reduced side effects. On the surface, this logic appears very sound; however, as we learn more about the pathophysiology of epilepsy, it is apparent that it is not a single ion channel or protein disease. It is a neuronal network disease, in which many intercellular and intracellular abnormalities have occurred to disturb network homeostasis (Faingold, 2004).

An appropriate illustration of this concept is that even though limiting voltage-gated sodium conductances is thought to be anticonvulsant, severe loss-of-function mutations of sodium channels paradoxically result in severe myoclonic epilepsy of childhood (i.e., Dravet syndrome) (Mulley et al., 2005). The consequences of these sodium channel mutations are best understood only in the context of the network. The mutant sodium channels are primarily expressed in inhibitory interneurons, resulting in reduced sodium current densities and action potential generation in these neurons (Ogiwara et al., 2007; Yu et al., 2006). Loss of sodium channel function in interneurons may decrease network inhibitory tone and effectively disinhibit large populations of principal cells, leading to a loss of network homeostasis and hyperexcitability. Because of this network context, AEDs that antagonize sodium channels exacerbate seizures in Dravet syndrome (Arzimanoglou, 2009).

Many current AEDs have multiple low-affinity interactions with overlapping targets, and it is most likely the combined effect of these low-affinity or subtle interactions that returns network homeostasis. Treatment with a high-affinity "clean" AED may actually be less efficacious, with more detrimental side effects. Moreover, given that nature tends to reuse binding motifs across diverse proteins, it is possible that designed high-affinity AEDs will also have low-affinity targets that contribute to any perceived efficacy. Accordingly, there is now a movement, at least in the neurodegenerative and Alzheimer's disease drug development fields, to intentionally create compounds that bind with low affinity to multiple targets (Stephenson et al., 2005; Youdim and Buccafusco, 2005). This method has been termed the "magic shotgun" approach and deserves serious consideration for the future of AED development (Bianchi et al., 2009). Of course, identifying the appropriate targets that provide the greatest efficacy and fewest side effects will always be the rate-limiting step and predicts a painstaking trial and error procedure; however, armed with the "magic bullet" and "magic shotgun" approaches, the future promises that some very exciting and novel AEDs are on the horizon.

## REFERENCES

Ambrósio, A.F., A.P. Silva, J.O. Malva, P. Soares-da-Silva, A.P. Carvalho, C.M. Carvalho. (1999). Carbamazepine inhibits L-type $Ca^{2+}$ channels in cultured rat hippocampal neurons stimulated with glutamate receptor agonists. *Neuropharmacology*, 38(9):1349–1359.

Angehagen, M., E. Ben-Menachem, R. Shank, L. Ronnback, E. Hansson. (2004). Topiramate modulation of kainate-induced calcium currents is inversely related to channel phosphorylation level. *J. Neurochem.*, 88(2):320–325.

Angehagen, M., L. Ronnback, E. Hansson, E. Ben-Menachem. (2005). Topiramate reduces AMPA-induced Ca(2+) transients and inhibits GluR1 subunit phosphorylation in astrocytes from primary cultures. *J. Neurochem.*, 94(4):1124–1130.

Arzimanoglou, A. (2009). Dravet syndrome: from electroclinical characteristics to molecular biology. *Epilepsia*, 50(Suppl. 8):3–9.

Berger, T., H.R. Lüscher. (2004). Associative somatodendritic interaction in layer V pyramidal neurons is not affected by the antiepileptic drug lamotrigine. *Eur. J. Neurosci.*, 20(6):1688–1693.

Bialer, M., S.I. Johannessen, H.J. Kupferberg, R.H. Levy, E. Perucca, T. Tomson. (2007). Progress report on new antiepileptic drugs: a summary of the Eighth Eilat Conference (EILAT VIII). *Epilepsy Res.*, 73(1):1–52.

Bian, F., Z. Li, J. Offord, M.D. Davis, J. McCormick, C.P. Taylor, L.C. Walker. (2006). Calcium channel alpha2-delta type 1 subunit is the major binding protein for pregabalin in neocortex, hippocampus, amygdale, and spinal cord: an *ex vivo* autoradiographic study in alpha2-delta type 1 genetically modified mice. *Brain Res.*, 1075(1):68–80.

Bianchi, M.T., J. Pathmanathan, S.S. Cash. (2009). From ion channels to complex networks: magic bullet versus magic shotgun approaches to anticonvulsant pharmacotherapy. *Med. Hypotheses*, 72(3):297–305.

Biervert, C., B.C. Schroeder, C. Kubisch, S.F. Berkovic, P. Propping, T.J. Jentsch, O.K. Steinlein. (1998). A potassium channel mutation in neonatal human epilepsy. *Science*, 279(5349):403–406.

Bouwman, B.M., P. Suffczynski, I.S. Midzyanovskaya, E. Maris, P.L. van den Broek, C.M. van Rijn. (2007). The effects of vigabatrin on spike and wave discharges in WAG/Rij rats. *Epilepsy Res.*, 76(1):34–40.

Broicher, T., T. Seidenbecher, P. Meuth, T. Munsch, S.G. Meuth, T. Kanyshkova, H.C. Pape, T. Budde. (2007). T-current related effects of antiepileptic drugs and a $Ca^{2+}$ channel antagonist on thalamic relay and local circuit interneurons in a rat model of absence epilepsy. *Neuropharmacology*, 53(3):431–446.

Brumberg, J.C., L.G. Nowak, D.A. McCormick. (2000). Ionic mechanisms underlying repetitive high-frequency burst firing in supragranular cortical neurons. *J. Neurosci.*, 20(13):4829–4843.

Carunchio, I., M. Pieri, M.T. Ciotti, F. Albo, C. Zona. (2007). Modulation of AMPA receptors in cultured cortical neurons induced by the antiepileptic drug levetiracetam. *Epilepsia*, 48(4):654–662.

Chang, H.R., C.C. Kuo. (2007a). Characterization of the gating conformational changes in the felbamate binding site in NMDA channels. *Biophys. J.*, 93(2):456–466.

Chang, H.R., C.C. Kuo. (2007b). Extracellular proton-modulated pore-blocking effect of the anticonvulsant felbamate on NMDA channels. *Biophys. J.*, 93(6):1981–1992.

Chang, H.R., C.C. Kuo. (2008). Molecular determinants of the anticonvulsant felbamate binding site in the *N*-methyl-D-aspartate receptor. *J. Med. Chem.*, 51(6):1534–1545.

Chao, T.I., C. Alzheimer (1995). Effects of phenytoin on the persistent $Na^+$ current of mammalian CNS neurones. *NeuroReport*, 6(13):1778–1780.

Chen, Y., J. Lu, H. Pan, Y. Zhang, H. Wu, K. Xu, X. Liu, Y. Jiang, X. Bao, Z. Yao, K. Ding, W.H. Lo, B. Qiang, P. Chan, Y. Shen, X. Wu. (2003). Association between genetic variation of CACNA1H and childhood absence epilepsy. *Ann. Neurol.*, 54(2):239–243.

Chioza, B., K. Everett, H. Aschauer, O. Brouwer, P. Callenbach et al. (2006). Evaluation of CACNA1H in European patients with childhood absence epilepsy. *Epilepsy Res.*, 69(2):177–181.

Coenen, A.M., E.H. Blezer, E.L. van Luijtelaar. (1995). Effects of the GABA-uptake inhibitor tiagabine on electroencephalogram, spike–wave discharges and behaviour of rats. *Epilepsy Res.*, 21(2):89–94.

Cooper, E.C., L.Y. Jan. (2003). M-channels: neurological diseases, neuromodulation, and drug development. *Arch. Neurol.*, 60(4):496–500.

Cope, D.W., S.W. Hughes, V. Crunelli. (2005). $GABA_A$ receptor-mediated tonic inhibition in thalamic neurons. *J. Neurosci.*, 25(50):11553–11563.

Coulter, D.A., J.R. Huguenard, D.A. Prince. (1989). Characterization of ethosuximide reduction of low-threshold calcium current in thalamic neurons. *Ann. Neurol.*, 25(6):582–593.

Crowder, K.M., J.M. Gunther, T.A. Jones, B.D. Hale, H.Z. Zhang, M.R. Peterson, R.H. Scheller, C. Chavkin, S.M. Bajjalieh. (1999). Abnormal neurotransmission in mice lacking synaptic vesicle protein 2A (SV2A). *Proc. Natl. Acad. Sci. U.S.A.*, 96(26):15268–15273.

Crunelli, V., N. Leresche. (2002). Childhood absence epilepsy: genes, channels, neurons and networks. *Nat. Rev. Neurosci.*, 3(5):371–382.

Custer, K.L., N.S. Austin, J.M. Sullivan, S.M. Bajjalieh. (2006). Synaptic vesicle protein 2 enhances release probability at quiescent synapses. *J. Neurosci.*, 26(4):1303–1313.

De Simone, G., A. Di Fiore, V. Menchise, C. Pedone, J. Antel, A. Casini, A. Scozzafava, M. Wurl, C.T. Supuran. (2005). Carbonic anhydrase inhibitors. Zonisamide is an effective inhibitor of the cytosolic isozyme II and mitochondrial isozyme. V. Solution and x-ray crystallographic studies. *Bioorg. Med. Chem. Lett.*, 15(9):2315–2320.

DeLorenzo, R.J., S. Sombati, D.A. Coulter. (2000). Effects of topiramate on sustained repetitive firing and spontaneous recurrent seizure discharges in cultured hippocampal neurons. *Epilepsia*, 41(Suppl. 1):S40–S44.

Derry, J.M., S.M. Dunn, M. Davies. (2004). Identification of a residue in the gamma-aminobutyric acid type A receptor alpha subunit that differentially affects diazepam-sensitive and -insensitive benzodiazepine site binding. *J. Neurochem.*, 88(6):1431–1438.

Devaux, J.J., K.A. Kleopa, E.C. Cooper, S.S. Scherer. (2004). KCNQ2 is a nodal K+ channel. *J. Neurosci.*, 24(5):1236–1244.

Díez-Sampedro, A., W.R. Silverman, J.F. Bautista, G.B. Richerson. (2006). Mechanism of increased open probability by a mutation of the BK channel. *J. Neurophysiol.*, 96(3):1507–1516.

Dodgson, S.J., R.P. Shank, B.E. Maryanoff. (2000). Topiramate as an inhibitor of carbonic anhydrase isoenzymes. *Epilepsia*, 41(Suppl. 1):S35–S39.

Dooley, D.J., C.P. Taylor, S. Donevan, D. Feltner. (2007). $Ca^{2+}$ channel alpha2delta ligands: novel modulators of neurotransmission. *Trends Pharmacol. Sci.*, 28(2):75–82.

Dreixler, J.C., J. Bian, Y. Cao, M.T. Roberts, J.D. Roizen, K.M. Houamed. (2000). Block of rat brain recombinant SK channels by tricyclic antidepressants and related compounds. *Eur. J. Pharmacol.*, 401(1):1–7.

Du, W., J.F. Bautista, H. Yang, A. Diez-Sampedro, S.A. You, L. Wang, P. Kotagal, H.O. Luders, J. Shi, J. Cui, G.B. Richerson, Q.K. Wang. (2005). Calcium-sensitive potassium channelopathy in human epilepsy and paroxysmal movement disorder. *Nat. Genet.*, 37(7):733–738.

Errington, A.C., T. Stohr, C. Heers, G. Lees. (2008). The investigational anticonvulsant lacosamide selectively enhances slow inactivation of voltage-gated sodium channels. *Mol. Pharmacol.*, 73(1):157–169.

Faingold, C.L. (2004). Emergent properties of CNS neuronal networks as targets for pharmacology: application to anticonvulsant drug action. *Prog. Neurobiol.*, 72(1):55–85.

ffrench-Mullen, J.M., J.L. Barker, M.A. Rogawski. (1993). Calcium current block by (–)-pentobarbital, phenobarbital, and CHEB but not (+)-pentobarbital in acutely isolated hippocampal CA1 neurons: comparison with effects on GABA-activated Cl- current. *J. Neurosci.*, 13(8):3211–3221.

Fohlmeister, J.F., W.J. Adelman, Jr., J.J. Brennan. (1984). Excitable channel currents and gating times in the presence of anticonvulsants ethosuximide and valproate. *J. Pharmacol. Exp. Ther.*, 230(1):75–81.

Gibbs, J.W., 3rd, S. Sombati, R.J. DeLorenzo, D.A. Coulter. (2000). Cellular actions of topiramate: blockade of kainate-evoked inward currents in cultured hippocampal neurons. *Epilepsia.* 41(Suppl. 1):S10–S16.

Gillard, M., P. Chatelain, B. Fuks. (2006). Binding characteristics of levetiracetam to synaptic vesicle protein 2A (SV2A) in human brain and in CHO cells expressing the human recombinant protein. *Eur. J. Pharmacol.*, 536(1–2):102–108.

Gomora, J.C., A.N. Daud, M. Weiergraber, E. Perez-Reyes. (2001). Block of cloned human T-type calcium channels by succinimide antiepileptic drugs. *Mol. Pharmacol.*, 60(5):1121–1132.

Gordey, M., T.M. DeLorey, R.W. Olsen. (2000). Differential sensitivity of recombinant GABA(A) receptors expressed in *Xenopus* oocytes to modulation by topiramate. *Epilepsia.* 41(Suppl. 1):S25–S29.

Granger, P., B. Biton, C. Faure, X. Vige, H. Depoortere, D. Graham, S.Z. Langer, B. Scatton, P. Avenet. (1995). Modulation of the gamma-aminobutyric acid type A receptor by the antiepileptic drugs carbamazepine and phenytoin. *Mol. Pharmacol.*, 47(6):1189–1196.

Gross, R.A., R.L. Macdonald. (1988). Barbiturates and nifedipine have different and selective effects on calcium currents of mouse DRG neurons in culture: a possible basis for differing clinical actions. The 1987 S. Weir Mitchell Award. *Neurology*, 38(3):443–451.

Grunze, H., R.W. Greene, H.J. Moller, T. Meyer, J. Walden. (1998a). Lamotrigine may limit pathological excitation in the hippocampus by modulating a transient potassium outward current. *Brain Res.*, 791(1–2):330–334.

Grunze, H., J. von Wegerer, R.W. Greene, J. Walden. (1998b). Modulation of calcium and potassium currents by lamotrigine. *Neuropsychobiology*, 38(3):131–138.

Gryder, D.S., M.A. Rogawski. (2003). Selective antagonism of GluR5 kainate-receptor-mediated synaptic currents by topiramate in rat basolateral amygdala neurons. *J. Neurosci.*, 23(18):7069–7074.

Hainsworth, A.H., N.C. McNaughton, A. Pereverzev, T. Schneider, A.D. Randall. (2003). Actions of sipatrigine, 202W92 and lamotrigine on R-type and T-type $Ca^{2+}$ channel currents. *Eur. J. Pharmacol.*, 467(1–3):77–80.

Hara, K., T. Sata. (2007). Inhibitory effect of gabapentin on N-methyl-D-aspartate receptors expressed in *Xenopus* oocytes. *Acta Anaesthesiol. Scand.*, 51(1):122–128.

Harty, T.P., M.A. Rogawski. (2000). Felbamate block of recombinant N-methyl-D-aspartate receptors: selectivity for the NR2B subunit. *Epilepsy Res.*, 39(1):47–55.

Herrero, A.I., N. Del Olmo, J.R. Gonzalez-Escalada, J.M. Solis. (2002). Two new actions of topiramate: inhibition of depolarizing GABA(A)-mediated responses and activation of a potassium conductance. *Neuropharmacology*, 42(2):210–220.

Hough, C.J., R.P. Irwin, X.M. Gao, M.A. Rogawski, D.M. Chuang. (1996). Carbamazepine inhibition of N-methyl-D-aspartate-evoked calcium influx in rat cerebellar granule cells. *J. Pharmacol. Exp. Ther.*, 276(1):143–149.

Huang, C.W., C.C. Huang, Y.C. Liu, S.N. Wu. (2004). Inhibitory effect of lamotrigine on A-type potassium current in hippocampal neuron-derived H19-7 cells. *Epilepsia*, 45(7):729–736.

Huang, C.W., C.C. Huang, S.N. Wu. (2007). Activation by zonisamide, a newer antiepileptic drug, of large-conductance calcium-activated potassium channel in differentiated hippocampal neuron-derived H19-7 cells. *J. Pharmacol. Exp. Ther.*, 321(1):98–106.

Huang, C.W., C.C. Huang, M.W. Lin, J.J. Tsai, S.N. Wu. (2008). The synergistic inhibitory actions of oxcarbazepine on voltage-gated sodium and potassium currents in differentiated NG108-15 neuronal cells and model neurons. *Int. J. Neuropsychopharmacol.*, 11(5):597–610.

Huguenard, J.R. (2002). Block of T-type Ca(2+) channels is an important action of succinimide antiabsence drugs. *Epilepsy Curr.*, 2(2):49–52.

Kaminski, R.M., M. Banerjee, M.A. Rogawski. (2004). Topiramate selectively protects against seizures induced by ATPA, a GluR5 kainate receptor agonist. *Neuropharmacology*, 46(8):1097–1104.

Kaminski, R.M., M. Gillard, K. Leclercq, E. Hanon, G. Lorent, D. Dassesse, A. Matagne, H. Klitgaard. (2009). Proepileptic phenotype of SV2A-deficient mice is associated with reduced anticonvulsant efficacy of levetiracetam. *Epilepsia*, 50(7):1729–1740.

Kawano, H., S. Sashihara, T. Mita, K. Ohno, M. Kawamura, K. Yoshii. (1994). Phenytoin, an antiepileptic drug, competitively blocked non-NMDA receptors produced by *Xenopus* oocytes. *Neurosci. Lett.*, 166(2):183–186.

Kemp, J.A., G.R. Marshall, E.H. Wong, G.N. Woodruff. (1987). The affinities, potencies and efficacies of some benzodiazepine-receptor agonists, antagonists and inverse-agonists at rat hippocampal GABA$_A$-receptors. *Br. J. Pharmacol.*, 91(3):601–608.

Kim, D., I. Song, S. Keum, T. Lee, M.J. Jeong, S.S. Kim, M.W. McEnery, H.S. Shin. (2001). Lack of the burst firing of thalamocortical relay neurons and resistance to absence seizures in mice lacking alpha(1G) T-type Ca(2+) channels. *Neuron*, 31(1):35–45.

Kleckner, N.W., J.C. Glazewski, C.C. Chen, T.D. Moscrip. (1999). Subtype-selective antagonism of N-methyl-D-aspartate receptors by felbamate: insights into the mechanism of action. *J. Pharmacol. Exp. Ther.*, 289(2):886–894.

Kuo, C.C. (1998). A common anticonvulsant binding site for phenytoin, carbamazepine, and lamotrigine in neuronal Na$^+$ channels. *Mol. Pharmacol.*, 54(4):712–721.

Kuo, C.C., B.P. Bean. (1994). Slow binding of phenytoin to inactivated sodium channels in rat hippocampal neurons. *Mol. Pharmacol.*, 46(4):716–725.

Kuo, C.C., L. Lu. (1997). Characterization of lamotrigine inhibition of Na$^+$ channels in rat hippocampal neurones. *Br. J. Pharmacol.*, 121(6):1231–1238.

Kuo, C.C., R.S. Chen, L. Lu, R.C. Chen. (1997). Carbamazepine inhibition of neuronal Na$^+$ currents: quantitative distinction from phenytoin and possible therapeutic implications. *Mol. Pharmacol.*, 51(6):1077–1083.

Kuo, C.C., B.J. Lin, H.R. Chang, C.P. Hsieh. (2004). Use-dependent inhibition of the N-methyl-D-aspartate currents by felbamate: a gating modifier with selective binding to the desensitized channels. *Mol. Pharmacol.*, 65(2):370–380.

Kuzmiski, J.B., W. Barr, G.W. Zamponi, B.A. MacVicar. (2005). Topiramate inhibits the initiation of plateau potentials in CA1 neurons by depressing R-type calcium channels. *Epilepsia*, 46(4):481–489.

Lacinová, L. (2004). Pharmacology of recombinant low-voltage activated calcium channels. *Curr. Drug Targets CNS Neurol. Disord.*, 3(2):105–111.

Lampe, H., H. Bigalke. (1990). Carbamazepine blocks NMDA-activated currents in cultured spinal cord neurons. *NeuroReport*, 1(1):26–28.

Lampl, I., P. Schwindt, W. Crill. (1998). Reduction of cortical pyramidal neuron excitability by the action of phenytoin on persistent Na$^+$ current. *J. Pharmacol. Exp. Ther.*, 284(1):228–237.

Lang, D.G., C.M. Wang, B.R. Cooper. (1993). Lamotrigine, phenytoin and carbamazepine interactions on the sodium current present in N4TG1 mouse neuroblastoma cells. *J. Pharmacol. Exp. Ther.*, 266(2):829–835.

Lee, C.Y., W.M. Fu, C.C. Chen, M.J. Su, H.H. Liou. (2008). Lamotrigine inhibits postsynaptic AMPA receptor and glutamate release in the dentate gyrus. *Epilepsia*, 49(5):888–897.

Leresche, N., H.R. Parri, G. Erdemli, A. Guyon, J.P. Turner, S.R. Williams, E. Asprodini, V. Crunelli. (1998). On the action of the anti-absence drug ethosuximide in the rat and cat thalamus. *J. Neurosci.*, 18(13):4842–4853.

Lorenz, S., A. Heils, J.M. Kasper, T. Sander. (2007). Allelic association of a truncation mutation of the KCNMB3 gene with idiopathic generalized epilepsy. *Am. J. Med. Genet. B. Neuropsychiatr. Genet.*, 144B(1):10–13.

Ludwig, A., T. Budde, J. Stieber, S. Moosmang, C. Wahl et al. (2003). Absence epilepsy and sinus dysrhythmia in mice lacking the pacemaker channel HCN2. *EMBO J.*, 22(2):216–224.

Lukyanetz, E.A., V.M. Shkryl, P.G. Kostyuk. (2002). Selective blockade of N-type calcium channels by levetiracetam. *Epilepsia*, 43(1):9–18.

Lynch, B.A., N. Lambeng, K. Nocka, P. Kensel-Hammes, S.M. Bajjalieh, A. Matagne, B. Fuks. (2004). The synaptic vesicle protein SV2A is the binding site for the antiepileptic drug levetiracetam. *Proc. Natl. Acad. Sci. U.S.A.*, 101(26):9861–9866.

Macdonald, R.L., C.J. Rogers, R.E. Twyman. (1989). Barbiturate regulation of kinetic properties of the $GABA_A$ receptor channel of mouse spinal neurones in culture. *J. Physiol.*, 417:483–500.

Madeja, M., D.G. Margineanu, A. Gorji, E. Siep, P. Boerrigter, H. Klitgaard, E.J. Speckmann. (2003). Reduction of voltage-operated potassium currents by levetiracetam: a novel antiepileptic mechanism of action? *Neuropharmacology*, 45(5):661–671.

Martella, G., C. Costa, A. Pisani, L.M. Cupini, G. Bernardi, P. Calabresi. (2008). Antiepileptic drugs on calcium currents recorded from cortical and PAG neurons: therapeutic implications for migraine. *Cephalalgia*, 28(12):1315–1326.

Matar, N., W. Jin, H. Wrubel, J. Hescheler, T. Schneider, M. Weiergraber. (2009). Zonisamide block of cloned human T-type voltage-gated calcium channels. *Epilepsy Res.*, 83(2–3):224–234.

Mazarati, A., J. Wu, D. Shin, Y.S. Kwon, R. Sankar. (2008). Antiepileptogenic and antiictogenic effects of retigabine under conditions of rapid kindling: an ontogenic study. *Epilepsia*, 49(10):1777–1786.

McCormick, D.A., D. Contreras. (2001). On the cellular and network bases of epileptic seizures. *Annu. Rev. Physiol.*, 63:815–846.

McLean, M.J., A.A. Bukhari, A.W. Wamil. (2000). Effects of topiramate on sodium-dependent action-potential firing by mouse spinal cord neurons in cell culture. *Epilepsia*, 41(Suppl. 1):S21–S24.

McNaughton, N.C., C.H. Davies, A. Randall. (2004). Inhibition of alpha(1E) Ca(2+) channels by carbonic anhydrase inhibitors. *J. Pharmacol. Sci.*, 95(2):240–247.

Meldrum, B.S., M.A. Rogawski. (2007). Molecular targets for antiepileptic drug development. *Neurotherapeutics*, 4(1):18–61.

Mody, I., R.A. Pearce. (2004). Diversity of inhibitory neurotransmission through GABA(A) receptors. *Trends Neurosci.*, 27(9):569–575.

Mulley, J.C., I.E. Scheffer, S. Petrou, L.M. Dibbens, S.F. Berkovic, L.A. Harkin. (2005). *SCN1A* mutations and epilepsy. *Hum. Mutat.*, 25(6):535–542.

Munsch, T., H.C. Pape. (1999). Upregulation of the hyperpolarization-activated cation current in rat thalamic relay neurones by acetazolamide. *J. Physiol.*, 519(Pt. 2):505–514.

Niespodziany, I., H. Klitgaard, D.G. Margineanu. (2001). Levetiracetam inhibits the high-voltage-activated Ca(2+) current in pyramidal neurones of rat hippocampal slices. *Neurosci. Lett.*, 306(1–2):5–8.

Niespodziany, I., H. Klitgaard, D.G. Margineanu. (2004). Is the persistent sodium current a specific target of anti-absence drugs? *NeuroReport*, 15(6):1049–1052.

Nobile, M., L. Lagostena. (1998). A discriminant block among $K^+$ channel types by phenytoin in neuroblastoma cells. *Br. J. Pharmacol.*, 124(8):1698–1702.

Nobile, M., P. Vercellino. (1997). Inhibition of delayed rectifier K+ channels by phenytoin in rat neuroblastoma cells. *Br. J. Pharmacol.*, 120(4):647–652.

Nusser, Z., I. Mody. (2002). Selective modulation of tonic and phasic inhibitions in dentate gyrus granule cells. *J. Neurophysiol.*, 87(5):2624–2628.

Ogiwara, I., H. Miyamoto, N. Morita, N. Atapour, E. Mazaki et al. (2007). Na(v)1.1 localizes to axons of parvalbumin-positive inhibitory interneurons: a circuit basis for epileptic seizures in mice carrying an *Scn1a* gene mutation. *J. Neurosci.*, 27(22):5903–5914.

Phillips, I., K.F. Martin, K.S. Thompson, D.J. Heal. (1997). Weak blockade of AMPA receptor-mediated depolarisations in the rat cortical wedge by phenytoin but not lamotrigine or carbamazepine. *Eur. J. Pharmacol.*, 337(2–3):189–195.

Pisani, A., A. Stefani, A. Siniscalchi, N.B. Mercuri, G. Bernardi, P. Calabresi. (1995). Electrophysiological actions of felbamate on rat striatal neurones. *Br. J. Pharmacol.*, 116(3):2053–2061.

Pisciotta, M., G. Prestipino. (2002). Anticonvulsant phenytoin affects voltage-gated potassium currents in cerebellar granule cells. *Brain Res.*, 941(1–2):53–61.

Poolos, N.P. (2005). The h-channel: a potential channelopathy in epilepsy? *Epilepsy Behav.*, 7(1):51–56.

Poolos, N.P., M. Migliore, D. Johnston. (2002). Pharmacological upregulation of h-channels reduces the excitability of pyramidal neuron dendrites. *Nat. Neurosci.*, 5(8):767–774.

Poulsen, C.F., T.A. Simeone, T.E. Maar, V. Smith-Swintosky, H.S. White, A. Schousboe. (2004). Modulation by topiramate of AMPA and kainate mediated calcium influx in cultured cerebral cortical, hippocampal and cerebellar neurons. *Neurochem. Res.*, 29(1):275–282.

Pritchett, D.B., H. Luddens, P.H. Seeburg. (1989). Type I and type II GABA$_A$-benzodiazepine receptors produced in transfected cells. *Science*, 245(4924):1389–1392.

Raol, Y.H., D.A. Lapides, J.G. Keating, A.R. Brooks-Kayal, E.C. Cooper. (2009). A KCNQ channel opener for experimental neonatal seizures and status epilepticus. *Ann. Neurol.*, 65(3):326–336.

Remy, S., B.W. Urban, C.E. Elger, H. Beck. (2003). Anticonvulsant pharmacology of voltage-gated Na⁺ channels in hippocampal neurons of control and chronically epileptic rats. *Eur. J. Neurosci.*, 17(12):2648–2658.

Rho, J.M., R. Sankar. (1999). The pharmacologic basis of antiepileptic drug action. *Epilepsia*, 40(11):1471–1483.

Rho, J.M., S.D. Donevan, M.A. Rogawski. (1994). Mechanism of action of the anticonvulsant felbamate: opposing effects on *N*-methyl-D-aspartate and gamma-aminobutyric acidA receptors. *Ann. Neurol.*, 35(2):229–234.

Rho, J.M., S.D. Donevan, M.A. Rogawski. (1996). Direct activation of GABA$_A$ receptors by barbiturates in cultured rat hippocampal neurons. *J. Physiol.*, 497(Pt. 2):509–522.

Rho, J.M., S.D. Donevan, M.A. Rogawski. (1997). Barbiturate-like actions of the propanediol dicarbamates felbamate and meprobamate. *J. Pharmacol. Exp. Ther.*, 280(3):1383–1391.

Rogawski, M.A. (2006). Diverse mechanisms of antiepileptic drugs in the development pipeline. *Epilepsy Res.*, 69(3):273–294.

Rogawski, M.A., W. Löscher. (2004). The neurobiology of antiepileptic drugs. *Nat. Rev. Neurosci.*, 5(7):553–564.

Sankar, R., G.L. Holmes. (2004). Mechanisms of action for the commonly used antiepileptic drugs: relevance to antiepileptic drug-associated neurobehavioral adverse effects. *J. Child Neurol.*, 19(Suppl. 1):S6–S14.

Schauf, C.L. (1987a). Anticonvulsants modify inactivation but not activation processes of sodium channels in *Myxicola* axons. *Can. J. Physiol. Pharmacol.*, 65(6):1220–1225.

Schauf, C.L. (1987b). Zonisamide enhances slow sodium inactivation in *Myxicola*. *Brain Res.*, 413(1):185–188.

Schmutz, M., F. Brugger, C. Gentsch, M.J. McLean, H.R. Olpe. (1994). Oxcarbazepine: preclinical anticonvulsant profile and putative mechanisms of action. *Epilepsia*, 35(Suppl. 5):S47–S50.

Schousboe, A., A. Sarup, O.M. Larsson, H.S. White. (2004). GABA transporters as drug targets for modulation of GABAergic activity. *Biochem. Pharmacol.*, 68(8):1557–1563.

Segal, M.M., A.F. Douglas. (1997). Late sodium channel openings underlying epileptiform activity are preferentially diminished by the anticonvulsant phenytoin. *J. Neurophysiol.*, 77(6):3021–3034.

Shah, M.M., M. Migliore, I. Valencia, E.C. Cooper, D.A. Brown. (2008). Functional significance of axonal Kv7 channels in hippocampal pyramidal neurons. *Proc. Natl. Acad. Sci. U.S.A.*, 105(22):7869–7874.

Shank, R.P., V.L. Smith-Swintosky, B.E. Maryanoff. (2008). Carbonic anhydrase inhibition: insight into the characteristics of zonisamide, topiramate, and the sulfamide cognate of topiramate. *J. Enzyme Inhib. Med. Chem.*, 23(2):271–276.

Sigel, E., R. Baur. (1988). Allosteric modulation by benzodiazepine receptor ligands of the GABA$_A$ receptor channel expressed in *Xenopus* oocytes. *J. Neurosci.*, 8(1):289–295.

Sigel, E., R. Baur, G. Trube, H. Mohler, P. Malherbe. (1990). The effect of subunit composition of rat brain GABA$_A$ receptors on channel function. *Neuron*, 5(5):703–711.

Simeone, T.A., S.D. Donevan, J.M. Rho. (2003). Molecular biology and ontogeny of gamma-aminobutyric acid (GABA) receptors in the mammalian central nervous system. *J. Child Neurol.*, 18(1):39–49.

Simeone, T.A., R.M. Sanchez, J.M. Rho. (2004). Molecular biology and ontogeny of glutamate receptors in the mammalian central nervous system. *J. Child Neurol.*, 19(5):343–361.

Simeone, T.A., J.F. Otto, K.S. Wilcox, H.S. White. (2006a). Felbamate is a subunit selective modulator of recombinant gamma-aminobutyric acid type A receptors expressed in *Xenopus* oocytes. *Eur. J. Pharmacol.*, 552(1–3):31–35.

Simeone, T.A., K.S. Wilcox, H.S. White. (2006b). Subunit selectivity of topiramate modulation of heteromeric GABA(A) receptors. *Neuropharmacology*, 50(7):845–857.

Singh, N.A., C. Charlier, D. Stauffer, B.R. DuPont, R.J. Leach et al. (1998). A novel potassium channel gene, KCNQ2, is mutated in an inherited epilepsy of newborns. *Nat. Genet.*, 18(1):25–29.

Skradski, S., H.S. White. (2000). Topiramate blocks kainate-evoked cobalt influx into cultured neurons. *Epilepsia*, 41(Suppl. 1):S45–S47.

Sohal, V.S., R. Keist, U. Rudolph, J.R. Huguenard. (2003). Dynamic GABA(A) receptor subtype-specific modulation of the synchrony and duration of thalamic oscillations. *J. Neurosci.*, 23(9):3649–3657.

Song, J.H., K. Nagata, C.S. Huang, J.Z. Yeh, T. Narahashi. (1996). Differential block of two types of sodium channels by anticonvulsants. *NeuroReport*, 7(18):3031–3036.

Spadoni, F., A.H. Hainsworth, N.B. Mercuri, L. Caputi, G. Martella, F. Lavaroni, G. Bernardi, A. Stefani. (2002). Lamotrigine derivatives and riluzole inhibit $I_{Na,P}$ in cortical neurons. *NeuroReport*, 13(9):1167–1170.

Stafstrom, C.E. (2007). Persistent sodium current and its role in epilepsy. *Epilepsy Curr.*, 7(1):15–22.

Stefani, A., P. Calabresi, A. Pisani, N.B. Mercuri, A. Siniscalchi, G. Bernardi. (1996a). Felbamate inhibits dihydropyridine-sensitive calcium channels in central neurons., *J. Pharmacol. Exp. Ther.*, 277(1):121–127.

Stefani, A., F. Spadoni, A. Siniscalchi, G. Bernardi. (1996b). Lamotrigine inhibits $Ca^{2+}$ currents in cortical neurons: functional implications. *Eur. J. Pharmacol.*, 307(1):113–116.

Stell, B.M., I. Mody. (2002). Receptors with different affinities mediate phasic and tonic GABA(A) conductances in hippocampal neurons. *J. Neurosci.*, 22(10):RC223.

Stephenson, V.C., R.A. Heyding, D.F. Weaver. (2005). The "promiscuous drug concept" with applications to Alzheimer's disease. *FEBS Lett.*, 579(6):1338–1342.

Subramaniam, S., J.M. Rho, L. Penix, S.D. Donevan, R.P. Fielding, M.A. Rogawski. (1995). Felbamate block of the *N*-methyl-D-aspartate receptor. *J. Pharmacol. Exp. Ther.*, 273(2):878–886.

Sun, G.C., T.R. Werkman, A. Battefeld, J.J. Clare, W.J. Wadman. (2007). Carbamazepine and topiramate modulation of transient and persistent sodium currents studied in HEK293 cells expressing the Na(v)1.3 alpha-subunit. *Epilepsia*, 48(4):774–782.

Surges, R., T.M. Freiman, T.J. Feuerstein. (2003). Gabapentin increases the hyperpolarization-activated cation current $I_h$ in rat CA1 pyramidal cells. *Epilepsia*, 44(2):150–156.

Surges, R., T.M. Freiman, T.J. Feuerstein. (2004). Input resistance is voltage dependent due to activation of $I_h$ channels in rat CA1 pyramidal cells. *J. Neurosci. Res.*, 76(4):475–480.

Taglialatela, M., E. Ongini, A.M. Brown, G. Di Renzo, L. Annunziato. (1996). Felbamate inhibits cloned voltage-dependent Na+ channels from human and rat brain. *Eur. J. Pharmacol.*, 316(2–3):373–377.

Taverna, S., M. Mantegazza, S. Franceschetti, G. Avanzini. (1998). Valproate selectively reduces the persistent fraction of Na+ current in neocortical neurons. *Epilepsy Res.*, 32(1–2):304–308.

Taverna, S., G. Sancini, M. Mantegazza, S. Franceschetti, G. Avanzini. (1999). Inhibition of transient and persistent Na+ current fractions by the new anticonvulsant topiramate. *J. Pharmacol. Exp. Ther.*, 288(3):960–968.

Taylor, C.P., T. Angelotti, E. Fauman. (2007). Pharmacology and mechanism of action of pregabalin: the calcium channel $\alpha_2$-$\delta$ (alpha2-delta) subunit as a target for antiepileptic drug discovery. *Epilepsy Res.*, 73(2):137–150.

Thompson, S.A., P.J. Whiting, K.A. Wafford. (1996). Barbiturate interactions at the human $GABA_A$ receptor: dependence on receptor subunit combination. *Br. J. Pharmacol.*, 117(3):521–527.

Thompson, S.M., B.H. Gähwiler. (1992). Effects of the GABA uptake inhibitor tiagabine on inhibitory synaptic potentials in rat hippocampal slice cultures. *J. Neurophysiol.*, 67(6):1698–1701.

Todorovic, S.M., C.J. Lingle. (1998). Pharmacological properties of T-type $Ca^{2+}$ current in adult rat sensory neurons: effects of anticonvulsant and anesthetic agents. *J. Neurophysiol.*, 79(1):240–252.

Tricarico, D., M. Barbieri, A. Mele, G. Carbonara, D.C. Camerino. (2004). Carbonic anhydrase inhibitors are specific openers of skeletal muscle BK channel of K+-deficient rats. *FASEB J.*, 18(6):760–761.

Twyman, R.E., C.J. Rogers, R.L. Macdonald. (1989). Differential regulation of gamma-aminobutyric acid receptor channels by diazepam and phenobarbital. *Ann. Neurol.*, 25(3):213–220.

Van den Berg, R.J., P. Kok, R.A. Voskuyl. (1993). Valproate and sodium currents in cultured hippocampal neurons. *Exp. Brain Res.*, 93(2):279–287.

Velísek, L., S.L. Moshe, W. Cammer. (1993a). Developmental changes in seizure susceptibility in carbonic anhydrase II-deficient mice and normal littermates. *Brain Res. Dev. Brain Res.*, 72(2):321–324.

Velísek, L., S.L. Moshe, S.G. Xu, W. Cammer. (1993b). Reduced susceptibility to seizures in carbonic anhydrase II deficient mutant mice. *Epilepsy Res.*, 14(2):115–121.

Vergnes, M., C. Marescaux, A. Micheletti, A. Depaulis, L. Rumbach, J.M. Warter. (1984). Enhancement of spike and wave discharges by GABAmimetic drugs in rats with spontaneous petit-mal-like epilepsy. *Neurosci. Lett.*, 44(1):91–94.

Wamil, A.W., M.J. McLean. (1993). Phenytoin blocks *N*-methyl-D-aspartate responses of mouse central neurons. *J. Pharmacol. Exp. Ther.*, 267(1):218–227.

Wang, S.J., C.C. Huang, K.S. Hsu, J.J. Tsai, P.W. Gean. (1996). Inhibition of N-type calcium currents by lam-otrigine in rat amygdalar neurones. *NeuroReport*, 7(18):3037–3040.

Wang, S.J., J.J. Tsai, P.W. Gean. (1998). Lamotrigine inhibits depolarization-evoked $Ca^{++}$ influx in dissociated amygdala neurons. *Synapse*, 29(4):355–362.

White, H.S., S.D. Brown, J.H. Woodhead, G.A. Skeen, H.H. Wolf. (1997). Topiramate enhances GABA-mediated chloride flux and GABA-evoked chloride currents in murine brain neurons and increases sei-zure threshold. *Epilepsy Res.*, 28(3):167–179.

White, H.S., S.D. Brown, J.H. Woodhead, G.A. Skeen, H.H. Wolf. (2000). Topiramate modulates GABA-evoked currents in murine cortical neurons by a nonbenzodiazepine mechanism. *Epilepsia*, 41(Suppl. 1):S17–S20.

White, H.S., M.D. Smith, K.S. Wilcox. (2007). Mechanisms of action of antiepileptic drugs. *Int. Rev. Neurobiol.*, 81:85–110.

Xie, X., T.J. Dale, V.H. John, H.L. Cater, T.C. Peakman, J.J. Clare. (2001). Electrophysiological and pharma-cological properties of the human brain type IIA $Na^+$ channel expressed in a stable mammalian cell line. *Pflügers Arch.*, 441(4):425–433.

Xu, T., S.M. Bajjalieh. (2001). SV2 modulates the size of the readily releasable pool of secretory vesicles. *Nat. Cell Biol.*, 3(8):691–698.

Ying, S.W., F. Jia, S.Y. Abbas, F. Hofmann, A. Ludwig, P.A. Goldstein. (2007). Dendritic HCN2 channels con-strain glutamate-driven excitability in reticular thalamic neurons. *J. Neurosci.*, 27(32):8719–8732.

Youdim, M.B., J.J. Buccafusco. (2005). Multi-functional drugs for various CNS targets in the treatment of neurodegenerative disorders. *Trends Pharmacol. Sci.*, 26(1):27–35.

Yu, F.H., M. Mantegazza, R.E. Westenbroek, C.A. Robbins, F. Kalume, K.A. Burton, W.J. Spain, G.S. McKnight, T. Scheuer, W.A. Catterall. (2006). Reduced sodium current in GABAergic interneurons in a mouse model of severe myoclonic epilepsy in infancy. *Nat. Neurosci.*, 9(9):1142–1149.

Yue, C., Y. Yaari. (2004). KCNQ/M channels control spike after depolarization and burst generation in hip-pocampal neurons. *J. Neurosci.*, 24(19):4614–4624.

Zhang, X., A.A. Velumian, O.T. Jones, P.L. Carlen. (2000). Modulation of high-voltage-activated calcium channels in dentate granule cells by topiramate. *Epilepsia*, 41(Suppl. 1):S52–S60.

Zona, C., M.T. Ciotti, M. Avoli. (1997). Topiramate attenuates voltage-gated sodium currents in rat cerebellar granule cells. *Neurosci. Lett.*, 231(3):123–126.

Zona, C., V. Tancredi, P. Longone, G. D'Arcangelo, M. D'Antuono, M. Manfredi, M. Avoli. (2002). Neocortical potassium currents are enhanced by the antiepileptic drug lamotrigine. *Epilepsia*, 43(7):685–690.

# 9 Epilepsy and Disease Modification: Animal Models for Novel Drug Discovery

*H. Steve White*

## CONTENTS

## INTRODUCTION

Approximately 1 to 3% of the general worldwide population suffers from epilepsy. For the most part, pharmacotherapy represents the mainstay of patient treatment. Vagus nerve stimulation represents a potential option for those patients who are not viable candidates for surgery or whose seizures cannot be controlled by existing anticonvulsant drugs. Advances in brain imaging and seizure mapping techniques have made surgery an option for patients wherein a resectable seizure focus can be identified. Unfortunately, despite the availability of these new therapeutic options, a significant fraction of the patients with epilepsy continue to live with uncontrolled seizures, often at the expense of significant drug-induced adverse effects. Clearly, there is a need for more efficacious therapies that will not infringe on a patient's quality of life.

As with any other classes of drugs, the discovery and development of new antiepileptic drugs (AEDs) relies heavily on the employment of animal models to demonstrate efficacy and safety prior to their introduction in human volunteers. Obviously, the more predictive an animal model for a particular seizure type or syndrome, the greater the likelihood that the investigational AED will demonstrate efficacy in human clinical trials. Herein lies one of the most often discussed issues in the current-day AED discovery process: *What is the most appropriate animal model to employ when attempting to screen for efficacy against human epilepsy?*

**TABLE 9.1**
**Proposed Characteristics for an Animal**
**Model of Human Epilepsy**

Pathology consistent with human epilepsy (e.g., cell loss,
   gliosis, neurogenesis, axonal and synaptic reorganization)
Species-appropriate latent period following initial insult
Chronic hyperexcitability
Expression of spontaneous seizures following latent period
Resistance to one or more AEDs

This chapter reviews briefly the different approaches employed in the AED discovery process. Furthermore, an attempt is made to address some of the issues surrounding the development of more appropriate models of pharmacoresistant epilepsy. Finally, attention will be paid to the employment of strategies being discussed that may lead to the identification and development of the truly novel drug that prevents the development of epilepsy: the antiepileptogenic compound. An extensive discussion of these issues is beyond the scope of this chapter. Where possible, the reader will be referred to pertinent reviews for a more detailed discussion.

## THE IDEAL MODEL SYSTEM

The ideal screening model should reflect a pathophysiology and phenomenology similar to those of human epilepsy. Based on our current understanding of the factors contributing to the development of temporal lobe epilepsy, a number of characteristics can be suggested (Table 9.1). For example, the ideal animal model would be expected to display spontaneous seizures within some time frame following an initiating insult. The latent period between initiating insult and expression of a behavioral seizure would be expected to be days to weeks in duration (which would be analogous to months to years in humans). The latent period should reflect a period of clinical quiescence, in which the animal does not display behavioral seizures. The latent period also defines the time frame in which functional and structural reorganization occurs as a consequence of a specific series of pathological events set in motion by the activation of a series of critical modulators. As such, the functional and structural changes that contribute to the expression of spontaneous seizures should also be consistent with those observed in human tissue. The ideal model of human temporal lobe epilepsy may also be characterized by additional neuroplastic changes that manifest as pharmacological resistance to existing anticonvulsant drugs or as neurobehavioral or cognitive impairment.

From a drug discovery perspective, the ideal model would provide an opportunity to assess efficacy against a pharmacoresistant seizure phenotype and permit the early evaluation of a drug's ability to modify the course of epilepsy following an epileptogenic insult. Because human epilepsy is a heterogeneous neurological disorder that encompasses many seizure phenotypes and syndromes, it is highly unlikely that any one animal model will ever predict the full therapeutic potential of an investigational AED. This necessitates the evaluation of an investigational AED in several syndrome-specific model systems.

## IDENTIFICATION OF ANTICONVULSANT ACTIVITY

The current era of AED discovery was ushered in by Merritt and Putnam (1937), who demonstrated the feasibility of using the maximal electroshock seizure (MES) model to identify the anticonvulsant potential of phenytoin. Subsequently, a number of animal models have been employed in the search for more efficacious and tolerable AEDs. In the early 1970s, the National Institute of Neurological Disorders and Stroke (NINDS) embarked on a mission to encourage basic research aimed at a greater understanding of the factors that contribute to the initiation, propagation, and amelioration of seizures. As part of this effort, the Anticonvulsant Drug Development (ADD) program was created to foster the development of new drugs for the treatment of epilepsy.

Since 1975, the ADD program has accessioned over 34,000 investigational anticonvulsant drugs from the academic community and the pharmaceutical industry. This ongoing effort has led to the identification and development of 12 new anticonvulsant drugs, including felbamate (1993), gabapentin (1993), lamotrigine (1994), fosphenytoin (1996), topiramate (1996), tiagabine (1997), levetiracetam (1999), zonisamide (2000), oxcarbazepine (2000), pregabalin (2005), rufinamide (2006), and lacosamide (2008). The introduction of these second- and third-generation anticonvulsant drugs has had a clear impact on the lives of many hundreds of thousands of epilepsy patients. Many have benefited from improved seizure control and fewer drug-induced adverse events. In addition, many of the newer drugs have demonstrated a more favorable pharmacokinetic profile. Despite the introduction of several new AEDs, however, it is estimated that the percentage of patients with partial epilepsy that remain pharmacoresistant to the existing AEDs has not changed significantly since the early 1970s and remains relatively stable, between 25 and 40% (Löscher, 2002). As such, this patient population has a substantial and unmet need for better therapies.

In addition to the advancements made at the therapeutic level, significant progress has also been made at the basic science level, as well. Results from basic research have led to greater insights into the biology of epilepsy at both molecular and genetic levels. Why, then, given our current level of understanding of the pathophysiology of epilepsy, the introduction of several new drugs, greater access and improvements in neurosurgery, and the introduction of the vagus nerve stimulator, does there still remain such a significant unmet clinical need? Furthermore, will it ever be possible to translate findings at the bench to an effective therapy for the treatment of therapy-resistant seizures—or, better yet, devise a therapy that will prevent or modify the development of epilepsy in the susceptible population? Addressing these questions will require appropriate model systems wherein proposed therapies can be systematically evaluated. Unfortunately, raising this as an issue leads to the suggestion that the model systems currently employed in the search for new therapies are inadequate. The results from a multitude of clinical trials and vast clinical experience would support the validity of the current approach. For example, as mentioned above, 12 new drugs have been identified and developed on the basis of preclinical findings using two easy-to-conduct animal seizure models: the maximal electroshock seizure (MES) and the subcutaneous pentylenetetrazol (scPTZ) tests (Table 9.2). Clearly, these models have been effective in identifying and characterizing several new and novel therapies for the symptomatic treatment of epilepsy. Moreover, the registration of the newer anticonvulsant drugs by the U.S. Food and Drug Administration (FDA) or other regulatory agencies requires demonstrated proof of efficacy in randomized, double-blind, placebo-controlled, add-on trials employing patients with refractory epilepsy. As such, it is difficult to argue that these newer AEDs are not pharmacologically different from the first-generation AEDs; however, it is important to keep in mind that efficacy required for regulatory purposes is defined by the ability of a drug to reduce the seizure frequency in 50% or more of the patient population and not by the ability of a drug to produce complete seizure freedom. Indeed, the seizure-free rate from most add-on clinical trials is less than 10%. Based on these findings, it becomes easier to appreciate why there still remains a substantial therapy-resistant patient population.

## DEFINING THE ANTICONVULSANT POTENTIAL OF AN INVESTIGATIONAL AED

### The MES and PTZ Tests

The MES and subcutaneous PTZ seizure models continue to represent the two most widely used animal seizure models employed in the search for new AEDs (White, 1997; White et al., 1995a, 2002). As mentioned above, Merritt and Putnam (1937) successfully employed the MES test in

**TABLE 9.2**
**Correlation between Anticonvulsant Efficacy and Clinical Utility of the Established and Second-Generation AEDs in Experimental Animal Models**

| | Clinical Seizure Type | | | |
| Experimental Model | Tonic and/or Clonic Generalized Seizures | Myoclonic/ Generalized Absence Seizures | Generalized Absence Seizures | Partial Seizures |
|---|---|---|---|---|
| MES (tonic extension)[a] | CBZ, PHT, VPA, PB, [FBM, GBP, LCM, LTG, PGB, RUF, TPM, ZNS] | — | — | — |
| Subcutaneous PTZ (clonic seizures)[a] | — | ESM, VPA, PB,[b] BZD, [FBM, GBP, PGB, RUF, TGB,[b] VGB[b]] | — | — |
| Spike–wave discharges[c] | — | — | ESM, VPA, BZD, [LTG, TPM, LVT] | — |
| Electrical kindling (focal seizures) | — | — | — | CBZ, PHT, VPA, PB, BZD, [FBM, GBP, LCM, LTG, TPM, TGB, ZNS, LVT, VGB] |
| Phenytoin-resistant kindled rat[d] | — | — | — | [LVT, GBP, TPM, FBM, LTG] |
| 6-Hz (44 mA)[e] | — | — | — | VPA, [LVT] |

[a]  Data summarized from White et al. (2002) and Bialer et al. (2007, 2009).

[b]  PB, TGB, and VGB block clonic seizures induced by sc PTZ but are inactive against generalized absence seizures and may exacerbate spike wave seizures.

[c]  Data summarized from GBL, GAERS, and *lh/lh* spike-wave models (Hosford et al., 1992, 1997; Marescaux and Vergnes, 1995; Snead, 1992).

[d]  Data summarized from Löscher (2002).

[e]  Data summarized from Barton et al. (2001).

*Note:*  [ ] indicates second-generation antiepileptic drugs.

*Abbreviations:*  BZD, benzodiazepines; CBZ, carbamazepine; ESM, ethosuximide; FBM, felbamate; GBP, gabapentin; LCM, lacosamide; LTG, lamotrigine; LVT, levetiracetam; PB, phenobarbital; PGB, pregabalin; PHT, phenytoin; RUF, rufinamide; TGB, tiagabine; TPM, topiramate; VGB, vigabatrin; VPA, valproic acid; ZNS, zonisamide.

a systematic screening program to identify phenytoin. This observation, when coupled with the subsequent success of phenytoin in the clinical management of generalized tonic–clonic seizures, provided the validation necessary to consider the MES test as a reasonable model of human generalized tonic–clonic seizures.

In 1944, Everett and Richards demonstrated that PTZ-induced seizures could be blocked by trimethadione and phenobarbital but not by phenytoin. A year later, Lennox demonstrated that trimethadione was effective in decreasing or preventing petit mal (absence) attacks in 50 patients but was ineffective or worsened grand mal attacks in 10 patients (Lennox, 1945). Trimethadione's success in the clinic and its ability to block threshold seizures induced by PTZ provided the necessary correlation to establish the PTZ test as a model of petit mal or generalized absence seizures. With these observations, the current era of AED screening using the MES and scPTZ tests was launched.

The MES and scPTZ tests are easily conducted with a minimal investment in equipment and technical expertise. They provide valuable data regarding the potential anticonvulsant activity of an investigational drug, and with one exception (i.e., levetiracetam) all of the currently available AEDs have been found to be active in one or both of these tests (Table 9.2). Furthermore, both tests are amenable to high-volume screening with widely available, relatively inexpensive normal rodents.

Although amenable to high-volume screening, the MES and scPTZ tests fail to meet any of the remaining criteria described earlier (Table 9.1); for example, there are now several examples wherein the pharmacological profile of an AED can be affected by the disease state. Because the MES and scPTZ tests are conducted in pathologically normal rodents, there is no guarantee that AEDs identified by these two tests will be equally effective in pathologically abnormal rodents. The best example to illustrate this point is levetiracetam. As mentioned above, the MES and scPTZ tests failed to identify the anticonvulsant activity of levetiracetam. Subsequent investigations demonstrated that levetiracetam was active in pathologically abnormal models of partial and primary generalized seizures (Gower et al., 1992, 1995; Klitgaard et al., 1998; Löscher and Honack, 1993; Löscher et al., 1998). In this regard, levetiracetam appears to represent the first truly novel AED to be identified in recent years. The identification and subsequent development of levetiracetam as an efficacious AED for the treatment of partial seizures underscore the need for flexibility when screening for efficacy and the need to incorporate levetiracetam-sensitive models into the early evaluation process.

One distinct disadvantage of the MES and scPTZ tests is that they do not possess a pharmacological profile consistent with therapy-resistant human epilepsy. The MES test, for example, is sensitive to all of the first-generation AEDs with demonstrated clinical efficacy in the treatment of generalized tonic–clonic seizures (phenytoin, carbamazepine, valproate, and phenobarbital). As summarized in Table 9.2, many of the AEDs brought to the market since 1993 (felbamate, gabapentin, lacosamide, lamotrigine, pregabalin, rufinamide, topiramate, and zonisamide) are also active in the MES test (Bialer et al., 2007, 2009). In contrast, the scPTZ test is sensitive to those first-generation AEDs used in the management of generalized absence seizures (ethosuximide, valproate, and the benzodiazepines). As mentioned above, this is due in part to the fact that these two models were developed and validated initially using the older, established AEDs: phenytoin, phenobarbital, and trimethadione (Kupferberg, 2001; White et al., 1995b). This particular validation process led to their selection on the basis that these two model systems displayed a pharmacological profile consistent with established medical practice at the time (for historical reviews and discussion, see Kupferberg, 2001; White et al., 1995b).

The lack of any demonstrable efficacy by tiagabine, vigabatrin, and levetiracetam in the MES test argues against its utility as a predictive model of partial seizures. Consistent with this conclusion is the observation that $N$-methyl-D-aspartate (NMDA) antagonists are very effective against tonic extension seizures induced by MES; however, they were found to be without benefit in patients with partial seizures (Meldrum, 1995).

Historically, positive results obtained in the scPTZ seizure test were considered suggestive of potential clinical utility against generalized absence seizures. Based on this argument, phenobarbital, gabapentin, pregabalin, and tiagabine, which are all effective in the scPTZ test, should all be effective against spike–wave seizures, and lamotrigine, which is inactive in the scPTZ test, should be inactive against spike–wave seizures; however, clinical experience has demonstrated that this is an invalid prediction. The barbiturates, gabapentin, pregabalin, and tiagabine all aggravate spike–wave seizure discharge, whereas lamotrigine has been found to be effective against absence seizures. Thus, the overall utility of the scPTZ test in predicting activity against human spike–wave seizures is limited. Before reaching any conclusions concerning potential clinical utility against spike–wave seizures, positive results in the scPTZ test should be corroborated by positive findings in other models of absence such as the γ-butyrolactone seizure test (Snead, 1992), the genetic absence epileptic rat of Strasbourg (Marescaux and Vergnes, 1995), the lethargic (*lh/lh*) mouse (Coenen and Van Luijtelaar, 2003; Hosford and Wang, 1997; Hosford et al., 1992), and the WAG/Rij rat (Coenen and Van Luijtelaar, 2003). Another important advantage of all four of these models is that they accurately predict the potentiation of spike–wave seizures by drugs that elevate γ-aminobutyric acid (GABA) concentrations (e.g., vigabatrin and tiagabine), drugs that directly activate the $GABA_B$ receptor, and the barbiturates.

One might ask, then, what, if any, benefit these two tests might provide when screening for novel AEDs. First, both tests provide some insight into the central nervous system (CNS) bioavailability of a particular investigational AED. Furthermore, both models are nonselective with respect to mechanism of action. As such, they are very well suited for the early evaluation of anticonvulsant activity because neither model assumes that the pharmacodynamic activity of a particular drug is dependent on its molecular mechanism of action. Finally, both model systems display clear and definable seizure endpoints and require minimal technical expertise. This, coupled with lack of dependence on a molecular mechanism, makes them ideally suited to screen large numbers of chemically diverse entities. Levetiracetam, however, taught the community that a lack of efficacy in either of these tests does not translate into lack of human efficacy. As such, there is no longer any reason to limit further screening on the basis of results obtained in the MES and scPTZ seizure tests. In fact, there is no *a priori* reason to assume that a novel AED will be active in the MES, scPTZ, or other acute seizure tests.

## THE KINDLED RAT

In recent years, the kindling model of partial epilepsy has been frequently employed in the AED development process (Table 9.2). Kindling refers to the process whereby there is a progressive increase in electrographic and behavioral seizure activity in response to repeated stimulation of a limbic brain region such as the amygdala or hippocampus with an initially subconvulsive current (Goddard et al., 1969). Furthermore, the kindling process is associated with a progressive increase in seizure severity and duration, a decrease in the focal seizure threshold, and neuronal degeneration in limbic brain regions that resemble human mesotemporal lobe epilepsy (Table 9.3). Both electrographic and behavioral components of the kindled seizure begin locally at the site of stimulation and quickly become secondarily generalized.

From a therapeutic perspective, the kindled rat model offers perhaps the best predictive value for efficacy against partial seizures; for example, it is the only model that adequately predicted the clinical utility of the first- and second-generation AEDs including tiagabine and vigabatrin (Table 9.2). Furthermore, the kindled rat is the only model that accurately predicted the lack of clinical efficacy of NMDA antagonists (Löscher and Honack, 1991). Given the predictive nature of the kindled rat, one might legitimately ask why this model or a similarly predictive chronic model is not utilized as a primary screen rather than a secondary screen for the early identification and evaluation of novel AEDs. The answer is primarily one of logistics. Any chronic animal model such as kindling is labor intensive and requires adequate facilities and resources to surgically implant the stimulating/recording electrode and to kindle and house sufficient rats over a chronic period of time. Furthermore, unlike the acute seizure models, the time required to conduct a drug study with a chronic model far exceeds the time required to conduct a similar study with the MES or scPTZ tests, thereby severely limiting the number of AEDs that can be screened in a timely manner. If, however, our goal is to provide novel therapeutic options that will meet the needs of the therapy-resistant patient population, we should no longer let these reasons be a barrier to testing of drugs in chronic epilepsy models.

## WHAT IS THE FUTURE OF AED DISCOVERY AND DEVELOPMENT?

### THERAPY-RESISTANT EPILEPSY

As discussed above, the models summarized in Table 9.2 have been successfully used to identify effective therapies for the treatment of human epilepsy. Moreover, those drugs found effective in one or more of these models can be expected to be clinically effective against partial, generalized,

**TABLE 9.3**
**Animal Models of Epileptogenesis**

| Experimental Model | Neuropathology Present? | Chronic Hyperexcitability? | Latent Period? | Spontaneous Seizures? |
|---|---|---|---|---|
| Kindling[a] | Yes | Yes | No | No[b] |
| Alumina hydroxide gel | Yes | Yes | Yes | Yes |
| Post-status epilepticus[c] (kainic acid, pilocarpine ± lithium, electrical models of self-sustaining status, cobalt–homocysteine, fluorothyl) | Yes | Yes | Yes (days, weeks) | Yes |
| Tetanus toxin[d] | Yes | Yes | Yes | Yes |
| Traumatic brain injury | | | | |
|   Cortical undercut[e] | Yes | Yes | Yes | Yes |
|   Ferric chloride[f] | Yes | Yes | Yes | Yes |
|   Fluid percussion[g] | Yes | Yes | Yes | Yes |
| Neonatal hypoxia–ischemia[h] | Yes | Yes | Yes | Yes |
| Neonatal hyperthermia[i] | No | Yes | Yes | Brief electrographic discharges |
| Cortical dysplasia models (MAM, freeze lesion, Otx$^{-/-}$)[j] | Yes | Decreased seizure threshold (MAM and freeze lesion models) | ? | Yes in Otx$^{-/-}$ but not other models |
| Encephalitis-induced epilepsy (Theiler's mouse encephalitis virus)[k] | Yes | Yes | Yes | Yes |
| Genetic models[l] | Model-specific | Yes | Model-specific | Model-specific |

[a]  Pitkanen and Sutula (2002); Stafstrom and Sutula (2005); Sutula et al. (1986, 1988).
[b]  Spontaneous seizures do develop after kindling stimulation.
[c]  Ben-Ari et al. (1979, 1980); Hellier et al. (1998); Turski et al. (1987, 1989).
[d]  Jefferys (1996); Jefferys et al. (1992); Jiang et al. (1998).
[e]  Sharpless and Halpern (1962); Echlin and Battista (1963); Prince and Jacobs (1998); Bush et al. (1999).
[f]  Willmore et al. (1978).
[g]  Pitkanen et al. (2009).
[h]  Jensen et al. (1995); Williams et al. (2004); Williams and Dudek (2007); Kadam and Dudek (2007).
[i]  Baram et al. (1997); Bender et al. (2004).
[j]  Avanzini et al. (2008).
[k]  Libbey et al. (2008).
[l]  Noebels et al. (2008).

*Source:*  Adapted from Sarkisian, M.R., Epilepsy Behav., 2, 201–216, 2001; Avanzini, G.G. et al., in Epilepsy: A Comprehensive Textbook (pp. 415–444), J. Engel et al., Eds., Lippincott Williams & Wilkins, Philadelphia, PA, pp. 415–444.

and secondarily generalized seizures. Unfortunately, in spite of their utility in screening for effective therapies for the symptomatic treatment of seizures, there still remains a substantial need for the identification of therapies for the patient with refractory seizures. In this respect, one might argue that the models summarized in Table 9.2 are not predictive of efficacy in those patients with refractory epilepsy. Thus, the identification and characterization of one or more model systems that would predict efficacy in the pharmacoresistant patient population would be a valuable asset to the epilepsy community. In addition to being useful for therapy development, the ability to segregate animals on the basis of their responsiveness or lack of sensitivity to a given AED would prove to be (1) useful for attempting to understand the molecular mechanisms underlying pharmacoresistance, (2) an asset for those studies designed to assess whether it is possible to reverse drug resistance, and (3) useful for identification of surrogate markers that might be able to predict which patient will remit and become pharmacoresistant. Unfortunately, a model will only become clinically validated at the time that a drug is found to markedly reduce the incidence of therapy-resistant epilepsy in patients with epilepsy. In the meantime, drugs in development for the treatment of epilepsy should be evaluated in as many proposed models of pharmacoresistance as feasible. At the time that a new drug is introduced into the market that substantially reduces the percentage of patients with refractory epilepsy, it is hoped that we will be able to identify the one or more model systems retrospectively that would predict efficacy against refractory seizures.

At the present time, a number of potentially interesting model systems of therapy resistance are available. Before briefly discussing the salient features of those models that have emerged in recent years, it is first important to define AED pharmacoresistance from a model-development perspective. The participants of two National Institutes of Health (NIH)/National Institute of Neurological Disorders and Stroke (NINDS)/American Epilepsy Society (AES)-sponsored workshops on animal models held in 2001 and 2002 agreed that any proposed model of pharmacoresistant epilepsy should meet certain criteria. For example, human experience dictates that pharmacoresistance can be defined as persistent seizure activity that does not respond to monotherapy at tolerable doses with at least two current AEDs (Stables et al., 2002, 2003). Thus, the demonstrated resistance to two or more of the first-line AEDs employed in the treatment of partial epilepsy was considered pivotal. Ideally, it can be hoped that any proposed model will lead to the identification of a new therapy that will ultimately be highly effective in humans resistant to existing AEDs.

All of the following display a phenotype consistent with pharmacoresistant epilepsy: the phenytoin (PHT)-resistant kindled rat (Löscher, 2002; Löscher et al., 2000); the lamotrigine (LTG)-resistant kindled rat (Postma et al., 2000; Srivastava and White, 2005; Srivastava et al., 2003, 2004); the carbamazepine (CBZ)-resistant kindled rat (Nissinen et al., 2000); the 6-Hz psychomotor seizure model of partial epilepsy (Barton et al., 2001; Brown et al., 1953); post-status epilepticus models of temporal lobe epilepsy (Ben-Ari et al., 1980; Brandt et al., 2004; Glien et al., 2002; Grabenstatter and Dudek, 2005; Grabenstatter et al., 2005; Hellier et al., 1998; Leite and Cavalheiro, 1995; Turski et al., 1987; 1989; van Vliet et al., 2006); the methylazoxymethanol acetate (MAM) *in utero* model of nodular heterotopias (Smyth et al., 2002); and the *in vitro* low-magnesium, entorhinal–hippocampal slice preparation (Armand et al., 2000). Each of these models possesses some utility when attempting to differentiate the anticonvulsant profile of an investigational drug from existing antiepileptic drugs. This is not to imply that other approaches using *in vitro* systems are of any less value; the reader is referred to Dichter and Pollard (2006) and Heinemann et al. (2006) for reviews and references. The remainder of this section briefly discusses some of the salient features of the LTG-resistant kindled rat, the 6-Hz psychomotor seizure model, and the post-status recurrent seizure models as they relate to their utility for early differentiation of new therapies.

## The Lamotrigine-Resistant Kindled Rat Model

The LTG-resistant kindled rat model of partial epilepsy was first described by Postma and colleagues (2000). Unlike the PHT-resistant kindled rat, resistance to LTG is induced when a rat is exposed to low-dose LTG during the kindling acquisition phase; thus, exposure to a low dose (5 mg/kg) of LTG during kindling development leads to reduced efficacy of LTG when administered to the fully kindled rat (Postma et al., 2000). A similar phenomenon has been observed for CBZ (Weiss and Post, 1991). Perhaps more important is the observation that LTG-resistant rats are also refractory to CBZ, PHT, and topiramate (TPM; Topamax®) but not valproic acid (VPA) or the investigational *KCNQ2* activator retigabine (Srivastava and White, 2005; Srivastava et al., 2003, 2004) or the investigational AED carisbamate (White et al., 2006).

## The 6-Hz Psychomotor Seizure Model of Partial Epilepsy

Although the high-frequency, short-duration stimulation employed in the MES test has become a standard for screening AEDs, it is only one of several electroshock paradigms initially developed in the 1940s and 1950s (Swinyard, 1972). An alternative stimulation paradigm is the low-frequency (6 Hz), long-duration (3 seconds) corneal stimulation model, which was stated to produce "psychic" or '"psychomotor" seizures (Table 9.2). Instead of the tonic extension seizure characteristic of the MES test, the 6-Hz seizure was reported to involve a minimal, clonic phase followed by stereotyped automatisms reminiscent of human complex partial seizures (Brown et al., 1953; Toman, 1951; Toman et al., 1952).

At the time of its initial description, the authors were attempting to validate the 6-Hz model as a screening test for partial seizures; however, the pharmacological profile was not consistent with clinical practice (Brown et al., 1953). Phenytoin, for example, was found to be inactive in the 6-Hz seizure test. This observation led the authors to suggest that it was no more predictive of clinical utility than the other models available at the time (i.e., the MES and scPTZ tests). Subsequent investigations confirmed the relative insensitivity of the 6-Hz test to phenytoin and extended this observation to include carbamazepine, lamotrigine, and topiramate (Barton et al., 2001). The relative resistance of some patients to phenytoin and other AEDs in today's clinical setting and the lack of sensitivity of the MES and scPTZ to levetiracetam prompted studies to reevaluate the 6-Hz seizure test as a potential screen for therapy-resistant epilepsy (Barton et al., 2001). Subsequent investigations demonstrated that levetiracetam affords protection against the 6-Hz seizure at a stimulus intensity at which other AEDs display little to no efficacy (Table 9.2). In addition to levetiracetam, the investigational AEDs retigabine (Bialer et al., 2009) and carisbamate (White et al., 2006) are both effective against 6-Hz seizures at doses that are devoid of toxicity. These observations clearly demonstrate the potential utility of this model as a screen for differentiating investigational AEDs from marketed AEDs.

## Post-Status Epilepticus Models of Temporal Lobe Epilepsy

Post-status epilepticus models of refractory epilepsy are beginning to emerge as important tools in the differentiation of investigational AEDs. These models of pharmacoresistant epilepsy differ significantly from the AED-resistant kindled rat and the 6-Hz seizure model in that seizures are spontaneously evolving and not evoked. This adds yet another level of complexity to the pharmacological evaluation of an investigational drug but yields potentially important information about its relative efficacy compared to existing AEDs.

The development and characterization of this model system for pharmacological testing resulted from a focused effort of the epilepsy community to identify clinically relevant models of chronic epilepsy (Stables et al., 2002, 2003). In many respects, the post-status models described thus far fulfill one very important characteristic of the ideal model system: spontaneous recurrent seizures (SRSs) following a species-appropriate latent period. The post-status epileptic rat is particularly useful in that this model allows the investigator the opportunity to evaluate the efficacy of a given treatment on seizure frequency, seizure type (i.e., focal or generalized), and the liability for tolerance development following chronic treatment. Unfortunately, by their very nature, drug trials in rats with spontaneous seizures take on another level of complexity. They are extremely laborious and time consuming and require a greater level of technical expertise. As such, there have only been a few pharmacological studies conducted to date (Brandt et al., 2004; Glien et al., 2002; Grabenstatter and Dudek, 2005; Grabenstatter et al., 2005; Leite and Cavalheiro, 1995; van Vliet et al., 2006). Having said this, the advantages that this model provides for differentiating a given compound from the established AEDs are well worth the investment.

All of these models are being used with increasing frequency in the search for novel antiepileptic drugs. They can play an important role in efforts to differentiate investigational drugs from existing AEDs. Furthermore, it is important to note that the use of post-status recurrent seizure models has led to the development of novel drug-testing protocols in animals that more closely resemble human clinical protocols. It is also important to note that each of these models provides a biological system that will likely lead to a greater understanding of the mechanisms underlying pharmacoresistant epilepsy. As such, they can be used to test novel approaches designed to overcome or reverse therapy resistance and to perhaps identify appropriate surrogate markers of pharmacoresistance. One can envision the day when we will be able to identify the patient at risk for developing therapy-resistant epilepsy and institute a prophylactic therapy that prevents the emergence of pharmacoresistance.

Unfortunately, it will take the successful clinical development of a drug with demonstrated clinical efficacy in the management of refractory epilepsy before any one of these (or other) model systems will be clinically validated. Nonetheless, this should not preclude the community at large from continuing the search for a more effective therapy using these and any other viable models that are currently available. In fact, until there is a validated model, it becomes even more important to characterize and incorporate several of the available models into the drug discovery process while at the same time continuing to identify new models of refractory epilepsy.

## DISEASE MODIFICATION: MOVING BEYOND THE SYMPTOMATIC TREATMENT OF EPILEPSY

Since the first White House-initiated Citizen's United for Research in Epilepsy (CURE) conference in March 2000, a substantial interest has been placed on finding a "cure" for epilepsy. The multifaceted approach that has been supported by the NIH, CURE, the Epilepsy Therapy Development Project, the American Epilepsy Society, the Epilepsy Foundation of America, and the pharmaceutical industry has clearly led to:

- A greater understanding of the pathophysiology of epilepsy at the molecular and genetic level
- The search for surrogate markers of the epileptogenic process
- The development and utilization of model systems (animal and human) for evaluating novel therapeutic approaches
- Ongoing discussions and collaborations within the scientific, regulatory, and pharmaceutical communities

The need to find a therapy that would prevent or delay the development of epilepsy in the susceptible individual is evident, and many of the required tools have been developed in recent years. Nonetheless, many practical challenges and hurdles must be overcome before a cure can be realized.

*Epileptogenesis* is broadly defined as the process by which brain function is altered in such a way that there is a propensity for recurrent, spontaneous seizure (RSS) activity. The ultimate identification of a therapy that could prevent the development of epilepsy in a person with a known genetic mutation or following a brain injury (traumatic brain injury, stroke, infection, complex febrile seizures, cortical malformation) would clearly represent a quantum leap forward from current treatment options. Perhaps a more attainable goal in the near term is to define a therapy that might possess disease-modifying properties—for example, slowing the progression or decreasing the severity of seizures, preventing the cognitive decline and other comorbidities associated with epilepsy such as depression, or preventing the emergence of pharmacoresistance. Such a therapy would offer substantial improvement over currently available therapies.

Prior to embarking on a clinical trial, there should be some evidence that a given therapy is effective in one or more animal models of epileptogenesis. Unlike the animal models routinely employed in the search for novel anticonvulsant compounds, the animal models of epileptogenesis that are currently available have not been validated clinically. As with models for therapy-resistant epilepsy, clinical validation will not be provided until the first truly antiepileptic or disease-modifying therapy has been found to be effective in an appropriately designed clinical trial. Having said this, the community should not be discouraged from pursing this approach but should clearly be aware of the limitations of the existing models and employ caution when designing preclinical studies and interpreting the results obtained.

A number of experimental models could be used to test a potential disease-modifying or anti-epileptogenic therapy. These include a number of acquired epilepsy models (Table 9.3) and an expanding number of genetic mutants that possess a known mutation found to be associated with a particular genetic form of human epilepsy (for a review and references, see Noebels et al., 2008). Each of the acquired animal models listed in Table 9.3 involves an initial insult that is followed by a variable latent period that often culminates in the evolution of recurrent, spontaneous seizure activity or measurable hyperexcitability some time later. The rodent models of acquired epilepsy most often used are the kindling models (Pitkanen and Sutula, 2002; Stafstrom and Sutula, 2005; Sutula et al., 1986, 1988), status epilepticus models (Ben-Ari et al., 1979, 1980; Hellier et al., 1998; Turski et al., 1987, 1989), and traumatic brain injury models (Bush et al., 1999; Echlin and Battista, 1963; Pitkanen et al., 2009; Prince and Jacobs, 1998; Sharpless and Halpern, 1962; Willmore et al., 1978). Each of these displays varying degrees of cell loss, synaptic reorganization, network hyper-excitability, and a latent period that is followed by the expression of an altered seizure threshold or development of spontaneous seizures. As such, they fulfill some important characteristics of a model of acquired partial epilepsy. Furthermore, they offer something unique for the investigator interested in evaluating a proposed antiepileptogenic therapy.

A number of other models of epileptogenesis have been developed over the years that display many phenotypic, histological, and electrophysiological similarities with their human counterparts. These include, but are not limited to, the tetanus toxin model of partial epilepsy (Jefferys, 1996; Jefferys et al., 1992; Jiang et al., 1998), neonatal hypoxia–ischemia model of hypoxic–ischemic injury (Jensen, 1995; Kadam and Dudek, 2007; Williams and Dudek, 2007; Williams et al., 2004), neonatal hyperthermia model of febrile seizures (Baram et al., 1997; Bender et al., 2004), models of cortical dysplasia (for a review and references, see Avanzini et al., 2008), and a model of viral encephalitis-induced epilepsy (Libbey et al., 2008).

Conducting an intervention study with any one of the available models has its limitations (Dudek, 2009); for example, there is a substantial degree of variability among individual animals in the latent period between their initial insult and the development of spontaneous seizures. The variability in the latent period is only one of the many issues that must be considered when interpreting results from an intervention study. Before declaring that a therapy is antiepileptogenic in a particular model, the investigator is almost certainly obligated to conduct months of continuous video-electroencephalography (EEG) recording to ensure that a seizure was not missed due to a lapse in monitoring or due to the fact that seizures in animals, like humans, often cluster and could be easily missed by intermittent monitoring.

Another important issue that has to be considered when interpreting results from a positive intervention study is the possibility that the therapy itself decreased the insult-induced seizure severity or duration (Dudek, 2009). This is not to imply that such a therapy would not be important; however, it becomes difficult under these conditions to separate the acute anticonvulsant effect of a therapy from its purported antiepileptogenic effects.

Attempts have been made to prevent the development of epilepsy in one or more of the available models; unfortunately, the results from these attempts have been somewhat disappointing or non-conclusive given the caveats surrounding the study design or seizure monitoring protocol employed. Nonetheless, each of these studies has been extremely useful in setting the stage for further investigations and providing useful information regarding potential mechanisms underlying epileptogenesis for further hypothesis testing. Also, the results obtained have been encouraging and suggest that early treatment after an insult may offer some therapeutic benefit in the way of disease modification. Application of a cannabinoid receptor 1 (CB1) antagonist during *in vivo* febrile seizures, for example, blocked the cannabinoid-receptor-mediated, depolarization-induced suppression of inhibition (DSI), prevented seizure-induced upregulation of CB1 receptors, and prevented the development of long-term limbic hyperexcitability (Chen et al., 2007). In a separate study, Pitkanen and colleagues (2004) found that chronic treatment with the alpha-2 receptor antagonist atipamezole decreased the severity of epilepsy and provided some apparent neuroprotection in a post-status epilepticus rat model. These results suggest that chronic treatment with atipamezole is not antiepileptogenic but possesses disease-modifying potential in that the epilepsy that does develop is milder and not progressive.

In addition to the models of acquired epilepsy, a number of genetic mouse models developed in recent years recapitulate many of the features of human genetic epilepsy resulting from a specific defect in a particular receptor or voltage-gated ion channel. These genetic models have provided the community with a wealth of information regarding the pathophysiology of epilepsy at the molecular and genetic levels. Two excellent examples of animal models that have emerged from our knowledge of human genetic mutations are the various mouse models that contain mutations in Nav1.1 (Yu et al., 2006) and Kv7.2/7.3 (Singh et al., 2008) genes. For the most part, these mice recapitulate many of the phenotypic and pharmacologic features of human severe myoclonic epilepsy of infancy (SMEI) and benign familial neonatal convulsions (BFNCs), respectively. The extent to which these and other emerging genetic mouse models will lead to the identification and validation of novel drug targets has yet to be fully appreciated.

As mentioned above, the WAG/Rij rat, like the GAERS rat, is an excellent model of human absence epilepsy. It displays an age-dependent expression of spike–wave seizures that increases over age (Coenen and Van Luijtelaar, 2003). In a novel experimental approach, Blumenfeld and colleagues (2008) demonstrated that early treatment with ethosuximide from postnatal day 21 through 5 months of age blocked changes in the expression of the ion channels Nav1.1, Nav1.6, and HCN1, normally associated with the onset and expression of spike–wave seizures in the WAG/Rij rat model of absence epilepsy. Furthermore, they observed that early treatment was associated with a prolonged suppression of seizures long after ethosuximide treatment was discontinued. Again, this study does not show that ethosuximide prevented the development of epilepsy; however, it provides supportive information suggesting that early intervention with ethosuximide has the potential for disease modification.

Many animal models of epileptogenesis that have been described in recent years recapitulate several of the features of human epilepsy. The use of these models for anticonvulsant drug discovery efforts may lead to novel treatment strategies, better efficacy, less toxic profiles, and unique therapeutic targets for refractory patients; however, the holy grail of epilepsy research is the identification of a novel therapy that prevents or cures epilepsy. Animal models of epileptogenesis will no doubt play a fundamental role in attaining each of these goals.

## SUMMARY

The current approach to AED discovery is effective for identifying drugs that are useful for the treatment of systematic seizures. Unfortunately, the utility of the existing AEDs is limited by their marginal effectiveness against refractory seizures. Some models of pharmacoresistance that have been developed in recent years display some degree of therapy resistance to several of the available AEDs. Whether efficacy in these resistant models will translate into improved treatment for the therapy-resistant patient population has yet to be determined by clinical practice. Nonetheless, their utility should not be dismissed but rather endorsed as part of a comprehensive screen to thoroughly evaluate the anticonvulsant potential of an investigational AED.

A greater understanding of the pathophysiology of acquired epilepsy at both molecular and genetic levels may lead to the development of a new therapeutic approach that reaches beyond the symptomatic treatment of epilepsy to modify the progression or, dare we suggest, prevent the development of epilepsy in the susceptible patient. The realization of such a possibility will necessitate a change in our current AED discovery approach. Significant progress in our understanding of the factors that contribute to human epileptogenesis has been made in recent years. These advances have in large part been made possible through the development and characterization of genetic-, insult-, and age-specific animal models of epileptogenesis and secondary neuronal hyperexcitability. The true validation of a given model of epileptogenesis requires the development of an effective therapy that prevents or delays the development of epilepsy or hyperexcitability in the human condition and whose activity was predicted by preclinical testing. As we gain additional insight into the molecular biology and genetics of the acquired and genetic epilepsies, the day may come when new therapies that target the epileptogenic process will be developed.

# REFERENCES

Armand, V., C. Rundfeldt, U. Heinemann. (2000). Effects of retigabine (D-23129) on different patterns of epileptiform activity induced by low magnesium in rat entorhinal cortex hippocampal slices. *Epilepsia*, 41:28–33.

Avanzini, G.G., D.M. Treiman, J. Engel. (2008). Animal models of acquired epilepsies and status epilepticus. In *Epilepsy: A Comprehensive Textbook* (pp. 415–444), J. Engel, T.A. Pedley, J. Aicardi, M.A. Dichter, and S. Moshé, Eds. Philadelphia, PA: Lippincott Williams & Wilkins.

Baram, T.Z., A. Gerth, L. Schultz. (1997). Febrile seizures: an appropriate-aged model suitable for long-term studies. *Brain Res. Dev. Brain Res.*, 98:265–270.

Barton, M.E., B.D. Klein, H.H. Wolf, H.S. White. (2001). Pharmacological characterization of the 6-Hz psychomotor seizure model of partial epilepsy. *Epilepsy Res.*, 47:217–227.

Ben-Ari, Y., J. Lagowska, E. Tremblay, G. Le Gal La Salle. (1979). A new model of focal status epilepticus: intra-amygdaloid application of kainic acid elicits repetitive secondarily generalized convulsive seizures. *Brain Res.*, 163:176–179.

Ben-Ari, Y., E. Tremblay, O.P. Ottersen. (1980). Injections of kainic acid into the amygdaloid complex of the rat: an electrographic, clinical and histological study in relation to the pathology of epilepsy. *Neuroscience*, 5:515–528.

Bender, R.A., C. Dube, T.Z. Baram. (2004). Febrile seizures and mechanisms of epileptogenesis: insights from an animal model. *Adv. Exp. Med. Biol.*, 548:213–225.

Bialer, M., S.I. Johannessen, H.J. Kupferberg, R.H. Levy, E. Perucca, T. Tomson. (2007). Progress report on new antiepileptic drugs: a summary of the Eighth Eilat Conference (Eilat VIII). *Epilepsy Res.*, 73(1):1–52.

Bialer, M., S.I. Johannessen, R.H. Levy, E. Perucca, T. Tomson, H.S. White. (2009). Progress report on new antiepileptic drugs: a summary of the Ninth Eilat Conference (Eilat IX). *Epilepsy Res.*, 83(1):1–43.

Blumenfeld, H., J.P. Klein, U. Schridde, M. Vestal, T. Rice et al. (2008). Early treatment suppresses the development of spike–wave epilepsy in a rat model. *Epilepsia*, 49:400–409.

Brandt, C., H.A. Volk, W. Löscher. (2004). Striking differences in individual anticonvulsant response to phenobarbital in rats with spontaneous seizures after status epilepticus. *Epilepsia*, 45:1488–1497.

Brown, W.C., D.O. Schiffman, E.A. Swinyard, L.S. Goodman. (1953). Comparative assay of antiepileptic drugs by psychomotor seizure test and minimal electroshock threshold test. *J. Pharmacol. Exp. Ther.*, 107:273–283.

Bush, P.C., D.A. Prince, K.D. Miller. (1999). Increased pyramidal excitability and NMDA conductance can explain posttraumatic epileptogenesis without disinhibition: a model. *J. Neurophysiol.*, 82:1748–1758.

Chen, K., A. Neu, A.L. Howard, C. Foldy, J. Echegoyen, L. Hilgenberg, M. Smith, K. Mackie, I. Soltesz. (2007). Prevention of plasticity of endocannabinoid signaling inhibits persistent limbic hyperexcitability caused by developmental seizures. *J. Neurosci.*, 27:46–58.

Coenen, A.M., E.L. Van Luijtelaar. (2003). Genetic animal models for absence epilepsy: a review of the WAG/Rij strain of rats. *Behav. Genet.*, 33:635–655.

Dichter, M.A., J. Pollared. (2006). Cell culture models for studying epilepsy. In *Models of Seizures and Epilepsy* (pp. 23–34), A. Pitkanen, P.A. Schwartzkroin, and S.L. Moshe, Eds. New York: Elsevier.

Dudek, F.E. (2009). Commentary: a skeptical view of experimental gene therapy to block epileptogenesis. *Neurotherapeutics*, 6:319–322.

Echlin, F.A., A. Battista. (1963). Epileptiform seizures from chronic isolated cortex. *Arch. Neurol.*, 9:154–170.

Everett, G.M., R.K. Richards. (1944). Comparative anticonvulsive action of 3,5,5-trimethyloxazolidine-2,4-dione (tridione), dilantin, and phenobarbital. *J. Pharmacol. Exp. Ther.*, 81(4):402–407.

Glien, M., C. Brandt, H. Potschka, W. Löscher. (2002). Effects of the novel antiepileptic drug levetiracetam on spontaneous recurrent seizures in the rat pilocarpine model of temporal lobe epilepsy. *Epilepsia*, 43:350–357.

Goddard, G.V., D.C. McIntyre, C.K. Leech. (1969). A permanent change in brain function resulting from daily electrical stimulation. *Exp. Neurol.*, 25:295–330.

Gower, A.J., M. Noyer, R. Verloes, J. Gobert, E. Wulfert. (1992). UCB l059, a novel anti-convulsant drug: pharmacological profile in animals. *Eur. J. Pharmacol.*, 222:193–203.

Gower, A.J., E. Hirsch, A. Boehrer, M. Noyer, C. Marescaux. (1995). Effects of levetiracetam, a novel antiepileptic drug, on convulsant activity in two genetic rat models of epilepsy. *Epilepsy Res.*, 22:207–213.

Grabenstatter, H.L., F.E. Dudek. (2005). The effect of carbamazepine on spontaneous seizures in freely-behaving rats with kainite-induced epilepsy. *Epilepsia*, 46(Suppl. 8):287.

Grabenstatter, H.L., D.J. Ferraro, P.A. Williams, P.L. Chapman, F.E. Dudek. (2005). Use of chronic epilepsy models in antiepileptic drug discovery: the effect of topiramate on spontaneous motor seizures in rats with kainate-induced epilepsy. *Epilepsia*, 46:8–14.

Heinemann, U., O. Kann, S. Schuchmann. (2006). An overview of *in vitro* seizure models in acute and organotypic slices. In *Models of Seizures and Epilepsy* (pp. 35–44), A. Pitkanen, P.A. Schwartzkroin, and S.L. Moshe, Eds. New York: Elsevier.

Hellier, J.L., P.R. Patrylo, P.S. Buckmaster, F.E. Dudek. (1998). Recurrent spontaneous motor seizures after repeated low-dose systemic treatment with kainate: assessment of a rat model of temporal lobe epilepsy. *Epilepsy Res.*, 31:73–84.

Hosford, D.A., Y. Wang. (1997). Utility of the lethargic (*lh/lh*) mouse model of absence seizures in predicting the effects of lamotrigine, vigabatrin, tiagabine, gabapentin, and topiramate against human absence seizures. *Epilepsia*, 38:408–414.

Hosford, D.A., S. Clark, Z. Cao, W.A. Wilson, Jr., F.H. Lin, R.A. Morrisett, A. Huin. (1992). The role of $GABA_B$ receptor activation in absence seizures of lethargic (lh/lh) mice. *Science*, 257:398–401.

Jefferys, J.G. (1996). Chronic epileptic foci induced by intracranial tetanus toxin. *Epilepsy Res. Suppl.*, 12:111–117.

Jefferys, J.G., B.J. Evans, S.A. Hughes, S.F. Williams. (1992). Neuropathology of the chronic epileptic syndrome induced by intrahippocampal tetanus toxin in rat: preservation of pyramidal cells and incidence of dark cells. *Neuropathol. Appl. Neurobiol.*, 18:53–70.

Jensen, F.E. (1995). An animal model of hypoxia-induced perinatal seizures. *Ital. J. Neurol. Sci.*, 16:59–68.

Jiang, M., C.L. Lee, K.L. Smith, J.W. Swann. (1998). Spine loss and other persistent alterations of hippocampal pyramidal cell dendrites in a model of early-onset epilepsy. *J. Neurosci.*, 18:8356–8368.

Kadam, S.D., F.E. Dudek. (2007). Neuropathogical features of a rat model for perinatal hypoxic–ischemic encephalopathy with associated epilepsy. *J. Comp. Neurol.*, 505:716–737.

Klitgaard, H., A. Matagne, J. Gobert, E. Wulfert. (1998). Evidence for a unique profile of levetiracetam in rodent models of seizures and epilepsy. *Eur. J. Pharmacol.*, 353:191–206.

Kupferberg, H. (2001). Animal models used in the screening of antiepileptic drugs. *Epilepsia*, 42(Suppl. 4):7–12.

Leite, J.P., E.A. Cavalheiro. (1995). Effects of conventional antiepileptic drugs in a model of spontaneous recurrent seizures in rats. *Epilepsy Res.*, 20:93–104.

Lennox, W.G. (1945). The petit mal epilepsies: their treatment with Tridone. *JAMA*, 129:1069–1074.

Libbey, J.E., N.J. Kirkman, M.C. Smith, T. Tanaka, K.S. Wilcox, H.S. White, R.S. Fujinami. (2008). Seizures following picornavirus infection. *Epilepsia*, 49:1066–1074.

Löscher, W. (2002). Animal models of epilepsy for the development of antiepileptogenic and disease-modifying drugs: a comparison of the pharmacology of kindling and post-status epilepticus models of temporal lobe epilepsy. *Epilepsy Res.*, 50:105–123.

Löscher, W., D. Honack. (1991). Responses to NMDA receptor antagonists altered by epileptogenesis. *Trends Pharmacol. Sci.*, 12:52.

Löscher, W., D. Honack. (1993). Profile of ucb l059, a novel anticonvulsant drug, in models of partial and generalized epilepsy in mice and rats. *Eur. J. Pharmacol.*, 232:147–158.

Löscher, W., D. Honack, C. Rundfeldt. (1998). Antiepileptogenic effects of the novel anticonvulsant levetiracetam (UCB l059) in the kindling model of temporal lobe epilepsy. *J. Pharmacol. Exp. Ther.*, 284:474–479.

Löscher, W., E. Reissmuller, U. Ebert. (2000). Anticonvulsant efficacy of gabapentin and levetiracetam in phenytoin-resistant kindled rats. *Epilepsy Res.*, 40:63–77.

Marescaux, C., M. Vergnes. (1995). Genetic absence epilepsy in rats from Strasbourg (GAERS). *Ital. J. Neurol. Sci.*, 16:113–118.

Meldrum, B.S. (1995). Neurotransmission in epilepsy. *Epilepsia*, 36:S30–S35.

Nissinen, J., T. Halonen, E. Koivisto, A. Pitkanen. (2000). A new model of chronic temporal lobe epilepsy induced by electrical stimulation of the amygdala in rat. *Epilepsy Res.*, 38:177–205.

Noebels, J.L., D.M. Treiman, J. Engel. (2008). Genetic models of epilepsy. In *Epilepsy: A Comprehensive Textbook* (pp. 445–455), J. Engel, T.A. Pedley, J. Aicardi, M.A. Dichter, and S. Moshé, Eds. Philadelphia, PA: Lippincott Williams & Wilkins.

Pitkanen, A., T.P. Sutula. (2002). Is epilepsy a progressive disorder? Prospects for new therapeutic approaches in temporal-lobe epilepsy. *Lancet Neurol.*, 1:173–181.

Pitkanen, A., S. Narkilahti, Z. Bezvenyuk, A. Haapalinna, J. Nissinen. (2004). Atipamezole, an alpha(2)-adrenoceptor antagonist, has disease modifying effects on epileptogenesis in rats. *Epilepsy Res.*, 61:119–140.

Pitkanen, A., R.J. Immonen, O.H. Grohn, I. Kharatishvili. (2009). From traumatic brain injury to posttraumatic epilepsy: what animal models tell us about the process and treatment options. *Epilepsia*, 50(Suppl. 2):21–29.

Postma, T., E. Krupp, X.L. Li, R.M. Post, S.R. Weiss. (2000). Lamotrigine treatment during amygdala-kindled seizure development fails to inhibit seizures and diminishes subsequent anticonvulsant efficacy. *Epilepsia*, 41:1514–1521.

Prince, D.A., K. Jacobs. (1998). Inhibitory function in two models of chronic epileptogenesis. *Epilepsy Res.*, 32:83–92.

Putnam, T.J., H.H. Merritt. (1937). Experimental determination of the anticonvulsant properties of some phenyl derivatives. *Science*, 85:525–526.

Sarkisian, M.R. (2001). Overview of the current animal models for human seizure and epileptic disorders. *Epilepsy Behav.*, 2:201–216.

Sharpless, S.K., L.M. Halpern. (1962). The electrical excitability of chronically isolated cortex studied by means of permanently implanted electrodes. *Electroencephalogr. Clin. Neurophysiol.*, 14:244–255.

Singh, N.A., J.F. Otto, E.J. Dahle, C. Pappas, J.D. Leslie, A. Vilaythong, J.L. Noebels, H.S. White, K.S. Wilcox, M.F. Leppert. (2008). Mouse models of human *KCNQ2* and *KCNQ3* mutations for benign familial neonatal convulsions show seizures and neuronal plasticity without synaptic reorganization. *J. Physiol.*, 586:3405–3423.

Smyth, M.D., N.M. Barbaro, S.C. Baraban. (2002). Effects of antiepileptic drugs on induced epileptiform activity in a rat model of dysplasia. *Epilepsy Res.*, 50:251–264.

Snead, O.C., 3rd. (1992). Pharmacological models of generalized absence seizures in rodents. *J. Neural Transm.*, 35(Suppl.):7–19.

Srivastava, A., H.S. White. (2005). Retigabine decreases behavioral and electrographic seizures in the lamotrigine-resistant amygdale kindled rat model of pharmacoresistant epilepsy. *Epilepsia*, 46(Suppl. 8):217–218.

Srivastava, A., J.H. Woodhead, H.S. White. (2003). Effect of lamotrigine, carbamazepine and sodium valproate on lamotrigine-resistant kindled rats. *Epilepsia*, 44(Suppl. 9):42.

Srivastava, A., M.R. Franklin, B.S. Palmer, H.S. White. (2004). Carbamazepine, but not valproate, displays pharmaco-resistance in lamotrigine-resistant amygdala kindled rats. *Epilepsia*, 45(Suppl. 7):12.

Stables, J.P., E.H. Bertram, H.S. White, D.A. Coulter, M.A. Dichter, M.P. Jacobs, W. Löscher, D.H. Lowenstein, S.L. Moshe, J.L. Noebels, M. Davis. (2002). Models for epilepsy and epileptogenesis: report from the NIH workshop, Bethesda, Maryland. *Epilepsia*, 43:1410–1420.

Stables, J.P., E. Bertram, F.E. Dudek, G. Holmes, G. Mathern, A. Pitkanen, H.S. White. (2003). Therapy discovery for pharmacoresistant epilepsy and for disease-modifying therapeutics: summary of the NIH/NINDS/AES Models II workshop. *Epilepsia*, 44:1472–1478.

Stafstrom, C.E., T.P. Sutula. (2005). Models of epilepsy in the developing and adult brain: implications for neuroprotection. *Epilepsy Behav.*, 7(Suppl. 3):S18–S24.

Sutula, T., C. Harrison, O. Steward. (1986). Chronic epileptogenesis induced by kindling of the entorhinal cortex: the role of the dentate gyrus. *Brain Res.*, 385:291–299.

Sutula, T., X.X. He, J. Cavazos, G. Scott. (1988). Synaptic reorganization in the hippocampus induced by abnormal functional activity. *Science*, 239:1147–1150.

Swinyard, E.A. (1972). Electrically induced convulsions. In *Experimental Models of Epilepsy* (pp. 443–458), D.B. Purpura, J.K. Penry, D. Tower, D.M. Woodbury, and R. Walter, Eds. New York: Raven Press.

Toman, J.E. (1951). Neuropharmacologic considerations in psychic seizures. *Neurology*, 1:444–460.

Toman, J.E., G.M. Everett, R.K. Richards. (1952). The search for new drugs against epilepsy. *Tex. Rep. Biol. Med.*, 10:96–104.

Turski, L., E.A. Cavalheiro, S.J. Czuczwar, W.A. Turski, Z. Kleinrok. (1987). The seizures induced by pilocarpine: behavioral, electroencephalographic and neuropathological studies in rodents. *Pol. J. Pharmacol. Pharm.*, 39:545–555.

Turski, L., C. Ikonomidou, W.A. Turski, Z.A. Bortolotto, E.A. Cavalheiro. (1989). Review: cholinergic mechanisms and epileptogenesis. The seizures induced by pilocarpine: a novel experimental model of intractable epilepsy. *Synapse*, 3:154–171.

van Vliet, E.A., R. van Schaik, P.M. Edelbroek, S. Redeker, E. Aronica, W.J. Wadman, N. Marchi, A. Vezzani, J.A. Gorter. (2006). Inhibition of the multidrug transporter P-glycoprotein improves seizure control in phenytoin-treated chronic epileptic rats. *Epilepsia*, 47:672–680.

Weiss, S.R., R.M. Post. (1991). Development and reversal of contingent inefficacy and tolerance to the anticonvulsant effects of carbamazepine. *Epilepsia*, 32:140–145.

White, H.S. (1997). Mechanisms of antiepileptic drugs. In *Epilepsies II* (pp. 1–30), R. Porter and D. Chadwick, Eds. Boston, MA: Butterworth-Heinemann.

White, H.S., J.H. Woodhead, M.R. Franklin, E.A. Swinyard, H.H. Wolf. (1995a). General principles: experimental selection, quantification, and evaluation of antiepileptic drugs. In *Antiepileptic Drugs*, 4th ed. (pp. 99–110), R.H. Levy, R.H. Mattson, and B.S. Meldrum, Eds. New York: Raven Press.

White, H.S., M. Johnson, H.H. Wolf, H.J. Kupferberg. (1995b). The early identification of anticonvulsant activity: role of the maximal electroshock and subcutaneous pentylenetetrazol seizure models. *Ital. J. Neurol. Sci.*, 16:73–77.

White, H.S., J.H. Woodhead, K.S. Wilcox, J.P. Stables, H.J. Kupferberg, H.H. Wolf. (2002). Discovery and preclinical development of antiepileptic drugs. In *Antiepileptic Drugs* (pp. 36–48), R.H. Levy, R.H. Mattson, B. Meldrum, and E. Perucca, Eds. Philadelphia, PA: Lippincott Williams & Wilkins.

White, H.S., A. Srivastava, B. Klein, B. Zhao, Y. Moon Choi, R. Gordon, S.J. Lee. (2006). The novel investigational neuromodulator RWJ 333369 displays a broad-spectrum anticonvulsant profile in rodent seizure and epilepsy models. *Epilepsia*, 47(Suppl. 4):200.

Williams, P.A., F.E. Dudek. (2007). A chronic histopathological and electrophysiological analysis of a rodent hypoxic–ischemic brain injury model and its use as a model of epilepsy. *Neuroscience*, 149:943–961.

Williams, P.A., P. Dou, F.E. Dudek. (2004). Epilepsy and synaptic reorganization in a perinatal rat model of hypoxia–ischemia. *Epilepsia*, 45:1210–1218.

Willmore, L.J., G.W. Sypert, J.V. Munson, R.W. Hurd. (1978). Chronic focal epileptiform discharges induced by injection of iron into rat and cat cortex. *Science*, 200:1501–1503.

Yu, F.H., M. Mantegazza, R.E. Westenbroek, C.A. Robbins, F. Kalume, K.A. Burton, W.J. Spain, G.S. McKnight, T. Scheuer, W.A. Catterall. (2006). Reduced sodium current in GABAergic interneurons in a mouse model of severe myoclonic epilepsy in infancy. *Nat. Neurosci.*, 9:1142–1149.

# 10 Pharmacogenetics of AED Development

*Russell J. Buono, Thomas N. Ferraro,*
*Dennis J. Dlugos, and Karen S. Wilcox*

## CONTENTS

## INTRODUCTION

Numerous examples can be found throughout medicine of differences in individual patient responses to medications. These differences can involve efficacy, potency, and adverse reactions. This is also true for drugs used to treat epilepsy, and recent work in pharmacogenetics has begun to unravel the myriad contributing factors to individual responses to antiepileptic drug (AED) therapy. A number of genetic variations are now known to influence both the pharmacokinetics and pharmacodynamics of AEDs. In addition, changes in the regulation of neuronal and glial gene expression following seizure activity can also influence the response to drugs. Finally, the problem of refractory, or pharmacoresistant, epilepsy may very well also have its roots in pharmacogenetics. In this chapter, we highlight some of the most recent findings in patients and animal models that are relevant to individual variations in response to antiepileptic drugs, how these findings might ultimately impact AED development and personalized therapy, and what challenges remain for the effective use of pharmacogenetic information in the development of treatments for epilepsy.

## PHARMACOKINETICS OF AEDS

Following administration, medications are absorbed into the body, distributed to various tissues and organs, metabolized, and finally excreted (ADME). This pharmacokinetic dispensation of compounds determines drug concentration at target sites and leads to both therapeutic and adverse responses. This is certainly true for AEDs, and each clinically available AED has its own unique ADME profile, the descriptions of which are beyond the scope of this chapter. However, several well-known pharmacogenetic studies have elucidated proteins responsible for aspects of ADME that are known to impact the clinical use of AEDs and the potential for adverse side effects.

Most notable in this regard is the well-documented cytochrome P450 polymorphisms that influence phenytoin (PHT), mephenytoin, and, to a lesser extent, phenobarbital metabolism (Anderson, 2008). Metabolism of these standard AEDs occurs primarily via the cytochrome P450 enzymes CYP2C9 and CYP2C19. These two isozymes are known to have genetic polymorphisms that influence their activity. CYP2C9 has four known polymorphisms, three of which result in a low rate of metabolism of PHT: CYP2C9*2, CYP2C9*3, and the rare CYP2C9*4 (Browne and LeDuc, 2002). In patients with these polymorphisms, administration of commonly used doses of PHT can result in toxicity due to the high concentrations of plasma PHT that are achieved. PHT concentrations can be elevated further if there is also a concomitant polymorphism in CYP2C19 (Brandolese et al., 2001).

Although genetic screening can be used to determine the risks inherent in treatment with PHT, the low incidence of these polymorphisms in most ethnic populations and their relative effect on the overall clinical response suggest that such genetic testing would not be cost efficient or clinically justified at this time. In addition, these particular liver enzymes are not intimately involved in the metabolism of newer AEDs. Although this information provides clear evidence that ADME can be influenced by genetics, the clinical significance is somewhat limited.

The other major class of proteins associated with pharmacokinetic differences in AEDs that may contribute to drug efficacy are the drug transporters. As these molecules are discussed in more detail in the context of drug resistance in Chapter 6, we will only highlight some of the salient features of the potential role of polymorphisms of these proteins with respect to the distribution of AEDs. The hypothesis that multidrug transporter proteins might contribute to drug resistance in epilepsy was born out of the findings in cancer research showing that they are responsible in some circumstances for the refractoriness of malignant cells to the cytotoxic effects of chemotherapeutic drugs. The adenosine triphosphate (ATP)-binding cassette (ABC) family of transporters has received much attention over the last several years. These proteins are expressed in endothelial cells that comprise the blood–brain barrier and mediate the transport of xenobiotics away from the brain. The ABC genes encode P-glycoprotein (P-gp), and polymorphisms in the *ABCB1* gene have been associated with altered levels of P-gp and altered transport of AEDs (Siddiqui et al., 2003). Following this initial report, however, subsequent studies have been unable to find a connection between the *ABCB1* polymorphisms and drug resistance (Sills et al., 2005; Tan et al., 2004). A recent meta-analysis combining data from all published studies concluded that variation in the *ABCB1* transporter is not associated with AED resistance, suggesting that polymorphisms in *ABCB1* may not be predictive of refractory epilepsy or be useful in identifying appropriate AED therapies (Bournissen et al., 2009). Furthermore, it is not clear the extent to which AEDs even serve as substrates for P-gp transport. Finally, most AEDs are fairly lipophilic; therefore, entry into the central nervous system (CNS) for most drugs proceeds passively according to mass action. Medical refractoriness to an AED is often seen despite overt signs of CNS toxicity, suggesting that achievement of relevant tissue levels is unlikely to be the major cause of therapy failure. Thus, evidence in support of a role for *ABCB1* polymorphisms in altering brain levels of AEDs, at least at the present time, is not very strong.

## ADVERSE EVENTS

Adverse events are a common problem with all classes of drugs. Most drugs will exhibit toxicity if their levels build to high enough concentrations and organ systems are often differentially more sensitive depending on the pharmacokinetic and pharmacodynamic properties of the drugs in question. AEDs impact several organ systems, including the CNS, liver, and skin. The use of genomic information to predict adverse events induced by AEDs is not yet a common clinical practice, but one line of research in the field of epilepsy has fueled excitement and demonstrates the potential usefulness of applied pharmacogenetics. Patients who harbor a specific human leukocyte antigen (HLA1502B) are at a very high risk (>90%) for developing Stevens–Johnson syndrome, a severe and potentially fatal adverse reaction involving the skin when treated with carbamazepine (CBZ) (Man et al., 2007). HLA1502B is a common allele found in persons of Han Chinese ancestry, and the Food and Drug Administration (FDA) has issued a "black box" warning to inform physicians and patients of the risk involved in taking CBZ in the context of an HLA1502B allele. The FDA warning recommends genotyping this marker in patients if Han Chinese ancestry is known or suspected. This finding is certainly of clinical importance. Not only does this work demonstrate that a single marker can lead to a diagnostic test with high clinical validity and reliability, but, perhaps even more importantly, the discovery was also initially made with less than 100 individuals with Stevens–Johnson syndrome. These data suggest that adverse events can be linked to genetic variation without necessarily studying thousands or tens of thousands of patients.

Adverse events are known for other AEDs, and the genetic contributions are the subject of much research. PHT, CBZ, and valproic acid (VPA) can all lead to liver toxicity if serum concentrations reach high levels due to poor metabolic rates (Franciotta et al., 2009). This can happen in patients given standard doses of drugs but who harbor CYP variations that lower their ability to metabolize these AEDs. Thus, toxic blood levels may rise due to genetic variation that affects the pharmacokinetics of metabolism. The variations that account for the majority of this effect are still not fully identified, and, until they are, testing of the currently known markers cannot yield enough information to inform dosing strategies. Vigabatrin has been shown to be associated with an adverse event in the retina leading to constriction of the visual fields in epilepsy patients treated with this drug (Conway et al., 2008). Finally, there is good evidence that VPA can cause birth defects *in utero* when mothers are treated with this AED (Sankar, 2007). Furthermore, there is a persistent adverse effect on learning and cognitive skills during early childhood in children exposed to VPA *in utero* who have no overt structural birth defects (Meador et al., 2009). Thus, more research is being focused on identifying genetic variations that could be used as biomarkers to predict adverse events from AED treatment. In the future, more testing should become available to discern those individuals at greatest risk for AED-induced adverse events.

## PHARMACODYNAMICS OF AEDS

The mechanisms of AED actions comprise the pharmacodynamic response. Just as changes in proteins underlying pharmacokinetics can alter the potency, efficacy, and adverse effects of drugs in individual patients, genetic differences in the targets of drugs also play an important role in the ability of an AED to provide symptomatic relief from seizures. In addition, we are beginning to learn that gene expression changes that occur in target neurons, and maybe even in glial cells as a consequence of seizures *per se*, may also alter the pharmacodynamic profile of an AED. These recent findings, many in animal models of epilepsy, suggest that drug development should not be confined exclusively to acute seizure models. Instead, the use of a variety of chronic epilepsy models should be incorporated into drug screening activities, as they may provide important clues as to the efficacy of novel drugs in patients with epilepsy. This section reviews what is currently known about the pharmacogenetics of some of the primary AED targets and also discusses the consequences of *acquired pharmacogenomics*, which refers to changes in gene expression that occur as a consequence of either seizure activity or CNS insults and lead to the development of epilepsy and concomitantly affect the density or distribution of drug targets. We discuss several examples of acquired pharmacogenomics, with consideration of how changes in drug target expression can influence individual AED therapy.

## Sodium Channels

Commonly used AEDs for the treatment of both generalized and partial seizures include PHT, CBZ, and lamotrigine (LTG), and the primary mechanism of action of these compounds is blockade of voltage-gated sodium channels. It was recently determined that the maximal doses of either PHT or CBZ in patients with epilepsy were found to be associated with a polymorphism in the *SCN1A* gene (Tate et al., 2005), and the PHT maximal dose is associated with CYP2C9 polymorphisms. As described below, mutations in the *SCN1A* gene are also now known to underlie a number of epilepsy syndromes. The polymorphism that was associated with the maximum doses of PHT and CBZ was found to be at the IVS5-91 G>A locus, with the minor allele (G) being associated with 10 to 15% reductions in the maximal tolerated doses of PHT and 15 to 20% reductions in the maximal tolerated doses of CBZ (Tate et al., 2005). The maximum PHT dose varied from 326 to 373 mg per day, and the maximum CBZ doses varied from 1083 and 1313 mg per day, depending on the *SCN1A* genotype (Ferraro et al., 2006; Tate et al., 2005). The report by Tate et al. (2006) demonstrated a genomic influence on PHT dosing on both pharmacokinetic and pharmacodynamic bases. These results suggest potential therapeutic strategies, as the genotype could dictate the titration schedule for new patients, thus leading to a more rapid seizure control for the patient with acceptable tolerability.

As mentioned above, a number of mutations in the *SCN1A* subunit of the sodium channel have been found to result in generalized epilepsy with febrile seizures plus (GEFS+) and Dravet syndrome (Mullen and Scheffer, 2009). Approximately 95% of the over 200 mutations in *SCN1A* that are associated with Dravet syndrome appear to have occurred *de novo*; therefore, there is often no familial component in this severe form of epilepsy and yet the finding that it is due to sodium channelopathies is an important diagnostic tool (Berkovic et al., 2006). Recent work has determined that, in mouse models of some of the mutations that result in Dravet syndrome, hyperexcitability is most likely due to decreased expression of sodium channels in interneurons. These findings not only explain the paradoxical finding that loss-of-function mutations in sodium channels result in hyperexcitability and seizure activity but also provide a rationale for the unusual finding that LTG may exacerbate myoclonic seizures in this population (Ogiwara et al., 2007; Yu et al., 2006). This work certainly highlights the important role in epilepsy research for novel animal models that express the same mutations as human patients. Experiments in such animal models will no doubt lead to important insights into the role of sodium channels, not only regarding the pathophysiology of the disease process but also the selection of appropriate AED therapy.

Genotyping is now viewed as an important component of the process used for diagnosis of Dravet syndrome and certainly could dictate therapeutic decisions. For example, rapid identification of the genetics underlying Dravet's syndrome could lead to more aggressive therapeutic management of febrile seizures and more informed counseling regarding prognosis (Delgado-Escueta and Bourgeois, 2008). Detection of mutations early in the evaluation of new patients could also rule out other possibilities for the observed seizures, thus making the diagnosis more accurate. For example, Berkovic et al. (2006) found in a retrospective study that, for a substantial portion of cases where the initial diagnosis of seizures was thought to be due to vaccination-induced encephalopathy, 11 out of 14 children were subsequently found to actually have *SCN1A* mutations. It should be noted that mutations in *SCN1A* have been identified in unaffected family members of Dravet patients as well and that not all Dravet patients have an identified mutation in *SCN1A*.

The ability of AEDs to effectively control seizures depends on the degree to which these drugs may enter the CNS (pharmacokinetics) and, once distributed appropriately, the degree to which these drugs act on selective targets (pharmacodynamics). Important experiments performed by Beck and colleagues have led to the hypothesis that pharmacoresistant epilepsy may, in some cases, be due to the fact that the drug targets have in some way been altered (Remy et al., 2003). In these experiments, the authors obtained resected tissue from patients with either CBZ-sensitive or CBZ-resistant seizures and recorded sodium currents from acutely dissociated granule cells of the dentate gyrus. In stark contrast to the cells obtained from CBZ-sensitive patients, cells obtained from the CBZ-resistant patients did *not* exhibit a CBZ-induced, use-dependent block of sodium currents. In essence, neurons from these patients were "pharmacoresistant." Furthermore, electographic seizure-like activity induced in brain slices obtained from the CBZ-resistant patients was also found to be CBZ resistant, whereas bursting in slices from CBZ-sensitive patients was significantly reduced following CBZ administration. Importantly, these *in vitro* experiments eliminate any possibility that drug transporters or metabolism may contribute to the concentration of the drug that is ultimately delivered to the target tissue (Remy et al., 2003). The investigators concluded from this work that there is a loss of sensitivity to CBZ in sodium channels which may in fact lead to the lack of CBZ efficacy in these patients. It is not known if the CBZ-resistant patients possessed sodium channel mutations that altered the CBZ binding site or if the disease process itself altered expression of sodium channel subunits. Nevertheless, this work has led to the acceptance of the idea that pharmacoresistant epilepsy may result as a consequence of altered drug targets. There is no doubt that future pharmacogenomic data obtained from both animal models and humans will provide insight into these processes and will help guide future drug development as well as individualized AED treatment.

## POTASSIUM CHANNELS

### *KCNQ2* and the M-Channel

The M-type potassium current regulates resting membrane potential and spike frequency adaptation in the neuron. Mutations in two subunits that comprise the M-channel, *KCNQ2* (Kv7.2) and *KCNQ3* (Kv7.3), result in the seizure disorder benign familial neonatal convulsions (BFNC) (Charlier et al., 1998; Singh et al., 1998). To date, over 60 different mutations in *KCNQ2* and 4 different mutations in *KCNQ3* have been identified in families with BFNC. Interestingly, the primary mechanism of action of the novel AED retigabine (RGB) is via a shift in the activation curve for $I_{K(M)}$ to a more hyperpolarized potential (Wickenden et al., 2000). Thus, in recent years, there has been substantial enthusiasm for consideration of the M-channel as a potential and novel therapeutic target.

To study the contribution of the M-channel to seizures, a number of animal models have been developed. The *Szt1* mouse strain has a spontaneous 300-kb deletion on the C57BL/6 genetic background that includes the genomic DNA encoding the *KCNQ2* C-terminus, thus resulting in a *Kcnq2* haploinsufficiency (Yang et al., 2003). Such haploinsufficiencies are the likely outcome of many of the identified mutations in this gene in human BFNC patients. Initial work with this strain of mice demonstrated that the heterozygotes (i.e., one functional *Kcnq2* allele) have a substantial reduction in seizure threshold in several different corneal stimulation models (Otto et al., 2004; Yang et al., 2003). Interestingly, these mice also displayed a significantly decreased sensitivity to RGB in the 6-Hz model of psychomotor seizures. Electrophysiology experiments were then performed on CA1 neurons in the *in vitro* hippocampal brain slices obtained from wild-type and *Szt1* mice to test the hypothesis that this deletion alters $I_{K(M)}$ function and, importantly, pharmacology. CA1 neurons in *Szt1* mice were found to have a decreased $I_{K(M)}$ amplitude and current density, and action potential accommodation was compromised compared to that of wild-type C57BL/6 littermates. Most interesting, however, was that significant differences in $I_{K(M)}$ pharmacology in *Szt1* CA1 neurons were observed. These differences had intriguing parallels to the *in vivo* work, including a highly significant decrease in sensitivity to RGB (Figure 10.1) (Otto et al., 2006). These results suggest that AEDs that target the M-channel (e.g., retigabine, flupertine) may be less effective in patients with underlying *KCNQ2* mutations that result in a haploinsufficiency. It is not yet known if many epilepsy patients outside of BFNC patients have underlying mutations or insufficiencies in *KCNQ2* expression; however, this may turn out to be an important gene to evaluate when determining the likelihood of a patient's beneficial therapeutic response to AEDs that target the M-channel.

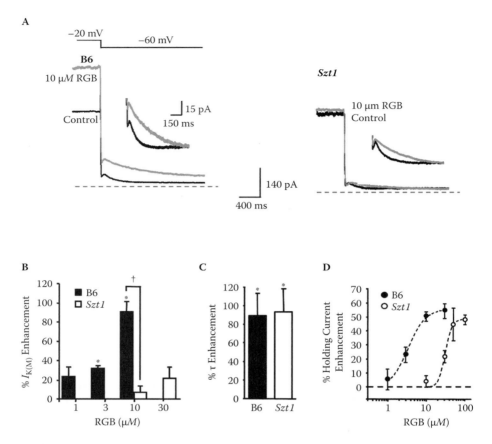

**FIGURE 10.1** *Szt1* confers decreased sensitivity to the $I_{(KM)}$-enhancing properties of retigabine (RGB). (A) Sample traces in control (black) and 10-μ*M* RGB (gray) illustrate that *Szt1* CA1 neurons exhibit decreased sensitivity to the $I_{(KM)}$-enhancing effects of RGB. (B) *Szt1* slices exhibit dramatically decreased sensitivity to the $I_{(KM)}$-amplitude-enhancing effects of RGB. Although 3- and 10-μ*M* RGB significantly enhance $I_{(KM)}$ amplitude in B6 slices, $I_{(KM)}$ amplitude is not enhanced by RGB at any concentration in *Szt1* slices. The effects of 10-μ*M* RGB differ significantly between B6 and *Szt1* slices ($^\dagger p < 0.0005$, two-way ANOVA, drug effect vs. mouse genotype comparison; $^* p < 0.05$). Notice, however, that RGB does prolong $I_{(KM)}$ deactivation kinetics in *Szt1* slices (A, inset, C). (C) RGB significantly and similarly prolongs the $I_{(KM)}$ of deactivation in CA1 neurons of B6 and *Szt1* mice ($^* p < 0.02$). (D) Dose–response curves show that *Szt1* slices are ~10-fold less sensitive to the effects of RGB on holding current at –20 mV (*Szt1* $EC_{50}$, $31.2 \pm 0.4$ μ*M*; B6 $EC_{50}$, $3.5 \pm 0.4$ μ*M*). The RGB EC shift in *Szt1* CA1 neurons was significant at $p \le 0.05$. (From Otto, J.F. et al., *J. Neurosci.*, 26(7), 2053–2059, 2006. With permission.)

### *KCNJ10*

*KCNJ10* encodes a subunit for an inward-rectifying potassium channel (KIR4.1) that is highly expressed in glial cells in the CNS as well as in cardiac, kidney, and other tissues. In the brain, KIR channels function in glial cells to buffer extracellular potassium after action potentials, thereby acting as a potassium sink allowing neurons to return to their resting potential. Variation in the mouse *KCNJ10* was first linked to seizure susceptibility by utilizing quantitative trait loci mapping experiments (Ferraro et al., 2004). Shortly thereafter variation in the human *KCNJ10* showed a positive association with common forms of human epilepsy (Buono et al., 2004). These findings have since been replicated in two separate cohorts, confirming a role for *KCNJ10* variation in common forms of human epilepsy (Heuser et al., 2008; Lenzen et al., 2005). Recently, two separate groups have identified rare coding region mutations in the *KCNJ10* gene that are linked to familial forms of epilepsy with ataxia, sensorineural deafness, and kidney tubulopathy (Bockenhauer et al., 2009; Scholl et al., 2009). Thus, although it appears that common haplotypes of *KCNJ10* are associated with common forms of human epilepsy, these haplotypes may be concealing rare, more deleterious mutations that can underlie rare familial forms of epilepsy involving additional syndromic phenotypes.

These data exemplify the notion that epilepsy and other complex trait diseases in humans are likely to be caused by a combination of common genetic variants that increase susceptibility as well as rare mutations in critical genes. These rare mutations, in combination with genetic background modifiers, can cause disease. The same is likely to hold true for pharmacogenetic influences on drug response. Thus, it is reasonable to suggest that the efficacy of and tolerability to AEDs are likely influenced by common variations across the human genome as well as by rare mutations in key pharmacokinetic and pharmacodynamic genes.

## GABA$_A$ Receptors

Inhibitory neurotransmission in the CNS is mediated to a great extent by the γ-aminobutyric acid (GABA) system. The neurotransmitter GABA activates three types of receptors: GABA$_A$, GABA$_B$, and GABA$_C$. Receptors in brain regions that are thought to be involved in epilepsy include primarily the ionotropic GABA$_A$ and metabotropic GABA$_B$ receptors. The GABA$_A$ receptor is a ligand-gated ion channel that is permeable to chloride and is comprised of five transmembrane-spanning subunits ($\alpha_{1-6}$, $\beta_{1-3}$, $\gamma_{1-3}$, $\delta$, $\varepsilon$, and $\rho$) (Reid et al., 2009). The most commonly expressed synaptic GABA$_A$ receptors contain two α, two β, and one γ subunit, and the types of subunits that are assembled determine the properties of the receptors, such as pharmacology, kinetics, and conductance. It has long been recognized that pharmacologically induced reductions in GABAergic signaling can induce seizures, and AEDs such as phenobarbital and diazepam work primarily by increasing current flow through the GABA$_A$ receptor. Perhaps it was not too surprising, then, to find that mutations in the GABA$_A$ receptor underlie or are associated with several clinical forms of epilepsy.

Mutations that are associated with epilepsy have been found in the *GABRG2* gene, which encodes the $\gamma_2$ subunit, and in the *GABRA1* gene, which encodes the $\alpha_1$ subunit. Mutations in the $\gamma_2$ subunit result in GEFS+, childhood absence epilepsy (CAE), and febrile seizures; a mutation in the $\alpha_1$ subunit is associated with juvenile myoclonic epilepsy (JME) (Cossette et al., 2002; Reid et al., 2009) and a sporadic form of CAE (Maljevic et al., 2006). Many of these mutations reduce the extent of GABA$_A$-mediated inhibition, although a variety of molecular mechanisms, such as impaired trafficking of the receptors to the membrane and altered kinetics, have been identified. Interestingly, it was found that trafficking of receptors that contain $\gamma_2$ mutations that are associated with febrile seizures (R43Q, K289M, and Q351X) are highly temperature sensitive. This is consistent with the clinical observation that febrile seizures are triggered when patients have a high fever and suggests that GABA receptor expression at the neuronal membrane in these patients may be reduced, thus altering neuronal excitability during fevers (Kang et al., 2006). Although *in vitro* expression systems have aided our understanding of the effect of these mutations on many properties of the GABA$_A$ receptor, it will be necessary to use transgenic animal models with the mutations knocked in to determine how the changes in inhibition are manifested at the network level and to help predict appropriate therapeutic strategies for the patients harboring these mutations.

Acquired changes in GABA$_A$ receptor subunit expression are thought to play a role in the process of epileptogenesis following status epilepticus (SE) (Brooks-Kayal et al., 1998, 1999). Granule cells of the adult dentate gyrus primarily express GABA$_A$ receptors containing the $\alpha_1$ subunit; however, following pilocarpine-induced status epilepticus in rats, granule cells dramatically increase expression of the $\alpha_4$ subunit and decrease expression of the $\alpha_1$ subunit (Brooks-Kayal et al., 1998). This has functional ramifications for the hippocampus, as $\alpha_4$-containing receptors have a reduced sensitivity for benzodiazepines and an increased sensitivity to zinc blockade. Thus, these changes may contribute to hyperexcitability and the development of epilepsy following SE.

Recent work by Raol et al. (2006), capitalized on these findings to trick the dentate gyrus into expressing $\alpha_1$ subunits and thereby test the hypothesis that increased expression of $\alpha_1$ subunits of the $GABA_A$ receptor could block epileptogenesis following SE. In their study, the authors injected into the dentate gyrus an adeno-associated virus (AAV) designed to express a bicistronic RNA encoding the $\alpha_1$ subunit under the control of the promoter region of the $GABA_A$ $\alpha_4$ receptor subunit (which has been shown to be upregulated following SE) (Brooks-Kayal et al., 2009; Raol et al., 2006). Compared to control animals, rats that had received the AAV-$\alpha_1$ injection prior to SE had a dramatic increase in $\alpha_1$ expression and a delayed onset of epilepsy. Treatment of patients at risk for the development of epilepsy in such a manner may not be practical, but this proof-of-principle experiment suggests that increased expression of $\alpha_1$ subunits may delay the onset of epilepsy, and administration of drugs that address the regulatory mechanism of gene expression may be a viable approach to prevent the development of epilepsy in at-risk individuals (Brooks-Kayal et al., 2009).

These same changes in GABA receptor subunits have also been observed in granule cells obtained from resected hippocampus from temporal lobe epilepsy (TLE) patients (Brooks-Kayal et al., 1999). Interestingly, this change in GABA receptor subunit expression may also partially explain why topiramate, a commonly prescribed drug for the treatment of partial seizures, may be an effective treatment for some TLE patients. In addition to possessing many mechanisms of action, including blockade of kainate receptors and sodium channels, topiramate has also been shown to enhance GABA-mediated currents when receptors contain the $\alpha_4$ subunit (Simeone et al., 2006). Thus, as we have seen before, insights gained from animal models of epilepsy can have important translational value, leading to insight into ways to improve treatment with AEDs.

## NICOTINIC RECEPTORS

The first genetic mutation to be identified as a cause for epilepsy was found in a subunit of the nicotinic acetylcholine receptor (Steinlein et al., 1995). The nicotinic acetylcholine receptor is a pentameric receptor-gated ion channel that is permeable to cations and is assembled as either a homomer or heteromer from a selection of 17 different subunits ($\alpha_{1-10}$, $\beta_{1-4}$, $\delta$, $\epsilon$, and $\gamma$) (Reid et al., 2009). A gain-of-function mutation in the $\alpha_4$ subunit, which is encoded by the *CHRNA4* gene, causes autosomal dominant nocturnal frontal lobe epilepsy (ADNFLE). Patients with ADNFLE exhibit focal seizures in the frontal lobe during sleep and usually respond well to treatment with CBZ. Since the first mutation was identified almost 15 years ago, three additional mutations in the $\alpha_4$ subunit have also been found to result in ADNFLE. In addition, several mutations in the $\beta_2$ subunit and a single mutation in $\alpha_2$, encoded by the *CHRNB2* and *CHRNA2* genes, respectively, have also been found to be gain-of-function mutations that also result in ADNFLE (Reid et al., 2009). Although discovery of the genetic cause of this epilepsy has not changed the way it is treated, basic research regarding the structure and function of receptors comprised of subunits containing these receptors has provided insight into the pharmacodynamic effects of CBZ in the treatment of this disorder (Bertrand et al., 2002; Löscher et al., 2009; Picard et al., 1999). As discussed above, CBZ acts primarily at sodium channels to exert its AED effects; however, CBZ is also known to block nicotinic acetylcholine receptors, and in nicotinic receptors expressing the common pore mutations found in ADNFLE there is a threefold greater sensitivity for CBZ (Löscher et al., 2009).

Recent work with knockin mice with the same mutations as human patients has provided additional insight into this gain-of-function disorder (Klaassen et al., 2006). Electrophysiology recordings in brain slices obtained from the knockin mice, which have spontaneous seizures, have demonstrated that there is a substantial increase in the nicotine-evoked synaptic release of GABA at inhibitory synapses onto cortical pyramidal cells (Klaassen et al., 2006; Mann and Mody, 2008). This surprising finding suggests that enhanced inhibitory transmission in the cortex of these animals underlies their seizures. This hypothesis was supported by the finding that low concentrations of GABA receptor antagonists can actually block the spontaneous seizures that are observed in these animals (Klaassen et al., 2006). It is anticipated that future work with these important new mouse strains will provide important insights into this seizure disorder.

## Calcium Channels

The first evidence that voltage-gated calcium channels might be implicated in epilepsy came from the finding that the AED ethosuximide (ESM) blocked low-threshold, T-type calcium channels in thalamic neurons (Coulter et al., 1989). ESM very effectively blocks absence seizures, characterized by 3-Hz spike and wave electroencephalography (EEG) activity, but is largely ineffective against generalized tonic–clonic or partial seizures. As the field of epilepsy genetics has progressed over the last 15 years, several mutations in voltage-gated calcium channels have been identified in both human epilepsies and mouse models of epilepsy. As is the case with other voltage-gated ion channels, multiple genes encode a wide variety of calcium channel subunits. Calcium channels, like sodium channels, are comprised of an $\alpha$ subunit, which defines the main properties of the channel, and auxiliary $\beta$ subunits, which modify the kinetics of the channel. To date, while sporadic mutations in genes that encode the P/Q types of high-voltage-gated calcium channels have been found in a few families, the largest number of epilepsy-related mutations that have been identified are in the *CACNA1H* gene (Reid et al., 2009). The *CACN1H* gene encodes the $Ca_v3.2$ subunit of the low-voltage-gated T-type calcium channel. Over 30 different mutations have been identified in patients with generalized epilepsy, with many of the patients exhibiting absence seizures (Heron et al., 2007; Reid et al., 2009). However, it is not clear that all mutations that have been identified actually underlie the seizures, and they may be viewed instead as seizure susceptibility genes.

Similar to the case for $GABA_A$ receptors that showed a subunit change in gene expression following SE, expression of the $Ca_v3.2$ subunit has been observed to increase in CA1 pyramidal cells of the hippocampus following pilocarpine-induced SE in both rats and mice. This increase in expression of T-type calcium channels in pyramidal cells causes a switch from regular firing to intrinsic bursting in these cell types, resulting in hyperexcitability and possibly contributing to the process of epileptogenesis (Becker et al., 2008; Sanabria et al., 2001; Su et al., 2002). These results emphasize the notion that protein targets of AEDs are altered in the epileptic brain and reinforce the necessity of incorporating a variety of both experimental and genetic models of epilepsy early into the development phase of novel AEDs.

In addition to the changes in GABA receptor and $Ca^{2+}$ channel expression that occur following SE in animal models of TLE, a number of other voltage-gated and receptor-gated ion channels have been found to have altered expression following a variety of epilepsy-inducing insults. Furthermore, these changes are also thought to contribute to hyperexcitability of the hippocampus or cortex and must be considered when evaluating neuronal changes that may be involved in both epileptogenesis and TLE. These channels include, but certainly are not limited to, the A-type potassium channel and the hyperpolarization-activated cyclic nucleotide-gated (HCN) channel (Bernard et al., 2004; Chen et al., 2001; Jung et al., 2007).

## CHALLENGES IN THE ERA OF PHARMACOGENETICS

This is indeed an exciting time in epilepsy research. The molecular era has ushered in vast new insights into the genetic causes of many forms of epilepsy, as well as the molecular basis of both effective and noneffective treatments for specific seizure disorders. The ability to engineer novel mouse models that contain the exact mutations as those possessed by patients will dramatically increase our ability to develop new syndrome-specific AEDs, develop new treatment approaches such as viral-mediated expression of important genes, and, finally, to develop new treatment regimens to give hope to both the pharmacoresistant epilepsy patient and the patient at risk for the development of epilepsy. In addition, sophisticated animal models of epilepsy, coupled with advances in molecular biology, electrophysiology, imaging, and computational neuroscience, will continue to provide the research community with important insights into how subtle changes at the molecular level are manifested as seizures at the complex circuit level. Significant challenges in therapeutics still remain, however, as evidenced most strongly by the fact that, despite the overwhelming accomplishments within the fields of clinical and basic epilepsy research, between 30 and 40% of patients still do not have their seizures controlled by existing AEDs or other therapies, and this proportion has not changed over recent decades.

A major challenge in the field is the clinical diagnosis of epilepsy itself. Clinicians debate classification schemes and diagnostic criteria for various seizure disorders. Even the definition of epilepsy itself can be controversial. A unified classification scheme would be beneficial to genetic research, which relies on homogeneity of samples to increase the power to detect important risk factors. Another limitation is the lack of consensus regarding standard design for genetic association studies and the statistical significance levels and supporting functional data required to identify and confirm risk alleles. It appears likely that complex disease traits are caused by a combination of common variations across the genome acting in concert with rare mutations in key genes and contributions from environmental cues. Whole genome association studies involving very large patient populations may identify common susceptibility alleles for any given phenotype but will miss rare mutations that arise *de novo*. Similarly, those seeking rare mutations in candidate genes as a cause for phenotype will miss the common variations in the genetic background that act to modify phenotype with respect to penetrance and severity. Regarding pharmacogenetics, a major challenge remains in the definition of drug resistance. Each individual patient has a unique seizure frequency and severity associated with his or her illness, and it is difficult to standardize a definition of drug resistance based on this heterogeneity. Many of these issues have been reviewed in detail (Ferraro et al., 2006).

Another significant clinical challenge facing the research community today is the very limited information available regarding the pharmacogenetics of the newer AED drugs. Indeed, we are still in the beginning stages of determining the full spectrum of proteins that contribute to the pharmacokinetics and pharmacodynamics of AEDs. Although this chapter has highlighted some of the information regarding a few of the genes that are known to be important for AED pharmacogenetics, Table 10.1 points out many additional classes of proteins that need to be considered for a comprehensive understanding. This challenge may seem daunting, but there is considerable excitement in the field as new resources are brought to bear on the problem; for example, insights from the International HapMap project are likely to facilitate our understanding of common haplotype blocks in human disease and response to drugs. Whole genome association studies in epilepsy are underway, and these data will identify common variations associated with disease as well as those associated with responses to AEDs. In addition, new technology designed to sequence large amounts of DNA within a rapid time frame will allow for rare mutations in critical genes to be discovered. Finally, an emerging perspective is that genomics, proteomics, and metabolomics will be successfully combined to identify predictive pharmacogenetic biomarkers that will aid in the development of novel therapeutics as well as novel approaches to individualized AED treatment. The complexities inherent in both epilepsy and its treatment are substantial, but the progress that has been made in genetics and in basic research over the last 15 years since the first epilepsy gene was identified (Steinlein et al., 1995) leaves us optimistic that better treatments for individuals with epilepsy will continue to be developed over the next 15 years.

## ACKNOWLEDGMENTS

This work was supported by NS40554 [TNF], NS40396 [RJB], NS05399 and NS45803 [DJD], and NS62419 [KSW].

**TABLE 10.1**
**Candidate Gene Classes Relevant to AED Pharmacogenetic Studies**

| Gene/Protein Class | Putatively Related AED |
| --- | --- |
| Albumin | CBZ, PB, PHT, VPA |
| $\alpha_1$ Acid glycoprotein | CBZ, PHT |
| ATP-binding cassette transporters | CBZ, LTG, PB, PHT, VPA |
| Monocarboxylate carriers | VPA |
| CYP monooxygenases | CBZ, FBM, OXC, PHT, TGB, VPA, ZNS |
| UDP-glucuronyltransferases | CBZ, FMB, LTG, TGB, VPA |
| Epoxide hydrolase | CBZ |
| Glutathione-S-transferases | CBZ, FBM, VPA |
| Sulfatases | ZNS |
| Sodium channel subunits | CBZ, FBM, LTG, OXC, PHT, TPM, VPA, ZNS |
| Calcium channel subunits | CBZ, ETX, FBM, GPT, LTG, LVT, OXC, PB, TPM, VPA, ZNS |
| Potassium channel subunits | CBZ, OXC |
| GABA receptor subunits | BZ, FBM, LVT, PB, TPM |
| GABA-transaminase | GPT, VGB, VPA |
| GABA transporters | GPT, TGB |
| Synaptic vesicle protein | LVT |
| Succinnic semialdehyde dehydrogenase | GPT, VPA |
| Glutamate receptor subunits | FBM, PB, TPM |
| Carbonic anhydrase | TPM, ZNS |

*Abbreviations:* AED, antiepileptic drug; ATP, adenosine triphosphate; BZ, benzodiazepine; CBZ, carbamazepine; CYP, cytochrome P450; ETX, ethosuximide; FBM, felbamate; GABA, γ-aminobutyric acid; GPT, gabapentin; LTG, lamotrigine; LVT, levetiracetam; OXC, oxcarbazepine; PB, phenobarbital; PHT, phenytoin; TGB, tiagabine; TPM, topiramate; UDP, uridine diphosphate; VPA, valproic acid; ZNS, zonisamide.

*Source:* Adapted from Buono, R.J. et al., *Pharmacogenomics*, 7(1), 89–103, 2006. With permission of Future Medicine Ltd.

# REFERENCES

Anderson, G.D. (2008). Pharmacokinetic, pharmacodynamic, and pharmacogenetic targeted therapy of antiepileptic drugs. *Ther. Drug Monit.*, 30:173–180.

Becker, A.J., J. Pitsch, D. Sochivko, T. Opitz, M. Staniek, C.C. Chen, K.P. Campbell, S. Schoch, Y. Yaari, H. Beck. (2008). Transcriptional upregulation of CAV3.2 mediates epileptogenesis in the pilocarpine model of epilepsy. *J. Neurosci.*, 28:13341–13353.

Berkovic, S.F., L. Harkin, J.M. McMahon, J.T. Pelekanos, S.M. Zuberi, E.C. Wirrell, D.S. Gill, X. Iona, J.C. Mulley, I.E. Scheffer. (2006). *De novo* mutations of the sodium channel gene *SCN1A* in alleged vaccine encephalopathy: a retrospective study. *Lancet Neurol.*, 5:488–492.

Bernard, C., A. Anderson, A. Becker, N.P. Poolos, H. Beck, D. Johnston. (2004). Acquired dendritic channelopathy in temporal lobe epilepsy. *Science*, 305:532–535.

Bertrand, D., F. Picard, S. Le Hellard, S. Weiland, I. Favre, H. Phillips, S. Bertrand, S.F. Berkovic, A. Malafosse, J. Mulley. (2002). How mutations in the nAChRs can cause ADNFLE epilepsy. *Epilepsia*, 43(Suppl. 5):112–122.

Bockenhauer, D., S. Feather, H.C. Stanescu, S. Bandulik, A.A. Zdebik et al. (2009). Epilepsy, ataxia, sensorineural deafness, tubulopathy, and *KCNJ10* mutations. *N. Engl. J. Med.*, 360:1960–1970.

Bournissen, F.G., M.E. Moretti, D.N. Juurlink, G. Koren, M. Walker, Y. Finkelstein. (2009). Polymorphism of the MDR1/ABCB1 C3435T drug transporter and resistance to anticonvulsant drugs: a meta-analysis. *Epilepsia*, 50:898–903.

Brandolese, R., M.G. Scordo, E. Spina, M. Gusella, R. Padrini. (2001). Severe phenytoin intoxication in a subject homozygous for CYP2C9*3. *Clin. Pharmacol. Ther.*, 70:391–394.

Brooks-Kayal, A.R., M.D. Shumate, H. Jin, T.Y. Rikhter, D.A. Coulter. (1998). Selective changes in single cell
    GABA(A) receptor subunit expression and function in temporal lobe epilepsy. *Nat. Med.*, 4:1166–1172.
Brooks-Kayal, A.R., M.D. Shumate, H. Jin, D.D. Lin, T.Y. Rikhter, K.L. Holloway, D.A. Coulter. (1999).
    Human neuronal gamma-aminobutyric acid(A) receptors: coordinated subunit mRNA expression and
    functional correlates in individual dentate granule cells. *J. Neurosci.*, 19:8312–8318.
Brooks-Kayal, A.R., Y.H. Raol, S.J. Russek. (2009). Alteration of epileptogenesis genes. *Neurotherapeutics*,
    6:312–318.
Browne, T.R., B. LeDuc. (2002). Phenytoin and other hydantoins: chemistry and biotransformation. In
    *Antiepileptic Drugs*, 5th ed. (Section X, pp. 565–580), R.H. Levy, R.H. Mattson, B.S. Meldrum, and E.
    Perucca, Eds. Philadelphia, PA: Lippincott Williams & Wilkins.
Buono, R.J., F.W. Lohoff, T. Sander, M.R. Sperling, M.J. O'Connor, D.J. Dlugos, S.G. Ryan, G.T. Golden,
    H. Zhao, T.M. Scattergood, W.H. Berrettini, T.N. Ferraro. (2004). Association between variation in the
    human *KCNJ10* potassium ion channel gene and seizure susceptibility. *Epilepsy Res.*, 58:175–183.
Charlier, C., N.A. Singh, S.G. Ryan, T.B. Lewis, B.E. Reus, R.J. Leach, M. Leppert. (1998). A pore mutation in
    a novel KQT-like potassium channel gene in an idiopathic epilepsy family. *Nat. Genet.*, 18:53–55.
Chen, K., I. Aradi, N. Thon, M. Eghbal-Ahmadi, T.Z. Baram, I. Soltesz. (2001). Persistently modified H-channels
    after complex febrile seizures convert the seizure-induced enhancement of inhibition to hyperexcitability.
    *Nat. Med.*, 7:331–337.
Conway, M., R.P. Cubbidge, S.L. Hosking. (2008). Visual field severity indices demonstrate dose-dependent
    visual loss from vigabatrin therapy. *Epilepsia*, 49:108–116.
Cossette, P., L. Liu, K. Brisebois, H. Dong, A. Lortie, M. Vanasse, J.M. Saint-Hilaire, L. Carmant, A. Verner,
    W.Y. Lu, Y.T. Wang, G.A. Rouleau. (2002). Mutation of $GABA_{A1}$ in an autosomal dominant form of
    juvenile myoclonic epilepsy. *Nat. Genet.*, 31:184–189.
Coulter, D.A., J.R. Huguenard, D.A. Prince. (1989). Characterization of ethosuximide reduction of low-thresh-
    old calcium current in thalamic neurons. *Ann. Neurol.*, 25:582–593.
Delgado-Escueta, A.V., B.F. Bourgeois. (2008). Debate: Does genetic information in humans help us treat
    patients? PRO—genetic information in humans helps us treat patients. CON—genetic information does
    not help at all. *Epilepsia*, 49(Suppl. 9):13–24.
Ferraro, T.N., G.T. Golden, G.G. Smith, J.F. Martin, F.W. Lohoff, T.A. Gieringer, D. Zamboni, C.L. Schwebel,
    D.M. Press, S.O. Kratzer, H. Zhao, W.H. Berrettini, R.J. Buono. (2004). Fine mapping of a seizure sus-
    ceptibility locus on mouse chromosome 1: nomination of *Kcnj10* as a causative gene. *Mamm. Genome*,
    15:239–251.
Ferraro, T.N., D.J. Dlugos, R.J. Buono. (2006). Challenges and opportunities in the application of pharmaco-
    genetics to antiepileptic drug therapy. *Pharmacogenomics*, 7:89–103.
Franciotta, D., P. Kwan, E. Perucca. (2009). Genetic basis for idiosyncratic reactions to antiepileptic drugs.
    *Curr. Opin. Neurol.*, 22:144–149.
Heron, S.E., H. Khosravani, D. Varela, C. Bladen, T.C. Williams, M.R. Newman, I.E. Scheffer, S.F. Berkovic,
    J.C. Mulley, G.W. Zamponi. (2007). Extended spectrum of idiopathic generalized epilepsies associated
    with CACNA1H functional variants. *Ann. Neurol.*, 62:560–568.
Heuser K, E. Nagelhus, E. Tauboll, U. Indahl, P. Berg, S. Nakken, S. Lien, L. Gjerstad, O. Ottersen. (2008). Do
    patients with temporal lobe epilepsy and childhood febrile seizures constitute a genetic subgroup? Results
    from an association study of aquaporin-4 and potassium channel genes. *Epilepsia*, 49(Suppl. 7):344.
Jung, S., T.D. Jones, J.N. Lugo, Jr., A.H. Sheerin, J.W. Miller, R. D'Ambrosio, A.E. Anderson, N.P. Poolos.
    (2007). Progressive dendritic HCN channelopathy during epileptogenesis in the rat pilocarpine model of
    epilepsy. *J. Neurosci.*, 27:13012–13021.
Kang, J.Q., W. Shen, R.L. Macdonald. (2006). Why does fever trigger febrile seizures? $GABA_A$ receptor
    gamma2 subunit mutations associated with idiopathic generalized epilepsies have temperature-depen-
    dent trafficking deficiencies. *J. Neurosci.*, 26:2590–2597.
Klaassen, A., J. Glykys, J. Maguire, C. Labarca, I. Mody, J. Boulter. (2006). Seizures and enhanced cortical
    GABAergic inhibition in two mouse models of human autosomal dominant nocturnal frontal lobe epi-
    lepsy. *Proc. Natl. Acad. Sci. U.S.A.*, 103:19152–19157.
Lenzen, K.P., A. Heils, S. Lorenz, A. Hempelmann, S. Hofels, F.W. Lohoff, B. Schmitz, T. Sander. (2005).
    Supportive evidence for an allelic association of the human *KCNJ10* potassium channel gene with idio-
    pathic generalized epilepsy. *Epilepsy Res.*, 63:113–118.
Löscher, W., U. Klotz, F. Zimprich, D. Schmidt. (2009). The clinical impact of pharmacogenetics on the treat-
    ment of epilepsy. *Epilepsia*, 50:1–23.

Maljevic, S., K. Krampfl, J. Cobilanschi, N. Tilgen, S. Beyer, Y.G. Weber, F. Schlesinger, D. Ursu, W. Melzer, P. Cossette, J. Bufler, H. Lerche, A. Heils. (2006). A mutation in the GABA(A) receptor alpha(1)-subunit is associated with absence epilepsy. *Ann. Neurol.*, 59:983–987.

Man, C.B., P. Kwan, L. Baum, E. Yu, K.M. Lau, A.S. Cheng, M.H. Ng. (2007). Association between HLA-B*1502 allele and antiepileptic drug-induced cutaneous reactions in Han Chinese. *Epilepsia*, 48:1015–1018.

Mann, E.O., I. Mody. (2008). The multifaceted role of inhibition in epilepsy: seizure-genesis through excessive GABAergic inhibition in autosomal dominant nocturnal frontal lobe epilepsy. *Curr. Opin. Neurol.*, 21:155–160.

Meador, K.J., G.A. Baker, N. Browning, J. Clayton-Smith, D.T. Combs-Cantrell, M. Cohen, L.A. Kalayjian, A. Kanner, J.D. Liporace, P.B. Pennell, M. Privitera, D.W. Loring. (2009). Cognitive function at 3 years of age after fetal exposure to antiepileptic drugs. *N. Engl. J. Med.*, 360:1597–1605.

Mullen, S.A., I.E. Scheffer. (2009). Translational research in epilepsy genetics: sodium channels in man to interneuronopathy in mouse. *Arch. Neurol.*, 66:21–26.

Ogiwara, I., H. Miyamoto, N. Morita, N. Atapour, E. Mazaki, I. Inoue, T. Takeuchi, S. Itohara, Y. Yanagawa, K. Obata, T. Furuichi, T.K. Hensch, K. Yamakawa. (2007). Na(v)1.1 localizes to axons of parvalbumin-positive inhibitory interneurons: a circuit basis for epileptic seizures in mice carrying an *Scn1a* gene mutation. *J. Neurosci.*, 27:5903–5914.

Otto, J.F., Y. Yang, W.N. Frankel, K.S. Wilcox, H.S. White. (2004). Mice carrying the *Szt1* mutation exhibit increased seizure susceptibility and altered sensitivity to compounds acting at the M-channel. *Epilepsia*, 45:1009–1016.

Otto, J.F., Y. Yang, W.N. Frankel, H.S. White, K.S. Wilcox. (2006). A spontaneous mutation involving *KCNQ2* (Kv7.2) reduces M-current density and spike frequency adaptation in mouse CA1 neurons. *J. Neurosci.*, 26:2053–2059.

Picard, F., S. Bertrand, O.K. Steinlein, D. Bertrand. (1999). Mutated nicotinic receptors responsible for autosomal dominant nocturnal frontal lobe epilepsy are more sensitive to carbamazepine. *Epilepsia*, 40:1198–1209.

Raol, Y.H., I.V. Lund, S. Bandyopadhyay, G. Zhang, D.S. Roberts, J.H. Wolfe, S.J. Russek, A.R. Brooks-Kayal. (2006). Enhancing GABA(A) receptor alpha 1 subunit levels in hippocampal dentate gyrus inhibits epilepsy development in an animal model of temporal lobe epilepsy. *J. Neurosci.*, 26:11342–11346.

Reid, C.A., S.F. Berkovic, S. Petrou. (2009). Mechanisms of human inherited epilepsies. *Prog. Neurobiol.*, 87:41–57.

Remy, S., S. Gabriel, B.W. Urban, D. Dietrich, T.N. Lehmann, C.E. Elger, U. Heinemann, H. Beck. (2003). A novel mechanism underlying drug resistance in chronic epilepsy. *Ann. Neurol.*, 53:469–479.

Sanabria, E.R., H. Su, Y. Yaari. (2001). Initiation of network bursts by Ca$^{2+}$-dependent intrinsic bursting in the rat pilocarpine model of temporal lobe epilepsy. *J. Physiol.*, 532:205–216.

Sankar, R. (2007). Teratogenicity of antiepileptic drugs: role of drug metabolism and pharmacogenomics. *Acta. Neurol. Scand.*, 116:65–71.

Scholl, U.I., M. Choi, T. Liu, V.T. Ramaekers, M.G. Hausler, J. Grimmer, S.W. Tobe, A. Farhi, C. Nelson-Williams, R.P. Lifton. (2009). Seizures, sensorineural deafness, ataxia, mental retardation, and electrolyte imbalance (SeSAME syndrome) caused by mutations in *KCNJ10*. *Proc. Natl. Acad. Sci. U.S.A.*, 106:5842–5847.

Siddiqui, A., R. Kerb, M.E. Weale, U. Brinkmann, A. Smith, D.B. Goldstein, N.W. Wood, S.M. Sisodiya. (2003). Association of multidrug resistance in epilepsy with a polymorphism in the drug-transporter gene *ABCB1*. *N. Engl. J. Med.*, 348:1442–1448.

Sills, G.J., R. Mohanraj, E. Butler, S. McCrindle, L. Collier, E.A. Wilson, M.J. Brodie. (2005). Lack of association between the C3435T polymorphism in the human multidrug resistance (*MDR1*) gene and response to antiepileptic drug treatment. *Epilepsia*, 46:643–647.

Simeone, T.A., K.S. Wilcox, H.S. White. (2006). Subunit selectivity of topiramate modulation of heteromeric GABA(A) receptors. *Neuropharmacology*, 50:845–857.

Singh, N.A., C. Charlier, D. Stauffer, B.R. DuPont, R.J. Leach, R. Melis, G.M. Ronen, I. Bjerre, T. Quattlebaum, J.V. Murphy, M.L. McHarg, D. Gagnon, T.O. Rosales, A. Peiffer, V.E. Anderson, M. Leppert. (1998). A novel potassium channel gene, *KCNQ2*, is mutated in an inherited epilepsy of newborns. *Nat. Genet.*, 18:25–29.

Steinlein, O.K., J.C. Mulley, P. Propping, R.H. Wallace, H.A. Phillips, G.R. Sutherland, I.E. Scheffer, S.F. Berkovic. (1995). A missense mutation in the neuronal nicotinic acetylcholine receptor alpha 4 subunit is associated with autosomal dominant nocturnal frontal lobe epilepsy. *Nat. Genet.*, 11:201–203.

Su, H., D. Sochivko, A. Becker, J. Chen, Y. Jiang, Y. Yaari, H. Beck. (2002). Upregulation of a T-type $Ca^{2+}$ channel causes a long-lasting modification of neuronal firing mode after status epilepticus. *J. Neurosci.*, 22:3645–3655.

Tan, N.C., S.E. Heron, I.E. Scheffer, J.T. Pelekanos, J.M. McMahon, D.F. Vears, J.C. Mulley, S.F. Berkovic. (2004). Failure to confirm association of a polymorphism in *ABCB1* with multidrug-resistant epilepsy. *Neurology*, 63:1090–1092.

Tate, S.K., C. Depondt, S.M. Sisodiya, G.L. Cavalleri, S. Schorge, N. Soranzo, M. Thom, A. Sen, S.D. Shorvon, J.W. Sander, N.W. Wood, D.B. Goldstein. (2005). Genetic predictors of the maximum doses patients receive during clinical use of the anti-epileptic drugs carbamazepine and phenytoin. *Proc. Natl. Acad. Sci. U.S.A.*, 102:5507–5512.

Wickenden, A.D., W. Yu, A. Zou, T. Jegla, P.K. Wagoner. (2000). Retigabine, a novel anti-convulsant, enhances activation of KCNQ2/Q3 potassium channels. *Mol. Pharmacol.*, 58:591–600.

Yang, Y., B.J. Beyer, J.F. Otto, T.P. O'Brien, V.A. Letts, H.S. White, W.N. Frankel. (2003). Spontaneous deletion of epilepsy gene orthologs in a mutant mouse with a low electroconvulsive threshold. *Hum. Mol. Genet.*, 12:975–984.

Yu, F.H., M. Mantegazza, R.E. Westenbroek, C.A. Robbins, F. Kalume, K.A. Burton, W.J. Spain, G.S. McKnight, T. Scheuer, W.A. Catterall. (2006). Reduced sodium current in GABAergic interneurons in a mouse model of severe myoclonic epilepsy in infancy. *Nat. Neurosci.*, 9:1142–1149.

# 11 Neuroprotective Strategies Using Antiepileptic Drugs

*Patrick G. Sullivan and Ryan D. Readnower*

## CONTENTS

## INTRODUCTION

Seizures induce neuronal damage/death via a variety of mechanisms, including glutamate-induced excitotoxicity, oxidative stress, and mitochondrial dysfunction. Antiepileptic drugs (AEDs) have long been used to control seizure activity; however, interest in the neuroprotective actions of AEDs is growing. During seizure activity, elevated levels of excitatory amino acids (EAAs), such as glutamate, activate $\alpha$-amino-3-hydroxy-5-methyl-4-isoxazoleproprionic acid (AMPA) receptors, which cause depolarization. This increase in EAA (glutamate surge) can also lead to the activation of voltage-dependent $Na^+$ channels, leading to further depolarization resulting from increased $Na^+$ entry into the cell and osmotic swelling of the neuron as $Cl^-$ passively enters the cell, causing an influx of water.

Aberrant depolarization can also result in the removal of the $Mg^{2+}$ block from $N$-methyl-D-aspartate (NMDA) receptors, allowing them to be activated by glutamate. Activation of NMDA receptors and group I metabotropic glutamate receptors allows $Ca^{2+}$ influx into the neuron, subsequently activating downstream mediators of programmed cell death (Sullivan, 2005; Walker, 2007). This disruption of $Ca^{2+}$ homeostasis is believed to underlie glutamate-mediated toxicity. Based on several distinct animal models of excitotoxicity and injury, it has become apparent that several antiepileptic drugs may possess neuroprotective properties (Trojnar et al., 2002). Although their sites of action may be very distinct, generally AEDs displaying neuroprotective properties inhibit components of the excitotoxic cascade (Sullivan, 2005).

Several of the AEDs shown to be neuroprotective are known to enhance GABA-meditated inhibition at the synapse, including barbiturates, benzodiazepines, vigabatrin, tiagabine, felbamate, and topiramate. Others, such as phenytoin, carbamazepine, oxcarbazepine, lamotrigine, zonisamide, and valproate, are able to inactivate voltage-dependent $Na^+$ channels. Lamotrigine, felbamate, and valproate may act as $Ca^{2+}$ channel blockers, and AEDs such as felbamate, gabapentin, lamotrigine, tiagabine, topiramate, and valproate all exert antiseizure and neuroprotective actions at numerous sites (or combinations of sites), including inhibition of $Na^+$ and $Ca^{2+}$ channels, enhancing GABA-mediated inhibition, and acting as antagonists of the AMPA receptors (for a full review of AED mechanisms, see Chapter 8 in this volume; for reviews of neuroprotection by AEDs, see Sullivan, 2005; Trojnar et al., 2002). Zonisamide, in addition to inhibition of $Na^+$ channels, may also act as an antioxidant, which could enhance its neuroprotective properties.

The complex pharmacology of many of these AEDs results in a blanket-type approach being used in the treatment of epilepsy which has proven beneficial in reducing seizure severity; however, this approach has proven controversial with regard to the proposed neuroprotective benefits of AEDs. Valproate, tiagabine, and even benzodiazepines, for example, have been shown to be neurotoxic and can exacerbate neuronal damage in some paradigms, particularly in immature animal models of CNS injury (Table 11.1) (Olney et al., 2002; Trojnar et al., 2002). Furthermore, in several cases, neuroprotection is limited unless the dosage is raised considerably above the clinical range, and efficacy can vary greatly (even becoming neurotoxic) depending on treatment initiation and the therapeutic window that is targeted. In addition, concerns about adverse neurobehavioral effects have been raised and demonstrated in several animal models of epilepsy (Sankar and Holmes, 2004). Even though the precise mechanisms underlying these adverse effects are unknown, they may hinder the potential for using AEDs as neuroprotective agents. However, if one moves downstream of the receptor-mediated events currently utilized by AEDs, it becomes apparent that $Ca^{2+}$-mediated mitochondrial dysfunction may be a prime target for neuroprotective interventions.

In short, the overall pharmacological target for AEDs is to decrease neuronal excitation (Landmark, 2007). By decreasing neuronal excitation, AEDs may exert their neuroprotective mechanisms of action through inhibition of the excitotoxic cascade. It is important to note that not all AEDs are neuroprotective in every paradigm, and some have been shown to be neurotoxic, especially in immature animal models of injury (Olney et al., 2002; Trojnar et al., 2002). This chapter focuses on reviewing the main neuroprotective mechanisms of action for AEDs, first discussing AEDs that act at the GABAergic synapse followed by those functioning at the glutamatergic synapse (Figure 11.1; see also figures in Chapter 8 in this volume). Given the emergent role of $Ca^{2+}$-mediated mitochondrial dysfunction in epilepsy, we begin by covering the pivotal role mitochondria can play in neuronal cell survival or death during excitotoxic insults through their regulation of both energy metabolism and apoptotic pathways.

## EPILEPTIC MITOCHONDRIAL DYSFUNCTION

Mitochondria primarily function to regulate energy metabolism and serve as high-capacity $Ca^{2+}$ sinks to maintain neuronal $Ca^{2+}$ homeostasis. Mitochondria accomplish these functions by pumping protons across the inner mitochondrial membrane via the electron transport system (ETS), thus establishing a mitochondrial membrane potential ($\Delta\psi$). $\Delta\psi$ is used to drive $Ca^{2+}$ into the mitochondrial

**TABLE 11.1**
**AEDs and Neuroprotection**

| Site of Action | Drug | Neuroprotection (Model) | | |
|---|---|---|---|---|
| | | Ischemia/Trauma | Excitotoxicity | Neurotoxicity |
| GABAergic synapse | Benzodiazepine | ✓ | ✓ | ✓ |
| | Phenobarbital | ✓ | ✓ | ✓ |
| | Valproic acid | ✓ | Yes/no | ✓ |
| | Vigabatrin | ✓ | ✓ | ✓ |
| | Tiagabine | ✓ | ✓ | ✓ |
| Glutamatergic synapse | Topiramate | Maybe | ✓ | ? |
| | Lamotrigine | ✓ | ✓ | ? |
| | Carbamazepine | ✓ | Yes/no | ✓ |
| | Felbamate | ✓ | ✓ | ? |

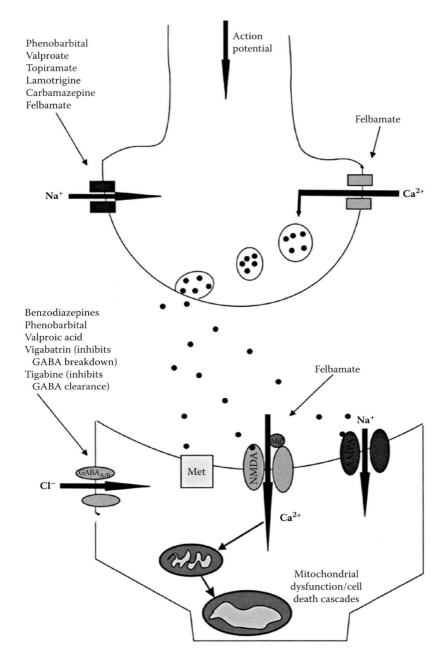

**FIGURE 11.1** The excitotoxic cascade and the potential or purposed targets of several antiepileptic drugs (AEDs). Extended periods of excitation can lead to chronic depolarization and the release of neurotransmitters such as glutamate (Glu) causing influxes of $Ca^{2+}$ into the dendrite via NMDA receptor activation. If unchecked, this $Ca^{2+}$ influx into neurons can result in a subsequent loss of neuronal $Ca^{2+}$ homeostasis. This disruption of $Ca^{2+}$ homeostasis is believed to be a defining factor of glutamate-mediated toxicity. Several AEDs may possess neuroprotective properties against excitotoxicity and neuronal cell death. Importantly, the AEDs that display the most neuroprotection generally have multiple sites of actions/targets that can reduce neuronal $Ca^{2+}$ loss of homeostasis. *Abbreviations:* AMPA, α-amino-3-hydroxy-5-methyl-4-isoxazoleproprionic acid; Met, metabotropic glutamate receptor; NMDA, *N*-methyl-D-aspartate. Please see the text for detailed information.

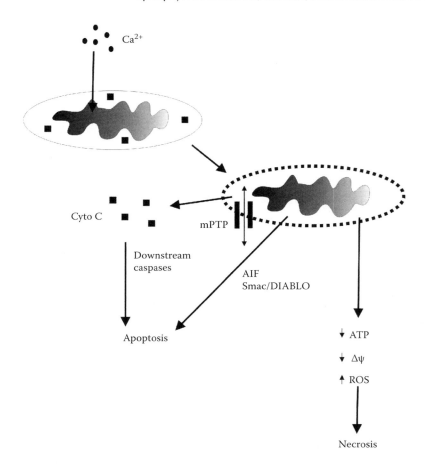

**FIGURE 11.2**  Highly simplified schematic diagram illustrating the pivotal role of mitochondria in apoptotic and necrotic neuronal cell death following excitotoxic insult. The excitatory amino acid glutamate increases $Ca^{2+}$ influx into neurons. Mitochondria sequester the $Ca^{2+}$ in an effort to maintain cytosolic $Ca^{2+}$ homeostasis. Excessive mitochondrial $Ca^{2+}$ cycling/loading can result in mitochondrial permeability that can be mediated by the mitochondrial permeability transition pore (mPTP) opening or via increased permeability of the outer membrane without mPTP opening. In either case, pro-apoptotic proteins such as Smac/DIABLO, apoptosis-inducing factor (AIF), and cytochrome $c$ (Cyto C) can be released from the mitochondria, leading to activation of downstream caspases and apoptosis. $Ca^{2+}$-induced mPT results in increased conductance of the inner membrane, mitochondrial swelling, loss of ATP production due to decreased membrane potential ($\Delta\psi$), and increased reactive oxygen species (ROS) production. These events promote mPT and can lead to necrosis depending on the extent and severity of the insult or injury.

matrix through membrane channels and phosphorylate adenosine diphosphate (ADP) through utilization of complex V (ATP synthase) to produce adenosine triphosphate (ATP). During excitotoxic periods, excessive mitochondrial $Ca^{2+}$ uptake causes increased reactive oxygen species (ROS) production, inhibits ATP production, induces mitochondrial permeability transition pore (MPTP), and promotes the release of cytochrome $c$ (Figure 11.2) (Sullivan, 2005).

Mitochondrial dysfunction has been implicated in a variety of neurological disorders, such as Alzheimer's disease, Parkinson's disease, stroke, and traumatic brain injury (Beal, 1998). Prolonged epileptic seizures have been shown to increase lipid peroxidation, induce oxidative DNA damage, and cause mitochondrial dysfunction (Bruce and Baudry, 1995; Kudin et al., 2002; Liang et al., 2000). Disruption of mitochondrial $Ca^{2+}$ homeostasis during seizures or excitotoxic insults may be an upstream mediator of neuronal cell death or damage associated with epilepsy.

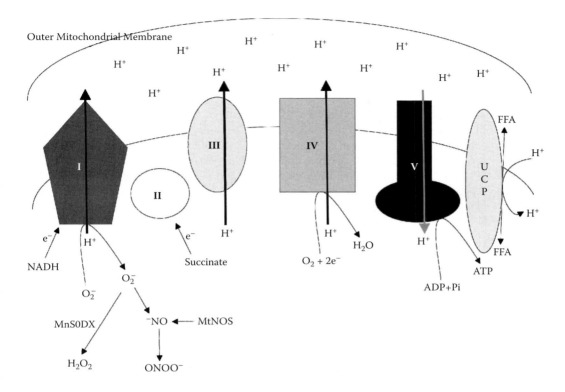

**FIGURE 11.3** The relationship between mitochondrial uncoupling proteins (UCPs), the ketogenic diet (KD), and reactive oxygen species formation. Schematic illustrates the relationship of the electron transport system (ETS) to ATP production and reactive oxygen species formation. As electrons are transported down the ETS, complexes I, III, and IV translocate protons from the matrix, generating an electrochemical gradient. This store of energy can then be coupled to ATP production as protons move down the gradient and back into the matrix via $F_0F_1$ ATP synthase. Superoxide production is a byproduct of electron transport and occurs when electrons escape or slip from the complexes and reduce $O_2$ due to a backup of electrons in the ETS. Neuronal UCPs (UCP-2) can short-circuit the system, leading to increased electron flow and subsequent decreased superoxide production. Although the exact UCP-mediated mechanisms underlying the increased proton flow back across the inner membrane are unknown, it is well established that free fatty acids (FFAs) do activate and modulate UCPs. One possible mechanism involves FFAs inducing proton channels in the UCP, whereas another postulates that protonated FFAs flip-flop across the membrane using UCP to transport the unprotonated FFA.

Because AEDs are capable of attenuating some components of the excitotoxic cascade associated with seizure, they may exert their neuroprotective effects by inhibiting $Ca^{2+}$-mediated mitochondrial dysfunction.

The role of mitochondrial dysfunction in epilepsy pathology was further elucidated in recent studies examining the effects of the ketogenic diet and endogenous mitochondrial uncoupling proteins (UCPs). UCPs uncouple the pumping of protons out of the matrix via the ETS from the flow of protons through ATP synthase and thus reduce $\Delta\psi$ (Figure 11.3). This reduction in $\Delta\psi$ would be expected to decrease mitochondrial $Ca^{2+}$ uptake and ROS production, as both are $\Delta\psi$ dependent. UCPs use free fatty acids to translocate protons from the inner membrane space into the mitochondrial matrix. It has been hypothesized that by increasing dietary fat (i.e., ketogenic diet) it could be expected that UCP expression or activity would increase followed by a subsequent decrease in mitochondrial ROS production. Recent studies have shown that, through dietary modulation using the ketogenic diet, UCP expression or activity is increased and mitochondrial ROS production is decreased (Sullivan et al., 2004). The ketogenic diet has been shown to reduce neuronal injury in a kainic-acid-induced injury model, further supporting the key role that mitochondria play in epilepsy pathology (Sullivan et al., 2004).

## THE GABAERGIC SYNAPSE

Benzodiazepines act by binding to the $GABA_A$ receptor subtype, inducing an increase in the frequency of opening-associated $Cl^-$ channels and resulting in membrane hyperpolarization (O'Brien, 2005; Pitkanen et al., 2005). This class of drugs represents over 50 family members, including diazepam, lorazepam, midazolam, and clonazepam. Certain benzodiazepines, such as diazepam, have been shown to decrease the likelihood of future seizures and are neuroprotective if administered shortly after induction of status epilepticus (Deckers et al., 2000). Importantly, benzodiazepine neuroprotection appears to be dependent on neuronal development; their administration has shown neuroprotection in adult animals but has shown a dose-dependent induction of cell death in immature animals (Sankar and Holmes, 2004; Sullivan, 2005; Trojnar et al., 2002).

Phenobarbital increases the affinity of GABA for $GABA_A$ receptors and blocks sodium channels (Deckers et al., 2000; Trojnar et al., 2002). Phenobarbital has been shown to be neuroprotective and to reduce future seizures when administered before or during seizure but not after induction of seizure (Brown-Croyts et al., 2000). Additionally, phenobarbital has been shown to be a free-radical scavenger, enhancing its neuroprotective actions (Bashkatova et al., 2003). Neurotoxicity of phenobarbital has been shown in immature animal models (Olney et al., 2002) where mitochondrial degeneration and deficits in hippocampal-based behavior measurements have been observed (Beal, 1998; Trojnar et al., 2002).

Vigabatrin, a structural GABA analog, is a newer AED that was designed to be an irreversible inhibitor of GABA transaminase. Through inhibition of the enzyme that degrades GABA, vigabatrin has been shown to increase extracellular GABA concentrations *in vivo* (Verhoeff et al., 1999). Vigabatrin has shown neuroprotection in several different models when administered early after seizure induction (André et al., 2001; Jolkkonen et al., 1996). Tiagabine, a newer AED, is designed to inhibit glial/neuronal uptake of GABA and thereby delay clearance of GABA from the synaptic cleft (Krauss et al., 2003). The administration of tiagabine after status epilepticus has not been shown to be efficacious in preventing seizure or providing neuroprotection (Verhoeff et al., 1999).

## THE GLUTAMATERGIC SYNAPSE

Several AEDs have mechanisms of action that modulate $Na^+$ channel conductance (Bruce and Baudry, 1995; O'Brien, 2005). Phenytoin, carbamazepine, oxcarbazepine, lamotrigine, felbamate, topiramate, zonisamide, and rufinamide all inhibit voltage-gated $Na^+$ channels which, in turn, decreases the frequency of generated action potentials (Bialer et al., 2002; Kuo, 1998; Liu et al., 2003). It is conceivable that the neuroprotective effects of topiramate, a special case, may result from a combination of its actions that involve the non-NMDA glutamatergic receptors (AMPA–kainate subtypes) and some subtypes of GABA receptors. Topiramate may be especially useful as a neuroprotectant in the developing brain (Cha et al., 2002; Liu et al., 2004; Sfaello et al., 2005). Administration of topiramate has been shown to inhibit neuronal degeneration in the hippocampus and to improve mitochondrial bioenergetics following seizure (Kudin et al., 2004; Rigoulot et al., 2004).

Felbamate has several mechanisms of action, including NMDA receptor antagonism and inhibition of both voltage-gated sodium and calcium channels (Kuo et al., 2004; Rogawski and Löscher, 2004) as well as enhancement of GABA-mediated chloride currents (Rho et al., 1994). Toxicity associated with felbamate administration has limited its clinical use, but neuroprotective effects of felbamate have been demonstrated in different models of brain injury (Wallis and Panizzon, 1995; Wasterlain et al., 1993). Fluorofelbamate, a nontoxic analog of felbamate, has proved neuroprotective in ischemia models (Bruce and Baudry, 1995).

## CONCLUSION

In summary, although several AEDs have been shown to have neuroprotective properties, their usefulness for preventing excitotoxic cell death is limited by adverse effects ranging from unwanted neurobehavioral changes to exacerbating neuronal damage. AEDs are used to treat and prevent seizures by decreasing neuronal excitability. Recently, AEDs have been shown to be neuroprotective in multiple injury models (Acharya et al., 2008). By preventing components of the excitotoxic cascade, AEDs likely inhibit $Ca^{2+}$-mediated mitochondrial dysfunction and thus exert their neuroprotective mechanisms of action (Sullivan, 2005). A pivotal role in neuronal cell survival and death following neuronal injury is played by mitochondria. They serve as the powerhouse of the cell by maintaining ratios of ATP to ADP that thermodynamically favor the hydrolysis of ATP to ADP + Pi; yet, a byproduct of this process is the generation of ROS. Proton-pumping by components of the ETS generates $\Delta\psi$, which can then be used to phosphorylate ADP to ATP or to sequester $Ca^{2+}$ out of the cytosol into the mitochondrial matrix. This allows mitochondria to act as a cellular $Ca^{2+}$ sink and to be in phase with changes in cytosolic $Ca^{2+}$ levels. Epileptic hyperexcitability is known to cause calcium-mediated mitochondrial dysfunction subsequently leading to neuronal injury or death.

Following epileptic seizure activity there is a well-documented increase in excitotoxicity that leads to significant increases in cytosolic $Ca^{2+}$, ROS production, oxidative damage, and ultimately cell death. *In vitro* studies have demonstrated that this cascade of events is dependent on mitochondrial $Ca^{2+}$ cycling and that a reduction in $\Delta\psi$ is sufficient to reduce excitotoxic cell death. Additional support for these findings comes from experiments demonstrating that overexpression of endogenous mitochondrial uncouplers (UCPs) reduces EAA-mediated ROS production and cell death, whereas dietary manipulations that reduce UCP levels increase ROS production and susceptibility to excitotoxic insults. Thus, the fine line between cell survival and cell death relies on mitochondrial integrity and ultimately the state of $\Delta\psi$. This scenario predicts neuronal survival following seizure activity can be enhanced by reducing mitochondrial $\Delta\psi$.

## REFERENCES

Acharya, M.M., B. Hattiangady, A.K. Shetty. (2008). Progress in neuroprotective strategies for preventing epilepsy. *Prog. Neurobiol.*, 84(4):363–404.

André, V., A. Ferrandon, C. Marescaux, A. Nehlig. (2001). Vigabatrin protects against hippocampal damage but is not antiepileptogenic in the lithium–pilocarpine model of temporal lobe epilepsy. *Epilepsy Res.*, 47(1–2):99–117.

Bashkatova, V., V. Narkevich, G. Vitskova, A. Vanin. (2003). The influence of anticonvulsant and antioxidant drugs on nitric oxide level and lipid peroxidation in the rat brain during penthylenetetrazole-induced epileptiform model seizures. *Prog. Neuropsychopharmacol. Biol. Psychiatry*, 27(3):487–492.

Beal, M.F. (1998). Mitochondrial dysfunction in neurodegenerative diseases. *Biochim. Biophys. Acta*, 1366(1–2):211–223.

Bialer, M., S.I. Johannessen, H.J. Kupferberg, R.H. Levy, P. Loiseau, E. Perucca. (2002). Progress report on new antiepileptic drugs: a summary of the Sixth Eilat Conference (EILAT VI). *Epilepsy Res.*, 51(1–2):31–71.

Brown-Croyts, L.M., P.W. Caton, D.T. Radecki, S.L. McPherson. (2000). Phenobarbital pre-treatment prevents kainic acid-induced impairments in acquisition learning. *Life Sci.*, 67(6):643–650.

Bruce, A.J., M. Baudry. (1995). Oxygen free radicals in rat limbic structures after kainate-induced seizures. *Free Radic. Biol. Med.*, 18(6):993–1002.

Cha, B.H., D.C. Silveira, X. Liu, Y. Hu, G.L. Holmes. (2002). Effect of topiramate following recurrent and prolonged seizures during early development. *Epilepsy Res.*, 51(3):217–232.

Czapinski, P., B. Blaszczyk, S.J. Czuczwar. (2005). Mechanisms of action of antiepileptic drugs. *Curr. Top. Med. Chem.*, 5(1):3–14.

Deckers, C.L., S.J. Czuczwar, Y.A. Hekster, A. Keyser, H. Kubova, H. Meinardi, P.N. Patsalos, W.O. Renier, C.M. Van Rijn. (2000). Selection of antiepileptic drug polytherapy based on mechanisms of action: the evidence reviewed. *Epilepsia*, 41(11):1364–1374.

Jolkkonen, J., T. Halonen, E. Jolkkonen, J. Nissinen, A. Pitkanen. (1996). Seizure-induced damage to the hippocampus is prevented by modulation of the GABAergic system. *NeuroReport*, 7(12):2031–2035.

Krauss, G.L., M.A. Johnson, S. Sheth, N.R. Miller. (2003). A controlled study comparing visual function in patients treated with vigabatrin and tiagabine. *J. Neurol. Neurosurg. Psychiatry*, 74(3):339–343.

Kudin, A.P., T.A. Kudina, J. Seyfried, S. Vielhaber, H. Beck, C.E. Elger, W.S. Kunz. (2002). Seizure-dependent modulation of mitochondrial oxidative phosphorylation in rat hippocampus. *Eur. J. Neurosci.*, 15(7):1105–1114.

Kudin, A.P., G. Debska-Vielhaber, S. Vielhaber, C.E. Elger, W.S. Kunz. (2004). The mechanism of neuroprotection by topiramate in an animal model of epilepsy. *Epilepsia*, 45(12):1478–1487.

Kuo, C.C. (1998). A common anticonvulsant binding site for phenytoin, carbamazepine, and lamotrigine in neuronal $Na^+$ channels. *Mol. Pharmacol.*, 54(4):712–721.

Kuo, C.C., B.J. Lin, H.R. Chang, C.P. Hsieh. (2004). Use-dependent inhibition of the *N*-methyl-D-aspartate currents by felbamate: a gating modifier with selective binding to the desensitized channels. *Mol. Pharmacol.*, 65(2):370–380.

Landmark, C.J. (2007). Targets for antiepileptic drugs in the synapse. *Med. Sci. Monit.*, 13(1):RA1–RA7.

Liang, L.P., Y.S. Ho, M. Patel. (2000). Mitochondrial superoxide production in kainate-induced hippocampal damage. *Neuroscience*, 101(3):563–570.

Liu G., V. Yarov-Yarovoy, J. Nobbs, J.J. Clare, T. Scheuer, W.A. Catterall. (2003). Differential interactions of lamotrigine and related drugs with transmembrane segment IVS6 of voltage-gated sodium channels. *Neuropharmacology*, 44(3):413–422.

Liu, Y., J.D. Barks, G. Xu, F.S. Silverstein. (2004). Topiramate extends the therapeutic window for hypothermia-mediated neuroprotection after stroke in neonatal rats. *Stroke*, 35(6):1460–1465.

O'Brien, C.P. (2005). Benzodiazepine use, abuse, and dependence. *J. Clin. Psychiatry*, 66(Suppl. 2):28–33.

Olney, J.W., D.F. Wozniak, V. Jevtovic-Todorovic, N.B. Farber, P. Bittigau, C. Ikonomidou. (2002). Drug-induced apoptotic neurodegeneration in the developing brain. *Brain Pathol.*, 12(4):488–498.

Pitkanen, A., I. Kharatishvili, K. Narkilahti, K. Lukasiuk, J. Nissinen. (2005). Administration of diazepam during status epilepticus reduces development and severity of epilepsy in rat. *Epilepsy Res.*, 63(1):27–42.

Rho, J.M., S.D. Donevan, M.A. Rogawski. (1994). Mechanism of action of the anticonvulsant felbamate: opposing effects on *N*-methyl-D-aspartate and gamma-aminobutyric acidA receptors. *Ann. Neurol.*, 35(2):229–234.

Rigoulot, M.A., E. Koning, A. Ferrandon, A. Nehlig. (2004). Neuroprotective properties of topiramate in the lithium–pilocarpine model of epilepsy. *J. Pharmacol. Exp. Ther.*, 308(2):787–795.

Rogawski, M.A., W. Löscher. (2004). The neurobiology of antiepileptic drugs for the treatment of nonepileptic conditions. *Nat. Med.*, 10(7):685–692.

Sankar, R., G.L. Holmes. (2004). Mechanisms of action for the commonly used antiepileptic drugs: relevance to antiepileptic drug-associated neurobehavioral adverse effects. *J. Child. Neurol.*, 19(Suppl. 1):S6–S14.

Sfaello, I., O. Baud, A. Arzimanoglou, P. Gressens. (2005). Topiramate prevents excitotoxic damage in the newborn rodent brain. *Neurobiol. Dis.*, 20(3):837–848.

Sullivan, P.G. (2005). Interventions with neuroprotective agents: novel targets and opportunities. *Epilepsy Behav.*, 7(Suppl. 3):S12–S17.

Sullivan, P.G., N.A. Rippy, K. Dorenbos, R.C. Concepcion, A.K. Agarwal, J.M. Rho. (2004). The ketogenic diet increases mitochondrial uncoupling protein levels and activity. *Ann. Neurol.*, 55(4):576–580.

Trojnar, M.K., R. Malek, M. Chroscinska, S. Nowak, B. Blaszczyk, S.J. Czuczwar. (2002). Neuroprotective effects of antiepileptic drugs. *Pol. J. Pharmacol.*, 54(6):557–566.

Verhoeff, N.P., O.A. Petroff, F. Hyder, S.S. Zoghbi, N. Fujita, N. Rajeevan, D.L. Rothman, J.P. Seibyl, R.H. Mattson, R.B. Innis. (1999). Effects of vigabatrin on the GABAergic system as determined by [[123]I] iomazenil SPECT and GABA MRS. *Epilepsia*, 40(10):1433–1438.

Walker, M. (2007). Neuroprotection in epilepsy. *Epilepsia*, 48(Suppl. 8):66–68.

Wallis, R.A., K.L. Panizzon. (1995). Felbamate neuroprotection against CA1 traumatic neuronal injury. *Eur. J. Pharmacol.*, 294(2–3):475–482.

Wasterlain, C.G., L.M. Adams, P.H. Schwartz, H. Hattori, R.D. Sofia, J.K. Wichmann. (1993). Posthypoxic treatment with felbamate is neuroprotective in a rat model of hypoxia–ischemia. *Neurology*, 43(11):2303–2310.

# 12 Posttraumatic Seizures and Epileptogenesis: Good and Bad Plasticity

*Christopher C. Giza*

## CONTENTS

## INTRODUCTION

Few things in medicine are predictable, and the clinical condition of epilepsy is one diagnosis that is fraught with this unpredictability—who will get it, when will it start, when will a seizure occur, when are medications necessary or not necessary? However, amid this uncertainty, some things are well known. Traumatic brain injury (TBI) is a definite risk factor for epilepsy (Annegers et al., 1998;

Jennett, 1975a; Lowenstein, 2009). Increased TBI severity brings with it a higher risk of subsequent seizures and epilepsy (Annegers et al., 1998). Seizures can occur early after TBI, and these seizures appear to be different from those occurring later after TBI, both in terms of clinical characteristics and underlying pathophysiology. Perhaps the most intriguing characteristic of seizures associated with brain trauma is the fact that the initiating event in the process of developing seizures is overt. Thus, the temporal pattern of this process can be studied, with an eye to developing both a scientific understanding of the neurobiological mechanisms and, perhaps equally important, specific mechanism-based therapeutic interventions that may alter the course of epileptogenesis, perhaps even before the first seizure ever occurs.

## DEFINITIONS

*Traumatic brain injury* refers to any biomechanically induced damage to the brain, either temporary or permanent. This has traditionally been thought of as having primarily a structural component, although more recent investigations have raised the likelihood that neural dysfunction, even transient, may also play a major role. In fact, dysfunction in the absence of overt structural damage is particularly relevant in mild TBI/concussion and in the manifestation of cumulative damage from repetitive mild injuries as occurs in contact sports and in military-associated mild TBI. The definition of TBI severity varies from study to study, although general convention holds that TBI can be categorized as mild, moderate, or severe (based upon either the Glasgow Coma Scale or other clinical parameters, including loss of consciousness and posttraumatic amnesia). Importantly, there is growing recognition that TBI is not a single entity but a series of distinct processes sharing a biomechanical origin but demonstrating substantial variations in pathophysiology, clinical symptomatology, potential complications, and effective therapeutic interventions (Saatman et al., 2008).

*Posttraumatic seizures* (PTSs) and *posttraumatic epilepsy* (PTE) share this quandary of classification and terminology. Any seizure occurring as a result of TBI can be referred to as a PTS, although this relationship may become faint if there is a long delayed interval between the TBI and the first PTS. The PTS literature, however, uses many different definitions that require careful consideration. By convention, *early posttraumatic seizures* (EPTSs) are those seizures occurring within the first 7 days after injury, and *late posttraumatic seizures* (LPTSs) are those occurring later. Most authorities now also make the distinction of *immediate posttraumatic seizures* (IPTSs)—namely, those that occur within the first 24 hours of injury and which may have important differences in underlying physiological origin and in prognosis. A further subdivision of IPTS is the syndrome of an *impact seizure*, which represents a hyperacute, usually generalized, seizure with rapid postictal recovery. Impact seizures tend to be reported more often in younger patients, have a benign prognosis, and are often excluded in studies of PTS epidemiology. Another very important distinction in the literature is the designation of PTE. Many older studies of PTS classify any posttraumatic seizure, early or late, single or multiple, as "posttraumatic epilepsy"; however, in keeping with more current epilepsy nomenclature, the moniker PTE should be reserved for recurrent, otherwise unprovoked LPTSs (first occurring >7 days post-TBI).

## EPIDEMIOLOGY

Traumatic brain injury is the most common cause of death and disability among those under the age of 45 years, with peaks in incidence in infancy and adolescence/young adulthood. A third peak occurs in senescence (CDC, 1999; Keenan and Bratton, 2006; Langlois et al., 2006; Rutland-Brown et al., 2006). There is a substantial gender inequity in TBI, with a male/female ratio of 2 or 3 to 1. Mechanisms of TBI vary widely and by age: inflicted TBI (child abuse), falls, motor vehicle accidents, sports concussions, assaults, gunshot wounds (and other penetrating TBI), and military TBI, to name a few of the most common categories (Keenan and Bratton, 2006). Although moderate to severe TBI is widely recognized as causing lasting cognitive, motor, and behavioral impairments, these higher levels of TBI severity account for only 15 to 25% of all TBI. The vast majority of TBI is considered "mild," with only transient deficits followed by full recovery, but there is now increased awareness that persistent deficits can be seen in a subset of mild injuries and that recurrent mild TBI very likely carries a risk of chronic neurocognitive impairment (Collins et al., 1999; Guskiewicz et al., 2005). Given recent conflicts in Iraq and Afghanistan, as well as the increased threat of global terrorism, all of which are associated with the use of improvised explosive devices (IEDs), a new type of TBI is being recognized: blast injury. Military TBI in past conflicts has been highly associated with the development of PTS and PTE, but the subgroup at highest risk were those individuals who experienced penetrating wounds to the brain (Aarabi, 1990; Caveness et al., 1962; Lowenstein, 2009). With improved protective head gear and the change in the type of warfare, the numbers of closed head injuries, particularly those classified as "blast" injuries, appear to be rising in incidence (Warden, 2006). The diagnosis, pathophysiology, and long-term prognosis (and potential risk of complications such as epilepsy) from these types of injuries have only very recently been investigated and clarified.

Epilepsy resulting from TBI is an important clinical entity because it constitutes up to 20% of all symptomatic epilepsies and 6% of all epilepsies (Hauser et al., 1993). Among acquired etiologies for epilepsy, severe TBI is in the top 3% with regard to relative risk (29-fold), exceeded only by subarachnoid hemorrhage (34-fold) and brain tumor (40-fold) (Herman, 2002). Furthermore, PTE may be difficult to treat and is often associated with additional neurocognitive sequelae of the initial TBI. An important therapeutic consideration for PTE is that there is almost always a delay (latent period) between the acute TBI and the later development of recurrent unprovoked seizures. With improved understanding of the underlying neurobiological mechanisms of posttraumatic epileptogenesis, this may provide a window for effective "preventive" treatment.

The risk of PTE is clearly dependent upon TBI severity, with incidence rates as high as 50% for survivors of penetrating brain injury. The rates for PTE in individuals with mild, moderate, or severe nonpenetrating TBI are 0.7%, 1.2%, and 16.7%, respectively, 5 years after the injury (Annegers et al., 1980, 1998). Increased risk of late PTE was associated with several other factors related to the acute injury, including cerebral contusion, subdural hematoma, depressed skull fracture, and cerebral edema (Chiaretti et al., 2000; Frey, 2003; Kieslich and Jacobi, 1995; Temkin, 2003). The occurrence of EPTSs also corresponds to a higher rate of LPTSs/PTE in many studies, although the prevention of EPTSs by acute treatment with anticonvulsants does not appear to mitigate this risk (Temkin et al., 1990, 1999). This implies that perhaps early and late seizures are markers of some other pathophysiological phenomena, but not causally related, or perhaps that EPTSs are not being accurately quantified (for example, by inclusion of benign variants such as impact seizures or missing subclinical EPTS/status). Finally, younger age appears to be a risk factor in the development of PTS (Chiaretti et al., 2000; Ratan et al., 1999). This factor undoubtedly adds complexity to our understanding of the process of posttraumatic epileptogenesis but may also provide important clues to unraveling these mechanisms.

## PATHOPHYSIOLOGY OF TRAUMATIC BRAIN INJURY

An understanding of PTS and the process of posttraumatic epileptogenesis begins with the pathophysiology of acute TBI. The ionic and neurometabolic cascade of events occurring after TBI has been increasingly well defined, both experimentally and clinically. Biomechanical brain injury is known to massively disturb the ionic equilibrium of the neuron, triggering indiscriminate ionic flux, neurotransmitter release, and metabolic stress. Whether these processes result in only transient dysfunction or lead to cellular demise is likely dependent upon a host of factors, including severity of the initial injury, presence of secondary insults, regional distinctions in the brain, age at injury, premorbid function, and genetic makeup, to name a few.

## NEUROTRANSMITTER RELEASE AND IONIC DISEQUILIBRIUM

The initial physiological event after TBI is a widespread ionic disequilibrium, of which excessive potassium efflux is the hallmark (Bullock et al., 1998; Katayama et al., 1990). This can occur due to membrane poration as well as through opening of ion channels (e.g., NMDA receptors, AMPA receptors, voltage-gated sodium channels) (Faden et al., 1989; Giza and Hovda, 2001). Indiscriminate depolarization can result in pathological levels of glutamate release, creating a vicious cycle of ongoing ionic stress (Bullock et al., 1998; Katayama et al., 1990, 1995). Adenosine triphosphate (ATP)-driven membrane ionic pumps work overtime in an attempt to restore homeostasis but require significant energy supplies to do so. This acute increase of metabolic demand in a setting of widespread electrochemical disturbance then sets the stage for a cellular energy crisis in neural tissues (Bergsneider et al., 1997; Reinert et al., 2000; Sunami et al., 1989; Vespa et al., 2005; Yoshino et al., 1991). Periodic cortical depolarizations resembling spreading depression have been described following both experimental and human TBI, and these events appear to be associated with increased metabolic stress (Fabricius et al., 2008; Kubota et al., 1989; Strong et al., 2002).

## METABOLIC–BLOOD FLOW COUPLING

In normal neural tissue, increases in metabolic demand are met by increases in substrate delivery. Acutely following TBI, this coupling between cellular metabolism and blood flow becomes disconnected, and at the same time there is a period of markedly increased glucose uptake (Yoshino et al., 1991). Disturbances in cerebral flow (Kelly et al., 1997; Martin et al., 1995; Muizelaar et al., 1989) and cerebral autoregulation (Kelly et al., 1996; Vavilala et al., 2006) have been reported after TBI. Reduced cerebral blood flow (CBF) in a setting of high metabolic demand can result in worsened metabolic state and may provide the impetus for early and delayed cell death. In addition to the spreading depression-like state mentioned above, other settings where uncoupling may be of particular concern would include regional blood flow disturbances in pericontusional areas (Bergsneider et al., 1997), relative CBF reduction in a developing brain (which has metabolic needs higher than adult) (Adelson et al., 1997), and secondary insults that worsen this mismatch (e.g., reduced cerebral perfusion pressure, seizures, hyperthermia) (Vespa et al., 1998, 2005).

## MITOCHONDRIAL DYSFUNCTION AND OXIDATIVE STRESS

Calcium flux is another major component of ionic disequilibrium after TBI (Fineman et al., 1993; McIntosh et al., 1998; Osteen et al., 2001) and contributes to metabolic dysfunction by altering biochemical pathways and interfering with mitochondrial oxidative metabolism (Lifshitz et al., 2004; Robertson et al., 2006; Sullivan et al., 2005). Mitochondrial dysfunction may manifest as both reduced capability to generate ATP as well as an increase in the creation of free radicals (Lewen et al., 2000; Singh et al., 2006). Other sources of oxidative stress and free-radical generation may be enzyme activation, heme products from intracranial bleeding, and activation of inflammatory processes. Another potentially important contributor to oxidative stress is the stage of neural development, as some experimental work suggests that the immature nervous system may be less capable of upregulating certain free-radical scavenging mechanisms and thus be more susceptible to this type of injury (Chang et al., 2005; Fan et al., 2003).

## EXCITOTOXICITY AND APOPTOSIS

Excessive influx of calcium and reduction of ATP both conspire to acutely tip the balance of homeostasis toward cellular compromise and even death. Calcium flux triggers a widespread cascade of intracellular signaling that also includes increased activity of caspases and calpains and leads to both acute cell death and ongoing apoptosis (McIntosh et al., 1998; Raghupathi, 2004).

One important factor that modulates the direction of calcium signaling is the $N$-methyl-D-aspartate receptor (NMDAR). As one of the critical gateways for calcium flux after TBI, it is not only the amount of receptors but also the subunit composition and location of the NMDARs that govern the net outcome of this calcium flux (Hardingham and Bading, 2003; Osteen et al., 2004). NMDARs containing predominantly the NR2A subunit increase with development, result in a more discriminating NMDAR-mediated signal, and are preferentially located synaptically (Cull-Candy et al., 2001; Flint et al., 1997; Quinlan et al., 1999). NMDAR composed of NR2B subunits are generally more sensitive to glutamate and show both synaptic and extrasynaptic localization. These distinctions appear to be of great importance, as a growing body of evidence indicates that indiscriminate activation of NR2B-containing (predominantly extrasynaptic) NMDARs favors pathological stimulation of the neuron and activation of cell death pathways (Hardingham et al., 2002). Conversely, more "physiological" activation of neurons via NR2A-containing (predominantly synaptic) NMDARs triggers expression of molecules associated with cell survival and improved neuroplasticity (Deridder et al., 2006; Hardingham, 2006; Liu et al., 2007).

Sensitivity to excitotoxicity and apoptosis also appears to be highly age dependent. In neonatal animal models (younger than postnatal day 7 in rodents), TBI appears to induce greater cell death than in adults but primarily through apoptotic mechanisms (Bittigau et al., 1999). In weanlings and young juveniles, both apoptotic and necrotic cell death are relatively less than in the mature animals (Adelson et al., 2001; Bittigau et al., 1999; Gurkoff et al., 2006; Prins et al., 1996). Maturational changes in NMDAR subunit composition occur normally throughout this time range, with NR2A expression rapidly increasing after P10 and peaking between P25 and P35, shortly after weaning.

A final consideration with respect to early TBI-induced cell death is that the consequences of such damage are related to the cell types being affected. Among neurons, hilar interneurons appear to be particularly susceptible to TBI-induced damage (Lowenstein et al., 1992). Their death appears to underlie an excitatory–inhibitory imbalance that begins and evolves after injury. Non-neuronal cell types are also affected, with dysfunction and death of astrocytes (Floyd and Lyeth, 2007; Zhao et al., 2003) and oligodendrocytes (Beer et al., 2000; Fukuda et al., 1996) being important contributors to the response to TBI and also likely playing a role in post-injury network excitability and plasticity.

## PLASTICITY

Following TBI, the neuroplastic response of the brain is affected by many concurrent mechanisms, including the development of focal lesions, diffuse cell loss, axonal disconnection, functional changes in neurotransmission, neurogenesis, and sprouting. There is evidence for both enhancement and suppression of plasticity after TBI which complicates our understanding of processes that may be involved in post-injury epileptogenesis.

For example, brain-derived neurotrophic factor (BDNF) is upregulated in multiple brain regions after TBI in several experimental models (Griesbach et al., 2002; Hicks et al., 1997). Conversely, experience-dependent upregulation of BDNF expression following voluntary exercise shows diminished responsiveness early after TBI that eventually recovers as a function of time and injury severity (Griesbach et al., 2004, 2007).

Rearing in an enriched environment (EE) has generally been shown to have beneficial effects on functional outcome after adult TBI and other acute injuries (Johansson and Belichenko, 2002; Passineau et al., 2001; Wagner et al., 2002). However, the effects of EE rearing are greatly attenuated early after TBI in the immature animal (Fineman et al., 2000), although elements of this responsiveness seem to recover over time (Giza et al., 2005).

In the developing brain, the neuroplastic response to injury is very dependent upon the injury model and the outcome being measured. In conjunction with the EE data above, an injury-induced NMDAR subunit change (reduction of NR2A) has been reported in the first week after P19 TBI (Giza et al., 2006). Quantification of dendritic arbors after P19 TBI followed by EE has shown both

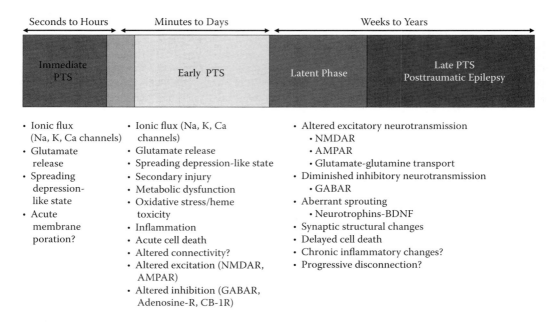

| Seconds to Hours | Minutes to Days | Weeks to Years |
|---|---|---|
| Immediate PTS | Early PTS | Latent Phase     Late PTS Posttraumatic Epilepsy |

- Ionic flux (Na, K, Ca channels)
- Glutamate release
- Spreading depression-like state
- Acute membrane poration?

- Ionic flux (Na, K, Ca channels)
- Glutamate release
- Spreading depression-like state
- Secondary injury
- Metabolic dysfunction
- Oxidative stress/heme toxicity
- Inflammation
- Acute cell death
- Altered connectivity?
- Altered excitation (NMDAR, AMPAR)
- Altered inhibition (GABAR, Adenosine-R, CB-1R)

- Altered excitatory neurotransmission
  - NMDAR
  - AMPAR
  - Glutamate-glutamine transport
- Diminished inhibitory neurotransmission
  - GABAR
- Aberrant sprouting
  - Neurotrophins-BDNF
- Synaptic structural changes
- Delayed cell death
- Chronic inflammatory changes?
- Progressive disconnection?

**FIGURE 12.1** Time course of post-injury responses, clinical seizure syndromes, and underlying pathophysiological mechanisms.

a lack of dendritic elaboration (ipsilateral to injury) and an enhancement of arborization (contralateral to injury) (Ip et al., 2002). However, using a weight-drop model in P21 mice, both anatomical and behavioral measures also show the occurrence of later-appearing deficits (Pullela et al., 2006).

These are just a few examples of altered neuroplasticity after TBI; however, mechanisms just like these (alterations in neurotrophins, axonal sprouting, dendritic arborization, changes in neurotransmitter receptors) are also involved in the process of epileptogenesis. This leads to a complicated goal of trying to restore "good" plasticity while inhibiting "bad" plasticity and is one of the major challenges facing investigators in this arena.

## ACUTE POSTTRAUMATIC SEIZURES

Acute posttraumatic seizures have clinically been divided into immediate posttraumatic seizures (IPTSs; those occurring within 24 hours of TBI) and early posttraumatic seizures (EPTSs; those occurring <7 days after TBI). This is a simple temporal distinction that is clinically straightforward; however, this subdivision may also have implications relevant to underlying cellular and physiological mechanisms. In general, EPTSs are thought to be worrisome occurrences—potential signs of an acute cerebral perturbation, be it hemorrhage or secondary ischemia or edema. EPTSs have clinically been linked to a higher risk of LTPSs and to worse global outcomes; thus, there has been a whole series of clinical investigations attempting to prevent EPTS. Although 24 hours is somewhat of an arbitrary breakpoint, it is also known that some very early seizures (that would be classified as IPTSs or, by some, as impact seizures) do not appear to be associated with any long-term sequelae. This raises the possibility that there are two separate syndromes of acute PTS that have different underlying pathophysiology and therefore different clinical significance (Figure 12.1). We will investigate the underlying mechanisms of acute seizures with these distinctions in mind (Figure 12.2).

A second important consideration is the likelihood that studies that identified only clinical EPTSs after TBI may underestimate the true incidence of EPTSs by missing subclinical seizures. The use of continuous electroencephalography (cEEG) monitoring in neurological/neurosurgical

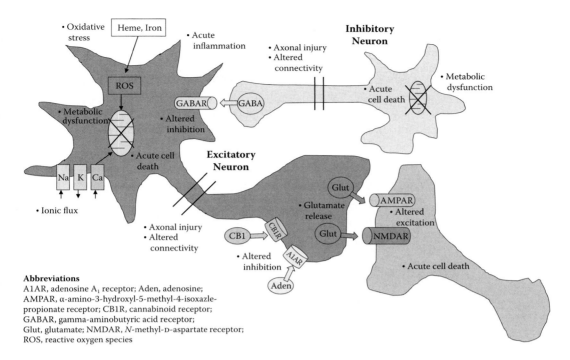

**Abbreviations**
A1AR, adenosine A$_1$ receptor; Aden, adenosine;
AMPAR, α-amino-3-hydroxyl-5-methyl-4-isoxazle-
propionate receptor; CB1R, cannabinoid receptor;
GABAR, gamma-aminobutyric acid receptor;
Glut, glutamate; NMDAR, N-methyl-D-aspartate receptor;
ROS, reactive oxygen species

**FIGURE 12.2** Summary of potential mechanisms of early posttraumatic seizures discussed in the text. Predominant acute mechanisms include uncontrolled ionic flux and excitatory neurotransmitter release. Excessive energy demands may occur as a result of the primary injury or due to secondary injuries (such as hypoxia, hypotension, or seizure activity). These metabolic perturbations, along with oxidative stress, may go on to trigger acute cell death in vulnerable populations.

intensive-care settings is becoming increasingly common. cEEG provides valuable additional information regarding neurological status, including the detection of subclinical seizures, distinction of non-epileptic spells, ability to monitor sedation, management of level of burst suppression, recognition of cerebral ischemia, and prognostication (Hirsch, 2004). Overall, 18% of neurocritical care patients were found to have subclinical seizures, and 10% of these were in subclinical status epilepticus (Claassen, 2004). Some important factors associated with seizures on cEEG included coma at time of EEG initiation, convulsive seizures prior to monitoring, and age less than 18 years. Prior history of epilepsy and presence of periodic epileptiform discharges on EEG were also implicated. Furthermore, the timing of monitoring is very important. In noncomatose patients, 95% had their first seizure within the first 24 hours of monitoring. Interestingly, of patients comatose at the onset of cEEG, only 80% had their first seizure within 24 hours, and this value increased to 87% by 48 hours, suggesting that a longer period of monitoring is valuable for neuroICU patients presenting with altered mental status or coma (Claassen, 2004).

Subclinical seizures have also been described using cEEG specifically in patients following TBI. In a cohort of 94 adult patients with moderate–severe TBI, 22% had EPTSs. Of these, seizure activity was subclinical in 52% (11/21) and only detectable through cEEG monitoring (Vespa et al., 1999). Another report (n = 70) found seizures in 33% of moderate–severe TBI patients and epileptiform discharges in an additional 16% (Ronne-Engstrom and Winkler, 2006). A more recent study of 47 TBI patients did not find any clinical or subclinical seizure activity during cEEG monitoring (Olivecrona et al., 2009). Factors that may have contributed to this difference from the Vespa et al. study include the fact that cEEG monitoring began later in the Olivecrona et al. study (50 hours vs. 9.6 hours). Also, a 5-lead EEG montage was used by Olivecrona et al. vs. 12 by Vespa et al. and 11 by Ronne-Engstrom and Winkler (2006). The TBI protocol utilized by Olivecrona et al. included continuous sedation with midazolam/thiopental with no awakening tests for several days. Vespa et al. used phenytoin prophylaxis and followed levels, but no antiepileptic drug (AED) prophylaxis was used by either Olivecrona et al. or Ronne-Engstrom et al. (Olivecrona et al., 2009; Ronne-Engstrom and Winkler, 2006; Vespa et al., 1999).

Overall, these studies suggest that PTSs can occur frequently in the neurocritical care population and specifically in TBI patients. EEG monitoring is helpful to identify subclinical seizures and underlying epileptiform activity in common ICU situations. The absence of cEEG monitoring may have had confounding effects on previous longer term outcome studies, particularly studies of anticonvulsant prophylaxis.

## EXCITATORY NEUROTRANSMITTER FLUX

Indiscriminate release of excitatory neurotransmitters, such as glutamate, can result in widespread depolarization of neurons after TBI. This mechanism may theoretically lead to seizure activity and potentiate metabolic stresses on the injured brain. Early seizures can occur in response to excessive levels of extracellular glutamate, and human microdialysis studies in patients with severe TBI have shown some relationship between high glutamate in the extracellular space and electrographic seizure activity (Vespa et al., 1998, 2007). Furthermore, post-TBI seizure activity has been correlated with physiological biomarkers of neural distress, such as elevations of the lactate/pyruvate ratio (LPR), reductions of extracellular glucose, and increased intracranial pressure temporally related to ongoing ictal activity (Vespa et al., 2007).

Animal models confirm increased brain glutamate after experimental TBI (Katayama et al., 1990), although little animal work has been done to determine specific relationships between acute physiological parameters and early seizures. One observational study that used EEG monitoring for 2 hours after controlled cortical impact (CCI) detected bilateral seizures in 82% of adult rats. Acute microdialysis measurements showed significant spikes in excitatory amino acids (particularly aspartate and glutamate) in the immediate post-injury period, which corresponded to the electrographic onset of bilateral seizure activity (Nilsson et al., 1994).

Other endogenous modulators of excitatory neurotransmitter release that have been studied experimentally include adenosine, which exerts an inhibitory effect on presynaptic glutamate release via the $A_1$ adenosine receptor ($A_1AR$) (Emptage et al., 2001; Malva et al., 2003). Using CCI in adult mice, $A_1AR$ knockouts showed substantially greater incidence of both clinical and electrographic seizure activity than wild-type or $A_1AR$ heterozygotes (Kochanek et al., 2006). Acute histopathological assessments using H&E and Fluoro-Jade® B did not reveal any qualitative differences in cellular injury, but rates of status epilepticus and mortality were significantly higher in the $A_1AR$ knockouts. The use of animal models to study acute post-TBI seizures is an area of investigation that is ripe for expanded inquiry and has the potential to shed light onto cellular mechanisms underlying IPTS and EPTS.

## IONIC DISTURBANCES AND SPREADING DEPRESSION

Glutamate release and biomechanical stretching of cellular membranes trigger acute ionic flux after TBI. Potassium efflux into the extracellular space has been measured in animal models using microdialysis (Katayama et al., 1995). This leads to a spreading depression-like state characterized by periodic DC shifts in experimental animals. Recently, cortical electrode arrays have detected similar electrophysiological phenomena after severe TBI in humans (Hartings et al., 2009; Strong et al., 2002), and electrographic seizure activity is associated with these events (Fabricius et al., 2008). Concomitant microdialysis in the vicinity of cortical electrodes has detected neurochemical signals of metabolic distress, including a wave of reduced tissue glucose and an elevation of the lactate/pyruvate ratio occurring with the spreading depression (Parkin et al., 2005).

In experimental models, investigators have characterized factors relevant to post-injury ionic perturbations and cellular excitability. Glial dysfunction has been reported to alter reuptake of extracellular potassium and result in abnormal neuronal excitability (D'Ambrosio et al., 1999), although the relationship between this elevated extracellular potassium and glial dysfunction has been debated (Santhakumar et al., 2003). Much as is seen after fluid percussion injury (FPI), substantial potassium efflux has been measured following experimental intraparenchymal hemorrhage in cats (Hubschmann and Kornhauser, 1983) and during focal seizures induced by penicillin application (Moody et al., 1974).

## BLOOD PRODUCTS, OXIDATIVE STRESS, AND FREE RADICALS

Clinically, in addition to injury severity, the presence of blood in contact with neuropil appears to be an independent predictor of PTE. EPTSs occur more often in those with subdural hematoma, subarachnoid hemorrhage, and cerebral contusions (D'Alessandro et al., 1988; Frey, 2003; Jennett, 1975b). Blood products such as heme and transferrin can release iron into the brain parenchyma, and this can result in generation of toxic free-radical intermediates. Furthermore, patients with epilepsy have a higher likelihood of having haptoglobin 2-2, the haptoglobin phenotype least effective in binding hemoglobin (Sadrzadeh et al., 2004), suggesting at least some epilepsy risk in patients less able to bind up iron-containing proteins (also see below under genetic predisposition and Table 12.2). Other types of hemorrhagic brain injuries in humans are also associated with development of seizure activity, including hemorrhagic cerebral infarction, aneurysmal subarachnoid hemorrhage, and spontaneous intraparenchymal hemorrhage.

Experimentally, a model of posttraumatic seizures and epilepsy has been characterized after cortical injection of iron salts in rats and cats. Within 72 hours of ferrous or ferric chloride subpial injection, electrocorticography showed epileptiform discharges that became associated with behavioral seizures between 2 and 5 days after injection (Willmore et al., 1978a,b). These early seizures almost always resulted in chronic recurrent seizures (94% of injected rats continued to have seizures after 12 weeks), suggesting that toxic effects from metallic ions in blood products can contribute to the genesis of PTS/PTE.

## SECONDARY INJURY

It is well known that other processes such as hypotension, hypoxia, and ischemia can worsen outcome after traumatic brain injury. The Traumatic Coma Data Bank showed that a single episode of hypotension or hypoxia was associated with worse global outcomes in a moderate–severe TBI patient population (Chesnut et al., 1993). Seizures were not specifically studied as a form of secondary injury in this analysis. In many experimental TBI models, hypotension or hypoxia are also known to worsen outcome (Jenkins et al., 1989; Matsushita et al., 2001), although the timing of the secondary injury is critical, and at some time intervals a phenomenon of preconditioning has been reported (Liu et al., 1992; Perez-Pinzon et al., 1999).

Early posttraumatic seizures may be not only a marker of initial injury severity but also a cause of secondary injury due to increased metabolic and ionic stresses. Acute electrographic seizures seen after severe TBI in humans were associated not only with increased intracranial pressure but also with an elevated lactate/pyruvate ratio and elevated glutamate, indicative of underlying neurometabolic stress (Vespa et al., 2007). Low extracellular glucose has been associated with worse long-term outcome (Vespa et al., 2003) and showed a strong correlation with spreading depression-like events in pericontusional cortex as measured by electrocorticography (Parkin et al., 2005). Given the current understanding of TBI pathophysiology as a neurometabolic cascade, these changes may logically raise concern for ongoing problems in cerebral energy metabolism and the potential for secondary injury.

Clinically, both pediatric and adult patients with EPTSs have worse long-term outcomes (Chiaretti et al., 2000; Corkin et al., 1984; Ong et al., 1996; Walker and Blumer, 1989). Most series also describe a higher risk for LPTSs/PTE in individuals who had EPTSs and that the occurrence of PTE worsens long-term outcomes. Although several large prospective studies of anticonvulsant prophylaxis have shown the efficacy of conventional anticonvulsants in preventing EPTS, none has been effective in stopping the development of PTE (Beghi, 2003; Temkin et al., 1990, 1999). This suggests that the mechanisms of EPTSs and LPTSs are distinct and diverge at some early point.

Seizures as a secondary form of injury after trauma have only been studied in a limited fashion using experimental models. It has long been known that acute seizures are observed after TBI using FPI (Kharatishvili et al., 2006), CCI (Nilsson et al., 1994), and weight-drop models (Golarai et al., 2001); however, precise characterization of these seizures has only focused on their acute characteristics and less so on their potential long-term consequences. Electrophysiological monitoring for acute seizures showed high rates of EPTSs in rats immediately after controlled cortical impact (82%, 14/17), but the presence of EPTSs was not studied in relationship to any chronic histological or functional outcomes (Nilsson et al., 1994).

In terms of specific examination of seizures as secondary injuries, there is one report of lateral FPI followed by kainate-induced status epilepticus in adult rats resulting in worse histopathological outcome (Zanier et al., 2003). In this study, adult rats underwent a moderate lateral FPI followed 1 hour later by injection of either saline or a sublethal dose of kainate, which induced acute generalized seizure activity in 6 of 8 FPI+kainate animals but none of the FPI-only or sham+kainate animals. The group with the combined injury (FPI+kainate) showed about 70% cell loss in CA3/CA4, which was significantly greater than all of the other groups. This study also showed no regional breakdown of the blood–brain barrier to explain this cellular vulnerability; in fact, using 2-deoxy-glucose autoradiography, there was an increase in glucose uptake in the vulnerable regions, leading the authors to hypothesize that the increase in neuronal energy-requiring activity due to seizures after TBI was the underlying cause of the selective cell death.

A second study using lithium–pilocarpine (PC)-induced status epilepticus as a secondary injury after lateral FPI in the immature rat (postnatal day 19) showed different results (Gurkoff et al., 2009). In this study, administration of PC 1 hour or 6 hours after lateral FPI actually resulted in less acute cell death in hippocampus than PC without preceding FPI. This apparent protection was lost in a second experiment where long-term outcomes were examined at 20 weeks. At the late time point, there were no significant differences in hippocampal cell death, pentylenetetrazole seizure threshold, or mossy fiber sprouting between the FPI+PC and the '-only groups; however, significant differences in these measures were noted between chronic epileptic and nonepileptic animals. In the chronic PC group, 50% had observed seizures, and seizures were seen in 75% of the FPI+PC group (N.S.). These studies, like other experimental studies of post-TBI secondary injury, demonstrate that seizures may worsen acute outcomes and that other variables may affect this post-injury response, such as timing of secondary injury, age at injury, and age at followup.

## EPILEPTOGENESIS AND POSTTRAUMATIC EPILEPSY

Although EPTSs may be considered provoked seizures and appear to occur in response to acute physiological perturbations such as glutamate release, ionic imbalance, metabolic distress, toxic blood products, oxidative damage, and other disturbances, LPTSs and PTE are characterized by recurrent spontaneous seizures temporally distant from the acute injury (Figure 12.1). The underlying mechanisms by which LPTSs and PTE occur are thus certainly mediated by different mechanisms, and the underlying processes that set the stage for LPTSs and PTE may also be distinct from those causing EPTS (Figure 12.3).

The time course of epileptogenesis is typically thought of as having three phases: (1) acute injury, (2) latent period, and (3) spontaneous recurrent seizures. Although some of the acute pathophysiology triggering EPTSs may overlap the acute signals that initiate the pathway of epileptogenesis, the

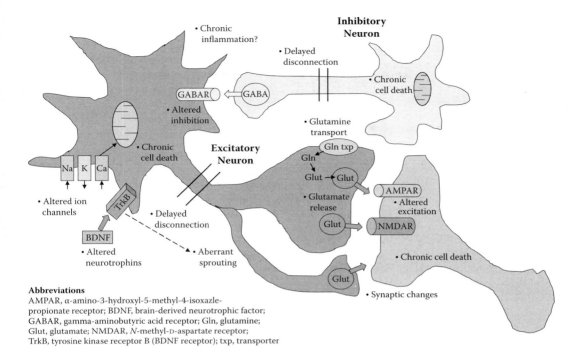

**Abbreviations**
AMPAR, α-amino-3-hydroxyl-5-methyl-4-isoxazle-
propionate receptor; BDNF, brain-derived neurotrophic factor;
GABAR, gamma-aminobutyric acid receptor; Gln, glutamine;
Glut, glutamate; NMDAR, *N*-methyl-D-aspartate receptor;
TrkB, tyrosine kinase receptor B (BDNF receptor); txp, transporter

**FIGURE 12.3** Summary of potential mechanisms of epileptogenesis discussed in the text. Predominant mechanisms in this phase include chronic changes in ion channels and neurotransmitter receptors/transporters. Additionally, alterations in expression of neurotrophins may promote aberrant sprouting and synaptic changes that strengthen abnormal excitatory connections. Chronic neuronal death and disconnection may still occur and may be exacerbated by ongoing seizure activity.

strategy of preventing EPTSs has proven insufficient to prevent PTE. Therefore, strategies to avoid the long-term disability and lost functionality due to PTE currently focus on treating PTE after it develops. In the setting of TBI, however, the timing of the inciting event is clearly known. This raises the intriguing possibility that strategies targeted toward abnormal cellular and network responses in either the acute or latent phases may actually stop or prevent the progression to epilepsy.

## EXCITATORY–INHIBITORY IMBALANCE

Normal network function in both the immature and mature brain requires a balance between excitatory and inhibitory neurotransmission, and this balance appears to be altered by acute brain injuries, particularly TBI. Initially, the imbalance is due to damaged or dysfunctional connections and neurons, but as the brain recovers both structural and functional amendments occur that may either ultimately restore normalcy or promote lasting but aberrant change.

Electrophysiological recordings have demonstrated impaired hippocampal CA1 long-term potentiation (LTP) in the first 2 weeks post-injury (D'Ambrosio et al., 1998; Miyazaki et al., 1992; Sick et al., 1998). Interestingly, normal long-term depression (LTD) was reported at the same time points when LTP induction was diminished (D'Ambrosio et al., 1998). Although LTP induction was restored by 8 weeks post-injury, chronic deficits in LTP maintenance developed at this time after mild, moderate, and severe FPI (Sanders et al., 2000). These results demonstrate altered hippocampal plasticity in both the acute and latent periods of epileptogenesis. The relationship of these specific changes to the development of chronic PTE is unclear, although some studies have suggested a perturbation in NMDA-mediated glutamatergic neurotransmission in the first post-injury week. Specific reductions in hippocampal NR2A subunits have been reported after developmental lateral FPI (Giza et al., 2006; Li et al., 2005). These results are consistent with the finding that LTP is mediated by activation of NR2A-containing NMDAR, whereas LTD requires NR2B (Liu et al., 2004), and further substantiate an important role for NMDAR subunit change in post-injury plasticity. AMPA receptor dysfunction, changes in dendritic morphology, and reduced α-CaMKII expression have also been reported as potential mechanisms contributing to impaired hippocampal LTP after TBI in mice (Schwarzbach et al., 2006). Although this literature pertains to the effect of injury on altered excitability, as measured by LTP/LTD, a direct relationship between these changes and epileptogenesis has yet to be established.

Molecular changes in glutamate receptors have been implicated in altered excitatory activity after TBI, but changes in neurotransmitter transport may also be important. System A glutamine transporters (sodium-coupled neutral amino acid transporters SNAT1 and SNAT2) are upregulated in a cortical undercut model of PTE. In the presence of physiologically elevated glutamate levels, these preparations are associated with increased epileptiform potentials and spreading depression-like events (Tani et al., 2007).

At least some of these changes in excitatory–inhibitory balance may be mediated through changes in intracellular calcium homeostasis. In a PC-induced model of acquired epilepsy, increases in intracellular calcium flux were identified in the acute post-injury, latent, and recurrent seizure phases of epileptogenesis (Raza et al., 2004). These changes were not seen in PC-treated animals that did not undergo status, PC-treated animals that did experience status but did not develop spontaneous epilepsy, or PC-treated animals that experienced status but had been pretreated with the NMDAR antagonist MK-801. This is concordant with some of the mechanistic data reported after experimental TBI, above.

Using a weight-drop model in P11–P15 rats, CA1 but not CA3 pyramidal cells showed an increase in the frequency of spontaneous action potentials at 3 hours and 24 hours after injury and recovered by 72 hours. CA1 interneurons, like CA3 pyramidal cells, did not show major changes in spontaneous action potentials post-injury (Griesemer and Mautes, 2007).

Hippocampal mossy cells have also been studied after FPI, albeit in immature (P13–P15) rats. Basic input–output functions of these cells did not show major changes 1 week after injury, but more detailed investigation revealed alterations in sodium, potassium, and hyperpolarized-activated cation currents (Howard et al., 2007). When these parameters were entered into a computational model of single perforant path stimulation, each change individually showed the potential to significantly affect network excitability. The opposing nature of these changes contributed to the overall lack of difference in current–firing frequency and current–voltage relationships in this study. These data illustrate how overall network function is dependent upon a series of parameters whose alterations may individually cause substantial electrical change but in total can result in a new homeostatic balance of excitation and inhibition.

In an interesting and somewhat paradoxical report, very early (<2 minutes) administration of a cannabinoid type-1 (CB1) antagonist after lateral FPI reduced long-term (6-week) susceptibility to kainate-induced seizures (Echegoyen et al., 2009). Endogenous cannabinoids acting via the CB1 receptor play a role in controlling acute seizures (Monory et al., 2006; Wallace et al., 2003), whereas CB1 antagonists have been shown to promote seizure activity (Chen et al., 2007; Monory et al., 2006). In this scenario, then, it is hypothesized that the early CB1 activation may represent an initiating step on the path of epileptogenesis.

Investigations such as these demonstrate the level of complexity inherent to understanding post-injury changes in both intrinsic neuronal excitability as well as functional alterations in the properties of the entire network. Important factors affecting the post-injury response include cell type being studied, electrophysiological parameter investigated, and time post-injury. Injury model and severity may also play an important role in the degree and type of neuronal dysfunction vs. neuronal death. Furthermore, these studies of post-TBI electrophysiology are reported in a broad age range, from P11 up to adulthood, but the age-at-injury variable has not been systematically studied (Table 12.1).

**TABLE 12.1**
**Age-Dependent Experimental Studies of Posttraumatic Excitability**

| Age/Strain | Injury | Experimental Variable | Dependent Variable | Result | Refs. |
|---|---|---|---|---|---|
| P11–P15/Wistar | WD | Injury | Whole-cell recordings | PID2–24: CA1 (not CA3) ↑ spontaneous action potential frequency | Griesemer and Mautes (2007) |
| P13–P15/Wistar | LFPI | Injury | Whole-cell recordings | PID5–8: ↑ $I_{Na+}$; ↓ inward $I_{A(K+)}$; ↑ $I_h$ | Howard et al. (2007) |
| P16–P18/Sprague | CCI | Injury | Seizure threshold, MRI | PID16–24: ↔ in threshold; PID42–47: ↓ in threshold; MRI results variable | Statler et al. (2008) |
| P19/Sprague | LFPI | Injury ± LiPC SE | Seizures, histology, PTZ threshold | PID1: LFPI+LiPC leads to ↓ hippocampal cell death than LiPC alone, but PIW20 LFPI+LiPC and LiPC only show similar cell death, PTZ threshold, and mossy fiber sprouting | Gurkoff et al. (2009) |
| P21–P22/Wistar | LFPI | CB1 antagonist | KA seizure susceptibility | CB1 antagonist < 2 min p TBI prevented late PIW6 KA seizure susceptibility | Echegoyen et al. (2009) |
| P21–P25/Sprague | Cortical undercut | Glutamine | Cortical slice FPs | Glutamine ↑ FPs; effect is blocked by MeAIB (SNAT1/2 inhibitor) | Tani et al. (2007) |
| P26–P31/Sprague | LFPI | Injury | Hippocampal slice FPs, K+, GFAP | PID2: CA3 glia ↑ GFAP, ↑ extracellular K+, ↓ inward K+ currents | D'Ambrosio et al. (1999) |
| P32–P35 Sprague | LFPI | Injury | EEG and behavioral seizures; cortical slice FPs; histology | PIW2: 60% of rats had seizures; PIW10: 92% had seizures; ↑ ipsilateral cortex afterdischarges; ↑ ipsilateral cortex GFAP immunohistochemistry | D'Ambrosio et al. (2004) |
| P33–P35 Sprague | LFPI | Injury | EEG and behavioral seizures; histology | PIW2: 90% focal cortical seizures; PIW17–18: 10% focal cortical, 40% spreading cortical, 50% limbic seizures | D'Ambrosio et al. (2005) |
| P33–35 Sprague | LFPI | Injury | EEG (surface and ECoG) and behavioral seizures | PIW2–28: seizures defined differently (from 0.8–15 sec on ECoG to 1–5 sec on scalp leads) showed PTE latencies of increasing duration | D'Ambrosio et al. (2009) |

*Abbreviations:* CCI, controlled cortical impact; ECoG, electrocorticography; EEG, electroencephalography; FPs, field potentials; GFAP, glial fibrillary acidic protein; KA, kainic acid; LFPI, lateral fluid percussion injury; LiPC, lithium and pilocarpine; MRI, magnetic resonance imaging; PID, post-injury day; PIW, post-injury week; PTZ, pentylenetetrazole; SE, status epilepticus; SNAT, sodium-coupled neutral amino acid transporters; TBI, traumatic brain injury; WD, weight drop.

## REGIONAL CELL DEATH

Several processes have been postulated to interact with or cause the excitatory–inhibitory imbalance that appears to underlie epileptogenesis. Selective cell death of vulnerable cell populations is one such mechanism. Hippocampal neuronal death is most often implicated in posttraumatic epileptogenesis, although direct death of cortical neurons has also been investigated in more severe injuries that include cortical contusions.

For patients with acquired epilepsy, trauma has been identified as the likely etiology in a substantial proportion (Herman, 2002; Lowenstein, 2009). There is conflicting evidence for the association between an etiological diagnosis of PTE and the pathological picture of mesial temporal sclerosis, with some studies reporting a significant correlation (Marks et al., 1995; Mathern et al., 1994), while others dispute this (Diaz-Arrastia et al., 2000; Hudak et al., 2004). Nonetheless, the histopathology of mesial temporal sclerosis is clearly seen in cases of intractable temporal lobe epilepsy, including those with a posttraumatic etiology.

Experimental TBI models have shown particular vulnerability of hippocampal neurons, including hilar inhibitory interneurons (Hicks et al., 1993; Lowenstein et al., 1992; Toth et al., 1997). Following lateral FPI, a severity-dependent loss of hilar neurons was seen both ipsilaterally and to a lesser extent contralaterally. Dentate granule, CA1, and CA3 pyramidal cell counts did not decrease in these same animals. Field potentials in response to perforant path stimulation in the dentate granule cells showed multiple population spikes indicative of underlying hyperexcitability (Lowenstein et al., 1992). Weight drops onto somatosensory cortex result in both a cortical lesion and selective loss of hippocampal CA3 and hilar neurons (Golarai et al., 2001). In this study, the functional consequences of this injury included a greater susceptibility for pentylenetetrazole-induced seizures 15 weeks post-TBI and increased excitability of dentate granule neurons to perforant path stimulation measured from 2 to 15 weeks post-injury. Diffusion MRI showed reduced diffusivity in ipsilateral hippocampus within 3 hours of lateral FPI that corresponded with later development of hyperexcitability and mossy fiber sprouting (Kharatishvili et al., 2007). Hippocampal diffusion measures then progressively increased and were significantly higher relative to sham when measured anywhere from 2 to 11 months post-TBI, consistent with ongoing hippocampal cell death.

## INFLAMMATION AND GLIOSIS

Following TBI, inflammatory cascades are triggered in response to acute excitotoxic cell death, blood–brain barrier breakdown, release of toxic blood products, and activation of second-messenger signaling systems (Chang et al., 2005; Kadhim et al., 2008). As discussed earlier, the presence of intracranial hemorrhage is a clinical risk factor for both EPTSs and PTE (Jennett, 1975b; Kieslich and Jacobi, 1995; Temkin, 2003). One of the earliest models of PTE involved direct application of iron onto the cortex (Willmore et al., 1978a,b). In nontraumatic epilepsy models, the presence of inflammation can significantly modulate seizure-induced injury. As an example, the administration of lipopolysaccharide, a potent trigger of systemic inflammation, resulted in significantly greater neuronal injury after lithium–pilocarpine-induced status epilepticus in the developing rat (Auvin et al., 2007).

It is well known that reactive gliosis occurs following many types of TBI. Early involvement of glial cells in the injury response has been described (Floyd and Lyeth, 2007; Hill et al., 1996; Hill-Felberg et al., 1999). Glial scarring may be indicative of underlying cell death and has been associated with the development of neocortical PTE. In a rat model (P32–P35) of rostral–parasagittal FPI, electrocorticography demonstrated seizure foci in the vicinity of the cortical injury site in 92% of injured animals within 2 to 8 weeks. Unilateral cortical field recordings showed after-discharges, and the presence of PTE was associated with a cloud of glial fibrillary acidic protein (GFAP) immunoreactivity that extended from the site of injury into the temporal neocortex (D'Ambrosio et al.,

2004). In a follow-up study with a longer duration of post-TBI monitoring, progressive asymmetry was noted in both neocortex and hippocampus, and the epilepsy syndrome appeared to evolve from primarily cortical to hippocampal (D'Ambrosio et al., 2005). What role the astrocyte reaction played in this evolution, or whether it was just an innocent bystander indicative of ongoing neuronal death, has not yet been determined. It is also important to understand that, in general, the glial response to TBI may also serve critical roles in limiting the extent of injury and promoting recovery. In transgenic models where reactive glia may be selectively ablated after injury, outcomes are substantially worse, although specific observation for seizures or epilepsy was not performed (Myer et al., 2006).

## ABERRANT SPROUTING AND RECURRENT CONNECTIVITY

One of the hallmarks of plasticity is the ability to grow new connections in response to activity. This can be a mechanism for recovery of function but has also been implicated in the process of epileptogenesis, particularly with the aberrant sprouting and development of collateral excitatory connections. Within the hippocampus, mossy fiber sprouting has been implicated in epileptogenesis (Scharfman, 2002; Sutula, 2001). Although not necessarily required for the development of epilepsy (Armitage et al., 1998; Mohapel et al., 2000), sprouting of these intrahippocampal fibers has been correlated with the development of spontaneous recurrent seizures in many nontraumatic epilepsy models (Cavazos, 1991). Sprouting and recurrent connectivity may also occur in other regions and contribute to enhance the excitability and synchronization of neural firing patterns that herald the development of epilepsy.

Mossy fiber sprouting is a marker for aberrant recurrent excitatory structural plasticity and has been described in human cases of temporal lobe epilepsy (Mathern et al., 1996; Sutula et al., 1989). Mossy fiber sprouting and seizure susceptibility have also been studied after TBI in experimental models, making a more substantial link between the traumatic injury response and development of epilepsy. Mossy fiber sprouting is seen after both FPI and contusive weight-drop experimental TBI, is generally more prominent in the hippocampus ipsilateral to injury, and is associated with signs of hyperexcitability (Golarai et al., 2001; Gurkoff et al., 2009; Kharatishvili et al., 2006).

Axonal sprouting and generation of new connections may be a final common pathway of enhanced excitatory activity or excitatory–inhibitory imbalance that may be mediated through many of the mechanisms already discussed in this chapter. These include selective death of inhibitory neurons, changes in ionic currents in injured cells, altered neurotransmitter receptor composition, and injury-induced alterations of neurotrophic factor expression. TBI itself has been shown to result in altered patterns of neuronal connectivity. Using diffusion tensor imaging (DTI), many clinical studies have reported global and regional changes in white-matter fractional anisotropy that imply changes in connectivity after human TBI (Kumar et al., 2009; Wilde et al., 2008). In a recent study of mild TBI/concussion, focal slowing seen using magnetoencephalography overlies areas of abnormal DTI signal, making a correlation between abnormal cortical connectivity and electrophysiological dysfunction (Huang et al., 2009). Advanced imaging of neural connectivity and functional activation will almost certainly provide future data leading to a better understanding of this aspect of epileptogenesis in patients.

In an animal model of entorhinal cortex lesion combined with TBI, the carefully orchestrated pattern of perforant path regrowth is substantially altered after fluid percussion injury (Prins et al., 2003). Abnormal synaptic ultrastructure and cognitive impairment were associated with the abnormal plasticity in this model, although development of seizures was not monitored. Other studies of TBI have shown abnormal cortical connectivity and dendritic branching following fluid percussion injury at P19, demonstrating that injury alone alters connectivity but not specifically looking at epileptogenesis (Ip et al., 2002).

## OTHER CONSIDERATIONS

### GENETIC PREDISPOSITION

A more recent consideration with regard to posttraumatic episodic neurological disturbances (such as seizure and migraine) is the contribution of genetic factors (Table 12.2). Mutations in the *CACNA1A* calcium channel subunit gene have been associated with hemiplegic migraine attacks and coma induced by delayed cerebral edema following mild TBI (Kors et al., 2001). This same mutation has also been associated with childhood seizures (Chan et al., 2008; Kors et al., 2001) and raises the question as to whether individuals with this mutation (or in other known mutations of ion channel genes associated with epilepsy and migraine) have a lower threshold for posttraumatic paroxysmal electrical disturbances.

Another gene associated with worse neurocognitive outcomes after TBI, apolipoprotein E4 (*apoE4*), has shown an independent correlation with risk for LPTSs. In the study by Diaz-Arrastia et al. (2003), 10/29 apoE4+ patients had late seizures, as compared with 11/77 apoE4– individuals (RR = 2.41, $p < 0.03$). No significant demographic or injury severity differences between the apoE4+ and apoE4– groups were found. Haptoglobin 2-2 (Hp 2-2) is an isoform of a hemoglobin-binding protein that may also be involved in controlling inflammatory responses to hemorrhagic events. Although not yet studied in relation to TBI or specifically PTSs, Hp 2-2 was found more frequently in an epileptic cohort (67%) than normal controls (35%, $p < 0.001$) (Sadrzadeh et al., 2004). In an animal model, knockout of the $A_1AR$ gene resulted in lethal status epilepticus following experimental TBI (Kochanek et al., 2006). A human correlate to this has not yet been described.

These reports invite the study of PTS in transgenic animal models for these types of genes. They also suggest that obtaining genetic information on patients who experience TBI, perhaps even mild TBI, may offer insight into the risks of seizure outcomes for these patients.

### AGE AT INJURY

Inherent differences in connectivity, excitability, and neuroplasticity exist throughout brain maturation. These ongoing changes undoubtedly play a role in the response to TBI and the subsequent development of PTSs. In neonates, $GABA_A$ receptor (GABAR) activation may trigger excitation due to differences in chloride transport (Ben Ari et al., 1997). This has implications for seizures occurring in this early age range, which in humans may include infant victims of inflicted TBI. This type of TBI has a high association with PTS (Ewing-Cobbs et al., 1998).

Other neurotransmitter receptors also undergo molecular changes during development that may influence their potential contribution to the genesis of both EPTSs and LTPSs/PTE. NMDAR subunit composition favors the NR2B subunit early in life, and NR2B-containing NMDARs are known to be more sensitive to glutamate and allow more calcium flux than their NR2A-containing counterparts (Cull-Candy et al., 2001). Conversely, activation of NR2A-containing NMDARs appears to promote the expression of antiapoptotic and proplasticity mechanisms (Hardingham and Bading, 2003). NMDAR expression is altered after experimental TBI and shows different patterns when the injury occurs during development vs. in adulthood (Giza et al., 2006; Kumar et al., 2002; Osteen et al., 2004). These changes have been associated with alterations in electrophysiology, plasticity, and recovery, but whether or how these subunit changes are related to post-TBI epileptogenesis remains to be studied.

α-Amino-3-hydroxy-5-methyl-4-isoxazoleproprionic acid receptors (AMPARs) lacking the GluR2 subunit are expressed in greater amounts during development and after brain injuries, and GluR2(–) AMPARs are calcium permeable (Gorter et al., 1997; Pellegrini-Giampietro et al., 1997). This type of molecular change thus has potential relevance for vulnerability to excitotoxicity as well as for responsiveness to post-injury activation. This dichotomy of adverse pathological stimulation and beneficial physiological activation is a common theme in development and recovery from injury and offers a major challenge to future strategies to prevent post-TBI epileptogenesis without adversely affecting mechanisms of recovery.

**TABLE 12.2**
**Genes and Posttraumatic Seizures**

| Gene | Mechanism | Comparison | Association with Human PTS | Refs. |
| --- | --- | --- | --- | --- |
| apoE4 | Lipoprotein; associated with worse TBI outcomes | Moderate–severe neurosurgery TBI admissions ± PTS | RR = 2.41 of PTS; $p = 0.03$ | Diaz-Arrastia et al. (2003) |
| CACNA1A | Calcium channel subunit; associated with FHM, post-TBI edema | Case/series reports | Cases of post-TBI edema and coma; PTS | Kors et al. (2001); Chan et al. (2008) |
| Haptoglobulin 2-2 | Hemoglobin binding protein; antiinflammatory | Epileptics vs. normal controls | 67% epilepsy patients vs. 35% controls; $p < 0.001$ | Sadrzadeh et al. (2004) |
| $A_1AR$? | Presynaptic inhibition of neurotransmitter release | Associated with post-TBI status epilepticus in knockout mice | No human PTS studies yet | Kochanek et al. (2006) |
| Other channelopathies? | Ionic disequilibrium? Altered neurotransmission? | Many have been implicated in epilepsy | No human PTS studies yet | Reid et al. (2009) |
| NMDA receptor subunits? | Altered neurotransmission? | In humans, increased NMDAR subunits in nonhippocampal sclerosis, decreased NR2A in hippocampal sclerosis cases | NMDAR changes seen in epileptic patients vs. nonepileptic autopsy cases | Mathern et al. (1999) |
| BDNF? | Altered plasticity? | In humans, Val/Met variant associated with cognitive impairment and risk of psychiatric disorders | No human PTS studies yet | Egan et al. (2003) |

*Abbreviations:* $A_1AR$, $A_1$ adenosine receptor; BDNF, brain-derived neurotrophic factor; FHM, familial hemiplegic migraine; NMDA, $N$-methyl-D-aspartate; NMDAR, $N$-methyl-D-aspartate receptor; PTS, posttraumatic seizure; RR, risk ratio; TBI, traumatic brain injury.

Connectivity is also changing throughout brain maturation, and ontogenetic profiles of cerebral structures demonstrate continued development of frontal lobe white matter well into the early 20s in humans (Giedd, 2008). Experimental studies of TBI using the midline FPI model have shown increased vulnerability of unmyelinated fibers compared to myelinated fibers in the corpus callosum. The functional assessment of these fibers was performed using *ex vivo* electrophysiology of compound action potentials (Reeves et al., 2005). Vulnerability of unmyelinated axons is certainly relevant to development, and disrupting the ongoing process of myelination by TBI may influence both normal recovery as well as aberrant plasticity responses such as those associated with epileptogenesis.

Injury during development can have long-term sequelae that are different from the acute consequences of injury, and the interaction between post-injury pathophysiology and ongoing brain maturation can result in both recovery from acute deficits as well as later appearance of impairments. Experimentally, developmental TBI has been shown to lead to late-appearing cognitive dysfunction that becomes manifest weeks after the initial injury (Fineman et al., 2000; Pullela et al., 2006). As discussed above, fluid percussion injury in juvenile rats leads to PTE weeks or months after injury (D'Ambrosio et al., 2004), and a recent study showed that controlled cortical impact injury to pre-weanling rats results in reduced minimal clonic seizure thresholds measurable in adulthood, but not when tested during adolescence (Statler et al., 2008). In another study, FPI injury in P19 rats followed by PC-induced status epilepticus appeared to protect against acute seizure-induced injury, but this early neuroprotection did not result in any long-term benefits (as measured by spontaneous seizures, mossy fiber sprouting, pentylenetetrazole threshold, and hippocampal neuronal death) (Table 12.1) (Gurkoff et al., 2009).

The appropriate treatments for developmental TBI-induced seizures may also be different from PTS in adults. Using a lithium–pilocarpine status epilepticus model at P15 and P28, Suchomelova et al. (2006) were able to dissociate anticonvulsant and antiepileptogenic effects at different developmental stages. In this study, acute treatment with topiramate had little effect on acute status duration in P15 rats; however, it almost completely prevented the later development of spontaneous recurrent seizures, even if the treatment were delayed 70 minutes after the onset of the initial status epilepticus. In P28 animals, topiramate had better anticonvulsant effects but much less in the way of an antiepileptogenic effect. Diazepam was effective at stopping early seizures and, if administered early (<20 minutes) after onset of status, was also associated with reduced later epilepsy (Suchomelova et al., 2006). In the next section, we will also discuss some of the potential developmental toxicities of many standard anesthetics and anticonvulsants that would argue for judicious, rather than indiscriminate, use of these agents in EPTS prophylaxis (Bittigau et al., 2002; Kaindl et al., 2006).

## GOOD VS. BAD PLASTICITY AND FUNCTIONAL RECOVERY

One of the major challenges to controlling epileptogenesis is that many of the general molecular responses that appear to underlie functional plasticity and sprouting may be similar or identical to those that result in aberrant connectivity and PTE. Thus, interventions that may mitigate epileptogenesis may also have deleterious effects on normal recovery.

Increased expression of neurotrophins such as BDNF has been implicated in enhanced function, in both non-injury and injury settings. Naïve or control animals that upregulate BDNF through voluntary exercise or other environmental enhancements may also show increased dendritic arborization and better neurocognitive performance (Griesbach et al., 2002, 2004; Ip et al., 2002; Neeper et al., 1995). In fact, when exercise-induced upregulation of BDNF is impaired after brain injury, there is a concomitant absence or even reversal of behavioral improvement (Griesbach et al., 2004). Blockade of BDNF pathways using selective inhibitors can block downstream molecular responses and diminish the cognitive enhancements often associated with increased BDNF expression (Vaynman et al., 2004). However, excessive BDNF may actually be associated with seizures or epilepsy, and, in other models, blockade of BDNF pathways can serve as antiepileptogenic therapy (Koyama et al., 2004; Roberts et al., 2006).

Treatments to prevent EPTSs typically involve agents that block excitatory neurotransmission or promote inhibition. Although EPTSs have been associated with a higher risk of developing PTE, effective prophylaxis of EPTSs has not yet been shown to reduce this risk (Temkin et al., 1990, 1999). In fact, numerous preclinical studies, particularly in immature rodents, suggest that treatment with anticonvulsants in therapeutic doses results in substantially increased apoptosis in both naïve and injured animals (Bittigau et al., 2002; Kaindl et al., 2006) and that these treatments result in later electrophysiological and behavioral impairments (Jevtovic-Todorovic et al., 2003).

Direct comparisons of anticonvulsant or sedative-induced developmental cell death or cognitive impairments are difficult to obtain clinically; however, at least two important clinical studies provide supportive data. When children with febrile convulsions were studied after receiving either phenobarbital or no preventive therapy, those receiving phenobarbital demonstrated a significantly lower IQ on cognitive testing (Farwell et al., 1990). This reduction improved after being taken off of phenobarbital. More recently, review of a population-based birth cohort showed that exposure to anesthetics before age 4 was associated with an increased risk of being diagnosed with a learning disability (Wilder et al., 2009). As mentioned above, these studies, in conjunction with the substantial preclinical literature, warrant clear indications and careful selection when using anticonvulsants in infants and children.

## RECURRENT MILD INJURY

One clinical scenario with theoretical relevance to this discussion is the setting of mild recurrent TBI, as occurs in both contact sports and also in military activities. There are no definitive series demonstrating that recurrent sports concussions predispose to the development of PTE, although a recent, large population-based study showed an increased risk of epilepsy after even mild TBI (Christensen et al., 2009). Most published opinions support the judicious participation of those with epilepsy in athletic endeavors (Fountain and May, 2003), and there are no peer-reviewed studies that have explicitly examined the risk of late epilepsy after recurrent mild TBI. This is a challenging area for future investigation, and the timing and recovery from each individual mild TBI may be important variables that increase or diminish underlying physiological processes relevant to epileptogenesis. Certainly, deleterious changes in many other neurological variables (cognitive functions, time of onset of dementia, and behavioral disturbances) are being associated with an accumulation of mild concussions (Collins et al., 1999; Guskiewicz et al., 2005; Matser et al., 1999).

Despite the relative lack of evidence for a cumulative effect of concussions on the development of PTE, there is a growing concern of this potential risk in the setting of military TBI. The past literature of military TBI shows significant risks for PTE, but this has generally been after severe, and usually penetrating, TBI (Caveness et al., 1962; Salazar et al., 1985; Walker and Blumer, 1989). Modern military TBI is more commonly associated with diffuse concussive injury (Warden, 2006), due in part to better head protective gear and to the increased use of improvised explosive devices. There is lack of consensus on the best way to document mild TBI occurring in the setting of military deployment (Hoge et al., 2008). Nonetheless, the seizure-related sequelae of military TBI remain an area of important focus in the future and may shed additional light onto the mechanisms and risks for development of PTE.

## CONCLUSIONS

Seizures associated with traumatic brain injury may be classified into three categories, which differ in terms of pathophysiology, clinical characteristics, and outcome. Impact seizures or immediate posttraumatic seizures (IPTSs) generally occur very early after biomechanical injury. The mechanisms underlying this entity have not been elucidated in detail, but they are rapidly reversible and do not generally portend any long-term impairments. Early posttraumatic seizures (EPTSs) are provoked seizures occurring within the first week post-injury in response to primary brain injury mechanisms (e.g., glutamate release, ionic disequilibrium, metabolic crisis, intracranial bleeding, cell death, oxidative stress) or secondary injuries (e.g., hypoxia, hypotension, cerebral edema, infection). These seizures are related to higher mortality, an increased risk for late epilepsy, and worse global outcomes. Late posttraumatic seizures (LPTSs) are spontaneous seizures that generally occur after 7 days, and recurrent LPTSs constitute posttraumatic epilepsy (PTE). Mechanisms underlying PTE are related to excitatory–inhibitory imbalance, which can be mediated through selective death or dysfunction of inhibitory neurons, changes in neurotransmission, alterations of cellular ion channels, inflammation and glial reaction, and aberrant sprouting and altered neuronal connectivity. Based upon these distinctions, therapies proven effective for prophylaxis of EPTS do not mitigate the risk of later development of PTE, and a better understanding of the mechanisms of epileptogenesis is needed to identify novel targets to prevent PTE.

The process of becoming epileptic, namely epileptogenesis, occurs in three stages: acute injury, latent period, and recurrent spontaneous seizures. Current anticonvulsants may be effective for PTE once it develops, but it may be possible to intervene earlier, during the latent period, for patients in high-risk groups. This would require better understanding of the processes active in the early injury and latent periods and likely involve neuroplastic mechanisms that may be in common with normal recovery of function. Thus, effective antiepileptogenic interventions would have to be temporally targeted as well as mechanistically selective to minimize interference with "good plasticity."

Finally, within individuals, many factors interact to determine the response to injury, likelihood of seizures, and potential for maximal long-term outcome. Injury factors play an important role, such as the severity of injury and the presence of bleeding or secondary damage; however, a host of other variables may also be important, including genetic predisposition, age at injury, post-injury environment and plasticity, and injury recurrence.

## ACKNOWLEDGMENTS

Special thanks to Rick Staba and Jason Lerner, for their timely review and comments on this manuscript, and to Raman Sankar, Dan Arndt, and David Hovda, for countless discussions on the topic. The author would also like to acknowledge support from NIH NS057420, NS027544, NS02197, the Child Neurology Foundation/Winokur Family Foundation, the Thrasher Research Foundation, and the UCLA Brain Injury Research Center

## REFERENCES

Aarabi, B. (1990). Surgical outcome in 435 patients who sustained missile head wounds during the Iran–Iraq War. *Neurosurgery*, 27:692–695.

Adelson, P.D., B. Clyde, P.M. Kochanek, S.R. Wisniewski, D.W. Marion, H. Yonas. (1997). Cerebrovascular response in infants and young children following severe traumatic brain injury: a preliminary report. *Pediatr. Neurosurg.*, 26:200–207.

Adelson, P.D., L.W. Jenkins, R.L. Hamilton, P. Robichaud, M.P. Tran, P.M. Kochanek. (2001). Histopathologic response of the immature rat to diffuse traumatic brain injury. *J. Neurotrauma*, 18:967–976.

Annegers, J.F., J.D. Grabow, R.V. Groover, E.R. Laws, Jr., L.R. Elveback, L.T. Kurland. (1980). Seizures after head trauma: a population study. *Neurology*, 30:683–689.

Annegers, J.F., W.A. Hauser, S.P. Coan, W.A. Rocca. (1998). A population-based study of seizures after traumatic brain injuries. *N. Engl. J. Med.*, 338:20–24.

Armitage, L.L., P. Mohapel, E.M. Jenkins, D.K. Hannesson, M.E. Corcoran. (1998). Dissociation between mossy fiber sprouting and rapid kindling with low-frequency stimulation of the amygdala. *Brain Res.*, 781:37–44.

Auvin, S., D. Shin, A. Mazarati, J. Nakagawa, J. Miyamoto, R. Sankar. (2007). Inflammation exacerbates seizure-induced injury in the immature brain. *Epilepsia*, 48(Suppl. 5):27–34.

Beer, R., G. Franz, A. Srinivasan, R.L. Hayes, B.R. Pike, J.K. Newcomb, X. Zhao, E. Schmutzhard, W. Poewe, A. Kampfl. (2000). Temporal profile and cell subtype distribution of activated caspase-3 following experimental traumatic brain injury. *J. Neurochem.*, 75:1264–1273.

Beghi, E. (2003). Overview of studies to prevent posttraumatic epilepsy. *Epilepsia*, 44(Suppl. 10):21–26.

Ben-Ari, Y., R. Khazipov, X. Leinekugel, O. Caillard, J.L. Gaiarsa. (1997). GABA$_A$, NMDA and AMPA receptors: a developmentally regulated 'menage a trois.' *Trends Neurosci.*, 20:523–529.

Bergsneider, M., D.A. Hovda, E. Shalmon, D.F. Kelly, P.M. Vespa, N.A. Martin, M.E. Phelps, D.L. McArthur, M.J. Caron, J.F. Kraus, D.P. Becker. (1997). Cerebral hyperglycolysis following severe traumatic brain injury in humans: a positron emission tomography study. *J. Neurosurg.*, 86:241–251.

Bittigau, P., M. Sifringer, D. Pohl, D. Stadthaus, M. Ishimaru, H. Shimizu, M. Ikeda, D. Lang, A. Speer, J.W. Olney, C. Ikonomidou. (1999). Apoptotic neurodegeneration following trauma is markedly enhanced in the immature brain. *Ann. Neurol.*, 45:724–735.

Bittigau, P., M. Sifringer, K. Genz, E. Reith, D. Pospischil, S. Govindarajalu, M. Dzietko, S. Pesditschek, I. Mai, K. Dikranian, J.W. Olney, C. Ikonomidou. (2002). Antiepileptic drugs and apoptotic neurodegeneration in the developing brain. *Proc. Natl. Acad. Sci. U.S.A.*, 99:15089–15094.

Bullock, R., A. Zauner, J.J. Woodward, J. Myseros, S.C. Choi, J.D. Ward, A. Marmarou, H.F. Young. (1998). Factors affecting excitatory amino acid release following severe human head injury. *J. Neurosurg.*, 89:507–518.

Cavazos, J.E., G. Golarai, T.P. Sutula. (1991). Mossy fiber synaptic reorganization induced by kindling: time course of development, progression, and permanence. *J. Neurosci.*, 11:2795–2803.

Caveness, W.F., A.E. Walker, P.B. Ascroft. (1962). Incidence of posttraumatic epilepsy in Korean veterans as compared with those from World War I and World War II. *J. Neurosurg.*, 19:122–129.

CDC. (1999). *Traumatic Brain Injury in the United States: A Report to Congress*. Atlanta, GA: Centers for Disease Control and Prevention.

Chan, Y.C., J.M. Burgunder, E. Wilder-Smith, S.E. Chew, K.M. Lam-Mok-Sing, V. Sharma, B.K. Ong. (2008). Electroencephalographic changes and seizures in familial hemiplegic migraine patients with the *CACNA1A* gene S218L mutation. *J. Clin. Neurosci.*, 15:891–894.

Chang, E.F., C.P. Claus, H.J. Vreman, R.J. Wong, L.J. Noble-Haeusslein. (2005). Heme regulation in traumatic brain injury: relevance to the adult and developing brain. *J. Cereb. Blood Flow Metab.*, 25:1401–1417.

Chen, K., A. Neu, A.L. Howard, C. Foldy, J. Echegoyen, L. Hilgenberg, M. Smith, K. Mackie, I. Soltesz. (2007). Prevention of plasticity of endocannabinoid signaling inhibits persistent limbic hyperexcitability caused by developmental seizures. *J. Neurosci.*, 27:46–58.

Chesnut, R.M., L.F. Marshall, M.R. Klauber, B.A. Blunt, N. Baldwin, H.M. Eisenberg, J.A. Jane, A. Marmarou, M.A. Foulkes. (1993). The role of secondary brain injury in determining outcome from severe head injury. *J. Trauma*, 34:216–222.

Chiaretti, A., R. De Benedictis, G. Polidori, M. Piastra, A. Iannelli, C. Di Rocco. (2000). Early post-traumatic seizures in children with head injury. *Childs Nerv. Syst.*, 16:862–866.

Christensen, J., M.G. Pedersen, C.B. Pedersen, P. Sidenius, J. Olsen, M. Vestergaard. (2009). Long-term risk of epilepsy after traumatic brain injury in children and young adults: a population-based cohort study. *Lancet*, 373:1105–1110.

Claassen, J., S.A. Mayer, R.G. Kowalski, R.G. Emerson, L.J. Hirsch. (2004). Detection of electrographic seizures with continuous EEG monitoring in critically ill patients. *Neurology*, 62:1743–1748.

Collins, M.W., S.H. Grindel, M.R. Lovell, D.E. Dede, D.J. Moser, B.R. Phalin, S. Nogle, M. Wasik, D. Cordry, K.M. Daugherty, S.F. Sears, G. Nicolette, P. Indelicato, D.B. McKeag. (1999). Relationship between concussion and neuropsychological performance in college football players. *JAMA*, 282:964–970.

Corkin, S., E.V. Sullivan, F.A. Carr. (1984). Prognostic factors for life expectancy after penetrating head injury. *Arch. Neurol.*, 41:975–977.

Cull-Candy, S., S. Brickley, M. Farrant. (2001). NMDA receptor subunits: diversity, development and disease. *Curr. Opin. Neurobiol.*, 11:327–335.

D'Alessandro, R., R. Ferrara, G. Benassi, P.L. Lenzi, L. Sabattini. (1988). Computed tomographic scans in posttraumatic epilepsy. *Arch. Neurol.*, 45:42–43.

D'Ambrosio, R., D.O. Maris, M.S. Grady, H.R. Winn, D. Janigro. (1998). Selective loss of hippocampal long-term potentiation, but not depression, following fluid percussion injury. *Brain Res.*, 786:64–79.

D'Ambrosio, R., D.O. Maris, M.S. Grady, H.R. Winn, D. Janigro. (1999). Impaired K(+) homeostasis and altered electrophysiological properties of post-traumatic hippocampal glia. *J. Neurosci.*, 19:8152–8162.

D'Ambrosio, R., J.P. Fairbanks, J.S. Fender, D.E. Born, D.L. Doyle, J.W. Miller. (2004). Post-traumatic epilepsy following fluid percussion injury in the rat. *Brain*, 127:304–314.

D'Ambrosio, R., J.S. Fender, J.P. Fairbanks, E.A. Simon, D.E. Born, D.L. Doyle, J.W. Miller. (2005). Progression from frontal–parietal to mesial–temporal epilepsy after fluid percussion injury in the rat. *Brain*, 128:174–188.

D'Ambrosio, R., S. Hakimian, T. Stewart, D.R. Verley, J.S. Fender, C.L. Eastman, A.H. Sheerin, P. Gupta, R. Diaz-Arrastia, J. Ojemann, J.W. Miller. (2009). Functional definition of seizure provides new insight into post-traumatic epileptogenesis. *Brain*, 132:2805–2821.

DeRidder, M.N., M.J. Simon, R. Siman, Y.P. Auberson, R. Raghupathi, D.F. Meaney. (2006). Traumatic mechanical injury to the hippocampus *in vitro* causes regional caspase-3 and calpain activation that is influenced by NMDA receptor subunit composition. *Neurobiol. Dis.*, 22:165–176.

Diaz-Arrastia, R., M.A. Agostini, A.B. Frol, B. Mickey, J. Fleckenstein, E. Bigio, P.C. Van Ness. (2000). Neurophysiologic and neuroradiologic features of intractable epilepsy after traumatic brain injury in adults. *Arch. Neurol.*, 57:1611–1616.

Diaz-Arrastia, R., Y. Gong, S. Fair, K.D. Scott, M.C. Garcia, M.C. Carlile, M.A. Agostini, P.C. Van Ness. (2003). Increased risk of late posttraumatic seizures associated with inheritance of APOE epsilon4 allele. *Arch. Neurol.*, 60:818–822.

Echegoyen, J., C. Armstrong, R.J. Morgan, I. Soltesz. (2009). Single application of a CB1 receptor antagonist rapidly following head injury prevents long-term hyperexcitability in a rat model. *Epilepsy Res.*, 85:123–127.

Egan, M.F., M. Kojima, J.H. Callicott, T.E. Goldberg, B.S. Kolachana et al. (2003). The BDNF val66met polymorphism affects activity-dependent secretion of BDNF and human memory and hippocampal function. *Cell*, 112(2):257–269.

Emptage, N.J., C.A. Reid, A. Fine. (2001). Calcium stores in hippocampal synaptic boutons mediate short-term plasticity, store-operated $Ca^{2+}$ entry, and spontaneous transmitter release. *Neuron*, 29:197–208.

Ewing-Cobbs, L., L. Kramer, M. Prasad, D.N. Canales, P.T. Louis, J.M. Fletcher, H. Vollero, S.H. Landry, K. Cheung. (1998). Neuroimaging, physical, and developmental findings after inflicted and noninflicted traumatic brain injury in young children. *Pediatrics*, 102:300–307.

Fabricius, M., S. Fuhr, L. Willumsen, J.P. Dreier, R. Bhatia, M.G. Boutelle, J.A. Hartings, R. Bullock, A.J. Strong, M. Lauritzen. (2008). Association of seizures with cortical spreading depression and peri-infarct depolarisations in the acutely injured human brain. *Clin. Neurophysiol.*, 119:1973–1984.

Faden, A.I., P. Demediuk, S.S. Panter, R. Vink. (1989). The role of excitatory amino acids and NMDA receptors in traumatic brain injury. *Science*, 244:798–800.

Fan, P., T. Yamauchi, L.J. Noble, D.M. Ferriero. (2003). Age-dependent differences in glutathione peroxidase activity after traumatic brain injury. *J. Neurotrauma*, 20:437–445.

Farwell, J.R., Y.J. Lee, D.G. Hirtz, S.I. Sulzbacher, J.H. Ellenberg, K.B. Nelson. (1990). Phenobarbital for febrile seizures: effects on intelligence and on seizure recurrence. *N. Engl. J. Med.*, 322:364–369.

Fineman, I., D.A. Hovda, M. Smith, A. Yoshino, D.P. Becker. (1993). Concussive brain injury is associated with a prolonged accumulation of calcium: a 45Ca autoradiographic study. *Brain Res.*, 624:94–102.

Fineman, I., C.C. Giza, B.V. Nahed, S.M. Lee, D.A. Hovda. (2000). Inhibition of neocortical plasticity during development by a moderate concussive brain injury. *J. Neurotrauma*, 17:739–749.

Flint, A.C., U.S. Maisch, J.H. Weishaupt, A.R. Kriegstein, H. Monyer. (1997). NR2A subunit expression shortens NMDA receptor synaptic currents in developing neocortex. *J. Neurosci.*, 17:2469–2476.

Floyd, C.L., B.G. Lyeth. (2007). Astroglia: important mediators of traumatic brain injury. *Prog. Brain Res.*, 161:61–79.

Fountain, N.B., A.C. May. (2003). Epilepsy and athletics. *Clin. Sports Med.*, 22:605–616, x–xi.

Frey, L.C. (2003). Epidemiology of posttraumatic epilepsy: a critical review. *Epilepsia*, 44(Suppl. 10):11–17.

Fukuda, K., J.D. Richmon, M. Sato, F.R. Sharp, S.S. Panter, L.J. Noble. (1996). Induction of heme oxygenase-1 (HO-1) in glia after traumatic brain injury. *Brain Res.*, 736:68–75.

Giedd, J.N. (2008). The teen brain: insights from neuroimaging. *J. Adolesc. Health*, 42:335–343.

Giza, C.C., D.A. Hovda. (2001). The neurometabolic cascade of concussion. *J. Athl. Train.*, 36:228–235.

Giza, C.C., G.S. Griesbach, D.A. Hovda. (2005). Experience-dependent behavioral plasticity is disturbed following traumatic injury to the immature brain. *Behav. Brain Res.*, 157:11–22.

Giza, C.C., N.S. Maria, D.A. Hovda. (2006). *N*-methyl-D-aspartate receptor subunit changes after traumatic injury to the developing brain. *J. Neurotrauma*, 23:950–961.

Golarai, G., A.C. Greenwood, D.M. Feeney, J.A. Connor. (2001). Physiological and structural evidence for hippocampal involvement in persistent seizure susceptibility after traumatic brain injury. *J. Neurosci.*, 21:8523–8537.

Gorter, J.A., J.J. Petrozzino, E.M. Aronica, D.M. Rosenbaum, T. Opitz, M.V. Bennett, J.A. Connor, R.S. Zukin. (1997). Global ischemia induces downregulation of Glur2 mRNA and increases AMPA receptor-mediated $Ca^{2+}$ influx in hippocampal CA1 neurons of gerbil. *J. Neurosci.*, 17:6179–6188.

Griesbach, G.S., D.A. Hovda, R. Molteni, F. Gomez-Pinilla. (2002). Alterations in BDNF and synapsin I within the occipital cortex and hippocampus after mild traumatic brain injury in the developing rat: reflections of injury-induced neuroplasticity. *J. Neurotrauma*, 19:803–814.

Griesbach, G.S., D.A. Hovda, R. Molteni, A. Wu, F. Gomez-Pinilla. (2004). Voluntary exercise following traumatic brain injury: brain-derived neurotrophic factor upregulation and recovery of function. *Neuroscience*, 125:129–139.

Griesbach, G.S., F. Gomez-Pinilla, D.A. Hovda. (2007). Time window for voluntary exercise-induced increases in hippocampal neuroplasticity molecules after traumatic brain injury is severity dependent. *J. Neurotrauma*, 24:1161–1171.

Griesemer, D., A.M. Mautes. (2007). Closed head injury causes hyperexcitability in rat hippocampal CA1 but not in CA3 pyramidal cells. *J. Neurotrauma*, 24:1823–1832.

Gurkoff, G.G., C.C. Giza, D.A. Hovda. (2006). Lateral fluid percussion injury in the developing rat causes an acute, mild behavioral dysfunction in the absence of significant cell death. *Brain Res.*, 1077:24–36.

Gurkoff, G.G., C.C. Giza, D. Shin, S. Auvin, R. Sankar, D.A. Hovda. (2009). Acute neuroprotection to pilocarpine-induced seizures is not sustained after traumatic brain injury in the developing rat. *Neuroscience*, 164(2):862–876.

Guskiewicz, K.M., S.W. Marshall, J. Bailes, M. McCrea, R.C. Cantu, C. Randolph, B.D. Jordan. (2005). Association between recurrent concussion and late-life cognitive impairment in retired professional football players. *Neurosurgery*, 57:719–726.

Hardingham, G.E. (2006). Pro-survival signalling from the NMDA receptor. *Biochem. Soc. Trans.*, 34:936–938.

Hardingham, G.E., H. Bading. (2003). The yin and yang of NMDA receptor signalling. *Trends Neurosci.*, 26:81–89.

Hardingham, G.E., Y. Fukunaga, H. Bading. (2002). Extrasynaptic NMDARs oppose synaptic NMDARs by triggering CREB shut-off and cell death pathways. *Nat. Neurosci.*, 5:405–414.

Hartings, J.A., A.J. Strong, M. Fabricius, A. Manning, R. Bhatia, J.P. Dreier, A.T. Mazzeo, F.C. Tortella, M.R. Bullock, Cooperative Study of Brain Injury Depolarizations. (2009). Spreading depressions and late secondary insults after traumatic brain injury. *J. Neurotrauma*, 26(11):1857–1866.

Hauser, W.A., J.F. Annegers, L.T. Kurland. (1993). Incidence of epilepsy and unprovoked seizures in Rochester, Minnesota: 1935–1984. *Epilepsia*, 34:453–468.

Herman, S.T. (2002). Epilepsy after brain insult: targeting epileptogenesis. *Neurology*, 59:S21–S26.

Hicks, R.R., D.H. Smith, D.H. Lowenstein, R. Saint Marie, T.K. McIntosh. (1993). Mild experimental brain injury in the rat induces cognitive deficits associated with regional neuronal loss in the hippocampus. *J. Neurotrauma*, 10:405–414.

Hicks, R.R., S. Numan, H.S. Dhillon, M.R. Prasad, K.B. Seroogy. (1997). Alterations in BDNF and NT-3 mRNAs in rat hippocampus after experimental brain trauma. *Brain Res. Mol. Brain Res.*, 48:401–406.

Hill, S.J., E. Barbarese, T.K. McIntosh. (1996). Regional heterogeneity in the response of astrocytes following traumatic brain injury in the adult rat. *J. Neuropathol. Exp. Neurol.*, 55:1221–1229.

Hill-Felberg, S.J., T.K. McIntosh, D.L. Oliver, R. Raghupathi, E. Barbarese. (1999). Concurrent loss and proliferation of astrocytes following lateral fluid percussion brain injury in the adult rat. *J. Neurosci. Res.*, 57:271–279.

Hirsch, L.J. (2004). Continuous EEG monitoring in the intensive care unit: an overview. *J. Clin. Neurophysiol.*, 21:332–340.

Hoge, C.W., D. McGurk, J.L. Thomas, A.L. Cox, C.C. Engel, C.A. Castro. (2008). Mild traumatic brain injury in U.S. soldiers returning from Iraq. *N. Engl. J. Med.*, 358:453–463.

Howard, A.L., A. Neu, R.J. Morgan, J.C. Echegoyen, I. Soltesz. (2007). Opposing modifications in intrinsic currents and synaptic inputs in post-traumatic mossy cells: evidence for single-cell homeostasis in a hyperexcitable network. *J. Neurophysiol.*, 97:2394–2409.

Huang, M., R.J. Theilmann, A. Robb, A. Angeles, S. Nichols, A. Drake, J. Dandrea, M. Levy, M. Holland, T. Song, S. Ge, E. Hwang, K. Yoo, L. Cui, D.G. Baker, D. Trauner, R. Coimbra, R.R. Lee. (2009). Integrated imaging approach with MEG and DTI to detect mild traumatic brain injury in military and civilian patients. *J. Neurotrauma*, 26(8):1213–1226.

Hubschmann, O.R., D. Kornhauser. (1983). Effects of intraparenchymal hemorrhage on extracellular cortical potassium in experimental head trauma. *J. Neurosurg.*, 59:289–293.

Hudak, A.M., K. Trivedi, C.R. Harper, K. Booker, R.R. Caesar, M. Agostini, P.C. Van Ness, R. Diaz-Arrastia. (2004). Evaluation of seizure-like episodes in survivors of moderate and severe traumatic brain injury. *J. Head Trauma Rehabil.*, 19:290–295.

Ip, E.Y., C.C. Giza, G.S. Griesbach, D.A. Hovda. (2002). Effects of enriched environment and fluid percussion injury on dendritic arborization within the cerebral cortex of the developing rat. *J. Neurotrauma*, 19:573–585.

Jenkins, L.W., K. Moszynski, B.G. Lyeth, W. Lewelt, D.S. DeWitt, A. Allen, C.E. Dixon, J.T. Povlishock, T.J. Majewski, G.L. Clifton et al. (1989). Increased vulnerability of the mildly traumatized rat brain to cerebral ischemia: the use of controlled secondary ischemia as a research tool to identify common or different mechanisms contributing to mechanical and ischemic brain injury. *Brain Res.*, 477:211–224.

Jennett, B. (1975a). *Epilepsy After Non-Missile Head Injuries*. Chicago, IL: William Heinemann.

Jennett, B. (1975b). Epilepsy and acute traumatic intracranial haematoma. *J. Neurol. Neurosurg. Psychiatry*, 38:378–381.

Jevtovic-Todorovic, V., R.E. Hartman, Y. Izumi, N.D. Benshoff, K. Dikranian, C.F. Zorumski, J.W. Olney, D.F. Wozniak. (2003). Early exposure to common anesthetic agents causes widespread neurodegeneration in the developing rat brain and persistent learning deficits. *J. Neurosci.*, 23:876–882.

Johansson, B.B., P.V. Belichenko. (2002). Neuronal plasticity and dendritic spines: effect of environmental enrichment on intact and postischemic rat brain. *J. Cereb. Blood Flow Metab.*, 22:89–96.

Kadhim, H.J., J. Duchateau, G. Sebire. (2008). Cytokines and brain injury: invited review. *J. Intensive Care Med.*, 23:236–249.

Kaindl, A.M., S. Asimiadou, D. Manthey, M.V. Hagen, L. Turski, C. Ikonomidou. (2006). Antiepileptic drugs and the developing brain. *Cell Mol. Life Sci.*, 63:399–413.

Katayama, Y., D.P. Becker, T. Tamura, D.A. Hovda. (1990). Massive increases in extracellular potassium and the indiscriminate release of glutamate following concussive brain injury. *J. Neurosurg.*, 73:889–900.

Katayama, Y., T. Maeda, M. Koshinaga, T. Kawamata, T. Tsubokawa. (1995). Role of excitatory amino acid-mediated ionic fluxes in traumatic brain injury. *Brain Pathol.*, 5:427–435.

Keenan, H.T., S.L. Bratton. (2006). Epidemiology and outcomes of pediatric traumatic brain injury. *Dev. Neurosci.*, 28:256–263.

Kelly, D.F., R.K. Kordestani, N.A. Martin, T. Nguyen, D.A. Hovda, M. Bergsneider, D.L. McArthur, D.P. Becker. (1996). Hyperemia following traumatic brain injury: relationship to intracranial hypertension and outcome. *J. Neurosurg.*, 85:762–771.

Kelly, D.F., N.A. Martin, R. Kordestani, G. Counelis, D.A. Hovda, M. Bergsneider, D.Q. McBride, E. Shalmon, D. Herman, D.P. Becker. (1997). Cerebral blood flow as a predictor of outcome following traumatic brain injury. *J. Neurosurg.*, 86:633–641.

Kharatishvili, I., J.P. Nissinen, T.K. McIntosh, A. Pitkanen. (2006). A model of posttraumatic epilepsy induced by lateral fluid percussion brain injury in rats. *Neuroscience*, 140:685–697.

Kharatishvili, I., R. Immonen, O. Grohn, A. Pitkanen. (2007). Quantitative diffusion MRI of hippocampus as a surrogate marker for post-traumatic epileptogenesis. *Brain*, 130:3155–3168.

Kieslich, M., G. Jacobi. (1995). Incidence and risk factors of post-traumatic epilepsy in childhood. *Lancet*, 345(8943):187.

Kochanek, P.M., V.A. Vagni, K.L. Janesko, C.B. Washington, P.K. Crumrine, R.H. Garman, L.W. Jenkins, R.S. Clark, G.E. Homanics, C.E. Dixon, J. Schnermann, E.K. Jackson. (2006). Adenosine A$_1$ receptor knockout mice develop lethal status epilepticus after experimental traumatic brain injury. *J. Cereb. Blood Flow Metab.*, 26:565–575.

Kors, E.E., G.M. Terwindt, F.L. Vermeulen, R.B. Fitzsimons, P.E. Jardine, P. Heywood, S. Love, A.M. van den Maagdenberg, J. Haan, R.R. Frants, M.D. Ferrari. (2001). Delayed cerebral edema and fatal coma after minor head trauma: role of the *CACNA1A* calcium channel subunit gene and relationship with familial hemiplegic migraine. *Ann. Neurol.*, 49:753–760.

Koyama, R., M.K. Yamada, S. Fujisawa, R. Katoh-Semba, Y. Matsuki, Y. Ikegaya. (2004). Brain-derived neurotrophic factor induces hyperexcitable reentrant circuits in the dentate gyrus. *J. Neurosci.*, 24:7215–7224.

Kubota, M., T. Nakamura, K. Sunami, Y. Ozawa, H. Namba, A. Yamaura, H. Makino. (1989). Changes of local cerebral glucose utilization, DC potential and extracellular potassium concentration in experimental head injury of varying severity. *Neurosurg. Rev.*, 12(Suppl. 1):393–399.

Kumar, A., L. Zou, X. Yuan, Y. Long, K. Yang. (2002). *N*-methyl-D-aspartate receptors: transient loss of NR1/NR2A/NR2B subunits after traumatic brain injury in a rodent model. *J. Neurosci. Res.*, 67:781–786.

Kumar, R., M. Husain, R.K. Gupta, K.M. Hasan, M. Haris, A.K. Agarwal, C.M. Pandey, P.A. Narayana. (2009). Serial changes in the white matter diffusion tensor imaging metrics in moderate traumatic brain injury and correlation with neuro-cognitive function. *J. Neurotrauma*, 26:481–495.

Langlois, J.A., W. Rutland-Brown, M.M. Wald. (2006). The epidemiology and impact of traumatic brain injury: a brief overview. *J. Head Trauma Rehabil.*, 21:375–378.

Lewen, A., P. Matz, P.H. Chan. (2000). Free radical pathways in CNS injury. *J. Neurotrauma*, 17:871–890.

Li, Q., I. Spigelman, D.A. Hovda, C.C. Giza. (2005). Decreased NMDA receptor mediated synaptic currents in CA1 neurons following fluid percussion injury in developing rats. *J. Neurotrauma*, 20(10):1249.

Lifshitz, J., P.G. Sullivan, D.A. Hovda, T. Wieloch, T.K. McIntosh. (2004). Mitochondrial damage and dysfunction in traumatic brain injury. *Mitochondrion*, 4:705–713.

Liu, L., T.P. Wong, M.F. Pozza, K. Lingenhoehl, Y. Wang, M. Sheng, Y.P. Auberson, Y.T. Wang. (2004). Role of NMDA receptor subtypes in governing the direction of hippocampal synaptic plasticity. *Science*, 304:1021–1024.

Liu, Y., H. Kato, N. Nakata, K. Kogure. (1992). Protection of rat hippocampus against ischemic neuronal damage by pretreatment with sublethal ischemia. *Brain Res.*, 586:121–124.

Liu, Y., T.P. Wong, M. Aarts, A. Rooyakkers, L. Liu, T.W. Lai, D.C. Wu, J. Lu, M. Tymianski, A.M. Craig, Y.T. Wang. (2007). NMDA receptor subunits have differential roles in mediating excitotoxic neuronal death both *in vitro* and *in vivo*. *J. Neurosci.*, 27:2846–2857.

Lowenstein, D.H. (2009). Epilepsy after head injury: an overview. *Epilepsia*, 50(Suppl. 2):4–9.

Lowenstein, D.H., M.J. Thomas, D.H. Smith, T.K. McIntosh. (1992). Selective vulnerability of dentate hilar neurons following traumatic brain injury: a potential mechanistic link between head trauma and disorders of the hippocampus. *J. Neurosci.*, 12:4846–4853.

Malva, J.O., A.P. Silva, R.A. Cunha. (2003). Presynaptic modulation controlling neuronal excitability and epileptogenesis: role of kainate, adenosine and neuropeptide Y receptors. *Neurochem. Res.*, 28:1501–1515.

Marks, D.A., J. Kim, D.D. Spencer, S.S. Spencer. (1995). Seizure localization and pathology following head injury in patients with uncontrolled epilepsy. *Neurology*, 45:2051–2057.

Martin, N.A., C. Doberstein, M. Alexander, R. Khanna, H. Benalcazar, G. Alsina, C. Zane, D. McBride, D. Kelly, D. Hovda et al. (1995). Posttraumatic cerebral arterial spasm. *J. Neurotrauma*, 12:897–901.

Mathern, G.W., T.L. Babb, B.G. Vickrey, M. Melendez, J.K. Pretorius. (1994). Traumatic compared to nontraumatic clinical-pathologic associations in temporal lobe epilepsy. *Epilepsy Res.*, 19:129–139.

Mathern, G.W., T.L. Babb, P.S. Mischel, H.V. Vinters, J.K. Pretorius, J.P. Leite, W.J. Peacock. (1996). Childhood generalized and mesial temporal epilepsies demonstrate different amounts and patterns of hippocampal neuron loss and mossy fibre synaptic reorganization. *Brain*, 119(Pt. 3):965–987.

Mathern, G.W., J.K. Pretorius, D. Mendoza, J.P. Leite, L. Chimelli, D.E. Born, I. Fried, J.A. Assirati, G.A. Ojemann, P.D. Adelson, L.D. Cahan, H.I. Kornblum. (1999). Hippocampal *N*-methyl-D-aspartate receptor subunit mRNA levels in temporal lobe epilepsy patients. *Ann. Neurol.*, 46(3):343–358.

Matser, E.J., A.G. Kessels, M.D. Lezak, B.D. Jordan, J. Troost. (1999). Neuropsychological impairment in amateur soccer players. *JAMA*, 282:971–973.

Matsushita, Y., H.M. Bramlett, J.W. Kuluz, O. Alonso, W.D. Dietrich. (2001). Delayed hemorrhagic hypotension exacerbates the hemodynamic and histopathologic consequences of traumatic brain injury in rats. *J. Cereb. Blood Flow Metab.*, 21:847–856.

McIntosh, T.K., K.E. Saatman, R. Raghupathi, D.I. Graham, D.H. Smith, V.M. Lee, J.Q. Trojanowski. (1998). The Dorothy Russell Memorial Lecture. The molecular and cellular sequelae of experimental traumatic brain injury: pathogenetic mechanisms. *Neuropathol. Appl. Neurobiol.*, 24:251–267.

Miyazaki, S., Y. Katayama, B.G. Lyeth, L.W. Jenkins, D.S. DeWitt, S.J. Goldberg, P.G. Newlon, R.L. Hayes. (1992). Enduring suppression of hippocampal long-term potentiation following traumatic brain injury in rat. *Brain Res.*, 585:335–339.

Mohapel, P., L.L. Armitage, T.H. Gilbert, D.K. Hannesson, G.C. Teskey, M.E. Corcoran. (2000). Mossy fiber sprouting is dissociated from kindling of generalized seizures in the guinea-pig. *NeuroReport*, 11:2897–2901.

Monory, K., F. Massa, M. Egertova, M. Eder, H. Blaudzun et al. (2006). The endocannabinoid system controls key epileptogenic circuits in the hippocampus. *Neuron*, 51:455–466.

Moody, W.J., K.J. Futamachi, D.A. Prince. (1974). Extracellular potassium activity during epileptogenesis. *Exp. Neurol.*, 42:248–263.

Muizelaar, J.P., A. Marmarou, A.A. DeSalles, J.D. Ward, R.S. Zimmerman, Z. Li, S.C. Choi, H.F. Young. (1989). Cerebral blood flow and metabolism in severely head-injured children. Part 1. Relationship with GCS score, outcome, ICP, and PVI. *J. Neurosurg.*, 71:63–71.

Myer, D.J., G.G. Gurkoff, S.M. Lee, D.A. Hovda, M.V. Sofroniew. (2006). Essential protective roles of reactive astrocytes in traumatic brain injury. *Brain*, 129:2761–2772.

Neeper, S.A., F. Gomez-Pinilla, J. Choi, C. Cotman. (1995). Exercise and brain neurotrophins. *Nature*, 373:109.

Nilsson, P., E. Ronne-Engstrom, R. Flink, U. Ungerstedt, H. Carlson, L. Hillered. (1994). Epileptic seizure activity in the acute phase following cortical impact trauma in rat. *Brain Res.*, 637:227–232.

Olivecrona, M., B. Zetterlund, M. Rodling-Wahlstrom, S. Naredi, L.O. Koskinen. (2009). Absence of electroencephalographic seizure activity in patients treated for head injury with an intracranial pressure-targeted therapy. *J. Neurosurg.*, 110:300–305.

Ong, L.C., M.K. Dhillon, B.M. Selladurai, A. Maimunah, M.S. Lye. (1996). Early post-traumatic seizures in children: clinical and radiological aspects of injury. *J. Paediatr. Child Health*, 32:173–176.

Osteen, C.L., A.H. Moore, M.L. Prins, D.A. Hovda. (2001). Age-dependency of 45calcium accumulation following lateral fluid percussion: acute and delayed patterns. *J. Neurotrauma*, 18:141–162.

Osteen, C.L., C.C. Giza, D.A. Hovda. (2004). Injury-induced alterations in *N*-methyl-D-aspartate receptor subunit composition contribute to prolonged 45calcium accumulation following lateral fluid percussion. *Neuroscience*, 128:305–322.

Parkin, M., S. Hopwood, D.A. Jones, P. Hashemi, H. Landolt, M. Fabricius, M. Lauritzen, M.G. Boutelle, A.J. Strong. (2005). Dynamic changes in brain glucose and lactate in pericontusional areas of the human cerebral cortex, monitored with rapid sampling on-line microdialysis: relationship with depolarisation-like events. *J. Cereb. Blood Flow Metab.*, 25:402–413.

Passineau, M.J., E.J. Green, W.D. Dietrich. (2001). Therapeutic effects of environmental enrichment on cognitive function and tissue integrity following severe traumatic brain injury in rats. *Exp. Neurol.*, 168:373–384.

Pellegrini-Giampietro, D.E., J.A. Gorter, M.V. Bennett, R.S. Zukin. (1997). The GluR2 (GluR-B) hypothesis: Ca(2+)-permeable AMPA receptors in neurological disorders. *Trends Neurosci.*, 20:464–470.

Perez-Pinzon, M.A., O. Alonso, S. Kraydieh, W.D. Dietrich. (1999). Induction of tolerance against traumatic brain injury by ischemic preconditioning. *NeuroReport*, 10:2951–2954.

Prins, M.L., S.M. Lee, C.L. Cheng, D.P. Becker, D.A. Hovda. (1996). Fluid percussion brain injury in the developing and adult rat: a comparative study of mortality, morphology, intracranial pressure and mean arterial blood pressure. *Brain Res. Dev. Brain Res.*, 95:272–282.

Prins, M.L., J.T. Povlishock, L.L. Phillips. (2003). The effects of combined fluid percussion traumatic brain injury and unilateral entorhinal deafferentation on the juvenile rat brain. *Brain Res. Dev. Brain Res.*, 140:93–104.

Pullela, R., J. Raber, T. Pfankuch, D.M. Ferriero, C.P. Claus, S.E. Koh, T. Yamauchi, R. Rola, J.R. Fike, L.J. Noble-Haeusslein. (2006). Traumatic injury to the immature brain results in progressive neuronal loss, hyperactivity and delayed cognitive impairments. *Dev. Neurosci.*, 28:396–409.

Quinlan, E.M., B.D. Philpot, R.L. Huganir, M.F. Bear. (1999). Rapid, experience-dependent expression of synaptic NMDA receptors in visual cortex *in vivo*. *Nat. Neurosci.*, 2:352–357.

Raghupathi, R. (2004). Cell death mechanisms following traumatic brain injury. *Brain Pathol.*, 14:215–222.

Ratan, S.K., R. Kulshreshtha, R.M. Pandey. (1999). Predictors of posttraumatic convulsions in head-injured children. *Pediatr. Neurosurg.*, 30:127–131.

Raza, M., R.E. Blair, S. Sombati, D.S. Carter, L.S. Deshpande, R.J. DeLorenzo. (2004). Evidence that injury-induced changes in hippocampal neuronal calcium dynamics during epileptogenesis cause acquired epilepsy. *Proc. Natl. Acad. Sci. U.S.A.*, 101:17522–17527.

Reeves, T.M., L.L. Phillips, J.T. Povlishock. (2005). Myelinated and unmyelinated axons of the corpus callosum differ in vulnerability and functional recovery following traumatic brain injury. *Exp. Neurol.*, 196:126–137.

Reid, C.A., S.F. Berkovic, S. Petrou. (2009). Mechanisms of human inherited epilepsies. *Prog. Neurobiol.*, 87(1):41–57.

Reinert, M., B. Hoelper, E. Doppenberg, A. Zauner, R. Bullock. (2000). Substrate delivery and ionic balance disturbance after severe human head injury. *Acta Neurochir. Suppl.*, 76:439–444.

Roberts, D.S., Y. Hu, I.V. Lund, A.R. Brooks-Kayal, S.J. Russek. (2006). Brain-derived neurotrophic factor (BDNF)-induced synthesis of early growth response factor 3 (Egr3) controls the levels of type A GABA receptor alpha 4 subunits in hippocampal neurons. *J. Biol. Chem.*, 281:29431–29435.

Robertson, C.L., L. Soane, Z.T. Siegel, G. Fiskum. (2006). The potential role of mitochondria in pediatric traumatic brain injury. *Dev. Neurosci.*, 28:432–446.

Ronne-Engstrom, E., T. Winkler. (2006). Continuous EEG monitoring in patients with traumatic brain injury reveals a high incidence of epileptiform activity. *Acta Neurol. Scand.*, 114:47–53.

Rutland-Brown, W., J.A. Langlois, K.E. Thomas, Y.L. Xi. (2006). Incidence of traumatic brain injury in the United States, 2003. *J. Head Trauma Rehabil.*, 21:544–548.

Saatman, K.E., A.C. Duhaime, R. Bullock, A.I. Maas, A. Valadka, G.T. Manley. (2008). Classification of traumatic brain injury for targeted therapies. *J. Neurotrauma.*, 25:719–738.

Sadrzadeh, S.M., Y. Saffari, J. Bozorgmehr. (2004). Haptoglobin phenotypes in epilepsy. *Clin. Chem.*, 50:1095–1097.

Salazar, A.M., B. Jabbari, S.C. Vance, J. Grafman, D. Amin, J.D. Dillon. (1985). Epilepsy after penetrating head injury. I. Clinical correlates: a report of the Vietnam Head Injury Study. *Neurology*, 35:1406–1414.

Sanders, M.J., T.J. Sick, M.A. Perez-Pinzon, W.D. Dietrich, E.J. Green. (2000). Chronic failure in the maintenance of long-term potentiation following fluid percussion injury in the rat. *Brain Res.*, 861:69–76.

Santhakumar, V., J. Voipio, K. Kaila, I. Soltesz. (2003). Post-traumatic hyperexcitability is not caused by impaired buffering of extracellular potassium. *J. Neurosci.*, 23:5865–5876.

Scharfman, H.E. (2002). Epilepsy as an example of neural plasticity. *Neuroscientist*, 8:154–173.

Schwarzbach, E., D.P. Bonislawski, G. Xiong, A.S. Cohen. (2006). Mechanisms underlying the inability to induce area CA1 LTP in the mouse after traumatic brain injury. *Hippocampus*, 16:541–550.

Sick, T.J., M.A. Perez-Pinzon, Z.Z. Feng. (1998). Impaired expression of long-term potentiation in hippocampal slices 4 and 48 h following mild fluid-percussion brain injury *in vivo*. *Brain Res.*, 785:287–292.

Singh, I.N., P.G. Sullivan, Y. Deng, L.H. Mbye, E.D. Hall. (2006). Time course of post-traumatic mitochondrial oxidative damage and dysfunction in a mouse model of focal traumatic brain injury: implications for neuroprotective therapy. *J. Cereb. Blood Flow Metab.*, 26:1407–1418.

Statler, K.D., S. Swank, T. Abildskov, E.D. Bigler, H.S. White. (2008). Traumatic brain injury during development reduces minimal clonic seizure thresholds at maturity. *Epilepsy Res.*, 80:163–170.

Strong, A.J., M. Fabricius, M.G. Boutelle, S.J. Hibbins, S.E. Hopwood, R. Jones, M.C. Parkin, M. Lauritzen. (2002). Spreading and synchronous depressions of cortical activity in acutely injured human brain. *Stroke*, 33:2738–2743.

Suchomelova, L., R.A. Baldwin, H. Kubova, K.W. Thompson, R. Sankar, C.G. Wasterlain. (2006). Treatment of experimental status epilepticus in immature rats: dissociation between anticonvulsant and antiepileptogenic effects. *Pediatr. Res.*, 59:237–243.

Sullivan, P.G., A.G. Rabchevsky, P.C. Waldmeier, J.E. Springer. (2005). Mitochondrial permeability transition in CNS trauma: cause or effect of neuronal cell death? *J. Neurosci. Res.*, 79:231–239.

Sunami, K., T. Nakamura, Y. Ozawa, M. Kubota, H. Namba, A. Yamaura. (1989). Hypermetabolic state following experimental head injury. *Neurosurg. Rev.*, 12(Suppl. 1):400–411.

Sutula, T. (2001). Secondary epileptogenesis, kindling, and intractable epilepsy: a reappraisal from the perspective of neural plasticity. *Int. Rev. Neurobiol.*, 45:355–386.

Sutula, T.P., G. Cascino, J. Cavazos, I. Parada, L. Ramirez. (1989). Mossy fiber synaptic reorganization in the epileptic human temporal lobe. *Ann. Neurol.*, 26:321–330.

Tani, H., A.E. Bandrowski, I. Parada, M. Wynn, J.R. Huguenard, D.A. Prince, R.J. Reimer. (2007). Modulation of epileptiform activity by glutamine and system A transport in a model of post-traumatic epilepsy. *Neurobiol. Dis.*, 25:230–238.

Temkin, N.R. (2003). Risk factors for posttraumatic seizures in adults. *Epilepsia*, 44(Suppl. 10):18–20.

Temkin, N.R., S.S. Dikmen, A.J. Wilensky, J. Keihm, S. Chabal, H.R. Winn. (1990). A randomized, double-blind study of phenytoin for the prevention of post-traumatic seizures., *N. Engl. J. Med.*, 323:497–502.

Temkin, N.R., S.S. Dikmen, G.D. Anderson, A.J. Wilensky, M.D. Holmes, W. Cohen, D.W. Newell, P. Nelson, A. Awan, H.R. Winn. (1999). Valproate therapy for prevention of posttraumatic seizures: a randomized trial. *J. Neurosurg.*, 91:593–600.

Toth, Z., G.S. Hollrigel, T. Gorcs, I. Soltesz. (1997). Instantaneous perturbation of dentate interneuronal networks by a pressure wave-transient delivered to the neocortex. *J. Neurosci.*, 17:8106–8117.

Vavilala, M.S., S. Muangman, N. Tontisirin, D. Fisk, C. Roscigno, P. Mitchell, C. Kirkness, J.J. Zimmerman, R. Chesnut, A.M. Lam. (2006). Impaired cerebral autoregulation and 6-month outcome in children with severe traumatic brain injury: preliminary findings. *Dev. Neurosci.*, 28:348–353.

Vaynman, S., Z. Ying, F. Gomez-Pinilla. (2004). Hippocampal BDNF mediates the efficacy of exercise on synaptic plasticity and cognition. *Eur. J. Neurosci.*, 20:2580–2590.

Vespa, P.M. (2009). Nonconvulsive seizures in sepsis: what you can't see may hurt you. *Crit. Care Med.*, 37:2132–2133.

Vespa, P.M., M. Prins, E. Ronne-Engstrom, M. Caron, E. Shalmon, D.A. Hovda, N.A. Martin, D.P. Becker. (1998). Increase in extracellular glutamate caused by reduced cerebral perfusion pressure and seizures after human traumatic brain injury: a microdialysis study. *J. Neurosurg.*, 89:971–982.

Vespa, P.M., M.R. Nuwer, V. Nenov, E. Ronne-Engstrom, D.A. Hovda, M. Bergsneider, D.F. Kelly, N.A. Martin, D.P. Becker. (1999). Increased incidence and impact of nonconvulsive and convulsive seizures after traumatic brain injury as detected by continuous electroencephalographic monitoring. *J. Neurosurg.*, 91:750–760.

Vespa, P.M., D. McArthur, K. O'Phelan, T. Glenn, M. Etchepare, D. Kelly, M. Bergsneider, N.A. Martin, D.A. Hovda. (2003). Persistently low extracellular glucose correlates with poor outcome 6 months after human traumatic brain injury despite a lack of increased lactate: a microdialysis study. *J. Cereb. Blood Flow Metab.*, 23:865–877.

Vespa, P.M., M. Bergsneider, N. Hattori, H.M. Wu, S.C. Huang, N.A. Martin, T.C. Glenn, D.L. McArthur, D.A. Hovda. (2005). Metabolic crisis without brain ischemia is common after traumatic brain injury: a combined microdialysis and positron emission tomography study. *J. Cereb. Blood Flow Metab.*, 25:763–774.

Vespa, P.M., C. Miller, D. McArthur, M. Eliseo, M. Etchepare, D. Hirt, T.C. Glenn, N. Martin, D. Hovda. (2008). Nonconvulsive electrographic seizures after traumatic brain injury result in a delayed, prolonged increase in intracranial pressure and metabolic crisis. *Crit. Care Med.*, 36(7):2218–2219.

Wagner, A.K., A.E. Kline, J. Sokoloski, R.D. Zafonte, E. Capulong, C.E. Dixon. (2002). Intervention with environmental enrichment after experimental brain trauma enhances cognitive recovery in male but not female rats. *Neurosci. Lett.*, 334:165–168.

Walker, A.E., D. Blumer. (1989). The fate of World War II veterans with posttraumatic seizures. *Arch. Neurol.*, 46:23–26.

Wallace, M.J., R.E. Blair, K.W. Falenski, B.R. Martin, R.J. DeLorenzo. (2003). The endogenous cannabinoid system regulates seizure frequency and duration in a model of temporal lobe epilepsy. *J. Pharmacol. Exp. Ther.*, 307:129–137.

Warden, D. (2006). Military TBI during the Iraq and Afghanistan wars. *J. Head Trauma Rehabil.*, 21:398–402.

Wilde, E.A., S.R. McCauley, J.V. Hunter, E.D. Bigler, Z. Chu, Z.J. Wang, G.R. Hanten, M. Troyanskaya, R. Yallampalli, X. Li, J. Chia, H.S. Levin. (2008). Diffusion tensor imaging of acute mild traumatic brain injury in adolescents. *Neurology*, 70:948–955.

Wilder, R.T., R.P. Flick, J. Sprung, S.K. Katusic, W.J. Barbaresi, C. Mickelson, S.J. Gleich, D.R. Schroeder, A.L. Weaver, D.O. Warner. (2009). Early exposure to anesthesia and learning disabilities in a population-based birth cohort. *Anesthesiology*, 110:796–804.

Willmore, L.J., G.W. Sypert, J.B. Munson. (1978a). Recurrent seizures induced by cortical iron injection: a model of posttraumatic epilepsy. *Ann. Neurol.*, 4:329–336.

Willmore, L.J., G.W. Sypert, J.V. Munson, R.W. Hurd. (1978b). Chronic focal epileptiform discharges induced by injection of iron into rat and cat cortex. *Science*, 200:1501–1503.

Yoshino, A., D.A. Hovda, T. Kawamata, Y. Katayama, D.P. Becker. (1991). Dynamic changes in local cerebral glucose utilization following cerebral conclusion in rats: evidence of a hyper- and subsequent hypometabolic state. *Brain Res.*, 561:106–119.

Zanier, E.R., S.M. Lee, P.M. Vespa, C.C. Giza, D.A. Hovda. (2003). Increased hippocampal CA3 vulnerability to low-level kainic acid following lateral fluid percussion injury. *J. Neurotrauma*, 20:409–420.

Zhao, X., A. Ahram, R.F. Berman, J.P. Muizelaar, B.G. Lyeth. (2003). Early loss of astrocytes after experimental traumatic brain injury. *Glia*, 44:140–152.

# 13 Possible Roles of Nonsynaptic Mechanisms in Synchronization of Epileptic Seizures: Potential Antiepileptic Targets?

*F. Edward Dudek, Li-Rong Shao, and John E. Rash*

## CONTENTS

## INTRODUCTION

This review describes nonsynaptic mechanisms that likely play an important role in synchronization of neuronal electrical activity during seizures and may contribute to chronic epilepsy. Both electrotonic coupling through gap junctions and electrical field effects (i.e., ephaptic interactions) via currents in the extracellular space can synchronize action potentials and are probably important in the high-frequency oscillations (i.e., "fast ripples") seen at seizure onset in humans with temporal lobe epilepsy and in animal models. Robust seizure-like activity can be generated in hippocampal slices after active chemical transmission is blocked with either low-calcium solutions or with γ-aminobutyric acid (GABA)- and glutamate-receptor antagonists. Gap junction blockers and treatments that increase the size of the extracellular space suppress or block this type of seizure-like activity. Research in this area may lead to approaches that block seizures without negatively impacting normal brain function, as compared to drugs that alter neurotransmitter systems (e.g., glutamatergic or GABAergic mechanisms). Relatively few data are available regarding the hypothesis that nonsynaptic mechanisms underlie or contribute to chronic epileptogenesis (i.e., spontaneous recurrent seizures from genetic mutations or after brain injury), and additional experiments are required to test this hypothesis.

## WHICH "NONSYNAPTIC" MECHANISMS PLAY A ROLE IN SEIZURES?

Nearly all basic research on the mechanisms of generation and propagation of experimental seizures and on the underlying process of epileptogenesis (i.e., the occurrence of spontaneous recurrent seizures) has focused on the roles of chemical synaptic transmission, specifically glutamatergic and GABAergic neurotransmission, or on voltage-gated ion channels. Experimental research on nonsynaptic mechanisms, particularly the synchronization of electrical activity during epileptic seizures, has a long and controversial history. Over 25 years ago, evidence that electrical mechanisms of neuronal interaction play an important role in synchronization during epileptiform activity was presented (Dudek et al., 1983, 1986), and similar studies on ionic mechanisms had also been published (Heinemann et al., 1986; Lux et al., 1986). The widely used expression *nonsynaptic* refers to those cellular mechanisms that are independent of active chemical synaptic transmission. Electrotonic coupling through gap junctions, which many workers would consider to be *electrical synapses* (Bennett, 1972, 1977, 1997; Bennett and Zukin, 2004), is one such mechanism. Electrical field effects (i.e., ephaptic interactions), which are mediated by current flow in the extracellular space, are another nonsynaptic mechanism. These two electrical mechanisms are distinctly different: electrotonic coupling involves a specialized membrane structure (i.e., gap junctions, intercellular channels composed of connexin proteins), whereas electrical field effects arise because of specific types of neuronal orientation and the volume of the extracellular space (i.e., tightly packed neurons arranged in parallel are susceptible to the extracellular effects of activity-induced electrical fields). Activity-dependent shifts in the intracellular and extracellular concentration of ions are a third nonsynaptic mechanism that likely contributes to epileptic seizures. The intense and synchronous electrical activity that defines an epileptic seizure involves transmembrane ionic currents, which redistribute ions such as potassium and calcium. Increases in the concentration of extracellular potassium and decreases in extracellular calcium, both of which increase membrane excitability, have a slow synchronizing effect on a neuronal network. This chapter summarizes: (1) evidence for the presence of these mechanisms in the hippocampus, (2) the hypothetical roles of these mechanisms in synchronization of the electrical activity of neurons during epileptic seizures, and (3) the reasons why these mechanisms are potential targets for clinical therapies. Nonsynaptic interactions among neocortical neurons are also likely to be important, although the lower packing density in the neocortex would be expected to reduce the importance of electrical field effects (i.e., ephaptic transmission) in the synchronization of electrical activity of neocortical pyramidal cells.

### Electrotonic Junctions

Considerable experimental support for the hypothesis that some vertebrate neurons, including those in several subcortical areas in mammals (e.g., mesenchephalic nucleus, lateral vestibular nucleus, inferior olive, suprachiasmatic nucleus, and olfactory bulb), are electrotonically coupled through gap junctions has been available for several years, and some of the evidence was obtained decades ago (for reviews, see Bennett, 1977, 1997; Bennett and Pereda, 2006; Bennett and Zukin, 2004; Connors and Long, 2004; Dudek et al., 1983, 1986). During the 1980s, several studies suggested that hippocampal pyramidal cells and dentate granule cells are also electrotonically coupled through gap junctions (MacVicar and Dudek, 1980, 1981, 1982; Schmalbruch and Jahnsen, 1981; Taylor and Dudek, 1982a).

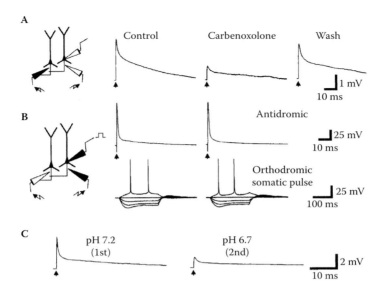

**FIGURE 13.1**    Bath application of carbenoxolone blocked partial spikes in response to antidromic stimulation of CA1 pyramidal cells. (A) Application of the gap junction blocker carbenoxolone (100 μ*M*) reversibly reduced the amplitude of the spikelets evoked through extracellular stimulation at a distant site (filled electrode). (B) Carbenoxolone did not affect the antidromically evoked action potential (i.e., direct stimulation of the axon; filled electrode) or the somatically evoked action potentials (i.e., through the recording electrode). (C) Similar data were obtained from a neuron first recorded in whole-cell configuration with the standard pipette solution (pH = 7.2) and then subsequently with another pipette with pH = 6.7. The partial spikes were greatly reduced when the pipette contained a solution with the lower pH. Subsequent recordings with two other patch pipettes as a control showed no significant alteration in the amplitude or waveform of the spikelets. (From Schmitz, D. et al., *Neuron*, 31(5), 831–840, 2001. With permission.)

Intracellular injections of low-molecular-weight tracers revealed that intracellular acidification reduces tracer coupling (MacVicar and Jahnsen, 1985), and alkalinization increases tracer coupling (Church and Baimbridge, 1991). Although this concept remains controversial, several aspects of the experimental evidence have been replicated and extended with other techniques (for reviews, see Carlen et al., 1996; Dudek et al., 1998; Jefferys, 1995). In particular, whole-cell recordings with patch pipettes in hippocampal slices have shown that antidromic stimulation triggers "spikelets" in CA1 pyramidal cells, and gap junction blockers reduce the amplitude of these events without directly altering evoked action potentials (Figure 13.1) (Schmitz et al., 2001). Furthermore, tracer injections under direct visual observation confirmed the presence of tracer coupling between CA1 pyramidal cells. The site of apparent junctional coupling appeared to be axo–axonal, which is consistent with computer modeling studies (Traub et al., 1999, 2001, 2002). More recently, dual intracellular recording experiments have demonstrated the presence of direct electrotonic coupling between neighboring pyramidal cells, which confirmed and extended in CA1 the results of MacVicar and Dudek (1981) in CA3 (Figure 13.2). Additional experiments also confirmed that action potentials in one CA1 pyramidal cell caused spikelets or fast prepotentials in a coupled pyramidal cell, which promoted the generation of synchronous action potentials in the coupled cell (Figure 13.3). Thus, a considerable body of electrophysiological data from several laboratories suggests that hippocampal pyramidal cells and dentate granule cells show homotypic coupling. In addition, homotypic coupling occurs between interneurons.

Early observations of freeze-fracture replicas of putative hippocampal pyramidal cells (Schmalbruch and Jahnsen, 1981) and dentate granule cells (MacVicar and Dudek, 1982) suggested that gap junctions were located on the somata and dendrites of these neurons, but after reanalysis of the published data—based on more specific criteria for cell-type identification—those gap junctions appear to have

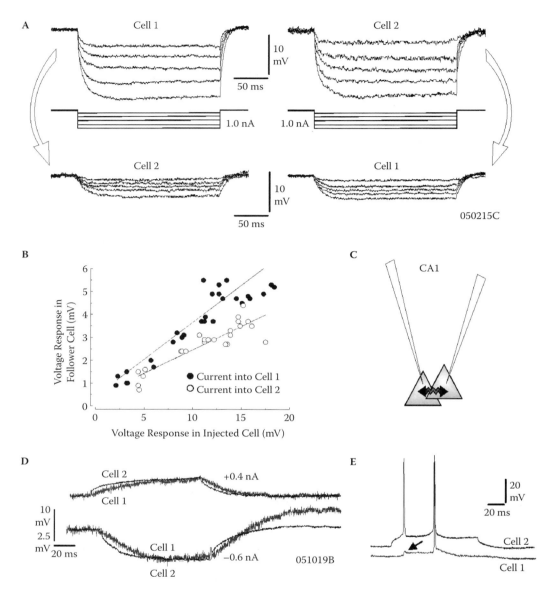

**FIGURE 13.2** Dual intracellular recordings from two pairs of electrically coupled CA1 pyramidal cells. (A) Current pulses injected into cell 1 (left) or cell 2 (right) evoked voltage responses in both cells, but the response in the follower cell was smaller. (B) Amplitude of voltage responses to current injected into the follower cell plotted against the amplitude of voltage responses in the injected cell demonstrated linear relationships with slopes of 0.32 + 0.03 (correlation = 0.88, cell 1 to cell 2) and 0.21 + 0.02 (correlation = 0.90). (C) Diagram illustrating the recording arrangement. (D) In another coupled pair, voltage responses in the follower cell (cell 1) are superimposed at a higher gain (×4) on voltage responses in the injected cell (cell 2) to illustrate the slower rate of rise of voltage in the follower cell. (E) In the same pair, a larger pulse evoked two action potentials in cell 2. The first action potential elicited a "spikelet" in cell 1. The spikelet elicited by the second action potential in cell 2 evoked an action potential in cell 1. (From Mercer, A. et al., *Brain Cell Biol.*, 35(1), 13–27, 2006. With permission.)

been on oligodendrocytes (Rash et al., 1997). More recent freeze-fracture data have confirmed the presence of neuronal gap junctions in the hippocampus, but cell identity (i.e., principal neurons vs. interneurons) has been unclear (Dudek et al., 1998). Thin-section ultrastructural studies, however,

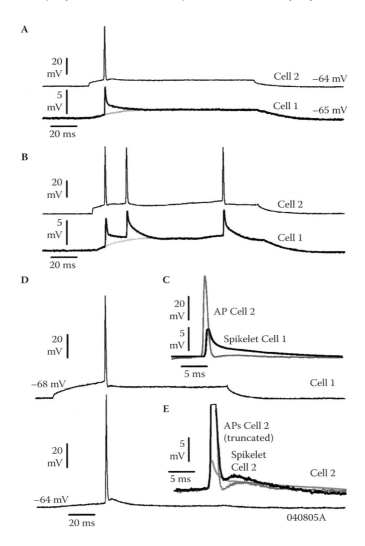

**FIGURE 13.3**  Action potentials in one CA1 pyramidal cell activate "spikelets" in a coupled follower cell which can elicit action potentials in that cell. (A, B) Suprathreshold current pulses injected into cell 2 generated action potentials that activated spikelets in cell 1. (C) The action potential recorded in cell 2 and the spikelet recorded in cell 1 were superimposed (with different gains), illustrating the delay in activation of the spikelet and its slower time course. The shape of the spikelet was obtained by subtracting from the original records the spike artifact and the underlying current-activated depolarization. (D) An action potential generated in cell 1 elicits a spikelet in cell 2 that activates an action potential. (E) The cell 2 action potential activated by a spikelet (illustrated in D), an action potential activated by direct current injection (as in A), and the spikelet in cell 2 are superimposed to illustrate the effect that the underlying spikelet with its prolonged time course has on the shape of the spike after hyperpolarization. (From Mercer, A. et al., *Brain Cell Biol.*, 35(1), 13–27, 2006. With permission.)

have clearly shown gap junctions between hippocampal interneurons (Kosaka, 1983; Kosaka and Hama, 1985). Furthermore, a recent study using thin-section electron microscopy (Figure 13.4) and freeze-fracture immunogold labeling (FRIL) techniques (Figure 13.5) has revealed gap junctions on mossy fiber axons projecting to the CA3 area (Hamzei-Sichani et al., 2007). That study also showed a gap junction on a dendritic spine projecting into the mossy fiber bundle, suggesting that it was on a CA3 pyramidal cell, as interneurons typically do not have spines, and the few that have spines

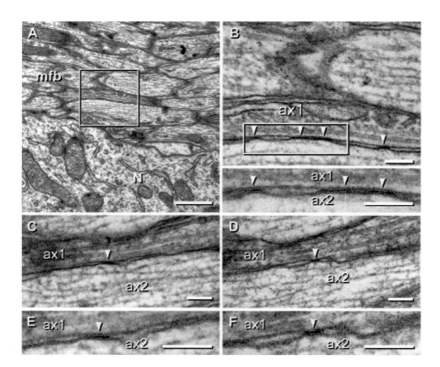

**FIGURE 13.4** Electron micrographs of six closely spaced gap junctions between two mossy fiber axons, cut longitudinally in CA3 stratum lucidum. (A) Low-magnification electron micrograph of stratum lucidum showing bundle of MF axons (mfb); N, CA3 pyramidal neuron cell body. (B) High-magnification electron micrograph of the region within the box marked in A; the arrowheads mark gap junctions between two axons, ax1 and ax2. A higher magnification electron micrograph of the three gap junctions within the box is shown as an inset. (C, D) Electron micrographs of two sections (50-nm thick) adjacent to the section shown in B, with two additional presumptive gap junctions between the same pair of axons. Arrowheads mark the site of each gap junction between the two axons, ax1 and ax2. (E, F) A higher magnification electron micrographs of the gap junctions shown in C and D. Scale bars: A, 500 nm; B–F, 100 nm. (From Hamzei-Sichani, F. et al., *Proc. Natl. Acad. Sci. U.S.A.*, 104(30), 12548–12553, 2007. With permission.)

do not project them into the mossy fiber axon bundles (Hamzei-Sichani et al., 2007) Therefore, several lines of evidence have suggested that some hippocampal pyramidal cells, granule cells, and associated interneurons show homotypic electrotonic coupling through gap junctions, but more rigorous ultrastructural data with immunogold labeling of specific connexins and identified neurons are needed. Furthermore, the relative numbers and sizes of gap junctions for different cell types in the hippocampus are essentially unknown. The available evidence, however, strongly suggests that gap junctions on principal neurons in the hippocampus should be low in number (or have a small number of connexons), so that any one pyramidal or granule cell is normally coupled to only one or a few other neurons (for reviews, see Dudek et al., 1983, 1986; Traub et al., 2002). Several lines of evidence suggest that nonsynaptic synchronization of epileptiform events (Roper et al., 1993) and electrotonic coupling through gap junctions are more prominent in the immature hippocampus and cortex (Rorig and Sutor, 1996), but the issue of developmental differences will not be considered in this chapter.

## ELECTRICAL FIELD EFFECTS

Electrical field effects or ephaptic interactions depend on the orientation and density of neurons within a population. For over half a century, electrophysiologists have known that the hippocampus generates large field potentials during synchronous activity because of the tight packing and parallel

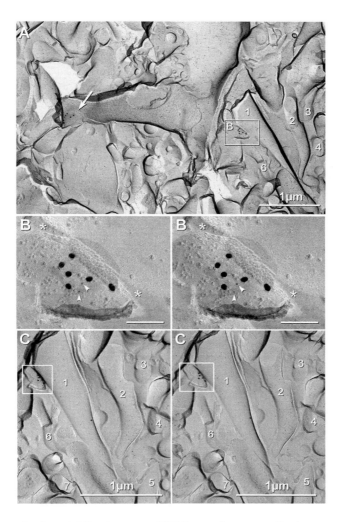

**FIGURE 13.5** **(See color insert following page 458.)** Freeze-fracture replica immunogold labeled (FRIL) electron micrographs of a small-diameter plaque gap junction on a MF axon and a dendritic gap junction in the CA3c stratum lucidum labeled with anti-Cx36 immunogold beads. (A) Low-magnification FRIL electron micrograph of an axonal gap junction (box B). Numbers 1 to 4 represent MF axons in the stratum lucidum of the CA3c field or the rat dorsal hippocampus. The white arrow points to a dendritic gap junction labeled with anti-Cx36 immunogold beads. (B) High-magnification stereoscopic electron micrographs of the plaque gap junction (red overlay), which consisted of ≈100 connexons labeled by six 18-nm gold beads and two 6-nm gold beads (arrowheads). Slightly displaced gold beads reflect the length of the double antibody bridge, slight clumping of immunogold, and displacement of lipids and proteins during SDS washing (Kamasawa et al., 2006). Asterisks mark the extracellular space, which narrows to ≈3 nm at the gap junction contact point. Axon E-face has few intramembrane particles, whereas the P-face of the unidentified but coupled cell process contains densely packed particles around the gap junction. (C) High-magnification stereoscopic electron micrographs of the bundle of MF axons (numbers 1 to 6) tilted ≈45° with respect to the plane shown in A to show a clearer view of the interior of axon 1. The area inscribed by the box contains the steeply tilted axonal gap junction. Scale bars: A, 1000 nm; B, 100 nm; C, 1000 nm. (From Hamzei-Sichani, F. et al., *Proc. Natl. Acad. Sci. U.S.A.*, 104(30), 12548–12553, 2007. With permission.)

orientation of the hippocampal pyramidal cells and dentate granule cells. The extracellular space in the hippocampus is comparatively small (McBain et al., 1990), and this leads to high extracellular resistance, which effectively channels current in the extracellular space across the membranes of hippocampal pyramidal cells and dentate granule cells (Dudek et al., 1986; Taylor and Dudek, 1984a,b). This spatial arrangement augments the amplitude of the extracellular field potentials and their associated currents. The technique of transmembrane recording (i.e., intracellular minus focal extracellular) revealed that hippocampal population spikes generate an intracellular depolarization that can be recorded in the inactive neurons (Taylor and Dudek, 1982b, 1984a,b; reviewed in Dudek et al., 1986; Jefferys, 1995). These field-effect depolarizations are roughly half of the amplitude of the population spike (i.e., a 10-mV population spike generates a 5-mV depolarization). Moreover, these electrical field effects also may act in concert with electrotonic coupling through gap junctions to synchronize the action potentials of hippocampal pyramidal cells (Taylor and Dudek, 1982b; Traub et al., 1985a,b).

## IONIC MECHANISMS

Intense synchronous neuronal activity, as occurs during electrographic seizures, is known to cause increases in the concentration of extracellular potassium and decreases in the concentration of extracellular calcium (for reviews, see Heinemann et al., 1986; Lux et al., 1986). Numerous studies with ion-selective electrodes have corroborated earlier findings that major shifts in the concentration of extracellular potassium and calcium occur during seizures. Furthermore, experiments with hippocampal slices have shown that changes in the concentration of these ions alter the susceptibility to seizure activity (see below). These ionic mechanisms would generally be expected to have a powerful but slower effect than the electrical interactions described above.

## ARE NONSYNAPTIC MECHANISMS IMPORTANT IN SEIZURES?

Pharmacological agents that block $GABA_A$ receptors generally induce epileptiform activity, and chronic alterations in GABAergic systems (e.g., loss of inhibitory GABAergic interneurons) have been proposed to contribute to epileptogenesis. Excitatory glutamatergic synaptic mechanisms also probably contribute importantly to seizure generation, and synaptic reorganization of these pathways has been hypothesized to be a central mechanism in epileptogenesis. Nonetheless, experiments from numerous laboratories have clearly shown that robust seizure-like activity can occur in hippocampal slices after active chemical synaptic transmission has been blocked pre- and postsynaptically with ionic and pharmacological treatments, respectively (for reviews, see below and Carlen et al., 1996, 2000; Dudek et al., 1998; Jefferys, 1995). These experiments, which will be summarized below, support the hypothesis that active chemical synapses are not necessary for the synchronization of neuronal activity that characterizes seizure activity and that nonsynaptic mechanisms alone can synchronize neurons during seizures under a variety of conditions.

## LOW-CALCIUM SEIZURE MODELS

### Initial Experiments in the CA1 Area

In the early 1980s, three research groups showed nearly simultaneously that solutions containing little or no extracellular calcium (which effectively blocks active chemical synaptic transmission) induce hyperexcitability and promote seizure-like activity. Taylor and Dudek (1982b) showed that low-calcium solutions containing manganese, which demonstrably blocked excitatory postsynaptic potentials (EPSPs) in CA1 pyramidal cells from Schaffer collateral stimulation, led to a progressive increase in excitability that was manifest as long, antidromically evoked afterdischarges and prolonged spontaneous bursts (Figure 13.6). Simultaneous intracellular and extracellular recordings showed that intracellular action potentials were synchronous with extracellular population spikes and that partial spikes (i.e., fast prepotentials, or spikelets) also occurred synchronously with small

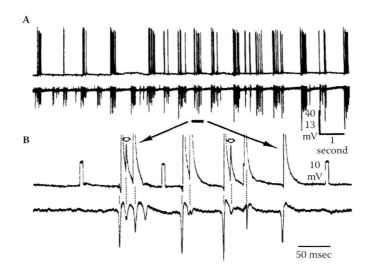

**FIGURE 13.6**   Simultaneous intracellular and extracellular recording of spontaneous bursts of synchronous action potentials in the CA1 area in a low-calcium solution containing manganese to block active chemical synaptic transmission. This solution was shown to block active chemical transmission. (A) Spontaneous bursts of action potentials recorded intracellularly (upper trace) occurred synchronously with the bursts of population spikes recorded extracellularly (lower trace). (B) Expansion of the time scale (see bar in A) showed that the action potentials and population spikes occurred synchronously (see the vertical dashed lines). Fast depolarizations that were subthreshold (see open arrows) and resembled fast prepotentials (or spikelets) were also recorded synchronously with the population spikes. (From Taylor, C.P. and Dudek, F.E., *Science*, 218(4574), 810–812, 1982. With permission.)

population spikes. Furthermore, differential transmembrane recording revealed that extracellular population spikes generated field-effect depolarizations (Taylor and Dudek, 1982b, 1984a,b). This work showed that: (1) synchronous spontaneous bursts could occur in low-calcium solutions without active chemical transmission, (2) partial spikes (suggestive of electrotonic coupling) were synchronous with population spikes, and (3) synchronization involved electrical field effects (i.e., ephaptic interactions). Jefferys and Haas (1982) showed nearly simultaneously that spontaneous bursts of population spikes could occur in low-calcium solutions that blocked chemical transmission (Haas and Jefferys, 1984). Konnerth et al. (1984, 1986) clearly showed with ion-selective electrodes that these prolonged nonsynaptic bursts were associated with an increase in the concentration of extracellular potassium and that the propagation of these events coincided with shifts in the concentration of extracellular ions.

Spontaneous release of transmitter (i.e., miniature postsynaptic potentials) occurs in low-calcium solutions. To confirm that spontaneous miniature postsynaptic potentials were not required for low-calcium bursting, experiments were conducted in a medium containing antagonists to γ-aminobutyric acid type A (GABA$_A$), $N$-methyl-D-aspartate (NMDA), and α-amino-3-hydroxy-5-methyl-4-isoxazoleproprionic acid (AMPA) receptors, such as bicuculline, 2-amino-5-phospho-nopentanoic acid (AP-5), and 6,7-dinitroquinoxaline-2,3-dione (DNQX), which alone were shown to block fast evoked chemical postsynaptic potentials in hippocampal slices (Dudek et al., 1990). The observation that epileptiform activity still occurred in the presence of the GABA$_A$-, NMDA-, and AMPA-receptor antagonists indicated that spontaneous spike-independent release of fast amino acid transmitters (i.e., miniature postsynaptic potentials) is not required for nonsynaptic bursting. These results have been replicated and extended by several other laboratories (for reviews, see, for example, Carlen et al., 1996, 2000; Dudek et al., 1998; Jefferys, 1995).

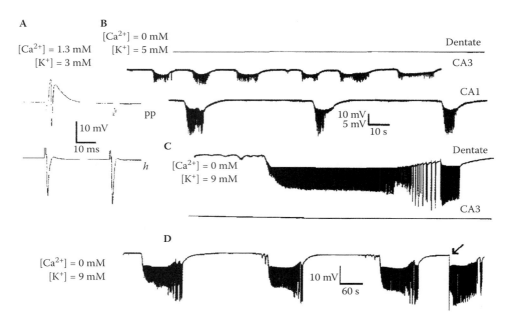

**FIGURE 13.7**   Effect of deletion of $Ca^{2+}$ from the bath on burst firing in CA3, CA1, and the dentate gyrus. (A) Average of 5 responses recorded to single stimuli delivered to the perforant path (above) or hilus (below). Left: Control conditions, 1.3-mM extracellular calcium and 3-mM extracellular potassium. Right: Same slice after 1-hr perfusion with medium containing 0-added extracellular calcium and 3-mM extracellular potassium. Note that the perforant path response was lost, whereas the hilar (antidromic) response was not significantly changed. (B) Spontaneous activity in the dentate gyrus, CA3, and CA1 areas in 0-added extracellular calcium and 5-mM extracellular potassium. Both pyramidal cell regions showed spontaneous bursts that were more prolonged than those seen in 0.5-mM extracellular calcium. The dentate gyrus showed no spontaneous bursts. Vertical calibration bar is 10 mV for the dentate and 5 mV for CA3 and CA1. (C) At 9-mM extracellular potassium with 0-added extracellular calcium, spontaneous field bursts occurred in the dentate that were more prolonged than those occurring in 0.5-mM extracellular calcium. CA3 and CA1 (not shown) no longer had spontaneous bursts. (D) Spontaneous field bursts in the dentate in 0-added extracellular calcium and 9-mM extracellular potassium. Arrow indicates a burst resulting from a single hilar stimulus. (From Schweitzer, J.S. et al., *J. Neurophysiol.*, 68(6), 2016–2025, 1992. With permission.)

## Other Hippocampal Regions

One important concern regarding the experiments described above on the CA1 area of hippocampal slices is whether neocortical regions, or even other hippocampal areas, show nonsynaptic mechanisms of synchronization during epileptiform activity. Normally, CA1 pyramidal cells are more tightly packed than other pyramidal cells in hippocampus and neocortex, and their resting potentials are relatively close to threshold; therefore, CA1 pyramidal cells would seem particularly susceptible to low-calcium solutions, which essentially lower spike threshold in addition to blocking spike-mediated release of neurotransmitters. Dentate granule cells are also tightly packed but generally have a more negative resting potential (Fricke and Prince, 1984); they were also shown to be susceptible to low-calcium solutions (Figure 13.7), particularly when extracellular potassium was elevated to higher levels (Roper et al., 1992; Schweitzer et al., 1992). Based on histological observations, CA3 pyramidal cells would be expected to be less susceptible to electrical field and ionic interactions because of their lower packing density, but CA3 pyramidal cells also show low-calcium bursting (Figure 13.7) (Schweitzer et al., 1992).

Studies using both field-potential and multiple-unit recordings have revealed that these field bursts are preceded by an increase in spontaneous action potential activity and that this phenomenon can also occur in the medial entorhinal cortex, a cortical region that is not as laminated or densely packed as the hippocampus (Patrylo et al., 1996). These experiments also showed that spontaneous multiple-unit activity occurred when CA1 pyramidal cells generated slow negative shifts in the field potential without superimposed population spikes, which are likely the result of ionic interactions. Medial entorhinal cortex, with its decreased lamination and packing density, showed slow negative shifts in the extracellular field potential with coincident multiple-unit activity but did not show population spikes. These experiments indicate that all of the major regions of the hippocampus, and even some of the parahippocampal regions, like the entorhinal cortex, demonstrate low-calcium bursting. Therefore, nonsynaptic mechanisms appear to involve: (1) ionic mechanisms that affect both membrane potential and the action potential threshold and mediate slow synchronization responsible for negative field-potential shifts, and (2) faster electrical mechanisms that generate spontaneous population spikes.

## NONSYNAPTIC SEIZURE ACTIVITY IN NORMAL CONCENTRATIONS OF EXTRACELLULAR CALCIUM

One potential criticism of the experiments described above is that the extracellular ionic concentrations are abnormal and thus may not reflect what occurs in epileptic brain. As reviewed above, however, a variety of studies have shown that seizure activity causes an increase in extracellular potassium and a decrease in the concentration of extracellular calcium (Heinemann et al., 1986; Lux et al., 1986). Therefore, these levels of extracellular ions at least partially mimic the changes that would be expected to occur during a seizure. Furthermore, other studies have used physiological calcium concentrations to evaluate the potential role of nonsynaptic mechanisms in the generation and synchronization of seizure-like activity in a more normal ionic environment. Synchronous bursts could be generated in the presence of $GABA_A$-, NMDA-, and AMPA-receptor antagonists when calcium was maintained at more physiological concentrations and if potassium was also raised to higher levels (Figure 13.8) (Patrylo et al., 1994; Schweitzer et al., 1992). Thus, prolonged bursts can occur in normal calcium concentrations if excitability is raised, even when chemical synaptic transmission is blocked pharmacologically (for additional information on this issue, see Bikson et al., 2002; Xiong and Stringer, 2001). It is important to note that, in plasma, about half of the calcium is bound to protein; therefore, the free, ionized calcium concentration is much lower than the normal concentration and is even lower in cerebrospinal fluid (CSF). This factor should be considered when defining a "normal" extracellular calcium concentration in the solution for *in vitro* slice experiments. These studies suggest that ionic and electrical mechanisms, such as electrotonic coupling through gap junctions or electrical field effects, contribute importantly to seizure generation under a wide range of ionic conditions that include relatively normal concentrations of extracellular calcium (Jensen and Yaari, 1997). Thus, these nonsynaptic mechanisms, particularly when operating together, are powerful and can synchronize hyperexcitable cortical neurons independently of chemical transmission. In normal physiological solutions with chemical transmission intact, recurrent excitation and intrinsic spike-timing mechanisms also play an important role in synchronizing action potentials and generating fast ripples and high-frequency oscillations (Dzhala and Staley, 2004).

## EXPERIMENTAL APPROACHES TO BLOCK NONSYNAPTIC MECHANISMS OF SEIZURE GENERATION AND SYNCHRONIZATION

The concept that nonsynaptic interactions between neurons may play an important role in seizure activity is consistent with suggestions that these mechanisms may represent targets for clinical intervention. At present, the pharmacological mechanisms of antiepileptic drugs (AEDs) are thought to involve primarily either voltage-gated ion channels or ligand-activated channels. Drugs that would target neuronal gap junctions or modulate the size of the extracellular space (and thus influence electrical field effects and activity-induced ion shifts) are two additional strategies that would potentially represent novel ways to control seizures. In support of this, tonabersat (SB-220453) is a novel benzoylaminobenzopyran compound with potent anticonvulsant activity in preclinical models that blocks neuronal gap junctions and is currently in clinical development (Bialer et al., 2009).

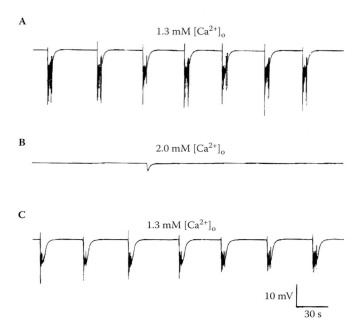

**FIGURE 13.8** Prolonged bursts of population spikes were recorded in 1.3-m$M$ extracellular calcium when extracellular potassium was raised to 12 m$M$, even when GABA$_A$, NMDA, and AMPA receptors were blocked pharmacologically. To eliminate fast chemical synaptic transmission while maintaining extracellular calcium at physiological levels, GABA$_A$, NMDA, and AMPA receptors were blocked with 30-$\mu M$ bicuculline, 50-$\mu M$ AP-5, and 50-$\mu M$ DNQX, respectively. (A) Repetitive bursts of population spikes were recorded in normal (1.3-m$M$) extracellular calcium and high potassium (12-m$M$). (B) The spontaneous bursts of population spikes were blocked by 2.0-m$M$ extracellular calcium. (C) The burst recovered when the extracellular calcium was returned to 1.3 mM. (From Patrylo, P.R. et al., *Neuroscience*, 61(1), 13–19, 1994. With permission.)

## ELECTROTONIC JUNCTIONS

The experimental approaches to evaluate whether gap junctions mediate electrotonic coupling have largely included the use of generalized gap junction blockers in brain-slice experiments; that is, the gap junction blockers would suppress coupling not only between interneurons and principal neurons but also between glial cells (e.g., astrocytes are extensively coupled) and even cells in other organs (e.g., heart). Studies aimed at providing evidence that CA1 pyramidal cells are electrically coupled by axonal gap junctions showed that antidromic stimulation evoked spikelets, or partial action potentials. Bath application of 100-$\mu M$ carbenoxolone, a relatively specific blocker of diverse types of gap junctions (Schmitz et al., 2001), reduced the amplitude of these events (Figure 13.1). Several studies using different acute seizure models in hippocampal or neocortical slices have provided evidence that gap junction blockers, including carbenoxolone, halothane, quinine, and octanol, depress or even block epileptiform activity (Gajda et al., 2005; Gigout et al., 2006a,b; He et al., 2009; Kohling et al., 2001; Margineanu and Klitgaard, 2001; Ross et al., 2000). A critical issue when using these pharmacological agents is to determine whether the gap junction blocker alters membrane excitability, which would also influence seizure susceptibility. These studies have provided different lines of evidence to suggest that excitability was unaltered and, therefore, that changes in excitability did not account for the blocking effect of these agents on epileptiform activity. Although these studies support the hypothesis that gap junctions between neurons, including both pyramidal cells and interneurons, play a role in the synchronization of electrical activity during epileptiform events, additional experiments are warranted to ensure that slight decreases in excitability do not contribute to depression of seizure activity. For example, recordings that show a loss of *synchrony during electrical activity* after application of the gap junction blocker would be stronger evidence that the gap junction blockers specifically blocked gap junctional transmission and did not elevate spike threshold. Future studies need to test more extensively the specificity of particular pharmacological agents on *neuronal* gap junctions, because an added complication for the above experiments is the extensive gap junctional coupling known to occur between glial cells, particularly astrocytes. An important conceptual difficulty with gap junction blockers as potential AEDs, in addition to the problem with affecting glial cells, is that gap junctions connect many other types of cells in a variety of tissues, including heart, liver, and intestine. Compared to these other tissues, neurons display a relative *lack* of gap junctions, even in neuronal structures where electrotonic coupling is thought to be present. Thus, treatments that block all classes of gap junctions would be expected to have widespread physiological effects, including functional disruption of critical organs such as the heart. A large body of data, however, has begun to identify which connexins are in gap junctions from different tissues, and neuronal gap junctions appear to have neuron-specific connexins (Rash et al., 2001), including Cx36 and Cx45 (Li et al., 2008; Rash et al., 2005).

Mefloquine has been suggested to be a highly specific gap junction blocker; in particular, mefloquine is an antimalarial drug that has been reported to specifically block Cx36-containing neuronal gap junctions with little or no effect on either glial or other non-neuronal gap junctions (Cruikshank et al., 2004). Mefloquine appears to suppress tremor in a harmaline-treated mouse model (Martin and Handforth, 2006), possibly blocking gap junctions in the inferior olive; however, few, if any, studies have analyzed the electrophysiological effects of mefloquine on seizure-like activity in brain slice preparations. *In vivo* studies suggest that mefloquine actually induces seizures, possibly by suppressing GABAergic transmission (Amabeoku and Farmer, 2005). Clinically, mefloquine may cause seizures in susceptible patients (Reuther et al., 2003; Wooltorton, 2002). Because most human exposure to mefloquine is for treatment of malaria, and malaria is often associated with seizures (Mishra and Newton, 2009), it is difficult to know the degree to which the mefloquine or the malaria caused the seizures in these reports on humans. Thus, these reports suggest that mefloquine is unlikely to be a specific blocker of neuronal gap junctions; furthermore, even if gap junctions contribute to synchronization of neuronal activity during seizures, mefloquine is unlikely to have potential as an antiepileptic drug. Based on the overall body of data with mefloquine, further development of neuron-specific gap junction blockers is necessary.

## ELECTRICAL AND IONIC MECHANISMS INVOLVING EXTRACELLULAR SPACE

Pharmacological mechanisms directed at electrical field effects or changes in the concentration of extracellular ions may be more difficult to implement for selective control of seizures than agents directed at gap junctions. A series of classic studies showed that seizure-like activity causes neuronal swelling that is manifest as a decrease in the size of the extracellular space (Dietzel et al., 1980; for a review, see Lux et al., 1986). Using potassium-induced electrographic seizures in the CA1 area of hippocampal slices, Traynelis and Dingledine (1989) showed that hyperosmotic solutions suppressed electrographic seizure activity, and this was associated with a decrease in extracellular resistance that arose from an increase in the volume of the extracellular space. Nonsynaptic mechanisms were considered likely to be responsible for these effects of hyperosmotic solutions. Experiments with low-calcium solutions and pharmacological treatments that blocked synaptic transmission both pre- and postsynaptically showed that hyperosmolar solutions containing membrane-impermeant solutes could block nonsynaptic seizure-like activity (Figure 13.9), whereas hyperosmolar solutions with membrane-permeant solutes did not (Dudek et al., 1990; Roper et al., 1992). Those slices that did not generate low-calcium bursts were treated with hypo-osmolar solutions, which would be expected to cause cell swelling; these hypo-osmolar solutions induced synchronous burst discharges (Figure 13.9; see also Heinemann et al., 1986). Therefore, hyper-osmolar solutions, such as mannitol, appear to depress seizures by causing cell shrinkage and an increase in

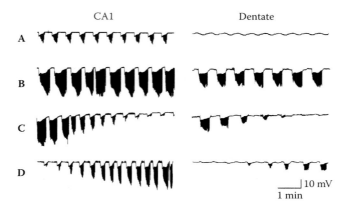

**FIGURE 13.9**   Nonsynaptic seizure-like activity in CA1 and the dentate gyrus was sensitive to changes in the osmolality of the extracellular medium. (A) Spontaneous burst discharges were present in CA1 but not the dentate gyrus in low-calcium solution. (B) The CA1 bursts increased in amplitude and duration and spontaneous bursts began to occur in the dentate gyrus after the osmolality of the extracellular medium was reduced from 308 mOsm to 254 mOsm. (C) Bursts were blocked in both the CA1 area and the dentate gyrus when mannitol (20-m$M$) was added to the extracellular fluid to increase the osmolality. (D) Bursts began to occur again when the extracellular fluid was diluted. (From Roper, S.N. et al., *Ann. Neurol.*, 31(1), 81–85, 1992. With permission.)

the volume of the extracellular space, which would be expected to reduce electrical field effects and dampen the effects of activity-induced changes in ion concentrations. Conversely, dilute media (i.e., hypo-osmolar solutions that cause cell swelling) promote seizure activity. These results support the hypothesis that nonsynaptic mechanisms involving the extracellular space, in particular electrical-field effects and shifts in the extracellular concentration of potassium, contribute to synchronization of epileptiform activity (Andrew et al., 1989). Similar to the need for studies of gap junction blockers, additional experiments are needed to demonstrate that the effects of these changes in osmolality are specifically on cell volume and the size of the extracellular space.

Other approaches have been used to test the extracellular-volume hypothesis. Hochman et al. (1995) showed that furosemide, a loop diuretic that would be expected to reduce activity-dependent increases in cell volume, depressed several different types of seizure activity *in vivo* and *in vitro*. The effect of furosemide was observed in low-calcium solutions, further suggesting that nonsynaptic mechanisms involving changes in extracellular space could hypothetically be used to manage clinical seizures. Furosemide, similar to the gap junction blockers, however, has additional mechanisms of action. Transient electrical fields can block or disrupt epileptiform activity (Bikson et al., 2001; Richardson and O'Reilley, 1995; Weinstein, 2001), and this effect could be due to a disruption of electrical field effects. Nonsynaptic epileptiform activity can cross a knife-cut lesion that would eliminate gap-junctional communication and reduce electrical field effects, which supports the hypothesis of an important role for ionic interactions in the synchronization and propagation of slow epileptiform activity (Lian et al., 2001). These experiments suggest that nonsynaptic mechanisms are potential clinical targets for future AEDs.

## ARE NONSYNAPTIC MECHANISMS ALTERED DURING EPILEPTOGENESIS?

Most hypotheses concerning the cellular mechanisms of chronic epileptogenesis, at least in regard to temporal lobe epilepsy, have focused on synaptic mechanisms, and few studies have tested hypotheses concerning nonsynaptic mechanisms of *chronic epileptogenesis* (vs. seizure generation and synchronization). This is important because one must separate the concepts of blocking seizures through a mechanism that is *not* part of the epileptogenic mechanism from one that is. If epileptogenesis were to depend, at least in part, on alterations in nonsynaptic mechanisms and if a selective therapeutic intervention strategy could be found, this might provide a way to depress epileptic seizures with minimal side effects.

## ELECTROTONIC COUPLING AND GAP JUNCTIONS

Whereas numerous studies have addressed the issue of whether gap junctions between neurons may contribute to synchronization of epileptiform events, relatively few experiments have been directed at the hypothesis that epileptogenesis is associated with an increase in neuronal gap junctions. Gap junction blockers (carbenoxolone and meclofenamic acid; all-*trans*-retinoic acid) have recently been reported to suppress seizures in animal models of chronic epilepsy (tetanus toxin and amygdala kindling models, respectively) but do not provide evidence that neuronal gap junctions are part of the epileptogenic process. However, they do suggest (given the caveats raised earlier) that neuronal gap junctions are a hypothetical mechanism of neuronal synchronization during electrographic seizures (Nilsen et al., 2006; Sayyah et al., 2007).

Colling et al. (1996), however, reported increased dye coupling between rat hippocampal CA1 pyramidal cells in the tetanus toxin model of epilepsy. Those data suggest that the increased seizure susceptibility associated with this model may be due in part to increased gap junctional coupling between CA1 pyramidal cells, which would suggest that the data of Nilsen et al. (2006) could reflect an uncoupling of new gap junctions between CA1 pyramidal neurons and other cortical neurons. Additional experiments, however, are needed to address this issue in other models of temporal lobe epilepsy and in other cortical areas.

Gap junctions among glial cells could also be altered in epileptogenesis. Gap junctions between astrocytes are thought to participate in spatial buffering of extracellular potassium and other ions. A decrease in gap junctional communication might be expected to impede this process. Lee et al. (1995) isolated astrocytes from human surgical tissue and reported an *increase* in gap junctional coupling among astrocytes, based on the propagation of intracellular calcium oscillations and fluorescence recovery after photobleaching. A subsequent analysis of functional coupling in gliomas using fluorescence recovery after photobleaching showed reduced coupling among cultured astrocytes obtained from human surgical tissue (i.e., gliomas and mesial temporal sclerosis) (Soroceanu et al., 2001). Using a freeze-lesion model of cortical dysplasia, Bordey et al. (2001) found *reduced* dye coupling among proliferative astrocytes at the site of injury and *enhanced* coupling among astrocytes in a hyperexcitable zone near the cortical injury. Finally, Xu et al. (2009) have recently reported that gap junction coupling is impaired between astrocytes in a mouse model of tuberous sclerosis complex, and this in turn is associated with altered spatial buffering of extracellular potassium. These studies suggest altered gap junctional coupling among the glia around a focal injury, in an area of mesial temporal sclerosis, or in a model of tuberous sclerosis. Additional experiments are needed to test these hypotheses. It is also unclear whether these changes are compensatory or that they contribute to chronic seizures.

Numerous studies over the last decade have suggested an increase in the expression of different connexins, particularly connexins Cx43 and Cx32, in human tissue due to surgical treatment for temporal lobe epilepsy or in animal models of this disorder. Many of these studies have suggested that this may reflect an increase in neuronal coupling, but evidence from Rash et al. (2001) using freeze-fracture immunogold labeling of specific connexins on gap junctions of identified cells in the central and peripheral nervous system indicates (at least in normal brain) that Cx43 is specific for astrocytes and Cx32 for oligodendrocytes. The data concerning these two connexins in various animal models and in tissue from human surgery are unclear on whether the gliosis in epileptic lesions is associated with normal, reduced, or enhanced expression of connexin proteins and gap junctions in glia (compare Aronica et al., 2001; Elisevich et al., 1997, 1998; Fonseca et al., 2002; Khurgel and Ivy, 1996; Li et al., 2001; Naus, 1991; Sohl et al., 2000).

Recently, McCracken and Roberts (2006) reported that a single, evoked after-discharge causes a significant *decrease* in expression of Cx36 in adult rat dorsal hippocampus, and the effect was specific for dorsal (not ventral) hippocampus and for Cx36, but not Cx26, Cx32, or Cx43. These data suggest that neuronal coupling may be a dynamic function of seizure history. A study using immunostaining of surgically resected human tissue from patients undergoing amygdalo–hippocampectomy for treatment of intractable seizures found an increase in expression of Cx43, a decrease of Cx32, and preservation of Cx36 (Collignon et al., 2006), which suggests preferential alterations in glial vs. neuronal gap junctions. Studies on human tissue, however, are difficult to perform and interpret, and additional studies on animal models with more appropriate controls may be able to address this issue in a syndrome-specific manner.

## Ionic and Electrical Field Mechanisms

Several hypothetical mechanisms could be responsible for an enhanced contribution of ionic or electrical field effects in epileptogenesis (Schwartzkroin, 1998). For example, a decrease in extracellular space, as might be expected to occur with gliosis and seizure-induced cell swelling, could increase seizure susceptibility similar to the application of dilute media. A study on resected human epileptic tissue showed that increased numbers of water-transport channels (aquaporin-4) may more effectively transport extracellular water (and $K^+$) into or out of glial cells and contribute to an altered extracellular space and seizure susceptibility (Lee et al., 2004). However, neuronal loss, as is known to occur in mesial temporal sclerosis, would reduce neuronal density and the amplitude of the field potentials, which would thus reduce electrical field effects. Also, an increase in astrocytes associated with mesial temporal sclerosis might better buffer changes in the concentration of extracellular ions associated with seizure activity, so it is difficult to interpret how altered aquaporin-4 channels would affect seizure susceptibility, and this hypothesis needs to be tested more directly. Mice lacking aquaporin-4 water channels have been reported to have increased seizure duration and slowed potassium kinetics (Binder et al., 2006; Hsu et al., 2007).

In conclusion, little or no direct evidence is presently available concerning the hypothesis that changes in ionic mechanisms or electrical field effects are *responsible for* or contribute to chronic epileptogenesis. This lack of evidence, however, should not be interpreted as evidence *against* such mechanisms contributing to chronic epileptogenesis, but only the need for directly relevant research.

## CONCLUSION

This chapter reviewed the different types of nonsynaptic mechanisms that may contribute to seizure generation and synchronization. Electrotonic coupling through gap junctions and/or electrical field effects via currents in the extracellular space appear to synchronize action potentials and be responsible for the fast oscillations (i.e., "fast ripples") seen at seizure onset in animal models of temporal lobe epilepsy and in human patients with this disorder (Bragin et al., 2002; Draguhn et al., 1998; Traub et al., 2001, 2002). Seizure-like activity after active chemical transmission is blocked in hippocampal slices (i.e., with low-calcium solutions or with normal calcium levels, but fast chemical neurotransmission blocked pharmacologically) represents a possible experimental model for testing AEDs targeted to nonsynaptic mechanisms. Several laboratories have conducted experiments aimed at blocking this and other types of seizure activity with gap junction blockers or with treatments that increase the size of the extracellular space, and the data so far suggest that either mechanism can be effective, although the specificity of these treatments is a concern. Future research in this area could identify AEDs that act specifically on nonsynaptic mechanisms and hypothetically do not impact negatively on normal brain function. Relatively few data suggest that the mechanisms of *chronic epileptogenesis* (vs. synchronization of neuronal seizures) involve nonsynaptic mechanisms, but further experiments are required to test this hypothesis.

# REFERENCES

Amabeoku, G.J., C.C. Farmer. (2005). Gamma-aminobutyric acid and mefloquine-induced seizures in mice. *Prog. Neuropsychopharmacol. Biol. Psychiatry*, 29(6):917–921.

Andrew, R.D., M. Fagan, B.A. Ballyk, A.S. Rosen. (1989). Seizure susceptibility and the osmotic state. *Brain Res.*, 498(1):175–180.

Aronica, E., J.A. Gorter, G.H. Jansen, S. Leenstra, B. Yankaya, D. Troost. (2001). Expression of connexin 43 and connexin 32 gap-junction proteins in epilepsy-associated brain tumors and in the perilesional epileptic cortex. *Acta Neuropathol.*, 101(5):449–459.

Bennett, M.V.L. (1972). A comparison of electrically and chemically mediated transmission. In *Structure and Function of Synapses* (pp. 221–256), G.D. Pappas and D.P. Purpura, Eds. New York: Raven Press.

Bennett, M.V.L. (1977). Electrical transmission: a functional analysis and comparison with chemical transmission. In *Handbook of Physiology*. Section I. *The Nervous System*. Vol. I. *Cellular Biology of Neurons*, Part I (pp. 357–416), E.R. Kandel, Ed. Bethesda, MD: American Physiological Society.

Bennett, M.V. (1997). Gap junctions as electrical synapses. *J. Neurocytol.*, 26(6):349–366.

Bennett, M.V., A. Pereda. (2006). Pyramid power: principal cells of the hippocampus unite! *Brain Cell Biol.*, 35(1):5–11.

Bennett, M.V., R.S. Zukin. (2004). Electrical coupling and neuronal synchronization in the mammalian brain. *Neuron*, 41(4):495–511.

Bialer, M., S.I. Johannessen, R.H. Levy, E. Perucca, T. Tomson, H.S. White. (2009). Progress report on new antiepileptic drugs: a summary of the Ninth Eilat Conference (EILAT IX). *Epilepsy Res.*, 83(1):1–43.

Bikson, M., J. Lian, P.J. Hahn, W.C. Stacey, C. Sciortino, D.M. Durand. (2001). Suppression of epileptiform activity by high frequency sinusoidal fields in rat hippocampal slices. *J. Physiol.*, 531(Pt. 1):181–191.

Bikson, M., S.C. Baraban, D.M. Durand. (2002). Conditions sufficient for nonsynaptic epileptogenesis in the CA1 region of hippocampal slices. *J. Neurophysiol.*, 87(1):62–71.

Binder, D.K., X. Yao, Z. Zador, T.J. Sick, A.S. Verkman, G.T. Manley. (2006). Increased seizure duration and slowed potassium kinetics in mice lacking aquaporin-4 water channels. *Glia*, 53(6):631–636.

Bordey, A., S.A. Lyons, J.J. Hablitz, H. Sontheimer. (2001). Electrophysiological characteristics of reactive astrocytes in experimental cortical dysplasia. *J. Neurophysiol.*, 85(4):1719–1731.

Bragin, A., I. Mody, C.L. Wilson, J. Engel, Jr. (2002). Local generation of fast ripples in epileptic brain. *J. Neurosci.*, 22(5):2012–2021.

Carlen, P.L., J.L. Perez-Valazquez, T.A. Valiante, S.S. Jahromi, B.K. Bardakjian. (1996). Electric coupling in epileptogenesis. In *Gap Junctions in the Nervous System* (pp. 289–299), D.C. Spray and R. Dermietzel, Eds. Austin, TX: R.G. Landes.

Carlen, P.L., F. Skinner, L. Zhang, C. Naus, M. Kushnir, J.L. Perez Velazquez. (2000). The role of gap junctions in seizures. *Brain Res. Brain Res. Rev.*, 32(1):235–241.

Church, J., K.G. Baimbridge. (1991). Exposure to high-pH medium increases the incidence and extent of dye coupling between rat hippocampal CA1 pyramidal neurons *in vitro*. *J. Neurosci.*, 11(10):3289–3295.

Collignon, F., N.M. Wetjen, A.A. Cohen-Gadol, G.D. Cascino, J. Parisi, F.B. Meyer, W.R. Marsh, P. Roche, S.D. Weigand. (2006). Altered expression of connexin subtypes in mesial temporal lobe epilepsy in humans. *J. Neurosurg.*, 105(1):77–87.

Colling, S.B., W.D. Man, A. Draguhn, J.G. Jefferys. (1996). Dendritic shrinkage and dye-coupling between rat hippocampal CA1 pyramidal cells in the tetanus toxin model of epilepsy. *Brain Res.*, 741(1–2):38–43.

Connors, B.W., M.A. Long. (2004). Electrical synapses in the mammalian brain. *Annu. Rev. Neurosci.*, 27:393–418.

Cruikshank, S.J., M. Hopperstad, M. Younger, B.W. Connors, D.C. Spray, M. Srinivas. (2004). Potent block of Cx36 and Cx50 gap junction channels by mefloquine. *Proc. Natl. Acad. Sci. U.S.A.*, 101(33):12364–12369.

Dietzel, I., U. Heinemann, G. Hofmeier, H.D. Lux. (1980). Transient changes in the size of the extracellular space in the sensorimotor cortex of cats in relation to stimulus-induced changes in potassium concentration. *Exp. Brain Res.*, 40(4):432–439.

Draguhn, A., R.D. Traub, D. Schmitz, J.G. Jefferys. (1998). Electrical coupling underlies high-frequency oscillations in the hippocampus *in vitro*. *Nature*, 394(6689):189–192.

Dudek, F.E., R.D. Andrew, B.A. MacVicar, R.W. Snow, C.P. Taylor. (1983). Recent evidence for and possible significance of gap junctions and electrotonic synapses in the mammalian brain. In *Basic Mechanisms of Neuronal Hyperexcitability* (pp. 31–73), H.H. Jasper and N.M. van Gelder, Eds. New York: Alan R. Liss.

Dudek, F.E., R.W. Snow, C.P. Taylor. (1986). Role of electrical interactions in synchronization of epileptiform bursts. *Adv. Neurol.*, 44:593–617.

Dudek, F.E., A. Obenaus, J.G. Tasker. (1990). Osmolality-induced changes in extracellular volume alter epileptiform bursts independent of chemical synapses in the rat: importance of non-synaptic mechanisms in hippocampal epileptogenesis. *Neurosci. Lett.*, 120(2):267–270.

Dudek, F.E., T. Yasumura, J.E. Rash. (1998). "Non-synaptic" mechanisms in seizures and epileptogenesis. *Cell Biol. Int.*, 22(11–12):793–805.

Dzhala, V.I., K.J. Staley. (2004). Mechanisms of fast ripples in the hippocampus. *J. Neurosci.*, 24(40):8896–8906.

Elisevich, K., S.A. Rempel, B.J. Smith, K. Edvardsen. (1997). Hippocampal connexin 43 expression in human complex partial seizure disorder. *Exp. Neurol.*, 145(1):154–164.

Elisevich, K., S.A. Rempel, B. Smith, K. Hirst. (1998). Temporal profile of connexin 43 mRNA expression in a tetanus toxin-induced seizure disorder. *Mol. Chem. Neuropathol.*, 35(1–3):23–37.

Fonseca, C.G., C.R. Green, L.F. Nicholson. (2002). Upregulation in astrocytic connexin 43 gap junction levels may exacerbate generalized seizures in mesial temporal lobe epilepsy. *Brain Res.*, 929(1):105–116.

Fricke, R.A., D.A. Prince. (1984). Electrophysiology of dentate gyrus granule cells. *J. Neurophysiol.*, 51(2):195–209.

Gajda, Z., Z. Szupera, G. Blazso, M. Szente. (2005). Quinine, a blocker of neuronal Cx36 channels, suppresses seizure activity in rat neocortex *in vivo*. *Epilepsia*, 46(10):1581–1591.

Gigout, S., J. Louvel, H. Kawasaki, M. D'Antuono, V. Armand, I. Kurcewicz, A. Olivier, J. Laschet, B. Turak, B. Devaux, R. Pumain, M. Avoli. (2006a). Effects of gap junction blockers on human neocortical synchronization. *Neurobiol. Dis.*, 22(3):496–508.

Gigout, S., J. Louvel, R. Pumain. (2006b). Effects *in vitro* and *in vivo* of a gap junction blocker on epileptiform activities in a genetic model of absence epilepsy. *Epilepsy Res.*, 69(1):15–29.

Haas, H.L., J.G. Jefferys. (1984). Low-calcium field burst discharges of CA1 pyramidal neurones in rat hippocampal slices. *J. Physiol.*, 354:185–201.

Hamzei-Sichani, F., N. Kamasawa, W.G. Janssen, T. Yasumura, K.G. Davidson, P.R. Hof, S.L. Wearne, M.G. Stewart, S.R. Young, M.A. Whittington, J.E. Rash, R.D. Traub. (2007). Gap junctions on hippocampal mossy fiber axons demonstrated by thin-section electron microscopy and freeze-fracture replica immunogold labeling. *Proc. Natl. Acad. Sci. U.S.A.*, 104(30):12548–12553.

He, J., H.L. Hsiang, C. Wu, S. Mylvagnanam, P.L. Carlen, L. Zhang. (2009). Cellular mechanisms of cobalt-induced hippocampal epileptiform discharges. *Epilepsia*, 50(1):99–115.

Heinemann, U., A. Konnerth, R. Pumain, W.J. Wadman. (1986). Extracellular calcium and potassium concentration changes in chronic epileptic brain tissue. *Adv. Neurol.*, 44:641–661.

Hochman, D.W., S.C. Baraban, J.W. Owens, P.A. Schwartzkroin. (1995). Dissociation of synchronization and excitability in furosemide blockade of epileptiform activity. *Science*, 270(5233):99–102.

Hsu, M.S., D.J. Lee, D.K. Binder. (2007). Potential role of the glial water channel aquaporin-4 in epilepsy. *Neuron Glia Biol.*, 3(4):287–297.

Jefferys, J.G. (1995). Nonsynaptic modulation of neuronal activity in the brain: electric currents and extracellular ions. *Physiol. Rev.*, 75(4):689–723.

Jefferys, J.G., H.L. Haas. (1982). Synchronized bursting of CA1 hippocampal pyramidal cells in the absence of synaptic transmission. *Nature*, 300(5891):448–450.

Jensen, M.S., Y. Yaari. (1997). Role of intrinsic burst firing, potassium accumulation, and electrical coupling in the elevated potassium model of hippocampal epilepsy. *J. Neurophysiol.*, 77(3):1224–1233.

Kamasawa, N., C.S. Furman, K.G. Davidson, J.A. Sampson, A.R. Magnie, B.R. Gebhardt, M. Kamasawa, T. Yasumura, J.R. Zumbrunnen, G.E. Pickard, J.I. Nagy, J.E. Rash. (2006). Abundance and ultrastructural diversity of neuronal gap junctions in the OFF and ON sublaminae of the inner plexiform layer of rat and mouse retina. *Neuroscience*, 142(4):1093–1117.

Khurgel, M., G.O. Ivy. (1996). Astrocytes in kindling: relevance to epileptogenesis. *Epilepsy Res.*, 26(1):163–175.

Kohling, R., S.J. Gladwell, E. Bracci, M. Vreugdenhil, J.G. Jefferys. (2001). Prolonged epileptiform bursting induced by 0-Mg(2+) in rat hippocampal slices depends on gap junctional coupling. *Neuroscience*, 105(3):579–587.

Konnerth, A., U. Heinemann, Y. Yaari. (1984). Slow transmission of neural activity in hippocampal area CA1 in absence of active chemical synapses. *Nature*, 307(5946):69–71.

Konnerth, A., U. Heinemann, Y. Yaari. (1986). Nonsynaptic epileptogenesis in the mammalian hippocampus *in vitro*. I. Development of seizurelike activity in low extracellular calcium. *J. Neurophysiol.*, 56(2):409–423.

Kosaka, T. (1983). Neuronal gap junctions in the polymorph layer of the rat dentate gyrus. *Brain Res.*, 277(2):347–351.

Kosaka, T., K. Hama. (1985). Gap junctions between non-pyramidal cell dendrites in the rat hippocampus (CA1 and CA3 regions): a combined Golgi–electron microscopy study. *J. Comp. Neurol.*, 231(2):150–161.

Lee, S.H., S. Magge, D.D. Spencer, H. Sontheimer, A.H. Cornell-Bell. (1995). Human epileptic astrocytes exhibit increased gap junction coupling. *Glia*, 15(2):195–202.

Lee, T.S., T. Eid, S. Mane, J.H. Kim, D.D. Spencer, O.P. Ottersen, N.C. de Lanerolle. (2004). Aquaporin-4 is increased in the sclerotic hippocampus in human temporal lobe epilepsy. *Acta Neuropathol.*, 108(6):493–502.

Li, J., H. Shen, C.C. Naus, L. Zhang, P.L. Carlen. (2001). Upregulation of gap junction connexin 32 with epileptiform activity in the isolated mouse hippocampus. *Neuroscience*, 105(3):589–598.

Li, X., N. Kamasawa, C. Ciolofan, C.O. Olson, S. Lu, K.G. Davidson, T. Yasumura, R. Shigemoto, J.E. Rash, J.I. Nagy. (2008). Connexin45-containing neuronal gap junctions in rodent retina also contain connexin36 in both apposing hemiplaques, forming bihomotypic gap junctions, with scaffolding contributed by zonula occludens-1. *J. Neurosci.*, 28(39):9769–9789.

Lian, J., M. Bikson, J. Shuai, D.M. Durand. (2001). Propagation of non-synaptic epileptiform activity across a lesion in rat hippocampal slices. *J. Physiol.*, 537(Pt. 1):191–199.

Lux, H.D., U. Heinemann, I. Dietzel. (1986). Ionic changes and alterations in the size of the extracellular space during epileptic activity. *Adv. Neurol.*, 44:619–639.

MacVicar, B.A., F.E. Dudek. (1980). Dye-coupling between CA3 pyramidal cells in slices of rat hippocampus. *Brain Res.*, 196(2):494–497.

MacVicar, B.A., F.E. Dudek. (1981). Electrotonic coupling between pyramidal cells: a direct demonstration in rat hippocampal slices. *Science*, 213(4509):782–785.

MacVicar, B.A., F.E. Dudek. (1982). Electrotonic coupling between granule cells of rat dentate gyrus: physiological and anatomical evidence. *J. Neurophysiol.*, 47(4):579–592.

MacVicar, B.A., H. Jahnsen. (1985). Uncoupling of CA3 pyramidal neurons by propionate. *Brain Res.*, 330(1):141–145.

Margineanu, D.G., H. Klitgaard. (2001). Can gap-junction blockade preferentially inhibit neuronal hypersynchrony vs. excitability? *Neuropharmacology*, 41(3):377–383.

Martin, F.C., A. Handforth. (2006). Carbenoxolone and mefloquine suppress tremor in the harmaline mouse model of essential tremor. *Mov. Disord.*, 21(10):1641–1649.

McBain, C.J., S.F. Traynelis, R. Dingledine. (1990). Regional variation of extracellular space in the hippocampus. *Science*, 249(4969):674–677.

McCracken, C.B., D.C. Roberts. (2006). A single evoked afterdischarge produces rapid time-dependent changes in connexin36 protein expression in adult rat dorsal hippocampus. *Neurosci. Lett.*, 405(1–2):84–88.

Mercer, A., A.P. Bannister, A.M. Thomson. (2006). Electrical coupling between pyramidal cells in adult cortical regions. *Brain Cell Biol.*, 35(1):13–27.

Mishra, S.K., C.R. Newton. (2009). Diagnosis and management of the neurological complications of falciparum malaria. *Nat. Rev. Neurol.*, 5(4):189–198.

Naus, C.C., J.F. Bechberger, D.L. Paul. (1991). Gap junction gene expression in human seizure disorder. *Exp. Neurol.*, 111(2):198–203.

Nilsen, K.E., A.R. Kelso, H.R. Cock. (2006). Antiepileptic effect of gap-junction blockers in a rat model of refractory focal cortical epilepsy. *Epilepsia*, 47(7):1169–1175.

Patrylo, P.R., J.S. Schweitzer, F.E. Dudek. (1994). Potassium-dependent prolonged field bursts in the dentate gyrus: effects of extracellular calcium and amino acid receptor antagonists. *Neuroscience*, 61(1):13–19.

Patrylo, P.R., A.J. Kuhn, J.S. Schweitzer, F.E. Dudek. (1996). Multiple-unit recordings during slow field-potential shifts in low-$[Ca^{2+}]0$ solutions in rat hippocampal and cortical slices. *Neuroscience*, 74(1):107–118.

Rash, J.E., H.S. Duffy, F.E. Dudek, B.L. Bilhartz, L.R. Whalen, T. Yasumura. (1997). Grid-mapped freeze-fracture analysis of gap junctions in gray and white matter of adult rat central nervous system, with evidence for a "panglial syncytium" that is not coupled to neurons. *J. Comp. Neurol.*, 388(2):265–292.

Rash, J.E., T. Yasumura, F.E. Dudek, J.I. Nagy. (2001). Cell-specific expression of connexins and evidence of restricted gap junctional coupling between glial cells and between neurons. *J. Neurosci.*, 21(6):1983–2000.

Rash, J.E., K.G.V. Davidson, N. Kamasawa, T. Yasumura, M. Kamasawa, C. Zhang, R. Michaels, D. Restrepo, O.P. Ottersen, C.O. Olson, J.I. Nagy. (2005). Ultrastructural localization of connexins (Cx36, Cx43, Cx45), glutamate receptors and aquaporin-4 in rodent olfactory mucosa, olfactory nerve and olfactory bulb. *J. Neurocytol.*, 34:307–341.

Reuther, L.O., S.T. Pedersen, A.M. Ronn. (2003). Drug-induced seizures. *Ugeskr. Laeger.*, 165(14):1447–1451.

Richardson, T.L., C.N. O'Reilly. (1995). Epileptiform activity in the dentate gyrus during low-calcium perfusion and exposure to transient electric fields. *J. Neurophysiol.*, 74(1):388–399.

Roper, S.N., A. Obenaus, F.E. Dudek. (1992). Osmolality and nonsynaptic epileptiform bursts in rat CA1 and dentate gyrus. *Ann. Neurol.*, 31(1):81–85.

Roper, S.N., A. Obenaus, F.E. Dudek. (1993). Increased propensity for nonsynaptic epileptiform activity in immature rat hippocampus and dentate gyrus. *J. Neurophysiol.*, 70(2):857–862.

Rorig, B., B. Sutor. (1996). Regulation of gap junction coupling in the developing neocortex. *Mol. Neurobiol.*, 12(3):225–249.

Ross, F.M., P. Gwyn, D. Spanswick, S.N. Davies. (2000). Carbenoxolone depresses spontaneous epileptiform activity in the CA1 region of rat hippocampal slices. *Neuroscience*, 100(4):789–796.

Sayyah, M., M. Rezaie, S. Haghighi, A. Amanzadeh. (2007). Intra-amygdala all-*trans* retinoic acid inhibits amygdala-kindled seizures in rats. *Epilepsy Res.*, 75(2–3):97–103.

Schmalbruch, H., H. Jahnsen. (1981). Gap junctions on CA3 pyramidal cells of guinea pig hippocampus shown by freeze-fracture. *Brain Res.*, 217(1):175–178.

Schmitz, D., S. Schuchmann, A. Fisahn, A. Draguhn, E.H. Buhl, E. Petrasch-Parwez, R. Dermietzel, U. Heinemann, R.D. Traub. (2001). Axo-axonal coupling. a novel mechanism for ultrafast neuronal communication. *Neuron*, 31(5):831–840.

Schwartzkroin, P.A., S.C. Baraban, D.W. Hochman. (1998). Osmolarity, ionic flux, and changes in brain excitability. *Epilepsy Res.*, 32(1–2):275–285.

Schweitzer, J.S., P.R. Patrylo, F.E. Dudek. (1992). Prolonged field bursts in the dentate gyrus: dependence on low calcium, high potassium, and nonsynaptic mechanisms. *J. Neurophysiol.*, 68(6):2016–2025.

Sohl, G., M. Guldenagel, H. Beck, B. Teubner, O. Traub, R. Gutierrez, U. Heinemann, K. Willecke. (2000). Expression of connexin genes in hippocampus of kainate-treated and kindled rats under conditions of experimental epilepsy. *Brain Res. Mol. Brain Res.*, 83(1–2):44–51.

Soroceanu, L., T.J. Manning, Jr., H. Sontheimer. (2001). Reduced expression of connexin-43 and functional gap junction coupling in human gliomas. *Glia*, 33(2):107–117.

Taylor, C.P., F.E. Dudek. (1982a). A physiological test for electrotonic coupling between CA1 pyramidal cells in rat hippocampal slices. *Brain Res.*, 235(2):351–357.

Taylor, C.P., F.E. Dudek. (1982b). Synchronous neural afterdischarges in rat hippocampal slices without active chemical synapses. *Science*, 218(4574):810–812.

Taylor, C.P., F.E. Dudek. (1984a). Excitation of hippocampal pyramidal cells by an electrical field effect. *J. Neurophysiol.*, 52(1):126–142.

Taylor, C.P., F.E. Dudek. (1984b). Synchronization without active chemical synapses during hippocampal afterdischarges. *J. Neurophysiol.*, 52(1):143–155.

Traub, R.D., F.E. Dudek, R.W. Snow, W.D. Knowles. (1985a). Computer simulations indicate that electrical field effects contribute to the shape of the epileptiform field potential. *Neuroscience*, 15(4):947–958.

Traub, R.D., F.E. Dudek, C.P. Taylor, W.D. Knowles. (1985b). Simulation of hippocampal afterdischarges synchronized by electrical interactions. *Neuroscience*, 14(4):1033–1038.

Traub, R.D., D. Schmitz, J.G. Jefferys, A. Draguhn. (1999). High-frequency population oscillations are predicted to occur in hippocampal pyramidal neuronal networks interconnected by axoaxonal gap junctions. *Neuroscience*, 92(2):407–426.

Traub, R.D., M.A. Whittington, E.H. Buhl, F.E. LeBeau, A. Bibbig, S. Boyd, H. Cross, T. Baldeweg. (2001). A possible role for gap junctions in generation of very fast EEG oscillations preceding the onset of, and perhaps initiating, seizures. *Epilepsia*, 42(2):153–170.

Traub, R.D., A. Draguhn, M.A. Whittington, T. Baldeweg, A. Bibbig, E.H. Buhl, D. Schmitz. (2002). Axonal gap junctions between principal neurons: a novel source of network oscillations, and perhaps epileptogenesis. *Rev. Neurosci.*, 13(1):1–30.

Traynelis, S.F., R. Dingledine. (1989). Role of extracellular space in hyperosmotic suppression of potassium-induced electrographic seizures. *J. Neurophysiol.*, 61(5):927–938.

Weinstein, S. (2001). The anticonvulsant effect of electrical fields. *Curr. Neurol. Neurosci. Rep.*, 1(2):155–161.

Wooltorton, E. (2002). Mefloquine: contraindicated in patients with mood, psychotic or seizure disorders. *CMAJ*, 167(10):1147.

Xiong, Z.Q., J.L. Stringer. (2001). Prolonged bursts occur in normal calcium in hippocampal slices after raising excitability and blocking synaptic transmission. *J. Neurophysiol.*, 86(5):2625–2628.

Xu, L., L.H. Zeng, M. Wong. (2009). Impaired astrocytic gap junction coupling and potassium buffering in a mouse model of tuberous sclerosis complex. *Neurobiol. Dis.*, 34(2):291–299.

# 14 A Mechanistic Approach to Antiepileptic Drug Interactions

*Gail D. Anderson*

## CONTENTS

# INTRODUCTION

Antiepileptic drugs (AEDs) are associated with a wide range of pharmacologic interactions. Although monotherapy with AEDs is preferred, patients with multiple seizure types or those with refractory epilepsy generally require various combinations of AEDs. AEDs are also used to treat other non-epilepsy disorders, including affective disorders, migraine, and pain. Long-term treatment is common, and concurrent treatment is often prescribed by multiple caregivers. Thus, there is a significant risk of an AED-associated drug interaction. Drug interactions can occur by pharmacokinetic or pharmacodynamic mechanisms. Pharmacodynamic interactions occur when the pharmacology of one agent alters the pharmacology or effect of the other drug without altering the plasma concentration. Theoretically, the interaction can occur at the receptor or site of action or indirectly by affecting other physiological mechanisms. For example, classic signs of carbamazepine neurotoxicity (diplopia, dizziness, ataxia) were reported when lamotrigine was added to carbamazepine therapy (Warner et al., 1992; Wolf, 1992). In addition, a case report describes four patients who experienced intolerable carbamazepine-related adverse effects when levetiracetam was added to their therapy without an alteration in carbamazepine or carbamazepine-epoxide concentrations (Sisodiya et al., 2002). However, the most commonly occurring AED interactions are pharmacokinetic interactions, where one drug alters the plasma concentrations of another. Pharmacokinetic interactions include hepatic enzyme induction and inhibition and protein-binding displacement.

# MECHANISMS OF DRUG INTERACTIONS

## HEPATIC ENZYME INDUCTION

Hepatic enzyme induction is generally the result of an increase in the amount of enzyme protein. In most cases, enzyme induction results in an increase in the rate of metabolism of the affected drug, a decrease in the serum concentration of a parent drug, and possibly a loss of clinical efficacy. If the affected drug has an active metabolite, induction can result in increased metabolite concentrations and potentially an increase in the therapeutic effect and toxicity of the drug. Enzyme induction causes major effects on a limited number of extensively metabolized drugs (>75% metabolized) with a low therapeutic index. For the drugs listed in Table 14.1, the addition or removal of an inducer could result in loss of efficacy or toxicity if plasma concentrations are not adjusted. Dosage adjustments of approximately 50 to 100%, with careful clinical monitoring, may be required.

The time required for induction depends on both the time to reach steady state of the inducing agent and the rate of synthesis of new enzymes. The amount of enzyme induction is proportional to the dose of the inducing agent. This has been shown for phenytoin (Perucca et al., 1984), phenobarbital (Perucca et al., 1984), and carbamazepine (Tomson et al., 1989). In contrast to the dose relationship, induction is not strictly additive when patients are receiving multiple inducers (Patsalos et al., 1988; Perucca et al., 1984). In a group of patients receiving both carbamazepine and phenytoin, the elimination half-life of antipyrine was determined (as a marker used to assess oxidative metabolism) before and after discontinuation of either carbamazepine or phenytoin (Patsalos et al., 1988). In patients in whom carbamazepine was discontinued, the antipyrine half-life did not change. However, when phenytoin was discontinued instead of carbamazepine, the antipyrine half-life increased. Assuming no change in the volume of distribution, this suggests that phenytoin has stronger inducing properties than carbamazepine and that induction is not additive.

Because induction is a gradual process, allowing time for gradual increases in the dose of the affected drug is required. The time course of deinduction is dependent on the rate of degradation of the enzyme and the time required to eliminate the inducing drug. For AEDs, the rate-limiting step in deinduction is generally dependent on the elimination of the inducing drug. When the inducer is removed, plasma concentrations of the affected drug will increase. Serious adverse events can occur if the dose of the affected drug is not reduced. The magnitude and timing of these interactions are critical to allow clinicians to adjust doses in such a way as to maintain therapeutic effect and avoid toxicity.

## TABLE 14.1
## Drugs for Which Addition or Discontinuation of a Hepatic Enzyme Inducer Could Cause Clinically Significant Effects

**Analgesics**
Alfentanyl
Fentanyl
Methadone
Morphine

**Antidepressant drugs**[a]
Amitriptyline
Amoxapine
Clomipramine
Desipramine
Doxepine
Imipramine
Nortriptyline
Protriptyline
Trimipramine

**Antiepileptic drugs**
Carbamazepine
Clobazam
Ethosuximide
Felbamate
Lamotrigine
Phenytoin
Tiagabine
Topiramate
Valproate
Zonisamide

**Anti-infectious agents**
Voriconazole
Itraconazole
Ketoconazole
Mebendazole

**Antipsychotic agents**
Aripiprazole
Clozapine
Haloperidol
Quetiapine
Risperidone

**Antiviral agents**
Amprenavir
Atazanavir
Darunavir
Delavirdine
Efavirenz
Indinavir
Nelfinavir
Ritonavir
Saquinavir
Tipranavir
Zidovudine

**Benzodiazepines**
Alprazolam
Clonazepam
Diazepam
Lorazepam
Midazolam
Triazolam

**Calcium-channel blockers**
Amlodipine
Belpridil
Diltiazem
Felodipine
Isradipine
Nicardipine
Nifedipine
Nimodipine
Nisodipine
Nitrendipine
Verapamil

**Cardioactive drugs**
Amiodarone
Disopyramide
Procainamide
Propranolol
Quinidine

**Corticosteroids**
Betamethasone
Cortisone
Dexamethasone
Hydrocortisone
Methylprednisolone
Prednisolone
Prednisone
Triamcinolone

**HMG-CoA reductase inhibitors (statins)**
Atorvastatin
Lovastatin
Simvastatin

**Immunosuppressants**
Cyclosporine
Sirolimus
Tacrolimus

**Oral anticoagulants**
Dicumarol
Warfarin

**Oral contraceptives**
Conjugated estrogens
Ethinyl estradiol
Levonorgestrel
Norethindrone

**Miscellaneous**
Cyclophosphamide[b]
Thiotepa[b]
Theophylline
Vincristine

[a]  Many antidepressants have active metabolites; therefore, the effect of enzyme induction on efficacy is unpredictable.

[b]  Increased exposure to active metabolite is associated with increased toxicity.

## HEPATIC ENZYME INHIBITION

Hepatic enzyme inhibition usually occurs because of competition at the enzyme site and results in a decrease in the rate of metabolism of the affected drug. Clinically, this is associated with an increased plasma concentration of the affected drug and potentially an increased pharmacologic response. The onset of the interaction is frequently rapid, and the extent of the interaction is highly variable. The initial effects of hepatic enzyme inhibition usually occur within 24 hours of addition of the inhibitor, but the time to maximal inhibition will depend on the time required to achieve steady state of both the affected drug and the inhibiting drug.

Metabolic reactions are catalyzed by the cytochrome P450 (CYP) and UDP glucuronosyltransferase (UGT) enzymes. CYP enzymes are a family of multiple enzymes; the individual isozymes comprise three major families (CYP1, CYP2, and CYP3). Seven primary isozymes are involved in the hepatic metabolism of most drugs: CYP1A2, CYP2A6, CYP2C8/9, CYP2C19, CYP2D6, CYP2E1, and CYP3A4 (Wrighton and Stevens, 1992). The most abundant isozyme, CYP3A4, which accounts for approximately 30% of the total hepatic CYP, has the broadest substrate specificity and is involved in the metabolism of more than 50% of all drugs. The UGTs are family of enzymes that catalyze the transfer of a glucuronic acid moiety from a donor cosubstrate, uridine 5′-diphosphoglucuronic acid (UDPGA), to an aglycone (Clark and Burchell, 1994). The UGT1 family members have been shown to be capable of glucuronidating a wide range of drugs, xenobiotics, and endobiotics. The UGT2 family of isoforms has long been considered to be more involved in the glucuronidation of endobiotics, including steroids and bile acids. UGT2 isoenzymes seem to favor these types of compounds as substrates but can glucuronidate other chemical types, as well.

Recent knowledge of the specific CYP and UGT isozymes involved in the metabolism of AEDs allows prediction of potential inhibition and induction interactions (Table 14.2). The extent of the drug interaction is more difficult to predict than the type of interaction. A large number of patient and drug factors will influence the extent of inhibition. The extent of inhibition is dependent on the dose of the inhibitor, as the majority of inhibition interactions are competitive. Intersubject variability in the expression of the CYP and UGT isozymes will influence the fraction of the dose associated with each metabolic pathway that is inhibited. The expression of the isozymes is dependent on both genetic and environmental influences, including concurrent diseases.

## PROTEIN-BINDING DISPLACEMENT

Drug interactions related to protein binding result from the displacement of one drug with less affinity for the protein by another drug with greater affinity. Clinically significant interactions occur only with highly protein-bound drugs (>90%). For highly protein-bound drugs that are primarily eliminated by low hepatic extraction metabolism, protein-binding displacement causes a decrease in total plasma concentrations of the displaced drug and no change in the unbound drug. Transient increases in unbound drug can be associated with acute toxicity. For most AEDs, total plasma concentrations are used for clinical monitoring. Interpretation of total concentrations in the context of protein-binding interactions will result in dosing adjustments that will possibly lead to AED toxicity. In the case of AEDs, phenytoin and valproate are the only drugs involved in clinically important protein-binding interactions, as described below.

# ANTIEPILEPTIC DRUG INTERACTIONS

## CARBAMAZEPINE

### Effect of Carbamazepine on Other Drugs

Clinically, carbamazepine has been shown to induce the metabolism of drugs that are metabolized by the CYP1A2, CYP2C, and CYP3A isozymes and UGTs (Anderson, 1998). More recently, expression profiling studies of carbamazepine upregulated CYP1A, CYP2A, CYP2B, CYP2C, and CYP3A

**TABLE 14.2**

**Inducers and Inhibitors of CYP Isozymes and UGTs Involved in Antiepileptic Drug Metabolism**

| Isozyme | AED Substrate | Inducers | Inhibitors[a] |
|---------|---------------|----------|--------------|
| CYP1A2 | Carbamazepine | Carbamazepine<br>Phenobarbital<br>Phenytoin | Fluvoxamine |
| CYP2C8 | Brivaracetam<br>Carbamazepine | Phenobarbital | — |
| CYP2C9 | Carbamazepine<br>Phenobarbital<br>Phenytoin<br>Valproate | Carbamazepine<br>Phenobarbital<br>Phenytoin | Amiodarone<br>Cimetidine<br>Fluconazole<br>Miconazole<br>Propoxyphene<br>Sulfaphenazole<br>Valproate |
| CYP2C19 | Clobazam<br>N-Desmethylclobazam<br>Diazepam<br>N-Desmethyldiazepam<br>Phenobarbital<br>Phenytoin<br>Valproate | Carbamazepine<br>Phenobarbital<br>Phenytoin | Felbamate<br>Fluoxetine<br>Fluvoxamine<br>Isoniazid<br>Ticlopidine |
| CYP3A4 | Carbamazepine<br>Clobazam<br>Diazepam<br>Ethosuximide<br>Tiagabine[a] | Carbamazepine<br>Felbamate<br>Oxcarbazepine<br>Phenobarbital<br>Phenytoin<br>Topiramate | Amprenavir<br>Atazanavir<br>Cimetidine<br>Clarithromycin<br>Danazol<br>Darunavir<br>Diltiazem<br>Erythromycin<br>Fluconazole<br>Grapefruit juice<br>Indinavir<br>Isoniazid<br>Itraconazole<br>Ketoconazole<br>Miconazole<br>Nelfinavir<br>Propoxyphene<br>Quinupristin<br>Ritonavir<br>Saquinavir<br>Telithromycin<br>Troleandomycin<br>Verapamil |
| UGTs | Carisbamate<br>Lamotrigine<br>Lorazepam<br>Retigabine<br>Valproate | Carbamazepine<br>Eratopenem<br>Imipenem<br>Lamotrigine<br>Meropenem<br>Oral contraceptives<br>Oxcarbazepine<br>Panipenem<br>Phenobarbital<br>Phenytoin | Valproate |

[a] These drugs have been shown to be inhibitors of the various CYP and UGT isozymes. Not all interactions have been demonstrated for all drugs. Caution should be used with concurrent therapy of known inhibitors and inducers.

subfamilies of enzymes; UGT1A; glutathione *S*-transferase 1A and Z1; and sulfotransferase 1A1, as well as several drug transporters (Oscarson et al., 2006). In addition to inducing the metabolism of other drugs, carbamazepine induces its own metabolism. The plasma clearance of carbamazepine more than doubles during the initial weeks of therapy (Geradin et al., 1976). Auto-induction of carbamazepine occurs via induction of CYP3A4-catalyzed metabolism to carbamazepine epoxide (CBZ-epoxide), the active metabolite of carbamazepine. In pharmacokinetic modeling studies, carbamazepine demonstrated an induction half-life of 4 days (Geradin et al., 1976). The majority of the induction occurs within 1 week of initiation of carbamazepine and should be completed within approximately 3 weeks. The time course of the deinduction is approximately the same (Schaffler et al., 1994). Using midazolam and caffeine as probe substrates for CYP3A4 and CYP1A2, the half-lives of induction by carbamazepine were 70 and 105 hours, respectively (Magnusson et al., 2008). Reductions in plasma concentrations of other AEDs and non-AEDs will occur when carbamazepine is added to therapy (Table 14.1).

### Effect of Other Drugs on Carbamazepine

Carbamazepine is extensively metabolized, with less than 1% excreted unchanged in the urine. CBZ-epoxide is the predominant metabolite, accounting for approximately 25% of the dose in monotherapy and 50% in polytherapy with other inducing AEDs. In *in vitro* human liver microsomes, carbamazepine is a substrate for CYP3A4 (major) with minor metabolism by CYP1A2 and CYP2C8. CBZ-epoxide is pharmacologically active and contributes to the therapeutic effects of carbamazepine as well as its neurotoxicity. When considering the effects of other drugs on carbamazepine, one must consider the effects on the active metabolite, which is rarely measured in practice. Clinically, the patients may show significant signs of carbamazepine toxicity with what might appear to be small increases in carbamazepine plasma concentrations. For example, valproate inhibits epoxide hydrolase, the enzyme that catalyzes the metabolism of CBZ-epoxide (Rambeck et al., 1990). Increases in serum concentrations of CBZ-epoxide are seen with either an increase or no change in carbamazepine concentrations when valproate is added to carbamazepine therapy. Increased plasma concentrations of carbamazepine resulting in carbamazepine toxicity have been reported with several drugs that are potent inhibitors of CYP3A4 activity. Cases of carbamazepine toxicity—because of increased carbamazepine plasma concentrations—have been reported for propoxyphene, danazol, nicotinamide, the macrolide antibiotics (erythromycin, clarithromycin, and troleandomycin), and calcium channel blockers (verapamil and diltiazem) (Anderson, 1998; Perucca, 2006). Other known CYP3A4 inhibitors, including several antiviral agents, are listed in Table 14.2. Co-administration with carbamazepine often results in reciprocal interactions; that is, carbamazepine decreases the plasma concentrations and efficacy of the CYP3A4 inhibitor (Table 14.1), and the CYP3A4 inhibitor increases the plasma concentrations of carbamazepine and results in carbamazepine toxicity.

### Herbal/Carbamazepine Interactions

St. John's wort, an inducer of CYP3A metabolism (Roby et al., 2000), demonstrated no significant effect on carbamazepine pharmacokinetics (Burstein et al., 2000). Mechanistically, after the initial autoinduction of CYP3A4 by carbamazepine occurs in patients, St. John's wort is not able to further increase the CYP3A4 induction of carbamazepine. A study in rabbits evaluated the potential pharmacokinetic interaction of carbamazepine and Mentat, an herbal medication used in India for psychiatric conditions. Mentat significantly increased the plasma concentrations of carbamazepine, suggesting a possible inhibitor effect (Tripathi et al., 2000).

### Clobazam

### Effect of Clobazam on Other Drugs

In studies of retrospectively collected plasma concentrations, clobazam had no significant effect on the concentration/dose ratio of carbamazepine, phenobarbital, or phenytoin (Sennoune et al., 1992; Theis et al., 1997). A significant decrease in valproate concentrations was found in one study (Theis et al., 1997), but not the other (Sennoune et al., 1992). In a small group of patients receiving cloba-zam plus carbamazepine, carbamazepine concentrations did not differ. However, concentrations of CBZ-epoxide and its sequential metabolite, *trans*-CBZ-diol, were moderately increased compared to concentrations obtained in patients receiving carbamazepine monotherapy. This suggests that cloba-zam either induces carbamazepine metabolism or inhibits epoxide-hydrolase-catalyzed CBZ-epoxide metabolism. Clobazam-induced carbamazepine toxicity associated with significantly increased car-bamazepine and CBZ-epoxide concentrations was reported in a patient receiving carbamazepine and topiramate. Similarly, there is a case report of three patients with phenytoin toxicity associated with increased phenytoin concentrations (Zifkin et al., 1991). The concentration/dose ratio of lamotrigine is significantly lower with co-administration of clobazam compared to monotherapy (Reimers et al., 2007). For both the carbamazepine and phenytoin reports, there was a significant delay in the time course of the interaction, implicating accumulation of the active metabolite of clobazam, *N*-desmethylclobazam (*N*-clobazam), as a possible mechanism. Because the interaction does not consistently occur in all patients, the interaction may be dependent on the CYP2C19 genotype. The *N*-desmethylclobazam/clobazam ratio is sixfold higher in patients homozygous for the CYP2C19 mutant allele (Kosaki et al., 2004) which could be responsible for the selective occurrence of the interaction.

### Effect of Other Drugs on Clobazam

Clobazam is eliminated predominately by hepatic metabolism to multiple metabolites. The primary metabolite, *N*-clobazam, is active and accumulates to approximately eightfold higher serum concen-trations than clobazam after multiple dosing. Clobazam is metabolized by CYP2C19 and CYP3A4. CYP2C19 is the major enzyme involved in *N*-clobazam hydroxylation (Giraud et al., 2004). The ratio of *N*-clobazam to clobazam was significantly higher in patients receiving phenobarbital, phe-nytoin, and carbamazepine (Sennoune et al., 1992; Theis et al., 1997). The concentration/dose ratio of *N*-clobazam and the ratio of *N*-clobazam to clobazam are significantly higher in patients receiving carbamazepine, phenytoin, phenobarbital, or concurrent felbamate (Contin et al., 1999), plus induc-ing AEDs, compared to those receiving inducing AEDs alone. Stiripentol inhibits the hydroxylation of *N*-clobazam, and clobazam doses must be reduced by half (Chiron et al., 2000).

### CLONAZEPAM

### Effect of Clonazepam on Other Drugs

Using a population analysis of routine serum concentrations, Yukawa et al. (2001) found that clon-azepam significantly reduced the clearance of carbamazepine by 22% and increased carbamazepine concentrations. In contrast, the concentration/dose ratio of lamotrigine is significantly lower with co-administration of clonazepam compared to monotherapy (Reimers et al., 2007), suggesting induction of UGT metabolism.

### Effect of Other Drugs on Clonazepam

Clonazepam is extensively metabolized to inactive metabolites by CYP3A4 with less than 1% excreted unchanged in the urine. Carbamazepine increases clonazepam clearance and decreases clonazepam concentrations by 20 to 30% (Lai et al., 1978; Yukawa et al., 2001). Phenobarbital or phenytoin treatment increased the clearance of a single dose of clonazepam by approximately 20% and 50%, respectively (Khoo et al., 1980).

### ETHOSUXIMIDE

### Effect of Ethosuximide on Other Drugs

Therapeutic doses of ethosuximide do not cause induction of enzymes in humans, and ethosux-imide has not been associated with inhibition of other drugs (Gilbert et al., 1974). A report by Salke-Kellermann et al. (1997) suggested that ethosuximide administration may reduce valproate concentrations by approximately one third; however, the mechanism is unknown. Overall, ethosuximide has a low capacity to alter the pharmacokinetics of other drugs.

## Effect of Other Drugs on Ethosuximide

Ethosuximide is eliminated primarily by hepatic metabolism: major metabolism by CYP3A4 and minor metabolism by CYP2E1, with 20% of a given dose excreted unchanged. Valproate (Bauer et al., 1982) and isoniazid (van Wieringen and Vrijlandt, 1983) have been reported to inhibit the metabolism of ethosuximide and cause small increases in ethosuximide plasma concentration; however, a dosage adjustment of ethosuximide is usually not required.

## Felbamate

### Effect of Felbamate on Other Drugs

Felbamate is both an inhibitor of drugs metabolized by CYP2C19 and β-oxidation and an inducer of drugs metabolized by CYP3A4 (Hachad et al., 2002). An *in vitro* study in human liver microsomes demonstrated that, of the CYPs evaluated, only CYP2C19-metabolized drugs are inhibited (Glue et al., 1997). This is consistent with the clinical experience with felbamate. Felbamate reduces the concentrations of carbamazepine but increases CBZ-epoxide concentrations, which may require a decrease in carbamazepine dose when felbamate is added (Albani et al., 1991). Felbamate also significantly decreases the concentrations of gestodene, the estrogen component in low-dose oral contraceptives (Saano et al., 1995). Phenytoin, phenobarbital, and valproate doses may need to be reduced when felbamate therapy is initiated due to inhibition by felbamate of CYP2C19 (phenytoin, phenobarbital) and β-oxidation (valproate) (Hachad et al., 2002).

### Effect of Other Drugs on Felbamate

Felbamate is eliminated by both renal excretion of unchanged drug (50%), and hepatic metabolism via UGT as a glucuronide conjugate (20%), and as a substrate for CYP3A4 (20%) and CYP2E1. Plasma concentrations of felbamate are decreased by carbamazepine, phenytoin, and phenobarbital (Hachad et al., 2002). Erythromycin (a CYP3A4 inhibitor) did not result in a significant increase in felbamate plasma concentrations (Sachdeo et al., 1998). Due to the small percent of felbamate metabolized by CYP3A4, this is not unexpected. A retrospective evaluation of felbamate plasma concentrations found that dose-normalized concentrations of felbamate with concurrent therapy with gabapentin were 37% higher compared to patients receiving monotherapy (Hussein et al., 1996). A pharmacokinetic study of felbamate and gabapentin has not yet been conducted to confirm the retrospective observations.

## Gabapentin

### Effect of Gabapentin on Other Drugs

Gabapentin is not an inducer of CYP- or UGT-dependent metabolism (Anderson, 1998) and does not significantly affect the elimination of the commonly used AEDs, oral contraceptives, or lithium. Gabapentin may decrease the plasma clearance of felbamate (see above); however, the mechanism is unclear (Hussein et al., 1996).

### Effect of Other Drugs on Gabapentin

Gabapentin is eliminated almost completely unchanged by the kidneys. The elimination of gabapentin is not affected by co-administration with any other drugs. Administration of aluminum hydroxide and magnesium hydroxide (Maalox® TC) simultaneously and within 2 hours of a single dose of gabapentin reduced gabapentin bioavailability by approximately 10 to 20% (Busch et al., 1992). For this reason, separating the administration of antacids and gabapentin by more than 2 hours is recommended.

## Lamotrigine

### Effect of Lamotrigine on Other Drugs

Lamotrigine is an inducer of UGT. Lamotrigine auto-induces its own metabolism (Yau et al., 1992) and decreases valproate plasma concentrations by approximately 25% when co-administered with valproate (Anderson et al., 1996). Levonorgestrel concentrations are reduced approximately 20% and ethinyl estradiol concentrations are unchanged when lamotrigine is co-administered with the combined oral contraceptive (Sidhu et al., 2006). In this study, there was no change in the serum progesterone concentrations, suggesting that suppression of ovulation was maintained. Carbamazepine toxicity presents with only small changes in carbamazepine and CBZ-epoxide concentrations when lamotrigine is added, suggesting that the interaction with lamotrigine may be pharmacodynamic rather than pharmacokinetic (Besag et al., 1998). Similarly, there is an increase in adverse events with co-therapy of lamotrigine and oxcarbazepine without an effect on the pharmacokinetics of either drug (Theis et al., 2005). Lamotrigine co-therapy resulted in a modest increase in olanzapine concentrations with no effect on concentrations of clozapine, risperidone, and their active metabolites (Spina et al., 2006).

### Effect of Other Drugs on Lamotrigine

Lamotrigine is extensively metabolized to two $N$-glucuronide metabolites via a reaction catalyzed by UGT1A4, with only minor renal elimination. Lamotrigine plasma concentrations are reduced by carbamazepine, phenytoin, and phenobarbital (Eriksson et al., 1996; May et al., 1996). In a retrospective analysis, dose-corrected plasma concentrations of lamotrigine in patients receiving methsuximide and oxcarbazepine were significantly lower compared to concentrations in patients receiving lamotrigine monotherapy (May et al., 1999). The inducing properties of oxcarbazepine were less (29%) compared to carbamazepine (54%); therefore, replacement of carbamazepine with oxcarbazepine should result in an increase in lamotrigine plasma concentrations, and a reduction in lamotrigine dose may be necessary (May et al., 1999). Methsuximide is not a known inducer of UGTs; however, a similar decrease in lamotrigine plasma concentrations was found in a pharmacokinetic study involving 16 patients.

Valproate significantly increases lamotrigine plasma concentrations, presumably by competitively inhibiting lamotrigine glucuronidation. The magnitude of the inhibition does not appear to be dependent on the valproate dose or concentration (Gidal et al., 2000). The maximum inhibition of lamotrigine plasma clearance at the lowest measured valproate plasma concentration is consistent with the approximately 20-fold higher molar concentrations of valproate used compared to lamotrigine. In addition, the lamotrigine–valproate combination has been associated with rash (Guberman et al., 1999). The incidence of rash in children is significantly higher than that in adults. In some cases, the rash has been potentially life threatening (e.g., Stevens–Johnson syndrome, toxic epidermal necrolysis). Patients should be carefully counseled to report any rash or other symptom of hypersensitivity (e.g., fever, lymphadenopathy) to their clinician. Because the lamotrigine–valproate combination therapy has been reported to have significant efficacy in refractory patients with atypical absence and complex partial seizures (Brodie and Yuen, 1997), the combination is not necessarily contraindicated in adult patients. By decreasing the initial starting dose and the rate of dose escalation of lamotrigine, the incidence of the rash is significantly decreased (Guberman et al., 1999).

Lamotrigine concentrations are decreased by ethinyl-estradiol-containing oral contraceptives; progestin-only contraceptives had no effect on lamotrigine concentration (Reimers et al., 2005). Co-administration of lamotrigine with the combined oral contraceptive results in almost a doubling of lamotrigine concentrations during the first week after the oral contraceptive is stopped (Christensen et al., 2007). Lopinavir/ritonavir decreases lamotrigine concentrations by 50%, presumably due to induction of UGT metabolism (van der Lee et al., 2006). A case report of two patients receiving sertraline and lamotrigine co-therapy reported that sertraline significantly increased plasma concentrations of lamotrigine (Kaufman and Gerner, 1998).

## LEVETIRACETAM

### Effect of Levetiracetam on Other Drugs

Levetiracetam is not an inducer of CYP- or UGT-dependent metabolism and does not appear to alter the plasma concentrations of the other AEDs measured in clinical studies (Gidal et al., 2005; Otoul et al., 2007a). *In vitro* evaluation of potential drug interactions with levetiracetam on 11 different drug-metabolizing enzyme activities demonstrated that levetiracetam is unlikely to inhibit the metabolism of other drugs by CYPs, UGTs, or epoxide hydrolase (Nicolas et al., 1999). Levetiracetam does not affect the pharmacokinetics of warfarin, oral contraceptives (Patsalos, 2000), or digoxin (Levy et al., 2001). A case report describes four patients who experienced intolerable carbamazepine-related adverse effects when levetiracetam was added to their therapy without an alteration in carbamazepine or carbamazepine epoxide concentrations (Sisodiya et al., 2002). In another case report, increased topiramate adverse events (decreased appetite, weight loss, nervousness) occurred after addition of levetiracetam in four children without a change in topiramate plasma concentrations (Glauser et al., 2002).

### Effect of Other Drugs on Levetiracetam

Levetiracetam is eliminated predominately by renal excretion of unchanged drug (~2/3) and by hydrolysis of the acetamide group, a reaction catalyzed by amidases, an enzyme that is present in a number of tissues. Concentrations of levetiracetam are lower in patients receiving enzyme-inducing drugs and slightly higher in patients also receiving valproate; however, dosage adjustments are not needed (Perucca et al., 2003). Digoxin does not alter the pharmacokinetics of levetiracetam (Levy et al., 2001).

## OXCARBAZEPINE

### Effect of Oxcarbazepine on Other Drugs

Unlike carbamazepine, oxcarbazepine is not a broad-spectrum inducer of CYP enzymes; however, it does induce the metabolism of drugs that are catalyzed by CYP3A4 and UGT (May et al., 2003). The extent of induction is on the average 46% higher with carbamazepine than with oxcarbazepine (Andreasen et al., 2007). Oral contraceptives need to be used cautiously in patients receiving any AED that causes induction of CYP3A4. Two weeks of oxcarbazepine co-therapy with an oral contraceptive resulted in a significant decrease in the plasma concentrations of ethinyl estradiol and levonorgestrel (Fattore et al., 1999). Consistent with CYP3A4 induction, a case report found that oxcarbazepine decreased cyclosporine plasma concentrations (Rosche et al., 2001). Felodipine plasma concentrations were significantly reduced by co-administration of oxcarbazepine but less so than with carbamazepine (Zaccara et al., 1993). Oxcarbazepine does not alter the pharmacokinetics of lamotrigine (Theis et al., 2005). Both phenobarbital and phenytoin plasma concentrations may increase when oxcarbazepine is added in a dose-dependent manner. This is consistent with inhibition of CYP2C19. Oxcarbazepine does not alter the pharmacokinetics of warfarin (May et al., 2003), risperidone, or olanzepine (Rosaria Muscatello et al., 2005).

### Effect of Other Drugs on Oxcarbazepine and 10-Monohydroxy Metabolites

Oxcarbazepine is a prodrug that is rapidly converted to 10-monohydroxy metabolites (MHDs) on oral administration, a reaction catalyzed by cytosolic arylketone reductase. MHD is predominately excreted unchanged in the urine or conjugated by UGT before being excreted, with only minor oxidation metabolism to dihydroxy derivative (DHD). The conversion of oxcarbazepine to MHD appears to be a non-inducible pathway (Kumps and Wurth, 1990). Phenobarbital and phenytoin decrease the plasma concentration of MHD, presumably due to induction of UGT (Kumps and Wurth, 1990). Valproate slightly decreases the elimination of MHD, most likely due to inhibition of

UGT-catalyzed metabolism (Tartara et al., 1993). Verapamil co-administration resulted in a small decrease in MHD (Kramer et al., 1991). Cimetidine, erythromycin, and propoxyphene did not affect the pharmacokinetics of oxcarbazepine (May et al., 2003).

## PHENOBARBITAL

### Effect of Phenobarbital on Other Drugs

Phenobarbital induces the metabolism of drugs metabolized by the CYP1A, CYP2A6, CYP2B, and CYP3A families and by UGT. The time course of induction and deinduction is primarily dependent on the long elimination half-life of phenobarbital. Induction usually begins in approximately 1 week, with maximal induction occurring 2 to 3 weeks after phenobarbital therapy is initiated (Ohnhaus et al., 1989). The time course of the deinduction will follow a similar course, as phenobarbital plasma concentrations decline over 2 to 3 weeks following removal of the drug (Dossing et al., 1983). Decreases in the plasma concentrations of other AEDs and non-AEDs will occur when phenobarbital is added to therapy (Table 14.1).

### Effect of Other Drugs on Phenobarbital

Phenobarbital is eliminated by both renal excretion of unchanged drug and hepatic metabolism. The two primary metabolites of phenobarbital are parahydroxyphenobarbital (PbOH) and phenobarbital N-glucoside. CYP2C9 plays a major role in the formation of PbOH, with minor metabolism by CYP2C19 and CYP2E1. The diversity of the elimination pathways of phenobarbital and the low protein binding (50%) minimize the effects of other drugs on phenobarbital. For example, inhibitors of CYP2C9 should alter the formation of PbOH; however, because the fraction of the dose of phenobarbital metabolized by CYP2C9 to PbOH is low (20%), CYP2C9 inhibitors do not cause clinically significant increases in phenobarbital plasma concentrations. Valproate causes the only clinically significant increase in phenobarbital plasma concentrations because of its broad spectrum of enzyme inhibition. Valproate inhibits both major metabolic pathways of phenobarbital: the formation of PbOH and phenobarbital N-glucoside (Bernus et al., 1994).

## PHENYTOIN

### Effect of Phenytoin on Other Drugs

Phenytoin increases the metabolism of drugs that are metabolized by the CYP2C and CYP3A families and UGT enzymes. Maximal induction or deinduction occurs approximately 1 to 2 weeks after initiation of phenytoin therapy, corresponding to the estimated time to steady-state phenytoin concentrations (Fleishaker et al., 1995). Theoretically, deinduction would require a similar period of time. Decreases in the plasma concentrations of other AEDs and non-AEDs can occur when phenytoin is added to therapy (Table 14.1).

Oral contraceptives should be used cautiously in patients receiving any AED that causes induction of CYP3A4 (Table 14.3). Phenytoin decreases the area under the concentration time curve (AUC) of ethinyl estradiol and levonorgestrel (Crawford et al., 1990). A number of unplanned pregnancies have occurred in women taking phenytoin or other inducing AEDs (Janz and Schmidt, 1974; Kenyon, 1972). Breakthrough bleeding and spotting may be a sign of inadequate contraception, and an alternative form of birth control may be advisable.

Although phenytoin is considered an inducer of hepatic metabolism, competition by phenytoin at the site of metabolism can occur with other drugs metabolized by the same CYP (Table 14.2). Both phenytoin and phenobarbital are metabolized by CYP2C9. Phenobarbital plasma concentrations reportedly decrease when phenytoin is discontinued in some patients (Morselli et al., 1971); therefore, combined use of phenytoin and phenobarbital may decrease the concentrations of both drugs, or, in some patients, phenobarbital concentrations and phenytoin concentrations may increase.

**TABLE 14.3**
**Induction and Inhibition Effect of Antiepileptic Drugs on Hepatic Enzymes**

| Antiepileptic Drug | Effect | Enzyme Involved |
|---|---|---|
| Carbamazepine | Inducer | CYP1A2, CYP2C, CYP3A, UGT |
| Ethosuximide | No effect | — |
| Felbamate | Inducer | CYP3A4 |
| | Inhibitor | CYP2C19, β-oxidation |
| Gabapentin | No effect | — |
| Lamotrigine | Inducer | UGT |
| | Inhibitor | CYP2C19 |
| Levetiracetam | No effect | — |
| Oxcarbazepine | Inducer | CYP3A4, UGT |
| | Inhibitor | CYP2C19 |
| Phenobarbital/primidone | Inducer | CYP1A, CYP2A6, CYP2B, CYP2C, CYP3A, UGT |
| Phenytoin | Inducer | CYP2C, CYP3A, UGT |
| Pregabalin | No effect | — |
| Tiagabine | No effect | — |
| Topiramate | Inducer | β-Oxidation |
| | Inhibitor | CYP2C19 |
| Valproate | Inhibitor | CYP2C9, UGT, epoxide hydrolase |
| Vigabatrin | No effect | — |
| Zonisamide | No effect | — |

Initiation of phenytoin in a patient receiving warfarin requires special consideration. There have been case reports of a hypoprothrombinemic response after phenytoin was given to patients receiving chronic warfarin therapy (Nappi, 1979; Panegyres and Rischbieth, 1991). Two proposed mechanisms could account for this response. First, when phenytoin therapy is started in a patient stabilized on warfarin therapy, phenytoin may displace warfarin from albumin-binding sites and cause a transient increase in warfarin effect (Nappi, 1979; Panegyres and Rischbieth, 1991). Second, phenytoin initially may competitively inhibit the metabolism of warfarin because both phenytoin and $S$-warfarin are CYP2C9 substrates (Bajpai et al., 1996). After the initial increased effect of $S$-warfarin, plasma concentrations may decline within 1 to 2 weeks after phenytoin is added because of CYP219 induction. Therefore, an initial decrease and then increase in warfarin dose may be necessary in order to maintain the desired anticoagulation effect.

### Effect of Other Drugs on Phenytoin

Phenytoin is eliminated predominately by CYP2C9- and CYP2C19-dependent hepatic metabolism. CYP2C9 and CYP2C19 are polymorphically distributed. Genetic mutations in CYP2C9 result in significantly greater impairment of phenytoin clearance than those of CYP2C19. Clinically significant increases in phenytoin plasma concentrations have been demonstrated for inhibitors of CYP2C9 and CYP2C19 enzymes (Table 14.2). Some commonly used CYP2C9 inhibitors are amiodarone, fluconazole (Blum et al., 1991), miconazole, propoxyphene, sulfaphenazole, and valproate. CYP2C19 inhibitors include felbamate, omeprazole, cimetidine, fluoxetine, fluvoxamine, isonizaid, and ticlopidine (Anderson, 1998).

When carbamazepine and phenytoin are given concurrently, the serum concentrations of both drugs may decrease. Similar to the phenytoin–phenobarbital interactions, phenytoin concentrations may increase in some patients when carbamazepine is added; however, the mechanism is unclear (Browne et al., 1988).

The effect of valproate on phenytoin is a combination of a protein-binding displacement and enzyme inhibition (Levy and Koch, 1982). The interactions result in a disruption of the relationship between unbound and total phenytoin concentrations. Total phenytoin concentrations can increase, decrease, or not change when valproate is added; however, the unbound phenytoin concentrations will increase. Ideally, unbound phenytoin concentrations should be monitored in a patient receiving both valproate and phenytoin.

### Herbal/Phenytoin Drug Interactions

There is one report of two patients receiving phenytoin who lost seizure control after Shankapulshipi, an Ayurvedic preparation used for the treatment of epilepsy, was added. A follow-up study in rats found that co-administration resulted in a 50% decrease in plasma phenytoin concentration (Dandekar et al., 1992). A study in rats evaluated the effect of a traditional Chinese medicine, Paeoniae Radix, used in the treatment of epilepsy in Asian countries on phenytoin pharmacokinetics. Paeoniae Radix co-therapy resulted in no significant changes in phenytoin pharmacokinetics (Chen et al., 2001). A study in rabbits evaluated the potential pharmacokinetic interaction of phenytoin and Mentat, an herbal medication used in India for psychiatric disorders. Mentat significantly increased the plasma concentrations of phenytoin, suggesting a possible inhibitor effect (Tripathi et al., 2000). The Chinese herb *Angelica dahurica* significantly inhibited the metabolism of tolbutamide (a CYP2C9 substrate) after intravenous administration in rats (Ishihara et al., 2000). *Ginkgo biloba* also significantly inhibits the metabolism of tolbutamide (Uchida et al., 2006). These studies suggest a significant potential interaction between herbal preparations and phenytoin; however, clinical data are still lacking.

## PREGABALIN

### Effect of Pregabalin on Other Drugs

Steady-state plasma concentrations of carbamazepine, lamotrigine, phenobarbital, phenytoin, topiramate, and valproate were not altered when pregabalin was evaluated as add-on therapy in clinical efficacy/safety studies (Bockbrader et al., 2001). Therefore, pregabalin is neither an inducer nor an inhibitor of metabolic enzymes.

### Effect of Other Drugs on Pregabalin

Pregabalin is eliminated by renal excretion of unchanged drug with little or no metabolism and is not bound to plasma proteins. Clinically significant drug interactions are not expected (Bialer et al., 2001); however, a retrospective analysis of pregabalin concentrations suggests that enzyme-inducing AEDs can decrease pregabalin concentrations by 20 to 30% (May et al., 2007).

## PRIMIDONE

### Effect of Primidone on Other Drugs

Primidone therapy results in the formation of phenobarbital, one of its active metabolites; therefore, the interactions described above for phenobarbital also relate to administration of primidone.

### Effect of Other Drugs on Primidone

Fifty percent of primidone is eliminated unchanged via urinary excretion. The remainder is metabolized by CYP to two active metabolites, phenobarbital and phenylethylmalonamide (PEMA). The CYPs involved have not been identified. A decreased ratio of primidone to phenobarbital occurs with concurrent administration of the CYP inducers phenytoin (Sato et al., 1992) and carbamazepine (Callaghan et al., 1977). Valproate inhibits the formation and elimination of phenobarbital, resulting in variable effects on the primidone/phenobarbital ratio (Bruni, 1981). Isoniazid inhibits the formation of phenobarbital, causing an increased primidone/phenobarbital ratio (Sutton and Kupferberg, 1975). As both primidone and phenobarbital are active, the clinical significance of these interactions is unclear.

## STIRIPENTOL

### Effect of Stiripentol on Other Drugs

Stiripentol is a potent inhibitor of the majority of the CYPs involved in drug metabolism. *In vitro* studies have demonstrated inhibition of CYP1A2, CYP2C9, CYP2C19, CYP2D6, and CYP3A4 (Tran et al., 1997). Clinically, stiripentol has been shown to significantly reduce the clearance of many AEDs, including carbamazepine (Kerr et al., 1991), phenytoin, phenobarbital (Levy et al., 1984), and the clearance of the active metabolite of clobazam, *N*-clobazam (Chiron et al., 2000). There is a small, but not clinically significant, increase in valproate plasma concentrations (Levy et al., 1990). As stiripentol inhibits all the major CYPs involved in drug metabolism, caution should be used when stiripentol is co-administered with any drug eliminated predominately by CYP-catalyzed metabolism.

### Effect of Other Drugs on Stiripentol

Stiripentol is eliminated extensively by CYP and UGT hepatic metabolism to 13 metabolites and displays Michaelis–Menton nonlinear pharmacokinetics within the range of plasma concentrations found clinically. Plasma concentrations of stiripentol are significantly decreased in patients receiving concurrent inducing AEDs (Chiron, 2007).

## TIAGABINE

### Effect of Tiagabine on Other Drugs

Tiagabine appears to be neither an inducer nor an inhibitor of CYP or UGT enzymes. Tiagabine did not affect the plasma concentrations or pharmacokinetics of carbamazepine, valproate, phenytoin, triazolam, theophylline, warfarin, digoxin, cimetidine, oral contraceptives, or ethanol (Hachad et al., 2002). Tiagabine co-administration resulted in a slight, but clinically insignificant reduction in valproate plasma concentrations (~10%) (Gustavson et al., 1998).

### Effect of Other Drugs on Tiagabine

Tiagabine is extensively metabolized, with less than 2% excreted unchanged in the urine. CYP3A4 has been identified as the primary isozyme responsible for metabolism of tiagabine to 5-oxo-tiagabine (~22% of the dose). In patients with epilepsy treated with enzyme-inducing AEDs, tiagabine clearance is significantly increased (Hachad et al., 2002). *In vitro* inhibitors of CYP3A4 (erythromycin, ketoconazole) significantly decreased the metabolism of tiagabine (Bopp et al., 1995). However, in a study in normal subjects, erythromycin did not significantly alter the clearance of tiagabine (Thomsen et al., 1998) due to the small fraction of the dose metabolized by CYP3A4. Temazepam, a CYP3A4 substrate, did not alter the pharmacokinetics of tiagabine (Richens et al., 1998).

## TOPIRAMATE

### Effect of Topiramate on Other Drugs

Topiramate does not significantly affect the plasma concentrations of carbamazepine, lamotrigine, phenobarbital, primidone, or valproate (Bialer et al., 2004). Topiramate at 200 mg/day has minimal effects on lithium, haloperidol, propranolol, and nortriptyline pharmacokinetics (Bialer et al., 2004). Topiramate doses of 50 to 200 mg/day do not interact with oral contraceptives containing ethinyl estradiol and norethindrone (Doose et al., 2003). Reduced ethinyl estradiol plasma concentrations occur with topiramate at doses greater than 200 mg/day, suggesting the need for higher doses of oral contraceptives (Rosenfeld et al., 1997). In some patients, topiramate increases the metabolism of phenytoin by 25%, possibly by inhibiting CYP2C19. The intersubject variability in the phenytoin–topiramate interaction may reflect the intersubject variability in the fraction of phenytoin metabolized by the polymorphically distributed CYP2C9 and CYP2C19 (Bajpai et al., 1996).

The inhibition spectrum of topiramate was evaluated in human liver microsomes (Levy et al., 1995). Topiramate significantly inhibited the model substrate of CYP2C19 and had no effect on any of the other CYPs evaluated. The *in vitro* study was consistent with the inhibitory effect of topiramate on phenytoin and lack of inhibitory effect on the other AEDs.

### Effect of Other Drugs on Topiramate

Topiramate is eliminated by hepatic metabolism and renal excretion of unchanged drug (40%). Concurrent use of enzyme-inducing drugs decreases topiramate plasma concentrations by approximately 40 to 50% (Britzi et al., 2005). Valproate does not affect the pharmacokinetics of topiramate (Mimrod et al., 2005).

## VALPROATE

### Effect of Valproate on Other Drugs

Valproate is a broad-spectrum inhibitor of hepatic metabolism, including inhibition of CYP2C9, UGT, and epoxide hydrolase, and is a weak inhibitor of CYP1A2 (Anderson, 1998). Valproate does not inhibit cyclosporine or oral contraceptives (Crawford et al., 1986), suggesting a lack of inhibition of CYP3A-metabolized drugs. Valproate increases the plasma concentrations of phenobarbital and phenytoin, presumably via inhibition of CYP2C9 (phenytoin, phenobarbital) and *N*-glucosidation (phenobarbital) (Perucca, 2006). Valproate inhibits the glucuronide conjugation of lamotrigine, lorazepam, and zidovudine, resulting in significant increases in plasma concentrations of the affected drugs (Anderson, 1998; Perucca, 2006). A case report of elevated clomipramine and desmethylclomipramine plasma concentrations after addition of valproate suggests that valproate inhibited the parallel glucuronidation pathway of clomipramine elimination (Fehr et al., 2000).

Compared to a group of patients receiving clozapine monotherapy, patients receiving clozapine with valproate had higher clozapine plasma concentrations and lower plasma concentrations of desmethylclozapine, an active metabolite. The interaction should not be clinically significant (Facciola et al., 1999). Valproate is highly protein bound to albumin (>90%). Because of the high molar concentrations of valproate obtained clinically, valproate displaces other AEDs (phenytoin, carbamazepine, and diazepam) from albumin-binding sites, which is only clinically significant for the therapeutic drug monitoring of phenytoin (Levy and Koch, 1982).

### Effect of Other Drugs on Valproate

Valproate predominately undergoes hepatic metabolism, with less than 5% of the dose excreted unchanged in the urine. Major metabolism occurs by UGT-catalyzed glucuronide conjugation (UGT1A6, UGT1A9, and UGT2B7) and β-oxidation with minor CYP-dependent metabolism via CYP2C9 and CYP2C19. Valproate concentrations are decreased in the presence of enzyme-inducing AEDs (i.e., carbamazepine, phenobarbital, and phenytoin). The carbapenem antibiotics, meropenem, panipenem, eratopenem, and imipenem, significantly decrease valproate concentrations by an average of 34% (Mori et al., 2007). The effect of meropenem on valproate concentrations in 39 patients led to a electroclinical deterioration in over half of the patients (Spriet et al., 2007). Due to its lack of enzyme-inducing properties, the use of valproate in women taking oral contraceptives has been recommended in the past; however, a recent study has found that total and unbound valproate concentrations are decreased, on average, 18% and 30%, respectively, when women are receiving the combined oral contraceptives (Galimberti et al., 2006). Clonazepam (Yukawa et al., 2003) and felbamate

(Anderson, 1998; Hachad et al., 2002) significantly inhibit valproate metabolism and increase valproate concentrations. There are case reports of clinically significant decreases in valproate concentrations with co-administration of cisplatin (Ikeda et al., 2005) and efavirenz (Saraga et al., 2006).

## Herbal/Valproate Drug Interactions

A study in healthy subjects evaluating the effect of a traditional Chinese medicine, Paeoniae Radix, used in Asian countries for the treatment of epilepsy, found that it had no effect on the pharmacokinetics of valproate (Chen et al., 2000).

## VIGABATRIN

### Effect of Vigabatrin on Other Drugs

Vigabatrin had historically been available in many countries except the United States. In mid-2009, vigabatrin finally received approval for use in patients with infantile spasms and refractory complex partial seizures in adults. Vigabatrin is neither an inducer (Bartoli et al., 1997) nor inhibitor of drug metabolism and is not associated with any significant drug interactions, with one exception (Rey et al., 1992). A clinically significant decrease of 25 to 40% in phenytoin plasma concentrations occurs in approximately one third of the patients receiving both drugs. The mechanism has not yet been elucidated.

### Effect of Other Drugs on Vigabatrin

Vigabatrin is eliminated almost completely unchanged by the kidneys (Haegele et al., 1988).There have been no reports of significant interactions affecting the plasma concentrations of vigabatrin. The lack of protein binding of vigabatrin makes protein-binding interactions unlikely.

## ZONISAMIDE

### Effect of Zonisamide on Other Drugs

Zonisamide does not significantly affect the plasma concentrations of lamotrigine, phenytoin, phenobarbital, primidone, or valproate (Hachad et al., 2002; Schmidt et al., 1993). Results of clinical studies of the effect of zonisamide on carbamazepine have been conflicting, with no effect, increases in carbamazepine concentrations, and decreases in the carbamazepine epoxide/carbamazepine ratio reported (Hachad et al., 2002).

### Effect of Other Drugs on Zonisamide

Zonisamide is eliminated by a combination of renal excretion of unchanged drug (~35%), metabolism via $N$-acetylation (~15%) and reduction to 2-sulfamoyolacetylphenol (50%). Studies using expressed human CYPs demonstrated that CYP3A4, CYP3A5, and CYP2C19 are all capable of catalyzing zonisamide reduction. The intrinsic clearance of CYP2C19 and CYP3A5 was low compared to CYP3A4, indicating that the relative contribution of CYP2C19 and CYP3A5 may be low *in vivo* (Nakasa et al., 1998). Zonisamide plasma clearance is increased in the presence of the enzyme-inducing AEDs carbamazepine, phenytoin, and phenobarbital (Sackellares et al., 1985).

## LACOSAMIDE

### Effect of Lacosamide on Other Drugs

Based on *in vitro* studies, lacosamide has a low potential to induce or inhibit CYP enzymes. Lacosamide does not alter the concentrations of carbamazepine, digoxin, levetiracetam, lamotrigine, metformin, omeprazole, topiramate, valproate, or the combined oral contraceptive containing ethinyl estradiol and levonorgestrel (Doty et al., 2007).

## Effect of Other Drugs on Lacosamide

Lacosamide is eliminated primarily by renal excretion of unchanged drug and minor metabolism to an *O*-desmethyl metabolite. In clinical studies, lacosamide concentrations were not altered by co-administration of carbamazepine, digoxin, levetiracetam, lamotrigine, metformin, omeprazole, topiramate, valproate, or the combined oral contraceptive containing ethinyl estradiol and levonorgestrel (Doty et al., 2007).

### RUFINAMIDE

### Effect of Rufinamide on Other Drugs

*In vitro* studies in human liver microsomes have demonstrated that rufinamide does not inhibit model substrates of CYP1A2, CYP2A6, CYP2C9, CYP2C19, CYP2D6, CYP2E1, CYP3A4/5, or CYP4A9/11 (Kapeghian et al., 1996). In a population pharmacokinetic analysis of safety and efficacy trials (Perucca et al., 2008), rufinamide did not affect the clearance of topiramate or valproate. Rufinamide increased the clearance of carbamazepine and lamotrigine and slightly decreased the clearance of phenobarbital and phenytoin (Perucca et al., 2008); however, all the changes were less than 20% and unlikely to be clinically significant. Rufinamide did not affect the trough plasma concentrations of any of the co-administered AEDs including carbamazepine, clobazam, clonazepam, phenobarbital, phenytoin, primidone, oxcarbazepine, and valproate (Bialer et al., 2001).

### Effect of Other Drugs on Rufinamide

Rufinamide is extensively metabolized, with less than 2% of the dose excreted in the urine as unchanged drug. The primary metabolic pathway is a hydrolysis of the carboxylamide group to an inactive metabolite, which is subsequently excreted in the urine (Bialer et al., 2001). In a population pharmacokinetic analysis, concurrent therapy with CYP-enzyme-inducing AEDs increased rufinamide oral clearance by approximately 25%. Valproate reduced the oral clearance of rufinamide by 22% (Perucca et al., 2008). The inhibition effect of valproate on rufinamide in children was significantly greater than in adults (Perucca et al., 2008).

## ANTIEPILEPTIC DRUG INTERACTIONS OF DRUGS IN DEVELOPMENT

By the time a new AED is released to market, some information regarding the elimination of the drug as well as a potential drug interaction spectrum is usually known. Clinical studies may or may not be available in the literature at the time of initial release.

### BRIVARACETAM

### Effect of Brivaracetam on Other Drugs

Brivaracetam is not an inducer of CYP- or UGT-dependent metabolism and does not appear to alter the plasma concentrations carbamazepine, lamotrigine, levetiracetam, oxcarbazepine, or topiramate in clinical studies (Otoul et al., 2007b). Carbamazepine epoxide concentrations were moderately increased at the highest brivaracetam dose studied (150 mg/day). The CBZ-epoxide/carbamazepine ratio was 65% higher than in the placebo group. No difference in the ratio was found with brivaracetam doses of 5, 20, or 50 mg/day (Otoul et al., 2007b). Dose-normalized concentrations of phenytoin concentrations were 25% higher in patients receiving brivaracetam in safety/efficacy studies.

### Effect of Other Drugs on Brivaracetam

Brivaracetam is eliminated almost entirely by metabolism with minimal renal excretion. Major metabolism occurs via hydrolysis of the acetamide group and CYP2C8-mediated hydroxylation.

A population pharmacokinetic evaluation of brivaracetam found that concurrent administration of enzyme-inducing AEDs increased the oral clearance of brivaracetam, which would predict a 30% average decrease in brivaracetam steady-state concentrations (Lacroix et al., 2007).

## CARISBAMATE

### Effect of Carisbamate on Other Drugs

Carisbamate did not alter the pharmacokinetics of tolbutamide (CYP2C9), desipramine (CYP2D6), or midazolam (CYP3A4), suggesting a low potential to induce or inhibit CYP enzymes (Novak et al., 2007). Concurrent administration of carisbamate resulted in 15% and 20% decreases in the AUCs for valproate and lamotrigine, respectively. These data would be consistent with a modest induction of UGT enzymes (Chien et al., 2007). No significant change in levels of carbamazepine or ethinyl estradiol and norethindrone (combined oral contraceptive) was observed (Novak et al., 2007).

### Effect of Other Drugs on Carisbamate

Carisbamate is extensively metabolized through UGT-catalyzed $O$-glucuronidation and hydrolysis of the carbamate ester (Novak et al., 2007). Co-administration of carbamazepine or oral contraceptives decreases carisbamate concentration by 36% and 20 to 30%, respectively (Novak et al., 2007).

## RETIGABINE

### Effect of Retigabine on Other Drugs

*In vitro*, retigabine does not interact with valproate, lamotrigine, imipramine, propofol, or probe substrates of UGT isozymes (Borlak et al., 2006). In a study of healthy subjects, retigabine increased the clearance of lamotrigine by 22% and decreased lamotrigine concentrations by 18% (Hermann et al., 2003), suggesting that retigabine is a modest inducer of UGT metabolism. Retigabine does not alter the plasma concentrations of ethinyl estradiol, norgestrel, carbamazepine, phenytoin, phenobarbital, valproate, or topiramate (Bialer et al., 2009).

### Effect of Other Drugs on Retigabine

Retigabine is eliminated predominately by metabolism via $N$-glucuronidation, catalyzed by UGT1A1 and UGT1A9 (Borlak et al., 2006), and acetylation, with no evidence of CYP-dependent oxidation pathways (Hempel et al., 1999). Plasma concentrations of retigabine are decreased approximately 30% when co-administered with phenytoin or carbamazepine (Bialer et al., 2009).

## CONCLUSIONS

Drug interactions in patients receiving AEDs are a common complication of therapy. AED induction of the metabolism of drugs with narrow therapeutic ranges will lead to patients receiving suboptimal doses of concurrent therapy if standard doses are not adjusted. By understanding the time course of induction and deinduction, serious adverse events as well as loss of efficacy can be avoided. Inhibitory drug interactions are the leading cause of clinically significant drug interactions due to their often rapid onset and unpredictable side effects. Interactions involving inhibition of the AEDs often result in clinical toxicity. Increased knowledge regarding the specific isozymes of CYPs and UGTs involved in the metabolism of the AEDs is enabling the qualitative prediction of interactions for both the new AEDs as well as interactions of new non-AEDs. Recent correlations between findings of *in vitro* studies and clinical interaction studies have demonstrated the value of *in vitro* studies in predicting and directing clinical interaction studies.

## REFERENCES

Albani, F., W.H. Theodore, P. Washington, O. Devinsky, E. Bromfield, R.J. Porter, F.J. Nice. (1991). Effect of felbamate on plasma levels of carbamazepine and its metabolites. *Epilepsia*, 32:130–132.

Anderson, G.D. (1998). A mechanistic approach to antiepileptic drug interactions. *Ann. Pharmacother.*, 32:554–563.

Anderson, G.D., M.K. Yau, B.E. Gidal, S.J. Harris, R.H. Levy, A.A. Lai, K.B. Wolf, W.A. Wargin, A.T. Dren. (1996). Bidirectional interaction of valproate and lamotrigine in healthy subjects. *Clin. Pharmacol. Ther.*, 60:145–156.

Andreasen, A.H., K. Brosen, P. Damkier. (2007). A comparative pharmacokinetic study in healthy volunteers of the effect of carbamazepine and oxcarbazepine on CYP3A4. *Epilepsia*, 48:490–496.

Bajpai, M., L.K. Roskos, D.D. Shen, R.H. Levy. (1996). Roles of cytochrome P4502C9 and cytochrome P4502C19 in the stereoselective metabolism of phenytoin to its major metabolite. *Drug Metab. Dispos.*, 24:1401–1403.

Bartoli, A., G. Gatti, G. Cipolla, N. Barzaghi, G. Veliz, C. Fattore, J. Mumford, E. Perucca. (1997). A double-blind, placebo-controlled study on the effect of vigabatrin on *in vivo* parameters of hepatic microsomal enzyme induction and on the kinetics of steroid oral contraceptives in healthy female volunteers. *Epilepsia*, 38:702–707.

Bauer, L.A., C. Harris, A.J. Wilensky, V.A. Raisys, R.H. Levy. (1982). Ethosuximide kinetics: possible interaction with valproic acid. *Clin. Pharmacol. Ther.*, 31:741–745.

Bernus, I., R.G. Dickinson, W.D. Hooper, M.J. Eadie. (1994). Inhibition of phenobarbitone *N*-glucosidation by valproate. *Br. J. Clin. Pharmacol.*, 38:411–416.

Besag, F.M., D.J. Berry, F. Pool, J.E. Newbery, B. Subel. (1998). Carbamazepine toxicity with lamotrigine: pharmacokinetic or pharmacodynamic interaction? *Epilepsia*, 39:183–187.

Bialer, M., S.I. Johannessen, H.J. Kupferberg, R.H. Levy, P. Loiseau, E. Perucca. (2001). Progress report on new antiepileptic drugs: a summary of the Fifth Eilat Conference (EILAT V). *Epilepsy Res.*, 43:11–58.

Bialer, M., D.R. Doose, B. Murthy, C. Curtin, S.S. Wang, R.E. Twyman, S. Schwabe. (2004). Pharmacokinetic interactions of topiramate. *Clin. Pharmacokinet.*, 43:763–780.

Bialer, M., S.I. Johannessen, R.H. Levy, E. Perucca, T. Tomson, H.S. White. (2009). Progress report on new antiepileptic drugs: a summary of the Ninth Eilat Conference (ELIAT IX). *Epilepsy Res.*, 83:1–43.

Blum, R.A., J.H. Wilton, D.M. Hilligoss, M.J. Gardner, E.B. Henry, N.J. Harrison, J.J. Schentag. (1991). Effect of fluconazole on the disposition of phenytoin. *Clin. Pharmacol. Ther.*, 49:420–425.

Bockbrader, H.N., P.J. Burger, B.W. Corrigan A.R. Kugler, L.E. Knapp, E.A. Garofalo, R.L. Lalonde. (2001). Population pharmacokinetic (PK) analysis of commonly prescribed antiepileptic drugs (AEDs) coadminstered with pregabalin (PGB) in adult patients with refractory partial seizures. *Epilepsia*, 42(Suppl. 7):84.

Bopp, B.A., G.D. Nequist, A.D. Rodrigues. (1995). Role of the cytochrome P450 3A subfamily in the metabolism of [$^{14}$C] tiagabine by human hepatic microsomes. *Epilepsia*, 36(Suppl. 3):S158–S159.

Borlak, J., A. Gasparic, M. Locher, H. Schupke, R. Hermann. (2006). *N*-Glucuronidation of the antiepileptic drug retigabine: results from studies with human volunteers, heterologously expressed human UGTs, human liver, kidney, and liver microsomal membranes of Crigler–Najjar type II. *Metabolism*, 55:711–721.

Britzi, M., E. Perucca, S. Soback, R.H. Levy, C. Fattore, F. Crema, G. Gatti, D.R. Doose, B.E. Maryanoff, M. Bialer. (2005). Pharmacokinetic and metabolic investigation of topiramate disposition in healthy subjects in the absence and in the presence of enzyme induction by carbamazepine. *Epilepsia*, 46:378–384.

Brodie, M.J., A.W. Yuen. (1997). Lamotrigine substitution study: evidence for synergism with sodium valproate? 105 study group. *Epilepsy Res.*, 26:423–432.

Browne, T.R., G.K. Szabo, J.E. Evans, B.A. Evans, D.J. Greenblatt, M.A. Mikati. (1988). Carbamazepine increases phenytoin serum concentration and reduces phenytoin clearance. *Neurology*, 38:1146–1150.

Bruni, J. (1981). Valproic acid and plasma levels of primidone and derived phenobarbital. *Can. J. Neurol. Sci.*, 8:91–92.

Burstein, A.H., R.L. Horton, T. Dunn, R.M. Alfaro, S.C. Piscitelli, W. Theodore. (2000). Lack of effect of St. John's wort on carbamazepine pharmacokinetics in healthy volunteers. *Clin. Pharmacol. Ther.*, 68:605–612.

Busch, J.A., L.L. Radulovic, H.N. Bockbrader, B.A. Underwood, A.J. Sedman, T. Chang. (1992). Effect of Maalox TC on single-dose pharmacokinetics of gabapentin capsules in healthy subjects. *Pharm. Res.*, 9(Suppl. 2):S315.

Callaghan, N., M. Feely, F. Duggan, M. O'Callaghan, J. Seldrup. (1977). The effect of anticonvulsant drugs which induce liver microsomal enzymes on derived and ingested phenobarbitone levels. *Acta Neurol. Scand.*, 56:1–6.

Chen, L.C., M.H. Chou, M.F. Lin, L.L. Yang. (2000). Lack of pharmacokinetic interaction between valproic acid and a traditional Chinese medicine, Paeoniae Radix, in healthy volunteers. *J. Clin. Pharm. Ther.*, 25:453–459.

Chen, L.C., M.H. Chou, M.F. Lin, L.L. Yang. (2001). Effects of Paeoniae Radix, a traditional Chinese medicine, on the pharmacokinetics of phenytoin. *J. Clin. Pharm. Ther.*, 26:271–278.

Chien, S., C. Yao, A. Mertens, T. Verhaeghe, B. Solanki, D.R. Doose, G. Novak, M. Bialer. (2007). An interaction study between the new antiepileptic and CNS drug carisbamate (RWJ-333369) and lamotrigine and valproic acid. *Epilepsia*, 48:1328–1338.

Chiron, C. (2007). Stiripentol. *Neurotherapeutics*, 4:123–125.

Chiron, C., M.C. Marchand, A. Tran, E. Rey, P. d'Athis, J. Vincent, O. Dulac, G. Pons. (2000). Stiripentol in severe myoclonic epilepsy in infancy: a randomised placebo-controlled syndrome-dedicated trial. STICLO study group. *Lancet*, 356:1638–1642.

Christensen, J., V. Petrenaite, J. Atterman, P. Sidenius, I. Ohman, T. Tomson, A. Sabers. (2007). Oral contraceptives induce lamotrigine metabolism: evidence from a double-blind, placebo-controlled trial. *Epilepsia*, 48:484–489.

Clark, D.J., B. Burchell. (1994). The uridine diphosphate glucuronosyltransferase multigene family: function and regulation. In *Handbook of Experimental Pharmacology* (Vol. 112, pp. 3–43), C. Frederic and F.C. Kaufsan, Eds. New York: Springer-Verlag.

Contin, M., R. Riva, F. Albani, A.A. Baruzzi. (1999). Effect of felbamate on clobazam and its metabolite kinetics in patients with epilepsy. *Ther. Drug Monit.*, 21:604–608.

Crawford, P., D. Chadwick, P. Cleland, J. Tjia, A. Cowie, D.J. Back, M.L. Orme. (1986). The lack of effect of sodium valproate on the pharmacokinetics of oral contraceptive steroids. *Contraception*, 33:23–29.

Crawford, P., D.J. Chadwick, C. Martin, J. Tjia, D.J. Back, M. Orme. (1990). The interaction of phenytoin and carbamazepine with combined oral contraceptive steroids. *Br. J. Clin. Pharmacol.*, 30:892–896.

Dandekar, U.P., R.S. Chandra, S.S. Dalvi, M.V. Joshi, P.C. Gokhale, A.V. Sharma, P.U. Shah, N.A. Kshirsagar. (1992). Analysis of a clinically important interaction between phenytoin and Shankhapushpi, an Ayurvedic preparation. *J. Ethnopharmacol.*, 35:285–288.

Doose, D.R., S.S. Wang, M. Padmanabhan, S. Schwabe, D. Jacobs, M. Bialer. (2003). Effect of topiramate or carbamazepine on the pharmacokinetics of an oral contraceptive containing norethindrone and ethinyl estradiol in healthy obese and nonobese female subjects. *Epilepsia*, 44:540–549.

Dossing, M., H. Pilsgaard, B. Rasmussen, H.E. Poulsen. (1983). Time course of phenobarbital and cimetidine mediated changes in hepatic drug metabolism. *Eur. J. Clin. Pharmacol.*, 25:215–222.

Doty, P., G.D. Rudd, T. Stoehr, D. Thomas. (2007). Lacosamide. *Neurotherapeutics*, 4:145–148.

Eriksson, A.S., K. Hoppu, A. Nergardh, L. Boreus. (1996). Pharmacokinetic interactions between lamotrigine and other antiepileptic drugs in children with intractable epilepsy. *Epilepsia*, 37:769–773.

Facciola, G., A. Avenoso, M.G. Scordo, A.G. Madia, A. Ventimiglia, E. Perucca, E. Spina. (1999). Small effects of valproic acid on the plasma concentrations of clozapine and its major metabolites in patients with schizophrenic or affective disorders. *Ther. Drug Monit.*, 21:341–345.

Fattore, C., G. Cipolla, G. Gatti, G.L. Limido, Y. Sturm, C. Bernasconi, E. Perucca. (1999). Induction of ethinylestradiol and levonorgestrel metabolism by oxcarbazepine in healthy women. *Epilepsia*, 40:783–787.

Fehr, C., G. Grunder, C. Hiemke, N. Dahmen. (2000). Increase in serum clomipramine concentrations caused by valproate. *J. Clin. Psychopharmacol.*, 20:493–494.

Fleishaker, J.C., L.K. Pearson, G.R. Peters. (1995). Phenytoin causes a rapid increase in 6 beta-hydroxycortisol urinary excretion in humans: a putative measure of CYP3A induction. *J. Pharm. Sci.*, 84:292–294.

Galimberti, C.A., I. Mazzucchelli, C. Arbasino, M.P. Canevini, C. Fattore, E. Perucca. (2006). Increased apparent oral clearance of valproic acid during intake of combined contraceptive steroids in women with epilepsy. *Epilepsia*, 47:1569–1572.

Geradin, A.P., F.V. Abadie, J.A. Campestrini, W. Theobald. (1976). Pharmacokinetics of carbamazepine in normal humans after single and repeated oral doses. *J. Pharmacokinet. Biopharm.*, 4:521–535.

Gidal, B.E., G.D. Anderson, P.R. Rutecki, R. Shaw, A. Lanning. (2000). Lack of an effect of valproate concentration on lamotrigine pharmacokinetics in developmentally disabled patients with epilepsy. *Epilepsy Res.*, 42:23–31.

Gidal, B.E., E. Baltes, C. Otoul, E. Perucca. (2005). Effect of levetiracetam on the pharmacokinetics of adjunctive antiepileptic drugs: a pooled analysis of data from randomized clinical trials. *Epilepsy Res.*, 64:1–11.

Gilbert, J.C., A.K. Scott, D.B. Galloway, J.C. Petrie. (1974). Ethosuximide: liver enzyme induction and D-glucaric acid excretion. *Br. J. Clin. Pharmacol.*, 1:249–252.

Giraud, C., A. Tran, E. Rey, J. Vincent, J.M. Treluyer, G. Pons. (2004). *In vitro* characterization of clobazam metabolism by recombinant cytochrome P450 enzymes: importance of CYP2C19. *Drug Metab. Dispos.*, 32:1279–1286.

Glauser, T.A., J.M. Pellock, E.M. Bebin, N.B. Fountain, F.J. Ritter, C.M. Jensen, W.D. Shields. (2002). Efficacy and safety of levetiracetam in children with partial seizures: an open-label trial. *Epilepsia*, 43:518–524.

Glue, P., C.R. Banfield, J.L. Perhach, G.G. Mather, J.K. Racha, R.H. Levy. (1997). Pharmacokinetic interactions with felbamate. *In vitro–in vivo* correlation. *Clin. Pharmacokinet.*, 33:214–224.

Guberman, A.H., F.M. Besag, M.J. Brodie, J.M. Dooley, M.S. Duchowny, J.M. Pellock, A. Richens, R.S. Stern, E. Trevathan. (1999). Lamotrigine-associated rash: risk/benefit considerations in adults and children. *Epilepsia*, 40:985–991.

Gustavson, L.E., K.W. Sommerville, S.W. Boellner, G.F. Witt, H.J. Guenther, G.R. Granneman. (1998). Lack of a clinically significant pharmacokinetic drug interaction between tiagabine and valproate. *Am. J. Ther.*, 5:73–79.

Hachad, H., I. Ragueneau-Majlessi, R.H. Levy. (2002). New antiepileptic drugs: review on drug interactions. *Ther. Drug Monit.*, 24:91–103.

Haegele, K.D., N.D. Huebert, M. Ebel, G.P. Tell, P.J. Schechter. (1988). Pharmacokinetics of vigabatrin: implications of creatinine clearance. *Clin. Pharmacol. Ther.*, 44:558–565.

Hempel, R., H. Schupke, P.J. McNeilly, K. Heinecke, C. Kronbach, C. Grunwald, G. Zimmermann, C. Griesinger, J. Engel, T. Kronbach. (1999). Metabolism of retigabine (D-23129), a novel anticonvulsant. *Drug Metab. Dispos.*, 27:613–622.

Hermann, R., N.G. Knebel, G. Niebch, L. Richards, J. Borlak, M. Locher. (2003). Pharmacokinetic interaction between retigabine and lamotrigine in healthy subjects. *Eur. J. Clin. Pharmacol.*, 58:795–802.

Hussein, G., A.S. Troupin, G. Montouris. (1996). Gabapentin interaction with felbamate. *Neurology*, 47:1106.

Ikeda, H., T. Murakami, M. Takano, T. Usui, K. Kihira. (2005). Pharmacokinetic interaction on valproic acid and recurrence of epileptic seizures during chemotherapy in an epileptic patient. *Br. J. Clin. Pharmacol.*, 59:593–597.

Ishihara, K., H. Kushida, M. Yuzurihara, Y. Wakui, T. Yanagisawa, H. Kamei, S. Ohmori, M. Kitada. (2000). Interaction of drugs and Chinese herbs: pharmacokinetic changes of tolbutamide and diazepam cause by extract of *Angelica dahurica*. *J. Pharm. Pharmacol.*, 52:1023–1029.

Janz, D., D. Schmidt. (1974). Anti-epileptic drugs and failure of oral contraceptives [letter]. *Lancet*, 1:1113.

Kapeghian, J.C., A. Madan, A. Parkinson, S.L. Tripp, A. Probst. (1996). Evaluation of rufinamide, a novel anticonvulsant for potential drug interactions *in vitro*. *Epilepsia*, 37(Suppl. 5):26.

Kaufman, K.R., R. Gerner. (1998). Lamotrigine toxicity secondary to sertraline. *Seizure*, 7:163–165.

Kenyon, I.E. (1972). Unplanned pregnancy in an epileptic. *Br. Med. J.*, 1:686–687.

Kerr, B.M., J.M. Martinez-Lage, C. Viteri, J. Tor, A.C. Eddy, R.H. Levy. (1991). Carbamazepine dose requirements during stiripentol therapy: influence of cytochrome P-450 inhibition by stiripentol. *Epilepsia*, 32:267–274.

Khoo, K.C., J. Mendels, M. Rothbart, W.A. Garland, W.A. Colburn, B.H. Min, R. Lucek, J.J. Carbone, H.G. Boxenbaum, S.A. Kaplan. (1980). Influence of phenytoin and phenobarbital on the disposition of a single oral dose of clonazepam. *Clin. Pharmacol. Ther.*, 28:368–375.

Kosaki, K., K. Tamura, R. Sato, H. Samejima, Y. Tanigawara, T. Takahashi. (2004). A major influence of CYP2C19 genotype on the steady-state concentration of *N*-desmethylclobazam. *Brain Dev.*, 26:530–534.

Kramer, G., B. Tettenborn, G. Flesch. (1991). Oxcarbazepine–verapamil drug interaction in healthy volunteers. *Epilepsia*, 32(Suppl. 1):70.

Kumps, A., C. Wurth. (1990). Oxcarbazepine disposition: preliminary observations in patients. *Biopharm. Drug Dispos.*, 11:365–370.

Lacroix, B., P. Von Rosenstiel, M.L. Sargentini-Maier. (2007). Population pharmacokinetics of brivaraetam in patients with partial epilepsy. *Epilepsia*, 48(Suppl. 6):333–334.

Lai, A.A., R.H. Levy, R.E. Cutler. (1978). Time-course of interaction between carbamazepine and clonazepam in normal man. *Clin. Pharmacol. Ther.*, 24:316–323.

Levy, R.H., K.M. Koch. (1982). Drug interactions with valproic acid. *Drugs*, 24:543–556.

Levy, R.H., P. Loiseau, M. Guyot, H.M. Blehaut, J. Tor, T.A. Moreland. (1984). Stiripentol kinetics in epilepsy: nonlinearity and interactions. *Clin. Pharmacol Ther.*, 36:661–669.

Levy, R.H., A.W. Rettenmeier, G.D. Anderson, A.J. Wilensky, P.N. Friel, T.A. Baillie, A. Acheampong, J. Tor, M. Guyot, P. Loiseau. (1990). Effects of polytherapy with phenytoin, carbamazepine, and stiripentol on formation of 4-ene-valproate, a hepatotoxic metabolite of valproic acid. *Clin. Pharmacol. Ther.*, 48:225–235.

Levy, R.H., F. Bishop, A.J. Streeter, W.F. Trage, K.L. Kunzd, K.T. Thummel, G.G. Mather. (1995). Explanation and prediction of drug interactions with topiramate using CYP450 inhibition spectrum. *Epilepsia*, 36(Suppl. 4):47.

Levy, R.H., I. Ragueneau-Majlessi, E. Baltes. (2001). Repeated administration of the novel antiepileptic agent levetiracetam does not alter digoxin pharmacokinetics and pharmacodynamics in healthy volunteers. *Epilepsy Res.*, 46:93–99.

Magnusson, M.O., M.L. Dahl, J. Cederberg, M.O. Karlsson, R. Sandstrom. (2008). Pharmacodynamics of carbamazepine-mediated induction of CYP3A4, CYP1A2, and PGP as assessed by probe substrates midazolam, caffeine, and digoxin. *Clin. Pharmacol Ther.*, 84:52–62.

May, T.W., B. Rambeck, U. Jurgens. (1996). Serum concentrations of lamotrigine in epileptic patients: the influence of dose and comedication. *Ther. Drug Monit.*, 18:523–531.

May, T.W., B. Rambeck, U. Jurgens. (1999). Influence of oxcarbazepine and methsuximide on lamotrigine concentrations in epileptic patients with and without valproic acid comedication: results of a retrospective study. *Ther. Drug Monit.*, 21:175–181.

May, T.W., E. Korn-Merker, B. Rambeck. (2003). Clinical pharmacokinetics of oxcarbazepine. *Clin. Pharmacokinet.*, 42:1023–1042.

May, T.W., B. Rambeck, R. Neb, U. Jurgens. (2007). Serum concentrations of pregabalin in patients with epilepsy: the influence of dose, age, and comedication. *Ther. Drug Monit.*, 29:789–794.

Mimrod, D., L.M. Specchio, M. Britzi, E. Perucca, N. Specchio, A. La Neve, S. Soback, R.H. Levy, G. Gatti, D.R. Doose, B.E. Maryanoff, M. Bialer. (2005). A comparative study of the effect of carbamazepine and valproic acid on the pharmacokinetics and metabolic profile of topiramate at steady state in patients with epilepsy. *Epilepsia*, 46:1046–1054.

Mori, H., K. Takahashi, T. Mizutani. (2007). Interaction between valproic acid and carbapenem antibiotics. *Drug Metab. Rev.*, 39(4):647–657.

Morselli, P.L., M. Rizzo, S. Garattini. (1971). Interaction between phenobarbital and diphenylhydantoin in animals and in epileptic patients., *Ann. N.Y. Acad. Sci.*, 179:88–107.

Nakasa, H., H. Nakamura, S. Ono, M. Tsutsui, M. Kiuchi, S. Ohmori, M. Kitada. (1998). Prediction of drug–drug interactions of zonisamide metabolism in humans from *in vitro* data. *Eur. J. Clin. Pharmacol.*, 54:177–183.

Nappi, J.M. (1979). Warfarin and phenytoin interaction. *Ann. Intern. Med.*, 90:852.

Nicolas, J.M., P. Collart, B. Gerin, G. Mather, W. Trager, R. Levy, J. Roba. (1999). *In vitro* evaluation of potential drug interactions with levetiracetam, a new antiepileptic agent. *Drug Metab. Dispos.*, 27:250–254.

Novak, G.P., M. Kelley, P. Zannikos, B. Klein. (2007). Carisbamate (RWJ-333369). *Neurotherapeutics*, 4:106–109.

Ohnhaus, E.E., A.M. Breckenridge, B.K. Park. (1989). Urinary excretion of 6 beta-hydroxycortisol and the time course measurement of enzyme induction in man. *Eur. J. Clin. Pharmacol.*, 36:39–46.

Oscarson, M., U.M. Zanger, O.F. Rifki, K. Klein, M. Eichelbaum, U.A. Meyer. (2006). Transcriptional profiling of genes induced in the livers of patients treated with carbamazepine. *Clin. Pharmacol. Ther.*, 80:440–456.

Otoul, C., H. De Smedt, A. Stockis. (2007a). Lack of pharmacokinetic interaction of levetiracetam on carbamazepine, valproic acid, topiramate, and lamotrigine in children with epilepsy. *Epilepsia*, 48:2111–2115.

Otoul, C., P. von Rosenstiel, A. Stockis. (2007b). Evaluation of the pharmacokinetic interaction of brivaracetam on other antiepileptic drugs in adults with partial-onset seizures. *Epilepsia*, 48(Suppl. 6):334.

Panegyres, P.K., R.H. Rischbieth. (1991). Fatal phenytoin–warfarin interaction. *Postgrad. Med. J.*, 67:98.

Patsalos, P.N. (2000). Pharmacokinetic profile of levetiracetam: toward ideal characteristics. *Pharmacol. Ther.*, 85:77–85.

Patsalos, P.N., J.S. Duncan, S.D. Shorvon. (1988). Effect of the removal of individual antiepileptic drugs on antipyrine kinetics, in patients taking polytherapy. *Br. J. Clin. Pharmacol.*, 26:253–259.

Perucca, E. (2006). Clinically relevant drug interactions with antiepileptic drugs. *Br. J. Clin. Pharmacol.*, 61:246–255.

Perucca, E., A. Hedges, K.A. Makki, M. Ruprah, J.F. Wilson, A. Richens. (1984). A comparative study of the relative enzyme inducing properties of anticonvulsant drugs in epileptic patients. *Br. J. Clin. Pharmacol.*, 18:401–410.

Perucca, E., B.E. Gidal, E. Baltes. (2003). Effects of antiepileptic comedication on levetiracetam pharmacokinetics: a pooled analysis of data from randomized adjunctive therapy trials. *Epilepsy Res.*, 53:47–56.

Perucca, E., J. Cloyd, D. Critchley, E. Fuseau. (2008). Rufinamide: clinical pharmacokinetics and concentration–response relationships in patients with epilepsy. *Epilepsia*, 49:1123–1141.

Rambeck, B., A. Salke-Treumann, T. May, H.E. Boenigk. (1990). Valproic acid-induced carbamazepine-10,11-epoxide toxicity in children and adolescents. *Eur. Neurol.*, 30:79–83.

Reimers, A., G. Helde, E. Brodtkorb. (2005). Ethinyl estradiol, not progestogens, reduces lamotrigine serum concentrations. *Epilepsia*, 46:1414–1417.

Reimers, A., E. Skogvoll, J.K. Sund, O. Spigset. (2007). Lamotrigine in children and adolescents: the impact of age on its serum concentrations and on the extent of drug interactions. *Eur. J. Clin. Pharmacol.*, 63:687–692.

Rey, E., G. Pons, G. Olive. Vigabatrin. (1992). Clinical pharmacokinetics. *Clin. Pharmacokinet.*, 23:267–278.

Richens, A., R.W. Marshall, J. Dirach, J.A. Jansen, S. Snel, P.C. Pedersen. (1998). Absence of interaction between tiagabine, a new antiepileptic drug, and the benzodiazepine triazolam. *Drug Metabol. Drug Interact.*, 14:159–177.

Roby, C.A., G.D. Anderson, E. Kantor, D.A. Dryer, A.H. Burstein. (2000). St John's wort: effect on CYP3A4 activity. *Clin. Pharmacol. Ther.*, 67:451–457.

Rosaria Muscatello, M., M. Pacetti, M. Cacciola, D. La Torre, R. Zoccali, C. D'Arrigo, G. Migliardi, E. Spina. (2005). Plasma concentrations of risperidone and olanzapine during coadministration with oxcarbazepine. *Epilepsia*, 46:771–774.

Rosche, J., W. Froscher, D. Abendroth, J. Liebel. (2001). Possible oxcarbazepine interaction with cyclosporine serum levels: a single case study. *Clin. Neuropharmacol.*, 24:113–116.

Rosenfeld, W.E., D.R. Doose, S.A. Walker, R.K. Nayak. (1997). Effect of topiramate on the pharmacokinetics of an oral contraceptive containing norethindrone and ethinyl estradiol in patients with epilepsy. *Epilepsia*, 38:317–323.

Saano, V., P. Glue, C.R. Banfield, P. Reidenberg, R.D. Colucci, J.W. Meehan, P. Haring, E. Radwanski, A. Nomeir, C.C. Lin et al. (1995). Effects of felbamate on the pharmacokinetics of a low-dose combination oral contraceptive. *Clin. Pharmacol. Ther.*, 58:523–531.

Sachdeo, R.C., S. Narang-Sachdeo, P.A. Montgomery, R.C. Shumaker, J.L. Perhach, W.H. Lyness, A. Rosenberg. (1998). Evaluation of the potential interaction between felbamate and erythromycin in patients with epilepsy. *J. Clin. Pharmacol.*, 38:184–190.

Sackellares, J.C., P.D. Donofrio, J.G. Wagner, B. Abou-Khalil, S. Berent, K. Aasved-Hoyt. (1985). Pilot study of zonisamide (1,2-benzisoxazole-3-methanesulfonamide) in patients with refractory partial seizures. *Epilepsia*, 26:206–211.

Salke-Kellermann, R.A., T. May, H.E. Boenigk. (1997). Influence of ethosuximide on valproic acid serum concentrations. *Epilepsy Res.*, 26:345–349.

Saraga, M., M. Preisig, D.F. Zullino. (2006). Reduced valproate plasma levels possible after introduction of efavirenz in a bipolar patient. *Bipolar Disord.*, 8:415–417.

Sato, J., Y. Sekizawa, A. Yoshida, E. Owada, N. Sakuta, M. Yoshihara, T. Goto, Y. Kobayashi, K. Ito. (1992). Single-dose kinetics of primidone in human subjects: effect of phenytoin on formation and elimination of active metabolites of primidone, phenobarbital and phenylethylmalonamide. *J. Pharmacobiodyn.*, 15:467–472.

Schaffler, L., B.F. Bourgeois, H.O. Luders. (1994). Rapid reversibility of autoinduction of carbamazepine metabolism after temporary discontinuation. *Epilepsia*, 35:195–198.

Schmidt, D., R. Jacob, P. Loiseau, E. Deisenhammer, D. Klinger, A. Despland, M. Egli, G. Bauer, E. Stenzel, V. Blankenhorn. (1993). Zonisamide for add-on treatment of refractory partial epilepsy: a European double-blind trial. *Epilepsy Res.*, 15:67–73.

Sennoune, S., E. Mesdjian, J. Bonneton, P. Genton, C. Dravet, J. Roger. (1992). Interactions between clobazam and standard antiepileptic drugs in patients with epilepsy. *Ther. Drug Monit.*, 14:269–274.

Sidhu, J., S. Job, S. Singh, R. Philipson. (2006). The pharmacokinetic and pharmacodynamic consequences of the co-administration of lamotrigine and a combined oral contraceptive in healthy female subjects. *Br. J. Clin. Pharmacol.*, 61:191–199.

Sisodiya, S.M., J.W. Sander, P.N. Patsalos. (2002). Carbamazepine toxicity during combination therapy with levetiracetam: a pharmacodynamic interaction. *Epilepsy Res.*, 48:217–219.

Spina, E., C. D'Arrigo, G. Migliardi, V. Santoro, M.R. Muscatello, U. Mico, G. D'Amico, E. Perucca. (2006). Effect of adjunctive lamotrigine treatment on the plasma concentrations of clozapine, risperidone and olanzapine in patients with schizophrenia or bipolar disorder. *Ther. Drug Monit.*, 28:599–602.

Spriet, I., J. Goyens, W. Meersseman, A. Wilmer, L. Willems, W. Van Paesschen. (2007). Interaction between valproate and meropenem: a retrospective study. *Ann. Pharmacother.*, 41:1130–1136.

Sutton, G., H.J. Kupferberg. (1975). Isoniazid as an inhibitor of primidone metabolism. *Neurology*, 25:1179–1181.

Tartara, A., C.A. Galimberti, R. Manni, R. Morini, G. Limido, G. Gatti, A. Bartoli, G. Strada, E. Perucca. (1993). The pharmacokinetics of oxcarbazepine and its active metabolite 10-hydroxy-carbazepine in healthy subjects and in epileptic patients treated with phenobarbitone or valproic acid. *Br. J. Clin. Pharmacol.*, 36:366–368.

Theis, J.G., G. Koren, R. Daneman, A.L. Sherwin, E. Menzano, M. Cortez, P. Hwang. (1997). Interactions of clobazam with conventional antiepileptics in children. *J. Child Neurol.*, 12:208–213.

Theis, J.G., J. Sidhu, J. Palmer, S. Job, J. Bullman, J. Ascher. (2005). Lack of pharmacokinetic interaction between oxcarbazepine and lamotrigine. *Neuropsychopharmacology*, 30:2269–2274.

Thomsen, M.S., L. Groes, H. Agerso, T. Kruse. (1998). Lack of pharmacokinetic interaction between tiagabine and erythromycin. *J. Clin. Pharmacol.*, 38:1051–1056.

Tomson, T., J.O. Svensson, P. Hilton-Brown. (1989). Relationship of intraindividual dose to plasma concentration of carbamazepine: indication of dose-dependent induction of metabolism. *Ther. Drug Monit.*, 11:533–539.

Tran, A., E. Rey, G. Pons, M. Rousseau, P. d'Athis, G. Olive, G.G. Mather, F.E. Bishop, C.J. Wurden, R. Labroo, W.F. Trager, K.L. Kunze, K.E. Thummel, J.C. Vincent, J.M. Gillardin, F. Lepage, R.H. Levy. (1997). Influence of stiripentol on cytochrome P450-mediated metabolic pathways in humans: *in vitro* and *in vivo* comparison and calculation of *in vivo* inhibition constants. *Clin. Pharmacol. Ther.*, 62:490–504.

Tripathi, M., R. Sundaram, M. Rafiq, M.V. Venkataranganna, S. Gopumadhavan, S.K. Mitra. (2000). Pharmacokinetic interactions of Mentat with carbamazepine and phenytoin. *Eur. J. Drug Metab. Pharmacokinet.*, 25:223–226.

Uchida, S., H. Yamada, X.D. Li, S. Maruyama, Y. Ohmori, T. Oki, H. Watanabe, K. Umegaki, K. Ohashi, S. Yamada. (2006). Effects of *Ginkgo biloba* extract on pharmacokinetics and pharmacodynamics of tolbutamide and midazolam in healthy volunteers. *J. Clin. Pharmacol.*, 46:1290–1298.

van der Lee, M.J., L. Dawood, H.J. ter Hofstede, M.J. de Graaff-Teulen, E.W. van Ewijk-Beneken Kolmer, N. Caliskan-Yassen, P.P. Koopmans, D.M. Burger. (2006). Lopinavir/ritonavir reduces lamotrigine plasma concentrations in healthy subjects. *Clin. Pharmacol. Ther.*, 80:159–168.

van Wieringen, A., C.M. Vrijlandt. (1983). Ethosuximide intoxication caused by interaction with isoniazid. *Neurology*, 33:1227–1228.

Warner, T., P.N. Patsalos, M. Prevett, A.A. Elyas, J.S. Duncan. (1992). Lamotrigine-induced carbamazepine toxicity: an interaction with carbamazepine-10,11-epoxide. *Epilepsy Res.*, 11:147–150.

Wolf, P. (1992). Lamotrigine: preliminary clinical observations on pharmacokinetics and interactions with traditional antiepileptic drugs. *J. Epilepsy*, 5:73.

Wrighton, S.A., J.C. Stevens. (1992). The human hepatic cytochromes P450 involved in drug metabolism. *Crit. Rev. Toxicol.*, 22:1–21.

Yau, M.K., M.A. Adams, W.A. Wargin, A.A. Lai. (1992). A single dose and steady state pharmacokinetic study of lamotrigine in healthy male volunteers. In *Proceedings of the Third International Cleveland Clinic–Bethel Epilepsy Symposium on Antiepileptic Drug Pharmacology*, June 16–20, Cleveland, OH.

Yukawa, E., T. Nonaka, M. Yukawa, S. Ohdo, S. Higuchi, T. Kuroda, Y. Goto. (2001). Pharmacoepidemiologic investigation of a clonazepam–carbamazepine interaction by mixed effect modeling using routine clinical pharmacokinetic data in Japanese patients. *J. Clin. Psychopharmacol.*, 21:588–593.

Yukawa, E., T. Nonaka, M. Yukawa, S. Higuchi, T. Kuroda, Y. Goto. (2003). Pharmacoepidemiologic investigation of a clonazepam–valproic acid interaction by mixed effect modeling using routine clinical pharmacokinetic data in Japanese patients. *J. Clin. Pharm. Ther.*, 28:497–504.

Zaccara, G., P.F. Gangemi, L. Bendoni, G.P. Menge, S. Schwabe, G.C. Monza. (1993). Influence of single and repeated doses of oxcarbazepine on the pharmacokinetic profile of felodipine. *Ther. Drug Monit.*, 15:39–42.

Zifkin, B., A. Sherwin, F. Andermann. (1991). Phenytoin toxicity due to interaction with clobazam. *Neurology*, 41:313–314.

# Section III

---

*Epilepsy Surgery*

# 15 Advances in Structural and Functional Neuroimaging: How Are These Guiding Epilepsy Surgery?

*Hyunmi Kim, A. LeBron Paige, and Robert C. Knowlton*

## CONTENTS

## INTRODUCTION

Epilepsy surgery has been revolutionized by modern neuroimaging. Prior to the advent of computed tomography (CT) and early magnetic resonance imaging (MRI), the task of localizing seizures was, for the most part, a *needle-in-the-haystack* search based on electrophysiological localization methods with recording electrodes typically requiring placement in or directly on the brain. Even with the tomographic imaging that became widely available in the late 1970s and early 1980s, the majority of pathologic lesions serving as epileptogenic substrates or pathology remained undetected (cryptogenic).

The main goal of epilepsy neuroimaging advances has been to increasingly identify the location and extent of epileptogenic tissue not detected in previous scan techniques. One major impact of these advances has been to allow more patients to avoid costly, invasive intracranial electro-encephalography (IC-EEG) investigation. Additionally, based on the premise of better diagnostic accuracy, the goal of advanced epilepsy imaging is also to improve the selection of surgical candidates: decreased inclusion of candidates with a likely low yield for successful surgery and increased inclusion of those that may have been otherwise incorrectly excluded. Finally, more than just sorting patients, increased accuracy of epilepsy localization, either directly or indirectly, can improve the overall proportion of seizure-free outcomes from surgery. Thus, the role of advanced neuroimaging techniques is to allow the pursuit of both optimal patient selection and highest probability of cure regardless of whether IC-EEG is included in the treatment process.

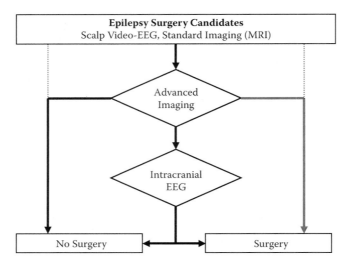

**FIGURE 15.1** Position of advanced imaging impact on epilepsy surgery. Advances in neuroimaging methods have the potential to affect all outcomes of surgical decision-making. New imaging information that informs clinicians that a patient would not be a good candidate for surgery may allow that patient to avoid any further evaluation and possibly even a failed surgery. Even if new localization information from imaging does not allow the patient to avoid intracranial EEG (IC-EEG) electrodes, it can direct optimal electrode placement, an effect that can avoid the result of false localization—either with respect to nonlocalized recordings when they otherwise would have been localized (leading to inappropriate denial of surgery) or incorrect localization (leading to a failed surgical outcome). Additional information about the location or extent of epileptogenic substrate that would not be available without advanced imaging techniques may also affect the net total of seizure-free outcomes.

Figure 15.1 shows the impact that advanced imaging methods can have on the evaluation and treatment of epilepsy surgical candidates. As a sorting tool alone, justification of imaging tests is probably limited to the possibility of decreasing evaluation costs by excluding patients who have a low yield of surgical success. Test specificity, however, would have to be extremely high to minimize the risk of excluding patients who could benefit from the overall dramatic effect of seizure surgery cure vs. continued cumulative consequences of lifelong medically intractable epilepsy; otherwise, decision analysis would suggest that all possible surgical candidates be treated even if the cure rate was relatively small. The only way for test usage to overcome this strong "treat-all" tendency would be to provide a sufficient degree of improvement in seizure-free outcome rate.

## HISTORY OF NEUROIMAGING IMPACT ON EPILEPSY SURGERY

The first major modern neuroimaging advance to widely reduce IC-EEG was the development of FDG–PET (Engel et al., 1982a). This novel modality for relatively high-resolution tomographic imaging of brain metabolism began to influence presurgical epilepsy evaluation by demonstrating focal functional defects of glucose uptake in regions associated with epileptogenic tissue that did not appear abnormal on conventional structural imaging, including early MRI (Engel et al., 1982b). This was mainly the case in mesial temporal lobe epilepsy (MTLE), for both the most common surgery for intractable epilepsy and, in the majority of cases, surgery without an obvious structural lesion. FDG–PET began to be applied to epilepsy surgery prior to the introduction of MRI, which became more widely available in the late 1980s (Abou et al., 1987; Engel, 1984; Mazziotta and Engel, 1984; Shimizu et al., 1985; Sperling et al., 1986; Stefan et al., 1987; Theodore et al., 1986). In 1989, the University of California, Los Angeles (UCLA) group showed that focal ictal EEG patterns along with concordant relative focal temporal lobe (TL) hypometabolism reliably localized MTLE cases and that IC-EEG recordings (at that time, performed in nearly all evaluations and often referred to as phase II evaluations) provided no additional information (Engel et al., 1990). Thus, UCLA and other centers with FDG–PET began utilizing the concept of "skip candidates," a presurgical evaluation strategy that allowed patients to proceed directly to surgery, skipping IC-EEG or phase II evaluations.

Not long after the discovery of the value of FDG–PET in MTLE, it was demonstrated that the technique also revealed cryptogenic epileptogenic substrates in neonates and children with catastrophic partial epilepsies, including infantile spasms (Chugani et al., 1988). The role of FDG–PET was remarkable in these cases because it provoked the discussion and ultimate decision to offer radical resections (hemispherectomy or large temporal–parietal–occipital resections), which turned out to afford a seizure-free outcome and nearly normal development in patients who were otherwise doomed to severe mental retardation due to uncontrolled seizures in a critical window of brain development. The concept elucidated was that, as long as the epileptogenic pathology and relentless seizures affecting the entire brain remained, potential normal development of nonpathological brain was precluded. Indeed, in spite of major initial deficits from the radical resections (e.g., hemiparesis), the children made remarkable gains in brain development. Prior to FDG–PET revealing a focal or lateralized epileptogenic focus, these patients were not considered surgical candidates because EEG suggested a global dysfunction of brain and generalized epilepsy.

In the late 1980s, MRI began to advance to the point that subtle abnormalities of brain structure could be detected. Most importantly, this included MRI findings indicative of hippocampal sclerosis (Berkovic et al., 1991; Jackson et al., 1990), the most common pathology of MTLE. Detection of relative hippocampal atrophy or prolonged $T_2$ signal concordant with ictal scalp EEG had an impact analogous to that of PET with respect to skipping IC-EEG (Sperling, 1990). Because of the wide availability of MRI, however, the magnitude of the impact has been far greater than that of FDG–PET. Additionally, both hippocampal volumes (HVs) and $T_2$ relaxation times ($T_2$ mapping) began to be performed in imaging laboratories to objectively detect very subtle or questionable abnormalities, further optimizing the sensitivity of MRI for this clinically critical brain structure in epilepsy (Jack et al., 1990; Jackson et al., 1993b; Watson et al., 1992). Shortly thereafter, correlations were demonstrated between HV and $T_2$ mapping measures and clinical parameters, including surgical histopathology (hippocampal neuronal cell counts), prediction of seizure-free outcome, and memory dysfunction (Cascino et al., 1991; Jackson et al., 1993a; Kuzniecky et al., 1993a; Loring et al., 1993; Martin et al., 2001; Trenerry et al., 1993). Arguably, the introduction of high-quality MRI capable of detecting evidence for hippocampal sclerosis and other previously cryptogenic epileptogenic lesions has had the greatest impact of any investigative tool on the surgical treatment of epilepsy.

Magnetic resonance imaging advances in the last two decades have continued to improve candidate selection and localization accuracy. The next most significant contribution of MRI was detection and delineation of malformations of cortical development (MCD) (Barkovich and Kuzniecky, 1996; Barkovich and Maroldo, 1993; Kuzniecky and Barkovich, 1996). This capability has provided for an entirely new *in vivo* imaging paradigm for classification of human brain development abnormalities in addition to an advancement in detection of what was before not defined from an imaging standpoint (Barkovich, 1996). What remains a challenge is determining which exact part of an extensive neuroimaging abnormality is responsible for seizure onset and to what extent abnormal cortical tissue remains undetected. For this challenge and others, the need for further improvement in structural imaging remains.

Functional imaging modalities have also evolved vastly over the past couple of decades. The methodologies range from those that measure biochemical molecular concentrations with magnetic resonance spectroscopy (MRS), to tomographic imaging of blood flow during seizures with ictal single photon emission computed tomography (SPECT), to mapping of neurotransmitter receptor ligands with positron emission tomography (PET), to localizing sources of electrophysiological disturbances of cerebral activity with combined EEG and functional MRI, to determining cortical connectivity derangements with diffusion weighted and diffusion tensor imaging (DWI/DTI). It is clear that advances in these *in vivo* functional imaging methods will aid the large challenge that remains with structural imaging, particularly determining the significance of many ambiguous or subtle and questionable structural abnormalities, as well as detecting functional abnormalities associated either indirectly or directly with epileptogenic substrates that remain invisible (cryptogenic) even on state-of-the-art high-resolution MRI. This chapter attempts to provide the reader with knowledge of both established and still developing advances in neuroimaging—both structural and functional—that have shown (or have the potential) to increase accuracy of epilepsy localization such that they can aid epilepsy surgery patient selection and treatment.

## STRUCTURAL IMAGING

### MAGNETIC RESONANCE IMAGING

Epilepsy MRI advances begin with enhancement in imaging quality. The main approaches to this are directed to increasing the signal-to-noise ratio (SNR) and, if possible, spatial resolution. Three different approaches that accomplish this task include the use of surface coils, image averaging, and high field strength imaging. Each method has its merits and limitations with regard to clinical applications.

Phase-array surface coils can increase the SNR at the surface of cortical mantle as much as sixfold vs. conventional quadrature head coils (Grant et al., 1997; Maravilla, 1997). The increase in signal allows for better contrast or can be traded for an increase in resolution without compromise of the signal. The increase in spatial resolution is truly only significant for two-dimensional (2D) imaging; with an increase in three-dimensional (3D) resolution, the drop-off in signal is proportional to the cube of the decrease in voxel size, an amount that cannot be overcome by even a sixfold increase in signal. Furthermore, with surface coils SNR decreases with distance from the coil; for example, at a depth of the hippocampus, the SNR increase is only 1.67-fold (Hayes et al., 1993). Although this variation of signal strength as a function of distance away from the coils is one of the limitations of surface coils, the signal can be normalized by algorithms that correct for the variation such that signal intensity is uniform throughout the image (Wald et al., 1995). Other limitations associated with surface coils include limited brain coverage, potential motion problems, and an overall increase in scanning time. Still, the improvement in contrast and fine detail discrimination can be valuable in the assessment of very subtle cortical dysgenesis abnormalities and hippocampal microstructural disturbances. In reality, the greatest limitation of surface coils is the fact that they are not routinely available.

Image averaging is a novel method to improve SNR that can be performed without any special hardware (Holmes et al., 1998). Improvement in image quality from increased SNR, mainly better gray–white matter contrast, is present throughout the brain with no trade other for deep vs. superficial tissue. The method requires the acquisition of several scans that are then coregistered and averaged. Gain in SNR is equal to the square root of the number of scans acquired. The same increase could be obtained in the sum of *n* scans if patients could hold their head motionless for more than several minutes (usually even the most motivated subjects cannot keep their head perfectly still more than ten minutes). The optimal ways to take advantage of image averaging remain unclear. In the clinical setting, time in the scanner is of utmost importance. A preliminary report on the application of image averaging in patients with partial epilepsy and subtle or ambiguous lesions on conventional MRI showed a benefit in defining the suspected lesions of interest. Greatly improved contrast allowed better delineation of very subtle abnormalities at the gray–white matter junction, yielding greater confidence in the interpretation (Knowlton and Kjos, 1999). Figure 15.2 shows the improvement in image quality, mainly due to better contrast, in a patient with a small area of cortical migration disturbance that is not confidently visualized on the unaveraged native image. Realistically only a few additional scans of several minutes' duration can be allowed. A practical number of scans is four (only three additional) to yield a twofold increase in SNR. The greatest

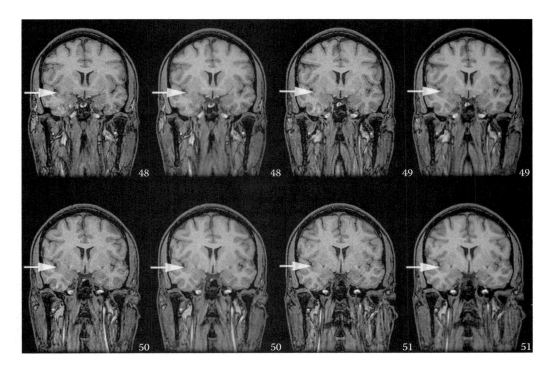

**FIGURE 15.2** Image averaging to improve signal to noise. A subtle area of abnormal cortical migration present in the right temporal stem is poorly visualized on the native single-acquisition MRI scan on the left. The lesion is more clearly defined in the averaged image on the right ($T_1$-weighted gradient echo volume sequence × 4).

drawback for averaging is not the extra time required for the exam, but rather the fact that the image processing is not automated. Ultimately, for routine clinical use, the registration and averaging must be done automatically immediately after acquisition of scans.

The most widespread approach to improve MR image quality is to take advantage of higher field magnets: 3 to 7 tesla (T). This approach provides the most effective method for increasing SNR with the fewest limitations. High-field imaging does not require special coils with limited brain coverage nor does it require any postprocessing or extended time in the magnet. In the range of 1.5 to 7 T, the increase in SNR is approximately linear. As with the other methods, by decreasing voxel size, the improvement in SNR can be traded for increased spatial resolution. Additionally, the dynamic range in $T_1$ contrast can be increased using new pulse sequences that exploit the SNR advantage relative to standard field strength images. One issue that remains with high-field MRI is the availability of such scanners, but, as this is written, it appears that 3-T instruments are in routine use at most major medical centers in the United States, Europe, and Australia. Still, higher field scanners are not available to most of the world's epilepsy population. Although it is clear that higher field strength provides a better quality image compared to 1.5-T scanners, it has not been shown in comparative efficacy studies that 3-T images increase diagnostic utility. One study that compared hippocampal and amygdala volumes between 1.5-T and 3.0-T MRI demonstrated no differences in measures (Scorzin et al., 2008).

When optimal MRI acquisition is completed, postprocessing techniques can be exploited to extract more information from the anatomical data that may not be apparent on standard planar tomographic visualization. Multiplanar reconstruction, curvilinear reformatting, surface reconstruction, and voxel-based morphometry (VBM) are either developed or emerging techniques that have been shown to benefit MRI interpretation, especially in epilepsy imaging, which often requires detection of very subtle abnormalities.

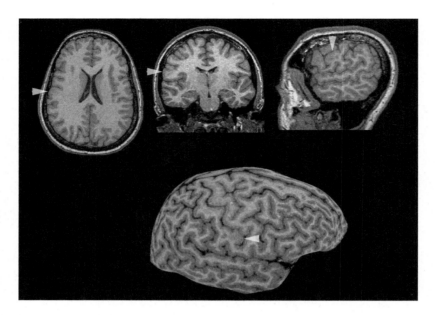

**FIGURE 15.3** Curvilinear multiplanar reformatting (CMPR). Cryptogenic cortical dysplasia (upper images) are standard orthogonal multiplanar reformatted transverse, coronal, and sagittal slices. The lesion (white arrows) on these images cannot be unequivocally interpreted to show focal cortical gray mantle thickening or blurring of the gray matter–white matter junction because most of the gyri are not orthogonally sliced. The lower CMPR image definitively reveals the lesion, for which histopathologic examination confirmed the diagnosis of Taylor Type II cortical dysplasia. (Courtesy of Alexandre Bastos, Department of Neurology and Neurosurgery and Brain Imaging Center, Montreal Neurological Hospital and Institute, McGill University, Montreal, Quebec, Canada.)

Subtle focal lesions, particularly focal malformations of cortical development that involve thickening of the cortex and blurring of gray–white matter interfaces, can sometimes be best detected with multiplanar reformatting of 3D high-resolution MRI (Barkovich et al., 1995). However, the complex gyral anatomy causes oblique slicing of much of the cortical gray matter ribbon. Interactive online arbitrary reslicing by neuroimagers helps to address this problem somewhat; however, it is extremely time consuming to attempt correction of all oblique slicing of the entire brain cortex. Reslicing in an increasingly deep concentric fashion with serial curved slices using the recently developed method curvilinear multiplanar reformatting (CMPR) solves the problem (Bastos et al., 1999). This type of slicing results in an approximately perpendicular orientation of the slices in relation to inward-folding gyri. As a result, artificial cortical thickening and partial volume averaging are minimized or even eliminated in many regions. Also, the method preserves topographical landmarks that are apparent from 3D reconstructions. In a report of applying this technique to five patients in whom conventional 2D and 3D MRI analysis, including multiplanar reformatting, was initially considered normal, five patients were detected to have focal cortical dysplasia with CMPR (Bastos et al., 1999). In four of these patients, histological diagnosis confirmed the finding. Figure 15.3 provides an example of CMPR in a patient with a questionable MCD in the right lateral frontal lobe region. With CMPR focal cortical thickening and blurring of the gray–white matter interface makes the potential abnormality unequivocal. CMPR can be seen to offer a true novel advance in image postprocessing to reduce the number of patients considered to have normal neuroanatomical imaging from even recent state-of-the-art MRI acquisitions.

Three-dimensional surface reconstruction of high-resolution MRI datasets allows a unique perspective for analysis of MCDs that cannot be appreciated by visual analysis of 2D tomographic slicing. Despite the variability and complex gyral patterns that exist from person to person, abnormal patterns can still be easily detected, especially when the abnormalities are unilateral. Symmetrical images that preserve tomographic landmarks allow valid comparison between homologous gyri. Three-dimensional surface visualization also correlates best to what is actually seen in the operating room. Reflection of deeper abnormalities to the surface can also be performed such that surgeons can plan an accurate and exact strategy even beyond that of a better visual approach (Chabrerie et al., 1997). The most logical use of 3D surface reconstructions is integrating the reconstructed volume into frameless stereotaxy systems.

Voxel-based morphometry (VBM) (Ashburner and Friston, 2000) in its simplest form is a voxel-by-voxel-based comparison of local concentrations of gray matter between a patient (or patient group) and a normal control group using, for example, statistical parametric mapping (SPM) (Acton and Friston, 1998). VBM represents a morphometric feature analysis beyond that of simple volume measurements, such as hippocampal volumetry. It can detect structural abnormalities that are overlooked with visual inspection. Sisodiya and colleagues (1995) have shown that VBM can detect much more extensive structural disturbances outside visually identified focal lesions of cortical dysplasia. The questions of interest involve the nature of the volumetric disturbance detected by VBM in these cases, and what impact the presence or absence of widespread abnormalities has on the success of surgical resection of only the visible lesion. At least in patients with hippocampal sclerosis, the answer to the latter question appears to be that patients that have cryptic extrahippocampal structural disturbances detected by VBM are less likely to be seizure free than those who do not have such abnormalities outside of the typical resection volume (Sisodiya et al., 1997).

Voxel-based morphometry does not bias particular structures and includes assessment of the whole brain. Taking further advantage of nonlinear registration techniques, VBM can be used to identify differences in relative position of structures (deformation-based morphometry) or differences in local shape of brain structures (tensor-based morphometry) (Ashburner and Friston, 2000). These methods, however, are not computationally practical at this time for examining small local differences from one subject compared to a group, such as a group of normals. In the future, it can be envisioned that, within a multivariate framework, VBM-based feature analysis could include disturbances of local indices of gyrification and image gradients. In particular, the corrections computed for registration between a patient to an appropriate normal control group can be used to discern differences in gyral anatomy that are beyond the range of normal variation in regions of interest or the entire brain (Ashburner and Friston, 2000).

## DIFFUSION TENSOR IMAGING

Diffusion tensor imaging (DTI) is a novel magnetic resonance imaging technique that provides quantitative information on axonal organization by measuring the random motion of water molecules (diffusion). In brain tissue, water diffusion is restricted by boundaries such as cell membranes or myelin sheaths. In sections of brain white matter with strongly aligned axons, molecules preferentially move parallel to the orientation of the axons rather than perpendicular to them (high anisotropy). The measurement of the molecular displacement in different directions gives indirect information about the microstructural architecture of the brain (Duncan, 2002; Mori and Zhang, 2006). Diffusion tensor imaging measures both anisotropy (direction) and diffusivity (magnitude). Anisotropy is influenced microscopically by intraaxonal organization, density of fiber and neuroglial cell packing, degree of myelination and individual fiber diameter (Beaulieu et al., 1996; Chenevert et al., 1990; Melhem et al., 2002; Pierpaoli et al., 1996). There are three possibilities that would lead to lower anisotropy: an increase in transverse (short axes) diffusivity (type 1 anisotropy loss), a decrease in axial (the longest axis) diffusivity (type 2 anisotropy loss), and a combination of the two (type 3 anisotropy loss). Demyelination often leads to type 1 loss, and axonal injury leads to type 2 anisotropy loss (Song et al., 2002). An explanation for these results is that demyelination leads to less restriction (diffusion barrier) in transverse diffusivity and that axonal injury causes a disarray of axons that reduces axial diffusivity. Trace and mean diffusivity are considered to reflect cellular density and extracellular volume (Gass et al., 2001). Increased diffusivity is likely correlated with neuronal loss and gliosis (Duncan, 2002).

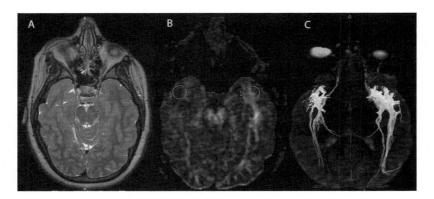

**FIGURE 15.4** **(See color insert following page 458.)** Fiber tractography depicting a reduction of the subcortical fibers and a connection between the white matter subadjacent to the cortical mantle and deep white matter in a malformation of cortical development (pathology; focal cortical dysplasia type IIa). (A) Decreased white matter volume in right temporal pole with poor gray matter–white matter differentiation. (B) Regions of interest drawn on colored fractional anisotropy (FA) map. (C) Relatively scant and poorly arborized white matter in the right temporal pole.

Neuronal migration occurs as a radial growth pattern from the ependymal portion of the embryonic brain to the cortex by the guidance of a glial fiber system. As development of the gray and white matter is closely linked, developmental malformations can affect both the cortex and the underlying white matter (Barth, 1987). Cortical lesions that occur during development include malformations of cortical development (MCD), such as cortical dysplasia, tuberous sclerosis complex (TSC), and developmental tumors, and they have the potential for generating aberrant fiber pathways due to a disruption of normal white matter formation.

Diffusion tensor imaging–fiber tractography (FT) can be used to evaluate the integrity of the white matter adjacent to developmental lesions by depicting an aberrant course of the underlying white matter tract. DTI–FT also localizes the lesion by showing a decreased connection between the deep white matter and the cortex and decreased volume of fiber bundles in the affected cortex. Lee et al. (2004) showed that semiquantitative analysis of fiber bundles adjacent to dysplastic cortex showed a significant mean fractional anisotropy (FA) reduction in the longitudinal fibers of the affected hemisphere in comparison to the normal contralateral region, and that fiber tractography localized the lesion by showing a decreased connection between the deep white matter and the cortex in seven cases in this study who received surgical resection. Figure 15.4 shows these findings in a patient with type II focal cortical dysplasia who was seizure free after epilepsy surgery. Chandra et al. (2006) reported that, in patients with tuberous sclerosis complex, larger volumes of hypometabolism relative to tuber size using FDG–PET/MRI coregistration and higher apparent diffusion coefficient (ADC) values in subtuber white matter had improved sensitivity, specificity, and accuracy compared with larger tuber volumes by structural MRI and FA values. The use of this noninvasive clinical approach to select tubers and adjoining cortex for surgical removal gave TSC patients a greater chance to be seizure free.

Neurons communicate with each other via axons in neural networks. The structural mapping of such networks during health and disease states is essential for understanding brain function; however, our understanding of brain structural connectivity is surprisingly limited, due in part to the lack of noninvasive methodologies to study axonal anatomy. In temporal lobe epilepsy, the most common localization-related epilepsy, pathologic changes have been reported beyond the temporal lobe with extension to the contralateral limbic system, thalami, and distant extratemporal lobe structures such as internal capsule, external capsule, and corpus callosum (Arfanakis et al., 2002; Concha et al., 2006; Gross et al., 2006; Hermann et al., 2003; Kim et al., 2008; Kimiwada et al.,

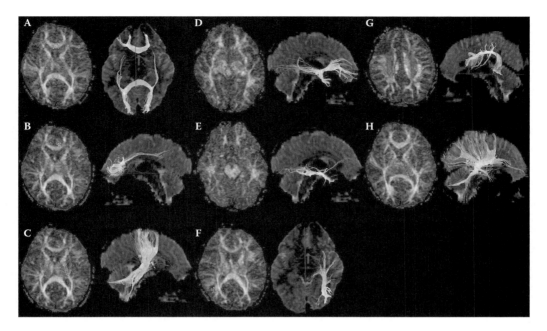

**FIGURE 15.5** **(See color insert following page 458.)** Anatomic regions of interest and fiber tractography. (A) Corpus callosum, (B) cingulum, (C) posterior limb of internal capsule: corticospinal tract, (D) inferior fronto-occipital fasciculus, (E) inferior longitudinal fasciculus, (F) posterior corona radiata, (G) superior longitudinal fasciculus, and (H) thalamocortical projection.

2006; Thivard et al., 2005). These results suggest that the epileptogenic zone, through abnormal neuronal firing and hyperexcitability, produces both functional and structural changes in a neuronal network along seizure propagation pathways.

Diffusion tensor imaging–fiber tractography is also an anatomic imaging tool that can delineate the extent and direction of fiber tracts. White matter tracts have been classified into five functional categories: (1) brainstem tracts, (2) projection fibers (cortex–subcortex connections), (3) association fibers (cortex–cortex connections), (4) limbic system tracts, and (5) commissural fibers (right–left hemispheric connections). Projection fibers include corticospinal tracts (motor function), thalamocortical projections (sensory function), and the visual pathways. Association fibers include cortical U fibers and four well-documented association fiber bundles: superior longitudinal fasciculus, inferior longitudinal fasciculus, inferior fronto-occipital fasciculus, and uncinate fasciculus (Figure 15.5) (Jellison et al., 2004; Wakana et al., 2004).

Diffusion tensor imaging–fiber tractography has been used to spare the eloquent function in tumor surgery trajectory. Corticospinal white matter, arcuate fascicule, and optic radiation transmit somatomotor, conductive language, and visual function, respectively. In temporal lobe epilepsy the optic radiations have been of most interest as an area of risk. Contralateral superior homonymous visual field defects (VFDs) occur in approximately 10% of patients after anterior temporal lobectomy, presumably caused by disruption of Meyer's loop. The anterior extent of Meyer's loop is not well localized on conventional imaging, and the extent of a postoperative VFD cannot be accurately predicted by conventional MRI or from the extent of the resection performed. The mean distance between the most anterior part of Meyer's loop and the temporal pole shows considerable variability by DTI–FT, with reports measuring it between 44 mm (Nilsson et al., 2007) and 37 mm (Yamamoto et al., 2005). Meyer's loop did not reach the tip of the temporal horn in any subject. A disruption in Meyer's loop could be demonstrated in a patient with quadrantanopia after temporal lobe resection. Even resection less than 31.3 mm from the tip resulted in visual field defects in 50% of operated patients, indicating that the anterior extent of the optic nerve radiations runs far more rostral than the tip of the temporal horn of the lateral ventricle (Krolak-Salmon et al., 2000). Power et al. (2005) suggested that that the use of DTI tractography preoperatively to image the anterior boundary of Meyer's loop will enable us to identify those patients at greatest risk of VFDs.

Both DTI and DTI–FT are useful techniques for evaluating the etiology and consequences of epilepsy. DTI and DTI–FT provide additional information to localize the epileptogenic lesion or cortex in cases of conventional MRI-negative focal epilepsy and to define functional neural network and its change or disturbance secondary to epilepsy. DTI–FT is helpful for mapping the eloquent white matter tract for minimizing postoperative neurologic deficits, including motor, language, and visual function.

## FUNCTIONAL IMAGING

### Positron Emission Tomography

Positron emission tomography (PET) is the most established functional imaging modality in the evaluation of patients with epilepsy. Its largest clinical application involves 2-[$^{18}$F]fluoro-2-deoxy-D-glucose (FDG)–PET. Imaging the tomographic distribution of glucose uptake in the brain with FDG–PET is equated with imaging cerebral metabolism. Ictal scans can be useful; however, the long duration for steady-state uptake of glucose (on the order of many minutes compared to partial seizures, which are typically less than a couple of minutes) often leads to scans that contain a mixture of interictal, ictal, and postictal states that is difficult to interpret. Presurgical epilepsy scans are typically performed in the interictal state, with the goal of detecting relative focal areas of decreased metabolism (hypometabolism), which are presumed to reflect focal functional disturbances of cerebral activity associated with the epileptogenic substrate. What is still remarkable about FDG–PET is that the cause of hypometabolism in and near epileptogenic regions of brain remains unclear. What is known is that it is not simply related to the presence of a structural lesion or frequency of epileptiform discharges as recorded by EEG.

From a purely imaging standpoint, FDG–PET provides a remarkable depiction of *in vivo* glucose metabolism. It offers relatively high resolution due to the fact that positron-emitting isotopes emit two photons after a positron annihilates with an electron. The distance of travel between a positron and an electron annihilation is very small (on the order of a millimeter or less). There is a relatively good, intrinsic SNR. With state-of-the-art cameras and collimators, images can now be obtained with full width at half maximum in-plane resolution as small as 2 to 3 mm (Valk et al., 1990).

The FDG–PET technique has been proven most valuable in the evaluation of medically refractory epilepsy surgery candidates with clinically suspected temporal lobe epilepsy. Because the availability of PET arose before high-resolution structural imaging, it was the only modality for years that might show an abnormality of brain imaging in surgical epilepsy candidates. Sensitivity for detecting relative temporal lobe hypometabolism with FDG–PET in MTLE ranges between 80 and 90% (Gaillard et al., 1995; Knowlton et al., 1997; Ryvlin et al., 1992, 1998; Swartz et al., 1992; Valk et al., 1993). Much of the variability in sensitivity reflects heterogeneity of the epilepsy more than the differences in quality or specifications of the PET camera (Henry et al., 1993a). Specificity for delineating the exact location and extent of the epileptogenic zone is considered significantly less than sensitivity. This is partly due to the more diffuse or regional relative hypometabolism seen in the temporal lobe, even with additional involvement of extratemporal, basal ganglia, and thalamic regions (Henry et al., 1993c). Still, the classic pattern of relative hypometabolism in MTLE (without lesions other than mesial temporal sclerosis) involves mesial temporal, temporal polar, and anterior lateral temporal neocortical regions (Hajek et al., 1993; Henry et al., 1993c).

When structural lesions are present, the presence of hypometabolism is effectively 100%, although it is commonly distributed over a larger area than the lesion itself (Ryvlin et al., 1992; Theodore et al., 1986). The question arises as to whether the distribution of hypometabolism beyond the lesion reflects a functional disturbance directly related to the epileptogenic zone. If not, no clinical utility of FDG–PET would exist for patients with focal epileptogenic lesions including hippocampal sclerosis. Hippocampal sclerosis is the most common epileptogenic pathology in epilepsy surgery candidates, and it can be very reliably detected with MRI (Jackson et al., 1993a). Initially it was believed that the more extensive anterior temporal hypometabolism seen in patients with hippocampal sclerosis was a secondary functional disturbance, leading to the conclusion that FDG–PET was nonspecific with respect to localization of the epileptogenic zone and therefore should not guide epilepsy surgery without other supportive evidence for localization. In temporal lobe epilepsy, this long-held belief has since been questioned by investigators who have shown that the region of seizure onsets, as defined by intracranial EEG recordings in patients with classic evidence of hippocampal sclerosis, frequently involves the temporal polar region as much as the hippocampus (Ryvlin et al., 1997). Thus, an important question arises as to whether the extent of temporal lobe hypometabolism should influence the surgical decision with regard to a standard anterior temporal lobectomy vs. a selective amygdalohippocampectomy.

The clinical value of FDG–PET in neocortical epilepsy is less clear. Earlier studies reported only modest sensitivity (da Silva et al., 1997; Henry et al., 1991; Lee et al., 1994; Swartz et al., 1989). More recently, a few notable larger retrospective studies have addressed our gap in knowledge regarding FDG–PET in neocortical epilepsy (Ryvlin et al., 1998; Won et al., 1999). They show, perhaps surprisingly given early reports, a relatively high diagnostic sensitivity (around 50%), with even normal MRI cases having a sensitivity ranging between 32 and 44%. Most important, it should be emphasized that focal defects in metabolism can be found in patients that have cryptogenic lesions. Specifically, some focal cortical dysplasias that still cannot be detected with MRI can be detected with FDG–PET. The question that remains is that of specificity. A study correlating neocortical focal hypometabolism with intracranial subdural grid EEG recordings found that not only is the distribution of hypometabolism often greater in extent but it also is more likely to be greatest on the margin of the region of seizure onset (Juhasz et al., 2000b). More recently, a multivariate analysis has revealed concordant localized FDG–PET to be an independent predictor of seizure-free outcome (Yun et al., 2006). The findings from a another recent prospective appear to support the same interpretations that localized FDG–PET has both a diagnostic value (sensitivity, 59%; specificity, 79%) and independent predictive value for seizure-free outcome (Knowlton et al., 2008).

These persistent issues with FDG–PET have elicited investigations into the use of flumazenil–PET (FMZ–PET) to see if gamma-aminobutyric acid type A receptor (GABA$_A$) binding disturbances are more focal and specific than hypometabolism. Imaging GABA receptor binding was a logical choice for advanced functional imaging of epilepsy with PET due to the well-established relationship of disturbed GABAergic function in epileptogenic tissue. Indeed, early studies of temporal lobe epilepsy (small series with highly selected patients with unilateral hippocampal sclerosis) revealed that disturbances of GABA receptor binding were highly sensitive and more localized to epileptogenic mesial temporal-only tissue, in contrast to the more diffuse anterior temporal hypometabolism seen with FDG–PET (Burdette et al., 1995; Henry et al., 1993b; Koepp et al., 1997; Savic et al., 1993).

The clinical utility of FMZ–PET was investigated in a large series of patients with all types of partial epilepsy. Somewhat surprisingly, this study concluded that FMZ–PET was not more clinically useful than FDG–PET (Ryvlin et al., 1998) and that it could be falsely lateralizing in temporal lobe epilepsy (Ryvlin et al., 1999). However, its value was noted for patients with bilateral hippocampal abnormalities on MRI. Also, the region of overlap of hypometabolism and decreased GABAergic function was highly specific for the epileptogenic zone as defined by intracranial ictal EEG. Countering this study is work from the group at Wayne State University comparing FMZ– and FDG–PET in children mostly with extratemporal lobe epilepsy. They reported that FMZ–PET had greater specificity. In examining lesional epilepsy, they found that the distribution of abnormal FMZ–PET in the perilesional region correlates with electrocorticographically defined epileptogenic tissue, the distribution of which is commonly eccentric in position with respect to the lesion (Juhasz et al., 2000a). They concluded that FMZ–PET offered the possibility of accurately

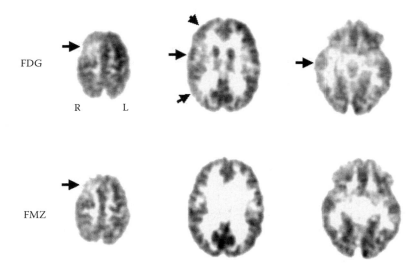

**FIGURE 15.6** FDG vs. FMZ PET. (Top) FDG–PET images show widespread right frontal–parietal–temporal relative hypometabolism in this patient with frontal lobe onset-only seizures. (Bottom) FMZ–PET images show decreased benzodiazepine receptor binding that is confined to the right frontal lobe, where this patient's seizures were confirmed to arise as defined by intracranial ictal EEG. (Courtesy of Csaba Juhasz, PET Center, Children's Hospital of Michigan, Department of Pediatrics, Wayne State University School of Medicine.)

identifying perilesional epileptogenic cortex, which should be removed in addition to the lesion for optimal surgical outcome. In another study comparing PET abnormalities to intracranial EEG, it was concluded that FMZ–PET, although not sufficient to preclude intracranial EEG recording, should be an effective tool for guiding placement of intracranial electrodes (Muzik et al., 2000). Figure 15.6 provides an example of more localized abnormal GABA receptor binding as compared to widespread hypometabolism in a patient with cryptogenic frontal lobe epilepsy. Finally, in a more direct study with respect to outcome, the main reason for surgical failure was residual epileptogenic tissue (Juhasz et al., 2001). The potential delineation by FMZ–PET of epileptogenic regions in lesional as well as nonlesional cases, even in patients with normal MRI, should be able to positively affect surgical outcome. Indeed, it was found that a better outcome occurred when a greater extent of FMZ–PET disturbance was included in the resection. In contrast, the extent of preoperative or nonresected cortex that was abnormal on FDG–PET was not related to surgical outcome.

## SINGLE PHOTON EMISSION COMPUTED TOMOGRAPHY

Epileptic seizures are typically rare events that occur in a random fashion. This fact constrains most clinically useful functional imaging modalities to more practical targets, those derangements present in the interictal state. Single photon emission computed tomography (SPECT) is the only imaging modality suited to revealing changes in brain activity present at the time of an epileptic seizure. This technique quantifies regional cerebral blood flow, relying on its known proportional linkage to synaptic activity, for seizure localization. At the heart of the practical advantage of SPECT is the technetium-99m ($^{99m}$Tc)-based radiopharmaceutical, which, in the case of ictal SPECT, is injected intravenously at the first sign of seizure activity. Once injected, these small, neutral, and lipophilic molecules pass rapidly from the blood into cerebrovascular endothelial cells with high first-pass blood extraction. The amount of radiotracer entering an endothelial cell depends on delivery, which is based on local cerebral blood flow, which in turn is proportional to focal synaptic activity. Once inside the cerebrovascular endothelial cells, radiotracer retention is facilitated by conversion to a hydrophilic form via intracellular glutathione or esterase systems. This unique combination of high extraction rate, low redistribution, and rapid washout from non-brain tissues produces a latent radioactive image or *snapshot* of the cerebral blood flow associated with seizure onset. In practical terms, its stable retention and 6-hour radioisotope half-life give the clinical team a window of at least 2 hours after injection during which the patient may be stabilized, transported, and scanned. As such, ictal SPECT has demonstrated its ability to provide reliable novel localizing information in the presurgical evaluation of partial epilepsies of all types (Bilgin et al., 2008; Burneo et al., 2006; Ho et al., 1994; Koh et al., 1999; Kuzniecky et al., 1993b; Marks et al., 1992; Newton et al., 1995; Shen et al., 1990; Spanaki et al., 1999; Sturm et al., 2000).

Two SPECT radiotracers are available commercially: technetium-99m–hexamethylpropylene amine oxime ($^{99m}$Tc-HMPAO) and technetium-99m–ethyl cysteinate diethylester ($^{99m}$Tc-ECD). Although they are comparable in most respects, the two exhibit minor differences in mechanisms of action, performance, and practicality. $^{99m}$Tc-HMPAO (Ceretec™) was the first to see commercial availability and has been used in the vast majority of ictal or peri-ictal SPECT studies in epilepsy. It has a high extraction rate, and a newer formulation containing cobalt chloride overcomes prior *in vitro* instability that required reconstitution just prior to injection (Lee et al., 2002). The more recently available $^{99m}$Tc-ECD (Neurolite®) is stable *in vitro* and has a high renal clearance rate, which lowers initial uptake in extracerebral tissues; however, because of a more complicated mechanism of transmembrane passage, $^{99m}$Tc-ECD exhibits a somewhat lower brain extraction than $^{99m}$Tc-HMPAO. The intracellular transformation of both radiotracers to lipophilic species enhances their retention and is accomplished by cytosolic glutathione or esterase in the case of $^{99m}$Tc-HMPAO or $^{99m}$Tc-ECD, respectively (Jacquier-Sarlin et al., 1996).

The same defining features that make ictal SPECT unique among presurgical epilepsy imaging modalities also represent its greatest challenge—the fact that it is an ictal study. A primary limitation is associated with the difficult logistics of successfully infusing the isotope as early as possible after seizure onset. Prompt ictal injection reduces the chance of registering ambiguous ictal blood flow changes and is therefore critically important, particularly with the rapidly spreading seizures of neocortical epilepsy. Compounding this urgency for early injection is the fact that patients with neocortical epilepsy are the most likely to benefit from the additional localizing information of high-quality ictal SPECT. The results of late ictal injection may be particularly misleading if a brain region remote from the seizure onset zone takes over as the main generator of seizure activity. Unfortunately, dedication to the highest quality injection is associated with the high cost of having healthcare staff sitting at the patient's bedside waiting to inject at the first sign of seizure onset. Some centers have devised and implemented automated or semiautomated injection systems with the help of basic engineering groups, but these have not been widely adopted (Feichtinger et al., 2007; Van Paesschen et al., 2000). Additional logistical considerations include the need for onsite facilities and expertise to produce the short-half-life $^{99m}$Tc isotope ($T_{1/2} < 6$ hours), as well as providing the appropriate transportation, storage, and training for use of radioactive pharmaceuticals in the epilepsy monitoring unit (Burneo et al., 2007).

An advance that has dramatically improved the clinical utility of ictal SPECT is digital image subtraction of the patient's interictal scan from the ictal scan (Zubal et al., 1995). Subtraction ictal SPECT and coregistration to MRI (SISCOM) has been shown to enhance the detection of blood flow differences as they vary in the region of seizure onset between interictal hypoperfusion and ictal hyperperfusion states (O'Brien et al., 1998). SISCOM consists of spatial coregistration of ictal and interictal scans, followed by normalization of pixel intensities to the mean intensity, and finally voxel-by-voxel subtraction of the two transformed images to produce a difference image. Thresholding this image enhances interpretability by displaying only difference voxels that are statistically significant (i.e., greater than two standard deviations above zero). Finally, this relatively low-resolution subtraction image is coregistered to the patient's own MRI, thereby incorporating anatomical landmarks that allow precise localization of significant features. Studies in patients with intractable partial epilepsy have demonstrated improved sensitivity and specificity of seizure

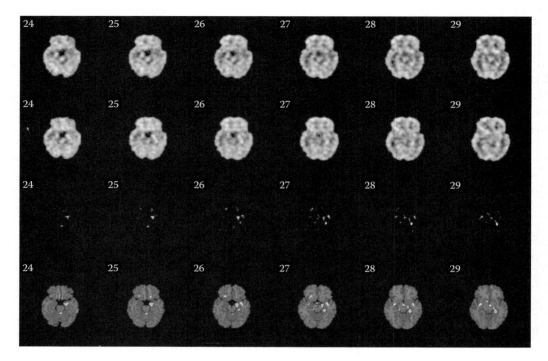

**FIGURE 15.7** Subtraction ictal SPECT and coregistration to MRI. Third row shows the pixel data remaining after precise SPECT–SPECT coregistration, pixel intensity normalization, and subtraction of interictal (top row) from ictal (second row) of cerebral blood flow images. Bottom row images with coregistered MRI reveal the anatomic localization of increased blood flow associated with a complex partial seizure in this patient with left mesial temporal lobe epilepsy. Localization is remarkably confined to the left amygdalo–hippocampal region, where the seizure was confirmed to arise and remain with little spread with recording by scalp-sphenoidal EEG electrodes.

localization, better interobserver agreement, and better concordance with EEG and MRI localization than traditional side-by-side interpretation of ictal and interictal scans (Ahnlide et al., 2007; Cascino et al., 2004; Lewis et al., 2000; Tan et al., 2008; Wetjen et al., 2006).

Figure 15.7 demonstrates the concept of perceived normal blood flow in an ictal scan that truly is a relative increase from that which is below normal from the scan in the interictal state. The figure further shows the precise localization of this relative increase in blood flow to that which is remarkably confined to the hippocampus, the small specific region of brain that was exclusively involved as the region of seizure activity.

Another direction of research that promises to yield more objective ictal SPECT results is based on the neuroimaging analysis technique of statistical parametric mapping (SPM) (Chang et al., 2002; Friston, 1994; Knowlton et al., 2004b). As with SISCOM, this technique relies on coregistration of SPECT datasets to a standard space and normalization of intensity; however, once transformed, SPM uses the general linear model to generate a voxel-wise statistical representation of the group of SPECT scans. This three-dimensional set of statistical parameters can then be used to test statistical hypotheses about these groups in a voxel-to-corresponding-voxel fashion. Clinical studies typically compare the patient's ictal–interictal SPECT difference image to pooled and parameterized difference images from a group of healthy normal subjects. The original difference image from each normal subject is generated by subtracting two interictal scans obtained on separate days. Individual voxel intensities in the resulting patient-normal comparison image, or Z-map, each represent a Z-score. This quantity shows how statistically significant a change in the patient's interictal to ictal perfusion is when compared to the variance of the corresponding voxel in the normal group's parameterized data. Thresholding to display only those voxels with high significance changes, greater than a user-specified $p$ value (e.g., $p < 0.001$), adds additional clarity. Optionally, the second-order threshold of cluster size is used to eliminate physiologically irrelevant small clusters. It is implemented by displaying only significant voxels that are also in contiguous groups larger than a specified parameter (e.g., $k = 300$ voxels). Finally, coregistration of this image to the patient's MRI allows precise localization of significant ictal activity. A growing number of studies have demonstrated that the SPM approach provides objective and accurate analysis of ictal–interictal SPECT data in patients with temporal lobe and extratemporal epilepsy (Kim et al., 2007; Knowlton et al., 2004a; Lee et al., 2000; McNally et al., 2005; Tae et al., 2005; Van Paesschen et al., 2003).

Studies utilizing this statistical approach continue to advance our understanding of the dynamics of seizure onset and spread by allowing group analyses of patients with specific seizure types. For example, SPM-based studies of temporal lobe seizures have demonstrated not only temporal lobe hyperperfusion but also vast areas of ictal hypoperfusion in the ipsilateral frontal and parietal lobes. This finding is thought to explain a patient's inability to understand or interact with his or her environment during a complex partial seizure of temporal lobe origin (Blumenfeld et al., 2004; Dupont et al., 2009). Other areas of exploration include patients with juvenile myoclonic epilepsy (JME) and other idiopathic generalized epilepsies, pediatric patients with temporal and frontal lobe epilepsy, and those patients undergoing electroconvulsive shock therapy (Enev et al., 2007; Joo et al., 2008; Tae et al., 2007). SPM has also lent itself to multimodality studies examining SPECT/PET ratio images, which have been shown to add seizure lateralizing value over PET alone (Buch et al., 2008; Zubal et al., 2000).

## MAGNETIC RESONANCE SPECTROSCOPY

Magnetic resonance spectroscopy (MRS) is the use of nuclear magnetic resonance to perform *in vivo* metabolite and macromolecule measurements. Protein- and phosphorus-based molecules have been the main chemicals of interest in the study of epilepsy with MRS. Studies use either a single-voxel technique or simultaneous acquisition of an array of voxels or regions using chemical shift imaging. When the latter is used, the technique may be labeled *magnetic resonance spectroscopic imaging* (MRSI).

Proton spectroscopy is primarily based on the detection of $N$-acetyl (NA) compounds (primarily $N$-acetylaspartate [NAA]), creatine, choline, lactate, and glutamate. NA compounds are of particular interest because NAA is found specifically in neurons or their processes in the fully developed central nervous system (Urenjak et al., 1992). The cellular biochemical function of NAA is unknown, although there is some suggestion that it is involved in cell volume regulation and electrolyte homeostasis (Davies et al., 1998; McIntosh and Cooper, 1965; Tsai and Coyle, 1995). Measurement of lactate and glutamate provides information about cellular energy status and localized information about excitatory amino acid concentration respectively. These are valuable tools that have yet to be fully exploited in studies directed toward further understanding of mechanisms underlying focal cellular biochemical disturbances associated with epileptogenic tissue.

Phosphorus-based MRS techniques detect phosphocreatine (PCr), adenosine triphosphate (ATP), inorganic phosphate (Pi), pH from the chemical shift of Pi, free magnesium ($Mg^{2+}$) from the chemical shift of ATP, phosphomonoesters (PMEs), and phosphodiesters (PDEs). PCr, ATP, Pi, and pH provide information concerning bioenergetics, and PDE and PME provide information concerning lipid metabolism.

Proton MRS and MRSI studies of temporal lobe epilepsy have demonstrated a remarkable sensitivity for the detection of relative depletion of NA, both ipsilateral and contralateral to the side of predominant seizure onsets (Cendes et al., 1997b; Connelly et al., 1998; Cross et al., 1996; Ende et al., 1995; Hetherington et al., 1995; Hugg et al., 1993; Knowlton et al., 1997; Kuzniecky et al., 1998; Ng et al., 1994; Vainio et al., 1994). The sensitivity for detecting an ipsilateral depletion of NA is relatively high, ranging from 85 to 100%. Most interesting is the high detection of contralateral abnormalities, with sensitivity ranging between 30 and 50%. Relative decreases in NA are measured in either absolute concentration or as a ratio of NA to creatine (Cr), choline (Cho), or the sum of Cr and Cho. This high sensitivity for detecting abnormalities of NA is seen with both single-voxel and spectroscopic imaging techniques. NA abnormalities are also seen, whether or not hippocampal atrophy or other evidence for mesial temporal sclerosis (a condition defined by hippocampal neuronal cell loss) exists (Cendes et al., 1997b; Connelly et al., 1998; Knowlton et al., 1997; Woermann et al., 1999). As an imaging tool, it is hoped that the predominantly lateralized NA abnormalities can help provide information regarding the site of epilepsy and surgical decision-making. The latter includes the possibility that the degree of disturbance may predict seizure-free outcome as well as postoperative neurological memory function (Capizzano et al., 2002; Incisa et al., 1995; Martin et al., 1999, 2001; Sawrie et al., 2000).

As with other tomographic imaging modalities MRS and MRSI studies are generally performed interictally for the obvious reason that it is difficult to make a patient have a seizure while inside the magnet. Preliminary reports of a couple of patients with partial seizures of temporal lobe origin that occurred in the magnet showed an increase in lactate, but no NA changes (Cendes et al., 1997a). Further, a report of postictal MRSI in patients with temporal lobe epilepsy showed no changes (Maton et al., 2001). Also to be emphasized, and in contradiction to what was initially believed about the concentration of NAA reflecting neuronal cell density, is the evidence that decreases in NA concentrations are reversible (Rango et al., 1995). This has been demonstrated in the evaluation of patients who have and have not become seizure free after anterior temporal lobectomy (Cendes et al., 1997a; Hugg et al., 1996). In patients who became seizure free after surgical treatment, it was shown that their abnormally decreased NA concentrations in the residual ipsilateral and contralateral temporal lobe (prior to surgery) increased or approached the normal range. Thus arises the important concept that MRS imaging of NA may be utilized more than just as a sensitive tool for *in vivo* detection of microscopic neuronal cell loss, as it reflects functional neuronal disturbance. As such it may provide much more predictive information of cognitive dysfunction and even recovery following temporal lobe surgery (Sawrie et al., 2000).

Relatively little work has been published on spectroscopic imaging of extratemporal partial epilepsy (Garcia et al., 1995; Krsek et al., 2007; Stanley et al., 1998). Two groups looked at proton MRSI in the evaluation of patients with malformations of cortical development (Kuzniecky et al., 1997; Li et al., 1998). These studies examined small numbers of patients with focal cortical dysplasia, polymicrogyria, and subcortical heterotopia with proton MRSI protocols. Interestingly, only the cortical dysplasias showed abnormal decreases in NA compounds. Both studies also showed abnormally low NA in neighboring cortex beyond the MRI-defined lesions, an important finding that holds promise for proton MRSI in providing clinically valuable information about the true extent of the epileptogenic zone in patients with cortical dysplasia. The surgical outcome of patients with cortical dysplasias is lower than that of other epileptogenic lesions, and the most common reason for failure is inadequacy of resection as proven by a relatively high success rate with second surgeries in such cases. Dysplastic cortex can be cryptic, but findings of a recent study have provided renewed hope for the role of MRS in revealing metabolite abnormalities in normal MRI neocortical epilepsies (Krsek et al., 2007). In spite of all the potential for proton MRSI based on NA measures, for it to be exploited as a more routine presurgical epilepsy imaging tool the development of fast-acquisition whole-slice and multislice or whole-brain techniques will be required.

In contrast to proton MRS, the regions of interest studied by $^{31}$P-MRSI are rather large due to the small magnitude of signal for phosphorus-based molecules compared to those that are proton based. Thus, most studies have generally confined interpretation of findings to lobar disturbances of $^{31}$P metabolites. Common findings from the two main groups that have published on this work in partial epilepsy are an elevation of inorganic phosphate and a decrease in the phosphocreatine-to-inorganic phosphate ratio (Hugg et al., 1992; Kuzniecky et al., 1992). Controversy exists as to whether an elevation in pH occurs in the epileptogenic lobar region (Chu et al., 1996; Laxer et al., 1992). With a high field magnet (4.1 T), a decrease in the ratio of phosphocreatine to inorganic phosphate was consistently found in temporal lobe regions ipsilateral to the seizure onset (Chu et al., 1998). This abnormality was specific to temporal lobe areas with no relative disturbances in extratemporal regions. One study evaluated patients with frontal lobe epilepsy (Garcia et al., 1994). This included a small number of patients, but in seven of the eight cases studied, a decrease in phosphomonoesters was seen. In contrast to temporal lobe epilepsy, there was no significant change in inorganic phosphate. It should be stated that $^{31}$P-MRSI is still a developing technology and remains far from clinical application for localization or lateralization of either temporal or extratemporal lobe epilepsy.

Overall, active research into the potential roles of MRS and MRSI in clinical applications for epilepsy localization for surgery has diminished. Obstacles to clinical utility will require great improvements in efficient acquisition and better resolution. More likely, advances in MRS for epilepsy imaging will have to be directed toward further understanding the nature and significance of NA disturbances and their correlation with measures of changes in cellular bioenergetics (with both PET and MRS).

## FUNCTIONAL MAGNETIC RESONANCE IMAGING

By taking advantage of blood oxygen level dependence (BOLD), functional magnetic resonance imaging (fMRI) can image neuronal activation (Ogawa et al., 1990). Increased relative neuronal activity is linked with a localized relative increase in cerebral blood flow (CBF) (Fox and Raichle, 1986). The increase in CBF is greater in magnitude relative to the increase in oxygen consumption; therefore, the ratio of increased oxyhemoglobin to deoxyhemoglobin causes a change in paramagnetic effect on $T_2$ relaxation time. This is due to the fact that deoxyhemoglobin is paramagnetic while oxyhemoglobin is not. The magnitude of signal change is small, on the order of 1 to 5%, with imaging at 1.5 T.

Applications of fMRI to epilepsy include mapping of interictal or ictal epileptiform disturbances of cerebral activity recorded with EEG and localization of important cortical function (especially memory and language). PET using [$^{15}$O] is also capable of imaging blood-flow changes associated with stimulus-induced cortical activation, in some ways better than fMRI. However, its requirement for repeated radioactive exposure and rare availability due to the expense of an onsite cyclotron relegate its use mostly to research study of brain function.

Imaging of epileptiform disturbances of cerebral activity by fMRI is a novel application still in development (Krakow et al., 2000; Seeck et al., 1998). The first hurdle was the need to develop techniques such that artifact-free and safe EEG recording could be performed in the MRI environment (Goldman et al., 2000; Krakow et al., 2000). The second hurdle was the low signal change associated with epileptiform discharges. It would be expected that large numbers of discharge-triggered acquisitions would be needed to gain adequate signal to noise. This latter problem has been helped by the increasing availability of 3-T magnets. An initial study of its application to localization of EEG spikes in patients with partial epilepsy showed that 12 of 24 patients had successful fMRI localization that was concordant with EEG-defined seizure onset and associated structural lesions. In 10 patients who did not show significant fMRI activation, spike amplitudes were observed to be much smaller compared to those with positive fMRI results. The potential for this technology as a non-invasive tool in presurgical epilepsy localization is large because of high spatial resolution compared to EEG. Yet, it is important to remember that spikes may not represent localization of seizure onsets. A case report confirms the ability of EEG-triggered fMRI to successfully localize the region of cortical activity involved during a seizure (Lazeyras et al., 2000); however, capturing seizures in MRI units is not logistically applicable to most patients.

Somatotopic mapping of sensory and motor function was demonstrated early in investigations of fMRI (Jack et al., 1994; Puce, 1995; Rao et al., 1993, 1995; Yetkin et al., 1995; Yousry et al., 1995). For epileptogenic lesions neighboring the primary motor or sensory cortex, prior knowledge of critical motor and sensory function has obvious implications for the optimal extent of surgical resection with minimal neurological morbidity (Conesa et al., 1999). Conventionally, identifying primary motor and sensory cortex is based on identification of the central sulcus on anatomic structural imaging (Kido et al., 1980; Sobel et al., 1993); however, this can be difficult in pathological conditions (e.g., tumors and strokes) which may distort normal landmarks. Additionally, in other types of lesions (dysplasia, developmental tumors, trauma, and ischemia), function may be shifted from that expected on normal anatomy. Over the past several years, numerous groups have shown successful somatotopic mapping in the setting of various pathologies with fMRI, and in many centers it is being used routinely for clinical mapping of function (Achten et al., 1999; Chapman et al., 1995; Maldjian et al., 1996; Mueller et al., 1996; Righini et al., 1996; Schad et al., 1996). Most paradigms use simple movement, which robustly activates cortical areas M-I, S-I, and S-II.

Mapping of languages is of particular interest in epilepsy surgery due to its elective nature and because it can commonly occur in the dominant hemisphere. Language involves very complex special sensory systems that entail distributed networks of neuronal populations, including regions in the nondominant hemisphere. Paradigms to activate language cortex have to deal with this complex distribution of neuronal activation. The best results have been with word generation or fluency tasks (listening, reading, repeating, and object naming tasks). Fluency tasks involve mainly word retrieval in response to a verbal cue. Semantic verbal fluency paradigms have demonstrated frontal activation asymmetry that correlates well with the intracarotid amobarbital injection test for determination of language dominance lateralization (the Wada exam) (Bahn et al., 1997; Benson et al., 1999; Binder et al., 1996; Gaillard et al., 2000; Lehericy et al., 2000; Yetkin et al., 1998). Of note, temporal lobe areas of activation did not correlate with the Wada exam. Other approaches that have been successful involve word-pairing comprehension and semantic categorization tasks. One of the difficult issues in designing paradigms for activation of language involves the development of nonlinguistic tasks to act as controls such that primary auditory, attention, working memory, and motor aspects of language can be removed.

Although fMRI also provides relatively detailed maps of intrahemispheric activation, this degree of localization is not a reality yet for assisting surgical planning. Regions not activated by one task may be activated by another. Just because a region is not activated does not mean it lacks important function. Conversely, because a region is activated does not mean it is critical for function. Thus, at this point, application of fMRI for language mapping in epilepsy should be confined to an attempt at lateralizing dominance that may ultimately replace the Wada test. Even with respect to this application, it should be emphasized that only a few atypical or cross-dominant subjects have been studied.

Memory functional lateralization and localization has obvious application for epilepsy surgery. This is mainly due to the fact that the most common procedure is temporal lobe resections, including much of the hippocampus. The initial goals are to determine the function of memory structures in pathological vs. nonpathological temporal lobes such that the risk of amnestic complications can be avoided and asymmetry of memory function may allow epilepsy lateralization. Early studies show promise for fMRI detection of lateralized asymmetries in memory activation (Detre et al., 1998; Killgore et al., 2000). Also, as with the Wada test, functional assessment of memory may provide information with respect to lateralization of temporal lobe epilepsy as well as predictive power for seizure-free outcome (Binder et al., 2000). However, with regard to more specific localization, as with language, memory function is activated as a distributed network; therefore, some regions may be critical for function while others may not. Most importantly, some paradigms may not activate all regions that are critical. Nevertheless, among all MR-based imaging advances, fMRI has the greatest potential clinical application in neuroimaging, and without doubt fMRI will increasingly impact the non-invasive presurgical evaluation of epilepsy more than any other imaging tool in the near future.

## CONCLUSIONS

Many of the advanced methods of both structural and functional imaging developments reviewed in this chapter have been demonstrated to have scientific value and interest in regard to better understanding epilepsy. The likelihood of clinical value is compelling, and many are even used in clinical practice. There is, however, a paucity of evidence for actual clinical utility. The onus is on clinical researchers investigating these and other rapidly advancing epilepsy-imaging tools used for epilepsy surgery to provide clinical validation toward three applications: (1) identifying appropriate candidates not captured with standard tests, (2) optimizing the selection of candidates and avoiding the exclusion of patients that can be successfully treated with surgery, and (3) demonstrating that the tests affect outcome. Without this evidence, it will become increasingly difficult to justify clinical application and exploit possible improvement and change in practice..

# REFERENCES

Abou, K.B., Siegel, G.J., Sackellares, J.C. et al. (1987). Positron emission tomography studies of cerebral glucose metabolism in chronic partial epilepsy. *Ann. Neurol.*, 22, 480–486.

Achten, E., Jackson, G.D., Cameron, J.A. et al. (1999). Presurgical evaluation of the motor hand area with functional MR imaging in patients with tumors and dysplastic lesions. *Radiology*, 210, 529–538.

Acton, P.D., Friston, K.J. (1998). Statistical parametric mapping in functional neuroimaging: beyond PET and fMRI activation studies. *Eur. J. Nucl. Med.*, 25, 663–667.

Ahnlide, J.A., Rosen, I., Linden-Mickelsson Tech P., Kallen, K. (2007). Does SISCOM contribute to favorable seizure outcome after epilepsy surgery? *Epilepsia*, 48, 579–588.

Arfanakis, K., Hermann, B.P., Rogers, B.P. et al. (2002). Diffusion tensor MRI in temporal lobe epilepsy. *Magn. Reson. Imaging*, 20, 511–519.

Ashburner, J., Friston, K.J. (2000). Voxel-based morphometry: the methods. *NeuroImage*, 11, 805–821.

Bahn, M.M., Lin, W., Silbergeld, D.L. et al. (1997). Localization of language cortices by functional MR imaging compared with intracarotid amobarbital hemispheric sedation. *Am. J. Roentgenol.*, 169, 575–579.

Barkovich, A.J. (1996). Malformations of neocortical development: magnetic resonance imaging correlates. *Curr. Opin. Neurol.*, 9, 118–121.

Barkovich, A.J., Kuzniecky, R.I. (1996). Neuroimaging of focal malformations of cortical development. *J. Clin. Neurophysiol.*, 13, 481–494.

Barkovich, A.J., Maroldo, T.V. (1993). Magnetic resonance imaging of normal and abnormal brain development. *Top. Magn. Reson. Imaging*, 5, 96–122.

Barkovich, A.J., Rowley, H.A., Andermann, F. (1995). MR in partial epilepsy: value of high-resolution volumetric techniques. *AJNR Am. J. Neuroradiol.*, 16, 339–343.

Barth, P.G. (1987). Disorders of neuronal migration. *Can. J. Neurol. Sci.*, 14, 1–16.

Bastos, A.C., Comeau, R.M., Andermann, F. et al. (1999). Diagnosis of subtle focal dysplastic lesions: curvilinear reformatting from three-dimensional magnetic resonance imaging. *Ann. Neurol.*, 46, 88–94.

Beaulieu, C., Does, M.D., Snyder, R.E., Allen, P.S. (1996). Changes in water diffusion due to Wallerian degeneration in peripheral nerve. *Magn. Reson. Med.*, 36, 627–631.

Benson, R., FitzGerald, D., LeSueur, L. et al. (1999). Language dominance determined by whole brain functional MRI in patients with brain lesions. *Neurology*, 52, 798–809.

Berkovic, S.F., Andermann, F., Olivier, A. et al. (1991). Hippocampal sclerosis in temporal lobe epilepsy demonstrated by magnetic resonance imaging. *Ann. Neurol.*, 29, 175–182.

Bilgin, O., Vollmar, C., Peraud, A. et al. (2008). Ictal SPECT in Sturge–Weber syndrome. *Epilepsy Res.*, 78, 240–243.

Binder, J.R., Swanson, S.J., Hammeke, T.A. et al. (1996). Determination of language dominance using functional MRI: a comparison with the Wada test. *Neurology*, 46, 978–984.

Binder, J.R., Bellgowan, P., Swanson, S. et al. (2000). FMRI activation asymmetry predicts side of seizure focus in temporal lobe epilepsy. *NeuroImage*, 11, 155.

Blumenfeld, H., McNally, KA., Vanderhill, SD. et al. (2004). Positive and negative network correlations in temporal lobe epilepsy. *Cereb. Cortex*, 14, 892–902.

Buch, K., Blumenfeld, H., Spencer, S. et al. (2008). Evaluating the accuracy of perfusion/metabolism (SPET/PET) ratio in seizure localization. *Eur. J. Nucl. Med. Mol. Imaging*, 35, 579–588.

Burdette, D.E., Sakurai, S.Y., Henry, T.R. et al. (1995). Temporal lobe central benzodiazepine binding in unilateral mesial temporal lobe epilepsy. *Neurology*, 45, 934–941.

Burneo, J.G., Hamilton, M., Vezina, W., Parrent, A. (2006). Utility of Ictal SPECT in the presurgical evaluation of Rasmussen's encephalitis. *Can. J. Neurol. Sci.*, 33, 107–110.

Burneo, J.G., Vezina, W., Romsa, J. et al. (2007). Evaluating the development of a SPECT protocol in a Canadian epilepsy unit. *Can. J. Neurol. Sci.*, 34, 225–229.

Capizzano, A.A., Vermathen, P., Laxer, K.D. et al. (2002). Multisection proton MR spectroscopy for mesial temporal lobe epilepsy. *AJNR Am. J. Neuroradiol.*, 23, 1359–1368.

Cascino, G.D., Jack, C.R., Parisi, J.E. et al. (1991). Magnetic resonance imaging-based volume studies in temporal lobe epilepsy: pathological correlations. *Ann. Neurol.*, 30, 31–36.

Cascino, G.D., Buchhalter, J.R., Mullan, B.P., So, E.L. (2004). Ictal SPECT in nonlesional extratemporal epilepsy. *Epilepsia*, 45(Suppl. 4), 32–34.

Cendes, F., Andermann, F., Dubeau, F. et al. (1997a). Normalization of neuronal metabolic dysfunction after surgery for temporal lobe epilepsy: evidence from proton MR spectroscopic imaging. *Neurology*, 49, 1525–1533.

Cendes, F., Caramanos, Z., Andermann, F. et al. (1997b). Proton magnetic resonance spectroscopic imaging and magnetic resonance imaging volumetry in the lateralization of temporal lobe epilepsy: a series of 100 patients. *Ann. Neurol.*, 42, 737–746.

Chabrerie, A., Ozlen, F., Nakajima, S. et al. (1997). Three-dimensional reconstruction and surgical navigation in pediatric epilepsy surgery. *Pediatr. Neurosurg.*, 27, 304–310.

Chandra, P.S., Salamon, N., Huang, J. et al. (2006). FDG–PET/MRI coregistration and diffusion–tensor imaging distinguish epileptogenic tubers and cortex in patients with tuberous sclerosis complex: a preliminary report. *Epilepsia*, 47, 1543–1549.

Chang, D.J., Zubal, I.G., Gottschalk, C. et al. (2002). Comparison of statistical parametric mapping and SPECT difference imaging in patients with temporal lobe epilepsy. *Epilepsia*, 43, 68–74.

Chapman, P.H., Buchbinder, B.R., Cosgrove, G.R., Jiang, H.J. (1995). Functional magnetic resonance imaging for cortical mapping in pediatric neurosurgery. *Pediatr. Neurosurg.*, 23, 122–126.

Chenevert, T.L., Brunberg, J.A., Pipe, J.G. (1990). Anisotropic diffusion in human white matter: demonstration with MR techniques *in vivo*. *Radiology*, 177, 401–405.

Chu, W.J., Hetherington, H.P., Kuzniecky, R.J. et al. (1996). Is the intracellular pH different from normal in the epileptic focus of patients with temporal lobe epilepsy? A $^{31}P$ NMR study. *Neurology*, 47, 756–760.

Chu, W.J., Hetherington, H.P., Kuzniecky, R.I. et al. (1998). Lateralization of human temporal lobe epilepsy by $^{31}P$ NMR spectroscopic imaging at 4.1 T. *Neurology*, 51, 472–479.

Chugani, H.T., Shewmon, D.A., Peacock, W.J. et al. (1988). Surgical treatment of intractable neonatal-onset seizures: the role of positron emission tomography. *Neurology*, 38, 1178–1188.

Concha, L., Gross, D.W., Wheatley, B.M., Beaulieu, C. (2006). Diffusion tensor imaging of time-dependent axonal and myelin degradation after corpus callosotomy in epilepsy patients. *NeuroImage*, 32, 1090–1099.

Conesa, G., Pujol, J., Deus, J. et al. (1999). EPI functional MRI: a useful tool for preoperative Rolandic fissure localization. *Front. Radiat. Ther. Oncol.*, 33, 23–27.

Connelly, A., Van Paesschen, W., Porter, D.A. et al. (1998). Proton magnetic resonance spectroscopy in MRI-negative temporal lobe epilepsy. *Neurology*, 51, 61–66.

Cross, J.H., Connelly, A., Jackson, G.D. et al. (1996). Proton magnetic resonance spectroscopy in children with temporal lobe epilepsy. *Ann. Neurol.*, 39, 107–113.

da Silva, E.A., Chugani, D.C., Muzik, O., Chugani, H.T. (1997). Identification of frontal lobe epileptic foci in children using positron emission tomography. *Epilepsia*, 38, 1198–1208.

Davies, S., Gotoh, M., Richards, D., Obrenovitch, T. (1998). Hypoosmolarity induces an increase in extracellular *N*-acetylaspartate concentration in rat striatum. *Neurochem. Res.*, 23, 1021–1025.

Detre, J., Maccotta, L., King, D. et al. (1998). Functional MRI lateralization of memory in temporal lobe epilepsy. *Neurology*, 50, 926–932.

Duncan, J.S. (2002). Neuroimaging methods to evaluate the etiology and consequences of epilepsy. *Epilepsy Res.*, 50, 131–140.

Dupont, P., Zaknun, J.J., Maes, A. et al. (2009). Dynamic perfusion patterns in temporal lobe epilepsy. *Eur. J. Nucl. Med. Mol. Imaging*, 36(5), 823–830.

Ende, G., Laxer, K.D., Knowlton, R.C. et al. (1995). Quantitative $^{1}H$-SI shows bilateral metabolite changes in unilateral TLE patients with and without hippocampal atrophy. In *SMR Third Meeting Proceedings*, Nice, France, August 19–25.

Enev, M., McNally, K.A., Varghese, G. et al. (2007). Imaging onset and propagation of ECT-induced seizures. *Epilepsia*, 48, 238–244.

Engel, J.J. (1984). The use of positron emission tomographic scanning in epilepsy. *Ann. Neurol.*, S180–S191.

Engel, J.J., Kuhl, D.E., Phelps, M.E., Crandall, P.H. (1982a). Comparative localization of epileptic foci in partial epilepsy by PCT and EEG. *Ann. Neurol.*, 12, 529–537.

Engel, J.J., Kuhl, D.E., Phelps, M.E., Mazziotta, J.C. (1982b). Interictal cerebral glucose metabolism in partial epilepsy and its relation to EEG changes. *Ann. Neurol.*, 12, 510–517.

Engel, J.J., Henry, T.R., Risinger, M.W. et al. (1990). Presurgical evaluation for partial epilepsy: relative contributions of chronic depth-electrode recordings versus FDG–PET and scalp–sphenoidal ictal EEG. *Neurology*, 40, 1670–1677.

Feichtinger, M., Eder, H., Holl, A. et al. (2007). Automatic and remote controlled ictal SPECT injection for seizure focus localization by use of a commercial contrast agent application pump. *Epilepsia*, 48, 1409–1413.

Fox, P., Raichle, M. (1986). Focal physiological uncoupling of cerebral blood flow and oxidative metabolism during somatosensory stimulation in human subjects. *Proc. Natl. Acad. Sci. USA*, 83, 1140–1144.

Friston, K. (1994). Statistical parametric mapping. In *Functional Neuroimaging*, Thatcher, R.W., Hallet, M., Zeffiro, T., Roy John, E., Huerta, M., Eds. (p. 91). San Diego, CA: Academic Press.

Gaillard, W., Hertz-Pannier, L., Mott, S. et al. (2000). Functional anatomy of cognitive development: fMRI of verbal fluency in children and adults. *Neurology*, 54, 180–185.

Gaillard, W., Bhatia, S., Bookheimer, SY. et al. (1995). FDG–PET and volumetric MRI in the evaluation of patients with partial epilepsy. *Neurology*, 45, 123–126.

Garcia, P.A., Laxer, K.D. et al. (1994). Phosphorus magnetic resonance spectroscopic imaging in patients with frontal lobe epilepsy. *Ann. Neurol.*, 35, 217–221.

Garcia, P.A., Laxer, K.D., van der Grond, J. et al. (1995). Proton magnetic resonance spectroscopic imaging in patients with frontal lobe epilepsy. *Ann. Neurol.*, 37, 279–281.

Gass, A., Niendorf, T., Hirsch, J.G. (2001). Acute and chronic changes of the apparent diffusion coefficient in neurological disorders: biophysical mechanisms and possible underlying histopathology. *J. Neurol. Sci.*, 186(Suppl. 1):S15–S23.

Goldman, R., Stern, J., Engel, J.J., Cohen, M. (2000). Acquiring simultaneous EEG and functional MRI. *Clin. Neurophysiol.*, 111, 1974–1980.

Grant, P.E., Barkovich, A.J., Wald, L.L. et al. (1997). High-resolution surface-coil MR of cortical lesions in medically refractory epilepsy: a prospective study. *AJNR Am. J. Neuroradiol.*, 18, 291–301.

Gross, D.W., Concha, L., Beaulieu, C. (2006). Extratemporal white matter abnormalities in mesial temporal lobe epilepsy demonstrated with diffusion tensor imaging. *Epilepsia*, 47, 1360–1363.

Hajek, M., Antonini, A., Leenders, K.L., Wieser, H.G. (1993). Mesiobasal versus lateral temporal lobe epilepsy: metabolic differences in the temporal lobe shown by interictal $^{18}$F-FDG positron emission tomography [see comments]. *Neurology*, 43, 79–86.

Hayes, C.E., Tsuruda, J.S., Mathis, C.M. (1993). Temporal lobes: surface MR coil phased-array imaging. *Radiology*, 189, 918–920.

Henry, T.R., Sutherling, W.W., Engel, J.J. et al. (1991). Interictal cerebral metabolism in partial epilepsies of neocortical origin. *Epilepsy Res.*, 10, 174–182.

Henry, T.R., Engel, J.J., Mazziotta, J.C. (1993a). Clinical evaluation of interictal fluorine-18–fluorodeoxyglucose PET in partial epilepsy. *J. Nucl. Med.*, 34, 1892–1898.

Henry, T.R., Frey, K.A., Sackellares, J.C. et al. (1993b). *In vivo* cerebral metabolism and central benzodiazepine-receptor binding in temporal lobe epilepsy. *Neurology*, 43, 1998–2006.

Henry, T.R., Mazziotta, J.C., Engel, J., Jr. (1993c). Interictal metabolic anatomy of mesial temporal lobe epilepsy. *Arch. Neurol.*, 50, 582–589.

Hermann, B., Hansen, R., Seidenberg, M. et al. (2003). Neurodevelopmental vulnerability of the corpus callosum to childhood onset localization-related epilepsy. *NeuroImage*, 18, 284–292.

Hetherington, H., Kuzniecky, R., Pan, J. et al. (1995). Proton nuclear magnetic resonance spectroscopic imaging of human temporal lobe epilepsy at 4.1 T. *Ann. Neurol.*, 38, 396–404.

Ho, S.S., Berkovic, S.F., Newton, M.R. et al. (1994). Parietal lobe epilepsy: clinical features and seizure localization by ictal SPECT. *Neurology*, 44, 2277–2284.

Holmes, C.J., Hoge, R., Collins, L. et al. (1998). Enhancement of MR images using registration for signal averaging. *J. Comput. Assist. Tomogr.*, 22, 324–333.

Hugg, J., Laxer, K., Matson, G. et al. (1992). Lateralization of human focal epilepsy by $^{31}$P magnetic resonance spectroscopic imaging. *Neurology*, 42, 2001–2018.

Hugg, J., Laxer, K., Matson, G. et al. (1993). Neuron loss localizes human temporal lobe epilepsy by *in vivo* proton magnetic resonance spectroscopic imaging. *Ann. Neurol.*, 34, 488–794.

Hugg, J., Kuzniecky, R.I., Gilliam, F.G. et al. (1996). Normalization of contralateral metabolic function following temporal lobectomy demonstrated by $^1$H magnetic resonance spectroscopic imaging. *Ann. Neurol.*, 40, 236–239.

Incisa della Rocchetta, A., Gadian, D.G. et al. (1995). Verbal memory impairment after right temporal lobe surgery: role of contralateral damage as revealed by $^1$H magnetic resonance spectroscopy and $T_2$ relaxometry. *Neurology*, 45, 797–802.

Jack, C.R., Thompson, R.M., Butts, R.K. et al. (1994). Sensory motor cortex: correlation of presurgical mapping with functional MR imaging and invasive cortical mapping. *Radiology*, 190, 85–92.

Jack, J.R., Sharbrough, F.W., Twomey, C.K. et al. (1990). Temporal lobe seizures: lateralization with MR volume measurements of the hippocampal formation. *Radiology*, 175, 423–429.

Jackson, G.D., Berkovic, S.F., Tress, B.M. et al. (1990). Hippocampal sclerosis can be reliably detected by magnetic resonance imaging. *Neurology*, 40, 1869–1875.

Jackson, G.D., Berkovic, S.F., Duncan, J.S., Connelly, A. (1993a). Optimizing the diagnosis of hippocampal sclerosis using MR imaging. *AJNR Am. J. Neuroradiol.*, 14, 753–762.

Jackson, G.D., Connelly, A., Duncan, J.S. et al. (1993b). Detection of hippocampal pathology in intractable partial epilepsy: increased sensitivity with quantitative magnetic resonance $T_2$ relaxometry. *Neurology*, 43, 1793–1799.

Jacquier-Sarlin, M.R., Polla, B.S., Slosman, D.O. (1996). Cellular basis of ECD brain retention. *J. Nucl. Med.*, 37, 1694–1697.

Jellison, B.J., Field, A.S., Medow, J. et al. (2004). Diffusion tensor imaging of cerebral white matter: a pictorial review of physics, fiber tract anatomy, and tumor imaging patterns. *AJNR Am. J. Neuroradiol.*, 25, 356–369.

Joo, E.Y., Tae, W.S., Hong, S.B. (2008). Cerebral blood flow abnormality in patients with idiopathic generalized epilepsy. *J. Neurol.*, 255, 520–525.

Juhasz, C., Chugani, D.C., Muzik, O. et al. (2000a). Electroclinical correlates of flumazenil and fluorodeoxyglucose PET abnormalities in lesional epilepsy. *Neurology*, 55, 825–835.

Juhasz, C., Watson, C., Chugani, D. et al. (2000b). Epileptogenicity of 'metabolic borderzones' in human neocortical epilepsy. Neurology, 54 (Suppl 3), 107.

Juhasz, C., Chugani, D.C., Muzik, O. et al. (2001). Relationship of flumazenil and glucose PET abnormalities to neocortical epilepsy surgery outcome. *Neurology*, 56, 1650–1658.

Kido, D., LeMay, M., Levinson, A., Benson, W. (1980). Computed tomographic localization of the precentral sulcus. *Radiology*, 135, 373–377.

Killgore, W.D., Casasanto, D.J., Yurgelun-Todd, D.A. et al. (2000). Functional activation of the left amygdala and hippocampus during associative encoding. *NeuroReport*, 11, 2259–2263.

Kim, H., Piao, Z., Liu, P. et al. (2008). Secondary white matter degeneration of the corpus callosum in patients with intractable temporal lobe epilepsy: a diffusion tensor imaging study. *Epilepsy Res.*, 81, 136–142.

Kim, J.H., Im, K.C., Kim, J.S. et al. (2007). Ictal hyperperfusion patterns in relation to ictal scalp EEG patterns in patients with unilateral hippocampal sclerosis: a SPECT study. *Epilepsia*, 48, 270–277.

Kimiwada, T., Juhasz, C., Makki, M. et al. (2006). Hippocampal and thalamic diffusion abnormalities in children with temporal lobe epilepsy. *Epilepsia*, 47, 167–175.

Knowlton, R.C., Kjos, B. (1999). Enhanced visualization of anatomy and epileptogenic lesions using MR image registration and signal averaging. *Epilepsia*, 40(Suppl. 7), 182.

Knowlton, R.C., Laxer, K.D., Ende, G. et al. (1997). Presurgical multimodality neuroimaging in electroencephalographic lateralized temporal lobe epilepsy. *Ann. Neurol.*, 42, 829–837.

Knowlton, R.C., Lawn, N.D., Mountz, J.M. et al. (2004a). Ictal single-photon emission computed tomography imaging in extra temporal lobe epilepsy using statistical parametric mapping. *J. Neuroimaging*, 14, 324–330.

Knowlton, R.C., Lawn, N.D., Mountz, J.M., Kuznicky, R.I. (2004b). Ictal SPECT analysis in epilepsy: subtraction and statistical parametric mapping techniques. *Neurology*, 63, 10–15.

Knowlton, R.C., Elgavish, R.A., Bartolucci, A. et al. (2008). Functional imaging. II. Prediction of epilepsy surgery outcome. *Ann. Neurol.*, 64, 35–41.

Koepp, M.J., Richardson, M.P., Labbe, C. et al. (1997). [11]C-flumazenil PET, volumetric MRI, and quantitative pathology in mesial temporal lobe epilepsy. *Neurology*, 49, 764–773.

Koh, S., Jayakar, P., Resnick, T. et al. (1999). The localizing value of ictal SPECT in children with tuberous sclerosis complex and refractory partial epilepsy. *Epileptic Disord.*, 1, 41–46.

Krakow, K., Allen, P., Lemieux, L. et al. (2000). Methodology: EEG-correlated fMRI. *Adv. Neurol.*, 38, 187–201.

Krolak-Salmon, P., Guenot, M., Tiliket, C. et al. (2000). Anatomy of optic nerve radiations as assessed by static perimetry and MRI after tailored temporal lobectomy. *Br. J. Ophthalmol.*, 84, 884–889.

Krsek, P., Hajek, M., Dezortova, M. et al. (2007). (1)H MR spectroscopic imaging in patients with MRI-negative extratemporal epilepsy: correlation with ictal onset zone and histopathology. *Eur. Radiol.*, 17, 2126–2135.

Kuzniecky, R., Barkovich, A.J. (1996). Pathogenesis and pathology of focal malformations of cortical development and epilepsy. *J. Clin. Neurophysiol.*, 13, 468–480.

Kuzniecky, R., Elgavish, G.A., Hetherington, H.P. et al. (1992). *In vivo* $^{31}$P nuclear magnetic resonance spectroscopy of human temporal lobe epilepsy. *Neurology*, 42, 1586–1590.

Kuzniecky, R., Burgard, S., Faught, E. et al. (1993a). Predictive value of magnetic resonance imaging in temporal lobe epilepsy surgery. *Arch. Neurol.*, 50, 65–69.

Kuzniecky, R., Mountz, J.M., Thomas, F. (1993b). Ictal $^{99m}$Tc HM–PAO brain single-photon emission computed tomography in electroencephalographic nonlocalizable partial seizures. *J. Neuroimaging*, 3, 100–102.

Kuzniecky, R., Hetherington, H., Pan, J. et al. (1997). Proton spectroscopic imaging at 4.1 tesla in patients with malformations of cortical development and epilepsy. *Neurology*, 48, 1018–1024.

Kuzniecky, R., Hugg, J.W., Hetherington, H. et al. (1998). Relative utility of $^1$H spectroscopic imaging and hippocampal volumetry in the lateralization of mesial temporal lobe epilepsy. *Neurology*, 51, 66–71.

Laxer, K.D., Hubesch, B., Sappey-Marinier, D., Weiner, M.W. (1992). Increased pH and inorganic phosphate in temporal seizure foci demonstrated by [$^{31}$P]MRS. *Epilepsia*, 33, 618–623.

Lazeyras, F., Blanke, O., Zimine, I. et al. (2000). MRI, $^1$H-MRS, and functional MRI during and after prolonged nonconvulsive seizures activity. *Neurology*, 55, 1677–1682.

Lee, D.S., Lee, S.K., Kim, Y.K. et al. (2002). Superiority of HMPAO ictal SPECT to ECD ictal SPECT in localizing the epileptogenic zone. *Epilepsia*, 43, 263–269.

Lee, J.D., Kim, H.J., Lee, B.I. et al. (2000). Evaluation of ictal brain SPET using statistical parametric mapping in temporal lobe epilepsy. *Eur. J. Nucl. Med.*, 27, 1658–1665.

Lee, N., Radtke, R.A., Gray, L. et al. (1994). Neuronal migration disorders: positron emission tomography correlations. *Ann. Neurol.*, 35, 290–297.

Lee, S.K., Kim, D.I., Mori, S. et al. (2004). Diffusion tensor MRI visualizes decreased subcortical fiber connectivity in focal cortical dysplasia. *NeuroImage*, 22, 1826–1829.

Lehericy S., Cohen L., Bazin B. et al. (2000). Functional MR evaluation of temporal and frontal language dominance compared with the Wada test. *Neurology*, 54, 1625–1633.

Lewis, P.J., Siegel, A., Siegel, A.M. et al. (2000). Does performing image registration and subtraction in ictal brain SPECT help localize neocortical seizures? *J. Nucl. Med.*, 41, 1619–1626.

Li, L.M., Cendes, F., Bastos, A.C. et al. (1998). Neuronal metabolic dysfunction in patients with cortical developmental malformations: a proton magnetic resonance spectroscopic imaging study. *Neurology*, 50, 755–759.

Loring, D.W., Murro, A.M., Meador, K.J. et al. (1993). Wada memory testing and hippocampal volume measurements in the evaluation for temporal lobectomy. *Neurology*, 43, 1789–1793.

Maldjian, J., Atlas, S.W., Howard, R.S., 2nd. et al. (1996). Functional magnetic resonance imaging of regional brain activity in patients with intracerebral arteriovenous malformations before surgical or endovascular therapy. *J. Neurosurgery*, 84, 477–483.

Maravilla, K. (1997). Advances in MR imaging and MR spectroscopy evaluation of patients with temporal lobe epilepsy. *Chonnam J. Med. Sci.*, 10, 162–170.

Marks, D.A., Katz, A., Hoffer, P., Spencer, S.S. (1992). Localization of extratemporal epileptic foci during ictal single photon emission computed tomography. *Ann. Neurol.*, 31, 250–255.

Martin, R.C., Sawrie, S., Hugg, J. et al. (1999). Cognitive correlates of $^1$H MRSI-detected hippocampal abnormalities in temporal lobe epilepsy. *Neurology*, 53, 2052–2058.

Martin, R.C., Sawrie, S.M., Knowlton, R.C. et al. (2001). Bilateral hippocampal atrophy: consequences to verbal memory following temporal lobectomy. *Neurology*, 57, 597–604.

Maton, B., Londono, A., Sawrie, S. et al. (2001). Postictal stability of proton magnetic resonance spectroscopy imaging ($^1$H-MRSI) ratios in temporal lobe epilepsy. *Neurology*, 42, 651–659.

Mazziotta, J.C., Engel, J.J. (1984). The use and impact of positron computed tomography scanning in epilepsy. *Epilepsia*, S86–S104.

McIntosh, J., Cooper, J. (1965). Studies on the function of *N*-acetyl-aspartic acid in brain. *J. Neurochem.*, 12, 825–835.

McNally, K.A., Paige, A.L., Varghese, G. et al. (2005). Localizing value of ictal–interictal SPECT analyzed by SPM (ISAS). *Epilepsia*, 46, 1450–1464.

Melhem, E.R., Mori, S., Mukundan, G. et al. (2002). Diffusion tensor MR imaging of the brain and white matter tractography. *Am. J. Roentgenol.*, 178, 3–16.

Mori, S., Zhang, J. (2006). Principles of diffusion tensor imaging and its applications to basic neuroscience research. *Neuron*, 51, 527–539.

Mueller, W.M., Yetkin, F.Z., Hammeke, T.A. et al. (1996). Functional magnetic resonance imaging mapping of the motor cortex in patients with cerebral tumors. *Neurosurgery*, 39, 515–521.

Muzik, O., da Silva, E.A., Juhasz, C. et al. (2000). Intracranial EEG versus flumazenil and glucose PET in children with extratemporal lobe epilepsy. *Neurology*, 54, 171–179.

Newton, M.R., Berkovic, S.F., Austin, M.C. et al. (1995). SPECT in the localisation of extratemporal and temporal seizure foci. *J. Neurol. Neurosurg. Psychiatry*, 59, 26–30.

Ng, T.C., Comair, Y.G., Xue, M. et al. (1994). Temporal lobe epilepsy: presurgical localization with proton chemical shift imaging. *Radiology*, 193, 465–472.

Nilsson, D., Starck, G., Ljungberg, M. et al. (2007). Intersubject variability in the anterior extent of the optic radiation assessed by tractography. *Epilepsy Res.*, 77, 11–16.

O'Brien, T.J., So, E.L., Mullan, B.P. et al. (1998). Subtraction ictal SPECT co-registered to MRI improves clinical usefulness of SPECT in localizing the surgical seizure focus. *Neurology*, 50, 445–454.

Ogawa, S., Lee, T.M., Kay, A.R., Tank, D.W. (1990). Brain magnetic resonance imaging with contrast dependent on blood oxygenation. *Proc. Natl. Acad. Sci. U.S.A.*, 87:9868–9872.

Pierpaoli, C., Jezzard, P., Basser, P.J. et al. (1996). Diffusion tensor MR imaging of the human brain. *Radiology*, 201, 637–648.

Powell, H.W., Parker, G.J., Alexander, D.C. et al. (2005). MR tractography predicts visual field defects following temporal lobe resection. *Neurology*, 65, 596–599.

Puce, A. (1995). Comparative assessment of sensorimotor function using functional magnetic resonance imaging and electrophysiological methods. *J. Clin. Neurophysiol.*, 12, 450–459.

Rango, M., Spangnoli, D., Tomei, G. et al. (1995). Central nervous system trans-synaptic effects of acute axonal injury: a ¹H magnetic resonance spectroscopy study. *Magn. Reson. Med.*, 33, 595–600.

Rao, S.M., Binder, J.R., Bandettini, P.A. et al. (1993). Functional magnetic resonance imaging of complex human movements. *Neurology*, 43, 2311–2318.

Rao, S.M., Binder, J.R., Hammeke, T.A. et al. (1995). Somatotopic mapping of the human primary motor cortex with functional magnetic resonance imaging. *Neurology*, 45, 919–924.

Righini, A., de Divitiis, O., Prinster, A. et al. (1996). Functional MRI: primary motor cortex localization in patients with brain tumors. *J. Comput. Assist. Tomogr.*, (20, 702–708.

Ryvlin, P., Philippon, B., Cinotti, L. et al. (1992). Functional neuroimaging strategy in temporal lobe epilepsy: a comparative study of ¹⁸FDG–PET and ⁹⁹ᵐTc–HMPAO–SPECT. *Ann. Neurol.*, 31, 650–656.

Ryvlin, P., Guenot, M., Isnard, J. et al. (1997). The role of the temporopolar cortex in temporal lobe epilepsy. *Stereotact. Funct. Neurosurg.*, 67, 143.

Ryvlin, P., Bouvard, S., Le Bars, D. et al. (1998). Clinical utility of flumazenil–PET versus and [¹⁸F]fluorodeoxyglucose–PET and MRI in refractory partial epilepsy: a prospective study in 100 patients. *Brain*, 121, 2067–2081.

Ryvlin, P., Bouvard, S., Le Bars, D., Mauguiere, F. (1999). Transient and falsely lateralizing flumazenil-PET asymmetries in temporal lobe epilepsy. *Neurology*, 53, 1882–1885.

Savic, I., Ingvar, M., Stone, E.S. (1993). Comparison of [¹¹C]flumazenil and [¹⁸F]FDG as PET markers of epileptic foci. *J. Neurol. Neurosurg. Psychiatry*, 56, 615–621.

Sawrie, S.M., Martin, R.C., Gilliam, F.G. et al. (2000). Visual confrontation naming and hippocampal function: a neural network study using quantitative (1)H magnetic resonance spectroscopy. *Brain*, 123, 770–780.

Schad, L.R., Bock, M., Baudendistel, K. et al. (1996). Improved target volume definition in radiosurgery of arteriovenous malformations by stereotactic correlation of MRA, MRI, blood bolus tagging, and functional MRI. *Eur. Radiology*, 6, 38–45.

Scorzin, J.E., Kaaden, S., Quesada, C.M. et al. (2008). Volume determination of amygdala and hippocampus at 1.5 and 3.0T MRI in temporal lobe epilepsy. *Epilepsy Res.*, 82, 29–37.

Seeck, M., Lazeyras, F., Michel, C. et al. (1998). Non-invasive epileptic focus localization using EEG -triggered functional MRI and electromagnetic tomography. *Electroencephalogr. Clin. Neurophysiol.*, 106, 508–512.

Shen, W., Lee, B.I., Park, H.M. et al. (1990). HIPDM–SPECT brain imaging in the presurgical evaluation of patients with intractable seizures. *J. Nucl. Med.*, 31, 1280–1284.

Shimizu, H., Ishijima, B., Iio, M. (1985). [Diagnosis of temporal lobe epilepsy by positron emission tomography]. *No To Shinkei*, 37, 507–512.

Sisodiya, S.M., Free, S.L., Stevens, J.M. et al. (1995). Widespread cerebral structural changes in patients with cortical dysgenesis and epilepsy. *Brain*, 118, 1039–1050.

Sisodiya, S.M., Moran, N., Free, S.L. et al. (1997). Correlation of widespread preoperative magnetic resonance imaging changes with unsuccessful surgery for hippocampal sclerosis. *Ann. Neurol.*, 41, 490–496.

Sobel, D.F., Gallen, C.C., Schwartz, B.J., Waltz, T.A., Copeland, B., Yamada, S., Hirschkoff, E.C., Bloom, F.E. (1993). Locating the central sulcus: comparison of MR anatomic and magnetoencephalographic functional methods. *AJNR Am. J. Neuroradiol.*, 14:915–925.

Song, S.K., Sun, S.W., Ramsbottom, M.J. et al. (2002). Dysmyelination revealed through MRI as increased radial (but unchanged axial) diffusion of water. *NeuroImage*, 17, 1429–1436.

Spanaki, M.V., Spencer, S.S., Corsi, M. et al. (1999). Sensitivity and specificity of quantitative difference SPECT analysis in seizure localization. *J. Nucl. Med.*, 40, 730–736.

Sperling, M.R. (1990). Neuroimaging in epilepsy: contribution of MRI, PET and SPECT. *Semin. Neurol.*, 10, 349–356.

Sperling, M.R., Wilson, G., Engel, J.J. et al. (1986). Magnetic resonance imaging in intractable partial epilepsy: correlative studies. *Ann. Neurol.*, 20, 57–62.

Stanley, J.A., Cendes, F., Dubeau, F. et al. (1998). Proton magnetic resonance spectroscopic imaging in patients with extratemporal epilepsy. *Epilepsia*, 39, 267–273.

Stefan, H., Pawlik, G., Bocher, S.H. et al. (1987). Functional and morphological abnormalities in temporal lobe epilepsy: a comparison of interictal and ictal EEG, CT, MRI, SPECT and PET. *J. Neurol.*, 234, 377–384.

Sturm, J.W., Newton, M.R., Chinvarun, Y. et al. (2000). Ictal SPECT and interictal PET in the localization of occipital lobe epilepsy. *Epilepsia*, 41, 463–466.

Swartz, B.E., Halgren, E., Delgado, E.A. et al. (1989). Neuroimaging in patients with seizures of probable frontal lobe origin. *Epilepsia*, 30, 547–558.

Swartz, B.E., Tomiyasu, U., Delgado, E.A. et al. (1992). Neuroimaging in temporal lobe epilepsy: test sensitivity and relationships to pathology and postoperative outcome. *Epilepsia*, 33, 624–634.

Tae, W.S., Joo, E.Y., Kim, J.H. et al. (2005). Cerebral perfusion changes in mesial temporal lobe epilepsy: SPM analysis of ictal and interictal SPECT. *NeuroImage*, 24, 101–110.

Tae, W.S., Joo, E.Y., Han, S.J. et al. (2007). CBF changes in drug naive juvenile myoclonic epilepsy patients. *J. Neurol.*, 254, 1073–1080.

Tan, K.M., Britton, J.W., Buchhalter, J.R. et al. (2008). Influence of subtraction ictal SPECT on surgical management in focal epilepsy of indeterminate localization: a prospective study. *Epilepsy Res.*, 82(2-3), 190–193.

Theodore, W.H., Holmes, M.D., Dorwart, R.H. et al. (1986). Complex partial seizures: cerebral structure and cerebral function. *Epilepsia*, 27, 576–582.

Thivard, L., Lehericy, S., Krainik, A. et al. (2005). Diffusion tensor imaging in medial temporal lobe epilepsy with hippocampal sclerosis. *NeuroImage*, 28, 682–690.

Trenerry, M.R., Jack, C.J., Ivnik, R.J. et al. (1993). MRI hippocampal volumes and memory function before and after temporal lobectomy. *Neurology*, 43, 1800–1805.

Tsai, G., Coyle, J.T. (1995). *N*-acetylaspartate in neuropsychiatric disorders. *Prog. Neurobiol.*, 46, 531–540.

Urenjak, J., Williams, S.R., Gadian, D.G., Noble, M. (1992). Specific expression of *N*-acetylaspartate in neurons, oligodendrocyte-type-2 astrocyte progenitors, and immature oligodendrocytes *in vitro*. *J. Neurochem.*, 59, 55–61.

Vainio, P., Usenius, JP., Vapalahti, M. et al. (1994). Reduced *N*-acetylaspartate concentration in temporal lobe epilepsy by quantitative $^1$H MRS *in vivo*. *NeuroReport*, 5, 1733–1736.

Valk, P.E., Jagust, W.J., Derenzo, S.E. et al. (1990). Clinical evaluation of a high-resolution (2.6-mm) positron emission tomography. *Radiology*, 176, 783–790.

Valk, P.E., Laxer, K.D., Barbaro, N.M. et al. (1993). High-resolution (2.6-mm) PET in partial complex epilepsy associated with mesial temporal sclerosis [see comments]. *Radiology*, 186, 55–58.

Van Paesschen, W., Dupont, P., Van Heerden, B. et al. (2000). Self-injection ictal SPECT during partial seizures. *Neurology*, 54, 1994–1997.

Van Paesschen, W., Dupont, P., Van Driel, G. et al. (2003). SPECT perfusion changes during complex partial seizures in patients with hippocampal sclerosis. *Brain*, 126, 1103–1111.

Wakana, S., Jiang, H., Nagae-Poetscher, LM. et al. (2004). Fiber tract-based atlas of human white matter anatomy. *Radiology*, 230, 77–87.

Wald, L.L., Carvajal, L., Moyher, S.E. et al. (1995). Phased array detectors and an automated intensity-correction algorithm for high-resolution MR imaging of the human brain. *Magn. Reson. Med.*, 34, 433–439.

Watson, C., Andermann, F., Gloor, P. et al. (1992). Anatomic basis of amygdaloid and hippocampal volume measurement by magnetic resonance imaging. *Neurology*, 42, 1743–1750.

Wetjen, N.M., Cascino, G.D., Fessler, A.J. et al. (2006). Subtraction ictal single-photon emission computed tomography coregistered to magnetic resonance imaging in evaluating the need for repeated epilepsy surgery. *J. Neurosurg.*, 105, 71–76.

Woermann, F.G., McLean, M.A., Bartlett, P.A. et al. (1999). Short echo time single-voxel $^1$H magnetic resonance spectroscopy in magnetic resonance imaging–negative temporal lobe epilepsy: different biochemical profile compared with hippocampal sclerosis. *Ann. Neurol.*, 45, 369–376.

Won, H.J., Chang, K.H., Cheon, J.E. et al. (1999). Comparison of MR imaging with PET and ictal SPECT in 118 patients with intractable epilepsy. *AJNR Am. J. Neuroradiol.*, 20, 593–599.

Yamamoto, T., Yamada, K., Nishimura, T., Kinoshita, S. (2005). Tractography to depict three layers of visual field trajectories to the calcarine gyri. *Am. J. Ophthalmol.*, 140, 781–785.

Yetkin, F.Z., Mueller, W.M., Hammeke, T.A. et al. (1995). Functional magnetic resonance imaging mapping of the sensorimotor cortex with tactile stimulation. *Neurosurgery*, 36, 921–925.

Yetkin, F.Z., Swanson, S., Fischer, M. et al. (1998). Functional MR of frontal lobe activation: comparison with Wada language results. *AJNR Am. J. Neuroradiol.*, 19, 1095–1098.

Yousry, T.A., Schmid, U.D., Jassoy, A.G. et al. (1995). Topography of the cortical motor hand area: prospective study with functional MR imaging and direct motor mapping at surgery. *Radiology*, 195, 23–29.

Yun, C.H., Lee, S.K., Lee, S.Y. et al. (2006). Prognostic factors in neocortical epilepsy surgery: multivariate analysis. *Epilepsia*, 47, 574–579.

Zubal, I.G., Avery, R.A., Stokking, R. et al. (2000). Ratio-images calculated from interictal positron emission tomography and single-photon emission computed tomography for quantification of the uncoupling of brain metabolism and perfusion in epilepsy. *Epilepsia*, 41, 1560–1566.

Zubal, I.G., Spencer, S.S., Imam, K. et al. (1995). Difference images calculated from ictal and interictal technetium-99m–HMPAO SPECT scans of epilepsy. *J. Nucl. Med.*, 36, 684–689.

# 16 Magnetoencephalography in Clinical Epilepsy: A Translational Viewpoint

*Jerry J. Shih and Michael P. Weisend*

## CONTENTS

## INTRODUCTION

Biomagnetism is the study of magnetic fields generated in biological systems. The study of biomagnetism began when Baule and McFee (1963) measured magnetic signals generated by a human heart. Magnetoencephalography (MEG) is a discipline in the field of biomagnetism that investigates magnetic fields originating in brain cells. MEG was officially born in 1968 when David Cohen (1972) demonstrated the magnetic field equivalent of the human posterior-dominant alpha rhythm. Although many of the initial proposed clinical applications for biomagnetism have been abandoned (studies of lung, liver, abdomen, eyeball), the use of MEG in research and clinical practice has not only taken root but continues to increase.

Investigators from a number of scientific disciplines currently use biomagnetometers as their principal research tool, and MEG is used clinically in Europe, Asia, and the United States. MEG has been used in clinical research studies on schizophrenia, depression, stroke, myoclonus, dyslexia, auditory processing, visual processing, language lateralization, and localization, to name a select few (Brancucci et al., 2008; Hirano et al., 2008; Huang et al., 2004; Kahkonen et al., 2007; Luo and Poeppel, 2007; Merrifield et al., 2007; Nahum et al., 2009; Okazaki et al., 2008; Reulbach et al., 2007; Salmelin, 2007; Silen et al., 2002; Stephen et al., 2006; Tarkiainen et al., 2003). MEG is clinically approved in the United States to map and localize human somatosensory and motor cortex; however, the main clinical application for MEG is in the characterization and localization of the epileptogenic focus.

In this chapter, we discuss the pretranslational studies that underlie knowledge of the neurophysiologic basis of MEG. Also described are the methodology and analysis techniques used in MEG studies in epilepsy. Finally, we explore the seminal translational studies on human epilepsy and conclude with a posttranslational look at ongoing epilepsy projects and some thoughts for future studies.

## NEUROPHYSIOLOGIC BASIS OF MEG

The physiological signal measured in MEG is the magnetic field that wraps around intracellular current flow in the dendrites of neurons. In a hippocampal slice preparation, Wu and Okada (1999) demonstrated that stimulation of the apical dendrites of hippocampal pyramidal cells produced an electric and magnetic field consisting of an initial spike based on the influx of sodium followed by a slow component that was based on potassium currents. Coordinated postsynaptic potentials in thousands of the long apical dendrites of neocortical pyramidal cells are thought to be the primary source of the MEG signal (Murakami and Okada, 2006). As current flows from distal dendrites toward the somas of the neurons, a magnetic field rotates around the dendrites. The rotation of the magnetic field obeys the "right-hand rule," meaning that current flow in the direction of the thumb on the right hand is accompanied by a magnetic field that wraps around the current in the same direction in which the fingers naturally curl. Thus, activity measured with MEG, like electroencephalography (EEG), has a positive and a negative pole. In EEG, the current is flowing between the positive and negative poles. In contrast, the negative pole in MEG represents the locations where the magnetic field is rotating into the head while the positive pole represents the location where the magnetic field is rotating out of the head. The direction of current flow is perpendicular to a line drawn between the positive and negative magnetic poles, with the negative pole on the right and the positive pole on the left of the source.

The advantage of MEG is that the signal passes through the scalp unimpeded, whereas the EEG is distorted by resistance across the scalp layers (Barth et al., 1986; Kaufman et al., 1981). Careful comparisons of the EEG and MEG signals in animal experiments established that the skull is essentially transparent to the magnetic fields measured in MEG, but not the electrical potentials measured with EEG. MEG measurements from above the skull and from above the dura, with the skull removed, are similar in time course, amplitude, and spatial distribution (Okada et al., 1999b). In contrast, the EEG measured from the skull surface and the cortical surface show different amplitudes, time courses, and spatial contours (Okada et al., 1999a). In general, this results in a higher signal-to-noise ratio for MEG than for EEG, as a great number of sources are measured with MEG but EEG records only the sources that can overcome the impedance of the tissue intervening between the brain and the electrode.

The physical nature of the signal remains the same, but the configuration of the MEG sensor array can alter the appearance of the signal much like choosing different montages in EEG. The superconducting quantum interference device (SQUID) is the heart of the MEG system. There are three basic arrangements of SQUIDs in MEG systems today: magnetometers, axial gradiometers, and planar gradiometers. Magnetometers consist of a single loop of wire in which the magnetic field generated by the brain induces current flow that is amplified to produce the MEG recording. Magnetometers are highly sensitive to magnetic sources that are both local to (e.g., brain activity, muscle activity from eye blinks, movement) and distant from (e.g., passing vehicles, opening and closing doors, computers, light fixtures) the sensors. The high sensitivity of magnetometers is desirable, but the merging of signals from many sources can be problematic. Elegant signal processing methods have been developed to aid in resolving brain activity from data recorded with magnetometers. Thus, the mixing of many signals in the recording should not be considered a significant drawback. Axial gradiometers consist of two pickup coils that are vertically displaced along the axis of the sensor. Planar gradiometers also have two pickup coils, but the wires are displaced horizontally. Gradiometers cancel noise from distant sources while retaining high sensitivity to sources that are close to the sensor array. Thus, the raw signals are cleaner and more readily interpretable in their raw form than those recorded with magnetometers. Raw signals from both magnetometers and axial gradiometers show the MEG signal to be largest at the magnetic poles, whereas planar gradiometers show the signal to be largest on sensors directly over the source.

Magnetoencephalography is exquisitely sensitive to activity that is in the walls of the gyri. Activity in the apical dendrites of pyramidal cells in the gyral walls produces electrical currents that flow parallel to the surface of the skull and generate magnetic fields that are tangential to the skull surface. The crowns of the gyri have currents in the pyramidal neurons that flow roughly perpendicular to the skull surface. MEG is less sensitive to magnetic fields that rotate around currents perpendicular to the skull surface because a large fraction of the magnetic field rotation remains inside the head. When this current is perfectly radial to the sensor array or to the head model (head models will be discussed later in this chapter), the magnetic field will theoretically be completely undetected. The numbers of sources that are undetectable in the cortical mantle are generally considered to be small (Hillebrand and Barnes, 2002); however, in any MEG recording the possibility of "silent" sources cannot be ruled out. When a source is neither perfectly radial nor perfectly tangential it will be measured at partial strength because a tilt in the axis of a source relative to the sensor array effectively increases the distance between the source and location on the sensor array where it is measured.

Distance of the current source from the sensor array is an important consideration in the measurement of brain activity with MEG. This fact has resulted in spirited debate about the sensitivity of MEG to deep sources such as the hippocampus. Herein, no specific reference to hippocampus is made; however, the conditions that influence the detection of deep sources are noted. First, a deep source must generate strong signals, as the strength of the magnetic field falls off exponentially with distance (Williamson and Kaufman, 1981). The mean amplitude of an epileptic spike recorded with planar gradiometers over a series of 50 sequential cases was 600 nAm (Weisend et al., 2002). Studies that recorded simultaneous MEG and electrocorticography suggest that patches of activated temporal neocortex can be between 3 and 8 cm$^2$ (Baumgartner et al., 2000; Mikuni et al., 1997; Shigeto et al., 2002). These demonstrate that sources giving rise to MEG signals can be strong. Second, the geometry of the source cannot be closed. The asymmetry in the dendritic tree of the neocortical pyramidal cell is what makes it an ideal generator for MEG signals. In contrast, the more symmetric shape of dendritic trees in many subcortical nuclei is not optimal for the generation of magnetic fields. In radially symmetrical neurons, uniform activation of the dendritic tree causes the cancellation of magnetic fields; however, MEG signal could be expected even from neurons with radially symmetric dendritic trees when postsynaptic activation is asymmetrical (Murakami and Okada, 2006). Third, the signal is spread across many sensors with deeper sources; thus, when looking for deep sources the same criterion for detection of a superficial source (i.e., focality) cannot be used. Last, the accuracy of source localization decreases with distance from the sensor array (Hillebrand and Barnes, 2002; Jerbi et al., 2004). Superficial sources can be localized in MEG with subcentimeter accuracy. The same level of localization accuracy is not possible with deep sources. This is problematic in that, although a source might be measured, it could be difficult to accurately localize. With respect to deep sources and MEG, there is no *a priori* reason to reject their detection; however, one should interpret deep sources with caution.

Studies with a human skull phantom confirm an increase in the accuracy of source localizations with MEG as compared to EEG. The mean localization error across 32 dipolar sources was 3 mm with MEG, whereas the error for EEG was 7 to 8 mm (Leahy et al., 1998). This indeed shows that the localization accuracy of MEG is superior, although this experiment also illustrates that the advantage is small when a complex head model, necessary for EEG localization, is carefully calculated. Multiple studies using simulated and measured data illustrate that source analysis with combined MEG and EEG data is superior to that with either one alone (Babiloni et al., 2001; Baillet et al., 1999; Cohen and Cuffin, 1987; Huang et al., 2007; Huizenga et al., 2001; Liu et al., 2002; Yoshinaga et al., 2002). These studies suggest that MEG and EEG are best considered complementary measures that can synergize to produce superior results in the localization of brain activity (Figure 16.1).

**Interictal Epileptiform Activity**

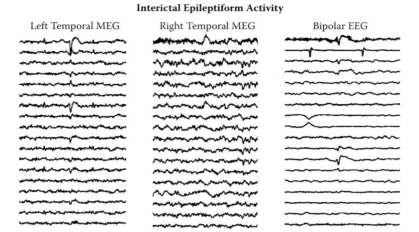

Left Temporal MEG            Right Temporal MEG                Bipolar EEG

**FIGURE 16.1**    Interictal epileptiform activity shown in simultaneously recorded magnetoencephalography (MEG) and electroencephalography (EEG). The left column shows approximately 2 sec of MEG activity recorded during from the left temporal sensors during an interictal spike and wave event that localized to the left anterior temporal lobe. The center column shows activity recorded concurrently from the right temporal sensors. The right column shows data recorded at the same time from EEG recorded in a double banana montage. The top EEG tracing is electro-oculography (EOG), and the second tracing is electrocardiography (EKG). Note the similarity in the timing and shape of the spike recorded with both MEG and EEG.

## METHODOLOGY OF MEG STUDIES IN EPILEPSY

### The MEG Exam

Magnetoencephalography studies in patients with epilepsy are most frequently ordered as part of the presurgical evaluation of medically refractory seizure patients. The goal of the presurgical MEG exam is twofold: (1) localize eloquent cortices to be preserved, and (2) localize epileptogenic tissue to be removed. Somatosensory, motor, auditory, and visual cortices are routinely localized in evoked response paradigms. Language localization and lateralization procedures have also been developed and validated (Bower et al., 2004; Kober et al., 2001; Papanicolaou et al., 2005). Epileptiform events are routinely detected in and localized from continuous MEG recordings (Paetau et al., 1990; Sato, 1990; Sutherling et al., 2001). Abnormal low-frequency activity that is often associated with lesions of many types is also localizable with MEG (Lewine et al., 2007; Vieth et al., 1996). The efforts to localize both function and dysfunction with MEG are frequently at odds; for example, spikes are most frequently detected when the patient is falling asleep, but the performance of tasks that target language localization or lateralization is typically poor in a very sleepy patient. It is optimal if the continuous recordings for detection and localization of spikes are long but that may not leave time in a busy schedule for recording a complete battery of evoked responses to map cortical functions. A set of typical exams for presurgical evaluation with approximate timing, stimuli for evoked response paradigms, data collection/analysis parameters, and sample waveforms is provided in Table 16.1.

Our typical clinical epilepsy exam lasts for approximately 3 hours including setup (30 minutes), MEG recording (2 hours), and magnetic resonance imaging (MRI) recording (30 minutes). After the patient is positioned in the MEG instrument, a 1-minute noise run is recorded. The noise run is conducted prior to attaching any other apparatus to the patient so any object creating magnetic noise can be quickly removed. Electrodes are placed on the patient to record the activity of the eyes and heart. Head position indicator coils are affixed to the patient for co-registration with MRI data,

**TABLE 16.1**
**Examples of Exams, Data Collection and Analysis Parameters, and Waveforms**

| Exam | Length | Stimuli | Data Collection Parameters | Filter Settings for Analysis |
|---|---|---|---|---|
| Somatosensory, hand | 7.5 min per hand | Electrical, median nerve | SR = 1 kHz<br>BW = 1–330 Hz<br>Nave = 300<br>ISI = 1.5 ± .5 sec | BW = 10–150 Hz<br>NF = 60 ± 5 Hz<br>SL = 17–25 msec |
| Somatosensory, foot | 7.5 min per foot | Electrical, tibial nerve | SR = 1 kHz<br>BW = 1–330 Hz<br>Nave = 300<br>ISI = 1.5 ± .5 sec | BW = 5–100 Hz<br>NF = 60 ± 5 Hz<br>SL = 25–45 msec |
| Somatosensory, face | 10 min per site on the lip | Pneumatic, lip | SR = 1 kHz<br>BW = 1–330 Hz<br>Nave = 300<br>ISI = 2 ± .5 sec | BW = 1–100 Hz<br>NF = 60 ± 5 Hz<br>SL = 20–35 msec |
| Auditory | 5 min per ear | Pure tones or clicks | SR = 400 Hz<br>BW = .1–120 Hz<br>Nave = 200<br>ISI = 1.5 ± .5 sec | BW = 2–40 Hz<br>NF = NA<br>SL = 80–120 msec |
| Visual | 5 min per visual field | Reversing checkerboard | SR = 400 Hz<br>BW = .1–120 Hz<br>Nave = 200<br>ISI = 1.5 ± .5 sec | BW = 1–50 Hz<br>NF = NA<br>SL = 65–95 msec |
| Language, receptive | 17 min | Dichotic listening | SR = 400 Hz<br>BW = .1–120 Hz<br>Nave = 200<br>ISI = 5 ± .5 sec | BW = 0.5–40 Hz<br>NF = NA<br>SL = 250–400 msec |
| Language, productive | 30 min | Covert naming | SR = 400 Hz<br>BW = .1–120 Hz<br>Nave = 200<br>ISI = 1.5 ± .5 sec | BW = 0.5–50 Hz<br>NF = NA<br>SL = 200–750 msec |
| Motor, hand | 5 min per finger | Index finger pointing | SR = 400 Hz<br>BW = .1–120 Hz<br>Nave = 200<br>ISI = 1.5 ± .5 sec | BW = .1–20 Hz<br>NF = NA<br>SL = –50 to –10 msec |
| Spontaneous spikes | 40 min | 4 × 10-min recordings | 1 ictal or 10 interictal events for valid exam | BW = 6–70 Hz<br>NF = 60 ± 5 Hz<br>SL = peak ± 35 msec |
| Spontaneous slow-wave | 40 min | 4 × 10-min recordings | 5 rhythmic low-frequency bursts for valid exam | BW = 1–8 Hz<br>NF = NA<br>SL = 1000 msec |

*Abbreviations:* BW, bandwidth; ISI, interstimulus interval; Nave, number of averages; NF, notch filter; SL, epoch for localization of source activity; SR, sampling rate.

and a digital head shape is recorded. These two steps require roughly 15 minutes. If simultaneous EEG is to be recorded with MEG this adds 15 minutes for a 19-channel cap and up to 60 minutes for high-density EEG.

The MEG exam itself can last for 2 hours or more. When looking for epileptiform events, we collect at least 40 minutes of spontaneous data. This leaves 1 hour and 20 minutes for evoked response paradigms. Because the patient is often sleep deprived, we collect evoked response data first. We typically allow several minutes between exams to give the patient instructions or reposition a patient who has moved from an optimal recording position. To stay within the 2-hour exam window, no more than 1 hour of exams should be selected from Table 16.1. Language and motor function are often of primary interest to the surgeon and thus are almost always part of the MEG exam. That leaves time for only one other modality. In temporal lobe patients, auditory evoked testing is the most frequent exam. In patients with neocortical epilepsy, visual or somatosensory exams are usually performed.

In MEG, each patient's electrophysiological data are co-registered with the individual's MRI. Three things must be considered when collecting the MRI. First, the MRI dataset should be collected in a three-dimensional (3D) volume. Most MEG source-reconstruction algorithms require a dataset with data for every voxel in the source space. Moreover, export of the MRI dataset with embedded source localizations from MEG into a format for image-guided surgery requires 3D volume MRI data. Second, the MRI must have a wide enough field of view to capture the data from both ears and the complete nose of the patient. Co-registration of the MRI data with the MEG data is accomplished by matching the positions of known, fiducial points on or near the ears and the digitized head shape with the external skin surface from the MRI data. Failure to capture the full shape of the head, including the ears, will result in co-registration errors and inaccurate source localizations. Third, the MRI dataset should be collected with symmetrical voxels. Most automated MRI segmentation routines (i.e., FreeSurfer, BrainSuite) assume symmetrical voxels and can produce inaccurate results when this assumption is violated.

## MEG Data Analysis and Source Localization

Several basic issues that must be considered in the analysis of MEG data include artifact rejection, head modeling, and solutions to the inverse problem. Appropriate artifact rejection is essential for clear interpretation of MEG data.

Frequent artifacts in MEG data include eyeblinks, respiration, movement, heart beat, moving doors, or even vehicles passing nearby. Artifact rejection in MEG data can take various forms. Filtering is one form of artifact rejection. A high-pass filter rejects low-frequency artifacts such as DC drift in the sensor array that are not related to brain activity. Trial rejection is another straightforward method of artifact rejection. Trial-based artifact rejection can be accomplished either manually or in an automated fashion. Manual trial rejection requires a human user to examine each epoch and decide if it is artifact free or not. This process is obviously subjective and runs the risk of being investigator unique. Automated trial-based rejection can be carried out by specifying a threshold in either the amplitude or magnitude of the signal and excluding all trials from further analysis when this threshold is exceeded. Although less subjective than manual trial rejection, the threshold is arbitrary and can also be investigator unique. Manual and automated trial rejection strategies share a common problem. If the criterion for rejection is too strict, there may be too few trials after artifact rejection to accomplish meaningful analysis. If the criterion is not sufficiently stringent, the data will be too noisy for meaningful analysis. For these reasons, several alternatives to trial-based artifact rejection have been developed.

One well-established method for artifact rejection in MEG is signal-space projection (SSP) (Tesche et al., 1995). All artifacts show a specific pattern of activity on the sensor array. In SSP, this pattern of activity is subtracted from the signal on a channel-by-channel basis whenever it occurs. SSP works very well for common artifacts such as eye blinks, heart beats, and static background activity; however, if the subject moves, the pattern of activity on the sensor array changes and SSP fails until it can be recalculated for the new position of the patient. Another well-established method for artifact rejection is independent component analysis (ICA) (Ikeda and Toyama, 2000;

Rong and Contreras-Vidal, 2006). In ICA, statistically independent signal components are derived. The different components in the signal can be examined, the components identified as noise can be removed, and the remaining components reassembled to produce an artifact-free signal. ICA is preferably done in the raw recordings and has proven to be an exceptional de-noising technique; however, it can be very computationally demanding, and there are multiple ICA methods that have yet to be evaluated in the analysis of clinical MEG data.

Another statistical technique for rejecting artifacts is variance-based sample rejection (VBSR). In calculating the evoked response, all samples surrounding the presentation of the stimulus are aligned in time and averaged across trials. In VBSR, the standard deviation of each sample in the evoked response is calculated, and when a sample falls outside the range of the mean plus or minus two standard deviations the sample is rejected from the analysis. In this way, aberrant samples, not trials or signal components, are rejected. This method carries the advantage over ICA of being computationally inexpensive, but, similarly to ICA, it has not been tested on clinical data. One last method of artifact rejection is signal-space separation (SSS; also referred to as tSSS). SSS specifies the volume from which sources will be accepted (Taulu et al., 2004). Any sources outside this volume are minimized in the signal space. This elegant method of artifact rejection is unfortunately vendor specific and is not in widespread use at the current time. These computationally elegant and brute force approaches to cleaning MEG and EEG data deal very effectively with the common artifacts.

The head model is another issue that must be considered in the processing of MEG data. The terms *lead field*, *head model*, and *forward model* are sometimes used interchangeably; however, there are important distinctions. The lead field refers to the view of a sensor to many sources, whereas the forward model determines how a source appears to many sensors. The forward model consists of two parts: the head model and the sensor model (Hamalainen, 1992). The head model portion of the forward model is typically different for MEG and EEG experiments. Because the scalp and skull are transparent to magnetic but not electrical activity, sophisticated multiple shell head models are necessary for accurate source localization from EEG data, whereas simpler, single shell models can be used for MEG. The head model can be standardized across patients (i.e., a spherical model) or individualized for each patient (i.e., a realistic head model). The spherical head model is simple and computationally efficient. Boundary element models (BEMs) or realistic head models incorporate cortical geometry information and can increase the accuracy of source localization but are computationally intensive. The multiple overlapping spheres approach (Huang et al., 1999) is a compromise between the spherical and realistic head models; it is computationally efficient and has excellent source reconstruction results. The forward model is used to project a putative source in the brain onto the sensor array for matching to the observed signal.

There are many types of source localization algorithms that can be used with MEG data. Source localization requires a solution to the inverse problem; however, there is no unique solution to the inverse problem from MEG or EEG recordings if no assumptions are made about the sources. When reasonable assumptions are made about the source configuration, MEG can produce results that are accurate within millimeters when compared to intracranial recordings. Broadly, the source localization techniques can be broken down into three approaches: equivalent current dipole fitting, beamforming, and cortical imaging (Ermer et al., 2001). The dipole fitting approach assumes that the observed magnetic field can be explained by one or more point sources of electrical activity. The dipole fitting approach begins with an initial guess for the locations of the activity. The sources are projected through the forward model onto the sensor array and compared to the recorded signal, and an error term is calculated for the difference between the predicted and observed values at each sensor. The position and orientation of the dipole sources are adjusted, and the error term is recalculated. This process continues iteratively until some threshold for minimal improvement in the match between predicted and observed signals is reached. The dipole fitting approach is simple and robust in many types of experiments, but the accuracy of the final solution is highly influenced by the number of dipoles that are modeled as well as the starting locations of these sources. The multi-start spatial–temporal downhill simplex algorithm (MSST) eliminates the effects of spatial bias in the starting location of the dipole sources by starting dipoles in random locations over many replications of the same source localization (Huang et al., 1998). In the end, these sparse solutions require the judgment of an expert user to choose the best among many possible solutions.

The beamforming approach to source localization takes several forms: multiple signal classification (MUSIC), synthetic aperture magnetometry (SAM), and linearly constrained minimum variance (LCMV) beamformer. Generally, beamformers share the common feature of point-by-point scanning the brain volume across a set of equivalent current dipoles in fixed locations (Mosher et al., 1992; Van Veen et al., 1997; Vrba and Robinson, 2001). The strength of the dipole is specified at each location based on the contribution of the source to the activity observed on the sensor array. In the beamforming approach, the correspondence between the predicted and observed magnetic field is a product of the forward model and a weighting matrix calculated for each dipole location based on the lead field of each MEG sensor. The weighting matrix may or may not be normalized with respect to background noise. Beamforming approaches provide a computationally efficient approach to source localization that eliminates the biases based on starting location and number of sources in dipole-fitting algorithms; however, correlated and distributed sources can be problematic for beamformers.

Imaging methods for source reconstruction also come in several varieties: $L_1$ and $L_2$ minimum norm approaches, LORETA, FOCUSS, and VESTAL (Gorodnitsky et al., 1995; Hamalainen and Ilmoniemi, 1994; Wagner et al., 2004). Generally, the imaging approaches place dipoles in fixed locations at fixed orientations. Like beamforming approaches, the inverse problem is to assign values to the dipole at each location. The locations are typically constrained to a tessellation of the cortical surface that has been extracted from a magnetic resonance image. The orientations of the cortical dipoles are often constrained to be perpendicular to the cortical surface, as this models the orientation of the apical dendrites in the cortical columns. Unlike beamforming, there is no point-by-point scanning of the source space. Instead, the values of all dipoles are optimized to minimize the error between the predicted and observed MEG signal. The weights of the dipoles can be adjusted until the difference between the observed and expected approaches zero; however, the imaging approaches are optimized for distributed sources models. In fact, even when the sources are known to be focal, they may be displayed in the source image as distributed. Thus, the imaging methods are fairly low resolution and, like beamformers, hugely underdetermined. This makes additional assumptions (i.e., the maximum strength of any dipole) necessary in order to arrive at a stable solution. Imaging solutions tend to be memory intensive but computationally fast.

Each of these solutions to the inverse problem has strengths and weaknesses. Although most MEG laboratories use the equivalent current dipole method for clinical studies, no inverse solution approach has been accepted as convincingly superior. In general, the data analysis approach should reflect our understanding of the underlying magnetic field source and may be individualized.

## SEMINAL TRANSLATIONAL STUDIES

### ADULT EPILEPSY STUDIES

Although the human alpha rhythm was first measured in 1972, technical difficulties had to be resolved before pathologic epileptic activity in patients could be reliably detected. In one of the earliest ictal MEG studies, Sutherling et al. (1987) captured 63 complex partial seizures in 4 patients with a one-channel magnetometer and demonstrated that MEG signals had a morphology and frequency similar to those for EEG. The MEG data resolved an ambiguity in scalp EEG in one patient, highlighting the potential of MEG in providing added value to EEG. The UCLA team then demonstrated, in a truly seminal study (Sutherling et al., 1988a), that the magnetic field localization of epileptic interictal spikes agreed with intracranial electrode localization of EEG spikes. Given the theoretical advantage of MEG over scalp EEG for spatial resolution, the finding that MEG spikes colocalize to spikes detected by intracranial EEG electrodes strongly suggested a clinical utility of MEG.

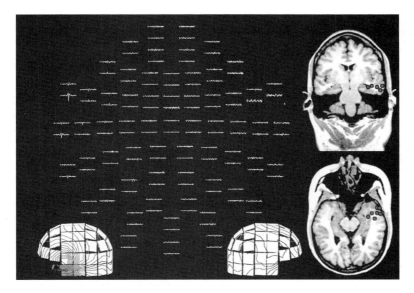

**FIGURE 16.2    (See color insert following page 458.)** Source localizations from left temporal spikes with a sample spike on the sensor array. The white tracings are approximately 750 msec of magnetoencephalography (MEG) data recorded on a 122-channel whole-head biomagnetometer (Elekta Neuromag®; Helsinki, Finland) filtered from 6 to 70 Hz. In this display, the nose would be at the top of the image with true left and right. The left anterior temporal sensors show an interictal spike. At the bottom of the figure are images showing the rotation of the magnetic field in a contour plot overlaid onto the sensor helmet (represented with white squares). At top and bottom right are the source localizations of several similar interictal events on the patient's MRI images. The top image is a coronal slice through the hippocampi; the bottom slice shows an oblique axial cut through the lengths of the hippocampi.

In a large surgical series, Smith et al. (1995) evaluated 40 patients with neocortical or extratemporal lobe epilepsy using MEG to localize interictal spikes. EEG and MEG spikes were localized to the same region in 21 patients, and 13 patients underwent resection of the brain region with localized EEG and MEG spiking. Of the 13 patients, 8 became seizure free. Of 4 patients with discordant MEG and EEG spike localizations who also underwent resective surgery of the EEG-identified region, none became seizure free. This study illustrated several points. The most important finding was that *interictal* MEG spikes may have localizing value in delineating the epileptogenic region (zone of tissue necessary to initiate and sustain epileptic seizures and thus the removal of which would render a patient seizure free). Electroencephalographers have long recognized that *interictal* EEG spikes obtained noninvasively on scalp recording may localize to regions outside of the epileptogenic zone. Therefore, EEG localization of the epileptogenic zone relied almost exclusively on localization of the *ictal* onset zone either with scalp EEG or intracranial EEG. The possibility that interictal spikes as detected by MEG may accurately localize the seizure onset zone was somewhat of a paradigm shift for epileptologists trained on the dictum that only ictal data were truly localizing (Figure 16.2). Another possibility raised by the Smith et al. study was that MEG may be of prognostic value in predicting seizure freedom after surgery.

Approximately 2 years later, Knowlton et al. (1997) found MEG to localize spike sources in 11 of 12 patients with normal or nonlocalizing MRI results. MEG localizations were localized to the epileptogenic zone as ultimately defined by the gold standard of intracranial EEG. The investigators concluded that MEG sometimes provided noninvasive localization data not available with magnetic resonance imaging or scalp ictal EEG. This finding was confirmed by Wheless et al. (1999) in a prospective study on 58 patients with refractory partial epilepsy. The study compared the relative efficacy and contribution of interictal scalp EEG, ictal scalp EEG, ictal intracranial EEG, MRI, and

MEG in identifying the epileptogenic zone. The main outcome measure was seizure freedom after surgery. MEG was found to be second only to intracranial ictal EEG in predicting the epileptogenic zone for patients with an excellent surgical outcome.

In one of the largest series of MEG subjects, Stefan et al. (2003) summarized their experiences with 455 epilepsy patients who underwent MEG study. The investigators found that MEG detected epileptic activity in about 70% of patients. Of the 131 patients who proceeded on to surgical resection, MEG localized the correct lobe in 89%. They concluded that MEG supplied additional information compared to that obtained with their standard presurgical evaluation in 35% of patients, and "crucial" information that significantly altered management in 10% of patients. In a prospective observational study, Knowlton et al. (2008) showed that MEG had a positive predictive value of 78% in predicting seizure-free surgical outcome.

## Pediatric Epilepsy Studies

In one of the first pediatric MEG studies, Paetau et al. (1994) used a multichannel MEG system to record spike activity in children and adolescents. The MEG spike foci in three of their patients who underwent resective surgery agreed with the intracranial EEG localization. These data demonstrated that MEG may be useful in the presurgical evaluation of pediatric epilepsy patients. Over the past 10 years, the Toronto group has made significant contributions to understanding how MEG may help evaluate pediatric epilepsy patients. Minassian et al. (1999) performed MEG in 11 children with intractable, nonlesional, extratemporal focal epilepsy. These children underwent intracranial EEG monitoring and subsequent surgical resection. In 10 of 11 children, the MEG-localized spike zone corresponded to the ictal onset zone as defined by intracranial EEG. Nine of the 11 patients were either seizure free or had a >90% reduction in seizures at 24-month follow-up. These results suggested that MEG may provide valuable noninvasive presurgical data on nonlesional pediatric epilepsy patients. RamachandranNair et al. (2007) extended the previous study by evaluating for factors that predict postsurgical seizure freedom. The investigators examined scalp EEG, MEG, and intracranial EEG data on 22 children with medically refractory epilepsy and normal or nonfocal MRI findings. All children with seizure freedom had an MEG spike cluster in the final resection area, suggesting that MEG accurately detected the epileptogenic region. None of the children with bilateral or scattered MEG spikes achieved seizure freedom after resection based on EEG data, suggesting that diffuse MEG spike sources portend a lower likelihood of postsurgical seizure freedom. To a large degree, the results of the pediatric epilepsy studies mirror those of adults in demonstrating the clinical utility of MEG in providing additional information in a subset of medically refractory seizure patients.

## Mapping Functional Cortex

Because of its high spatial and temporal resolution, MEG can also localize human functional cortices. In 1988, the UCLA group (Sutherling et al., 1988b) mapped the somatosensory cortex in three epilepsy patients by using MEG, EEG, combined MEG/EEG, and electrocorticography. The average distance of localizations from the central sulcus was 4 mm for MEG, 3 mm for combined MEG/EEG, and 3 mm for electrocorticography. Over the next 5 years, hand, lip, and face regions on somatosensory cortex were mapped with MEG (Baumgartner et al., 1991a,b; Orrison et al., 1990; Walter et al., 1992). The mapping of somatosensory cortex is a routine part of epilepsy studies in most clinical MEG centers, and some neurosurgeons also utilize this test for presurgical planning in tumor patients with lesions near the central sulcus.

One of the earliest indications that MEG is useful in the study of brain mechanisms underlying speech perception was Hari's (1991) study, which showed that sounds generated sustained magnetic fields that localized to auditory cortex. Consonant and vowel combinations evoked stronger magnetic responses compared to pure tones, suggesting activation of language cortex when word stimuli are presented. Subsequently, several groups delineated the areas of brain activated by various types of language stimuli (Eulitz et al., 1994; Gross et al., 1998; Kuriki et al., 1999; Martin et al., 1993; Simos et al., 1998). Papanicolaou et al. (1999) introduced an MEG language paradigm to identify the language-dominant hemisphere and specific language-related regions within the dominant hemisphere. Later, Papanicolaou et al. (2004) evaluated the sensitivity and specificity of their language paradigm in 100 surgical epilepsy candidates who also underwent Wada testing. MEG showed lateralized language responses in 98% of the patients and was concordant with Wada in 87%. Most of the discordance was secondary to MEG showing bihemispheric activation with language, whereas Wada showed unilateral hemispheric dominance. The results of this study suggest that MEG is effective in identifying hemispheric dominance for language in epilepsy patients. An interesting but unproven possibility highlighted by this study is that, although the Wada test is considered the gold standard for determining hemispheric dominance for language, MEG may actually be more sensitive in detecting language cortex.

## POSTTRANSLATIONAL STUDIES

Despite large case series from multiple centers suggesting the clinical utility of MEG in the presurgical evaluation of epilepsy, Lau et al. (2008) concluded in a systematic review of MEG papers from 1987 to 2006 that there is insufficient evidence to support the relationship between the use of MEG in surgical planning and seizure-free outcome after epilepsy surgery. One can debate the merits of the review's analysis and the inclusion of two discordant studies, but there is no question that the ability of MEG to improve postsurgical seizure-free outcome has not been evaluated in a randomized, double-blind, controlled fashion. Two points bear emphasis. First, epilepsy surgery as a treatment for medically refractory medial temporal lobe epilepsy has been accepted by epileptologists since the 1960s, but only within the last decade has a randomized, double-blind, controlled trial demonstrated the superiority of anterior temporal lobectomy over continued medical therapy in this patient population (Wiebe et al., 2001). Two, positron emission tomography (PET) and ictal single photon emission computed tomography (SPECT), are routinely used tests for presurgical epilepsy evaluations that have not been demonstrated to be useful in any randomized, double-blind, controlled study. A randomized study would provide additional support, although the preponderance of evidence argues for a role for MEG in the presurgical evaluation of select epilepsy patients.

A major focus of some investigators is the development of a memory paradigm to lateralize verbal and visual memory. Given that language lateralization can be reliably determined by MEG, the ability of MEG to reliably lateralize memory may obviate the need for the Wada test. The technical and logistical challenges of performing this invasive study are well known to epileptologists, neuroradiologists, and neuropsychologists; therefore, any noninvasive study that replaces the Wada test would be welcomed by many clinicians. Tendolkar et al. (2000) found that magnetic evoked fields underlying recognition memory localized to the right medial temporal, right inferior frontal, and left inferior parietal cortices of three young healthy volunteers. Several groups around the world are actively performing memory research using MEG. A number of memory tests activate specific brain regions in some patients (Dhond et al., 2005; Duzel et al., 2003, 2004; Hanlon et al., 2003; Martin et al., 2007; Staresina et al., 2005). Unfortunately, no memory paradigm for MEG has been developed with enough sensitivity and specificity to be clinically applicable to a wide range of epilepsy patients.

Magnetoencephalography may also have a role as a confirmatory test for the diagnosis of epilepsy, especially frontal lobe epilepsy. Approximately 10 to 30% of epilepsy patients have normal EEGs, even with repeat testing (Salinsky et al., 1987); however, we have demonstrated that MEG can detect epileptic spikes not well seen on EEG (Shih et al., 2003). In a study of 36 patients with temporal lobe epilepsy and interictal spiking, Lin et al. (2003) reported that 22% of the spikes were seen only with MEG, and almost 10% were seen exclusively by EEG. These results demonstrate not only that MEG and EEG are complementary techniques but also that MEG is of added value to routine EEG in certain patients. The EEG literature documents that frontal lobe epilepsy patients often present a diagnostic challenge because their outpatient EEGs are frequently normal, and the clinical semiology of the epileptic event may incorporate behaviors often associated with non-epileptic events. In a modeling study, de Jongh et al. (2005) found that the signal-to-noise ratio of MEG was significantly higher than EEG in the frontal head regions, thus indicating MEG may be especially useful in detecting frontal lobe spikes. A preliminary study of 24 patients suggested MEG is more successful for screening and localizing frontal lobe epilepsy than EEG (Ossenblok et al., 2007). Further studies are needed to confirm these findings and better define the role of MEG in the diagnosis of epilepsy.

Multimodality imaging incorporating MEG has been used in basic and patient-oriented research (Eswaran et al., 2005; Gonsalves et al., 2005; Im et al., 2005; Ishibashi et al., 2002; Lewine et al., 2007; Peyron et al., 2002; Ramon et al., 2004; Shih and Weisend, 2005; Shih et al., 2004). Although many have used MEG as part of a battery of independent imaging modalities, we and others have used information from one technique to constrain and enhance the spatial or temporal resolution of another technique. Im et al. (2005) applied a functional magnetic resonance imaging (fMRI)-constrained imaging method to the datasets of language experiments and found that some MEG sources may be eliminated by the introduction of the fMRI weighting. We demonstrated in a group of 20 patients with nonlesional temporal lobe epilepsy that focal areas of epileptiform discharges as determined by MEG were associated with metabolic abnormalities detected by proton magnetic resonance spectroscopy (Shih et al., 2004). We have also demonstrated similar findings in patients with frontal lobe epilepsy (Shih and Weisend, 2005). Further studies are needed to determine if MEG-guided proton magnetic resonance spectroscopy improves upon current tests used to localize the epileptogenic zone.

Many EEG/MEG analysis algorithms remain experimental and are not in widespread clinical use. The current difficulty limiting widespread use of MEG in epilepsy is that MEG manufacturers provide tools for only basic or very narrowly focused analyses. Some commercial software packages such as BESA® and Curry can process data from any of the major whole-head MEG instruments; however, even these commercial packages require an expert user in both MEG and the software to properly implement myriad settings. This limits accessibility to MEG analysis by the typical neurologist who cannot focus exclusively on MEG. Some basic science research laboratories have developed data processing and source localization algorithms, but these packages are not widely distributed to clinical users and may have limited clinical applicability. The implementation of noncommercial software packages such as BrainStorm, Nutmeg, MNE, FOCUSS, and VESTAL is very often machine or site specific, thus limiting the general utility of the algorithm. One recent study addressed this problem. MNE and BrainStorm now have generalized data import routines that can handle the data from any of the major whole-head MEG machines and have demonstrated similar results across machines when the same task is performed in the same individuals (Weisend et al., 2007). However, these analysis routines still require an expert user for implementation. The proliferation of algorithms for data analysis provides the diversity necessary for innovative data analysis solutions but has impeded progress in the widespread acceptance of MEG as a clinical imaging tool. The plethora of data analysis tools, among which the differences are highlighted far more often than the similarities, has led to the impression that the analysis of clinical MEG data yields unreliable results, even though the opposite is true (Figure 16.3). These analysis methods share much more in common than is often highlighted by the advocates of any particular algorithm. A generalized analysis tool for MEG analysis with reasonable assumptions that can be used by clinical MEG laboratories to understand brain function and dysfunction would be a significant advance in the field of clinical MEG.

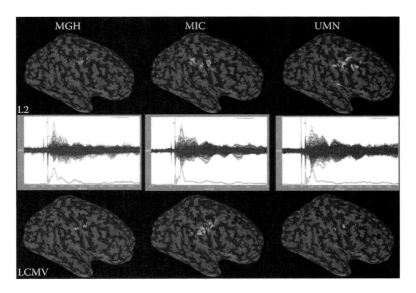

**FIGURE 16.3 (See color insert following page 458.)** Correspondence of median nerve localizations across different magnetoencephalography (MEG) machines and source localizations. The columns in each figure show data recorded at different sites with different MEG instruments. The top row in the figure shows source localizations obtained using BrainStorm software on inflated cortical surfaces extracted from magnetic resonance (MR) images using the BrainSuite software package and the L2 minimum norm algorithm. The center row shows the recorded wave forms (black) and the source time course for the same voxels in the post central gyrus, and the bottom row depicts localization of the same data using the LCMV beamforming algorithm. These data were recorded from the same subject during visits to Massachusetts General Hospital (MGH), the Mind Research Network Imaging Center (MIC), and the University of Minnesota (UMN). The data at MGH were recorded with an Elekta Neuromag® 306-channel vector view instrument. Data from MIC were recorded with a VSM MedTech OMEGA 275-channel instrument. A 248-channel 4D neuroimaging system was used to record data at UMN. The same region was identified in the post central gyrus across all sites and algorithms.

## REFERENCES

Babiloni, F., F. Carducci, F. Cincotti, C. Del Gratta, V. Pizzella, G.L. Romani, P.M. Rossini, F. Tecchio, C. Babiloni. (2001). Linear inverse source estimate of combined EEG and MEG data related to voluntary movements. *Hum. Brain Mapp.*, 14:197–209.

Baillet, S., L. Garnero, G. Marin, J.P. Hugonin. (1999). Combined MEG and EEG source imaging by minimization of mutual information. *IEEE Trans. Biomed. Eng.*, 46:522–534.

Barth, D.S., W. Sutherling, J. Broffman, J. Beatty. (1986). Magnetic localization of a dipolar current source implanted in a sphere and a human cranium. *Electroencephalogr. Clin. Neurophysiol.*, 63:260–273.

Baule, G., R. McFee. (1963). Detection of the magnetic field of the heart. *Am. Heart J.*, 66:95–96.

Baumgartner, C., A. Doppelbauer, L. Deecke, D.S. Barth, J. Zeitlhofer, G. Lindinger, W.W. Sutherling. (1991a). Neuromagnetic investigation of somatotopy of human hand somatosensory cortex. *Exp. Brain Res.*, 87:641–648.

Baumgartner, C., A. Doppelbauer, W.W. Sutherling, J. Zeitlhofer, G. Lindinger, C. Lind, L. Deecke. (1991b). Human somatosensory cortical finger representation as studied by combined neuromagnetic and neuroelectric measurements. *Neurosci. Lett.*, 134:103–108.

Baumgartner, C., E. Pataraia, G. Lindinger, L. Deecke. (2000). Neuromagnetic recordings in temporal lobe epilepsy. *J. Clin. Neurophysiol.*, 17:177–189.

Bowyer, S.M., J.E. Moran, K.M. Mason, J.E. Constantinou, B.J. Smith, G.L. Barkley, N. Tepley. (2004). MEG localization of language-specific cortex utilizing MR-FOCUSS. *Neurology*, 62:2247–2255.

Brancucci, A., S.D. Penna, C. Babiloni, F. Vecchio, P. Capotosto, D. Rossi, R. Franciotti, K. Torquati, V. Pizzella, P.M. Rossini, G.L. Romani. (2008). Neuromagnetic functional coupling during dichotic listening of speech sounds. *Hum. Brain Mapp.*, 29:253–264.

Cohen, D. (1972). Magnetoencephalography: detection of the brain's electrical activity with a superconducting magnetometer. *Science*, 175:664–666.

Cohen, D., B.N. Cuffin. (1987). A method for combining MEG and EEG to determine the sources. *Phys. Med. Biol.*, 32:85–89.

de Jongh, A., J.C. de Munck, S.I. Goncalves, P. Ossenblok. (2005). Differences in MEG/EEG epileptic spike yields explained by regional differences in signal-to-noise ratios. *J. Clin. Neurophysiol.*, 22:153–158.

Dhond, R.P., T. Witzel, A.M. Dale, E. Halgren. (2005). Spatiotemporal brain maps of delayed word repetition and recognition. *Neuroimage*, 28:293–304.

Duzel, E., R. Habib, B. Schott, A. Schoenfeld, N. Lobaugh, A.R. McIntosh, M. Scholz, H.J. Heinze. (2003). A multivariate, spatiotemporal analysis of electromagnetic time–frequency data of recognition memory. *Neuroimage*, 18:185–197.

Duzel, E., R. Habib, S. Guderian, H.J. Heinze. (2004). Four types of novelty–familiarity responses in associative recognition memory of humans. *Eur. J. Neurosci.*, 19:1408–1416.

Ermer, J.J., J.C. Mosher, S. Baillet, R.M. Leah. (2001). Rapidly recomputable EEG forward models for realistic head shapes. *Phys. Med. Biol.*, 46:1265–1281.

Eswaran, H., C.L. Lowery, J.D. Wilson, P. Murphy, H. Preissl. (2005). Fetal magnetoencephalography: a multimodal approach. *Brain Res. Dev. Brain Res.*, 154:57–62.

Eulitz, C., T. Elbert, P. Bartenstein, C. Weiller, S.P. Muller, C. Pantev. (1994). Comparison of magnetic and metabolic brain activity during a verb generation task. *NeuroReport*, 6:97–100.

Gonsalves, B.D., I. Kahn, T. Curran, K.A. Norman, A.D. Wagner. (2005). Memory strength and repetition suppression: multimodal imaging of medial temporal cortical contributions to recognition. *Neuron*, 47:751–761.

Gorodnitsky, I.F., J.S. George, B.D. Rao. (1995). Neuromagnetic source imaging with FOCUSS: a recursive weighted minimum norm algorithm. *Electroencephalogr. Clin. Neurophysiol.*, 95:231–251.

Gross, J., A.A. Ioannides, J. Dammers, B. Maess, A.D. Friederici, H.W. Muller-Gartner. (1998). Magnetic field tomography analysis of continuous speech. *Brain Topogr.*, 10:273–281.

Hamalainen, M.S. (1992). Magnetoencephalography: a tool for functional brain imaging. *Brain Topogr.*, 5:95–102.

Hamalainen, M.S., R.J. Ilmoniemi. (1994). Interpreting magnetic fields of the brain: minimum norm estimates. *Med. Biol. Eng. Comput.*, 32:35–42.

Hanlon, F.M., M.P. Weisend, M. Huang, R.R. Lee, S.N. Moses, K.M. Paulson, R.J. Thoma, G.A. Miller, J.M. Canive. (2003). A non-invasive method for observing hippocampal function. *NeuroReport*, 14:1957–1960.

Hari, R. (1991). Activation of the human auditory cortex by speech sounds. *Acta Otolaryngol. Suppl.*, 491:132–138.

Hillebrand, A., G.R. Barnes. (2002). A quantitative assessment of the sensitivity of whole-head MEG to activity in the adult human cortex. *Neuroimage*, 16:638–650.

Hirano, S., Y. Hirano, T. Maekawa, C. Obayashi, N. Oribe, T. Kuroki, S. Kanba, T. Onitsuka. (2008). Abnormal neural oscillatory activity to speech sounds in schizophrenia: a magnetoencephalography study. *J. Neurosci.*, 28:4897–4903.

Huang, M.X., C.J. Aine, S. Supek, E. Best, D. Ranken, E.R. Flynn. (1998). Multi-start downhill simplex method for spatio-temporal source localization in magnetoencephalography. *Electroencephalogr. Clin. Neurophysiol.*, 108:32–44.

Huang, M.X., J.C. Mosher, R.M. Leahy. (1999). A sensor-weighted overlapping-sphere head model and exhaustive head model comparison for MEG. *Phys. Med. Biol.*, 44:423–440.

Huang, M.X., L.E. Davis, C. Aine, M. Weisend, D. Harrington, R. Christner, J. Stephen, J.C. Edgar, M. Herman, J. Meyer, K. Paulson, K. Martin, R.R. Lee. (2004). MEG response to median nerve stimulation correlates with recovery of sensory and motor function after stroke. *Clin. Neurophysiol.*, 115:820–833.

Huang, M.X., A.M. Dale, T. Song, E. Halgren, D.L. Harrington, I. Podgorny, J.M. Canive, S. Lewis, R.R. Lee. (2006). Vector-based spatial-temporal minimum $L_1$-norm solution for MEG. *Neuroimage*, 31:1025–1037.

Huang, M.X., T. Song, D.J. Hagler, Jr., I. Podgorny, V. Jousmaki, L. Cui, K. Gaa, D.L. Harrington, A.M. Dale, R.R. Lee, J. Elman, E. Halgren. (2007). A novel integrated MEG and EEG analysis method for dipolar sources. *Neuroimage*, 37:731–748.

Huizenga, H.M., T.L. van Zuijen, D.J. Heslenfeld, P.C. Molenaar. (2001). Simultaneous MEG and EEG source analysis. *Phys. Med. Biol.*, 46:1737–1751.

Ikeda, S., K. Toyama. (2000). Independent component analysis for noisy data: MEG data analysis. *Neural Netw.*, 13:1063–1074.

Im, C.H., H.K. Jung, N. Fujimaki. (2005). fMRI-constrained MEG source imaging and consideration of fMRI invisible sources. *Hum. Brain Mapp.*, 26:110–118.

Ishibashi, H., P.G. Simos, J.W. Wheless, J.E. Baumgartner, H.L. Kim, R.N. Davis, W. Zhang, A.C. Papanicolaou. (2002). Multimodality functional imaging evaluation in a patient with Rasmussen's encephalitis. *Brain Dev.*, 24:239–244.

Jerbi, K., S. Baillet, J.C. Mosher, G. Nolte, L. Garnero, R.M. Leahy. (2004). Localization of realistic cortical activity in MEG using current multipoles. *Neuroimage*, 22:779–793.

Kahkonen, S., H. Yamashita, H. Rytsala, K. Suominen, J. Ahveninen, E. Isometsa. (2007). Dysfunction in early auditory processing in major depressive disorder revealed by combined MEG and EEG. *J. Psychiatry Neurosci.*, 32:316–322.

Kaufman, L., Y. Okada, D. Brenner, S.J. Williamson. (1981). On the relation between somatic evoked potentials and fields. *Int. J. Neurosci.*, 15:223–239.

Knowlton, R.C., K.D. Laxer, M.J. Aminoff, T.P. Roberts, S.T. Wong, H.A. Rowley. (1997). Magnetoencephalography in partial epilepsy: clinical yield and localization accuracy. *Ann. Neurol.*, 42:622–631.

Knowlton, R.C., R.A. Elgavish, A. Bartolucci, B. Ojha, N. Limdi, J. Blount, J.G. Burneo, L. Ver Hoef, L. Paige, E. Faught, P. Kankirawatana, K. Riley, R. Kuzniecky. (2008). Functional imaging. II. Prediction of epilepsy surgery outcome. *Ann. Neurol.*, 64:35–41.

Kober, H., M. Moller, C. Nimsky, J. Vieth, R. Fahlbusch, O. Ganslandt. (2001). New approach to localize speech relevant brain areas and hemispheric dominance using spatially filtered magnetoencephalography. *Hum. Brain Mapp.*, 14:236–250.

Kuriki, S., T. Mori, Y. Hirata. (1999). Motor planning center for speech articulation in the normal human brain. *NeuroReport*, 10:765–769.

Lau, M., D. Yam, J.G. Burneo. (2008). A systematic review on MEG and its use in the presurgical evaluation of localization-related epilepsy. *Epilepsy Res.*, 79:97–104.

Leahy, R.M., J.C. Mosher, M.E. Spencer, M.X. Huang, J.D. Lewine. (1998). A study of dipole localization accuracy for MEG and EEG using a human skull phantom. *Electroencephalogr. Clin. Neurophysiol.*, 107:159–173.

Lewine, J.D., J.T. Davis, E.D. Bigler, R. Thoma, D. Hill, M. Funke, J.H. Sloan, S. Hall, W.W. Orrison. (2007). Objective documentation of traumatic brain injury subsequent to mild head trauma: multimodal brain imaging with MEG, SPECT, and MRI. *J. Head Trauma Rehabil.*, 22:141–155.

Lin, Y.Y., Y.H. Shih, J.C. Hsieh, H.Y. Yu, C.H. Yiu, T.T. Wong, T.C. Yeh, S.Y. Kwan, L.T. Ho, D.J. Yen, Z.A. Wu, M.S. Chang. (2003). Magnetoencephalographic yield of interictal spikes in temporal lobe epilepsy: comparison with scalp EEG recordings. *Neuroimage*, 19:1115–1126.

Liu, A.K., A.M. Dale, J.W. Belliveau. (2002). Monte Carlo simulation studies of EEG and MEG localization accuracy. *Hum. Brain Mapp.*, 16:47–62.

Luo, H., D. Poeppel. (2007). Phase patterns of neuronal responses reliably discriminate speech in human auditory cortex. *Neuron*, 54:1001–1010.

Martin, N.A., J. Beatty, R.A. Johnson, M.L. Collaer, F. Vinuela, D.P. Becker, M.R. Nuwer. (1993). Magnetoencephalographic localization of a language processing cortical area adjacent to a cerebral arteriovenous malformation: case report. *J. Neurosurg.*, 79:584–588.

Martin, T., M.A. McDaniel, M.J. Guynn, J.M. Houck, C.C. Woodruff, J.P. Bish, S.N. Moses, D. Kicic, C.D. Tesche. (2007). Brain regions and their dynamics in prospective memory retrieval: a MEG study. *Int. J. Psychophysiol.*, 64:247–258.

Merrifield, W.S., P.G. Simos, A.C. Papanicolaou, L.M. Philpott, W.W. Sutherling. (2007). Hemispheric language dominance in magnetoencephalography: sensitivity, specificity, and data reduction techniques. *Epilepsy Behav.*, 10:120–128.

Mikuni, N., T. Nagamine, A. Ikeda, K. Terada, W. Taki, J. Kimura, H. Kikuchi, H. Shibasaki. (1997). Simultaneous recording of epileptiform discharges by MEG and subdural electrodes in temporal lobe epilepsy. *Neuroimage*, 5:298–306.

Minassian, B.A., H. Otsubo, S. Weiss, I. Elliott, J.T. Rutka, O.C. Snead 3rd. (1999). Magnetoencephalographic localization in pediatric epilepsy surgery: comparison with invasive intracranial electroencephalography. *Ann. Neurol.*, 46:627–633.

Mosher, J.C., P.S. Lewis, R.M. Leahy. (1992). Multiple dipole modeling and localization from spatio-temporal MEG data. *IEEE Trans. Biomed. Eng.*, 39:541–557.

Murakami, S., Y. Okada. (2006). Contributions of principal neocortical neurons to magnetoencephalography and electroencephalography signals. *J. Physiol.*, 575:925–936.

Nahum, M., H. Renvall, M. Ahissar. (2009). Dynamics of cortical responses to tone pairs in relation to task difficulty: a MEG study. *Hum. Brain Mapp.*, 30:1592–1604.

Okada, Y., A. Lahteenmaki, C. Xu. (1999a). Comparison of MEG and EEG on the basis of somatic evoked responses elicited by stimulation of the snout in the juvenile swine. *Clin. Neurophysiol.*, 110:214–229.

Okada, Y.C., A. Lahteenmaki, C. Xu. (1999b). Experimental analysis of distortion of magnetoencephalography signals by the skull. *Clin. Neurophysiol.*, 110:230–238.

Okazaki, Y., A. Abrahamyan, C.J. Stevens, A.A. Ioannides. (2008). The timing of face selectivity and attentional modulation in visual processing. *Neuroscience*, 152:1130–1144.

Orrison, W.W., L.E. Davis, G.W. Sullivan, F.A. Mettler, Jr., E.R. Flynn. (1990). Anatomic localization of cerebral cortical function by magnetoencephalography combined with MR imaging and CT. *AJNR Am. J. Neuroradiol.*, 11:713–716.

Ossenblok, P., J.C. de Munck, J.C. et al. (2007). Magnetoencephalography is more successful for screening and localizing frontal lobe epilepsy than electroencephalography. *Epilepsia*, 48:2139–2149.

Paetau, R., M. Kajola, R. Hari. (1990). Magnetoencephalography in the study of epilepsy. *Neurophysiol. Clin.*, 20:169–187.

Paetau, R., M. Hamalainen, R. Hari, M. Kajola, J. Karhu, T.A. Larsen, E. Lindahl, O. Salonen. (1994). Magnetoencephalographic evaluation of children and adolescents with intractable epilepsy. *Epilepsia*, 35:275–284.

Papanicolaou, A.C., P.G. Simos, J.I. Breier, G. Zouridakis, L.J. Willmore, J.W. Wheless, J.E. Constantinou, W.W. Maggio, W.B. Gormley. (1999). Magnetoencephalographic mapping of the language-specific cortex. *J. Neurosurg.*, 90:85–93.

Papanicolaou, A.C., P.G. Simos, E.M. Castillo, J.I. Breier, S. Sarkari, E. Pataraia, R.L. Billingsley, S. Buchanan, J. Wheless, V. Maggio, W.W. Maggio. (2004). Magnetocephalography: a noninvasive alternative to the Wada procedure. *J. Neurosurg.*, 100:867–876.

Papanicolaou, A.C., P.G. Simos, E.M. Castillo. (2005). MEG localization of language-specific cortex utilizing MR-FOCUSS. *Neurology*, 64:765.

Peyron, R., M. Frot, F. Schneider, L. Garcia-Larrea, P. Mertens, F.G. Barral, M. Sindou, B. Laurent, F. Mauguiere. (2002). Role of operculoinsular cortices in human pain processing: converging evidence from PET, fMRI, dipole modeling, and intracerebral recordings of evoked potentials. *Neuroimage*, 17:1336–1346.

RamachandranNair, R., H. Otsubo, M.M. Shroff, A. Ochi, S.K. Weiss, J.T. Rutka, O.C. Snead 3rd. (2007). MEG predicts outcome following surgery for intractable epilepsy in children with normal or nonfocal MRI findings. *Epilepsia*, 48:149–157.

Ramon, C., J. Haueisen, T. Richards, K. Maravilla. (2004). Multimodal imaging of somatosensory evoked cortical activity. *Neurol. Clin. Neurophysiol.*, 2004:96.

Reulbach, U., S. Bleich, C. Maihofner, J. Kornhuber, W. Sperling. (2007). Specific and unspecific auditory hallucinations in patients with schizophrenia: a magnetoencephalographic study. *Neuropsychobiology*, 55:89–95.

Rong, F., J.L. Contreras-Vidal. (2006). Magnetoencephalographic artifact identification and automatic removal based on independent component analysis and categorization approaches. *J. Neurosci. Methods*, 157:337–354.

Salinsky, M., R. Kanter, R.M. Dasheiff. (1987). Effectiveness of multiple EEGs in supporting the diagnosis of epilepsy: an operational curve. *Epilepsia*, 28:331–334.

Salmelin, R. (2007). Clinical neurophysiology of language: the MEG approach. *Clin. Neurophysiol.*, 118:237–254.

Sato, S. (1990). Epilepsy research: NIH experience. *Adv. Neurol.*, 54:223–230.

Shigeto, H., T. Morioka, K. Hisada, S. Nishio, H. Ishibashi, D. Kira, S. Tobimatsu, M. Kato. (2002). Feasibility and limitations of magnetoencephalographic detection of epileptic discharges: simultaneous recording of magnetic fields and electrocorticography. *Neurol. Res.*, 24:531–536.

Shih, J.J., M.P. Weisend. (2005). Magnetoencephalography-guided MR spectroscopy show areas of metabolite abnormalities in frontal lobe epilepsy. *Epilepsia*, 46(Suppl. 7):75–76.

Shih, J.J., D. Devier, E. Bryniarski, M.P. Weisend. (2003). Focal MEG montages can detect frontal lobe spikes not seen in routine EEG recordings. *Epilepsia*, 44(Suppl. 9):252–253.

Shih, J.J., M.P. Weisend, J. Lewine, J. Sanders, J. Dermon, R. Lee. (2004). Areas of interictal spiking are associated with metabolic dysfunction in MRI-negative temporal lobe epilepsy. *Epilepsia*, 45:223–229.

Silen, T., N. Forss, S. Salenius, T. Karjalainen, R. Hari. (2002). Oscillatory cortical drive to isometrically contracting muscle in Unverricht–Lundborg type progressive myoclonus epilepsy (ULD). *Clin. Neurophysiol.*, 113:1973–1979.

Simos, P.G., J.I. Breier, G. Zouridakis, A.C. Papanicolaou. (1998). Identification of language-specific brain activity using magnetoencephalography. *J. Clin. Exp. Neuropsychol.*, 20:706–722.

Smith, J.R., B.J. Schwartz, C. Gallen, W. Orrison, J. Lewine, A.M. Murro, D.W. King, Y.D. Park. (1995). Multichannel magnetoencephalography in ablative seizure surgery outside the anteromesial temporal lobe. *Stereotact. Funct. Neurosurg.*, 65:81–85.

Staresina, B.P., H. Bauer, L. Deecke, P. Walla. (2005). Magnetoencephalographic correlates of different levels in subjective recognition memory. *Neuroimage*, 27:83–94.

Stefan, H., C. Hummel, G. Scheler, A. Genow, K. Druschky, C. Tilz, M. Kaltenhauser, R. Hopfengartner, M. Buchfelder, J. Romstock. (2003). Magnetic brain source imaging of focal epileptic activity: a synopsis of 455 cases. *Brain*, 126:2396–2405.

Stephen, J.M., D.F. Ranken, C.J. Aine. (2006). Frequency-following and connectivity of different visual areas in response to contrast-reversal stimulation. *Brain Topogr.*, 18:257–272.

Sutherling, W.W., P.H. Crandall, J. Engel, Jr., T.M. Darcey, L.D. Cahan, S. Barth. (1987). The magnetic field of complex partial seizures agrees with intracranial localizations. *Ann. Neurol.*, 21:548–558.

Sutherling, W.W., P.H. Crandall, L.D. Cahan, D.S. Barth. (1988a). The magnetic field of epileptic spikes agrees with intracranial localizations in complex partial epilepsy. *Neurology*, 38:778–786.

Sutherling, W.W., P.H. Crandall, T.M. Darcey, D.P. Becker, M.F. Levesque, D.S. Barth. (1988b). The magnetic and electric fields agree with intracranial localizations of somatosensory cortex. *Neurology*, 38:1705–1714.

Sutherling, W.W., M. Akhtari, A.N. Mamelak, J. Mosher, D. Arthur, S. Sands, P. Weiss, N. Lopez, M. DiMauro, E. Flynn, R. Leah. (2001). Dipole localization of human induced focal afterdischarge seizure in simultaneous magnetoencephalography and electrocorticography. *Brain Topogr.*, 14:101–116.

Tarkiainen, A., P. Helenius, R. Salmelin. (2003). Category-specific occipitotemporal activation during face perception in dyslexic individuals: an MEG study. *Neuroimage*, 19:1194–1204.

Taulu, S., J. Simola, M. Kajola. (2004). MEG recordings of DC fields using the signal space separation method (SSS). *Neurol. Clin. Neurophysiol.*, 2004:35.

Tendolkar, I., M. Rugg, J. Fell, H. Vogt, M. Scholz, H. Hinrichs, H.J. Heinze. (2000). A magnetoencephalographic study of brain activity related to recognition memory in healthy young human subjects. *Neurosci. Lett.*, 280:69–72.

Tesche, C.D., M.A. Uusitalo, R.J. Ilmoniemi, M. Huotilainen, M. Kajola, O. Salonen. (1995). Signal-space projections of MEG data characterize both distributed and well-localized neuronal sources. *Electroencephalogr. Clin. Neurophysiol.*, 95:189–200.

Van Veen, B.D., W. van Drongelen, M. Yuchtman, A. Suzuki. (1997). Localization of brain electrical activity via linearly constrained minimum variance spatial filtering. *IEEE Trans. Biomed. Eng.*, 44:867–880.

Vieth, J.B., H. Kober, P. Grummich. (1996). Sources of spontaneous slow waves associated with brain lesions, localized by using the MEG. *Brain Topogr.*, 8:215–221.

Vrba, J., S.E. Robinson. (2001). Signal processing in magnetoencephalography. *Methods*, 25:249–271.

Wagner, M., M. Fuchs, J. Kastner. (2004). Evaluation of sLORETA in the presence of noise and multiple sources. *Brain Topogr.*, 16:277–280.

Walter, H., R. Kristeva, U. Knorr, G. Schlaug, Y. Huang, H. Steinmetz, B. Nebeling, H. Herzog, R.J. Seitz. (1992). Individual somatotopy of primary sensorimotor cortex revealed by intermodal matching of MEG, PET, and MRI. *Brain Topogr.*, 5:183–187.

Weisend M.P., F.M. Hanlon, S.N. Moses, M.X. Huang, R.R. Lee. (2002). Simulations of hippocampal and nearby temporal lobe activity in magnetoencephalography. In *Proceedings of the 13th International Conference on Biomagnetism*, Jena, Germany, October 10–14.

Weisend, M.P., F.M. Hanlon, R. Montaño, S.P. Ahlfors, A.C. Leuthold, D. Pantazis, J.C. Mosher, A.P. Georgopoulos, M.S. Hamalainen, C.J. Aine. (2007). Paving the way for cross-site pooling of magnetoencephalography (MEG) data. *Int. Congress Ser.*, 1300:615–618.

Wheless, J.W., L.J. Willmore, J.I. Breier, M. Kataki, J.R. Smith, D.W. King, K.J. Meador, Y.D. Park, D.W. Loring, G.L. Clifton, J. Baumgartner, A.B. Thomas, J.E. Constantinou, A.C. Papanicolaou. (1999). A comparison of magnetoencephalography, MRI, and V-EEG in patients evaluated for epilepsy surgery. *Epilepsia*, 40:931–941.

Wiebe, S., W.T. Blume, J.P. Girvin, M. Eliasziw. (2001). A randomized, controlled trial of surgery for temporal-lobe epilepsy. *N. Engl. J. Med.*, 345:311–318.

Williamson, S.J., L. Kaufman. (1981). Magnetic fields of the cerebral cortex. In *Biomagnetism* (pp. 353–402), S.N. Erne, H.D. Hahlbohm, and H. Lubbig, Eds. Berlin: Walter de Gruyter.

Wu, J., Y.C. Okada. (1999). Roles of a potassium afterhyperpolarization current in generating neuromagnetic fields and field potentials in longitudinal CA3 slices of the guinea-pig. *Clin. Neurophysiol.*, 110:1858–1867.

Yoshinaga, H., T. Nakahori, Y. Ohtsuka, E. Oka, Y. Kitamura, H. Kiriyama, K. Kinugasa, K. Miyamoto, T. Hoshida. (2002). Benefit of simultaneous recording of EEG and MEG in dipole localization. *Epilepsia*, 43:924–928.

# 17 What Have We Learned from Resective Surgery in Pediatric Patients with Cortical Dysplasia?

*Carlos Cepeda and Gary W. Mathern*

## CONTENTS

## INTRODUCTION

The feasibility of maintaining *in vitro* human brain slices from surgical tissue samples opened the door to electrophysiologic and mechanistic studies of seizure generation. Pioneering figures include David Prince, Phil Schwartzkroin, and others (Prince and Wong, 1981; Schwartzkroin and Knowles, 1984; Schwartzkroin and Prince, 1976). Although somewhat disappointing at first due to technical limitations, the functional investigation of epileptogenic tissue is now a burgeoning field. In recent years, the number of cellular electrophysiological studies on human tissue resected for the treatment of pharmacoresistant epilepsy has grown considerably. Current studies encompass, among others, tissue obtained from patients with temporal lobe epilepsy (TLE), cortical dysplasia (CD), hypothalamic hamartomas (HHs), and tuberous sclerosis complex (TSC). These studies require a collaborative approach with careful cooperation of the experimental design and data interpretation between the clinical and basic science teams.

"Why do you study human tissue?" This is a question we frequently hear when we describe the results of our correlative clinicopathologic research studies of children with CD. Our answer is that there are many valuable reasons to spend the time, energy, and resources to perform these studies. Using human brain tissue removed at epilepsy surgery is the only opportunity to directly study the pathology that generates the seizures. Whereas animal models of epilepsy offer more controlled experimental manipulation, they never completely replicate the human clinical and pathological conditions. Thus, mechanisms developed in model systems still have to be verified in humans. Basic science studies of human brain tissue also allow us the opportunity to uncover mechanisms that help explain the epileptogenesis of CD tissue. Sometimes the discovered mechanisms are similar to those found in animal models, but oftentimes they are not. This helps to elucidate mechanisms that might be critical in designing new treatments for patients. Finally, the study of human brain tissue offers us the opportunity to try out novel pharmacological therapies before clinical trials. New non-invasive treatments might reduce the need for resective operations. Hence, in our opinion, the study of human brain tissue is an essential research arm to understand the human epilepsies that are treated surgically.

Our laboratory at the University of California, Los Angeles (UCLA) has been studying brain tissue samples from pediatric epilepsy surgery patients for the past 20 years. Cases have mainly included CD, TSC, hemimegalencephaly, Rasmussen's encephalitis, tumors, stroke, and hippocampal sclerosis. As it would be impossible to cover the multiple facets of research on human brain tissue, we are going to limit this chapter to our correlative functional and morphological studies of CD tissue samples and comment briefly on other epileptic conditions as they pertain to these studies.

## DEFINITION AND CLASSIFICATIONS

Taylor and associates defined focal CD as a histopathologic abnormality characterized by cortical dyslamination, misoriented pyramidal neurons, and the presence of dysmorphic cells, in particular cytomegalic neurons and balloon cells (Taylor et al., 1971). The potential role of these abnormal cells in CD epileptogenesis had long been hypothesized (Kerfoot et al., 1999; Schwartzkroin and Walsh, 2000; Spreafico et al., 1998); however, a functional characterization of these cells had to be deferred until technological advances in optical and recording techniques allowed visualization of individual cells in slices. Thanks to the introduction in the early 1990s of infrared (IR) video microscopy in tissue slices, the principal features of abnormal cells in CD could finally be examined.

Cortical dysplasia is not a homogeneous pathology, and several types have been described. In a first approximation, the neuropathology group at UCLA, directed by Harry Vinters, divided CD into mild and severe forms. Mild CD is characterized by dyslamination and cell misorientation, whereas severe CD includes the presence of dysmorphic cytomegalic neurons and balloon cells (Cepeda et al., 2003; Mischel et al., 1995). More recently, Palmini et al. (2004) proposed a similar classification of CD that includes Type I, where only architectural abnormalities of the cortex are found, and more severe cases where, in addition to cortical dyslamination, the sample contains abnormal cells; Type IIA (with cytomegalic neurons); and Type IIB (with balloon cells). For the sake of simplicity, in this chapter, we will use the terms mild CD or Palmini Type I and severe CD or Palmini Type IIA/B interchangeably.

It should be noted that even within the same patient, areas removed at the time of surgery have different degrees of histopathologic abnormality. In our studies, tissue is typically separated into most abnormal (MA) and least abnormal (LA) based on degree of abnormality in electrocorticography (ECoG), magnetic resonance imaging (MRI), and positron emission tomography (PET) studies. Thus, in our analyses, we first establish a within-subject comparison (LA vs. MA areas); subsequently, data from CD cases with different degrees of pathology (mild Type I vs. severe Type II) are compared against non-CD cases. Areas with the highest degree of ECoG abnormalities correlate with worse MRI and histopathologic findings (Cepeda et al., 2005a). It is in the MA samples where the likelihood of finding dysmorphic cells is highest.

## CLINICAL CHARACTERISTICS OF EPILEPSY SURGERY PATIENTS WITH CD

Patients with CD, as a group, have a younger age at seizure onset and greater seizure frequencies compared with other etiologies in epilepsy surgery cases. In pediatric surgery patients, CD is the most common pathology (Harvey et al., 2008; Mathern et al., 1999). By comparison, in adult epilepsy surgery patients, it is the third most common etiology behind hippocampal sclerosis and tumors (Becker et al., 2006). Mild CD is associated with seizures arising more frequently from the temporal lobe (Fauser et al., 2004; Tassi et al., 2002; Widdess-Walsh et al., 2005), whereas patients with severe CD are more likely to have multilobar and hemispheric abnormalities involving extratemporal regions. Likewise, a history of infantile spasms and status epilepticus is more common in patients with Type II compared with Type I CD (Koh et al., 2005; Krsek et al., 2008).

There are no distinctive interictal or ictal scalp EEG features that are exclusively associated with CD in patients with refractory epilepsy. The EEG can show interictal background slowing, interictal spikes and polyspikes, and ictal events with electrographic characteristics similar to other patients with intractable epilepsy undergoing presurgical evaluation (Noachtar et al., 2008). However, there are several features on structural MRI that specifically identify areas of CD (Raybaud et al., 2006; Tassi et al., 2002; Yagishita et al., 1997). These MRI findings include increased thickness of the cortical gray matter often with abnormal gyral patterns, blurring of the cortical–white matter junction, and increased $T_2$ and fluid-attenuated inversion recovery (FLAIR) signal intensity in the subcortical white matter (Figure 17.1). When present, the constellation of MRI findings is diagnostic for CD in patients with intractable epilepsy; however, MRI findings can often be subtle and difficult to detect, especially for patients with mild CD and those with very small severe CD lesions. In about one third of patients, structural MRI can be normal or nonspecific for the diagnosis of CD (Lerner et al., 2009). As expected, normal MRI scans are reported more frequently in patients with Type I compared with Type II CD.

2-[$^{18}$F]fluoro-2-deoxy-D-glucose (FDG)–PET has been shown to be one of the more sensitive techniques in identifying areas of CD (Chugani et al., 1990). Contemporary studies indicate that FDG–PET will show interictal hypometabolism localized to areas of CD in approximately 75% of patients. To increase the sensitivity of FDG–PET, the UCLA group recently modified the technique whereby pseudo-colored FDG–PET images representing escalating levels of hypometabolism were overlaid onto the structural MRI (Burri et al., 2008). This method identified CD in 98% of patients, including a substantial proportion of patients with mild CD with previously interpreted normal structural MRI scans (Salamon et al., 2008). Hence, FDG–PET can be a very useful tool in detecting CD in patients undergoing presurgical evaluation for their intractable epilepsy, especially if the MRI is normal or nonspecific for CD.

In the UCLA series, 37% of patients with CD had hemispheric or multilobar operations, and most of those were in patients with Type II CD (Lerner et al., 2009). In adult cohorts, up to 80% of patients were reported to have Type I CD (Kim et al., 2008). After surgery, 82% of patients were seizure free at 1 to 2 years of follow-up in the UCLA series (Salamon et al., 2008). This is an improvement from studies published from 1971 to 1999 that reported that 38 to 40% of patients with CD were seizure free after surgery (Sisodiya, 2000).

## PRINCIPAL HISTOPATHOLOGIC FINDINGS IN CD

As mentioned earlier, the key histopathologic criteria that define CD are cortical neuronal disorganization, dyslamination, and the presence of abnormal dysmorphic cells (Mischel et al., 1995; Taylor et al., 1971). Cortical disorganization and dyslamination consist of an irregular arrangement of

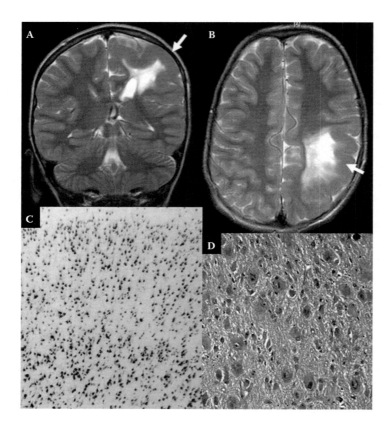

**FIGURE 17.1** Clinicopathologic example of severe Type II CD. This 6-year-old began having intractable epilepsy at 2 years of age. MRI revealed an area of CD characterized by thickened cortex with blurring of the gray–white matter interface and increased $T_2$ signal in the subcortical white matter (A and B; arrows). Histopathology revealed disorganized cortex (C) that at higher magnification showed a mix of dysmorphic cytomegalic neurons and balloon cells (D). Panel C is stained with NeuN, and panel D with H&E.

pyramidal neurons, loss of normal cortical laminar organization, irregular clustering, and neurons showing abnormal polarity and misoriented apical dendrites. As discussed later, these histopathologic features are consistent with incomplete development of the cerebral cortex. Other histopathologic features of CD include excessive heterotopic neurons in the white matter and molecular layer, as well as marginal and nodular glioneuronal heterotopia and polymicrogyria, probably resulting from overproduction of late-generated neurons (Andres et al., 2005). This type of polymicrogyria in CD patients generally appears as thickened cortex on MRI (Salamon et al., 2006).

## METHODOLOGICAL ASPECTS OF HUMAN CD STUDIES

Early work by Avoli and associates in mostly adult surgical patients showed that human CD tissue is sensitive to epileptogenic agents such as 4-aminopyridine (4-AP), a K⁺ channel blocker that also enhances neurotransmitter release (Avoli et al., 1999; Mattia et al., 1995). However, the evoked ictal-like events in the slice could not be correlated with the presence of abnormal cells, as these studies used blind recording techniques and neuronal staining was not attempted. Other studies also failed to find abnormal cells or aberrant synaptic transmission in tissue samples resected from children with intractable epilepsy, even though intracellular labels were used (Dudek et al., 1995; Tasker et al., 1996; Wuarin et al., 1990). In all likelihood, this was due to the fact that the patient population in these studies was heterogeneous and included multiple pathologies, many without CD. Using Lucifer yellow, an intracellular fluorescent dye, we demonstrated increased dye coupling, an indication of the presence of gap junctions, in young cases of pediatric epilepsy that included some CD cases (Cepeda et al., 1993). Furthermore, some pyramidal neurons looked abnormally shaped, but it was difficult to correlate cell morphology with electrophysiological abnormalities.

With the introduction of IR video microscopy in combination with differential interference contrast (DIC) optics (Dodt and Zieglgänsberger, 1994), visualization of individual neurons in slices became possible, especially as we focused on cases of CD. In a first communication, we demonstrated the feasibility of identifying pyramidal and nonpyramidal neurons in slices from children with catastrophic epilepsy (Cepeda et al., 1999). Soon thereafter, we were able to record from abnormal cells in severe CD and TSC cases (Cepeda et al., 2003; Mathern et al., 2000), where balloon cells and cytomegalic neurons are common histopathologic elements. One limitation of IR–DIC video microscopy is that the health of the cells and the possibility of maintaining the slices for long periods of time are best in the young tissue samples. In our cohort, many cases are less than 2 years, which permits electrophysiological recordings under optimal conditions.

Obviously, there are limitations inherent to the use of human tissue samples for understanding mechanisms of epileptogenesis (Köhling and Avoli, 2006), most importantly the scarcity of "real" control tissue, the chronic use of antiepileptic drugs (AEDs), and restrictions in sampling. In spite of these limitations, in our tissue samples we have been able to find the multiple classes of abnormal cells described by anatomical studies. In our slice studies we routinely record basic membrane properties (capacitance, input resistance, and time constant), voltage-gated currents, and spontaneous or evoked excitatory and inhibitory synaptic activities both in current- and voltage-clamp. Drugs are applied in the bath and in some cases iontophoretically. In addition, some slices are used for acute cell dissociation using enzymatic treatment. Cell dissociation allows better space-clamp control and a more rigorous measurement of voltage- and ligand-gated currents.

## MORPHOLOGIC AND ELECTROPHYSIOLOGIC CHARACTERIZATION OF ABNORMAL CELL TYPES IN PEDIATRIC CD

Figure 17.2 depicts various camera lucida drawings of abnormal cells in CD which were recorded electrophysiologically and filled with biocytin. The following subsections highlight important features of the six major types of abnormal CD cells.

### NORMAL-SIZED MISORIENTED PYRAMIDAL NEURONS

Besides cortical dyslamination, a hallmark of CD is the presence of misoriented cortical pyramidal neurons. Misorientation is manifested by a rotation of the cell body and deviation of the apical dendrite from the normal perpendicular orientation with respect to the pial surface. In some pyramidal cells misorientation is very subtle, but in other cases it is more noticeable, and complete inversion of the soma is not uncommonly seen. In nearly all of our CD tissue samples we encountered misoriented pyramidal neurons. Despite their abnormal orientation, cell membrane capacitance, input resistance, and time constant were very similar to those of normal appearing and oriented cortical pyramidal neurons (Cepeda et al., 2003). Also, action-potential firing induced by membrane depolarization did not reveal signs of cellular hyperexcitability. This indicates that misoriented neurons probably do not contribute significantly to the intrinsic epileptogenicity of CD tissue. This is in line with the finding that a mouse model of cortical migration disorder, the *Reeler* mouse, does not display spontaneous seizures (Caviness, 1976; Schwartzkroin and Walsh, 2000), although *Reeler* mice have lowered thresholds for induction of epileptic activity (Patrylo et al., 2006).

**Abnormal Cells in Pediatric CD**

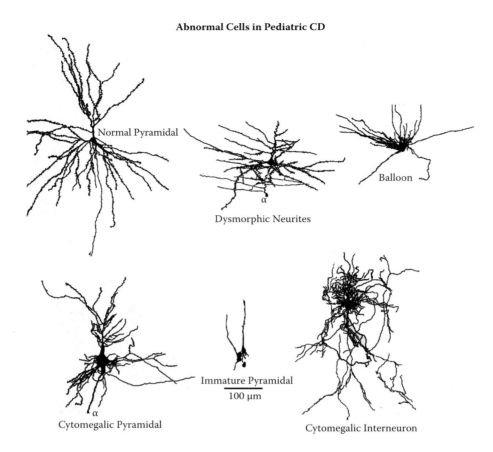

**FIGURE 17.2**  Camera lucida drawings of abnormal cells recorded electrophysiologically and filled with biocytin. For comparison, a normal pyramidal neuron is also illustrated (top left). All cells were recorded in Type II CD, except for the balloon cell, which was from a TSC case. $\alpha$ indicates the axon. (Drawings made by Dr. Robin S. Fisher and Marea K. Boyan. Adapted from Cepeda, C. et al., *J. Neurosci. Res.*, 72, 472–486, 2003. With permission of Wiley-Liss, Inc., a subsidiary of John Wiley & Sons, Inc.)

## PYRAMIDAL NEURONS WITH DYSMORPHIC NEURITES

Another abnormal morphologic finding in CD tissue is the presence of normal-sized pyramidal neurons with dysmorphic or dystrophic neurites. These neurons have normal somata, but the apical or basilar dendrites and sometimes the axon display segments that are tortuous. Many cells with dysmorphic dendrites also show misorientation. Dystrophic neurites are seen in many neurologic diseases, including temporal lobe epilepsy (Porter et al., 2003); however, the occurrence of dystrophic neurites appears to be relatively specific to CD (Judkins et al., 2006). The role of dystrophic neurites in epileptogenesis is currently unknown. Our single-cell electrophysiological recordings did not reveal significant differences in basic membrane properties or signs of cellular hyperexcitability compared with normal-appearing pyramidal neurons (Cepeda et al., 2003).

## BALLOON CELLS

Abnormal balloon cells are found in severe CD tissue from ~40% of pediatric surgery cases and are also very abundant in tubers from children with TSC. These are probably the most bizarre cells present in severe CD tissue and, from an electrophysiological perspective, they represent a paradox. Morphologically, balloon cells are reminiscent of fibrillary astrocytes. They lack dendritic spines, and an axon is not visible. They usually display abundant, tortuous processes that extend for several hundred microns and become progressively thinner with greater distances away from the soma. Immunocytochemical studies reveal that balloon cells may display neuronal and glial markers alone or in combination (Farrell et al., 1992; Mathern et al., 2000). This suggests that balloon cells do not commit to a neuronal or glial phenotype and remain undifferentiated as suggested by other studies (Lamparello et al., 2007; Ying et al., 2005).

For some time, it was believed that balloon cells could be the culprits of epileptogenicity in CD (Schwartzkroin and Walsh, 2000); however, our electrophysiological recordings demonstrated that these cells are unable to generate sodium or calcium inward currents and do not depolarize when excitatory amino acids are applied (Cepeda et al., 2003, 2005a). In more recent studies, balloon cells were found not to be completely inert, as they produced minor membrane oscillations, similar to those generated by glial cells, when 4-AP was applied to the slice. The role of balloon cells in the generation of epileptic activity in severe CD is still an open question, but based on the fact that these cells are unable to generate action potentials their role in epileptogenesis is probably only minor.

## CYTOMEGALIC PYRAMIDAL NEURONS

The presence of cytomegalic pyramidal neurons is another hallmark of severe CD and occurs in ~60% of pediatric CD cases. The cytoplasm of these cells often contains neurofibrillay tangle-like cytoplasmic inclusions, irregular clumping of the Nissl substance around the nucleus, and cytoplasmic vacuolization (Hildebrandt et al., 2005; Vinters, 2002). These cells are often two to three times larger than nearby normal-appearing pyramidal neurons and also display very thick processes, including the axon hillock. Dendritic spines are present and can be abundant in some segments whereas in others they are scarce. Cytomegalic pyramidal neurons have abnormal membrane properties, some of which could be considered as contributing to epileptogenesis. Due to their large size, the membrane capacitance and time constants are significantly larger compared to normal pyramidal neurons. In contrast, their input resistance is very low. The resting membrane potential of cytomegalic pyramidal neurons is relatively hyperpolarized (Cepeda et al., 2005b). These features indicate that, due to their large size and reduced input resistance, it takes more synaptic inputs for these cells to reach firing threshold. However, once depolarized, one of the most distinct features of cytomegalic pyramidal neurons is their capacity to generate repetitive calcium oscillations, suggesting upregulation of calcium channels (Cepeda et al., 2003). Indeed, large calcium current densities were observed in cytomegalic pyramidal neurons using dissociated cell preparations. We also demonstrated that cytomegalic pyramidal neurons display reduced $Mg^{2+}$ sensitivity (André et al., 2004). This finding implies that $N$-methyl-D-aspartate (NMDA) receptors can be activated at more hyperpolarized membrane potentials compared with neurons with normal $Mg^{2+}$ sensitivity, which can also lead to cellular hyperexcitability.

## CYTOMEGALIC GABAERGIC INTERNEURONS

In a proportion of severe CD cases, particularly in children with hemimegalencephaly, we recently described another type of abnormal cell, the cytomegalic interneuron (André et al., 2007). Cytomegalic interneurons were found in areas of greatest CD abnormality and in close proximity to cytomegalic pyramidal neurons. Cytomegalic interneurons are several times larger than normal interneurons and express GABAergic markers such as glutamic acid decarboxylase (GAD), the synthesizing enzyme for GABA, along with the calcium-binding proteins calbindin and parvalbumin, which usually localize to inhibitory neurons. Cytomegalic interneurons are generally basket-like in shape and display increased numbers of dendrites compared with normal interneurons. Dysmorphic interneurons display signs of cellular hyperexcitability such as non-inactivating sodium spikes when a depolarizing ramp voltage command is applied. An important feature of a subpopulation of cytomegalic interneurons is that these are the only cells in which we have observed spontaneous paroxysmal depolarizing shifts leading to action potential firing in the *in vitro* slice preparation. Thus, cytomegalic interneurons could play an important role in the generation of epileptic activity in the minority of patients that display these cells in the CD tissue sample.

### IMMATURE PYRAMIDAL NEURONS

Immature-looking pyramidal neurons can be observed in mild and severe CD tissue from patients generally younger than 2 years of age. They tend to occur in clusters, similar to the clones occurring during early neuroblast migration (Cepeda et al., 2003, 2007). Their morphological and electro-physiological properties are typical of immature pyramidal neurons. They have incipient dendrites and spines, very high membrane input resistance, low capacitance, fast time constants, and small inward currents due to their small size. Although rigorously speaking these are not abnormal cells, their presence in postnatal tissue was unexpected as we have not observed similar cells in patients with non-CD pathologies operated at similar ages.

## THE PATHOGENESIS OF CD

The cause of CD is still debated. In a series of studies, we have compared CD surgical cases with autopsy or non-CD cases for changes in MRI-assessed cerebral cortical and white-matter volumes and cortical neuronal densities (Andres et al., 2005; Chandra et al., 2007; Salamon et al., 2006). Total MRI brain volumes were similar between non-CD and CD cases. Compared with autopsy cases, NeuN cell densities in CD were increased in the molecular layer, upper gray matter, and white matter. The presence of higher neuronal densities in upper cortical layers is consistent with the hypothesis that CD pathogenesis involves increased neurogenesis in the late, not early, phases of neurogenesis. In addition, more neurons in the molecular layer and white matter supports the concept that CD pathogenesis also involves incomplete programmed cell death in the remnant cells occupying the preplate and subplate regions (Mathern et al., 2007).

The mechanisms that induce cortical dyslamination and pyramidal cell misorientation are unknown. However, important clues have been provided by the isolation of genes related to neuronal guidance and final positioning in the cortical plate such as *Reelin* (D'Arcangelo et al., 1995). The glycoprotein Reelin appears early during embryogenesis (by gestational week 11 in humans) and is expressed prominently in the marginal zone by the Cajal–Retzius neurons as well as by subpial granular cells (Meyer and Goffinet, 1998). Reelin promotes neuronal migration towards the cortical surface, and this has been proposed to occur via a differential gradient of extracellular Reelin, so a low level promotes upward migration whereas a high level induces cell detachment from the radial glia and migration arrest (D'Arcangelo, 2006). Interestingly, the number of Cajal–Retzius neurons appears increased in some types of CD, suggesting a role in migration abnormalities (Garbelli et al., 2001; Mischel et al., 1995; Thom et al., 2003).

What causes dystrophic dendritic and axonal growth? The observation that many misoriented neurons also showed dystrophic neurites suggests common mechanisms. In fact, some of the mechanisms involved in axonal growth are also implicated in cell migration (Ridley et al., 2003). Intrinsic and extrinsic signals aid pathfinding by neurites. The most important signals include calcium (Gomez and Zheng, 2006), the surface receptor Notch1 (Redmond et al., 2000; Sestan et al., 1999), semaphorins (Polleux et al., 1998), netrins (Round and Stein, 2007), and integrins (Clegg et al., 2003). The growth of apical dendrites toward the pial surface is regulated by chemoattractant molecules located in the marginal zone; semaphorin 3A, a chemorepellant for cortical axons, is also a chemoattractant for apical dendrites (Polleux et al., 2000).

The origin of balloon cells in severe CD and in TSC remains an enigma. Understanding their origin could provide important clues about the timing of CD pathogenesis. From the outset these cells have been difficult to classify due to their similarities to both neurons and glia. Although they may express neuronal markers, the absence of an axon casts doubts about their neuronal pheno-type. The first clues about the origin of these atypical cells came from anatomic studies suggesting they were undifferentiated cells (Huttenlocher and Heydemann, 1984; Probst and Ohnacker, 1977) that, in a sense, mimicked features of neuroembryonic development (Trombley and Mirra, 1981). Stronger evidence came from immunocytochemical and molecular biological studies demonstrating that balloon cells express both neuronal and glial markers (Englund et al., 2005; Farrell et al., 1992; Mathern et al., 2000), as well as a wide variety of embryological markers characteristic of undif-ferentiated progenitors (Crino et al., 1996, 1997; Fukutani et al., 1992; Hirose et al., 1995; Thom et al., 2005, 2007; Ying et al., 2005). More specifically, we and others have proposed that balloon cells may be remnants of radial glial progenitor cells (Cepeda et al., 2006; Lamparello et al., 2007) or from the minority of embryonic progenitors that co-express neuronal and glial markers described by Zecevic (2004). Why balloon cells do not fully differentiate into astrocytes remains unknown; however, a recent study found that many balloon cells fail to complete the cell cycle, suggesting that cell arrest occurs at an early point of the $G_1$ phase (Thom et al., 2007).

The origin of cytomegalic pyramidal neurons and interneurons is another riveting question. One possibility is that these cells are a normal constituent of cortical development but do not undergo programmed cell death, a normal process common in mid- and late-gestational periods (Rakic and Zecevic, 2000). The fact that cytomegalic neurons resemble cells previously described in the human subplate (Mrzljiak et al., 1988) supports this hypothesis (Andres et al., 2005; Cepeda et al., 2003, 2006). Alterations in the mTOR pathway, which controls cell growth, could also explain the presence of cytomegalic neurons. For example, using microarray analysis of gene expression and immunohistochemistry it has been shown that some proteins in this pathway, such as the ribosomal proteins S6, eIF4G, and Akt, are hyperphosphorylated in cytomegalic neurons from human CD tis-sue (Ljungberg et al., 2006). If this is correct, these findings open avenues for novel treatments by targeting the mTOR pathway (Franz et al., 2006; Ruegg et al., 2007).

The presence of cytomegalic interneurons in severe CD cases is particularly intriguing. These cells are unique in that they can generate spontaneous depolarizing shifts, making them suitable as generators of synchronous activity. It is interesting that in cell cultures from embryonic cerebral rat cortex a distinct subpopulation of large GABAergic neurons, which constitute only a small minority of cells, are a key element in the generation of synchronous oscillatory network activity (Voigt et al., 2001). In addition, these basket-like neurons are derived from the primordial plexiform layer and reside in the subplate at the time of birth, placing them in an ideal position to distribute activity in the local networks (Voigt et al., 2001).

## SYNAPTIC CIRCUITS IN PEDIATRIC CD

Spontaneous synaptic activity, mostly due to activation of glutamatergic and GABAergic receptors, is an accurate index of synaptic inputs onto the cell. In our studies, we routinely record sponta-neous synaptic currents at two holding potentials (–70 and +10 mV) to isolate glutamatergic and GABAergic inputs, respectively. In addition, selective blockers of these receptors allow pharma-cological isolation. Using these methods we found that some abnormal cells also display abnormal synaptic activity. Although misoriented and pyramidal neurons with dysmorphic neurites do not show obvious changes in spontaneous synaptic activity, balloon cells are unique in that no synaptic activity is observed (Cepeda et al., 2003, 2005a). This finding supports observations that these cells do not form synaptic contacts with neighboring neurons (Alonso-Nanclares et al., 2005; Hirose et al., 1995; Lippa et al., 1993). Cytomegalic pyramidal neurons are very diverse in terms of the type and abundance of synaptic inputs. Whereas some cells display abundant synaptic activity, others display very little activity. Further, in some cytomegalic neurons the majority inputs are GABAergic. In the case of immature neurons, the most remarkable feature is the presence of abundant, rhyth-mic spontaneous GABAergic synaptic events (Cepeda et al., 2003, 2006). In current-clamp record-ings, GABAergic synaptic events are depolarizing and may induce action potentials that can be blocked with bicuculline, a GABA$_A$ receptor antagonist, suggesting that some local circuits in CD are immature (Cepeda et al., 2007) and supporting the finding that GABA can be excitatory in early development (Ben-Ari, 2006; Cherubini et al., 1991).

These observations led us to ask what types of inputs are present in normal-appearing pyramidal neurons in CD tissue. Recordings of spontaneous synaptic activity in these neurons indicated that, in CD tissue, GABAergic activity is not reduced compared to relatively normal tissue. In fact, in severe CD it is the glutamatergic activity that is reduced (Cepeda et al., 2005b). Paradoxically, immunohistological studies using calcium-binding proteins, indirect markers of inhibitory cells, have consistently found that the number of GABAergic interneurons is decreased in CD tissue. How can this be reconciled with the observation of relatively high spontaneous synaptic activity in pediatric CD tissue? One possibility is that aberrant synaptic organization from the remaining GABA interneurons occurs in CD tissue, leading to compensation for reduced number of cells. In human temporal lobe epilepsy, calbindin-positive cells in the hilus become hypertrophic to compensate for the loss of interneurons (Magloczky et al., 2000). Similarly, in focal CD and microdysgenesis hypertrophic calbindin- and pavalbumin-positive neurons have been reported (Ferrer et al., 1992; Thom et al., 2003). The demonstration of extensive but aberrant GABA synaptic organization in CD (e.g., the hypertrophic parvalbumin-positive basket formations surrounding cytomegalic pyramidal neurons) (Alonso-Nanclares et al., 2005; Ferrer et al., 1992; Garbelli et al., 1999; Spreafico et al., 1998) reinforces the idea of compensatory mechanisms in CD. Another possibility is that, in severe CD cases, cytomegalic interneurons provide rich innervation onto cytomegalic, normal, and immature pyramidal neurons or that synaptic inputs could arise from the molecular layer that contains GABAergic neurons during cortical development.

## DIFFERENTIAL CHANGES OF GABA$_A$ RECEPTORS IN TYPE I AND TYPE II CD

Reductions in GABA$_A$ receptor function have been invoked to explain epileptogenesis in CD tissue (Spreafico et al., 1998). However, with the discovery that, during development, GABA actions change from depolarizing to hyperpolarizing (Ben-Ari, 2006), the mechanisms by which GABA$_A$ receptors are involved in the generation of epileptic discharges in CD have to be reevaluated (D'Antuono et al., 2004). In a recent study, our group demonstrated differential alterations in GABA$_A$ receptor function in Type I and Type II CD (André et al., 2008). In Type II CD, peak currents induced by GABA were larger in cytomegalic pyramidal neurons; however, GABA peak current densities and the modulatory effect of zolpidem were smaller in Type II CD, specifically in cytomegalic pyramidal neurons. Changes in zolpidem and zinc sensitivities supported the concept that GABA$_A$ receptor subunit composition is probably altered in Type II CD, with responses consistent with decreases in the relative proportion of $\alpha_1$ and increases in $\gamma_2$ subunits. Anatomically, there were fewer GAD cells in Type II CD tissue containing cytomegalic pyramidal neurons, but the remaining GAD cells were larger than in non-CD, Type I CD and Type II CD tissue samples without cytomegalic pyramidal neurons. In addition, cytomegalic pyramidal neurons were densely surrounded by terminals expressing GABAergic markers, consistent with hyperinnervation of these abnormal cells. In contrast, the tissue from Type I CD showed no loss of GABAergic cells, but functionally there was decreased GABA sensitivity compared with non-CD, indicating that GABA function is likely decreased in Type I CD. The implications of these findings are that patients with Type I and Type II CD will respond differently to AEDs acting on GABA$_A$ receptors and that cytomegalic neurons have features similar to immature neurons with prolonged GABA$_A$ receptor open-channel times (André et al., 2008).

## WHAT MAKES CD TISSUE EPILEPTOGENIC?

Hypotheses attempting to explain the intrinsic epileptogenicity of CD tissue range from those proposing that abnormal dysmorphic neurons can generate epileptic activity (Kerfoot et al., 1999; Schwartzkroin and Walsh, 2000) to those suggesting a more prominent involvement of aberrant circuits including abnormal NMDA receptors (André et al., 2004; Ying et al., 1999) or reduced $GABA_A$ receptor-mediated synaptic activity (Calcagnotto et al., 2005; Spreafico et al., 1998). A combination of those factors is perhaps more likely an explanation. Among the diverse classes of abnormal cells described, we believe that cytomegalic pyramidal neurons and interneurons along with immature neurons have membrane properties that could contribute to the mechanisms involved in seizure genesis, in contrast to balloon cells, which are unable to generate action potentials and lack synaptic connections and axonal processes. Cytomegalic interneurons have the potential to generate spontaneous depolarizing shifts, making them good candidates as generators of synchronous activity. If GABAergic basket formations surrounding cytomegalic pyramidal neurons depolarize these cells, upregulated calcium channels could generate synchronous oscillations that spread to surrounding normal-appearing pyramidal neurons. This could serve as an amplifier for the synchronization of cortical networks and the propagation of epileptic activity to other cortical regions. Immature pyramidal neurons are also candidates for hyperexcitable elements in younger, severe forms of CD. High membrane input resistances can amplify incoming signals. Immature properties also include depolarizing GABA actions (Cepeda et al., 2007), an observation supported by immunocytochemical studies showing altered expression of chloride transporters NKCC1 and reduced KCC2 in CD tissue (Aronica et al., 2007; Munakata et al., 2007). However, it is important to note that there are many CD patients with intractable epilepsy in which the tissue specimen does *not* contain cytomegalic neurons. These cases display only mild CD without dysmorphic cells. Thus, mechanisms of epileptogenesis are probably multifactorial, and understanding how cells interact with each other to induce paroxysmal activity in CD will require additional studies.

## COMPARISON WITH OTHER PATHOLOGIES: IS CORTICAL DYSMATURITY A COMMON THREAD IN EPILEPTOGENESIS?

The idea that dysmaturity, or a return to immaturity, could explain the proclivity of neuronal tissue to produce paroxysmal activity is not new in the epilepsy field. In human TLE, elegant slice electrophysiology studies by Cohen et al. (2002) demonstrated spontaneous, rhythmic activity similar to interictal discharges in the subiculum. This activity was mediated by glutamate and GABA and encompassed a network of both subicular interneurons and a subgroup of pyramidal cells. A substantial percentage of these cells displayed immature $GABA_A$ receptor properties, such as depolarizing activity, which was suggested to play a role in the generation of interictal activity. In fact, these changes in cellular and synaptic function caused by deafferentation were suggested to reflect a pathological replay of developmental mechanisms (Cohen et al., 2003). Furthermore, GABA currents elicited in *Xenopus* oocytes microinjected with cell membranes isolated from the subiculum of TLE patients had a more depolarized reversal potential compared with the hippocampus proper or the neocortex, and quantitative reverse transcription–polymerase chain reaction (RT–PCR) analyses revealed an upregulation of NKCC1 and a downregulation of KCC2 mRNA (Palma et al., 2006).

Recent electrophysiologic and immunocytochemic studies in hypothalamic hamartomas (HHs) tissue demonstrated that, compared to small neurons, large neurons exhibited more positive chloride equilibrium potentials, higher intracellular chloride concentrations, lower KCC2 expression, and an immature phenotype, consistent with $GABA_A$ receptor-mediated excitation (Wu et al., 2008). In addition, the $GABA_A$ receptor agonist muscimol induced membrane depolarization and a rise in intracellular calcium in large but not small neurons (Kim et al., 2008). These results suggested that $GABA_A$ receptor-mediated excitation and subsequent activation of L-type calcium channels may contribute to seizure genesis in HH tissue.

In human CD, it was also shown that epileptiform synchronization leading to *in vitro* ictal activity is initiated by a synchronizing mechanism that paradoxically relies on $GABA_A$ receptor activation that causes significant increases in $K^+$ concentration (D'Antuono et al., 2004). In addition, spontaneous synchronous discharges occurring in CD tissue can be abolished by NMDA receptor antagonists and slowed down by blockers of gap junctions (Gigout et al., 2006). Electrotonic coupling via gap junctions is generally increased in immature brains (Cepeda et al., 1993), and this is probably another sign that dysmaturity is a common feature of epileptogenic tissue.

Based on our initial studies, we proposed that CD pathogenesis probably involves partial failure of events occurring during later phases of corticogenesis resulting in incomplete cerebral development (Andres et al., 2005; Cepeda et al., 2006). The timing of these events during cortical development would explain the different forms of CD. Developmental alterations during the late second or early third trimester would account for severe CD with numerous dysmorphic and cytomegalic cells (Type II), while events occurring closer to birth after the subplate has nearly degenerated would explain mild CD (Type I). As a consequence, subplate and radial glial degeneration and transformation would be prevented, giving the appearance of abnormal dysmorphic cells in the postnatal human brain. In addition, failure of late cortical maturation could explain the presence of thickened, abnormally placed secondary gyri with indistinct cortical gray–white matter junctions in postnatal CD tissue. However, it is important to acknowledge that these observations do not exclude that elements of severe CD pathogenesis might begin much earlier during neurogenesis and involve the generation of the preplate itself. Such early defects may be of particular importance in large-scale congenital dysplasias associated with severe epileptogenesis such as hemimegalencephaly (Salamon et al., 2006).

The presence of undifferentiated cells, retained embryonic neurons, immature pyramidal neurons, predominance of GABAergic synaptic activity, and depolarizing actions of GABA all point to the idea that in severe CD we are probably dealing with a problem of tissue and cellular dysmaturity (Cepeda et al., 2006). Interestingly, the ECoG in severe CD is characterized by high-amplitude slow waves that resemble the EEG patterns of preterm infants such as the *tracé alternant* (Khazipov and Luhmann, 2006). Slow EEG transients decrease toward full term in association with an increase in the expression of KCC2 (Vanhatalo et al., 2005). GABAergic markers, including the vesicular GABA transporter, have been observed in the subplate of human embryos as early as gestation week 11.5. Consequently, it has been proposed that GABA receptor-mediated synaptic activity occurs at this early stage of cerebral cortical development (Bayatti et al., 2008) and could play an important role in brain development (Owens and Kriegstein, 2002). Thus, many of the observations reported in our studies of children with CD are congruent with the idea that in severe CD the pre- and subplate or some of their elements do not dissolve before birth and retain immature features.

## IMPLICATIONS FOR TREATMENT

The differences between mild and severe CD in terms of the timing and severity of the cortical lesion indicate that different therapeutic approaches may be necessary to control seizures. For example, if GABA activity is not significantly affected in pediatric CD, medications used to upregulate $GABA_A$ receptor function may not be useful and could be detrimental in severe CD tissue if GABA is depolarizing onto cytomegalic and immature neurons. Interestingly, in a rodent model of neonatal seizures it was shown recently that, although phenobarbital alone fails to abolish seizure activity, when combined with bumetanide, which blocks the NKCC1 transporter, it becomes a more effective seizure suppressor (Dzhala et al., 2008). Furthermore, differences in $GABA_A$ receptor-mediated pyramidal cell responses in Type I and Type II CD also have important implications for treatment. Changes in zolpidem and zinc sensitivities suggest that cytomegalic neurons have altered $GABA_A$ receptor subunit composition, similar to immature neurons with prolonged $GABA_A$ receptor open-channel times. These findings support the hypothesis that patients with Type I and Type II CD will respond differently to GABA receptor-mediated AEDs. It is thus clear that a better understanding of the basic mechanisms of epileptogenesis in pediatric CD will further our therapeutic perspective. In other words, basic science studies of tissue removed for epilepsy neurosurgery in children with CD have important value in our understanding of the pathogenesis and epileptogenesis of this neurologic condition that cannot be obtained from studies of animal models. It is through studies of the human tissue that we hope to develop novel treatments that will improve the lives of children with epilepsy from CD.

## ACKNOWLEDGMENTS

We would like to thank the young patients and their parents for allowing use of surgically resected tissue samples for research purposes. We also thank Véronique M. André, Raymond S. Hurst, Jorge Flores-Hernández, Elizabeth Hernández-Echeagaray, Nanping Wu, Irene Yamazaki, Marea K. Boylan, and Robin S. Fisher for their assistance in electrophysiological data collection and morphological analyses. This research would not be possible without the invaluable collaboration of Harry V. Vinters, MD, and the section of Neuropathology at UCLA. The expertise and dedication of the UCLA Hospital Pediatric Neurology Staff are also greatly appreciated. This work was supported by NIH grant NS 38992 from the NINDS.

## REFERENCES

Alonso-Nanclares, L., R. Garbelli, R.G. Sola, J. Pastor, L. Tassi, R. Spreafico, J. DeFelipe. (2005). Microanatomy of the dysplastic neocortex from epileptic patients. *Brain*, 128:158–173.

André, V.M., J. Flores-Hernandez, C. Cepeda, A.J. Starling, S. Nguyen, M.K. Lobo, H.V. Vinters, M.S. Levine, G.W. Mathern. (2004). NMDA receptor alterations in neurons from pediatric cortical dysplasia tissue. *Cereb. Cortex*, 14:634–646.

André, V.M., N. Wu, I. Yamazaki, S.T. Nguyen, R.S. Fisher, H.V. Vinters, G.W. Mathern, M.S. Levine, C. Cepeda. (2007). Cytomegalic interneurons: a new abnormal cell type in severe pediatric cortical dysplasia. *J. Neuropathol. Exp. Neurol.*, 66:491–504.

André, V.M., C. Cepeda, H.V. Vinters, M. Huynh, G.W. Mathern, M.S. Levine. (2008). Pyramidal cell responses to gamma-aminobutyric acid differ in type I and type II cortical dysplasia. *J. Neurosci. Res.*, 86:3151–3162.

Andres, M., V.M. André, S. Nguyen, N. Salamon, C. Cepeda, M.S. Levine, J.P. Leite, L. Neder, H.V. Vinters, G.W. Mathern. (2005). Human cortical dysplasia and epilepsy: an ontogenetic hypothesis based on volumetric MRI and NeuN neuronal density and size measurements. *Cereb. Cortex*, 15:194–210.

Aronica, E., K. Boer, S. Redeker, W.G. Spliet, P.C. van Rijen, D. Troost, J.A. Gorter. (2007). Differential expression patterns of chloride transporters, Na$^+$–K$^+$–2Cl$^-$-cotransporter and K$^+$–Cl$^-$-cotransporter, in epilepsy-associated malformations of cortical development. *Neuroscience*, 145:185–196.

Avoli, M., A. Bernasconi, D. Mattia, A. Olivier, G.G. Hwa. (1999). Epileptiform discharges in the human dysplastic neocortex: *in vitro* physiology and pharmacology. *Ann. Neurol.*, 46:816–826.

Bayatti, N., J.A. Moss, L. Sun, P. Ambrose, J.F. Ward, S. Lindsay, G.J. Clowry. (2008). A molecular neuroanatomical study of the developing human neocortex from 8 to 17 postconceptional weeks revealing the early differentiation of the subplate and subventricular zone. *Cereb. Cortex*, 18:1536–1548.

Becker, A.J., I. Blumcke, H. Urbach, V. Hans, M. Majores. (2006). Molecular neuropathology of epilepsy-associated glioneuronal malformations. *J. Neuropathol. Exp. Neurol.*, 65:99–108.

Ben-Ari, Y. (2006). Basic developmental rules and their implications for epilepsy in the immature brain. *Epileptic Disord.*, 8:91–102.

Burri, R.J., B. Rangaswamy, L. Kostakoglu, B. Hoch, E.M. Genden, P.M. Som, J. Kao. (2008). Correlation of positron emission tomography standard uptake value and pathologic specimen size in cancer of the head and neck. *Int. J. Radiat. Oncol. Biol. Phys.*, 71:682–688.

Calcagnotto, M.E., M.F. Paredes, T. Tihan, N.M. Barbaro, S.C. Baraban. (2005). Dysfunction of synaptic inhibition in epilepsy associated with focal cortical dysplasia. *J. Neurosci.*, 25:9649–9657.

Caviness, V.S., Jr. (1976). Patterns of cell and fiber distribution in the neocortex of the *Reeler* mutant mouse. *J. Comp. Neurol.*, 170:435–447.

Cepeda, C., J.P. Walsh, W. Peacock, N.A. Buchwald, M.S. Levine. (1993). Dye-coupling in human neocortical tissue resected from children with intractable epilepsy. *Cereb. Cortex*, 3:95–107.

Cepeda, C., Z. Li, H.C. Cromwell, K.L. Altemus, C.A. Crawford, E.A. Nansen, M.A. Ariano, D.R. Sibley, W.J. Peacock, G.W. Mathern, M.S. Levine. (1999). Electrophysiological and morphological analyses of cortical neurons obtained from children with catastrophic epilepsy: dopamine receptor modulation of glutamatergic responses. *Dev. Neurosci.*, 21:223–235.

Cepeda, C., R.S. Hurst, J. Flores-Hernandez, E. Hernandez-Echeagaray, G.J. Klapstein, M.K. Boylan, C.R. Calvert, E.L. Jocoy, O.K. Nguyen, V.M. André, H.V. Vinters, M.A. Ariano, M.S. Levine, G.W. Mathern. (2003). Morphological and electrophysiological characterization of abnormal cell types in pediatric cortical dysplasia. *J. Neurosci. Res.*, 72:472–486.

Cepeda, C., V.M. André, J. Flores-Hernandez, O.K. Nguyen, N. Wu, G.J. Klapstein, S. Nguyen, S. Koh, H.V. Vinters, M.S. Levine, G.W. Mathern. (2005a). Pediatric cortical dysplasia: correlations between neuroimaging, electrophysiology and location of cytomegalic neurons and balloon cells and glutamate/GABA synaptic circuits. *Dev. Neurosci.*, 27:59–76.

Cepeda, C., V.M. André, H.V. Vinters, M.S. Levine, G.W. Mathern. (2005b). Are cytomegalic neurons and balloon cells generators of epileptic activity in pediatric cortical dysplasia? *Epilepsia*, 46(Suppl. 5):82–88.

Cepeda, C., V.M. André, M.S. Levine, N. Salamon, H. Miyata, H.V. Vinters, G.W. Mathern. (2006). Epileptogenesis in pediatric cortical dysplasia: the dysmature cerebral developmental hypothesis. *Epilepsy Behav.*, 9:219–235.

Cepeda, C., V.M. André, N. Wu, I. Yamazaki, B. Uzgil, H.V. Vinters, M.S. Levine, G.W. Mathern. (2007). Immature neurons and GABA networks may contribute to epileptogenesis in pediatric cortical dysplasia. *Epilepsia*, 48(Suppl. 5):79–85.

Chandra, P.S., N. Salamon, S.T. Nguyen, J.W. Chang, M.N. Huynh, C. Cepeda, J.P. Leite, L. Neder, S. Koh, H.V. Vinters, G.W. Mathern. (2007). Infantile spasm-associated microencephaly in tuberous sclerosis complex and cortical dysplasia. *Neurology*, 68:438–445.

Cherubini, E., J.L. Gaiarsa, Y. Ben-Ari. (1991). GABA: an excitatory transmitter in early postnatal life. *Trends Neurosci.*, 14:515–519.

Chugani, H.T., W.D. Shields, D.A. Shewmon, D.M. Olson, M.E. Phelps, W.J. Peacock. (1990). Infantile spasms. I. PET identifies focal cortical dysgenesis in cryptogenic cases for surgical treatment. *Ann. Neurol.*, 27:406–413.

Clegg, D.O., K.L. Wingerd, S.T. Hikita, E.C. Tolhurst. (2003). Integrins in the development, function and dysfunction of the nervous system. *Front. Biosci.*, 8:d723–d750.

Cohen, I., V. Navarro, S. Clemenceau, M. Baulac, R. Miles. (2002). On the origin of interictal activity in human temporal lobe epilepsy *in vitro*. *Science*, 298:1418–1421.

Cohen, I., V. Navarro, C. Le Duigou, R. Miles. (2003). Mesial temporal lobe epilepsy: a pathological replay of developmental mechanisms? *Biol. Cell.*, 95:329–333.

Crino, P.B., J.Q. Trojanowski, M.A. Dichter, J. Eberwine. (1996). Embryonic neuronal markers in tuberous sclerosis: single-cell molecular pathology. *Proc. Natl. Acad. Sci. U.S.A.*, 93:14152–14157.

Crino, P.B., J.Q. Trojanowski, J. Eberwine. (1997). Internexin, MAP1B, and nestin in cortical dysplasia as markers of developmental maturity. *Acta Neuropathol.*, 93:619–627.

D'Antuono, M., J. Louvel, R. Köhling, D. Mattia, A. Bernasconi, A. Olivier, B. Turak, A. Devaux, R. Pumain, M. Avoli. (2004). GABA$_A$ receptor-dependent synchronization leads to ictogenesis in the human dysplastic cortex. *Brain*, 127:1626–1640.

D'Arcangelo, G. (2006). *Reelin* mouse mutants as models of cortical development disorders. *Epilepsy Behav.*, 8:81–90.

D'Arcangelo, G., G.G. Miao, S.C. Chen, H.D. Soares, J.I. Morgan, T. Curran. (1995). A protein related to extracellular matrix proteins deleted in the mouse mutant *Reeler*. *Nature*, 374:719–723.

Dodt, H.U., W. Zieglgänsberger. (1994). Infrared videomicroscopy: a new look at neuronal structure and function. *Trends Neurosci.*, 17:453–458.

Dudek, F.E., J.P. Wuarin, J.G. Tasker, Y.I. Kim, W.J. Peacock. (1995). Neurophysiology of neocortical slices resected from children undergoing surgical treatment for epilepsy. *J. Neurosci. Methods*, 59:49–58.

Dzhala, V.I., A.C. Brumback, K.J. Staley. (2008). Bumetanide enhances phenobarbital efficacy in a neonatal seizure model. *Ann. Neurol.*, 63:222–235.

Englund, C., R.D. Folkerth, D. Born, J.M. Lacy, R.F. Hevner. (2005). Aberrant neuronal–glial differentiation in Taylor-type focal cortical dysplasia (type IIA/B). *Acta Neuropathol.*, 109:519–533.

Farrell, M.A., M.J. DeRosa, J.G. Curran, D.L. Secor, M.E. Cornford, Y.G. Comair, W.J. Peacock, W.D. Shields, H.V. Vinters. (1992). Neuropathologic findings in cortical resections (including hemispherectomies) performed for the treatment of intractable childhood epilepsy. *Acta Neuropathol.*, 83:246–259.

Fauser, S., A. Schulze-Bonhage, J. Honegger, H. Carmona, H.J. Huppertz, G. Pantazis, S. Rona, T. Bast, K. Strobl, B.J. Steinhoff, R. Korinthenberg, D. Rating, B. Volk, J. Zentner. (2004). Focal cortical dysplasias: surgical outcome in 67 patients in relation to histological subtypes and dual pathology. *Brain*, 127:2406–2418.

Ferrer, I., M. Pineda, M. Tallada, B. Oliver, A. Russi, L. Oller, R. Noboa, M.J. Zujar, S. Alcantara. (1992). Abnormal local-circuit neurons in epilepsia partialis continua associated with focal cortical dysplasia. *Acta Neuropathol.*, 83:647–652.

Franz, D.N., J. Leonard, C. Tudor, G. Chuck, M. Care, G. Sethuraman, A. Dinopoulos, G. Thomas, K.R. Crone. (2006). Rapamycin causes regression of astrocytomas in tuberous sclerosis complex. *Ann. Neurol.*, 59:490–498.

Fukutani, Y., M. Yasuda, M. Saitoh, S. Kyoya, K. Kobayashi, K. Miyazu, I. Nakamura. (1992). An autopsy case of tuberous sclerosis. Histological and immunohistochemical study. *Histol. Histopathol.*, 7:709–714.

Garbelli, R., C. Munari, S. De Biasi, L. Vitellaro-Zuccarello, C. Galli, M. Bramerio, R. Mai, G. Battaglia, R. Spreafico. (1999). Taylor's cortical dysplasia: a confocal and ultrastructural immunohistochemical study. *Brain Pathol.*, 9:445–461.

Garbelli, R., C. Frassoni, A. Ferrario, L. Tassi, M. Bramerio, R. Spreafico. (2001). Cajal–Retzius cell density as marker of type of focal cortical dysplasia. *NeuroReport*, 12:2767–2771.

Gigout, S., J. Louvel, H. Kawasaki, M. D'Antuono, V. Armand, I. Kurcewicz, A. Olivier, J. Laschet, B. Turak, B. Devaux, R. Pumain, M. Avoli. (2006). Effects of gap junction blockers on human neocortical synchronization. *Neurobiol. Dis.*, 22:496–508.

Gleeson, J.G. (2001). Neuronal migration disorders. *Ment. Retard. Dev. Disabil. Res. Rev.*, 7:167–171.

Gomez, T.M., J.Q. Zheng. (2006). The molecular basis for calcium-dependent axon pathfinding. *Nat. Rev. Neurosci.*, 7:115–125.

Harvey, A.S., J.H. Cross, S. Shinnar, B.W. Mathern. (2008). Defining the spectrum of international practice in pediatric epilepsy surgery patients. *Epilepsia*, 49:146–155.

Hildebrandt, M., T. Pieper, P. Winkler, D. Kolodziejczyk, H. Holthausen, I. Blumcke. (2005). Neuropathological spectrum of cortical dysplasia in children with severe focal epilepsies. *Acta Neuropathol.*, 110:1–11.

Hirose, T., B.W. Scheithauer, M.B. Lopes, H.A. Gerber, H.J. Altermatt, M.J. Hukee, S.R. VandenBerg, J.C. Charlesworth. (1995). Tuber and subependymal giant cell astrocytoma associated with tuberous sclerosis: an immunohistochemical, ultrastructural, and immunoelectron and microscopic study. *Acta Neuropathol.*, 90:387–399.

Huttenlocher, P.R., P.T. Heydemann. (1984). Fine structure of cortical tubers in tuberous sclerosis: a Golgi study. *Ann. Neurol.*, 16:595–602.

Judkins, A.R., B.E. Porter, N. Cook, R.R. Clancy, A.C. Duhaime, J.A. Golden. (2006). Dystrophic neuritic processes in epileptic cortex. *Epilepsy Res.*, 70:49–58.

Kerfoot, C., H.V. Vinters, G.W. Mathern. (1999). Cerebral cortical dysplasia: giant neurons show potential for increased excitation and axonal plasticity. *Dev. Neurosci.*, 21:260–270.

Khazipov, R., H.J. Luhmann. (2006). Early patterns of electrical activity in the developing cerebral cortex of humans and rodents. *Trends Neurosci.*, 29:414–418.

Kim, D.W., S.K. Lee, K. Chu, K.I. Park, S.Y. Lee, C.H. Lee, C.K. Chung, G. Choe, J.Y. Kim. (2009). Predictors of surgical outcome and pathologic considerations in focal cortical dysplasia. *Neurology*, 72:211–216.

Kim, D.Y., K.A. Fenoglio, T.A. Simeone, S.W. Coons, J. Wu, Y. Chang, J.F. Kerrigan, J.M. Rho. (2008). GABA$_A$ receptor-mediated activation of L-type calcium channels induces neuronal excitation in surgically resected human hypothalamic hamartomas. *Epilepsia*, 49:861–871.

Koh, S., G.W. Mathern, G. Glasser, J.Y. Wu, W.D. Shields, R. Jonas, S. Yudovin, C. Cepeda, N. Salamon, H.V. Vinters, R. Sankar. (2005). Status epilepticus and frequent seizures: incidence and clinical characteristics in pediatric epilepsy surgery patients. *Epilepsia*, 46:1950–1954.

Köhling, R., M. Avoli. (2006). Methodological approaches to exploring epileptic disorders in the human brain *in vitro*. *J. Neurosci. Methods*, 155:1–19.

Krsek, P., B. Maton, B. Korman, E. Pacheco-Jacome, P. Jayakar, C. Dunoyer, G. Rey, G. Morrison, J. Ragheb, H.V. Vinters, T. Resnick, M. Duchowny. (2008). Different features of histopathological subtypes of pediatric focal cortical dysplasia. *Ann. Neurol.*, 63:758–769.

Lamparello, P., M. Baybis, J. Pollard, E.M. Hol, D.D. Eisenstat, E. Aronica, P.B. Crino. (2007). Developmental lineage of cell types in cortical dysplasia with balloon cells. *Brain*, 130:2267–2276.

Lerner, J.T., N. Salamon, J.S. Hauptman, T.R. Velasco, M. Hemb, J.Y. Wu, R. Sankar, W. Donald Shields, J. Engel, Jr., I. Fried, C. Cepeda, V.M. André, M.S. Levine, H. Miyata, W.H. Yong, H.V. Vinters, G.W. Mathern. (2009). Assessment and surgical outcomes for mild type I and severe type II cortical dysplasia: a critical review and the UCLA experience. *Epilepsia*, 50:1310–1335.

Lippa, C.F., D. Pearson, T.W. Smith. (1993). Cortical tubers demonstrate reduced immunoreactivity for synapsin I, *Acta Neuropathol.*, 85:449–451.

Ljungberg, M.C., M.B. Bhattacharjee, Y. Lu, D.L. Armstrong, D. Yoshor, J.W. Swann, M. Sheldon, G. D'Arcangelo. (2006). Activation of mammalian target of rapamycin in cytomegalic neurons of human cortical dysplasia. *Ann. Neurol.*, 60:420–429.

Magloczky, Z., L. Wittner, Z. Borhegyi, P. Halasz, J. Vajda, S. Czirjak, T.F. Freund. (2000). Changes in the distribution and connectivity of interneurons in the epileptic human dentate gyrus. *Neuroscience*, 96:7–25.

Mathern, G.W., C.C. Giza, S. Yudovin, H.V. Vinters, W.J. Peacock, D.A. Shewmon, W.D. Shields. (1999). Postoperative seizure control and antiepileptic drug use in pediatric epilepsy surgery patients: the UCLA experience, 1986–1997. *Epilepsia*, 40:1740–1749.

Mathern, G.W., C. Cepeda, R.S. Hurst, J. Flores-Hernandez, D. Mendoza, M.S. Levine. (2000). Neurons recorded from pediatric epilepsy surgery patients with cortical dysplasia. *Epilepsia*, 41(Suppl. 6):S162–S167.

Mathern, G.W., M. Andres, N. Salamon, P.S. Chandra, V.M. André, C. Cepeda, M.S. Levine, J.P. Leite, L. Neder, H.V. Vinters. (2007). A hypothesis regarding the pathogenesis and epileptogenesis of pediatric cortical dysplasia and hemimegalencephaly based on MRI cerebral volumes and NeuN cortical cell densities. *Epilepsia*, 48(Suppl. 5):74–78.

Mattia, D., A. Olivier, M. Avoli. (1995). Seizure-like discharges recorded in human dysplastic neocortex maintained *in vitro*. *Neurology*, 45:1391–1395.

Meyer, G., A.M. Goffinet. (1998). Prenatal development of Reelin–immunoreactive neurons in the human neocortex. *J. Comp. Neurol.*, 397:29–40.

Mischel, P.S., L.P. Nguyen, H.V. Vinters. (1995). Cerebral cortical dysplasia associated with pediatric epilepsy. Review of neuropathologic features and proposal for a grading system. *J. Neuropathol. Exp. Neurol.*, 54:137–153.

Mrzljiak, L., H.B. Uylings, I. Kostovic, C.G. Van Eden. (1988). Prenatal development of neurons in the human prefrontal cortex. I. A qualitative Golgi study. *J. Comp. Neurol.*, 271:355–386.

Munakata, M., M. Watanabe, T. Otsuki, H. Nakama, K. Arima, M. Itoh, J. Nabekura, K. Iinuma, S. Tsuchiya. (2007). Altered distribution of KCC2 in cortical dysplasia in patients with intractable epilepsy. *Epilepsia*, 48:837–844.

Noachtar, S., O. Bilgin, J. Remi, N. Chang, I. Midi, C. Vollmar, B. Feddersen. (2008). Interictal regional polyspikes in noninvasive EEG suggest cortical dysplasia as etiology of focal epilepsies. *Epilepsia*, 49:1011–1017.

Owens, D.F., A.R. Kriegstein. (2002). Is there more to GABA than synaptic inhibition? *Nat. Rev. Neurosci.*, 3:715–727.

Palma, E., M. Amici, F. Sobrero, G. Spinelli, S. Di Angelantonio, D. Ragozzino, A. Mascia, C. Scoppetta, V. Esposito, R. Miledi, F. Eusebi. (2006). Anomalous levels of Cl⁻ transporters in the hippocampal subiculum from temporal lobe epilepsy patients make GABA excitatory. *Proc. Natl. Acad. Sci. U.S.A.*, 103:8465–8468.

Palmini, A., I. Najm, G. Avanzini, T. Babb, R. Guerrini, N. Foldvary-Schaefer, G. Jackson, H.O. Luders, R. Prayson, R. Spreafico, H.V. Vinters. (2004). Terminology and classification of the cortical dysplasias. *Neurology*, 62:S2–S8.

Patrylo, P.R., R.A. Browning, S. Cranick. (2006). Reeler homozygous mice exhibit enhanced susceptibility to epileptiform activity. *Epilepsia*, 47:257–266.

Polleux, F., R.J. Giger, D.D. Ginty, A.L. Kolodkin, A. Ghosh. (1998). Patterning of cortical efferent projections by semaphorin–neuropilin interactions. *Science*, 282:1904–1906.

Polleux, F., T. Morrow, A. Ghosh. (2000). Semaphorin 3A is a chemoattractant for cortical apical dendrites. *Nature*, 404:567–573.

Porter, B.E., A.R. Judkins, R.R. Clancy, A. Duhaime, D.J. Dlugos, J.A. Golden. (2003). Dysplasia: a common finding in intractable pediatric temporal lobe epilepsy. *Neurology*, 61:365–368.

Prince, D.A., R.K. Wong. (1981). Human epileptic neurons studied *in vitro*. *Brain Res.*, 210:323–333.

Probst, A., H. Ohnacker. (1977). Tuberous sclerosis in a premature infant [authors' transl.]. *Acta Neuropathol.*, 40:157–161.

Rakic, S., N. Zecevic. (2000). Programmed cell death in the developing human telencephalon. *Eur. J. Neurosci.*, 12:2721–2734.

Raybaud, C., M. Shroff, J.T. Rutka, S.H. Chuang. (2006). Imaging surgical epilepsy in children. *Childs Nerv. Syst.*, 22:786–809.

Redmond, L., S.R. Oh, C. Hicks, G. Weinmaster, A. Ghosh. (2000). Nuclear Notch1 signaling and the regulation of dendritic development. *Nat. Neurosci.*, 3:30–40.

Ridley, A.J., M.A. Schwartz, K. Burridge, R.A. Firtel, M.H. Ginsberg, G. Borisy, J.T. Parsons, A.R. Horwitz. (2003). Cell migration: integrating signals from front to back. *Science*, 302:1704–1709.

Round, J., E. Stein. (2007). Netrin signaling leading to directed growth cone steering. *Curr. Opin. Neurobiol.*, 17:15–21.

Ruegg, S., M. Baybis, H. Juul, M. Dichter, P.B. Crino. (2007). Effects of rapamycin on gene expression, morphology, and electrophysiological properties of rat hippocampal neurons. *Epilepsy Res.*, 77:85–92.

Salamon, N., M. Andres, D.J. Chute, S.T. Nguyen, J.W. Chang, M.N. Huynh, P.S. Chandra, V.M. André, C. Cepeda, M.S. Levine, J.P. Leite, L. Neder, H.V. Vinters, G.W. Mathern. (2006). Contralateral hemimicrencephaly and clinical–pathological correlations in children with hemimegalencephaly. *Brain*, 129:352–365.

Salamon, N., J. Kung, S.J. Shaw, J. Koo, S. Koh, J.Y. Wu, J.T. Lerner, R. Sankar, W.D. Shields, J. Engel, Jr., I. Fried, H. Miyata, W.H. Yong, H.V. Vinters, G.W. Mathern. (2008). FDG–PET/MRI coregistration improves detection of cortical dysplasia in patients with epilepsy. *Neurology*, 71:1594–1601.

Schwartzkroin, P.A., W.D. Knowles. (1984). Intracellular study of human epileptic cortex: *in vitro* maintenance of epileptiform activity? *Science*, 223:709–712.

Schwartzkroin, P.A., D.A. Prince. (1976). Microphysiology of human cerebral cortex studied *in vitro*, *Brain Res.*, 115:497–500.

Schwartzkroin, P.A., C.A. Walsh. (2000). Cortical malformations and epilepsy. *Ment. Retard. Dev. Disabil. Res. Rev.*, 6:268–280.

Sestan, N., S. Artavanis-Tsakonas, P. Rakic. (1999). Contact-dependent inhibition of cortical neurite growth mediated by notch signaling. *Science*, 286:741–746.

Sisodiya, S.M. (2000). Surgery for malformations of cortical development causing epilepsy. *Brain*, 123(Pt. 6):1075–1091.

Spreafico, R., G. Battaglia, P. Arcelli, F. Andermann, F. Dubeau, A. Palmini, A. Olivier, J.G. Villemure, D. Tampieri, G. Avanzini, M. Avoli. (1998). Cortical dysplasia: an immunocytochemical study of three patients. *Neurology*, 50:27–36.

Tasker, J.G., N.W. Hoffman, Y.I. Kim, R.S. Fisher, W.J. Peacock, F.E. Dudek. (1996). Electrical properties of neocortical neurons in slices from children with intractable epilepsy. *J. Neurophysiol.*, 75:931–939.

Tassi, L., N. Colombo, R. Garbelli, S. Francione, G. Lo Russo, R. Mai, F. Cardinale, M. Cossu, A. Ferrario, C. Galli, M. Bramerio, A. Citterio, R. Spreafico. (2002). Focal cortical dysplasia: neuropathological subtypes, EEG, neuroimaging and surgical outcome. *Brain*, 125:1719–1732.

Taylor, D.C., M.A. Falconer, C.J. Bruton, J.A. Corsellis. (1971). Focal dysplasia of the cerebral cortex in epilepsy. *J. Neurol. Neurosurg. Psychiatry*, 34:369–387.

Thom, M., B.N. Harding, W.R. Lin, L. Martinian, H. Cross, S.M. Sisodiya. (2003). Cajal–Retzius cells, inhibitory interneuronal populations and neuropeptide Y expression in focal cortical dysplasia and microdysgenesis. *Acta. Neuropathol.*, 105:561–569.

Thom, M., L. Martinian, S.M. Sisodiya, J.H. Cross, G. Williams, K. Stoeber, W. Harkness, B.N. Harding. (2005). Mcm2 labelling of balloon cells in focal cortical dysplasia. *Neuropathol. Appl. Neurobiol.*, 31:580–588.

Thom, M., L. Martinian, A. Sen, W. Squier, B.N. Harding, J.H. Cross, W. Harkness, A. McEvoy, S.M. Sisodiya. (2007). An investigation of the expression of $G_1$-phase cell cycle proteins in focal cortical dysplasia type IIB. *J. Neuropathol. Exp. Neurol.*, 66:1045–1055.

Trombley, I.K., S.S. Mirra. (1981). Ultrastructure of tuberous sclerosis: cortical tuber and subependymal tumor. *Ann. Neurol.*, 9:174–181.

Vanhatalo, S., J.M. Palva, S. Andersson, C. Rivera, J. Voipio, K. Kaila. (2005). Slow endogenous activity transients and developmental expression of $K^+$–$Cl^-$ cotransporter 2 in the immature human cortex. *Eur. J. Neurosci.*, 22:2799–2804.

Vinters, H.V. (2002). Histopathology of brain tissue from patients with infantile spasms. *Int. Rev. Neurobiol.*, 49:63–76.

Voigt, T., T. Opitz, A.D. de Lima. (2001). Synchronous oscillatory activity in immature cortical network is driven by GABAergic preplate neurons. *J. Neurosci.*, 21:8895–8905.

Widdess-Walsh, P., C. Kellinghaus, L. Jeha, P. Kotagal, R. Prayson, W. Bingaman et al. (2005) Electro-clinical and imaging characteristics of focal cortical dysplasia: correlation with pathological subtypes. *Epilepsy Res.*, 67:25–33.

Wu, J., J. Dechon, F. Xue, G. Li, K. Ellsworth, M. Gao, Q. Liu, K. Yang, C. Zheng, P. He, J. Tu, D.Y. Kim, J.M. Rho, H. Rekate, J.F. Kerrigan, Y. Chang. (2008). $GABA_A$ receptor-mediated excitation in dissociated neurons from human hypothalamic hamartomas. *Exp. Neurol.*, 213:397–404.

Wuarin, J.P., Y. I. Kim, C. Cepeda, J.G. Tasker, J.P. Walsh, W.J. Peacock, N.A. Buchwald, F.E. Dudek. (1990). Synaptic transmission in human neocortex removed for treatment of intractable epilepsy in children. *Ann. Neurol.*, 28:503–511.

Yagishita, A., N. Arai, T. Maehara, H. Shimizu, A.M. Tokumaru, M. Oda. (1997). Focal cortical dysplasia: appearance on MR images. *Radiology*, 203:553–559.

Ying, Z., T.L. Babb, N. Mikuni, I. Najm, J. Drazba, W. Bingaman. (1999). Selective coexpression of NMDAR2A/B and NMDAR1 subunit proteins in dysplastic neurons of human epileptic cortex. *Exp. Neurol.*, 159:409–418.

Ying, Z., J. Gonzalez-Martinez, C. Tilelli, W. Bingaman, I. Najm. (2005). Expression of neural stem cell surface marker CD133 in balloon cells of human focal cortical dysplasia. *Epilepsia*, 46:1716–1723.

Zecevic, N. (2004). Specific characteristic of radial glia in the human fetal telencephalon. *Glia*, 48:27–35.

# 18 Nonsurgical Ablation for Epilepsy

*Erich O. Richter, Marina V. Abramova,*
*William A. Friedman, and Steven N. Roper*

## CONTENTS

## INTRODUCTION

Stereotactic radiosurgery is a method of ablating pathologic tissue by delivering precise convergent radiation in a noninvasive manner. Radiosurgery requires the use of ionizing radiation (IR), a form of high-energy electromagnetic radiation, to exert its effect on tissues. Ionizing refers to the ability of the radiation to excite and eject electrons away from atoms by overcoming the binding energy holding the electrons to the nucleus. The two types of radiation used in these applications are photon beams and particle beams. Photon beam systems can be further divided based on the use of gamma rays and x-rays or based on targeting methods.

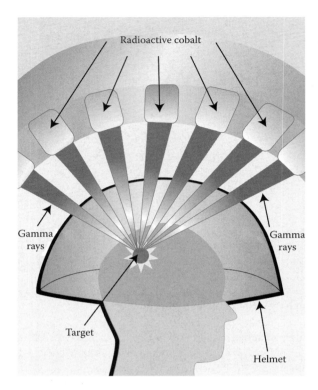

**FIGURE 18.1**   Gamma knife collimator helmet diagram demonstrating the 201 cobalt sources focused at the beam's isocenter. (Courtesy of Erich O. Richter, MD, and Ellie D. Guillot, Coordinator of Information, Neuroscience Center of Excellence, Louisiana State University Health Sciences Center.)

## TYPES OF RADIOSURGERY

### Photon Beam

### Gamma Knife

The first radiosurgical device was called the gamma knife (GK), developed in 1951 by Lars Leksell at the Karolinska Institute in Stockholm, Sweden. It is based on gamma rays from a cobalt source. Gamma rays are produced by radioactive decay of a heavy metal atom (e.g., cobalt) undergoing fission. During fission, fragmentation of the nucleus leads to a change in energy state of the nuclear particles, leading to photon emission. Depending on the radionuclide, a specific amount of energy is emitted which determines the frequency of the photon. For a single atom of cobalt-60, two levels of energy can be produced: 1173 KeV and 1332 KeV. The helmet-shaped collimator of the gamma knife unit has 201 ports through which photons from the cobalt-60 source may pass (Figure 18.1). The gamma rays converge at the center of the sphere, termed the *isocenter*. Image-based software allows for targeting of multiple isocenters to conform to complex target shapes.

### LINAC

Another method of producing photon beam irradiation is a linear accelerator (LINAC)-based system. LINAC systems (Figure 18.2) utilize high-power electromagnetic fields, pushing a stream of electrons to collide with a metal target which results in x-ray release. Targeting is achieved by a collimator traveling in multiple noncoplanar arcs relative to a head frame, yielding an isocentric dose distribution similar to that of the GK. It is possible, as with GK, to use multiple isocenters to

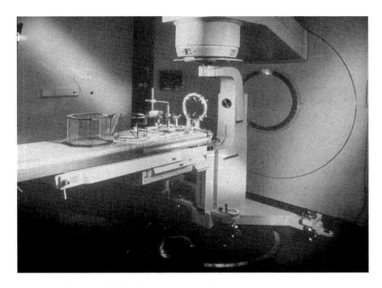

**FIGURE 18.2**   Linear accelerator (LINAC) and the head frames used to position the patient for treatment. (Courtesy of David A. Peace, MS, CMI.)

conform to an irregularly shaped target. When patient selection has been compared between GK and LINAC systems in institutions with access to both modalities, no significant differences are seen (Stern et al., 2008).

### Frameless Image-Guided Stereotactic Radiosurgery

Several newer systems can deliver photon-based radiosurgical treatment plans without the use of the head frame. The CyberKnife® system, for example, incorporates a small linear accelerator on a robotic arm. Two orthogonal x-ray cameras take frequent pictures and match the patients' intraprocedure x-rays with predicted reconstructions from preprocedure CT scans to determine patient position and guide treatment delivery (Figure 18.3).

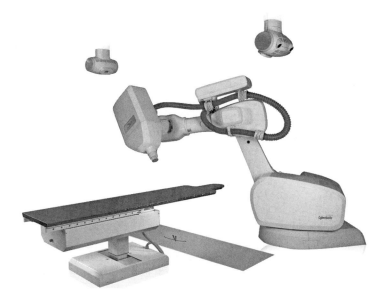

**FIGURE 18.3**   CyberKnife® frameless image-guided system. (Courtesy of Accuray, Inc.; Sunnyvale, CA.)

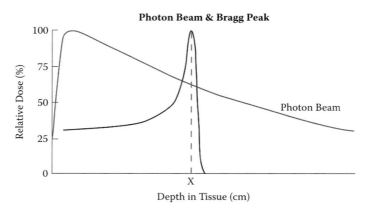

The depth, X, of maximum energy delivery is determined by the energy of the proton beam.

**FIGURE 18.4** Proton beam and Bragg peak. (Courtesy of Erich O. Richter, MD, and Ellie D. Guillot, Coordinator of Information, Neuroscience Center of Excellence, Louisiana State University Health Sciences Center.)

## PARTICLE BEAM

Charged particles such as protons can be used in radiosurgery as well. These particles are able to create high peak energy at the tissue target with minimal energy delivered to the entry path and none to the exit path. A cyclotron utilizes a perpendicular magnetic field to accelerate the particles in a spiral manner as they pass through the gap between a pair of electrodes. As the beam passes through tissue, most of the energy is deposited at a depth known as the Bragg peak (Figure 18.4), which is a function of initial proton energy. By modifying the energy parameters of accelerated protons, it is possible to create a desired depth of penetration. The risk of dose distribution beyond the target is minimal.

## BIOLOGICAL EFFECT OF IONIZING RADIATION

Radiation absorbed dose, the amount of energy absorbed per unit mass of tissue, is expressed in gray (Gy). One Gy is equal to 1 joule of energy absorbed per kilogram of tissue. In older literature, radiation dose was specified in rads (100 rads = 1 Gy). The effect on tissue of the energy deposited by radiation depends on the relative biological effectiveness (RBE) of the radiation. The RBE is 1.0 for γ-rays and x-rays and 1.1 for protons.

Molecular effects of radiation include chromosomal deletions and inversions (Kavathas et al., 1980; Thomas et al., 1998; Urlaub et al., 1986; You et al., 1997). The main underlying mechanism of deletions is DNA breaks, which can be measured by pulsed-field gel electrophoresis (Prise et al., 2001). Single-strand breaks (SSBs) are quite common; double-strand breaks (DSBs) are less common.

"Clustered" DNA damage, areas where both SSB and DSB occur, can lead to cell death. Both the yield and the spatial distribution of DSB are influenced by radiation quality (Prise et al., 2001). This is due to the interaction of particle tracks with the higher order chromosomal structures within cell nuclei. The frequency distributions of fragments induced by charged particles are shifted toward smaller sizes compared to those induced by comparable doses of photon irradiation (Belli et al., 2001). DSBs are two times greater for proton beam radiation than equivalently dosed γ-ray treatment. Notably, the most important process for the survival of a tumor cell following IR is the repair of DNA double-strand breaks. DNA damage from IR is mediated through the formation of free radicals and peroxides and by direct damage from charged particles. The hydroxyl radical (OH⁻) is an example of one such molecule formed by the effects of IR on cellular water. The passage of an energetic charged particle through a cell produces a region of dense ionization along its track. The ionization of water and other cell components can damage DNA molecules and is both time and dose dependent (Gajdusek et al., 2001; Raicu et al., 1993).

Cellular adaptive responses to radiation exposure include the following:

1. Alterations in cellular signal transduction are triggered by radiation (Szumiel, 2008). Radiation-induced activation of the enzyme secreted clusterin (sCLU) may lead to a tumor cell adaptive response, specifically an increased radioresistance (Klokov et al., 2004).
2. DNA repair potential occurs via nonhomologous end joining (Sankaranarayanan, 2006; Sankaranarayanan and Wassom, 2005). Cells with a functional DNA repair system may show increased radiation sensitivity due to deficiencies in specific kinases essential for repair activation (Szumiel, 2008).
3. Activation of a different form of clusterin, nuclear clusterin (nCLU), may trigger programmed cell death (Leskov et al., 2001).
4. P53-dependent radiation induces apoptosis (Mayberg et al., 2000; Wood and Youle, 1995).
5. Autocrine self-destruction via reactive oxygen species causes cell death (O'Connor and Mayberg, 2000).

Cells may undergo apoptosis or necrosis, dedifferentiation, or an alteration of phenotypic expression. Vascular changes may result in ischemic effects at sites remote from the area of direct injury. The local inflammatory response appears to be initiated by endothelial damage. The dose of radiation, time since treatment, and degree of tissue radiosensitivity determine the overall effect of IR on a particular region in the brain. A physiologic effect on tissue can be achieved in the absence of radiation-induced necrosis (Chen et al., 2001; Regis et al., 1996).

## THERAPEUTIC EFFECT OF CEREBRAL IRRADIATION ON EPILEPSY IN ANIMAL MODELS

Focal cerebral irradiation has successfully reduced seizures in certain animal models of epilepsy (Kitchen, 1995). Models of epilepsy, seizure outcome, and behavioral effects are summarized in Table 18.1. Several papers have reported an antiepileptic effect of radiation at subnecrotic doses (absence of coagulative necrosis) with preservation of normal tissue histology (Maesawa et al., 2000; Mori et al., 2000), demonstrating the potential of stereotactic radiosurgery to treat the underlying tissue pathophysiology while preserving function. Regis et al. (1996), using comparative radioisozyme labeling for cholinergic and GABAergic systems in rat striatum, found that single-fraction gamma irradiation transiently reduced levels of glutamic acid decarboxylase, choline acetyl transferase, and excitatory amino acids while increasing levels of glycine. These neurochemical changes may play a role in the antiepileptic effects associated with IR.

## CLINICAL RESULTS OF RADIOSURGERY FOR EPILEPSY BY DISORDER

### ARTERIOVENOUS MALFORMATIONS

Arteriovenous malformations (AVMs) are congenital vascular abnormalities consisting of a pathological nidus, rather than a normal capillary bed, connecting the arterial and venous systems. Seizures are the second most common presenting symptom (Esteves et al., 2008; Hadjipanayis et al., 2001; Kida et al., 2000; Moreno-Jimenez et al., 2007; Schlienger et al., 2000; Sirin et al., 2006). Whereas

**TABLE 18.1**
**Animal Models of Epilepsy**

| Study | Animal Model | Radiation Source | Seizure Outcome |
|---|---|---|---|
| Gaffey et al. (1981) | Alumina-cream-induced epilepsy in cats | Proton beam radiation | 7 days post-radiation, EEG normalized |
| Barcia-Salorio et al. (1987) | Cobalt-induced cat epilepsy model | Gamma irradiation | 3 weeks post-radiation, animals ceased having behavioral manifestations of seizures; by 6 months, EEG recordings returned to normal |
| Sun et al. (1998) | Chronic-stimulus-induced electrographic seizures | LINAC irradiation | 3 months post-radiation, increased seizure threshold and decreased duration of after-discharges |
| Mori et al. (2000) | Kainic-acid-induced hippocampal epilepsy in rats | Gamma knife | Seizure control achieved by rats treated with 20, 40, and 60 Gy without complete tissue destruction |
| Chen et al. (2001) | Chronic spontaneous limbic epilepsy in rats | Gamma knife | 10 months post-radiation, reductions in both the frequency and duration of spontaneous seizures; functional increase in seizure threshold of hippocampal neurons radiated at 40 Gy |

most studies focus on hemorrhage prevention, many have also reported seizure control rates after stereotactic radiosurgery as well (Table 18.2). Seizure control ranges from 33 to 85% (Eisenschenk et al., 1998; Falkson et al., 1997; Gerszten et al., 1996; Hadjipanayis et al., 2001; Heikkinen et al., 1989; Hoh et al., 2002; Kida et al., 2000; Kjellberg et al., 1983; Kurita et al., 1998; Lim et al., 2006; Lundsford et al., 1991; Schauble et al., 2004; Sirin et al., 2006; Steiner et al., 1992; Sutcliffe et al., 1992).

Several characteristics of AVMs may lead to epileptogenesis. The low resistance of the nidus shunts blood away from normal brain parenchyma. This process, known as a *steal phenomenon*, is characteristic for AVMs and results in hypoperfusion of adjacent normal cortex (Fleetwood and Steinberg, 2002; Kawaguchi et al., 2002; Takeshita et al., 1994). The resulting ischemic zone can lead to seizure activity (Kraemer and Awad, 1994; Schauble et al., 2004). Perilesional neurochemical changes consistent with abnormal balance between excitatory and inhibitory signals have also been observed. For example, immunoreactivity for glutamate decarboxylase, GABA receptors, and N-methyl-D-aspartate receptors is decreased adjacent to AVMs (Wolf et al., 1996). Hemosiderin deposits associated with previous hemorrhage may be responsible for epileptogenic activity (Leblanc et al., 1983; Murphy, 1985; Yeh et al., 1990). In some reports, the angioarchitecture of AVMs correlates with seizure activity. For example, superficial supratentorial AVMs in the region of the middle cerebral artery with the cortical location of the feeders are prone to induce seizures, as are those in the frontal lobe. This may be because these AVMs are more likely to cause a steal phenomenon or maybe for other reasons (Eisenschenk et al., 1998; Lasjaunias et al., 2008; Murphy, 1985; Turjman et al., 1995a,b). Finally, in a phenomenon known as *secondary epileptogenesis* (Yeh and Privitera, 1991; Yeh et al., 1990), arteriovenous lesions may produce epileptogenic foci at distant sites. These foci generate seizures independent of the AVM.

The effect of stereotactic radiosurgery on AVMs is well described. The pathological changes induced by radiosurgery ultimately lead to obliteration of the nidus. The time course of obliteration may take up to 3 years, a process termed the *latency period*. During the latency period, endothelial cell damage and proliferation, massive degeneration of vessels, hyaline thickening (Figure 18.5), partial or complete vaso-occlusion by thrombi, smooth muscle neoproliferation, smooth muscle re-endothelization, and proliferation of subendothelial myofibroblasts occur (Eisenschenk et al., 1998;

**TABLE 18.2**
**Seizure Outcome in Patients with Arteriovenous Malformations after Stereotactic Radiosurgery**

| Study | Number of Patients Available for Follow-Up | Treatment Methods | Seizure-Free Patients[a] (%) | Improvement[b] (%) |
|---|---|---|---|---|
| Kjellberg et al. (1983) | 24 | Proton beam | 33 | 46 |
| Heikkinen et al. (1989) | 29 | Proton beam | 55 | Not specified |
| Lunsford et al. (1991) | 43 | Gamma knife | — | 51[c] |
| Steiner et al. (1992) | 59 | Gamma knife | 19 | 51 |
| Sutcliffe et al. (1992) | 48 | Gamma knife | 38 | 22 |
| Gerszten et al. (1996) | 13 | Gamma knife | 85 | 15 |
| Falkson et al. (1997) | 16 | LINAC | 63 | 31 |
| Eisenschenk et al. (1998) | 32 | LINAC | 59 | 19 |
| Kurita et al. (1998) | 35 | Gamma knife | 80 | 11 |
| Hoh et al. (2002) | 110 | Surgical resection, proton beam therapy, or embolization | 66 | 11 |
| Schauble et al. (2004) | 51 | Gamma knife | 51 | 78 |
| Hadjipanayis et al. (2001) | 27 | Gamma knife | 63 | 37 |
| Kida et al. (2000) | 79 | Gamma knife | Group A, 71; Group B, 38[d] | Group A, 21; Group B, 25[d] |
| Lim et al. (2006) | 43 | Gamma knife | 54 | 23 |
| Sirin et al. (2006)[e] | 13 | Gamma knife | 62 | 23 |

[a] Described as Engel class I.
[b] Described as Engel classes II and III.
[c] Reported as improvement or elimination of seizures.
[d] Patients in group A had seizures as the presenting symptom; patients in group B presented with seizures secondary to hemorrhagic episode.
[e] Radiosurgery was performed for large arteriovenous malformations (>15 mL); 62% of patients were reported as stable on medication following radiosurgery.

Szeifert et al., 2007; Tu et al., 2006). Small-volume AVMs are more likely to be completely obliterated in the first 3 years (Hadjipanayis et al., 2001; Moreno-Jimenez et al., 2007). Reported rates of complete obliteration by angiography range from 36 to 70% (Hadjipanayis et al., 2001; Kurita et al., 2000; Maruyama et al., 2004; Moreno-Jimenez et al., 2007; Ross et al., 2000; Sirin et al., 2006).

Whether complete obliteration of cerebral AVMs is necessary or sufficient for seizure cessation is still controversial. LINAC stereotactic radiosurgery used to treat 33 patients with AVMs at the University of Florida had a 59% seizure-free outcome. Interestingly, 40% of the seizure-free patients in this study did not have angiographic occlusion of their AVMs. Other studies have reported similar results (Eisenschenk et al., 1998; Falkson et al., 1997; Gerszten et al., 1996; Kida et al., 2000; Kurita et al., 1998; Schauble et al., 2004; Steiner et al., 1992). It also remains unclear whether AVM size is strongly correlated with success of AVM radiosurgery for seizure control (Pollock et al., 1998). Some studies have shown an effect of AVM size on seizure freedom after radiosurgery (Schauble et al., 2004), whereas others have not (Hoh et al., 2002; Kida et al., 2000). Factors that these studies have implicated as potentially affecting seizure freedom after stereotactic radiosurgery are summarized in Table 18.3. Large AVMs (more than 15 mL) rarely respond to a single treatment with doses that are safe to the surrounding parenchyma but may be treated with staged radiosurgery (Sirin et al., 2006). This requires several courses of radiation (16 Gy) separated by several years, each of which shrinks the AVM and may ultimately lead to complete obliteration (Sirin et al., 2006).

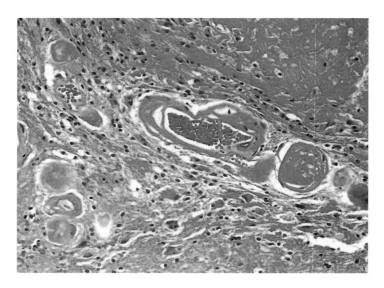

**FIGURE 18.5    (See color insert following page 458.)** Radiation-induced tissue changes in the brain. Evident in this histological preparation is pronounced hyalinization and occlusion of vessels, gliosis, and inflammatory response. (Courtesy of Anthony Yachnis, MD.)

Although stereotactic radiosurgery may be effective for seizure control in patients with AVMs, there may also be complications and side effects. Risks include:

1. Postradiosurgical hemorrhage—The latency period may place patients who have undergone stereotactic radiosurgery at an increased risk of hemorrhage because the damaged vessels remain exposed to arterial pressure. Radiosurgery eventually eliminates the high pressure gradient by obliterating the nidus and restores normal hemodynamics, ultimately eliminating the ischemia which may underlie the seizures. The question of whether radiosurgery increases or decreases the risk of hemorrhage in patients who have not achieved complete obliteration remains unclear. Some studies provide evidence that radiosurgery reduces the hemorrhage rate even in the absence of complete obliteration (Karlsson et al., 2001; Maruyama et al., 2004, 2005), others report an unchanged hemorrhage rate (Friedman et al., 1996; Pollock et al., 1996), and still others show an increased risk of hemorrhage (Fabrikant et al., 1992; Steinberg et al., 1990).

2. Radiation-induced complications—A dose greater than 12 Gy and brain-stem location appear to be positive predictive factors for postradiosurgical neurological deficits (Flickinger, 1989; Flickinger et al., 1997, 1998; Friedman et al., 2003).

---

**TABLE 18.3**

**Some Factors Implicated in Seizure Freedom Rates after Stereotactic Radiosurgery for Arteriovenous Malformations**

| Less Likely to Be Seizure Free | More Likely to Be Seizure Free |
| --- | --- |
| Seizure in the setting of hemorrhage (Kida et al., 2000) | Small arteriovenous malformation size, median 2.9 cm at 3 years' follow-up (Schauble et al., 2004) |
| | Present with generalized tonic–clonic seizure (Schauble et al., 2004) |
| | Present with less than 5 seizures (Kurita et al., 1998) |
| | Epilepsy duration less than 6 months (Kurita et al., 1998) |

3. Increased seizure activity or new seizure onset (Flickinger et al., 1999; Gerszten et al., 1996; Kurita et al., 1998; Lunsford et al., 1991; Schlienger et al., 2000; Steiner et al., 1992).
4. Delayed cyst formation (Chang et al., 2004; Flickinger et al., 1999; Izawa et al., 2007; Pan et al., 2005; Pollock and Brown, 2001)—Embolization and radiation-induced brain edema have been reported to lead to an increased incidence (Pan et al., 2005).
5. One report of *de novo* paranidal aneurysm formation following stereotactic radiosurgery due to changes in flow dynamics, with increased intraluminal pressure within the proximal arteries (Huang et al., 2001).
6. Radiosurgery-induced tumor growth (Berman et al., 2007; Crocker et al., 2007; Sheehan et al., 2006).

It should be noted that the best results in the literature are not with radiosurgery alone but by employing a multimodal, multidisciplinary, team-based approach to these complex lesions. It is probably best for most of these lesions to be referred to centers where expertise is available in microsurgery, endovascular embolization, and radiosurgery.

## CAVERNOUS MALFORMATIONS

Cavernous malformations (CMs) account for 5 to 10% of all central nervous system vascular malformations (McCormick and Nofzinger, 1966; New et al., 1986). They are most easily recognized by their characteristic appearance on magnetic resonance imaging and are not seen on angiography (Ide et al., 2000; Sage and Blumbergs, 2001). These lesions are composed of abnormally dilated, thin-walled, vascular channels with no intervening brain parenchyma (Wong et al., 2000; Zabramski et al., 1999) and are associated with seizures (Huang et al., 2006; Kim et al., 2005; Lee et al., 2008a; Liscak et al., 2005; Liu et al., 2005; Shih and Pan, 2005). As many as 40% of patients with supratentorial CMs may have medically refractory epilepsy (Casazza et al., 1996). Mesiotemporal location is associated with more severe epilepsy (Casazza et al., 1996). Patients with epilepsy from CMs will rarely remit without treatment (Kondziolka et al., 1995a).

Stereotactic radiosurgery can be effective for seizure control (Table 18.4) (Amin-Hanjani et al., 1998; Huang et al., 2006; Kim et al., 2005; Liscak et al., 2005; Liu et al., 2005; Regis et al., 2000b; Shih et al., 2005). Prognostic factors for seizure control are summarized in Table 18.5 (Regis et al., 2000b). There is no consensus regarding the optimal peripheral dose of radiation. Some studies suggest that a dose of less than 15 Gy is optimal for seizure control (Kim et al., 2005; Shih et al., 2005). Stereotactic radiosurgery for CMs is associated with a relatively high complication rate (Table 18.6). These complications include:

1. Although some researchers have demonstrated decreased hemorrhage rate after stereotactic radiosurgery, bleeding remains significant for 2 to 3 years after treatment (Hasegawa et al., 2002; Kim et al., 2005; Liscak et al., 2005; Pollock et al., 2000). Reported annual risk of hemorrhage after radiosurgery ranges from 8.8 to 12.3% (Amin-Hanjani et al., 1998; Hasegawa et al., 2002; Pollock et al., 2000) with a tendency toward a decrease in risk of rebleeding 2 to 3 years later (Chang et al., 1998; Hasegawa et al., 2002; Kondziolka et al., 1995b). Hemorrhage rate may be related to dose (Shih et al., 2005).
2. Radiation-induced toxicity is associated with higher marginal doses (Hasegawa et al., 2002; Kim et al., 2005; Liscak et al., 2005; Pollock et al., 2000). This appears to be likely in patients with increased radiosensitivity of skin fibroblasts *in vitro* (Malone et al., 2000).
3. One study reports a glioblastoma multiforme development in a patient treated with a marginal dose of 10 Gy 13 years following stereotactic radiosurgery (Salvati et al., 2003).
4. There has been one case report of delayed cyst formation 93 months after stereotactic radiosurgery (Casazza et al., 1996).

## TABLE 18.4
### Seizure Outcome after Stereotactic Radiosurgery in Patients with Cavernous Malformations

| Study | No of Patients | Treatment Modality | Marginal Dose (Gy) | Follow-Up | Seizure-Free N/N (%) | Improvement (%) | No Change (%) |
|---|---|---|---|---|---|---|---|
| Amin-Hanjani et al. (1998) | 18 | Proton beam | 15 | 5.4 years | 1/18 (6%) | 38 | 0 |
| Regis et al. (2000b) | 49 | Gamma knife | 19.17 | 24 months | 26/49 (53%) | 20 | 26 |
| Kim et al. (2005) | 12 | Gamma knife | 14.55 | 27 months | 9/12 (75%) | — | 25 |
| Liscak et al. (2005) | 44 | Gamma knife | 16 | 48 months | 0/44 | 45 | 0 |
| Liu et al. (2005) | 43 | Gamma knife | Not specified | 50 months | 12/43 (28%) | 56 | 0 |
| Shih et al. (2005) | 16 | Gamma knife | 13.3 | 53 months | 4/16 (25%) | Not specified | Not specified |
| Huang et al. (2006) | 13 | LINAC | 16 | 60 months | 8/13 (62%) | 23 | 15 |

## TABLE 18.5
### Predictive Factors for Seizure Outcome in Patients with Cavernous Malformations after Stereotactic Radiosurgery

| Factors That Improve Seizure Outcome | Factors That Worsen Seizure Outcome |
|---|---|
| Mesiotemporal location | Simple partial seizures |
| Complex partial seizures | Lateral temporal cortex location of cavernous malformation |

*Source:* Adapted from Regis, J. et al., *Neurosurgery*, 47, 1091–1097, 2000.

## TABLE 18.6
### Complications after Stereotactic Radiosurgery in Patients with Cavernous Malformations

| Study | No. of Patients | Dose (Gy) | Brain Edema N (%) | Neurologic Deficit[b] N/N (%) | Annual Hemorrhage Rate (%) | Death[c] N (%) |
|---|---|---|---|---|---|---|
| Pollock et al. (2000)[a] | 17 | 18 | Not specified | 10/17 (59%) | 8.8 | 0 |
| Tsien et al. (2001) | 20 | 25 | 3 (15) | 1/20 (5%) | 3.2 | 1 (5%) |
| Hasegawa et al. (2002) | 82 | 16.2 | Not specified | 11/82 (13.4%) | 12.3[d] | 0 |
| Liscak et al. (2005) | 107 | 16 | 30 (27)[d] | 22/107 (20.5%) | 1.6 | 2 (1.8%) |
| Kim et al. (2005) | 42 | 14.55 | 11 (26.2) | 3/42 (7.1%) | 1 (2.3) | 0 |
| Huang et al. (2006) | 30 | 16 | 2 (6.7) | 0–30 | 0.67 | 0 |

[a]  Reported for the first 2 years after stereotactic radiosurgery.
[b]  Reported as both transient and persistent morbidity.
[c]  Death was due to rebleeding.
[d]  Magnetic resonance imaging was performed in the 110 patients.

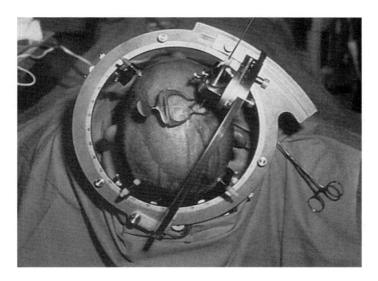

**FIGURE 18.6**  Patient positioned for stereotactic radiofrequency lesioning (for pallidotomy). Pictured here is a standard stereotactic head-ring with lesioning electrode visible in upper right. (Courtesy of David A. Peace, MS, CMI.)

For comparison, surgical excision of the CM has success rates for seizure control ranging from 20 to 80% (Acciarri et al., 1995; Cappabianca et al., 1997; Cohen et al., 1995; Dodick et al., 1994; Giulioni et al., 1995; Schroeder et al., 1996; Shih and Pan, 2005). In cases of good concordance between the electroclinical data and the location of the CM, complete lesionectomy led to the disappearance of seizures in 86% of patients with chronic forms of epilepsy (Casazza et al., 1996). In studies that directly compare the two modalities, craniotomy appeared to have a better seizure control rate (79 to 87%) than stereotactic radiosurgery (25 to 64%) (Hsu et al., 2007; Shih and Pan, 2005).

In conclusion, for surgically accessible symptomatic cavernous malformations, microsurgery or observation remains the treatment of choice, especially for those who present with hemorrhage. The use of stereotactic radiosurgery for intracranial CMs is considered to be highly controversial at best. For patients presenting with seizures and multiple hemorrhages with a deep-seated cavernous malformation, stereotactic radiosurgery may be considered a viable alternative in some cases.

## Brain Tumors

A great deal of literature exists for the oncologic response of tumors to stereotactic radiosurgery, but the data on seizure presentation and resolution are less prolific (Davey et al., 2007; Fahrig et al., 2007; Franzin et al., 2008; Germanwala et al., 2008; Kondziolka et al., 2008; Kong et al., 2008; Lee et al., 2008; Lekovic et al., 2007; Lo et al., 2008; Mathieu et al., 2007, 2008; Matsunaga et al., 2007; Muacevic et al., 2008; Nataf et al., 2008; Swinson and Friedman, 2008; Yuan et al., 2008). Seizures occur in 15 to 95% of all primary brain tumors on presentation (DeAngelis, 2001). They are more frequently associated with low-grade gliomas (65 to 95%) than malignant gliomas (15 to 25%). Conventional fractionated radiation therapy for low-grade gliomas has yielded limited success with seizure control. Among patients treated with conventional radiotherapy for gliomas, seizure-free outcome was achieved in 20 to 75%, with seizure relapse when growth progressed (Chalifoux and Elisevich, 1996; Goldring et al., 1986; Rogers et al., 1993) (Figure 18.6).

Radiosurgery is an effective treatment modality to control seizures in this group of patients. Studies have demonstrated the seizure response rate to be 52 to 100% (Kwon et al., 2006; Schoggl et al., 1998; Schrottner et al., 2002; Whang and Kwon, 1996) (Table 18.7). Factors that potentially predict favorable seizure outcome are summarized in Table 18.8 (Schrottner et al., 1998). Complications

**TABLE 18.7**
**Seizure Control Rates with Stereotactic Radiosurgery, by Tumor Histology**

| Study | Tumor Type | No. of Patients | Marginal Dose (Gy) | Seizure Free N/N (%) |
|---|---|---|---|---|
| Whang and Kwon (1996) | Not specified | 23 | 23 | 12/23 (52%) |
| Schoggl et al. (1998) | Metastatic lesion from the renal cell carcinoma | 5 | 22 | 5/5 (100%) |
| Schrottner et al. (2002) | Low-grade astrocytomas, gangliogliomas, and cavernomas | 19 | 17 | 11/19 (58%) |
| Kwon et al. (2006) | Dysembryoplastic neuroepithelial tumor | 1 | Not specified | 1/1 (100%) |

**TABLE 18.8**
**Factors That Predict Seizure Control in Patients with Brain Tumors**

**Favorable Predictive Factors for Seizure Outcome**

Radiation dose more than 10 Gy outside the tumor margin
Temporal location of lesion
Seizure duration less than 2.5 years

*Source:* Adapted from Schrottner, O. et al., *Stereotact. Funct. Neurosurg.*, 70(Suppl. 1), 50–56, 1998.

after stereotactic radiosurgery include new seizure onset, edema, necrosis, hemorrhagic event, new recurrent lesion, transient worsening of paresis, and tumor expansion (Chang and Adler, Jr., 1997; Curry et al., 2005; Hillard et al., 2003; Jensen et al., 2005; Lavine et al., 1999; Pan et al., 1998; Schindler et al., 2006; Shenouda et al., 1997; Singh et al., 2000).

## MESIAL TEMPORAL LOBE EPILEPSY

Mesial temporal lobe epilepsy (MTLE) is associated with a good response to traditional surgical methods (Jutila et al., 2002; McIntosh et al., 2004; Spencer et al., 2005; Wieser et al., 2003; Yoon et al., 2003). MTLE is gradually progressive, often with multiple epileptogenic sites (Bartolomei et al., 2008a). Imaging studies often show atrophy of the hippocampus, entorhinal cortex, thalamus, and cerebellar structures, as well as cortical and corpus callosum thinning (Carne et al., 2007; Lin et al., 2007; McDonald et al., 2008a,b; Meade et al., 2008; Weber et al., 2007). Criteria for selecting the appropriate candidates for stereotactic radiosurgery are summarized in Table 18.9. The optimal dose is 21 to 25 Gy (Bartolomei et al., 2008b; Hoggard et al., 2008; Regis et al., 2002, 2004; Rheims et al., 2008). A lower dose of 15 Gy failed to reliably produce seizure control (Cmelak et al., 2001; Kawai et al., 2001; Srikijvilaikul et al., 2004). Seizure outcomes are summarized in Table 18.10. Seizure episodes tend to increase during the latency period, which usually lasts 12 to 18 months (Bartolomei et al., 2008b; Regis et al., 2004). Factors that play an important role in stereotactic radiosurgery outcome include:

1. Complexity of targeting requires good conformality and complex multiisocentric technology (Regis et al., 2004) (Figure 18.7).
2. Patients with lesions solely in the mesiotemporal lobe respond better than patients with lesions extending beyond the mesiotemporal structures (Rheims et al., 2008).
3. Failure to maintain appropriate antiepileptic drug therapy is directly correlated with seizure relapse (Bartolomei et al., 2008b).

Complications after stereotactic radiosurgery are summarized below and in Table 18.11.

**TABLE 18.9**

**Criteria for Selection of Patients with Mesial Temporal Lobe Epilepsy for Stereotactic Radiosurgery**

**Best Candidates for Stereotactic Radiosurgery**

History of drug-resistant epilepsy
Hippocampal atrophy
No history of psychiatric disorder
No space-occupying lesions

*Source:* Adapted from Regis, J. et al., *Epilepsia*, 45, 504–515, 2004.

**TABLE 18.10**

**Seizure Outcome after Stereotactic Radiosurgery for Mesial Temporal Lobe Epilepsy[a]**

| Study | No. of Patients | Follow-Up[c] | Dose[d] (Gy) | Seizure-Free (%) | Seizure Cessation Delay |
|---|---|---|---|---|---|
| Heikkinen et al. (1992) | 1 | 27 months | 10 | 100 | 7 months |
| Barcia-Salorio et al. (1993)[b] | 11 | 18 months | 10–20 | 36 | Not specified |
| Kurita et al. (2001) | 1 | 24 months | 18 | 100 | Not specified |
| Kawai et al. (2001) | 2 | 23 months | 18 | 0 | — |
| Cmelak et al. (2001) | 1 | 12 months | 15 | 0 | — |
| Srikijvilaikul et al. (2004) | 5 | 20 months | 20 | 0 | — |
| Regis et al. (2004) | 20 | 24 months | 24 ± 1 | 65 | 16 months |
| Prayson and Yoder (2007) | 4 | 20 months | 20 | 0 | — |
| Rheims et al. (2008) | 15 | 60 ± 22 months | 21.1 ± 2.6 | 47 | 20 months |
| Bartolomei et al. (2008b) | 15 | 8 years | 24 | 60 | 12 months |
| Hoggard et al. (2008) | 8 | 42 months | 25 | 38 | 18–24 months |

[a] All studies used the gamma knife.

[b] Betatron was used in two patients.

[c] Reported as a mean or range.

[d] Reported as a marginal dose that corresponds to the 50% isodose.

**TABLE 18.11**

**Complications after Stereotactic Radiosurgery in Patients with Temporal Lobe Epilepsy[a]**

| Study | No. of Patients | Visual Deficit N/N (%) Hemianopsia | Asymptomatic Quadrantanopsia | Memory Decline N/N (%) | Death N/N (%) |
|---|---|---|---|---|---|
| Hoggard et al. (2008)[b] | 8 | Not specified | Not specified | 0/8 | 0 |
| Bartolomei et al. (2008a) | 15 | 1/15 (6%) | 8/15 (53%) | Not specified | 0 |
| Prayson and Yoder (2007) | 4 | Not specified | Not specified | Not specified | 1/4 (25%) |
| Srikijvilaikul et al. (2004) | 5 | Not specified | Not specified | 1/5 (20%) | 2/5 (40%) |
| McDonald et al. (2004) | 3 | Not specified | Not specified | 3/3 (100%) | 0 |
| Regis et al. (2004)[c] | 20 | 1/20 (5%) | 8/20 (40%) | 0 | 0 |

[a] Absolute number and percentage.

[b] Of 8 patients, 2 (25%) developed mild dysphasia that responded well to steroids.

[c] One patient died of a myocardial infarction during the study.

1. Verbal memory decline is a common complication after conventional surgery and occurs more frequently in patients with lesions in the left temporal lobe (Bell and Davies, 1998; McIntosh et al., 2004; Spencer et al., 2005; Yoon et al., 2003). This has been reported after stereotactic radiosurgery as well (McDonald et al., 2004; Srikijvilaikul et al., 2004). In contrast, others have shown no memory decline in patients after stereotactic radiosurgery (Regis et al., 2004). Some have concluded that memory decline is due to a refusal to take steroids after stereotactic radiosurgery, although steroids reduce posttreatment edema. One study demonstrates significant improvement in confrontational naming at 1 year following GK (McDonald et al., 2004).
2. Visual deficits are most commonly due to the injury of the optic radiation (Regis et al., 2004). Optic neuropathy can occur in patients that have received a radiation dose of more than 10 Gy to their visual apparatus (Leber et al., 1995, 1998). Visual field deficit in post-resection patients undergoing formal field testing after anterior temporal lobectomy is significantly higher (69 to 83%) (Egan et al., 2000; Guenot et al., 1999; Jensen and Seedorff, 1976).
3. Sudden unexpected death in epilepsy (SUDEP) may occur during the latent period and has been reported, from 2 weeks to 13 months following radiosurgery in several studies (Bartolomei et al., 2008b; Prayson and Yoder, 2007; Srikijvilaikul et al., 2004).

## EXTRATEMPORAL IDIOPATHIC FOCAL EPILEPSY

The treatment of extratemporal idiopathic focal epilepsy with IR has been reported (Barcia-Salorio et al., 1994; Smith et al., 2000; Stefan et al., 1998). The largest clinical study published to date involved the treatment of 11 patients with medically intractable extratemporal epilepsy (Barcia-Salorio et al., 1994). All of the patients had invasive subdural electrode monitoring through burr holes, followed by a second monitoring admission with depth electrodes around the focus to define its margins for the stereotactic radiosurgery treatment plan. Treatment outcome is summarized in Table 18.12.

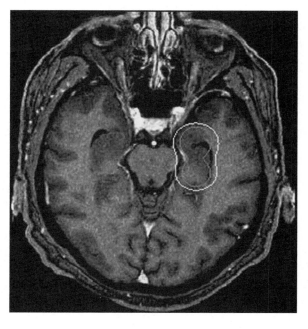

**FIGURE 18.7**    MRI-based LINAC dose plan for stereotactic radiosurgery treatment of mesial temporal lobe epilepsy. Inner circle represents prescription dose, and outer circle is the 50% isodose line. (Courtesy of William Friedman, MD.)

**TABLE 18.12**
**Seizure Outcome after Stereotactic Radiosurgery**
**in Patients with Extratemporal Focal Epilepsy**

| Study | No. of Patients | Dose | Follow-Up (months) | Seizure Outcome |
|---|---|---|---|---|
| Barcia-Salorio et al. (1994) | 11 | 10–20 Gy | 75 | 36% seizure free, 45% improved |
| Stefan et al. (1998) | 1 | 7 daily fractions, 3 Gy | 9 | Seizure free |

## DROP ATTACKS

Callosotomy has been successfully used to treat medically refractory epilepsy, particularly drop attacks (Cendes et al., 1993; Fuiks et al., 1991; Jenssen et al., 2006; Oguni et al., 1991; Rahimi et al., 2007; Reutens et al., 1993; Shimizu, 2005). Due to its significant complications (Polkey, 2003; Rossi et al., 1996; Sass et al., 1990; Shimizu, 2005; Spencer, 1988), stereotactic radiosurgery has been investigated (Table 18.13). It appears comparable to conventional open callosotomy in seizure control with a lower complication rate (Feichtinger et al., 2006).

## HYPOTHALAMIC HAMARTOMAS

Hypothalamic hamartomas (HHs) are benign lesions that can cause seizures, precocious puberty, and mental retardation. They consist of neuronal and glial tissue that resembles tissue in other parts of the brain. Sessile hamartomas (Boyko et al., 1991) as well as intrahypothalamic hamartomas (Arita et al., 1999) can cause seizure activity. Regis et al. (2007) proposed a classification that correlates with seizure severity and is useful for deciding surgical strategy.

Seizure-free outcome ranges from 37 to 67% (Table 18.14) (Akai et al., 2002; Arita et al., 1998; Dunoyer et al., 2002; Mathieu et al., 2006; Regis et al., 2000a, 2006; Selch et al., 2005; Shim et al., 2008; Unger et al., 2002). One study suggests that a dose of more than 17 Gy improves seizure-free outcome (Regis et al., 2000a), but the close relationship of HHs with the surrounding optic pathways must be taken into account. Recommended thresholds are 10 Gy for the optic tract and 8 Gy for the chiasm (Regis et al., 2006). Typical postradiosurgery evolution has been well described (Regis et al., 2006). At first, patients experience a global improvement of both seizure frequency and severity accompanied by an improvement in behavior. During the following 2 to 6 months, seizure frequency

**TABLE 18.13**
**Seizure Outcome after Stereotactic Callosotomy in Patients with Refractory Epilepsy**

| Study | No. of Patients | Treatment Modality | Seizure-Free (%) | Outcome Drop Attacks | Outcome Other Seizures |
|---|---|---|---|---|---|
| Pendl et al. (1999) | 3 | Gamma knife | None | Significant reduction | Significant improvement of seizure frequency and severity |
| Feichtinger et al. (2006) | 8 | Gamma knife | 38 | Reduction in 2 (25%) | 100% reduction in 2 (50%)[a]; significant reduction in 2 (50% and 60%) |
| Celis et al. (2007) | 1 | LINAC | — | 84% reduction | Improvement of generalized tonic–clonic seizures |

[a] Four patients presented with additional generalized tonic–clonic seizures.

**TABLE 18.14**
**Seizure Outcome after Stereotactic Radiosurgery in Patients with Hypothalamic Hamartoma**

| Study | Number of Patients | Age of Onset | Treatment Modality | Marginal Dose (Gy) | Maximal Dose to Optic Pathway (Gy) | Follow-Up | Seizure-Free Outcome N (%) | Improvement |
|---|---|---|---|---|---|---|---|---|
| Regis et al. (2000a) | 8 | 4.9 years | Gamma knife | 15.25 | 10 | 28 months | 4 (50%) | 4 (50%) |
| Dunoyer et al. (2002) | 2 | 21 months | Gamma knife | 11–14 (median, 12.5) | <8, < 6.5 | 26–32 months (median, 29) | 1 (50%) | 1 (50%) with >90% seizure reduction |
| Unger et al. (2002) | 4 | Not specified | Gamma knife | 12–14 | Not specified | 12–68 months | Not specified | Engel class IIa, 3 (75%); Engel class IIIa, 1 (25%) |
| Akai et al. (2002) | 1 | 3 years | LINAC | 25 | <8 | 10 months | Not specified | Improvement 10 months after surgery (resection has been performed after stereotactic radiosurgery) |
| Selch et al. (2005) | 3 | Not specified | LINAC | 16 | Not specified | 9–17 months (median, 13) | 2 (67%) | Engel class IIb, 1 (33%) |
| Mathieu et al. (2006) | 4 | Not specified | Gamma knife | 17.5 | Not specified | 22 | Not specified | Engel class II, 2 (50%) out of 3 (75%) |
| Regis et al. (2006) | 27 | Not specified | Gamma knife | 17 | 6.4 (median) | >3 years | 10 (37%) | 6 (22%) |
| Shim et al. (2008) | 4 | Not specified | Gamma knife | 9–13 (median, 11.5) | <7 | 25–54 months (median, 42.2) | Not specified | Engel class III, 2 (50%); Engel class IV, 2 (50%) |

increases, becoming comparable to the preoperative state; however, neurophysiologic parameters such as the electroencephalogram, behavior, and sleep patterns gradually normalize. The third period is characterized by a marked increase in seizure frequency for up to a month. In the fourth period, the seizures cease. No studies have shown permanent neurologic deficits following radiosurgery for HHs. Transient poikilothermia and transient increase in seizures have been reported (Regis et al., 2006).

## CONCLUSION

Stereotactic radiosurgery can produce seizure control with low complication rates and is particularly attractive in lesional epilepsy. When traditional surgical intervention is considered too risky, stereotactic radiosurgery may be a viable alternative. The potential risks and benefits have been summarized here for several disorders. Much about radiosurgery for epilepsy remains to be studied. Larger, randomized, blinded studies with longer follow-up will be important in the future.

## REFERENCES

Acciarri, N., M. Giulioni, R. Padovani, E. Galassi, G. Gaist. (1995). Surgical management of cerebral cavernous angiomas causing epilepsy. *J. Neurosurg. Sci.*, 39:13–20.

Akai, T., K. Okamoto, H. Iizuka, H. Kakinuma, T. Nojima. (2002). Treatments of hamartoma with neuroendoscopic surgery and stereotactic radiosurgery: a case report. *Minim. Invasive Neurosurg.*, 45:235–239.

Amin-Hanjani, S., C.S. Ogilvy, G.J. Candia, S. Lyons, P.H. Chapman. (1998). Stereotactic radiosurgery for cavernous malformations: Kjellberg's experience with proton beam therapy in 98 cases at the Harvard Cyclotron. *Neurosurgery*, 42:1229–1238.

Arita, K., K. Kurisu, K. Iida, R. Hanaya, T. Akimitsu, S. Hibino, B. Pant, M. Hamasaki, S. Shinagawa. (1998). Subsidence of seizure induced by stereotactic radiation in a patient with hypothalamic hamartoma. Case report. *J. Neurosurg.*, 89:645–648.

Arita, K., F. Ikawa, K. Kurisu, M. Sumida, K. Harada, T. Uozumi, S. Monden, J. Yoshida, Y. Nishi. (1999). The relationship between magnetic resonance imaging findings and clinical manifestations of hypothalamic hamartoma. *J. Neurosurg.*, 91:212–220.

Barcia-Salorio, J.L., V. Vanaclocha, M. Cerdá, J. Ciudad, L. López-Gómez. (1987). Response of experimental epileptic focus to focal ionizing radiation. *Appl. Neurophysiol.*, 50(1–6):359–364.

Barcia-Salorio, J.L., J.A. Barcia, P. Roldán, G. Hernández, L. López-Gómez. (1993). Radiosurgery of epilepsy. *Acta Neurochir. Suppl. (Wien)*, 58:195–197.

Barcia-Salorio, J.L., J.A. Barcia, G. Hernandez, L. Lopez-Gomez. (1994). Radiosurgery of epilepsy: long-term results. *Acta Neurochir.*, 62(Suppl.):111–113.

Bartolomei, F., P. Chauvel, F. Wendling. (2008a). Epileptogenicity of brain structures in human temporal lobe epilepsy: a quantified study from intracerebral EEG. *Brain*, 131:1818–1830.

Bartolomei, F., M. Hayashi, M. Tamura, M. Rey, C. Fischer, P. Chauvel, J. Regis. (2008b). Long-term efficacy of gamma knife radiosurgery in mesial temporal lobe epilepsy. *Neurology*, 70:1658–1663.

Bell, B.D., K.G. Davies. (1998). Anterior temporal lobectomy, hippocampal sclerosis, and memory: recent neuropsychological findings. *Neuropsychol. Rev.*, 8:25–41.

Belli, M., R. Cherubini, M. Dalla Vecchia, V. Dini, G. Esposito, G. Moschini, O. Sapora, C. Signoretti, G. Simone, E. Sorrentino, M.A. Tabocchini. (2001). DNA fragmentation in mammalian cells exposed to various light ions. *Adv. Space Res.*, 27:393–399.

Berman, E.L., T.N. Eade, D. Brown, M. Weaver, J. Glass, G. Zorman, S.J. Feigenberg. (2007). Radiation-induced tumor after stereotactic radiosurgery for an arteriovenous malformation: case report. *Neurosurgery*, 61:E1099.

Boyko, O.B., J.T. Curnes, W.J. Oakes, P.C. Burger. (1991). Hamartomas of the tuber cinereum: CT, MR, and pathologic findings. *AJNR Am. J. Neuroradiol.*, 12:309–314.

Cappabianca, P., A. Alfieri, F. Maiuri, G. Mariniello, S. Cirillo, E. de Divitiis. (1997). Supratentorial cavernous malformations and epilepsy: seizure outcome after lesionectomy on a series of 35 patients. *Clin. Neurol. Neurosurg.*, 99:179–183.

Carne, R.P., T.J. O'Brien, C.J. Kilpatrick, L.R. Macgregor, L. Litewka, R.J. Hicks, M.J. Cook. (2007). "MRI-negative PET-positive" temporal lobe epilepsy (TLE) and mesial TLE differ with quantitative MRI and PET: a case control study. *BMC Neurol.*, 7:16.

Casazza, M., G. Broggi, A. Franzini, G. Avanzini, R. Spreafico, M. Bracchi, M.C. Valentini. (1996). Supratentorial cavernous angiomas and epileptic seizures: preoperative course and postoperative outcome. *Neurosurgery*, 39:26–32; discussion, 34.

Celis, M.A., S. Moreno-Jiménez, J.M. Lárraga-Gutiérrez, M.A. Alonso-Vanegas, O.A. García-Garduño, I.E. Martínez-Juárez, M.C. Fernández-Gónzalez. (2007). Corpus callosotomy using conformal stereotactic radiosurgery. *Childs Nerv. Syst.*, 23(8):917–920.

Cendes, F., P.C. Ragazzo, V. da Costa, L.F. Martins. (1993). Corpus callosotomy in treatment of medically resistant epilepsy: preliminary results in a pediatric population. *Epilepsia*, 34:910–917.

Chalifoux, R., K. Elisevich. (1996). Effect of ionizing radiation on partial seizures attributable to malignant cerebral tumors. *Stereotact. Funct. Neurosurg.*, 67:169–182.

Chang, S.D., J.R. Adler, Jr. (1997). Treatment of cranial base meningiomas with linear accelerator radiosurgery. *Neurosurgery*, 41:1019–1027.

Chang, S.D., R.P. Levy, J.R. Adler, Jr., D.P. Martin, P.R. Krakovitz, G.K. Steinberg. (1998). Stereotactic radiosurgery of angiographically occult vascular malformations: 14-year experience. *Neurosurgery*, 43:213–221.

Chang, T.C., H. Shirato, H. Aoyama, S. Ushikoshi, N. Kato, S. Kuroda, T. Ishikawa, K. Houkin, Y. Iwasaki, K. Miyasaka. (2004). Stereotactic irradiation for intracranial arteriovenous malformation using stereotactic radiosurgery or hypofractionated stereotactic radiotherapy. *Int. J. Radiat. Oncol. Biol. Phys.*, 60:861–870.

Chen, Z.F., T. Kamiryo, S.L. Henson, H. Yamamoto, E.H. Bertram, F. Schottler, F. Patel, L. Steiner, D. Prasad, N.F. Kassell, S. Shareghis, K.S. Lee. (2001). Anticonvulsant effects of gamma surgery in a model of chronic spontaneous limbic epilepsy in rats. *J. Neurosurg.*, 94:270–280.

Cmelak, A.J., B. Abou-Khalil, P.E. Konrad, D. Duggan, R.J. Maciunas. (2001). Low-dose stereotactic radiosurgery is inadequate for medically intractable mesial temporal lobe epilepsy: a case report. *Seizure*, 10:442–446.

Cohen, D.S., G.P. Zubay, R.R. Goodman. (1995). Seizure outcome after lesionectomy for cavernous malformations. *J. Neurosurg.*, 83:237–242.

Crocker, M., R. deSouza, P. Epaliyanage, I. Bodi, N. Deasy, R. Selway. (2007). Masson's tumour in the right parietal lobe after stereotactic radiosurgery for cerebellar AVM: case report and review. *Clin. Neurol. Neurosurg.*, 109:811–815.

Curry, W.T., Jr., G.R. Cosgrove, F.H. Hochberg, J. Loeffler, N.T. Zervas. (2005). Stereotactic interstitial radiosurgery for cerebral metastases. *J. Neurosurg.*, 103:630–635.

Davey, P., M.L. Schwartz, D. Scora, S. Gardner, P.F. O'Brien. (2007). Fractionated (split dose) radiosurgery in patients with recurrent brain metastases: implications for survival. *Br. J. Neurosurg.*, 21:491–495.

DeAngelis, L.M. (2001). Brain tumors. *N. Engl. J. Med.*, 344:114–123.

Dodick, D.W., G.D. Cascino, F.B. Meyer. (1994). Vascular malformations and intractable epilepsy: outcome after surgical treatment. *Mayo. Clin. Proc.*, 69:741–745.

Dunoyer, C., J. Ragheb, T. Resnick, L. Alvarez, P. Jayakar, N. Altman, A. Wolf, M. Duchowny. (2002). The use of stereotactic radiosurgery to treat intractable childhood partial epilepsy. *Epilepsia*, 43:292–300.

Egan, R.A., W.T. Shults, N. So, K. Burchiel, J.X. Kellogg, M. Salinsky. (2000). Visual field deficits in conventional anterior temporal lobectomy versus amygdalohippocampectomy. *Neurology*, 55:1818–1822.

Eisenschenk, S., R.L. Gilmore, W.A. Friedman, R.A. Henchey. (1998). The effect of LINAC stereotactic radiosurgery on epilepsy associated with arteriovenous malformations. *Stereotact. Funct. Neurosurg.*, 71:51–61.

Esteves, S.C., W. Nadalin, R.L. Piske, S. Benabou, E. Souza, A.C. Oliveira. (2008). Radiosurgery with a linear accelerator in cerebral arteriovenous malformations. *Rev. Assoc. Med. Bras.*, 54:167–172.

Fabrikant, J.I., R.P. Levy, G.K. Steinberg, M.H. Phillips, K.A. Frankel, J.T. Lyman, M.P. Marks, G.D. Silverberg. (1992). Charged-particle radiosurgery for intracranial vascular malformations. *Neurosurg. Clin. N. Am.*, 3:99–139.

Fahrig, A., O. Ganslandt, U. Lambrecht, G. Grabenbauer, G. Kleinert, R. Sauer, K. Hamm. (2007). Hypofractionated stereotactic radiotherapy for brain metastases: results from three different dose concepts. *Strahlenther. Onkol.*, 183:625–630.

Falkson, C.B., K.B. Chakrabarti, D. Doughty, P.N. Plowman. (1997). Stereotactic multiple arc radiotherapy. III. Influence of treatment of arteriovenous malformations on associated epilepsy. *Br. J. Neurosurg.*, 11:12–15.

Feichtinger, M., O. Schrottner, H. Eder, H. Holthausen, T. Pieper, F. Unger, A. Holl, L. Gruber, E. Korner, E. Trinka, F. Fazekas, E. Ott. (2006). Efficacy and safety of radiosurgical callosotomy: a retrospective analysis. *Epilepsia*, 47:1184–1191.

Fleetwood, I.G., G.K. Steinberg. (2002). Arteriovenous malformations. *Lancet*, 359:863–873.

Flickinger, J.C. (1989). An integrated logistic formula for prediction of complications from radiosurgery. *Int. J. Radiat. Oncol. Biol. Phys.*, 17:879–885.

Flickinger, J.C., D. Kondziolka, B.E. Pollock, A.H. Maitz, L.D. Lunsford. (1997). Complications from arteriovenous malformation radiosurgery: multivariate analysis and risk modeling. *Int. J. Radiat. Oncol. Biol. Phys.*, 38:485–490.

Flickinger, J.C., D. Kondziolka, A.H. Maitz, L.D. Lunsford. (1998). Analysis of neurological sequelae from radiosurgery of arteriovenous malformations: how location affects outcome. *Int. J. Radiat. Oncol. Biol. Phys.*, 40:273–278.

Flickinger, J.C., D. Kondziolka, L.D. Lunsford, B.E. Pollock, M. Yamamoto, D.A. Gorman, P.J. Schomberg, P. Sneed, D. Larson, V. Smith, M.W. McDermott, L. Miyawaki, J. Chilton, R.A. Morantz, B. Young, H. Jokura, R. Liscak. (1999). A multi-institutional analysis of complication outcomes after arteriovenous malformation radiosurgery. *Int. J. Radiat. Oncol. Biol. Phys.*, 44:67–74.

Franzin, A., A. Vimercati, P. Picozzi, C. Serra, S. Snider, L. Gioia, C. Ferrari da Passano, A. Bolognesi, M. Giovanelli. (2008). Stereotactic drainage and gamma knife radiosurgery of cystic brain metastasis. *J. Neurosurg.*, 109:259–267.

Friedman, W.A., D.L. Blatt, F.J. Bova, J.M. Buatti, W.M. Mendenhall, P.S. Kubilis. (1996). The risk of hemorrhage after radiosurgery for arteriovenous malformations. *J. Neurosurg.*, 84:912–919.

Friedman, W.A., F.J. Bova, S. Bollampally, P. Bradshaw. (2003). Analysis of factors predictive of success or complications in arteriovenous malformation radiosurgery. *Neurosurgery*, 52:296–308.

Fuiks, K.S., A.R. Wyler, B.P. Hermann, G. Somes. (1991). Seizure outcome from anterior and complete corpus callosotomy. *J. Neurosurg.*, 74:573–578.

Gaffey, C.T., V.J. Montoya, J.T. Lyman, J. Howard. (1981). Restriction of the spread of epileptic discharges in cats by means of Bragg peak, intracranial irradiation. *Int. J. Appl. Radiat. Isot.*, 32(11):779–784.

Gajdusek, C., K. Onoda, S. London, M. Johnson, R. Morrison, M. Mayberg. (2001). Early molecular changes in irradiated aortic endothelium. *J. Cell. Physiol.*, 188:8–23.

Germanwala, A.V., J.C. Mai, N.D. Tomycz, A. Niranjan, J.C. Flickinger, D. Kondziolka, L.D. Lunsford. (2008). Boost gamma knife surgery during multimodality management of adult medulloblastoma. *J. Neurosurg.*, 108:204–209.

Gerszten, P.C., P.D. Adelson, D. Kondziolka, J.C. Flickinger, L.D. Lunsford. (1996). Seizure outcome in children treated for arteriovenous malformations using gamma knife radiosurgery. *Pediatr. Neurosurg.*, 24:139–144.

Giulioni, M., N. Acciarri, R. Padovani, E. Galassi. (1995). Results of surgery in children with cerebral cavernous angiomas causing epilepsy. *Br. J. Neurosurg.*, 9:135–141.

Goldring, S., K.M. Rich, S. Picker. (1986). Experience with gliomas in patients presenting with a chronic seizure disorder. *Clin. Neurosurg.*, 33:15–42.

Guenot, M., P. Krolak-Salmon, P. Mertens, J. Isnard, P. Ryvlin, C. Fischer, A. Vighetto, F. Mauguiere, M. Sindou. (1999). MRI assessment of the anatomy of optic radiations after temporal lobe epilepsy surgery. *Stereotact. Funct. Neurosurg.*, 73:84–87.

Hadjipanayis, C.G., E.I. Levy, A. Niranjan, A.D. Firlik, D. Kondziolka, J.C. Flickinger, L.D. Lunsford. (2001). Stereotactic radiosurgery for motor cortex region arteriovenous malformations. *Neurosurgery*, 48:70–77.

Hasegawa, T., J. McInerney, D. Kondziolka, J.Y. Lee, J.C. Flickinger, L.D. Lunsford. (2002). Long-term results after stereotactic radiosurgery for patients with cavernous malformations. *Neurosurgery*, 50:1190–1198.

Heikkinen, E.R., B. Konnov, L. Melnikov, N. Yalynych, N. Zubkov Yu, A. Garmashov Yu, V.A. Pak. (1989). Relief of epilepsy by radiosurgery of cerebral arteriovenous malformations. *Stereotact. Funct. Neurosurg.*, 53:157–166.

Heikkinen, E.R., M.I. Heikkinen, K. Sotaniemi. (1992). Stereotactic radiotherapy instead of conventional epilepsy surgery: a case report. *Acta Neurochir. (Wien)*,119(1–4):159–160.

Hillard, V.H., L.L. Shih, S. Chin, C.R. Moorthy, D.L. Benzil. (2003). Safety of multiple stereotactic radiosurgery treatments for multiple brain lesions. *J. Neurooncol.*, 63:271–278.

Hoggard, N., I.D. Wilkinson, P.D. Griffiths, P. Vaughan, A.A. Kemeny, J.G. Rowe. (2008). The clinical course after stereotactic radiosurgical amygdalohippocampectomy with neuroradiological correlates. *Neurosurgery*, 62:336–346.

Hoh, B.L., P.H. Chapman, J.S. Loeffler, B.S. Carter, C.S. Ogilvy. (2002). Results of multimodality treatment for 141 patients with brain arteriovenous malformations and seizures: factors associated with seizure incidence and seizure outcomes. *Neurosurgery*, 51:303–311.

Hsu, P.W., C.N. Chang, C.K. Tseng, K.C. Wei, C.C. Wang, C.C. Chuang, Y.C. Huang. (2007). Treatment of epileptogenic cavernomas: surgery versus radiosurgery. *Cerebrovasc. Dis.*, 24:116–121.

Huang, P.P., T. Kamiryo, P.K. Nelson. (2001). *De novo* aneurysm formation after stereotactic radiosurgery of a residual arteriovenous malformation: case report. *AJNR Am. J. Neuroradiol.*, 22:1346–1348.

Huang, Y.C., C.K. Tseng, C.N. Chang, K.C. Wei, C.C. Liao, P. W. Hsu. (2006). LINAC radiosurgery for intracranial cavernous malformation: 10-year experience. *Clin. Neurol. Neurosurg.*, 108:750–756.

Husain, A.M., M. Mendez, A.H. Friedman. (2001). Intractable epilepsy following radiosurgery for arteriovenous malformation. *J. Neurosurg.*, 95:888–892.

Ide, C., B. De Coene, V. Baudrez. (2000). MR features of cavernous angioma. *JBR-BTR*, 83:320.

Izawa, M., M. Chernov, M. Hayashi, K. Nakaya, S. Kamikawa, K. Kato, T. Higa, H. Ujiie, H. Kasuya, T. Kawamata, Y. Okada, O. Kubo, H. Iseki, T. Hori, K. Takakura. (2007). Management and prognosis of cysts developed on long-term follow-up after gamma knife radiosurgery for intracranial arteriovenous malformations. *Surg. Neurol.*, 68:400–406.

Jensen, I., H.H. Seedorff. (1976). Temporal lobe epilepsy and neuro-ophthalmology: ophthalmological findings in 74 temporal lobe resected patients. *Acta Ophthalmol.*, 54:827–841.

Jensen, A.W., P.D. Brown, B.E. Pollock, S.L. Stafford, M.J. Link, Y.I. Garces, R.L. Foote, D.A. Gorman, P.J. Schomberg. (2005). Gamma knife radiosurgery of radiation-induced intracranial tumors: local control, outcomes, and complications. *Int. J. Radiat. Oncol. Biol. Phys.*, 62:32–37.

Jenssen, S., M.R. Sperling, J.I. Tracy, M. Nei, L. Joyce, G. David, M. O'Connor. (2006). Corpus callosotomy in refractory idiopathic generalized epilepsy. *Seizure*, 15:621–629.

Jutila, L., A. Immonen, E. Mervaala, J. Partanen, K. Partanen, M. Puranen, R. Kalviainen, I. Alafuzoff, H. Hurskainen, M. Vapalahti, A. Ylinen. (2002). Long-term outcome of temporal lobe epilepsy surgery: analyses of 140 consecutive patients. *J. Neurol. Neurosurg. Psychiatry*, 73:486–494.

Karlsson, B., I. Lax, M. Soderman. (2001). Risk for hemorrhage during the 2-year latency period following gamma knife radiosurgery for arteriovenous malformations. *Int. J. Radiat. Oncol. Biol. Phys.*, 49:1045–1051.

Kavathas, P., F.H. Bach, R. DeMars. (1980). Gamma ray-induced loss of expression of HLA and glyoxalase I alleles in lymphoblastoid cells. *Proc. Natl. Acad. Sci. U.S.A.*, 77:4251–4255.

Kawaguchi, S., T. Sakaki, R. Uranishi. (2002). Color Doppler flow imaging of the superior ophthalmic vein in dural arteriovenous fistulas. *Stroke*, 33:2009–2013.

Kawai, K., I. Suzuki, H. Kurita, M. Shin, N. Arai, T. Kirino. (2001). Failure of low-dose radiosurgery to control temporal lobe epilepsy. *J. Neurosurg.*, 95:883–887.

Kida, Y., T. Kobayashi, T. Tanaka, Y. Mori, T. Hasegawa, T. Kondoh. (2000). Seizure control after radiosurgery on cerebral arteriovenous malformations. *J. Clin. Neurosci.*, 7(Suppl. 1):6–9.

Kim, M.S., S.Y. Pyo, Y.G. Jeong, S.I. Lee, Y.T. Jung, J.H. Sim. (2005). Gamma knife surgery for intracranial cavernous hemangioma. *J. Neurosurg.*, 102(Suppl.):102–106.

Kitchen, N. (1995). Experimental and clinical studies on the putative therapeutic efficacy of cerebral irradiation (radiotherapy) in epilepsy. *Epilepsy Res.*, 20:1–10.

Kjellberg, R.N., T. Hanamura, K.R. Davis, S.L. Lyons, R.D. Adams. (1983). Bragg-peak proton-beam therapy for arteriovenous malformations of the brain. *N. Engl. J. Med.*, 309:269–274.

Klokov, D., T. Criswell, K.S. Leskov, S. Araki, L. Mayo, D.A. Boothman. (2004). IR-inducible clusterin gene expression: a protein with potential roles in ionizing radiation-induced adaptive responses, genomic instability, and bystander effects. *Mutat. Res.*, 568:97–110.

Kondziolka, D., L.D. Lunsford, J.C. Flickinger, J.R. Kestle. (1995a). Reduction of hemorrhage risk after stereotactic radiosurgery for cavernous malformations. *J. Neurosurg.*, 83:825–831.

Kondziolka, D., L.D. Lunsford, J.R. Kestle. (1995b). The natural history of cerebral cavernous malformations. *J. Neurosurg.*, 83:820–824.

Kondziolka, D., D. Mathieu, L.D. Lunsford, J.J. Martin, R. Madhok, A. Niranjan, J.C. Flickinger. (2008). Radiosurgery as definitive management of intracranial meningiomas. *Neurosurgery*, 62:53–60.

Kong, D.S., J.I. Lee, K. Park, J.H. Kim, D.H. Lim, D.H. Nam. (2008). Efficacy of stereotactic radiosurgery as a salvage treatment for recurrent malignant gliomas. *Cancer*, 112:2046–2051.

Kraemer, D.L., I.A. Awad. (1994). Vascular malformations and epilepsy: clinical considerations and basic mechanisms. *Epilepsia*, 35(Suppl. 6):S30–S43.

Kurita, H., S. Kawamoto, I. Suzuki, T. Sasaki, M. Tago, A. Terahara, T. Kirino. (1998). Control of epilepsy associated with cerebral arteriovenous malformations after radiosurgery. *J. Neurol. Neurosurg. Psychiatry*, 65:648–655.

Kurita, H., S. Kawamoto, T. Sasaki, M. Shin, M. Tago, A. Terahara, K. Ueki, T. Kirino. (2000). Results of radiosurgery for brain stem arteriovenous malformations. *J. Neurol. Neurosurg. Psychiatry*, 68:563–570.

Kurita, H., I. Suzuki, M. Shin, K. Kawai, M. Tago, T. Momose, T. Kirino. (2001). Successful radiosurgical treatment of lesional epilepsy of mesial temporal origin. *Minim. Invasive Neurosurg.*, 44(1):43–46.

Kwon, K.H., J.I. Lee, S.C. Hong, D.W. Seo, S.B. Hong. (2006). Gamma knife radiosurgery for epilepsy related to dysembryoplastic neuroepithelial tumor. *Stereotact. Funct. Neurosurg.*, 84:243–247.

Lasjaunias, P.L., P. Landrieu, G. Rodesch, H. Alvarez, A. Ozanne, S. Holmin, W.Y. Zhao, S. Geibprasert, D. Ducreux, T. Krings. (2008). Cerebral proliferative angiopathy: clinical and angiographic description of an entity different from cerebral AVMs. *Stroke*, 39:878–885.

Lavine, S.D., Z. Petrovich, A.A. Cohen-Gadol, L.S. Masri, D.L. Morton, S.J. O'Day, R. Essner, V. Zelman, C. Yu, G. Luxton, M.L. Apuzzo. (1999). Gamma knife radiosurgery for metastatic melanoma: an analysis of survival, outcome, and complications. *Neurosurgery*, 44:59–64; discussion, 66.

Leber, K.A., J. Bergloff, G. Langmann, M. Mokry, O. Schrottner, G. Pendl. (1995). Radiation sensitivity of visual and oculomotor pathways. *Stereotact. Funct. Neurosurg.*, 64(Suppl. 1):233–238.

Leber, K.A., J. Bergloff, G. Pendl. (1998). Dose–response tolerance of the visual pathways and cranial nerves of the cavernous sinus to stereotactic radiosurgery. *J. Neurosurg.*, 88:43–50.

Leblanc, R., W. Feindel, R. Ethier. (1983). Epilepsy from cerebral arteriovenous malformations. *Can. J. Neurol. Sci.*, 10:91–95.

Lee, J.W., D.S. Kim, K.W. Shim, J.H. Chang, S.K. Huh, Y.G. Park, J.U. Choi. (2008a). Management of intracranial cavernous malformation in pediatric patients. *Childs. Nerv. Syst.*, 24:321–327.

Lee, Y.K., N.H. Park, J.W. Kim, Y.S. Song, S.B. Kang, H.P. Lee. (2008b). Gamma-knife radiosurgery as an optimal treatment modality for brain metastases from epithelial ovarian cancer. *Gynecol. Oncol.*, 108:505–509.

Lekovic, G.P., L.F. Gonzalez, A.G. Shetter, R.W. Porter, K.A. Smith, D. Brachman, R.F. Spetzler. (2007). Role of gamma knife surgery in the management of pineal region tumors. *Neurosurg. Focus*, 23:E12.

Leskov, K.S., T. Criswell, S. Antonio, J. Li, C.R. Yang, T.J. Kinsella, D.A. Boothman. (2001). When x-ray-inducible proteins meet DNA double strand break repair. *Semin. Radiat. Oncol.*, 11:352–372.

Lim, Y.J., C.Y. Lee, J.S. Koh, T.S. Kim, G.K. Kim, B.A. Rhee. (2006). Seizure control of gamma knife radiosurgery for non-hemorrhagic arteriovenous malformations. *Acta Neurochir.*, 99(Suppl.):97–101.

Lin, J.J., N. Salamon, A.D. Lee, R.A. Dutton, J.A. Geaga, K.M. Hayashi, E. Luders, A.W. Toga, J. Engel, Jr., P.M. Thompson. (2007). Reduced neocortical thickness and complexity mapped in mesial temporal lobe epilepsy with hippocampal sclerosis. *Cereb. Cortex*, 17:2007–2018.

Liscak, R., V. Vladyka, G. Simonova, J. Vymazal, J. Novotny, Jr. (2005). Gamma knife surgery of brain cavernous hemangiomas. *J. Neurosurg.*, 102(Suppl.):207–213.

Liu, A.L., C.C. Wang, K. Dai. (2005). Gamma knife radiosurgery for cavernous malformations. *Zhongguo Yi Xue Ke Xue Yuan Xue Bao*, 27:18–21.

Lo, S.S., A.J. Fakiris, R. Abdulrahman, M.A. Henderson, E.L. Chang, J.H. Suh, R.D. Timmerman. (2008). Role of stereotactic radiosurgery and fractionated stereotactic radiotherapy in pediatric brain tumors. *Expert Rev. Neurother.*, 8:121–132.

Lunsford, L.D., D. Kondziolka, J.C. Flickinger, D.J. Bissonette, C.A. Jungreis, A.H. Maitz, J.A. Horton, R.J. Coffey. (1991). Stereotactic radiosurgery for arteriovenous malformations of the brain. *J. Neurosurg.*, 75:512–524.

Maesawa, S., D. Kondziolka, C.E. Dixon, J. Balzer, W. Fellows, L.D. Lunsford. (2000). Subnecrotic stereotactic radiosurgery controlling epilepsy produced by kainic acid injection in rats. *J. Neurosurg.*, 93:1033–1040.

Malone, S., G.P. Raaphorst, R. Gray, A. Girard, G. Alsbeih. (2000). Enhanced *in vitro* radiosensitivity of skin fibroblasts in two patients developing brain necrosis following AVM radiosurgery: a new risk factor with potential for a predictive assay. *Int. J. Radiat. Oncol. Biol. Phys.*, 47:185–189.

Maruyama, K., D. Kondziolka, A. Niranjan, J.C. Flickinger, L.D. Lunsford. (2004). Stereotactic radiosurgery for brainstem arteriovenous malformations: factors affecting outcome. *J. Neurosurg.*, 100:407–413.

Maruyama, K., N. Kawahara, M. Shin, M. Tago, J. Kishimoto, H. Kurita, S. Kawamoto, A. Morita, T. Kirino. (2005). The risk of hemorrhage after radiosurgery for cerebral arteriovenous malformations. *N. Engl. J. Med.*, 352:146–153.

Mathieu, D., D. Kondziolka, A. Niranjan, J. Flickinger, L.D. Lunsford. (2006). Gamma knife radiosurgery for refractory epilepsy caused by hypothalamic hamartomas. *Stereotact. Funct. Neurosurg.*, 84:82–87.

Mathieu, D., D. Kondziolka, P.B. Cooper, J.C. Flickinger, A. Niranjan, S. Agarwala, J. Kirkwood, L.D. Lunsford. (2007). Gamma knife radiosurgery for malignant melanoma brain metastases. *Clin. Neurosurg.*, 54:241–247.

Mathieu, D., D. Kondziolka, J.C. Flickinger, D. Fortin, B. Kenny, K. Michaud, S. Mongia, A. Niranjan, L.D. Lunsford. (2008). Tumor bed radiosurgery after resection of cerebral metastases. *Neurosurgery*, 62:817–824.

Matsunaga, S., T. Shuto, S. Inomori, H. Fujino, I. Yamamoto. (2007). Gamma knife radiosurgery for intracranial haemangioblastomas. *Acta Neurochir. (Wien)*, 149:1007–1013.

Mayberg, M.R., S. London, J. Rasey, C. Gajdusek. (2000). Inhibition of rat smooth muscle proliferation by radiation after arterial injury: temporal characteristics *in vivo* and *in vitro*. *Radiat. Res.*, 153:153–163.

McCormick, W.F., J.D. Nofzinger. (1966). "Cryptic" vascular malformations of the central nervous system. *J. Neurosurg.*, 24:865–875.

McDonald, C.R., M.A. Norman, E. Tecoma, J. Alksne, V. Iragui. (2004). Neuropsychological change following gamma knife surgery in patients with left temporal lobe epilepsy: a review of three cases. *Epilepsy Behav.*, 5:949–957.

McDonald, C.R., D.J. Hagler, Jr., M.E. Ahmadi, E. Tecoma, V. Iragui, A.M. Dale, E. Halgren. (2008a). Subcortical and cerebellar atrophy in mesial temporal lobe epilepsy revealed by automatic segmentation. *Epilepsy Res.*, 79:130–138.

McDonald, C.R., D.J. Hagler, Jr., M.E. Ahmadi, E. Tecoma, V. Iragui, L. Gharapetian, A.M. Dale, E. Halgren. (2008b). Regional neocortical thinning in mesial temporal lobe epilepsy. *Epilepsia*, 49:794–803.

McIntosh, A.M., R.M. Kalnins, L.A. Mitchell, G.C. Fabinyi, R.S. Briellmann, S.F. Berkovic. (2004). Temporal lobectomy: long-term seizure outcome, late recurrence and risks for seizure recurrence. *Brain*, 127:2018–2030.

Meade, C.E., S.C. Bowden, G. Whelan, M.J. Cook. (2008). Rhinal cortex asymmetries in patients with mesial temporal sclerosis. *Seizure*, 17:234–246.

Moreno-Jimenez, S., M.A. Celis, J.M. Larraga-Gutierrez, J. de Jesus Suarez-Campos, A. Garcia-Garduno, M. Hernandez-Bojorquez. (2007). Intracranial arteriovenous malformations treated with linear accelerator-based conformal radiosurgery: clinical outcome and prediction of obliteration. *Surg. Neurol.*, 67:487–492.

Mori, Y., D. Kondziolka, J. Balzer, W. Fellows, J.C. Flickinger, L.D. Lunsford, K.R. Thulborn. (2000). Effects of stereotactic radiosurgery on an animal model of hippocampal epilepsy. *Neurosurgery*, 46:157–168.

Muacevic, A., B. Wowra, A. Siefert, J.C. Tonn, H.J. Steiger, F.W. Kreth. (2008). Microsurgery plus whole brain irradiation versus gamma knife surgery alone for treatment of single metastases to the brain: a randomized controlled multicentre phase III trial. *J. Neurooncol.*, 87:299–307.

Murphy, M.J. (1985). Long-term follow-up of seizures associated with cerebral arteriovenous malformations: results of therapy. *Arch. Neurol.*, 42:477–479.

Nataf, F., M. Schlienger, Z. Liu, J.N. Foulquier, B. Gres, A. Orthuon, J.M. Vannetzel, B. Escudier, J.F. Meder, F.X. Roux, E. Touboul. (2008). Radiosurgery with or without a 2-mm margin for 93 single brain metastases. *Int. J. Radiat. Oncol. Biol. Phys.*, 70:766–772.

New, P.F., R.G. Ojemann, K.R. Davis, B.R. Rosen, R. Heros, R.N. Kjellberg, R.D. Adams, E.P. Richardson. (1986). MR and CT of occult vascular malformations of the brain. *AJR Am. J. Roentgenol.*, 147:985–993.

O'Connor, M.M., M.R. Mayberg. (2000). Effects of radiation on cerebral vasculature: a review. *Neurosurgery*, 46:138–151.

Oguni, H., A. Olivier, F. Andermann, J. Comair. (1991). Anterior callosotomy in the treatment of medically intractable epilepsies: a study of 43 patients with a mean follow-up of 39 months. *Ann. Neurol.*, 30:357–364.

Pan, H.C., J. Sheehan, M. Stroila, M. Steiner, L. Steiner. (2005). Late cyst formation following gamma knife surgery of arteriovenous malformations. *J. Neurosurg.*, 102(Suppl.):124–127.

Pan, L., E.M. Wang, B.J. Wang, L.F. Zhou, N. Zhang, P.W. Cai, J.Z. Da. (1998). Gamma knife radiosurgery for hemangioblastomas. *Stereotact. Funct. Neurosurg.*, 70(Suppl. 1):179–186.

Pendl, G., H.G. Eder, O. Schroettner, K.A. Leber. (1999). Corpus callosotomy with radiosurgery. *Neurosurgery*, 45(2):303–308.

Polkey, C.E. (2003). Alternative surgical procedures to help drug-resistant epilepsy: a review. *Epileptic Disord.*, 5:63–75.

Pollock, B.E., R.D. Brown, Jr. (2001). Management of cysts arising after radiosurgery to treat intracranial arteriovenous malformations. *Neurosurgery*, 49:259–265.

Pollock, B.E., J.C. Flickinger, L.D. Lunsford, D.J. Bissonette, D. Kondziolka. (1996). Hemorrhage risk after stereotactic radiosurgery of cerebral arteriovenous malformations. *Neurosurgery*, 38:652–661.

Pollock, B.E., J.C. Flickinger, L.D. Lunsford, A. Maitz, D. Kondziolka. (1998). Factors associated with successful arteriovenous malformation radiosurgery. *Neurosurgery*, 42:1239–1247.

Pollock, B.E., Y.I. Garces, S.L. Stafford, R.L. Foote, P.J. Schomberg, M.J. Link. (2000). Stereotactic radiosurgery for cavernous malformations. *J. Neurosurg.*, 93:987–991.

Prayson, R.A., B.J. Yoder. (2007). Clinicopathologic findings in mesial temporal sclerosis treated with gamma knife radiotherapy. *Ann. Diagn. Pathol.*, 11:22–26.

Prise, K.M., M. Pinto, H.C. Newman, B.D. Michael. (2001). A review of studies of ionizing radiation-induced double-strand break clustering. *Radiat. Res.*, 156:572–576.

Rahimi, S.Y., Y.D. Park, M.R. Witcher, K.H. Lee, M. Marrufo, M.R. Lee. (2007). Corpus callosotomy for treatment of pediatric epilepsy in the modern era. *Pediatr. Neurosurg.*, 43:202–208.

Raicu, M., A. Vral, H. Thierens, L. De Ridder. (1993). Radiation damage to endothelial cells *in vitro*, as judged by the micronucleus assay. *Mutagenesis*, 8:335–339.

Regis, J., L. Kerkerian-Legoff, M. Rey, M. Vial, D. Porcheron, A. Nieoullon, J.C. Peragut. (1996). First biochemical evidence of differential functional effects following gamma knife surgery. *Stereotact. Funct. Neurosurg.*, 66(Suppl. 1):29–38.

Regis, J., F. Bartolomei, B. de Toffol, P. Genton, T. Kobayashi, Y. Mori, K. Takakura, T. Hori, H. Inoue, O. Schrottner, G. Pendl, A. Wolf, K. Arita, P. Chauvel. (2000a). Gamma knife surgery for epilepsy related to hypothalamic hamartomas. *Neurosurgery*, 47:1343–1352.

Regis, J., F. Bartolomei, Y. Kida, T. Kobayashi, V. Vladyka, R. Liscak, D. Forster, A. Kemeny, O. Schrottner, G. Pendl. (2000b). Radiosurgery for epilepsy associated with cavernous malformation: retrospective study in 49 patients. *Neurosurgery*, 47:1091–1097.

Regis, J., F. Bartolomei, M. Hayashi, P. Chauvel. (2002). What role for radiosurgery in mesial temporal lobe epilepsy? *Zentralbl. Neurochir.*, 63:101–105.

Regis, J., M. Rey, F. Bartolomei, V. Vladyka, R. Liscak, O. Schrottner, G. Pendl. (2004). Gamma knife surgery in mesial temporal lobe epilepsy: a prospective multicenter study. *Epilepsia*, 45:504–515.

Regis, J., D. Scavarda, M. Tamura, M. Nagayi, N. Villeneuve, F. Bartolomei, T. Brue, D. Dafonseca, P. Chauvel. (2006). Epilepsy related to hypothalamic hamartomas: surgical management with special reference to gamma knife surgery. *Childs. Nerv. Syst.*, 22:881–895.

Regis, J., D. Scavarda, M. Tamura, N. Villeneuve, F. Bartolomei, T. Brue, I. Morange, D. Dafonseca, P. Chauvel. (2007). Gamma knife surgery for epilepsy related to hypothalamic hamartomas. *Semin. Pediatr. Neurol.*, 14:73–79.

Reutens, D.C., A.M. Bye, I.J. Hopkins, A. Danks, E. Somerville, J. Walsh, A. Bleasel, R. Ouvrier, R.A. MacKenzie, J.I. Manson. (1993). Corpus callosotomy for intractable epilepsy: seizure outcome and prognostic factors. *Epilepsia*, 34:904–909.

Rheims, S., C. Fischer, P. Ryvlin, J. Isnard, M. Guenot, M. Tamura, J. Regis, F. Mauguiere. (2008). Long-term outcome of gamma-knife surgery in temporal lobe epilepsy. *Epilepsy Res.*, 80:23–29.

Rogers, L.R., H.H. Morris, K. Lupica. (1993). Effect of cranial irradiation on seizure frequency in adults with low-grade astrocytoma and medically intractable epilepsy. *Neurology*, 43:1599–1601.

Ross, D.A., H.M. Sandler, J.M. Balter, J.A. Hayman, J. Deveikis, D.L. Auer. (2000). Stereotactic radiosurgery of cerebral arteriovenous malformations with a multileaf collimator and a single isocenter. *Neurosurgery*, 47:123–130.

Rossi, G.F., G. Colicchio, E. Marchese, A. Pompucci. (1996). Callosotomy for severe epilepsies with generalized seizures: outcome and prognostic factors. *Acta Neurochir. (Wien)*, 138:221–227.

Sage, M.R., P.C. Blumbergs. (2001). Cavernous haemangiomas (angiomas) of the brain. *Australas. Radiol.*, 45:247–256.

Salvati, M., A. Frati, N. Russo, E. Caroli, F.M. Polli, G. Minniti, R. Delfini. (2003). Radiation-induced gliomas: report of 10 cases and review of the literature. *Surg. Neurol.*, 60:60–67.

Sankaranarayanan, K. (2006). Estimation of the genetic risks of exposure to ionizing radiation in humans: current status and emerging perspectives [review]. *J. Radiat. Res. (Tokyo)*, 47(Suppl. B):B57–B66.

Sankaranarayanan, K., J.S. Wassom. (2005). Ionizing radiation and genetic risks. XIV. Potential research directions in the post-genome era based on knowledge of repair of radiation-induced DNA double-strand breaks in mammalian somatic cells and the origin of deletions associated with human genomic disorders. *Mutat. Res.*, 578:333–370.

Sass, K.J., R.A. Novelly, D.D. Spencer, S.S. Spencer. (1990). Postcallosotomy language impairments in patients with crossed cerebral dominance. *J. Neurosurg.*, 72:85–90.

Schauble, B., G.D. Cascino, B.E. Pollock, D.A. Gorman, S. Weigand, A.A. Cohen-Gadol, R.L. McClelland. (2004). Seizure outcomes after stereotactic radiosurgery for cerebral arteriovenous malformations. *Neurology*, 63:683–687.

Schindler, K., E.R. Christ, T. Mindermann, H.G. Wieser. (2006). Transient MR changes and symptomatic epilepsy following gamma knife treatment of a residual GH-secreting pituitary adenoma in the cavernous sinus. *Acta Neurochir. (Wien)*, 148:903–908.

Schlienger, M., D. Atlan, D. Lefkopoulos, L. Merienne, E. Touboul, O. Missir, F. Nataf, H. Mammar, K. Platoni, P. Grandjean, J.N. Foulquier, J. Huart, C. Oppenheim, J.F. Meder, E. Houdart, J.J. Merland. (2000). LINAC radiosurgery for cerebral arteriovenous malformations: results in 169 patients. *Int. J. Radiat. Oncol. Biol. Phys.*, 46:1135–1142.

Schoggl, A., K. Kitz, A. Ertl, K. Dieckmann, W. Saringer, W.T. Koos. (1998). Gamma-knife radiosurgery for brain metastases of renal cell carcinoma: results in 23 patients. *Acta Neurochir. (Wien)*, 140:549–555.

Schroeder, H.W., M.R. Gabb, U. Runge. (1996). Supratentorial cavernous angiomas and epileptic seizures: preoperative course and postoperative outcome. *Neurosurgery*, 39:1271.

Schrottner, O., H.G. Eder, F. Unger, K. Feichtinger, G. Pendl. (1998). Radiosurgery in lesional epilepsy: brain tumors. *Stereotact. Funct. Neurosurg.*, 70(Suppl. 1):50–56.

Schrottner, O., F. Unger, H.G. Eder, M. Feichtinger, G. Pendl. (2002). Gamma-knife radiosurgery of mesiotem-poral tumour epilepsy observations and long-term results. *Acta Neurochir. Suppl.*, 84:49–55.

Selch, M.T., A. Gorgulho, C. Mattozo, T.D. Solberg, C. Cabatan-Awang, A.A. DeSalles. (2005). Linear accel-erator stereotactic radiosurgery for the treatment of gelastic seizures due to hypothalamic hamartoma. *Minim. Invasive Neurosurg.*, 48:310–314.

Sheehan, J., C.P. Yen, L. Steiner. (2006). Gamma knife surgery-induced meningioma: report of two cases and review of the literature. *J. Neurosurg.*, 105:325–329.

Shenouda, G., L. Souhami, E.B. Podgorsak, J.P. Bahary, J.G. Villemure, J.L. Caron, G. Mohr. (1997). Radiosurgery and accelerated radiotherapy for patients with glioblastoma. *Can. J. Neurol. Sci.*, 24:110–115.

Shih, Y.H., D.H. Pan. (2005). Management of supratentorial cavernous malformations: craniotomy versus gamma knife radiosurgery. *Clin. Neurol. Neurosurg.*, 107:108–112.

Shim, K.W., J.H. Chang, Y.G. Park, H.D. Kim, J.U. Choi, D.S. Kim. (2008). Treatment modality for intractable epilepsy in hypothalamic hamartomatous lesions. *Neurosurgery*, 62:847–856.

Shimizu, H. (2005). Our experience with pediatric epilepsy surgery focusing on corpus callosotomy and hemi-spherotomy. *Epilepsia*, 46(Suppl. 1):30–31.

Singh, V.P., S. Kansai, S. Vaishya, P.K. Julka, V.S. Mehta. (2000). Early complications following gamma knife radiosurgery for intracranial meningiomas. *J. Neurosurg.*, 93(Suppl. 3):57–61.

Sirin, S., D. Kondziolka, A. Niranjan, J.C. Flickinger, A.H. Maitz, L.D. Lunsford. (2006). Prospective staged volume radiosurgery for large arteriovenous malformations: indications and outcomes in otherwise untreatable patients. *Neurosurgery*, 58:17–27.

Smith, J.R., D.W. King, Y.D. Park, M.R. Lee, G.P. Lee, P.D. Jenkins. (2000). Magnetic source imaging guid-ance of gamma knife radiosurgery for the treatment of epilepsy. *J. Neurosurg.*, 93(Suppl. 3):136–140.

Spencer, S.S. (1988). Corpus callosum section and other disconnection procedures for medically intractable epilepsy. *Epilepsia*, 29(Suppl. 2):S85–S99.

Spencer, S.S., A.T. Berg, B.G. Vickrey, M.R. Sperling, C.W. Bazil, S. Shinnar, J.T. Langfitt, T.S. Walczak, S.V. Pacia. (2005). Predicting long-term seizure outcome after resective epilepsy surgery: the multicenter study. *Neurology*, 65:912–918.

Srikijvilaikul, T., I. Najm, N. Foldvary-Schaefer, T. Lineweaver, J.H. Suh, W.E. Bingaman. (2004). Failure of gamma knife radiosurgery for mesial temporal lobe epilepsy: report of five cases. *Neurosurgery*, 54:1395–1404.

Stefan, H., C. Hummel, G.G. Grabenbauer, R.G. Muller, S. Robeck, W. Hofmann, M. Buchfelder. (1998). Successful treatment of focal epilepsy by fractionated stereotactic radiotherapy. *Eur. Neurol.*, 39:248–250.

Steinberg, G.K., J.I. Fabrikant, M.P. Marks, R.P. Levy, K.A. Frankel, M.H. Phillips, L.M. Shuer, G.D. Silverberg. (1990). Stereotactic heavy-charged-particle Bragg-peak radiation for intracranial arteriovenous malfor-mations. *N. Engl. J. Med.*, 323:96–101.

Steiner, L., C. Lindquist, J.R. Adler, J.C. Torner, W. Alves, M. Steiner. (1992). Clinical outcome of radiosurgery for cerebral arteriovenous malformations. *J. Neurosurg.*, 77:1–8.

Stern, R.L., J.R. Perks, C.T. Pappas, J.E. Boggan, A.Y. Chen. (2008). The option of LINAC-based radiosurgery in a gamma knife radiosurgery center. *Clin. Neurol. Neurosurg.*, 110:968–972.

Sun, B., A.A. DeSalles, P.M. Medin, T.D. Solberg, B. Hoebel, M. Felder-Allen, S.E. Krahl, R.F. Ackermann. (1998). Reduction of hippocampal-kindled seizure activity in rats by stereotactic radiosurgery. *Exp. Neurol.*, 154(2):691–695.

Sutcliffe, J.C., D.M. Forster, L. Walton, P.S. Dias, A.A. Kemeny. (1992). Untoward clinical effects after stereot-actic radiosurgery for intracranial arteriovenous malformations. *Br. J. Neurosurg.*, 6:177–185.

Swinson, B.M., W.A. Friedman. (2008). Linear accelerator stereotactic radiosurgery for metastatic brain tumors: 17 years of experience at the University of Florida. *Neurosurgery*, 62:1018–1032.

Szeifert, G.T., W.R. Timperley, D.M. Forster, A.A. Kemeny. (2007). Histopathological changes in cerebral arte-riovenous malformations following gamma knife radiosurgery. *Prog. Neurol. Surg.*, 20:212–219.

Szumiel, I. (2008). Intrinsic radiation sensitivity: cellular signaling is the key. *Radiat. Res.*, 169:249–258.

Takeshita, G., H. Toyama, K. Nakane, M. Nomura, H. Osawa, Y. Ogura, K. Katada, A. Takeuchi, S. Koga, Y. Kato. (1994). Evaluation of regional cerebral blood flow changes on perifocal brain tissue SPECT before and after removal of arteriovenous malformations. *Nucl. Med. Commun.*, 15:461–468.

Thomas, J.W., C. LaMantia, T. Magnuson. (1998). X-ray-induced mutations in mouse embryonic stem cells. *Proc. Natl. Acad. Sci. U.S.A.*, 95:1114–1119.

Tsien, C., L. Souhami, A. Sadikot, A. Olivier, R. del Carpio-O'Donovan, R. Corns, H. Patrocinio, W. Parker, E. Podgorsak. (2001). Stereotactic radiosurgery in the management of angiographically occult vascular malformations. *Int. J. Radiat. Oncol. Biol. Phys.*, 50(1):133–138.

Tu, J., M.A. Stoodley, M.K. Morgan, K.P. Storer. (2006). Responses of arteriovenous malformations to radiosurgery: ultrastructural changes. *Neurosurgery*, 58:749–758.

Turjman, F., T.F. Massoud, J.W. Sayre, F. Vinuela, G. Guglielmi, G. Duckwiler. (1995a). Epilepsy associated with cerebral arteriovenous malformations: a multivariate analysis of angioarchitectural characteristics. *AJNR Am. J. Neuroradiol.*, 16:345–350.

Turjman, F., T.F. Massoud, F. Vinuela, J.W. Sayre, G. Guglielmi, G. Duckwiler. (1995b). Correlation of the angioarchitectural features of cerebral arteriovenous malformations with clinical presentation of hemorrhage. *Neurosurgery*, 37:856–862.

Unger, F., O. Schrottner, M. Feichtinger, G. Bone, K. Haselsberger, B. Sutter. (2002). Stereotactic radiosurgery for hypothalamic hamartomas. *Acta Neurochir. Suppl.*, 84:57–63.

Urlaub, G., P.J. Mitchell, E. Kas, L.A. Chasin, V.L. Funanage, T.T. Myoda, J. Hamlin. (1986). Effect of gamma rays at the dihydrofolate reductase locus: deletions and inversions. *Somat. Cell. Mol. Genet.*, 12:555–566.

Weber, B., E. Luders, J. Faber, S. Richter, C.M. Quesada, H. Urbach, P.M. Thompson, A.W. Toga, C.E. Elger, C. Helmstaedter. (2007). Distinct regional atrophy in the corpus callosum of patients with temporal lobe epilepsy. *Brain*, 130:3149–3154.

Whang, C.J., Y. Kwon. (1996). Long-term follow-up of stereotactic gamma knife radiosurgery in epilepsy. *Stereotact. Funct. Neurosurg.*, 66(Suppl. 1):349–356.

Wieser, H.G., M. Ortega, A. Friedman, Y. Yonekawa. (2003). Long-term seizure outcomes following amygdalohippocampectomy. *J. Neurosurg.*, 98:751–763.

Wolf, H.K., D. Roos, I. Blumcke, T. Pietsch, O.D. Wiestler. (1996). Perilesional neurochemical changes in focal epilepsies. *Acta Neuropathol.*, 91:376–384.

Wong, J.H., I.A. Awad, J.H. Kim. (2000). Ultrastructural pathological features of cerebrovascular malformations: a preliminary report. *Neurosurgery*, 46:1454–1459.

Wood, K.A., R.J. Youle. (1995). The role of free radicals and p53 in neuron apoptosis *in vivo*. *J. Neurosci.*, 15:5851–5857.

Yeh, H.S., M.D. Privitera. (1991). Secondary epileptogenesis in cerebral arteriovenous malformations. *Arch. Neurol.*, 48:1122–1124.

Yeh, H.S., S. Kashiwagi, J.M. Tew, Jr., T.S. Berger. (1990). Surgical management of epilepsy associated with cerebral arteriovenous malformations. *J. Neurosurg.*, 72:216–223.

Yoon, H.H., H.L. Kwon, R.H. Mattson, D.D. Spencer, S.S. Spencer. (2003). Long-term seizure outcome in patients initially seizure-free after resective epilepsy surgery. *Neurology*, 61:445–450.

You, Y., R. Bergstrom, M. Klemm, B. Lederman, H. Nelson, C. Ticknor, R. Jaenisch, J. Schimenti. (1997). Chromosomal deletion complexes in mice by radiation of embryonic stem cells. *Nat. Genet.*, 15:285–288.

Yuan, J., J.Z. Wang, S. Lo, J.C. Grecula, M. Ammirati, J.F. Montebello, H. Zhang, N. Gupta, W.T. Yuh, N.A. Mayr. (2008). Hypofractionation regimens for stereotactic radiotherapy for large brain tumors. *Int. J. Radiat. Oncol. Biol. Phys.*, 72:390–397.

Zabramski, J.M., J.S. Henn, S. Coons. (1999). Pathology of cerebral vascular malformations. *Neurosurg. Clin. N. Am.*, 10:395–410.

# 19 Deep Brain Stimulation for Epilepsy

*Francisco Velasco, Ana Luisa Velasco, Luisa Rocha,
Marcos Velasco, Jose D. Carrillo-Ruiz, Manoela Cuellar,
Guillermo Castro, Fiacro Jimenez, and David Trejo*

## CONTENTS

## INTRODUCTION

Electrical stimulation of central nervous system structures for the treatment of chronic neurologic disorders started over 40 years ago (Velasco, 2000). Electric current is delivered through electrodes placed in specific areas and generated by internalized pulse generators (IPGs), powered by high-performance batteries and connected to electrodes by extension cables. There are two types of electrodes. One consists of 4 to 12 rounded or square platinum contacts and wires embedded in a plate of silicon used for cortical, cerebellar, and spinal cord stimulation. The other type is tetrapolar; 1.5-mm-length individual platinum rings mounted on silicon tubing and separated from each other by a distance of 0.5 to 1.5 mm. This type of electrode is stereotactically implanted into deep brain structures, and the process is referred to as *deep brain stimulation* (DBS). IPGs are usually implanted in a subcutaneous subclavicular or abdominal, surgically created pouch, and an extension cable is passed from this place through a subcutaneous tunnel to meet the proximal (or extracranial) end of the electrode. IPGs are programmed through an external device that uses radiofrequency to read and adjust an IPG program. Pulse frequency (Hz), amplitude (V or mA), and duration (µs) are readily adjusted within a range approved for safety. The possibility of using several combinations of contacts acting as cathodes or anodes in bipolar stimulation or cathodes in monopolar stimulation provides the clinician with multiple options of extending or restricting the stimulated area to optimize the effect on individual cases.

Stimulation can be continuous or cycling. Cycling may be set to deliver current for fixed periods of time, or the patient can use a programmer to turn the current on or off as needed.

Correct placement of electrodes is essential for successful treatment and to avoid undesirable side effects; therefore, implanting electrodes has become a multidisciplinary procedure where neurosurgeons share responsibilities with imaging specialists, neurophysiologists, neurologists, and biomedical engineers. The use of stereotaxis coupled with neuronavigation systems, where virtual trajectories for electrodes designed on the basis of magnetic resonance imaging (MRI) fusion, computed tomography (CT), functional MRI (fMRI), and stereotactic atlases of human brain, increases the precision and safety of surgical procedures. Transoperative confirmation of the correct placement using evoked responses, microelectrode recording, and deep brain stimulation mapping is used routinely in stereotactic procedures. Sometimes, a therapeutic trial of stimulation with temporarily externalized electrodes is necessary before deciding on internalization of the stimulation system. Of course, imaging confirmation of electrode location is also imperative and nowadays is often carried out with intraoperative MRI facilities.

Although the mechanisms of action of electrical stimulation are still far from being fully understood, certain procedures have been adopted that have proven to work consistently. For example, low-frequency stimulation is performed to enhance the neuronal or fiber activity of stimulated structures, whereas high-frequency stimulation often mimics the effects of lesions of stimulated targets. This lesion-like effect is reversible, as it tends to disappear when stimulators are turned off. Recently, this dual effect of electric current on neurologic symptoms, related to the frequency of the delivered current, has been demonstrated for tremor amplitude (Birdno and Grill, 2008). Because electric current may enhance or inhibit neural activity depending on the parameters of stimulation and the stimulated site, there is a tendency to refer to the technique as *neuromodulation*, because stimulation may mistakenly be considered to be equivalent to excitation.

# NEUROBIOLOGICAL BACKGROUND FOR TARGETS PROPOSED FOR SEIZURE CONTROL

## CEREBELLAR CORTEX

In 1973, Irving Cooper published the first report on neuromodulation in the treatment of epilepsy (Cooper et al., 1973). The proposal was based on experimental observations indicating that low-frequency stimulation (LFS) (10 Hz) of paravermian cerebellar cortex decreased paroxysmal electroencephalography (EEG) discharges and seizures induced by electric shocks (Moruzzi, 1941) or cobalt powder (Dow et al., 1962) applied on the motor cortex in cats. Iawata and Snider (1959) extended the studies to the hippocampus and demonstrated that cerebellar stimulation (CS) could stop seizures and prolonged afterdischarges induced by hippocampal electrical stimulation.

Chronic CS in monkeys indicated that current intensity with charge densities below 7.4 $\mu$C/cm$^2$/phase did not induce neuronal damage in the stimulated area (Brown et al., 1977). Such charge density is 4 to 5 times that required to induce cerebellar efferent activity in monkeys (Babb et al., 1977). On the other hand, in animal models of epilepsy, the anticonvulsant effect induced by CS is closely related to stimulation parameters (Ebner et al., 1980). Moreover, electrical stimulation on experimental epileptic foci in monkeys within the range of 3.0 to 4.0 $\mu$C/cm$^2$/phase may decrease rather than increase epileptic activity in contrast with density charges above 10.0 $\mu$C/cm$^2$/phase. Finally, stimulation of the vermis and intermediate cortex on the superomedial cerebellar surface is more effective at controlling epileptic activity in animals, especially that originating in limbic areas (Laxer et al., 1980). Under normal conditions, paravermian cortex inhibits deep brain cerebellar nuclei activity, which facilitates both cortical excitability and the spinal cord monosynaptic reflex. CS of this cortex has been successfully used to treat epilepsy and spasticity (Cooper, 1976).

## CENTROMEDIAN AND ANTERIOR THALAMIC NUCLEI

Intralaminar and midline thalamic nuclei participate in the genesis and propagation of epileptic seizures, according to clinical (Penfield and Jasper, 1954; Velasco et al., 1993a) and experimental (Hunter and Jasper, 1949; Miller and Ferrendelli, 1990; Pollen et al., 1963) observations. They are also anatomically linked with brain-stem structures involved in seizure onset (Velasco and Velasco, 1990). Although the controversy regarding the cortical vs. subcortical origin of epileptic attacks remains unsolved, there is a general consensus that thalamocortical interactions are essential in the development and propagation of many types of seizures (Avoli et al., 1983; Gloor et al., 1977; Pollen et al., 1963; Velasco et al., 1982). High-frequency stimulation (HFS) in the centromedial thalamic nucleus (CM), which is part of the intralaminar thalamic nuclei, tends to interfere with seizure propagation (Velasco et al., 1987).

On the other hand, pentylenetetrazol (PTZ)-induced seizures in rats and guinea pigs were accompanied by hypermetabolism of the anterior hypothalamus, mammillary bodies, mammillothalamic tract, and anterior thalamic nucleus (which is part of the intralaminar nuclei), as seen by autoradiography (Mirski and Ferrendelli, 1986). HFS of the anterior thalamic nucleus (AN) interrupts PTZ-induced generalized EEG seizures in rats (Mirski et al., 1997). Therefore, AN stimulation was proposed to treat epileptic seizures in humans. This target would have the advantage over CM stimulation by interfering with propagation of epileptic attacks originating in the mesial temporal lobes and propagating via the fornix–mammillothalamic–cingulate circuit (Fisher, 2004).

## HIPPOCAMPUS, PARAHIPPOCAMPAL CORTEX, AND CORTICAL ELOQUENT AREAS

The kindling model of epilepsy consists of progressive enhancement of neuronal activity by delivering electric current of low amplitude and duration in consecutive daily sessions. If, during the development of kindling, continuous low-amplitude (15 μA) direct current (DC) is applied through the same electrode, kindling is disrupted and the threshold for afterdischarges increases. Disruption of epileptogenesis is known as *quenching* (Weiss et al., 1995).

A trial of electrical stimulation of the epileptic focus was conducted on a group of patients who were candidates for anterior temporal lobectomy for seizure control and were being studied with intracranial subdural or intrahippocampal electrodes to define the place and extension of the epileptic zone. All patients had discontinued anticonvulsant medication to allow the occurrence of spontaneous seizures and determine the site of ictal activity onset. Bipolar HFS at amplitudes from 150 to 300 μA were delivered at the epileptic foci over a period of 15 days. Significant decreases in interictal EEG spike activity and arrest of spontaneous seizures occurred after 5 to 7 days of stimulation (Velasco et al., 2000b). Patients were thereafter operated upon; the surgical specimens were studied morphologically and for different neurotransmitter receptor content by autoradiography. Patients with better antiepileptic response to HFS had higher γ-aminobutyric acid (GABA) concentrations and lower rates of cell loss in the hippocampus and parahippocampus (mesial temporal sclerosis). Quantification of receptor binding in those patients revealed lower mu, delta, and nociceptin-binding, whereas $GABA_A$ and benzodiazepine receptor binding were not modified. Results suggest that HFS efficacy is associated with high GABA tissue content and low opioid receptor binding (Cuellar-Herrera et al., 2004). These findings extrapolate to extratemporal foci (Ondarza et al., 2002).

On the other hand, HFS at the hippocampus or parahippocampus over 15 days increased the threshold for afterdischarges tenfold, which, in contrast to the prestimulation period, did not propagate to other areas. Moreover, single photon emission computed tomography (SPECT) studies after stimulation showed decreased regional cerebral blood flow (rCBF) at the stimulation site. These findings suggest that HFS induces neuronal inhibition (Velasco et al., 2004).

## CAUDATE NUCLEUS

Stimulation of the caudate nucleus (Cd) decreases EEG spike activity in hippocampal epileptic foci induced by local penicillin injection in cats and alumina cream in monkeys (Oakley and Ojemann, 1982). In humans with electrodes implanted in the Cd, LFS induced decreases in cortical paroxysmal EEG discharges, whereas HFS increased them (Sramka et al., 1980).

## SUBTHALAMIC NUCLEUS

Substantia nigra reticulata (SNr) has been recognized as "a key site of GABAergic drugs for the control of generalized seizures" (Iadarola and Gale, 1982). Enhancement of SNr GABAergic activity has anticonvulsant effect in several models of experimental epilepsy (Frye et al., 1983; Garant and Gale, 1986; Gonzalez and Hettinger, 1984; McNamara et al., 1984). On the other hand, SNr and globus pallidum internus (GPi) are under the control of subthalamic nucleus (STN) afferents. STNs in monkeys exert an inhibitory influence on SNr, directly or indirectly through GPi. HFS of the STN has been used in the control of motor symptoms of Parkinson's disease and mimics the effects of lesions of the same structure (Limousin et al., 1998). HFS of STN in rats excites SNr and GPi neurons which may be the mechanism by which it mediates anticonvulsant effects (Benazzouz et al., 1995).

## POSTERIOR HYPOTHALAMUS AND MAMMILLOTHALAMIC TRACT

High-frequency stimulation of the posterior hypothalamus increases the threshold for PTZ-induced seizures in rats (Mirski and Fisher, 1994), and, as mentioned above, PTZ increases the metabolic activity of the posterior hypothalamus, mammillothalamic tract, and anterior thalamus (Mirski and Ferrendelli, 1986). Recent attempts have been made to control mesial-temporal-initiated seizures through hypothalamic stimulation. Figure 19.1 summarizes the neuronal circuits involved in seizure generation and propagation that have been used as targets for electrical stimulation.

# CLINICAL TRIALS

## CEREBELLAR STIMULATION

Cooper et al. (1973) reported a series of 34 patients with epilepsy being treated with CS who showed a decrease of over 50% in the number of seizures in 18. Between 1976 and 1992, about 200 patients were reported to have been treated for seizure control using dorsal paravermian cerebellar stimulation (Cooper et al., 1976; Davis and Emmonds, 1992; Levy and Auchterloni, 1979). Results in open-label studies were excellent in 27% of the patients, who demonstrated a greater than 80% decrease in seizures, with many becoming seizure free. Improvement was greater than 50% in 53% of the patients, and only 20% were nonresponders.

The technique of implanting electrodes is relatively simple in that two burr holes are placed in the occipital bone just below the transverse sinus and 2.5 cm at either side of the midline. Four contact plate electrodes are inserted through the burr holes under fluoroscopic control, to the left of each side of the vermis (Figure 19.2A). The electrodes are connected to IPGs placed in the abdominal wall, and cycling LFS at 10 to 20 Hz is delivered.

Two double-blind controlled studies reported no differences in seizure occurrence with the stimulators on or off, thereby casting doubt on the efficacy of CS (Van Buren et al., 1978; Wright et al., 1984). However, in a recent study, double blinding was used at the beginning of the period of stimulation to avoid any anticonvulsant residual effect that may result from chronic electrical stimulation (see below). During the double-blind period, a significant difference was observed between those cases being stimulated and the ones that remained off stimulation. Generalized

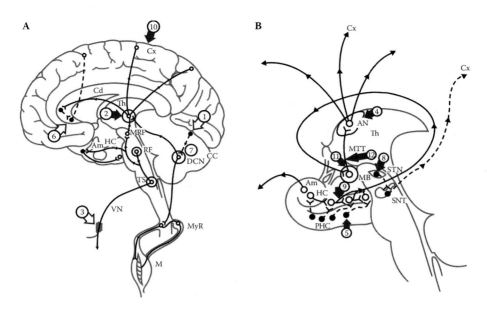

**FIGURE 19.1** Diagrammatic representation of neuronal circuits involved in genesis and propagation of seizures that have been targeted with neuromodulation. Clear circles and continuous lines ending in synaptic terminals indicate excitatory circuits for seizure genesis. Dark circles and discontinuous lines ending on synaptic terminals indicate inhibitory circuitry. Clear arrows point to the places stimulated with low-frequency pulses (excitatory). Dark arrows point to the places stimulated with high-frequency pulses (inhibitory). (1) Low-frequency stimulation (LFS) of cerebellar cortex (CC) inhibits deep cerebellar nuclei (DCN) facilitation of cortical excitability (Cx) and spinal cord monosynaptic reflex (MyR). (2) High-frequency stimulation (HFS) of centromedial thalamic (CM) nucleus interferes with the propagation of cortical and subcortically initiated seizures. (3) Vagus nerve (VN) LFS antidominantly activates the solitary tract (TS) nucleus, which in turn connects with the raphe nucleus (RF), mesencephalic reticular formation (MRF), thalamus (Th), and temporal lobe amygdala (Am). (4) Anterior thalamic nuclei (AN) HFS interferes with seizures propagated through the thalamus and in particular with mesial temporal lobe-initiated seizures through mammillary body (MB) and mammillothalamic tract (MTT). (5) HFS of parahippocampal cortex (PHC) inhibits hippocampal neurons through a GABA receptor-mediated mechanism. (6) LFS of the caudate nucleus (Cd) inhibits cortical excitability. (7) DCN–HFS inhibits cortical excitability through the same mechanism as CS. (8) Subthalamic nucleus (STN) HFS stimulation inhibits local neurons and liberates substantia nigra reticulata (SNr) to exert GABA-mediated inhibition on cortical excitability. (9) HFS of the hippocampus (HC) inhibits mesial temporal lobe-initiated seizures. (10) HFS cortical stimulation inhibits epileptic foci.

tonic–clonic seizures (GTCSs) and tonic seizures (TSs) decreased 76% and 68%, respectively, after a 2-year follow-up period (Velasco et al., 2005). In another report of CS, stimulation was directed to deep cerebellar nuclei that, as mentioned above, facilitate cortical excitability; therefore, HFS (>50 Hz) is used to inhibit neuronal activity in those nuclei with similar anticonvulsive results (Chkhenkeli et al., 2004).

## CENTROMEDIAN AND ANTERIOR THALAMIC NUCLEI STIMULATION

In 1985, there were reports of seizure control by stimulating the mesothalamic reticular area, which corresponds to the centromedial thalamic nucleus (CM) and upper part of the posterior subthalamus (Andy and Jurko, 1985; 1986). Centromedial thalamic stimulation (CMS) using HFS tends to interfere with propagation of seizures initiating in the cortex and with other seizures probably originating in subcortical structures or through corticothalamic interplay (Miller and Ferrendelli, 1990; Velasco et al., 1993a). Doubts about initially promising results (Velasco et al., 1987) were raised by

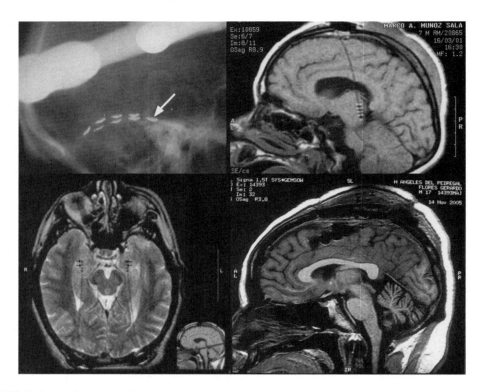

**FIGURE 19.2**    (A) Intraoperative x-ray showing the supracerebellar parasagittal placement of tetrapolar plate electrodes. (B) Placement of deep brain stimulation (DBS) tetrapolar electrodes in the CM parvocellular subnucleus. (C) Bilateral DBS in the hippocampus to treat difficult-to-control, temporal-lobe-initiated seizures with independent bilateral foci. (D) Plate electrode placed in supplementary motor cortex to treat nonlesional seizures initiated in this area.

a controlled, double-blind study in which no differences in seizure number were found between on and off periods (Fisher et al., 1992). This was confirmed by another double-blind study by Velasco et al. (2000a).

In both protocols, however, the double-blind maneuver was established after a period of 6 to 9 months of the stimulation being on, when seizures had already decreased to 80% of the baseline level and did not increase during the off stimulation period. This raised the suspicion that chronic CMS had left a residual anticonvulsive effect that outlasted the stimulation period by several months. Figure 19.3 shows the follow-up of patients treated with CMS for many years; IPG battery depletion is accompanied by a slight increase in the number of seizures that represent no more than 20% of the pre-CMS number. Seizures decreased again as soon as the batteries were replaced and stimulation restarted. No fluctuation in the number of seizures was seen while stimulation was maintained. This indicates that CMS is responsible for seizure control.

So far, over 62 patients have been treated with CMS, which appears to be safe and effective in controlling GTCSs, atypical absences (AAs), and partial motor seizures (>80%) and less effective in complex partial (CxP) and adversive (Adv) seizures (±50%) (Andy and Jurko, 1985; Velasco et al., 1993b, 2000a, 2002). Recently, we have restricted CMS to cases with GTCSs and AAs, particularly those with Lennox–Gastaut syndrome who demonstrated an overall seizure reduction of 83%, with 15.4% being seizure free (Velasco et al., 2006b).

Centromedial thalamic stimulation has been performed in the cycling mode, with 1 minute on in one side, a 4-minute interval, 1 minute on in the opposite side, a 4-minute interval, etc., 24 hours a day. Placement of electrodes is stereotactically oriented to leave the tip of tetrapolar DBS electrodes

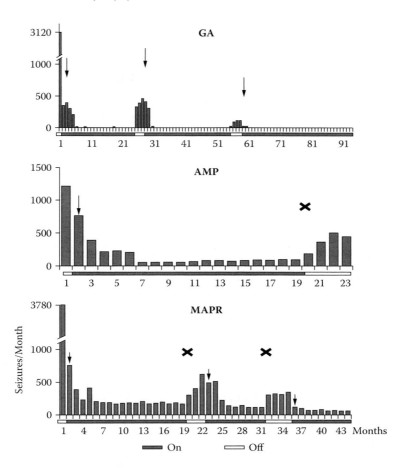

**FIGURE 19.3** (A) Long-term follow up of two cases with Lennox–Gastaut syndrome being treated by centromedial thalamic stimulation (GA and MAPR). The initial (baseline) number of seizures (thousands per month) decreased progressively to reach maximum decrement in 5 to 6 months and remain stable until the battery power of the internalized pulse generator (IPG) depletes. It then increases to reach about 20% of the baseline value. When stimulation restarts after replacing the IPGs (arrows), the number of seizures decreases rapidly to reach maximal improvement. (B) Graph representing the follow-up of a patient (AMP) that had to be explanted due to skin erosions. Seizures increased slowly to reach 30% baseline level after several months off stimulation. (From Velasco, A.L. et al., *Epilepsia*, 47(7):1203–1212, 2006. With permission.)

in contact with the posterior commissure at a 60° angle with regard to the anterior–posterior commissure (AC–PC) (Figure 19.2B). Correct positioning of electrodes within the most anterior and lateral part of the nucleus is accomplished by eliciting typical cortical recruiting responses using 6- to 8-Hz bipolar stimulation between adjacent DBS contacts (Velasco et al., 1997).

In the 1980s, stimulating the anterior thalamic nucleus (AN) of a patient was reported to provide good control of GTC and AA seizures (Upton et al., 1985). At that time, however, targeting of thalamic nuclei other than the ventrolateral and ventroposterior was rudimentary.

Preliminary reports of anterior thalamic nucleus stimulation (ANS) for seizure control provides evidence for the safety of the method; however, seizure control varies from 30 to 76% for GTC and CxP seizures and 62% for partial sensory seizures (Kerrigan et al., 2004; Lim et al., 2007). A group of 100 patients is currently being evaluated in a controlled, double-blind, multicenter protocol of ANS for seizure control. Targeting AN uses recruiting responses as well as microelectrode recordings to determine the entrance into the dorsal thalamus from the lateral ventricle.

## Caudate Stimulation

Low-frequency stimulation of the ventral part of the caudate nucleus was observed to decrease spontaneous epileptic cortical activity in humans (Chkhenkeli, 1978), leading to a clinical trial of chronic caudate nucleus stimulation (CdS). In a group of 38 patients followed for an average of 15 months, GTC seizures decreased >80% and CxP seizures decreased >50% (Chkhenkeli et al., 2004). LFS (4 to 5 Hz) at 200 μsec and 4.0 to 6.0 V in cycling mode, 5 minutes on/5 minutes off, were used to induce excitation of caudate neurons.

## Hippocampus Stimulation

The clinical, neurophysiological, imaging, and neurochemical evidence that HFS inhibits epileptogenic foci in the hippocampus and parahippocampus (Velasco et al., 2000b) was followed by a clinical trial using chronic stimulation with internalized systems. The first report was a case with bilateral independent hippocampal foci, as evidenced by depth EEG recordings. Simultaneous continuous stimulation in both hippocampi at 3.5 V, 130 Hz, 450 μsec, resulted in an immediate seizure reduction, and the patient became seizure free after 1 year on stimulation and has remained so (Velasco et al., 2004).

Two recent publications report promising results. One was an 18-month follow up of 9 patients with significant decreases in CxP and secondary GTC seizures, with an overall improvement of 70%; 3 patients were seizure free (Velasco et al., 2007). In that report, there was a clear difference in response between cases with normal MRI and those with mesial temporal sclerosis. Although the former had an immediate response and 3/5 patients became seizure free, the ones with mesial temporal sclerosis had a delayed response, and the decrement in seizures was less marked, with none becoming seizure free. The other report (Boon et al., 2007) included 10 patients with follow-up from 12 to 52 months. One patient was seizure free, and the degree of improvement in the others varied from >90% (2 patients) to >50% (5 patients) to >30% (2 patients), with one nonresponder.

The difference between results in these two reports may be that, in the former study, exploratory eight-contact electrodes were used to define precisely the location and extension of epileptogenic area, while in the other study electrodes were directed to the most anterior part of the hippocampus, which often, but not always, corresponds to the epileptogenic zone.

Another trial of hippocampal stimulation for patients with mesial temporal lobe sclerosis and epilepsy reported a significant but modest decrease in seizures (Tellez-Zenteno et al., 2006). In this study, epileptic foci were stimulated with 1.0 to 3.0 μC/cm²/phase, which is considerably less than the charge density used in other trials. Patients were followed for 6 months. As mentioned previously, such patients have a delayed response, possibly due to the fact that sclerotic tissue with higher impedance requires larger charge densities than nonsclerotic tissue.

Electrodes have also been implanted through a transoccipital approach, placing all electrode contacts along the hippocampal axis. Implantation, in this case, is guided by EEG recordings (Figure 19.1C) and stimulation mode is cycling with a program similar to CMS. It is important to remark that even bilateral simultaneous HCS has not induced decreases in memory, as evidenced by neuropsychological testing.

## Extratemporal Foci

Epileptic foci located in eloquent areas and in particular those without evidence of cortical lesions or dysplasias have been considered poor candidates for ablative procedures. The antiepileptic effect induced with low-amplitude HFS in both hippocampus and parahippocampal cortex, without interfering with primary functions, led to stimulation of eloquent areas. So far, six epileptic foci have been stimulated in five patients (four in primary motor cortex for epilepsia partialis continua and two in supplementary motor cortex). Those cases have been followed for over one year with a >80% reduction in GTC and partial motor seizures and >50% reduction in partial sensory seizures. No deleterious effects have been detected in motor or sensory functions. Cortical areas are stimulated through tetrapolar plate electrodes, and the placement is oriented by fMRI and brain mapping (Figure 19.1D) (Velasco et al., 2006a, 2008). In one case report, subdural stimulation of five epileptic foci was used to successfully treat a life-threatening status epilepticus (Schreder et al., 2006).

A protocol of stimulation of epileptic foci using stimulating devices coupled with detectors of EEG signals that may anticipate the onset of seizures is now under evaluation. EEG signals would trigger stimulation at the precise moment that a seizure occurs (Lesser et al., 1999). One still unsolved problem with these devices is the lack of specific EEG signals to anticipate seizure occurrence. In fact, either desynchronization or synchronization may occur spontaneously in normal conditions.

Experience shows that the antiepileptic effect induced by electrical stimulation takes time to reach its maximal effect. For hippocampal stimulation, it can take weeks; for centromedial thalamic stimulation, it takes months; and for cerebellar stimulation and vagus nerve stimulation it takes years.

## SUBTHALAMIC NUCLEUS SIMULATION

Subthalamic nucleus simulation (STNS) was proposed as an alternative for difficult-to-control seizures in diverse epileptic syndromes. It has been tested in small groups of patients in France and the United States (Chabardes et al., 2002; Handforth et al., 2006). All protocols included double-blind procedures and demonstrated seizure decrements related to STNS; however, the degree of improvement varied considerably from one study to the other. Seizures decreased from 30% (Handforth et al., 2006) to over 80% (Chabardes et al., 2002), with an average of 64%. Complex partial, partial motor, and myoclonic seizures responded best to STNS (Chabardes et al., 2008). Targeting the STN is well established from its extensive use in treating Parkinson's disease and couples imaging fusion (CT–MRI and ventriculography) with an anatomical atlas, using microelectrode recording and microstimulation as coadjutors in the procedure.

Figure 19.1 indicates the locations where the circuits for genesis and propagation of seizures have been targeted with neuromodulation. Table 19.1 summarizes the trials that better define the efficacy and stimulation parameters used for each target. In summary, we have 35 years of experience in the use of neuromodulation for seizure control. So far, the technique has been safe.

As far as efficacy, most open-label studies have reported significant decreases in seizure occurrence. Double-blind studies have questioned some of the results; however, it has been proven that chronic electrical stimulation induces changes in neurotransmitters that may extend the anticonvulsant effect beyond the period of on stimulation. When stimulation is discontinued, it takes months to years for the seizures to increase, and they may never reach the initial baseline level. Protocols must incorporate double blinding at the onset of stimulation period and avoid crossovers.

Most important is the evidence that, with the stimulation parameters used for therapeutic purposes, electrical stimulation does not deteriorate normal function. On the contrary, it may improve neuropsychological performance; this is true even when bilateral symmetric epileptic foci in eloquent areas are being stimulated. Therefore, neuromodulation of bilateral foci has a definite advantage over bilateral ablative procedures.

**TABLE 19.1**
**Targets Used for Neuromodulation in the Treatment of Epilepsy**

| Target | No. of Patients | Stimulation | | Effect on Different Seizure Types | | | | | | | Follow-Up (months) | Refs. |
|---|---|---|---|---|---|---|---|---|---|---|---|---|
| | | Parameters | Mode | GTC | AA | CxP | PM | PS | GEL | MYO | | |
| CC | 34 | 20 Hz 360 µsec 5.0 V | Cycling, 10 min on, 30 min off | 0–++ | 0–++ | ++ | NA | NA | NA | ++ | 12 | Cooper et al. (1973) |
| | 27 | 10–20 Hz 2.0 µC/cm² Phase | Cycling, 4 min on, 4 min off | 0–++ | ++ | 0–+++ | 0–++ | ++ | NA | +++ | 14–53 | Cooper et al. (1976) |
| | 6 | 10 Hz Phase width NA 2.0–4.0V | Cycling, 8–60 min on-off period | +–+++ | +++ | 0–++ | NA | NA | NA | NA | 6–18 | Levy and Auchterlonie (1979) |
| | 6 | 9.0–2.0 µC/cm² Phase 10 Hz 500 µsec | Cycling, NA | ++ | 0–+++ | 0–+++ | NA | NA | NA | ++ | 99.6–204 | Davis and Emmonds (1992) |
| | 5 | 2.0 µC/cm² Phase 10 Hz 450 µsec | Cycling, 4 min on, 4 min off | +++ | ++ | NA | ++ | NA | NA | NA | 24 | Velasco et al. (2005) |
| DCN | 11 | 50 Hz 200 µsec 4.0–6.0 V | Cycling, 5 min on, 5 min off | ++ | NA | ++ | 0 | NA | NA | NA | 15 | Chkhenkeli et al. (2004) |
| CM | 5 | 60 Hz 450 µsec 3.0–5.0 V | Cycling, 1 min on, 4 min off | +++ | ++ | ++ | NA | NA | NA | NA | 12 | Velasco et al. (1987) |
| | 5 | 60 Hz 450 µsec 3.5 V | Cycling, 1 min on, 4 min off | ++ | + | + | NA | NA | NA | NA | 12 | Fisher et al. (1993) |
| | 49 | 130 Hz 450 µsec 3.5 V | Cycling, 1 min on, 4 min off | +++ | +++ | ++ | +++ | NA | NA | NA | 41.2 | Velasco et al. (2002) |

| Group | N | Parameters | Mode | | | | | | | | Follow-up | Reference |
|---|---|---|---|---|---|---|---|---|---|---|---|---|
| | 5 | 20–130 Hz<br>200 μsec<br>6.0 V | Cycling, 5 min on, 5 min off | NA | +++ | NA | ++ | NA | NA | NA | 15 | Chkhenkeli et al. (2004) |
| | 13 | 130 Hz<br>450 μsec<br>3.0–5.0 V | Cycling, 1 min on, 4 min off | NA | NA | ++ | +++ | +++ | NA | NA | 12 | Velasco et al. (2006) |
| AN | 5 | 5–10 Hz<br>90–200 μsec<br>5.0–10.0 V | Cycling, NA | ++ | NA | ++ | NA | ++ | NA | NA | 12 | Kerrigan et al. (2004) |
| | 4 | 90–110 Hz<br>60–90 μsec<br>4.5 V | Cycling | +++ | NA | +++ | NA | +++ | NA | NA | 43.8 | Lim et al. (2006) |
| Cd | 38 | 4–5 Hz<br>200 μsec<br>6.0–8.0 V | Cycling, 10 min on, 15 min off | +++ | 0 | ++ | NA | +++ | NA | NA | 15 | Chkhenkeli et al. (2004) |
| HC | 10 | 130 Hz<br>450 μsec<br>3.0 V | Continuous | +++ | NA | +++ | NA | +++ | NA | NA | 0.5<br>Subacute | Velasco et al. (2000) |
| | 10 | 130 Hz<br>450 μsec<br>4.0–5.0 V | Continuous | ++++ | NA | ++++ | NA | ++++ | NA | NA | 31 | Boon et al. (2007) |
| | 9 | 130 Hz<br>450 μsec<br>3.0–4.0 V | Cycling, 1 min on, 4 min off | ++ ++++ | NA | ++ ++++ | NA | ++ ++++ | NA | NA | 18 | Velasco et al. (2007) |
| Cx | 1 | 0.5 Hz | Cycling, 30 min on, every 24 hr | +++ | NA | NA | ++ | +++ | NA | NA | 1 week | Schrader et al. (2006) |
| | 1 | 130 Hz<br>450 μsec<br>3.0 V | Cycling, 1 min on, 4 min off | +++ | ++ | + | +++ | +++ | NA | NA | 9 | Velasco et al. (2006) |
| | 3 | 130 Hz<br>450 μsec<br>3.0 V | Cycling, 1 min on, 4 min off | +++ | +++ | NA | ++ | +++ | NA | NA | 12 | Velasco et al. (2008) |

**(continued)**

**TABLE 19.1 (cont.)**
**Targets Used for Neuromodulation in the Treatment of Epilepsy**

| Target | No. of Patients | Stimulation | | Effect on Different Seizure Types | | | | | | | Follow-Up (months) | Refs. |
|---|---|---|---|---|---|---|---|---|---|---|---|---|
| | | Parameters | Mode | GTC | AA | CxP | PM | PS | GEL | MYO | | |
| AN | 5 | 5–10 Hz 90–200 μsec 5.0–10.0 V | Cycling, NA | ++ | NA | ++ | NA | ++ | NA | NA | 12 | Kerrigan et al. (2004) |
| | 4 | 90–110 Hz 60–90 μsec 4.5 V | Cycling | +++ | NA | +++ | NA | NA | NA | NA | 43.8 | Lim et al. (2006) |
| Cd | 38 | 4–5 Hz 200 μsec 6.0–8.0 V | Cycling, 10 min on, 15 min off | +++ | NA | ++ | 0 | NA | NA | NA | 15 | Chkhenkeli et al. (2004) |
| HC | 10 | 130 Hz 450 μsec 3.0 V | Continuous | +++ | NA | +++ | NA | NA | NA | NA | 0.5 Subacute | Velasco et al. (2000) |
| | 10 | 130 Hz 450 μsec 4.0–5.0 V | Continuous | ++++ | NA | ++++ | NA | NA | NA | NA | 31 | Boon et al. (2007) |
| | 9 | 130 Hz 450 μsec 3.0–4.0 V | Cycling, 1 min on, 4 min off | +++++ | NA | +++++ | NA | NA | NA | NA | 18 | Velasco et al. (2007) |

| | n | Stimulation parameters | Stimulation mode | | | | | | | | Follow-up | Reference |
|---|---|---|---|---|---|---|---|---|---|---|---|---|
| Cx | 1 | 0.5 Hz | Cycling, 30 min on, every 24 hr | +++ | ++ | NA | NA | NA | NA | NA | 1 week | Schrader et al. (2006) |
| | 1 | 130 Hz 450 μsec 3.0 V | Cycling, 1 min on, 4 min off | +++ | +++ | + | ++ | NA | NA | NA | 9 | Velasco et al. (2006) |
| | 3 | 130 Hz 450 μsec 3.0 V | Cycling, 1 min on, 4 min off | +++ | NA | NA | +++ | ++ | NA | NA | 12 | Velasco et al. (2008) |
| STN | 5 | 130 Hz 60 μsec Below side effects | Continuous | NA | NA | ++ | ++ | ++ | NA | NA | 12 | Chabardes et al. (2002) |
| | 2 | 130 Hz 60 μsec 3.5 V | Continuous | + | + | + | NA | NA | NA | NA | 9 | Handfoth et al. (2006) |
| | 3 | 130 Hz 60 μsec Below side effects | Continuous | ++ | NA | NA | +++ | + | NA | ++ | 25–31 | Chabardes et al. (2008) |

*Note:* AN, anterior thalamic nucleus; CC, cerebellar cortex; Cd, caudate nucleus; CM, centromedian thalamic nucleus; Cx, extratemporal foci (eloquent areas); DCN, deep cerebellar nuclei; HC, hippocampus; STN, subthalamic nucleus. Reports that included the number of patients, stimulation parameters, and stimulation mode (cycling vs. continuous) were selected. They also specified the type of seizures treated and the percentage improvement: AA, atypical absences; CxP, complex partial; GTC, generalized tonic–clonic; GEL, gelastic; myo, myoclonic; NA, not available; O, no improvement; PM, partial motor; PS, partial sensory: +, 30–50%; ++, 50–80%; +++, >80%. Follow-up is provided in months unless otherwise specified. Reports are cited in the Reference section.

## REFERENCES

Andy, O.J., M. Jurko. (1985). Absence attacks controlled by thalamic stimulation. *Appl. Neurophysiol.*, 48:423–426.

Andy, O.J., M.F. Jurko. (1986). Seizure control by mesothalamic reticular stimulation. *Clin. Electroencephalogr.*, 17:52–60.

Avoli, M., P. Gloor, G. Kostopoulos, J. Gotman. (1983). An analysis of penicillin-induced generalized spike and wave discharges using simultaneous recordings of cortical and thalamic single neurons. *J. Neurophysiol.*, 50:819–837.

Babb, T.L., H.V. Soper, J.P. Lieb, W.J. Brown, C.A. Ottino, P.H. Crandall. (1977). Electrophysiological studies of long-term electrical stimulation of the cerebellum in monkeys. *J. Neurosurg.*, 47:353–365.

Benazzouz, A., B. Piallat, P. Pollak, A.L. Benabid. (1995). Responses of substantia nigra pars reticulata and globus pallidus complex to high frequency stimulation of the subthalamic nucleus in rats: electrophysiological data. *Neurosci. Lett.*, 189:77–80.

Birdno, M.J., W.M. Grill. (2008). Mechanisms of deep brain stimulation in movement disorders as revealed by changes in stimulus frequency. *Neurotherapeutics*, 5:14–25.

Boon, P., K. Vonck, V. De Herdt, A. Van Dycke, M. Goethals, L. Goossens, M. Van Zandijcke, T. De Smedt, I. Dewaele, R. Achten, W. Wadman, F. Dewaele, J. Caemaert, D. Van Roost. (2007). Deep brain stimulation in patients with refractory temporal lobe epilepsy. *Epilepsia*, 48:1551–1560.

Brown, W.J., T.L. Babb, H.V. Soper, J.P. Lieb, C.A. Ottino, P.H. Crandall. (1977). Tissue reactions to long-term electrical stimulation of the cerebellum in monkeys. *J. Neurosurg.*, 47:366–379.

Chabardes, S., P. Kahane, L. Minotti, A. Koudsie, E. Hirsch, A.L. Benabid. (2002). Deep brain stimulation in epilepsy with particular reference to the subthalamic nucleus. *Epileptic Disord.*, 4(Suppl. 3):S83–S93.

Chabardes, S., L. Minotti, S. Chassagnon, B. Piallat, N. Torres, E. Seigneuret, L. Vercueil, R. Carron, E. Hirsch, P. Kahane, A.L. Benabid. (2008). Basal ganglia deep-brain stimulation for treatment of drug-resistant epilepsy: review and current data. *Neurochirurgie*, 54:436–440.

Chkhenkeli, S.A. (1978). Inhibitory influences of caudate stimulation of epileptic activity of human amygdala and hippocampus during temporal lobe epilepsy. *Physiol. Hum. Anim.*, 4:406–411.

Chkhenkeli, S.A., M. Sramka, G.S. Lortkipanidze, T.N. Rakviashvili, E. Bregvadze, G.E. Magalashvili, T. Gagoshidze, I.S. Chkhenkeli. (2004). Electrophysiological effects and clinical results of direct brain stimulation for intractable epilepsy. *Clin. Neurol. Neurosurg.*, 106:318–329.

Cooper, I.S. (1978). *Cerebellar Stimulation in Man*. New York: Raven Press.

Cooper, I.S., I. Amin, S. Gilman. (1973). The effect of chronic cerebellar stimulation upon epilepsy in man. *Trans. Am. Neurol. Assoc.*, 98:192–196.

Cooper, I.S., I. Amin, M. Riklan, J.M. Waltz, T.P. Poon. (1976). Chronic cerebellar stimulation in epilepsy: clinical and anatomical studies. *Arch. Neurol.*, 33:559–570.

Cuellar-Herrera, M., M. Velasco, F. Velasco, A.L. Velasco, F. Jimenez, S. Orozco, M. Briones, L. Rocha. (2004). Evaluation of GABA system and cell damage in parahippocampus of patients with temporal lobe epilepsy showing antiepileptic effects after subacute electrical stimulation. *Epilepsia*, 45:459–466.

Davis, R., S.E. Emmonds. (1992). Cerebellar stimulation for seizure control: 17-year study. *Stereotact. Funct. Neurosurg.*, 58:200–208.

Dow, R.S., A. Fernandez-Guardiola, E. Manni. (1962). The influence of the cerebellum on experimental epilepsy. *Electroencephalogr. Clin. Neurophysiol.*, 14:383–398.

Ebner, T.J., H. Bantli, J.R. Bloedel. (1980). Effects of cerebellar stimulation on unitary activity within a chronic epileptic focus in a primate. *Electroencephalogr. Clin. Neurophysiol.*, 49:585–599.

Fisher, R.S. (2004). Anterior thalamic nucleus stimulation: issues in study design. In *Deep Brain Stimulation and Epilepsy* (pp. 305–319), H.O. Lüders, Ed. London: Martin Dunitz.

Fisher, R.S., S. Uematsu, G.L. Krauss, B.J. Cysyk, R. McPherson, R.P. Lesser, B. Gordon, P. Schwerdt, M. Rise. (1992). Placebo-controlled pilot study of centromedian thalamic stimulation in treatment of intractable seizures. *Epilepsia*, 33:841–851.

Frye, G.D., T.J. McCown, G.R. Breese. (1983). Characterization of susceptibility to audiogenic seizures in ethanol-dependent rats after microinjection of gamma-aminobutyric acid (GABA) agonists into the inferior colliculus, substantia nigra or medial septum., *J. Pharmacol. Exp. Ther.*, 227:663–670.

Garant, D., K. Gale. (1986). Intranigral muscimol attenuates electrographic signs of seizure activity induced by intravenous bicuculline in rats. *Eur. J. Pharmacol.*, 124:365–369.

Gloor, P., L.F. Quesney, H. Zumstein. (1977). Pathophysiology of generalized penicillin epilepsy in the cat: the role of cortical and subcortical structures. II. Topical application of penicillin to the cerebral cortex and to subcortical structures. *Electroencephalogr. Clin. Neurophysiol.*, 43:79–94.

Gonzalez, L.P., M.K. Hettinger. (1984). Intranigral muscimol suppresses ethanol withdrawal seizures. *Brain Res.*, 298:163–166.

Handforth, A., A.A. DeSalles, S.E. Krahl. (2006). Deep brain stimulation of the subthalamic nucleus as adjunct treatment for refractory epilepsy. *Epilepsia*, 47:1239–1241.

Hunter, J., H.H. Jasper. (1949). Effects of thalamic stimulation in unanaesthetised animals; the arrest reaction and petit mal-like seizures, activation patterns and generalized convulsions. *Electroencephalogr. Clin. Neurophysiol.*, 1:305–324.

Iadarola, M.J., K. Gale. (1982). Substantia nigra: site of anticonvulsant activity mediated by gamma-aminobutyric acid. *Science*, 218:1237–1240.

Iatawa, K., R.S. Snider. (1959). Cerebello–hippocampal influences on the electroencephalogram. *Appl. Neurophysiol.*, 11:439–446.

Kerrigan, J.F., B. Litt, R.S. Fisher, S. Cranstoun, J.A. French, D.E. Blum, M. Dichter, A. Shetter, G. Baltuch, J. Jaggi, S. Krone, M. Brodie, M. Rise, N. Graves. (2004). Electrical stimulation of the anterior nucleus of the thalamus for the treatment of intractable epilepsy. *Epilepsia*, 45:346–354.

Laxer, K.D., L.T. Robertson, R.M. Julien, R.S. Dow. (1980). Phenytoin: relationship between cerebellar function and epileptic discharges. *Adv. Neurol.*, 27:415–427.

Lesser, R.P., S.H. Kim, L. Beyderman, D.L. Miglioretti, W.R. Webber, M. Bare, B. Cysyk, G. Krauss, B. Gordon. (1999). Brief bursts of pulse stimulation terminate afterdischarges caused by cortical stimulation. *Neurology*, 53:2073–2081.

Levy, L.F., W.C. Auchterlonie. (1979). Chronic cerebellar stimulation in the treatment of epilepsy. *Epilepsia*, 20:235–245.

Lim, S.N., S.T. Lee, Y.T. Tsai, I.A. Chen, P.H. Tu, J.L. Chen, H.W. Chang, Y.C. Su, T. Wu. (2007). Electrical stimulation of the anterior nucleus of the thalamus for intractable epilepsy: a long-term follow-up study. *Epilepsia*, 48:342–347.

Limousin, P., P. Krack, P. Pollak, A. Benazzouz, C. Ardouin, D. Hoffmann, A.L. Benabid. (1998). Electrical stimulation of the subthalamic nucleus in advanced Parkinson's disease. *N. Engl. J. Med.*, 339:1105–1111.

McNamara, J.O., M.T. Galloway, L.C. Rigsbee, C. Shin. (1984). Evidence implicating substantia nigra in regulation of kindled seizure threshold. *J. Neurosci.*, 4:2410–2417.

Miller, J.W., J.A. Ferrendelli. (1990). The central medial nucleus: thalamic site of seizure regulation. *Brain Res.*, 508:297–300.

Mirski, M.A., J.A. Ferrendelli. (1986). Selective metabolic activation of the mammillary bodies and their connections during ethosuximide-induced suppression of pentylenetetrazol seizures. *Epilepsia*, 27:194–203.

Mirski, M.A., R.S. Fisher. (1994). Electrical stimulation of the mammillary nuclei increases seizure threshold to pentylenetetrazol in rats. *Epilepsia*. 35:1309–1316.

Mirski, M.A., L.A. Rossell, J.B. Terry, R.S. Fisher. (1997). Anticonvulsant effect of anterior thalamic high frequency electrical stimulation in the rat. *Epilepsy Res.*, 28:89–100.

Moruzzi, G. (1941). Sui rapporti fra cervelleto e corteccia cerebrale: azione d'impulse cerebellari sulle attivita motriciti provocate della stimulazione farandica o chimica del giro sigmoldeo nel gato. *Arch. Fisiol.*, 41:157–182.

Oakley, J.C., G.A. Ojemann. (1982). Effects of chronic stimulation of the caudate nucleus on a preexisting alumina seizure focus. *Exp. Neurol.*, 75:360–367.

Ondarza, R., D. Trejo-Martinez, R. Corona-Amezcua, M. Briones, L. Rocha. (2002). Evaluation of opioid peptide and muscarinic receptors in human epileptogenic neocortex: an autoradiography study. *Epilepsia*, 43(Suppl. 5):230–234.

Penfield W., H. Jasper. (1954). *Epilepsy and the Functional Anatomy of the Human Brain*. Boston, MA: Little Brown, pp. 566–596.

Pollen, D.A., P. Perot, K.H. Reid. (1963). Experimental bilateral wave and spike from thalamic stimulation in relation to level of arousal. *Electroencephalogr. Clin. Neurophysiol.*, 15:1017–1028.

Rocha, L., M. Cuellar-Herrera, M. Velasco, F. Velasco, A.L. Velasco, F. Jimenez, S. Orozco-Suarez, A. Borsodi. (2007). Opioid receptor binding in parahippocampus of patients with temporal lobe epilepsy: its association with the antiepileptic effects of subacute electrical stimulation. *Seizure*, 16:645–652.

Schrader, L.M., J.M. Stern, C.L. Wilson, T.A. Fields, N. Salamon, M.R. Nuwer, P.M. Vespa, I. Fried. (2006). Low frequency electrical stimulation through subdural electrodes in a case of refractory status epilepticus. *Clin. Neurophysiol.*, 117:781–788.

Sramka, M., G. Fritz, D. Gajdosova, P. Nadvornik. (1980). Central stimulation treatment of epilepsy. *Acta Neurochir. Suppl. (Wien)*, 30:183–187.

Tellez-Zenteno, J.F., R.S. McLachlan, A. Parrent, C.S. Kubu, S. Wiebe. (2006). Hippocampal electrical stimulation in mesial temporal lobe epilepsy. *Neurology*, 66:1490–1494.

Upton, A.R., I.S. Cooper, M. Springman, I. Amin. (1985). Suppression of seizures and psychosis of limbic system origin by chronic stimulation of anterior nucleus of the thalamus. *Int. J. Neurol.*, 19–20:223–230.

Van Buren, J.M., J.H. Wood, J. Oakley, F. Hambrecht. (1978). Preliminary evaluation of cerebellar stimulation by double-blind stimulation and biological criteria in the treatment of epilepsy. *J. Neurosurg.*, 48:407–416.

Velasco, A.L., F. Velasco, F. Jimenez, M. Velasco. (2006a). Interfering with the genesis and propagation of epileptic seizures by neuromodulation. In *Biomedical Engineering Fundamentals*, 3rd ed. (pp. 36.1–36.12), J.D. Bronzino, Ed. London: Taylor & Francis.

Velasco, A.L., F. Velasco, F. Jimenez, M. Velasco, G. Castro, J.D. Carrillo-Ruiz, G. Fanghanel, B. Boleaga. (2006b). Neuromodulation of the centromedian thalamic nuclei in the treatment of generalized seizures and the improvement of the quality of life in patients with Lennox–Gastaut syndrome. *Epilepsia*, 47:1203–1212.

Velasco, A.L., F. Velasco, M. Velasco, D. Trejo, G. Castro, J.D. Carrillo-Ruiz. (2007). Electrical stimulation of the hippocampal epileptic foci for seizure control: a double-blind, long-term follow-up study. *Epilepsia*, 48:1895–1903.

Velasco, A.L., F. Velasco, M. Velasco, B. Boleaga, M. Kuri, F. Jimenez, J.M. Nuñez. (2008). Neuromodulation: current trends interfering with epileptic seizures. In *Neuroengineering* (pp. 3-1–3-13), D.J. Di Lorenzo and J.D. Bronzino, Eds. Boca Raton, FL: CRC Press.

Velasco, F. (2000). Neuromodulation: an overview. *Arch. Med. Res.*, 31:232–236.

Velasco F., M. Velasco. (1990). Mesencephalic structures and tonic clonic generalized seizures. In *Generalized Epilepsy* (pp. 368–384), M. Avoli, P. Gloor, G. Kustopoulus, and R. Naquet, Eds. Boston: Birkhouser.

Velasco, F., M. Velasco, R. Romo. (1982). Specific and non-specific multiple unit activities during pentylenetetrazol seizures in animals with mesencephalic transections. *Electroencephalogr. Clin. Neurophysiol.*, 53:289–297.

Velasco, F., M. Velasco, C. Ogarrio, G. Fanghanel. (1987). Electrical stimulation of the centromedian thalamic nucleus in the treatment of convulsive seizures: a preliminary report. *Epilepsia*, 28:421–430.

Velasco, F., M. Velasco, I. Marquez, G. Velasco. (1993a). Role of the centromedian thalamic nucleus in the genesis, propagation and arrest of epileptic activity: an electrophysiological study in man. *Acta Neurochir. Suppl. (Wien)*, 58:201–204.

Velasco, F., M. Velasco, A.L. Velasco, F. Jimenez. (1993b). Effect of chronic electrical stimulation of the centromedian thalamic nuclei on various intractable seizure patterns. I. Clinical seizures and paroxysmal EEG activity. *Epilepsia*, 34:1052–1064.

Velasco, F., M. Velasco, F. Jimenez, A.L. Velasco, F. Brito, M. Rise, J.D. Carrillo-Ruiz. (2000a). Predictors in the treatment of difficult-to-control seizures by electrical stimulation of the centromedian thalamic nucleus. *Neurosurgery*, 47:295–305.

Velasco, F., M. Velasco, F. Jimenez, A.L. Velasco, B. Rojas, M.L. Perez. (2002). Centromedian nucleus stimulation for epilepsy: clinical, electroencephalographic, and behavioral observations. *Thalamus & Related Systems*, 1:387.

Velasco F., A.L. Velasco, M. Velasco, J. Rocha, D. Menes. (2004). Electrical neuromodulation of the epileptic focus in cases of temporal lobe seizures. In *Deep Brain Stimulation and Epilepsy* (pp. 285–298). H.O. Lüders, Ed. London: Martin Dunitz.

Velasco, F., J.D. Carrillo-Ruiz, F. Brito, M. Velasco, A.L. Velasco, I. Marquez, R. Davis. (2005). Double-blind, randomized controlled pilot study of bilateral cerebellar stimulation for treatment of intractable motor seizures. *Epilepsia*, 46:1071–1081.

Velasco, M., F. Velasco, A.L. Velasco, F. Brito, F. Jimenez, I. Marquez, B. Rojas. (1997). Electrocortical and behavioral responses produced by acute electrical stimulation of the human centromedian thalamic nucleus. *Electroencephalogr. Clin. Neurophysiol.*, 102:461–471.

Velasco, M., F. Velasco, A.L. Velasco, B. Boleaga, F. Jimenez, F. Brito, I. Marquez. (2000b). Subacute electrical stimulation of the hippocampus blocks intractable temporal lobe seizures and paroxysmal EEG activities. *Epilepsia*, 41:158–169.

Weiss, S.R., X.L. Li, J.B. Rosen, H. Li, T. Heynen, R.M. Post. (1995). Quenching: inhibition of development and expression of amygdala kindled seizures with low frequency stimulation. *NeuroReport*, 6:2171–2176.

Wright, G.D., D.L. McLellan, J.G. Brice. (1984). A double-blind trial of chronic cerebellar stimulation in twelve patients with severe epilepsy. *J. Neurol. Neurosurg Psychiatry*, 47:769–774.

# 20 Optical Imaging Techniques in Pediatric Neocortical Epilepsy

*Daryl W. Hochman and Michael M. Haglund*

## CONTENTS

## OPTICAL IMAGING TECHNIQUES IN NEOCORTICAL EPILEPSY

One of the most common forms of epilepsy affecting pediatric patients is neocortical (Obeid et al., 2009a,b). The majority of these pediatric patients require lengthy work-ups, which include grid arrays and strip electrodes, to localize the neocortical focus (Obeid et al., 2009a,b; Sheth, 2000). The additional monitoring for epileptogenic zone localization requires second craniotomies, with the additional risk of infection and complications. Newer imaging techniques may assist epileptologists and epilepsy surgeons in the localization of these neocortical epileptic foci (Snead and Nelson, 1993). Intraoperative techniques such as stereotactic magnetic resonance imaging (MRI) and ultrasound can localize anatomic malformations and larger neuronal migration disorders but still do not have the ability to look at the gold-standard, real-time epileptiform activity. One possible substitute for electrical recordings from the neocortical surface is optical imaging. In this chapter, we will go over the basics of optical imaging and give tangible examples of how this technique may be able to localize epileptiform activity and possibly even neocortical epileptic foci and epileptic onset zones. Basic fundamental work on the mechanisms underlying the optical intrinsic signal, and the paradigms to remove noise from the acquired images, will move this field forward to the ultimate goal of using optical imaging in real time to localize epileptic foci and allow epileptologists and epilepsy surgeons to do smaller resections and avoid unnecessary surgeries and their inherent risks.

## POTENTIAL APPLICATIONS OF OPTICAL IMAGING OF INTRINSIC SIGNALS

Changes in the level of neuronal activity can alter the way in which visible light is absorbed and scattered by brain tissue (Haglund and Hochman, 2004, 2007). Increases in action potential firing within organized networks of cortical neurons elicit transient optical changes that are localized in the tissue to the areas undergoing increased activity. Because these optical changes arise from the tissue itself and do not involve the application of dyes or contrast-enhancing agents, they are often referred to as *intrinsic optical signals* (IOSs). IOS imaging involves attaching a sensitive digital camera to an operating microscope, illuminating the cortical surface with light at appropriate wavelengths, and acquiring a series of images that can then be processed and analyzed using digital imaging techniques and statistical modeling. This relatively straightforward technique makes it possible to acquire movies showing dynamic patterns of the IOS resulting from action potential firing. IOS imaging has been used as an experimental technique in the laboratory since the 1980s to study the functional organization of sensory cortex in nonhuman primates (Grinvald et al., 1988). It has been most often applied in the laboratory for studying visual cortex, where it is possible to map functional units such as ocular dominance and orientation selectivity columns (Snead and Nelson, 1993). In recent years, IOS imaging has been adapted to the operating room, where it has been shown to be capable of mapping language, motor, and sensory cortex, as well as the spread of epileptic activity in human subjects (Cannestra et al., 2001; Haglund and Hochman, 2004; Haglund et al., 1992; Sato et al., 2002 2005; Schwartz et al., 2004; Zhao et al., 2007).

High-resolution IOS imaging has several important limitations. These include its restriction to mapping activity on the exposed cortical surface and that its temporal and spatial relationships to neuronal activity are still not completely understood. In spite of these limitations, IOS imaging has several major advantages giving it the potential to be a widely useful tool in the operating room for online mapping of functional and epileptic tissue. Its main advantages are its ability to safely map patterns of cortical activity with micron-level resolution and its relative low cost. Because IOS imaging uses non-ionizing light and does not require the use of potentially phototoxic optical dyes, it can safely map the cortex without any possibility of causing tissue damage. Additionally, an IOS imaging system is a fraction of the cost and far simpler and more portable than other imaging technologies, such as functional MRI and positron emission tomography (PET).

More complete understanding and interpretation of the activity-evoked optical changes in the cortex are required for IOS imaging to become a practical and useful clinical tool. An important issue that needs to be better understood is how accurately optical changes at various wavelengths localize normal and epileptic neuronal activity. Resolving this issue requires better elucidating the hemodynamic changes that are responsible for generating the activity-evoked optical changes in the neocortex. The remainder of this chapter focuses on this issue.

## RELATIONSHIPS BETWEEN THE OPTICAL SIGNAL, CEREBRAL HEMODYNAMICS, AND CORTICAL ACTIVITY

Changes in the intrinsic optical properties of cortical tissue are thought to arise from activity-evoked changes in blood volume and blood oxygenation (Haglund and Hochman, 2007). Action potential firing by cortical neurons is known to elicit a transient redistribution of blood within the active tissue, resulting in significant and localized increases in blood flow and blood volume (Roland, 1997). Signaling molecules, such as adenosine and nitric oxide, are released by neurons during action potential firing and diffuse through the extracellular space to cause nearby pial arterioles to dilate (Iadecola, 2004). It is known that the smallest pial arterioles with diameters less than 100 µm are the major mediators of this activity-evoked redistribution of blood within the neocortex (Mchedlishvili, 1987). Poiseuille's equation predicts that blood flow is related to the fourth power of vessel diameter (Fung, 1997); hence, small changes in neuronal activity are transformed into large increases in blood flow and volume through dilation of the pial arterioles, whose diameters can increase by 30% above their resting baseline size. The dramatic increase in flow velocity means that the transit time of hemoglobin molecules within the vicinity of the active tissue is significantly reduced, resulting in an increase in the amount of oxygenated hemoglobin in that portion of the venous network receiving the blood from the active tissue (Roland, 1997). Blood-oxygen-level-dependent (BOLD) fMRI, for example, relies on the detection of this activity-evoked increase in oxyhemoglobin.

Intrinsic optical signal imaging takes advantage of the fact that oxyhemoglobin ($HbO_2$) and deoxyhemoglobin (Hb) absorb light throughout the visible spectrum differently from each other (Figure 20.1A). Their absorption spectra show that at certain wavelengths, such as 535 nm (green light) and 800 nm (near-infrared light), both $HbO_2$ and Hb absorb exactly the same amount of light, so optical imaging at these wavelengths would be insensitive to changes in the oxygen content of blood. In contrast, optical imaging at 660 nm (red light) would be expected to be sensitive to changes in blood oxygenation, as $HbO_2$ and Hb are maximally distinguishable at this wavelength.

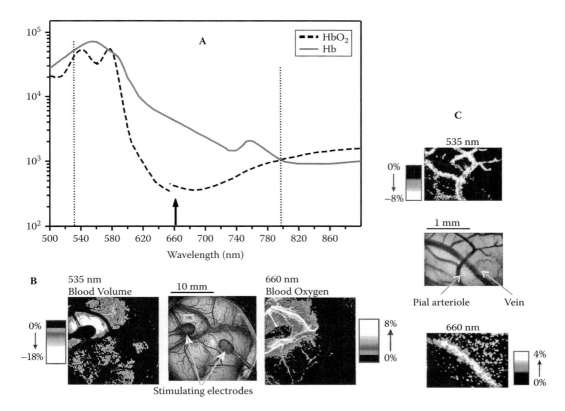

**FIGURE 20.1** (A) Plots of the optical absorption spectra for oxyhemoglobin (HbO$_2$) and deoxyhemoglobin (Hb). Plots are shown of the amount of light absorbed by HbO$_2$ (dashed black line) and Hb (solid gray line) across the optical spectrum, plotted on a logarithmic scale in units of molar extinction coefficient ($M^{-1}cm^{-1}$) on the y-axis vs. wavelength (nm) on the x-axis. HbO$_2$ and Hb both absorb exactly the same amount of light at certain wavelengths, such as 535 nm and 800 nm (vertical dotted lines); these wavelengths of equal optical absorption for both light-absorbing species are known as *isosbestic points*. Imaging at isosbestic points is useful for mapping changes in blood volume independent of changes in blood oxygenation. At 660 nm (vertical arrow), the difference between HbO$_2$ and Hb is maximal; hence, imaging at this wavelength is maximally sensitive to changes in blood oxygenation. (B) Low-magnification optical imaging of electrical-stimulation-evoked activity in monkey cortex. The center image shows the position of the bipolar stimulating electrode on the cortical surface. The cortex was stimulated for 4 seconds with a current (6 mA; 60-Hz biphasic) that was just below the afterdischarge threshold. The leftmost image shows a map of the blood volume changes acquired with 535-nm light. As is typical, the largest changes occurs predominantly around one of the two stimulating electrodes. The largest optical changes at this wavelength appear to occur diffusely throughout the tissue surrounding the electrode and are absent in the larger blood vessels. The right-most image shows the blood oxygenation map acquired with 660-nm light. At this wavelength, the largest changes occur in the larger veins nearest the stimulating electrodes. (C) High-magnification optical imaging of electrical stimulation-evoked cortical activity. These images show highly magnified changes occurring within the small square region drawn above the left electrode in the middle figure of B). The middle figure shows the microvascular network within the field of view, where the pial arteriole is approximately 30 μm in diameter. During stimulation, the pial arterial dilated approximately 30% above its resting diameter, mediating the blood-volume change within that region of tissue. The top image shows blood-volume changes that were acquired with 535-nm light, demonstrating that the optical changes were restricted to the dilating pial arteriole. The bottom image shows that the blood oxygenation images acquired with 660-nm light were restricted to the small veins. (The graphs in Part A were generated from publicly available data compiled by Dr. Scott Prahl, Oregon Medical Laser Center, http://omlc.ogi.edu/spectra/hemoglobin/summary.html, and were tabulated from data by Dr. W.B. Gratzer, Medical Research Council Labs, Holly Hill, London, and Dr. N. Kollias, Wellman Laboratories, Harvard Medical School, Boston. Parts B and C were modified from Haglund, M.M. and Hochman, D.W., *Epilepsia*, 48(Suppl. 4), 65–74, 2007. With permission from Wiley-Blackwell.)

Several predictions follow from the facts described in the preceding section. First, it would be expected that, during activation, the cortex would absorb more light when illuminated with 535-nm light (i.e., become darker). This follows from the fact that dilating microscopic pial arterioles results in an increase in the number of light-absorbing hemoglobin molecules within a volume of tissue, so more light would be expected to be absorbed at 535 nm, a wavelength that is insensitive to changes in blood oxygenation. In contrast, it would be expected that the veins receiving blood from active tissue would absorb less light, or become brighter under illumination at 660 nm. This follows from the facts that there would be an increase in $HbO_2$ in the venous network, and $HbO_2$ absorbs less than Hb at this wavelength (Figure 20.1A).

The imaging hypotheses above have been tested and validated in monkey neocortex (Haglund and Hochman, 2007) (Figure 20.1B,C). IOS imaging was performed in those studies at sufficiently high magnification to reveal the optical changes occurring within the microscopic pial arterioles and surrounding venous network during electrical stimulation of the cortical surface. It can be seen that the optical changes at 535 nm are restricted to the dilating pial arterioles that absorb more light following cortical activation. At 660 nm, the optical changes are restricted to the nearby veins that become brighter at the same time (Figure 20.1C). These experiments suggest the optical changes at 535 nm are restricted to the dilating pial arterioles nearest to the neurons that fire action potentials; hence, IOS imaging at this wavelength would be expected to generate maps that reliably localize cortical areas of increased activity. Optical changes at 660 nm are localized to the venous network that receives blood from active cortical areas and is more highly oxygenated during increased neuronal activity. Because the more oxygenated blood may flow into vessels that may be located at various distances from the site of activation, it would be expected that IOS imaging at 660 nm would be less reliable for localizing active cortical areas. In summary, these recent studies suggest that IOS imaging at 535 nm, which reflects activity-evoked changes in blood volume, may be an accurate and reliable method for intraoperative cortical mapping.

Intraoperative IOS imaging studies on human subjects further support the notion that IOS imaging of blood volume changes at 535 nm can be used to reliably map functional and epileptic activity (Haglund and Hochman, 2004). We have recently carried out a more extensive series of careful studies on the use of 535-nm IOS imaging for localizing cortical activity. One experimental method for investigating that ability of IOS imaging for mapping epileptic activity involves electrical stimulation of the cortex at various currents that are either below or at the threshold for eliciting afterdischarge (AD) activity, which is electrophysiological and similar to ictal discharges (Figure 20.2). Through stimulation (60-Hz, biphasic) of the cortex for 4-second durations with an Ojemann Cortical Stimulator (Integra™; Plainsboro, NJ), stimulation currents can be observed that are just sufficient to reliably and repeatedly elicit ADs. Stimulation at a current that is just 1 mA below this afterdischarge threshold will fail to elicit any ADs. Figure 20.2A shows data from one such experiment in which 6 mA was found to be the afterdischarge threshold. In these experiments, it has been found that afterdischarge activity generates optical changes at 535 nm that are of significantly greater magnitude and of longer duration than optical changes elicited by neuronal activity that is below the afterdischarge threshold. Further, the optical changes are observed to spread into cortical areas that are more distant from the stimulating electrodes during the afterdischarge activity. These observations suggest that epileptic activity may have a distinctly different "optical signature" from nonepileptic activity. Intraoperative IOS imaging in combination with electrical stimulation may thus provide a reliable means to map epileptogenic cortical tissue.

## MAKING IOS IMAGING A PRACTICAL CLINICAL TOOL

Intrinsic optical signal imaging is currently an experiment technique that will require further development before it can become a practical method for intraoperative cortical mapping. Necessary steps include: (1) further elucidation of the accuracy and reliability for localizing epileptic and functional activity, and (2) further development of data processing and analysis techniques to provide a

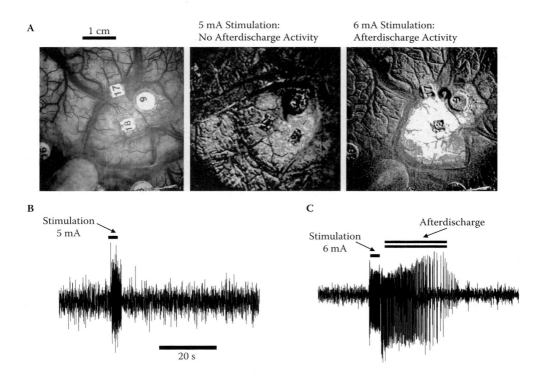

**FIGURE 20.2**    Optical imaging of electrical-stimulation-evoked afterdischarge (ictal) activity in the human cortex. This figure compares blood-volume changes elicited by a stimulation current that was just below the afterdischarge threshold (5 mA) to those elicited by a current at the afterdischarge threshold (6 mA) that elicited approximately 30 seconds of discharge activity. The left-most image in Part A shows the position of the bipolar electrodes on the cortical surface that are labeled "17" and "18" and the surface EEG recording electrode labeled "9." The middle image shows the blood-volume changes imaged at 535 nm resulting from 4 seconds of stimulation in which no afterdischarge activity was elicited. The right-most image in Part A shows the pattern of blood-volume changes occurring during the afterdischarge activity following 4 seconds of stimulation with 6-mA current. Note the increase in spread and magnitude of the optical changes. Parts B and C show the simultaneously acquired EEG recordings at 5 mA and 6 mA, respectively. The afterdischarge activity in Part C is indicated by the double horizontal lines above the EEG recording.

meaningful representation of optical imaging data to the neurosurgeon. This first step will require further human and animal studies aimed at carefully correlating optical changes to normal and epileptic electrophysiological activities. The second step is necessary, as intraoperatively acquired optical data is extremely noisy, contaminated with artifacts from heartbeat, respiration, and random movements in the case of awake subjects. Data analysis methods need to be developed that remove the noise from the optical data while preserving the underlying activity-evoked signal. Ideally, these techniques will allow for a statistical analysis, so that (for example) optical maps can be generated that represent areas of optical changes with a well-defined statistical confidence level. Preliminary work in this area shows that statistical modeling of optical imaging data with dynamic linear models (DLMs) may be a promising approach to addressing this issue (Hochman et al., 2006).

Although still an experimental tool, optical imaging has the opportunity to become an intraoperative technique that could help in localizing neocortical epileptic foci. The advances made by several groups thus far have allowed optical imaging to localize interictal spikes, epileptogenic zones, and even the spread of neocortical epilepsy. Many of the techniques described in this chapter show a pathway forward for intraoperative optical imaging. Several key questions remain surrounding real-time data analysis and the ability to only visualize epileptic foci on the exposed cortical surface.

However, localization of epileptic foci, minimizing second craniotomies for invasive monitoring, and possibly smaller resections of key epileptogenic zones and their pathways of spread across the cortex may be accomplished using this technique.

## REFERENCES

Cannestra, A.F., N. Pouratian, S.Y. Bookheimer, N.A. Martin, D.P. Beckerand, A.W. Toga. (2001). Temporal spatial differences observed by functional MRI and human intraoperative optical imaging. *Cereb. Cortex*, 11:773–782.

Fung, Y.C. (1997). *Biomechanics: Circulation*. New York: Springer.

Grinvald, A., R.D. Frostig, E. Lieke, R. Hildesheim. (1988). Optical imaging of neuronal activity. *Physiol. Rev.*, 68:1285–1366.

Haglund, M.M., D.W. Hochman. (2004). Optical imaging of epileptiform activity in human neocortex. *Epilepsia*, 45(Suppl. 4):43–47.

Haglund, M.M., D.W. Hochman. (2007). Imaging of intrinsic optical signals in primate cortex during epileptiform activity. *Epilepsia*, 48(Suppl. 4):65–74.

Haglund, M.M., G.A. Ojemann, D.W. Hochman. (1992). Optical imaging of epileptiform and functional activity in human cerebral cortex. *Nature*, 358:668–671.

Hochman, D.W., M. Lavine, M.M. Haglund. (2006). Dynamic linear modeling of optical brain data. In *Proceedings of the Statistical Analysis of Neuronal Data Workshop (SAND3)*, May 11–13, Philadelphia, PA.

Iadecola, C. (2004). Neurovascular regulation in the normal brain and in Alzheimer's disease. *Nat. Rev. Neurosci.*, 5:347–360.

Mchedlishvili, G. (1987). *Arterial Behavior and Blood Circulation in the Brain*. New York: Plenum Press.

Obeid, M., E. Wyllie, A.C. Rahi, M.A. Mikati. (2009a). Approach to pediatric epilepsy surgery: state of the art. Part I. General principles and presurgical workup. *Eur. J. Paediatr. Neurol.*, 13:102–114.

Obeid, M., E. Wyllie, A.C. Rahi, M.A. Mikati. (2009b). Approach to pediatric epilepsy surgery: state of the art. Part II. Approach to specific epilepsy syndromes and etiologies. *Eur. J. Paediatr. Neurol.*, 13:115–127.

Roland, P.E. (1997). *Brain Activation*. New York: Wiley-Liss.

Sato, K., T. Nariai, S. Sasaki, I. Yazawa, H. Mochida, N. Miyakawa, Y. Momose-Sato, K. Kamino, Y. Ohta, K. Hirakawa, K. Ohno. (2002). Intraoperative intrinsic optical imaging of neuronal activity from subdivisions of the human primary somatosensory cortex. *Cereb. Cortex*, 12:269–280.

Sato, K., T. Nariai, Y. Tanaka, T. Maehara, N. Miyakawa, S. Sasaki, Y. Momose-Sato, K. Ohno. (2005). Functional representation of the finger and face in the human somatosensory cortex: intraoperative intrinsic optical imaging. *NeuroImage*, 25:1292–1301.

Schwartz, T.H., L.M. Chen, R.M. Friedman, D.D. Spencer, A.W. Roe. (2004). Intraoperative optical imaging of human face cortical topography: a case study. *NeuroReport*, 15:1527–1531.

Sheth, R.D. (2000). Intractable pediatric epilepsy: presurgical evaluation. *Semin. Pediatr. Neurol.*, 7:158–165.

Snead III, O.C., M.D. Nelson, Jr. (1993). PET does not eliminate need for extraoperative, intracranial monitoring in pediatric epilepsy surgery. *Pediatr. Neurol.*, 9:409–411.

Xiao, Y., A. Casti, J. Xiao, E. Kaplan. (2007). Hue maps in primate striate cortex. *NeuroImage*, 35:771–786.

Zhao, M., M. Suh, H. Ma, C. Perry, A. Geneslaw, T.H. Schwartz. (2007). Focal increases in perfusion and decreases in hemoglobin oxygenation precede seizure onset in spontaneous human epilepsy. *Epilepsia*, 48:2059–2067.

# 21 High-Frequency Oscillations in Epileptic Brain

*S. Matt Stead, Mark Bower, and Greg Worrell*

## CONTENTS

## INTRODUCTION

Neuronal oscillations recorded from the human brain span a range of spatial and temporal scales that extend far beyond traditional clinical electroencephalography (EEG). Extracellular local field potentials (LFPs) recorded directly from human cortex range from near DC to ultra-fast oscillations (~0 to 1000 Hz) (Figure 21.1). The mechanisms generating these oscillations are varied, and the correlation with mechanism, function, or pathology is challenging. Despite the lack of mechanistic specificity for neuronal oscillations, EEG research has a rich, successful history of identifying neural correlates of physiological and pathological brain function.

Until recently, human EEG was primarily focused on neuronal oscillations within the *Berger bands* (1 to 25 Hz), named in honor of Hans Berger, who first reported on human EEG (Berger, 1929; Gloor, 1969, 1975). This relatively narrow clinical EEG bandwidth is largely a technical legacy and at least partly associated with the challenge of recording and storing wide-bandwidth EEG data. In addition to the technical challenge, the EEG spectral power is dominated by activity in the Berger frequency bands. At high frequency, the EEG power has been reported to fall off rapidly, as either $f^{-1}$ (Linkenkaer-Hansen et al., 2001; Worrell et al., 2002) or $f^{-2}$ (Bédard et al., 2006; Milstein et al., 2009). This is most dramatic when recording from the scalp but is also evident from intracranial EEG (iEEG). Recent studies, however, have begun to establish the importance of neural activity occurring outside the Berger bands in animal and human electrophysiology (Bragin et al., 1999a, 2002a,b; Curio, 2000a; Jacobs et al., 2008; Jirsch et al., 2006; Traub et al., 2001; Vanhatalo et al., 2005; Worrell et al., 2004, 2008) (Figure 21.1).

Studies from both animals and humans describe gamma-frequency (25 to 80 Hz) oscillations, which are involved in sensory binding (Singer, 1999; Uhlhaas and Singer, 2006); ripple-frequency (80 to 200 Hz) oscillations, which may be important for memory consolidation (Buzsaki and Draguhn, 2004; Buzsaki et al., 1992; Lisman and Idiart, 1995); fast ripples (200 to 600 Hz), which may serve as markers for epileptic networks (Bragin et al., 1999a,b; 2002a; Dzhala and Staley, 2004); and ultra-fast (~600 Hz) oscillations associated with somatosensory evoked response (Curio,

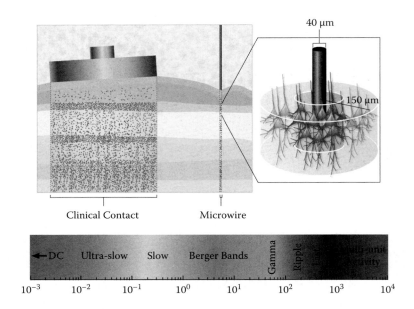

**FIGURE 21.1** **(See color insert following page 458.)** Spatiotemporal scales: the large volume sampled by millimeter-scale clinical electrodes (~$10^5$ neurons) vs. a 40-μm microwire. Cortex is organized into columns of neurons that are ~400 to 600 μm in diameter (~$10^4$ neurons in each column). Bottom figure shows the frequency range of neuronal network oscillations (DC, 1000 Hz).

2000a,b). In addition to these normal physiological high-frequency oscillations (HFOs), studies in animals and humans have shown that, in some cases, HFOs may be pathological and involved in seizure generation (Bikson et al., 2003; Grenier et al., 2003; Timofeev and Steriade, 2004) and epileptogenesis (Bragin et al., 2000; Engel et al., 2009). High ratios of fast ripples to ripple discharges seem to be associated with histopathologic changes of hippocampal sclerosis in human temporal lobe epilepsy (Staba et al., 2007).

The generation of spontaneous seizures is the defining characteristic of epilepsy. However, it was recognized early in the development of EEG that abnormal brain activity also occurred at times removed from seizures (i.e., interictal epileptiform spikes and sharp waves) (Neidermeyer and Da Silva, 2005). These interictal signatures of epileptic brain are transient local field potentials created by the synchronous discharge of large neuronal populations and are highly specific for brain that generates spontaneous seizures (Ayala et al., 1973). Similar to interictal epileptiform spikes, researchers have recently speculated that some interictal high-frequency oscillations are pathological and may be signatures (i.e., electrophysiological biomarkers) of epileptic brain (Bragin et al., 1999a,b; Engel et al., 2009; Le Van Quyen et al., 2006; Worrell et al., 2004), epileptogenesis (Engel et al., 2009), and seizure generation (Timofeev and Steriade, 2004; Worrell et al., 2004).

Although epileptiform spikes are highly specific for the epileptic brain, it has proven more difficult to classify pathological HFO because normal physiological oscillations occur in the same frequency range (Engel et al., 2009; Le Van Quyen et al., 2006). In addition, HFOs are brief, low-amplitude transients that currently require iEEG recordings. Because these studies are limited to patients with medically resistant partial epilepsy, the specificity of the results remains an open question. Whether HFOs recorded in epileptic brain are generated by unique pathological mechanisms or represent an aberration of normal physiological oscillations is not clear. There are currently no established criteria for distinguishing physiological from pathological HFOs. Important questions that remain unanswered include: (1) Are there distinctly pathological HFOs in epileptic brain that can be differentiated from physiological HFOs? (2) Are HFOs mechanistically involved in the generation of focal seizures? (3) Are pathological HFOs biomarkers for epileptic brain and epileptogenesis? In this chapter, some of the technical issues will be reviewed relating to recording wide-bandwidth electrophysiology and research investigating HFOs in human epileptic brain.

## LARGE-SCALE HUMAN ELECTROPHYSIOLOGY: ELECTRODES, ACQUISITION, STORAGE, AND MINING

Despite advances in digital electronics and computing that have revolutionized animal electrophysiology (Buzsaki, 2004), clinical iEEG continues to primarily utilize narrow-bandwidth (0.1 to 100 Hz) recordings from widely spaced (5 to 10 mm), large-diameter (>1 mm$^2$) macroelectrodes. However, wide-bandwidth recordings from microwires and even clinical macroelectrodes in humans have begun to redefine the physiological bandwidth of human brain activity (Engel et al., 2009; Vanhatalo et al., 2005). Remaining fundamental questions about the optimal spatial and spectral resolution for clinical iEEG are now receiving significant attention.

### Spatial Resolution of iEEG

Despite the knowledge that neocortex is spatially organized on the scale of the cortical column (0.3 to 0.6 mm), clinical iEEG uses large (>1 mm$^2$), widely spaced (5 to 10 mm) macroelectrodes. The cortical column has been proposed as a possible anatomic substrate for the initiation of epileptic activity (Gabor et al., 1979; Reichenthal and Hocherman, 1977, 1979), but currently there is no direct experimental evidence (Chesi and Stone, 1998; Schwartz, 2003). Although the role of submillimeter-scale neuronal assemblies in the generation of seizures is not known, the fact that fast-ripple oscillations (250 to 1000 Hz) are localized to submillimeter volumes (<1 mm$^3$) in human hippocampus (Bragin et al., 1999a, 2002b; Staba et al., 2002; Worrell et al., 2008) and animal models (Bragin et al., 2002a) supports the hypothesis that the functional organization of the epileptic brain also extends to submillimeter scales.

### Hybrid Electrodes

Intracranial EEG is associated with a modest but real risk of morbidity and mortality to the patient (Hamer et al., 2002; Van Gompel et al., 2008b). For this reason, iEEG can only be obtained in the appropriate clinical setting. In carefully designed studies, experimental electrode systems can be investigated, but their safety must be critically reviewed. The level of invasiveness of experimental intracranial electrodes varies considerably. Studies investigating HFOs in humans have utilized standard clinical subdural or depth macroelectrodes (Fisher et al., 1992; Jacobs et al., 2008; Jirsch et al., 2006; Traub et al., 2001; Urrestarazu et al., 2006; Worrell et al., 2004; 2008), hybrid electrode systems (Figure 21.2) (Bragin et al., 2002a,b; Worrell et al., 2008), or more invasive electrode systems that penetrate brain tissue and are independent of the clinical electrode system (Schevon et al., 2008; Ulbert et al., 2004). The hybrid electrode systems take advantage of the clinical need for iEEG by using modified clinical electrodes (Van Gompel et al., 2008a). Hybrid subdural electrodes are modifications of standard clinical subdural electrodes with additional microwire arrays (Figure 21.2). Clinical depth electrodes with microwires exiting the tip of the depth have been used extensively by the University of California, Los Angeles, group with excellent safety (Jerome Engel, pers. comm.). Using a similar depth electrode design with additional microwires embedded in the electrode shaft, we have also achieved excellent safety (Van Gompel et al., 2008a). The complication rate (infections, subdural and epidural hematomas, status epilepticus) was 4.2% (1 of 24 cases with a complication of subdural hematoma), which did not differ from our previously reported (Van Gompel et al., 2008b) complication rate of 6.6% from 198 patients implanted with standard clinical subdural intracranial electrodes.

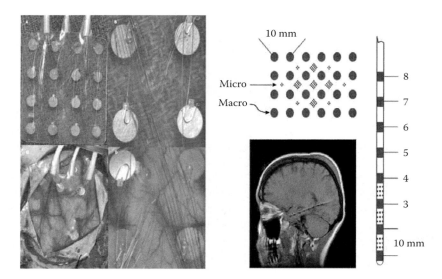

**FIGURE 21.2**   (Left) Photographic montage of hybrid subdural grid containing clinical 16 macroelectrodes (4 mm) and 112 microelectrodes (0.040 mm). (Right) Schematic of hybrid subdural grid and depth electrodes; MRI of hippocampal hybrid depth implant (below). (From Brinkmann, B.H. et al., *J. Neurosci. Methods*, 180(1), 185–192, 2009. With permission.)

## ACQUISITION, STORAGE, AND MINING

Wide-bandwidth electrophysiology from a large number of microwires required to record submillimeter regions of brain has largely remained limited to animal research (Buzsaki, 2004). Recently, these advances have been applied to human intracranial EEG, and the approach has shown potential for clinical utility (Bragin et al., 1999a,b; Brinkmann et al., 2009). The acquisition of wide-bandwidth iEEG (DC, 1000 Hz) and single neuron action potentials from high-impedance microwires is not possible with current clinical iEEG systems (Table 21.1). However, a few clinical iEEG acquisition systems do have wide-frequency bandwidth capability suitable for recording gamma, ripple, and fast-ripple oscillations. Because of the technical challenges associated with wide bandwidth from high-impedance microelectrodes, the progress in determining the optimal spatial and temporal recording resolution for localization of epileptic brain and the generation of seizures has been slow. Nonetheless, significant progress has been made in the past decade, and studies from a number of groups have demonstrated the feasibility and potential clinical importance (Engel et al., 2009).

In addition to the technical challenge of large-scale recording, the transfer, storage, and analysis of multiple-terabyte datasets remain significant challenges but are within the capability of commercially available systems (Brinkmann et al., 2009) (Figure 21.3). Identifying epileptiform transients in continuous multiday iEEG recordings presents a significant challenge. Human experts (i.e. electroencephalographers) are considered the gold standard for marking epileptiform activity, but it is not feasible to routinely label entire iEEG records manually. Objective and accurate labeling of epileptiform activity in large-scale recordings requires automated detectors. The automated mining and analysis of large-scale datasets and HFO detectors represent an area of active research (Gardner et al., 2007a,b; Worrell et al., 2008).

Advances in automated analysis applied to EEG event detection include techniques from many areas of signal processing and machine learning. In addition, more recently the literature makes clear how difficult it is to reliably detect interictal spikes and seizure (Gardner et al., 2007b; Saab and Gotman, 2005; Wilson and Emerson, 2002; Wilson et al., 2003). The detection and labeling of interictal and ictal epileptiform activity in EEG records can be broadly categorized into three different approaches:

**TABLE 21.1**
**EEG Systems**

| | Neuralynx (Animal) | Xltek® (Human) | Stellate (Human) | Nicolet® (Human) | NeuroScan (Human) |
|---|---|---|---|---|---|
| Bandwidth sampling rate | 0–9,000 Hz | 0.1–500 Hz | 0.1–500 Hz | 0.1–100 Hz | DC, 3500 Hz |
| | 32,556 Hz | 512–2000 Hz | 2000 Hz | 200 Hz | 20,000 Hz |
| A/D resolution | 24 bit | 16 bit | 16 bit | 16 bit | 24 bit |
| Input impedance (Z) | — | 106 | 106 | 106 | 109 |
| DC-capable amplifiers | Yes | No | No | No | Yes |
| High-impedance electrodes | Yes | No | No | No | No |
| Number of channels | 32–1024 | 128 | 128 | 128 | 256 |

1. Expert manual EEG review is considered the gold standard but is not feasible for large-scale datasets. In addition, there is more variability between human experts than is widely acknowledged.
2. High-sensitivity automated detection combined with expert review is the primary approach for automated analysis and can significantly reduce the data volume to review; however, the approach generally does not provide detector specificity.

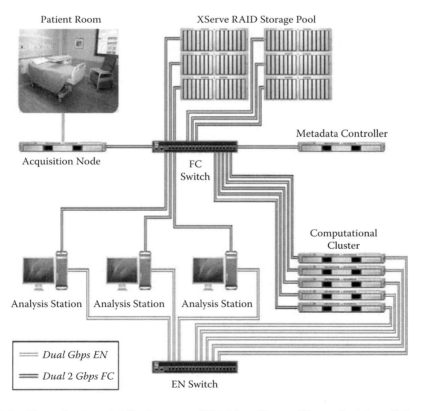

**FIGURE 21.3** **(See color insert following page 458.)** Mayo Human Electrophysiology Suite acquisition using scalable Neuralynx Cheetah. The system streams data from the patient's ICU room to the acquisition node via a dedicated dual-gigabit Ethernet. Data are stored in a 70-terabyte storage pool and are accessed via a fiber channel Service Area Network. Large-scale analysis is performed on a dedicated computational cluster. (From Brinkmann, B.H. et al., *J. Neurosci. Methods*, 180(1), 185–192, 2009. With permission.)

3. Fully automated detection and labeling of epileptiform events require high-specificity and high-sensitivity detectors. The approach is highly efficient if detectors can be realized and likely would be required for real-time device applications.

Existing methods for automated HFO detection implicitly model the events as short-duration, high-frequency transients added to background iEEG (Gardner et al., 2007b; Staba et al., 2002). The detection of HFO events can generally be separated into two stages: preprocessing and event detection. During preprocessing, data are bandpass filtered to restrict the range of frequencies under consideration. Additional preprocessing steps include spectral equalization (i.e., whitening) (Shiro and Itzhak, 1982), and both manual and automated artifact removal (Hu et al., 2007). The detection of HFOs usually consists of calculating a signal feature (e.g., spectral energy or line length) (Esteller et al., 2005; Firpi et al., 2007; Gardner et al., 2007b; Staba et al., 2002) and applying a threshold to select statistically significant events. Further refinement in classification of HFO events can be made by sequentially applying additional classifiers (e.g., HFO duration or number of oscillations cycles) (Gardner et al., 2007b; Staba et al., 2002). This approach can be extended to include a cascade of classifiers to sieve the data down to candidate events (Viola and Jones, 2001).

Gardner et al. (2007a,b) have reported that line length (LL) is a more robust feature than spectral energy for false-positive detections produced by large-amplitude spikes and artifacts and that the distributions of short-time energy and LL are not Gaussian distributed. Therefore, thresholds based on parametric statistics are not optimal. Most importantly, thresholds based on global statistics perform poorly (often even on short datasets), because iEEG is often highly nonstationary. We show significant improvements in HFO detection using data-driven nonparametric adaptive thresholds (Gardner et al., 2007a,b). While our threshold selection method requires subjective specification of the threshold based on cumulative density function percentile, we observe that this method yields more uniform performance across patients than specifying a global fixed threshold based on a cutoff of mean plus standard deviation.

## PHYSIOLOGIC AND PATHOLOGIC HIGH-FREQUENCY OSCILLATIONS

The physiological importance of gamma-frequency HFO activity has received significant attention (Singer, 1999; Uhlhaas and Singer, 2006), and the experimental data support that synchronization of gamma HFOs plays a fundamental role in binding of neural activity. Microwire recordings from the hippocampus of freely behaving rats support an important physiological role for ripple frequency oscillations in memory consolidation (Buzsaki, 1989, 1992; 1998; Jadhav and Frank, 2009) (Figure 21.4). Studies from neocortex have identified sigma frequency bursts (~600 Hz) that appear to be involved in the sensory stimulation processing (Curio, 2000a,b).

In contrast to gamma, ripple, and sigma frequency HFOs, hippocampal fast-ripple oscillations are reported to be distinct pathological oscillations. In experimental rat models of epileptogenesis, fast-ripple oscillations were only recorded in the hippocampus of rats that ultimately developed spontaneous seizures (Bragin et al., 2000; Engel et al., 2009). Similarly, fast-ripple oscillations were strongly lateralized to seizure-generating hippocampus, rare in the contralateral hippocampus and absent in control animals. In addition to pathological fast-ripple oscillations, the presence of ripple-frequency HFOs was identified only in dentate gyrus of epileptic hippocampus and not the dentate gyrus of normal controls (Bragin et al., 1999a). These results from rat models support the hypothesis that fast ripples are biomarkers of epilepsy, the process of epileptogenesis, and epilepsy. The presence of ripples in the dentate of epileptic hippocampus is an interesting example of a particular HFO that is pathological because of anatomic location.

Similar to the studies in rats, microwire recordings from human medial temporal lobe show ripple frequency oscillations in both epileptogenic and nonepileptogenic hippocampus, but fast-ripple oscillations primarily occur in epileptogenic hippocampus (Bragin et al., 2002a,b; Worrell et al., 2008), and also have been observed using standard clinical macroelectrodes (Jacobs et al.,

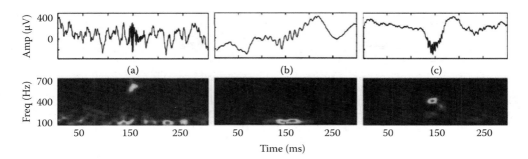

**FIGURE 21.4 (See color insert following page 458.)** Representative high-frequency oscillations (HFOs); each plot shows two views of HFO activity over a 300-msec epoch. (Top) Unfiltered EEG with an HFO event centered at 150 msec. (Bottom) Spectrogram (2.6-msec window). Note that HFOs are primarily characterized by a sharp spectral mode in the fast ripple (a, c) or ripple frequency range (b).

2008; Worrell et al., 2008). Unlike recordings from rat models, the precise location (hippocampal subfield and dentate) of the depth electrodes in human iEEG is not known; therefore, whether ripple oscillations are present in normal or epileptic human dentate is not known. In fact, while initial studies in humans reported a reduction in ripple oscillations in epileptic hippocampus compared to contralateral hippocampus, we have recently reported evidence that both fast ripples and ripples are increased in human epileptic hippocampus (Worrell et al., 2008) (Figure 21.5). The origin of these differences is unclear but may be related to the underlying hippocampal pathology of the patients investigated. The local field potentials recorded by microwire electrodes and macroelectrodes are qualitatively different (Worrell et al., 2008). We have speculated that this is due to the significant difference in electrode surface area ($10^{-3}$ mm$^2$ vs. 9.3 mm$^2$), and the physical properties of neuronal assemblies generating these signals. High-frequency oscillations tend to be localized to smaller regions of brain (Bragin et al., 2002a; Worrell et al., 2008), and the spatial averaging of local field potentials by the large surface area electrodes would effectively obscure them under sample focal HFO activity. The fact that we find that ripple HFOs are increased in the seizure onset zone (SOZ), rather than decreased, is consistent with reports from nonprimates describing an increase in ripples prior to seizures (Dzhala and Staley, 2003; Grenier et al., 2003) and supports the hypothesis that ripple-frequency HFO may be involved in the generation of seizures (Timofeev and Steriade, 2004).

## Seizure Onset Patterns

Spontaneous seizures recorded from the hippocampus of pilocarpine rat models show an increase in ripple and fast-ripple oscillations prior to and at the onset of seizures (Dzhala and Staley, 2003). In a series of elegant *in vivo* experiments by Greiner et al. (2001, 2003) in cat neocortex, ripple frequency oscillations were directly associated with generation of seizures. In cats under ketamine–xylazine anesthesia, spontaneous and electrically induced epileptiform discharges were recorded using simultaneous intracellular and extracellular recordings. In these experiments, the onset of seizure was associated with an increase in the amplitude of LFP ripple oscillations. The amplitude of ripple frequency LFPs was shown to have a direct effect on the transmembrane potential of local neurons, and the authors speculated about a possible mechanistic role for ripple oscillations in the genesis of seizures. During paroxysmal epileptiform population events, the intracellular recorded action potentials of neurons exhibited firing that was phase locked to very fast field potential oscillations or to ripple fluctuations (80 to 200 Hz). The authors speculated that the spontaneous field potential oscillations influence neuronal behavior, bringing nearby neurons to threshold, and, because neuronal activity is the generator of these field potential oscillations, a positive feedback loop may be

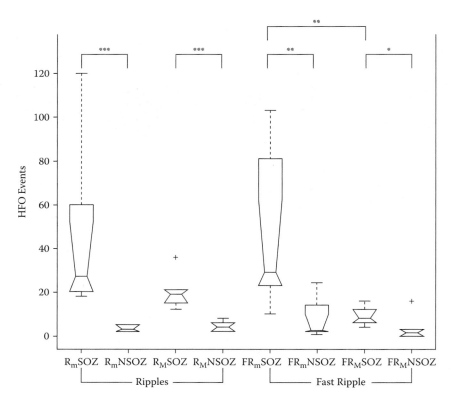

**FIGURE 21.5** Kurskal–Wallis applied to high-frequency oscillation (HFO) (ripple/fast ripple), electrode type (microwire/macro), and brain region (SOZ/non-SOZ). Box plots and the results from *post hoc* analysis using Wilcoxon rank sum (\*\*\**p* < 0.002, \*\**p* < 0.01, \**p* < 0.05). The number of microwire ripple ($R_m$) and fast ripple ($FR_m$) HFOs are increased in the SOZ compared to non-SOZ (NSOZ). The number of macroelectrode ripple ($R_M$) and fast ripple ($FR_M$) HFOs are increased in the SOZ compared to NSOZ. (From Plecko, B. et al., *Hum. Mutat.*, 28, 19–26, 2007. With permission.)

involved in the generation of neocortical ripples and the initiation and spread of seizures (Timofeev and Steriade, 2004). They propose that beyond a certain threshold of ripple HFO amplitude the neurons are entrained into the ripple frequency seizure discharge (Timofeev and Steriade, 2004).

Intracranial electroencephalography recordings in patients undergoing evaluation for epilepsy surgery show that partial seizures originating in hippocampus and temporal and extratemporal neocortex often begin with low-amplitude, high-frequency oscillations (Figure 21.6). The ranges of frequencies reported vary but they are primarily in the gamma-frequency range: 30 to 80 Hz (Alarcon et al., 1995), 40 to 120 Hz (Fisher et al., 1992), 60 to 100 Hz (Wetjen et al., 2009; Worrell et al., 2004), 70 to 90 Hz (Traub et al., 2001), 80 to 110 Hz (Allen et al., 1992), and 100 to 500 Hz (Jirsch et al., 2006).

Currently no conclusive studies have shown that HFOs are precursor events in human partial epilepsy, but, similar to the above-mentioned studies in rat pilocarpine model (Dzhala and Staley, 2003), gamma-frequency HFOs are increased in some patients prior to seizure (Worrell et al., 2004). These results suggest that HFOs within the seizure onset zone may be useful for identifying periods of increased predisposition to clinical seizures (Engel et al., 2009; Le Van Quyen et al., 2006; Worrell et al., 2004, 2008). Thus, it is fair to say that there is rapidly accumulating evidence that high-frequency oscillations may be a functional signature of epileptogenic brain and play a role in seizure generation. Whether there are clinical advantages to recording iEEG on submillimeter spatial scales over a wide spectral bandwidth is yet to be determined, but the preliminary data are encouraging.

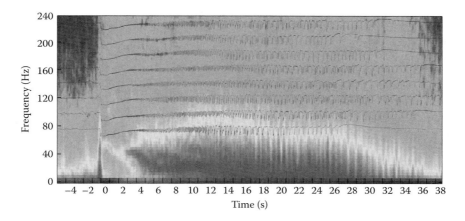

**FIGURE 21.6** **(See color insert following page 458.)** Average of eight seizures recorded from the hippocampus of a patient with medically resistant partial epilepsy. The figure show the time traces of eight of the patient's habitual seizures aligned so the seizure onset is at 0 seconds. The average time-frequency spectrogram is shown in the background. The seizures all show a characteristic slow wave transient that is associated with a high-frequency oscillation (~110 Hz) at seizure onset.

# ACKNOWLEDGMENTS

This work was supported in part by funding from the Epilepsy Therapy and Development Program, CURE Foundation, and National Institutes of Health (R01#NS 06939).

# REFERENCES

Alarcon, G., C.D. Binnie, R.D. Elwes, C.E. Polkey. (1995). Power spectrum and intracranial EEG patterns at seizure onset in partial epilepsy. *Electroencephalogr. Clin. Neurophysiol.*, 94(5):326–337.

Allen, P.J., D.R. Fish, S.J. Smith. (1992). Very high-frequency rhythmic activity during SEEG suppression in frontal lobe epilepsy. *Electroencephalogr. Clin. Neurophysiol.*, 82(2):155–159.

Ayala, G.F., M. Dichter, R.J. Gumnit, H. Matsumoto, W.A. Spencer. (1973). Genesis of epileptic interictal spikes: new knowledge of cortical feedback systems suggests a neurophysiological explanation of brief paroxysms. *Brain Res.*, 52:1–17.

Berger, H. (1929). Über das Elektrenkephalogramm des Menschen. *I. Mitteilung Arch. Psychiatr. Nervenkr.*, 87:527–570.

Bédard, C., H. Kroger, A. Destexhe. (2006). Does the $1/f$ frequency scaling of brain signals reflect self-organized critical states? *Phys. Rev. Lett.*, 97(11):118102.

Bikson, M., J.E. Fox, J.G. Jefferys. (2003). Neuronal aggregate formation underlies spatiotemporal dynamics of nonsynaptic seizure initiation. *J. Neurophysiol.*, 89(4):2330–2333.

Bragin, A., J. Engel, Jr., C.L. Wilson, I. Fried, G. Buzsaki. (1999a). High-frequency oscillations in human brain. *Hippocampus*, 9(2):137–142.

Bragin, A., J. Engel, Jr., C.L. Wilson, I. Fried, G.W. Mathern. (1999b). Hippocampal and entorhinal cortex high-frequency oscillations (100–500 Hz) in human epileptic brain and in kainic acid-treated rats with chronic seizures. *Epilepsia*, 40(2):127–137.

Bragin, A., C.L. Wilson, J. Engel, Jr. (2000). Chronic epileptogenesis requires development of a network of pathologically interconnected neuron clusters: a hypothesis. *Epilepsia*, 41(Suppl. 6):S144–S152.

Bragin, A., I. Mody, C.L. Wilson, J. Engel, Jr. (2002a). Local generation of fast ripples in epileptic brain. *J. Neurosci.*, 22(5):2012–2021.

Bragin, A., C.L. Wilson, R.J. Staba, M. Reddick, I. Fried, J. Engel, Jr. (2002b). Interictal high-frequency oscillations (80–500 Hz) in the human epileptic brain: entorhinal cortex. *Ann. Neurol.*, 52(4):407–415.

Brinkmann, B.H., M.R. Bower, K.A. Stengel, G.A. Worrell, M. Stead. (2009). Large-scale electrophysiology: acquisition, compression, encryption, and storage of big data. *J. Neurosci. Methods*, 180(1):185–192.

Buzsaki, G. (1989). Two-stage model of memory trace formation: a role for "noisy" brain states. *Neuroscience*, 31(3):551–570.

Buzsaki, G. (1998). Memory consolidation during sleep: a neurophysiological perspective. *J. Sleep. Res.*, 7(Suppl. 1):17–23.

Buzsaki, G. (2004). Large-scale recording of neuronal ensembles. *Nat. Neurosci.*, 7(5):446–451.

Buzsaki, G., A. Draguhn. (2004). Neuronal oscillations in cortical networks. *Science*, 304(5679):1926–1929.

Buzsaki, G., Z. Horvath, R. Urioste, J. Hetke, K. Wise. (1992). High-frequency network oscillation in the hippocampus. *Science*, 256(5059):1025–1027.

Chesi, A.J., T.W. Stone. (1998). Epileptiform activity in supragranular and infragranular blocks of mouse neocortex. *Epilepsy Res.*, 31(1):29–38.

Curio, G. (2000a). Ain't no rhythm fast enough: EEG bands beyond beta. *J. Clin. Neurophysiol.*, 17(4):339–340.

Curio, G. (2000b). Linking 600-Hz "spikelike" EEG/MEG wavelets ("sigma-bursts") to cellular substrates: concepts and caveats. *J. Clin. Neurophysiol.*, 17(4):377–396.

Dzhala, V.I., K.J. Staley. (2003). Transition from interictal to ictal activity in limbic networks *in vitro*. *J. Neurosci.*, 23(21):7873–7880.

Dzhala, V.I., K.J. Staley. (2004). Mechanisms of fast ripples in the hippocampus. *J. Neurosci.*, 24(40):8896–8906.

Engel, Jr., J., A. Bragin, R. Staba, I. Mody. (2009). High-frequency oscillations: what is normal and what is not? *Epilepsia*, 50(4):598–604.

Esteller, R., J. Echauz, M. D'Alessandro, G. Worrell, S. Cranstoun, G. Vachtsevanos, B. Litt. (2005). Continuous energy variation during the seizure cycle: towards an on-line accumulated energy. *Clin. Neurophysiol.*, 116(3):517–526.

Firpi, H., O. Smart, G. Worrell, E. Marsh, D. Dlugos, B. Litt. (2007). High-frequency oscillations detected in epileptic networks using swarmed neural-network features. *Ann. Biomed. Eng.*, 35(9):1573–1584.

Fisher, R.S., W.R. Webber, R.P. Lesser, S. Arroyo, S. Uematsu. (1992). High-frequency EEG activity at the start of seizures. *J. Clin. Neurophysiol.*, 9(3):441–448.

Gabor, A.J., R.P. Scobey, C.J. Wehrli. (1979). Relationship of epileptogenicity to cortical organization. *J. Neurophysiol.*, 42(6):1609–1625.

Gardner, A.B., A. Krieger, G. Vachtsevanos, B. Litt. (2007a). One-class novelty detection for seizure analysis from intracranial EEG. *J. Mach. Learn. Res.*, 7, 1025–1044.

Gardner, A.B., G.A. Worrell, E. Marsh, D. Dlugos, B. Litt. (2007b). Human and automated detection of high-frequency oscillations in clinical intracranial EEG recordings. *Clin. Neurophysiol.*, 118(5):1134–1143.

Gloor, P. (1969). The work of Hans Berger. *Electroencephalogr. Clin. Neurophysiol.*, 27(7):649.

Gloor, P. (1975). Contributions of electroencephalography and electrocorticography to the neurosurgical treatment of the epilepsies. *Adv. Neurol.*, 8:59–105.

Grenier, F., I. Timofeev, M. Steriade. (2001). Focal synchronization of ripples (80–200 Hz) in neocortex and their neuronal correlates. *J. Neurophysiol.*, 86(4):1884–1898.

Grenier, F., I. Timofeev, M. Steriade. (2003). Neocortical very fast oscillations (ripples, 80–200 Hz) during seizures: intracellular correlates. *J. Neurophysiol.*, 89(2):841–852.

Hamer, H.M., H.H. Morris, E.J. Mascha, M.T. Karafa, W.E. Bingaman, M.D. Bej, R.C. Burgess, D.S. Dinner, N.R. Foldvary, J.F. Hahn, P. Kotagal, I. Najm, E. Wyllie, H.O. Luders. (2002). Complications of invasive video-EEG monitoring with subdural grid electrodes. *Neurology*, 58(1):97–103.

Hu, S., M. Stead, G.A. Worrell. (2007). Automatic identification and removal of scalp reference signal for intracranial EEGs based on independent component analysis. *IEEE Trans. Biomed. Eng.*, 54(9):1560–1572.

Jacobs, J., P. LeVan, R. Chander, J. Hall, F. Dubeau, J. Gotman. (2008). Interictal high-frequency oscillations (80–500 Hz) are an indicator of seizure onset areas independent of spikes in the human epileptic brain. *Epilepsia*. 49(11):1893–1907.

Jadhav, S.P., L.M. Frank. (2009). Reactivating memories for consolidation. *Neuron*, 62(6):745–746.

Jirsch, J.D., E. Urrestarazu, P. LeVan, A. Olivier, F. Dubeau, J. Gotman. (2006). High-frequency oscillations during human focal seizures. *Brain*, 129(Pt. 6):1593–1608.

Le Van Quyen, M., I. Khalilov, Y. Ben-Ari. (2006). The dark side of high-frequency oscillations in the developing brain. *Trends Neurosci.*, 29(7):419–427.

Linkenkaer-Hansen, K., V.V. Nikouline, J.M. Palva, R.J. Ilmoniemi. (2001). Long-range temporal correlations and scaling behavior in human brain oscillations. *J. Neurosci.*, 21(4):1370–1377.

Lisman, J.E., M.A. Idiart. (1995). Storage of 7 ± 2 short-term memories in oscillatory subcycles. *Science*, 267(5203):1512–1515.

Milstein, J., F. Mormann, I. Fried, C. Koch. (2009). Neuronal shot noise and Brownian $1/f^2$ behavior in the local field potential. *PLoS One*, 4(2):e4338.

Neidermeyer, E., F.L. Da Silva. (2005). *Electroencephalography: Basic Principals, Clinical Applications, and Related Fields*, Philadelphia, PA: Lippincott & Wilkins.

Reichenthal, E., S. Hocherman. (1977). The critical cortical area for development of penicillin-induced epilepsy. *Electroencephalogr. Clin. Neurophysiol.*, 42(2):248–251.

Reichenthal, E., S. Hocherman. (1979). A critical epileptic area in the cat's cortex and its relation to the cortical columns. *Electroencephalogr. Clin. Neurophysiol.*, 47:147–152.

Saab, M.E., J. Gotman. (2005). A system to detect the onset of epileptic seizures in scalp EEG. *Clin. Neurophysiol.*, 116(2):427–442.

Schevon, C.A., S.K. Ng, J. Cappell, R.R. Goodman, G. McKhann, Jr., A. Waziri, A. Branner, A. Sosunov, C.E. Schroeder, R.G. Emerson. (2008). Microphysiology of epileptiform activity in human neocortex. *J. Clin. Neurophysiol.*, 25(6):321–330.

Schwartz, T.H. (2003). Optical imaging of epileptiform events in visual cortex in response to patterned photic stimulation. *Cereb. Cortex*, 13(12):1287–1298.

Shiro, U., A. Itzhak. (1982). Digital low-pass differentiation for biological signal processing. *IEEE Trans. Biomed. Eng.*, 29:686–693.

Singer, W. (1999). Neuronal synchrony: a versatile code for the definition of relations? *Neuron*, 24(1):49–65, 111–125.

Staba, R.J., C.L. Wilson, A. Bragin, I. Fried, J. Engel, Jr. (2002). Quantitative analysis of high-frequency oscillations (80–500 Hz) recorded in human epileptic hippocampus and entorhinal cortex. *J. Neurophysiol.*, 88(4):1743–1752.

Staba, R.J., L. Frighetto, E.J. Behnke, G.W. Mathern, T. Fields, A. Bragin, J. Ogren, I. Fried, C.L. Wilson, J. Engel, Jr. (2007). Increased fast ripple to ripple ratios correlate with reduced hippocampal volumes and neuron loss in temporal lobe epilepsy patients. *Epilepsia*, 48(11):2130–2138.

Timofeev, I., M. Steriade. (2004). Neocortical seizures: initiation, development and cessation. *Neuroscience*, 123(2):299–336.

Traub, R.D., M.A. Whittington, E.H. Buhl, F.E. LeBeau, A. Bibbig, S. Boyd, H. Cross, T. Baldeweg. (2001). A possible role for gap junctions in generation of very fast EEG oscillations preceding the onset of, and perhaps initiating, seizures. *Epilepsia*, 42(2):153–170.

Uhlhaas, P.J., W. Singer. (2006). Neural synchrony in brain disorders: relevance for cognitive dysfunctions and pathophysiology. *Neuron*, 52(1):155–168.

Ulbert, I., Z. Magloczky, L. Eross, S. Czirjak, J. Vajda, L. Bognar, S. Toth, Z. Szabo, P. Halasz, D. Fabo, E. Halgren, T.F. Freund, G. Karmos. (2004). *In vivo* laminar electrophysiology co-registered with histology in the hippocampus of patients with temporal lobe epilepsy. *Exp. Neurol.*, 187(2):310–318.

Urrestarazu, E., J.D. Jirsch, P. LeVan, J. Hall, M. Avoli, F. Dubeau, J. Gotman. (2006). High-frequency intracerebral EEG activity (100–500 Hz) following interictal spikes. *Epilepsia*, 47(9):1465–1476.

Van Gompel, J.J., S.M. Stead, C. Giannini, F.B. Meyer, W.R. Marsh, T. Fountain, E. So, A. Cohen-Gadol, K.H. Lee, G.A. Worrell. (2008a). Phase I trial: safety and feasibility of intracranial electroencephalography using hybrid subdural electrodes containing macro- and microelectrode arrays. *Neurosurg. Focus*, 25(3):E23.

Van Gompel, J.J., G.A. Worrell, M.L. Bell, T.A. Patrick, G.D. Cascino, C. Raffel, W.R. Marsh, F.B. Meyer. (2008b). Intracranial electroencephalography with subdural grid electrodes: techniques, complications, and outcomes. *Neurosurgery*, 63(3):498–506.

Vanhatalo, S., J. Voipio, K. Kaila. (2005). Full-band EEG (fbEEG): a new standard for clinical electroencephalography. *Clin. EEG Neurosci.*, 36(4):311–317.

Viola, P., M. Jones. (2001). Rapid object detection using a boosted cascade of simple features. In *Proceedings of the IEEE Computer Society Conference on Computer Vision and Pattern Recognition (CVPR 2001)*, December 8–14, Kauai, HI, p. 511.

Wetjen, N.M., W.R. Marsh, F.B. Meyer, G.D. Cascino, E. So, J.W. Britton, S.M. Stead, G.A. Worrell. (2009). Intracranial electroencephalography seizure onset patterns and surgical outcomes in nonlesional extratemporal epilepsy. *J. Neurosurg.*, 110(6):1147–1152.

Wilson, S.B., R. Emerson. (2002). Spike detection: a review and comparison of algorithms. *Clin. Neurophysiol.*, 113(12):1873–1881.

Wilson, S.B., M.L. Scheuer, C. Plummer, B. Young, S. Pacia. (2003). Seizure detection: correlation of human experts. *Clin. Neurophysiol.*, 114(11):2156–2164.

Worrell, G.A., S.D. Cranstoun, J. Echauz, B. Litt. (2002). Evidence for self-organized criticality in human epileptic hippocampus. *NeuroReport*, 13(16):2017–2021.

Worrell, G.A., L. Parish, S.D. Cranstoun, R. Jonas, G. Baltuch, B. Litt. (2004). High-frequency oscillations and seizure generation in neocortical epilepsy. *Brain*, 127(Pt. 7):1496–1506.

Worrell, G.A., A.B. Gardner, S.M. Stead, S. Hu, S. Goerss, G.J. Cascino, F.B. Meyer, R. Marsh, B. Litt. (2008). High-frequency oscillations in human temporal lobe: simultaneous microwire and clinical macroelectrode recordings. *Brain*, 131(Pt. 4):928–937.

# Section IV

## Alternative Therapies

# 22 Special Treatments in Epilepsy

*Kevin Chapman and James W. Wheless*

## CONTENTS

## INTRODUCTION

Seizure disorders represent a frequently occurring neurologic problem. Antiepileptic drugs (AEDs) are the primary treatment modality and provide good seizure control in most patients; however, more than 25% of children and adults with seizure disorders either have intractable seizures or suffer significant adverse effects secondary to medications. A limited number will benefit from surgical therapy. The shortcomings of antiepileptic drug therapy have allowed alternative treatments to emerge. In this chapter, three specific and unique treatments of special interest to the epileptologist are reviewed (pyridoxine, acetazolamide, and magnesium sulfate).

# PYRIDOXINE (VITAMIN B$_6$)

## HISTORY

The relationship between vitamin B$_6$ and convulsions has attracted general interest, especially in light of the role of vitamin B$_6$ in γ-aminobutyric acid (GABA) metabolism (Figure 22.1 and Table 22.1) (Jakobs et al., 1993; Minns, 1980). Vitamin B$_6$ deficiency (Snyderman et al., 1950) and dependency (Hunt et al., 1954) were described first about 50 years ago and are well known among the vitamin B$_6$-related convulsions. Hansson and Hagberg (1968) reported the efficacy of high-dose pyridoxine in some epileptic children. These seizures were called *vitamin B$_6$-responsive seizures* (Ekelund et al., 1969; Hansson and Hagberg, 1968), indicating they were either vitamin B$_6$ dependent or deficient in etiology.

## MECHANISM OF ACTION

Pyridoxine (vitamin B$_6$) is absorbed in dietary form and converted to its active form, pyridoxal 5′-phosphate (PLP), through the action of a kinase and oxidase (Figure 22.2). PLP is the coenzyme for glutamate decarboxylase (GAD) and GABA transaminase (GABA-T), the enzymes that regulate GABA in the central nervous system. Rats treated with semicarbazide demonstrated inhibition of glutamic acid decarboxylase and a subsequent decrease in GABA levels, as well as development of seizures. This sequence was competitively reversed by pyridoxine, demonstrating its critical role in GABA metabolism (Killam and Bain, 1957) and its presumed effect in treating pyridoxine-deficient seizures. Mice lacking tissue nonspecific alkaline phosphatases develop seizures due to defective metabolism of PLP and have decreased brain GABA levels (Waymire et al., 1995). This mutant seizure phenotype can be rescued by the administration of pyridoxine. Additionally, PLP plays a crucial role in the central nervous system as a cofactor for over 100 enzymatic reactions. These pyridoxine-dependent enzymes are involved in the metabolism of amino acids and neurotransmitters (dopamine, serotonin, GABA, and norepinephrine), as well as sphingolipids and polyamines (Dakshinamurti et al., 1990). There exists experimental evidence for vitamin B$_6$ acting as a modulator of steroid activity, a mechanism that may explain its efficacy in treating infantile spasms (Cake et al., 1978).

Initially, the cause of pyridoxine-dependent seizures (PDSs) was felt to be a defect in GAD that would reduce pyridoxal phosphate (PLP) binding and enzyme activity. This would result in low GABA and high glutamate levels, a combination that favors seizures. This was supported by the finding that neonatal seizures due to pyridoxine deficiency or dependency have low cerebrospinal fluid (CSF) GABA levels, which improve after treatment with vitamin B$_6$ (Kurlemann et al., 1987, 1991, 1992). Baumeister et al. (1994) examined CSF levels of glutamate and GABA in a patient with

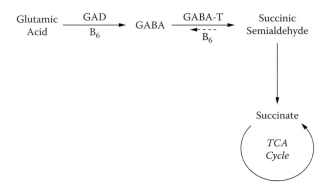

**FIGURE 22.1** GABA metabolism; longer arrows indicate major direction of metabolism. GABA, γ-aminobutyric acid; GABA-T, γ-aminobutyric acid transaminase; GAD, glutamic acid decarboxylase.

**TABLE 22.1**
**Pyridoxine Associated Seizure Disorders**

| Condition | B₆ Level | Age of Onset | Family History | Response to AEDs | Response of Seizures to B₆ | Seizure Recurrence after Stopping B₆ | Interictal EEG | Duration of Treatment | Genetics |
|---|---|---|---|---|---|---|---|---|---|
| I. B₆ deficiency | Decreased | 1 to 4 months | Negative | Poor | Good, physiologic dose (1.0 to 5.0 mg) | (−) | Normal | Brief (single dose) | None known, all acquired |
| II. B₆ dependency | Normal | Neonatal (siblings with similar problem or SE in infancy) | Positive | Poor | Good, pharmacologic dose (10 to 50 mg/day) | (+) | Abnormal | Indefinite | *ALDH7A1* gene (PNPO) |
| III. B₆ responsiveness | Normal | 2 to 14 months | Negative | Poor | Good, pharmacologic dose (20 to 400 mg/kg/day) | (−) | Abnormal | Brief (weeks to months) | None known |

*Note:* AEDs, antiepileptic drugs; SE, status epilepticus.

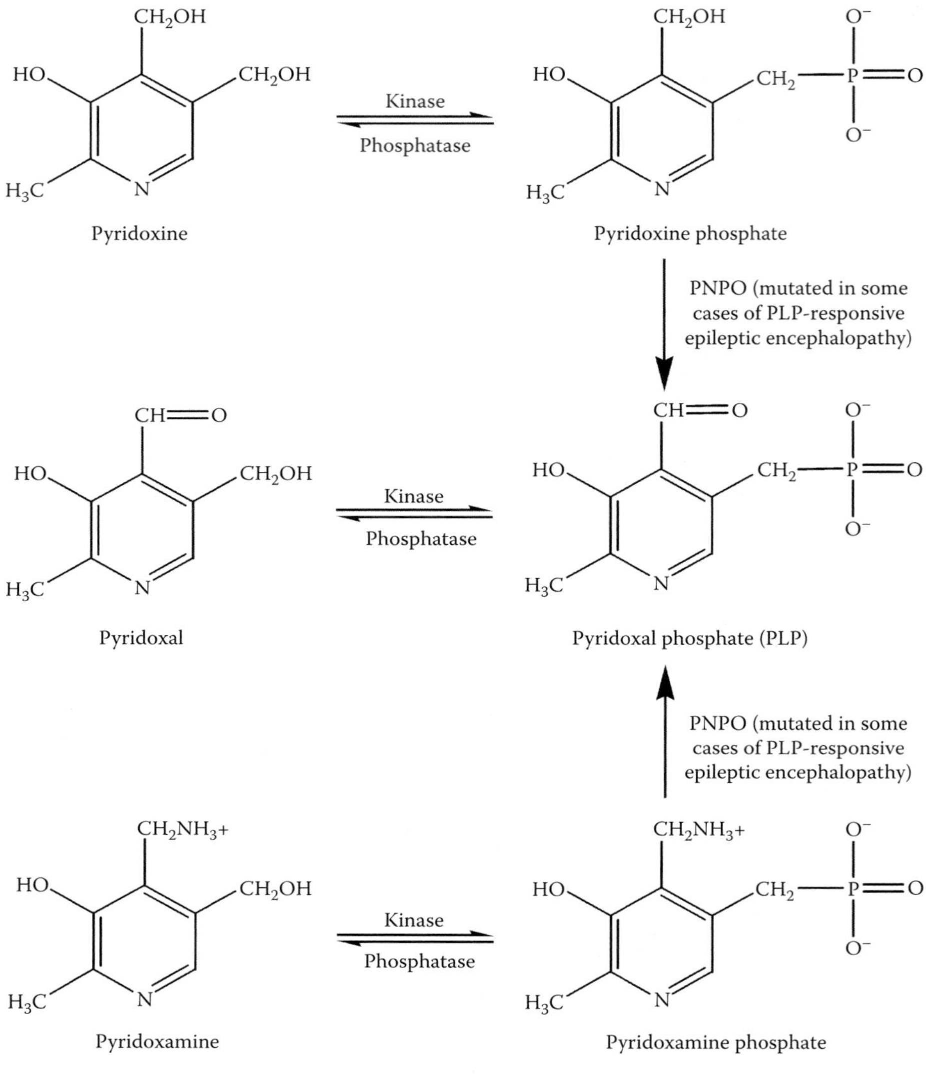

**FIGURE 22.2**    Vitamin B$_6$ metabolism. PNPO, pyridoxamine 5′-phosphate oxidase.

pyridoxine dependency while on and off vitamin B$_6$ treatment. GABA levels were normal, but off vitamin B$_6$ the glutamate level was markedly elevated and required higher doses of pyridoxine to normalize than did the electroencephalography (EEG). The imbalance of excitatory (glutamate) to inhibitory (GABA) neurotransmitters may explain the seizures and the cognitive difficulties. A child with vitamin B$_6$-dependent seizures had reduced GAD and GABA levels in the frontal and occipital cortices (Lott et al., 1978). A boy with pyridoxine-responsive West syndrome had CSF GABA levels examined before and during treatment with high-dose vitamin B$_6$ (Kurlemann et al., 1997). The low GABA levels normalized with treatment, the hypsarrhythmia resolved, and vitamin B$_6$ was withdrawn. The authors hypothesized that pyridoxine-responsive West syndrome may be due to a developmental change of GABA metabolism that responds to high-dose pyridoxine therapy. This was presumed to be due to abnormal cofactor binding (Jakobs et al., 1993); however, genetic linkage studies did not implicate mutations in GAD genes as the primary causative factor in pyridoxine-dependent seizures (Battaglioli et al., 2000; Cormier-Daire et al., 2000; Kure et al., 1998).

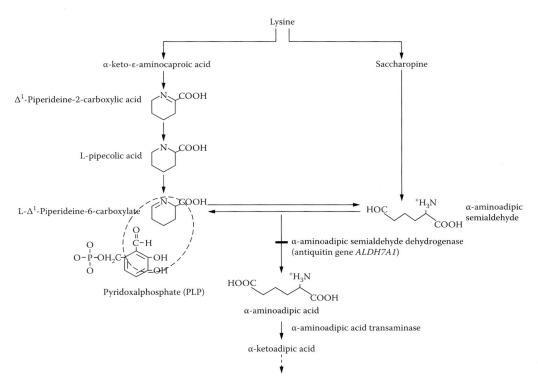

**FIGURE 22.3** Description of pathway leading to inactivation of PLP. (From Plecko, B. et al., *Hum. Mutat.*, 28, 19–26, 2007. With permission.)

Reports of pyridoxine dependency occurring in multiple family members suggested that this was a genetic condition, with a recessive pattern of inheritance (Bankier et al., 1983; Baxter et al., 1996; Coker, 1992; Cormier-Daire et al., 2000; Garty et al., 1962; Jardim et al., 1994; Mikati et al., 1991; Nabbout et al., 1999; Scriver, 1960; Waldinger and Berg, 1963). Results of genetic linkage analysis in four families with consanguineous parents identified the locus for pyridoxine-dependent epilepsy to chromosome 5q31 (Cormier-Daire et al., 2000). Reports of elevated pipecolic acid in the CSF of several patients with PDSs led to the identification of a pathogenic mutation in the antiquitin gene, a $\Delta^1$-piperideine-6-carboxylase (P6C)–$\alpha$-aminoadipic semialdehyde ($\alpha$-AASA) dehydrogenase deficiency. This enzyme was found to be encoded by *ALDH7A1* and mapped to chromosome 5q31, the gene locus for PDSs (Mills et al., 2006).

Mutations in antiquitin inhibit the metabolism of pipecolic acid and $\alpha$-AASA, causing elevations in CSF and plasma levels. $\alpha$-AASA dehydrogenase deficiency causes seizures because elevated P6C (Figure 22.3) inactivates PLP through a Knoevenagel condensation reaction. This inactivation leads to a decrease in PLP, thereby reducing this important enzymatic cofactor. Supplementation of pyridoxine can overcome this condensation reaction and restore the PLP pool available for the formation of GABA and other enzymatic reactions. There is also speculation that the highly reactive aldehyde groups of $\alpha$-AASA may interact with cellular proteins causing neurotoxicity (Plecko et al., 2007).

It should be pointed out, however, that genetic heterogeneity has been described with PDSs (Bennett et al., 2005; Striano et al., 2009), and not all patients with PDSs have antiquitin mutations. Therefore, even with the availability of reliable biomarkers, early consideration of a pyridoxine trial is still critically important in an infant with pharmacoresistant epilepsy of unknown etiology. Interestingly, patients with hyperprolinemia type II also have PDSs. These patients produce excess L-$\Delta^1$-pyrroline-5-carboxylate (P5C) that inactivates PLP through a Knoevenagel condensation reaction. This inactivation leads to a similar decrease in the PLP pool for enzymatic reactions (Farrant et al., 2001).

## PYRIDOXINE DEFICIENCY

The requirement for pyridoxine was first noted in the 1930s and 1940s after removal from the diet caused convulsions and a hypochromic, microcytic anemia in chicks, swine, calves, and ducks (Snyderman et al., 1953; Tower et al., 1956). However, little was known about human requirements until Snyderman et al. (1950) gave two infants a pyridoxine-deficient diet. One developed seizures, which were promptly relieved by the intravenous administration of 50 mg of pyridoxine. Subsequently, between 1951 and 1954, a considerable number of infants were reported to have developed convulsions that were abruptly alleviated by administering 5 to 10 mg of pyridoxine (Bessey et al., 1954). In each instance, the child had received a commercial food formula with inadequate amounts of vitamin $B_6$ (Bessey et al., 1954, 1957; Coursin, 1954, 1955; Molony and Parmelee, 1954). These observations led to a report in the *Journal of the American Medical Association* (Report of the Council on Pharmacy and Chemistry, 1951) warning that convulsions in infancy could be produced by pyridoxine deficiency. Prompt treatment was associated with immediate cessation of symptoms and no residual findings (Coursin, 1955; Scriver and Hutchison, 1963), although if prompt treatment is not given the prognosis for mental development is poor. In recent years, isoniazid poisoning has caused a relative pyridoxine deficiency by inhibiting the activity of brain PLP. In this clinical setting, children present in refractory convulsive status epilepticus which is responsive to intravenous pyridoxine administration.

## PYRIDOXINE-DEPENDENT SEIZURES

Pyridoxine dependency is a rare cause of generalized seizures in children that was first reported about 50 years ago (Hunt et al., 1954). It is an autosomal recessive disorder with variable penetrance that typically presents in neonates as generalized seizures or status epilepticus unresponsive to standard antiepileptic drug therapy; however, seizures stop immediately following parenteral pyridoxine. Lifelong therapy with pyridoxine is required to prevent seizure recurrence. Long-term follow-up suggests that the prognosis for complete seizure control is excellent, with patients experiencing occasional seizures during illness or noncompliance (Basura et al., 2009; Mikati et al., 1991); however, death may occur in untreated patients (Gospe, 2006)

Knowledge about this condition comes from case reports and small clinical series; however, from these there appears to be a consistent constellation of features. The typical presentation is that of neonatal seizures beginning before birth or in the first two days of life that are intractable to antiepileptic drugs, and when pyridoxine is administered they stop completely (Baxter et al., 1996, 2001; Crowell and Roach, 1983; Garty et al., 1962; Gordon, 1997; Gospe, 1998, 2006; Haenggeli et al., 1991; Hunt et al., 1954; Scriver, 1960; Waldinger and Berg, 1963). The seizures are frequent and brief. All seizure types have been described, and multiple seizure types are usual, particularly clonic and generalized tonic (Baxter et al., 1996; Crowell and Roach 1983; Haenggeli et al., 1991; Mikati et al., 1991).

Atypical (later onset or onset beyond infancy) presentations do not follow this classic pattern and may occur in up to 35% of affected children (Bankier et al., 1983; Baxter, 1999; Coker, 1992; Goutieres and Aicardi, 1985; Krishnamoorthy, 1983; Mikati et al., 1991; Shih et al., 1996; Tanaka et al., 1992). Children may present up to 3 years of age, have initial seizures occurring only during febrile illnesses, or experience bouts of status epilepticus; the initial seizures may be partially or completely controlled with standard AEDs but later become difficult to control (Coker, 1992; Ekelund et al., 1969; Goutieres and Aicardi, 1985). Prolonged seizure-free intervals (up to 6 months) may occur without pyridoxine. Rarely, high-dose pyridoxine may be required; however, in others, small doses (1 mg) can control the seizures, allowing multivitamin supplements to confuse the diagnosis (Grillo et al., 2001).

The EEG findings in untreated patients have not been specifically described. The EEG can be normal before seizure onset. In typical (i.e., neonatal) pyridoxine-dependent seizures, the background may be normal or diffusely slow, but normal sleep patterns are lacking (Nabbout et al.,

1999). Superimposed paroxysmal features include focal or multifocal spikes or sharp waves, or burst suppression patterns (Bass et al., 1996; Crowell and Roach, 1983; Mikati et al., 1991; Nabbout et al., 1999; Waldinger and Berg, 1963). In a small number of late-onset patients, the interictal EEG is normal, and high-voltage slow activity is less prominent.

Early-onset seizures are not the only clinical manifestation of this condition. Over half of the infants present with an encephalopathy that is not secondary to the seizure activity. This is characterized by hyperalertness, irritability, tremulousness, abnormal cry, feeding difficulties and vomiting, and an exaggerated startle response that can trigger seizures (Haenggeli et al., 1991; Nabbout et al., 1999; Scriver, 1960). Meconium staining, hypotonia, and difficulties at birth can lead to a misdiagnosis of hypoxic–ischemic encephalopathy (Baxter, 1999). In contrast, most children with the late-onset form are normal or mildly delayed before seizure onset. Mothers of those with early-onset pyridoxine-dependent seizures may experience abnormal episodic fetal movements, which begin at 20 gestation weeks, last 15 to 20 minutes, and occur one or more times a day (Baxter et al., 1996; Nabbout et al., 1999). The use of pyridoxine in pregnancy is advantageous to babies with pyridoxine-dependent seizures. In addition to treating the seizures, the neurodevelopmental outcome is normal in cases in which mothers take pyridoxine during pregnancy.

Administration of a therapeutic trial of pyridoxine is generally the first step in identifying patients with pyridoxine-dependent seizures. When performing a trial of pyridoxine, the doses used vary from 5 to 100 mg, although 100 mg has commonly been used in more recent reports. Both oral and intravenous routes are effective, but the latter yields a more rapid onset. The EEG reveals an initial disappearance of spikes followed by slow activity (Mikati et al., 1991). Rarely, the pyridoxine trial may cause severe electrocerebral suppression lasting hours (Bass et al., 1996; Mikati et al., 1991; Tanaka et al., 1992). Patients with a short-lived response or no initial response and then a response to subsequent doses have been reported (Bass et al., 1996). As a result, Gospe (1998) recommended that an initial dose of 100 mg be given; if no improvement is seen, this can be repeated every 10 minutes to a total of 500 mg. In late-onset forms, a dose of 100 to 200 mg is given daily for one week. The seizure control should improve by the second day, but complete control may take the entire week. A lack of response to this regimen or seizure recurrence while receiving pyridoxine probably excludes the diagnosis. The administration of pyridoxine stops the seizures, and typically the child will sleep for hours and remain hypotonic and less responsive. This is primarily seen in the typical form.

A recent report indicates that folinic-acid-responsive seizures are identical to pyridoxine-dependent epilepsy (Gallagher et al., 2009). This has several important clinical implications. First, patients previously diagnosed with folinic-acid-responsive seizures should also be treated with pyridoxine. Second, this report confirms the necessity of an oral pyridoxine trial in those children who do not respond to intravenous pyridoxine. Third, folinic acid therapy may be of added benefit in some children with α-AASA dehydrogenase deficiency. As a result, new treatment recommendations have been developed (Figure 22.4) to include both pyridoxine and folinic acid in neonates and infants with epileptic encephalopathy (if unknown etiology).

The diagnosis of pyridoxine-dependent seizures was previously based on confirming that the seizures stopped after pyridoxine administration and recurred when it was withdrawn (Baxter, 2001; Gospe, 2006). With the identification of the *ALDH7A1* mutation, assays for α-AASA and pipecolic acid can be used to screen patients for PDSs, without the need for the withdrawal test. Urinary determinants of α-AASA have been suggested as the easiest and best method for screening patients with PDS. Although pipecolic levels are elevated, they may only be mildly elevated in PDS and thus are best evaluated using the more invasive testing of CSF. In addition, pipecolic acid may be elevated in peroxisomal disorders, confounding diagnosis (Struys and Jakobs, 2007).

As noted previously, there are reports of patients with PDSs who do not have the *ALDH7A1* mutation, and confirmation of the diagnosis is demonstrated by seizure recurrence with pyridoxine withdrawal and resolution of seizures when pyridoxine is restarted. After stopping pyridoxine, in typical cases the clinical sequence consists of a prodrome of altered behavior and irritability, with

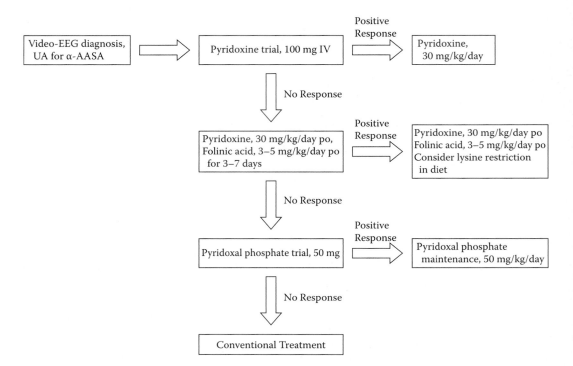

**FIGURE 22.4**  Proposed treatment algorithm for the treatment of intractable cryptogenic neonatal seizures.

seizure recurrence typically within 5 days. If the trial is performed later in infancy or childhood, it may take 6 weeks or more before seizures recur (Gospe, 1998; Jakobs et al., 1993; Yoshikawa et al., 1999). The best age to perform a withdrawal trial is not known. Some authors recommend the preschool years, with the parents having 100 mg of liquid pyridoxine on hand to use as a rescue agent. Intercurrent illnesses have been reported to precipitate seizures, and increasing the daily dose by 100 mg for the duration of the illness has been recommended (Baxter et al., 1996; Tanaka et al., 1992). Using the previous clinical criteria (prior to identification of the *ALDH7A1* mutation) epidemiologic studies demonstrated a prevalence of 1:20,000 to 1:700,000 (Baxter et al., 1996, 2001, 1999; Ebinger et al., 1999).

Maintenance doses of 50 to 100 mg/day have typically been proposed; however, a recent report suggests that, although the seizures may respond to these doses, higher doses (15 to 18 mg/kg/day, up to 500 mg/day) may improve the cognitive outcome, and the daily dose should be based on the latter (Baxter, 2001; Baxter et al., 1996; Tanaka et al., 1992). This dose is unlikely to produce a sensory neuropathy, a finding confirmed using similar doses for the treatment of homocystinuria (Baxter, 2001; Mpofu et al., 1991).

The prognosis is determined by several factors. Untreated patients die, usually in status epilepticus, typically within the first days to months of life (Haenggeli et al., 1991). In treated patients, a poorer prognosis has been associated with onset in the neonatal period, initial responsiveness to routine AEDs, persistently abnormal EEG background, and a delay in treatment of over one week (Mikati et al., 1991). Antenatal pyridoxine might positively influence outcome, and the maintenance dose may have some influence (Baxter, 2001; Nabbout et al., 1999). Specific difficulties are often encountered even with early treatment. Limited formal neuropsychologic testing has been reported. Baxter noted normal nonverbal skills, although they were below average, and impairment of expressive verbal skills even with early treatment (Baxter, 2001; Baxter et al., 1996). Additionally, some children show a marked articulatory dysphasia. These findings are less pronounced in children with late-onset forms. Neuroimaging studies, primarily magnetic resonance imaging (MRI), show abnormalities of white matter and brain structure (Baxter et al., 1996; Grillo et al., 2001; Ito et al., 1984; Jardim et al., 1994; Tanaka et al., 1992). The most common finding is hypoplasia of the posterior portion of the corpus callosum (Yoshikawa et al., 1999). Also reported is cerebellar hypoplasia with a mega cisterna magna and rarely hydrocephalus.

## Pyridoxine-Responsive Seizures (Infantile Spasms)

The finding that CSF GABA levels were significantly lower in children with infantile spasms than in controls led to initial trials with vitamin $B_6$ (Ito et al., 1984). Initial reports suggested efficacy in some patients and led to larger open-label prospective trials (Blennow and Starck, 1986; French et al., 1965; Ito, 1998; Ito et al., 1991; Ohtsuka et al., 1987, 2000; Pietz et al., 1993). These children had a variable age of onset of seizures; no signs or symptoms of vitamin $B_6$ deficiency or cryptogenic or symptomatic seizures; and a response to high-dose pyridoxine without the need for indefinite treatment (hence, the term *vitamin $B_6$-responsive seizures*). Typically, spasms were completely suppressed, and the EEG normalized or dramatically improved within 3 to 4 weeks of initiation of vitamin $B_6$ therapy. Cryptogenic patients responded best, but symptomatic patients with various etiologies also responded. Additionally, follow-up studies have suggested that their mental and seizure outcomes were more favorable in children with vitamin $B_6$-responsive West syndrome than those with vitamin $B_6$-nonresponsive West syndrome, regardless of etiology (Ohtsuka et al., 2000). Pyridoxine therapy can be discontinued after one year of age if the child's EEG does not exhibit epileptiform discharges.

Clinical features and laboratory data cannot predict which children with infantile spasms will respond to pyridoxine. Accordingly, some authors suggest that high-dose (200 to 400 mg/day) vitamin $B_6$ treatment for 2 weeks should be tried in all cases of West syndrome.

Unfortunately, no standard treatment protocols have evaluated initial and subsequent drug therapy for infantile spasms. Various surveys of the treatment of infantile spasms have been performed over the last several years (Appleton, 1996; Bobele and Bodensteiner, 1994; Ito, 1998; Ito et al., 2000; Ohashi et al., 1995). The disparities in treatment protocols in populations that have no known pharmacogenomic differences to account for differing drug responses further support the need for comparative sequential treatment protocols in West syndrome.

## Pyridoxal Phosphate

Pyridoxal 5′-phosphate is the functional coenzyme form of vitamin $B_6$ (pyridoxine) and is required for multiple enzymatic reactions, including the formation of neurotransmitters, sphingolipids, and amino acids (Dakshinamurti et al., 1990). Case reports identified neonates with an epileptic encephalopathy that failed to respond to pyridoxine but responded to PLP (Clayton et al., 2003; Kuo and Wang, 2002). Abnormalities in the level of homovanillic acid, 5-hydroxyindoleacetic acid, and 3-methoxytyrosine in CSF and elevation of plasma glycine and threonine were consistent with reduced function in PLP-dependent enzymes. These abnormalities suggested that PLP supplementation may be an effective treatment (Clayton et al., 2003). Previous reports had demonstrated the utility of PLP in the treatment of infantile spasms, either alone or as adjunctive treatment (Ohtsuka et al., 1987; Takuma and Seki, 1996). A single case report of an adult with intractable status epilepticus who responded to PLP was attributed to low pyridoxine levels associated with his malnourished state (Nakagawa et al., 1997).

Further research of families with PLP-responsive epileptic encephalopathy revealed mutations in the *PNPO* gene on chromosome 17q21.2 encoding pyridox(am)ine 5′-phosphate oxidase (PNPO) (Mills et al., 2005). The five patients described in this study had similar clinical phenotypes: premature birth, acidosis (4 of 5 patients), seizures within the first 12 hours of life, and a burst-suppression pattern on EEG. Biochemical evaluation of CSF and plasma demonstrated similar abnormalities as described above (Bagci et al., 2008). Reported treatment doses with PLP ranged from 40 to 50 mg followed by maintenance doses between 30 and 60 mg/kg/day (Bagci et al., 2008; Kuo and Wang, 2002; Mills et al., 2005). One patient became apneic and hypotonic associated with an isoelectric pattern on his EEG that lasted for 4 days (Mills et al., 2005). Unfortunately, PLP is not available as an intravenous preparation in the United States and is only available as an unlicensed supplement, making initiation challenging.

A trial of pyridoxal phosphate should be considered in patients with intractable neonatal epileptic encephalopathy who fail a trial of pyridoxine, especially when associated with abnormalities in CSF neurotransmitters or elevated glycine and threonine in plasma (Figure 22.4). Further research into this supplement may expand the clinical spectrum in which it may be an effective treatment.

## Adverse Events

The adverse event profile of vitamin $B_6$ is only partially established, as many of the authors do not mention side effects and all of the studies are open label. Intravenous $B_6$ administration has been associated with apnea, lethargy, pallor, decreased responsiveness, and hypotonia (Haenggeli et al., 1991; Mikati et al., 1991; Nabbout et al., 1999; Tanaka et al., 1992; Waldinger and Berg, 1963). These reactions may occur immediately and persist for hours.

Rarely, these same side effects have followed the initial oral dose (Grillo et al., 2001; Kroll, 1985) or intramuscular administration (Bankier et al., 1983; Garty et al., 1962). These symptoms are usually mild but exceptionally have resulted in intubation and mechanical ventilation (Heeley et al., 1978). The pathophysiology of these events is felt to be a burst of GABA release from accumulated substrate and exaggeration of its physiological effect (Kroll, 1985). As a result, some authors have suggested that it would be prudent to administer the first dose in an environment with full resuscitation facilities at hand (Kroll, 1985).

Loss of appetite (71%), periods of restlessness and crying (59%), vomiting (47%), apathy (29%), and hemorrhagic gastritis (6%) have been reported during high-dose vitamin $B_6$ therapy for infantile spasms (Pietz et al., 1993). Laboratory abnormalities reported during therapy for infantile spasms consist of rare elevations of serum transaminases (AST, ALT) with doses over 400 mg/day, rapidly recovering by decreasing the dose (Ohtsuka et al., 1987).

Long-term pyridoxine use can produce a sensory neuropathy that has been documented experimentally in rats, guinea pigs, dogs (Antopol and Tarlov, 1942; Krinke et al., 1981; Xu et al., 1989) and humans (Berger et al., 1992) and described clinically in humans (Dalton and Dalton, 1987; Schaumburg et al., 1983). Depending on the daily dose and duration of administration, distal neuropathy, neuronopathy, and diffuse axonopathy have all been described after experimental pyridoxine intoxication. The dose required to produce toxicity to the dorsal root ganglia, with subsequent degeneration of the peripheral sensory nerves, was initially reported to be 2000 to 6000 mg/day (Schaumburg et al., 1983). However, later reports indicated daily doses of 200 to 500 mg could be neurotoxic in adults (Berger and Schaumburg, 1984; Parry and Bredesen, 1985). In all cases, the effects were dose dependent and at least partially reversible upon discontinuation of pyridoxine.

Prospective studies of adults taking 100 to 300 mg/day of pyridoxine showed no clinical or electrophysiologic evidence of neuropathy (Bernstein and Lobitz, 1988; DelTredici et al., 1985). Vitamin $B_6$-associated neuropathy appears to be very rare in children. Only a single report evaluated and documented unchanged sensory nerve conduction velocities in three children treated with high-dose vitamin $B_6$ for infantile spasms (Blennow and Starck, 1986). Only one case of sensory neuropathy associated with high-dose (2000 mg/day), long-term pyridoxine therapy (for pyridoxine-dependent epilepsy) has been reported (McLachlan and Brown, 1995). A single child had a paradoxical increase in neonatal seizures after intravenous treatment with pyridoxine (Hammen et al., 1998).

$$CO_2 \; + \; H_2O \;\xleftrightarrow{\text{Carbonic anhydrase}}\; H_2CO_3 \xleftrightarrow{\hspace{2cm}} H^+ \; + \; HCO_3^-$$

**FIGURE 22.5** Interconversion of carbon dioxide ($CO_2$) and water to bicarbonate ($HCO_3^-$) and $H^+$ via the action of carbonic anhydrase.

## ACETAZOLAMIDE

### HISTORY

Acetazolamide ($N$-[5-(aminosulfonyl)–1,3,4-thiadiazol-2-yl]acetamide) is an unsubstantiated sulfonamide that was synthesized in 1950 (Roblin and Clapp, 1950) and first used to treat epilepsy in 1952 (Bergstom et al., 1952). Acetazolamide inhibits the enzyme carbonic anhydrase (CA) (Figure 22.5), with subsequent generation of carbon dioxide and water. Carbonic anhydrase activity was first discovered in the early 1930s in red blood cells and has subsequently been found in the central nervous system, eyes, adrenal cortex, pancreas, and gastric mucosa. Early open-label reports confirmed the effectiveness of acetazolamide in most seizure types (Ansell and Clark, 1956; Forsythe et al., 1981; Golla and Hodge, 1957; Lim et al., 2001; Millichap, 1956; Oles et al., 1989; Reiss and Oles, 1996; Resor and Resor, 1990); however, this was limited by the development of tolerance (Ansell and Clark, 1956; Lombroso and Forxythe, 1960; Lombroso et al., 1956; Millichap, 1956; Woodbury, 1980). There have been no prospective, controlled clinical trials of acetazolamide to establish its efficacy against specific seizure types or in epilepsy syndromes. As a result, the use of acetazolamide in epilepsy is empiric and guidelines for its use are lacking.

### PHARMACOKINETICS

Acetazolamide is absorbed mainly in the duodenum and upper jejunum. Absorption is rapid, with peak levels achieved within 2 to 4 hours after oral ingestion of 5 to 10 mg/kg. At higher doses, the absorption is erratic. The time-release formulation reaches a plateau at 3.5 hours and is maintained for 10 hours. Acetazolamide is 90 to 95% protein bound. The percentage of protein binding is dose dependent. Acetazolamide diffuses into tissue water as the free form, where it binds to and inhibits CA. In the tissue, its concentration is higher than in plasma. The highest concentrations are found in tissue that contains the highest amounts of CA (i.e., erythrocytes, glia). After 24 hours, 90% of the drug is bound to CA. The volume of distribution is 0.2 liters/kg. Acetazolamide is completely excreted in the urine without metabolism. The renal half-life is 10 to 15 hours, but the enzyme–inhibitor complex has a slow dissociation constant; thus, acetazolamide has an effective half-life of several days (Woodbury, 1980). In children, acetazolamide doses range from 8 to 36 mg/kg/day (Lombroso and Forxythe, 1960; Millichap, 1956), typically divided into two or three doses. Acetazolamide is usually initiated at one third of the total daily dose and is increased by this same amount weekly, as needed. In adults, the usual dose varies from 10 to 20 mg/kg/day (Resor et al., 1995), with occasional benefit at doses up to 30 mg/kg/day. In some patients, increasing the dose overcomes the tolerance developed at lower doses. Those patients with compromised renal function may need a reduced dose because the ability to clear unbound acetazolamide is directly related to creatinine clearance. Plasma acetazolamide levels are less than erythrocyte concentrations, and in most studies no relationship between plasma levels and efficacy has been found.

### MECHANISM OF ACTION

Among the 14 CA isoenzymes that have been identified are 3 CA-related proteins that lack the zinc-binding capability necessary for enzymatic reaction (Thiry et al., 2007). The carbonic anhydrase II isoenzyme is localized in the cytoplasm of oligodendrocytes and astrocytes, the myelin sheath, and choroid plexus (Sapirstein et al., 1984). The function of CA-related proteins remains unknown but may relate to neural cell development (Taniuchi et al., 2002).

Carbonic anhydrase belongs to a family of zinc metal enzymes involved in the transfer of $H^+$, $HCO_3^-$, or both from neurons to glial cells. Interruption of this process by acetazolamide leads to $CO_2$ accumulation in neurons. Development of tolerance to the anticonvulsant effect of acetazolamide is the primary limitation to its therapeutic use. This results from the induction of both increased amount and activity of carbonic anhydrase in glial cells (Woodbury, 1980).

Acetazolamide inhibits CA in the brain by more than 99%, leading to carbon dioxide accumulation, which appears to cause the anticonvulsant effect. This is the only known mechanism of action of this drug and is believed to mediate its activity (Woodbury, 1980).

The accumulation of $CO_2$ is sufficient to prevent the tonic extensor component of maximal electroshock-induced seizures (a measure of spread of seizure activity and efficacy against generalized tonic–clonic seizures) (Millichap, 1957, 1958; Millichap et al., 1958; Woodbury and Esplin, 1959; Woodbury and Karler, 1960). At higher doses, acetazolamide inhibits erythrocyte CA, causing further $CO_2$ accumulation, which may contribute to its anticonvulsant effect. The degree of inhibition of brain CA correlates with the anticonvulsant effect of acetazolamide in mice (Anderson et al., 1986; Gray et al., 1957; Gray and Rauth, 1967; Millichap et al., 1955). Nephrectomy does not alter the anticonvulsant activity, proving it is not secondary to the systemic acidosis produced from renal CA inhibition (Millichap et al., 1955). Doses of acetazolamide greater than those that produce complete CA inhibition do not increase the anticonvulsant activity. Mutant mice, deficient in carbonic anhydrase II enzyme, have a reduced seizure susceptibility to flurothyl-, pentylenetetrazol-, and audiogenic-induced seizures (Velisek et al., 1993). These data provide supporting evidence of the role of CA in seizure generation. The increased extracellular concentration of protons may block the $N$-methyl-D-aspartate receptor (Traynelis and Cull-Candy, 1990). In animal models, acetazolamide has been shown to protect against seizures induced by various methods—audiogenic seizures (Engstrom et al., 1986; Gray et al., 1957), pentylenetetrazol- and picrotoxin-induced seizures (Resor et al., 1995), and maximal electroshock (Millichap, 1957, 1958; Millichap et al., 1958; Woodbury and Esplin, 1959; Woodbury and Karler, 1960)—suggesting a broad spectrum of clinical efficacy in the treatment of human epilepsy.

## CLINICAL USE

Acetazolamide has been used over the last 50 years to treat a variety of seizure types or epilepsy syndromes. Unfortunately, trial designs using modern methods have not been performed. Most patients are given acetazolamide as adjunctive therapy, after failing standard AEDs (Bergstom et al., 1952; Lombroso and Forxythe, 1960; Lombroso et al., 1956; Millichap, 1956; Resor and Resor, 1990). Many of the early studies were performed before modern seizure classification, and all but one (Millichap, 1956) was open label; however, given these limitations, all but one (Livingston, 1956) suggested that, at least in the short term, acetazolamide was effective in the treatment of various types of epilepsy. Over time, most patients appear to develop tolerance to its antiepileptic effect. This could be observed after 2 to 8 weeks (Bergstom et al., 1952; Millichap, 1956), but one study (Reiss and Oles, 1996) showed no loss of seizure control over 2 to 3 years, which raises doubts about the role of tolerance in these patients.

### Partial Seizures

Forsythe et al. (1981) performed an open-label, long-term trial to assess the effectiveness of adjunctive therapy with acetazolamide in the treatment of 54 children with carbamazepine-resistant epilepsy. Patients who responded were followed for at least 2 years. Fifty percent (7 of 14) of children with temporal lobe epilepsy (i.e., complex partial seizures) were seizure free for 2 years, 67% (6 of 9) at the end of 3 years, and 100% (2 of 2) at the end of 4 years.

Eight years later, Oles et al. (1989), performed a similar study, using adjunctive acetazolamide therapy in 48 refractory partial seizure patients. The patients, ages 6 to 64 years (average age, 28 years), were on monotherapy with carbamazepine. This was an open-label, retrospective review.

Twenty-one patients (44%) had a greater than 50% reduction in seizure frequency, and 3 were seizure free for more than 3 months. The mean time on treatment was 9.6 months. With this length of follow-up, tolerance was not a problem.

## Generalized Seizures

Lombroso and Forxythe (1960) reported a series of 277 patients treated with acetazolamide, of which 91 had absence seizures; 80 of the patients were less than 12 years old (i.e., most had childhood-onset absence epilepsy). All had failed at least one AED, and acetazolamide monotherapy was used for 48 of them. This is the easiest group to judge the effect of acetazolamide alone. At the end of 3 months, 9 (19%) had at least a 99% seizure reduction, and 20 (42%) had a 90% or more seizure reduction. At 1 year, 8 (17%) of the patients continued to be at least 99% absence seizure free, while 12 (25%) continued to have at least a 90% reduction in seizure frequency.

In the same series reported by Lombroso and Forxythe (1960), 19 patients had only generalized tonic–clonic (GTC) seizures; however, most were under the age of 12 years and had failed prior AEDs, making it unlikely that they had primary generalized epilepsy. In 15 of the patients, acetazolamide was used as adjunctive therapy. At the end of 3 months, 7 (37%) experienced seizures decreased by at least 99%, and 12 (67%) had at least a 90% improvement; however, by 2 years only two patients continued to have at least a 99% improvement. The same authors noted that acetazolamide helped decrease GTC seizures in patients who also had absence or myoclonic seizures, often associated with photic sensitivity and generalized spike and slow-wave discharges. This suggests that acetazolamide might be useful in the treatment of GTC seizures seen in primary generalized epilepsy. Resor and Resor (1990) reported the use of chronic acetazolamide monotherapy in 51 patients with juvenile myoclonic epilepsy (JME) unresponsive to valproate. Among the 51 patients, they could isolate the effect of acetazolamide on the GTC seizures of JME in 31; 14 patients (45%) were seizure free or had a single provoked GTC seizure (1 due to sleep deprivation and 1 due to antihistamines) on acetazolamide monotherapy for 10 to 70 months (average, 41 months). Acetazolamide appeared less effective in controlling myoclonus. Initially, all 31 patients reported complete control of myoclonus; however, only 4 remained seizure free, although the myoclonus was improved in all. This suggests that, even when acetazolamide completely controls the GTC seizures of JME, most patients develop a degree of tolerance to its antimyoclonic effect.

## Catamenial Epilepsy

Tolerance to the anticonvulsant effect of acetazolamide limits its use as a daily preventative medication; however, alternate-day or cyclical dosing has been suggested to reduce the development of tolerance. It has been hypothesized that intermittent dosing would allow this drug to be a useful adjunct in the treatment of catamenial seizures, without development of tolerance.

Ansell and Clark (1956) reported the initial use of acetazolamide to treat catamenial epilepsy. The drug was initiated 2 days prior to menses and a benefit was seen in 2 patients over a 3-month period. Lim et al. (2001) retrospectively analyzed the efficacy, safety profile, and tolerability of acetazolamide in catamenial epilepsy in the modern era. Eleven women took the drug continuously and 9 intermittently; 8 out of the 20 women (40%) reported that their catamenial seizures decreased by 50% or more, with no significant difference in effectiveness being noted between continuous and intermittent dosing. Response rates were similar in generalized or focal epilepsy. A loss of efficacy over 6 to 24 months was reported in 3 (15%) women. This study suggests that acetazolamide is a useful adjunct in the treatment of catamenial epilepsy, but a double-blind, prospective study is needed.

## SIDE EFFECTS

Acetazolamide is a well-tolerated AED. Side effects are mostly mild or transient. All of the adverse effects are thought to be related to inhibition of CA, with the exception of hypersensitivity reactions. There are only rare reports of sulfonamide-type hypersensitivity reactions to acetazolamide, and cross-sensitivity to acetazolamide is uncommon even in patients with a history of allergy to sulfa drugs (Stock, 1990).

Dyspepsia occurs in at least 90% of patients (Resor et al., 1995). Acetazolamide eliminates the tingle of carbonated beverages, giving them a flat taste. The effect is specific for CA inhibitors and is localized to the taste buds, which contain CA. Anorexia is common and, in many patients, a pleasant surprise. Renal nephrolithiasis has been reported, although the incidence varies from 2 to 43% in adults and appears much lower in children (Resor et al., 1995). Encouraging fluid intake and, when indicated, specific therapy based on the stone composition can minimize recurrence. The most common adverse effects are transient, mild polyuria, hypokalemia, paresthesias, irritability, anorexia, headache, nausea, diarrhea, and drowsiness (Reiss and Oles, 1996).

Acetazolamide causes a metabolic acidosis by inhibition of renal CA in the proximal tubular epithelium with resultant bicarbonate diuresis and loss of sodium and potassium. A normal anion gap hypochloremic acidosis may be found on routine serum chemistries, although this is usually asymptomatic; however, Futagi et al. (1996) reported growth suppression in children receiving acetazolamide as adjunctive therapy. They speculated this was related to the metabolic acidosis. Rarely, acetazolamide-associated blood dyscrasias (thrombocytopenia and aplastic anemia) have occurred (Reiss and Oles, 1996). Most of these occurred in patients being treated for glaucoma; in one study, the median age of patients who developed aplastic anemia was 71 years (Keisu et al., 1990). To date, aplastic anemia has not been associated with the use of acetazolamide in epilepsy. If periodic blood counts are checked, it appears sufficient to do this for the first 6 months to detect hematologic reactions to acetazolamide. Acetazolamide is a teratogen in animals, and in some species acetazolamide has caused forelimb abnormalities. In humans, this has not been reported.

## MAGNESIUM

### HISTORY

Magnesium is second only to potassium in abundance as an intracellular cation. Tetany and convulsions, subsequent to experimental magnesium deficiency, were first shown to occur in rats in 1932 (Kruse et al., 1932) and subsequently in other animals (Barron et al., 1949; Blaxter et al., 1954; Orent et al., 1932; Roine et al., 1949). Hirschfielder (1934) first reported the association of seizures and magnesium depletion in humans; however, scientific documentation of the relationship between convulsions and hypomagnesemia in humans was not possible until the development of the multichannel flame spectrometer (Vallee et al., 1960). This allowed simple, accurate, and rapid measurement of magnesium levels; correlation with clinical symptoms; and reversal of symptoms upon treatment with parenteral magnesium sulfate.

As early as 1906, magnesium sulfate was injected intrathecally to prevent eclamptic seizures (Alton and Lincoln, 1925; Horn, 1906). Subsequently, reports of intramuscular magnesium sulfate controlling convulsions associated with tetanus led to a similar regimen being used in 1926 to prevent recurrent seizures in women with eclampsia (Dorsett, 1926). Lazard, following up on his initial observations (Lazard, 1925), administered magnesium sulfate intravenously to over 500 women with preeclampsia and eclampsia at Los Angeles General Hospital (Lazard, 1933). All of these early studies used small doses of magnesium. Later authors used larger doses given intramuscularly (Eastman and Steptoe, 1945), but Pritchard (1955) and later Zuspan (1966) popularized the current protocol of intramuscular and intravenous treatment.

### MECHANISM OF ACTION

The pathophysiology of hypomagnesemic seizures and the beneficial role that magnesium plays in the treatment of preeclampsia and eclampsia are not clearly known. First, we will review the proposed pathophysiology in hypomagnesemic seizures. Hypomagnesemia may diminish the normal magnesium block of the N-methyl-D-aspartate (NMDA) subtype of glutamate receptors, triggering neuronal depolarization (Avoli et al., 1987; Mody et al., 1987). The NMDA glutamate receptor is a voltage-dependent, ligand-gated channel that modulates flux of $Ca^{2+}$ and $Na^+$. Magnesium ordinarily inhibits cation flux by means of a voltage-dependent block of the ion channel, preventing it from contributing to the excitatory postsynaptic potential (Dingledine and McBain, 1994; Mayer et al., 1984; Nowak et al., 1984). At the resting membrane potential (about $-70$ mV), ambient extracellular concentrations of magnesium block the NMDA-receptor ion channel and prevent current flow, even when the glutamate and glycine binding sites are occupied (Goldman and Finkbeiner, 1988). It has been hypothesized that hypomagnesemia removes the inhibitory influence this electrolyte has on glutamate-activated depolarizations, increasing neuronal excitability and seizure susceptibility. Supportive of this hypothesis is the observation that low levels of extracellular magnesium induce epileptiform activity in the rat hippocampal slice model (Dreier and Heinemann, 1991; Herron et al., 1985; Mody et al., 1987; Walther et al., 1986). These epileptiform events closely resemble the electrographic pattern associated with tonic–clonic seizures.

Although excellent results have been obtained in treating preeclampsia and eclampsia with magnesium sulfate, the physiologic basis for the use of large doses in modern obstetrics remains unclear. However, it is suspected that magnesium is superior to traditional AEDs because mechanisms other than anticonvulsant properties enhance its therapeutic benefit for women with preeclampsia. The animal and human experiments that have been performed to verify or refute its anticonvulsant action are reviewed next, followed by a discussion of other possible mechanisms.

First, investigations in the intact animal have demonstrated that the blood–brain barrier offers considerable resistance to changes in the brain and cerebrospinal fluid magnesium concentrations (Hilmy and Somjen, 1968). As a result, intravenous administration of magnesium raises the magnesium content of brain tissue only slightly. Other authors, however, have argued that, because acute convulsions may cause significant alterations in the blood–brain barrier permeability, additional research on convulsing animals is necessary to understand the effect of magnesium. Borges and Gucer (1978) demonstrated diminished epileptiform activity and seizures in animals given intravenous magnesium sulfate. The epileptiform activity was induced by topical application of penicillin G to the motor cortex in anesthetized cats and dogs and in awake primates. This animal model is felt to be predictive of efficacy against partial seizures in humans. Magnesium was able to directly suppress neuronal burst firing and interictal EEG spike generation at serum levels below those producing paralysis; however, Koontz and Reid (1985) subsequently disputed these findings. They studied the effect of parental magnesium sulfate on established penicillin-induced seizure foci in anesthetized cats. Control animals were infused with normal saline. Analysis of the electroencephalogram recordings demonstrated no significant difference in epileptic spike activity between the control and the magnesium sulfate groups. In both groups, maximum spike frequency occurred about 30 minutes after application of penicillin and declined rapidly thereafter. The discrepancy between the two studies was felt to be due to a lack of controls by Borges and Gucer (1978).

An investigation of thresholds for electroshock convulsions (felt to be predictive of efficacy in partial seizures with or without secondary generalization) and pentylenetetrazol-induced seizures (felt to predictive of efficacy in human absence seizures) in mice failed to show significant attenuation of the electrographic seizures, despite serum levels of magnesium sufficient to cause neuromuscular blockade in the mice (Krauss et al., 1989).

Disruption of the blood–brain barrier may be important for the entry and efficacy of magnesium. Hallak et al. (1982) demonstrated that sustained, elevated, serum magnesium concentrations increased magnesium concentrations in the cortex and hippocampus of rats, and induction of hippocampal seizure activity further elevated CSF magnesium concentrations. Additionally, the threshold for hippocampal seizure in the magnesium sulfate treatment group was increased significantly more (34% vs. 3.5%) than the control group. Once in the extracellular space of the brain, magnesium has been shown to antagonize depolarizations and seizures mediated by the NMDA subtype of glutamate receptor (Cotton et al., 1992, 1993; Mayer et al., 1984). These studies demonstrate that magnesium does have a central inhibitory action and that this action may be at least partially mediated by suppression of the NMDA receptor. This counters the long-standing argument that magnesium sulfate has no central anticonvulsant activity; however, these same authors then compared the effect of magnesium sulfate vs. phenytoin for seizure prevention in amygdala-kindled rats (Standley et al., 1994). Although kindled seizures are unrelated to eclamptic seizures, this study was performed to provide information regarding the effect of magnesium sulfate in a model of partial seizure activity. Phenytoin significantly reduced seizure duration, duration of postictal depression, and behavioral seizure stage. Magnesium sulfate had no effect on any of the seizure parameters assessed.

This result may be explained by the magnesium being more effective at reducing seizure activity when it acts as an NMDA antagonist. The outcome may reflect a difference in NMDA receptor distribution (i.e., relatively few in the amygdala) and subunit combination. Alternatively, this could be explained because excitotoxicity in hippocampal and cortical neurons depends only slightly on extracellular magnesium (Lipton and Rosenberg, 1994). The magnesium block of the ligand-gated calcium channel is voltage dependent and is present at physiologic levels of magnesium. During a seizure (i.e., continuous depolarization), elevated serum magnesium levels may not effectively block this channel.

Magnesium sulfate was given by intravenous infusion to treat a human with myoclonic status epilepticus (Fisher et al., 1988). Over 3 days, the serum magnesium was increased from 1.5 to 14.2 mEq/L. CSF magnesium was elevated by the infusion to only 3.5 mgEq/L (normal is 2 to 2.5 mEq/L), reflecting the difficulty in elevating CSF levels. Despite magnesium-related neuromuscular blockade and cessation of visible myoclonus, the electroencephalogram revealed ongoing epileptiform activity at the baseline frequency. Although magnesium was ineffective in this case of myoclonic epilepsy, this may not bear directly on the controversy over the use of magnesium sulfate in eclampsia, a condition with a different pathophysiology.

How magnesium sulfate acts as an anticonvulsant in preeclampsia remains controversial. Some believe this agent acts primarily by neuromuscular blockade, but magnesium also has a central depressant effect. The abnormal electroencephalograms commonly found in preeclampsia or eclampsia may not be altered by therapeutic blood levels of magnesium (Sibai et al., 1985). In addition, eclamptic seizures are occasionally seen in patients with confirmed therapeutic magnesium serum levels. Recent studies suggest that independently of any possible anticonvulsant efficacy magnesium may have several other potentially beneficial mechanisms of action.

Watson et al. (1986) investigated the action of magnesium sulfate on prostaglandin I2 (prostacyclin). This is based on the observation that in advanced stages preeclampsia is characterized pathologically by microvascular occlusions consisting of platelet and fibrin thrombi. Prostaglandin production may be relatively deficient during preeclampsia. Magnesium sulfate amplified release of prostacyclin and enhanced the anti-aggregatory effect of intact human umbilical vein endothelial cells using serum levels achieved in treating preeclampsia. These results provide a physiologic basis for the use of large amounts of magnesium sulfate in preeclampsia.

Another mechanism proposed for the action of magnesium sulfate in preeclampsia includes opposition of calcium-dependent arterial vasoconstriction by antagonizing the increase in the intracellular calcium concentration caused by ischemia. Belfort and Moise (1992) evaluated the effect of intravenous magnesium sulfate (or placebo) on maternal brain blood flow in preeclampsia using Doppler ultrasonography. They showed that magnesium vasodilates the smaller diameter intracranial vessels distal to the middle cerebral artery, and they hypothesized that the drug may exert its main effect in the prophylaxis and treatment of eclampsia by relieving cerebral ischemia. This improved perfusion would prevent cell damage, cerebral edema, and possibly convulsions. The effect of magnesium sulfate may be multiple in eclamptic patients, including a depressant effect on the central nervous system, induction of hypotension, and changes in the renin–angiotensin system. These multiple sites of action may explain its efficacy, rather than viewing it as purely anticonvulsant. It has been argued that an eclamptic seizure is like any other seizure and therefore should be treated as such, but the efficacy of magnesium sulfate suggests that this assumption may not be correct. Ultimately, once the exact cascade of pathophysiologic events in preeclampsia is understood, then specific, targeted therapy will be advocated.

## CLINICAL EFFICACY

### Hypomagnesemia

Symptomatic hypomagnesemia is rare in childhood (Fishman, 1965; Tsau et al., 1998). Hypomagnesemia occurs as a result of: (1) abnormal gastrointestinal absorption, (2) renal wasting of a genetic or acquired origin, and (3) primary hypomagnesemia (Ahsan et al., 1998; Bettinelli et al., 1992; Dudin and Teebi, 1987; Evans et al., 1981; Martin, 1969; Muldowney et al., 1970; Nuytten et al., 1991; Rude and Singer, 1981). Neurologic disturbances constitute the most prominent clinical manifestations. These include confusional states, focal or generalized tonic–clonic seizures, and increased neuromuscular irritability with hyperreflexia, tremors, tachycardia, and myoclonic jerks (Avoli et al., 1987; Fishman, 1965; Mody et al., 1987). Phenomenologically, the magnesium-deficiency tetany syndrome is indistinguishable from hypocalcemic injury. Additionally, the neurologic manifestations of hypomagnesemia most frequently occur in association with hypocalcemia. Serum chemistries, magnesium, calcium, and albumin levels are necessary for differentiation. Familial hypomagnesemia due to mutations in the *TRPM6* gene can present with tetany, muscle spasms, and seizures that are responsive to magnesium supplementation (Schlingmann et al., 2002). Hypomagnesemic seizures are usually generalized tonic–clonic, although partial seizures have been reported (Ahsan et al., 1998; Chutkow and Meyers, 1968; Fishman, 1965; Nuytten et al., 1991; Rude and Singer, 1981; Tsau et al., 1998; Vallee et al., 1960). Seizures usually are associated with serum magnesium levels less than 1.0 mg/dL (Chutkow and Meyers, 1968; Rude and Singer, 1981), but seizures have occurred at 1.4 mg/dL (Fishman, 1965). Two groups identified mutations in *KCNJ10* as a cause of epilepsy, ataxia, sensorineural deafness, mental retardation, and tubulopathy associated with hypomagnesemia and hypocalcemia (Bockenhauer et al., 2009; Scholl et al., 2009).

Parenteral or oral preparations of magnesium sulfate can correct hypomagnesemic seizures. For oral administration, an isotonic, 4% solution is prepared, avoiding an osmotic cathartic effect. No established therapeutic regimen exists in the pediatric literature. Bhasker et al. (1999) used magnesium sulfate (10 mg/kg intramuscularly) to stop ongoing seizures and then a maintenance dose of 130 mg/kg/day given in three divided doses. To treat neonatal hypomagnesemic convulsions, 200 mg/kg is given intravenously (IV) or by intramuscular (IM) administration (Morgan et al., 1999). To acutely correct hypomagnesemia, *The Harriet Lane Handbook* recommends magnesium sulfate 25 to 50 mg/kg IV or IM every 4 to 6 hours, followed by a maintenance oral dose of 30 to 60 mg/kg/day (Siberry and Iannone, 2000).

### Preeclampsia/Eclampsia

Eclampsia is one of the dreaded complications of pregnancy as it carries a high risk of morbidity and mortality for the baby and mother. Preeclampsia and eclampsia represent a continuum of an acute hypertensive disorder peculiar to women (Ramin, 1999). Preeclampsia is characterized by gestational hypertension in association with proteinuria, generalized edema, or both, and it occurs after 20 weeks gestation. Other causes of hypertension and seizures must be excluded (Kaplan, 1999). Eclampsia is defined by the appearance of seizures or coma. Yearly, about 200,000 maternal deaths are attributed to eclampsia worldwide, and it remains the leading cause of maternal death in the United States (Kaplan, 1999).

Preeclampsia is a multisystem disorder (Bolte, 2001). One mechanism is thought to be vasospasm, with a secondary hypertensive encephalopathy and failure of cerebral blood pressure autoregulation. The exact cause of eclamptic seizures remains unclear. A number of pathological changes, including breakdown of the blood–brain barrier, vascular spasm, ischemia, and hypertensive encephalopathy, have been reported. The neuropathologic hallmarks of eclampsia are petechial hemorrhages and microinfarcts in cerebral cortex, subcortical hemorrhages, cerebral edema, small subarachnoid hemorrhage, and multiple 3- to 5-mm hemorrhages in the corona radiata, caudate nucleus, or brain stem (Donaldson, 1988; Kenny and Baker, 1999; Thomas, 1998).

The treatment of preeclampsia is directed at decreasing elevated blood pressure, preventing seizures, preventing brain edema, and delivering a viable baby (Donaldson, 1988). The treatment of seizures and eclampsia has been a subject of controversy for obstetricians and neurologists (Dinsdale, 1988; Donaldson, 1986; Kaplan et al., 1988, Sankar and Licht, 1995). Numerous, open-label retrospective studies have documented the efficacy of magnesium sulfate in the treatment of severe preeclampsia or eclamptic seizures (Pritchard, 1955; Pritchard et al., 1984; Sibai et al., 1984). Recently, prospective, randomized, large trials in the United States and the United Kingdom have compared magnesium sulfate to phenytoin or diazepam as therapy for the prevention of eclampsia (Lucas et al., 1995) or for the management of eclampsia (Eclampsia Trial Collaborative Group, 1995; Sawhney et al., 1999). Magnesium sulfate was found to be superior to phenytoin for the prevention of eclampsia in hypertensive pregnant women (Lucas et al., 1995). The findings of this study were complemented by the Eclampsia Trial Collaborative Group (1995). They reported that women receiving magnesium were half as likely to have recurrent seizures as women receiving diazepam and one third as likely as women receiving phenytoin. Both studies reinforced the finding of seizure occurrence being associated with substantially increased maternal morbidity and established the scientific validity of magnesium sulfate as a treatment regimen.

Subsequently, a case-controlled evaluation of the efficacy of magnesium sulfate in preeclampsia (Abi-Said et al., 1997) and Cochrane reviews of the efficacy of magnesium sulfate in eclampsia (Duley and Henderson-Smart, 2001a,b) have been performed. All concluded that there is compelling evidence in favor of magnesium sulfate for the treatment of eclampsia. Eclampsia appears to be distinguished from other forms of seizures in that it is better controlled by magnesium sulfate than by conventional AEDs (i.e., diazepam or phenytoin). This knowledge may offer opportunities to explore the pathogenesis of eclampsia.

## SIDE EFFECTS

Serious toxic symptoms can develop with magnesium overdosage, especially if renal function is compromised. Signs and symptoms of magnesium toxicity are dose dependent, typically occurring at magnesium concentrations greater than 4 mEq/L (the normal serum $Mg^{2+}$ concentration is 1.5 to 2.5 mEq/L). Depression of blood pressure is observed with serum levels of 3 to 5 mEq/L (Rude and Singer, 1981). Loss of knee and ankle deep-tendon reflexes occurs at magnesium levels of 8 to 10 mEq/L. Somnolence and slurred speech may also be seen. Generalized muscular paralysis, including respiratory muscles, occurs when the levels rise above 10 mEq/L and coma at levels between 12 to 15 mEq/L. Cardiac arrest can occur at levels of 15 mEq/L. It is imperative to monitor the patient's clinical status, urine output, and serum levels of magnesium periodically. Magnesium toxicity should be identified promptly and remedial measures instituted. Administration of 10 to 20 cc of 10% calcium gluconate intravenously is helpful in combating the respiratory and cardiac toxicity of magnesium. Loop diuretics may hasten the urinary excretion when renal function is normal. In the setting of renal failure, dialysis may be required.

Hypermagnesemia causes a myasthenic syndrome and blocks transmission in sympathetic ganglion (Donaldson, 1988). Administration of magnesium sulfate to a woman with myasthenia gravis will precipitate a myasthenic crisis. Hypermagnesemia potentiates neuromuscular blocking agents during anesthesia; hence, reduced doses must be used in toxemia patients treated with magnesium sulfate.

In preeclampsia, the load of magnesium transferred to the fetus depends on peak maternal levels and the duration of maternal hypermagnesemia. A serum level should be performed in the neonate and appropriate supportive care provided. Rarely, it may take up to 1 week for neonatal magnesium levels to become normal. Fetal morbidity has been shown to be reduced in randomized studies comparing magnesium sulfate to either phenytoin or benzodiazepines (Anthony et al., 1996; Duley and Henderson-Smart, 2001a,b; Eclampsia Trial Collaborative Group, 1995; Lucas et al., 1995). The risk of cerebral palsy is lower among babies born to mothers given magnesium sulfate for eclampsia when compared to those not given magnesium sulfate (Nelson and Grether, 1995). Some studies have suggested a lower rate of cerebral palsy in preterm infants whose mothers were given magnesium sulfate prior to delivery (Crowther et al., 2003; Marret et al., 2008; Rouse et al., 2008).

## CONCLUSION

Of these three agents—vitamin $B_6$, acetazolamide, and magnesium—the last has had the most thorough clinical evaluation to date. For all three, there is a clear need for future double-blind, controlled trials to establish their roles in the treatment of epilepsy using the criteria established for evidence-based medicine. In the interim, however, magnesium sulfate appears to have a unique and important role in the treatment of preeclampsia. In women with an established seizure disorder, a traditional AED will also be a necessary part of their treatment regimen.

Vitamin $B_6$-dependent seizures should be considered in all infants and children with intractable epilepsy that begins prior to age 3 years. They should have an appropriate trial of vitamin $B_6$; if a patient responds to a challenge, evaluation of urinary α-AASA as a screening tool for the *ALDH7A1* gene mutation should be considered. For those patients without the mutation, vitamin $B_6$ should be withdrawn and reinitiated at a convenient time to document seizure recurrence and clinical response, respectively. It is important to firmly establish the diagnosis of pyridoxine-dependent seizures and to convince the family of an affected patient that lifelong supplementation is required, as it is a treatable disorder. Further, a controlled trial of vitamin $B_6$ in the treatment of infantile spasms is necessary to define a logical treatment sequence for this difficult to treat seizure disorder. Acetazolamide continues to be a useful adjunctive AED, although its use may be limited now by the emergence of two other agents possessing carbonic anhydrase activity (i.e., zonisamide and topiramate). Acetazolamide's niche may ultimately be in the treatment of catamenial epilepsy; however, double-blind, prospective studies are needed to define the exact role of intermittent acetazolamide therapy.

All of the three agents reviewed in this chapter provide special insights into the complex neurochemistry and pathophysiology of epilepsy. This fact alone, combined with their potential therapeutic utility, justifies further investigation and understanding of their precise roles in the treatment of special populations with epilepsy.

## REFERENCES

Abi-Said, D., J.F. Annegers, D. Combs-Cantrell, R. Suki, R.F. Frankowski, L.J. Willmore. (1997). A case-control evaluation of treatment efficacy: the example of magnesium sulfate prophylaxis against eclampsia in patients with preeclampsia. *J. Clin. Epidemiol.*, 50:419–423.

Ahsan, S.K., S. al-Swoyan, M. Hanif, M. Ahmad. (1998). Hypomagnesemia and clinical implications in children and neonates. *Indian J. Med Sci.*, 52:541–547.

Alton, B.H., G.C. Lincoln. (1925). The control of eclampsia convulsions by intraspinal injections of magnesium sulphate. *Am. J. Obstet. Gynecol.*, 9:167–177.

Anderson, R.E., R.A. Howard, D.M. Woodbury. (1986). Correlation between effects of acute acetazolamide administration to mice on electroshock seizure threshold and maximal electroshock seizure pattern, and on carbonic anhydrase activity in subcellular fractions of brain. *Epilepsia*, 27:504–509.

Ansell, B., E. Clark. (1956). Acetazolamide in treatment of epilepsy. *Br. Med. J.*, 1:650–654.

Anthony, J., R.B. Johanson, L. Duley. (1996). Role of magnesium sulfate in seizure prevention in patients with eclampsia and pre-eclampsia. *Drug Saf.*, 15:188–199.

Antopol, W., I.M. Tarlov. (1942). Experimental study of the effects produced by large doses of vitamin $B_6$. *J. Neuropathol. Exp. Neurol.*, 1:330–336.

Appleton, R.E. (1996). The treatment of infantile spasms by paediatric neurologists in the U.K. and Ireland. *Dev. Med. Child Neurol.*, 38:278–279.

Avoli, M., J. Louvel, R. Pumain, A. Olivier. (1987). Seizure-like discharges induced by lowering $[Mg^{2+}]_0$ in the human epileptogenic neocortex maintained *in vitro*. *Brain Res.*, 417:199–203.

Bagci, S., J. Zschocke, G.F. Hoffmann, T. Bast, J. Klepper, A. Muller, A. Heep, P. Bartmann, A.R. Franz. (2008). Pyridoxal phosphate-dependent neonatal epileptic encephalopathy. *Arch. Dis. Child. Fetal Neonatal Ed.*, 93:F151–F152.

Bankier, A., M. Turner, I.J. Hopkins. (1983). Pyridoxine dependent seizures: a wider clinical spectrum. *Arch. Dis. Child.*, 58:415–418.

Barron, G.P., S.O. Brown, P.B. Pearson. (1949). Histological manifestations of a magnesium deficiency in the rat and rabbit. *Proc. Soc. Exp. Biol. Med.*, 70:220–223.

Bass, N.E., E. Wyllie, B. Cohen, S.A. Joseph. (1996). Pyridoxine-dependent epilepsy: the need for repeated pyridoxine trials and the risk of severe electrocerebral suppression with intravenous pyridoxine infusion. *J. Child Neurol.*, 11:422–424.

Basura, G.J., S.P. Hagland, A.M. Wiltse, S.M. Gospe, Jr. (2009). Clinical features and the management of pyridoxine-dependent and pyridoxine-responsive seizures: review of 63 North American cases submitted to a patient registry. *Eur. J. Pediatr.*, 168:697–704.

Battaglioli, G., D.R. Rosen, S.M. Gospe, Jr., D.L. Martin. (2000). Glutamate decarboxylase is not genetically linked to pyridoxine-dependent seizures. *Neurology*, 55:309–311.

Baumeister, F.A., W. Gsell, Y.S. Shin, J. Egger. (1994). Glutamate in pyridoxine-dependent epilepsy: neurotoxic glutamate concentration in the cerebrospinal fluid and its normalization by pyridoxine. *Pediatrics*, 94:318–321.

Baxter, P. (1999). Epidemiology of pyridoxine dependent and pyridoxine responsive seizures in the UK. *Arch. Dis. Child.*, 81:431–433.

Baxter, P. (2001). Pyridoxine-dependent and pyridoxine-responsive seizures. *Dev. Med. Child Neurol.*, 43:416–420.

Baxter, P., P. Griffiths, T. Kelly, D. Gardner-Medwin. (1996). Pyridoxine-dependent seizures: demographic, clinical, MRI and psychometric features, and effect of dose on intelligence quotient. *Dev. Med. Child Neurol.*, 38:998–1006.

Belfort, M.A., K.J. Moise, Jr. (1992). Effect of magnesium sulfate on maternal brain blood flow in preeclampsia: a randomized, placebo-controlled study. *Am. J. Obstet. Gynecol.*, 167:661–666.

Bennett, C.L., H.M. Huynh, P.F. Chance, I.A. Glass, S.M. Gospe, Jr. (2005). Genetic heterogeneity for autosomal recessive pyridoxine-dependent seizures. *Neurogenetics*, 6:143–149.

Berger, A.R., H.H. Schaumburg. (1984). More on neuropathy from pyridoxine abuse. *N. Engl. J. Med.*, 311:986–987.

Berger, A.R., H.H. Schaumburg, C. Schroeder, S. Apfel, R. Reynolds. (1992). Dose response, coasting, and differential fiber vulnerability in human toxic neuropathy: a prospective study of pyridoxine neurotoxicity. *Neurology*, 42:1367–1370.

Bergstom, W.H., R.F. Carzoli, C. Lombroso, D.T. Davidson, W.M. Wallace. (1952). Observations on the metabolic and clinical effects of carbonic-anhydrase inhibitors in epileptics. *AMA Am. J. Dis. Child.*, 84:771–772.

Bernstein, A.L., C.S. Lobitz. (1988). A clinical and electrophysiologic study of the treatment of painful diabetic neuropathies with pyridoxine. In *Clinical and Physiological Application of Vitamin $B_6$* (pp. 415–423), J.E. Leklem and R.D. Reynolds, Eds. New York: Alan R. Liss.

Bessey, O.A., D.J.D. Adam, D.R. Bussey, A.E. Hansen. (1954). Vitamin $B_6$ requirements in infants. *Fed. Proc.*, 13:451.

Bessey, O.A., D.J. Adam, A.E. Hansen. (1957). Intake of vitamin $B_6$ and infantile convulsions: a first approximation of requirements of pyridoxine in infants. *Pediatrics*, 20:33–44.

Bettinelli, A., M.G. Bianchetti, E. Girardin, A. Caringella, M. Cecconi, A.C. Appiani et al. (1992). Use of calcium excretion values to distinguish two forms of primary renal tubular hypokalemic alkalosis: Bartter and Gitelman syndromes. *J. Pediatr.*, 120:38–43.

Bhasker, B., P. Raghupathy, T.M. Nair, S.R. Ahmed, V. deSilva, B.C. Bhuyan, S.M. Al Khusaiby. (1999). External hydrocephalus in primary hypomagnesaemia: a new finding. *Arch. Dis. Child.*, 81:505–507.

Blaxter, K.L., J.A. Rook, A.M. Macdonald. (1954). Experimental magnesium deficiency in calves. I. Clinical and pathological observations. *J. Comp. Pathol.*, 64:157–175.

Blennow, G., L. Starck. (1986). High dose $B_6$ treatment in infantile spasms. *Neuropediatrics*, 17:7–10.

Bobele, G.B., J.B. Bodensteiner. (1994). The treatment of infantile spasms by child neurologists. *J. Child Neurol.*, 9:432–435.

Bockenhauer, D., S. Feather, H.C. Stanescu, S. Bandulik, A.A. Zdebik, M. Reichold et al. (2009). Epilepsy, ataxia, sensorineural deafness, tubulopathy, and KCNJ10 mutations. *N. Engl. J. Med.*, 360:1960–1970.

Bolte, A.C., H.P. van Geijn, G.A. Dekker. (2001). Management and monitoring of severe preeclampsia. *Eur. J. Obstet. Gynecol. Reprod. Biol.*, 96:8–20.

Borges, L.F., G. Gucer. (1978). Effect of magnesium on epileptic foci. *Epilepsia*, 19:81–91.

Cake, M.H., D.M. DiSorbo, G. Litwack. (1978). Effect of pyridoxal phosphate on the DNA binding site of activated hepatic glucocorticoid receptor. *J. Biol. Chem.*, 253:4886–4891.

Chutkow, J.G., S. Meyers. (1968). Chemical changes in the cerebrospinal fluid and brain in magnesium deficiency. *Neurology*, 18:963–974.

Clayton, P.T., R.A. Surtees, C. DeVile, K. Hyland, S.J. Heales. (2003). Neonatal epileptic encephalopathy. *Lancet*, 361:1614.

Coker, S.B. (1992). Postneonatal vitamin $B_6$-dependent epilepsy. *Pediatrics*, 90:221–223.

Cormier-Daire, V., N. Dagoneau, R. Nabbout, L. Burglen, C. Penet, C. Soufflet, I. Desguerre, A. Munnich, O. Dulac. (2000). A gene for pyridoxine-dependent epilepsy maps to chromosome 5q31. *Am. J. Hum. Genet.*, 67:991–993.

Cotton, D.B., C.A. Janusz, R.F. Berman. (1992). Anticonvulsant effects of magnesium sulfate on hippocampal seizures: therapeutic implications in preeclampsia-eclampsia. *Am. J. Obstet. Gynecol.*, 166:1127–1136.

Cotton, D.B., M. Hallak, C. Janusz, S.M. Irtenkauf, R.F. Berman. (1993). Central anticonvulsant effects of magnesium sulfate on *N*-methyl-D-aspartate-induced seizures. *Am. J. Obstet. Gynecol.*, 168:974–978.

Coursin, D.B. (1954). Convulsive seizures in infants with pyridoxine-deficient diet. *J. Am. Med. Assoc.*, 154:406–408.

Coursin, D.B. (1955). Vitamin $B_6$ deficiency in infants; a follow-up study. *AMA Am. J. Dis. Child.*, 90:344–348.

Crowell, G.F., E.S. Roach. (1983). Pyridoxine-dependent seizures. *Am. Fam. Physician*, 27:183–187.

Crowther, C.A., J.E. Hiller, L.W. Doyle, R.R. Haslam. (2003). Effect of magnesium sulfate given for neuroprotection before preterm birth: a randomized controlled trial. *JAMA*, 290:2669–2676.

Dakshinamurti, K., C.S. Paulose, M. Viswanathan, Y.L. Siow, S.K. Sharma, B. Bolster. (1990). Neurobiology of pyridoxine. *Ann. N.Y. Acad. Sci.*, 585:128–144.

Dalton, K., M.J. Dalton. (1987). Characteristics of pyridoxine overdose neuropathy syndrome. *Acta Neurol. Scand.*, 76:8–11.

DelTredici, A.M., A.L. Berstein, K. Chinn. (1985). Carpal tunnel syndrome and vitamin $B_6$ therapy. In *Vitamin $B_6$: Its Role in Health and Disease* (pp. 459–462), R.D. Reynolds and J.E. Leklem, Eds. New York: Alan R. Liss.

Dingledine, R., C.J. McBain. (1994). Excitatory: amino acid neurotransmitters. In *Basic Neurochemistry* (pp. 367–387), C.J. Siegel, Ed. New York: Raven Press.

Dinsdale, H.B. (1988). Does magnesium sulfate treat eclamptic seizures? Yes. *Arch. Neurol.*, 45:1360–1361.

Donaldson, J.O. (1986). Does magnesium sulfate treat eclamptic convulsions? *Clin. Neuropharmacol.*, 9:37–45.

Donaldson, J.O. (1988). Eclampsia. In *Major Problems in Neurology*. Vol. 19. *Neurology of Pregnancy*, 2nd ed. (pp. 269–310), J.O. Donaldson, Ed. Philadelphia, PA: W.B. Saunders.

Dorsett, L. (1926). The intramuscular injection of magnesium sulphate for the control of convulsions in eclampsia. *Am. J. Obstet. Gynecol.*, 11:227–231.

Dreier, J.P., U. Heinemann. (1991). Regional and time dependent variations of low $Mg^{2+}$ induced epileptiform activity in rat temporal cortex slices. *Exp. Brain Res.*, 87:581–596.

Dudin, K.I., A.S. Teebi. (1987). Primary hypomagnesaemia: a case report and literature review. *Eur. J. Pediatr.*, 146:303–305.

Duley, L., D. Henderson-Smart. (2001a). Magnesium sulphate versus diazepam for eclampsia. *Cochrane Database of Systematic Review*, 4:CD000127 (DOI: 10.1002/14651858.CD000127).

Duley, L., D. Henderson-Smart. (2001b). Magnesium sulphate versus phenytoin for eclampsia. *Cochrane Database of Systematic Review*, 4:CD000128 (DOI: 10.1002/14651858.CD000128).

Eastman, N.J., P.P. Steptoe. (1945). The management of pre-eclampsia. *Can. Med. Assoc. J.*, 52:562–568.

Ebinger, M., C. Schultze, S. Konig. (1999). Demographics and diagnosis of pyridoxine-dependent seizures. *J. Pediatr.*, 134:795–796.

Eclampsia Trial Collaborative Group. (1995). Which anticonvulsant for women with eclampsia? Evidence from the Collaborative Eclampsia Trial. *Lancet*, 345:1455–1463.

Ekelund, H., I. Gamstorp, W. Von Studnitz. (1969). Apparent response of impaired mental development, minor motor epilepsy and ataxia to pyridoxine. *Acta Paediatr. Scand.*, 58:572–576.

Engstrom, F.L., H.S. White, J.W. Kemp, D.M. Woodbury. (1986). Acute and chronic acetazolamide administration in DBA and C57 mice: effects of age. *Epilepsia*, 27:19–26.

Evans, R.A., J.N. Carter, C.R. George, R.S. Walls, R.C. Newland, G.D. McDonnell, J.R. Lawrence. (1981). The congenital "magnesium-losing kidney": report of two patients. *Q. J. Med.*, 50:39–52.

Farrant, R.D., V. Walker, G.A. Mills, J.M. Mellor, G.J. Langley. (2001). Pyridoxal phosphate de-activation by pyrroline-5-carboxylic acid: increased risk of vitamin $B_6$ deficiency and seizures in hyperprolinemia type II. *J. Biol. Chem.*, 276:15107–15116.

Fisher, R.S., P.W. Kaplan, A. Krumholz, R.P. Lesser, S.A. Rosen, M.R. Wolff. (1988). Failure of high-dose intravenous magnesium sulfate to control myoclonic status epilepticus. *Clin. Neuropharmacol.*, 11:537–544.

Fishman, R.A. (1965). Neurological aspects of magnesium metabolism. *Arch. Neurol.*, 12:562–569.

Forsythe, W.I., J.R. Owens, C. Toothill. (1981). Effectiveness of acetazolamide in the treatment of carbamazepine-resistant epilepsy in children. *Dev. Med. Child Neurol.*, 23:761–769.

French, J.H., B.B. Grueter, R. Druckman, D. O'Brien. (1965). Pyridoxine and infantile myoclonic seizures. *Neurology*, 15:101–113.

Futagi, Y., K. Otani, J. Abe. (1996). Growth suppression in children receiving acetazolamide with antiepileptic drugs. *Pediatr. Neurol.*, 15:323–326.

Gallagher, R.C., J.L. Van Hove, G. Scharer, K. Hyland, B. Plecko, P.J. Waters, S. Mercimek-Mahmutoglu, S. Stockler-Ipsiroglu, G.S. Salomons, E.H. Rosenberg, E.A. Struys, C. Jakobs. (2009). Folinic acid-responsive seizures are identical to pyridoxine-dependent epilepsy. *Ann. Neurol.*, 65:550–556.

Garty, R., Z. Yonis, J. Braham, K. Steinitz. (1962). Pyridoxine-dependent convulsions in an infant. *Arch. Dis. Child.*, 37:21–24.

Goldman, R.S., S.M. Finkbeiner. (1988). Therapeutic use of magnesium sulfate in selected cases of cerebral ischemia and seizure. *N. Engl. J. Med.*, 319:1224–1225.

Golla, F.L., R.S. Hodge. (1957). Control of petit mal by acetazolamide. *J. Ment. Sci.*, 103:214–217.

Gordon, N. (1997). Pyridoxine dependency: an update. *Dev. Med. Child Neurol.*, 39:63–65.

Gospe, Jr., S.M. (1998). Current perspectives on pyridoxine-dependent seizures. *J. Pediatr.*, 132:919–923.

Gospe, Jr., S.M. (2006). Pyridoxine-dependent seizures: new genetic and biochemical clues to help with diagnosis and treatment. *Curr. Opin. Neurol.*, 19:148–153.

Goutieres, F., J. Aicardi. (1985). Atypical presentations of pyridoxine-dependent seizures: a treatable cause of intractable epilepsy in infants. *Ann. Neurol.*, 17:117–120.

Gray, W.D., C.E. Rauh. (1967). The anticonvulsant action of inhibitors of carbonic anhydrase: site and mode of action in rats and mice. *J. Pharmacol. Exp. Ther.*, 156:383–396.

Gray, W.D., T.H. Maren, G.M. Sisson, F.H. Smith. (1957). Carbonic anhydrase inhibition. VII. Carbonic anhydrase inhibition and anticonvulsant effect. *J. Pharmacol. Exp. Ther.*, 121:160–170.

Grillo, E., R.J. da Silva, J.H. Barbato, Jr. (2001). Pyridoxine-dependent seizures responding to extremely low-dose pyridoxine. *Dev. Med. Child Neurol.*, 43:413–415.

Haenggeli, C.A., E. Girardin, L. Paunier. (1991). Pyridoxine-dependent seizures: clinical and therapeutic aspects. *Eur. J. Pediatr.*, 150:452–455.

Hallak, M., R.F. Berman, S.M. Irtenkauf, M.I. Evans, D.B. Cotton. (1992). Peripheral magnesium sulfate enters the brain and increases the threshold for hippocampal seizures in rats. *Am. J. Obstet. Gynecol.*, 167:1605–1610.

Hammen, A., B. Wagner, M. Berkhoff, F. Donati. (1998). A paradoxical rise of neonatal seizures after treatment with vitamin $B_6$. *Eur. J. Paediatr. Neurol.*, 2:319–322.

Hansson, O., B. Hagberg. (1968). Effect of pyridoxine treatment in children with epilepsy. *Acta Soc. Med. Ups.*, 73:35–43.

Heeley, A., R.J. Pugh, B.E. Clayton, J. Shepherd, J. Wilson. (1978). Pyridoxol metabolism in vitamin $B_6$-responsive convulsions of early infancy. *Arch. Dis. Child.*, 53:794–802.

Herron, C.E., R.A. Lester, E.J. Coan, G.L. Collingridge. (1985). Intracellular demonstration of an *N*-methyl-D-aspartate receptor mediated component of synaptic transmission in the rat hippocampus. *Neurosci. Lett.*, 60:19–23.

Hilmy, M.I., G.G. Somjen. (1968). Distribution and tissue uptake of magnesium related to its pharmacological effects. *Am. J. Physiol.*, 214:406–413.

Hirschfelder, A.D. (1934). Clinical manifestations of high and low plasma magnesium; dangers of Epson salt purgation in nephritis. *JAMA*, 102:1138–1141.

Horn, E. (1906). To tilfaelde af eclampsia gravidarum behandlet med sulfas magnesicus injeceret: rygmarvens subarachnoidalrum. *Medicinsk Rev.*, 32:264–272.

Hunt, Jr., A.D., J. Stokes, Jr., W.W. McCrory, H.H. Stroud. (1954). Pyridoxine dependency: report of a case of intractable convulsions in an infant controlled by pyridoxine. *Pediatrics*, 13:140–145.

Ito, M. (1998). Antiepileptic drug treatment of West syndrome. *Epilepsia*, 39(Suppl. 5):38–41.

Ito, M., H. Mikawa, T. Taniguchi. (1984). Cerebrospinal fluid GABA levels in children with infantile spasms. *Neurology*, 34:235–238.

Ito, M., T. Okuno, H. Hattori, T. Fujii, H. Mikawa. (1991). Vitamin B$_6$ and valproic acid in treatment of infantile spasms. *Pediatr. Neurol.*, 7:91–96.

Ito, M., T. Seki, Y. Takuma. (2000). Current therapy for West syndrome in Japan. *J. Child Neurol.*, 15:424–428.

Jakobs, C., J. Jaeken, K.M. Gibson. (1993). Inherited disorders of GABA metabolism. *J. Inherit. Metab. Dis.*, 16:704–715.

Jardim, L.B., R.F. Pires, C.E. Martins, C.R. Vargas, J. Vizioli, F.A. Kliemann, R. Giugliani. (1994). Pyridoxine-dependent seizures associated with white matter abnormalities. *Neuropediatrics*, 25:259–261.

Kaplan, P.W. (1999). Neurologic issues in eclampsia. *Rev. Neurol. (Paris)*, 155:335–341.

Kaplan, P.W., R.P. Lesser, R.S. Fisher, J.T. Repke, D.F. Hanley. (1988). No, magnesium sulfate should not be used in treating eclamptic seizures. *Arch. Neurol.*, 45:1361–1364.

Keisu, M., B.E. Wiholm, A. Ost, O. Mortimer. (1990). Acetazolamide-associated aplastic anaemia. *J. Intern. Med.*, 228:627–632.

Kenny, L., P.N. Baker. (1999). Maternal pathophysiology in pre-eclampsia. *Baillieres Best Pract. Res. Clin. Obstet. Gynaecol.*, 13:59–75.

Killam, K.F., J.A. Bain. (1957). Convulsant hydrazides. I. *In vitro* and *in vivo* inhibition of vitamin B$_6$ enzymes by convulsant hydrazides. *J. Pharmacol. Exp. Ther.*, 119:255–262.

Koontz, W.L., K.H. Reid. (1985). Effect of parenteral magnesium sulfate on penicillin-induced seizure foci in anesthetized cats. *Am. J. Obstet. Gynecol.*, 153:96–99.

Krauss, G.L., P. Kaplan, R.S. Fisher. (1989). Parenteral magnesium sulfate fails to control electroshock and pentylenetetrazol seizures in mice. *Epilepsy Res.*, 4:201–206.

Krinke, G., H.H. Schaumburg, P.S. Spencer, J. Suter, P. Thomann, R. Hess. (1981). Pyridoxine megavitaminosis produces degeneration of peripheral sensory neurons (sensory neuronopathy) in the dog. *Neurotoxicology*, 2:13–24.

Krishnamoorthy, K.S. (1983). Pyridoxine-dependency seizure: report of a rare presentation. *Ann. Neurol.*, 13:103–104.

Kroll, J.S. (1985). Pyridoxine for neonatal seizures: an unexpected danger. *Dev. Med. Child Neurol.*, 27:377–379.

Kruse, H.D., E.R. Orent, E.V. McCollum. (1932). Studies on magnesium deficiency in animals. I. Symptomatology resulting from magnesium deficiency. *J. Biol. Chem.*, 96:519–539.

Kuo, M.F., H.S. Wang. (2002). Pyridoxal phosphate-responsive epilepsy with resistance to pyridoxine. *Pediatr. Neurol.*, 26:146–147.

Kure, S., Y. Sakata, S. Miyabayashi, K. Takahashi, T. Shinka, Y. Matsubara, H. Hoshino, K. Narisawa. (1998). Mutation and polymorphic marker analyses of 65K- and 67K-glutamate decarboxylase genes in two families with pyridoxine-dependent epilepsy. *J. Hum. Genet.*, 43:128–131.

Kurlemann, G., W. Löscher, H.C. Dominick, G.D. Palm. (1987). Disappearance of neonatal seizures and low CSF GABA levels after treatment with vitamin B$_6$. *Epilepsy Res.*, 1:152–154.

Kurlemann, G., E.M. Menges, D.G. Palm. (1991). Low level of GABA in CSF in vitamin B$_6$-dependent seizures. *Dev. Med. Child. Neurol.*, 33:749–750.

Kurlemann, G., R. Ziegler, M. Gruneberg, T. Bomelburg, K. Ullrich, D.G. Palm. (1992). Disturbance of GABA metabolism in pyridoxine-dependent seizures. *Neuropediatrics*, 23:257–259.

Kurlemann, G., T. Deufel, G. Schuierer. (1997). Pyridoxine-responsive West syndrome and gamma-aminobutyric acid. *Eur. J. Pediatr.*, 156:158–159.

Lazard, E.M. (1925). A preliminary report on the intravenous use of magnesium sulphate in puerperal eclampsia. *Am. J. Obstet. Gynecol.*, 9:178–188.

Lazard, E.M. (1933). An analysis of 575 cases of eclamptic and preeclamptic toxemias treated by intravenous injections of magnesium sulphate. *Am. J. Obstet. Gynecol.*, 26:647–656.

Lim, L.L., N. Foldvary, E. Mascha, J. Lee. (2001). Acetazolamide in women with catamenial epilepsy. *Epilepsia*, 42:746–749.

Lipton, S.A., P.A. Rosenberg. (1994). Excitatory amino acids as a final common pathway for neurologic disorders. *N. Engl. J. Med.*, 330:613–622.

Livingston, S., D. Peterson, L. Boks. (1956). Ineffectiveness of diamox in the treatment of childhood epilepsy. *Pediatrics*, 17:541.

Lombroso, C.T., I. Forsythe. (1960). A long-term follow-up of acetazolamide (diamox) in the treatment of epilepsy. *Epilepsia*, 1:493–500.

Lombroso, C.T., D.T. Davidson, Jr., M.L. Grossi-Bianchi. (1956). Further evaluation of acetazolamide (diamox) in treatment of epilepsy. *J. Am. Med. Assoc.*, 160:268–272.

Lott, I.T., T. Coulombe, R.V. Di Paolo, E.P. Richardson, Jr., H.L. Levy. (1978). Vitamin $B_6$-dependent seizures: pathology and chemical findings in brain. *Neurology*, 28:47–54.

Lucas, M.J., K.J. Leveno, F.G. Cunningham. (1995). A comparison of magnesium sulfate with phenytoin for the prevention of eclampsia. *N. Engl. J. Med.*, 333:201–205.

Marret, S., L. Marpeau, C. Follet-Bouhamed, G. Cambonie, D. Astruc, B. Delaporte, H. Bruel, B. Guillois, D. Pinquier, V. Zupan-Simunek, J. Benichou. (2008). Effect of magnesium sulphate on mortality and neurologic morbidity of the very-preterm newborn (of less than 33 weeks) with two-year neurological outcome: results of the prospective PREMAG trial. *Gynecol. Obstet. Fertil.*, 36:278–288.

Martin, H.E. (1969). Clinical magnesium deficiency. *Ann. N.Y. Acad. Sci.*, 162:891–900.

Mayer, M.L., G.L. Westbrook, P.B. Guthrie. (1984). Voltage-dependent block by $Mg^{2+}$ of NMDA responses in spinal cord neurones. *Nature*, 309:261–263.

McLachlan, R.S., W.F. Brown. (1995). Pyridoxine dependent epilepsy with iatrogenic sensory neuronopathy. *Can. J. Neurol. Sci.*, 22:50–51.

Mikati, M.A., E. Trevathan, K.S. Krishnamoorthy, C.T. Lombroso. (1991). Pyridoxine-dependent epilepsy: EEG investigations and long-term follow-up. *Electroencephalogr. Clin. Neurophysiol.*, 78:215–221.

Millichap, J.G. (1956). Anticonvulsant action of diamox in children. *Neurology*, 6:552–559.

Millichap, J.G. (1957). Development of seizure patterns in newborn animals; significance of brain carbonic anhydrase. *Proc. Soc. Exp. Biol. Med.*, 96:125–129.

Millichap, J.G. (1958). Seizure patterns in young animals; significance of brain carbonic anhydrase, Part II. *Proc. Soc. Exp. Biol. Med.*, 97:606–611.

Millichap, J.G., D.M. Woodbury, L.S. Goodman. (1955). Mechanism of the anticonvulsant action of acetazoleamide, a carbonic anhydrase inhibitor. *J. Pharmacol. Exp. Ther.*, 115:251–258.

Millichap, J.G., M. Balter, P. Hernandez. (1958). Development of susceptibility to seizures in young animals. III. Brain water, electrolyte and acid–base metabolism. *Proc. Soc. Exp. Biol. Med.*, 99:6–11.

Mills, P.B., R.A. Surtees, M.P. Champion, C.E. Beesley, N. Dalton, P.J. Scambler, S.J. Heales, A. Briddon, I. Scheimberg, G.F. Hoffmann, J. Zschocke, P.T. Clayton. (2005). Neonatal epileptic encephalopathy caused by mutations in the PNPO gene encoding pyridox(am)ine 5′-phosphate oxidase. *Hum. Mol. Genet.*, 14:1077–1086.

Mills, P.B., E. Struys, C. Jakobs, B. Plecko, P. Baxter, M. Baumgartner, M.A. Willemsen, H. Omran, U. Tacke, B. Uhlenberg, B. Weschke, P.T. Clayton. (2006). Mutations in antiquitin in individuals with pyridoxine-dependent seizures. *Nat. Med.*, 12:307–309.

Minns, R. (1980). Vitamin $B_6$ deficiency and dependency. *Dev. Med. Child. Neurol.*, 22:795–799.

Mody, I., J.D. Lambert, U. Heinemann. (1987). Low extracellular magnesium induces epileptiform activity and spreading depression in rat hippocampal slices. *J. Neurophysiol.*, 57:869–888.

Molony, C.J., A.H. Parmelee. (1954). Convulsions in young infants as a result of pyridoxine (vitamin $B_6$) deficiency. *J. Am. Med. Assoc.*, 154:405–406.

Morgan, J.D., M.J. Painter. (1999). Neonatal seizures. In *Pediatric Neurology: Principles and Practice* (pp. 183–190), K.F. Swaiman and S. Ashwal, Eds. St. Louis, MO: Mosby.

Mpofu, C., S.M. Alani, C. Whitehouse, B. Fowler, J.E. Wraith. (1991). No sensory neuropathy during pyridoxine treatment in homocystinuria. *Arch. Dis. Child.*, 66:1081–1082.

Muldowney, F.P., T.J. McKenna, L.H. Kyle, R. Freaney, M. Swan. (1970). Parathormone-like effect of magnesium replenishment in steatorrhea. *N. Engl. J. Med.*, 282:61–68.

Nabbout, R., C. Soufflet, P. Plouin, O. Dulac. (1999). Pyridoxine dependent epilepsy: a suggestive electroclinical pattern. *Arch. Dis. Child. Fetal. Neonatal. Ed.*, 81:F125–F129.

Nakagawa, E., T. Tanaka, M. Ohno, T. Yamano, M. Shimada. (1997). Efficacy of pyridoxal phosphate in treating an adult with intractable status epilepticus. *Neurology*, 48:1468–1469.

Nelson, K.B., J.K. Grether. (1995). Can magnesium sulfate reduce the risk of cerebral palsy in very low birthweight infants? *Pediatrics*, 95:263–269.

Nowak, L., P. Bregestovski, P. Ascher, A. Herbet, A. Prochiantz. (1984). Magnesium gates glutamate-activated channels in mouse central neurones. *Nature*, 307:462–465.

Nuytten, D., J. Van Hees, A. Meulemans, H. Carton. (1991). Magnesium deficiency as a cause of acute intractable seizures. *J. Neurol.*, 238:262–264.

Ohashi, N., K. Watanabe, K. Aso. (1995). The treatment of West syndrome by child neurologists in Japan. *Psychiatry Clin. Neurosci.*, 49:S244–S245.

Ohtsuka, Y., M. Matsuda, T. Ogino, K. Kobayashi, S. Ohtahara. (1987). Treatment of the West syndrome with high-dose pyridoxal phosphate. *Brain. Dev.*, 9:418–421.

Ohtsuka, Y., T. Ogino, T. Asano, J. Hattori, H. Ohta, E. Oka. (2000). Long-term follow-up of vitamin B(6)-responsive West syndrome. *Pediatr. Neurol.*, 23:202–206.

Oles, K.S., J.K. Penry, D.L. Cole, G. Howard. (1989). Use of acetazolamide as an adjunct to carbamazepine in refractory partial seizures. *Epilepsia*, 30:74–78.

Orent, E.R., H.D. Kruse, E.V. McCollum. (1932). Studies on magnesium deficiency in animals. II. Species variation in symptomatology of magnesium deprivation. *Am. J. Physiol.*, 101:454–461.

Parry, G.J., D.E. Bredesen. (1985). Sensory neuropathy with low-dose pyridoxine. *Neurology*, 35:1466–1468.

Pettit, R.E. (1987). Pyridoxine dependency seizures: report of a case with unusual features. *J. Child Neurol.*, 2:38–40.

Pietz, J., C. Benninger, H. Schafer, D. Sontheimer, G. Mittermaier, D. Rating. (1993). Treatment of infantile spasms with high-dosage vitamin $B_6$. *Epilepsia*, 34:757–763.

Plecko, B., K. Paul, E. Paschke, S. Stoeckler-Ipsiroglu, E. Struys, C. Jakobs et al. (2007). Biochemical and molecular characterization of 18 patients with pyridoxine-dependent epilepsy and mutations of the antiquitin (*ALDH7A1*) gene. *Hum. Mutat.*, 28:19–26.

Pritchard, J.A. (1955). The use of the magnesium ion in the management of eclamptogenic toxemias. *Surg. Gynecol. Obstet.*, 100:131–140.

Pritchard, J.A., F.G. Cunningham, S.A. Pritchard. (1984). The Parkland Memorial Hospital protocol for treatment of eclampsia: evaluation of 245 cases. *Am. J. Obstet. Gynecol.*, 148:951–963.

Ramin, K.D. (1999). The prevention and management of eclampsia. *Obstet. Gynecol. Clin. North. Am.*, 26:ix, 489–503.

Reiss, W.G., K.S. Oles. (1996). Acetazolamide in the treatment of seizures. *Ann. Pharmacother.*, 30:514–519.

Report of the Council on Pharmacy and Chemistry. (1951). Pyridoxine hydrochloride (vitamin $B_6$): a status report. *J. Am. Med. Assoc.*, 147:322–325.

Resor, Jr., S.R., L.D. Resor. (1990). Chronic acetazolamide monotherapy in the treatment of juvenile myoclonic epilepsy. *Neurology*, 40:1677–1681.

Resor, Jr., S.R., L.D. Resor, D.M. Woodbury, J.W. Kemp. (1995). Other antiepileptic drugs: acetazolamide. In *Antiepileptic Drugs*, 4th ed. (pp. 969–985), R.H. Levy, R.H. Mattson, and B.S. Meldrum, Eds. New York: Raven Press.

Roblin, R.O., J.W. Clapp. (1950). The preparation of heterocyclic sulfonamides. *J. Am. Chem. Soc.*, 72:4890–4892.

Roine, P., A.N. Booth et al. (1949). Importance of potassium and magnesium in nutrition of the guinea pig. *Proc. Soc. Exp. Biol. Med.*, 71:90.

Rouse, D.J., D.G. Hirtz, E. Thom, M.W. Varner, C.Y. Spong, B.M. Mercer et al. (2008). A randomized, controlled trial of magnesium sulfate for the prevention of cerebral palsy. *N. Engl. J. Med.*, 359:895–905.

Rude, R.K., F.R. Singer. (1981). Magnesium deficiency and excess. *Annu. Rev. Med.*, 32:245–259.

Sankar, R., E.A. Licht. (1995). Magnesium sulfate versus phenytoin for the prevention of eclampsia. *N. Engl. J. Med.*, 333:1638; author reply, 9.

Sapirstein, V.S., P. Strocchi, J.M. Gilbert. (1984). Properties and functions of brain carbonic anhydrase. In *Biology and Chemistry of Carbonic Anhydrases* (pp. 481–493), R.E. Tashian and D. Hewitt-Emmett, Eds. New York: New York Academy of Sciences.

Sawhney, H., I.M. Sawhney, R. Mandal, Subramanyam, K. Vasishta. (1999). Efficacy of magnesium sulphate and phenytoin in the management of eclampsia. *J. Obstet. Gynaecol. Res.*, 25:333–338.

Schaumburg, H., J. Kaplan, A. Windebank, N. Vick, S. Rasmus, D. Pleasure, M.J. Brown. (1983). Sensory neuropathy from pyridoxine abuse: a new megavitamin syndrome. *N. Engl. J. Med.*, 309:445–448.

Schlingmann, K.P., S. Weber, M. Peters, L. Niemann Nejsum, H. Vitzthum, K. Klingel et al. (2002). Hypomagnesemia with secondary hypocalcemia is caused by mutations in *TRPM6*, a new member of the *TRPM* gene family. *Nat. Genet.*, 31:166–70.

Scholl, U.I., M. Choi, T. Liu, V.T. Ramaekers, M.G. Hausler, J. Grimmer, S.W. Tobe, A. Farhi, C. Nelson-Williams, R.P. Lifton. (2009). Seizures, sensorineural deafness, ataxia, mental retardation, and electrolyte imbalance (SeSAME syndrome) caused by mutations in KCNJ10. *Proc. Natl. Acad. Sci. U.S.A.*, 106:5842–5847.

Scriver, C.R. (1960). Vitamin $B_6$-dependency and infantile convulsions. *Pediatrics*, 26:62–74.

Scriver, C.R., J.H. Hutchison. (1963). The vitamin $B_6$ deficiency syndrome in human infancy: biochemical and clinical observations. *Pediatrics*, 31:240–250.

Shih, J.J., H. Kornblum, D.A. Shewmon. (1996). Global brain dysfunction in an infant with pyridoxine dependency: evaluation with EEG, evoked potentials, MRI, and PET. *Neurology*, 47:824–826.

Sibai, B.M., J.A. Spinnato, D.L. Watson, G.A. Hill, G.D. Anderson. (1984). Pregnancy outcome in 303 cases with severe preeclampsia. *Obstet. Gynecol.*, 64:319–325.

Sibai, B.M., J.A. Spinnato, D.L. Watson, J.A. Lewis, G.D. Anderson. (1985). Eclampsia. IV. Neurological findings and future outcome. *Am. J. Obstet. Gynecol.*, 152:184–192.

Siberry, G.K., R. Iannone. (2000). *The Harriet Lane Handbook*, 15th ed. St. Louis, MO: Mosby, 762 pp.

Snyderman, S.E., R. Carretero, L.E. Holt. (1950). Pyridoxine deficiency in the human being. *Fed. Proc.*, 9:372–373.

Snyderman, S.E., L.E. Holt, Jr., R. Carretero, K. Jacobs. (1953). Pyridoxine deficiency in the human infant. *J. Clin. Nutr.*, 1:200–207.

Standley, C.A., S.M. Irtenkauf, L. Stewart, B. Mason, D.B. Cotton. (1994). Magnesium sulfate versus phenytoin for seizure prevention in amygdala-kindled rats. *Am. J. Obstet. Gynecol.*, 171:948–951.

Stock, J.G. (1990). Sulfonamide hypersensitivity and acetazolamide. *Arch. Ophthalmol.*, 108:634–635.

Striano, P., S. Battaglia, L. Giordano, G. Capovilla, F. Beccaria, E.A. Struys, G.S. Salomons, C. Jakobs. (2009). Two novel *ALDH7A1* (antiquitin) splicing mutations associated with pyridoxine-dependent seizures. *Epilepsia*, 50:933–936.

Struys, E.A., C. Jakobs. (2007). Alpha-aminoadipic semialdehyde is the biomarker for pyridoxine dependent epilepsy caused by alpha-aminoadipic semialdehyde dehydrogenase deficiency. *Mol. Genet. Metab.*, 91:405.

Takuma, Y., T. Seki. (1996). Combination therapy of infantile spasms with high-dose pyridoxal phosphate and low-dose corticotropin. *J. Child Neurol.*, 11:35–40.

Tanaka, R., M. Okumura, J. Arima, S. Yamakura, T. Momoi. (1992). Pyridoxine-dependent seizures: report of a case with atypical clinical features and abnormal MRI scans. *J. Child Neurol.*, 7:24–28.

Taniuchi, K., I. Nishimori, T. Takeuchi, K. Fujikawa-Adachi, Y. Ohtsuki, S. Onishi. (2002). Developmental expression of carbonic anhydrase-related proteins VIII, X, and XI in the human brain. *Neuroscience*, 112:93–99.

Thiry, A., J.M. Dogne, C.T. Supuran, B. Masereel. (2007). Carbonic anhydrase inhibitors as anticonvulsant agents. *Curr. Top. Med. Chem.*, 7:855–864.

Thomas, S.V. (1998). Neurological aspects of eclampsia. *J. Neurol. Sci.*, 155:37–43.

Tower, D.B. (1956). Neurochemical aspects of pyridoxine metabolism and function. *Am. J. Clin. Nutr.*, 4:329–345.

Traynelis, S.F., S.G. Cull-Candy. (1990). Proton inhibition of $N$-methyl-D-aspartate receptors in cerebellar neurons. *Nature*, 345:347–350.

Tsau, Y.K., W.Y. Tsai, F.L. Lu, W.S. Tsai, C.H. Chen. (1998). Symptomatic hypomagnesemia in children. *Zhonghua Min Guo Xiao Er Ke Yi Xue Hui Za Zhi*, 39:393–397.

Vallee, B.L., W.E. Wacker, D.D. Ulmer. (1960). The magnesium-deficiency tetany syndrome in man. *N. Engl. J. Med.*, 262:155–161.

Velisek, L., S.L. Moshe, S.G. Su, W. Cammer. (1993). Reduced susceptibility to seizures in carbonic anhydrase II deficient mutant mice. *Epilepsy Res.*, 14(2):115–121.

Waldinger, C., R.B. Berg. (1963). Signs of pyridoxine dependency manifest at birth in siblings. *Pediatrics*, 32:161–168.

Walther, H., J.D. Lambert, R.S. Jones, U. Heinemann, B. Hamon. (1986). Epileptiform activity in combined slices of the hippocampus, subiculum and entorhinal cortex during perfusion with low magnesium medium. *Neurosci. Lett.*, 69:156–161.

Watson, K.V., C.F. Moldow, P.L. Ogburn, H.S. Jacob. (1986). Magnesium sulfate: rationale for its use in preeclampsia. *Proc. Natl. Acad. Sci. U.S.A.*, 83:1075–1078.

Waymire, K.G., J.D. Mahuren, J.M. Jaje, T.R. Guilarte, S.P. Coburn, G.R. MacGregor. (1995). Mice lacking tissue non-specific alkaline phosphatase die from seizures due to defective metabolism of vitamin B-6. *Nat. Genet.*, 11:45–51.

Woodbury, D.M. (1980). Carbonic anhydrase inhibitors. In *Advances in Neurology*. Vol. 27. *Antiepileptic Drugs: Mechanisms of Action* (pp. 617–634), G.H. Glaser, J.K. Penry, and D.M. Woodbury, Eds. New York: Raven Press.

Woodbury, D.M., D.W. Esplin. (1959). Neuropharmacology and neurochemistry of anticonvulsant drugs. *Res. Publ. Assoc. Res. Nerv. Ment. Dis.*, 37:24–56.

Woodbury, D.M., R. Karler. (1960). The role of carbon dioxide in the nervous system. *Anesthesiology*, 21:686–703.

Xu, Y., J.T. Sladky, M.J. Brown. (1989). Dose-dependent expression of neuronopathy after experimental pyridoxine intoxication. *Neurology*, 39:1077–1083.

Yoshikawa, H., T. Abe, Y. Oda. (1999). Pyridoxine-dependent seizures in an older child. *J. Child Neurol.*, 14:687–690.

Zuspan, F.P. (1966). Treatment of severe preeclampsia and eclampsia. *Clin. Obstet. Gynecol.*, 9:954–972.

# 23 Herbs and Botanicals

*Steven C. Schachter*

## CONTENTS

## INTRODUCTION

Botanicals and herbs have been used for centuries by persons with epilepsy from many cultures around the world. Currently, herbal therapies are often tried by patients in developing as well as developed countries for control of seizures, to reduce side effects from antiepileptic drugs (AEDs), or for general health maintenance. The physicians who prescribe these patients their AEDs are rarely told about the herbal therapies. Well-designed clinical trials of herbal therapies in patients with epilepsy are scarce, and methodological issues prevent any firm conclusions about their efficacy or safety in this population. Further, anecdotal evidence suggests that some botanicals and herbs may be proconvulsant or alter AED metabolism (Samuels et al., 2008). In spite of these limitations, further preclinical evaluation of herbs and botanicals and their constituent compounds using validated scientific methods is warranted based on numerous anecdotal observations of clinical benefit in patients with epilepsy and published reports showing mechanisms of action relevant to epilepsy or anticonvulsant effects in animal models of epilepsy. This chapter highlights the use of herbal therapies for epilepsy, outlines the U.S. Food and Drug Administration's role in regulating herbal products, and discusses safety issues, specific herbal therapies that have been used and evaluated for epilepsy, and the author's tradition-to-bench-to-bedside approach to herbal therapy research for epilepsy.

## UNMET MEDICAL NEED AND HERBAL THERAPIES

Despite the availability of many AEDs, nearly one in three patients with epilepsy who have access to AEDs continue to have seizures and a similar proportion experience unacceptable AED-related side effects (Brodie, 2005). In addition, the large majority of people with epilepsy around the world are not under treatment with AEDs, largely because of their lack of access to physicians, the costs of AEDs, and cultural attitudes toward modern treatments (Meinardi et al., 2001).

For thousands of years, people with epilepsy have used a variety of herbs and botanicals, hereafter referred to as "herbal therapies" (no clinical benefit is implied or suggested by this term). Today, herbal therapies are among the most commonly used forms of complementary and alternative medical (CAM) therapies by patients. The National Institutes of Health–National Center of Complementary and Alternative Medicine (http://nccam.nih.gov/) identifies CAM therapies as healthcare and medical practices that are not currently an integral part of conventional medicine, meaning the system of medical knowledge and practices taught in Western medical schools (for example, in the United States) and as practiced by Western-trained physicians, including neurologists.

Patients with a variety of chronic illnesses, including epilepsy, take herbal therapies for many reasons. For example, patients in developed countries may view herbal therapies as natural and time tested and therefore safe compared with "artificial" drugs, an attitude supported by recent reports of safety issues associated with widely prescribed Food and Drug Administration (FDA)-approved drugs. By contrast, persons in developing countries may have access to herbal therapies but not pharmaceuticals, due to both cultural and economic factors. Herbal traditions include traditional Chinese medicine (TCM), Ayurveda, and other culturally specific practices in which plant materials, processed or not, are ingested by persons with the intention of reducing symptoms or curing disease.

## EXTENT AND PATTERN OF USE

### DEVELOPED COUNTRIES

In developed countries, CAM treatments, including herbal therapies, are often used for general health maintenance or for chronic conditions such as pain or epilepsy that respond poorly or incompletely to standard medical treatments (Wahner-Roedler et al., 2005). Liow et al. (2007) surveyed 228 adult patients with epilepsy in the U.S. Midwest about their use of, perception of, and attitudes toward CAM therapies. Thirty-nine percent of the respondents reported using CAM therapies, and 25% reported use of CAM therapies specifically for seizure control. The most common forms of CAM were herbal therapies and other dietary supplements, prayer/spirituality, "mega" vitamins, chiropractic care, and stress management. Other surveys in the United States or United Kingdom suggest that up to one in three patients with epilepsy take herbal therapies and/or dietary supplements and that the majority do not inform their treating physicians (Easterford et al., 2005; Peebles et al., 2000; Sirven et al., 2003), including, for example, persons of South Asian origin living in the United Kingdom (Rhodes et al., 2008). In one survey, ginseng (*Panax* and *Eleutherococcus* species; less often, *Withania* species), St. John's wort (*Hypericum perforatum*), melatonin, gingko (*Ginkgo biloba*), garlic (*Allium sativum*), and black cohosh (*Actaea racemosa*) were most frequently used (Peebles et al., 2000). In another study, garlic, gingko, soy (*Glycine max*), melatonin, and kava (*Piper methysticum*) were most often taken (Sirven et al., 2003). Gingko is commonly taken by nursing home residents, including those with epilepsy, and the potential for proconvulsant effects of gingko products is cause for concern (Bressler, 2005; Harms et al., 2006; Kupiec and Raj, 2005).

Clinical experience suggests that patients may try some herbal therapies in order to reduce AED-related side effects or comorbid conditions—for example, valerian for insomnia, St. John's wort for depression, or gingko for memory disturbance. Therefore, by inquiring whether patients are taking herbal therapies, physicians may gain insight into nonseizure-related problems that patients may be experiencing.

### DEVELOPING COUNTRIES

Herbal therapies for epilepsy in developing countries have evolved over the centuries, often in conjunction with changing cultural beliefs, systems of health care, and levels of education (Elferink, 1999; Kim et al., 2006). For example, nearly 40% of persons responding to a 1988 survey in China said they would suggest that a friend with epilepsy ask for an herbal medicine doctor or seek acuncture (Lai et al., 1990). Herbal therapies in developing countries are generally available from traditional healers or practitioners, either instead of or in addition to medications prescribed by Western-trained physicians. It is not unusual in some regions for herbal therapies to be taken together with AEDs (Danesi and Adetunji, 1994), with combinations that can be tested experimentally in laboratory animals (Vattanajun et al., 2005), or after AEDs prescribed by physicians fail to control seizures (Tandon et al., 2002).

## REGULATORY ASPECTS

Governmental regulations concerning herbal therapies vary around the world. In the United States, prescription drugs are federally regulated by the Federal Food, Drug, and Cosmetic Act, but herbal therapies by the 1994 Dietary Supplement and Health Education Act (DSHEA), under the auspices of the FDA's Office of Nutritional Products, Labeling, and Dietary Supplements, which is also responsible for developing policies for dietary supplements. DSHEA defines a dietary supplement as a product taken by mouth that contains a dietary ingredient intended to supplement the diet. The dietary ingredients in these products may include vitamins, minerals, herbs or other botanicals, amino acids, and substances such as enzymes, organ tissues, glandular tissues, and metabolites (http://www.cfsan.fda.gov/~dms/ds-oview.html).

Under U.S. law, manufacturers of herbal therapies are responsible for the truthfulness of claims made on product labels and for controlling the quality of their products and verifying their safety. Manufacturers cannot claim effectiveness for a specific medical condition, such as epilepsy, but may claim an effect on a bodily part or function when accompanied by the following words: "This statement has not been evaluated by the FDA. This product is not intended to diagnose, treat, cure, or prevent any disease." Under DSHEA, no government agency is required to verify the labeling claims of dietary supplements or to assess their quality and safety. Furthermore, dietary supplements, including herbal therapies, are not required to be produced according to good manufacturing practices (GMPs), a legal requirement for pharmaceutical products. Consequently, herbal therapies could potentially be contaminated with, for example, microorganisms or pesticides; they may contain potentially toxic levels of heavy metals, as has been documented for some Ayurvedic products (Saper et al., 2004); or they may be adulterated with other herbs or drugs (Huang et al., 1997). Also, the purity and amount of herbal extract or standardized presumed active ingredient per unit may vary significantly within the same bottle or from batch to batch, or from one branded herbal therapy product to another, because of variable manufacturing processes. Patients and many of their treating physicians are generally unaware of the potential for wide variability in purity and in per-dose quantity of herbal therapies.

Two recent developments are changing the role of the FDA in overseeing the manufacturing and labeling of herbal therapies. First, in June 2007, the FDA issued a final rule, "Current Good Manufacturing Practice in Manufacturing, Packaging, Labeling, or Holding Operations for Dietary Supplements" (http://www.fda.gov/ohrms/dockets/98fr/cf0441.pdf), which significantly tightens manufacturing standards for dietary supplements, including herbal therapies, thereby ensuring their quality. Second, the FDA established a process to evaluate the safety and efficacy of herbal therapies (which they refer to as *botanical drug products*) in mitigating, treating, curing, or diagnosing a disease, which could potentially lead to FDA approval of herbal therapies for specific medical indications, including epilepsy. The first FDA approval of an herbal therapy under the new guidelines was in late 2006 for an extract of green tea as a topical treatment of genital warts caused by the human papillomavirus (Verigen; http://www.fda.gov/cder/rdmt/InternetNME06.htm).

## SAFETY ISSUES

Despite being generally regarded as natural and therefore judged to be safe by the public, herbal therapies, like pharmaceuticals, may cause serious or life-threatening side effects (Ernst, 2003); furthermore, the long-term safety profile of most herbal therapies, including those used for epilepsy, is unknown (Huxtable, 1990; Pearl et al., 2005). Numerous herbal therapies have been anecdotally reported to cause seizures, including in patients with epilepsy (Luciano and Spinella, 2005). These include anisatin (a component of Japanese star anise, or *Ilicium anisatum*), which is used in Spain and other countries to treat infant colic (Gil Campos et al., 2002; Johanns et al., 2002); gingko nuts (Miwa et al., 2001); essential oils (Spinella, 2001); evening primrose (*Oenothera* species) and borage (*Borago officinalis*) (Spinella, 2001); and the stimulant ephedra (ma huang, or *Ephedra sinica*) (Sirven, 2007). The extract of star fruit (*Averrhoa carambola*) may cause seizures in uremic patients (Neto et al., 2003) and is used to induce seizures experimentally (Carolino et al., 2005), as is the extract from *Catha edulis* (Khat), whose fresh young leaves are used recreationally by an estimated 5 million people in eastern Africa and the Arabian Peninsula (Oyungu et al., 2009). Likewise, an extract of a Chinese herbal therapy for schizophrenia (*Coriaria* lactone, which is made from the active parts of the plant *Loranthus* on *Coriaria sinica* Maxim) is the basis of a rat model for pharmacoresistant temporal lobe epilepsy (Wang et al., 2003).

The pharmacokinetic interactions between herbal therapies and drugs, including AEDs, have been inadequately studied. Available evidence suggests that St. John's wort (Obach, 2000), garlic, Echinacea (various *Echinacea* species), pine bark extract (*Pinus pinaster*, also known as pycnogel, Pygenol, or Pychnogenol), milk thistle (*Silybum* species), American hellebore (*Veratrum viride*), *Ginkgo biloba* (Bressler, 2005; Kupiec and Raj, 2005), mugwort (*Artemisia* species), and pipissewa (*Chimaphila umbellata*) affect the cytochrome P450 system and could therefore potentially affect serum concentrations of hepatically metabolized AEDs (Delgoda and Westlake, 2004), perhaps with fatal consequences (Kupiec and Raj, 2005).

## HERBAL THERAPIES FOR EPILEPSY

Historical evidence suggests that herbal therapies were used to treat convulsive seizures as early as 6000 B.C. in India with the origin of Ayurveda (Jain, 2005), and 3000 B.C. in China and in Peru; similarly, Africa and South America have rich traditions in herbal therapies, including for convulsions (Mahomed and Ojewole, 2006; Ojewole, 2008a,b).

The Ayurvedic literature contains treatises on epilepsy-like symptoms, causes, recognition, and treatment. Herbal and dietary therapies, which are recommended for external application, internal use, and topical use in the eyes and nose, include Brahmirasayan, Brahmighritham, Ashwagandha, old pure desi ghee, daily fresh juice of brahmi (*Centella asiatica* or *Bacopa* species, among others) with honey, garlic juice in oil, and powdered root of wild asparagus (*Asparagus racemosus*) with milk. Others are *Acacia nilotica* (syn. *Acacia arabica*), *Acorus calamus*, *Bacopa monnieri*, *Clitorea ternatea*, *Celastrus paniculatus*, *Convulvulus pluricaulis*, *Phyllanthus emblica* (syn. *Emblica officinalis*, mukta pishti (processed from pearls of *Mytilus magaritiferus* mussels), and *Whithania somnifera* (Chauhan et al., 1988). Recent scientific publications provide the scientific rationale for proceeding with controlled trials of some of these herbal therapies in patients with epilepsy (Kasture et al., 2002, 2008).

The oldest Chinese document on epilepsy, *The Yellow Emperor's Classic of Internal Medicine* (*Huang Di Nei Jing*), dates from 770 to 221 B.C. (Lai and Lai, 1991). Herbal therapies used to treat convulsive diseases in Asia in modern times include (Schachter et al., 2008):

- Chai-Hu-Long-Ku-Mu-Li-Tan (TW-001), a mixture of extracts from 13 herbal therapies
- *Gastrodia elata* (Tian Ma; gastrodia root)
- *Uncaria rhynchophylla* (cat's claw)
- *Menispermum dauricum* (moonseed)
- Shitei (kaki calyx; the calyx of *Diospyros kaki* persimmon)
- Shokyo (gingerroot; rhizome of *Zingiber officinale*)
- Choji (clove; flowerbud of *Syzygium aromaticum*; pharmaceutical name, caryophylli flos)

- Mixture of radish (*Raphanus sativus*) and pepper (*Piper* species, containing the alkaloid piperine)
- Qingyangshen (root of *Cynanchum otophyllum*)
- Kanbaku-taiso-to, a mixture of three herbal drugs—glycyrrhizae radix (licorice root; *Glycyrrhiza* species), tritici semen (wheat seed; *Triticum aestivum*), and zizyphi fructus (spiny jujube fruit; *Ziziphus spinosa*)
- Paeoniae radix (peony root; *Paeonia lactiflora*, syn. *P. albiflora*)
- Zheng Tai instant powder (a complex prescription of traditional Chinese medicine used for tonic–clonic seizures)

Several of these herbal therapies have been shown to have neuroprotective properties (Hsieh et al., 1999a, 2000; Kim et al., 2001), efficacy in animal models of epilepsy (Chiou et al., 1997; Hsieh et al., 1999b; Minami et al., 1999) and hippocampal slice models (Ameri et al., 1997), and effects on gene expression (Sugaya et al., 1999). These studies, however, generally do not specify the methods used to (1) authenticate the source plants, (2) produce extracts and fractions, (3) characterize the active ingredients, or (4) perform the preclinical evaluations (Schachter et al., 2008).

Recent investigations have addressed these limitations. For example, in their study of an extract of *Bacopa monnieri*, a traditional Ayurvedic therapy, on glutamate receptor binding and *GRIN1* gene expression (glutamate receptor, ionotropic, *N*-methyl-D-aspartate 1 gene; previously *NMDAR1*) in the hippocampus of rats following pilocarpine-induced seizures, Khan et al. (2008) described the source of the plant material, as well as the authentication and extraction methods. Furthermore, voucher specimens of the plant materials were prepared and stored, allowing for further characterization and replication of the findings.

In 2005, a comprehensive literature search identified 3 randomized controlled trials, 5 nonrandomized controlled trials, 6 case-control studies, and 57 observational studies, including case reports, of herbal therapies from East Asia for the treatment of epilepsy (Park et al., 2005). Over 135 different herbal extracts were used individually or in various combinations (formulas) in these studies, although rarely was the same herbal formula used in more than one study. The most frequently used plant (or animal) extracts were from the species *Pinellia ternata* (Ban Xia), *Arisaemi serratum* (syn. *A. japonicum*; Tian Nan Xing), *Acorus calamus* (Shi Chang Pu), *Gastrodia elata* (Tian Ma), *Buthus martensii* (Quan Xie; bark scorpion), *Wolfiporia extensa* (syn. *Poria cocos*; Fu Ling), *Bombyx mori* (Jiang Qing; silkmoth; the pharmaceutical name, bombyx batryticatus, indicates silkmoth infestation with a particular fungus), *Citrus reticulata* (Chen Pi), *Uncaria rhynchophylla* (Gou Teng), *Glycyrrhiza glabra* (Gan Cao), and *Salvia miltiorrhiza* (Dan Shen), as well as *Scolopendra subspinipes* (centipede), *Bupleurum falcatum*, *Succinum* (processed amber), *Paeonia lactifolia* (syn. *P. albiflora*), *Panax ginseng*, *Perichaeta communissma* (earthworm), and *Curcuma longa* (Schachter et al., 2008). The clinical studies were generally limited by methodological issues in study design, inadequate powering, insufficient categorization of seizure types and epilepsy syndromes, questionable choices of outcome measures and statistical methods, and lack of characterization of the herbal extracts. Only one clinical epilepsy study of an herbal extract has been published since 2005 in the English literature (Hijikata et al., 2006).

## HARVARD EPILEPSY BOTANICAL PROGRAM

Although properly controlled clinical evidence to support the use of specific herbal therapies for patients with epilepsy is lacking, available anecdotal and laboratory-based evidence suggests that further evaluation is warranted. A program was established at the Osher Research Center of Harvard Medical School to support preclinical evaluation of herbal therapies for epilepsy. The goals of the program are to (1) identify herbal therapies and compounds isolated from them that have promising activity in animal epilepsy models and relevant *in vitro* assays, (2) conduct the preclinical studies necessary to proceed with early stage clinical studies, and (3) plan and initiate these clinical studies.

In pursuit of these goals, collaborations were established with herbal experts and natural product chemists in the Far East, South America, and Africa in order to (1) identify herbal therapies for evaluation (based on clinical recommendations of herbal experts, electronic database searches, review of original text references, and published results of laboratory or clinical studies), and (2) test crude extracts and selected fractions of these herbal therapies, as well as pure compounds isolated from those fractions, in well-validated *in vitro* assays and animal models of epilepsy. Extracts are tested, in addition to the pure compounds, because they may contain active compounds that were too low in concentration to be isolated. The results from these studies provide a reasonable estimate of the anticonvulsant profile of a candidate substance and form the basis for selecting lead substances for further preclinical and subsequent clinical evaluation.

Although the most common reasons for selecting herbal therapies for preclinical evaluation have been either a tradition of use for seizures or known mechanisms of action that are relevant to epilepsy, epidemiological observations may also be helpful. For example, Salih and Mustafa (2008) reasoned that the lower prevalence of epilepsy among school children of Khartoum Province, Sudan, compared to that of Europe and North America, may be related to frequent dietary ingestion of broad beans (*Vicia faba*). The authors therefore prepared an extract of *Vicia faba* and found that it blocked seizures in the mouse strychnine model.

To date, more than 40 herbal extracts and extract-derived compounds isolated from Chinese, Japanese, and Indian herbal therapies have been studied by the Harvard program in animal models of epilepsy through the National Institute of Neurological Disorders and Stroke Anticonvulsant Screening Project (ASP), and many have also been evaluated using *in vitro* assays of neuronal receptor or ion channel function. Approximately two thirds of these extracts and extract-derived compounds show activity *in vitro*, *in vivo*, or both. An example is Huperzine A.

## HUPERZINE A

Based on its proposed action as a noncompetitive *N*-methyl-D-aspartate (NMDA) receptor antagonist (Zhang et al., 2002), Huperzine A was selected for further evaluation as a potential anticonvulsant. Huperzine A is a sesquiterpene *Lycopodium* alkaloid, typically isolated from Chinese club moss (*Huperzia serrata*); in Chinese folk medicine, it is known as *qian ceng ta*. The chemical nomenclature for Huperzine A is [5*R*-(5α,9β,11*E*)]-5-amino-11-ethlidene-5,6,9,10-tetrahydro-7-methyl-5,9-methanocycloocta [*b*] pyridin-2(1*H*)-one. The molecular formula is $C_{15}H_{18}N_2O$, and its molecular weight is 242.32.

Huperzine A has been traditionally used in China for the treatment of swelling, fever and inflammation, blood disorders, and schizophrenia (Zangara, 2003), and it is currently used in China for Alzheimer's disease. Huperzine A was classified as a dietary supplement by the FDA in 1997. It is available in health-food stores or via the Internet, labeled as a memory aid, and was recently studied in a multicenter NIH-sponsored trial for Alzheimer's disease at dosages up to 400 μg b.i.d. (http://clinicaltrials.gov/ct/show/NCT00083590?order=9).

Huperzine A (ADD 357133) was submitted to the ASP and found to be potently active against subcutaneously administered pentylenetetrazol-induced seizures and less so against maximal electroshock-induced seizures following oral administration to Swiss–Webster mice, with peak anticonvulsant activity at 1 hour (White et al., 2005). At doses of 1, 2, and 4 mg/kg, a maximum of 62.5% protection was observed. Impairment on the rotarod test was observed in 75 and 100% of mice tested at doses of 2 and 4 mg/kg, respectively. The $TD_{50}$ was 0.83 mg/kg.

In the 6-Hz model, $ED_{50}$ values for intraperitoneal Huperzine A were 0.28, 0.34, and 0.78 mg/kg for 22, 32, and 44 mA, respectively, suggesting a possible advantage over phenytoin, carbamazepine, lamotrigine, and topiramate, each of which displays limited efficacy in this model at doses devoid of behavioral toxicity (Schachter et al., 2007). The less than twofold ratio of dosages effective across the range of current strengths suggests a further possible advantage over other drugs that are active in this model, such as levetiracetam.

## CONCLUSION

Despite the widespread use of herbal therapies by patients with epilepsy, there is a striking lack of controlled evidence to support their use, and anecdotal reports suggest that some herbal therapies may pose a safety risk to this population. Absence of proof, however, is not necessarily proof of absence, and the centuries-old traditions of use of herbal therapies for epilepsy provide a reasonable basis for systematically proceeding with preclinical assessments using modern scientific methods. Indeed, preclinical work at Harvard and elsewhere (Bum et al., 2009), based on this approach, suggests that the study of herbal therapies and herbal-derived compounds may yield promising candidates for further clinical development.

Herbal therapies may, therefore, potentially yield new treatment options for patients whose seizures are uncontrolled despite available AEDs and may also represent inexpensive, culturally acceptable treatments for the millions of people around the world with untreated epilepsy.

## ACKNOWLEDGMENTS

This work was supported by grants from the Epilepsy Therapy Project, the Epilepsy Research Foundation, and the American Epilepsy Society. This chapter has been modified with permission from Schachter, S.C., Botanicals and herbs: a traditional approach to treating epilepsy, *Neurotherapeutics*, 6(2), 415–420, 2009.

## REFERENCES

Ameri, A., J. Gleitz, T. Peters. (1997). Bicuculline-induced epileptiform activity in rat hippocampal slices: suppression by *Aconitum* alkaloids. *Planta Med.*, 63:228–232.

Bressler, R. (2005). Herb–drug interactions: interactions between *Ginkgo biloba* and prescription medications. *Geriatrics*, 60:30–33.

Brodie, M.J. (2005). Diagnosing and predicting refractory epilepsy. *Acta Neurol. Scand. Suppl.*, 181:36–39.

Bum, E.N., G.S. Taiwe, L.A. Nkainsa, F.C. Moto, P.F. Seke Etet, I.R. Hiana, T. Bailabar, Rouyatou, P. Seyni, A. Rakotonirina, S.V. Rakotonirina. (2009). Validation of anticonvulsant and sedative activity of six medicinal plants. *Epilepsy Behav.*, 14:454–458.

Carolino, R.O., R.O. Beleboni, A.B. Pizzo, F.D. Vecchio, N. Garcia-Cairasco, M. Moyses-Neto, W.F. Santos, J. Coutinho-Netto. (2005). Convulsant activity and neurochemical alterations induced by a fraction obtained from fruit *Averrhoa carambola* (Oxalidaceae: Geraniales). *Neurochem. Int.*, 46:523–531.

Chauhan, A.K., M.P. Dobhal, B.C. Joshi. (1988). A review of medicinal plants showing anticonvulsant activity. *J. Ethnopharmacol.*, 22:11–23.

Chiou, L.C., J.Y. Ling, C.C. Chang. (1997). Chinese herb constituent beta-eudesmol alleviated the electroshock seizures in mice and electrographic seizures in rat hippocampal slices. *Neurosci. Lett.*, 231:171–174.

Danesi, M.A., J.B. Adetunji. (1994). Use of alternative medicine by patients with epilepsy: a survey of 265 epileptic patients in a developing country. *Epilepsia*, 35:344–351.

Delgoda, R., A.C. Westlake. (2004). Herbal interactions involving cytochrome p450 enzymes: a mini review. *Toxicol. Rev.*, 23:239–249.

Easterford, K., P. Clough, S. Comish, L. Lawton, S. Duncan. (2005). The use of complementary medicines and alternative practitioners in a cohort of patients with epilepsy. *Epilepsy Behav.*, 6:59–62.

Elferink, J.G. (1999). Epilepsy and its treatment in the ancient cultures of America. *Epilepsia*, 40:1041–1046.

Ernst, E. (2003). Serious psychiatric and neurological adverse effects of herbal medicines: a systematic review. *Acta Psychiatr. Scand.*, 108:83–91.

Gil Campos, M., J.L. Perez Navero, I. Ibarra De La Rosa. (2002). Convulsive status secondary to star anise poisoning in a neonate. *An. Esp. Pediatr.*, 57:366–368.

Harms, S.L., J. Garrard, P. Schwinghammer, L.E. Eberly, Y. Chang, I.E. Leppik. (2006). *Ginkgo biloba* use in nursing home elderly with epilepsy or seizure disorder. *Epilepsia*, 47:323–329.

Hijikata, Y., A. Yasuhara, Y. Yoshida, S. Sento. (2006). Traditional Chinese medicine treatment of epilepsy. *J. Altern. Complement Med.*, 12:673–677.

Hsieh, C.L., M.F. Chen, T.C. Li, S.C. Li, N.Y. Tang, C.T. Hsieh, C.Z. Pon, J.G. Lin. (1999a). Anticonvulsant effect of *Uncaria rhynchophylla* (MIQ) jack in rats with kainic acid-induced epileptic seizure. *Am. J. Chin. Med.*, 27:257–264.

Hsieh, C.L., N.Y. Tang, S.Y. Chiang, C.T. Hsieh, J.G. Lin. (1999b). Anticonvulsive and free radical scavenging actions of two herbs, *Uncaria rhynchophylla* (MIQ) jack and *Gastrodia elata* Bl., in kainic acid-treated rats. *Life Sci.*, 65:2071–282.

Hsieh, C.L., C.H. Chang, S.Y. Chiang, T.C. Li, N.Y. Tang, C.Z. Pon, C.T. Hsieh, J.G. Lin. (2000). Anticonvulsive and free radical scavenging activities of vanillyl alcohol in ferric chloride-induced epileptic seizures in Sprague–Dawley rats. *Life Sci.*, 67:1185–1195.

Huang, W.F., K.C. Wen, M.L. Hsiao. (1997). Adulteration by synthetic therapeutic substances of traditional Chinese medicines in Taiwan. *J. Clin. Pharmacol.*, 37:344–350.

Huxtable, R.J. (1990). The harmful potential of herbal and other plant products. *Drug Saf.*, 5(Suppl. 1):126–136.

Jain, S. (2005). Ayurveda: the ancient Indian system of medicine. In *Complementary and Alternative Therapies for Epilepsy* (pp. 123–128), O. Devinsky, S. Schachter, and S. Pacia, Eds. New York: Demos Medical Publishing.

Johanns, E.S., L.E. van der Kolk, H.M. van Gemert, A.E. Sijben, P.W. Peters, I. de Vries. (2002). An epidemic of epileptic seizures after consumption of herbal tea. *Ned. Tijdschr. Geneeskd.*, 146:813–816.

Kasture, V.S., V.K. Deshmukh, C.T. Chopde. (2002). Anxiolytic and anticonvulsive activity of *Sesbania grandiflora* leaves in experimental animals. *Phytother. Res.*, 16:455–460.

Khan, R., A. Krishnakumar, C.S. Paulose. (2008). Decreased glutamate receptor binding and NMDA R1 gene expression in hippocampus of pilocarpine-induced epileptic rats: neuroprotective role of *Bacopa monnieri* extract. *Epilepsy Behav.*, 12:54–60.

Kim, H.J., K.D. Moon, S.Y. Oh, S.P. Kim, S.R. Lee. (2001). Ether fraction of methanol extracts of *Gastrodia elata*, a traditional medicinal herb, protects against kainic acid-induced neuronal damage in the mouse hippocampus. *Neurosci. Lett.*, 314:65–68.

Kim, I.J., J.K. Kang, S.A. Lee. (2006). Factors contributing to the use of complementary and alternative medicine by people with epilepsy. *Epilepsy Behav.*, 8:620–624.

Kulkarni, S.K., K.K. Akula, A. Dhir. (2008).Effect of *Withania somnifera* Dunal root extract against pentylenetetrazol seizure threshold in mice: possible involvement of GABAergic system. *Indian J. Exp. Biol.*, 46:465–469.

Kupiec, T., V. Raj. (2005). Fatal seizures due to potential herb–drug interactions with *Ginkgo biloba*. *J. Anal. Toxicol.*, 29:755–758.

Lai, C.W., Y.H. Lai. (1991). History of epilepsy in Chinese traditional medicine. *Epilepsia*, 32:299–302.

Lai, C.W., X.S. Huang, Y.H. Lai, Z.Q. Zhang, G.J. Liu, M.Z. Yang. (1990). Survey of public awareness, understanding, and attitudes toward epilepsy in Henan Province, China. *Epilepsia*, 31:182–187.

Liow, K., E. Ablah, J.C. Nguyen, T. Sadler, D. Wolfe, K.D. Tran, L. Guo, T. Hoang. (2007). Pattern and frequency of use of complementary and alternative medicine among patients with epilepsy in the Midwestern United States. *Epilepsy Behav.*, 10:576–582.

Luciano, D. J., M. Spinella. (2005). Herbal treatment of epilepsy: phytotherapy. *Complementary and Alternative Therapies for Epilepsy* (pp. 143–155), O. Devinsky, S. Schachter, and S. Pacia, Eds. New York: Demos Medical Publishing.

Mahomed, I.M., J.A. Ojewole. (2006). Anticonvulsant activity of *Harpagophytum procumbens* DC [Pedaliaceae] secondary root aqueous extract in mice. *Brain Res. Bull.*, 69:57–62.

Meinardi, H., R.A. Scott, R. Reis, J.W. Sander. (2001). The treatment gap in epilepsy: the current situation and ways forward. *Epilepsia*, 42:136–149.

Minami, E., H. Shibata, Y. Nunoura, M. Nomoto, T. Fukuda. (1999). Efficacy of shitei-to, a traditional Chinese medicine formulation, against convulsions in mice. *Am. J. Chin. Med.*, 27:107–115.

Miwa, H., M. Iijima, S. Tanaka, Y. Mizuno. (2001). Generalized convulsions after consuming a large amount of gingko nuts. *Epilepsia*, 42:280–281.

Neto, M.M., J.A. da Costa, N. Garcia-Cairasco, J.C. Netto, B. Nakagawa, M. Dantas. (2003). Intoxication by star fruit (*Averrhoa carambola*) in 32 uraemic patients: treatment and outcome. *Nephrol. Dial. Transplant.*, 18:120–125.

Obach, R.S. (2000). Inhibition of human cytochrome P450 enzymes by constituents of St. John's wort, an herbal preparation used in the treatment of depression. *J. Pharmacol. Exp. Ther.*, 294:88–95.

Ojewole, J.A. (2008a). Anticonvulsant activity of *Hypoxis hemerocallidea* (Fisch. & C.A. Mey.) [Hypoxidaceae] corm ('African potato') aqueous extract in mice. *Phytother. Res.*, 22:91–96.

Ojewole, J.A. (2008b). Anticonvulsant property of *Sutherlandia frutescens* R. BR. (variety Incana E. MEY.) [Fabaceae] shoot aqueous extract. *Brain Res. Bull.*, 75:126–132.

Oyungu, E., P.G. Kioy, N.B. Patel. (2009). Proconvulsant effect of Khat (*Catha edulis*) in Sprague–Dawley rats. *J. Ethnopharmacol.*, 121:476–478.

Park, J., H. Wei, D. Lawhorn, S.C. Schachter. (2005). Herbal formulas in epilepsy: a systematic review. *Epilepsia*, 46(Suppl. 8):215.

Pearl, P.L., E.L. Robbins, H.D. Bennett, J.A. Conry. (2005). Use of complementary and alternative therapies in epilepsy: cause for concern. *Arch. Neurol.*, 62:1472–1475.

Peebles, C.T., J.W. McAuley, J. Roach, J.L. Moore, A.L. Reeves. (2000). Alternative medicine use by patients with epilepsy. *Epilepsy Behav.*, 1:74–77.

Rhodes, P.J., N. Small, H. Ismail, J.P. Wright. (2008). The use of biomedicine, complementary and alternative medicine, and ethnomedicine for the treatment of epilepsy among people of South Asian origin in the U.K. *BMC Complement. Altern. Med.*, 8:7.

Salih, M.A., A.A. Mustafa. (2008). A substance in broad beans (*Vicia faba*) is protective against experimentally induced convulsions in mice. *Epilepsy Behav.*, 12:25–29.

Samuels, N., Y. Finkelstein, S.R. Singer, M. Oberbaum. (2008). Herbal medicine and epilepsy: proconvulsive effects and interactions with antiepileptic drugs. *Epilepsia*, 49:373–380.

Saper, R.B., S.N. Kales, J. Paquin, M.J. Burns, D.M. Eisenberg, R.B. Davis, R.S. Phillips. (2004). Heavy metal content of ayurvedic herbal medicine products. *JAMA*, 292:2868–2873.

Schachter, S.C., H.S. White, L.J. Murphree, J. Stables. (2007). Huperzine A. *Epilepsy Res.*, 73:10–11.

Schachter, S.C., C. Acevedo, K.A. Acevedo et al. (2008). Complementary and alternative medical therapies. In *Epilepsy: A Comprehensive Textbook*, 2nd ed. (pp. 1407–1414), J. Engel and T.A. Pedley, Eds. Philadelphia, PA: Lippincott Williams & Wilkins.

Sirven, J.I. (2007). Alternative therapies for seizures: promises and dangers. *Semin. Neurol.*, 27:325–330.

Sirven, J.I., J.F. Drazkowski, R.S. Zimmerman, J.J. Bortz, D.L. Shulman, M. Macleish. (2003). Complementary/alternative medicine for epilepsy in Arizona. *Neurology*, 61:576–577.

Spinella, M. (2001). Herbal medicines and epilepsy: the potential for benefit and adverse effects. *Epilepsy Behav.*, 2:524–532.

Sugaya, E., N. Yuyama, K. Kajiwara, T. Tsuda, H. Ohguchi, K. Shimizu-Nishikawa, M. Kimura, A. Sugaya. (1999). Regulation of gene expression by herbal medicines: a new paradigm of gene therapy for multifocal abnormalities of genes. *Res. Commun. Mol. Pathol. Pharmacol.*, 106:171–180.

Tandon, M., S. Prabhakar, P. Pandhi. (2002). Pattern of use of complementary/alternative medicine (CAM) in epileptic patients in a tertiary care hospital in India. *Pharmacoepidemiol. Drug Saf.*, 11:457–463.

Vattanajun, A., H. Watanabe, M.H. Tantisira, B. Tantisira. (2005). Isobolographically additive anticonvulsant activity between *Centella asiatica*'s ethyl acetate fraction and some antiepileptic drugs. *J. Med. Assoc. Thai.*, 88(Suppl. 3):S131–S140.

Wahner-Roedler, D.L., P.L. Elkin, A. Vincent, J.M. Thompson, T.H. Oh, L.L. Loehrer, J.N. Mandrekar, B.A. Bauer. (2005). Use of complementary and alternative medical therapies by patients referred to a fibromyalgia treatment program at a tertiary care center. *Mayo Clin. Proc.*, 80:55–60.

Wang, Y., D. Zhou, B. Wang, H. Li, H. Chai, Q. Zhou, S. Zhang, H. Stefan. (2003). A kindling model of pharmacoresistant temporal lobe epilepsy in Sprague–Dawley rats induced by *Coriaria* lactone and its possible mechanism. *Epilepsia*, 44:475–488.

White, H.S., S. Schachter, D. Lee, J. Xiaoshen, D. Eisenberg. (2005). Anticonvulsant activity of Huperzine A, an alkaloid extract of Chinese club moss (*Huperzia serrata*). *Epilepsia*, 46(Suppl. 8):220.

Zangara, A. (2003). The psychopharmacology of Huperzine A: an alkaloid with cognitive enhancing and neuroprotective properties of interest in the treatment of Alzheimer's disease. *Pharmacol. Biochem. Behav.*, 75:675–686.

Zhang, Y.H., X.Y. Zhao, X.Q. Chen, Y. Wang, H.H. Yang, G.Y. Hu. (2002). Spermidine antagonizes the inhibitory effect of huperzine A on [$^3$H]dizocilpine (MK-801) binding in synaptic membrane of rat cerebral cortex. *Neurosci. Lett.*, 319:107–110.

# 24 The Ketogenic Diet: Scientific Principles Underlying Its Use

*Kristopher J. Bough and Carl E. Stafstrom*

## CONTENTS

## INTRODUCTION

The ketogenic diet (KD) is a high-fat, low-carbohydrate, low-protein diet that was initially formulated in the early 1920s as a treatment for intractable epilepsy (Wilder, 1921). Physicians in the 1920s and 1930s were well aware of the effectiveness of the diet; at that time, phenobarbital was the only other epilepsy treatment in widespread use. With the discovery of phenytoin in 1935, and other anticonvulsant medications over subsequent decades, the KD was used much less frequently; however, over the past 15 years, the KD has re-emerged as a viable alternative to standard anticonvulsants

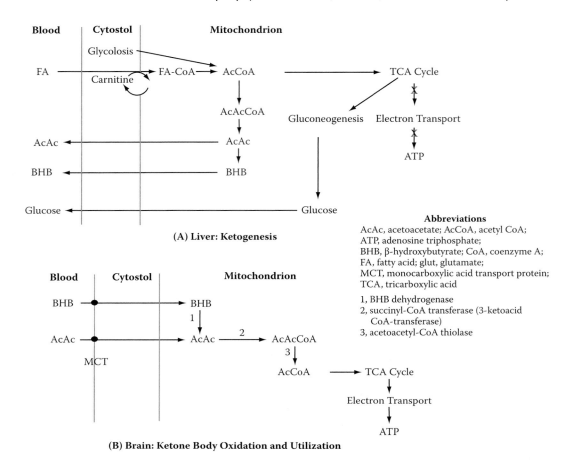

**FIGURE 24.1** Schematic summary of ketogenesis and ketone body utilization by the liver and brain. (A) Ketogenesis in the liver. Fatty acids from the circulation enter hepatocytes, then cross the inner mitochondrial membrane by either diffusion (short- and medium-chain fatty acids [FAs]) or via carnitine transport (long-chain FAs). Under conditions of fasting or the high-fat, low-carbohydrate ketogenic diet, carbohydrate substrate is lacking, so FAs are metabolized. Because oxaloacetate (a TCA intermediate) is diverted to gluconeogenesis, the TCA is not actively involved in energy generation (X'd arrows). Acetyl–CoA molecules are therefore not funneled into the TCA but instead are used for ketone body synthesis (ACA, BHB). The ketones are exported into the circulation, as the liver lacks the enzymes required to metabolize ACA and BHB. (B) In the brain, BHB and ACA enter neurons via the monocarboxylic acid transport system. In mitochondria, BHB and ACA are broken down (by enzymes 1, 2, and 3) into acetyl–CoA molecules that can then enter the TCA cycle for energy production. (C) TCA cycle and GABA shunt. In ketosis, α-KG is increased (De Vivo et al., 1978), and GABA synthesis from glutamate (via enzyme 4) is favored (Yudkoff et al., 2001). In addition, the GABA shunt, bypassing normal oxidative metabolism, facilitates GABA production.

in the treatment of refractory epilepsy, especially in children (Freeman et al., 2007b; Kossoff et al., 2009). The history of the KD, including its origin and the reasons for the recent resurgence in its use, is reviewed elsewhere in detail (Freeman et al., 2007a,b; Swink et al., 1997; Wheless, 2004).

As the KD has again found a place among epilepsy treatments, the time is now ripe to discover the scientific rationale underlying its use. Since its inception, little has changed in the formulation or administration of the KD. Much of the KD protocol is based on lore, passed on from generation to generation, rather than on rigorously tested principles. Although there is no longer any question that the KD works (Neal et al., 2008), to optimize its clinical usefulness its mechanism of action must

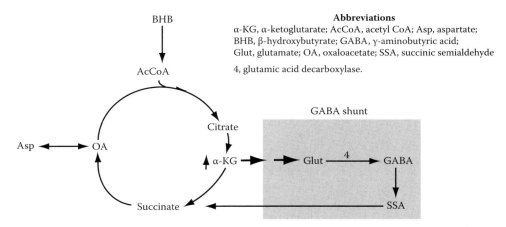

**Abbreviations**
α-KG, α-ketoglutarate; AcCoA, acetyl CoA; Asp, aspartate; BHB, β-hydroxybutyrate; GABA, γ-aminobutyric acid; Glut, glutamate; OA, oxaloacetate; SSA, succinic semialdehyde 4, glutamic acid decarboxylase.

**(C) Ketosis Favors an Increase in GABA Synthesis**

**FIGURE 24.1 (continued)**

be understood more clearly. At present, the KD remains "a treatment in search of an explanation" (Stafstrom and Spencer, 2000). Although some insights can be obtained through clinical studies, a full understanding of its mechanism requires animal experimentation. The optimal use of the KD in the future, including possible modifications of diet composition and utilization, will depend on new information based on sound experimental principles (Stafstrom, 2004).

In this chapter, we briefly review the clinical aspects of the KD, summarize up-to-date clinical and animal data that may underlie KD mechanisms, and synthesize this information into practical guidelines for clinicians using the diet. In so doing, it is our hope to stimulate future research to better understand the KD and its therapeutic implications.

## CLINICAL VARIATIONS OF KETOGENIC DIETS

As the KD has regained favor over the past 15 years, there has been increasing interest in alternative diets that are less restrictive but produce similar anticonvulsant efficacy. These less restrictive diets are intended to increase a patient's long-term compliance and avoid potential growth and developmental deficits. Toward this goal, several alternative KDs have been devised and evaluated. The classic KD and three alternative KDs are discussed below.

### CLASSIC KETOGENIC DIET

The classic KD was formulated to mimic the physiology of fasting. The term "ketogenic" refers to the physiological condition in which fats are preferentially metabolized to form ketone bodies in the liver. During treatment, ketones become a key supplemental source of metabolic fuel for the brain as glucose concentrations are diminished. Dietary fats are broken down in the liver into two-carbon acetyl–coenzyme A (CoA) molecules in a process called β-oxidation (Figure 24.1A). Acetyl–CoA molecules enter the tricarboxylic acid (TCA) cycle, producing energy via generation of adenosine triphosphate (ATP); however, under conditions of fasting or ketosis, acetyl–CoA cannot readily enter the TCA cycle because of low availability of the TCA rate-limiting substrate oxaloacetate and the rate-limiting enzyme α-ketoglutarate dehydrogenase (which have been diverted to gluconeogenesis) (Trauner, 1985). Instead, acetyl–CoA molecules are used to synthesize ketone bodies: acetone, β-hydroxybutyrate (BHB), and acetoacetate (ACA or AcAc). The liver lacks the enzymes necessary to break down ketone bodies, so ketone bodies are exported via the circulation to tissues where they can be utilized for energy, such as muscle and brain. Both fasting and the KD produce ketonemia, with elevated blood levels of ketone bodies, primarily BHB and ACA (McGarry and Foster, 1980; Saudek and Felig, 1976).

The classic KD is based on the consumption of long-chain saturated triglycerides (LCTs) with a ratio of fat to carbohydrate + protein of 4:1 by weight; that is, for every gram of combined protein and carbohydrate, 4 grams of fat are consumed. Each meal must be planned in accordance with this ratio. In the 4:1 KD, about 90% of daily calories are derived from fat sources. Adequate amounts of protein are provided to ensure patient growth (i.e., about 1 gram per kg body weight); somewhat more or less protein may be required in infants and adolescents, respectively. Other KD ratios are sometimes employed, depending on the patient's age, metabolic needs, or response to the diet (e.g., 3:1). It is important to emphasize that the KD is not a fad diet but rather a rigid medical regimen requiring strict adherence for optimal effect. Thus, the role of a dietitian experienced with the KD is critical.

Calorie restriction (CR), like diet ratio, is also a key aspect of KD efficacy. Calorie intake is typically reduced by 25% from the recommended daily allowance (RDA) during diet treatment and is commonly used to fine tune the diet. Because the consumption of excess calories can lead to the endogenous production of glucose (i.e., gluconeogenesis), the degree of CR helps ensure that ketosis is maintained throughout and the diet is maximally efficacious.

## ALTERNATIVE KETOGENIC DIETS

### The Modified Atkins Diet

The Atkins diet, like the KD, has long been known to induce a state of ketosis. Within the last few years, it was suggested that the Atkins diet could induce seizure protection similar to that produced by the KD. Like the classic KD, the modified Atkins diet is primarily comprised of fats and consists of approximately 60% fat, 30% protein, and 10% carbohydrate by weight (Kang et al., 2007; Kossoff et al., 2006). Carbohydrate intake is generally limited at first to approximately 10 g/day; however, patients can increase their carbohydrate intake over time to 15 or even 20 g/day as long as they stay within an overall maximum intake of ~10% carbohydrates (by weight). Unlike the classic KD, the modified Atkins diet does not limit fluid intake, does not restrict calories, and allows for greater protein consumption (Kossoff et al., 2007).

Several clinical studies have evaluated the anticonvulsant actions of the Atkins diet. An initial small study involved 6 subjects (5 children, 1 adult) with various types of epilepsy (Kossoff et al., 2003). Five of the 6 patients became ketotic, and 3 of the 5 children exhibited a >90% reduction in seizure frequency with a concurrent reduction in antiepileptic medications. A second prospective investigation involved a cohort of 20 children with various types of epilepsy (Kossoff et al., 2006); after 6 months of dietary treatment, 13 of the subjects (65%) exhibited a >50% reduction in seizure frequency, whereas 7 (35%) exhibited a >90% improvement. Additional data were collected in a trial involving 14 children with intractable seizures (Kang et al., 2007). Of the 7 patients remaining on the diet after 6 months, 5 exhibited a >50% reduction in seizures, 3 of which (21%) became seizure free. Adults have also been studied (Kossoff et al., 2008b). Although several of the adult subjects did not complete the trial, intent-to-treat analysis showed that, after 6 months, 33% of subjects had a >50% reduction in seizures.

Overall, the modified Atkins diet was well tolerated, and few adverse effects were observed. These data indicate that the modified Atkins diet can effectively increase the resistance to a wide variety of seizures.

### The Low-Glycemic-Index Treatment

Compared to the classic KD and the modified Atkins diet described above, the low-glycemic-index (LGI) treatment is aimed at providing even greater dietary flexibility. Like other KDs, this diet is also based upon fats (Ludwig, 2007). It consists of 60 to 70% fat, 20 to 30% protein, and 10%

carbohydrate by weight. The LGI treatment differs from the other dietary regimens in that alternative sources of carbohydrates are preferentially consumed; only carbohydrates that are slowly digested and absorbed (i.e., glycemic index values < 50) are used. Patients are free to consume as much as 40 to 60 grams of carbohydrates per day, as long as they consume a maximum carbohydrate intake of ~10%. The LGI treatment is typically administered to patients who have demonstrated some response to the classic KD but have had difficulty adhering to its more restrictive protocol.

Currently, only one clinical study has investigated the LGI diet as a possible treatment for epilepsy (Pfeifer and Thiele, 2005). This study investigated 20 subjects (18 children and 2 adults greater than 18 years old) with either generalized or partial complex epilepsy. Results were encouraging. Ten of the 18 children (56%) exhibited a >90% reduction in seizures, while 4 of the 18 children experienced a >50% reduction. Interestingly, seizure frequency in the 2 adults did not change during the first 4 weeks of treatment.

## The Medium-Chain Triglyceride Diet

A third example of an alternative KD is the medium-chain triglyceride (MCT) diet. Like most KDs, the MCT diet is chiefly composed of fats, with a limited amount of protein and carbohydrates. What makes this diet unique is that MCT-based KDs tend to induce more pronounced ketonemia for a given amount of fat than do classic KDs that utilize primarily long-chain triglycerides (Huttenlocher, 1976). Accordingly, the MCT diet ratio does not typically exceed 3:1, allowing for greater dietary flexibility (Carroll and Koenigsberger, 1998; Huttenlocher et al., 1971).

Because some studies have positively correlated ketonemia with seizure protection (Huttenlocher, 1976), it has been thought that MCT-based diets might also be more efficacious than the classic KD regimen. However, the MCT diet does not appear to induce greater seizure protection in humans (Huttenlocher, 1976; Sills et al., 1986; Trauner, 1985) or rodents (Thavendiranathan et al., 2000). In one study, 29% (5 of 17) of the patients had total seizure control, 29% exhibited at least some reduction in seizures, 2 showed no benefit, and 3 others could not comply with dietary therapy due to adverse events such as vomiting and diarrhea (Trauner, 1985). This represents a major drawback of the MCT diet. Unlike other alternative KDs, experience with the MCT diet has been consistently associated with unpleasant gastrointestinal side effects.

Overall, more studies of alternative dietary regimens are warranted. Evidence accumulated thus far suggests that the efficacy of alternative KDs is remarkably similar to the classic KD. The major benefit of the alternative KDs is the added flexibility. Although none of these alternative diets is as well defined as the classic KD protocol and their routine use is probably premature, alternative KDs have the potential for wider clinical use.

## ADMINISTRATION OF THE CLASSIC KD

The KD is implemented gradually over a period of about 5 days (Freeman et al., 2007a). It has been successfully started both in the hospital and on an outpatient basis (Vaisleib et al., 2004). Most often, the child is fasted for the initial 24 to 48 hours. Seizures often decrease, even during the fasting phase (Kang et al., 2007; Kossoff et al., 2008a), an observation that may have mechanistic implications. Once fasting has produced sufficient ketosis, the KD is advanced in one-third increments of the full diet over 3 days. During the fasting and initiation periods of the KD, patients are routinely screened for acute metabolic side effects such as hypoglycemia, vomiting, and dehydration (Kang et al., 2004).

The KD protocol must be strictly administered. Individual ingredients of the diet must be carefully weighed each day, and each meal must conform to the prescribed ratio. If carbohydrates are reintroduced (e.g., sneaking a cookie or piece of candy), ketosis can be lost and breakthrough seizures can occur rapidly, sometimes within an hour (Freeman et al., 2007a; Huttenlocher, 1976).

The requirement of strict compliance, the difficulty of preparation, and the narrow food choices have limited more frequent use of the KD, especially in children capable of making their own food choices. For these reasons, the KD has most commonly been used as a last-resort treatment for intractable epilepsies behind multiple, conventional antiepileptic medications. The effectiveness of the KD as a first- or second-line treatment in epilepsy is currently unknown.

The long-term goal is to maintain the KD for 2 years, although this has been an arbitrary target. Some children obtain long-lasting benefit from much shorter use, while others remain on the KD for several years. Compliance with the KD has generally been related to its effectiveness. Patients who remain on the diet for a year or longer experience good seizure control and tend to be free of significant side effects (Freeman et al., 2007a). Reasons for earlier discontinuation include the lack of KD effectiveness, difficulty in its preparation, and intercurrent illnesses that interfere with its administration. These three factors are interrelated; that is, parents find the diet more restrictive to administer if seizures are not well controlled (Hemingway et al., 2001).

The KD is the treatment of choice for certain inborn errors of metabolism, such as deficiencies of the glucose transporter GLUT1 (Boles et al., 1999; De Vivo et al., 1991; Rauchenzauner et al., 2008) and pyruvate dehydrogenase (PDH) (Wexler et al., 1997; Wijburg et al., 1992). Both disorders result in cerebral energy failure and epilepsy. In GLUT1 deficiency, the protein required for the transport of glucose into the central nervous system is dysfunctional, and in this situation ketone bodies generated by the diet can provide an alternative source of energy. Children with GLUT1 deficiency often present with seizures and developmental delay; treatment with the KD helps to limit seizures and improve developmental outcome. In PDH deficiency, the KD provides an alternative source for brain acetyl–CoA and enhances TCA cycle activity (Nordli and De Vivo, 1997).

## CLINICAL QUESTIONS CONCERNING KETOGENIC DIET EFFICACY

### WHAT IS THE OVERALL EFFICACY OF THE KD?

The KD is a time-tested anticonvulsant treatment. Its efficacy and administration of the KD continue to be optimized both here in the United States and abroad (Henderson et al., 2006; Keene, 2006; Kossoff and McGrogan, 2005; Neal et al., 2008). Despite early skepticism about whether a high-fat diet could improve seizure control, especially in refractory patients, there is now little doubt as to the effectiveness of the KD. The success rate of the KD in controlling refractory seizures is at least as good as, and often better than, most of the new antiepileptic medications (Lefevre and Aronson, 2000). In general, half of all patients treated with the KD will exhibit at least a 50% reduction in seizure frequency, as documented in both U.S. and international series (Coppola et al., 2002; Freeman et al., 1998; Helmholz and Keith, 1930; Kaytal et al., 2000; Kinsman et al., 1992; Livingston, 1972; Magrath et al., 2000; Neal et al., 2008; Peterman, 1925; Vining et al., 1998). Gender and seizure type do not typically affect outcome (Freeman et al., 1998; Schwartz et al., 1989; Vining et al., 1998), but some epilepsy types may respond preferentially. Interestingly, the success rate of the KD is similar now to when it was first introduced (Kinsman et al., 1992). Perhaps most importantly, the KD can often be discontinued without a concomitant loss of seizure control (Freeman et al., 2007a; Hemingway et al., 2001). This observation suggests that the KD may be both anticonvulsant (stops seizures) and antiepileptic (mitigates the development of the epileptic state).

Treatment with a KD has several advantages over more traditional pharmacotherapy. The KD effectively controls multiple types of epilepsy and works when other medications fail. At diet onset, a typical patient will have been treated with as many as seven antiepileptic drugs (AEDs) (Freeman et al., 1998). With KD treatment, AEDs can usually be reduced or even discontinued, providing a substantial cost savings and decrease in side effects from polytherapy (Wheless, 1995). Some reports cite improved behavioral function and mental capacity (Helmholz and Keith, 1930; Kinsman et al., 1992; Livingston, 1972; Peterman, 1925). A prospective study documented improved attention span, developmental quotient, and social functioning in a cohort of children on the KD for one year (Pulsifer et al., 2001). These behavioral improvements can sometimes be as important to a family as better seizure control. The need for fewer sedative anticonvulsants may account for some of this cognitive improvement.

## Is the KD Equally Effective across All Types of Epilepsy?

Clinically, the KD has been used to treat a wide variety of seizure types. Early studies were limited by failure to adequately describe seizure types and epilepsy syndromes, at least in terms of modern classification schemes. Recent prospective reports have found no consistent correlation between efficacy and seizure type (Freeman et al., 1998; Schwartz et al., 1989; Vining et al., 1998). Nonetheless, some differences may exist. For example, Livingston (1972) stated that, despite the lack of "an absolute relationship between the type of epilepsy and … a satisfactory response to the KD," the best results were observed in children with myoclonic epilepsy. In a prospective evaluation of 106 consecutive children who stayed on the KD for 6 months, about half (43 to 60%) of all patients with myoclonic, atonic/drop, or atypical absence seizures exhibited a greater than 90% reduction in the number of seizures (Freeman et al., 1998). Seizures in children with Lennox–Gastaut syndrome (tonic, myoclonic, atypical absence) are notoriously difficult to control, and yet such children often benefit from the KD (Freeman et al., 2009). Patients that present with partial seizures (with or without secondary generalization) also respond to the KD. About 63% of children with partial seizures had a 50% or better seizure reduction (Freeman et al., 1998). In another study, 27% of children with partial seizures had a 50% or better seizure reduction (Maydell et al., 2001).

Despite widely discrepant methodologies, laboratory evidence, like clinical findings, indicates that the KD is effective against a variety of seizure types (Table 24.1). The correlation between human epilepsies and animal epilepsy models is inexact, based upon both the clinical semiology of the seizures and their responsiveness to anticonvulsant drugs (Holmes and Zhao, 2008). In general, seizures in animals induced by electroshock (causing tonic limb extension) represent a model of generalized tonic seizures, while the chemoconvulsant pentylenetetrazole (PTZ) models generalized myoclonic seizures. Kindling, kainic acid, and the genetic EL mouse mimic partial-onset seizures. As seen in Table 24.1, the KD can afford protection across multiple generalized and partial seizure models (Hartman et al., 2007). Some studies, however, have failed to show a protective effect of the KD, particularly against status epilepticus induced by kainic acid or maximal electroconvulsive shock (Bough et al., 2000a; Thavendiranathan et al., 2000).

How might observed differences in animal models translate to differences in KD clinical efficacy? The relative efficacy of the KD in some models over others suggests that different brain regions might be affected more profoundly than others, and this could relate to localized differences in the function of γ-aminobutyric acid (GABA) or other neurotransmitters, such as norepinephrine (Szot et al., 2001). It remains unclear, however, whether seizure etiology plays a role in KD effectiveness in patients. For instance, a genetic susceptibility (e.g., a channelopathy) might respond very differently to KD treatment than would an acquired pathology (e.g., mesial temporal sclerosis) or cortical dysplasia. There are scarce data about KD responsiveness in different etiological subgroups. To complicate matters, many children have multiple seizure etiologies and seizure types. An improved understanding of KD efficacy in relation to seizure type and its cognate region of epileptogenicity or pathology will help to select appropriate KD candidates and administer the KD more effectively. Such information may also provide clues to underlying mechanisms.

Despite differences in KD protocols in the various laboratory experiments, data collectively show that the KD can increase the resistance to multiple seizure types by increasing seizure threshold (either chemical or electroconvulsive) (Raffo et al., 2008). The response of most seizure types to the KD implies that the diet exerts a general suppressant effect on neuronal excitability. The enhanced response of some generalized seizure types suggests that KD efficacy may preferentially

**TABLE 24.1**
**Animal Models of the Ketogenic Diet—Seizure Types**

| Model | Analogous Human Seizure Type | Selected References |
|---|---|---|
| Electroshock (maximal and threshold models) | Generalized (tonic, tonic–clonic) | Appleton and De Vivo (1974); Bough et al. (2000a); Millichap et al. (1964); Nakazawa et al. (1983); Otani et al. (1984); Thavendiranathan et al. (2000); Uhlemann and Neims (1972) |
| Pentylenetetrazole | Generalized (myoclonic) | Bough and Eagles (1999); Bough et al. (2000a); Gasior et al. (2007); Likhodii et al. (2002); Raffo et al. (2008) |
| Bicuculline | Generalized (myoclonic, tonic–clonic) | Bough et al. (2002) |
| Picrotoxin | Generalized (myoclonic, tonic–clonic) | Bough et al. (2002) |
| Flurothyl | Generalized (myoclonic, tonic–clonic) | Rho et al. (1999a); Szot et al. (2001) |
| Audiogenic seizures (Fring's mice) | Generalized reflex seizures | Rho et al. (2002) |
| EL (genetic; stress/handling-induced) | Complex partial with secondary generalization | Mantis et al. (2004); Todorova et al. (2000) |
| Kindling | Complex partial with secondary generalization | Hori et al. (1997) |
| Kainic acid | Complex partial with secondary generalization | Kwon et al. (2008); Lian et al. (2007); Muller-Schwarze et al. (1999); Noh et al. (2003, 2006); Su et al. (2000) |
| Lithium–pilocarpine | Complex partial with secondary generalization | Zhao et al. (2004) |
| 6-Hz seizures | Complex partial | Hartman et al. (2008) |
| Genetic absence epilepsy (GAERS rats) | Absence | Melo et al. (2006) |

increase the resistance to seizure initiation while doing little to limit seizure propagation and secondary generalization. For now, we recommend that the KD be considered for any patient with medically refractory epilepsy.

## DOES AGE AFFECT KD EFFICACY?

Clinically, the KD is very effective in controlling intractable seizures in children. It is believed that the age at which the KD is begun determines responsiveness. In a large study of 1001 patients, the greatest success with the KD was in the 2- to 5-year-old age group (Livingston, 1972). Freeman reported similar results for this age group, concluding that the efficacy of the KD was particularly prominent in this age range because these children were able to produce the most pronounced levels of ketonemia and adhered to the exacting regimen of the diet most closely (Freeman et al., 1998). Prospective clinical studies report that 50 to 54% children ages 3 to 16 years exhibited a more than 50% reduction in the number of seizures after one year (Freeman et al., 1998; Schwartz et al., 1989; Vining et al., 1998).

Recent studies, however, attest to the effectiveness of the KD across a wider age range. In 32 infants less than 2 years of age, 55% exhibited a more than 50% reduction in the number of seizures (Nordli et al., 2001). There is some evidence that the KD is effective in adults as well (Barborka, 1930). Of 11 patients ages 19 to 45 years with refractory epilepsy, the KD produced a sustained ketosis and afforded a seizure reduction of greater than 50% in 6 of the patients (Sirven et al., 1999).

As mentioned above, compliance with the KD regimen is often linked with its efficacy. In 150 children, the relationship between age, compliance with the KD regimen (i.e., tolerance), and KD efficacy was examined (Freeman et al., 1998). Of all children less than 5 years old, 68% remained on the diet for one year, whereas 49% of children 6 to 16 years old were still on the diet after the same time period. The authors concluded that retention time on the diet was not principally affected by age at onset; rather, the most common reason why the diet was discontinued was lack of efficacy. If there was significant seizure control, the diet was more likely to be continued; however, it follows that if older children are less likely to be protected from seizures with a KD they are also less likely to continue the diet. Therefore, it is difficult to tease apart how these three inextricably linked variables are related—i.e., to separate behavioral and psychological aspects of compliance from the physiological variables of age and KD responsiveness.

Experimentally, compliance issues can be circumvented, and the effect of age becomes an independent, controllable variable. To date, only a few animal studies have directly investigated the effects of age at diet onset on KD efficacy. Despite various diet types, duration of diet treatments, and seizure testing methods, all of the studies concluded that animals started on the diet at a younger age exhibited a greater level of seizure protection than those that started the diet at an older age (Bough et al., 1999a,b; Otani et al., 1984; Rho et al., 1999a,b; Uhlemann and Neims, 1972). However, adult animals also exhibit an elevated seizure threshold (Appleton and De Vivo, 1974; Bough and Eagles, 1999; Hori et al., 1997). Therefore, although it may be concluded that the efficacy of the KD is somewhat better at early ages, there is no reason to withhold its use in appropriately selected adult patients.

## Is KD Efficacy Correlated with Ketonemia?

Ketonemia is an obvious consequence of the KD, and it would seem like a simple matter to determine the relationship between ketonemia and seizure control; however, this relationship has been difficult to establish. Several clinical studies have described positive correlations between ketonemia and seizure control, but it is not clear whether ketosis is *sufficient* to explain seizure control. For children maintained on the KD for 3 or 6 months, the threshold BHB level for seizure control was about 4 mmol/L (Gilbert et al., 2000); correlative values at earlier and later time points are not known.

Of the 1001 patients treated by Livingston (1972), only those that exhibited both pronounced ketonemia throughout treatment and appreciable loss of body weight during the fasting period responded favorably to the KD. Huttenlocher (1976) studied the relationship between BHB and seizure control using the MCT diet. Children with the highest plasma BHB concentrations had the greatest degree of seizure control. In addition, if ketosis was lost, seizures returned rapidly. A 4-year-old boy fed the KD developed a breakthrough seizure 45 minutes after ketosis was eliminated by intravenous glucose infusion (Huttenlocher, 1976). Others have reported similar results; even a slight deviation from the KD regimen (e.g., increased intake of carbohydrates) can result in loss of seizure control (Freeman, 2001). Because hypoglycemia is infrequently found in KD-treated patients, it is presumed that the anticonvulsant effects of the KD were mediated by the action of BHB, ACA, or both in the brain. Serum glucose is in the normal range in children on the KD, suggesting that the antiseizure effect is independent of glucose concentration and may be more related to the level of ketones or some other factor.

The possibility has been raised that acetone may be involved in the KD mechanism (Gasior et al., 2007). When the concentration of ketone bodies in the brain was studied by [1]H-magnetic resonance spectroscopy, there was a significant reduction of *N*-acetylaspartate and creatine concentrations in gray matter in KD-treated epilepsy patients compared to untreated patients with epilepsy and healthy controls (Seymour et al., 1999). There was also a significant concentration of acetone in KD-treated epilepsy patients vs. controls, but no detectable BHB or ACA levels. In rats, breath acetone has been successfully monitored, and its concentration correlates with systemic ketosis (Likhodii et al., 2002). Given the poor correlation of urinary ketones with serum ketone levels, the possibility that breath acetone could provide a reliable measure of serum ketones is intriguing and may be of future clinical value during KD administration.

The aforementioned clinical reports include patients of many different ages. Given that there is some evidence that age and ketonemia are negatively correlated (Bough et al., 1999b; Freeman et al., 1998; Schwartz et al., 1989), the effects of ketonemia may have been masked by the effects of age on KD efficacy (as discussed above). Indeed, experimental studies show that when age is controlled ketonemia is not positively correlated with the efficacy of the KD. In 270 rats maintained on KDs of various ratios (i.e., 1:1 to 9:1), rats fed diets of higher ketogenic diet ratios had greater ketonemia but failed to show a positive correlation to PTZ seizure threshold (Bough et al., 2000a). In another study, the same investigators found that neither BHB nor age significantly influenced PTZ seizure threshold. Instead, body weight and ketogenic ratio markedly affected seizure protection in rats (Bough et al., 1999a).

The effect of diet type and its relation to ketonemia was also studied in rats (Likhodii et al., 2000). For animals maintained on any of three dissimilar diets (i.e., utilizing different types of fats), ketone levels for the various dietary groups were significantly different from one another, while similar seizure protection was noted for all. Neither diet type nor ketonemia appeared to correlate with seizure thresholds.

Other data also indicated a lack of correlation between ketonemia and seizure control. Clinical (Dekaban, 1966; Dodson et al., 1976; Helmholz and Keith, 1930) and experimental results (Appleton and De Vivo, 1974) suggest that the KD does not become maximally efficacious for days to weeks after KD initiation, whereas ketonemia typically develops within hours after the dietary regimen has begun (Dekaban, 1966; Schwartz et al., 1989). This apparent temporal disconnection between the development of ketonemia and seizure protection suggests that ketonemia may be *necessary but not sufficient* for KD-induced seizure control (Stafstrom and Spencer, 2000; Vining, 1999). Hence, it seems increasingly clear that ketonemia is not necessarily the best predictor of KD success.

Of course, it is possible that the concentration of ketone bodies in the blood may not accurately reflect their concentration in the brain. Because ketone bodies are transported across the blood–brain barrier by a family of monocarboxylate transporter proteins (MCT-1 and MCT-2), transport could be saturated as high concentrations of ketone bodies approach $V_{max}$ (Pellerin et al., 1998). Moreover, the time that it takes for maximal seizure protection to develop could reflect the time that is necessary for the brain endothelial cells to upregulate the expression of these transporters and thereby increase the concentration of ketones in brain tissue. Whatever the case, it would be better to make correlative measurements of brain levels of ketones in future studies (Pan et al., 2000; Seymour et al., 1999) to most appropriately evaluate KD efficacy.

Overall, it seems clear that some threshold level of ketosis is required and must be maintained chronically for the KD to be maximally effective. The clinical goal should be to maintain ketonemia (as measured via plasma concentrations of BHB) with the assumption that brain ketone levels will be proportionate.

## Does Calorie Restriction Affect KD Efficacy?

Some degree of CR has been part of the KD regimen since its initiation. The scientific rationale underlying this practice is that CR enhances ketosis. The most radical form of CR—fasting—can also diminish seizure activity (De Vivo et al., 1975; Geyelin, 1921). But, can CR afford seizure protection by itself?

Typically, children on the KD are restricted to 75% of the recommended daily allowance for their ideal weight and height, but this is flexible and depends on the child's basal metabolic rate, age, and activity level. This level of restricted caloric intake allows linear growth to occur but prevents the child from gaining significant weight. There is clinical evidence that CR is not required for seizure protection, and KDs have been administered successfully using diets with and without CR (Vaisleib et al., 2004); however, CR facilitates ketosis and may have an antiseizure effect that is independent from ketosis.

In laboratory models of the KD, various degrees of CR have been investigated. Whether rats are fed KDs that are hypocaloric (Bough et al., 1999b), eucaloric (Appleton and De Vivo, 1974), "slightly more" than eucaloric (Likhodii et al., 2000), or even *ad libitum* (Hori et al., 1997), the KD has been shown to increase the resistance to seizures. Interestingly, CR by itself was also shown to significantly increase seizure threshold. Rats maintained on a normal CR diet exhibited a significantly elevated seizure threshold compared to animals fed the same normal diet *ad libitum* (Bough et al., 2000b; Greene et al., 2001). Similarly, rats maintained on a calorie-restricted KD exhibited a significantly elevated seizure threshold compared to animals fed a KD *ad libitum*. Seizure thresholds for the respective diet treatments could be ranked in the following manner: ketogenic CR > normal CR > ketogenic *ad libitum* > normal *ad libitum* (Bough et al., 2000b). These data suggest that CR alone can significantly increase the resistance to seizures, irrespective of diet type.

It is unknown why CR is so important to the KD effect. In general, it is thought that CR facilitates and maintains the metabolic state induced by KD, ensuring that gluconeogenesis remains suppressed during diet treatment. Although the potential adverse effects of restricting calories during a child's critical periods of growth and development must be weighed against the benefits of seizure control and improved long-term outcome, CR seems to be one of the most important variables in the successful administration of the diet. It is important to note that ingestion of excess calories of any type can lead to decreased KD efficacy.

## DOES DIET TYPE AFFECT KD EFFICACY?

A persistent question is whether variation in the lipid composition (i.e., type of fats) or diet ratio (i.e., quantity of fat) can be closely linked to KD efficacy. Typically, a KD includes fats of all chain lengths, most often long-chain, saturated fatty acids in the form of heavy cream and butter. Although fatty acids of varying chain lengths are likely to be included, few attempts have been made to delineate or quantify a specific type of fat or chain length to include in the diet.

One exception includes the MCT diet, as discussed above. Polyunsaturated fatty acids (PUFAs) are thought to play critical roles in brain development and in the modulation of cellular excitability (Stafstrom, 2001; Uauy et al., 1996; Vreugdenhil et al., 1996). More specifically, PUFAs such as arachidonic acid and docosahexaenoic acid have been shown to diminish the excitability of neurons (Xiao and Li, 1999) and cardiac myocytes (Leaf et al., 1999). This led to the premise that a KD supplemented with greater concentrations of PUFAs might enhance diet efficacy. In one clinical study in which the role of these fatty acids was examined, PUFAs increased significantly in children fed a KD and were positively correlated with improved seizure control (Fraser et al., 2003). In another clinical trial, five institutionalized patients with intractable epilepsy (Schlanger et al., 2002) were fed a diet enriched with omega-3 PUFAs; although they exhibited a reduced seizure frequency and intensity over a 6-month period, this study was not blinded and did not include a control group, and there was no wash-out period. In addition, other clinical trials (Bromfield et al., 2008; Yuen et al., 2005) and animal studies (Taha et al., 2006; Willis et al., 2009) failed to show a sustained beneficial anticonvulsant effect of dietary supplementation with PUFAs. Taken together, these findings raise doubts as to whether PUFAs can enhance the anticonvulsant action of the classical KD.

Experimental studies in rats have yielded similar findings. Young rats (postnatal day 20) were fed a variety of qualitatively different types of diets, including butter, flaxseed oil, MCT, a mixture of fat types, or a control diet (Likhodii et al., 2000). The butter, flaxseed oil, and MCT KDs all resulted in similar levels of seizure protection, although the mixed diet did not. In another study from the same laboratory, fatty acid profiles were similar in the brains of rats fed any of the above fat combinations (Dell et al., 2001). Thus, it can be concluded that diet type (i.e., the source of fats used in the KD) does not markedly affect seizure control as much as other variables such as the degree of CR or diet ratio.

## Do AEDs Complement KD Efficacy?

In the clinical setting, the KD is often used as the last resort for medically refractory epilepsy. Most patients beginning the KD have had exhaustive trials of AEDs. For example, in one large prospective study, the 150 children had previously been on an average of more than 6 medications (Freeman et al., 1998). If the KD was successful, many children were able to discontinue or at least reduce their AEDs, with a corresponding decrease in sedation and other side effects; however, few data are available as to which AED, if any, is best utilized in conjunction with the KD.

This question is difficult to decipher clinically, but it is approachable in the laboratory. Bough and colleagues placed rats on the KD plus various combinations of the standard anticonvulsants phenytoin and valproic acid (Bough and Eagles, 2001). Using the PTZ infusion threshold and maximal electroshock models, they showed that the best seizure protection was obtained with a combination of valproic acid and the KD, suggesting a possible synergistic effect. The clinical relevance of these observations remains to be determined, but many children beginning the KD are already on valproic acid. Caution should be exercised with this combination of treatments (Ballaban-Gil et al., 1998).

Once the KD is working, the timing of AED discontinuation is another decision that must be made without the benefit of solid clinical data. The goal is to maintain the KD for 2 years or until the patient is seizure free off of medications for one year (Freeman et al., 2007a). The best clinical advice is to ensure that the KD is working and then slowly wean AEDs, one at a time.

## What Are the Complications and Contraindications for the Use of the KD?

With this unusual dietary regimen, adverse effects can occur. Some side effects, such as constipation, are readily managed, while others are more serious. In some series, up to 10% of patients have developed complications (including hypoproteinemia and liver toxicity), the most severe of which occurred in children treated concurrently with valproic acid (Ballaban-Gil et al., 1998). Renal calculi develop in 5 to 10% of patients (Furth et al., 2000; Herzberg et al., 1990), but, fortunately, this problem in patients usually responds well to conservative treatment such as fluids and urine alkalinization. Most of the stones are urate or calcium oxalate; monitoring the urine calcium-to-creatinine ratio is a useful measure for when more aggressive measures are indicated. Growth must be monitored carefully (Couch et al., 1999), and the diets of children must be supplemented with essential vitamins and minerals (Freeman et al., 2007a). Other potential complications include pancreatitis (Stewart et al., 2001), cardiac abnormalities (Best et al., 2000), and platelet dysfunction (Berry-Kravis et al., 2001b). With careful monitoring, however, the majority of children do very well on the KD (Freeman et al., 2007a).

In addition to those complications, many of which are manageable, it is possible that some children will develop idiosyncratic reactions when placed on the KD. Physicians must be especially aware of metabolic disorders that may be worsened by the KD, such as porphyria, pyruvate carboxylase deficiency, carnitine deficiency, fatty acid oxidation disorders, and mitochondrial cytopathies (Wheless, 2001). In many of these disorders, which involve alteration in energy regulation, fasting can precipitate severe neurologic dysfunction. Because carnitine is required for the transport of long-chain fatty acids across the inner mitochondrial membrane, there is a theoretical concern that carnitine deficiency could be caused by or result from KD use (De Vivo et al., 1998). In the clinical setting, symptomatic carnitine deficiency is rarely seen with KD use (Berry-Kravis et al., 2001a). Before starting the KD, it would be prudent to obtain a metabolic screen, including urine amino and organic acids, serum amino acids, lactate, and carnitine profile (Wheless, 2001).

There has been some concern about adverse cognitive effects of the KD, despite the clinical experience being overwhelmingly positive in this regard. Specifically, adults on the KD for weight loss performed poorer on some neurocognitive tasks than did nontreated controls (Wing et al., 1995). Experimentally, however, kindled rats maintained on the KD performed similarly to controls in both Morris water maze and open field tests (Hori et al., 1997). These findings suggest that the KD does not adversely affect spatial learning and memory, nor does it alter an animal's response to novel environments. Other experiments confirm that the KD does not produce neurobehavioral deficits. Rats fed a variety of different KDs (3:1, 4:1, 5:1, etc.) performed similarly to *ad libitum* and calorie-restricted controls in a battery of behavioral tests (Bough et al., 2000b). In comparison to acute administration of VPA and phenytoin, the KD did not exhibit any neurobehavioral side effects, whereas adverse behavioral changes were seen with high doses of both VPA and phenytoin (alone or in combination with KD) (Bough and Eagles, 2001). Visuospatial memory in the water maze was impaired in rats on the KD, though spontaneous seizures were decreased significantly (Zhao et al., 2004). In that study, cognitive performance was likely confounded by a high ketogenic ratio (8:1) and poor weight gain, as results from the same laboratory previously demonstrated a lack of cognitive impairment (Hori et al., 1997). In another study, the KD was associated with a transient (<24-hour) decrease in locomotor activity in rats on open field testing (Murphy and Burnham, 2006). Thus, most but not all data support the conclusion that the KD does not induce significant adverse cognitive side effects.

Overall, both clinical and experimental findings indicate that the KD is worth considering for cases of intractable epilepsy, once metabolic contraindications are excluded.

## ARE THE EFFECTS OF THE KD ANTICONVULSANT, ANTIEPILEPTIC, OR BOTH?

The anticonvulsant profile of the KD is a broad one. Although research has focused on the acute effects of ketosis on seizure thresholds, questions with more realistic clinical relevance would be how the KD affects long-term seizure occurrence and whether it alters epileptogenesis (i.e., the process by which the brain becomes epileptic exhibiting spontaneous, recurrent seizures) (Gasior et al., 2006).

Clinically, there are anecdotal reports that some children maintained on the KD for 2 or more years can be gradually weaned off the diet and remain seizure free (Freeman et al., 2007a; Livingston, 1972). Evidence that the KD might be antiepileptogenic, as well as anticonvulsant, comes from the observation that children can be gradually weaned off the KD and a normal diet can be reintroduced without a concomitant loss of seizure control (Freeman et al., 2007a). It is unclear whether this is related to natural spontaneous remission or to some prolonged or permanent effect that outlasts KD treatment.

Experimentally, the KD may inhibit the development of spontaneous seizures in rats. Two days after kainic-acid-induced status epilepticus, rats were put on a 4:1 KD or a normal diet (Muller-Schwarze et al., 1999). KD-fed animals exhibited about 75% fewer spontaneous seizures over the ensuing 8 weeks. Compared to controls, KD-fed animals had about 50% less sprouting of hippocampal mossy fibers. These findings suggest that the KD can retard the development of both abnormally reorganized synaptic circuitry (mossy fiber sprouting) and functional sequalae (spontaneous seizures) that are typically seen following kainic-acid-induced status epilepticus. However, this protective effect was seen only if the KD was started within 2 weeks after status epilepticus, suggesting that there is a limited window of opportunity for KD effectiveness (Su et al., 2000).

The mechanism of neuroprotection might involve increased neurogenesis as a consequence of KD treatment. Mice treated with a KD following kainate-induced status epilepticus demonstrated increased bromodeoxyuridine (BrdU) labeling in progenitor neurons of the dentate gyrus (Kwon et al., 2008). In another study employing seizure-naïve rats, however, neurogenesis was not observed (Strandberg et al., 2008). Therefore, the role of neurogenesis in the action of the KD remains controversial.

Similar results were observed in other animal models of epileptogenesis. In the maximal dentate activation-model (a one-day kindling protocol), the rate of increase in electrographic seizure duration was markedly reduced in KD-fed animals compared to controls (Bough et al., 2003). In knockout mice lacking the Kv1.1 potassium channel $\alpha$ subunit, spontaneous seizures and mossy fiber sprouting occur during development (Rho et al., 1999b). Treatment of these mice with an experimental KD suppressed mossy fiber sprouting, thereby inhibiting the development of the epileptic state. In the kindling model, KD treatment delayed the progression of focally stimulated kindled seizures in rats (Hori et al., 1997). It is interesting to note, though, that these effects were only maintained transiently; the after-discharge threshold was elevated for only 2 weeks after treatment onset. In the EL mouse model of epilepsy, treatment with either the KD (Todorova et al., 2000) or CR alone (Greene et al., 2001) significantly delayed the onset of seizures (Todorova et al., 2000). Taken together, these data indicate that the KD not only inhibits seizure onset but may also slow the development of the epilepsy.

## WHAT IS THE MECHANISM OF ACTION?

Despite almost 90 years of clinical use, we still have not fully elucidated the underlying anticonvulsant mechanisms of the KD. To date, mechanistic investigations have employed a bewildering variety of methodologies, and generalizations have proven difficult (Stafstrom, 1999). Nonetheless, we summarize here four prominent theories. For more in-depth evaluation of potential anticonvulsant KD mechanisms and alternative theories, the reader is referred to recent reviews (Bough and Rho, 2007; Hartman et al., 2007; Kim and Rho, 2008; Schwartzkroin, 1999).

### DIRECT ACTIONS OF KETONE BODIES

Preclinical evidence has suggested that ketones may directly contribute to the anticonvulsant mechanism of the KD. Behavioral studies have shown that injection of either acetone or ACA can acutely increase the resistance to a wide variety of seizure types (e.g., maximal electroshock, pentylenetetrazole, amygdala kindling, audiogenic seizure-susceptible [Fring's] mice) (Gasior et al., 2007; Likhodii et al., 2003; Rho et al., 2002). Although it does not appear that ketones antagonize excitatory receptors (i.e., AMPA or NMDA) or activate inhibitory $GABA_A$ receptors directly (Donevan et al., 2003; Thio et al., 2000), recent evidence suggests that ACA and BHB may indirectly activate $K_{ATP}$ channels *in vitro*, at least in some neural structures (Ma et al., 2007). These data show that ketones directly diminish the spontaneous firing rate of neurons in the substantia nigra pars reticulata, a region of the brain known to be critical to seizure propagation (Iadarola and Gale, 1982).

### DIMINISHED GLYCOLYSIS AS A RESULT OF KD

Treatment with the KD rapidly induces a state of hypoglycemia and diminishes the rate of glycolysis, most likely as a result of the initial fasting period associated with KD onset. As ketones become the primary energy source for the brain, glucose, the substrate for glycolysis, is reduced. Restricted carbohydrate and calorie intake commonly produces a state of asymptomatic hypoglycemia in KD patients (Freeman et al., 2007a). In addition, there are concomitant increases in ATP (De Vivo et al., 1978; Otani et al., 1984; Pan et al., 1999) and citrate (Yudkoff et al., 2001) levels, which should further inhibit glycolysis in a feedback manner. In this way, the KD appears to inhibit glycolysis.

When glycolysis is pharmacologically inhibited, animals are protected from seizures. Injection of rats with the glycolytic inhibitor fructose-1,6-bisphosphate induced resistance to a variety of seizure types (Lian et al., 2007). Similar results were observed in experiments employing the glycolytic inhibitor 2-deoxyglucose (2-DG) (Garriga-Canut et al., 2006). 2-DG potently reduced the progression of kindled seizures and increased afterdischarge thresholds; in addition, there was a diminished expression of brain-derived neurotrophic factor (BDNF) and its primary receptor, TrkB—two proteins that are known to underlie key processes of hyperexcitability. 2-DG also reduced electrographic seizures induced by elevated extracellular potassium in hippocampal slices and in several *in vivo* seizure models, including 6-Hz stimulation in rats and audiogenic stimulation in Fring's mice (Stafstrom et al., 2009).

Evidence also suggests that a reduction in the glycolytic energy supply through calorie restriction alone can increase the resistance to seizures (Melo et al., 2006). For example, in one study of epileptic EL mice, mild CR (~15%) induced a state of hypoglycemia which, in turn, was positively correlated with diminished seizure frequency (Greene et al., 2001). Analogous results were reported in rats. Moderate CR (~10 to 20%) was shown to increase the electroconvulsive threshold and paired-pulse inhibition *in vivo* as compared to controls (Bough et al., 2003).

Together, these data indicate that a diminished rate of glycolysis—like that produced during KD treatment—could contribute to the anticonvulsant nature of the KD. More importantly, because the activation of TrkB receptors by BDNF has been demonstrated to be necessary for kindling and the development of epilepsy (He et al., 2004), KD-induced inhibition of glycolysis may also lead to more long-lasting, antiepileptic actions.

## ENHANCED ENERGY PRODUCTION AND RESERVES AFTER KD

It was theorized over 30 years ago that enhanced energy reserves might act to stabilize neurons and increase the resistance to seizure onset. Human imaging studies have shown that areas of hypometabolism can be linked to epileptic foci and sites of seizure generation and propagation (Henry and Votaw, 2004; Janigro, 1999), perhaps representing an uncoupling of metabolic supply from demand (Janigro, 1999). Both experimental (Kudin et al., 2002) and clinical (Antozzi et al., 1995; Kunz et al., 2000) studies have indicated that impaired oxidative phosphorylation contributes appreciably to the epilepsy phenotype. The KD, by comparison, increases energy production and reserves in the brain. This may occur, in part, after a metabolically triggered transcriptional activation of uncoupling proteins (UCPs). Uncoupling proteins are found within the inner mitochondria membrane; they uncouple electron transport from ATP production and effectively reduce the formation of reactive oxygen species (ROS). Sullivan et al. (2004) demonstrated that the KD induces the expression of mitochondrial UCPs and diminishes the generation of ROS in juvenile mice. It might be predicted that an increase in the expression of UCPs would diminish ATP production and reduce energy reserves; however, chronic induction of UCPs after KD treatment triggers a 46% increase in the number of mitochondrial profiles (Bough et al., 2006). Consistent with these data, microarray studies show that diet treatment coordinately upregulated 17 genes linked to oxidative phosphorylation (Bough et al., 2006; Noh et al., 2004). With KD treatment, there is a noted increase in energy reserves both experimentally (Bough et al., 2006; De Vivo et al., 1978; Nakazawa et al., 1983) and clinically (Pan et al., 1999). Ketone bodies have been shown to reduce ROS and decrease neurotoxicity in neocortical neurons by increasing the $NAD^+/NADH$ ratio (Maalouf et al., 2007).

Because synaptic transmission is highly dependent upon tissue energy levels (Attwell and Laughlin, 2001), it is not surprising that synaptic transmission in slices from KD-fed animals was observed to be significantly more resistant to a metabolic challenge. In electrophysiological experiments, field potentials could be measured in KD-treated tissue for approximately 60% longer than in control tissue (Bough et al., 2006). Furthermore, ketones prevent oxidative impairment of hippocampal long-term potentiation by inhibiting the activation of the serine–threonine protein phosphatase 2A (Maalouf and Rho, 2008). These findings suggest that the KD enhances the energy production capacity in brain tissue. This enhancement appears to boost oxidative phosphorylation and elevate energy reserves so that, in the face of a seizure-inducing stimulus, ATP-dependent metabolic pumps would be able to maintain ionic homeostasis and repolarize membrane potentials, prevent synaptic dysfunction, and increase the resistance to seizures.

## ENHANCED/ PROLONGED GABAERGIC INHIBITION AFTER KD

GABAergic interneurons, which already rest at more depolarized membrane potentials, must metabolically endure nonaccommodating bursts of neuronal firing or else network inhibition becomes compromised (Attwell and Laughlin, 2001). In cases of human temporal lobe epilepsy, a phosphocreatine (PCr)-to-ATP "energy storage" ratio has been inversely correlated with the recovery of the membrane potential following periods of bursting activity (Pan et al., 2008; Williamson et al., 2005). Because the creatine kinase enzyme is predominantly localized within GABAergic neurons (Ducray et al., 2007), it has been proposed that PCr and energy stores are especially important to the maintenance of GABAergic inhibitory output.

As discussed above, the KD increases energy production capacity and reserves in the brain. Thus, it follows that greater levels of ATP should enhance and/or prolong synaptic inhibition. There is some evidence for this hypothesis. In one study, rats maintained on a KD exhibited enhanced short-term paired-pulse inhibition compared to controls when recorded *in vivo*; because $GABA_A$ agonists also enhance short-term paired-pulse inhibition, this finding is consistent with a KD-induced enhancement of $GABA_A$-mediated inhibition (Bough et al., 2003). In another study, the ability of the KD to rescue succinic semialdehyde dehydrogenase (SSADH) deficiency was investigated (Nylen et al., 2008). SSADH deficiency is a heritable disorder of GABA deficiency leading to seizures and early death in mice. KD treatment restored spontaneous inhibitory synaptic currents to control levels and prolonged the lifespan in these genetically susceptible mice (Aldh5a1−/−). A clinical study yielded similar findings. Using transcranial magnetic stimulation (TMS), cortical inhibition was assessed in humans maintained on the KD for 2 weeks; short-latency cortical inhibition (which is thought to reflect $GABA_A$-mediated inhibition) was augmented after KD treatment (Cantello et al., 2007).

Consistent with this notion, the KD diet appears to modify amino acid metabolism and enhance GABA synthesis. Ketone bodies (Erecinska et al., 1996) and the KD (Yudkoff et al., 2001) both have been shown to limit the amount of oxaloacetate that is available for the aspartate–aminotransferase reaction in cultured cells and in brain tissue, respectively. Because the aspartate–aminotransferase enzyme reversibly catalyzes the conversion of [$\downarrow$ aspartate + oxoglutarate $\leftrightarrow$ oxaloacetate + $\uparrow$ glutamate], this means that there is an increase in the concentration of glutamate as oxaloacetate is metabolically utilized elsewhere. In turn, this makes glutamate more accessible to the glutamate decarboxylase (GAD) reaction in GABAergic interneurons, thereby increasing the production of GABA (Melo et al., 2006; Yudkoff et al., 2007). These findings suggest that metabolic adaptations to the KD functionally restore or enhance $GABA_A$-mediated synaptic inhibition to increase seizure resistance.

Overall, the KD appears to induce a unique physiologic and metabolic state that increases the resistance to seizures in individuals with refractory epilepsy. As the body rapidly moves into a state of ketosis during KD treatment, ketone bodies may directly limit neuronal hyperexcitability and seizure onset. A diminished rate of glycolysis may further limit hyperexcitability and, if sustained, could modify gene expression to mitigate the further development of epilepsy. The KD also appears to adaptively upregulate numerous energy metabolism genes, induce mitochondrial biogenesis, and elevate energy reserves. This boost in energy reserves may stabilize metabolic deficits within epileptic foci or compensate for transient failures of synaptic inhibition (i.e., GABAergic rundown), which would otherwise favor the initiation of seizure activity; this adaptation should increase in the resistance to a wide variety of seizure types.

## SUMMARY

We are only beginning to sort out the critical variables involved in understanding how the KD exerts its anticonvulsant, and perhaps antiepileptic, effects. Many questions remain. An improved understanding of the underlying mechanisms of KD action would not only aid in the development of novel anticonvulsant and antiepileptic treatments but also significantly enhance our basic understanding of how bioenergetics impact neuronal excitability and synaptic transmission. Using a combination of clinical and laboratory research, it is hoped that future studies will optimize the application of the KD, improve its overall effectiveness, and elucidate its underlying mechanisms of action. Until that time, however, the principles outlined in Table 24.2 and in a recent consensus statement (Kossoff et al., 2009) provide general clinical guidelines for using the KD.

**TABLE 24.2**
**Principles and Clinical Applications of the Ketogenic Diet**

| Principles | Clinical Applications |
|---|---|
| 1. Seizure types | The KD is worth trying in refractory epilepsy, regardless of seizure/epilepsy type, though it may work best in the generalized epilepsies. |
| 2. Optimal age range | The KD is worth trying at any age. |
| 3. Role of ketonemia | It is important to maintain ketosis in any patient on the KD; ketosis is necessary for seizure control, but the correlation is inexact. |
| 4. Fasting | Fasting allows a quicker attainment of ketosis. |
| 5. Calorie restriction | Most practitioners restrict calories to 75% of the recommended daily allowance, but this guideline is flexible, based on the patient's age, weight, and activity level. |
| 6. Fluid restriction | Fluids are usually limited to 75 cc/kg/day; the role of this mild dehydration in seizure control is unknown. |
| 7. Diet type | The classical 4:1 diet is preferred. A ratio > 4:1 might theoretically afford better seizure control, but this is often impractical. At this time, there is insufficient evidence to recommend specific fat types. |
| 8. Contraindications | Before KD initiation, it is recommended that screening be performed for metabolic disorders that may be exacerbated by the KD. |
| 9. Adjunctive medications | There are insufficient data to recommend specific adjunctive AEDs. There are no data to guide decisions as to when to discontinue AEDs. |

*Note:* Table has been derived from both clinical and laboratory studies.

# REFERENCES

Antozzi, C., S. Franceschetti, G. Filippini, B. Barbiroli, M. Savoiardo, F. Fiacchino, M. Rimoldi, R. Lodi, P. Zaniol, M. Zeviani. (1995). Epilepsia, partialis continua associated with NADH-coenzyme Q reductase deficiency. *J. Neurol. Sci.*, 129:152–161.

Appleton, D.B., D.C. De Vivo. (1974). An animal model for the ketogenic diet. *Epilepsia*, 15:211–217.

Attwell, D., S.B. Laughlin. (2001). An energy budget for signaling in the grey matter of the brain. *J. Cereb. Blood Flow Metab.*, 21:1133–1145.

Ballaban-Gil, K., C. Callahan, C. O'Dell, M. Pappo, S. Moshe, S. Shinnar. (1998). Complications of the ketogenic diet. *Epilepsia*, 39:744–748.

Barborka, C.J. (1930). Results of treatment by ketogenic diet in 100 cases. *Arch. Neurol. Psychiatr.*, 23:904–914.

Berry-Kravis, E., G. Booth, A.C. Sanchez, J. Woodbury-Kolb. (2001a). Carnitine levels and the ketogenic diet. *Epilepsia*, 42:1445–1451.

Berry-Kravis, E., B. Booth, A. Taylor, L.A. Valentino. (2001b). Bruising and the ketogenic diet: evidence for diet-induced changes in platelet function. *Ann. Neurol.*, 49:98–103.

Best, T.H., D.N. Franz, D.L. Gilbert, D.P. Nelson, M.R. Epstein. (2000). Cardiac complications in pediatric patients on the ketogenic diet. *Neurology*, 54:2328–2330.

Boles, R.G., M.R. Seashore, W.G. Mitchell, P.R. Kollros, S. Mofidi, E.J. Novotny. (1999). Glucose transporter type 1 deficiency: a study of two cases with video-EEG. *Eur. J. Pediatr.*, 158:978–983.

Bough, K.J., D.A. Eagles. (1999). A ketogenic diet increases the resistance to pentylenetetrazole-induced seizures in the rat. *Epilepsia*, 40:138–143.

Bough, K.J., D.A. Eagles. (2001). Comparison of the anticonvulsant efficacies and neurotoxic effects of valproic acid, phenytoin, and the ketogenic diet. *Epilepsia*, 42:1345–1353.

Bough, K.J., J.M. Rho. (2007). Anticonvulsant mechanisms of the ketogenic diet. *Epilepsia*, 48:43–58.

Bough, K.J., R. Chen, D.A. Eagles. (1999a). Path analysis shows that increasing ketogenic ratio, but not beta-hydroxybutyrate, elevates seizure threshold in the rat. *Dev. Neurosci.*, 21:400–406.

Bough, K.J., R. Valiyil, F.T. Han, D.A. Eagles. (1999b). Seizure resistance is dependent upon age and caloric restriction in rats fed a ketogenic diet. *Epilepsy Res.*, 35:21–28.

Bough, K.J., P.J. Matthews, D.A. Eagles. (2000a). A ketogenic diet has different effects upon seizures induced by maximal electroshock and by pentylenetetrazole infusion. *Epilepsy Res.*, 38:105–114.

Bough, K.J., S.G. Yao, D.A. Eagles. (2000b). Higher ketogenic diet ratios confer protection from seizures without neurotoxicity. *Epilepsy Res.*, 38:15–25.

Bough, K.J., K. Gudi, F.T. Han, A.H. Rathod, D.A. Eagles. (2002). An anticonvulsant profile of the ketogenic diet. *Epilepsy Res.*, 50:313–325.

Bough, K.J., P.A. Schwartzkroin, J.M. Rho. (2003). Calorie restriction and ketogenic diet diminish neuronal excitability in rat dentate gyrus *in vivo*. *Epilepsia*, 44:752–760.

Bough, K.J., J. Wetherington, B. Hassel, J.F. Pare, J.W. Gawryluk, J.G. Greene, R. Shaw, Y. Smith, J.D. Geiger, R.J. Dingledine. (2006). Mitochondrial biogenesis in the anticonvulsant mechanism of the ketogenic diet. *Ann. Neurol.*, 60:223–235.

Bromfield, E., B. Dworetzky, S. Hurwitz, Z. Eluri, L. Lane, S. Replansky, D. Mostofsky. (2008). A randomized clinical trial of polyunsaturated fatty acids for refractory epilepsy. *Epilepsy Behav.*, 12:187–190.

Cantello, R., C. Varrasi, R. Tarletti, M. Cecchin, F. D'Andrea, P. Veggiotti, G. Bellomo, F. Monaco. (2007). Ketogenic diet: electrophysiological effects on the normal human cortex. *Epilepsia*, 48:1756–1763.

Carroll, J., D. Koenigsberger. (1998). The ketogenic diet: a practical guide for caregivers. *J. Am. Diet. Assoc.*, 98:316–321.

Coppola, G., P. Veggiotti, R. Cusmai, S. Bertoli, S. Cardinali, C. Dionisi-Vici, M. Elia, M.L. Lispi, C. Sarnelli, A. Tagliabue, C. Toraldo, A. Pascotto. (2002). The ketogenic diet in children, adolescents and young adults with refractory epilepsy: an Italian multicentric experience. *Epilepsy Res.*, 48:221–227.

Couch, S., F. Schwarzman, J. Carroll, D. Koenigsberger, D.R. Nordli, Jr., R.J. Deckelbaum, A.R. DeFelice. (1999). Growth and nutritional outcomes of children treated with the ketogenic diet. *J. Am. Diet. Assoc.*, 99:1573–1575.

De Vivo, D.C., K.L. Malas, M.P. Leckie. (1975). Starvation and seizures. *Arch. Neurol.*, 32:755–760.

De Vivo, D.C., M.P. Leckie, J.S. Ferrendelli, D.B. McDougal. (1978). Chronic ketosis and cerebral metabolism. *Ann. Neurol.*, 3:331–337.

De Vivo, D.C., R.R. Trifiletti, R.I. Jacobson, G.M. Ronen, R.A. Behmand, S.I. Harik. (1991). Defective glucose transport across the blood–brain barrier as a cause of persistent hypoglycorrhachia, seizures, and developmental delay. *N. Engl. J. Med.*, 325:703–709.

De Vivo, D.C., T.P. Bohan, D.L. Coulter, F.E. Dreifuss, R.S. Greenwood, D.R. Nordli, Jr., W.D. Shields, C.E. Stafstrom, I. Tein. (1998). L-Carnitine supplementation in childhood epilepsy: current perspectives. *Epilepsia*, 39:1216–1225.

Dekaban, A.S. (1966). Plasma lipids in epileptic children treated with the high fat diet. *Arch. Neurol.*, 15:177–184.

Dell, C.A., S.S. Likhodii, K. Musa, M.A. Ryan, W.M. Burnham, S.C. Cunnane. (2001). Lipid and fatty acid profiles in rats consuming different high-fat ketogenic diets. *Lipids* 36:373–378.

Dodson, W.E., A.L. Prensky, D.C. De Vivo, S. Goldring, P.R. Dodge. (1976). Management of seizure disorders: selected aspects, Part II. *J. Pediatr.*, 89:695–703.

Donevan, S.D., H.S. White, G.D. Anderson, J.M. Rho. (2003). Voltage-dependent block of N-methyl-D-aspartate receptors by the novel anticonvulsant dibenzylamine, a bioactive constituent of L-(+)-beta-hydroxybutyrate. *Epilepsia*, 44:1274–1279.

Ducray, A.D., R. Qualls, U. Schlattner, R.H. Andres, E. Dreher, R.W. Seiler, T. Wallimann, H.R. Widmer. (2007). Creatine promotes the GABAergic phenotype in human fetal spinal cord cultures. *Brain Res.*, 1137:50–57.

Erecinska, M., D. Nelson, Y. Daikhin, M. Yudkoff. (1996). Regulation of GABA level in rat brain synaptosomes: fluxes through enzymes of the GABA shunt and effects of glutamate, calcium, and ketone bodies. *J. Neurochem.*, 67:2325–2334.

Fraser, D.D., S. Whiting, R.D. Andrew, E.A. Macdonald, K. Musa-Veloso, S.C. Cunnane. (2003). Elevated polyunsaturated fatty acids in blood serum obtained from children on the ketogenic diet. *Neurology*, 60:1026–1029.

Freeman, J.M. (2001). The ketogenic diet and epilepsy. *Nestle Nutr. Workshop Ser. Clin. Perf. Prog.*, 5:307–321.

Freeman, J.M., E. Vining, D. Pillas, P. Pyzik, J. Casey, M. Kelly. (1998). The efficacy of the ketogenic diet—1998: a prospective evaluation of intervention in 150 children. *Pediatrics*, 102:1358–1363.

Freeman, J.M., E.H. Kossoff, J.B. Freeman, M.T. Kelly. (2007a). *The Ketogenic Diet—A Treatment for Children and Others with Epilepsy*, 4th ed. New York: Demos Publications.

Freeman, J.M., E.H. Kossoff, A.L. Hartman. (2007b). The ketogenic diet: one decade later. *Pediatrics*, 119:535–543.

Freeman, J.M., E.P.G. Vining, E.H. Kossoff, P.L. Pyzik, X. Ye, S.N. Goodman. (2009). A blinded, crossover study of the efficacy of the ketogenic diet. *Epilepsia*, 50(2):322–325.

Furth, S.L., J.C. Casey, P.L. Pyzik, A.M. Neu, S.G. Docimo, E.P. Vining, J.M. Freeman, B.A. Fivush. (2000). Risk factors for urolithiasis in children on the ketogenic diet. *Pediatr. Nephrol.*, 15:125–128.

Garriga-Canut, M., B. Schoenike, R. Quazi, K. Bergendahl, T.J. Daley, R.M. Pfender, J.F. Morrison, J. Ockuly, C. Stafstrom, T. Sutula, A. Roopra. (2006). 2-Deoxy-D-glucose reduces epilepsy progression by NRSF-CtBP-dependent metabolic regulation of chromatin structure. *Nat. Neurosci.*, 9:1382–1387.

Gasior, M., M.A. Rogawski, A.L. Hartman. (2006). Neuroprotective and disease-modifying effects of the ketogenic diet. *Behav. Pharmacol.*, 17:431–439.

Gasior, M., A. French, M.T. Joy, R.S. Tang, A.L. Hartman, M.A. Rogawski. (2007). The anticonvulsant activity of acetone, the major ketone body in the ketogenic diet, is not dependent on its metabolite acetol, 1,2-propanediol, methylglyoxal, or pyruvic acid. *Epilepsia*, 48:793–800.

Geyelin, H. (1921). Fasting as a method for treating epilepsy. *Med. Rec.*, 99:1037–1039.

Gilbert, D.L., P.L. Pyzik, J.M. Freeman. (2000). The ketogenic diet: seizure control correlates better with serum beta-hydroxybutyrate than with urine ketones. *J. Child Neurol.*, 15:787–790.

Greene, A.E., M.T. Todorova, R. McGowan, T.N. Seyfried. (2001). Caloric restriction inhibits seizure susceptibility in epileptic EL mice by reducing blood glucose. *Epilepsia*, 42:1371–1378.

Hartman, A.L., M. Gasior, E.P. Vining, M.A. Rogawski. (2007). The neuropharmacology of the ketogenic diet. *Pediatr. Neurol.*, 36:281–292.

Hartman, A.L., M. Lyle, M.A. Rogawski, M. Gasior. (2008). Efficacy of the ketogenic diet in the 6-Hz seizure test. *Epilepsia*, 49:334–339.

He, X.P., R. Kotloski, S. Nef, B.W. Luikart, L.F. Parada, J.O. McNamara. (2004). Conditional deletion of TrkB but not BDNF prevents epileptogenesis in the kindling model. *Neuron*, 43:31–42.

Helmholz, H.F., H.M. Keith. (1930). Eight years' experience with the ketogenic diet in the treatment of epilepsy. *JAMA*, 95:707–709.

Hemingway, C., J.M. Freeman, D.J. Pillas, P.L. Pyzik. (2001). The ketogenic diet: a 3- to 6-year follow-up of 150 children enrolled prospectively. *Pediatrics*, 108:898–905.

Henderson, C.B., F.M. Filloux, S.C. Alder, J.L. Lyon, D.A. Caplin. (2006). Efficacy of the ketogenic diet as a treatment option for epilepsy: meta-analysis. *J. Child Neurol.*, 21:193–198.

Henry, T.R., J.R. Votaw. (2004). The role of positron emission tomography with [18F]-fluorodeoxyglucose in the evaluation of the epilepsies. *Neuroimaging Clin. N. Am.*, 14:517–535.

Herzberg, G.Z., B.A. Fivush, S.L. Kinsman, J.P. Gearhart. (1990). Urolithiasis associated with the ketogenic diet. *J. Pediatr.*, 117:742–745.

Holmes, G.L., Q. Zhao. (2008). Choosing the correct antiepileptic drugs: from animal studies to the clinic. *Pediatr. Neurol.*, 38:151–162.

Hori, A., P. Tandon, G.L. Holmes, C.E. Stafstrom. (1997). Ketogenic diet: effects on expression of kindled seizures and behavior in adult rats. *Epilepsia*, 38:750–758.

Huttenlocher, P.R. (1976). Ketonemia and seizures: metabolic and anticonvulsant effects of two ketogenic diets in childhood epilepsy. *Pediatr. Res.*, 10:536–540.

Huttenlocher, P.R., A.J. Wilbourn, J.M. Signore. (1971). Medium-chain triglycerides as a therapy for intractable childhood epilepsy. *Neurology*, 21:1097–1103.

Iadarola, M.J., K. Gale. (1982). Substantia nigra: site of anticonvulsant activity mediated by gamma-aminobutyric acid. *Science*, 218:1237–1240.

Janigro, D. (1999). Blood–brain barrier, ion homeostasis and epilepsy: possible implications towards the understanding of ketogenic diet mechanisms. *Epilepsy Res.*, 37:223–232.

Kang, H.C., D.E. Chung, D.W. Kim, H.D. Kim. (2004). Early- and late-onset complications of the ketogenic diet for intractable epilepsy. *Epilepsia*, 45:1116–1123.

Kang, H.C., H.S. Lee, S.J. You, D.C. Kang, T.S. Ko, H.D. Kim. (2007). Use of a modified Atkins diet in intractable childhood epilepsy. *Epilepsia*, 48:182–186.

Kaytal, N.G., A.N. Koehler, B. McGhee, C.M. Foley, P.K. Crumrine. (2000). The ketogenic diet in refractory epilepsy: the experience of Children's Hospital of Pittsburgh. *Clin. Pediatr.*, 39:153–159.

Keene, D.L. (2006). A systematic review of the use of the ketogenic diet in childhood epilepsy. *Pediatr. Neurol.* 35:1–5.

Kim, D.Y., J.M. Rho. (2008). The ketogenic diet and epilepsy. *Curr. Opin. Clin. Nutr. Metab. Care* 11:113–120.

Kinsman, S.L., E.P.G. Vining, S.A. Quaskey, D. Mellits, J.M. Freeman. (1992). Efficacy of the ketogenic diet for intractable seizure disorders: review of 58 cases. *Epilepsia*, 33:1132–1136.

Kossoff, E.H., J.R. McGrogan. (2005). Worldwide use of the ketogenic diet. *Epilepsia*, 46:280–289.

Kossoff, E.H., G.L. Krauss, J.R. McGrogan, J.M. Freeman. (2003). Efficacy of the Atkins diet as therapy for intractable epilepsy. *Neurology*, 61:1789–1791.

Kossoff, E.H., J.R. McGrogan, R.M. Bluml, D.J. Pillas, J.E. Rubenstein, E.P. Vining. (2006). A modified Atkins diet is effective for the treatment of intractable pediatric epilepsy. *Epilepsia*, 47:421–424.

Kossoff, E.H., Z. Turner, R.,M. Bluml, P.L. Pyzik, E.P. Vining. (2007). A randomized, crossover comparison of daily carbohydrate limits using the modified Atkins diet. *Epilepsy Behav.*, 10:432–436.

Kossoff, E.H., L.C. Laux, R. Blackford, P.F. Morrison, P.L. Pyzik, R.M. Hamdy, Z. Turner, D.R. Nordli, Jr. (2008a). When do seizures usually improve with the ketogenic diet? *Epilepsia*, 49:329–333.

Kossoff, E.H., H. Rowley, S.R. Sinha, E.P. Vining. (2008b). A prospective study of the modified Atkins diet for intractable epilepsy in adults. *Epilepsia*, 49:316–319.

Kossoff, E.H., B.A. Zupec-Kania, P.E. Amark, K.R. Ballaban-Gil, A.G. Christina Bergqvist, R. Blackford et al. (2009). Optimal clinical management of children receiving the ketogenic diet: recommendations of the International Ketogenic Study Group. *Epilepsia*, 50(2):304–317.

Kudin, A.P., T.A. Kudina, J. Seyfried, S. Vielhaber, H. Beck, C.E. Elger, W.S. Kunz. (2002). Seizure-dependent modulation of mitochondrial oxidative phosphorylation in rat hippocampus. *Eur. J. Neurosci.*, 15:1105–1114.

Kunz, W.S., A.P. Kudin, S. Vielhaber, I. Blumcke, W. Zuschratter, J. Schramm, H. Beck, C.E. Elger. (2000). Mitochondrial complex I deficiency in the epileptic focus of patients with temporal lobe epilepsy. *Ann. Neurol.*, 48:766–773.

Kwon, Y.S., S.W. Jeong, D.W. Kim, E.S. Choi, B.K. Son. (2008). Effects of the ketogenic diet on neurogenesis after kainic acid-induced seizures in mice. *Epilepsy Res.*, 78:186–194.

Leaf, A., J. Kang, Y.F. Xiao, G. Billman, R. Voskuyl. (1999). Functional and electrophysiologic effects of poly-unsaturated fatty acids on excitable tissues: heart and brain. *Prostagland. Leukotr. Essen. Fatty Acids*, 60:307–312.

Lefevre, F., N. Aronson. (2000). Ketogenic diet for the treatment of refractory epilepsy in children: a systematic review of efficacy. *Pediatrics*, 105(4):E46.

Lian, X.Y., F.A. Khan, J.L. Stringer. (2007). Fructose-1,6-bisphosphate has anticonvulsant activity in models of acute seizures in adult rats. *J. Neurosci.*, 27:12007–12011.

Likhodii, S.S., K. Musa, A. Mendonca, C. Dell, W.M. Burnham, S.C. Cunnane. (2000). Dietary fat, ketosis, and seizure resistance in rats on the ketogenic diet. *Epilepsia*, 41:1400–1410.

Likhodii, S.S., K. Musa, S.C. Cunnane. (2002). Breath acetone as a measure of systemic ketosis assessed in a rat model of the ketogenic diet. *Clin. Chem.*, 48:115–120.

Likhodii, S.S., I. Serbanescu, M.A. Cortez, P. Murphy, O.C. Snead, 3rd, W.M. Burnham. (2003). Anticonvulsant properties of acetone, a brain ketone elevated by the ketogenic diet. *Ann. Neurol.*, 54:219–226.

Livingston, S. (1972). Dietary treatment of epilepsy. In *Comprehensive Management of Epilepsy in Infancy, Childhood and Adolescence* (pp. 378–405), S. Livingston, Ed. Springfield, IL: Charles C Thomas.

Ludwig, D.S. (2007). Clinical update: the low-glycaemic-index diet. *Lancet*, 369: 890–892.

Ma, W., J. Berg, G. Yellen. (2007). Ketogenic diet metabolites reduce firing in central neurons by opening K(ATP) channels. *J. Neurosci.*, 27:3618–3625.

Maalouf, M., J.M. Rho. (2008). Oxidative impairment of hippocampal long-term potentiation involves activation of protein phosphatase 2A and is prevented by ketone bodies. *J. Neurosci. Res.*, 86: 3322–3330.

Maalouf, M., P.G. Sullivan, L. Davis, D.Y. Kim, J.M. Rho. (2007). Ketones inhibit mitochondrial production of reactive oxygen species production following glutamate excitotoxicity by increasing NADH oxidation. *Neuroscience*, 145:256–264.

Magrath, G., A. MacDonald, W. Whitehouse. (2000). Dietary practices and use of the ketogenic diet in the UK. *Seizure*, 9:128–130.

Mantis, J.G., N.A. Centeno, M.T. Todorova, R. McGowan, T.N. Seyfried. (2004). Management of multifactorial idiopathic epilepsy in EL mice with caloric restriction and the ketogenic diet: role of glucose and ketone bodies. *Nutr. Metab.*, 1:1–11.

Maydell, B.V., E. Wyllie, N. Akhtar, P. Kotagal, K. Powaski, K. Cook, A. Weinstock, A.D. Rothner. (2001). Efficacy of the ketogenic diet in focal versus generalized seizures. *Pediatr. Neurol.*, 25:208–212.

McGarry, J.D., D.W. Foster. (1980). Regulation of hepatic fatty acid oxidation and ketone body production. *Annu. Rev. Biochem.*, 49:395–420.

Melo, T.M., A. Nehlig, U. Sonnewald. (2006). Neuronal-glial interactions in rats fed a ketogenic diet. *Neurochem. Int.*, 48:498–507.

Millichap, J.G., J.D. Jones, B.P. Rudis. (1964). Mechanism of anticonvulsant action of the ketogenic diet. *Am. J. Dis. Child*, 107:593–604.

Muller-Schwarze, A.B., P. Tandon, Z. Liu, Y. Yang, G.L. Holmes, C.E. Stafstrom. (1999). Ketogenic diet reduces spontaneous seizures and mossy fiber sprouting in the kainic acid model. *NeuroReport*, 10:1517–1522.

Murphy, P., W.M. Burnham. (2006). The ketogenic diet causes a reversible decrease in activity level in Long–Evans rats. *Exp. Neurol.*, 201:84–99.

Nakazawa, M., S. Kodama, T. Matsuo. (1983). Effects of ketogenic diet on electroconvulsive threshold and brain contents of adenosine nucleotides. *Brain Dev.*, 5:375–380.

Neal, E.G., H. Chaffe, R.H. Schwartz, M.S. Lawson, N. Edwards, G. Fitzsimmons, A. Whitney, J.H. Cross. (2008). The ketogenic diet for the treatment of childhood epilepsy: a randomized controlled trial. *Lancet Neurol.*, 7:500–506.

Noh, H.S., Y.S. Kim, H.P. Lee, K.M. Chung, D.W. Kim, S.S. Kang, G.J. Cho, W.S. Choi. (2003). The protective effect of a ketogenic diet on kainic acid-induced hippocampal cell death in the male ICR mice. *Epilepsy Res.*, 53:119–128.

Noh, H.S., H.P. Lee, D.W. Kim, S.S. Kang, G.J. Cho, J.M. Rho, W.S. Choi. (2004). A cDNA microarray analysis of gene expression profiles in rat hippocampus following a ketogenic diet. *Molec. Brain Res.*, 129:80–87.

Noh, H.S., D.W. Kim, G.J. Cho, W.S. Choi, S.S. Kang. (2006). Increased nitric oxide caused by the ketogenic diet reduces the onset time of kainic acid-induced seizures in ICR mice. *Brain Res.*, 1075:193–200.

Nordli, Jr., D.R., D.C. De Vivo. (1997). The ketogenic diet revisited: back to the future. *Epilepsia*, 38:743–749.

Nordli, Jr., D.R., M.M. Kuroda, J. Carroll, D.Y. Koenigsberger, L.J. Hirsch, H.J. Bruner, W.T. Seidel, D.C. De Vivo. (2001). Experience with the ketogenic diet in infants. *Pediatrics*, 108:129–133.

Nylen, K., J.L. Velazquez, S.S. Likhodii, M.A. Cortez, L. Shen, Y. Leshchenko, K. Adeli, K.M. Gibson, W.M. Burnham, O.C. Snead, 3rd. (2008). A ketogenic diet rescues the murine succinic semialdehyde dehydrogenase deficient phenotype. *Exp. Neurol.*, 210:449–457.

Otani, K., A. Yamatodani, H. Wada, T. Mimaki, H. Yabuuchi. (1984). Effect of ketogenic diet on convulsive threshold and brain monoamine levels in young mice. *No To Hattatsu*, 16:196–204.

Pan, J.W., E. Bebin, W. Chu, H.P. Hetherington. (1999). Ketosis and epilepsy: $^{31}$P spectroscopic imaging at 4.1 T. *Epilepsia*, 40:703–707.

Pan, J.W., D.L. Rothman, K.L. Behar, D.T. Stein, H.P. Hetherington. (2000). Human brain beta-hydroxybutyrate and lactate increase in fasting-induced ketosis. *J. Cereb. Blood Flow Metab.*, 20:1502–1507.

Pan, J.W., I. Cavus, J. Kim, H.P. Hetherington, D.D. Spencer. (2008). Hippocampal extracellular GABA correlates with metabolism in human epilepsy. *Metab. Brain. Dis.*, 23:457–468.

Pellerin, L., G. Pellegri, J.L. Martin, P.J. Magistretti. (1998). Expression of monocarboxylate transporter mRNAs in mouse brain: support for a distinct role of lactate as an energy substrate for the neonatal vs. adult brain. *Proc. Nat. Acad. Sci. U.S.A.*, 95:3990–3995.

Peterman, M.G. (1925). The ketogenic diet in epilepsy. *JAMA*, 84:1979–1983.

Pfeifer, H.H., E.A. Thiele. (2005). Low-glycemic-index treatment: a liberalized ketogenic diet for treatment of intractable epilepsy. *Neurology*, 65:1810–1812.

Pulsifer, M., J. Gordon, E. Vining, J. Freeman. (2001). Effects of the ketogenic diet on development and behavior: preliminary report of a prospective study. *Dev. Med. Child Neurol.* 43:301–306.

Raffo, E., J. Francois, A. Ferrandon, E. Koning, A. Nehlig. (2008). Calorie-restricted ketogenic diet increases thresholds to all patterns of pentylenetetrazole-induced seizures: critical importance of electroclinical assessment. *Epilepsia*, 49:320–328.

Rauchenzauner, M., J. Klepper, B. Leiendecker, G. Luef, K. Rostasy, C. Ebenbichler. (2008). The ketogenic diet in children with GLUT1 deficiency syndrome and epilepsy. *J. Pediatr.*, 153:716–718.

Rho, J.M., D. Kim, C. Robbins, G.D. Anderson, P. Schwartzkroin. (1999a). Age-dependent differences in flurothyl seizure sensitivity in mice treated with a ketogenic diet. *Epilepsy Res.*, 37:233–240.

Rho, J.M., P. Szot, B.L. Tempel, P.A. Schwartzkroin. (1999b). Developmental seizure susceptibility of Kv1.1 potassium channel knockout mice. *Dev. Neurosci.*, 21:320–327.

Rho, J.M., G.D. Anderson, S.D. Donevan, H.S. White. (2002). Acetoacetate, acetone, and dibenzylamine (a contaminant in l-(+)-beta-hydroxybutyrate) exhibit direct anticonvulsant actions *in vivo*. *Epilepsia*, 43:358–361.

Saudek, C.D., P. Felig. (1976). The metabolic events of starvation. *Am. J. Med.*, 60:117–126.

Schlanger, S., M. Shinitzky, D. Yam. (2002). Diet enriched with omega-3 fatty acids alleviates convulsion symptoms in epilepsy patients. *Epilepsia*, 43:101989.

Schwartz, R.M., J. Eaton, B.D. Bower, A. Aynsley-Green. (1989). Ketogenic diets in the treatment of epilepsy: short-term clinical effects. *Dev. Med. Child Neurol.*, 31:145–151.

Schwartzkroin, P. (1999). Mechanisms underlying the anti-epileptic efficacy of the ketogenic diet. *Epilepsy Res.*, 37:171–180.

Seymour, K.J., S. Bluml, J. Sutherling, W. Sutherling, B.D. Ross. (1999). Identification of cerebral acetone by ¹H-MRS in patients with epilepsy controlled by ketogenic diet. *MAGMA*, 8:33–42.

Sills, M.A., W.I. Forsythe, D. Haidukewych, A. MacDonald, M. Robinson. (1986). The medium chain triglyceride diet and intractable epilepsy. *Arch. Dis. Child*, 61:1168–1172.

Sirven, J., B. Whedon, D. Caplan, J. Liborace, D. Glosser, J. O'Dwyer, M.R. Sperling. (1999). The ketogenic diet for intractable epilepsy in adults: preliminary results. *Epilepsia*, 40:1721–1726.

Stafstrom, C.E. (1999). Animal models of the ketogenic diet: what have we learned, what can we learn? *Epilepsy Res.*, 37:241–259.

Stafstrom, C.E. (2001). Effects of fatty acids and ketones on neuronal excitability: implications for epilepsy and its treatment. In *Fatty Acids—Physiological and Behavioral Functions* (pp. 273–290), D.I. Mostofsky, S. Yehuda, and N. Salem, Eds. Totowa, NJ: Humana Press.

Stafstrom, C.E. (2004). Dietary approaches to epilepsy treatment: old and new options on the menu. *Epilepsy Curr.*, 4:215–222.

Stafstrom, C.E., S. Spencer. (2000). The ketogenic diet: a therapy in search of an explanation. *Neurology*, 54:282–283.

Stafstrom, C.E., J.C. Ockuly, L. Murphree, M.T. Valley, A. Roopra, T.P. Sutula. (2009). Anticonvulsant and antiepileptic action of 2-deoxy-D-glucose in epilepsy models. *Ann. Neurol.*, 65:435–447.

Stewart, W.A., K. Gordon, P. Camfield. (2001). Acute pancreatitis causing death in a child on the ketogenic diet. *J. Child Neurol.*, 16:682.

Strandberg, J., D. Kondziella, T. Thorlin, F. Asztely. (2008). Ketogenic diet does not disturb neurogenesis in the dentate gyrus in rats. *NeuroReport*, 19:1235–1237.

Su, S.W., M.R. Cilio, Y. Sogawa, D. Silveira, G.L. Holmes, C.E. Stafstrom. (2000). Timing of ketogenic diet initiation in an experimental epilepsy model. *Dev. Brain Res.*, 125:131–138.

Sullivan, P.G., N.A. Rippy, K. Dorenbos, R.C. Concepcion, A.K. Agarwal, J.M. Rho. (2004). The ketogenic diet increases mitochondrial uncoupling protein levels and activity. *Ann. Neurol.*, 55:576–580.

Swink, T.D., E.P.G. Vining, J.M. Freeman. (1997). The ketogenic diet. *Adv. Pediatr.*, 44:297–329.

Szot, P., D. Weinshenker, J.M. Rho, T.W. Storey, P.A. Schwartzkroin. (2001). Norepinephrine is required for the anticonvulsant effect of the ketogenic diet. *Dev. Brain Res.*, 129:211–214.

Taha, A.Y., B.M. Baghiu, R. Lui, K. Nylen, D.W. Ma, W.M. Burnham. (2006). Lack of benefit of linoleic and alpha-linolenic polyunsaturated fatty acids on seizure latency, duration, severity or incidence in rats. *Epilepsy Res.*, 71:40–46.

Thavendiranathan, P., A. Mendonca, C. Dell, S.S. Likhodii, K. Musa, C. Iracleous, S.C. Cunnane, W.M. Burnham. (2000). The MCT ketogenic diet: effects on animal seizure models. *Exp. Neurol.*, 161:696–703.

Thio, L.L., M. Wong, K.A. Yamada. (2000). Ketone bodies do not directly alter excitatory or inhibitory hippocampal synaptic transmission. *Neurology*, 54:325–331.

Todorova, M.T., P. Tandon, R.A. Madore, C.E. Stafstrom, T.N. Seyfried. (2000). The ketogenic diet inhibits epileptogenesis in EL mice: a genetic model for idiopathic epilepsy. *Epilepsia*, 41:933–940.

Trauner, D.A. (1985). Medium-chain triglyceride (MCT) diet in intractable seizure disorders. *Neurology*, 35:237–238.

Uauy, R., P. Peirano, D. Hoffman, P. Mena, D. Birch, E. Birch. (1996). Role of essential fatty acids in the function of the developing nervous system. *Lipids*, 31(Suppl.):S167–S176.

Uhlemann, E.R., A.H. Neims. (1972). Anticonvulsant properties of the ketogenic diet in mice. *J. Pharm. Exp. Therap.*, 180:231–238.

Vaisleib, I.I., J.R. Buchhalter, M.L. Zupanc. (2004). Ketogenic diet: outpatient initiation, without fluid, or caloric restrictions. *Pediatr. Neurol.*, 31:198–202.

Vining, E.P.G. (1999). Clinical efficacy of the ketogenic diet. *Epilepsy Res.*, 37:181–190.

Vining, E.P.G., J.M. Freeman, K. Ballaban-Gil, C.S. Camfield, P.R. Camfield, G.L. Holmes, S. Shinnar, R. Shuman, E. Trevathan, J.W. Wheless. (1998). A multicenter study of the efficacy of the ketogenic diet. *Arch. Neurol.*, 55:1433–1437.

Vreugdenhil, M., C. Bruehl, R.A. Voskuyl, J.X. Kang, A. Leaf, W.J. Wadman. (1996). Polyunsaturated fatty acids modulate sodium and calcium currents in CA1 neurons. *Proc. Nat. Acad. Sci. U.S.A.*, 93:12559–12563.

Wexler, I.D., S.G. Hemalatha, J. McConnell, N.R. Buist, H.H. Dahl, S.A. Berry, S.D. Cederbaum, M.S. Patel, D.S. Kerr. (1997). Outcome of pyruvate dehydrogenase deficiency treated with ketogenic diets. *Neurology*, 49:1655–1661.

Wheless, J.W. (1995). The ketogenic diet: fa(c)t or fiction. *J. Child Neurol.*, 10:419–423.

Wheless, J.W. (2001). The ketogenic diet: an effective medical therapy with side effects. *J. Child Neurol.*, 16:633–635.

Wheless, J.W. (2004). History and origin of the ketogenic diet. In *Epilepsy and the Ketogenic Diet* (pp. 31–50), C.E. Stafstrom and J.M. Rho, Eds. Totowa, NJ: Humana Press.

Wijburg, F.A, P.G. Barth, L.A. Bindoff, M.A. Birch-Machin, J.F. van der Blij, W. Ruitenbeek, D.M. Turnbull, R.B. Schutgens. (1992). Leigh syndrome associated with a deficiency of the pyruvate dehydrogenase complex: results of treatment with a ketogenic diet. *Neuropediatrics*, 23:147–152.

Wilder, R.M. (1921). The effects of ketonemia on the course of epilepsy. *Mayo Clin. Proc.*, 2:307–308.

Williamson, A., P.R. Patrylo, J. Pan, D.D. Spencer, H. Hetherington. (2005). Correlations between granule cell physiology and bioenergetics in human temporal lobe epilepsy. *Brain*, 128(Pt. 5):1199–1208.

Willis, S., R. Samala, T.A. Rosenberger, K. Borges. (2009). Eicosapentaenoic and docosahexaenoic acids are not anticonvulsant or neuroprotective in acute mouse seizure models. *Epilepsia*, 50:138–142.

Wing, R., J. Vazquez, C. Ryan. (1995). Cognitive effects of ketogenic weight-reducing diets. *Int. J. Obesity*, 19:811–816.

Xiao, Y.-F., X. Li. (1999). Polyunsaturated fatty acids modify mouse hippocampal neuronal excitability during excitotoxic or convulsant stimulation. *Brain Res.*, 846:112–121.

Yudkoff, M., Y. Daikhin, T.M. Melo, I. Nissim, U. Sonnewald, I. Nissim. (2007). The ketogenic diet and brain metabolism of amino acids: relationship to the anticonvulsant effect. *Annu. Rev. Nutr.*, 27:415–430.

Yudkoff, M., Y. Daikhin, I. Nissim, A. Lazarow, I. Nissim. (2001). Ketogenic diet, amino acid metabolism, and seizure control. *J. Neurosci. Res.*, 66:931–940.

Yuen, A.W., J.W. Sander, D. Fluegel, P.N. Patsalos, G.S. Bell, T. Johnson, M.J. Koepp. (2005). Omega-3 fatty acid supplementation in patients with chronic epilepsy: a randomized trial. *Epilepsy Behav.*, 7:253–258.

Zhao, Q., C.E. Stafstrom, D.D. Fu, Y. Hu, G.L. Holmes. (2004). Detrimental effects of the ketogenic diet on cognitive function in rats. *Pediatr. Res.*, 55:498–506.

# 25 Vagus Nerve Stimulation

*Matthew M. Troester and Dean K. Naritoku*

## CONTENTS

## INTRODUCTION

The development of vagus nerve stimulation (VNS) for the treatment of epilepsy has demonstrated well the translation of concepts gained through laboratory research into a clinical therapy. Information gained from clinical research has, in turn, generated important basic research questions resulting in new directions for laboratory studies. In clinical practice, trains of current are applied intermittently to the left vagus nerve using a pacemaker-like device, the NeuroCybernetic Prosthesis (NCP; Cyberonics, Inc., Houston, TX). Randomized, double-blind controlled trials have established its efficacy for partial-onset seizures in human epilepsy, and open-label studies have suggested good efficacy for other seizure types. As will be discussed, laboratory studies have also indicated that modulation of the vagal system may be a therapeutic modality for other neurobiological disorders, including depression, pain, obesity, and memory loss.

## BACKGROUND

As of the summer of 2009, approximately 55,000 vagus nerve stimulators had been implanted in the United States, most (~51,000) as an adjunctive therapy for treatment-resistant epilepsy. The remainder have been implanted in patients with treatment-resistant depression (D. Wallace, pers. comm.). Although the first vagus nerve stimulator (the NCP) was implanted in humans in the late 1980s, the notion of treating epilepsy with some sort of peripheral stimulation has been around since antiquity. Traditional Chinese medicine teachings noted that the pressure of a thumbnail applied to a patient's philtrum may arrest seizures. Legends about physicians applying ligatures around affected limbs during a seizure (to abort the seizure or to perhaps prevent it) abound, and famous historical neurologists such as Brown-Séquard and Gowers thought them to be effective (Gowers, 1881; Odier, 1811). Gowers (1881) also believed that noxious stimuli could prevent seizures from spreading. Penry and Dreifuss (1969a,b) later described how sensory input could change the behavioral manifestations of absence seizures and noted that these seizures could be arrested by tactile stimulation.

Stimulation of the vagus nerve as a treatment modality for seizures was also described in the late 1800s by a lesser known American neurologist, Dr. James Corning, who believed that aberrant excessive blood flow was causal for seizures. He thought manual compression of the carotid artery

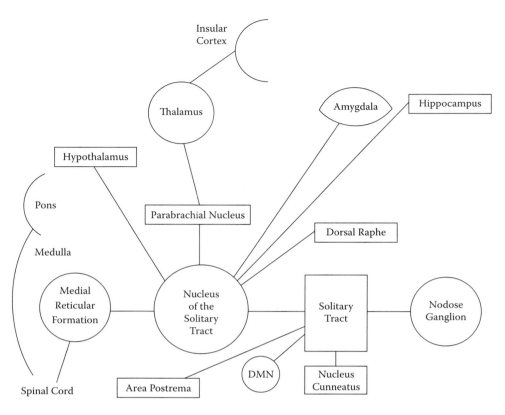

**FIGURE 25.1**    Schematic of efferent projections from the vagal system to the forebrain. The solitary nucleus is the primary afferent nucleus of the vagus nerve. It has widespread projections to the forebrain, either directly or indirectly through thalamic and brainstem monoaminergic nuclei. (From Rutecki, P., *Epilepsia*, 31(Suppl. 2), S1–S6, 1990. With permission.)

would mend this problem by decreasing the heart rate and thus the aberrant blood flow. He developed several devices (one that externally stimulated the vagus nerve with an electrical current) in an attempt to treat epilepsy and reported that they were effective (Corning, 1883; Lanska, 2002).

The vagus nerve is a large cranial nerve with nuclei in the medulla of the brain. It has a wide range of functions, including the transport of somatic and visceral afferent information (approximately 80% of vagus nerve fibers) from the heart, aorta, lungs, and gastrointestinal tract. Motor and visceral efferent information, which controls phonation, cardiac autonomic regulation, gastric function, and somatic and visceral sensation, comprises the other 20% of vagus nerve fibers. Although the vagal system is regarded as a brainstem system, it projects widely to the forebrain (Figure 25.1), suggesting that it has widespread influence on forebrain function. Of relevance to epilepsy and other neurological disorders are the direct projections from the solitary nucleus to the limbic system, via the amygdala, and to other regulatory structures in the brainstem, including the serotonergic (raphé nucleus) and noradrenergic (locus coeruleus) nuclei (Foley and DuBois, 1937; Rutecki, 1990; Schachter and Saper, 1998).

Neurophysiologic evidence that VNS may affect cortical function may be found in literature dating as far back as late 1930s, when Bailey and Bremer (1938) noted changes in the electrocorticogram of anesthetized cats upon stimulation of severed vagal nerve. They reported that in the cat VNS elicited synchronized activity in the orbital cortex. These findings were the first suggestion of a direct VNS influence on central nervous system function; Corning (1883) had used VNS for its indirect physiological effects. Although the electrographic changes in the cat were attributed to

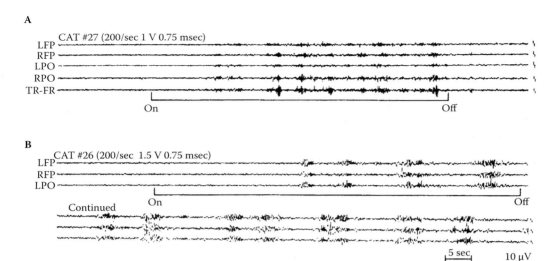

**FIGURE 25.2** EEG changes with vagus nerve stimulation. The tracings are from an anesthetized cat receiving vagal stimulation. Chase et al. (1967) demonstrated that in this preparation either hypersynchronization or desynchronization could be elicited, depending on the voltage and current used.

VNS-induced hypotension, Zanchetti et al. (1952) demonstrated this phenomenon could occur in the absence of systemic changes in cats who received spinal de-efferentation. Chase et al. (1966, 1967) demonstrated that different currents and frequencies could cause either electroencephalography (EEG) synchronization or desynchronization in anesthetized cats (Figure 25.2). Zabara (1992) postulated that, in light of the ability of VNS to desynchronize EEG activity, it could also arrest seizures, as they are presumably caused by hypersynchronous firing of neuronal populations. His initial studies demonstrated dramatic VNS-induced arrest of convulsive activity in canine seizures induced by strychnine (Figure 25.3). Later studies, however, found no acute desynchronization of electrocerebral activity during VNS in *awake* and freely moving rodents (McLachlan, 1993; Takaya et al., 1994) or during clinical VNS in humans (Hammond et al., 1992a). It is therefore possible that the EEG changes described in anesthetized animals may be dependent on anesthesia-induced unconsciousness and VNS-induced arousal.

A small study of the effects on primates treated with topical brain convulsants attempted to look at long-term effects of VNS on seizures (Lockard et al., 1990). This study suggested that VNS could suppress both ictal and interictal EEG abnormalities. Although the study was not of sufficient size to draw conclusions about efficacy, it was able to demonstrate the safety of chronic electrical stimulation. Long-term observational studies in persons with epilepsy and abnormal slowing of EEG background have shown improvement in the frequency of spikes and reduction in slowing in responders (Koo, 2001). Other authors have shown that VNS decreases interictal discharges in the short-term and may be an early predictive sign of VNS success (Kuba et al., 2002). More recent investigations have shown a progressive decrease in interictal discharges over 36 months in eight patients post vagus nerve stimulator placement (Wang et al., 2009).

Antiepileptic drug development has traditionally screened for potential therapies against standardized seizure models, which include administration of pentylenetetrazol (PTZ) or maximal electroshock (MES) (Krall et al., 1978) to mice or rats. Threshold convulsant doses of PTZ induce facial and forelimb convulsive seizures in rodents that are likely limbic in onset, whereas MES induces tonic seizures with hindlimb extension that probably originate in the brainstem. Woodbury and Woodbury (1990, 1991) demonstrated the ability of VNS to attenuate PTZ- and MES-induced seizures. Interestingly, when VNS was administered early in the PTZ-induced seizure, marked

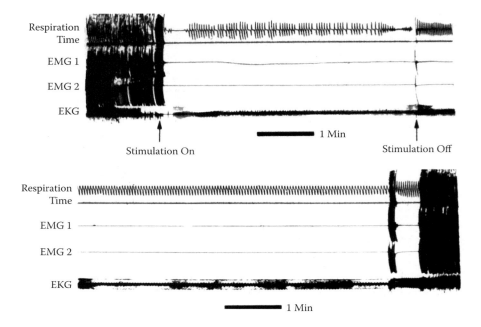

**FIGURE 25.3**    Abortion of strychnine-induced convulsions by vagus nerve stimulation. Zabara (1992) demonstrated the abortion of convulsive behavior using vagus nerve stimulation. The tracings show motor (EMG) patterns induced by strychnine. Note the abrupt arrest of convulsive activity with application of VNS.

abortive effects were observed, but if the stimulus was presented more than a few seconds after onset there was little effect on total seizure duration. This suggested that the abortive effect was most effective if presented early. Anecdotal reports from patients receiving VNS, as well as analysis of self-reported data from clinical trials (Naritoku et al., 1992b), also suggest that seizures may be terminated more readily if given early during the seizure.

Other basic research efforts have examined extensively the role of the vagal afferent system in modulating neural systems that have strong implications for its use in other neurologic disorders. VNS acutely attenuates nociceptive pain in animals (Randich and Gebhart, 1992) and appears to inhibit neuronal firing in at least some structures, including the nucleus of the trigeminal nerve (Bossut and Maixner, 1996; Fromm and Sun, 1993). An *in vivo* study demonstrated increases in slow afterhyperpolarization in cortical pyramidal cells of anesthetized rats (Zagon and Kemeny, 2000); however, the effect was maximal with very small stimulation current of 0.1 mA and declined with even slightly higher current of 0.5 mA, raising the question of whether it is a clinically relevant phenomenon.

The vagus nerve appears to mediate the effects of peripherally acting agents that modulate memory consolidation. In rats, functional inactivation or severing the vagus nerve blocks enhancement of memory consolidation normally induced by peripherally acting catecholamines, and electrical stimulation of the vagus nerve improves memory consolidation in rats given tasks (Clark et al., 1995). This effect appears to be mediated by the afferent vagal system, as inactivation of the vagus nerve proximal to the point of stimulation abolishes this effect (Clark et al., 1998).

## TRANSLATION TO CLINICAL PRACTICE

In translating laboratory findings to the clinical setting, some questions cannot be answered in advance and require a leap of faith in order to establish the proof of concept in humans. Key factors include the reliability of the animal model in predicting a response to the new therapy and conversion of dosing to humans. Physiological differences between species may result in differences in

| Model 100 | Model 101 | Model 102/102R | Model 103/104 |
|---|---|---|---|
| Thickness: 0.52" (13.2 mm) | 0.41" (10.3 mm) | 0.27" (6.9 mm) | 0.27" (6.9 mm) |
| Volume: 31 cc | 26 cc | 14/16 cc | 8 cc/10 cc |

**FIGURE 25.4** Human vagus nerve stimulator. Shown is the evolution of the commercially available stimulators approved by the FDA for human use. The earliest device is on the left, and the most recent device is pictured at the right. The device is a pacemaker-like object, with attached leads and spiral electrodes that attach to the vagus nerve. (Photograph courtesy of Cyberonics, Inc.; Houston, TX.)

efficacy and toxicity. Finally, interspecies differences raise questions of unseen adverse effects and safety issues. Thus, translation to a human study requires an initial best guess about the dosing or protocol that will demonstrate efficacy with minimal or acceptable adverse effects and allow resolution of outcomes between treatment groups. Unfortunately, the great expense associated with clinical studies, as well as limitations in research capital, often prohibit additional studies required to determine optimal intervention in humans.

At the time of initial human testing, the abortive effects of VNS on the acute seizure models of PTZ and MES were well known, but effects of VNS on other models of epilepsy were not established. An ongoing major criticism of PTZ and MES is that they produce symptomatic seizures, rather than true, spontaneous seizures that define the epileptic state, and the "goodness" of animal models of epilepsy has been the subject of review (Loscher, 2002). However, the utility of simple models such as PTZ and MES cannot be disputed. These screening methods boast a long track record of identifying successful antiepileptic drugs, thus validating their utility in predicting an antiepileptic effect.

Perhaps unique in medical device development, VNS therapy was subjected to rigorous clinical testing with blinded, randomized, controlled clinical trials utilized in modern antiepileptic drug development. An initial open-label trial of VNS for epilepsy was followed by two large multicenter randomized, double-blind active controlled trials for persons with medically intractable epilepsy with partial-onset seizures.

Although it was clear that VNS exerted an antiseizure effect in experimental models, it was not known what parameters would exert a therapeutic effect in human epilepsy and which parameters would be optimal. Animal studies focused on the abortive effects of VNS in acutely induced seizures. Thus, both induction of experimental seizures and timing of VNS were under direct control of the investigator. In clinical practice, accurate prediction of seizures to provide a closed-loop system was impractical with commercially available technology, thus prohibiting automatic application of the stimulus at seizure onset. The NCP, a commercial pacemaker-like device developed for VNS, was designed to provide intermittent trains of stimulation at programmable preset intervals (Figure 25.4). Other practical considerations included the duration of the nerve–electrode interface. Long-term loss of the interface, along with the potential difficulty in replacing the device, raised the question of whether successful therapy would be limited for mechanical reasons. Subsequent studies have shown a lack of nerve changes at postmortem examination (Cyberonics, data on file) and during electrode replacement (Espinosa et al., 1999).

Interestingly, currently established stimulation parameters in humans were selected primarily for practical reasons of patient comfort, safety of nerve stimulation, and battery life. Higher frequencies of 60 to 100 Hz were not tolerated in pilot clinical trials due to severe pain. More importantly, continuous chronic stimulation could lead to permanent nerve damage, and it was estimated that no more than a 50% duty cycle could be given using the currents and frequencies proposed (Agnew and McCreery, 1990). The high current consumption needed for continuous stimulation would make an implantable device less practical, because of the need for frequent battery replacement. Thus, the decision to provide lower frequency, intermittent stimulation on a fixed cycle addressed these pragmatic issues but represented a potential compromise of efficacy. At the time of clinical trial initiation, it was unclear if VNS would have an anticonvulsant effect if it did not occur during the time of the seizure. The stimulator device was designed so patients could trigger a stimulus on demand; unfortunately, in practice only a minority of patients can trigger the device before loss of consciousness.

The first descriptions of the implantable VNS Therapy™ appeared in the literature in the early 1990s (Terry et al., 1990, 1991). Around the same time, the initial results of the so-called EO1 and EO2 studies—both Phase I, single-blinded clinical trials—were released (Penry and Dean, 1990; Uthman et al., 1990; Wilder et al., 1991). These studies reported a reduction in seizure frequency and in some patients a reduction in severity. There was also a lag in the treatment effect noted. What followed were two randomized, "placebo"-controlled, double-blind studies: EO3 (an international multicenter trial) and EO5 (a multicenter trial in the United States). The word "placebo" is in quotes because some traditional thinkers did not feel the placebo group was really given "sham" stimulation. These double-blind VNS trials for complex partial epilepsy used intermittent electrical stimulation with 30-second trains of current: (1) high stimulation—30 Hz, 500-μsec pulse width, 5 minutes between trains, maximally tolerated current; and (2) low stimulation—1 Hz, 130-μsec pulse width, 180 minutes between trains, current set to perceived threshold. As patients can sense the stimulation, some degree of current stimulation had to be provided to the placebo, or low group, to protect the blinding. These low stimulations were believed to be less efficacious than the high group and to thus represent the best control possible in this clinical setting. Self-triggering of the device using a magnet was allowed in the high group to provide VNS on demand, whereas no stimulation occurred with triggering in the low group. This demonstrated a significantly greater reduction of seizures in the high group.

In both studies, the response rate (i.e., the percentage of patients achieving at least a 50% reduction in seizure frequency) was approximately 35% in patients receiving the higher stimulation parameters during the blinded phase (first 3 months) (Ben-Menachem et al., 1994; Handforth et al., 1998). Continued therapy during the open-label extension demonstrated a response in approximately 40% of the patients (35% at 1 year and up to 44% at 2 years, after which it reached a plateau). Furthermore, a steady improvement was seen over the following 12 months, suggesting a cumulative effect of VNS (DeGiorgio et al., 2000).

Other studies have followed, including a Swedish series showing a decrease in seizure frequency of again around 40% in patients treated up to 5 years (Ben-Menachem et al., 1999). Consistently, a Chinese study demonstrated an approximate 40% reduction in seizures after 18 months (Hui et al., 2004). An international study with an average follow-up of 33 months found a mean reduction in monthly seizure frequency of 55% (Vonck et al., 2004). A recent Czech study reviewed 90 patients after 5 years of VNS and found that at the last follow-up visit 64.4% of patients were responders; again, a cumulative effect was noted, with 44% responders at 1 year and 58.7% at 2 years. This group appears to lack the plateau effect seen in other study groups after 2 years (Kuba et al., 2009).

Although true prospective pediatric data are lacking, a retrospective review from the Cleveland Clinic also revealed a progressive seizure reduction of up to 74% at 36 months (Alexopoulos et al., 2006). A small retrospective series of VNS Therapy™ devices implanted in children less than 5 years of age (mean age at implantation, 20.5 months) revealed that most experienced a decrease in seizure frequency, and no patient became worse, suggesting some efficacy and safety in this fragile age group as well (Blount et al., 2006).

Most of the aforementioned data relate to those patients with localization-related epilepsy (e.g., complex partial seizures); however, retrospective studies and case series suggest VNS may also be useful for other seizure types and epilepsy syndromes. VNS is apparently effective for

Lennox–Gastaut syndrome (LGS) (Handforth et al., 1998; Hosain et al., 2000), which is characterized by poor response to medical therapy, multiple seizure types, and intellectual decline. Because of its benign nature, VNS has largely supplanted corpus callosotomy (in particular, for the "drop attack" type of seizure associated with LGS) as a surgical adjunct to medical therapy, although their efficacy is similar (Nei et al., 2006; You et al., 2008). Other series suggest efficacy against primarily generalized seizures. The EO4 study included 24 patients with generalized epilepsies. These patients had a seizure reduction of up to 46%, and generalized tonic seizures seemed most responsive (Labar et al., 1999). Several other authors have retrospectively described efficacy similar to localization-related epilepsy in treatment-resistant generalized epilepsy (Kostov et al., 2007; Ng and Devinsky, 2004; Quintana et al., 2002). In a prospective series out of Seattle, 16 generalized epilepsy patients showed a mean 43.3% reduction is seizure frequency (Holmes et al., 2004).

Overall efficacy may be better in patients with less intractable epilepsy. A postmarketing patient registry maintained by the manufacturer suggests that the response rate may be 50 to 60% in open-label clinical use (Labar, 2002). However, the possibility that differences in data acquisition in an open registry could be responsible for the improved numbers cannot be excluded.

Established functions of the vagal system initially raised apprehensions about the safety of VNS in humans. In rats, there is a robust slowing of heart rate with stimulation of the left vagus nerve, which results in a relative hypotension. The first stimulation of human vagus nerve was performed with full resuscitation equipment on standby in the operating room but was accomplished without complications. Surprisingly, no patients in either of the pivotal trials or in open-label trials experienced significant bradycardia or hypotension (Ramsay et al., 1994). During the second pivotal study, prolonged electrocardiogram (EKG) recording in humans using Holter monitoring was performed on each patient prior to and during VNS stimulation and showed a lack of EKG changes. However, in postmarketing experience, rare instances of cardiac asystole have been described (Asconape et al., 1999).

Other concerns included the potential for increased gastric acid secretion. Indeed, prior to the approval of histamine blockers for gastrointestinal ulcers, left vagotomy was one of the most commonly performed surgical procedures. Direct measurement of gastric acid production was performed on a small number of patients and showed no increases in acid production. Serum gastrin was measured in all patients in the second pivotal study, and no significant changes occurred (Handforth et al., 1998). These examples of lack of VNS-induced cardiac or gastric abnormalities underscore the unpredictable nature of translational medicine; our understanding of well-established physiologic mechanisms may not hold true in clinical applications, whether they are desired outcomes or adverse effects.

That is not to say, however, that VNS is without adverse effects. The most common complaints are a tingling sensation in the throat (thought to be due to secondary stimulation of the superior laryngeal nerve) and stimulation-related hoarseness thought to be due to stimulation of the recurrent laryngeal nerve (Charous et al., 1993; Claes and Jaco, 1986). Most common pediatric complaints include drooling (Pearl et al., 2008). VNS has been implicated for a multitude of other adverse effects, including Horner syndrome (Kim et al., 2001), epilepsy exacerbation (Koutroumanidis et al., 2000), psychosis (Gatzonis et al., 2000), unintentional adjustment of a magnetically controlled ventriculoperitoneal shunt valve (Guilfoyle et al., 2007), and chronic diarrhea (Sanossian and Haut, 2002). There is even a report of the VNS Therapy™ stimulator line being misidentified as intravenous access; there was no immediate damage to the stimulator, but it required replacement a year later due to infiltration of scar tissue into the wire (Pearl et al., 2008). VNS has also been implicated as causing obstructive sleep apnea (OSA) and as being an exacerbating factor of existing OSA, as well as directly causing difficulty with continuous positive airways pressure titration in the treatment of OSA (Ebben et al., 2008). That said, this is an extremely well-tolerated device as evidenced by the EO5 study in which 99% of patients reached study endpoints (Handforth et al., 1998).

## TRANSLATION TO THE LABORATORY

The demonstrated efficacy of VNS against partial-onset seizure in humans indicated that VNS alters seizure genesis in forebrain structures. This finding immediately raised the question of which brain structures and mechanisms mediate the anticonvulsant effects of the VNS. As previously discussed, few data existed to verify an electrophysiological effect in human forebrain. Although pretranslational work suggested EEG desynchronization as a possible VNS mechanism (Chase et al., 1967), VNS-induced EEG desynchrony was not initially demonstrated (Hammond et al., 1992a). In spite of this finding, other authors have hypothesized or suggested that EEG desynchronization is a mechanism of action for the antiseizure effect of VNS (Jaseja, 2004; Nemeroff et al., 2006). Marrosu et al. (2005) not only suggested the desynchronization notion but also further observed that VNS decreases synchronization in the theta frequency and increases the gamma power spectrum, ultimately concluding that the desynchronization coupled with an increase in gamma bands may be anticonvulsant.

Later studies designed to confirm the notion that VNS also evokes a cerebral response failed to reveal a VNS-evoked potential (Hammond et al., 1992b). Naritoku et al. (1992a) found modest prolongation of the subcortical–cortical (N20) potential in evoked potentials elicited from stimulation of median nerve in patients receiving VNS, suggesting a VNS-induced change in somatosensory neurotransmission, but visual and brainstem auditory evoked potentials were unchanged (Hammond et al., 1992b; Naritoku et al., 1992a). Later studies would show a reproducible VNS-evoked cerebral response, a so-called *vagus sensory evoked potential* (VSEP), suggesting that VNS can not only alter brain activity (as shown in above described work) but also evoke activity in particular brain regions (Fallgatter et al., 2003).

Other authors have suggested that the VNS effect has more to do with alteration of neurotransmitters such as γ-aminobutyric acid (GABA). Ben-Menachem et al. (1995) studied spinal fluid metabolites after chronic VNS and found an increase in GABA levels during VNS. This finding suggested that increased concentrations of GABA (the main central nervous system inhibitory neurotransmitter) may contribute to the VNS mechanism of action. Marrosu et al. (2003) used single photon emission computed tomography (SPECT) scans before and after VNS implantation to further explore VNS and GABA interactions. Based on the theory that concentrations of GABA are reduced in pathological specimens from individuals undergoing resective epilepsy surgery, these investigators found that the $GABA_A$ receptor density normalized after a year of chronic VNS and that this observation correlated with therapeutic response. Using transcranial magnetic stimulation, Di Lazzaro et al. (2004) further found that short-latency intracortical inhibition (SICI) is increased after 6 months of chronic intermittent VNS. This result failed to reach clinical significance ($p = 0.016$), but, because SICI is thought to represent $GABA_A$ intracortical inhibitory mechanisms, this may be an area for future mechanistic exploration of the VNS.

Interestingly, further evidence that VNS has a favorable influence on GABA was found in rats undergoing VNS after experimental brain injury; the GABAergic cortical cells of those rats exposed to VNS after the injury were protected from death (Neese et al., 2007). Expanding upon the theory that VNS affects GABAergic neurons, Valtschanoff et al. (1993) proposed that they contain nitric oxide synthase (NOS) and that release of GABA in the cortex is under some type of nitric oxide (NO)-dependent control (Ohkuma et al., 1996). Lyubashina and Panteleev (2009) later set out to determine whether the antiseizure effect of VNS is NO dependent. In a Wistar rat model, they demonstrated the antiseizure effect of VNS, as well as the reversal of this effect with inhibition of NOS. These results indicate that the effect of VNS may be partly dependent upon the presence of NO which may lead to GABA release from VNS activation of local NO-ergic mechanisms. Again, this assumption needs to be confirmed in other animal models of epilepsy.

Naritoku et al. (1995) identified VNS-induced changes in neural transmission in the forebrain by measuring *fos* expression in rats receiving VNS with parameters used for epilepsy therapy. In this study, VNS produced *fos* immunolabeling in amygdala, hypothalamus, cingulate cortex, and hypothalamus in the forebrain, suggesting increases in neuronal activity in these structures. In the hindbrain, labeling of the trigeminal sensory nucleus was observed, as well as the A5 nucleus and locus coeruleus, suggesting activation of the noradrenergic system. Although the role of monoamines in epilepsy remains unclear, both serotonin and norepinephrine regulate seizure thresholds in animal models (Jobe et al., 1986, 1999). The production of *fos*, which is an early-immediate gene product that promotes further gene transcription, raises the question of whether it may mediate sustained and long-term effects of VNS.

Metabolic mapping studies of the brain have yielded differing results that likely reflect the conditions under which the studies were performed. Autoradiographic studies in rat using radiolabeled 2-deoxyglucose compared sham controls and stimulated anesthetized rats receiving continuous stimulation for 1 hour using anticonvulsant parameters (Terry et al., 1996). Interestingly, there were no areas of increased uptake. VNS induced no changes in several brain regions; specifically, no reductions of glucose uptake were seen in several structures, most notably frontal cortex, hippocampus, and thalamus. In the brainstem, however, the reductions of glucose uptake observed in the solitary nucleus may be consistent with observations of Easton and colleagues (1996) that *inhibition* of the solitary nucleus with microinfusions of drugs is anticonvulsant.

An early positron emission tomography (PET) study by Ko et al. (1996) utilized patients who had already received VNS for several months. During acute VNS, they observed increases in blood flow in the ipsilateral thalamus, contralateral parietal cortex, and ipsilateral cerebellum when compared to interstimulation states; however, the design of the study did not account for long-term or sustained changes incurred by chronic VNS therapy. Another PET study by Henry et al. (1999) compared VNS-treated patients to their untreated baseline. Several areas of blood flow changes occurred during stimulation, but, most importantly, increases in thalamic blood flow correlated with VNS efficacy, suggesting that modulation of thalamic activity may be crucial to VNS effect (Henry et al., 1999). Interstimulation changes that may have occurred as a consequence of cumulative effects were not measured. A blood-oxygen-level-dependent functional magnetic resonance imaging (BOLD fMRI) study of five patients showed consistent activation in the frontal and occipital lobes and in the thalamus of the two patients with greatest seizure control. This suggested that BOLD fMRI activation can indeed be detected and that thalamic activation may be favorable (Liu et al., 2003).

The PET data from patients with treatment-resistant depression suggest that a decrease in glucose metabolism in the ventromedial prefrontal cortex appears to correlate with treatment response to VNS after one year of chronic stimulation in eight patients; no thalamic metabolism changes were noted (Pardo et al., 2008). The clinical data also raised important questions about the temporal aspects of VNS with respect to seizure control. By using intermittent 30-second stimulation at fixed intervals of 5 minutes, the probability that the stimulus would occur at the onset of a seizure (30:300, or 10%) was lower than the seizure efficacy observed. Although manual triggering of the device was possible, most patients did not routinely trigger the device. This raised the question of whether VNS exerts preventative (or antiepileptogenic), in addition to abortive, antiseizure effects.

The potential for sustained effects of VNS was addressed in laboratory studies by Takaya et al. (1996) utilizing the rodent PTZ model. Animals were pretreated with VNS, and stimulation was discontinued *prior* to PTZ administration (Figure 25.5). The findings verified that the anticonvulsant effects were not limited to the duration of stimulation; that is, VNS exhibited potential antiepileptogenic effects. Increasing the duration of VNS stimulation caused progressive increases in seizure latency, with reductions in total number of seizures, seizure duration, and seizure severity. These findings indicated that there may be a cumulative effect of VNS, at least under acute conditions. This notion is attractive in light of long-term findings in clinical trials that show progressive improvement over the first year of treatment.

Vagus nerve stimulation appears to retard electrical kindling of the amygdala in rodents and felines, further suggesting that it may have antiepileptogenic properties. Naritoku and Mikels (1997) demonstrated that the number of amygdalar stimulations required to reach Class 5 seizures was increased when VNS was administered prior to each kindling session. Furthermore, the threshold

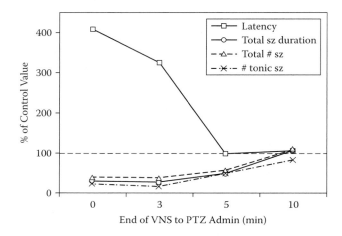

**FIGURE 25.5** Effects of vagus nerve stimulation duration on seizures. Pretreatment with VNS prior to pentylenetetrazol can block the induction of seizures, providing evidence for a sustained effect that persists after the end of stimulation. Shown in the diagram is the decay of the effect on seizure parameters. After 1 hour of VNS, the animals were treated with pentylenetetrazol immediately or 3 minutes, 5 minutes, or 10 minutes after the end of VNS. Note the return to control values by 10 minutes, suggesting that after an acute stimulation the effects diminish over that time period. (From Takaya, M. et al., *Epilepsia*, 37, 1111–1116, 1996. With permission.)

current required to trigger an afterdischarge was significantly reduced in control animals, but not in animals treated with VNS, even after reaching Class 5 seizures. VNS also reduced the duration of Class 5 seizures once kindling had already been achieved in rodents (Naritoku and Mikels, 1996). Fernandez-Guardiola et al. (1999) suggested that the effects of VNS on kindling, as well as its anticonvulsant effect, may relate to changes in sleep architecture. Malow et al. (2001) reported changes in sleep architecture in patients receiving VNS; however, changes in sleep cycles would not explain the efficacy of VNS on acutely induced seizures and therefore cannot explain fully the anticonvulsant and antiepileptogenic effect of the VNS. Finally, the fact that seizures recur after the battery has failed is a strong argument against an antiepileptogenic effect, and a single case report of long-lasting seizure control after VNS device removal (Labar and Ponticello, 2003) does not add much to the antiepileptogenic argument, either.

Laboratory studies have also focused on the potential modulation of the VNS effect by the mono-aminergic system. Krahl et al. (1998) found that either destroying the serotonergic nuclei or lesioning the locus coeruleus with focal infusions of 6-hydroxydopamine could reverse the effects of VNS on MES-induced seizures. The activation of monoaminergic systems by VNS may be an especially important mechanism for clinical application in depression and memory disorders.

There is a notion that stress and depression can lead to atrophy and cell loss in limbic brain structures and that antidepressant medications may reverse some of this atrophy and cell loss. This is the so-called neurotrophic hypothesis characterizing depression which others have termed a "neurogenesis" theory (Duman and Monteggia, 2006). Follesa et al. (2007) found that chronic VNS in rat brains increased the expression of brain-derived neurotrophic factor (BDNF) and fibroblast growth factor (FGF) in the hippocampus and cerebral cortex, and increased the concentration of norepinephrine in the prefrontal cortex. These findings suggest that VNS may indeed alter neurogenesis where these growth factors reverse the presumed depressive pathology. Others have found that short-term (i.e., 2 days) VNS treatment can create an increase of available progenitor cells in the adult rat dentate gyrus, suggesting that this may be a mechanism for stimulated neurogenesis which may be a part of the recovery process from the depressive state (Revesz et al., 2008).

Along similar lines is the notion that VNS affects long-term potentiation (LTP), thought to be a useful model of synaptic plasticity. In the hippocampus, LTP may represent the synaptic plasticity that underlies memory formation. Zuo et al. (2007) demonstrated that VNS exposure in rats at a moderate current or intensity facilitated induction of hippocampal LTP. They hypothesized that the VNS may modulate hippocampal memory-related plasticity primarily through neural messages conveyed by vagal afferents to the brain (Zuo et al., 2007). Future studies would require a demonstration of both improved memory function in the intact animal as well as the electrophysiological findings.

## AREAS TO BE EXPLORED IN LABORATORY AND CLINICAL SETTINGS

Several potential areas for future research have already been mentioned above. Fundamental research into the mechanisms of VNS action is vital for its neurobiological implications. VNS is unique in that it represents the first approved and validated therapy for brain disorders that rests on modulation of large neuronal networks by electrical stimulation. Understanding how these networks are activated and regulated by stimulation will undoubtedly yield information that may allow further exploitation of the vagal network for other clinical applications. Ultimately, information gained from such studies would lead to the development of paradigms for studying and utilizing other neural networks, as well.

There is a natural tendency to compare and contrast the potential mechanisms of VNS with other forms of neural stimulation, most notably deep brain stimulation (DBS). In current practice, VNS differs greatly from DBS. First, the stimulation paradigms are quite different. In the case of VNS, efficacy can be demonstrated using relatively lower frequencies of 20 to 30 Hz. In contrast, DBS uses a higher frequency of 100 Hz. Stimulation at 100 Hz may potentially induce a functional blockade of the nucleus, although this is currently under debate. At least some neurons may be able to fire at this rate. A stimulation frequency of 30 Hz clearly does not cause a conduction block of the peripheral vagus nerve, nor would it be expected to cause a conduction block of the afferent nuclei.

Other major differences between VNS and DBS are the anatomic targets for stimulation. At least for the present time, targets for DBS have been limited to brain nuclei directly involved with the seizure network, as identified by functional imaging of seizures (Mirski and Ferrendelli, 1986). This includes structures such as the anterior nucleus (Mirski et al., 1997) or centromedian nucleus of the thalamus (Velasco et al., 1993). In contrast, the primary afferent system for VNS lies remotely from the sites of seizure genesis in forebrain. Presumably, VNS exerts its action by modulation of epileptic foci through large neuronal networks within the brain.

As with many therapies for epilepsy, VNS may have potential applications in other diseases of the central nervous system. An early open-label pilot study showed a marked improvement of Hamilton Rating Scale for Depression scores in patients with medically intractable depression (Sackeim et al., 2001). A subsequent randomized controlled trial did not show a significant improvement during short-term stimulation because of a high degree of variability but did show some a significant response after one year (Rush et al., 2005a,b). More recent open-label studies have also shown a reduced severity of depression that increases over time, much like the antiseizure effect (Schlaepfer et al., 2008).

The findings of improved memory consolidation following VNS have been extended to human studies. A small randomized, double-blind controlled study performed in patients receiving VNS for intractable epilepsy evaluated the effects on consolidation of word-recognition memory (Clark et al., 1999). VNS or sham stimulation was given to patients immediately after they read a block of words. There was a significant improvement in recall in patients receiving 0.5-mA stimulation, but not at higher or lower current levels, thus supporting laboratory findings of both VNS enhancement of memory and an inverted "U" function of response related to the intensity of current. Another study showed a VNS-induced decline, rather than improvement, in memory functions in patients with epilepsy (Helmstaedter et al., 2001); however, in this study, the mean stimulation current was much higher and thus consistent with intensity-dependent findings identified in animal and human studies. A small open-label trial of VNS in patients with Alzheimer's disease showed improvements in 7 of 10 patients on the Alzheimer's Disease Assessment Scale–Cognitive Subscale and Mini-Mental State Exam, suggesting the possible utility of VNS for this disorder (Sjogren et al., 2002). Overall, although animal models seem to show a clear-cut memory-enhancing effect after vagus nerve stimulation, human studies, while promising, still require further investigation (Boon et al., 2006).

The use of VNS for pain management remains to be explored. Although extensive laboratory data support the use of VNS for pain, at present only a few small reports suggest that VNS may be useful for pain syndromes. An earlier study that measured pain perception in patients with epilepsy demonstrated reduced nociception during VNS (Kirchner et al., 2000). A small retrospective case series in patients who had epilepsy with migraine indicated that VNS may reduce migraine headaches (Hord et al., 2003). A recently completed trial of the VNS in patients with rheumatoid arthritis is pending publication, and a trial regarding VNS and fibromyalgia remains currently active (www.clinicaltrials.gov).

In the patient series of VNS for treatment-resistant depression, a subset of patients with obesity and depression who experienced weight loss was noted (Pardo et al., 2007). Using noncervical VNS, Camilleri et al. (2009) have shown that intermittent, intraabdominal vagal blocking may be associated with weight loss.

## CONCLUSIONS

Vagus nerve stimulation represents an excellent example of parallel development and interchange between clinical and basic science. Its value to neurobiological research extends beyond current applications for neurologic and psychiatric disorders. Understanding the complex interactions between the brainstem vagal system and the forebrain will undoubtedly identify neural mechanisms that may be leveraged for many brain disorders.

## REFERENCES

Agnew, W.F., D.B. McCreery. (1990). Considerations for safety with chronically implanted nerve electrodes. *Epilepsia*, 31(Suppl. 2):S27–S32.

Alexopoulos, A.V., P. Kotagal, T. Loddenkemper, J. Hammel, W.E. Bingaman. (2006). Long-term results with vagus nerve stimulation in children with pharmacoresistant epilepsy. *Seizure*, 15:491–503.

Asconape, J.J., D.D. Moore, D.P. Zipes, L.M. Hartman, W.H. Duffell, Jr. (1999). Bradycardia and asystole with the use of vagus nerve stimulation for the treatment of epilepsy: a rare complication of intraoperative device testing. *Epilepsia*, 40:1452–1454.

Baily, P., F.A. Bremer. (1938). Sensory cortical representation of the vagus nerve. *J. Neurophysiol.*, 1:405–412.

Ben-Menachem, E., R. Manon-Espaillat, R. Ristanovic, B.J. Wilder, H. Stefan, W. Mirza, W.B. Tarver, J.F. Wernicke. (1994). Vagus nerve stimulation for treatment of partial seizures. 1. A controlled study of effect on seizures. First International Vagus Nerve Stimulation Study Group. *Epilepsia*, 35:616–626.

Ben-Menachem, E., A. Hamberger, T. Hedner, E.J. Hammond, B.M. Uthman, J. Slater et al. (1995). Effects of vagus nerve stimulation on amino acids and other metabolites in the CSF of patients with partial seizures. *Epilepsy Res.*, 20:221–227.

Ben-Menachem, E., K. Hellstrom, C. Waldton, L.E. Augustinsson. (1999). Evaluation of refractory epilepsy treated with vagus nerve stimulation for up to 5 years. *Neurology*, 52:1265–1267.

Blount, J.P., R.S. Tubbs, P. Kankirawatana, S. Kiel, R. Knowlton, P.A. Grabb, M. Bebin. (2006). Vagus nerve stimulation in children less than 5 years old. *Childs Nerv. Syst.*, 22:1167–1169.

Boon, P., I. Moors, V. De Herdt, K. Vonck. (2006). Vagus nerve stimulation and cognition. *Seizure*, 15:259–263.

Bossut, D.F., W. Maixner. (1996). Effects of cardiac vagal afferent electrostimulation on the responses of trigeminal and trigeminothalamic neurons to noxious orofacial stimulation. *Pain*, 65:101–109.

Camilleri, M., J. Toouli, M.F. Herrera, L. Kow, J.P. Pantoja, C.J. Billington, K.S. Tweden, R.R. Wilson, F.G. Moody. (2009). Selection of electrical algorithms to treat obesity with intermittent vagal block using an implantable medical device. *Surg. Obes. Relat. Dis.*, 5:224–230.

Charous, S.J., G. Kempster, E. Manders, R. Ristanovic. (2001). The effect of vagal nerve stimulation on voice. *Laryngoscope*, 111:2028–2031.

Chase, M.H., M.B. Sterman, C.D. Clemente. (1966). Cortical and subcortical patterns of response to afferent vagal stimulation. *Exp. Neurol.*, 16:36–49.

Chase, M.H., Y. Nakamura, C.D. Clemente, M.B. Sterman. (1967). Afferent vagal stimulation: neurographic correlates of induced EEG synchronization and desynchronization. *Brain Res.*, 5:236–249.

Claes, J., P. Jaco. (1986). The nervus vagus. *Acta Otorhinolaryngol. Belg.*, 40:215–241.

Clark, K.B., S.E. Krahl, D.C. Smith, R.A. Jensen. (1995). Post-training unilateral vagal stimulation enhances retention performance in the rat. *Neurobiol. Learn. Mem.*, 63(3):213–216.

Clark, K.B., D.C. Smith, D.L. Hassert, R.A. Browning, D.K. Naritoku, R.A. Jensen. (1998). Posttraining electrical stimulation of vagal afferents with concomitant vagal efferent inactivation enhances memory storage processes in the rat. *Neurobiol. Learn Mem.*, 70:364–373.

Clark, K.B., D.K. Naritoku, D.C. Smith, R.A. Browning, R.A. Jensen. (1999). Enhanced recognition memory following vagus nerve stimulation in human subjects. *Nat. Neurosci.*, 2:94–98.

Corning, J.L. (1883). Considerations on pathology and therapeutics of epilepsy. *J. Nerv. Ment. Dis.*, 10:243–248.

DeGiorgio, C.M., S.C. Schachter, A. Handforth, M. Salinsky, J. Thompson, B. Uthman et al. (2000). Prospective long-term study of vagus nerve stimulation for the treatment of refractory seizures. *Epilepsia*, 41:1195–200.

Di Lazzaro, V., A. Oliviero, F. Pilato, E. Saturno, M. Dileone, M. Meglio, G. Colicchio, C. Barba, F. Papacci, P.A. Tonali. (2004). Effects of vagus nerve stimulation on cortical excitability in epileptic patients. *Neurology*, 62:2310–2312.

Duman, R.S., L.M. Monteggia. (2006). A neurotrophic model for stress-related mood disorders. *Biol. Psychiatry*, 59:1116–1127.

Easton, A., B. Walker, K. Gale. (1996). Influence of activity in nucleus tractus solitarius (NTS) on seizure manifestations. *Epilepsia*, 37(S5):68.

Ebben, M.R., N.K. Sethi, M. Conte, C.P. Pollak, D. Labar. (2008). Vagus nerve stimulation, sleep apnea, and CPAP titration. *J. Clin. Sleep Med.*, 4:471–473.

Espinosa, J., M.T. Aiello, D.K. Naritoku. (1999). Revision and removal of stimulating electrodes following long-term therapy with the vagus nerve stimulator. *Surg. Neurol.*, 51:659–664.

Fallgatter, A.J., B. Neuhauser, M.J. Herrmann, A.C. Ehlis, A. Wagener, P. Scheuerpflug, K. Reiners, P. Riederer. (2003). Far field potentials from the brain stem after transcutaneous vagus nerve stimulation. *J. Neural. Transm.*, 110:1437–1443.

Fernandez-Guardiola, A., A. Martinez, A. Valdes-Cruz, V.M. Magdaleno-Madrigal, D. Martinez, R. Fernandez-Mas. (1999). Vagus nerve prolonged stimulation in cats: effects on epileptogenesis (amygdala electrical kindling): behavioral and electrographic changes. *Epilepsia*, 40:822–829.

Foley, J., F. DuBois. (1937). Quantitative studies of the vagus nerve in the cat. I. The ratio of sensor and motor fibres. *J. Comp. Neurol.*, 67:49–67.

Follesa, P., F. Biggio, G. Gorini, S. Caria, G. Talani, L. Dazzi, M. Puligheddu, F. Marrosu, G. Biggio. (2007). Vagus nerve stimulation increases norepinephrine concentration and the gene expression of BDNF and BFGF in the rat brain. *Brain Res.*, 1179:28–34.

Fromm, G.H., K. Sun. (1993). How does vagal nerve stimulation prevent seizures? *Epilepsia*, 34(S6):52.

Gatzonis, S.D., E. Stamboulis, Siafakas, E. Angelopoulos, N. Georgaculias, E. Sigounas, A. Jekins. (2000). Acute psychosis and EEG normalisation after vagus nerve stimulation. *J. Neurol. Neurosurg. Psychiatry*, 69:278–279.

Gowers, W.R. (1881). *Epilepsy and Other Chronic Convulsive Diseases*. London: J. & A. Churchill.

Guilfoyle, M.R., H. Fernandes, S. Price. (2007). *In vivo* alteration of strata valve setting by vagus nerve stimulator-activating magnet. *Br. J. Neurosurg.*, 21:41–42.

Hammond, E.J., B.M. Uthman, S.A. Reid, B.J. Wilder. (1992a). Electrophysiological studies of cervical vagus nerve stimulation in humans. I. EEG effects. *Epilepsia*, 33:1013–1020.

Hammond, E.J., B.M. Uthman, S.A. Reid, B.J. Wilder. (1992b). Electrophysiologic studies of cervical vagus nerve stimulation in humans. II. Evoked potentials. *Epilepsia*, 33:1021–1028.

Handforth, A., C.M. DeGiorgio, S.C. Schachter, B.M. Uthman, D.K. Naritoku, E.S. Tecoma et al. (1998). Vagus nerve stimulation therapy for partial-onset seizures: a randomized active-control trial. *Neurology*, 51:48–55.

Helmstaedter, C., C. Hoppe, C.E. Elger. (2001). Memory alterations during acute high-intensity vagus nerve stimulation. *Epilepsy Res.*, 47:37–42.

Henry, T.R., J.R. Votaw, P.B. Pennell, C.M. Epstein, R.A. Bakay, T.L. Faber, S.T. Grafton, J.M. Hoffman. (1999). Acute blood flow changes and efficacy of vagus nerve stimulation in partial epilepsy. *Neurology*, 52:1166–1173.

Holmes, M.D., D.L. Silbergeld, D. Drouhard, A.J. Wilensky, L.M. Ojemann. (2004). Effect of vagus nerve stimulation on adults with pharmacoresistant generalized epilepsy syndromes. *Seizure*, 13:340–345.

Hord, E.D., M.S. Evans, B. Adamolekun, S. Mueed, D.K. Naritoku. (2003). Effect of vagus nerve stimulation (VNS) on migraine headaches. In *Proceedings of the 10th World Congress on Pain*, J. Dostrovsky, D. Carr, and M. Koltzenburg, Eds. Seattle, WA: International Association for the Study of Pain.

Hosain, S., B. Nikalov, C. Harden, M. Li, R. Fraser, D. Labar. (2000). Vagus nerve stimulation treatment for Lennox–Gastaut syndrome. *J. Child Neurol.*, 15:509–512.

Hui, A.C., J.M. Lam, K.S. Wong, R. Kay, W.S. Poon. (2004). Vagus nerve stimulation for refractory epilepsy: long term efficacy and side-effects. *Chin. Med. J. (Engl.)*, 117:58–61.

Jaseja, H. (2004). Vagal nerve stimulation technique: enhancing its efficacy and acceptability by augmentation with auto activation and deactivation mode of operation. *Med. Hypotheses.*, 63:76–79.

Jobe, P.C., J.W. Dailey, C.E. Reigel. (1986). Noradrenergic and serotonergic determinants of seizure susceptibility and severity in genetically epilepsy-prone rats. *Life Sci.*, 39:775–782.

Jobe, P.C., J.W. Dailey, J.F. Wernicke. (1999). A noradrenergic and serotonergic hypothesis of the linkage between epilepsy and affective disorders. *Crit. Rev. Neurobiol.*, 13:317–356.

Kim, W., R.R. Clancy, G.T. Liu. (2001). Horner syndrome associated with implantation of a vagus nerve stimulator. *Am. J. Ophthalmol.*, 131:383–384.

Kirchner, A., F. Birklein, H. Stefan, H.O. Handwerker. (2000). Left vagus nerve stimulation suppresses experimentally induced pain. *Neurology*, 55:1167–1171.

Ko, D., C. Heck, S. Grafton, M.L. Apuzzo, W.T. Couldwell, T. Chen, J.D. Day, V. Zelman, T. Smith, C.M. DeGiorgio. (1996). Vagus nerve stimulation activates central nervous system structures in epileptic patients during PET H$_2$(15)O blood flow imaging. *Neurosurgery*, 39:426–431.

Koo, B. (2001). EEG changes with vagus nerve stimulation. *J. Clin. Neurophysiol.*, 18:434–441.

Kostov, H., P.G. Larsson, G.K. Roste. (2007). Is vagus nerve stimulation a treatment option for patients with drug-resistant idiopathic generalized epilepsy? *Acta Neurol. Scand. Suppl.*, 187:55–58.

Koutroumanidis, M., M.J. Hennessy, C.D. Binnie, C.E. Polkey. (2000). Aggravation of partial epilepsy and emergence of new seizure type during treatment with VNS. *Neurology*, 55:892–893.

Krahl, S.E., K.B. Clark, D.C. Smith, R.A. Browning. (1998). Locus coeruleus lesions suppress the seizure-attenuating effects of vagus nerve stimulation. *Epilepsia*, 39:709–714.

Krall, R.L., J.K. Penry, B.G. White, H.J. Kupferberg, E.A. Swinyard. (1978). Antiepileptic drug development. II. Anticonvulsant drug screening. *Epilepsia*, 19:409–428.

Kuba, R., M. Guzaninova, M. Brazdil, Z. Novak, J. Christina, I. Rektor. (2002). Effect of vagal nerve stimulation on interictal epileptiform discharges: a scalp EEG study. *Epilepsia*, 43:1181–1188.

Kuba, R., M. Brazdil, M. Kalina, T. Prochazka, J. Hovorka, T. Nezadal, J. Hadac, K. Brozova, V. Sebronova, V. Komarek, P. Marusic, H. Oslejskova, J. Zarubova, I. Rektor. (2009). Vagus nerve stimulation: longitudinal follow-up of patients treated for 5 years. *Seizure*, 18:269–274.

Labar, D. (2002). Antiepileptic drug use during the first 12 months of vagus nerve stimulation therapy: a registry study. *Neurology*, 59:S38–S43.

Labar, D., L. Ponticello. (2003). Persistent antiepileptic effects after vagus nerve stimulation ends? *Neurology*, 61:1818.

Labar, D., J. Murphy, E. Tecoma. (1999). Vagus nerve stimulation for medication-resistant generalized epilepsy. E04 VNS Study Group. *Neurology*, 52:1510–1512.

Lanska, D.J. J.L. (2002). Corning and vagal nerve stimulation for seizures in the 1880s. *Neurology*, 58:452–459.

Liu, W.C., K. Mosier, A.J. Kalnin, D. Marks. (2003). BOLD fMRI activation induced by vagus nerve stimulation in seizure patients. *J. Neurol. Neurosurg. Psychiatry*, 74:811–813.

Lockard, J.S., W.C. Congdon, L.L. DuCharme. (1990). Feasibility and safety of vagal stimulation in monkey model. *Epilepsia*, 31(Suppl. 2):S20–S26.

Loscher, W. (2002). Animal models of drug-resistant epilepsy. *Novartis Found. Symp.*, 243:149–159; discussion 159–166, 180–185.

Lyubashina, O., S. Panteleev. (2009). Effects of cervical vagus nerve stimulation on amygdala-evoked responses of the medial prefrontal cortex neurons in rat. *Neurosci. Res.*, 65:122–125.

Malow, B.A., J. Edwards, M. Marzec, O. Sagher, D. Ross, G. Fromes. (2001). Vagus nerve stimulation reduces daytime sleepiness in epilepsy patients. *Neurology*, 57:879–884.

Marrosu, F., A. Serra, A. Maleci, M. Puligheddu, G. Biggio, M. Piga. (2003). Correlation between GABA(A) receptor density and vagus nerve stimulation in individuals with drug-resistant partial epilepsy. *Epilepsy Res.*, 55:59–70.

Marrosu, F., F. Santoni, M. Puligheddu, L. Barberini, A. Maleci, F. Ennas, M. Mascia, G. Zanetti, A. Tuveri, G. Biggio. (2005). Increase in 20–50 Hz (gamma frequencies) power spectrum and synchronization after chronic vagal nerve stimulation. *Clin. Neurophysiol.*, 116:2026–2036.

McLachlan, R.S. (1993). Suppression of interictal spikes and seizures by stimulation of the vagus nerve. *Epilepsia*, 34:918–923.

Mirski, M.A., J.A. Ferrendelli. (1986). Anterior thalamic mediation of generalized pentylenetetrazol seizures. *Brain Res.*, 399:212–223.

Mirski, M.A., L.A. Rossell, J.B. Terry, R.S. Fisher. (1997). Anticonvulsant effect of anterior thalamic high frequency electrical stimulation in the rat. *Epilepsy Res.*, 28:89–100.

Naritoku, D.K., J.A. Mikels. (1996). Vagus nerve stimulation (VNS) attenuates electrically kindled seizures. *Epilepsia*, 37(S5):75.

Naritoku, D.K., J.A. Mikels. (1997). Vagus nerve stimulation (VNS) is antiepileptogenic in the electrical kindling model. *Epilepsia*, 38(S3):220.

Naritoku, D.K., A. Morales, T.L. Pencek, D. Winkler. (1992a). Chronic vagus nerve stimulation increases the latency of the thalamocortical somatosensory evoked potential. *Pacing Clin. Electrophysiol.*, 15:1572–1578.

Naritoku, D., J. Willis, R. Manon-Espillat, NCP Study Group. (1992b). Patient-activated "therapeutic bursts" of stimulation delivered via the vagus nerve. *Epilepsia*, 33(S3):102.

Naritoku, D.K., W.J. Terry, R.H. Helfert. (1995). Regional induction of *fos* immunoreactivity in the brain by anticonvulsant stimulation of the vagus nerve. *Epilepsy Res.*, 22:53–62.

Neese, S.L., L.K. Sherill, A.A. Tan, R.W. Roosevelt, R.A. Browning, D.C. Smith, A. Duke, R.W. Clough. (2007). Vagus nerve stimulation may protect GABAergic neurons following traumatic brain injury in rats: an immunocytochemical study. *Brain Res.*, 1128:157–163.

Nei, M., M. O'Connor, J. Liporace, M.R. Sperling. (2006). Refractory generalized seizures: response to corpus callosotomy and vagal nerve stimulation. *Epilepsia*, 47:115–122.

Nemeroff, C.B., H.S. Mayberg, S.E. Krahl, J. McNamara, A. Frazer, T.R. Henry, M.S. George, D.S. Charney, S.K. Brannan. (2006). VNS therapy in treatment-resistant depression: clinical evidence and putative neurobiological mechanisms. *Neuropsychopharmacology*, 31:1345–1355.

Ng, M., O. Devinsky. (2004). Vagus nerve stimulation for refractory idiopathic generalised epilepsy. *Seizure*, 13:176–178.

Odier, L. (1811). *Manuel de Medicine Pratique*. Paris: J. Paschoud.

Ohkuma, S., M. Katsura, J.L. Guo, H. Narihara, T. Hasegawa, K. Kuriyama. (1996). Role of peroxynitrite in [$^3$H] gamma-aminobutyric acid release evoked by nitric oxide and its mechanism. *Eur. J. Pharmacol.*, 301:179–188.

Pardo, J.V., S.A. Sheikh, M.A. Kuskowski, C. Surerus-Johnson, M.C. Hagen, J.T. Lee, B.R. Rittberg, D.E. Adson. (2007). Weight loss during chronic, cervical vagus nerve stimulation in depressed patients with obesity: an observation. *Int. J. Obes. (Lond.)*, 31:1756–1759.

Pardo, J.V., S.A. Sheikh, G.C. Schwindt, J.T. Lee, M.A. Kuskowski, C. Surerus, S.M. Lewis, F.S. Abuzzahab, D.E. Adson, B.R. Rittberg. (2008). Chronic vagus nerve stimulation for treatment-resistant depression decreases resting ventromedial prefrontal glucose metabolism. *NeuroImage*, 42:879–889.

Pearl, P.L., J.A. Conry, A. Yaun, J.L. Taylor, A.M. Heffron, M. Sigman, T.N. Tsuchida, N.J. Elling, D.A. Bruce, W.D. Gaillard. (2008). Misidentification of vagus nerve stimulator for intravenous access and other major adverse events. *Pediatr. Neurol.*, 38:248–251.

Penry, J.K., F.E. Dreifuss. (1969a). Automatisms associated with the absence of petit mal epilepsy. *Arch. Neurol.*, 21:142–149.

Penry, J.K., J.C. Dean. (1990). Prevention of intractable partial seizures by intermittent vagal stimulation in humans: preliminary results. *Epilepsia*, 31(Suppl. 2):S40–S43.

Penry, J.K., F.E. Dreifuss. (1969b). A study of automatisms associated with the absence of petit mal. *Epilepsia*, 10:417–418.

Quintana, C., E.S. Tecoma, V.J. Iragui. (2002). Evidence that refractory partial onset and generalized epilepsy syndromes respond comparably to adjunctive vagus nerve stimulation therapy. *Epilepsia*, 43(Suppl. 7):S344–S345.

Ramsay, R.E., B.M. Uthman, L.E. Augustinsson, A.R. Upton, D. Naritoku, J. Willis, T. Treig, G. Barolat, J.F. Wernicke. (1994). Vagus nerve stimulation for treatment of partial seizures. 2. Safety, side effects, and tolerability. First International Vagus Nerve Stimulation Study Group. *Epilepsia*, 35:627–636.

Randich, A., G.F. Gebhart. (1992). Vagal afferent modulation of nociception. *Brain Res. Brain Res. Rev.*, 17:77–99.

Revesz, D., M. Tjernstrom, E. Ben-Menachem, T. Thorlin. (2008). Effects of vagus nerve stimulation on rat hippocampal progenitor proliferation. *Exp. Neurol.*, 214:259–265.

Rush, A.J., L.B. Marangell, H.A. Sackeim, M.S. George, S.K. Brannan, S.M. Davis et al. (2005a). Vagus nerve stimulation for treatment-resistant depression: a randomized, controlled acute phase trial. *Biol. Psychiatry*, 58:347–354.

Rush, A.J., H.A. Sackeim, L.B. Marangell, M.S. George, S.K. Brannan, S.M. Davis et al. (2005b). Effects of 12 months of vagus nerve stimulation in treatment-resistant depression: a naturalistic study. *Biol. Psychiatry*, 58:355–363.

Rutecki, P. (1990). Anatomical, physiological, and theoretical basis for the antiepileptic effect of vagus nerve stimulation. *Epilepsia*, 31(Suppl. 2):S1–S6.

Sackeim, H.A., A.J. Rush, M.S. George, L.B. Marangell, M.M. Husain, Z. Nahas, C.R. Johnson, S. Seidman, C. Giller, S. Haines, R.K. Simpson, Jr., R.R. Goodman. (2001). Vagus nerve stimulation (VNS) for treatment-resistant depression: efficacy, side effects, and predictors of outcome. *Neuropsychopharmacology*, 25:713–728.

Sanossian, N., S. Haut. (2002). Chronic diarrhea associated with vagal nerve stimulation. *Neurology*, 58:330.

Schachter, S.C., C.B. Saper. (1998). Vagus nerve stimulation. *Epilepsia*, 39:677–686.

Schlaepfer, T.E., C. Frick, A. Zobel, W. Maier, I. Heuser, M. Bajbouj et al. (2008). Vagus nerve stimulation for depression: efficacy and safety in a European study. *Psychol. Med.*, 38:651–661.

Sjogren, M.J., P.T. Hellstrom, M.A. Jonsson, M. Runnerstam, H.C. Silander, E. Ben-Menachem. (2002). Cognition-enhancing effect of vagus nerve stimulation in patients with Alzheimer's disease: a pilot study. *J. Clin. Psychiatry*, 63:972–980.

Takaya, M., W.J. Terry, D.K. Naritoku. (1994). Pretreatment with vagus nerve stimulation (VNS) results in a sustained anticonvulsant effect. *Epilepsia*, 35(S8):6.

Takaya, M., W.J. Terry, D.K. Naritoku. (1996). Vagus nerve stimulation induces a sustained anticonvulsant effect. *Epilepsia*, 37:1111–1116.

Terry, R., W.B. Tarver, J. Zabara. (1990). An implantable NeuroCybernetic prosthesis system. *Epilepsia*, 31(Suppl. 2):S33–S37.

Terry, R.S., W.B. Tarver, J. Zabara. (1991). The implantable NeuroCybernetic prosthesis system. *Pacing Clin. Electrophysiol.*, 14:86–93.

Terry, W.J., M. Takaya, D.K. Naritoku. (1996). Regional changes in brain glucose metabolism in rats following anticonvulsant stimulation of the vagus nerve. *Epilepsia*, 37(S5):117.

Uthman, B.M., B.J. Wilder, E.J. Hammond, S.A. Reid. (1990). Efficacy and safety of vagus nerve stimulation in patients with complex partial seizures. *Epilepsia*, 31(Suppl. 2):S44–S50.

Valtschanoff, J.G., R.J. Weinberg, V.N. Kharazia, H.H. Schmidt, M. Nakane, A. Rustioni. (1993). Neurons in rat cerebral cortex that synthesize nitric oxide: NADPH diaphorase histochemistry, NOS immunocytochemistry, and colocalization with GABA. *Neurosci. Lett.*, 157:157–161.

Velasco, F., M. Velasco, A.L. Velasco, F. Jimenez. (1993). Effect of chronic electrical stimulation of the centromedian thalamic nuclei on various intractable seizure patterns. I. Clinical seizures and paroxysmal EEG activity. *Epilepsia*, 34:1052–1064.

Vonck, K., V. Thadani, K. Gilbert, S. Dedeurwaerdere, L. De Groote, V. De Herdt et al. (2004). Vagus nerve stimulation for refractory epilepsy: a transatlantic experience. *J. Clin. Neurophysiol.*, 21:283–289.

Wang, H., X. Chen, Z. Lin, Z. Shao, B. Sun, H. Shen, L. Liu. (2009). Long-term effect of vagus nerve stimulation on interictal epileptiform discharges in refractory epilepsy. *J. Neurol. Sci.*, 284:96–102.

Wilder, B.J., B.M. Uthman, E.J. Hammond. (1991). Vagal stimulation for control of complex partial seizures in medically refractory epileptic patients. *Pacing Clin. Electrophysiol.*, 14:108–115.

Woodbury, D.M., J.W. Woodbury. (1990). Effects of vagal stimulation on experimentally induced seizures in rats. *Epilepsia*, 31(Suppl. 2):S7–S19.

Woodbury, J.W., D.M. Woodbury. (1991). Vagal stimulation reduces the severity of maximal electroshock seizures in intact rats: use of a cuff electrode for stimulating and recording. *Pacing Clin. Electrophysiol.*, 14:94–107.

You, S.J., H.C. Kang, T.S. Ko, H.D. Kim, M.S. Yum, Y.S. Hwang, J.K. Lee, D.S. Kim, S.K. Park. (2008). Comparison of corpus callosotomy and vagus nerve stimulation in children with Lennox–Gastaut syndrome. *Brain Dev.*, 30:195–199.

Zabara, J. (1992). Inhibition of experimental seizures in canines by repetitive vagal stimulation. *Epilepsia*, 33:1005–1012.

Zagon, A., A.A. Kemeny. (2000). Slow hyperpolarization in cortical neurons: a possible mechanism behind vagus nerve simulation therapy for refractory epilepsy? *Epilepsia*, 41:1382–1389.

Zanchetti, A., S.C. Wang, G. Moruzzi. (1952). The effect of vagal afferent stimulation on the EEG pattern of the cat. *Electroencephalogr. Clin. Neurophysiol.*, 4:357–361.

Zuo, Y., D.C. Smith, R.A. Jensen. (2007). Vagus nerve stimulation potentiates hippocampal LTP in freely moving rats. *Physiol. Behav.*, 90:583–589.

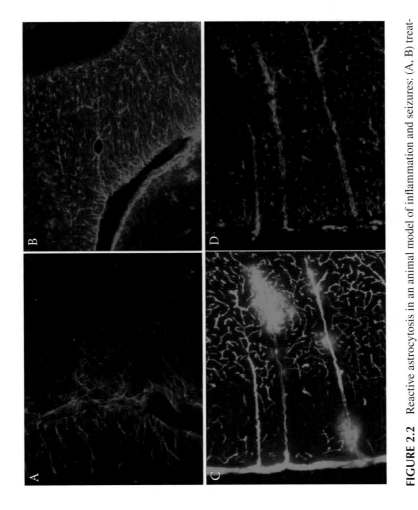

**FIGURE 2.2** Reactive astrocytosis in an animal model of inflammation and seizures: (A, B) treatment with lithium chloride; (C, D) treatment with pilocarpine. Note the enhanced GFAP staining (A, B, D) in specific regions and the leakage of a protein marker (C) in the proximity of gliosis. (From Marchi, N. et al., *Epilepsia*, 48(10), 1934–1946, 2007; Marchi, N. et al., *Neurobiol. Dis.*, 33, 171–271, 2009. With permission.)

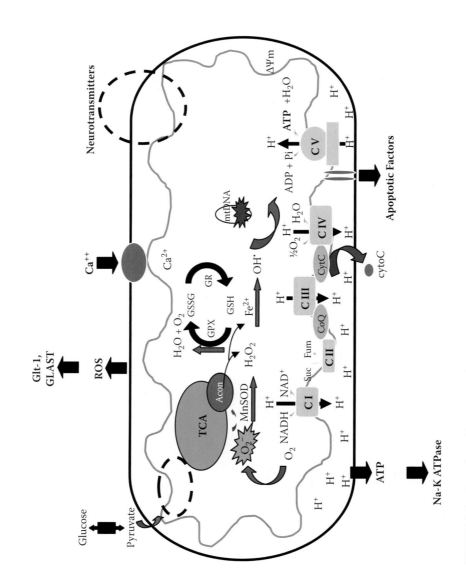

**FIGURE 3.1** Overview of important mitochondrial functions relevant to neuronal excitability.

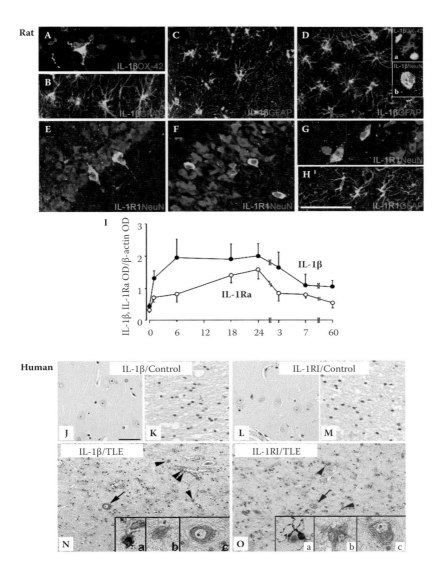

**FIGURE 4.1**  IL-1β system expression in the epileptogenic hippocampus of rats and temporal lobe epilepsy (TLE) patients. Panels A to H depict double–immunofluorescence micrographs showing the cellular localization of IL-1β (A–D) and IL-1R1 (E–H) in the rat hippocampus after status epilepticus (SE). During the acute phases of SE, IL-1β is expressed by reactive microglia cells (A) and astrocytes (B). Astrocytic upregulation persists during epileptogenesis (C) and in chronically epileptic tissue (D), where microglia (D, inset a) and neurons (D, inset b) also express the cytokine. IL-1R1 upregulation occurs both in neurons (E–G) and astrocytes (H). Panel I shows IL-1β and IL-1ra mRNA expression at different time points after electrically induced SE. Note the rapid increase in IL-1β levels, which are still higher than the control level 60 days after SE in rats with spontaneous seizures. IL-1ra, the endogenous IL-1R1 antagonist, is induced with a delayed time course and to a lesser extent as compared to IL-1β. Panels J to O present photomicrographs of immunohistochemical staining of IL-1β (J, K, N) and IL-1R1 (L, M, O) in the hippocampus of TLE patients. In epileptic specimens (N, O), note the strong activation of IL-1β (N) and IL-1R1 (O) in glial cells (N and O, insets a and b) and in neurons (N and O, inset c). In control tissue obtained at autopsy from patients without a history of seizures or other neurological diseases, no IL-1β (J, K) or IL-1R1 (L, M) staining was observed. (Adapted from De Simoni, M.G. et al., *Eur. J. Neurosci.*, 12(7), 2623–2633, 2000; Ravizza, T. et al., *Neurobiol. Dis.*, 29(1), 142–160, 2008.)

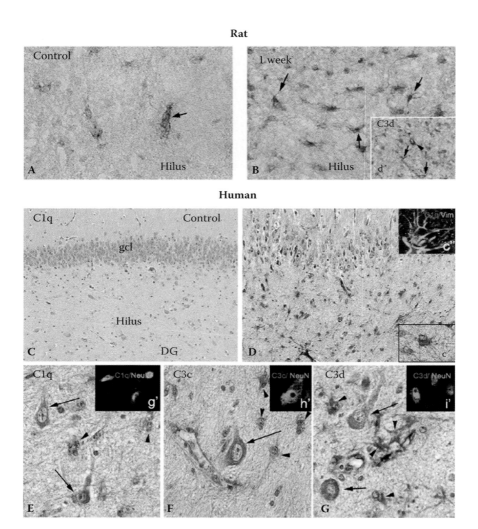

**FIGURE 4.2** Complement factor expression in epileptogenic rat hippocampus and in temporal lobe epilepsy (TLE) patients. Panels A and B depict C1q and C3d (inset in B) immunostaining in the rat hippocampus after status epilepticus (SE). Control hippocampus shows weak C1q immunoreactivity in the hilar region (A), whereas 1 week after SE onset a strong induction of C1q (B) and C3d (B, inset) was observed in glial cells. Panels C to G represent photomicrographs of immunohistochemical staining for C1q (C–E), C3c (F), and C3d (G) in the hippocampus of TLE patients. In control tissue, C1q immunostaining is not detectable (C), whereas in epileptic specimens a strong C1q signal was observed in reactive astrocytes (D and insets) and in neurons (E, inset). Panels F and G are photomicrographs showing C3c (F) and C3d (G) in epileptic tissue. C3c and C3d immunostaining was found in neurons (F and G, arrow and inset) and astrocytes (F and G, arrowheads). (Adapted from Aronica, E. et al., *Neurobiol. Dis.*, 26(3), 497–511, 2007.)

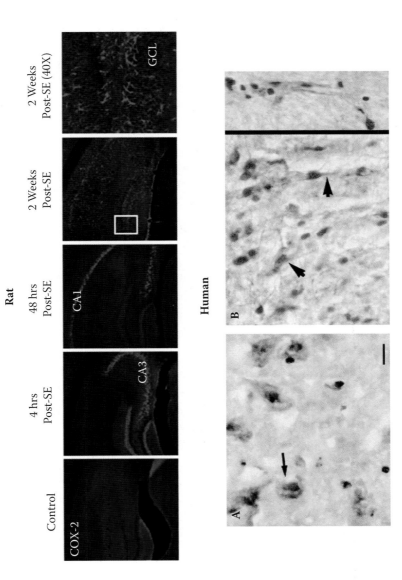

**FIGURE 4.3** COX-2 expression in the epileptogenic rat hippocampus and in TLE patients. Immunofluorescence micrographs depict COX-2 immunoreactivity at various time points after status epilepticus (SE) onset (4 hours to 2 weeks). Rapid neuronal upregulation of COX-2 expression was observed starting from 4 hours to 48 hours post-SE (second and third panels). In contrast, from 2 to 14 days after SE, COX-2 was observed in both astrocytes and microglia (fourth and fifth panels). Panels A and B are photomicrographs of immunohistochemical staining for COX-2 in the hippocampus of TLE patients. COX-2 immunostaining was found in neurons (A, arrow) and in astrocytes in TLE patients with hippocampal sclerosis (B, arrowheads). (Adapted from Desjardins, P. et al., *Neurochem. Int.*, 42(4), 299–303, 2003; Lee, B. et al., *Neurobiol. Dis.*, 25(1), 80–91, 2007.)

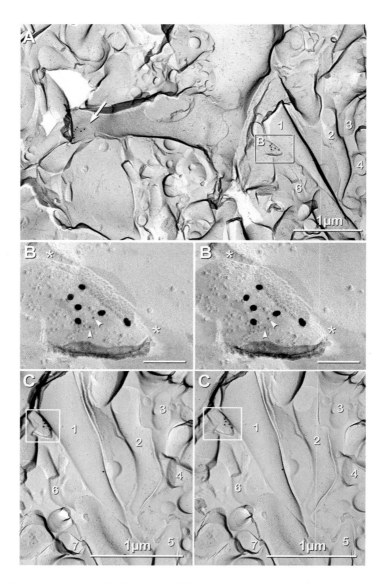

**FIGURE 13.5** Freeze-fracture replica immunogold labeled (FRIL) electron micrographs of a small-diameter plaque gap junction on a MF axon and a dendritic gap junction in the CA3c stratum lucidum labeled with anti-Cx36 immunogold beads. (A) Low-magnification FRIL electron micrograph of an axonal gap junction (box B). Numbers 1 to 4 represent MF axons in the stratum lucidum of the CA3c field or the rat dorsal hippocampus. The white arrow points to a dendritic gap junction labeled with anti-Cx36 immunogold beads. (B) High-magnification stereoscopic electron micrographs of the plaque gap junction (red overlay), which consisted of ≈100 connexons labeled by six 18-nm gold beads and two 6-nm gold beads (arrowheads). Slightly displaced gold beads reflect the length of the double antibody bridge, slight clumping of immunogold, and displacement of lipids and proteins during SDS washing (Kamasawa et al., 2006). Asterisks mark the extracellular space, which narrows to ≈3 nm at the gap junction contact point. Axon E-face has few intramembrane particles, whereas the P-face of the unidentified but coupled cell process contains densely packed particles around the gap junction. (C) High-magnification stereoscopic electron micrographs of the bundle of MF axons (numbers 1 to 6) tilted ≈45° with respect to the plane shown in A to show a clearer view of the interior of axon 1. The area inscribed by the box contains the steeply tilted axonal gap junction. Scale bars: A, 1000 nm; B, 100 nm; C, 1000 nm. (From Hamzei-Sichani, F. et al., *Proc. Natl. Acad. Sci. U.S.A.*, 104(30), 12548–12553, 2007. With permission.)

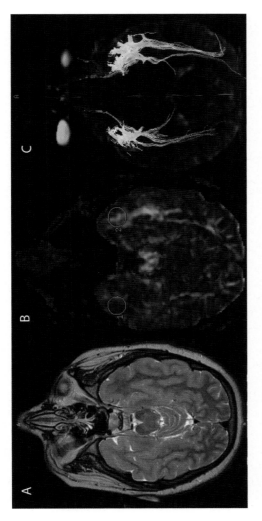

**FIGURE 15.4** Fiber tractography depicting a reduction of the subcortical fibers and a connection between the white matter subadjacent to the cortical mantle and deep white matter in a malformation of cortical development (pathology; focal cortical dysplasia type IIa). (A) Decreased white matter volume in right temporal pole with poor gray matter–white matter differentiation. (B) Regions of interest drawn on colored fractional anisotropy (FA) map. (C) Relatively scant and poorly arborized white matter in the right temporal pole.

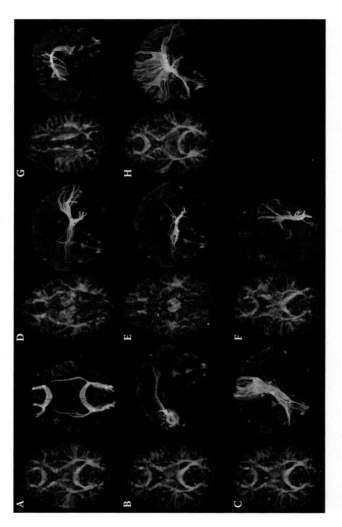

**FIGURE 15.5** Anatomic regions of interest and fiber tractography. (A) Corpus callosum, (B) cingulum, (C) posterior limb of internal capsule: corticospinal tract, (D) inferior fronto-occipital fasciculus, (E) inferior longitudinal fasciculus, (F) posterior corona radiata, (G) superior longitudinal fasciculus, and (H) thalamocortical projection.

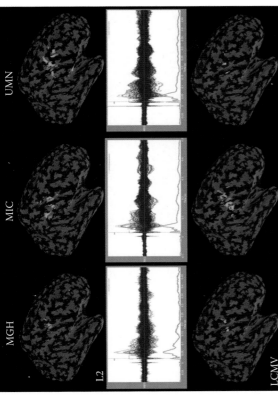

**FIGURE 16.3** Correspondence of median nerve localizations across different magnetoencephalography (MEG) machines and source localizations. The columns in each figure show data recorded at different sites with different MEG instruments. The top row in the figure shows source localizations obtained using BrainStorm software on inflated cortical surfaces extracted from magnetic resonance (MR) images using the BrainSuite software package and the L2 minimum norm algorithm. The center row shows the recorded wave forms (black) and the source time course for the same voxels in the post central gyrus, and the bottom row depicts localization of the same data using the LCMV beamforming algorithm. These data were recorded from the same subject during visits to Massachusetts General Hospital (MGH), the Mind Research Network Imaging Center (MIC), and the University of Minnesota (UMN). The data at MGH were recorded with a VSM MedTech OMEGA 275-channel instrument. Data from MIC were recorded with an Elekta Neuromag® 306-channel vector view view. Data from MIC were recorded with a VSM MedTech OMEGA 275-channel instrument. A 248-channel 4D neuroimaging system was used to record data at UMN. The same region was identified in the post central gyrus across all sites and algorithms.

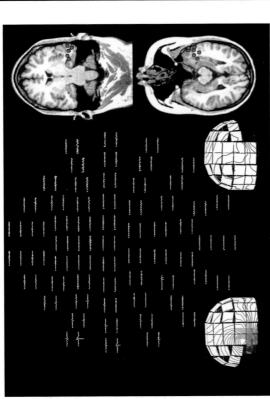

**FIGURE 16.2** Source localizations from left temporal spikes with a sample spike on the sensor array. The white tracings are approximately 750 msec of magnetoencephalography (MEG) data recorded on a 122-channel whole-head biomagnetometer (Elekta Neuromag®; Helsinki, Finland) filtered from 6 to 70 Hz. In this display, the nose would be at the top of the image with true left and right. The left anterior temporal sensors show an interictal spike. At the bottom of the figure are images showing the rotation of the magnetic field in a contour plot overlaid onto the sensor helmet (represented with white squares). At top and bottom right are the source localizations of several similar interictal events on the patient's MRI images. The top image is a coronal slice through the hippocampi; the bottom slice shows an oblique axial cut through the lengths of the hippocampi.

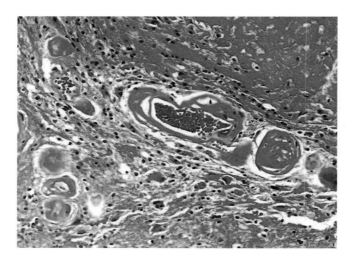

**FIGURE 18.5**  Radiation-induced tissue changes in the brain. Evident in this histological preparation is pronounced hyalinization and occlusion of vessels, gliosis, and inflammatory response. (Courtesy of Anthony Yachnis, MD.)

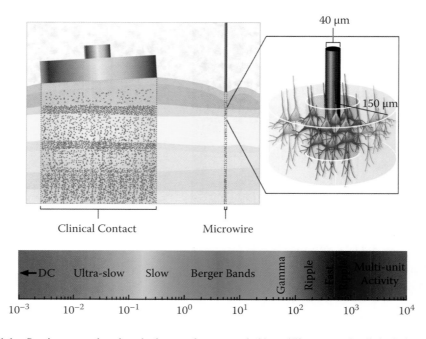

**FIGURE 21.1**  Spatiotemporal scales: the large volume sampled by millimeter-scale clinical electrodes (~$10^5$ neurons) vs. a 40-μm microwire. Cortex is organized into columns of neurons that are ~400 to 600 μm in diameter (~$10^4$ neurons in each column). Bottom figure shows the frequency range of neuronal network oscillations (DC, 1000 Hz).

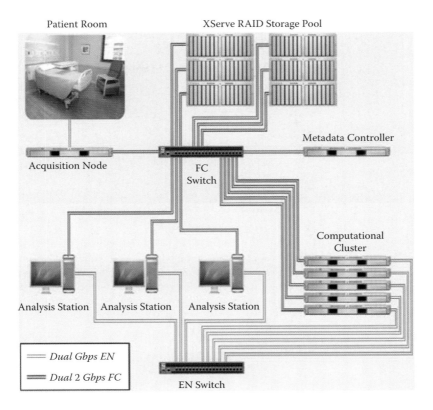

**FIGURE 21.3** Mayo Human Electrophysiology Suite acquisition using scalable Neuralynx Cheetah. The system streams data from the patient's ICU room to the acquisition node via a dedicated dual-gigabit Ethernet. Data are stored in a 70-terabyte storage pool and are accessed via a fiber channel Service Area Network. Large-scale analysis is performed on a dedicated computational cluster. (From Brinkmann, B.H. et al., *J. Neurosci. Methods*, 180(1), 185–192, 2009. With permission.)

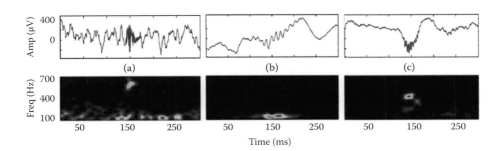

**FIGURE 21.4** Representative high-frequency oscillations (HFOs); each plot shows two views of HFO activity over a 300-msec epoch. (Top) Unfiltered EEG with an HFO event centered at 150 msec. (Bottom) Spectrogram (2.6-msec window). Note that HFOs are primarily characterized by a sharp spectral mode in the fast ripple (a, c) or ripple frequency range (b).

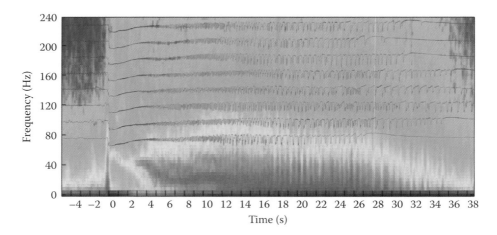

**FIGURE 21.6** Average of eight seizures recorded from the hippocampus of a patient with medically resistant partial epilepsy. The figure show the time traces of eight of the patient's habitual seizures aligned so the seizure onset is at 0 seconds. The average time-frequency spectrogram is shown in the background. The seizures all show a characteristic slow wave transient that is associated with a high-frequency oscillation (~110 Hz) at seizure onset.

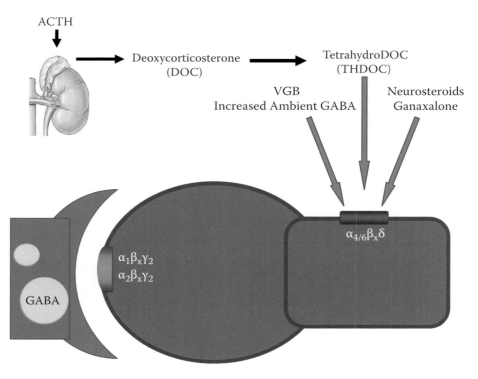

**FIGURE 26.1** A unifying view of the action of adrenocorticotropic hormone, vigabatrin, and ganaxalone on tonic inhibition mediated by extrasynaptic GABA receptors.

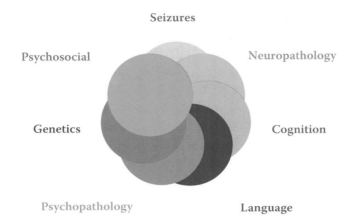

**FIGURE 30.1**  Comorbidity interactions with seizures and neuropathology, psychosocial, and genetic factors.

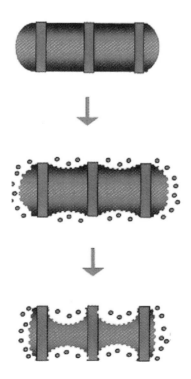

**FIGURE 31.2**  Schematic erosion of a RingCap® tablet. (From Rathbone, M.J. et al., Eds., *Modified-Release Drug Delivery Technology*, Marcel Dekker, New York, 2003. With permission.)

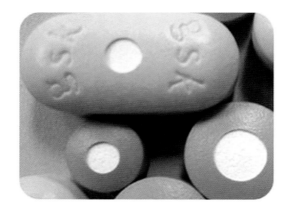

FIGURE 31.4 GlaxoSmithKline's DiffCORE® technology. (From http://www.gsk.com/infocus/diffcore-tablets.htm. © 2009 GlaxoSmithKline. Used with permission.)

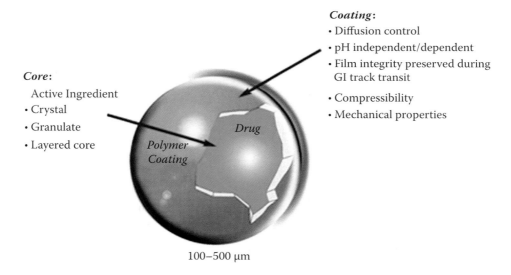

*Coating*:
• Diffusion control
• pH independent/dependent
• Film integrity preserved during GI track transit
• Compressibility
• Mechanical properties

*Core*:
    Active Ingredient
• Crystal
• Granulate
• Layered core

*Drug*

*Polymer Coating*

100–500 μm

FIGURE 31.5 Encapsulated microparticles used in the micropump technology. (From http://www.flamel.com/techAndProd/micropump.shtml.)

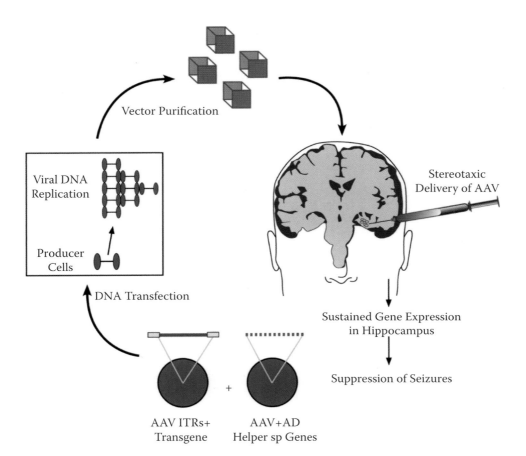

**FIGURE 33.3** General approach for adenovirus-associated gene therapy. Adeno-associated virus (AAV) is a small virus that infects humans and primate tissues. It is not known to cause disease and does not elicit a strong immune response, making it ideal for human gene therapy. Recombinant AAV (rAAV) is coinfected with a helper virus (AD) and a second AAV vector containing inverted terminal repeats (ITRs), which are required for efficient amplification of the AAV genome and for integration of the AAV DNA into the host cell genome. For human gene therapy, ITRs are required *in cis* next to the therapeutic gene. The AAV ITR transgene and AAV+AD genes are cotransfected into a producer cell line for viral DNA replication. The rAAV is purified and delivered by stereotaxic injection into the nervous system to achieve sustained gene expression.

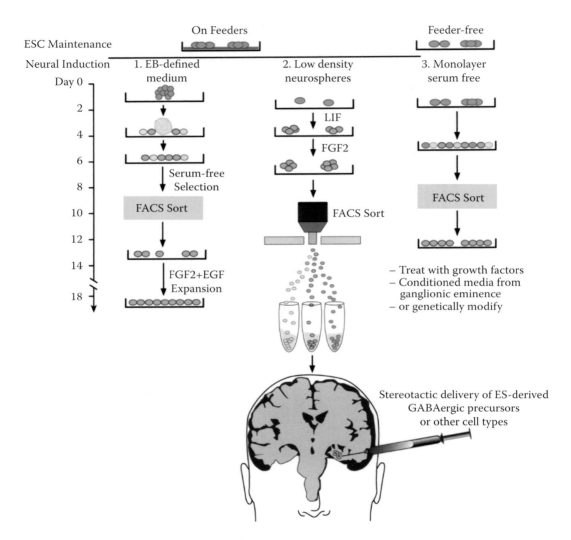

**FIGURE 33.5** Approaches for producing and purifying embryonic stem (ES) cell-derived neurons for stereotaxic delivery into limbic structures for treating temporal lobe epilepsy, based on studies in many labs. *Abbreviations:* EB, embryoid body; ESC, embryonic stem cell; FACS, fluorescence activated cell sorting; LIF, leukemia inhibitory factor. Figure is based on work in our laboratory and others (see, for example, review by Cai and Grabel, 2007).

# 26 Antiinflammatory Treatments for Seizure Syndromes and Epilepsy

*Stéphane Auvin and Raman Sankar*

## CONTENTS

## INTRODUCTION

There is a reciprocal relationship between seizures and inflammatory cytokines (see Chapter 4 in this volume). On the one hand, induced inflammation produces proconvulsant effects. On the other hand, seizure and status epilepticus (SE) in different experimental models enhanced the expression of proinflammatory cytokines such as interleukin-1β (IL-1β), interleukin-6 (IL-6), and tumor necrosis factor-α (TNF-α) (Vezzani and Granata, 2005). Moreover, preclinical data suggest the potent effects of various antiinflammatory drugs on seizure reduction. An understanding of the etiologic role played by the immune system in the genesis or modification of the human epilepsies is complicated by the paucity of data based on clinical research.

It is reasonable to propose a major role for inflammation in the intractable epilepsy associated with Rasmussen's encephalitis, and yet current treatment remains surgical. On the other hand, immunomodulatory therapy involving adrenocorticotropic hormone (ACTH) or corticosteroids is more convincingly established for the treatment of infantile spasms, a syndrome that is not generally attributed to an inflammatory etiology. Accumulating data also suggest that inflammation is involved in the occurrence of febrile seizure (FS); however, nonsteroidal antiinflammatory drugs (NSAIDs) failed to prevent FS recurrence in various clinical trials. Finally, there are accumulating data on immunomodulatory treatment of epilepsy from preclinical studies. We will discuss these data and the current limitations to initiate translational projects.

## ANTIINFLAMMATORY THERAPIES IN EPILEPTIC SYNDROMES

The clinical literature abounds with anecdotal reports on the efficacy of ACTH, corticosteroids, or intravenous immunoglobulin (IVIG) in a variety of refractory epilepsy syndromes, but almost always as treatments of last resort. No randomized, placebo-controlled, and blinded studies are available. The use of IVIG has all but replaced plasmapheresis. Indeed, this approach has generally been utilized either for syndromes that are known to respond poorly to antiepileptic drugs (AEDs) (e.g., infantile spasms, clinical syndromes involving electrical status epilepticus during slow-wave sleep such as the Landau–Kleffner syndrome) or in individuals with any type of seizure that has failed to respond adequately to AEDs. Klein and Livingston (1950) first reported the use of ACTH in children with intractable seizures; however, the pharmacology of these agents (ACTH, glucocorticoids, and IVIG) is so complex and diverse that we cannot be certain that the basis of their efficacy is their immunomodulatory ability. Experimental studies are needed to explain their underlying mechanisms.

A recent Cochrane review stated that no evidence exists for the efficacy or safety of corticosteroids in treating childhood epilepsies other than infantile spasms (Gayatri et al., 2007). Even in Rasmussen's encephalitis, the immunomodulatory approaches have been less than satisfactory and enduring, even though a strong role for inflammation is suspected in this syndrome. We briefly present our current state of knowledge regarding the use of immunotherapy in specific epilepsy syndromes.

### INFANTILE SPASMS

The strongest tradition to date for the use of immunomodulatory therapy exists for the use of ACTH in the catastrophic epilepsy syndrome of infancy. Even here, ACTH was deemed *probably effective* for the *short-term treatment* of infantile spasms by "evidence-based medicine" evaluative clinicians. These conclusions reflect the difficulty in undertaking the study of a relatively rare and age-specific, but not etiology-specific, disorder of considerable variability in the premorbid state (Hancock et al., 2008).

Previous studies have attempted to clarify whether ACTH treatment is superior to treatment with corticosteroids, if a higher dose of ACTH (150 IU/m$^2$) is superior to a lower dose (40 IU/m$^2$), and if an optimum duration of therapy can be determined. High-dose ACTH (150 IU/m$^2$) was superior to prednisone (3 mg/kg) in the cessation of infantile spasms, with a 100% reduction using ACTH vs. 59% reduction using prednisone. It was noted, however, that some patients who do not initially respond to ACTH may respond to prednisone and *vice versa* (Snead et al., 1983). Another study, double-blinded and placebo-controlled, compared ACTH (20 to 30 IU/day) to prednisone (2 mg/kg/day); however, at these dosages, no difference was observed in the cessation of infantile spasms and disappearance of hypsarrhythmia between ACTH and prednisone (Hrachovy et al., 1983).

The difference in dose may explain the significantly lower spasm-free rate of oral prednisone described in two previous studies in 1983 and 1996 (Baram et al., 1996; Hrachovy et al., 1983). Hrachovy et al. (1983) did not identify a statistical difference between ACTH and oral prednisone,

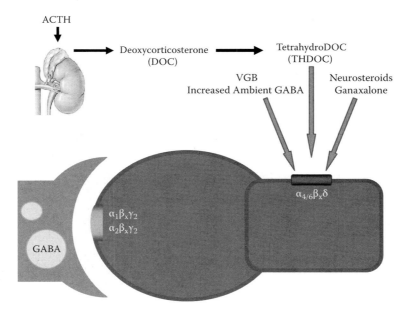

**FIGURE 26.1** **(See color insert following page 458.)** A unifying view of the action of adrenocorticotropic hormone, vigabatrin, and ganaxalone on tonic inhibition mediated by extrasynaptic GABA receptors.

but both were lower than other ACTH spasm-free reports. In both studies, prednisone was used at a dose of 2 mg/kg/day, which is half of the starting United Kingdom Infantile Spasms Study (UKISS) recommendation (Baram et al., 1996; Hrachovy et al., 1983; Lux et al., 2005).

Scattered reports describe the use if IVIG in the treatment of infantile spasms, but the data do not permit drawing conclusions in support of this treatment modality (Ariizumi et al., 1987; van Engelen et al., 1994). One study reported benefits from a single treatment with IVIG in children with refractory juvenile spasms (Bingel et al., 2003), but it is currently impossible to recognize IVIG as a valid therapy for infantile spasms.

## How Does ACTH Mitigate Infantile Spasms?

The mechanisms remain elusive. A major hypothesis has centered around the ability of ACTH to reduce the endogenous synthesis of corticotrophin-releasing hormone (CRH), which, when injected, produces severe seizures in immature rats (Brunson et al., 2001a,b). This theory does account for the time course of action of ACTH and the all-or-none response to treatment, and it is also supported by the findings of reduced ACTH and cortisol levels in the spinal fluids of patients with infantile spasms (Baram et al., 1992, 1995). Could ACTH be functioning, at least in part, by its effect on melanocortin receptors, independent of steroid release? Some peptides are ACTH fragments that can function as agonists at those receptors without mediating steroidogenesis, but no clinical trials with such compounds have been undertaken. Another hypothesis proposes that ACTH stimulates the synthesis of deoxycorticosterone (DOC), which is bioconverted to the tetrahydro derivative tetrahydrodeoxycorticosterone (THDOC), a potent agonist at the $GABA_A$ receptor site (Reddy and Rogawski, 2002; Rogawski and Reddy, 2002). This theory is consistent with the known ameliorative action on infantile spasms by vigabatrin and supports the current interest in ganaxalone, a neurosteroid derivative that is a highly selective agonist for the extrasynaptic $\delta$-subunit containing $GABA_A$ receptors and mediate tonic inhibition (Figure 26.1).

None of the above discussed mechanisms excludes the possibility that the administration of ACTH mitigates infantile spasms by modifying a wide range of cytokines (immunomodulation) which in turn may also influence brain excitability by interactions with glutamatergic and GABAergic mechanisms (see Chapter 4 in this volume). An explanation proposing a mechanism involving the modulation of cytokines can accommodate efficacy for both ACTH and glucocorticoids. Thus, both immune function related and other aspects of ACTH (and steroid) pharmacology may be important in this enduring treatment approach for infantile spasms.

## LENNOX–GASTAUT SYNDROME

The reports describing the use of ACTH (O'Regan and Brown, 1998; Yamatogi et al., 1979) and IVIG (van Engelen et al., 1994) in the treatment of Lennox–Gastaut syndrome (LGS) are extremely limited. In the study by Yamatogi and colleagues, all 45 children with LGS received ACTH in doses ranging from 10 IU/day for infants to 30 IU/day for older children. Of these children, 23 (51%) became seizure free for over 10 days; however, 78% of seizure-free children later relapsed (Yamatogi et al., 1979). In another study, 6 of 7 children with LGS and intractable seizures treated with ACTH (20 IU/day for 2 weeks followed by a gradual withdrawal over the succeeding weeks) showed an improvement in seizure control, EEG findings, and behavior (O'Regan and Brown, 1998). It is likely, however, that most specialists will preferentially consider rufinamide, felbamate, the ketogenic diet, or vagus nerve stimulation in medically refractory patients with this syndrome, besides valproic acid, topiramate, and lamotrigine. One report described a favorable response in 7 of 10 patients treated with prednisone (Sinclair, 2003), but it is not clear that the benefit was lasting. The use of corticosteroids in managing medically intractable LGS was likely more common before the availability of new-generation AEDs with demonstrated benefit in this syndrome. Anecdotal evidence describing the use of IVIG in LGS (Gross-Tsur et al., 1993; van Rijckevorsel-Harmant et al., 1986) indicates favorable effects, and the therapy is generally better tolerated than either ACTH or corticosteroids.

## SYNDROMES WITH ELECTRICAL STATUS EPILEPTICUS DURING SLOW-WAVE SLEEP: LANDAU–KLEFFNER SYNDROME AND CONTINUOUS SPIKE WAVES DURING SLOW-WAVE SLEEP

A spectrum of clinical disorders has been associated with the electroencephalogram (EEG) signature of electrical status epilepticus during slow-wave sleep (ESES); the clinical syndromes include the distinctive Landau–Kleffner syndrome (LKS) and the syndrome of continuous spike waves during slow-wave sleep (CSWS) (Galanopoulou et al., 2000; McVicar and Shinnar, 2004), the latter sometimes also referred to as Tassinari syndrome. LKS exhibits many clinical features that are typically associated with idiopathic epilepsy syndromes, whereas CSWS generally presents in a manner that suggests a symptomatic epilepsy syndrome (Van Hirtum-Das et al., 2006).

Clinical practice tends to approach the pharmacologic aspects of management of these syndromes in a more or less similar fashion, concentrating on the EEG trait of ESES. In the published literature, an immunologic approach seems to be more effective in LKS (Van Hirtum-Das et al., 2006). The treatment of LKS with corticosteroids in high doses for a protracted period was advocated by Marescaux and colleagues (1990). Sinclair and Snyder (2005) reported a favorable outcome in 9 of 10 patients with LKS who were treated with prednisone, which compares favorably to the surgical approach of multiple subpial transections (Grote et al., 1999; Morrell et al., 1995). Such a robust response has never been achieved in CSWS. One study reported two patients whose language functions recovered rapidly with the initiation of intravenous corticosteroid therapy, followed by conversion to oral therapy (Tsuru et al., 2000). It is uncertain, however, whether intravenous therapy adds much benefit to the more common oral treatment (Lerman et al., 1991; Marescaux et al., 1990; Sinclair and Snyder, 2005).

Successful therapy of LKS with IVIG was reported in isolated cases (Fayad et al., 1997; Lagae et al., 1998; Mikati and Saab, 2000). In a subsequent review, Mikati and Shamseddine (2005) stated that only 2 of 11 patients with LKS had responded adequately to treatment with IVIG.

No literature substantiates the specific use of ACTH, corticosteroids, or IVIG in ESES that presents as CSWS. The initial report by De Negri and colleagues (1995) advocating nightly diazepam treatment has been adopted by many clinicians, and abstracts have been published in support of this approach. No peer-reviewed publication has yet described this therapy and its outcome in a manner that will move it toward wider acceptance.

## How May IVIG Be Effective in Epileptic Disorders?

The concept that immunoglobulin treatment may have a beneficial effect on epileptic seizures in human epilepsy was first suggested after empirical observation (Pechadre et al., 1977). Although the role of the immune system in epilepsy is not completely clarified, the mechanism of action of IVIG in epilepsy is considered mainly immunological. IVIG is used as an immunomodulation therapy that acts by modulating the idiotypic antibody network and blocks Fc receptor–immunoglobulin interactions. Moreover, autoantibodies (Bartolomei et al., 1996; Majoie et al., 2006) have also been related to different forms of epileptic syndromes and to more complex neurological syndromes with prominent epilepsy such as Rasmussen's chronic encephalitis (Rogers et al., 1994). Nonimmunological mechanisms of action have also been suggested based on human epilepsy data. Evidence for a nonimmunological action of IVIG is related to the time course of the response to treatment of patients with different forms of epilepsy (either idiopathic or symptomatic); the often immediate response suggests a direct neuromodulating effect.

### RASMUSSEN'S ENCEPHALITIS

Rasmussen's syndrome probably represents the best example of an epilepsy syndrome caused by an inflammatory encephalopathy as shown by the histopathology of resected tissue. Nonviral autoimmune mechanisms were first suggested by pathologic evidence of cerebral vasculitis with immune complex deposition in a child with typical Rasmussen's encephalitis (RE) (Andrews et al., 1990; Prayson and Frater, 2002; Whitney et al., 1999). The report, suggesting a role for auto antibodies to the GluR3 subunit of the glutamate receptor (Rogers et al., 1994), initially resulted in much anticipation for the development of successful immunomodulating therapy. This initial enthusiasm has diminished considerably since then, with lack of consistent identification of such antibodies in many patients (Watson et al., 2004) and the less than gratifying sustained clinical response to steroids, plasmapheresis, or IVIG (Granata et al., 2003). Surgery remains the most effective treatment to date. Although hemispherotomy was the only treatment that stopped disease progression, immunomodulation could be considered as treatment until the surgical procedure is performed, as the available data suggest that it may slow disease progression (Bahi-Buisson et al., 2007).

### FEBRILE SEIZURE: INCONSISTENT EVIDENCE FOR INVOLVEMENT OF INFLAMMATION

The underlying mechanisms responsible for the occurrence of febrile seizures (FSs) remain incompletely understood. The association of several different factors seems to be involved in each patient, among which are an immature brain, an increase in body temperature, an inflammatory response, genetics, and a viral infection. Accumulating data suggest that inflammation has an important role for both the development and long-term consequences of FSs. In this regard, an association between polymorphisms involving the interleukin-1β–511T allele, prolonged febrile convulsions, and the emergence of temporal lobe epilepsy has been reported in the Japanese population (Kanemoto et al., 2000, 2003). This association has not been confirmed in other populations. These data have been challenged by other studies; however, a recent meta-analysis suggests a link between IL-1β gene polymorphism and temporal lobe epilepsy (TLE) (Kauffman et al., 2008).

The involvement of an inflammatory response in FS development or recurrence has not yet led to therapeutic advances leveraging this concept. The clinical trials using antiinflammatory drugs examined

**TABLE 26.1**
**Summary of Clinical Studies on Inflammation in Patients Who Experienced Febrile Seizure**

| Refs. | Studied Inflammatory Mediators | Methods |
|---|---|---|
| **Correlation between occurrence of FS and inflammation** | | |
| Tamai et al. (1983) | $PGF_{2\alpha}$ | CSF measurement |
| Loscher and Siemes (1988) | $PGE_2$ | CSF measurement |
| Helminen and Vesikari (1990) | IL-1β | Measurement of leukocyte production by LPS-induced stimulation |
| Straussberg et al. (2001) | IL-1β | Measurement of leukocyte production by LPS-induced stimulation |
| Tutuncuoglu et al. (2001) | IL-1β, $PGD_2$, $PGE_2$, $PGF_{2\alpha}$ | Serum and CSF measurement |
| Virta et al. (2002) | IL-1β/IL-1RA, IL-6 | Serum and CSF measurement |
| Haspolat et al. (2002) | IL-1β | CSF measurement |
| Masuyama et al. (2002) | IFN-α | Serum measurement (CF of *M. influenza*) |
| Matsuo et al. (2006) | IL-1β | Measurement of leukocyte production using TLR3 agonist stimulation |
| **Absence of correlation** | | |
| Lahat et al. (1997) | IL-1β | Serum and CSF measurement |
| Straussberg et al. (2001) | IL-6, IL-10 | Measurement of leukocyte production by LPS-induced stimulation |
| Tutuncuoglu et al. (2001) | IL-1α, TNF-α | Serum and CSF measurement |
| Virta et al. (2002) | IL-10, TNF-α | Serum and CSF measurement |
| Haspolat et al. (2002) | IL-1β, TNF-α/TNF-α | Serum and CSF measurement |
| Tomoum et al. (2007) | IL-1β | Serum measurement |

*Note:* CSF, cerebrospinal fluid; IFN, interferon; IL, interleukin; LPS, lipopolysaccharide; PG, prostaglandin; TLR3, Toll-like receptor 3; TNF, tumor necrosis factor.

whether nonsteroidal antiinflammatory drugs, frequently used for fever, are able to prevent FS occurrence. Two studies that examined the effect of ibuprofen on FS recurrences found little significant effect (Van Esch et al., 1995; van Stuijvenberg M. et al., 1998). One study, a randomized, double-blind, placebo-controlled trial involving 230 patients showed a significant reduction in temperature (0.7°C) after fever onset in the ibuprofen group compared with the placebo group if all 555 fever episodes were considered. In the fever episodes with a seizure recurrence, a similar temperature increase was shown in both groups, with no significant difference between the intention-to-treat and the per-protocol analysis (van Stuijvenberg M. et al., 1998). It is interesting to note that similar trials using acetaminophen also found no preventive effect on FS recurrence (Uhari et al., 1995; Van Esch et al., 1995).

This discrepancy between clinical trials and our current understanding of the underlying mechanisms of FS may be related to the drugs acting on the clinically significant pathway. Several studies pointed to the role of IL-1β. These studies were conducted in children as well as in experimental conditions, and, although clinical studies are numerous, they are often difficult to analyze because the methods and the control groups are very different from one study to another (Table 26.1).

The population of children that would enter any study of the role of inflammation in febrile convulsions comprises a genetically diverse group, possessing, as has already been demonstrated, a number of mutation in genes in coding for ion-channel subunits (Baulac et al., 2004). Thus, although inflammation may play a role, it is clearly not the only variable affecting this syndrome.

Experimental data highlight the role of IL-1β. Using hyperthermia induced-seizures in mice, Dubé et al. (2005) showed that an i.c.v. injection of 5 ng/μL IL-1β lowered the threshold for hyperthermic seizures in mouse pups and was even capable of producing seizures without a thermal provocation. This effect is related to the interaction of IL-1β with its receptor (Dubé et al., 2005). Using P14 rats treated with lipopolysaccharide (LPS) (200 μg/kg) to induce fever and a subconvulsant dose of kainic acid (1.75 mg/kg), it has been shown that excessive amounts of IL-1β may influence the genesis of FS (Heida and Pittman, 2005).

## PRECLINICAL DATA ON ANTIINFLAMMATORY DRUGS

Data coming from basic science suggest that other antiinflammatory drugs might be useful in epilepsy therapy in the future. We will discuss the available data for NSAID, rhIL-1Ra, cyclosporine A, tacrolimus, and polyunsaturated fatty acids.

### NONSTEROIDAL ANTIINFLAMMATORY DRUGS

Very few data regarding the potential anticonvulsive or antiepileptogenic proprieties of NSAIDs are available in humans. As previously discussed, clinical trials failed to demonstrate any effect on FS recurrence (Van Esch et al., 1995; van Stuijvenberg M. et al., 1998). NSAIDs act by blocking the synthesis of prostaglandins through the inhibition of cyclooxygenase (COX), the enzyme that converts arachidonic acid to cyclic endoperoxides, the precursors of prostaglandins. Three COX isoenzymes are known—COX-1, COX-2, and COX-3. COX-1 and COX-2 are expressed in different tissues at varying levels, and COX-3 is a splice variant of COX-1. COX-1 is considered a constitutive enzyme. COX-2 is normally undetectable in most normal tissues; it is an inducible enzyme, becoming abundant in cases of inflammatory response.

### COX-2 INHIBITORS SEEM TO HAVE ANTICONVULSANT PROPRIETIES IN SEIZURE-INDUCED ANIMAL MODELS

Pretreatment with COX inhibitors (aspirin, naproxen, nimesulide, or rofecoxib) dose-dependently showed protection against pentylenetetrazol (PTZ)-induced convulsions. COX-2 inhibitors were more effective as compared to nonselective COX inhibitors. Rofecoxib or nimesulide also enhanced the subprotective effect of diazepam showing GABAergic modulation of COX-2 inhibitors. COX-2 inhibitors also antagonized the effect of flumazenil against PTZ-induced convulsions, further confirming the involvement of a GABAergic mechanism (Dhir et al., 2006a). The possible involvement of the adenosinergic system was also suggested to explain the anticonvulsant effects of rofecoxib against the i.v. PTZ seizure threshold paradigm in mice (Akula et al., 2008). However, strong evidence of the involvement of the COX-2/prostaglandin $E_2$ ($PGE_2$) pathway has been recently demonstrated using i.c.v. administration of $PGE_2$ or anti-$PGE_2$ antibodies (Oliveira et al., 2008).

Similar results were reported using COX-2 inhibitors in other seizures models. COX-2 inhibitors given 45 minutes prior to an epileptic challenge prolonged the mean onset time of convulsions, decreased the duration of clonus, and decreased the percent mortality rate against both bicuculline- and picrotoxin-induced convulsions in mice. The effect of COX-2 inhibitors in treating maximal electroshock-induced seizures is controversial, with one group stating they were ineffective (Dhir et al., 2006b) and another describing protection against electroshock seizure (Shafiq et al., 2003).

More recently, it has been shown that rofecoxib potentiates the anticonvulsant effect of topiramate in PTZ-induced seizure models (Dhir et al., 2008). In response to this study, further investigations should be focused on the combination of various antiinflammatory drugs and antiepileptic drugs.

## COX-2 Inhibitors Decrease the Enhanced Role of Inflammation on Seizure Models

Inflammation induced by lipopolysaccharide (LPS) injection decreased the seizure threshold in a dose- and time-dependent manner in PTZ-induced seizures. Pretreatment of mice with cyclooxygenase inhibitor (piroxicam) completely reversed the proconvulsant effect of LPS (Sayyah et al., 2003). Although a specific COX-2 inhibitor (SC-58236) alone had no effect on PTZ-induced seizures, it did reverse the antiseizure activity observed 18 hours after LPS injection. In addition, SC-58236 partially acted against the proconvulsant changes observed 4 hours after LPS administration (Akarsu et al., 2006).

## NSAIDs Modulate Cerebral Excitability Induced by PTZ–Kindling Models

Chronic treatment with selective COX-2 inhibitors (rofecoxib for 15 days or nimesulide for 15 days) decreases the seizure score in PTZ-induced kindling in mice (Dhir et al., 2006c; Dhir et al., 2007).

## NSAIDs May Be Neuroprotective and Antiepileptogenic

In the lithium–pilocarpine seizure model, celecoxib attenuated the likelihood of developing spontaneous recurrent seizures, neuronal injury, and microglia activation in the hilus and CA1 (Jung et al., 2006). Antiinflammatory drugs with anti-COX activities were shown either to reduce or to exacerbate seizures induced by kainic acid or SE-induced cell injury. On the one hand, celecoxib administered before kainic acid treatment does not reduce kainic acid-induced neuronal loss in the hippocampus, but rats injected with kainic acid and treated with celecoxib (after the kainic injection) performed better in the spatial and nonspatial tasks than the group treated with kainic acid (Gobbo and O'Mara, 2004). Rofecoxib, a selective COX-2 inhibitor, selectively diminished cell loss in the hippocampus in rats after kainic acid injection. The same treatment selectively attenuated the number of terminal deoxynucleotidyl transferase dUTP nick end labeling (TUNEL)-positive cells in the hippocampus, whereas the cells of the thalamus, amygdala, and piriform cortex were not protected (Kunz and Oliw, 2001a). Nimesulide pretreatment augmented seizures and increased the mortality rate from approximately 10% to 69% after kainic acid injection. Cell loss did not seem to be reduced in animals treated with nimesulide 6 to 8 hours after kainic acid, but in the surviving animals receiving nimesulide before kainic acid injection a decrease in cell injury was observed (Kunz and Oliw, 2001b). On the other hand, others have shown that cyclooxygenase-2 selective inhibitors had no effect on cell loss and spontaneous recurrent seizures following SE (Holtman et al., 2009) and did not aggravate kainic acid-induced seizure and neuronal cell death in the hippocampus (Baik et al., 1999).

## Human Recombinant IL-1 Receptor Antagonist

The role of IL-1 in seizures and epilepsy has been extensively investigated. There is currently a consensus about the involvement of IL-1β in epilepsy. Three lines of evidence suggest a role for IL-1β brain tissue from epilepsy patients. Brain tissue from animal models shows increased IL-1β expression after seizures, and IL-1β has proconvulsive properties when applied exogenously.

The first study using the interleukin-1 receptor antagonist (IL-1Ra) in the epilepsy field was from the laboratory of Vezzani et al. (2000), who reported that IL-1Ra has an anticonvulsant action. IL-1Ra was administered by intrahippocampal injection, and it acts via the IL-1 receptor 1. More recently, it has been shown that animals pretreated with IL-1Ra exhibited significant reduction of SE onset and of blood–brain barrier damage (Marchi et al., 2009).

Recombinant human IL-1Ra (rhIL-1Ra) is now available for clinical use and is currently being used mainly in rheumatology (Mertens and Singh, 2009). Only a few studies are available in the neuroscience field. It is now established that rhIL-1Ra penetrates the human brain at experimentally therapeutic concentrations (Clark et al., 2008). A randomized, double-blind, placebo-controlled trial of rhIL-1Ra in patients with acute stroke was reported by Emsley et al. (2005). This Phase II trial suggests that rhIL-1Ra is safe and well tolerated in acute stroke. In addition, rhIL-1Ra exhibited biological activity that is relevant to the pathophysiology (decreasing neutrophil and total white cell counts, C-reactive protein, and IL-6) and clinical outcome of ischemic stroke.

## CYCLOSPORINE A AND TACROLIMUS

Cyclosporine is a cyclic peptide obtained from an extract of soil fungi. It is a powerful immuno-suppressant with a specific action on T-lymphocytes. Tacrolimus or FK506 is a macrolide isolated from the culture broth of a strain of *Streptomyces tsukubaensis* that has strong immunosuppressive activity *in vivo* and prevents the activation of T-lymphocytes in response to antigenic or mitogenic stimulation. Both cyclosporin and tacrolimus are known for their potent neurotoxicity in humans, producing posterior leukoencephalopathy. Seizure is then observed (Gaggero et al., 2006; Gleeson et al., 1998; Lavigne et al., 2004; Yoshida et al., 2003).

Despite some data describing the anticonvulsant properties of cyclosporine, more studies show that cyclosporine facilitates seizure rather than protecting against it. Two studies showed that cyclosporine decreases PTZ-induced seizures (Asanuma et al., 1994, 1995; Homayoun et al., 2002). In another study, cyclosporine ameliorated the course of SE induced by lithium–pilocarpine in adult rats (Setkowicz and Ciarach, 2007). Most of the experimental studies show that cyclosporin aggravates seizures in different models such as electroshock-induced seizures (Yamauchi et al., 2005), bicuculline-induced seizures (Shuto et al., 1999), lithium–pilocarpine-induced SE (Setkowicz et al., 2004), and *in vitro* brain slice preparation (Gorji et al., 2002). Two studies pointed out the potent role of the nitric oxidergic system in the therapeutic effects of cyclosporine on induced seizures, but they found opposing results (Fujisaki et al., 2002; Homayoun et al., 2002).

Regarding tacrolimus, experimental studies have not reported any anticonvulsant properties; however, concordant data suggest potent antiepileptogenic activity. In rat models of temporal lobe epilepsy, systemic injection of tacrolimus was responsible for protection against seizures, abnormal behaviors, and accompanying delayed neuronal damage in the hippocampus (Moriwaki et al., 1998; Nishimura et al., 2006). Similar results were reported in the lithium–pilocarpine model (Setkowicz and Ciarach, 2007; Setkowicz et al., 2004). In a kindling model, the kindling stage progression was reversibly blocked by the tacrolimus (Moia et al., 1994).

The discrepancy within basic science studies and between clinical data and experimental studies does not permit a clear view of the potential role of these drugs in the future of the epilepsy field. More studies about the potential antiepileptogenic proprieties are needed. The dose–response relationship of these drugs should be evaluated in human studies to determine if they can be used without the neurotoxicity observed following transplantation.

## POLYUNSATURATED FATTY ACIDS

At adequate physiologic levels, n-3 polyunsaturated fatty acids (PUFAs) decrease the production of inflammatory eicosanoids, cytokines, and reactive oxygen species, as well as the expression of adhesion molecules. PUFAs act both directly (through the replacement of arachidonic acid as an eicosanoid substrate and the inhibition of arachidonic acid metabolism) and indirectly (through the alteration of inflammatory gene expression through effects on transcription factor activation). Thus, n-3 PUFAs are potentially potent antiinflammatory agents (Calder, 1996a,b).

The literature suggests the antiepileptic properties of PUFAs, although this evidence emerges more from basic science (Porta et al., 2009; Taha et al., 2008, 2009) than from clinical trials (Bromfield et al., 2008; Yuen et al., 2005). Several mechanisms have been proposed to explain the effects of the n-3 PUFAs on seizure thresholds. One hypothesis suggests that PUFAs may act by modulating voltage-dependent ion channels (Young et al., 2000). This is the most commonly held hypothesis; however, a couple of other possibilities exist. Either n-3 PUFAs may act by binding to the peroxisome proliferator-activated receptors PPAR-$\alpha$ and PPAR-$\gamma$ or they may act by antagonizing neuroinflammation. These hypotheses are currently under investigation.

## CONCLUSIONS

There is a lack of convergence between our scientific advances thus far in linking inflammatory processes with epilepsy and our ability to translate those findings into therapeutic advances in humans. There is a paucity of scientific evidence to support the development of sound clinical guidelines for implementing immunomodulatory therapy, whereas the toxicities associated with extended ACTH or corticosteroid treatment are significant. Intravenous immunoglobulin is generally better tolerated, but evidence to support its placement in a therapeutic algorithm for any epilepsy syndrome is insufficient. Where the weight of evidence in support of efficacy for a treatment is somewhat stronger, it is still not clear whether it is the immunomodulating effect that is critical or some other aspect of the treatment pharmacology that is responsible for the observed benefits. Ultimately, clinical decisions are based on the empiric findings of efficacy and the demonstration of safety (or acceptable risk that is not out of proportion to the benefit). It is that equation that has relegated immune therapy to being a late treatment option in most epilepsy syndromes.

Data from basic science research are promising. The data on NSAIDs are still unclear, but they show some promise as anticonvulsants in seizure models. This has not been demonstrated in clinical practice. Some immunomodulatory drugs tantalize us as being the holy grails of antiepileptogenic therapy. In clinical practice, though, many of these agents (cyclosporine, tacrolimus) are either associated with the potential for encephalopathy or seizures as well as other risks to the patient in terms of infection and organ toxicity. Currently, PUFAs are probably the best candidates for use in clinical practice, but placebo-controlled, double-blinded clinical trials are needed for verification.

## REFERENCES

Akarsu, E.S., S. Ozdayi, E. Algan, F. Ulupinar. (2006). The neuronal excitability time-dependently changes after lipopolysaccharide administration in mice: possible role of cyclooxygenase-2 induction. *Epilepsy Res.*, 71:181–187.

Akula, K.K., A. Dhir, S.K. Kulkarni. (2008). Rofecoxib, a selective cyclooxygenase-2 (COX-2) inhibitor increases pentylenetetrazol seizure threshold in mice: possible involvement of adenosinergic mechanism. *Epilepsy Res.*, 78:60–70.

Andrews, J.M., J.A. Thompson, T.J. Pysher, M.L. Walker, M.E. Hammond. (1990). Chronic encephalitis, epilepsy, and cerebrovascular immune complex deposits. *Ann. Neurol.*, 28:88–90.

Ariizumi, M., K. Baba, S. Hibio, H. Shiihara, K. Michihiro, K. Ogawa, O. Okubo. (1987). Immunoglobulin therapy in the West syndrome. *Brain Dev.*, 9:422–425.

Asanuma, M., Y. Kondo, N. Ogawa, A. Mori. (1994). Effects of repeated immunosuppressant treatment on rat pentylenetetrazol-induced convulsion. *Jpn. J. Psychiatry Neurol.*, 48:281–284.

Asanuma, M., S. Nishibayashi, Y. Kondo, E. Iwata, M. Tsuda, N. Ogawa. (1995). Effects of single cyclosporin a pretreatment on pentylenetetrazol-induced convulsion and on TRE-binding activity in the rat brain. *Brain Res. Mol. Brain Res.*, 33:29–36.

Bahi-Buisson, N., V. Villanueva, C. Bulteau, O. Delalande, O. Dulac, C. Chiron, R. Nabbout. (2007). Long term response to steroid therapy in Rasmussen encephalitis. *Seizure*, 16:485–492.

Baik, E.J., E.J. Kim, S.H. Lee, C. Moon. (1999). Cyclooxygenase-2 selective inhibitors aggravate kainic acid induced seizure and neuronal cell death in the hippocampus. *Brain Res.*, 843:118–129.

Baram, T.Z., W.G. Mitchell, O.C. Snead, 3rd, E.J. Horton, M. Saito. (1992). Brain–adrenal axis hormones are altered in the CSF of infants with massive infantile spasms. *Neurology*, 42:1171–1175.

Baram, T.Z., W.G. Mitchell, R.A. Hanson, O.C. Snead, 3rd, E.J. Horton. (1995). Cerebrospinal fluid corticotropin and cortisol are reduced in infantile spasms. *Pediatr. Neurol.*, 13:108–110.

Baram, T.Z., W.G. Mitchell, A. Tournay, O.C. Snead, R.A. Hanson, E.J. Horton. (1996). High-dose corticotropin (ACTH) versus prednisone for infantile spasms: a prospective, randomized, blinded study. *Pediatrics*, 97:375–379.

Bartolomei, F., J. Boucraut, M. Barrie, J. Kok, C. Dravet, D. Viallat, D. Bernard, J.L. Gastaut. (1996). Cryptogenic partial epilepsies with anti-GM1 antibodies: a new form of immune-mediated epilepsy? *Epilepsia*, 37:922–926.

Baulac, S., I. Gourfinkel-An, R. Nabbout, G. Hubberfeld, J. Serratosa, E. LeGuern, M. (2004). Fever, genes, and epilepsy. *Lancet Neurol.*, 3:421–430.

Bingel, U., J.D. Pinter, M. Sotero de Menezes, J.M. Rho. (2003). Intravenous immunoglobulin as adjunctive therapy for juvenile spasms. *J. Child Neurol.*, 18:379–382.

Bromfield, E., B. Dworetzky, S. Hurwitz, Z. Eluri, L. Lane, S. Replansky, D. Mostofsky. (2008). A randomized trial of polyunsaturated fatty acids for refractory epilepsy. *Epilepsy Behav.*, 12:187–190.

Brunson, K.L., M. Eghbal-Ahmadi, T.Z. Baram. (2001a). How do the many etiologies of West syndrome lead to excitability and seizures? The corticotropin releasing hormone excess hypothesis. *Brain Dev.*, 23:533–538.

Brunson, K.L., N. Khan, M. Eghbal-Ahmadi, T.Z. Baram. (2001b). Corticotropin (ACTH) acts directly on amygdala neurons to down-regulate corticotropin-releasing hormone gene expression. *Ann. Neurol.*, 49:304–312.

Calder, P.C. (1996a). Can n-3 polyunsaturated fatty acids be used as immunomodulatory agents? *Biochem. Soc. Trans.*, 24:211–220.

Calder, P.C. (1996b). Effects of fatty acids and dietary lipids on cells of the immune system. *Proc. Nutr. Soc.*, 55:127–150.

Clark, S.R., C.J. McMahon, I. Gueorguieva et al. (2008). Interleukin-1 receptor antagonist penetrates human brain at experimentally therapeutic concentrations. *J. Cereb. Blood Flow Metab.*, 28:387–394.

De Negri, M., M.G. Baglietto, F.M. Battaglia, R. Gaggero, A. Pessagno, L. Recanati. (1995). Treatment of electrical status epilepticus by short diazepam (DZP) cycles after DZP rectal bolus test. *Brain Dev.*, 17:330–333.

Dhir, A., P.S. Naidu, S.K. Kulkarni. (2006a). Effect of cyclooxygenase inhibitors on pentylenetetrazol (PTZ)-induced convulsions: possible mechanism of action. *Prog. Neuropsychopharmacol. Biol. Psychiatry*, 30:1478–1485.

Dhir, A., P.S. Naidu, S.K. Kulkarni. (2006b). Effect of cyclooxygenase-2 (COX-2) inhibitors in various animal models (bicuculline, picrotoxin, maximal electroshock-induced convulsions) of epilepsy with possible mechanism of action. *Indian J. Exp. Biol.*, 44:286–291.

Dhir, A., P.S. Naidu, S.K. Kulkarni. (2006c). Effect of rofecoxib, a cyclo-oxygenase-2 inhibitor, on various biochemical parameters of brain associated with pentylenetetrazol-induced chemical kindling in mice. *Fundam. Clin. Pharmacol.*, 20:255–261.

Dhir, A., P.S. Naidu, S.K. Kulkarni. (2007). Neuroprotective effect of nimesulide, a preferential COX-2 inhibitor, against pentylenetetrazol (PTZ)-induced chemical kindling and associated biochemical parameters in mice. *Seizure*, 16:691–697.

Dhir, A., K.K. Akula, S.K. Kulkarni. (2008). Rofecoxib potentiates the anticonvulsant effect of topiramate. *Inflammopharmacology*, 16:83–86.

Dube, C., A. Vezzani, M. Behrens, T. Bartfai, T.Z. Baram. (2005). Interleukin-1beta contributes to the generation of experimental febrile seizures. *Ann. Neurol.*, 57:152–155.

Emsley, H.C., C.J. Smith, R.F. Georgiou et al. (2005). A randomised phase II study of interleukin-1 receptor antagonist in acute stroke patients. *J. Neurol. Neurosurg Psychiatry*, 76:1366–1372.

Fayad, M.N., R. Choueiri, M. Mikati. (1997). Landau–Kleffner syndrome: consistent response to repeated intravenous gamma-globulin doses: a case report. *Epilepsia*, 38:489–494.

Fujisaki, Y., A. Yamauchi, S. Dohgu et al. (2002). Cyclosporine A-increased nitric oxide production in the rat dorsal hippocampus mediates convulsions. *Life Sci.*, 72:549–556.

Gaggero, R., R. Haupt, M. Paola Fondelli, R. De Vescovi, A. Marino, E. Lanino, S. Dallorso, M. Faraci. (2006). Intractable epilepsy secondary to cyclosporine toxicity in children undergoing allogeneic hematopoietic bone marrow transplantation. *J. Child Neurol.*, 21:861–866.

Galanopoulou, A.S., A. Bojko, F. Lado, S.L. Moshe. (2000). The spectrum of neuropsychiatric abnormalities associated with electrical status epilepticus in sleep. *Brain Dev.*, 22:279–295.

Gayatri, N.A., C.D. Ferrie, H. Cross. (2007). Corticosteroids including ACTH for childhood epilepsy other than epileptic spasms. *Cochrane Database Syst. Rev.*, CD005222.

Gleeson, J.G., A.J. duPlessis, P.D. Barnes, J.J. Riviello, Jr. (1998). Cyclosporin A acute encephalopathy and seizure syndrome in childhood: clinical features and risk of seizure recurrence. *J. Child Neurol.*, 13:336–344.

Gobbo, O.L., S.M. O'Mara. (2004). Post-treatment, but not pre-treatment, with the selective cyclooxygenase-2 inhibitor celecoxib markedly enhances functional recovery from kainic acid-induced neurodegeneration. *Neuroscience*, 125:317–327.

Gorji, A., H.H. Scheld, E.J. Speckmann. (2002). Epileptogenic effect of cyclosporine in guinea-pig hippocampal slices. *Neuroscience*, 115:993–997.

Granata, T., L. Fusco, G. Gobbi et al. (2003). Experience with immunomodulatory treatments in Rasmussen's encephalitis. *Neurology*, 61:1807–1810.

Gross-Tsur, V., R.S. Shalev, E. Kazir, D. Engelhard, N. Amir. (1993). Intravenous high-dose gammaglobulins for intractable childhood epilepsy. *Acta Neurol. Scand.*, 88:204–209.

Grote, C.L., P. Van Slyke, J.A. Hoeppner. (1999). Language outcome following multiple subpial transection for Landau–Kleffner syndrome. *Brain*, 122(Pt. 3):561–566.

Hancock, E.C., J.P. Osborne, S.W. Edwards. (2008). Treatment of infantile spasms. *Cochrane Database Syst. Rev.*, CD001770.

Haspolat, S., E. Mihci, M. Coskun, S. Gumuslu, T. Ozben, O. Yegin. (2002). Interleukin-1beta, tumor necrosis factor-alpha, and nitrite levels in febrile seizures. *J. Child. Neurol.*, 17:749–751.

Heida, J.G., Q.J. Pittman. (2005). Causal links between brain cytokines and experimental febrile convulsions in the rat. *Epilepsia*, 46:1906–1913.

Helminen, M., T. Vesikari. (1990). Increased interleukin-1 (IL-1) production from LPS-stimulated peripheral blood monocytes in children with febrile convulsions. *Acta Paediatr. Scand.*, 79:810–816.

Holtman, L., E.A. van Vliet, R. van Schaik et al. (2009). Effects of SC58236, a selective COX-2 inhibitor, on epileptogenesis and spontaneous seizures in a rat model for temporal lobe epilepsy. *Epilepsy Res.*, 84:56–66.

Homayoun, H., S. Khavandgar, A.R. Dehpour. (2002). Anticonvulsant effects of cyclosporin A on pentylenetetrazole-induced seizure and kindling: modulation by nitricoxidergic system. *Brain Res.*, 939:1–10.

Hrachovy, R.A., J.D. Frost, Jr., P. Kellaway, T.E. Zion. (1983). Double-blind study of ACTH vs. prednisone therapy in infantile spasms. *J. Pediatr.*, 103:641–645.

Jung, K.H., K. Chu, S.T. Lee, J. Kim, D.I. Sinn, J.M. Kim, D.K. Park, J.J. Lee, S.U. Kim, M. Kim, S.K. Lee, J.K. Roh. (2006). Cyclooxygenase-2 inhibitor, celecoxib, inhibits the altered hippocampal neurogenesis with attenuation of spontaneous recurrent seizures following pilocarpine-induced status epilepticus. *Neurobiol. Dis.*, 23:237–246.

Kanemoto, K., J. Kawasaki, T. Miyamoto, H. Obayashi, M. Nishimura. (2000). Interleukin (IL)1beta, IL-1alpha, and IL-1 receptor antagonist gene polymorphisms in patients with temporal lobe epilepsy. *Ann. Neurol.*, 47:571–574.

Kanemoto, K., J. Kawasaki, S. Yuasa, T. Kumaki, O. Tomohiro, R. Kaji, M. Nishimura. (2003). Increased frequency of interleukin-1beta–511T allele in patients with temporal lobe epilepsy, hippocampal sclerosis, and prolonged febrile convulsion. *Epilepsia*, 44:796–799.

Kauffman, M.A., D.G. Moron, D. Consalvo, R. Bello, S. Kochen. (2008). Association study between interleukin 1 beta gene and epileptic disorders: a HuGe review and meta-analysis. *Genet. Med.*, 10:83–88.

Klein, R., S. Livingston. (1950). The effect of adrenocorticotropic hormone in epilepsy. *J. Pediatr.*, 37:733–742.

Kunz, T., E.H. Oliw. (2001a). The selective cyclooxygenase-2 inhibitor rofecoxib reduces kainate-induced cell death in the rat hippocampus. *Eur. J. Neurosci.*, 13:569–575.

Kunz, T., E.H. Oliw. (2001b). Nimesulide aggravates kainic acid-induced seizures in the rat. *Pharmacol. Toxicol.*, 88:271–276.

Lagae, L.G., J. Silberstein, P.L. Gillis, P.J. Casaer. (1998). Successful use of intravenous immunoglobulins in Landau–Kleffner syndrome. *Pediatr. Neurol.*, 18:165–168.

Lahat, E., M. Livne, J. Barr, Y. Katz. (1997). Interleukin-1beta levels in serum and cerebrospinal fluid of children with febrile seizures. *Pediatr. Neurol.*, 17:34–36.

Lavigne, C.M., D.A. Shrier, M. Ketkar, J.M. Powers. (2004). Tacrolimus leukoencephalopathy: a neuropathologic confirmation. *Neurology*, 63:1132–1133.

Lerman, P., T. Lerman-Sagie, S. Kivity. (1991). Effect of early corticosteroid therapy for Landau–Kleffner syndrome. *Dev. Med. Child Neurol.*, 33:257–260.

Loscher, W., H. Siemes. (1988). Increased concentration of prostaglandin E-2 in cerebrospinal fluid of children with febrile convulsions. *Epilepsia*, 29:307–310.

Lux, A.L., S.W. Edwards, E. Hancock, A.L. Johnson, C.R. Kennedy, R.W. Newton, F.J. O'Callaghan, C.M. Verity, J.P. Osborne. (2005). The United Kingdom Infantile Spasms Study (UKISS) comparing hormone treatment with vigabatrin on developmental and epilepsy outcomes to age 14 months: a multicentre randomised trial. *Lancet Neurol.*, 4:712–717.

Majoie, H.J., M. de Baets, W. Renier, B. Lang, A. Vincent. (2006). Antibodies to voltage-gated potassium and calcium channels in epilepsy. *Epilepsy Res.*, 71:135–141.

Marchi, N., G. Betto, V. Fazio, Q. Fan, C. Ghosh, A. Machado, D. Janigro. (2009). Blood–brain barrier damage and brain penetration of antiepileptic drugs: role of serum proteins and brain edema. *Epilepsia*, 50:664–677.

Marescaux, C., E. Hirsch, S. Finck et al. (1990). Landau–Kleffner syndrome: a pharmacologic study of five cases. *Epilepsia*, 31:768–777.

Masuyama, T., M. Matsuo, T. Ichimaru, K. Ishii, K. Tsuchiya, Y. Hamasaki. (2002). Possible contribution of interferon-alpha to febrile seizures in influenza. *Pediatr. Neurol.*, 27:289–292.

Matsuo, M., K. Sasaki, T. Ichimaru, S. Nakazato, Y. Hamasaki. (2006). Increased IL-1beta production from dsRNA-stimulated leukocytes in febrile seizures. *Pediatr. Neurol.*, 35:102–106.

McVicar, K.A., S. Shinnar. (2004). Landau–Kleffner syndrome, electrical status epilepticus in slow wave sleep, and language regression in children. *Ment. Retard. Dev. Disabil. Res. Rev.*, 10:144–149.

Mertens, M., J.A. Singh. (2009). Anakinra for rheumatoid arthritis. *Cochrane Database Syst. Rev.*, CD005121.

Mikati, M.A., R. Saab. (2000). Successful use of intravenous immunoglobulin as initial monotherapy in Landau–Kleffner syndrome. *Epilepsia*, 41:880–886.

Mikati, M.A., A.N. Shamseddine. (2005). Management of Landau–Kleffner syndrome. *Paediatr. Drugs.*, 7:377–389.

Moia, L.J., H. Matsui, G.A. de Barros et al. (1994). Immunosuppressants and calcineurin inhibitors, cyclosporin A and FK506, reversibly inhibit epileptogenesis in amygdaloid kindled rat. *Brain Res.*, 648:337–341.

Moriwaki, A., Y.F. Lu, K. Tomizawa, H. Matsui. (1998). An immunosuppressant, FK506, protects against neuronal dysfunction and death but has no effect on electrographic and behavioral activities induced by systemic kainate. *Neuroscience*, 86:855–865.

Morrell, F., W.W. Whisler, M.C. Smith, T.J. Hoeppner, L. de Toledo-Morrell, S.J. Pierre-Louis, A.M. Kanner, J.M. Buelow, R. Ristanovic, D. Bergen et al. (1995). Landau–Kleffner syndrome: treatment with subpial intracortical transection. *Brain*, 118(Pt. 6):1529–1546.

Nishimura, T., H. Imai, Y. Minabe, A. Sawa, N. Kato. (2006). Beneficial effects of FK506 for experimental temporal lobe epilepsy. *Neurosci. Res.*, 56:386–390.

Oliveira, M.S., A.F. Furian, L.F. Royes, M.R. Fighera, N.G. Fiorenza, M. Castelli, P. Machado, D. Bohrer, M. Veiga, J. Ferreira, E.A. Cavalheiro, C.F. Mello. (2008). Cyclooxygenase-2/PGE$_2$ pathway facilitates pentylenetetrazol-induced seizures. *Epilepsy Res.*, 79:14–21.

O'Regan, M.E., J.K. Brown. (1998). Is ACTH a key to understanding anticonvulsant action? *Dev. Med. Child Neurol.*, 40:82–89.

Pechadre, J.C., B. Sauvezie, C. Osier, J. Gibert. (1977). The treatment of epileptic encephalopathies with gamma globulin in children [authors' transl.]. *Rev. Electroencephalogr. Neurophysiol. Clin.*, 7:443–447.

Porta, N., B. Bourgois, C. Galabert, C. Lecointe, P. Cappy, R. Bordet, L. Vallee, S. Auvin. (2009). Anticonvulsant effects of linolenic acid are unrelated to brain phospholipid cell membrane compositions. *Epilepsia*, 50:65–71.

Prayson, R.A., J.L. Frater. (2002). Rasmussen encephalitis: a clinicopathologic and immunohistochemical study of seven patients. *Am. J. Clin. Pathol.*, 117:776–782.

Reddy, D.S., M.A. Rogawski. (2002). Stress-induced deoxycorticosterone-derived neurosteroids modulate GABA(A) receptor function and seizure susceptibility. *J. Neurosci.*, 22:3795–805.

Rogawski, M.A., D.S. Reddy. (2002). Neurosteroids and infantile spasms: the deoxycorticosterone hypothesis. *Int. Rev. Neurobiol.*, 49:199–219.

Rogers, S.W., P.I. Andrews, L.C. Gahring, T. Whisenand, K. Cauley, B. Crain, T.E. Hughes, S.F. Heinemann, J.O. McNamara. (1994). Autoantibodies to glutamate receptor GluR3 in Rasmussen's encephalitis. *Science*, 265:648–651.

Sayyah, M., M. Javad-Pour, M. Ghazi-Khansari. (2003). The bacterial endotoxin lipopolysaccharide enhances seizure susceptibility in mice: involvement of proinflammatory factors: nitric oxide and prostaglandins. *Neuroscience*, 122:1073–1080.

Setkowicz, Z., M. Ciarach. (2007). Neuroprotectants FK-506 and cyclosporin A ameliorate the course of pilocarpine-induced seizures. *Epilepsy Res.*, 73:151–155.

Setkowicz, Z., M. Ciarach, R. Guzik, K. Janeczko. (2004). Different effects of neuroprotectants FK-506 and cyclosporin A on susceptibility to pilocarpine-induced seizures in rats with brain injured at different developmental stages. *Epilepsy Res.*, 61:63–72.

Shafiq, N., S. Malhotra, P. Pandhi. (2003). Anticonvulsant action of celecoxib (alone and in combination with sub-threshold dose of phenytoin) in electroshock induced convulsion. *Methods Find. Exp. Clin. Pharmacol.*, 25:87–90.

Shuto, H., Y. Kataoka, K. Fujisaki, T. Nakao, M. Sueyasu, I. Miura, Y. Watanabe, M. Fujiwara, R. Oishi. (1999). Inhibition of GABA system involved in cyclosporine-induced convulsions. *Life Sci.*, 65:879–887.

Sinclair, D.B. (2003). Prednisone therapy in pediatric epilepsy. *Pediatr. Neurol.*, 28:194–198.

Sinclair, D.B., T.J. Snyder. (2005). Corticosteroids for the treatment of Landau–Kleffner syndrome and continuous spike-wave discharge during sleep. *Pediatr. Neurol.*, 32:300–306.

Snead, O.C., 3rd, J.W. Benton, G.J. Myers. (1983). ACTH and prednisone in childhood seizure disorders. *Neurology*, 33:966–970.

Straussberg, R., J. Amir, L. Harel, I. Punsky, H. Bessler. (2001). Pro- and anti-inflammatory cytokines in children with febrile convulsions. *Pediatr. Neurol.*, 24:49–53.

Taha, A.Y., P.S. Huot, S. Reza-Lopez et al. (2008). Seizure resistance in fat-1 transgenic mice endogenously synthesizing high levels of omega-3 polyunsaturated fatty acids. *J. Neurochem.*, 105:380–388.

Taha, A.Y., E. Filo, D.W. Ma, W. McIntyre Burnham. (2009). Dose-dependent anticonvulsant effects of linoleic and alpha-linolenic polyunsaturated fatty acids on pentylenetetrazol induced seizures in rats. *Epilepsia*, 50:72–82.

Tamai, I., T. Takei, K. Maekawa, H. Ohta. (1983). Prostaglandin $F_2$ alpha concentrations in the cerebrospinal fluid of children with febrile convulsions, epilepsy and meningitis. *Brain Dev.*, 5:357–362.

Tomoum, H.Y., Badawy, N.M., Mostafa, A.A., Harb, M.Y. (2007). Plasma interleukin-1beta levels in children with febrile seizures. *J. Child. Neurol.*, 22:689–692.

Tsuru, T., M. Mori, M. Mizuguchi, M.Y. Momoi. (2000). Effects of high-dose intravenous corticosteroid therapy in Landau–Kleffner syndrome. *Pediatr. Neurol.*, 22:145–147.

Tutuncuoglu, S., N. Kutukculer, L. Kepe, C. Coker, A. Berdeli, H. Tekgul. (2001). Proinflammatory cytokines, prostaglandins and zinc in febrile convulsions. *Pediatr. Int.*, 43:235–239.

Uhari, M., H. Rantala, L. Vainionpaa, R. Kurttila. (1995). Effect of acetaminophen and of low intermittent doses of diazepam on prevention of recurrences of febrile seizures. *J. Pediatr.*, 126:991–995.

van Engelen, B.G., W.O. Renier, C.M. Weemaes et al. (1994). High-dose intravenous immunoglobulin treatment in cryptogenic West and Lennox–Gastaut syndrome: an add-on study. *Eur. J. Pediatr.*, 153:762–769.

Van Esch, A., H.A. Van Steensel-Moll, E.W. Steyerberg, M. Offringa, J.D. Habbema, G. Derksen-Lubsen. (1995). Antipyretic efficacy of ibuprofen and acetaminophen in children with febrile seizures. *Arch. Pediatr. Adolesc. Med.*, 149:632–637.

Van Hirtum-Das, M., E.A. Licht, S. Koh, J.Y. Wu, W.D. Shields, R. Sankar. (2006). Children with ESES: variability in the syndrome. *Epilepsy Res.*, 70(Suppl. 1):S248–S258.

van Rijckevorsel-Harmant, K., M. Delire, M. Rucquoy-Ponsar. (1986). Treatment of idiopathic West and Lennox–Gastaut syndromes by intravenous administration of human polyvalent immunoglobulins. *Eur. Arch. Psychiatry Neurol. Sci.*, 236:119–122.

van Stuijvenberg, M., G. Derksen-Lubsen, E.W. Steyerberg, J.D. Habbema, H.A. Moll. (1998). Randomized, controlled trial of ibuprofen syrup administered during febrile illnesses to prevent febrile seizure recurrences. *Pediatrics*, 102:E51.

Vezzani, A., T. Granata. (2005). Brain inflammation in epilepsy: experimental and clinical evidence. *Epilepsia*, 46:1724–1743.

Vezzani, A., D. Moneta, M. Conti, C. Richichi, T. Ravizza, A. De Luigi, M.G. De Simoni, G. Sperk, S. Andell-Jonsson, J. Lundkvist, K. Iverfeldt, T. Bartfai. (2000). Powerful anticonvulsant action of IL-1 receptor antagonist on intracerebral injection and astrocytic overexpression in mice. *Proc. Natl. Acad. Sci. U.S.A.*, 97:11534–11539.

Virta, M., M. Hurme, M. Helminen. (2002). Increased plasma levels of pro- and anti-inflammatory cytokines in patients with febrile seizures. *Epilepsia*, 43:920–923.

Watson, R., Y. Jiang, I. Bermudez, L. Houlihan, L. Clover, K. McKnight, J.H. Cross, I.K. Hart, A. Roubertie, J. Valmier, Y. Hart, J. Palace, D. Beeson, A. Vincent, B. Lang. (2004). Absence of antibodies to glutamate receptor type 3 (GluR3) in Rasmussen encephalitis. *Neurology*, 63:43–50.

Whitney, K.D., P.I. Andrews, J.O. McNamara. (1999). Immunoglobulin G and complement immunoreactivity in the cerebral cortex of patients with Rasmussen's encephalitis. *Neurology*, 53:699–708.

Yamatogi, Y., Y. Ohtsuka, T. Ishida, N. Ichiba, S. Ishida, S. Miyake, E. Oka, S. Ohtahara. (1979). Treatment of the Lennox syndrome with ACTH: a clinical and electroencephalographic study. *Brain Dev.*, 1:267–276.

Yamauchi, A., H. Shuto, S. Dohgu, Y. Nakano, T. Egawa, Y. Kataoka. (2005). Cyclosporin A aggravates electroshock-induced convulsions in mice with a transient middle cerebral artery occlusion. *Cell Mol. Neurobiol.*, 25:923–928.

Yoshida, Y., H. Shimada, T. Mori, H. Yoshihara, M. Takeoka, T. Takahashi. (2003). FK506-associated limbic injury following umbilical cord blood transplantation. *Bone Marrow Transplant*, 32:523–525.

Young, C., P.W. Gean, L.C. Chiou, Y.Z. Shen. (2000). Docosahexaenoic acid inhibits synaptic transmission and epileptiform activity in the rat hippocampus. *Synapse*, 37:90–94.

Yuen, A.W., J.W. Sander, D. Fluegel, P.N. Patsalos, G.S. Bell, T. Johnson, M.J. Koepp. (2005). Omega-3 fatty acid supplementation in patients with chronic epilepsy: a randomized trial. *Epilepsy Behav.*, 7:253–258.

# Section V

## Other Modulators of the Epileptic State

# 27 Impact of Neuroendocrine Factors on Seizure Genesis and Treatment

*Pavel Klein and Jaromir Janousek*

## CONTENTS

## INTRODUCTION

Hormones affect seizures. They modulate the expression of individual seizures in patients with epilepsy, and they may also, in some situations, contribute to the development of an enduring state of spontaneous recurrent seizures (i.e., epileptogenesis). Modulation of seizure occurrence by hormonal changes in patients with epilepsy may be more common than is appreciated. Among the factors that precipitate seizures, stress is the most common, triggering seizures in 30 to 60% of patients with epilepsy. Among women, events related to the reproductive lifecycle are also commonly associated with changes in seizure manifestation and development. About 20% of women may develop epilepsy during the time around menarche ("perimenarche"); 15 to 30% of females with preexisting seizures experience seizure exacerbation during perimenarche. About a third of women with refractory epilepsy have seizure exacerbation in relation to their menstrual cycle (i.e., catamenial epilepsy), and seizure occurrence changes during pregnancy and perimenopause in 20 to 50% of women. These clinical effects are likely due, in part or completely, to the associated hormonal changes. Thus, hormonal changes may be a common cause of modulation (usually an increase) of seizure expression in patients with epilepsy. Understanding the hormonal changes involved may have important consequences for the treatment of patients with epilepsy and, possibly, for prevention of epilepsy in certain at-risk patient populations.

In this chapter, we discuss the effect of sex hormones and stress hormones on seizure expression and on epileptogenesis, as well as the therapeutic potential of hormonal manipulation on treatment of seizures. We first discuss the basic scientific data concerning hormonal influence on neuronal excitability and seizures. We then outline the clinical settings where hormonal modulation of seizure expression or development may be important. Finally, we review hormonal treatments of epilepsy. Because more is known about the role of hormones in precipitating individual seizures in patients with established epilepsy than about their role in epileptogenesis, we discuss the role of hormones in seizure provocation in more detail than their role in true epileptogenesis.

## HORMONAL EFFECTS ON NEURONAL EXCITABILITY AND SEIZURES: ANIMAL DATA

Gonadal and adrenal steroid hormones have a wide-ranging influence upon neuronal activity in the central nervous system (CNS). Steroidal effects in the CNS occur by both receptor-mediated (genomic) and membrane (nongenomic) mechanisms, with long and short latencies of action, respectively (Frye, 2008; McEwen, 1991; McEwen and Milner, 2007; Paul and Purdy, 1992; Scharfman and MacLusky, 2006; Verrotti et al., 2007). The effects of these hormones and related neuropeptides on neuronal excitability and in animal seizure models are discussed briefly below; a more detailed treatment can be found in the preceding chapter in this volume.

### GONADAL STEROIDS

In general, estrogens increase neuronal excitability and facilitate seizure occurrence and epileptogenesis. Progesterone has the opposite effect; it inhibits neuronal excitability, seizures, and epileptogenesis.

### ESTROGENS

In adult female rats, seizure threshold fluctuates during the estrous cycle. Susceptibility to seizures is highest during proestrus, when serum estrogen levels are highest (Terasawa and Timiras, 1968). Estrogens lower seizure threshold in different animal seizure models in both physiological and pharmacological doses and promote seizure kindling and epileptogenesis in animals in different brain regions, including the amygdala, the hippocampus, and the neocortex (Buterbaugh and Hudson, 1991; Frye, 2008; Hom and Buterbaugh, 1986; Logothetis and Harner, 1960; Marcus et al., 1966; Nicoletti et al., 1985; Scharfman and MacLusky, 2006; Woolley and Timiras, 1962). Interestingly, physiological doses of estradiol (E2) facilitate seizure kindling in both female and male rats (Hom and Buterbaugh, 1986; Nicoletti et al., 1985). Pharmacological doses of estrogens activate spike discharges and, in animals with preexisting cortical lesions, may induce seizures, including fatal status epilepticus (Logothetis and Harner, 1960; Marcus et al., 1966); however, ovariectomy in adult rats does not alter seizure threshold (Woolley and Timiras, 1962). Thus, lack of estrogen may not protect against seizures, while excess estrogen promotes them. Furthermore, recent evidence suggests that the effect of estradiol on seizures may not be uniformly proconvulsant but may, under certain circumstances, reduce seizure severity and protect against status-epilepticus-induced neuronal damage (Veliskova, 2007). The opposite effects (pro- vs. anticonvulsant) of estrogens may be seen depending on the dose used, treatment duration, hormonal status, epilepsy model, and mode of administration (Veliskova, 2007). The issue is further complicated by the fact that estrogen and progesterone affect each other both via intracellular receptor (genomic)-mediated and direct membrane-related mechanisms. Thus, for example, estrogen enhances the conversion of progesterone to allopregnanolone (AP), a potent positive gamma-aminobutyric acid type A ($GABA_A$) receptor modulator, and thus enhances the antiseizure effect of progesterone (Frye and Rhodes, 2005). Little is known about the interaction of factors that determine the overall effect on neuronal excitability and seizures.

Estrogens affect neuronal excitability by genomically dependent mechanisms; by nongenomic, direct effects on neuronal membrane; and by affecting neuronal plasticity and the number of excitatory synapses. The genomic mechanisms are mediated by cytosolic neuronal estrogen receptors (ERs). Estrogen receptors ($\alpha$, $\beta$) are present in neurons and glia in different parts of the CNS (Scharfman and MacLusky, 2006; Veliskova, 2007). In addition to the anterior and mesiobasal parts of the hypothalamus, where they modulate the activity of neurons involved in the control of endocrine and behavioral functions associated with reproduction, they are found in great density in the cortical and medial amygdaloid nuclei and in lesser numbers in the hippocampal pyramidal cell layer and the subiculum and the neocortex (Shughrue et al., 1997; Simerly et al., 1990). ER-containing neurons colocalize with other neurotransmitters, including GABA (Flugge et al., 1986; McEwen, 1991). By regulating the expression of genes affecting the activity, release, and postsynaptic action of different neurotransmitters and neuromodulators, estrogens may act to increase the excitability of neurons that concentrate estradiol. For example, E2 reduces GABA synthesis in the corticomedial amygdala by decreasing the activity of glutamic acid decarboxylase (Wallis and Luttge, 1980).

Estrogens also exert a direct excitatory effect at the neuronal membrane, where E2 augments both the $N$-methyl-D-aspartate (NMDA) and AMPA/kainate glutamate receptor activity (Moss and Gu, 1999; Scharfman and MacLusky, 2006; Smith, 1989; Veliskova, 2007). This enhances the resting discharge rates of neurons in a number of brain areas (Frye, 2008; Scharfman and MacLusky, 2006; Smith, 1989). In the hippocampus, for example, E2 increases excitability of CA1 pyramidal neurons and induces repetitive firing in response to Schaffer collateral stimulation (Kawakami et al., 1970; Wong and Moss, 1992). Finally, estrogens potentiate neuronal excitability by regulating neuronal plasticity and synaptogenesis. E2 increases the density of dendritic spines carrying excitatory, NMDA-receptor-containing synapses on hippocampal CA1 pyramidal neurons (Woolley and McEwen, 1993; Woolley et al., 1996). The density of the excitatory synapses fluctuates during the estrous cycle, being highest by about one third during the proestrus, when estrogen levels are highest, compared to diestrus, when they are low. It decreases markedly following oophorectomy (Woolley et al., 1996). The estrogen-driven increase in excitatory synapses results in enhanced excitatory input to the CA1 neurons, in neuronal excitability, and in synchronization of neuronal outflow that is critical to formation of paroxysmal depolarization shifts, seizure genesis, and propagation (Yankova et al., 2001).

## PROGESTINS

Progesterone depresses neuronal firing (Landgren et al., 1978, 1987; Smith et al., 1987), lessens epileptiform discharges, and elevates seizure threshold in animal models of epilepsy (Landgren et al., 1978, 1987), increasing seizure threshold in female but not male rats (Nicoletti et al., 1985; Woolley and Timiras, 1962a). The seizure threshold during the rat estrous cycle correlates with serum progesterone levels. It is highest during diestrus, when the serum progesterone level is highest, suggesting a protective effect of progesterone against seizures. Progesterone inhibits seizure occurrence in animals with existing seizures in models of epilepsy that include both limbic and neocortical localization-related epilepsy (Edwards et al., 1999; Landgren et al., 1987; Marcus et al., 1966; Woolley and Timiras, 1962a). Progesterone inhibits epileptogenesis in different animal models of epilepsy, including amygdaloid and hippocampal electrical kindling and kainate models of limbic epilepsy (Edwards et al., 1999; Frye and Scalise, 2000; Holmes and Weber, 1984; Nicoletti et al., 1985). Inhibition of kindling is particularly prominent in immature animals in which progesterone is strongly protective against seizure development, particularly against limbic seizure kindling (Edwards et al., 1999; Frye and Scalise, 2000; Holmes and Weber, 1984; Nicoletti et al., 1985; Woolley and Timiras, 1962a). The anticonvulsant effect of progesterone is largely mediated by its metabolite, 3α-hydroxy-5α-pregnan-20-one (AP) (Kokate et al., 1994, 1996); see discussion below.

Like estrogens, progesterone affects neuronal excitability by genomic mechanisms, by nongenomic membrane effects, and by influencing synaptic plasticity. Progesterone receptors are found in most ER-containing brain regions and in areas such as cerebral cortex that lack estrogen receptors (McEwen, 1983). As with E2, progesterone may influence, via genomic mechanisms, the enzymatic activity controlling the synthesis and release of various neurotransmitters and neuromodulators produced by progesterone receptor containing neurons (McEwen, 1991; McEwen and Milner, 2007). Chronic progesterone exposure decreases the number of dendritic spines and synapses on hippocampal CA1 pyramidal neurons, counteracting the stimulatory effects of E2 (Woolley and McEwen, 1993). However, the acute anticonvulsant effect of progesterone occurs independently of progesterone receptor-mediated genomic effects, as it persists in progesterone receptor knockout mice (Reddy et al., 2004). The direct membrane effects of progesterone are inhibitory (see below).

## NEUROACTIVE STEROIDS

Most of the membrane effects of progesterone are due to the action of its 3α-hydroxylated metabolite, 3α-hydroxy-5α-pregnan-20-one (AP) (Frye, 2008; Gee et al., 1995; Paul and Purdy, 1992; Scharfman and MacLusky, 2006). AP and the 3α,5α-hydroxylated natural metabolite of the mineralocorticoid deoxycorticosterone, allotetrahydrodeoxycorticosterone (alloTHDOC), are the two most potent of a number of endogenous neuroactive steroids with a direct membrane effect on neuronal excitability (Gee et al., 1995; Majewska et al., 1986; Paul and Purdy, 1992). AP (but not alloTHDOC) is devoid of hormonal effects and may, together with other related neuroactive steroids, be thought of as an endogenous regulator of brain excitability with anxiolytic, anticonvulsant, and sedative–hypnotic properties (Paul and Purdy, 1992; Reddy, 2004a).

Both AP and alloTHDOC hyperpolarize hippocampal and other neurons by potentiating GABA-mediated synaptic inhibition. They act as positive allosteric modulators of the GABA_A receptor (GABA_A R), interacting with a steroid-specific site near the receptor to facilitate inward chloride flux through the channel, thereby prolonging the inhibitory action of GABA on neurons (Hosie et al., 2006; Majewska et al., 1986; Paul and Purdy, 1992). AP is one of the most potent ligands of GABA_A receptors in the CNS, with affinities similar to the potent benzodiazepines, and approximately 1000 times higher than pentobarbital (Majewska et al., 1986; Paul and Purdy, 1992). The parent steroid, progesterone, enhances GABA-induced chloride currents only weakly and only in high concentrations (Paul and Purdy, 1992; Wu et al., 1990).

Plasma and brain levels of AP parallel those of progesterone in rats, and in normal women plasma levels of AP correlate with progesterone levels during the menstrual cycle and pregnancy (Paul and Purdy, 1992). However, brain activity of progesterone and AP is not dependent solely on ovarian and adrenal production, as they are both synthesized *de novo* in the brain (Cheney et al., 1995). Their synthesis is region specific and includes the cortex and the hippocampus (Cheney et al., 1995).

In animal seizure models and the 6-Hz model, AP, alloTHDOC, and a 3α-hydoxylated deoxy-corticosterone metabolite, dihydrodeoxycorticosterone (DHDOC), have potent anticonvulsant effects, including bicuculline-, metrazol-, picrotoxin-, pentylenetetrazol-, pilocarpine-, and kainic-acid-induced seizures (Gasior et al., 2000; Kaminski et al., 2004; Kokate et al., 1999a; Reddy and Rogawski; 2001; Reddy et al., 2001). AP is also effective against status epilepticus but is ineffective against electroshock and strychnine-induced seizures (Gee et al., 1995; Kokate et al., 1996).

The anticonvulsant properties of AP resemble those of the benzodiazepine clonazepam (Gee et al., 1995; Kokate et al., 1994, 1996); AP is less potent than clonazepam but may have lower relative toxicity. However, an important difference between the benzodiazepines and AP is that, although the anticonvulsant effect of benzodiazepines habituates, that of AP does not (Reddy and Rogawski, 2001). The anticonvulsant effect of AP is greater in female rats in diestrus (equivalent to human mid-luteal phase) than in estrus (equivalent to ovulation) or in the male. This may be due to increased functional sensitivity of the $GABA_AR$ complex to AP during diestrus. AP modulates the expression of a couple of $GABA_AR$ subunits (Maguire and Mody, 2007; Maguire et al., 2005). AP induces expression of the delta $GABA_AR$ subunit which increases tonic inhibitory currents at $GABA_A$ receptors and GABA inhibition. AP also modulates the expression of another $GABA_AR$ subunit, the $\alpha4$ $GABA_AR$ subunit. This subunit *blocks* the effects of benzodiazepines on $GABA_AR$. Its expression is reduced by AP and is increased when AP levels decline before menstruation (Smith et al., 1998, 2007). Thus, during diestrus, high levels of progesterone/AP are associated with high expression of the delta $GABA_AR$ subunit, potentiation of GABAergic tonic inhibition, reduced excitability, and decreased seizure propensity. During proestrus (premenstrual phase), progesterone/AP withdrawal induces expression of the $\alpha4$ $GABA_AR$ subunit, as well as reduced expression of the delta $GABA_AR$ subunit, reduced $GABA_AR$ activation, and increased seizure propensity. This effect can be triggered by blocking the conversion of progesterone to AP with the $\alpha$-reductase enzyme inhibitor finasteride.

In contrast, some of the sulfated neuroactive steroids have excitatory neuronal effects. They include pregnenolone sulfate and dehydroepiandrosterone sulfate (DHEAS), the naturally occurring sulfated esters of the progesterone precursor pregnenolone and of the progesterone metabolite DHEA (Kokate et al., 1999b; Paul and Purdy, 1992; Reddy and Kulkarni, 1998). These steroids increase neuronal firing when directly applied to neurons by negatively modulating $GABA_AR$ (Paul and Purdy, 1992) and by facilitating glutamate-induced excitation at the NMDA receptor (Irwin et al., 1992). In animals, pregnenolone sulfate and DHEAS have a dose-dependent proconvulsant effect when injected into the cerebral ventricles or directly into the hippocampus (Kokate at al., 1999b; Williamson et al., 2004), although possibly not when injected systemically, as the sulfated steroids do not cross the blood–brain barrier readily. However, like AP, pregnenolone sulfate is synthesized in the CNS by glia and is present in high concentrations in epileptogenic areas such as the hippocampus (Williamson et al., 2004). The proconvulsant effect is prevented by chronic pretreatment with progesterone (Reddy and Kulkarni, 1998).

## ANDROGENS

The effect of androgens on experimental seizures is uncertain, with both pro- and anticonvulsant findings described in animal models. This may be due to the fact that androgens may be metabolized to estrogens or to dihydroprogesterone (and thence to tetrahydroprogesterone, or AP), with different effects on neuronal excitability and seizures (Frye, 2006). One metabolite of testosterone, $5\alpha,3\alpha$-androstandione ($5\alpha,3\alpha$-A, or androsterone), like AP and THDOC, positively modulates $GABA_AR$ and has anticonvulsant properties in animal epilepsy models (Kaminski et al., 2005).

## ADRENAL STEROIDS

Progesterone, AP, THDOC, DHDOC, pregnenolone sulfate, and DHEAS are all adrenally produced steroids. Their neuroactive effects in relation to experimental epilepsy are discussed above and in the next chapter. Glucocorticoids (GCs) facilitate epileptiform discharges and seizures in animal models (Krugers et al., 1995; Lee et al., 1989; Marcus et al., 1966; Roberts and Keith, 1994; Roberts et al., 1993). Corticosterone and dexamethasone exacerbate seizures induced by kainic acid and pentylenetetrazol, as well as other types, whereas adrenalectomy attenuates them, as does administration of the steroid synthesis inhibitor aminoglutethimide and the mineralocorticoid antagonists spironolactone and RU-26752, in addition to chronic downregulation of the hypothalamo–pituitary–adrenal (HPA) axis with prolonged low-dose cortisone treatment (Krugers et al., 1995; Roberts and Keith, 1994; Roberts et al., 1993). Glucocorticoids (GCs) also facilitate the development of cocaine and alcohol withdrawal-caused seizures (Kling et al., 1993; Roberts et al., 1993).

Glucocorticoids are known to increase neuronal excitability (Birnstiel et al., 1995; Dafny et al., 1973; McEwen and Milner, 2007; Woodbury, 1952), but the mechanism has not been elucidated. Two types of GC receptors are found in the brain: (1) mineralocorticoid receptors (MRs), with a high affinity for corticosterone and aldosterone, and (2) glucocorticoid receptors (GRs), with a low affinity for corticosterone—10 times lower than for MRs (Joels et al., 2008a,b; Reul and de Kloet, 1985). Both receptor subtypes are present in the principal cells of the hippocampus (including the dentate gyrus) and modulate the excitability of these neurons. Normal basal levels of corticosterone occupy the majority of the MR receptors. Stimulation of these receptors reduces the activity of voltage-gated calcium channels and enhances long-term potentiation (LTP). Higher amounts of corticosterone activate glucocorticoid receptors. This facilitates voltage-gated calcium channels but inhibits LTP. During the diurnal surge of corticosterone and following stress, the mineralocorticoid and glucocorticoid receptors are both maximally occupied. Such high levels of glucocorticoid receptor activation markedly increase calcium currents and induce expression of NMDA receptor subunit 1 (NR1) mRNA in hippocampal neurons (Joels, 2008; Joels et al., 2008b; McEwen and Milner, 2007). It is these higher levels of corticosterone that are associated with a decrease in the action potential threshold (Birnstiel et al., 1995).

Glucocorticoids also affect neuronal survival and, like estrogens, plasticity (McEwen and Milner, 2007; Sapolsky et al., 1985; Woolley et al., 1990). Chronic glucocorticoid excess results in hippocampal pyramidal neuronal loss and in loss of dendritic spines and branching. Glucocorticoid elevation exacerbates hypoxia-, stress-, and seizure-induced CA3 pyramidal neuronal loss by potentiating glutamate (i.e., NMDA) receptor-dependent excitotoxicity and by limiting neuronal energy supply by inhibiting glucose uptake. Adrenalectomy has a protective effect in all these situations (McEwen and Milner, 2007; Sapolsky et al., 1985).

## CORTICOTROPIN-RELEASING HORMONE

Corticotropin-releasing hormone (CRH) is a neuropeptide that controls the release of adrenocorticotropin (ACTH). It is important in the endocrine and behavioral regulation of the stress response (Dunn and Berridge, 1990). It is found in the hypothalamus, where it acts as ACTH-releasing hormone, but also in the amygdala, hippocampus, and the cortex (De Souza et al., 1984), where it acts as an excitatory neurotransmitter or neuromodulator. CRH induces arousal and, in higher doses, epileptiform discharges and seizures. The seizures start in the amygdala and may spread to the hippocampus and the cortex (Aldenhoff et al., 1983; Baram et al., 1992; Ehlers et al., 1983; Marrosu et al., 1992; Weiss et al., 1986). In adult animals, seizures occur with a latency of a few hours (Ehlers et al., 1983), suggesting that the effect could be mediated by a CRH-induced increase in glucocorticoid activity rather than by CRH itself. In fact, CRH-induced seizures are potentiated by chronic glucocorticoid administration (Rosen et al., 1994). However, in neonatal rats, the convulsant effect occurs within 2 minutes and at a 200-fold lower dose, and seizures precede changes in serum corticosterone level. It may therefore be due to a direct effect of CRH itself (Baram et al., 1992). Tolerance develops to the seizure-inducing effect of CRH after 2 days (Weiss et al., 1986). This may be due to seizure-induced production of CRH-binding protein (CRF-BP) (Smith et al., 1997). The potentiating effect of CRH on non-CRH-induced seizures is long lasting (Weiss et al., 1986); thus, it could contribute to kindling and the development of seizures after an acute CNS insult such as febrile status epilepticus.

Corticotropin-releasing hormone has an excitatory effect on neurons. It increases neuronal firing in the hippocampus by decreasing the afterhyperpolarizations of CA1 and CA3 hippocampal pyramidal cells and of the dentate hilus granule cells (Aldenhoff et al., 1983; Smith et al., 1997). The mechanism of the excitatory effect is unknown.

Epileptiform discharges and seizures induced by CRH are suppressed by carbamazepine, verapamil, clonidine (Marrosu et al., 1992), a number of CRH nonselective peptide antagonists, and by the selective nonpeptide CRH receptor type 1 antagonist NBI 27914 (Baram et al., 1996). The CRH1 receptor mediates the excitatory actions of CRH in the developing rat brain (Baram et al., 1997). The selective CRH1 antagonist α-helical CRH elevates seizure threshold in an infant rat febrile seizure model.

Although CRH can induce limbic seizures, limbic seizures, in turn, activate hippocampal and amygdaloid CRH-producing neurons (Piekut et al., 1996; Smith et al., 1997). A positive feedback loop may thus exist between limbic seizures and CRH.

Like glucocorticoids, CRH exacerbates seizure-induced hippocampal neuronal loss (De Souza, 1984). Unlike other convulsants that do not cause neuronal loss in the neonate, CRH-induced status epilepticus in neonatal rats results in neuronal loss in the hippocampal CA3 region, the amygdala, and the pyriform cortex (Ribak and Baram, 1996). Possibly as a result of this, CRH lowers the animals' subsequent threshold to kainic-acid-induced seizures (Baram and Schultz, 1995). A CRH antagonist, astressin, protects against seizure-related hippocampal neuronal loss when injected intracerebroventricularly, either before or after seizures (Maecker et al., 1997).

## HORMONAL EFFECTS ON SEIZURES AND EPILEPTOGENESIS: CLINICAL DATA

Changes in reproductive hormonal status associated with menarche, menstrual cycles, pregnancy, and menopause may all affect clinical manifestations of seizures.

### MENARCHE

There is increased susceptibility to development of seizures around the time of menarche (i.e., perimenarche) (Gowers, 1893; Klein et al., 2003; Niijima and Wallace, 1989; Rosciszewska, 1975; Silveira and Guerreiro, 1991). However, this is not reflected in epidemiological studies, the majority of which do not show an increased incidence of seizures during the age associated with menarche or puberty (Hauser et al., 1991). This may be due to a number of reasons.

First, most epidemiological studies concentrate on age brackets, typically at 5-year intervals, rather than endocrine events such as puberty or menarche. Because the endocrine events may gradually evolve over many years and because the relevant age may differ among subjects, patient grouping by age may not detect these events. In one study where the age bracket was 1 year only, a rise in the incidence of seizures during the 15th year of life was seen (Shamansky and Glaser, 1979). Second, some seizures—for example, absence seizures and the primary idiopathic benign partial epilepsies of childhood—remit spontaneously during adolescence, whereas others, such as juvenile myoclonic epilepsy, begin during adolescence. These opposing effects might cancel each other out in studies of overall incidence and prevalence of epilepsy.

Third, puberty and sexual maturation are protracted, multifaceted endocrinological events, with different hormonal changes occurring during different stages of the 7- to 10-year-long process. This makes correlation with possible seizure onset (or remission) difficult without specifically directed prospective studies, which do not exist.

Several studies, however, have shown that, in women, seizures commonly start during perimenarche (Klein et al., 2003; Niijima and Wallace, 1989; Silveira and Guerreiro, 1991) or adolescence (Svalheim et al., 2006). In a questionnaire survey, 19% of 165 adult women with epilepsy reported seizure onset at menarche (Smith et al., 1997), including 33% of women with primary generalized epilepsy and 14% of women with partial epilepsy. This was confirmed in another study that compared seizure onset during menarche and perimenarche to seizure onset during other childhood periods. Of 76 consecutively evaluated women whose seizures began between the ages of 0.5 and 18 years, seizure onset occurred within 2 years of menarche in 35% of them. In 17%, seizures began during the year of menarche (Klein et al., 2003). Primary generalized and localization-related epilepsies were equally likely to start during perimenarche.

In addition, several studies have shown an increased risk of exacerbation of seizures during puberty. Rosciszewska (1975) showed in the 1970s that approximately one third of girls with preexisting seizures experience seizure exacerbation during puberty. Another third of girls showed improvement in seizure control, while the remaining third were not affected by sexual maturation. Seizure exacerbation is more likely to occur in girls with focal epilepsies, with refractory seizures, with evidence of CNS damage, and with delayed menarche. In more recent studies, 13 to 38% of girls experienced seizure exacerbation during perimenarche/puberty (Klein et al., 2003).

Changes in reproductive hormones may be responsible for these observations. Sexual maturation begins with adrenarche, usually between the ages of 8 and 10 years. It is initiated by a marked increase in the secretion of DHEAS and DHEA (Apter and Vihko, 1977). During perimenarche, which follows, ovarian secretion of estrogens precedes secretion of progesterone. Ovarian secretion of estrogens begins during thelarche (median age, 9.8 years) (Baram and Schultz, 1995) and increases steadily through menarche (median age, 12.8 years) until the onset of ovulation (Herman-Giddens et al., 1997). In contrast, ovarian secretion of progesterone does not begin until menstrual cycles become ovulatory, 12 to 18 months after menarche in the majority of girls (Speroff and Fritz, 2005), with a parallel increase in serum AP levels in late puberty (Genazzani et al., 2000). Thus, the DHEAS and estrogen secretion precedes that of progesterone by 2 to 6 and 2 to 4 years, respectively.

Localization-related epilepsy is commonly due to an underlying focal lesion. Patients with CNS insults such as severe head trauma, infections, complex febrile seizures, perinatal insults, and developmental disorders have an increased risk for developing seizures (Annegers et al., 1988, 1998; Hauser, 1994; Marks et al., 1992). Epilepsy may occur after a variable period that may last for years after the original insult (Marks et al., 1992). According to the "two-hit" hypothesis of epileptogenesis, reorganization of neuronal networks after the first insult may increase neuronal excitability, but not sufficiently to result in epilepsy (Dichter, 1997). A second event later in life may further lower seizure threshold and result in the appearance of epilepsy. Continued exposure of the brain during this time to the proconvulsant effects of estrogen and DHEAS, without the anticonvulsant effect of progesterone, may facilitate kindling of a preexisting CNS lesion into an epileptic focus and thus provide the second hit of epileptogenesis in susceptible females.

Animal studies support this hypothesis. Both DHEAS and estrogen promote epilepsy kindling, whereas progesterone retards it (Holmes and Weber, 1984; Hom et al., 1993; Reddy and Kulkarni, 1998; Terasawa and Timiras, 1968; Woolley and Timiras, 1962b). Female rats ovariectomized before puberty have a higher seizure threshold in adulthood compared to control animals (Hom et al., 1993; Reddy and Kulkarni, 1998; Woolley and Timiras, 1962b). The difference in seizure threshold between ovariectomized and intact animals only starts at the time of sexual maturation. Rats castrated after puberty do not differ from control animals (Terasawa and Timiras, 1968). This suggests that castration before puberty may be protective against later development of seizures. This is likely due to the estrogenic effect of maturing ovaries, as exposure of young female rats to even a single high dose of estrogen before maturation results in a permanent increase in neuronal excitability and a lowering of seizure threshold that persists well into adulthood (Heim, 1966). In contrast, administration of progesterone to rats shortly before sexual maturation is strongly protective against amygdaloid seizure kindling (Holmes and Weber, 1984).

The hormonal effects on epileptogenesis during sexual maturation could be due to the effects of gonadal hormones on neuronal plasticity and synaptogenesis. As noted earlier, in the adult female rat, the density of excitatory synapses on the apical dendrites of CA1 pyramidal neurons is increased by estrogen and reduced by progesterone (Woolley and McEwen, 1993; Woolley et al.,

1996). Ovariectomized animals treated with estradiol without progesterone, a situation analogous to human perimenarche, show a 30% increase in the density of the dendritic spines associated with excitatory synapses. These new synaptic connections form especially between previously unconnected hippocampal neurons (Yankova et al., 2001); thus, they facilitate synchronization of neuronal discharge and generation or propagation of seizures. The increase in excitatory synapses formation is opposed by progesterone (Woolley and McEwen, 1993). It is possible by promoting excitatory synaptogenesis and transmission that prolonged, continued exposure of the brain to DHEAS and estrogen without progesterone during adrenarche and perimenarche could facilitate kindling of a previously lesioned brain and provide the second hit of epileptogenesis in susceptible females.

The prolonged imbalance between the excitatory and inhibitory neuroactive steroids during perimenarche and the same pathophysiological mechanisms may also be responsible for the observed seizure exacerbation during perimenarche in girls with partial epilepsy.

## CATAMENIAL EPILEPSY

Catamenial (from the Greek *katamenia*: *kata*, "by"; *men*, "month") epilepsy refers to seizure exacerbation in relation to the menstrual cycle. Seizures do not occur randomly in the majority of men and women with epilepsy. They tend to cluster in over 50% of cases (Tauboll et al., 1991). In women, such clustering may relate to the menstrual cycle. This has been recognized since the 19th century (Gowers, 1893). Based on the analysis of both seizure frequency (average daily seizure frequency during four phases of the menstrual cycle) and the nature of menstrual cycles, Herzog et al. (1997) described three patterns of catamenial seizure exacerbation. The first pattern (C1) involves seizure exacerbation during the perimenstrual phase (menstrual cycle days –3 to 3, with the first day of menstrual blood flow being the first day of the cycle). The second pattern (C2) involves seizure exacerbation during periovulatory phase (days 10 to 13). Both C1 and C2 occur in women with normal ovulatory cycles. In the third pattern (C3), seizures are exacerbated during the entire second half (days 10 to 3) of anovulatory cycles.

Clinical determination of such exacerbation is made by daily seizure counts averaged during the different phases of the menstrual cycle (perimenstrual, day –3 to 3; follicular, day 4 to 10; periovulatory, day 11 to 16; and luteal, day 17 to –3) and the determination of midluteal (menstrual cycle day 20 to 22) serum progesterone levels to distinguish between normal and anovulatory menstrual cycles (>6 ng/mL in normal ovulatory cycles). Herzog et al. (Herzog, 2008; Herzog et al., 1997; 2004) proposed a definition of catamenial epilepsy based on the severity of seizure exacerbation: a twofold increase in average daily seizure frequency during the phases of exacerbation relative to the baseline phases. Using this definition, approximately one third of women with intractable partial epilepsy have catamenial seizure exacerbation. However, most studies of the frequency of catamenial seizures are based on populations of women with epilepsy seen in tertiary centers specializing in hormonal–seizure interaction (Herzog et al., 1997, 2004). The incidence of catamenial exacerbation is uncertain in the general population of women with epilepsy.

Menstrually related hormonal fluctuations of estrogen and progesterone likely underlie catamenial seizure exacerbation. Estrogens activate epileptiform discharges and may induce seizures in women with epilepsy (Logothetis et al., 1959). In contrast, progesterone given in doses resulting in luteal phase plasma levels suppresses epileptiform discharges (Backstrom et al., 1984). Accordingly, changes in reproductive hormonal status associated with menstrual cycles may affect the clinical manifestation of seizures (Laidlaw, 1956; Logothetis et al., 1959).

Both progesterone deficiency and estrogen excess relative to progesterone may contribute to the catamenial pattern of seizure exacerbation in both normal women and in women with menstrual irregularities (Herzog et al., 1997; Laidlaw, 1956). In rats, seizure propensity is greatest during proestrus, when the serum estrogen levels are highest, and lowest during diestrus, when the progesterone levels are highest (Woolley and Timiras, 1962b). In women with epilepsy, seizure frequency is similarly high during the periovulatory surge of serum estrogen levels and is low when serum progesterone levels are high during the luteal phase in normally ovulating women. The periovulatory seizure exacerbation may be related to the periovulatory surge in estrogen secretion. Seizure exacerbation pre- or perimenstrually may be related to the fall in serum progesterone levels (Herzog et al., 1997; Reddy and Rogawski, 2001). This may be due to the withdrawal of the progesterone metabolite AP. In an animal model of catamenial epilepsy, abrupt withdrawal of AP can be achieved biochemically with finasteride, a 5α-reductase inhibitor that blocks the conversion of progesterone to AP. Such withdrawal is associated with marked increased seizure susceptibility (Reddy, 2004b; Reddy et al., 2001). Few data are available on human serum levels of neurosteroids in relationship to catamenial seizure exacerbation. One study showed a reduced serum level of another progesterone $GABA_A$R-potentiating neurosteroid, THDOC, every 4 days throughout the menstrual cycle in women with catamenial epilepsy compared to controls. There was no difference in the concentrations of P4, pregnenolone, and AP (Tuveri et al., 2008). Although the mechanism underlying this effect is not clear, it has been suggested that the modulating effect of AP on $GABA_A$R may be responsible. Progesterone withdrawal is associated with a change in the expression of $GABA_A$R subunits and properties, a reduction in GABA inhibition, and an increase in neuronal excitability. Specifically, progesterone or AP withdrawal is associated with an increased expression of the α4 subunit of $GABA_A$R, which confers resistance of $GABA_A$R to GABA and to benzodiazepines (Smith et al., 1997, 2007) and with decreased expression of the delta $GABA_A$R subunit, which mediates the neurosteroid-induced tonic inhibitory current (Frye, 2008; Reddy, 2004; Smith et al., 2007).

In women with anovulatory cycles, seizure exacerbation is seen during the entire second half of the cycle (Backstrom et al., 1984; Herzog et al., 1997). During an anovulatory cycle, estrogen levels rise at the end of the follicular phase and stay elevated throughout the luteal phase until premenstrually, as in normally menstruating women, but there is little or no progesterone secretion. Thus, there is an estrogen/progesterone (E/P) imbalance, with a relative excess of estrogen throughout the entire second half of the menstrual cycle and associated seizure exacerbation (Backstrom, 1976). However, a recent observation suggests that the pattern of worsening of seizures throughout the menstrual cycle except in the early-mid-ovulatory phase (i.e., improvement in seizures during the follicular phase) may also occur in some women with ovulatory cycles (Herzog and Fowler, 2008).

Premenstrual exacerbation of seizures may also be related to a catamenial fluctuation in antiepileptic drug (AED) levels. Plasma levels of phenytoin and phenobarbital, for example, decline premenstrually (Rosciszewska et al., 1986; Shavit et al., 1984). These AEDs are metabolized by hepatic microsomal enzymes of the CYP3A subfamily which also metabolize gonadal steroids. Competition between the hormones and AEDs such as phenytoin results in greater phenytoin metabolism when gonadal serum levels decline premenstrually, accounting for the decline in AED level. A reduction of plasma lamotrigine levels may also occur during the luteal phase of the natural menstrual cycle (Herzog et al., 2009). Lamotrigine is metabolized by glucuronidation, with UDP–glucuronosyltransferase (UDP–GST) as the key enzyme. Gonadal steroids induce hepatic production of UDP–GST. Thus, when gonadal steroids are high, such as during periovulatory surge or during the luteal phase, lamotrigine levels may fall, and seizure breakthrough may occur (Herzog et al., 2009; Sabers et al., 2003). When gonadal plasma levels fall perimenstrually and remain low during the early-mid-follicular phase, lamotrigine levels may rise, and this could result in symptoms of toxicity. This phenomenon was first described in women taking estrogen-containing oral contraceptive (OC). In these women, plasma lamotrigine levels may drop by >50% during the active phase of the pill and rise during the inactive phase (Herzog et al., 2009; Sabers et al., 2003; Thorneycroft et al., 2006). The estrogens in the OC are primarily responsible for the effect. The fluctuation in levels may be associated with increased seizure frequency. Reduction of plasma levels during the active phase of OC treatment has also been reported with valproate (Herzog et al., 2009) and oxcarbazepine, both of which also undergo glucuronidation.

## REPRODUCTIVE ENDOCRINE DISORDERS AND THE DEVELOPMENT AND EXACERBATION OF EPILEPSY

Reproductive endocrine disorders may favor the development of temporal lobe epilepsy (TLE) in women. As noted above, women with anovulatory cycles have continued ovarian secretion of estrogen with failure of progesterone secretion during the luteal phase of the cycle. The resulting continuous exposure of the amygdala and the hippocampus to estrogen without the normal cyclical progesterone effects may promote interictal epileptiform activity and the possibility of kindling (Sharf et al., 1969).

Epilepsy itself is associated with an increased incidence of reproductive endocrine disorders. Approximately 20% of women with temporal lobe epilepsy suffer from polycystic ovarian syndrome, and another 5% each may suffer from hypothalamic hypogonadism and hyperprolactinemia (Herzog, 2008; Herzog et al., 1986). All of these conditions are associated with anovulatory menstrual cycles. One third of all menstrual cycles of women with epilepsy have been estimated to be anovulatory (Cummings et al., 1995). The chronically increased estrogen/progesterone ratio may promote seizure occurrence. Anovulatory cycles are associated with more frequent seizures (Backstrom, 1976; Mattson et al., 1981). Thus, a positive-feedback cycle may be envisaged, whereby TLE leads to endocrine reproductive disorders with failure of ovulation, which in turn further exacerbates the seizure disorder.

## MENOPAUSE

The effect of menopause on epilepsy has not been studied extensively (Harden, 2008). The term *menopause* refers to a complex process that encompasses menopause, cessation of all menstruation, and perimenopause, the preceding decline in reproductive endocrine function. Perimenopause often extends for several years. The decline in progesterone secretion occurs early in perimenopause, as ovulatory cycles change to anovulatory, an early endocrine change of the perimenopause. In contrast, estrogen secretion remains normal through most of perimenopause and may even increase episodically during it when, as a result of erratic follicular development, multiple follicles develop during some menstrual cycles (Burger et al., 1995; Santoro et al., 1996). Estrogen levels only drop consistently late in perimenopause, during the last few months before cessation of menses, as the follicle pool becomes exhausted. Thus, for a period of time lasting up to several years, there may be relative excess of serum estrogen levels compared to serum progesterone levels.

Based on the pattern of hormonal change, an evolving seizure pattern with seizure exacerbation during perimenopause might be expected: initial seizure exacerbation when progesterone secretion declines, but continued estrogen secretion followed by stabilization or improvement after menopause as estrogen secretion ceases. This pattern was, in fact, shown to occur in a recent questionnaire study, in which 64% of women experienced seizure exacerbation, and only 13% of women experienced seizure improvement during perimenopause. In contrast, 43% of women had seizure improvement during menopause, with only 31% experiencing seizure exacerbation (Harden et al., 1999). Hormone replacement therapy may further affect seizure expression during perimenopause and menopause. Of the women with seizure exacerbation during menopause, 63% were taking estrogen hormone replacement therapy (Harden et al., 1999). In a small but double-blind, randomized, placebo-controlled trial of the effect of hormone replacement therapy on seizure frequency in postmenopausal women with epilepsy taking stable doses of AEDs, Prempro® (premarin + medroxyprogesterone) dose-dependently increased seizure frequency and severity (Harden et al., 2006). Interestingly, partial epilepsy may begin during the climacteric, sometimes without an apparent cause. This occurred in 8 out of 61 (13%) menopausal women in one study (Abbasi et al., 1999). It is possible that the chronic exposure of the brain to estrogen without progesterone during the perimenopausal years could promote epileptogenesis in women with minor, occult CNS lesions in a way similar to the suggested epileptogenic effect of perimenarche. No studies have addressed this possibility.

## PREGNANCY

Pregnancy has variable effects on epilepsy. A small number of women experience seizures for the first time during pregnancy and have seizures only during pregnancy (Knight and Rhind, 1975). In about one third of women, seizures worsen during pregnancy (Harden and Sethi, 2008; Knight and Rhind, 1975; Schmidt et al., 1983). In about one sixth of women, seizures decrease in frequency. Seizure freedom in the 9 to 12 months before pregnancy is associated with seizure freedom during pregnancy (Harden and Sethi, 2008). Nonhormonal factors affecting seizure occurrence during pregnancy include altered medication compliance, AED absorption, body-space distribution, protein binding, liver metabolism, and renal clearance. Profound hormonal changes occur during pregnancy, including up to tenfold increases in estrogens, progesterone, cortisol, DHEAS, and other neuroactive hormones.

In the only study to date of the effects of hormones on seizures during pregnancy, no association was seen between serum progesterone and estrogen levels and estrogen/progesterone ratios and seizure frequency during pregnancy (Ramsay, 1987). Other hormones have not been examined. Levels of certain AEDs actually decline during pregnancy (Pennell et al., 2007). This is particularly true for lamotrigine, whose levels (total and free) decline steadily from the first trimester through the beginning of the third trimester, reaching a nadir at that time of >50% of their prepregnancy levels (Pennell et al., 2008). This is due to increased clearance of lamotrigine. The reduction in plasma levels is associated with increased seizure frequency. Levels return to those of prepregnancy during the puerperium, so toxicity may result if the lamotrigine dose is increased during pregnancy and not reduced after delivery. Similar observation has been made with levetiracetam, whose plasma levels also fall by >50% during pregnancy (Tomson et al., 2007; Westin et al., 2008). Because of these changes, AED levels should be monitored regularly (e.g., monthly) during pregnancy and puerperium.

## STRESS

Stress is probably the most common precipitant of seizures (Frucht et al., 2000; Joels et al., 2008b; Klein and Pezzullo, 2000; Mattson, 1991). Several surveys have shown that in 30 to 60% of epileptic patients seizure occurrence is related to stress, stressful events, or subjectively perceived change in stress levels (Frucht et al., 2000; Klein and Pezzullo, 2000; Mattson, 1991; Neugebauer et al., 1994; Spatt et al., 1998; Spector et al., 2000; Swinkels et al., 1998). In contrast, studies relating seizure diaries to life events or subjective distress have shown a more variable relationship of seizures to stress, with 14 to 58% of patients having an increase or, occasionally, decrease in seizure frequency with stress (Neugebauer et al., 1994; Swinkels et al., 1998; Temkin and Davis, 1984; Webster and Mawer, 1989). Several reports have documented seizure triggering in a controlled setting by exposure to emotionally stressful stimuli during a clinical interview or videotape presentation (Feldman and Paul, 1976; Groethuysen et al., 1957; Small et al., 1964; Stevens, 1959). Approximately one third of stress-related seizure facilitation may be due to seizure facilitation by concomitant sleep deprivation; the remainder of the effect is due to stress itself, unrelated to other stress-associated physiological and behavioral factors that might alter seizure threshold, such as medication compliance or alcohol or other substance abuse (Klein and Pezzullo, 2000).

The effect of stress on seizures is likely a direct effect of stress-related physiological changes on neuronal excitability and seizure threshold. Stress affects neuronal excitability and epileptiform discharges in both animals and in patients with epilepsy. In the monkey, spike discharges and seizure frequency increase under social stress (Lockard and Ward, 1980). In mice, brief restraint lowers seizure threshold to pentylenetetrazol- and electroshock-induced seizures, starting immediately after restraint. This effect is abrogated by adrenalectomy, suggesting that the increase in excitability is mediated by the adrenal gland (Swinyard et al., 1962). In contrast, more prolonged stress lasting 10 minutes may increase seizure threshold (Reddy and Rogawski, 2002; Woodbury, 1952). In human studies, a stress interview structured to elicit maximal emotional response activated interictal epileptiform spike activity within 30 seconds to minutes in 12 of 44 (27%) patients with epilepsy and in almost 50% among the patients with partial epilepsy (Small et al., 1964; Stevens, 1959). About 10% of patients developed a partial seizure during or following the interview. Reduction of epileptiform activity during stress was seen in 5% of patients. In one patient undergoing depth electrode recording, an emotional challenge triggered a seizure that started after 3 minutes in the left amygdala (Groethuysen et al., 1957).

Stress response includes the activation of the amygdalo–hypothalamo–pituitary–adrenal hormonal axis. Release of CRH leads to release of ACTH from the pituitary, which stimulates adrenal steroid secretion. As discussed earlier, CRH and a number of adrenal steroids affect neuronal excitability and seizures.

Corticotropin-releasing hormone acts as an excitatory neuromodulator. It increases neuronal firing in the hippocampus and induces epileptiform discharges and seizures in animals (Aldenhoff et al., 1983; Baram et al., 1992; Weiss et al., 1986). CRH-containing neurons are concentrated in the central nucleus of the amygdala, where they mediate the autonomic and behavioral aspects of the stress response (De Souza et al., 1984; Swanson et al., 1983). CRH-induced seizures start in the amygdala and spread to the hippocampus and cortex (Baram et al., 1992; Weiss et al., 1986). CRH neurons are activated under stress conditions (de Waal et al., 1995; Kalin et al., 1994). Thus, stress may activate CRH-containing neurons in the limbic system, and a CRH-mediated increase in neuronal excitability may result in seizure facilitation.

Glucocorticoids also increase neuronal excitability and facilitate epileptiform discharges and seizures in animal models, with a latency of hours (Klein and Herzog, 1997; Lee et al., 1989; Roberts and Keith, 1994; Woodbury, 1952). This effect is likely mediated by neuronal glucocorticoid receptors, as discussed previously. There is a high density of glucocorticoid receptors in the hippocampus, the amygdala, and the septum. Two types of glucocorticoid receptors are found in the brain. The glucocorticoid receptor has a low affinity for glucocorticoids and is only activated during times of high glucocorticoid levels, such as those seen during stress (Joels and de Kloet, 1994). Activation of this receptor might be important in lowering the seizure threshold at the time of stress.

Other adrenal neuroactive steroids also affect the seizure threshold. Pregnenolone sulfate (PS) and DHEAS both lower seizure threshold in animals (Heuser et al., 1965; Reddy and Rogawski, 2000). This is thought to occur by antagonism of $GABA_A$-mediated neuronal hyperpolarization and potentiation of NMDA receptor-mediated glutamatergic neuronal excitation (Paul and Purdy, 1992). In contrast, AP and the metabolites of the mineralocorticoids DHDOC and THDOC exert a potent anticonvulsant effect in a number of animal seizure models by enhancing $GABA_A$ receptor-mediated synaptic inhibition (Kokate et al., 1994; Majewska et al., 1986; Paul and Purdy, 1992; Reddy and Rogawski, 2002). Some of these neurosteroids (e.g., AP) are synthesized in the brain acutely following stress, independently of their adrenal synthesis (Cheney et al., 1995; Paul and Purdy, 1992). Withdrawal of this anticonvulsant effect could promote seizure activity when stress is terminated.

Finally, noradrenergic activity that is increased as part of the stress response exerts an anticonvulsant effect. The locus coeruleus is the principal source of ascending noradrenergic fibers. Activation of the locus coeruleus, which occurs in response to stress, elevates seizure threshold in animals and has been postulated as a possible important mechanism underlying the therapeutic anticonvulsant effect of vagus nerve stimulation (Naritoku et al., 1995).

Thus, stress induces both proconvulsant and anticonvulsant neuroendocrine and neuronal changes. These observations may explain why stressful events can be associated with both increased and decreased seizure frequency (Neugebauer et al., 1994; Swinkels et al., 1998). Stress-related seizure precipitation can occur with variable latency after change in stress level. The latency ranges from minutes through hours to days and may even include seizure precipitation during relief from stress (Klein and Pezzullo, 2000). This suggests that more than one pathophysiological mechanism may mediate stress-related seizure modulation. Seizure facilitation that occurs within minutes of a stressful event might result from activation of central CRH neurons, leading to augmentation of neuronal excitation, or from an acute increase in the CNS excitatory neurosteroids such as PS or DHEAS. Seizures occurring half an hour to hours after a stressful event could result from activation of the HPA axis and increased adrenal secretion of the proconvulsants cortisol, PS, and DHEAS. Increase in seizure frequency associated with chronic stress may be more difficult to explain but could be mediated by chronically elevated release of cortisol from the adrenal glands. Finally, seizure occurrence at the end of a stressful event (i.e., with termination of stress) might result from withdrawal of GABA-potentiating neurosteroids such as AP, DHDOC, and THDOC; acute secretion during the stressful event; or reduction of noradrenergic activity. The protective effect of stress reported in some patients might similarly be mediated by increased secretion of THDOC, DHDOC, or AP (Reddy and Rogawski, 2002).

If stress alters seizure threshold acutely, could sustained high levels of stress permanently increase excitability of selected neuronal tissue (e.g., in the limbic system) and promote epileptogenesis? In neonatal animals, acute CRH-induced status epilepticus exposure results in neuronal loss in the hippocampal CA3 region, amygdala, and the pyriform cortex (Ribak and Baram, 1996) and in a long-lasting potentiation of subsequent non-CRH induction of seizures (e.g., kainic acid), suggesting that this might be the case (Bender and Baram, 2007; Weiss et al., 1986). Clinical data are sparse in this respect. Several cases have been reported in which seizures starting shortly after sexual abuse (without apparent head trauma) were followed by seizure exacerbation during heightened symptoms of the posttraumatic stress disorder (PTSD) (Greig and Betts, 1992). There is an increased prevalence of PTSD among patients with intractable epilepsy: 37% vs. 5 to 10% in the general population (Rosenberg et al., 2000).

## HORMONAL TREATMENT IN EPILEPSY

Hormonal treatment of seizures may rationally be aimed at those endocrine aspects of seizures that act to cause, exacerbate, or ameliorate them. Endocrine intervention to prevent epileptogenesis has not been attempted to date. Several attempts have been made to use hormones for the treatment of established seizures. Progesterone may have an anticonvulsant effect, whereas estrogen, CRH, and cortisol may have proconvulsant effects. Thus, treatment with progesterone, estrogen antagonists, and medications that suppress the activity of the HPA axis could be useful adjunct treatment in the appropriate epilepsy patients.

### GONADAL-HORMONE-BASED TREATMENT

#### Progesterone

As noted previously, low progesterone levels or progesterone withdrawal may be a factor in the increased seizure frequency seen perimenstrually in women with catamenial epilepsy and normal ovulatory cycles and during the entire luteal phase of women with anovulatory cycles (Backstrom, 1976; Herzog et al., 1997; Laidlaw, 1956). Progesterone may be expected to benefit these women. Evaluation of progestational treatments has included synthetic progestins, natural progesterone, and a synthetic form of the neuroactive progesterone metabolite AP ganaxolone.

Oral forms of synthetic progestins have yielded inconsistent results. Few or no benefits have been noted in a number of studies (Mattson et al., 1984) with cyclically administered oral forms such as the testosterone-derived norethisterone (Dana-Haeri and Richens, 1983), although occasional benefits have been described in single case reports with its continuous use, as well as the use of the hydroxyprogesterone derivative medroxyprogesterone acetate (MPA) (Hall, 1977; Mattson et al., 1984).

Intramuscularly administered synthetic progestins may be more beneficial than oral forms (Mattson et al., 1984). In one open-label study of 13 women with refractory partial seizures and normal ovulatory cycles, a medroxyprogesterone dose large enough to induce amenorrhea (i.e., 120 to 150 mg every 6 to 12 weeks) resulted in a 40% seizure reduction (Mattson et al., 1984), from an average of 8 seizures per month to 5 seizures per month. It was unclear whether the effect was due to the direct anticonvulsant activity of MPA or to the hormonal consequences of the medroxy-progesterone-induced amenorrhea (i.e., loss of periovulatory estrogen surge). One patient who had absence rather than partial seizures did not improve. Potential side effects include depression, sedation, breakthrough vaginal bleeding, and lengthy delay in the return of regular menstrual cycles, the last side effect following only the depo-, not oral, medroxyprogesterone administration (Mattson et al., 1984).

Because the anticonvulsant effect of progesterone is likely due to the progesterone metabolite AP rather than due to progesterone itself, and because synthetic progestins are not metabolized to AP (Monaghan et al., 1999), natural progesterone may be a more rational and effective treatment than synthetic progestins (Herzog, 1986, 1995, 1999). In one open-label trial, 25 women with cata-menial exacerbation of complex partial seizures of temporal lobe origin with or without secondary generalization, 14 with inadequate luteal phase or anovulatory cycles, and 11 with normal cycles and perimenstrual seizure exacerbation received natural progesterone in doses sufficient to produce physiological luteal range progesterone serum levels between 5 and 25 ng/mL. Improvement was observed in 72% of the women, with a 55% decline in average seizure frequency, from 0.39 to 0.18 daily seizures (Herzog, 1995). Progesterone was administered as lozenges, 200 mg three times daily on days 23 to 25 of each menstrual cycle for perimenstrual exacerbation, and on days 15 to 25 of each menstrual cycle, with taper over days 26 to 28, for exacerbations lasting throughout the entire luteal phase. Similar findings had previously been found in a smaller study using progesterone vaginal suppositories (Herzog, 1986). Side effects included mild fatigue, depression, weight gain, fluid retention, and breakthrough bleeding. All side effects resolved promptly with dose reduction or medication withdrawal. The observations of the larger study were subsequently extended for 3 years, with continued improvement in seizure control at the same level and with no significant side effects during the 3 years of treatment (Herzog, 1999). An NIH-funded, multicenter, double-blind, randomized, placebo-controlled trial of progesterone add-on treatment in women with refractory localization-related epilepsy is currently in progress.

Natural progesterone is available as a soybean or yam extract in lozenge, micronized capsule, cream, and suppository form. The usual daily regimen to achieve physiological luteal range serum levels ranges from 100 to 200 mg three times daily (Herzog, 1995). Potential side effects may include sedation, depression, weight gain, breast tenderness, and breakthrough vaginal bleeding, all readily reversible upon discontinuation of the hormone or lowering of the dose.

## Neuroactive Steroids

The experimental data on the cellular effects of the GABA-mimetic neuroactive steroids such as AP and THDOC would suggest that these compounds may have a potent anticonvulsant effect, a suggestion borne out in animal studies (Kokate et al., 1994, 1996; Reddy and Rogawski, 2002).

Two older studies have investigated the antiepileptic potential of some neuroactive steroids. In one study, the 21-acetate form of deoxycorticosterone was found to be effective in 6 of 10 patients with refractory seizures (Aird, 1951). Its effect may have been due to its neuroactive steroid metabolite, THDOC (Reddy and Rogawski, 2002). In a second study, the synthetic neuroactive pregnane steroid anesthetic alphaxalone was effective in treating status epilepticus (Casaroli et al., 1980).

## Ganaxolone

The antiepileptic potential of AP has received considerable attention. AP and other related, naturally occurring neuroactive steroids with anticonvulsant properties produce motor impairment with a low therapeutic to toxic ratio, and they have very short half-lives (Kokate et al., 1994). Ganaxolone ($3\alpha$-hydroxy-$3\beta$-methyl-$5\alpha$-pregnan-20-one) is a $3\beta$-methylated synthetic analog of AP with positive allosteric modulatory effects on $GABA_A$ receptors comparable to AP (Nohria and Giller, 2007). It has a better therapeutic to toxic ratio in animal studies of pentylenetetrazol-, pilocarpine-, and bicuculline-induced seizures and a longer half-life than AP (Carter et al., 1997; Nohria and Giller, 2007). Ganaxolone has a neuroprotective activity in diverse rodent seizure models, including clonic seizures induced by pentylenetetrazol and bicuculline, the 6-Hz model, and amygdala-kindled seizures (Carter et al., 1997; Kaminski et al., 2004; Nohria and Giller, 2007; Reddy, 2004). Tolerance to the anticonvulsant effect, such as that seen with benzodiazepines, does not occur with ganaxolone in animal studies (Naritoku et al., 1995). In an open-label study of children and adolescents with refractory partial or generalized seizures, 6 of 15 patients had a greater than 50% seizure reduction, and 9 of 15 patients had mild to moderate side effects. There was no evidence of altered metabolism of other AEDs (Lechtenberg et al., 1996). In another open-label pediatric study of children ages 20 to 60 months with refractory seizures and a history of infantile spasms during the first year of life, approximately one third of patients improved (Dodson et al., 1997). A multicenter, double-blind, randomized, placebo-controlled trial of 50 adult patients with refractory partial complex seizures withdrawn from all AEDs as part of presurgical evaluation showed a trend toward efficacy (Laxer et al., 2000). Of ganaxolone-treated patients, 50% were withdrawn from the study because of seizures vs. 75% of placebo-treated patients ($p = 0.0795$). Side effects were similar between the ganaxolone and the placebo-treated groups. Unfortunately, the patient number was smaller, the study duration was shorter, and the response of the placebo group was better than in similar presurgical studies (i.e., studies of felbamate, gabapentin, and oxcarbazepine), likely accounting for the lack of statistical significance. The main adverse events have included somnolence and dizziness. Less frequently reported adverse effects included impaired coordination, unsteady gait, abnormal concentration, gastrointestinal disturbances, and malaise. In a more recent pilot open-label, dose-escalation study of the safety, tolerability, dose range, and potential efficacy of ganaxolone for the treatment of refractory epilepsy in pediatric and adolescent subjects with refractory epilepsy, subjects received an oral suspension of ganaxolone in a dose of 12 mg/kg t.i.d. for 8 weeks. Of the 15 patients enrolled, 4 (25%) had ≥50% reduction in seizure frequency, 2 of 15 (13%) had 25 to 50% reduction in seizure frequency, and 9 of 15 (60%) did not respond (Pieribone et al., 2007). One of the 15 subjects became essentially seizure free and remained so for ≥3.5 years of open-label extension ganaxolone administration. Ganaxolone was well tolerated. Somnolence was the most frequently reported adverse event (9 patients). Two Phase II studies of the safety and efficacy of add-on ganaxolone treatment in children with infantile spasms and in adults with refractory localization-related epilepsy are underway, and a pivotal Phase III add-on study of ganaxolone in adults with refractory localization-related epilepsy is being planned.

## Clomiphene

Clomiphene citrate is an estrogen analog with both estrogenic and antiestrogenic effects that are dose dependent. In clinical use, it acts primarily as an antiestrogen at the hypothalamic and pituitary level to stimulate gonadotropin secretion, ovulation, and fertility. It exerts an anticonvulsant effect in rats in a dose-related fashion (Frye, 2008). Remarkable reductions in seizure frequency have been reported in isolated cases involving both men and women (Herzog, 1988; Login, 1983). In one series of 12 women who had complex partial seizures and menstrual disorders (polycystic ovarian syndrome or inadequate luteal phase cycles) and who were given clomiphene, 10 improved, often dramatically, with an average 87% drop in seizure frequency (Herzog, 1988). Improvement in seizure frequency occurred in those women who had normalization of menstrual cycles and of luteal progesterone secretion. The only two women who did not improve continued to have menstrual abnormalities.

Thus, clomiphene may be a useful adjunct antiepileptic treatment in women with menstrual disorders. It is administered in dosages from 25 to 100 mg daily on days 5 to 9 of each menstrual cycle in women, and 25 to 50 mg daily or on alternate days in men. Unfortunately, side effects can be significant and include unwanted pregnancy, ovarian overstimulation syndrome, transient breast tenderness, and pelvic cramps. Furthermore, seizure frequency may increase during the enhanced preovulatory rise in serum estradiol levels in some women. Clomiphene treatment, therefore, should be restricted to situations where irregular anovulatory cycles cannot be readily normalized with cyclic progesterone use.

## Testosterone and Aromatase Inhibitors

We have reported improvement in seizures in a couple of patients treated for AED-related hyposexuality with testosterone and testolactone, an inhibitor of the enzyme aromatase, which inhibits conversion of estrogens and androgens (Herzog, 1998a). Treatment with testosterone alone was ineffective. The successful treatment was associated with normalization of previously elevated estradiol levels, suggesting that the anticonvulsant effect of testolactone may be due to reduction of the proconvulsant effects of estrogen. Newer aromatase inhibitors such as anastrazole and letrozole have been beneficial in a small open-label treatment of a small number of men with epilepsy and low testosterone levels (Harden and MacLusky, 2005). A larger study with a newer aromatase inhibitor is currently under way.

## LHRH Agonists

Lowering of estrogen levels by induction of menopause is expected to improve seizure control. Medical menopause can be achieved by chronic use of one of the long-acting luteinizing hormone-releasing hormone (LHRH) analogs. Long-term suppression of gonadotropin and ovarian secretion develops after an initial 3- to 4-week phase of reproductive endocrine stimulation. One patient with severe refractory seizures with perimenstrual exacerbation was reported to have improved markedly after treatment with the LHRH agonist goserelin (Haider and Barnett, 1991). One open-label study of 10 patients with catamenial seizures showed improvement in 8 of 10 patients, with adverse effects including hot flashes and headache occurring in 8 of 10 patients (Bauer et al., 1992). Seizure exacerbation during the first month of the stimulation phase may preclude their use in some cases. Moreover, the immediate and long-term effects of hypo-estrogenism need to be considered.

## Adrenal Steroid- and CRH-Based Treatments

Both ACTH and prednisone have been used extensively in the treatment of refractory infantile spasms (Baram et al., 1996; Mackay et al., 2004) and other primarily generalized seizures of childhood (Rogawski and Reddy, 2002; Snead, 1991). They have been successfully used in another childhood epileptic syndrome, Landau–Kleffner syndrome (i.e., acquired epileptic aphasia) (Snead, 1991), but not in other forms of partial epilepsy. In infantile spasms, low-dose ACTH and prednisone are equally efficacious, but a high dose of ACTH (150 IU/m$^2$/day) was found to be considerably more effective than prednisone during a 2-week-long course (Joels and de Kloet, 1994; Klein and Herzog, 1997). Longer usage of high-dose ACTH may be limited by severe side effects (Snead, 1991). The mechanism of ACTH anticonvulsant activity is uncertain. It may include suppression of CRH and its epileptogenic activity (Baram, 1993). There are also some data to suggest that ACTH may stimulate the production of endogenous steroids with anticonvulsant activity (Eneroth et al., 1972).

One study reported three patients with refractory seizures and hypercortisolemia whose seizures became controlled upon normalization of the cortisol levels by ketoconazole (Herzog et al., 1998b). Wider anticonvulsant potential of ketoconazole or other potential suppressors of cortisol synthesis, including low-dose prednisone and aminoglutethimide, and antagonists of the corticosteroid receptors, such as spironolactone (Roberts and Keith, 1994), have not been explored.

In animal studies, administration of a CRH receptor antagonist is protective against both seizures and seizure-related hippocampal neuronal loss (Maecker et al., 1997). The possibility of treating seizures and ictus-related excitatory neuronal injury by antagonizing CRH activity has been examined in a Phase I clinical study. The competitive CRH antagonist, α-helical CRH, failed to cross the blood–brain barrier and was ineffective in the treatment of infantile spasms in a small group of children (Baram et al., 1999). Oral small-molecule, nonpeptidal CRF receptor antagonists, such as the recently developed nonpeptidal CRF receptor 1 blocker NBI 27914, are potential candidates for testing in epilepsy, including, given the great sensitivity to the excitatory effect of CRH of developing CNS (Baram, et al., 1992), childhood epilepsies. Reduction of CRH activity by means of stimulating the activity of CRH-binding peptide may be another possible, as yet untested therapeutic avenue.

## REFERENCES

Abbasi, F., A. Krumholz, S.J. Kittner, P. Langenberg. (1999). Effects of menopause on seizures in women with epilepsy. *Epilepsia*, 40(2):205–210.

Aird, R.B., G.S. Gordan. (1951). Anticonvulsive properties of desoxycorticosterone. *J. Am. Med. Assoc.*, 145(10):715–719.

Aldenhoff, J.B., D.L. Gruol, J. Rivier, W. Vale, G.R. Siggins. (1983). Corticotropin releasing factor decreases postburst hyperpolarizations and excites hippocampal neurons. *Science*, 221(4613):875–877.

Annegers, J.F., W.A. Hauser, E. Beghi, A. Nicolosi, L.T. Kurland. (1988). The risk of unprovoked seizures after encephalitis and meningitis. *Neurology*, 38(9):1407–1410.

Annegers, J.F., W.A. Hauser, S.P. Coan, W.A. Rocca. (1998). A population-based study of seizures after traumatic brain injuries. *N. Engl. J. Med.*, 338(1):20–24.

Apter, D., R. Vihko. 1977. Serum pregnenolone, progesterone, 17-hydroxyprogesterone, testosterone and 5 alpha-dihydrotestosterone during female puberty. *J. Clin. Endocrinol. Metab.*, 45(5):1039–1048.

Backstrom, T. (1976). Epileptic seizures in women related to plasma estrogen and progesterone during the menstrual cycle. *Acta Neurol. Scand.*, 54(4):321–347.

Backstrom, T., B. Zetterlund, S. Blom, M. Romano. (1984). Effects of intravenous progesterone infusions on the epileptic discharge frequency in women with partial epilepsy. *Acta. Neurol. Scand.*, 69(4):240–248.

Baram, T.Z. (1993). Pathophysiology of massive infantile spasms: perspective on the putative role of the brain adrenal axis. *Ann. Neurol.*, 33(3):231–236.

Baram, T.Z., L. Schultz. (1995). ACTH does not control neonatal seizures induced by administration of exogenous corticotropin-releasing hormone. *Epilepsia*, 36(2):174–178.

Baram, T.Z., E. Hirsch, O.C. Snead, 3rd, L. Schultz. (1992). Corticotropin-releasing hormone-induced seizures in infant rats originate in the amygdala. *Ann. Neurol.*, 31(5):488–494.

Baram, T.Z., Y. Koutsoukos, L. Schultz, J. Rivier. (1996). The effect of 'Astressin,' a novel antagonist of corticotropin releasing hormone (CRH), on CRH-induced seizures in the infant rat: comparison with two other antagonists. *Mol. Psychiatry*, 1(3):223–226.

Baram, T.Z., D.T. Chalmers, C. Chen, Y. Koutsoukos, E.B. De Souza. (1997). The CRF₁ receptor mediates the excitatory actions of corticotropin releasing factor (CRF) in the developing rat brain: *in vivo* evidence using a novel, selective, non-peptide CRF receptor antagonist. *Brain Res.*, 770(1–2):89–95.

Baram, T.Z., W.G. Mitchell, K. Brunson, E. Haden. (1999). Infantile spasms: hypothesis-driven therapy and pilot human infant experiments using corticotropin-releasing hormone receptor antagonists. *Dev. Neurosci.*, 21(3–5):281–289.

Bauer, J., L. Wildt, D. Flugel, H. Stefan. (1992). The effect of a synthetic GnRH analogue on catamenial epilepsy: a study in ten patients. *J. Neurol.*, 239(5):284–286.

Bender, R.A., T.Z. Baram. (2007). Epileptogenesis in the developing brain: what can we learn from animal models? *Epilepsia*, 48(Suppl. 5):2–6.

Birnstiel, S., T.J. List, S.G. Beck. (1995). Chronic corticosterone treatment maintains synaptic activity of CA1 hippocampal pyramidal cells: acute high corticosterone administration increases action potential number. *Synapse*, 20(2):117–124.

Burger, H.G., E.C. Dudley, J.L. Hopper, J.M. Shelley, A. Green, A. Smith, L. Dennerstein, C. Morse. (1995). The endocrinology of the menopausal transition: a cross-sectional study of a population-based sample. *J. Clin. Endocrinol. Metab.*, 80(12):3537–3545.

Buterbaugh, G.G., G.M. Hudson. (1991). Estradiol replacement to female rats facilitates dorsal hippocampal but not ventral hippocampal kindled seizure acquisition. *Exp. Neurol.*, 111(1):55–64.

Carter, R.B., P.L. Wood, S. Wieland, J.E. Hawkinson, D. Belelli, J.J. Lambert, H.S. White, H.H. Wolf, S. Mirsadeghi, S.H. Tahir, M.B. Bolger, N.C. Lan, K.W. Gee. (1997). Characterization of the anticonvulsant properties of ganaxolone (CCD 1042; 3alpha-hydroxy-3beta-methyl-5alpha-pregnan-20-one), a selective, high-affinity, steroid modulator of the gamma-aminobutyric acid(A) receptor. *J. Pharmacol. Exp. Ther.*, 280(3):1284–1295.

Casaroli, D., C. Munari, G. Matteuzzi, L. Pacifico. (1980). Althesin in the treatment of status epilepticus. *Minerva Anestesiol.*, 46(2):129–140.

Cheney, D.L., D. Uzunov, E. Costa, A. Guidotti. (1995). Gas chromatographic-mass fragmentographic quantitation of 3 alpha-hydroxy-5 alpha-pregnan-20-one (allopregnanolone) and its precursors in blood and brain of adrenalectomized and castrated rats. *J. Neurosci.*, 15(6):4641–4650.

Cummings, L.N., L. Giudice, M.J. Morrell. (1995). Ovulatory function in epilepsy. *Epilepsia*, 36(4):355–359.

Dafny, N., M.I. Phillips, A.N. Taylor, S. Gilman. (1973). Dose effects of cortisol on single unit activity in hypothalamus, reticular formation and hippocampus of freely behaving rats correlated with plasma steroid levels. *Brain Res.*, 59:57–72.

Dana-Haeri, J., A. Richens. (1983). Effect of norethisterone on seizures associated with menstruation. *Epilepsia*, 24(3):377–381.

De Souza, E.B., M.H. Perrin, T.R. Insel, J. Rivier, W.W. Vale, M.J. Kuhar. (1984). Corticotropin-releasing factor receptors in rat forebrain: autoradiographic identification. *Science*, 224(4656):1449–1451.

de Waal, W.J., M. Torn, S.M. de Muinck Keizer-Schrama, R.S. Aarsen, S.L. Drop. (1995). Long term sequelae of sex steroid treatment in the management of constitutionally tall stature. *Arch. Dis. Child.*, 73(4):311–315.

Dichter, M.A. (1997). Basic mechanisms of epilepsy: targets for therapeutic intervention. *Epilepsia*, 38(Suppl. 9):2–6.

Dodson, W., B. Bourgeois, J. Kerrigan et al. (1997). An open label evaluation of safety and efficacy of ganaxolone in children with refractory seizures and history of infantile spasms. *Epilepsia*, 38(Suppl. 7):75.

Dunn, A.J., C.W. Berridge. (1990). Physiological and behavioral responses to corticotropin-releasing factor administration: is CRF a mediator of anxiety or stress responses? *Brain Res. Brain Res. Rev.*, 15(2):71–100.

Edwards, H.E., W.M. Burnham, A. Mendonca, D.A. Bowlby, N.J. MacLusky. (1999). Steroid hormones affect limbic afterdischarge thresholds and kindling rates in adult female rats. *Brain Res.*, 838(1–2):136–150.

Ehlers, C.L., S.J. Henriksen, M. Wang, J. Rivier, W. Vale, F.E. Bloom. (1983). Corticotropin releasing factor produces increases in brain excitability and convulsive seizures in rats. *Brain Res.*, 278(1–2):332–336.

Eneroth, P., H. Ferngren, J.A. Gustafsson, B. Ivemark, A. Stenberg. (1972). Excretion of steroid hormones in an anencephalic newborn infant. *Acta. Endocrinol. (Copenh.)*, 70(1):113–131.

Feldman, R.G., N.L. Paul. (1976). Identity of emotional triggers in epilepsy. *J. Nerv. Ment. Dis.*, 162(5):345–353.

Flugge, G., W.H. Oertel, W. Wuttke. (1986). Evidence for estrogen-receptive GABAergic neurons in the preoptic/anterior hypothalamic area of the rat brain. *Neuroendocrinology*, 43(1):1–5.

Frucht, M.M., M. Quigg, C. Schwaner, N.B. Fountain. (2000). Distribution of seizure precipitants among epilepsy syndromes. *Epilepsia*, 41(12):1534–1539.

Frye, C.A. (2006). Role of androgens in epilepsy. *Expert Rev. Neurother.*, 6(7):1061–1075.

Frye, C.A. (2008). Hormonal influences on seizures: basic neurobiology. *Int. Rev. Neurobiol.*, 83:7–77.

Frye, C.A., M.E. Rhodes. (2005). Estrogen-priming can enhance progesterone's anti-seizure effects in part by increasing hippocampal levels of allopregnanolone. *Pharmacol. Biochem. Behav.*, 81(4):907–916.

Frye, C.A., T.J. Scalise. (2000). Anti-seizure effects of progesterone and 3alpha,5alpha-THP in kainic acid and perforant pathway models of epilepsy. *Psychoneuroendocrinology*, 25(4):407–420.

Gasior, M., J.T. Ungard, M. Beekman, R.B. Carter, J.M. Witkin. (2000). Acute and chronic effects of the synthetic neuroactive steroid, ganaxolone, against the convulsive and lethal effects of pentylenetetrazol in seizure-kindled mice: comparison with diazepam and valproate. *Neuropharmacology*, 39(7):1184–1196.

Gee, K.W., L.D. McCauley, N.C. Lan. (1995). A putative receptor for neurosteroids on the $GABA_A$ receptor complex: the pharmacological properties and therapeutic potential of epalons. *Crit. Rev. Neurobiol.*, 9(2–3):207–227.

Genazzani, A.R., F. Bernardi, P. Monteleone, S. Luisi, M. Luisi. (2000). Neuropeptides, neurotransmitters, neurosteroids, and the onset of puberty. *Ann. N.Y. Acad. Sci.*, 900:1–9.

Gowers, A. (1893). *A Manual of Diseases of the Nervous System*. Philadelphia, PA: Blakinson, pp. 732–753.

Greig, E., T. Betts. (1992). Epileptic seizures induced by sexual abuse: pathogenic and pathoplastic factors. *Seizure*, 1(4):269–274.

Groethuysen, U.C., D.B. Robinson, C.H. Haylett, H.R. Estes, A.M. Johnson. (1957). Depth electrographic recording of a seizure during a structured interview; report of a case. *Psychosom. Med.* 19(5):353–362.

Haider, Y., D.B. Barnett. (1991). Catamenial epilepsy and goserelin. *Lancet*, 338(8781):1530.

Hall, S.M. (1977). Treatment of menstrual epilepsy with a progesterone-only oral contraceptive. *Epilepsia*, 18(2):235–236.

Harden, C.L. (2008). Issues for mature women with epilepsy. *Int. Rev. Neurobiol.*, 83:85–95.

Harden, C.L., N.J. MacLusky. (2005). Aromatase inhibitors as add-on treatment for men with epilepsy. *Expert Rev. Neurother.*, 5(1):123–127.

Harden, C.L., N.K. Sethi. (2008). Epileptic disorders in pregnancy: an overview. *Curr. Opin. Obstet. Gynecol.*, 20(6):557–562.

Harden, C.L., M.C. Pulver, L. Ravdin, A.R. Jacobs. (1999). The effect of menopause and perimenopause on the course of epilepsy. *Epilepsia*, 40(10):1402–1407.

Harden, C.L., A.G. Herzog, B.G. Nikolov, B.S. Koppel, P.J. Christos, K. Fowler, D.R. Labar, W.A. Hauser. (2006). Hormone replacement therapy in women with epilepsy: a randomized, double-blind, placebo-controlled study. *Epilepsia*, 47(9):1447–1451.

Hauser, W.A. (1994). The prevalence and incidence of convulsive disorders in children. *Epilepsia*, 35(Suppl. 2):1–6.

Hauser, W.A., J.F. Annegers, L.T. Kurland. (1991). Prevalence of epilepsy in Rochester, Minnesota: 1940–1980. *Epilepsia*, 32(4):429–445.

Heim, L.M. (1966). Effect of estradiol on brain maturation: dose and time response relationships. *Endocrinology*, 78(6):1130–1134.

Herman-Giddens, M.E., E.J. Slora, R.C. Wasserman, C.J. Bourdony, M.V. Bhapkar, G.G. Koch, C.M. Hasemeier. (1997). Secondary sexual characteristics and menses in young girls seen in office practice: a study from the Pediatric Research in Office Settings network. *Pediatrics*, 99(4):505–512.

Herzog, A.G. (1986). Intermittent progesterone therapy and frequency of complex partial seizures in women with menstrual disorders. *Neurology*, 36(12):1607–1610.

Herzog, A.G. (1988). Clomiphene therapy in epileptic women with menstrual disorders. *Neurology*, 38(3):432–434.

Herzog, A.G. (1995). Progesterone therapy in women with complex partial and secondary generalized seizures. *Neurology*, 45(9):1660–1662.

Herzog, A.G. (1999). Progesterone therapy in women with epilepsy: a 3-year follow-up. *Neurology*, 52(9):1917–1918.

Herzog, A.G. (2008a). Catamenial epilepsy: definition, prevalence pathophysiology and treatment. *Seizure*, 17(2):151–159.

Herzog, A.G. (2008b). Disorders of reproduction in patients with epilepsy: primary neurological mechanisms. *Seizure*, 17(2):101–110.

Herzog, A.G., K.M. Fowler. (2008). Sensitivity and specificity of the association between catamenial seizure patterns and ovulation. *Neurology*, 70(6):486–487.

Herzog, A.G., M.M. Seibel, D.L. Schomer, J.L. Vaitukaitis, N. Geschwind. (1986). Reproductive endocrine disorders in women with partial seizures of temporal lobe origin. *Arch Neurol.*, 43(4):341–346.

Herzog, A.G., P. Klein, B.J. Ransil. (1997). Three patterns of catamenial epilepsy. *Epilepsia*, 38(10):1082–1088.

Herzog, A.G., P. Klein, A.R. Jacobs. (1998a). Testosterone versus testosterone and testolactone in treating reproductive and sexual dysfunction in men with epilepsy and hypogonadism. *Neurology*, 50(3):782–784.

Herzog, A.G., A. Sotrel, M. Ronthal. (1998b). Reversible proximal myopathy in epilepsy related Cushing's syndrome. *J. Neurol. Neurosurg. Psychiatry*, 65(1):134.

Herzog, A.G., C.L. Harden, J. Liporace, P. Pennell, D.L. Schomer, M. Sperling, K. Fowler, B. Nikolov, S. Shuman, M. Newman. (2004). Frequency of catamenial seizure exacerbation in women with localization-related epilepsy. *Ann. Neurol.*, 56(3):431–434.

Herzog, A.G., A.S. Blum, E.L. Farina, X.E. Maestri, J. Newman, E. Garcia, K.B. Krishnamurthy, D.B. Hoch, S. Replansky, K.M. Fowler, S.D. Smithson, B.A. Dworetzky, E.B. Bromfield. (2009). Valproate and lamotrigine level variation with menstrual cycle phase and oral contraceptive use. *Neurology*, 72(10):911–914.

Heuser, G., G.M. Ling, N.A. Buchwald. (1965). Sedation or seizures as dose-dependent effects of steroids. *Arch Neurol.*, 13:195–203.

Holmes, G.L., D.A. Weber. (1984). The effect of progesterone on kindling: a developmental study. *Brain Res.*, 318(1):45–53.

Hom, A.C., G.G. Buterbaugh. (1986). Estrogen alters the acquisition of seizures kindled by repeated amygdala stimulation or pentylenetetrazol administration in ovariectomized female rats. *Epilepsia*, 27(2):103–108.

Hom, A.C., I.E. Leppik, C.A. Rask. (1993). Effects of estradiol and progesterone on seizure sensitivity in oophorectomized DBA/2J mice and C57/EL hybrid mice. *Neurology*, 43(1):198–204.

Hosie, A.M., M.E. Wilkins, H.M. da Silva, T.G. Smart. (2006). Endogenous neurosteroids regulate $GABA_A$ receptors through two discrete transmembrane sites. *Nature*, 444(7118):486–489.

Irwin, R.P., N.J. Maragakis, M.A. Rogawski, R.H. Purdy, D.H. Farb, S.M. Paul. (1992). Pregnenolone sulfate augments NMDA receptor mediated increases in intracellular $Ca^{2+}$ in cultured rat hippocampal neurons. *Neurosci. Lett.*, 141(1):30–34.

Joels, M. (2009). Stress, the hippocampus, and epilepsy. *Epilepsia*, 50(4):586–597.

Joels, M., E.R. de Kloet. (1994). Mineralocorticoid and glucocorticoid receptors in the brain: implications for ion permeability and transmitter systems. *Prog. Neurobiol.*, 43(1):1–36.

Joels, M., H. Karst, R. DeRijk, E.R. de Kloet. (2008a). The coming out of the brain mineralocorticoid receptor. *Trends Neurosci.*, 31(1):1–7.

Joels, M., H. Krugers, H. Karst. (2008b). Stress-induced changes in hippocampal function. *Prog. Brain Res.*, 167:3–15.

Kalin, N.H., L.K. Takahashi, F.L. Chen. (1994). Restraint stress increases corticotropin-releasing hormone mRNA content in the amygdala and paraventricular nucleus. *Brain Res.*, 656(1):182–186.

Kaminski, R.M., M.R. Livingood, M.A. Rogawski. (2004). Allopregnanolone analogs that positively modulate GABA receptors protect against partial seizures induced by 6-Hz electrical stimulation in mice. *Epilepsia*, 45(7):864–867.

Kaminski, R.M., H. Marini, W.J. Kim, M.A. Rogawski. (2005). Anticonvulsant activity of androsterone and etiocholanolone. *Epilepsia*, 46(6):819–827.

Kawakami, M., E. Terasawa, T. Ibuki. (1970). Changes in multiple unit activity of the brain during the estrous cycle. *Neuroendocrinology*, 6(1):30–48.

Klein, P., A. Herzog. (1997). Emerging applications of hormonal therapy of paroxysmal central nervous system disorders. *Expert Opin. Investig. Drugs*, 6(10):1337–1349.

Klein, P., J.C. Pezzullo. (2000). Effects of stress on localization-related epilepsy. *Epilepsia*, 41(Suppl. 7):112.

Klein, P., L.M. van Passel-Clark, J.C. Pezzullo. (2003). Onset of epilepsy at the time of menarche. *Neurology*, 60(3):495–497.

Kling, M.A., M.A. Smith, J.R. Glowa, D. Pluznik, J. Demas, M.D. DeBellis, P.W. Gold, J. Schulkin. (1993). Facilitation of cocaine kindling by glucocorticoids in rats. *Brain Res.*, 629(1):163–166.

Knight, A.H., E.G. Rhind. (1975). Epilepsy and pregnancy: a study of 153 pregnancies in 59 patients. *Epilepsia*, 16(1):99–110.

Kokate, T.G., B.E. Svensson, M.A. Rogawski. (1994). Anticonvulsant activity of neurosteroids: correlation with gamma-aminobutyric acid-evoked chloride current potentiation. *J. Pharmacol. Exp. Ther.*, 270(3):1223–1229.

Kokate, T.G., A.L. Cohen, E. Karp, M.A. Rogawski. (1996). Neuroactive steroids protect against pilocarpine- and kainic acid-induced limbic seizures and status epilepticus in mice. *Neuropharmacology*, 35(8):1049–1056.

Kokate, T.G., M.K. Banks, T. Magee, S. Yamaguchi, M.A. Rogawski. (1999a). Finasteride, a 5alpha-reductase inhibitor, blocks the anticonvulsant activity of progesterone in mice. *J. Pharmacol. Exp. Ther.*, 288(2):679–684.

Kokate, T.G., K.N. Juhng, R.D. Kirkby, J. Llamas, S. Yamaguchi, M.A. Rogawski. (1999b). Convulsant actions of the neurosteroid pregnenolone sulfate in mice. *Brain Res.*, 831(1–2):119–124.

Krugers, H.J., S. Knollema, R.H. Kemper, G.J. Ter Horst, J. Korf. (1995). Down-regulation of the hypothalamo–pituitary–adrenal axis reduces brain damage and number of seizures following hypoxia/ischaemia in rats. *Brain Res.*, 690(1):41–47.

Laidlaw, J. (1956). Catamenial epilepsy. *Lancet*, 271(6955):1235–1237.

Landgren, S., T. Backstrom, G. Kalistratov. (1978). The effect of progesterone on the spontaneous interictal spike evoked by the application of penicillin to the cat's cerebral cortex. *J. Neurol. Sci.*, 36(1):119–133.

Landgren, S., J. Aasly, T. Backstrom, B. Dubrovsky, E. Danielsson. (1987). The effect of progesterone and its metabolites on the interictal epileptiform discharge in the cat's cerebral cortex. *Acta. Physiol. Scand.*, 131(1):33–42.

Laxer, K., D. Blum, B.W. Abou-Khalil, M.J. Morrell, D.A. Lee, J.L. Data, E.P. Monaghan. (2000). Assessment of ganaxolone's anticonvulsant activity using a randomized, double-blind, presurgical trial design. Ganaxolone Presurgical Study Group. *Epilepsia*, 41(9):1187–1194.

Lechtenberg, R., R. Villeneuve, E.P. Monaghan, M.B. Densel, E. Rey, O. Dulac. (1996). An open label dose-escalation study to evaluate the safety and tolerability of ganaxolone in the treatment of refractory epilepsy in pediatric patients. *Epilepsia*, 38(Suppl. 5):204.

Lee, P.H., L. Grimes, J.S. Hong. (1989). Glucocorticoids potentiate kainic acid-induced seizures and wet dog shakes. *Brain Res.*, 480(1–2):322–325.

Lockard, J., A. Ward. (1980). *Epilepsy: A Window to the Brain Mechanisms*. New York: Raven Press.

Login, I.S., F.E. Dreifuss. (1983). Anticonvulsant activity of clomiphene. *Arch. Neurol.*, 40(8):525.

Logothetis, J., R. Harner. (1960). Electrocortical activation by estrogens. *Arch. Neurol.*, 3:290–297.

Logothetis, J., R. Harner, F. Morrell, F. Torres. (1959). The role of estrogens in catamenial exacerbation of epilepsy. *Neurology*, 9(5):352–360.

Mackay, M.T., S.K. Weiss, T. Adams-Webber, S. Ashwal, D. Stephens, K. Ballaban-Gill, T.Z. Baram, M. Duchowny, D. Hirtz, J.M. Pellock, W.D. Shields, S. Shinnar, E. Wyllie, O.C. Snead, 3rd. (2004). Practice parameter: medical treatment of infantile spasms: report of the American Academy of Neurology and the Child Neurology Society. *Neurology*, 62(10):1668–1681.

Maecker, H., A. Desai, R. Dash, J. Rivier, W. Vale, R. Sapolsky. (1997). Astressin, a novel and potent CRF antagonist, is neuroprotective in the hippocampus when administered after a seizure. *Brain Res.*, 744(1):166–170.

Maguire, J., I. Mody. (2007). Neurosteroid synthesis-mediated regulation of GABA(A) receptors: relevance to the ovarian cycle and stress. *J. Neurosci.*, 27(9):2155–2162.

Maguire, J.L., B.M. Stell, M. Rafizadeh, I. Mody. (2005). Ovarian cycle-linked changes in GABA(A) receptors mediating tonic inhibition alter seizure susceptibility and anxiety. *Nat. Neurosci.*, 8(6):797–804.

Majewska, M.D., N.L. Harrison, R.D. Schwartz, J.L. Barker, S.M. Paul. (1986). Steroid hormone metabolites are barbiturate-like modulators of the GABA receptor. *Science*, 232(4753):1004–1007.

Marcus, E.M., C.W. Watson, P.L. Goldman. (1966). Effects of steroids on cerebral electrical activity: epileptogenic effects of conjugated estrogens and related compounds in the cat and rabbit. *Arch. Neurol.*, 15(5):521–532.

Marks, D.A., J. Kim, D.D. Spencer, S.S. Spencer. (1992). Characteristics of intractable seizures following meningitis and encephalitis. *Neurology*, 42(8):1513–1518.

Marrosu, F., M. Giagheddu, G.L. Gessa, W. Fratta. (1992). Clonidine prevents corticotropin releasing factor-induced epileptogenic activity in rats. *Epilepsia*, 33(3):435–438.

Mattson, R.H. (1991). Emotional effects on seizure occurrence. *Adv. Neurol.*, 55:453–460.

Mattson R.H., J.M. Kramer, J.A. Cramer, B.V. Caldwell. (1981). Seizure frequency and the menstrual cycle: a clinical study. *Epilepsia*, 22:242.

Mattson, R.H., J.A. Cramer, B.V. Caldwell, B.C. Siconolfi. (1984). Treatment of seizures with medroxyprogesterone acetate: preliminary report. *Neurology*, 34(9):1255–1258.

McEwen, B.S. (1983). Progestin receptors in the brain and pituitary gland. In *Progesterone and Progestins* (pp. 59–76), M.C. Bardin, Ed. New York: Raven Press.

McEwen, B.S. (1991). Non-genomic and genomic effects of steroids on neural activity. *Trends Pharmacol. Sci.*, 12(4):141–147.

McEwen, B.S., T.A. Milner. (2007). Hippocampal formation: shedding light on the influence of sex and stress on the brain. *Brain Res. Rev.*, 55(2):343–355.

Monaghan, E.P., J.W. McAuley, J.L. Data. (1999). Ganaxolone: a novel positive allosteric modulator of the GABA(A) receptor complex for the treatment of epilepsy. *Expert. Opin. Investig. Drugs*, 8(10):1663–1671.

Moss, R.L., Q. Gu. (1999). Estrogen: mechanisms for a rapid action in CA1 hippocampal neurons. *Steroids*, 64(1–2):14–21.

Naritoku, D.K., W.J. Terry, R.H. Helfert. (1995). Regional induction of *fos* immunoreactivity in the brain by anticonvulsant stimulation of the vagus nerve. *Epilepsy Res.*, 22(1):53–62.

Neugebauer, R., M. Paik, W.A. Hauser, E. Nadel, I. Leppik, M. Susser. (1994). Stressful life events and seizure frequency in patients with epilepsy. *Epilepsia*, 35(2):336–343.

Nicoletti, F., C. Speciale, M.A. Sortino, G. Summa, G. Caruso, F. Patti, P.L. Canonico. (1985). Comparative effects of estradiol benzoate, the antiestrogen clomiphene citrate, and the progestin medroxyprogesterone acetate on kainic acid-induced seizures in male and female rats. *Epilepsia*, 26(3):252–257.

Niijima, S., W.G. Wallace. (1989). Effects of puberty on seizure frequency. *Dev. Med. Child Neurol.*, 32(2):174–180.

Nohria, V., E. Giller. (2007). Ganaxolone. *Neurotherapeutics*, 4(1):102–105.

Paul, S.M., R.H. Purdy. (1992). Neuroactive steroids. *FASEB J.*, 6(6):2311–2322.

Pennell, P.B., B.E. Gidal, A. Sabers, J. Gordon, E. Perucca. (2007). Pharmacology of antiepileptic drugs during pregnancy and lactation. *Epilepsy Behav.*, 11(3):263–269.

Pennell, P.B., L. Peng, D.J. Newport, J.C. Ritchie, A. Koganti, D.K. Holley, M. Newman, Z.N. Stowe. (2008). Lamotrigine in pregnancy: clearance, therapeutic drug monitoring, and seizure frequency. *Neurology*, 70(22, Pt. 2):2130–2136.

Piekut, D., B. Phipps, S. Pretel, C. Applegate. (1996). Effects of generalized convulsive seizures on corticotropin-releasing factor neuronal systems. *Brain Res.*, 743(1–2):63–69.

Pieribone, V.A., J. Tsai, C. Soufflet, E. Rey, K. Shaw, E. Giller, O. Dulac. (2007). Clinical evaluation of ganaxolone in pediatric and adolescent patients with refractory epilepsy. *Epilepsia*, 48(10):1870–1874.

Ramsay, R.E. (1987). Effect of hormones on seizure activity during pregnancy. *J. Clin. Neurophysiol.*, 4(1):23–25.

Reddy, D.S. (2004a). Pharmacology of catamenial epilepsy. *Methods Find Exp. Clin. Pharmacol.*, 26(7):547–561.

Reddy, D.S. (2004b). Role of neurosteroids in catamenial epilepsy. *Epilepsy Res.*, 62(2–3):99–118.

Reddy, D.S., S.K. Kulkarni. (1998). Proconvulsant effects of neurosteroids pregnenolone sulfate and dehydroepiandrosterone sulfate in mice. *Eur. J. Pharmacol.*, 345(1):55–59.

Reddy, D.S., M.A. Rogawski. (2000). Chronic treatment with the neuroactive steroid ganaxolone in the rat induces anticonvulsant tolerance to diazepam but not to itself. *J. Pharmacol. Exp. Ther.*, 295(3):1241–1248.

Reddy, D.S., M.A. Rogawski. (2001). Enhanced anticonvulsant activity of neuroactive steroids in a rat model of catamenial epilepsy. *Epilepsia*, 42(3):337–344.

Reddy, D.S., M.A. Rogawski. (2002). Stress-induced deoxycorticosterone-derived neurosteroids modulate GABA(A) receptor function and seizure susceptibility. *J. Neurosci.*, 22(9):3795–3805.

Reddy, D.S., H.Y. Kim, M.A. Rogawski. (2001). Neurosteroid withdrawal model of perimenstrual catamenial epilepsy. *Epilepsia*, 42(3):328–336.

Reddy, D.S., D.C. Castaneda, B.W. O'Malley, M.A. Rogawski. (2004). Anticonvulsant activity of progesterone and neurosteroids in progesterone receptor knockout mice. *J. Pharmacol. Exp. Ther.*, 310(1):230–239.

Reul, J.M., E.R. de Kloet. (1985). Two receptor systems for corticosterone in rat brain: microdistribution and differential occupation. *Endocrinology*, 117(6):2505–2511.

Ribak, C.E., T.Z. Baram. (1996). Selective death of hippocampal CA3 pyramidal cells with mossy fiber afferents after CRH-induced status epilepticus in infant rats. *Brain Res. Dev. Brain Res.*, 91(2):245–251.

Roberts, A.J., L.D. Keith. (1994). Mineralocorticoid receptors mediate the enhancing effects of corticosterone on convulsion susceptibility in mice. *J. Pharmacol. Exp. Ther.*, 270(2):505–511.

Roberts, A.J., J.C. Crabbe, L.D. Keith. (1993). Type I corticosteroid receptors modulate PTZ-induced convulsions of withdrawal seizure prone mice. *Brain Res.*, 626(1–2):143–148.

Rogawski, M.A., D.S. Reddy. (2002). Neurosteroids and infantile spasms: the deoxycorticosterone hypothesis. *Int. Rev. Neurobiol.*, 49:99–219.

Rosciszewska, D. (1975). The course of epilepsy in girls at the age of puberty. *Neurol. Neurochir. Pol.*, 9(5):597–602.

Rosciszewska, D., B. Buntner, I. Guz, L. Zawisza. (1986). Ovarian hormones, anticonvulsant drugs, and seizures during the menstrual cycle in women with epilepsy. *J. Neurol. Neurosurg. Psychiatry*, 49(1):47–51.

Rosen, J.B., S.K. Pishevar, S.R. Weiss, M.A. Smith, M.A. Kling, P.W. Gold, J. Schulkin. (1994). Glucocorticoid treatment increases the ability of CRH to induce seizures. *Neurosci. Lett.*, 174(1):113–116.

Rosenberg, H.J., S.D. Rosenberg, P.D. Williamson, G.L. Wolford, 2nd. (2000). A comparative study of trauma and posttraumatic stress disorder prevalence in epilepsy patients and psychogenic nonepileptic seizure patients. *Epilepsia*, 41(4):447–452.

Sabers, A., I. Ohman, J. Christensen, T. Tomson. (2003). Oral contraceptives reduce lamotrigine plasma levels. *Neurology*, 61(4):570–571.

Santoro, N., J.R. Brown, T. Adel, J.H. Skurnick. (1996). Characterization of reproductive hormonal dynamics in the perimenopause. *J. Clin. Endocrinol. Metab.*, 81(4):1495–1501.

Sapolsky, R.M., L.C. Krey, B.S. McEwen. (1985). Prolonged glucocorticoid exposure reduces hippocampal neuron number: implications for aging. *J. Neurosci.*, 5(5):1222–1227.

Scharfman, H.E., N.J. MacLusky. (2006). The influence of gonadal hormones on neuronal excitability, seizures, and epilepsy in the female. *Epilepsia*, 47(9):1423–1440.

Schmidt, D., R. Canger, G. Avanzini, D. Battino, C. Cusi, G. Beck-Mannagetta, S. Koch, D. Rating, D. Janz. (1983). Change of seizure frequency in pregnant epileptic women. *J. Neurol. Neurosurg. Psychiatry*, 46(8):751–755.

Shamansky, S.L., G.H. Glaser. (1979). Socioeconomic characteristics of childhood seizure disorders in the New Haven area: an epidemiologic study. *Epilepsia*, 20(5):457–474.

Sharf, M., B. Sharf, E. Bental, T. Kuzminsky. (1969). The electroencephalogram in the investigation of anovulation and its treatment by clomiphene. *Lancet*, 1(7598):750–753.

Shavit, G., P. Lerman, A.D. Korczyn, S. Kivity, M. Bechar, S. Gitter. (1984). Phenytoin pharmacokinetics in catamenial epilepsy. *Neurology*, 34(7):959–961.

Shughrue, P.J., M.V. Lane, I. Merchenthaler. (1997). Comparative distribution of estrogen receptor-alpha and -beta mRNA in the rat central nervous system. *J. Comp. Neurol.*, 388(4):507–525.

Silveira, D.C., C.A. Guerreiro. (1991). Beginning of epileptic seizures in menarche. *Arq. Neuropsiquiatr.*, 49(4):434–436.

Simerly, R.B., C. Chang, M. Muramatsu, L.W. Swanson. (1990). Distribution of androgen and estrogen receptor mRNA-containing cells in the rat brain: an *in situ* hybridization study. *J. Comp. Neurol.*, 294(1):76–95.

Small, J.G., J.R. Stevens, V. Milstein. (1964). Electro-clinical correlates of emotional activation of the electro-encephalogram. *J. Nerv. Ment. Dis.*, 138:146–155.

Smith, M.A., S.R. Weiss, R.L. Berry, L.X. Zhang, M. Clark, G. Massenburg, R.M. Post. (1997). Amygdala-kindled seizures increase the expression of corticotropin-releasing factor (CRF) and CRF-binding protein in GABAergic interneurons of the dentate hilus. *Brain Res.*, 745(1–2):248–256.

Smith, S.S. (1989). Estrogen administration increases neuronal responses to excitatory amino acids as a long-term effect. *Brain Res.*, 503(2):354–357.

Smith, S.S., B.D. Waterhouse, D.J. Woodward. (1987). Sex steroid effects on extrahypothalamic CNS. II. Progesterone, alone and in combination with estrogen, modulates cerebellar responses to amino acid neurotransmitters. *Brain Res.*, 422(1):52–62.

Smith, S.S., Q.H. Gong, F.C. Hsu, R.S. Markowitz, J.M. ffrench-Mullen, X. Li. (1998). GABA(A) receptor alpha4 subunit suppression prevents withdrawal properties of an endogenous steroid. *Nature*, 392(6679):926–930.

Smith, S.S., H. Shen, Q.H. Gong, X. Zhou. (2007). Neurosteroid regulation of GABA(A) receptors: focus on the alpha4 and delta subunits. *Pharmacol. Ther.*, 116(1):58–76.

Snead, O. (1991). ACTH and prednisone: use in seizure disorders other than infantile spasms. In *The Medical Treatment of Epilepsy* (pp. 445–448), S.R. Resor and H. Kutt, Eds. Raven Press: New York.

Spatt, J., G. Langbauer, B. Mamoli. (1998). Subjective perception of seizure precipitants: results of a questionnaire study. *Seizure*, 7(5):391–395.

Spector, S., C. Cull, L.H. Goldstein. (2000). Seizure precipitants and perceived self-control of seizures in adults with poorly-controlled epilepsy. *Epilepsy Res.*, 38(2–3):207–216.

Speroff, L., M.A. Fritz. (2005). *Clinical Gynecologic Endocrinology and Infertility*, 7th ed. Philadelphia, PA: Lippincott Williams & Wilkins.

Stevens, J.R. (1959). Emotional activation of the electroencephalogram in patients with convulsive disorders. *J. Nerv. Ment. Dis.*, 128(4):339–351.

Svalheim, S., E. Tauboll, T. Bjornenak, L.S. Roste, T. Morland, E.R. Saetre, L. Gjerstad. (2006). Onset of epilepsy and menarche: is there any relationship? *Seizure*, 15(8):571–575.

Swanson, L.W., P.E. Sawchenko, J. Rivier, W.W. Vale. (1983). Organization of ovine corticotropin-releasing factor immunoreactive cells and fibers in the rat brain: an immunohistochemical study. *Neuroendocrinology*, 36(3):165–186.

Swinkels, W.A., M. Engelsman, D.G. Kasteleijn-Nolst Trenite, M.G. Baal, G.J. de Haan, J. Oosting. (1998). Influence of an evacuation in February 1995 in the Netherlands on the seizure frequency in patients with epilepsy: a controlled study. *Epilepsia*, 39(11):1203–1207.

Swinyard, E.A., N. Radhakrishnan, L.S. Goodman. (1962). Effect of brief restraint on the convulsive threshold of mice. *J. Pharmacol. Exp. Ther.*, 138:37–42.

Tauboll, E., A. Lundervold, L. Gjerstad. (1991). Temporal distribution of seizures in epilepsy. *Epilepsy Res.*, 8(2):153–165.

Temkin, N.R., G.R. Davis. (1984). Stress as a risk factor for seizures among adults with epilepsy. *Epilepsia*, 25(4):450–456.

Terasawa, E., P.S. Timiras. (1968). Electrical activity during the estrous cycle of the rat: cyclic changes in limbic structures. *Endocrinology*, 83(2):207–216.

Thorneycroft, I., P. Klein, J. Simon. (2006). The impact of antiepileptic drug therapy on steroidal contraceptive efficacy. *Epilepsy Behav.*, 9(1):31–39.

Tomson, T., R. Palm, K. Kallen, E. Ben-Menachem, B. Soderfeldt, B. Danielsson, R. Johansson, G. Luef, I. Ohman. (2007). Pharmacokinetics of levetiracetam during pregnancy, delivery, in the neonatal period, and lactation. *Epilepsia*, 48(6):1111–1116.

Tuveri, A., A.M. Paoletti, M. Orru, G.B. Melis, M.F. Marotto, P. Zedda, F. Marrosu, C. Sogliano, C. Marra, G. Biggio, A. Concas. (2008). Reduced serum level of THDOC, an anticonvulsant steroid, in women with perimenstrual catamenial epilepsy. *Epilepsia*, 49(7):1221–1229.

Veliskova, J. (2007). Estrogens and epilepsy: why are we so excited? *Neuroscientist*, 13(1):77–88.

Verrotti, A., G. Latini, R. Manco, M. De Simone, F. Chiarelli. (2007). Influence of sex hormones on brain excitability and epilepsy. *J. Endocrinol. Invest.*, 30(9):797–803.

Wallis, C.J., W.G. Luttge. (1980). Influence of estrogen and progesterone on glutamic acid decarboxylase activity in discrete regions of rat brain. *J. Neurochem.*, 34(3):609–613.

Webster, A., G.E. Mawer. (1989). Seizure frequency and major life events in epilepsy. *Epilepsia*, 30(2):162–167.

Weiss, S.R., R.M. Post, P.W. Gold, G. Chrousos, T.L. Sullivan, D. Walker, A. Pert. (1986). CRF-induced seizures and behavior: interaction with amygdala kindling. *Brain Res.*, 372(2):345–351.

Westin, A.A., A. Reimers, G. Helde, K.O. Nakken, E. Brodtkorb. (2008). Serum concentration/dose ratio of levetiracetam before, during and after pregnancy. *Seizure*, 17(2):192–198.

Williamson, J., Z. Mtchedlishvili, J. Kapur. (2004). Characterization of the convulsant action of pregnenolone sulfate. *Neuropharmacology*, 46(6):856–864.

Wong, M., R.L. Moss. (1992). Long-term and short-term electrophysiological effects of estrogen on the synaptic properties of hippocampal CA1 neurons. *J. Neurosci.*, 12(8):3217–3225.

Woodbury, D.M. (1952). Effect of adrenocortical steroids and adrenocorticotrophic hormone on electroshock seizure threshold. *J. Pharmacol. Exp. Ther.*, 105(1):27–36.

Woolley, C.S., B.S. McEwen. (1993). Roles of estradiol and progesterone in regulation of hippocampal dendritic spine density during the estrous cycle in the rat. *J. Comp. Neurol.*, 336(2):293–306.

Woolley, C.S., E. Gould, B.S. McEwen. (1990). Exposure to excess glucocorticoids alters dendritic morphology of adult hippocampal pyramidal neurons. *Brain Res.*, 531(1–2):225–231.

Woolley, C.S., H.J. Wenzel, P.A. Schwartzkroin. (1996). Estradiol increases the frequency of multiple synapse boutons in the hippocampal CA1 region of the adult female rat. *J. Comp. Neurol.*, 373(1):108–117.

Woolley, D.E., P.S. Timiras. (1962a). Estrous and circadian periodicity and electroshock convulsions in rats. *Am. J. Physiol.*, 202:379–382.

Woolley, D.E., P.S. Timiras. (1962b). The gonad–brain relationship: effects of female sex hormones on electroshock convulsions in the rat. *Endocrinology*, 70:196–209.

Wu, F.S., T.T. Gibbs, D.H. Farb. (1990). Inverse modulation of gamma-aminobutyric acid- and glycine-induced currents by progesterone. *Mol. Pharmacol.*, 37(5):597–602.

Yankova, M., S.A. Hart, C.S. Woolley. (2001). Estrogen increases synaptic connectivity between single presynaptic inputs and multiple postsynaptic CA1 pyramidal cells: a serial electron-microscopic study. *Proc. Natl. Acad. Sci. U.S.A.*, 98(6):3525–3530.

# 28 Neurosteroid Replacement Therapy for Catamenial Epilepsy*

*Doodipala S. Reddy and Michael A. Rogawski*

## CONTENTS

## INTRODUCTION

The term "catamenial epilepsy" is used to describe the cyclical occurrence of seizure exacerbations during particular phases of the menstrual cycle in women with preexisting epilepsy (Newmark and Penry, 1980). The types of epilepsies and seizures that are susceptible to catamenial fluctuations have not been defined in detail. However, it seems that seizures in both partial epilepsies (such as mesial temporal lobe epilepsy) and certain primary generalized epilepsies (such as juvenile myoclonic epilepsy) can exhibit catamenial exacerbations (Agathonikou et al., 1997; Herzog et al., 1997; Panayiotopoulos et al., 1994). It has been recognized that the menstrual cycle can influence seizure susceptibility since antiquity. Today, because women with epilepsy severe enough to exhibit cyclical changes in seizure frequency are invariably treated with antiepileptic medications, catamenial epilepsy is now a specific form of intractable (pharmacoresistant) epilepsy. With drug treatment, some of these women experience a resolution of seizures except at certain times during the menstrual cycle; others do not respond to medications. In either case, the subjects have intractable seizures. The former situation is an example of state-dependent pharmacoresistance and may provide insight into mechanisms of drug intractability.

Catamenial seizure exacerbations affect as few as 10% or as many as 70% of women with epilepsy who are of reproductive age (Bazan et al., 2005; Duncan et al., 1993; Herzog et al., 2004; Tauboll et al., 1991). The large variation in prevalence is mainly due to definitional differences. Herzog et al. (1997) proposed a research definition that requires a twofold increase in average daily seizure frequency during a phase of exacerbation relative to the other phases. By this criterion, catamenial conditions are met by as many as one third of women with intractable partial epilepsy (Herzog et al., 2004). Herzog et al. (1997) defined three forms of catamenial epilepsy: (1) *perimenstrual* (C1, days −3 to 3),

---

* Adapted from D.S. Reddy and M.A. Rogawski, *Neurotherapeutics*, 6, 392-401, 2009. With permission.

(2) *periovulatory* (C2, days 10 to −13) in normal cycles, and (3) *luteal* (C3, days 10 to 3) in inadequate luteal phase cycles, where day 1 is the first day of menstrual flow and ovulation is presumed to occur 14 days before the subsequent onset of menses (day −14). The most common form is perimenstrual. In this article, treatment approaches for catamenial epilepsy are described and a rationale is provided for an investigational approach for the perimenstrual form involving exogenous administration of neurosteroids or neurosteroid analogs, referred to as *neurosteroid replacement therapy*.

## NONHORMONAL TREATMENTS FOR CATAMENIAL EPILEPSY

There are no validated therapeutic approaches to treat catamenial seizure exacerbations. A few anecdotal reports suggest that the carbonic anhydrase inhibitor acetazolamide may be effective. This was supported by a recent retrospective report in 20 women with temporal lobe (10), extratemporal (8), generalized (1), and unclassified (1) epilepsy; 30 to 40% of the patients reported improvement in the frequency and severity of perimenstrual seizure exacerbations while taking acetazolamide (Lim et al., 2001). Acetazolamide has a broad spectrum of efficacy, but its mechanism of action is not well understood. It is clear, however, that tolerance develops, which results in diminishing efficacy over time (Browne et al., 2008). This means that the drug can only be administered on an intermittent basis, which is appropriate for catamenial epilepsy but not for ordinary seizure prophylaxis and may explain why it has been adopted for use in the treatment of catamenial seizures. Still, there may be a rationale for its use in catamenial epilepsy. It is generally believed that the anticonvulsant effects of carbonic anhydrase inhibitors are produced by an accumulation of $CO_2$ in the brain, and it has been speculated that this somehow leads to reduced neuronal excitability (Rogawski and Porter, 1990). However, acetazolamide would also reduce intracellular bicarbonate. Outflux of bicarbonate through γ-aminobutyric acid type A (GABA$_A$) receptors has a depolarizing action that balances the hyperpolarization caused by influx of chloride (Staley et al., 1995). Although depolarizing GABA responses are characteristic of the immature brain, they also may occur pathologically in epileptic adult brain (Pathak et al., 2007). Reduction of intracellular bicarbonate has been shown to reduce depolarizing GABA responses (Bonnet and Bingmann, 1995). In addition, it can be expected that there would be enhancement of synaptic and tonic GABA$_A$ receptor inhibition (mediated by peri-synaptic/extrasynaptic GABA$_A$ receptors; see discussion below on the role of these receptors as a target for neurosteroids). Overall, carbonic anhydrase inhibition is expected to beneficially influence GABA inhibition and is therefore consistent with the hypothesis that altered GABA inhibition accounts for perimenstrual catamenial epilepsy and the strategy of enhancing GABA inhibition as a prophylactic approach.

Synaptic GABA$_A$ receptor-mediated inhibition can also be enhanced with benzodiazepines, which are powerful broad-spectrum anticonvulsants. As is the case for acetazolamide, tolerance develops to the anticonvulsant activity of benzodiazepines so they are of limited utility in seizure prophylaxis; however, they could theoretically be used on an intermittent basis for the treatment of catamenial seizures. In fact, the 1,5-benzodiazepine clobazam, administered intermittently, has been used to treat catamenial seizure exacerbations over long periods of time with good results (Feely and Gibson, 1984).

## HORMONAL BASIS OF PERIMENSTRUAL CATAMENIAL EPILEPSY

Catamenial seizure exacerbations have been attributed to a variety of mechanisms, including fluctuations in antiepileptic drug levels and changes in water and electrolyte balance. The best accepted hypothesis, however, posits that enhanced seizure susceptibility is related to physiological cyclic changes in ovarian hormone secretion (Bonuccelli et al., 1989; Reddy, 2004, 2007; Rogawski and Reddy, 2004; Scharfman and MacLusky, 2006). Whether there are abnormalities in hormonal dynamics that predispose one to catamenial epilepsy is not known. The periovulatory form of catamenial epilepsy is believed to be related to the midcycle increase in estrogen secretion that occurs at the time of ovulation and is relatively unopposed by progesterone until the early luteal phase. There is some evidence that estrogen has proconvulsant activity, although it may also be anticonvulsant under some circumstances (Buterbaugh, 1989; Scharfman and MacLusky, 2006; Veliskova, 2007). The molecular basis for these actions has not yet been defined. There is more substantial evidence that the perimenstrual form is due to withdrawal of progesterone that occurs at the time of menstruation. (An increase in the ratio of estrogen-to-progesterone levels during the perimenstrual period might also partly contribute to the development of perimenstrual seizure exacerbations.) The reproductive functions of progesterone are mediated via its interaction with intracellular progesterone receptors, expressed in A and B forms, arising from the same gene. The progesterone receptors are members of the nuclear receptor superfamily of transcription factors. In addition to being expressed in reproductive tissues where they control key female reproductive events such as ovulation, maintenance of pregnancy, and breast development, progesterone receptors are also expressed widely in the central nervous systems, including areas relevant to epilepsy such as the hippocampus, amygdala, and neocortex (Brinton et al., 2008). We have demonstrated, however, that the anticonvulsant effects of progesterone are not reduced in mice in which the progesterone receptor gene has been deleted by gene targeting (Reddy et al., 2004). This conclusively demonstrates that the seizure protection conferred by progesterone is not due to its interaction with progesterone receptors.

Progesterone has long been known to have anticonvulsant properties (Frye et al., 2002; Lonsdale and Burnham, 2003; Reddy et al., 2004; Rogawski and Reddy, 2004; Selye, 1942), and this stems from its conversion to the neurosteroid allopregnanolone (Kokate et al., 1999). Allopregnanolone is synthesized from progesterone by two sequential A-ring reductions catalyzed by $5\alpha$-reductase and $3\alpha$-hydroxysteroid oxidoreductase isoenzymes. During the menstrual cycle, circulating progesterone levels are low in the follicular phase but rise in the midluteal phase for about 10 to 11 days, before declining in the late luteal phase. Circulating allopregnanolone levels parallel those of its parent progesterone (Tuveri et al., 2008). The dynamics of *brain* allopregnanolone during the menstrual cycle have not been studied, although it is likely that local synthesis of $GABA_A$ receptor-modulating neurosteroids, including allopregnanolone, occurs in regions relevant to epilepsy such as the cortex, hippocampus, and amygdale (Agis-Balboa et al., 2006; Ebner et al., 2006; Mukai et al., 2008). Within these brain regions, neurosteroid synthetic enzymes are localized to glutamatergic principal neurons and not GABAergic inhibitory neurons; thus, neurosteroids may access $GABA_A$ receptors in the principal neurons by lateral membrane diffusion. The $GABA_A$ receptor-modulating neurosteroid allotetrahydrodeoxycorticosterone (THDOC), which is derived from the adrenal steroid deoxycorticosterone, also fluctuates during the menstrual cycle, with higher levels in the luteal phase (Tuveri et al., 2008). Overall, the serum levels of THDOC are lower than those of allopregnanolone (Hardoy et al., 2008), so it is likely to be less relevant to catamenial epilepsy, although it could contribute. One recent study suggested that women with catamenial epilepsy may have reduced levels of THDOC throughout the menstrual cycle (Tuveri et al., 2008); the significance of this finding is uncertain but may be responsible for an overall enhancement of seizure susceptibility.

Allopregnanolone is a potent, broad-spectrum anticonvulsant agent that protects against seizures in diverse animal models, including various chemoconvulsant models and the 6-Hz electroshock model, as well as in kindled animals (Kaminski et al., 2004; Rogawski and Reddy, 2004). It is noteworthy that both progesterone and allopregnanolone exacerbate seizures in animal models of absence epilepsy (Snead, 1998; van Luijtelaar et al., 2001). The proconvulsant effects of progesterone on absence seizures require conversion to neurosteroids, as is the case for the anticonvulsant actions of progesterone (van Luijtelaar et al., 2003). The anticonvulsant activity of allopregnanolone (and presumably also the proconvulsant activity in absence seizure models) is due to its action on $GABA_A$ receptors (Kokate et al., 1994). Allopregnanolone and related neurosteroids act as positive allosteric modulators of all $GABA_A$ receptor isoforms (Lambert et al., 2003). They are highly lipophilic molecules that are easily able to cross the blood–brain barrier and within the brain readily diffuse into cellular membranes. Indeed, they appear to act on $GABA_A$ receptors by entry into the plasma membrane, where they access specific sites on the receptors by lateral diffusion or from the cell interior (Akk et al., 2005). GABA$_A$ receptors, the main mediators of inhibition in the central nervous system, are pentameric protein complexes surrounding a central chloride-selective ion channel (Meldrum and Rogawski, 2007). The major isoforms consist of two $\alpha$, two $\beta$, and one $\gamma_2$ subunit, which are localized to synapses (and also can be expressed extrasynaptically). In some channels, $\delta$ substitutes for $\gamma_2$; these less abundant channels are not targeted to synapses and are believed to be principally perisynaptic (localized around the edges of synapses) and extrasynaptic (Glykys and Mody, 2007). The modulating effects of neurosteroids occur by binding to discrete sites within the transmembrane domains of the $\alpha$ and $\beta$ subunits that comprise the GABA$_A$ receptor (Hosie et al., 2007). Through homology modeling and the use of chimeras formed between $\alpha$ subunits and the neurosteroid-insensitive Rdl (*Drosophila* resistance to dieldrin) subunit, it has been determined that a highly conserved glutamine at position 241 in the M1 domain of the $\alpha$ subunit plays a key role in neurosteroid modulation (Hosie et al., 2009). Exposure to GABA$_A$ receptor-modulating neurosteroids enhances the open probability of the GABA$_A$ receptor chloride channel, so the mean open time is increased and the mean closed time is decreased (Akk et al., 2005). This increases the chloride current through the channel, ultimately resulting in a reduction of cellular excitability.

Neurosteroids, such as allopregnanolone and THDOC, are able to modulate most GABA$_A$ receptor isoforms. This distinguishes neurosteroids from benzodiazepines, which only act on GABA$_A$ receptors that (1) contain $\gamma_2$ subunits and (2) do not contain $\alpha_4$ or $\alpha_6$ subunits; however, GABA$_A$ receptors that contain the $\delta$ subunit are more sensitive to neurosteroid-induced potentiation of GABA responses (Belelli et al., 2002; Wohlfarth et al., 2002). The $\delta$ subunit does not contribute to the neurosteroid binding site but appears to confer enhanced transduction of neurosteroid action after the neurosteroid has bound to the receptor. GABA$_A$ receptors containing the $\delta$ subunit have a low degree of desensitization, and they are located perisynaptically/extrasynaptically (Glykys and Mody, 2007). These properties cause them to be prime candidates for mediating tonic GABA$_A$ receptor current activated by ambient concentrations of extracellular GABA. Tonic GABA$_A$ receptor current causes a steady inhibition of neurons and reduces their excitability. Glykys and Mody (2007) noted that GABA is a relatively low-efficacy agonist of $\delta$-containing GABA$_A$ receptors even though it binds with high affinity; therefore, neurosteroids have an opportunity to markedly enhance the current generated by $\delta$-containing GABA$_A$ receptors even in the presence of saturating GABA concentrations. During seizure activity, there is expected to be substantial release of GABA from active GABAergic interneurons that can interact with perisynaptic and extrasynaptic $\delta$-subunit-containing GABA$_A$ receptors. The powerful anticonvulsant activity of GABA$_A$ receptor modulating neurosteroids is likely to be due to their action on both synaptic and perisynaptic/extrasynaptic GABA$_A$ receptors; the effect on the latter type of GABA$_A$ receptor may be more significant given that $\delta$-subunit-containing receptors are more sensitive to neurosteroids.

Expression of the GABA$_A$ receptor subunit and the receptors composed of these subunits is not static in the cell; they undergo alterations that compensate for changes in the endogenous hormonal (neurosteroid) milieu and in response to exogenously administered pharmacological agents that modulate GABA$_A$ receptors, such as benzodiazepines or ethanol (Smith et al., 2007). The precise changes in brain GABA$_A$ receptor subunit expression occurring during the human menstrual cycle or in rat models of catamenial epilepsy have not been determined. It is now well recognized, however, that prolonged exposure to allopregnanolone in rats causes increased expression of the $\alpha_4$ GABA$_A$ receptor subunit in hippocampus, resulting in decreased benzodiazepine sensitivity of GABA$_A$ receptor currents (Gulinello et al., 2001). Although $\alpha_4$ can coassemble with $\gamma_2$ to form synaptic GABA$_A$ receptors, it preferentially coassembles with $\delta$ to form perisynaptic/extrasynaptic GABA$_A$ receptors (Sur et al., 1999). Allopregnanolone treatment of rats results in a transiently increased expression of the $\delta$ subunit in the hippocampus and an increased benzodiazepine-insensitive tonic current (Maguire et al., 2005; Shen et al., 2005). Progesterone also increases $\delta$ subunit expression, likely as a result of conversion to allopregnanolone (Shen et al., 2005). Interestingly, estrogen enhances the effect of progesterone (Shen et al., 2005). The basis of the effect of estrogen is not understood, but the phenomenon suggests that in physiological situations, such as in the pseudopregnancy model described below or during the human luteal phase where there are high circulating estrogen levels, estrogen enhances the plasticity of $GABA_A$ receptors. The relevance of the increased $\delta$ subunit expression for catamenial epilepsy is unclear, as $\delta$ subunit increases may be transitory and followed by *reduced* expression with chronic exposures as in pregnancy (Maguire and Mody, 2008) or in the prolonged luteal phase of the human menstrual cycle.

An important consequence of incorporating the normally low-abundance $\alpha_4$ subunit into synaptic $GABA_A$ receptors is that synaptic currents generated by these receptors have accelerated decay kinetics, resulting in less total charge transfer and reduced inhibition (Smith and Gong, 2005). $GABA_A$ receptor-modulating neurosteroids cause a prolongation of the decay of GABA-mediated synaptic currents (Belelli and Lambert, 2005); therefore, in the presence of high levels of allopregnanolone during the luteal phase, the acceleration due to $\alpha_4$ substitution is balanced. When neurosteroids are withdrawn at the time of menstruation, synaptic inhibition is diminished from normal, resulting in enhanced excitability, which, among other effects, predisposes to seizures.

Enhanced expression of $\alpha_4$ is associated with a compensatory reduction in $\alpha_1$ expression (Shen et al., 2005). *In vitro* studies have also indicated that withdrawal of neurosteroids is associated with the downregulation of $\alpha_1$ and $\gamma_2$ subunit expression (Mascia et al., 2002). Because the $\gamma_2$ subunit is required for synaptic clustering and targeting of $GABA_A$ receptors in dendrites, enhanced excitability may also result from a net reduction of synaptic $GABA_A$ receptors, although some studies have not detected the functional correlate of such receptor changes (i.e., reduced synaptic current) (Smith et al., 2007). It is noteworthy that periods of exposure to neurosteroids longer than 72 hours are associated with a return in $\alpha_4$ expression to control levels (Smith et al., 2007); therefore, the rise in $\alpha_4$ expression is transitory. Following prolonged periods of neurosteroid exposure, however, neurosteroid withdrawal results in a dramatic rebound rise in $\alpha_4$ expression (measured 24 hours after withdrawal) (Mascia et al., 2002; Smith et al., 2007). It is the *withdrawal* of neurosteroids around the time of menstruation rather than the prolonged exposure to neurosteroids during the menstrual cycle luteal phase that is likely to be most relevant to the enhanced excitability and greater seizure susceptibility in perimenstrual catamenial epilepsy. As noted above, chronic exposure to neurosteroids is also accompanied by the downregulation of the $\delta$ subunit expression and perisynaptic/extrasynaptic $GABA_A$ receptors (Maguire and Mody, 2008). This change is believed to be a compensatory mechanism, which would avoid excessive sedation caused by high neurosteroid levels acting on sensitive $\delta$-subunit-containing $GABA_A$ receptors. At the time of neurosteroid withdrawal, $\delta$ subunit expression rapidly recovers; however, if recovery is not sufficiently fast, there could be an enhancement of excitability due to a reduction in tonic inhibition mediated by perisynaptic/extrasynaptic $GABA_A$ receptors in the relative absence of neurosteroids.

Overall, the basis for the enhanced seizure susceptibility in perimenstrual catamenial epilepsy is multifaceted and may include: (1) withdrawal of the anticonvulsant effects of neurosteroids mediated through their action on $GABA_A$ receptors; (2) increased expression of $GABA_A$ receptor $\alpha_4$ subunits, especially following neurosteroid withdrawal, resulting in reduced GABA-mediated synaptic inhibition due to an acceleration in the decay of GABA-mediated synaptic currents; (3) overall reduction in $GABA_A$ receptor-mediated synaptic inhibition due to reduced expression of $GABA_A$ receptor subunits (such as $\alpha_1$ and $\gamma_2$) that compose synaptic $GABA_A$ receptors; and (4) reduced expression of extrasynaptic ($\delta$-subunit-containing) $GABA_A$ receptors with an insufficiently rapid recovery at the time of neurosteroid withdrawal. All of these factors are related to the high levels of circulating neurosteroids during the leuteal phase and the natural reduction or withdrawal of progesterone that occurs around the time of menstruation.

## RAT MODEL OF CATAMENIAL EPILEPSY

To better understand the biological basis of catamenial seizure exacerbations and to evaluate potential therapies, we developed a rodent model that is designed to simulate the hormonal changes that are believed to be relevant to perimenstrual catamenial epilepsy (Reddy et al., 2001). Rodents, which are the main mammalian species used for experimental epilepsy research, have a 4- to 5-day estrous cycle. Studies of fluctuations in seizure susceptibility during the estrous cycle have not provided results that are relevant to the 28-day human menstrual cycle (Finn and Gee, 1994); therefore, we attempted to simulate human hormonal menstrual changes in the rat. The basic requirement is a prolonged high level of progesterone followed by withdrawal to stimulate the fall in ovarian secretion that occurs at the time of menstruation. Abrupt discontinuation of progesterone after chronic treatment for 3 weeks leads to proconvulsant effects (Smith et al., 1998a). This can be considered a rudimentary model of catamenial epilepsy. A more physiological means of inducing elevated progesterone (and other hormonal changes of the human luteal phase such as increased estrogen) is to utilize a hormonal treatment regimen with pregnant mare's serum gonadotropin (PMSG) and human chorionic gonadotropin (hCG), as is commonly used to induce superovulation (Kim and Greenwald, 1986). This state has been referred to as *pseudopregnancy* (Smith et al., 1998b) because it is an anestrous condition (cessation of estrous cyclicity) associated with changes in the reproductive organs and mammary glands that simulate those occurring in pregnancy. During pseudopregnancy in the rat, which persists for 12 to 13 days, progesterone and estrogen levels are similar to those in the luteal phase of the human menstrual cycle. (Pseudopregnancy is ordinarily induced in female rats by vaginocervical stimulation, but gonadotropin treatment produces more reliable and robust sex steroid elevations.) Abrupt withdrawal from progesterone-derived neurosteroids, as is believed to occur around the time of menstruation in humans, can be induced either by ovariectomy or by injection of a 5α-reductase inhibitor that blocks neurosteroid synthesis (Kokate et al., 1999). Such neurosteroid withdrawal is associated with a marked enhancement in anxiety-like behaviors and seizure susceptibility (Reddy et al., 2001; Smith et al., 1998b). We have proposed that the period of heightened seizure susceptibility represents a model of human perimenstrual catamenial seizure exacerbations (Reddy et al., 2001) (Figure 28.1).

The pseudopregnancy neurosteroid withdrawal model has, until recently, been applied only with normal healthy rats induced to have seizures with pentylenetetrazol; however, Reddy and Zeng (2007) utilized the model in female rats that have experienced an episode of pilocarpine status epilepticus and who exhibit spontaneous recurrent seizures. As shown in Figure 28.2B, these epileptic animals exhibit a transient increase in the frequency of spontaneous seizures following neurosteroid withdrawal. The use of epileptic animals with spontaneous recurrent seizures has greater face validity for catamenial epilepsy than the pentylenetetrazol threshold method (Figure 28.2A). Although this new approach could possibly be more relevant to human epilepsy, it is substantially more labor intensive and is not well suited for extensive pharmacological studies.

## PHARMACOLOGY OF THE CATAMENIAL EPILEPSY MODEL

The neurosteroid withdrawal model of catamenial epilepsy has been used to investigate therapies for perimenstrual catamenial epilepsy (Reddy and Rogawski, 2000a, 2001). A key result is the observation that conventional antiepileptic drugs have reduced potency in protecting against seizures during the period of enhanced seizure susceptibility following neurosteroid withdrawal. These studies necessarily involved drugs effective against pentylenetetrazol seizures, including benzodiazepines and valproate, which were the type of seizures utilized with the model. The dose–response curves for seizure protection by these agents were shifted to the right in the period of enhanced seizure susceptibility following neurosteroid withdrawal, indicating reduced potency (see Figure 28.3B). Extrapolating these results to women with perimenstrual catamenial epilepsy suggests that seizure exacerbations may, at least in part, result from a state-dependent reduction in drug sensitivity. The model therefore represents a conditional form of antiepileptic drug pharmacoresistance of the antiepileptic drug target type (Remy and Beck, 2006). The increased relative expression of benzodiazepine-insensitive $GABA_A$ receptors at the time of neurosteroid withdrawal likely accounts for the reduced activity of benzodiazepines. The basis for the reduced activity of other anticonvulsant

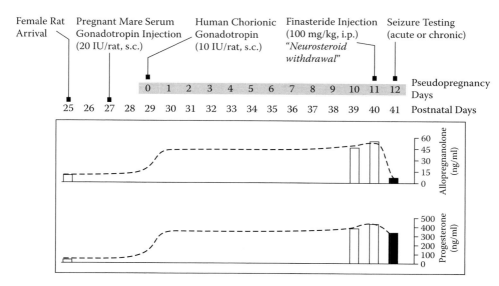

**FIGURE 28.1**    Rat neurosteroid withdrawal model of catamenial seizure exacerbations. Rats are treated sequentially with pregnant mare serum gonadotropin and human chorionic gonadotropin according to a protocol described by Kim and Greenwald (1986) which hyperstimulates the ovary to induce massive ovarian luteinization, resulting in prolonged elevated serum progesterone levels (described as a state of "pseudopregnancy"). On day 11 of pseudopregnancy, animals are treated with the 5α-reductase inhibitor finasteride, which blocks the conversion of progesterone to the neurosteroid allopregnanolone, resulting in "neurosteroid withdrawal." Acute seizure testing is performed on the day following finasteride administration. For studies of effects of neurosteroid withdrawal on spontaneous seizures (Figure 28.2B), the gonadotropin treatment was initiated 5 months after pilocarpine status epilepticus. Progesterone and allopregnanolone plasma levels are from Reddy et al. (2001); dashed lines schematically illustrate the time course of fluctuations in hormone levels and do not represent actual measurements except for the data points indicated with bars. The drop in allopregnanolone levels following finasteride treatment is significant, whereas the change in progesterone is not significant.

drugs is more difficult to define. In particular, there was a profound reduction in valproate potency, but it is difficult to speculate on the cause because the molecular mechanism of action of valproate is not well understood.

In contrast to the results with conventional anticonvulsant drugs, we unexpectedly found that neurosteroids, including allopregnanolone, THDOC, and their 5β-isomers, as well as the barbiturate phenobarbital (all of which are positive allosteric modulators of GABA$_A$ receptors), had enhanced activity in the perimenstrual catamenial epilepsy model (Reddy and Rogawski, 2001). Extensive studies were conducted with the neurosteroid analog ganaxolone (Figure 28.3) (Reddy and Rogawski, 2000b). As shown in Figure 28.3A, ganaxolone is the synthetic 3β-methyl derivative of allopregnanolone (Carter et al., 1997; Garofalo et al., 2008; Nohria and Giller, 2007; Reddy and Woodward, 2004). The 3β-methyl substituent minimizes back conversion by bidirectional 3α-hydroxysteroid oxidoreductase isoenzymes to the hormonally active 3-keto form. Hence, ganaxolone lacks hormonal activity, unlike the endogenous neurosteroid allopregnanolone, which may have hormonally active metabolites. Ganaxolone has pharmacological activity, including GABA$_A$ receptor modulating activity and anticonvulsant properties in animal models, that is similar to that of the parent allopregnanolone. Like naturally occurring neurosteroids, the anticonvulsant potency of ganaxolone was enhanced following neurosteroid withdrawal (Reddy and Rogawski, 2000a). Neurosteroid-withdrawn rats receiving a single dose of ganaxolone (7 mg/kg s.c.) exhibited markedly increased protective activity in the pentylenetetrazol threshold test than did control animals

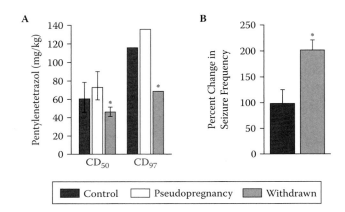

**FIGURE 28.2** Enhanced seizure susceptibility following neurosteroid withdrawal in an acute rat seizure model (pentylenetetrazol-induced clonic convulsions) and in a chronic epilepsy model (spontaneous recurrent seizures following pilocarpine status epilepticus). (A) Convulsant doses of pentylenetetrazol for 50% ($CD_{50}$) and 97% ($CD_{97}$) of rats in (1) control diestrous, (2) non-withdrawn (vehicle-treated) pseudopregnant, and (3) finasteride-treated (neurosteroid-withdrawn) pseudopregnant animals. The $CD_{50}$ and $CD_{97}$ values in withdrawn animals are significantly reduced compared with the values from control diestrous animals and from non-withdrawn pseudopregnant animals ($p < 0.05$), indicating greater seizure susceptibility. (Adapted from Reddy, D.S. et al., *Epilepsia*, 42, 328–336, 2001.) (B) Increased seizure frequency following neurosteroid withdrawal (catamenial seizure model) in epileptic animals 5 months after pilocarpine treatment. At the end of the 11 days of pseudopregnancy, rats on average exhibited about 2 seizures per day (= 100%; black bar). On the day after neurosteroid withdrawal (day 12), seizure frequency was 2-fold the prewithdrawal value (4 seizures per day; $p < 0.05$). The seizure frequency partially recovered to 2.8 seizures per day 2 days after withdrawal (day 13), which was within the baseline seizure frequency (2.6 to 3.0 seizures per day on days 1 and 2 of pseudopregnancy) (not shown) (D.S.R., unpublished).

(Reddy and Rogawski, 2008). Ganaxolone plasma and brain levels were not increased in neurosteroid-withdrawn animals; therefore, the enhanced potency of ganaxolone was not due to pharmacokinetic factors. In fact, brain ganaxolone levels were modestly reduced in withdrawn animals; the pharmacokinetic basis for this effect is uncertain. Based on the relationship between brain ganaxolone concentrations and the corresponding pentelenetetrazol thresholds, there was an approximately 70% elevation in threshold for any active brain ganaxolone concentration in the withdrawn animals compared with controls. The enhanced activity of ganaxolone was independent of brain concentration and must relate to alterations in pharmacodynamic activity. Although the molecular basis is still uncertain, it presumably results from alterations in $GABA_A$ receptor sensitivity. Specifically, it may relate to the relative increase in neurosteroid-sensitive δ-subunit containing $GABA_A$ receptors discussed previously. It is noteworthy that the motor toxicity of ganaxolone was not increased in neurosteroid withdrawn animals (Reddy and Rogawski, 2000a), indicating that the alterations in $GABA_A$ receptors that account for the potentiated anticonvulsant activity are not relevant to the general sedative activity of the steroid, which is the main side effect of neurosteroids and other $GABA_A$ receptor-positive modulators. Because of the selective increase in anticonvulsant potency, the therapeutic window (ratio of $TD_{50}$ for motor impairment and $ED_{50}$ for seizure protection in pentylenetetrazol test) of ganaxolone is increased from 1.6 in control rats to 6.4 in pseudopregnant neurosteroid-withdrawn animals (Reddy and Rogawski, 2000a).

In recent studies, the anticonvulsant activity of ganaxolone was examined in rats with spontaneous seizures that had experienced pilocarpine status epilepticus. As shown in Figure 28.3C, ganaxolone administered following neurosteroid withdrawal significantly reduced the occurrence of spontaneous seizures in these animals, further confirming its potential utility in the treatment of perimenstrual catamenial epilepsy.

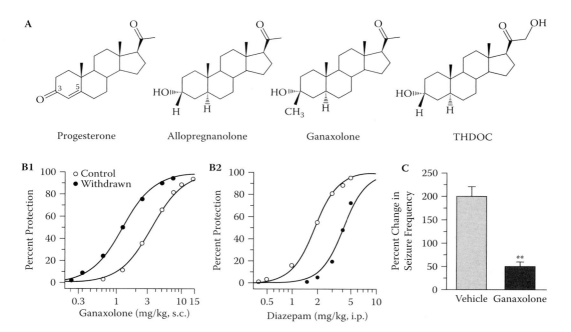

**FIGURE 28.3** (A) Structures of progesterone, the endogenous neurosteroids allopregnanolone and THDOC, and the synthetic allopregnanolone analog ganaxolone. Allopregnanolone is derived from progesterone by reduction at the 5- and 3-positions of the steroid A-ring. Ganaxolone differs from allopregnanolone by a 3β-methyl group. (B1) Anticonvulsant potency of ganaxolone is enhanced during the period of increased seizure susceptibility in the pseudopregnancy withdrawal model of catamenial epilepsy. Dose–response curves indicate ganaxolone protection against pentylenetetrazol-induced seizures in naïve female rats (control) and finasteride-treated pseudopregnant rats (withdrawn). The shift to the left indicates increased potency. (B2) Anticonvulsant potency of diazepam is reduced during the period of increased seizure susceptibility in the model of catamenial epilepsy. The dose–response curve for diazepam protection against pentylenetetrazol-induced seizures is shifted to the right in withdrawn animals, indicating reduced potency. In the withdrawn animals (B1 and B2), seizure testing with pentylenetetrazol was carried out on day 12 of pseudopregnancy (24 hours after finasteride administration). (Adapted from Reddy, D.S. and Rogawski, M.A., *J. Pharmacol. Exp. Ther.*, 295, 1241–1248, 2000.) (C) Antiseizure effects of ganaxolone (7 mg/kg, s.c.) against spontaneous seizures in epileptic rats following neurosteroid withdrawal (see Figure 28.1). Data values represent percent change in relation to average daily seizure frequency in control group. Ganaxolone treatment significantly ($p < 0.01$) reduced the frequency of spontaneous seizures compared with vehicle controls.

## NEUROSTEROID REPLACEMENT THERAPY

Our pharmacological results with the pseudopregnancy animal model of catamenial epilepsy suggest that neurosteroid replacement may be a useful approach to prevent catamenial seizure exacerbations. A neurosteroid or a neurosteroid-like compound such as ganaxolone could be administered in a pulse prior to menstruation and then either withdrawn or continuously administered throughout the month. Although intermittent administration at the time of increased seizure vulnerability is rational, continuous administration would avoid withdrawal of the therapeutic agent, which itself could predispose to seizures. This factor, as well as the practical difficulty many women experience predicting the time of their menstrual periods, suggests that continuous administration is preferred. The neuroactive steroid would be administered at low doses to avoid sedative side effects. Such low doses are expected to contribute little anticonvulsant activity during most of the menstrual cycle, so patients would require treatment with conventional antiepileptic medications. However, during the period of enhanced seizure susceptibility at the time of menstruation, the increased potency of the neurosteroid would confer protection against perimenstrual seizure exacerbations. It is noteworthy that anticonvulsant tolerance does not develop to ganaxolone and other $GABA_A$ receptor-modulating neurosteroids following chronic therapy in rodents (Kokate et al., 1998; Reddy and Rogawsky, 2000b), indicating that these agents, unlike benzodiazepines, may be suitable for chronic treatment. Dosing may have to be adjusted to account for alterations in metabolism occurring with chronic exposure.

To date, 961 human subjects (including 135 children) have been exposed to ganaxolone in safety and pharmacokinetic studies (260 subjects) and various clinical trials (701 subjects). The drug is well tolerated but causes reversible sedation and somnolence, which is dose limiting. Ganaxolone is metabolized by CYP3A to 16-hydroxyganaxolone, which is inactive, and then to various multihydroxylated metabolites. Despite this route of metabolism, the information available to date indicates that ganaxolone will have low drug–drug interactions; therefore, it could be administered easily with concomitant antiepileptic drugs.

In a preliminary uncontrolled, open-label study, ganaxolone was evaluated in two women with catamenial epilepsy who appeared to have a reduction in their catamenial seizures (McAuley et al., 2001). These women received oral ganaxolone (300 mg/day b.i.d.) starting on day 21 of the menstrual cycle and continuing through the third full day following the beginning of menstruation. Prospective clinical studies are warranted to validate these preliminary observations.

## CONCLUSIONS

Catamenial epilepsy is a multifaceted neuroendocrine condition. In perimenstrual catamenial epilepsy, the most common form, there is emerging evidence to suggest that neurosteroid withdrawal plays a key role in the seizure exacerbations that occur around the time of menstruation. There currently are no specific, approved treatments to prevent seizure exacerbations in this condition. Anecdotal evidence suggests that progesterone may be effective (Herzog, 1995, 1999), and a multicenter controlled trial is underway; however, progesterone is poorly absorbed orally and has a short half-life so it must be administered multiple times per day. The activity of progesterone is likely to be dependent upon conversion to allopregnanolone; consequently, individual variation in metabolism could cause variability in the response to therapy. An additional concern is that progesterone may be associated with undesired hormonal side effects. Neurosteroids and synthetic analogs such as ganaxolone might provide an effective approach for catamenial epilepsy therapy that is more reliable and does not expose patients to the risk of hormonal side effects. New oral formulations of ganaxolone are being developed with enhanced bioavailability and more consistent absorption. The treatment of catamenial epilepsy is a promising application for this investigational agent.

## ACKNOWLEDGMENTS AND DISCLOSURES

This chapter was originally presented at the 4th Workshop on New Horizons in the Development of Antiepileptic Drugs—Nontraditional Approaches to Treat Epilepsy, March 5–7, 2008, Dublin, Ireland. This work was supported by NIH grant NS052158 to DSR; MAR is a scientific founder and has served as consultant to Marinus Pharmaceuticals, Inc.

## REFERENCES

Agathonikou, A., M. Koutroumanidis, C.P. Panayiotopoulos. (1997). Fixation-off-sensitive epilepsy with absences and absence status: video–EEG documentation. *Neurology*, 48:231–234.

Agis-Balboa, R.C., G. Pinna, A. Zhubi, E. Maloku, M. Veldic, E. Costa, A. Guidotti. (2006). Characterization of brain neurons that express enzymes mediating neurosteroid biosynthesis. *Proc. Natl. Acad. Sci. U.S.A.*, 103:14602–14607.

Akk, G., H.J. Shu, C. Wang, J.H. Steinbach, C.F. Zorumski, D.F. Covey, S. Mennerick. (2005). Neurosteroid access to the $GABA_A$ receptor. *J. Neurosci.*, 25:11605–11613.

Bazan, A.C., M.A. Montenegro, F. Cendes, L.L. Min, C.A. Guerreiro. (2005). Menstrual cycle worsening of epileptic seizures in women with symptomatic focal epilepsy. *Arq. Neuropsiquiatr.*, 63:751–756.

Belelli, D., J.J. Lambert. (2005). Neurosteroids: endogenous regulators of the GABA(A) receptor. *Nat. Rev. Neurosci.*, 6:565–575.

Belelli, D., A. Casula, A. Ling, J.J. Lambert. (2002). The influence of subunit composition on the interaction of neurosteroids with GABA(A) receptors. *Neuropharmacology*, 43:651–661.

Bonnet, U., D. Bingmann. (1995). GABA$_A$-responses of CA3 neurones: contribution of bicarbonate and of Cl(–)-extrusion mechanisms. *NeuroReport*, 6:700–704.

Bonuccelli, U., G.B. Melis, A.M. Paoletti, P. Fioretti, L. Murri, A. Muratorio. (1989). Unbalanced progesterone and estradiol secretion in catamenial epilepsy. *Epilepsy Res.*, 3:100–106.

Brinton, R.D., R.F. Thompson, M.R. Foy, M. Baudry, J. Wang, C.E. Finch, T.E. Morgan, C.J. Pike, W.J. Mack, F.Z. Stanczyk, J. Nilsen. (2008). Progesterone receptors: form and function in brain. *Front. Neuroendocrinol.*, 29:313–339.

Browne, T.R., B.W. Leduc, M.D. Kosta-Rokosz, E.B. Bromfield, R.E. Ramsay, J. De Toledo. (2008). Trimethadione, paraldehyde, phenacemide, bromides, sulthiame, acetazolamide, and methsuximide. In *Epilepsy, A Comprehensive Textbook*, Vol. 1, 2nd ed. (pp. 1703–1719), J. Engel, Jr., T.A. Pedley, J. Aicardi, and M.A. Dichter, Eds. Philadelphia, PA: Lippincott Williams & Wilkins.

Buterbaugh, G.G. (1989). Estradiol replacement facilitates the acquisition of seizures kindled from the anterior neocortex in female rats. *Epilepsy Res.*, 4:207–215.

Carter, R.B., P.L. Wood, S. Wieland, J.E. Hawkinson, D. Belelli, J.J. Lambert et al. (1997). Characterization of the anticonvulsant properties of ganaxolone (CCD 1042; 3alpha-hydroxy-3beta-methyl-5alpha-pregnan-20-one), a selective, high-affinity, steroid modulator of the gamma-aminobutyric acid(A) receptor. *J. Pharmacol. Exp. Ther.*, 280:1284–1295.

Duncan, S., C.L. Read, M.J. Brodie. (1993). How common is catamenial epilepsy? *Epilepsia*, 34:827–831.

Ebner, M.J., D.I. Corol, H. Havlikova, J.W. Honour, J.P. Fry. (2006). Identification of neuroactive steroids and their precursors and metabolites in adult male rat brain. *Endocrinology*, 147:179–190.

Feely, M., J. Gibson. (1984). Intermittent clobazam for catamenial epilepsy: tolerance avoided. *J. Neurol. Neurosurg. Psychiatry*, 47:1279–1282.

Finn, D.A., K.W. Gee. (1994). The estrus cycle, sensitivity to convulsants and the anticonvulsant effect of a neuroactive steroid. *J. Pharmacol. Exp. Ther.*, 271:164–170.

Frye, C.A., M.E. Rhodes, A. Walf, J. Harney. (2002). Progesterone reduces pentylenetetrazol-induced ictal activity of wild-type mice but not those deficient in type I 5alpha-reductase. *Epilepsia*, 43(Suppl. 5):14–17.

Garofalo, E., J. Tsai, K. Shaw, M.A. Rogawski, V. Pieribone. (2008). Ganaxolone. In Progress report on new antiepileptic drugs: a summary of the Ninth Eilat Conference (EILAT IX), M. Bialer, S.I. Johannessen, R.H. Levy, E. Perucca, T. Tomson, and H.S. White, Eds. *Epilepsy Res.*, 83(1):1–43.

Glykys, J., I. Mody. (2007). Activation of GABA$_A$ receptors: views from outside the synaptic cleft. *Neuron*, 56:763–770.

Gulinello, M., Q.H. Gong, X. Li, S.S. Smith. (2001). Short-term exposure to a neuroactive steroid increases alpha4 GABA(A) receptor subunit levels in association with increased anxiety in the female rat. *Brain Res.*, 910:55–66.

Hardoy, M.C., C. Sardu, L. Dell'osso, M.G. Carta. (2008). The link between neurosteroids and syndromic/syndromal components of the mood spectrum disorders in women during the premenstrual phase. *Clin. Pract. Epidemol. Ment. Health*, 4:3.

Herzog, A.G. (1995). Progesterone therapy in women with complex partial and secondary generalized seizures. *Neurology*, 45:1660–1662.

Herzog, A.G. (1999). Progesterone therapy in women with epilepsy: a 3-year follow-up. *Neurology*, 52:1917–1918.

Herzog, A.G., P. Klein, B.J. Ransil. (1997). Three patterns of catamenial epilepsy. *Epilepsia*, 38:1082–1088.

Herzog, A.G., C.L. Harden, J. Liporace, P. Pennell, D.L. Schomer, M. Sperling, K. Fowler, B. Nikolov, S. Shuman, M. Newman. (2004). Frequency of catamenial seizure exacerbation in women with localization-related epilepsy. *Ann. Neurol.*, 56:431–434.

Hosie, A.M., M.E. Wilkins, T.G. Smart. (2007). Neurosteroid binding sites on GABA(A) receptors. *Pharmacol. Ther.*, 116:7–19.

Hosie, A.M., L. Clarke, H. da Silva, T.G. Smart. (2009). Conserved site for neurosteroid modulation of GABA$_A$ receptors. *Neuropharmacology*, 56:149–154.

Kaminski, R.M., M.R. Livingood, M.A. Rogawski. (2004). Allopregnanolone analogs that positively modulate GABA receptors protect against partial seizures induced by 6-Hz electrical stimulation in mice. *Epilepsia*, 45:864–867.

Kim, I., G.S. Greenwald. (1986). Steroidogenic effects of lipoproteins and 25-hydroxycholesterol on luteal and ovarian cells: a comparison of two pseudopregnant rat models. *Proc. Soc. Exp. Biol. Med.*, 181:242–248.

Kokate, T.G., B.E. Svensson, M.A. Rogawski. (1994). Anticonvulsant activity of neurosteroids: correlation with gamma-aminobutyric acid-evoked chloride current potentiation. *J. Pharmacol. Exp. Ther.*, 270:1223–1229.

Kokate, T.G., S. Yamaguchi, L.K. Pannell, U. Rajamani, D.M. Carroll, A.B. Grossman, M.A. Rogawski. (1998). Lack of anticonvulsant tolerance to the neuroactive steroid pregnanolone in mice. *J. Pharmacol. Exp. Ther.*, 287:553–558.

Kokate, T.G., M.K. Banks, T. Magee, S. Yamaguchi, M.A. Rogawski. (1999). Finasteride, a 5alpha-reductase inhibitor, blocks the anticonvulsant activity of progesterone in mice. *J. Pharmacol. Exp. Ther.*, 288:679–684.

Lambert, J.J., D. Belelli, D.R. Peden, A.W. Vardy, J.A. Peters. (2003). Neurosteroid modulation of GABA$_A$ receptors. *Prog. Neurobiol.*, 71:67–80.

Lim, L.L., N. Foldvary, E. Mascha, J. Lee. (2001). Acetazolamide in women with catamenial epilepsy. *Epilepsia*, 42:746–749.

Lonsdale, D., W.M. Burnham. (2003). The anticonvulsant effects of progesterone and 5alpha-dihydroprogesterone on amygdala-kindled seizures in rats. *Epilepsia*, 44:1494–1499.

Maguire, J., I. Mody. (2008). GABA(A)R plasticity during pregnancy: relevance to postpartum depression. *Neuron*, 59:207–213.

Maguire, J.L., B.M. Stell, M. Rafizadeh, I. Mody. (2005). Ovarian cycle-linked changes in GABA(A) receptors mediating tonic inhibition alter seizure susceptibility and anxiety. *Nat. Neurosci.*, 8:797–804.

Mascia, M.P., F. Biggio, L. Mancuso, S. Cabras, P.L. Cocco, G. Gorini, A. Manca, C. Marra, R.H. Purdy, P. Follesa, G. Biggio. (2002). Changes in GABA(A) receptor gene expression induced by withdrawal of, but not by long-term exposure to, ganaxolone in cultured rat cerebellar granule cells. *J. Pharmacol. Exp. Ther.*, 303:1014–1020.

McAuley, J.W., J.L. Moore, A.L. Reeves, J. Flyak, E.P. Monaghan, J. Data. (2001). A pilot study of the neurosteroid ganaxolone in catamenial epilepsy: clinical experience in two patients. *Epilepsia*, 42:85.

Meldrum, B.S., M.A. Rogawski. (2007). Molecular targets for antiepileptic drug development. *Neurotherapeutics*, 4:18–61.

Mukai, Y., T. Higashi, Y. Nagura, K. Shimada. (2008). Studies on neurosteroids. XXV. Influence of a 5alpha-reductase inhibitor, finasteride, on rat brain neurosteroid levels and metabolism. *Biol. Pharm. Bull.*, 31:1646–1650.

Newmark, M.E., J.K. Penry. (1980). Catamenial epilepsy: a review. *Epilepsia*, 21:281–300.

Nohria, V., E. Giller. (2007). Ganaxolone. *Neurotherapeutics*, 4:102–105.

Panayiotopoulos, C.P., T. Obeid, A.R. Tahan. (1994). Juvenile myoclonic epilepsy: a 5-year prospective study. *Epilepsia*, 35:285–296.

Pathak, H.R., F. Weissinger, M. Terunuma, G.C. Carlson, F.C. Hsu, S.J. Moss, D.A. Coulter. (2007). Disrupted dentate granule cell chloride regulation enhances synaptic excitability during development of temporal lobe epilepsy. *J. Neurosci.*, 27:14012–14022.

Reddy, D.S. (2004). Role of neurosteroids in catamenial epilepsy. *Epilepsy Res.*, 62:99–118.

Reddy, D.S. (2007). Perimenstrual catamenial epilepsy. *Women's Health*, 3(2):195–206.

Reddy, D.S., M.A. Rogawski. (2000a). Enhanced anticonvulsant activity of ganaxolone after neurosteroid withdrawal in a rat model of catamenial epilepsy. *J. Pharmacol. Exp. Ther.*, 294:909–915.

Reddy, D.S., M.A. Rogawski. (2000b). Chronic treatment with the neuroactive steroid ganaxolone in the rat induces anticonvulsant tolerance to diazepam but not to itself. *J. Pharmacol. Exp. Ther.*, 295:1241–1248.

Reddy, D.S., M.A. Rogawski. (2001). Enhanced anticonvulsant activity of neuroactive steroids in a rat model of catamenial epilepsy. *Epilepsia*, 42:337–344.

Reddy, D.S., M.A. Rogawski. (2008). Pharmacokinetic-pharmacodynamic comparison of ganaxolone in normal rats and in a rat model of catamenial epilepsy. *Epilepsia*, 49(Suppl. 7):371.

Reddy, D.S., R. Woodward. (2004). Ganaxolone: a prospective overview. *Drugs Future*, 29:227–242.

Reddy, D.S., Y.C. Zeng. (2007). Effect of neurosteroid withdrawal on spontaneous recurrent seizures in a rat model of catamenial epilepsy. *FASEB J.*, 21:885.14.

Reddy, D.S., H.Y. Kim, M.A. Rogawski. (2001). Neurosteroid withdrawal model of perimenstrual catamenial epilepsy. *Epilepsia*, 42:328–336.

Reddy, D.S., D.C. Castaneda, B.W. O'Malley, M.A. Rogawski. (2004). Anticonvulsant activity of progesterone and neurosteroids in progesterone receptor knockout mice. *J. Pharmacol. Exp. Ther.*, 310:230–239.

Remy, S., H. Beck. (2006). Molecular and cellular mechanisms of pharmacoresistance in epilepsy. *Brain*, 129:18–35.

Rogawski, M.A., R.J. Porter. (1990). Antiepileptic drugs: pharmacological mechanisms and clinical efficacy with consideration of promising developmental stage compounds. *Pharmacol. Rev.*, 42:223–286.

Rogawski, M.A., D.S. Reddy. (2004). Neurosteroids: endogenous modulators of seizure susceptibility. In *Epilepsy: Scientific Foundations of Clinical Practice* (pp. 319–355), J.M. Rho, R. Sankar, and J.E. Cavazos, Eds. New York: Marcel Dekker.

Scharfman, H.E., N.J. MacLusky. (2006). The influence of gonadal hormones on neuronal excitability, seizures, and epilepsy in the female. *Epilepsia*, 47:1423–1440.

Selye, H. (1942). Antagonism between anesthetic steroid hormones and pentamethylenetetrazol (Metrazol). *J. Lab. Clin. Med.*, 27:1051–1053.

Shen, H., Q.H. Gong, M. Yuan, S.S. Smith. (2005). Short-term steroid treatment increases delta $GABA_A$ receptor subunit expression in rat CA1 hippocampus: pharmacological and behavioral effects. *Neuropharmacology*, 49:573–586.

Smith, S.S., Q.H. Gong. (2005). Neurosteroid administration and withdrawal alter $GABA_A$ receptor kinetics in CA1 hippocampus of female rats. *J. Physiol.*, 564:421–436.

Smith, S.S., Q.H. Gong, F.C. Hsu, R.S. Markowitz, J.M. ffrench-Mullen, X. Li. (1998a). GABA(A) receptor alpha4 subunit suppression prevents withdrawal properties of an endogenous steroid. *Nature*, 392:926–930.

Smith, S.S., Q.H. Gong, X. Li, M.H. Moran, D. Bitran, C.A. Frye, F.C. Hsu. (1998b). Withdrawal from 3alpha-OH-5alpha-pregnan-20-one using a pseudopregnancy model alters the kinetics of hippocampal $GABA_A$-gated current and increases the $GABA_A$ receptor alpha4 subunit in association with increased anxiety. *J. Neurosci.*, 18:5275–5284.

Smith, S.S., H. Shen, Q.H. Gong, X. Zhou. (2007). Neurosteroid regulation of GABA(A) receptors: focus on the alpha4 and delta subunits. *Pharmacol. Ther.*, 116:58–76.

Snead, O.C., 3rd. (1998). Ganaxolone, a selective, high-affinity steroid modulator of the gamma-aminobutyric acid-A receptor, exacerbates seizures in animal models of absence. *Ann. Neurol.*, 44:688–691.

Staley, K.J., B.L. Soldo, W.R. Proctor. (1995). Ionic mechanisms of neuronal excitation by inhibitory $GABA_A$ receptors. *Science*, 269:977–981.

Sur, C., S.J. Farrar, J. Kerby, P.J. Whiting, J.R. Atack, R.M. McKernan. (1999). Preferential coassembly of alpha4 and delta subunits of the gamma-aminobutyric acidA receptor in rat thalamus. *Mol. Pharmacol.*, 56:110–115.

Tauboll, E., A. Lundervold, L. Gjerstad. (1991). Temporal distribution of seizures in epilepsy. *Epilepsy Res.*, 8:153–165.

Tuveri, A., A.M. Paoletti, M. Orru, G.B. Melis, M.F. Marotto, P. Zedda, F. Marrosu, C. Sogliano, C. Marra, G. Biggio, A. Concas. (2008). Reduced serum level of THDOC, an anticonvulsant steroid, in women with perimenstrual catamenial epilepsy. *Epilepsia*, 49:1221–1229.

van Luijtelaar, G., B. Budziszewska, L. Jaworska-Feil, J. Ellis, A. Coenen, W. Lason. (2001). The ovarian hormones and absence epilepsy: a long-term EEG study and pharmacological effects in a genetic absence epilepsy model. *Epilepsy Res.*, 46:225–239.

van Luijtelaar, G., B. Budziszewska, M. Tetich, W. Lason. (2003). Finasteride inhibits the progesterone-induced spike-wave discharges in a genetic model of absence epilepsy. *Pharmacol. Biochem. Behav.*, 75:889–894.

Veliskova, J. (2007). Estrogens and epilepsy: why are we so excited? *Neuroscientist*, 13:77–88.

Wohlfarth, K.M., M.T. Bianchi, R.L. Macdonald. (2002). Enhanced neurosteroid potentiation of ternary GABA(A) receptors containing the delta subunit. *J. Neurosci.*, 22:1541–1549.

# 29 Chronobiology and Sleep: Implications for Seizure Propensity

*Mark Quigg*

## CONTENTS

## INTRODUCTION

Circadian rhythms have marked effects on the expression of epilepsy, influencing the timing of occurrence of seizures and the characteristics of interictal epileptiform discharges (IIEDs). Circadian rhythms are not a singular entity but involve many systems within and external to the central nervous system. A circadian influence is not likely a single, unified mechanism; rather, circadian rhythms are better thought of as endogenously mediated, excitatory and inhibitory influences that vary with time of day and dynamically compete with other seizure precipitants to elevate or depress seizure threshold. Reviewed in this chapter are the basic concepts of chronobiology, the organization of the circadian timing system, the effects of epilepsy and seizures on circadian rhythms, and the influences of circadian rhythms on the timing of seizures.

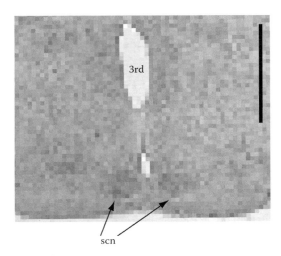

**FIGURE 29.1** Photomicrograph of the SCN in the anterior hypothalamus in the rat; 3rd = third ventricle, and bar = 100 μm.

## THE CIRCADIAN TIMING SYSTEM

A circadian (Latin: *circa*, "about"; *dies*, "day") rhythm is any self-sustained, behavioral or physiologic activity that spontaneously oscillates within a period of about 24 hours. The solar light–dark cycle is an important *zeitgeber* (German for "time giver"), an external time cue that entrains circadian rhythms to the local environment. Synchronization to the solar cycle is a widespread and highly conserved adaptation that confers advantages in survival and performance (DeCoursey et al., 1997; Ouyang et al., 1998; Recht et al., 1995).

A principal feature of circadian rhythms is that, once isolated from zeitgebers, endogenously maintained circadian rhythms persist and free-run over a period of about 24 hours, whereas exogenously maintained daily rhythms rapidly attenuate. Observation during constant environmental conditions, therefore, is the main test of whether a particular activity is truly circadian or is driven by external influences. Many behaviors or physiologic activities fulfill the strict criteria of circadian rhythms. In animal experiments, the rest–activity cycle and core body temperature are predominant because of ease of use and accuracy in long-term measurements. Perhaps more relevant to epilepsy, however, are the circadian cycles of sleep–wake and hormones. Circadian rhythms are not the only endogenous biological rhythms. Ultradian rhythms such as the 90- to 100-minute rapid eye movement (REM)–non-rapid eye movement (NREM) sleep cycle or the <1-minute cyclic alternating pattern of arousal (Parrino et al., 2000; Stevens et al., 1971), and the infradian menstrual cycle (Quigg et al., 2008) may also be important in precipitation of seizures or interictal epileptiform discharges (IIEDs).

The paired suprachiasmatic nuclei (SCN), located in the anterior hypothalamus, are the locus of the primary, mammalian biological clock (Figure 29.1). Ablation of the SCN attenuates circadian rhythms such as the cycles of rest–activity, body temperature, and corticosterone secretion (Eastman et al., 1984; Moore and Eicher, 1972; Stephan and Zucker, 1972). Transplantation of normal SCN cells into animals with genetically short circadian periods restores circadian rhythms with normal periods (Ralph et al., 1990). Circadian oscillation is a property of individual neurons; for example, neuronal cultures of SCN maintain sustained rhythmic activity (Welsh et al., 1995; Yamazaki et al., 2000).

Knowledge of the basic genetic mechanism of the clock has advanced rapidly in the last decade. The clock consists of a set of genes that form an autoregulatory cycle of transcription and translation (for review, see Takahashi et al., 2008). The clock, as originally described in *Drosophila*, consists of four core genetic components. Two genes—*clock* and *bmal1*—comprise the positive components of a regulatory loop. These genes activate the transcription of two other genes that form a negative portion of the feedback loop: *timeless* and *period*. Through interactions with their own genes, proteins encoded by *timeless* and *period* inhibit their continued transcription. With time, the proteins degrade (the rate of degradation is but one determinant of the period of the cycle), and their negative feedback effect weakens. Both *clock* and *bmal1* are then free to activate transcription again, and the 24-hour cycle resumes.

The primary neurotransmitter of SCN neurons is $\gamma$-aminobutyric acid (GABA). In the laboratory, circadian systems are susceptible to pharmacologic manipulation with agents active at the $GABA_A$ receptor (Ralph and Menaker, 1986). Other peptides and transmitters important in clock function or its efferent signaling include somatostatin, acetylcholine, neuropeptide Y, and vasopressin (Kalsbeek et al., 1996; Rusak and Bina, 1990).

The cyclic activity of the SCN and, in turn, the variety of circadian rhythms regulated by the SCN are synchronized to the solar light cycle. The SCN exert their rhythmic influence through diffuse projections throughout the hypothalamus (Buijs, 1996). Other efferents from the SCN, either directly or indirectly through hypothalamic connections, project to the thalamus and limbic system (Buijs, 1996; Peng et al., 1995).

Some consider the pineal body a second component of the circadian timing system. In mammals, the pineal body synthesizes and secretes melatonin during the nocturnal portion of the light–dark cycle in a pattern dependent upon SCN activity and on the suppressive influence of light exposure (Kalsbeek et al., 1999). The primary function of melatonin in mammals is to transmit information concerning light–dark cycles and day length to organize behavior dependent on seasonal functions. It may act directly on SCN neurons in a feedback loop to induce phase shifts and entrainment (Reppert et al., 1994).

The SCN are considered the primary circadian oscillators, but regions outside the SCN maintain rhythms that are either subordinate to or may contribute to circadian regulation. Genes with direct roles in circadian regulation are expressed in other diverse systems such as the retina, liver, lung, and skeletal muscle, as well as regions of the brain outside the SCN (Tosini, 2000; Yamazaki et al., 2000). Gene-chip experiments with *Drosophila* show that approximately 400 gene transcripts have significant circadian oscillations (Claridge-Chang et al., 2001), but the functions of rhythmic expression of these diverse genes, within and without the traditional circadian timing system, remain unclear.

The sleep–wake cycle is arguably the most evident of circadian rhythms and, in fact, the term is used synonymously with "circadian" in earlier reports of circadian effects in epilepsy. It is an oversimplification, though, as circadian timing and sleep–wake regulation are autonomous systems that mutually interact to regulate behavior (for review, see Saper et al., 2005). The temporal pattern of sleep–wakefulness can be represented by the following model (Figure 29.2) (Borbely, 1994): The homeostatic drive for sleep, Process S, gradually builds until reset by the next sleep episode. The 24-hour cycle of arousal, Process C, gradually increases the propensity for alertness as the day continues and declines late in the dark phase of the solar cycle. The patterns of sleep–wake states therefore reflect the dual influences of sleep debt and circadian organization. Sleep states are "circadian" in that the overall effect of the circadian clock is to organize sleep–wake states into daily cycles (Dijk and Czeisler, 1995). To date, experiments purporting to examine circadian mechanisms of seizure propensity have yet to adequately control for separate influences of sleep–wake state, circadian arousal, and sleep debt.

In addition to the above endogenous influences, sleep–wake cycles are uniquely susceptible to perturbation or interruption from exogenous influences, such as beepers and babies, as any medical resident or parent can attest. The self-sustained rhythm of arousal, however, continues without interruption, a property enabling *forced desynchronization protocols* in which sleep–wake are assigned to different periods and phases from underlying circadian cycles (Dijk and Czeisler, 1995).

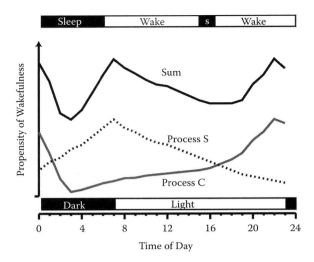

**FIGURE 29.2** Interactions of sleep homeostasis and the circadian rhythm of arousal on sleep–wake state. Sleep debt, represented by Process S, gradually accumulates following the last episode of sleep, resulting in a gradually decreasing level of alertness until reset by a major sleep episode. The circadian rhythm of arousal, Process C, counters sleep debt by gradually increasing its effect through the light phase of the cycle. The sum of Process S and Process C is a biphasic rhythm of propensity of wakefulness, organizing sleep–wake states into a major nocturnal episode of sleep and a shorter "nap" during the late afternoon. (Adapted from Borbely, A.A., in *Principles and Practice of Sleep Medicine*, 2nd ed., M. Kryger, T. Roth, and W. Dement, Eds., Saunders, Philadelphia, PA, 1994, pp. 309–320.)

Finally, a clear reason to differentiate the sleep–wake state from circadian timing is the observation that, among mammalian species, secondary circadian rhythms maintain different phase relationships to rhythms of the primary circadian timing system. Nocturnal animals, those active during the dark phase of the solar cycle, maintain rhythms of body temperature and cortisol secretion that peak at night. The opposite is true for diurnal animals, such as humans, with daily rhythms of activity, body temperate, and cortisol secretion that are maximal during the light phase. In contrast, primary rhythms of the circadian timing system—SCN activity and melatonin—remain in phase across all mammalian species. The activity of SCN neurons oscillates daily with the metabolic or electrical activity maintained during the daytime whether a species is diurnal or nocturnal (Inouye, 1996; Schwartz et al., 1983). Similarly, light suppresses melatonin production, so the daily oscillation of melatonin peaks during the night in both diurnal and nocturnal animals (Pang et al., 1993).

In summary, studies of circadian phenomena in epilepsy have historically—and understandably—simplified the complex interactions among primary and secondary endogenous circadian rhythms and exogenous stimuli. These complexities delineate interactions among seizures, epilepsy, and circadian phenomena.

## INTERACTIONS OF THE CIRCADIAN TIMING SYSTEM, EPILEPSY, AND SEIZURES

Three areas are important in the interactions between seizures, epilepsy, and circadian rhythms. The first is the effect of seizures and epilepsy upon circadian regulation. The second is the effect of circadian rhythms on epileptogenesis, the specific process of brain injury leading to the recurrence of spontaneous epileptic seizures. The third is the effect of primary and secondary circadian rhythms on the occurrence of IIEDs and seizures.

# EFFECTS OF SEIZURES OR EPILEPSY ON THE CIRCADIAN TIMING SYSTEM

Possible perturbations of the circadian timing system and its subordinate rhythms include attenuation or loss of normal circadian rhythmicity as a result of the chronic epileptic lesion, transient attenuation or perturbation of rhythms as an acute symptom of epileptic seizures, or more fundamental alterations of circadian timing induced by epileptic events.

## ATTENUATION OF CIRCADIAN RHYTHMICITY IN CHRONIC EPILEPSY

The epileptic state is a permanent alteration of neuronal excitability resulting in clinical seizures. The exact alteration varies with the pathogenesis of the specific epileptic syndrome. Because neuronal pathways connect the SCN to target regions that are responsible for individual rhythms, damage to those pathways or target regions could result in abnormalities manifested by changes in circadian amplitude or emergence of noncircadian rhythms.

One example of altered circadian rhythms in epilepsy is the nocturnal melatonin rhythm (Bazil et al., 2000; Rao et al., 1993; Schapel et al., 1995; Yalyn et al., 2006). Studies demonstrate inconsistent findings, with most patients treated with anticonvulsant medications having an interictal decrease in nocturnal melatonin (Bazil et al., 2000; Rao et al., 1993). A later study in patients treated with carbamazepine demonstrated no changes in melatonin secretion from controls (Yalyn et al., 2006). Another study documented increased nocturnal melatonin level in AED-naïve patients (Bazil et al., 2000). In a comparison of melatonin rhythms of photosensitive, epileptic children to those without photosensitivity, 2 out of 3 photosensitive children, compared to none of 11 epileptic children without photosensitivity, had attenuated nocturnal melatonin rhythms (Miyamoto et al., 1993). Although the above data suggest that some alterations in rhythms can result from the actions of anticonvulsant medications, attenuation of melatonin rhythmicity could represent a nonspecific response to injury, as evidenced by attenuation or loss of the circadian rhythms of body temperature, cortisol, or melatonin in patients with epileptic encephalopathy (Laakso et al., 1993), neuronal degenerative diseases (Heikkila et al., 1995), or brain injury (Olofsson et al., 2004).

Nevertheless, the basic machinery of the clock is robust; circadian rhythms emerge after incomplete destruction of bilateral SCN (Refinetti et al., 1994; Satinoff et al., 1982). Similarly, in a study of hypothalamic pathology in a rat model of limbic epilepsy, all animals maintain circadian rhythms of temperature (Quigg et al., 1999). Studies of patients with neurodegenerative diseases show that only those with the most severe and diffuse disease have clear circadian abnormalities (although most have disordered sleep) (Heikkila et al., 1995).

Along with the blunting of normal circadian rhythms, the epileptic lesion may also allow emergence of ultradian or infradian rhythms that may normally be suppressed by the primary circadian pacemaker. For example, circadian temperature rhythms of epileptic rats are more complex and polyphasic than normal rats, demonstrating ultradian rhythms with periods ranging from 2 to 16 hours (Quigg et al., 1999). Epileptic patients observed during a period of 72 hours off anticonvulsant medications have circadian temperature and cardiac rhythms that are more variable in the timing of temperature peaks, nadirs, and periods (Baust et al., 1976), consistent with either the effects of underlying ultradian rhythms or with instability of the circadian pacemaker. Ultradian rhythms can emerge in cases of specific brain lesions in experimental animals (Abrams and Hammel, 1965) and in nonspecific lesions attributable to brain injury in ICU patients (Dauch and Bauer, 1990). Immaturity in brain development is also associated with ultradian body temperature rhythms (Thomas, 1991). Studies of long-term seizure diaries show that some patients develop infradian patterns of seizure recurrence with periods ranging from 18 to 20 days (Binnie et al., 1984; Quigg and Straume, 2000), implying that an occult, underlying nonlunar rhythm is present in these subjects. In summary, these studies show that epileptic lesions largely spare overall circadian function, but circadian rhythms may be attenuated or impaired to the extent that ultradian or infradian patterns emerge.

## Seizures and Masking Effects

Many studies document postictal *masking* effects: transient, seizure-induced changes in normal function proportional to the physiologic severity of the seizure. Although the underlying rhythm is temporarily overcome, the masking stimulus has no long-lasting effect on period or phase of circadian rhythmicity. An illustration of masking is the transient disruption of the daily pattern of secretion of hormones following acute seizures (Meierkord et al., 1994; Meldrum, 1983). For example, severe, secondarily generalized partial seizures transiently elevate levels of noradrenaline, prolactin, vasopressin, and oxytocin (Meierkord et al., 1994). By contrast, simple partial seizures that spare the limbic system may fail to cause detectable changes in prolactin (Sperling et al., 1986). In this light, transient elevations in nocturnal melatonin seen immediately after partial-onset seizures can be interpreted as masking effects (Molina-Carballo et al., 2007).

## Seizures and Circadian Timing

Few studies have been designed to determine whether seizures affect clock function. A principal feature of circadian rhythms is that they can be entrained to the solar light–dark cycle. One specific alteration of circadian timing stems from shifts in phase necessary to synchronize organisms to the environmental light–dark cycle. For example, after transmeridian travel, humans reset their internal clocks by approximately one hour per day (Czeisler, 1995). Available studies disagree on the ability of seizures to perturb clock function. Most show a null effect on circadian phase or period. Early studies on wild rats that measured locomotor activity before and after electrically induced convulsions show transient, postictal masking of the rest–activity cycle. The subsequent phase of locomotor activity after recovery remains unchanged (Richter, 1965). Humans following electroconvulsive therapy (ECT) also demonstrate transient changes in amplitudes of daily temperature rhythms; shifts in phase are not observed (Szuba and Baxter, 1997). Similarly, in patients with spontaneous temporal lobe seizures captured during monitoring for epilepsy surgery, no clear effects on the phase of melatonin (Bazil et al., 2000; Molina-Carballo et al., 2007) or temperature rhythms (Bazil et al., 2000) are seen. The sampling resolution in these studies (e.g., every 3 hours in Bazil et al., 2000), however, is limited and might not be sensitive enough to detect small shifts in phase.

Other evidence suggests that seizures can shift the circadian phase. Adrenocorticotropic hormone (ACTH) and cortisol levels were monitored in depressed patients before and after antidepressant or electroconvulsive therapy; the timing of the peak of ACTH and cortisol levels shifted from phase-advanced to normal values following both treatments (Linkowski et al., 1987). Phase shifts in this study, however, may reflect the treatment of depression rather than the specific effect of a provoked seizure.

The above studies, however, are designed to show predictable, group effects. Rather than a consistent and physiologic shift in phase (as in the physiologic response to light stimuli), the circadian system may shift unpredictably in response to a pathologic stimulus. In a study of kindled rats, electrically provoked convulsions caused a variety of phase advances and delays in temperature rhythms (Quigg et al., 2001). Other studies that document alterations in entrainment to light stimuli in epileptic rats (Sanabria et al., 1996) and variable and inconsistent maintenance of body temperature and cardiac rhythms in humans (Baust et al., 1976) suggest that epilepsy is associated with an altered capacity for entrainment and synchronization.

In summary, there are conflicting studies regarding the effects of seizures on the circadian timing system. The focus on transient elevations or depressions of activity, rather than on its timing, has led to inadequate evaluation of interactions between circadian timing and epilepsy. Studies suggest, however, that time cues may be important in keeping epileptic subjects properly synchronized to environmental cycles following epileptic seizures.

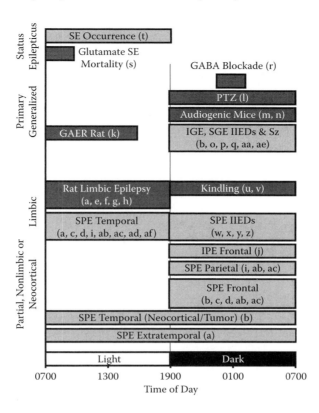

**FIGURE 29.3** Summary of 27 studies of the preferential phase of occurrence of seizures or IIEDs or, in some cases, phases of enhanced vulnerability to epileptogenic or ictogenic procedures, in a variety of human epileptic syndromes (light gray boxes) and in nocturnally active experimental models (dark gray boxes). In most studies, the times of day of phenomena are specified; in a few studies, as discussed in the text, sleep–wake state was made equivalent with the light–dark cycle by individual investigators. For comparison's sake, studies performed with point sampling (noon and midnight, for example) have been spread out to represent a whole photophase. Furthermore, slight variations in different light–dark schedules were ignored. Exceptions are those experiments with more frequent sampling across the clock that found effects confined to specific times of day. *Abbreviations:* GAER, genetic absence epilepsy rat; IIEDs, interictal epileptiform discharges; IPE/IGE, idiopathic partial/generalized epilepsy; PTZ, pentylenetetrazole; SE, status epilepticus; SGE, symptomatic generalized epilepsy; SPE, symptomatic partial epilepsy; Sz, seizure. *References:* (a) Quigg et al. (1998); (b) Janz (1974); (c) Herman et al. (2001); (d) Crespel et al. (1998); (e) Quigg et al. (2000); (f) Franceschi et al. (1984); (g) Bertram and Cornett (1994); (h) Hellier and Dudek (1999); (i) Quigg and Straume (2000); (j) Hayman et al. (1997); (k) Faradji et al. (2000); (l) Eidman et al. (1990); (m) Halberg et al. (1955); (n) Poirel (1991); (o) Sato et al. (1973); (p) Jovanivic (1967); (q) Stevens et al. (1972); (r) Naum et al. (2002); (s) Velasco et al. (1995); (t) Waterhouse et al. (1996); (u) Weiss et al. (1993); (v) Freeman (1980); (w) Lieb et al. (1980); (x) Malow et al. (1998); (y) Montplaisir et al. (1982 ); (z) Sammaritano et al. (1991); (aa) Degen and Degen (1991); (ab) Pavlova et al. (2004); (ac) Durazzo et al. (2008); (ad) Hofstra et al. (2009); (ae) Stewart et al. (2008); (af) Duckrow and Tcheng (2007).

## CIRCADIAN EFFECTS ON IIEDS, EPILEPTOGENIC LESIONS, AND SEIZURES

The following discussion is summarized in Figure 29.3. This schematic lays out the preferential phase of occurrence of seizures or IIEDs or, in some cases, phases of enhanced vulnerability to epileptogenic or ictogenic procedures, in a variety of human epileptic syndromes and in experimental models.

## EPILEPTOGENIC LESIONS

The effects of time of day on the process of epileptogenesis (distinct from circadian influences on ictogenesis) have not been thoroughly studied. Human evidence indirectly suggests that epileptogenic events vary in occurrence by time of day. Status epilepticus, for example, can be considered an epileptogenic event attended by neuronal injury (Meldrum and Brierley, 1973). Waterhouse et al. (1996) reported that a cohort of 1193 cases of status epilepticus in Richmond, Virginia, varied in occurrence by time of day in a unimodal distribution, with peak occurrence at 13:48 (pers. comm.). Even though this preliminary report did not explore outcome, one may speculate that the clinical impact of status epilepticus can vary with etiology, treatment opportunities, or physiologic effects, all of which may vary by time of day. By analogy, epidemiological studies of the outcome of stroke by time of day failed to demonstrate clear variables that distinguished strokes that occur in the morning (the peak time of stroke occurrence) vs. those that occurred at other times of day (Gur and Bornstein, 2000).

Clinical sequelae of status epilepticus may not necessarily occur in phase with epileptic injury. For example, rats treated with monosodium L-glutamate at 07:00 (early in the light phase) have convulsions and convulsive status epilepticus that occur more often and more severely than those treated at 15:00 (late in the light phase) or at 23:00 (middle of dark phase). Status epilepticus in the morning results in death in 70% of cases compared to none at other times (Feria-Velasco et al., 1995). Conversely, mice susceptible to generalized convulsions following audial stimulation experience a higher rate of fatal convulsions during the dark phase than during the light phase (Halberg et al., 1955). The process of kindling brings more direct evidence that epileptogenic events vary by time of day. In nocturnal animals, the dark phase of the daily cycle is particularly vulnerable to epileptogenic injury. Rats kindled by electrical stimulation of the amygdala require fewer stimulations to achieve full kindling and develop afterdischarges more quickly when stimulations are confined to dark phase of the daily cycle (Freeman, 1980; Weiss et al., 1993).

Thus, the differences in phases between the diurnally predominant occurrence and impact of status epilepticus and the nocturnally predominant effects of kindling suggest that these processes, in antiphase, are susceptible to different circadian influences.

## PHASIC OCCURRENCE OF INTERICTAL EPILEPTIFORM DISCHARGES

The relatively high rate of IIEDs (as opposed to ictal discharges) has facilitated their chronobiological study. Most studies concentrate on their occurrence at different sleep–wake states. Although examination of their occurrence is helpful in considering the specificity and sensitivity of EEG in diagnosis of epilepsy, the frequency of IIEDs, notably in partial epilepsies, does not clearly predict the occurrence of ictal discharges (Gotman and Koffler, 1989). Generalized epilepsies, because distinctions between interictal and ictal discharges can be difficult, nevertheless share similarities with partial epilepsies in the modulation of IIEDs by sleep–wake stages.

Most studies confirm that IIEDs in both partial and generalized epilepsies occur in patterns tied to regulation of sleep–wake states (Kellaway, 1985; Kellaway et al., 1980; Lieb et al., 1980; Malow et al., 1998; Montplaisir et al., 1982; Sammaritano et al., 1991; Stevens et al., 1971, 1972). IEDs in symptomatic partial epilepsies appear most frequently during non-REM sleep and are suppressed during REM sleep (Malow et al., 1998; Montplaisir et al., 1982; Sammaritano et al., 1991). Generalized epilepsies share similar phasic distributions of IIEDs. In the case of childhood absence epilepsy, a syndrome noted for frequent appearances of 3-Hz discharges during routine, daytime EEG, spike–waves are further activated by non-REM sleep (Sato et al., 1973). Slow spike–wave discharges in epileptic encephalopathy also increase in persistence during non-REM sleep (Degen and Degen, 1991). Similarly, difficult-to-localize syndromes such as Landau–Kleffner syndrome or electrical status epilepticus during slow-wave sleep (ESES) have IIEDs that are either exacerbated by or are limited to appearance during slow-wave sleep when they can become nearly continuous (Tassinari et al., 1992).

Long-term monitoring studies, performed in patients with a variety of epilepsy syndromes ranging from idiopathic partial (Rolandic epilepsy) to generalized polyspike wave epilepsy to temporal lobe epilepsy, show that IIEDs occur in 90- to 110-minute cycles, corresponding to the non-REM–REM ultradian cycles (Kellaway et al., 1980; Stevens et al., 1971, 1972). The ultradian pattern of IIED occurrence persists through the waking state (Stevens et al., 1971, 1972).

Another similarity shared among the epilepsies is that IIEDs are not only activated by non-REM sleep but also undergo changes in distribution. During non-REM sleep, focal spikes undergo generalization or appear bilaterally or contralaterally (Montplaisir et al., 1982; Sammaritano et al., 1991). Runs of 3-Hz discharges in absence epilepsy become disorganized and can appear as polyspike discharges (Sato, 1973). Conversely, the rare spikes during REM sleep are restricted in spatial distribution (Sammaritano et al., 1991).

No studies have clearly examined if IIED occurrence is tied to a particular circadian phase independent from sleep–wake state; however, certain components of the circadian timing system (CTS) have been evaluated in terms of diurnal differences in IIED occurrence. For example, hippocampal slices treated with hypomagnesia (Musshoff and Speckmann, 2003) to induce IIEDs demonstrate differential sensitivity to melatonin administration, with slices prepared during the day being susceptible to melatonin activation and slices at night being resistant. Slices rendered "epileptic" with bicuculline, however, were not activated by melatonin at any time. The findings suggest that epileptic effects of components of the CTS may differ by the epilepsy model. In humans, Kellaway and colleagues demonstrated that ultradian patterns of IIED occurrence could be modeled by interactions of sleep-associated ultradian and circadian rhythms (Kellaway, 1980). Potential studies to examine circadian effects independent of sleep state are complicated by the finding that sleep deprivation activates IIEDs in an independent fashion that supplements the activating effects of sleep state (Fountain et al., 1998). Future studies on daily patterns of IIEDs, therefore, may have to control for the duration from the last sleep episode. Thus, available evidence suggests that the nocturnal prevalence of IIEDs is circadian only in that the timing system organizes sleep–wake states into recognizable circadian and ultradian patterns of facilitation and inhibition.

## Phasic Occurrence of Seizures

Studies of the chronobiology of epilepsy reveal several important findings. Epileptic seizures recur at preferred times of day. Seizures in specific animal models of epilepsy occur in true, endogenously mediated circadian patterns, implying that other epileptic conditions are susceptible to endogenous circadian modulation. Finally, the phasic patterns of seizure occurrence vary with the pathophysiology of the underlying epileptic syndrome.

The tendency of epileptic seizures to occur in temporal patterns was first described over 100 years ago. Gowers (1885), in a study of 840 subjects, classified patients into three groups based on the daily distribution of "fits": *diurnal* (daytime onset, synonymous with wakefulness), *nocturnal* (nighttime onset, sleep), and *diffuse* (random onset). Diurnal fits comprised 43% of the sample, nocturnal 20%, and diffuse 37%. The day- or nighttime occurrence of the first fit in a cluster fairly accurately predicts that subsequent seizures would also occur within the preferred phase. Finally, transitions from sleep to wakefulness, especially in the morning, carry a relatively high risk of seizure occurrence, at least 5% of the sample. Later studies based on observations of subjects in the asylums of the period expanded upon Gowers' findings by plotting seizure occurrence as a function of time of day (Griffiths and Fox, 1938; Langdon-Down and Brain, 1929; Patry, 1931). A "diffuse" distribution was associated with worse severity and duration of epilepsy (Langdon-Down and Brain, 1929).

These early studies established that seizures vary by time of day. Evidence that daily variations are in fact circadian comes from case reports demonstrating that seizures, like circadian rhythms, can shift in phase given changes in zeitgebers. Halberg and Howard (1958) reported that, in a long-term seizure diary of an epileptic patient, not only did seizures occur in a daily pattern but the daily pattern also shifted in response to a phasic change in the subject's daily work schedule.

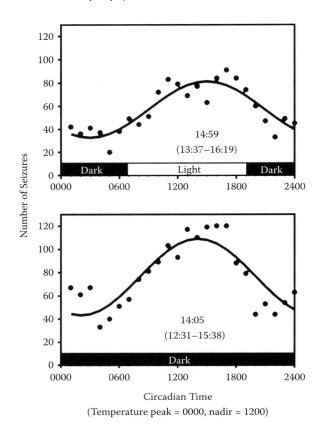

**FIGURE 29.4** Distribution of spontaneous limbic seizures in an animal model of limbic epilepsy. Seizures recur preferentially during the daytime when animals are maintained in a regular light–dark cycle. The same circadian distribution is maintained when animals are monitored in constant darkness and isolated from other zeitgebers. The persistent pattern of seizure recurrence therefore fulfills the strict definition of circadian rhythmicity. (From Patry, F., *Am. J. Psychiatry*, 10, 789–813, 1931.)

More recent studies carry advantages of confirmation of seizures with video–EEG and modern epilepsy syndrome classifications (Figure 29.3). In general, seizures arising from neocortical sites, limbic foci, and generalized epilepsies demonstrate marked differences in diurnal occurrence. None of these studies, however, has observed human seizures in "free-running" conditions.

Rigorous environment controls, however, can be imposed on animal models of epilepsy. In mice with audiogenic convulsions, changes in phases of the light–dark cycle provoke comparable changes in phases of convulsions (Poirel, 1991). Only one study thus far has conducted EEG monitoring under constant conditions to allow circadian rhythms to free-run (Figure 29.4) (Quigg et al., 2000). In this study, a rat model of mesial temporal lobe epilepsy was used that expresses spontaneous limbic seizures. During continuous EEG monitoring during a 12/12-hour light–dark cycle, nearly twice as many seizures occurred during the light phase than during the dark phase. During constant darkness, seizures continued to occur in a strong pattern when referenced to the free-running circadian rhythm of body temperature. Limbic seizures, therefore, fulfill the explicit criteria of circadian phenomena. It can be inferred that the daily patterns of spontaneous seizures noted in other animal and human studies are also susceptible to the endogenous modulation of the circadian timing system.

Figure 29.3 also compares the preferential phases of occurrence of a sample of animal models and human syndromes of partial and generalized epilepsies. The schematic highlights inconsistencies in phase within models of generalized epilepsies and models of limbic epilepsies. Animal models of generalized epilepsies (nocturnally active rats or mice) preferentially have greater susceptibility to seizures during the dark phase of the 24-hour light–dark cycle (Eidman et al., 1990; Halberg et al., 1955; Naum et al., 2002; Poirel, 1991; Stewart et al., 2008). These animals have seizures when pharmacologically or sensorially triggered. In contrast, the spontaneous spike–wave discharges of genetic absence epilepsy rats (GAERs) occur mainly during the light half of the cycle (Faradji et al., 2000).

Similarly, models of limbic epilepsy also vary in phase depending on experimental protocol. Kindling experiments that involve triggered afterdischarges or convulsions show greatest sensitivity during the dark (Freeman, 1980; Weiss et al., 1993), whereas spontaneous seizures in experimental models of limbic epilepsy occur preponderantly during the light half of the cycle (Bertram and Cornett, 1994; Cavalheiro et al., 1991; Hellier and Dudek, 1999; Quigg et al., 1998, 2000). One explanation of the internal inconsistencies within groups of generalized and limbic models, therefore, is that provoked seizures occur out of phase with spontaneous seizures. Arousal, handling, and transient light exposure can occur during provocation of seizures and can confound comparisons.

Experiments that measure spontaneous epileptic events, however, show clear differences between generalized, partial neocortical, and partial limbic epilepsies. In diurnally active humans, IIEDs of both generalized and partial epilepsies (discussed above), ictal discharges of generalized epilepsies (Degen and Degen, 1991; Janz, 1974; Sato et al., 1973; Stevens et al., 1972), and ictal discharges of neocortical epilepsies (Crespel et al., 1998; Durazzo et al., 2008; Hayman et al., 1997; Pavlova et al., 2004; Quigg and Straume, 2000) occur preponderantly during the dark half of the 24-hour cycle. In nocturnally active models of generalized epilepsies in which epileptic events occur spontaneously, ictal discharges appear more frequently during the light portion of the cycle in some reports (Faradji et al., 2000) and during the dark phase in another (Stewart et al., 2008). Therefore, models of generalized epilepsies and human generalized and cortically based epilepsies (with some exceptions) occur out of phase with each other. It follows, then, that circadian rhythms occurring antiphase between the two species can best modulate seizure occurrence in these syndromes. Sleep and other secondary circadian rhythms occur out of phase between diurnal and nocturnal species. In summary, generalized epileptic events favor the circadian half-cycle in which sleep occurs.

A similar deduction leads to an opposite conclusion in the case of limbic epilepsies. In a comparison between experimental limbic epilepsy in rats and mesial temporal lobe epilepsy in humans, both species have peak incidences of limbic seizures during the light portion of the 24-hour light–dark cycle (Quigg et al., 1998) (Figure 29.5). Rats, however, are nocturnal animals and have sleep–wake cycles 180° out of phase with those of diurnal humans. Rhythms of the primary circadian timing system occur in phase between diurnal and nocturnal animals. The finding that limbic seizures occur in phase among diurnal and nocturnal species implies a direct role of the circadian timing system on the modulation of seizure occurrence.

## Primary Circadian Influences

Only a few studies to date have examined direct involvement of the SCN or of the pineal body on seizure occurrence or seizure threshold. Indirect evidence, however, suggests that the circadian timing system may exert effects on spontaneous seizure occurrence. There are no studies addressing whether the circadian oscillation of the SCN directly affects epileptic seizures, although there is already evidence of direct connections from SCN to the limbic system (Buijs, 1996; Peng et al., 1995). Hippocampal cells in culture, although not self-sustaining, produce daily oscillations in electrical activity (Welsh et al., 1995).

Animal models generally show that melatonin may possess anticonvulsant properties, supported indirectly by studies showing suppression of nocturnal seizure rate in limbic epilepsies (Bertram and Cornett, 1994; Cavalheiro et al., 1991; Hellier and Dudek, 1999; Quigg et al., 1998, 2000) when melatonin secretion is high. Pinealectomy renders gerbils and parathyroidectomized rats susceptible to spontaneous convulsions, and melatonin administration confers partial protection against

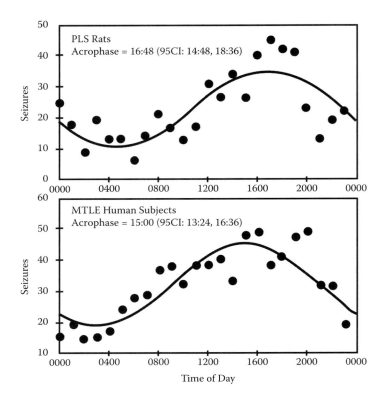

**FIGURE 29.5** Distribution of spontaneous limbic seizures from rats (above) rendered epileptic through continuous hippocampal stimulation and humans (below) with mesial temporal lobe epilepsy. Rats were maintained in a 12/12-hour light–dark cycle with lights on from 0700 to 1900. Humans were monitored in an epilepsy monitoring unit with a typical hospital schedule. Both had the occurrence of seizures confirmed with continuous video EEG. The occurrence of seizures during the light portion of the 24-hour light–dark cycle in both species confirms that rest–activity cycles are not the predominant mediator of the circadian pattern of seizure occurrence.

convulsions in pinealectomized gerbils (Champney and Peterson, 1993). In kindled animals, melatonin decreases the duration of afterdischarges (Albertson et al., 1981) and increases the threshold current required to trigger afterdischarges (Mevissen and Ebert, 1998). Melatonin decreases epileptiform activity recorded from human hippocampal slices, as well (Fauteck et al., 1995). Finally, adjunctive treatment with melatonin decreases seizures in humans (Fauteck et al., 1999; Molina-Carballo et al., 1997). In some situations, however, melatonin decreases (not increases) seizure thresholds (Yehuda and Mostofsky, 1993). In addition, pinealectomy reduces seizure susceptibility in some species, opposite of its proconvulsant effect (Champney and Peterson, 1993). Exogenous melatonin acts with much clearer efficacy in reducing seizures than does the endogenous cyclic increase in melatonin secretion (Champney et al., 1996; Mevissen and Ebert, 1998).

## SECONDARY CIRCADIAN INFLUENCES

### Sleep–Wake State

As discussed above, sleep, as organized by the circadian timing system, facilitates both the appearance and spread of IIEDs in many epilepsy syndromes and promotes seizure occurrence in generalized epilepsy syndromes. State-dependent changes in cortical excitability and thalamocortical synchronization play important roles in the occurrence of epileptic seizures or related paroxysmal activity in generalized epilepsies (Avoli and Gloor, 1981). The work of Steriade and colleagues demonstrated the progressive development from the hypersynchrony of non-REM sleep patterns to some forms of epileptic-like activities. Their work suggests that thalamocortical relay neurons are gradually recruited by cortical excitation into a hypersynchronous, epileptic state (Steriade and Contreras, 1995); however, thalamocortical interactions in promotion of seizures are not confined to the generalized epilepsies. Sleep spindles, mediated by GABAergic thalamocortical relay and reticular nucleus neurons (Steriade et al., 1994), have been recorded in the hippocampus as a finding associated with epilepsy (Malow et al., 1999). Growing experience in the functional anatomy of the limbic system suggests that interactions between the limbic system and thalamus comprise a network responsible for generation and maintenance of ictal discharges (Bertram et al., 1998).

## Hypothalamic–Pituitary–Adrenal Axis

The three components of the hypothalamic–pituitary–adrenal (HPA) axis—corticotropin-releasing hormone (CRH), adrenocorticotropic hormone (ACTH), and corticosterone—all form circadian rhythms that peak before and near the onset of locomotor activity (Atkinson and Waddell, 1997; Kwak et al., 1992; Moore and Eicher, 1972). These rhythms depend on SCN input for their synchronization (Buijs et al., 1997; Moore and Eicher, 1972). Contrary to the traditional top-down regulatory model, activity of the SCN directly affects pituitary (ACTH) as well as hypothalamic (CRH) activity (Buijs et al., 1997). All three components have been implicated as having roles in seizure threshold and epileptogenesis. Whereas CRH potentiates seizures (Baram et al., 1992; Ehlers et al., 1983), ACTH and corticosteroids are used as anticonvulsants in certain epileptic syndromes (Baram et al., 1996) (see Chapter 26, this volume).

Corticotropin-releasing hormone is particularly interesting with regard to the question of circadian modulation of seizures. CRH neurons are not confined to the hypothalamus. Although the main population of CRH neurons resides in the paraventricular nucleus (PVN), CRH neurons are also present in the limbic system, mainly located in the central nucleus of the amygdala, and are also found in CA1, CA3, and granule cell and hilar layers of the dentate (Gray and Bingaman, 1996; Merchenthaler, 1984; Swanson et al., 1983). CRH secretion not only undergoes circadian changes but is also a primary response to stress. Stressful stimuli immediately upregulate transcription of CRH in neurons of the PVN (Kovacs and Sawchenko, 1996). Because patients with epilepsy identify stress as an important precipitant of seizures (Frucht et al., 2000), CRH is uniquely suited to promote epileptic seizures provoked by both exogenous stimuli and endogenous rhythms. CRH is a potent proconvulsant, especially in the immature brain. Further, CRH is thought to amplify existing excitatory circuits to promote hyperexcitability (for review, see Baram and Hatalski, 1998); therefore, CRH released during illness or stress or in its daily cycle may facilitate epileptic seizures.

Adrenocorticotropic hormone and corticosteroids are used in the treatment of infantile spasms (Baram et al., 1996). Convulsions induced by CRH in infant rats do not cease with administration of ACTH and are independent of the effect of ACTH on glucocorticoid production, suggesting that downregulation of CRH synthesis may be one route of ACTH and steroid anticonvulsant action (Baram and Schultz, 1995). ACTH may have direct pharmacologic effects that vary with age. In infant rats, ACTH acts as a proconvulsant by reducing the threshold for electrically induced seizures. In adult rats, however, ACTH increases seizure threshold and slows down the rate of kindling (Pranzatelli, 1994).

Circadian aspects of neurosteroid regulation are just beginning to be elucidated. Estrogens potentiate interictal epileptiform discharges and lower seizure thresholds. Progesterone, in contrast, decreases interictal epileptiform discharges and increases seizure thresholds. The anticonvulsant effects of progesterone are probably mediated by its metabolites containing 3α-hydroxy-5. For example, 5α-pregnane-3α-ol-20-one (3α,5α-THP or alloprogesterone) is active at $GABA_A$ receptors, and there is strong evidence of it having anticonvulsant and neuroprotective effects (Frye, 1995). Androsterone acts differently on seizures provoked by inhibition of glutamic acid decarboxylase (thus preventing GABA synthesis), completely abolishing seizures at noon but being incompletely effective at midnight (Naum et al., 2002). The GABAergic system itself undergoes diurnal changes with peak activity at night (Cardinali and Golombek, 1998). By virtue of the interlinked metabolisms of corticosteroids and sex steroids, neurosteroids may have a role in the circadian modulation of seizures.

## Exogenous Rhythms

Daily and ultradian variations in anticonvulsant drugs can be important in patients with epilepsy because clusters of seizures may occur when anticonvulsants are at trough levels. Unfavorable drug timing in relationship to endogenous rhythms may complicate management of seizures. Carbamazepine, for example, has daily variations in auto-induction of its metabolism, showing more variability in total serum levels apparent during the day than during the night (Macphee et al., 1987). In contrast, valproate is absorbed less completely at night because of nocturnal changes in gastrointestinal physiology that affect disintegration of tablets (Cloyd, 1991). Daily differences in effectiveness of anticonvulsants have been demonstrated in the electroshock rat model. Low-dose diazepam is less effective between 1800 and 2200 hours, although the phasic effect attenuates with higher doses (Schmutz et al., 1990). In practice, however, most problems related to time-of-day effects of anticonvulsant medications stem from toxicities that are managed by adjustments in dose amounts or scheduling.

Finally, exogenous stress in the form of emotional stress (Frucht et al., 2000; Klein and van Passel, 2005) or physiologic arousal may exacerbate seizures, and daily recurring stress may contribute to an apparent schedule of seizure recurrence. For example, case studies show that changes in schedules (shift work) can cause a similar shift in the timing of seizures (Halberg and Howard, 1958). Avoidance therapy, in these cases, may aid in management of regularly recurring seizures.

## SUMMARY

This review has focused on the organization of the circadian timing system, its basic interactions with sleep–wake state, and its putative role in the modulation of epileptic seizures and IEDs. Future work is necessary to reveal dysfunction of circadian timing attributable to seizures and the mechanisms of circadian fluctuations in seizure propensity.

## REFERENCES

Abrams, R., H.T. Hammel. (1965). Cyclic variations in hypothalamic temperature in unanesthetized rats. *Am. J. Physiol.*, 208:698–702.

Albertson, T.E., S.L. Peterson, L.G. Stark. (1981). The anticonvulsant effects of diazepam and phenobarbital in prekindled and kindled seizures in rats. *Neuropharmacology*, 20:597–603.

Atkinson, H.C., B.J. Waddell. (1997). Circadian variation in basal plasma corticosterone and adrenocorticotropin in the rat: sexual dimorphism and changes across the estrous cycle. *Endocrinology*, 138:3842–3848.

Avoli, M., P. Gloor. (1981). The effects of transient functional depression of the thalamus on spindles and on bilateral synchronous epileptic discharges of feline generalized penicillin epilepsy. *Epilepsia*, 22:443–452.

Baram, T.Z., C.G. Hatalski. (1998). Neuropeptide-meditated excitability: a key triggering mechanism for seizure generation in the developing brain. *Trends Neurosci.*, 21:471–476.

Baram, T.Z., L. Schultz. (1995). ACTH does not control neonatal seizures induced by administration of exogenous corticotropin-releasing hormone. *Epilepsia*, 36:174–178.

Baram, T.Z., E. Hirsch, O.C. Snead, 3rd, L. Schultz. (1992). Corticotropin-releasing hormone-induced seizures in infant rats originate in the amygdala. *Ann. Neurol.*, 31:488–494.

Baram, T.Z., W.G. Mitchell, A. Tournay, O.C. Snead, R.A. Hanson, E.J. Horton. (1996). High-dose corticotropin (ACTH) versus prednisone for infantile spasms: a prospective, randomized, blinded study. *Pediatrics*, 97:375–379.

Baust, W., K. Irmscher, J. Jorg, T. Sommer. (1976). Studies on the circadian periodicity in patients with the awakening type of idiopathic epilepsy [author's transl.]. *J. Neurol.*, 213:283–294.

Bazil, C.W., D. Short, D. Crispin, W. Zheng. (2000). Patients with intractable epilepsy have low melatonin, which increases following seizures. *Neurology*, 55:1746–1748.

Bertram, E.H., J.F. Cornett. (1994). The evolution of a rat model of chronic spontaneous limbic seizures. *Brain Res.*, 661:157–162.

Bertram, E.H., D.X. Zhang, P. Mangan, N. Fountain, D. Rempe. (1998). Functional anatomy of limbic epilepsy: a proposal for central synchronization of a diffusely hyperexcitable network. *Epilepsy Res.*, 32:194–205.

Binnie, C.D., J.H. Aarts, M.A. Houtkooper, R. Laxminarayan, A. Martins da Silva, H. Meinardi, N. Nagelkerke, J. Overweg. (1984). Temporal characteristics of seizures and epileptiform discharges. *Electroencephalogr. Clin. Neurophysiol.*, 58:498–505.

Borbely, A.A. (1994). Sleep homeostasis and models of sleep regulation. In *Principles and Practice of Sleep Medicine*, 2nd ed. (pp. 309–320), M. Kryger, T. Roth, and W. Dement, Eds. Philadelphia, PA: Saunders.

Buijs, R.M. (1996). The anatomical basis for the expression of circadian rhythms: the efferent projections of the suprachiasmatic nucleus. In *Progress in Brain* Research (pp. 229–240), R.M. Buijs, A. Kalsheek, H.J. Romijn, C.M.A. Pennartz, and M. Mirmiran, Eds. Amsterdam: Elsevier.

Buijs, R.M., J. Wortel, J.J. Van Heerikhuize, A. Kalsbeek. (1997). Novel environment induced inhibition of corticosterone secretion: physiological evidence for a suprachiasmatic nucleus mediated neuronal hypo-thalamo–adrenal cortex pathway. *Brain Res.*, 758:229–236.

Cardinali, D.P., D.A. Golombek. (1998). The rhythmic GABAergic system. *Neurochem. Res.*, 23:607–614.

Cavalheiro, E.A., J.P. Leite, Z.A. Bortolotto, W.A. Turski, C. Ikonomidou, L. Turski. (1991). Long-term effects of pilocarpine in rats: structural damage of the brain triggers kindling and spontaneous recurrent seizures. *Epilepsia*, 32:778–782.

Champney, T.H., S.L. Peterson. (1993). Circadian, seasonal, pineal and melatonin influences on epilepsy. In *Melatonin: Physiological Effects, Biosynthesis and Clinical Applications* (pp. 478–494), R. Reiter and H. Yu, Eds. Boca Raton, FL: CRC Press.

Champney, T.H., W.H. Hanneman, M.E. Legare, K. Appel. (1996). Acute and chronic effects of melatonin as an anticonvulsant in male gerbils. *J. Pineal Res.*, 20:79–83.

Claridge-Chang, A., H. Wijnen, F. Naef, C. Boothroyd, N. Rajewsky, M.W. Young. (2001). Circadian regula-tion of gene expression systems in the *Drosophila* head. *Neuron*, 32:657–671.

Cloyd, J. (1991). Pharmacokinetic pitfalls of present antiepileptic medications. *Epilepsia*, 32(Suppl. 5):S53–S65.

Crespel, A., M. Baldy-Moulinier, P. Coubes. (1998). The relationship between sleep and epilepsy in frontal and temporal lobe epilepsies: practical and physiopathologic considerations. *Epilepsia*, 39:150–157.

Czeisler, C.A. (1995). The effect of light on the human circadian pacemaker. *Ciba Found Symp.*, 183:254–302.

Dauch, W.A., S. Bauer. (1990). Circadian rhythms in the body temperatures of intensive care patients with brain lesions. *J. Neurol. Neurosurg. Psychiatry*, 53:345–347.

DeCoursey, P.J., J.R. Krulas, G. Mele, D.C. Holley. (1997). Circadian performance of suprachiasmatic nuclei (SCN)-lesioned antelope ground squirrels in a desert enclosure. *Physiol. Behav.*, 62:1099–1108.

Degen, R., H.E. Degen. (1991). Sleep and sleep deprivation in epileptology. In *Epilepsy, Sleep, and Sleep Deprivation*, 2nd ed. (pp. 235–239), R. Degen and E.A. Rodin, Eds. Amsterdam: Elsevier.

Dijk, D.J., C.A. Czeisler. (1995). Contribution of the circadian pacemaker and the sleep homeostat to sleep propensity, sleep structure, electroencephalographic slow waves, and sleep spindle activity in humans. *J. Neurosci.*, 15:3526–3538.

Durazzo, T.S., S.S. Spencer, R.B. Duckrow, E.J. Novotny, D.D. Spencer, H.P. Zaveri. (2008). Temporal distri-butions of seizure occurrence from various epileptogenic regions. *Neurology*, 70:1265–1271.

Eastman, C.I., R.E. Mistlberger, A. Rechtschaffen. (1984). Suprachiasmatic nuclei lesions eliminate circadian temperature and sleep rhythms in the rat. *Physiol. Behav.*, 32:357–368.

Ehlers, C.L., S.J. Henriksen, M. Wang, J. Rivier, W. Vale, F.E. Bloom. (1983). Corticotropin releasing factor produces increases in brain excitability and convulsive seizures in rats. *Brain Res.*, 278:332–336.

Eidman, D.S., M.A. Benedito, J.R. Leite. (1990). Daily changes in pentylenetetrazol-induced convulsions and open-field behavior in rats. *Physiol. Behav.*, 47:853–856.

Faradji, H., C. Rousset, G. Debilly, M. Vergnes, R. Cespuglio. (2000). Sleep and epilepsy: a key role for nitric oxide? *Epilepsia*, 41:794–801.

Fauteck, J., J. Bockmann, T.M. Bockers, W. Wittkowski, R. Kohling, A. Lucke, H. Straub, E.J. Speckmann, I. Tuxhorn, P. Wolf et al. (1995). Melatonin reduces low-$Mg^{2+}$ epileptiform activity in human temporal slices. *Exp. Brain Res.*, 107:321–325.

Fauteck, J., H. Schmidt, A. Lerchl, G. Kurlemann, W. Wittkowski. (1999). Melatonin in epilepsy: first results of replacement therapy and first clinical results. *Biol. Signals Recept.*, 8:105–110.

Feria-Velasco, A., Y. Feria-Cuevas, R. Gutierrez-Padilla. (1995). Chronobiological variations in the convulsive effect of monosodium L-glutamate when administered to adult rats. *Arch. Med. Res.*, 26(Suppl.):S127–S132.

Fountain, N.B., J.S. Kim, S.I. Lee. (1998). Sleep deprivation activates epileptiform discharges independent of the activating effects of sleep. *J. Clin. Neurophysiol.*, 15:69–75.

Freeman, F.G. (1980). Development of kindled seizures and circadian rhythms. *Behav. Neural. Biol.*, 30:231–235.

Frucht, M.M., M. Quigg, C. Schwaner, N.B. Fountain. (2000). Distribution of seizure precipitants among epilepsy syndromes. *Epilepsia*, 41:1534–1539.

Frye, C.A. (1995). The neurosteroid 3alpha, 5alpha-THP has antiseizure and possible neuroprotective effects in an animal model of epilepsy. *Brain Res.*, 696:113–120.

Gotman, J., D.J. Koffler. (1989). Interictal spiking increases after seizures but does not after decrease in medication. *Electroencephalogr. Clin. Neurophysiol.*, 72:7–15.

Gowers, W. (1885). Course of epilepsy. In *Epilepsy and Other Chronic Convulsive Diseases: Their Causes, Symptoms and Treatment* (pp. 157–164), W. Gowers, Ed. New York: William Wood.

Gray, T.S., E.W. Bingaman. (1996). The amygdala: corticotropin-releasing factor, steroids, and stress. *Crit. Rev. Neurobiol.*, 10:155–168.

Griffiths, G.M., J.T. Fox. (1938). Rhythm in epilepsy. *Lancet*, 2:409–416.

Gur, A.Y., N.M. Bornstein. (2000). Are there any unique epidemiological and vascular risk factors for ischaemic strokes that occur in the morning hours? *Eur. J. Neurol.*, 7:179–181.

Halberg, F., R.B. Howard. (1958). 24-Hour periodicity and experimental medicine examples and interpretations. *Postgrad. Med.*, 24:349–358.

Halberg, F., J.J. Bittner, R.J. Gully, P.G. Albrecht, E.L. Brackney. (1955). 24-Hour periodicity and audiogenic convulsions in I mice of various ages. *Proc. Soc. Exp. Biol. Med.*, 88:169–173.

Hayman, M., I.E. Scheffer, Y. Chinvarun, S.U. Berlangieri, S.F. Berkovic. (1997). Autosomal dominant nocturnal frontal lobe epilepsy: demonstration of focal frontal onset and intrafamilial variation. *Neurology*, 49:969–975.

Heikkila, E., T.H. Hatonen, T. Telakivi, M.L. Laakso, H. Heiskala, T. Salmi, A. Alila, P. Santavuori. (1995). Circadian rhythm studies in neuronal ceroid-lipofuscinosis (NCL). *Am. J. Med. Genet.*, 57:229–234.

Hellier, J.L., F.E. Dudek. (1999). Spontaneous motor seizures of rats with kainate-induced epilepsy: effect of time of day and activity state. *Epilepsy Res.*, 35:47–57.

Inouye, S.I. (1996). Circadian rhythms of neuropeptides in the suprachiasmatic nucleus. In *Progress in Brain Research: Hypothalamic Integration of Circadian Rhythms* (pp. 75–90), R. Buijs, H. Kalsbeek, H. Romijn, C. Pennartz, and M. Mirmiriran, Eds. Amsterdam: Elsevier.

Janz, D. (1974). Epilepsy and the sleeping–waking cycle. *The Epilepsies* (pp. 457–490), O. Magnus and A. Lorentz De Haas, Eds. Amsterdam: North Holland.

Jovanivic, U.J. (1967). Das Schlafverhalten de Epileptiker I. Schlafdauer, Schlaftiefe und Besonderheiten der Schlafperiodik. *Deutch. Z. Nerven.*, 190:159.

Kalsbeek, A., R.M. Buijs. (1996). Rhythms of inhibitory and excitatory output from the circadian timing system as revealed by *in vivo* microdialysis. *Prog. Brain Res.*, 111:273–293.

Kalsbeek, A., R.A. Cutrera, J.J. Van Heerikhuize, J. Van Der Vliet, R.M. Buijs. (1999). GABA release from suprachiasmatic nucleus terminals is necessary for the light-induced inhibition of nocturnal melatonin release in the rat. *Neuroscience*, 91:453–461.

Kellaway, P. (1985). Sleep and epilepsy. *Epilepsia*, 26(Suppl. 1):S15–S30.

Kellaway, P., J.D. Frost, Jr., J.W. Crawley. (1980). Time modulation of spike-and-wave activity in generalized epilepsy. *Ann. Neurol.*, 8:491–500.

Klein, P., L. van Passel. (2005). Effect of stress related to the 9/11/2001 terror attack on seizures in patients with epilepsy. *Neurology*, 64:1815–1816.

Kovacs, K.J., P.E. Sawchenko. (1996). Regulation of stress-induced transcriptional changes in the hypothalamic neurosecretory neurons. *J. Mol. Neurosci.*, 7:125–133.

Kwak, S.P., E.A. Young, I. Morano, S.J. Watson, H. Akil. (1992). Diurnal corticotropin-releasing hormone mRNA variation in the hypothalamus exhibits a rhythm distinct from that of plasma corticosterone. *Neuroendocrinology*, 55:74–83.

Laakso, M.L., L. Leinonen, T. Hatonen, A. Alila, H. Heiskala. (1993). Melatonin, cortisol and body temperature rhythms in Lennox–Gastaut patients with or without circadian rhythm sleep disorders. *J. Neurol.*, 240:410–416.

Langdon-Down, M., W.R. Brain. (1929). Time of day in relation to convulsions in epilepsy. *Lancet*, 1:1029–1032.

Lieb, J.P., J.P. Joseph, J. Engel, Jr., J. Walker, P.H. Crandall. (1980). Sleep state and seizure foci related to depth spike activity in patients with temporal lobe epilepsy. *Electroencephalogr. Clin. Neurophysiol.*, 49:538–557.

Linkowski, P., J. Mendlewicz, M. Kerkhofs, R. Leclercq, J. Golstein, M. Brasseur, G. Copinschi, E. Van Cauter. (1987). 24-Hour profiles of adrenocorticotropin, cortisol, and growth hormone in major depressive illness: effect of antidepressant treatment. *J. Clin. Endocrinol. Metab.*, 65:141–152.

Macphee, G.J., E. Butler, M.J. Brodie. (1987). Intradose and circadian variation in circulating carbamazepine and its epoxide in epileptic patients: a consequence of autoinduction of metabolism. *Epilepsia*, 28:286–294.

Malow, B.A., X. Lin, R. Kushwaha, M.S. Aldrich. (1998). Interictal spiking increases with sleep depth in temporal lobe epilepsy. *Epilepsia*, 39:1309–1316.

Malow, B.A., P.R. Carney, R. Kushwaha, R.J. Bowes. (1999). Hippocampal sleep spindles revisited: physiologic or epileptic activity? *Clin. Neurophysiol.*, 110:687–693.

Meierkord, H., S. Shorvon, S.L. Lightman. (1994). Plasma concentrations of prolactin, noradrenaline, vasopressin and oxytocin during and after a prolonged epileptic seizure. *Acta Neurol. Scand.*, 90:73–77.

Meldrum, B.S. (1983). Endocrine consequences of status epilepticus. *Adv. Neurol.*, 34:399–403.

Meldrum, B.S., J.B. Brierley. (1973). Prolonged epileptic seizures in primates: ischemic cell change and its relation to ictal physiological events. *Arch. Neurol.*, 28:10–17.

Merchenthaler, I. (1984). Corticotropin releasing factor (CRF)-like immunoreactivity in the rat central nervous system: extrahypothalamic distribution. *Peptides*, 5(Suppl. 1):53–69.

Mevissen, M., U. Ebert. (1998). Anticonvulsant effects of melatonin in amygdala-kindled rats. *Neurosci. Lett.*, 257:13–16.

Miyamoto, A., M. Itoh, K. Hayashi, M. Hara, Y. Fukuyama. (1993). Diurnal secretion profile of melatonin in epileptic children with or without photosensitivity and an observation of altered circadian rhythm in a case of completely under dark living condition. *No To Hattatsu*, 25:405–411.

Molina-Carballo, A., A. Munoz-Hoyos, R.J. Reiter, M. Sanchez-Forte, F. Moreno-Madrid, M. Rufo-Campos, J.A. Molina-Font, D. Acuna-Castroviejo. (1997). Utility of high doses of melatonin as adjunctive anticonvulsant therapy in a child with severe myoclonic epilepsy: two years' experience. *J. Pineal Res.*, 23:97–105.

Molina-Carballo, A., A. Munoz-Hoyos, M. Sanchez-Forte, J. Uberos-Fernandez, F. Moreno-Madrid, D. Acuna-Castroviejo. (2007). Melatonin increases following convulsive seizures may be related to its anticonvulsant properties at physiological concentrations. *Neuropediatrics*, 38:122–125.

Montplaisir, J., M. Laverdiere, J.M. Saint-Hilaire. (1982). Sleep and focal epilepsy: contribution of depth recording. In *Sleep and Epilepsy* (pp. 301–314), M.B. Sterman, M.N. Shouse, and P. Passouant, Eds. New York: Academic Press.

Moore, R.Y., V.B. Eichler. (1972). Loss of a circadian adrenal corticosterone rhythm following suprachiasmatic lesions in the rat. *Brain Res.*, 42:201–206.

Musshoff, U., E.J. Speckmann. (2003). Diurnal actions of melatonin on epileptic activity in hippocampal slices of rats. *Life Sci.*, 73:2603–2610.

Naum, G., J. Cardozo, D.A. Golombek. (2002). Diurnal variation in the proconvulsant effect of 3-mercaptopropionic acid and the anticonvulsant effect of androsterone in the Syrian hamster. *Life Sci.*, 71:91–98.

Olofsson, K., C. Alling, D. Lundberg, C. Malmros. (2004). Abolished circadian rhythm of melatonin secretion in sedated and artificially ventilated intensive care patients. *Acta. Anaesthesiol. Scand.*, 48:679–684.

Ouyang, Y., C.R. Andersson, T. Kondo, S.S. Golden, C.H. Johnson. (1998). Resonating circadian clocks enhance fitness in cyanobacteria. *Proc. Natl. Acad. Sci. U.S.A.*, 95:8660–8664.

Pang, S., P. Lee, Y. Chan. (1993). Melatonin secretion and its rhythms in biological fluids. In *Melatonin: Biosynthesis, Physiological Effects, and Clinical Applications* (pp. 130–153), R. Reiter and H. Yu, Eds. Boca Raton, FL: CRC Press.

Parrino, L., A. Smerieri, M.C. Spaggiari, M.G. Terzano. (2000). Cyclic alternating pattern (CAP) and epilepsy during sleep: how a physiological rhythm modulates a pathological event. *Clin. Neurophysiol.*, 111(Suppl. 2):S39–S46.

Patry, F. (1931). The relationship of time of day, sleep, and other factors to the incidence of epileptic seizures. *Am. J. Psychiatry*, 10:789–813.

Pavlova, M.K., S.A. Shea, E.B. Bromfield. (2004). Day/night patterns of focal seizures. *Epilepsy Behav.*, 5:44–49.

Peng, Z.C., G. Grassi-Zucconi, M. Bentivoglio. (1995). Fos-related protein expression in the midline paraventricular nucleus of the rat thalamus: basal oscillation and relationship with limbic efferents. *Exp. Brain Res.*, 104:21–29.

Poirel, C. (1991). Circadian chronobiology of epilepsy: murine models of seizure susceptibility and theoretical perspectives for neurology. *Chronobiologia*, 18:49–69.

Pranzatelli, M.R. (1994). On the molecular mechanism of adrenocorticotrophic hormone in the CNS: neurotransmitters and receptors. *Exp. Neurol.*, 125:142–161.

Quigg, M., M. Straume. (2000). Dual epileptic foci in a single patient express distinct temporal patterns dependent on limbic versus nonlimbic brain location. *Ann. Neurol.*, 48:117–120.

Quigg, M., M. Straume, M. Menaker, E.H. Bertram, 3rd. (1998). Temporal distribution of partial seizures: comparison of an animal model with human partial epilepsy. *Ann. Neurol.*, 43:748–755.

Quigg, M., H. Clayburn, M. Straume, M. Menaker, E.H. Bertram, 3rd. (1999). Hypothalamic neuronal loss and altered circadian rhythm of temperature in a rat model of mesial temporal lobe epilepsy. *Epilepsia*, 40:1688–1696.

Quigg, M., H. Clayburn, M. Straume, M. Menaker, E.H. Bertram, 3rd. (2000). Effects of circadian regulation and rest-activity state on spontaneous seizures in a rat model of limbic epilepsy. *Epilepsia*, 41:502–509.

Quigg, M., M. Straume, T. Smith, M. Menaker, E.H. Bertram. (2001). Seizures induce phase shifts of rat circadian rhythms. *Brain Res.*, 913:165–169.

Quigg, M., K.M. Fowler, A.G. Herzog. (2008). Circalunar and ultralunar periodicities in women with partial seizures. *Epilepsia*, 49:1081–1085.

Rao, M.L., H. Stefan, J. Bauer, W. Burr. (1993). Hormonal changes in patients with partial epilepsy: attenuation of melatonin and prolactin circadian serum profiles. In *Chronopharmacology in Therapy of Epilepsies* (pp. 55–70), F.E Dreifuss, Ed. New York: Raven Press.

Ralph, M.R., M. Menaker. (1986). Effects of diazepam on circadian phase advances and delays. *Brain Res.*, 372:405–408.

Ralph, M.R., R.G. Foster, F.C. Davis, M. Menaker. (1990). Transplanted suprachiasmatic nucleus determines circadian period. *Science*, 247:975–978.

Recht, L.D., R.A. Lew, W.J. Schwartz. (1995). Baseball teams beaten by jet lag. *Nature*, 377:583.

Refinetti, R., C.M. Kaufman, M. Menaker. (1994). Complete suprachiasmatic lesions eliminate circadian rhythmicity of body temperature and locomotor activity in golden hamsters. *J. Comp. Physiol. A*, 175:223–232.

Reppert, S.M., D.R. Weaver, T. Ebisawa. (1994). Cloning and characterization of a mammalian melatonin receptor that mediates reproductive and circadian responses. *Neuron*, 13:1177–1185.

Richter, C.P. (1965). *Biological Clocks in Medicine and Psychiatry*. Springfield, IL: Charles C Thomas.

Rusak, B., K.G. Bina. (1990). Neurotransmitters in the mammalian circadian system. *Annu. Rev. Neurosci.*, 13:387–401.

Sammaritano, M., G.L. Gigli, J. Gotman. (1991). Interictal spiking during wakefulness and sleep and the localization of foci in temporal lobe epilepsy. *Neurology*, 41:290–297.

Sanabria, E.R., F.A. Scorza, Z.A. Bortolotto, L.S. Calderazzo-Filho, E.A. Cavalheiro. (1996). Disruption of light-induced c-Fos immunoreactivity in the suprachiasmatic nuclei of chronic epileptic rats. *Neurosci. Lett.*, 216:105–108.

Saper, C.B., T.E. Scammell, J. Lu. (2005). Hypothalamic regulation of sleep and circadian rhythms. *Nature*, 437:1257–1263.

Satinoff, E., J. Liran, R. Clapman. (1982). Aberrations of circadian body temperature rhythms in rats with medial preoptic lesions. *Am. J. Physiol.*, 242:R352–R357.

Sato, S., F.E. Dreifuss, J.K. Penry. (1973). The effect of sleep on spike-wave discharges in absence seizures. *Neurology*, 23:1335–1345.

Schapel, G.J., R.G. Beran, D.L. Kennaway, J. McLoughney, C.D. Matthews. (1995). Melatonin response in active epilepsy. *Epilepsia*, 36:75–78.

Schmutz M., J. Baud, A. Glatt, C. Portet, V. Baltzer. (1990). Chronopharmacological investigations on tonic–clonic seizures on rats and their suppression by carbamazepine, oxcarbazepine, and diazepam. In *Chronopharmacology in Therapy of the Epilepsies* (pp. 145–154), F. Dreifuss, H. Meinardi, and H. Stefan, Eds. New York: Raven Press.

Schwartz, W.J., S.M. Reppert, S.M. Eagan, M.C. Moore-Ede. (1983). *In vivo* metabolic activity of the suprachiasmatic nuclei: a comparative study. *Brain Res.*, 274:184–187.

Sperling, M.R., P.B. Pritchard, 3rd, J. Engel, Jr., C. Daniel, J. Sagel. (1986). Prolactin in partial epilepsy: an indicator of limbic seizures. *Ann. Neurol.*, 20:716–722.

Stephan, F.K., I. Zucker. (1972). Circadian rhythms in drinking behavior and locomotor activity of rats are eliminated by hypothalamic lesions. *Proc. Natl. Acad. Sci. U.S.A.*, 69:1583–1586.

Steriade, M., D. Contreras. (1995). Relations between cortical and thalamic cellular events during transition from sleep patterns to paroxysmal activity. *J. Neurosci.*, 15:623–642.

Steriade, M., D. Contreras, F. Amzica. (1994). Synchronized sleep oscillations and their paroxysmal developments. *Trends Neurosci.*, 17:199–208.

Stevens, J.R., H. Kodama, B. Lonsbury, L. Mills. (1971). Ultradian characteristics of spontaneous seizure discharges recorded by radio telemetry in man. *Electroencephalogr. Clin. Neurophysiol.*, 31:313–325.

Stevens, J.R., B.L. Lonsbury, S.L. Goel. (1972). Seizure occurrence and interspike interval: telemetered electroencephalogram studies. *Arch. Neurol.*, 26:409–419.

Stewart, L.S., K.J. Nylen, M.A. Persinger, M.A. Cortez, K.M. Gibson, O.C. Snead, 3rd. (2008). Circadian distribution of generalized tonic–clonic seizures associated with murine succinic semialdehyde dehydrogenase deficiency, a disorder of GABA metabolism. *Epilepsy Behav.*, 13:290–294.

Swanson, L.W., P.E. Sawchenko, J. Rivier, W.W. Vale. (1983). Organization of ovine corticotropin-releasing factor immunoreactive cells and fibers in the rat brain: an immunohistochemical study. *Neuroendocrinology*, 36:165–186.

Szuba, M.P., B.H. Guze, L.R. Baxter, Jr. (1997). Electroconvulsive therapy increases circadian amplitude and lowers core body temperature in depressed subjects. *Biol. Psychiatry*, 42:1130–1137.

Takahashi, J.S., H.K. Hong, C.H. Ko, E.L. McDearmon. (2008). The genetics of mammalian circadian order and disorder: implications for physiology and disease. *Nat. Rev. Genet.*, 9:764–775.

Tassinari, C.A., R. Michelucci, A. Forti, F. Salvi, R. Plasmati, G. Rubboli, M. Bureau, B. Dalla Bernardina, J. Roger. (1992). The electrical status epilepticus syndrome. *Epilepsy Res. Suppl.*, 6:111–115.

Thomas, K.A. (1991). The emergence of body temperature biorhythm in preterm infants. *Nurs. Res.*, 40:98–102.

Tosini, G. (2000). Melatonin circadian rhythm in the retina of mammals. *Chronobiol. Int.*, 17:599–612.

Velasco, M., A.E. Diaz-de-Leon, F. Brito, A.L. Velasco, F. Velasco. (1995). Sleep–epilepsy interactions in patients with intractable generalized tonic seizures and depth electrodes in the centro median thalamic nucleus. *Arch. Med. Res.*, 26(Suppl.):S117–S125.

Waterhouse, E., A. Towne, J. Boggs, L. Garnett, R. DeLorenzo. (1996). Circadian distribution of status epilepticus (abstract). *Epilepsia.* 37(Suppl. 5):137.

Weiss, G., K. Lucero, M. Fernandez, D. Karnaze, N. Castillo. (1993). The effect of adrenalectomy on the circadian variation in the rate of kindled seizure development. *Brain Res.*, 612:354–356.

Welsh, D.K., D.E. Logothetis, M. Meister, S.M. Reppert. (1995). Individual neurons dissociated from rat suprachiasmatic nucleus express independently phased circadian firing rhythms. *Neuron*, 14:697–706.

Yalyn, O., F. Arman, F. Erdogan, M. Kula. (2006). A comparison of the circadian rhythms and the levels of melatonin in patients with diurnal and nocturnal complex partial seizures. *Epilepsy Behav.*, 8:542–546.

Yamazaki, S., R. Numano, M. Abe, A. Hida, R. Takahashi, M. Ueda, G.D. Block, Y. Sakaki, M. Menaker, H. Tei. (2000). Resetting central and peripheral circadian oscillators in transgenic rats. *Science*, 288:682–685.

Yehuda, S., D.I. Mostofsky. (1993). Circadian effects of beta-endorphin, melatonin, DSIP, and amphetamine on pentylenetetrazol-induced seizures. *Peptides*, 14:203–205.

# 30 Pediatric Epilepsy: A Developmental Neuropsychiatric Disorder

*Rochelle Caplan*

## CONTENTS

## INTRODUCTION

Although pediatric epilepsy has been described as a window to brain–behavior relationships, there is increasing evidence suggesting that this is a developmental neuropsychiatric disorder. As such, in addition to seizures, pediatric epilepsy reflects and impacts brain development and the associated maturation of behavior, cognition, and language. Moreover, like seizures, the behavioral, cognitive, and linguistic comorbidities appear to be manifestations of the disorder.

It remains to be determined how epilepsy, psychosocial variables, and predisposing psychopathology interact and influence progression of this developmental neuropsychiatric disorder. More specifically, do the behavioral, cognitive, and linguistic comorbidities represent the effects on brain development of seizures, the underlying neuropathology, or both? In addition, how does the psychosocial response of the child and parents to having epilepsy interact with the biological factors underlying these comorbidities? Furthermore, what is the contribution of predisposing psychopathology in the child and parent to these comorbidities and to the psychosocial impact of the disorder?

To confirm that pediatric epilepsy is a neurodevelopmental disorder, this chapter first presents neuroimaging evidence indicating that pediatric epilepsy and seizures affect brain development. The subsequent review of the comorbidities examines the role of seizure variables and brain development on behavior, cognition, and language in pediatric epilepsy. Due to space limitations, this review focuses on comorbidities in children with epilepsy with average intelligence and not on those found in children with both epilepsy and intellectual disability.

The third section of the chapter integrates the previously reviewed findings to shed light on our understanding of the mechanisms of the comorbidities of pediatric epilepsy. It also underscores the understudied, complex interaction of the biology of the disorder (i.e., seizures and comorbidities) with psychosocial factors and predisposing psychopathology. The chapter concludes by emphasizing the importance of translational research to delineate the mechanisms underlying the interaction of the multiple and interrelated biological and psychosocial factors involved in epilepsy and its comorbidities.

## EPILEPSY AND BRAIN DEVELOPMENT: NEUROIMAGING EVIDENCE

Magnetic resonance imaging (MRI) studies have identified volumetric abnormalities in children with chronic, new onset (within 6 months), and recent onset (within 12 months) epilepsy. There is evidence for cortical and subcortical abnormalities in children with recent onset and chronic epilepsy. The cortical abnormalities include significantly smaller total and orbital frontal gyrus gray matter volumes in 41 children with idiopathic epilepsy with complex partial seizures (CPSs) and in 26 children with childhood absence epilepsy (CAE) compared to 41 normal children (Caplan et al., 2008a, 2009a; Daley et al. 2007). The children with CPSs also had increased white matter volume of the temporal lobe and a trend for smaller white matter volumes of the superior temporal gyrus (Caplan et al., 2008a). In contrast, a voxel-based morphometrics study of cortical volumes in 53 children with recent onset seizures revealed no significant differences compared to 50 healthy control subjects (Hermann et al., 2006). However, Pulsipher et al. (2009), using a semiautomated software package, found that 20 children with recent onset juvenile myoclonic epilepsy (JME) had significantly larger frontal lobe cerebral spinal fluid (CSF) volumes than the 51 healthy control subjects and 12 children with benign Rolandic epilepsy (BRE). In terms of subcortical structures, these new onset JME subjects had significantly smaller thalamic volumes than both the normal and BRE groups (Pulsipher et al., 2009).

The few extant studies on limbic structures have demonstrated smaller hippocampal volumes, particularly in the anterior hippocampus of 19 CPS and 11 CAE subjects (Daley et al., 2006); a 10% decrease in hippocampal volume of 231 children, including those with intellectual disability, different types of epilepsy, status epilepticus, and febrile seizures (Lawson et al., 2000); smaller hippocampus in 30 children with idiopathic partial epilepsy (5 with developmental disability) (Eroglu et al., 2008); and hippocampal asymmetry in 19 children with BRE (Lundberg et al., 2003). No significant differences have been found in the amygdala volume of 28 CPS and 30 normal children (Daley et al., 2008).

Regarding the relationship between brain volumes and seizure variables in children with chronic CPSs, early onset and longer duration of epilepsy were associated with decreased gray and white matter volumes of the orbital gyrus and white matter volumes of the temporal lobe, increased duration of the disorder with smaller posterior hippocampal volumes, a history of prolonged seizures and febrile convulsions with increased gray and white matter inferior frontal gyrus volumes, and left focal electroencephalography (EEG) findings with decreased total brain white matter volumes (Daley et al., 2007).

Similarly, total brain white matter volumes were significantly smaller in 37 adults with childhood onset (mean age, 7.8 years) of temporal lobe epilepsy compared to 16 patients with adult onset and 62 healthy controls (Hermann et al., 2002). Seizure variables were unrelated to frontotemporal volumes of children with CAE (Caplan et al., 2009a), as well as to anterior hippocampal volumes in children with CPSs (Daley et al., 2006) and in those with idiopathic partial epilepsy (Eroglu et al., 2008).

In summary, children with chronic epilepsy with average intelligence have structural volumetric abnormalities at the cortical and subcortical levels that are differentially associated with seizure variables by brain region, type of seizures, and chronicity. Data on normal brain development provide the context for understanding the implications of these findings. More specifically, gray and white matter volumes, cortical thickness, sulcal depth, and cortical complexity change with age in normal children (Barnea-Goraly et al., 2005; Lenroot and Giedd, 2006; Wilke et al., 2007). Gray matter volumes increase and then decrease with age as a U-shaped function in a posteroanterior direction, with dorsolateral prefrontal and superior temporal lobe gyri maturing last (Lenroot and Giedd, 2006) and the greatest decrease in gray matter volume occurring in the parietal lobe (Wilke et al., 2007). Normal myelination and increase in white matter volumes proceed in parallel in the parietal, temporal, and frontal lobes (Lenroot and Giedd, 2006).

The smaller cross-sectional gray matter volumes of the orbital frontal gyrus and temporal lobe in both CPSs (Daley et al., 2007) and CAE (Caplan et al., 2009a) compared to healthy control subjects suggest an impact of the disorder on brain development, particularly in regions involved in seizure onset and propagation in CAE (Holmes et al., 2004; Tucker et al., 2007) and seizure propagation in CPSs (Gloor, 1997). Furthermore, recent evidence for an increase in myelin-forming oligodendrocytes in the dentate gyrus described in stress conditions, such as seizures (Chetty et al., 2007), implies that the larger white matter volume of the temporal lobe in CPSs might also represent the effects of seizures on brain development. In fact, the association of early onset and longer duration of epilepsy with smaller gray matter orbital frontal volumes suggests that ongoing CPSs influence the normal development of the orbital frontal gyrus.

Yet, the underlying neuropathology might also influence the developmental course, irrespective of seizures, as suggested by the previously described frontotemporal and hippocampal volume findings in children with CAE, thalamic and frontal CSF volumes in recent onset JME, as well as hippocampal volumes in CPSs and idiopathic partial epilepsy. Thus, despite different imaging methodologies and sample sizes, as well as the cross-sectional nature of the previously reviewed studies, the relative role of seizures and the underlying neuropathology in brain development vary by type of epilepsy and its chronicity.

## COMORBIDITIES, SEIZURES, AND BRAIN DEVELOPMENT

### PSYCHOPATHOLOGY

Epidemiological (Davies et al., 2003; Hesdorffer et al., 2004, 2006; Hoie et al., 2006; Lossius et al., 2006) and community studies (for reviews, see Austin and Caplan, 2007; Rodenburg et al., 2005a,b) conducted over the past two decades clearly demonstrate a high rate of psychopathology in children with epilepsy who have average intelligence compared to children with chronic medical disorders not involving the brain and compared to healthy control subjects. In terms of type of psychopathology, these children met *Diagnostic and Statistical Manual of Mental Disorders*, 4th ed. (DSM-IV) criteria for a wide range of psychiatric disorders, including attention-deficit hyperactivity disorder (ADHD), affective-anxiety disorders, combined ADHD and affective/anxiety disorders, and psychosis (Austin et al., 2001, 2003; Caplan et al., 2004, 2008b; Dunn et al., 1999, 2003; Hermann et al., 2007; Jones et al., 2007; Oostrom et al., 2003). Cross-sectional studies identify ADHD as the most frequent diagnosis (for a review, see Dunn and Kronenberger, 2005).

Psychopathology measures, completed by parents and teachers, such as the Child Behavior Checklist (CBCL) (Achenbach, 1991), confirm clinically relevant internalizing behaviors, externalizing behaviors, and difficulties with attention, thinking, and social behavior compared to children with chronic medical illness and healthy children (for a review, see Rodenburg et al., 2005a,b). Self-report instruments also substantiate depression and anxiety (Caplan et al., 2005a; Dunn et al., 1999; Ettinger et al., 1998; Oguz et al., 2002).

Similar to adults with epilepsy, 20% of children with epilepsy have suicidal ideation, a rate that is significantly higher than in healthy control subjects (Caplan et al., 2005a). As found in the general population of youth (King et al., 2001; Strauss et al., 2000), suicidal ideation is associated with the presence of both affective/anxiety disorder diagnoses and ADHD rather than with depression or anxiety alone (Caplan et al., 2005a).

In terms of the role of seizures in the behavioral comorbidity of pediatric epilepsy, it is important to note that recent epidemiological studies identify ADHD (Hesdorffer et al., 2004) and depression (Hesdorffer et al., 2006) in individuals prior to the first unprovoked seizure. Community studies corroborate these epidemiological findings, with evidence for behavioral disturbances and psychiatric diagnoses before the first seizure in 32% of 224 children with new onset (Austin et al., 2001) and in 45% of 53 youth with recent onset seizures (Jones et al., 2007).

In addition, numerous cross-sectional studies and prospective community studies have examined the association of seizures with emotional and behavioral disturbances in children with epilepsy. Because seizure variables, such as age of onset, seizure frequency, duration of epilepsy, and antiepilepsy drug (AED) treatment, are interrelated, studies conducted on large samples of children with epilepsy are essential to tease out effects of individual factors related to the disorder (for a review, see Austin and Caplan, 2007). The findings of such studies support a relationship of behavioral disturbances with events that were probably seizures prior to the first recognized seizure and with recurrent seizures (Austin et al., 2001) in a new onset sample ($N = 224$) tested within 6 weeks of the first identified seizure. A new onset study on 51 subjects tested immediately after diagnosis of at least two unprovoked seizures (Oostrom et al., 2003) and a recent onset study on 53 youth within 12 months of their first seizure, however, did not confirm a relationship of behavior problems and psychopathology with seizure-related factors (Jones et al., 2007).

Findings regarding effects of the condition and psychopathology are also inconsistent in children with chronic epilepsy. Thus, studies on large chronic epilepsy samples that included youth with different types of epilepsy observed an association with seizure variables (Austin et al., 2000; Hermann et al., 1988; Schoenfeld et al., 1999), particularly in females (Austin et al., 2000). Similarly, the presence of a psychiatric diagnosis, particularly ADHD and anxiety disorders, and clinically relevant internalizing CBCL scores were also related to increased duration of epilepsy and seizure frequency in children with chronic CAE (Caplan et al., 2008b). In contrast, findings in a large sample of 101 children with idiopathic epilepsy with CPSs and temporal and frontal involvement did not confirm an association of seizure variables with psychiatric diagnoses, type of psychiatric diagnosis, suicidal ideation, CBCL scores, and self-report depression and anxiety scores (Caplan et al., 2004). Instead, verbal IQ predicted the psychopathology of these children.

Among seizure variables, the type of epilepsy does not appear to be related to the presence and range of psychopathology, other than psychosis (Caplan et al., 2004). Whereas psychosis appears to be associated only with CPSs (Caplan et al., 2004), children with CPSs (Caplan et al., 2004), CAE (Caplan et al., 2008b), BRE (Kavros et al., 2008), and JME (Gélisse et al., 2001) have ADHD, depression, and anxiety disorder diagnoses.

Similar to cross-sectional studies, follow-up studies of psychopathology also report an inconsistent association with seizure variables. Fifty-two children with recent onset seizures who had psychopathology continued to have psychiatric disturbances at the 2-year follow-up unrelated to the severity of their epilepsy (Hermann et al., 2008). Likewise, behavior problems continued at the baseline rate in 42 new onset patients at the 3.5-year follow-up (Oostrom et al., 2005).

In contrast, a 4-year prospective study of 115 children with chronic epilepsy demonstrated increased CBCL behavior problems, mainly in girls with worsening of seizures (Austin et al., 2000). Although there was no change in the rate of psychopathology over time in 65 children with chronic epilepsy, the presence of a psychiatric diagnosis 22 months after initial testing was related to both poor seizure control and male gender (Jones et al., 2008a). Similarly, a long-term outcome study of individuals with CAE followed from childhood through early adulthood demonstrated significantly worse behavioral outcome in the patients with continued poor seizure control compared to those with good seizure control (Wirrell, 2003).

Regarding the relationship of psychopathology with brain volumes, an ADHD diagnosis in children with new onset seizures was related to increased frontal lobe gray matter volumes and reduced brainstem volumes (Hermann et al., 2007). In children with chronic CPSs, the presence of a psychiatric diagnosis was related to small inferior frontal gyri volumes (Daley et al., 2007), but affective/

anxiety disorders were associated with increased posterior hippocampal volumes (Daley et al., 2006) and larger left amygdala volumes (Daley et al., 2008). In contrast to CPSs, no relationship between frontotemporal brain volumes and psychopathology in CAE has been found, even though children with CAE have similar rates and types of psychiatric diagnoses as children with CPSs (Caplan et al., 2004, 2008b). Involvement of frontal, temporal, and limbic brain regions in ADHD (Shaw et al., 2006), depression (Caetano et al., 2007), and anxiety disorders (Steingard et al., 2002) in children without epilepsy and in the onset, evolution, and spread of seizures (Ben-Ari et al., 2008; Meeren et al., 2005; Stafstrom, 2007), described in other sections of this book, might underlie the high rate of psychopathology in children with epilepsy.

In summary, the high rate of psychopathology prior to the first seizure, shortly after onset of seizures, and in chronic CPSs appears to be related to brain volumes rather than seizure severity. In contrast, the similar high psychopathology rate in children with CAE seems to reflect effects of ongoing seizures on brain development. Although prospective studies of children with new onset and chronic epilepsy demonstrate continued stable psychopathology rates over time, seizure control appears to play a role in the severity of psychopathology over time with increased behavioral disturbances in individuals with poor seizure control.

Thus, like seizures, psychopathology appears to be one of the manifestations of epilepsy. In addition, the behavioral component of the disorder in children is differentially related to seizures based on type of epilepsy and the duration of epilepsy. But, involvement of structural abnormalities in the psychopathology of pediatric epilepsy also supports the assumption that this is a developmental neuropsychiatric disorder. Notwithstanding the similar rate and spectrum of psychopathology in different types of epilepsy, the relative contribution of the underlying neuropathology and effects of ongoing seizures on the behavioral disturbances of children with epilepsy appear to vary across seizure types. Nonetheless, involvement of brain regions also implicated in psychiatric disorders unrelated to epilepsy implies a common diathesis to epilepsy and psychopathology in children with epilepsy.

## COGNITION

Unlike psychopathology, the intellectual functioning of children with epilepsy is consistently related to seizures. In terms of the type of epilepsy, epidemiological studies have demonstrated normal nonverbal intelligence in idiopathic epilepsies, benign Rolandic epilepsy (BRE), and absence and simple partial seizures (Hoie et al., 2005), but subnormal intelligence in epilepsy syndromes with symptomatic etiology and epileptic encephalopathy (Berg et al., 2008). Community studies have also found similar mean IQ scores in 101 children with CPSs (Caplan et al., 2004) and 69 with CAE (Caplan et al., 2008a). Likewise, youth with recent onset JME and those with BRE had average IQ scores (Pulsipher et al., 2009).

The profile of seizure variables associated with IQ differs by type of epilepsy. In a large sample of children with chronic CPSs, lower IQ scores were related to earlier age of onset, larger number of years with active seizures, higher seizure frequency, and AED polytherapy, as well as a history of prolonged seizures or febrile convulsions (Caplan et al., 2004; Schoenfeld et al., 1999). Similarly, in 68 children with chronic localization related epilepsy, most of whom had CPSs and a relatively low mean IQ of 87, age of onset was associated with the cognitive profile (alertness, memory, mental speed) (van Mil et al., 2008a). In contrast, the IQ scores of 69 children with CAE were associated with longer duration of epilepsy and AED treatment (any vs. no drugs) (Caplan et al., 2008a).

Seizure recurrence, found in 66.4% of 150 children with new onset seizures (Arthur et al., 2008), was related to teacher ratings of lower academic performance at the 24-month follow-up but not in a 12-month follow-up of 121 children (McNelis et al., 2007). Interestingly, a standardized school battery did not identify the impaired academic performance in these children; however, cognitive testing demonstrated dysfunction within 12 months of the first seizure in a smaller sample of 53 children with recent onset seizures compared to normal children (Hermann et al., 2006).

As for EEG findings, despite differences across studies of children with BRE (for a review, see Kavros et al., 2008), interictal spikes in the mid-temporal region were related to impaired attention but not memory (Northcott et al., 2007). EEG slowing rather than epileptic activity was associated with memory, not IQ, in a large sample of 37 children with new onset and 58 with chronic epilepsy (Koop et al., 2005). Worse memory scores and increased association of these scores with epileptic activity in the chronic sample also implied a possible long-term effect of continued seizures and duration of epilepsy on memory.

Prospective studies reported stable cognitive trajectories that were unrelated to seizures. In fact, parenting, as well as prediagnostic learning and behavioral histories, predicted both the cognitive and behavioral functioning of 42 children with new onset seizures and 30 gender matched classmates followed for 3.5 years with repeated testing 3, 12, and approximately 42 months after diagnosis (Oostrom et al., 2005). Similarly, a 4-year follow-up study of 45 children with chronic epilepsy (20 tested 3 times) found no drop in IQ (Aldenkamp et al., 1990). A cross-sectional study by the same group of researchers on duration of epilepsy and IQ in children with localization-related epilepsy ($N = 68$), generalized epilepsy ($N = 21$), and symptomatic epilepsy ($N = 24$) also demonstrated no time-related change in IQ (van Mil et al., 2008b). Furthermore, IQ mediated the effect of the epilepsy syndrome on educational underachievement in 126 children with epilepsy (Aldenkamp et al., 2005). These authors suggested that the underlying pathology determined both IQ and the epilepsy syndrome and that IQ level, like the epilepsy syndrome, reflects brain damage or dysfunction.

In terms of the association of IQ and brain development, a neuroradiological diagnosis of structural abnormalities in 14% of 239 children with a first recognized seizure 3 months prior to their participation in the study was associated with lower IQ, processing speed, executive/constructional ability, verbal memory, and learning factor scores (Byars et al., 2007). Similarly a volumetric study of 53 recent onset seizures demonstrated significantly decreased gray matter volume in the left parietal and occipital lobes of those with learning difficulties (Hermann et al., 2006).

Moreover, recent volumetric and morphometric studies identified dissimilar relationships between IQ and brain structure in children with both recent onset seizures and chronic epilepsy compared to normal children. Thus, 53 children with recent onset seizures did not exhibit the positive association between white matter cerebral volumes and IQ scores found in 50 healthy control subjects (Hermann et al., 2006). Similarly, whereas frontotemporal volumes were related to IQ in 42 normal children, they were positively associated with age, not IQ, in 41 children with CPSs (Daley et al., 2007) and in 26 with CAE (Caplan et al., 2009a).

Supporting the abnormal developmental trajectory of brain development in pediatric epilepsy, a morphometric study of 86 healthy youth revealed a positive correlation between cortical gray matter thickness and IQ among healthy children ages 6 to 8.5 years but a negative correlation, mainly in the frontal and temporal regions, among those 8.5 to 11.15 years old and 11.5 to 18 years old (Tosun et al., 2008). In contrast, a combined group of 68 children with recent onset and chronic CPSs demonstrated a predominantly negative correlation between intelligence and cortical thickness in the young epilepsy subgroup but a positive correlation in the late childhood and adolescent subgroups (Tosun et al., 2008). Of note, the association of IQ with morphometric measures was unrelated to seizure variables.

In summary, for children with chronic epilepsy, IQ is related to seizure variables with variable profiles across types of epilepsy. In chronic CPSs, all aspects of the condition, other than EEG findings, appear to be related to IQ. In chronic CAE, duration of epilepsy and AED treatment are associated with IQ; in BRE, EEG findings are correlated with IQ. In children with new onset seizures, recurrent seizures are related to teacher ratings of poor academic performance.

The following evidence, however, suggests that the neuropathology underlying epilepsy influences the intellectual functioning of these children. This includes cognitive dysfunction and its association with volumetric and neuroradiological abnormalities in new and recent onset seizures; abnormal association of IQ with brain volumes and cortical thickness that is unrelated to seizure variables in youth with recent onset seizures, chronic CPSs, and chronic CAE; and no change over time in IQ and learning problems in new onset, recent onset, and chronic epilepsy. Taken together, the reviewed findings suggest that the IQs of children with epilepsy reflect both the effects of ongoing seizures and the underlying neuropathology.

## LANGUAGE

Studies on large samples of children with epilepsy have demonstrated both basic (Caplan et al., 2004, 2006, 2008b; Fastenau et al., 2004; Henkin et al., 2003; Hermann et al., 2007, 2008; Northcott et al., 2005; Schoenfeld et al., 1999) and higher level linguistic deficits (Caplan et al., 2001, 2002) that might play an important role in problems these children have with learning, social skills, and behavior problems (Caplan et al., 2005b, 2006, 2009b).

### Basic Linguistic Deficits

The basic linguistic deficits of 182 children with chronic CPSs and CAE who were compared to 101 normal children included impaired use of syntax (i.e., how words combine to make sentences), semantics (i.e., meaning of words, expressions, and sentences), phonology (i.e., production and perception of speech sounds), and language-related skills, such as reading and writing (Caplan et al., 2009b). Although seizure variables were associated with linguistic deficits in the epilepsy subjects, the associations varied by both linguistic function and age. More specifically, among children with epilepsy who were 6.3 to 8.1 years old, longer duration of epilepsy was related to deficits in overall linguistic skills, including syntax, semantics, and phonology, whereas type of epilepsy (CAE > CPS) was associated with difficulties listening to language. In epilepsy subjects ages 9.1 to 11.7 years, however, there was a significant association of a history of prolonged seizures (>5 minutes) with impairment in overall language scores, including receptive and expressive syntax, as well as semantics skills. In adolescents (ages 13.0 to 15.2 years), poor seizure control was related to grammar, reading, and writing deficits, and longer duration of epilepsy to expressive language deficits.

These findings are particularly interesting given cross-sectional evidence for an age-related increase in the range and severity of basic linguistic deficits in youth with epilepsy (Caplan et al., 2009b). Although normal children have intact syntactic and semantic skills by age 5, these skills continue to develop and undergo acceleration during adolescence, with an increase in syntactic complexity, advanced use of grammar and vocabulary, and abstraction that can be attributed to the ongoing complex growth in thought, cognitive flexibility, and integration of knowledge with language (Berman and Nir-Sagiv, 2007; Nippold et al., 2005; Ravid, 2006).

This protracted normal development of language is paralleled by the previously described continued brain development with an age-related decrease in gray matter volume and thickness through adolescence in language-related areas of the temporal (i.e., superior temporal gyrus, Heschl's gyrus) and frontal lobes (inferior frontal gyrus) (for a review, see Lenroot and Giedd, 2006). Ongoing development of these brain regions in middle childhood and adolescence might, therefore, underlie the continued and changing linguistic vulnerability of youth with epilepsy to different aspects of the disorder over time.

Support for the role of abnormal brain development in the linguistic impairment of children with epilepsy also comes from recent evidence of different relationships between language skill and frontotemporal brain volumes in 69 children with epilepsy compared to 34 control subjects (Caplan et al., in press). Higher mean language scores were significantly associated with larger inferior frontal gyrus, temporal lobe, and posterior superior temporal gyrus gray matter volumes in the epilepsy group and in the children with epilepsy with average language scores. Increased total brain and dorsolateral prefrontal gray and white matter volumes, however, were associated with higher language scores in the healthy controls.

In terms of the importance of a history of prolonged seizures, the association of this variable with linguistic deficits, particularly semantic deficits in the previously described epilepsy subjects who were 9.1 to 11.7 years old, is reminiscent of the long-term effect of prolonged seizures in young animals on cognitive and learning deficits—even in the absence of evidence for neuropathological damage (Ben-Ari and Holmes, 2006; Sankar and Rho, 2007; Stafstrom, 2007). This, in turn, might reflect the age-related maturation of language-related brain regions.

Further underscoring this conclusion is the relationship between a history of prolonged seizures with smaller follow-up total and dorsolateral prefrontal gray matter volumes and reduced orbital frontal white matter volumes, brain regions involved in the linguistic deficits of children with epilepsy (Caplan et al., under review). Moreover, the association of increased duration of epilepsy with a wide range of language skills in young epilepsy subjects emphasizes that the continued condition impacts all aspects of ongoing language development in young children.

Studies of children with BRE, new onset seizures, and recent onset seizures support a role for the underlying neuropathology in the linguistic deficits of these children. The language deficits (phonological awareness and manipulation) of children with BRE were unrelated to seizure variables, including EEG, at both the baseline (Northcott et al., 2005) and at 12- to 60-month follow-up testing (Northcott et al., 2006). Moreover, the 14% of 239 children with new onset epilepsy with structural abnormalities on clinical MRI scans had lower language scores than the subjects without MRI abnormalities (Byars et al., 2007). The linguistic deficits of children with recent onset seizures were also unrelated to effects of the disorder at baseline and continued over 2 years, irrespective of the course of their seizures (Hermann et al., 2008).

## Higher Level Linguistic Deficits

The higher level linguistic or discourse deficits found in 93 children with chronic CPSs and 56 with chronic CAE included three components. When speaking, these children had difficulty organizing their thoughts (incoherence), using linguistic devices that tie together ideas across sentences (underuse of cohesion), as well as monitoring and repairing these communication breakdowns (underuse of self-initiated repair) (Caplan et al., 2001, 2002). A listener, therefore, has difficulty following who or what the children are talking about.

From the functional perspective, externalizing/disruptive behavior disorders, poor school performance, and problem social skills were associated with these discourse or social communication deficits in the CPS group (Caplan et al., 2006). Like basic linguistic skills, cross-sectional studies have demonstrated an association of these discourse deficits with seizure variables (Caplan et al., 2006), including a history of prolonged seizures, younger age of onset, higher seizure frequency, and AED polytherapy, albeit with different profiles for chronic CPSs and CAE.

In terms of the relationship with frontotemporal volumes, similar to basic linguistic deficits, higher level linguistic deficits were associated with the volumes of language-related regions (Caplan et al., 2008a). More specifically, the difficulty that children with epilepsy had organizing thoughts during conversation was significantly associated with smaller orbital frontal gray matter volumes, dorsolateral prefrontal white matter volumes, and superior temporal gyrus white matter volumes. Underuse of linguistic devices that tie together ideas across sentences was related to smaller orbital frontal gray matter volumes and increased Heschl's gyrus gray matter volumes. Poor monitoring and repair of these communication breakdowns were associated with smaller inferior frontal gray matter volumes, increased temporal lobe white matter volumes, and larger Heschl's gyrus gray matter volumes.

In summary, children with epilepsy have deficits in both basic and higher level linguistic skills with varying vulnerabilities at different ages for seizure variables. These linguistic impairments are also related to abnormal brain volumes in language-related regions in the temporal (superior temporal gyrus, Heschl's gyrus) and frontal lobe (orbital frontal gyrus, dorsolateral prefrontal gyrus, inferior frontal gyrus). Thus, this comorbidity, like psychopathology and cognition, appears to be integral to pediatric epilepsy and reflects the brain development of these children. From the developmental perspective, both the neuropathology underlying epilepsy and ongoing seizures might have a specific predilection for language-related brain regions due to the prolonged maturation of these brain regions from early childhood through adolescence.

## UNRAVELING THE MECHANISMS OF THE COMORBIDITIES

The findings of the studies reviewed in the previous sections confirm that pediatric epilepsy is a developmental neuropsychiatric disorder. They demonstrate abnormal brain development, as well as a high rate of problems with behavior, cognition, and language in this disorder. Cross-sectional studies suggest that, whereas brain volumes, cognition, and language are associated with the severity of seizures, the relationships with psychopathology are more complex. Thus, behavior problems are found prior to onset of seizures and are associated with seizure variables in some types of epilepsy (CAE) but not in others (CPSs, BRE, new onset seizures). Unlike cross-sectional studies, short-term follow-up studies over 2 to 5 years of the comorbidities in children with new onset, recent onset, and chronic pediatric epilepsy reveal a consistent trajectory related to seizure variables for psychopathology but not for cognition and language.

These findings suggest a complex and ongoing interaction of the underlying pathology, brain development, and seizures with the development of behavior, cognition, and language in children with epilepsy. Recent evidence for stress (Mazarati et al., 2009; Salzberg et al., 2007), depression (Jones et al., 2008a; Mazarati et al., 2008), anxiety (Bouilleret et al., 2009; Mesquita et al., 2006), and impaired cognition (Doucette et al., 2007; Gastens et al., 2008) in animal models of seizures, together with involvement of serotonin, noradrenaline, glutamate, and dopamine in seizures as well as depression, anxiety, schizophrenia, and ADHD, further support the biological basis for the psychopathology and learning difficulties in pediatric epilepsy.

However, the evidence of Hesdorffer and colleagues for ADHD (Hesdorffer et al., 2004), depression (Hesdorffer et al., 2006), and the co-occurrence of major depression, suicide attempt, and migraine with aura (Hesdorffer et al., 2007) before a first unprovoked seizure in a large number of subjects ages 10 years or older also implies that psychopathology *per se* (or the neuropathology underlying psychopathology) might increase the susceptibility to or beget seizures. Similarly, genetic absence epilepsy rats (GAERs), prior to the development of seizures, exhibit anxiety-like behaviors (Jones et al., 2008b).

These findings are particularly interesting because, as found in children with CPSs and affective/anxiety disorder diagnoses (Daley et al., 2008), the rats with anxiety-like behaviors also had increased volumes of the amygdala (Bouilleret et al., 2009). Furthermore, as previously mentioned, the behavior disturbances of children with epilepsy, ADHD, anxiety, and depression involve the same brain regions as for patients without epilepsy who have these psychiatric disorders.

This two-way diathesis for psychopathology and other comorbidities in children with epilepsy and for seizures in children with psychopathology does not occur in a vacuum because children live in families. Extensive research has documented the role of parenting, the family, socioeconomic status, and ethnicity in the behavioral (for a review, see Austin and Caplan, 2007), cognitive (Fastenau et al., 2008), and linguistic (Caplan et al., 2006) comorbidities of children with epilepsy. Recent animal studies reveal that maternal separation increases vulnerability for both anxiety and limbic epileptogenesis in rats with electrical kindling of seizures in the amygdala (Salzberg et al., 2007). In addition, environmental enrichment can reverse depression induced by seizures through transcriptional regulation of the serotonin receptor 5B in juvenile rats with kainite-induced seizures (Koh et al., 2007). These findings further highlight the need to consider the complex interaction of biological, including genetic and psychosocial, aspects in the comorbidities of pediatric epilepsy.

Understanding the mechanisms of the comorbidities of pediatric epilepsy is further complicated by the well-established genetic component for most psychiatric disorders (for a review, see Laporte et al., 2008) (Figure 30.1). Evidence for depression in mothers of children with epilepsy, particularly when their children have behavior problems (Shore et al., 2002), could imply both a genetic predisposition for psychopathology in the parent and child as well as the psychosocial impact of the disorder on the parents. Old (Hoare and Kerley, 1991) and recent (Siddarth et al., 2006) evidence for a high rate of psychiatric diagnoses in the siblings of children with chronic (not new) onset epilepsy

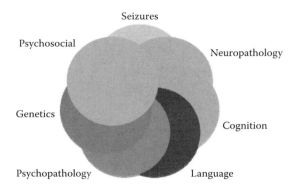

**FIGURE 30.1    (See color insert following page 458.)** Comorbidity interactions with seizures and neuropathology, psychosocial, and genetic factors.

and the association of sibling depression with increased duration of proband epilepsy (Siddarth et al., 2006) further underscores the psychosocial impact of the disorder on the family.

The role of family organization (Fastenau et al., 2004) and acculturation (Mitchell et al., 1991) in the achievement skills of large samples of children with chronic epilepsy and the role of parenting in the behavior and learning problems of children with new onset epilepsy (Oostrom et al., 2005) might represent both epilepsy-related psychosocial and genetic effects on parent behavior and psychopathology. Thus, parents with a genetic predisposition for depression or other types of psychopathology might have more difficulty parenting and coping (Dunn et al., 1999; Williams et al., 2003) with the stress of having a child with epilepsy and the associated comorbidities. In turn, these difficulties could increase behavior and learning problems in their children, both those with and without epilepsy.

The relative weighting of biological vs. psychosocial factors differs in individual children with epilepsy, their parents, and their families (Figure 30.2). For example, a larger biological loading in terms of severity of epilepsy might have fewer disruptive effects on parenting and family functioning, as well as on the behavior of the proband and siblings in the absence of a genetic predisposition for psychopathology. In contrast, economic and marital stressors might tip the scale so that relatively mild biological factors (in terms of severity of epilepsy) and underlying neuropathology result in significant comorbidities in the proband, parent, and siblings with a genetic predisposition to psychopathology.

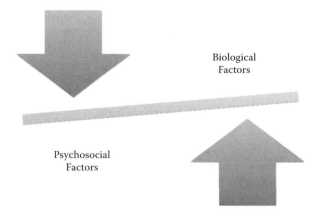

**FIGURE 30.2**    Schematic representation of variable weighting of biological and psychosocial factors in the comorbidities of pediatric epilepsy.

## WHERE TO NOW?

During the past decade, extensive progress has been made in identifying the neural circuits, histopathology, neurochemistry, and molecular mechanisms of epilepsy, as presented in the other chapters of this book. In parallel, behavioral researchers have begun to delineate the biological basis of the comorbidities of pediatric epilepsy and describe models for the impact of epilepsy on families, family resilience and coping, and parenting (for a review, see Austin and Caplan, 2007). This juncture highlights the importance of translational research that will bridge across the basic science, cognitive behavioral neuroscience, and psychosocial domains in order to tease out the complex interactions among the underlying neuropathology, brain development, seizures, comorbidities, genetic predispositions for epilepsy and its comorbidities, and the psychosocial impact of pediatric epilepsy. Advancing our knowledge of these complex mechanisms is essential for the design and operationalization of the comprehensive assessment and treatment required for this pediatric neuropsychiatric developmental disorder.

## REFERENCES

Achenbach, T. (1991). *Manual for the Child Behavior Checklist and Revised Child Behavior Profile.* Burlington: Department of Psychiatry, University of Vermont.

Aldenkamp, A., W.C.J. Alpherts, D. De Bruine-Seeder, M.J.A. Dekker. (1990). Test–retest variability in children with epilepsy—a comparison of WISC-R profiles. *Epilepsy Res.*, 7:165–172.

Aldenkamp, A, B. Weber, W.C. Overweg-Plandsoen, R. Reijs, S. van Mil. (2005). Educational underachievement in children with epilepsy: a model to predict the effects of epilepsy on educational achievement. *J. Child Neurol.*, 20:175–180.

Arthur, T., T.J. de Grauw, C.S. Johnson, S.M. Perkins, A. Kalnin, J.K. Austin, D.W. Dunn. (2008). Seizure recurrence risk following a first seizure in neurologically normal children. *Epilepsia*, 49:1950–1954.

Austin, J., R. Caplan. (2007). Behavioral and psychiatric comorbidities in pediatric epilepsy: toward an integrative model. *Epilepsia*, 48:1639–1651.

Austin, J., D.W. Dunn, G.A. Huster. (2000). Childhood epilepsy and asthma: changes in behavior problems related to gender and change in condition severity. *Epilepsia*, 41:615–623.

Austin, J., J. Harezlak, D.W. Dunn, G.A. Huster, D.F. Rose, W.T. Ambrosius. (2001). Behavior problems in children before first recognized seizures. *Pediatrics*, 107:115–122.

Austin, J., D.W. Dunn, H.M. Caffrey, S.M. Perkins, J. Harezlak, D. Rose. (2003). Recurrent seizures and behavior problems in children with first recognized seizures: a prospective study. *Epilepsia*, 43:1564–1573.

Barnea-Goraly, N., V. Menon, M. Eckert, L. Tamm, R. Bammer, A. Karchemskiy, C.C. Dant, A.L. Reiss. (2005). White matter development during childhood and adolescence: a cross-sectional diffusion tensor imaging study. *Cereb. Cortex*, 15:1848–1854.

Ben-Ari, Y., G.L. Holmes. (2006). Effects of seizures on developmental processes in the immature brain. *Lancet Neurol.*, 5:1055–1063.

Ben-Ari, Y., V. Crepel, A. Represa. (2008). Seizures beget seizures in temporal lobe epilepsies: the boomerang effects of newly formed aberrant kainatergic synapses. *Epilepsy Curr.*, 8:68–72.

Berg, A., J.T. Langfitt, F.M. Testa, S.R. Levy, F. Dimario, M. Westerveld, J. Kulas. (2008). Global cognitive function in children with epilepsy: a community-based study. *Epilepsia*, 49:608–614.

Berman, R., B. Nir-Sagiv. (2007). Comparing narrative and expository text construction across adolescence: a developmental paradox. *Discourse Processes*, 43:79–120.

Bouilleret, V., R.E. Hogan, D. Velakoulis, M.R. Salzberg, L. Wang, G.F. Egan, T.J. O'Brien, N.C. Jones. (2009). Morphometric abnormalities and hyperanxiety in genetically epileptic rats: a model of psychiatric comorbidity? *NeuroImage*, 45:267–274.

Byars, A., T.J. de Grauw, C.S. Johnson, P.S. Fastenau, S.M. Perkins, J.C. Egelhoff, A. Kalnin, D.W. Dunn, J.K. Austin. (2007). The association of MRI findings and neuropsychological functioning after the first recognized seizure. *Epilepsia*, 48:1067–1074.

Caetano, S., M. Fonseca, J.P. Hatch, R.L. Olvera, M. Nicoletti, K. Hunter, B. Lafer, S.R. Pliszka, J.C. Soares. (2007). Medial temporal lobe abnormalities in pediatric unipolar depression. *Neurosci. Lett.*, 427:142–147.

Caplan, R., D. Guthrie, S. Komo, S. Chayasirisobhon, W. Mitchell, W.D. Shields. (2001). Conversational repair in pediatric epilepsy. *Brain Lang.*, 78:82–93.

Caplan, R., D. Guthrie, S. Komo, P. Siddarth, S. Chayasirisobhon, H. Kornblum, R. Sankar. (2002). Social communication in pediatric epilepsy. *J. Child Psychol. Psychiatry*, 43:245–253.

Caplan, R., P. Siddarth, S. Gurbani, D. Ott, R. Sankar, W.D. Shields. (2004). Psychopathology and pediatric complex partial seizures: seizure-related, cognitive, and linguistic variables. *Epilepsia*, 45:1273–1286.

Caplan, R., P. Siddarth, S. Gurbani, R. Hanson, R. Sankar, W.D. Shields. (2005a). Depression and anxiety disorders in pediatric epilepsy. *Epilepsia*, 46:720–730.

Caplan, R., J. Sagun, P. Siddarth, S. Gurbani, S. Koh, R. Gowrinathan, R. Sankar. (2005b). Social competence in pediatric epilepsy: insights into underlying mechanisms. *Epilepsy Behav.*, 6:218–228.

Caplan, R., P. Siddarth, S. Gurbani, E. Lanphier, S. Koh, R. Sankar. (2006). Thought disorder: a developmental disability in pediatric epilepsy. *Epilepsy Behav.*, 8:726–735.

Caplan, R., J. Levitt, P. Siddarth, J. Taylor, M. Daley, K.N. Wu, S. Gurbani, W.D. Shields, R. Sankar. (2008a). Thought disorder and fronto-temporal volumes in pediatric epilepsy. *Epilepsy Behav.*, 13:593–599.

Caplan, R., P. Siddarth, L. Stahl, E. Lanphier, P. Vona, S. Gurbani, S. Koh, R. Sankar, W.D. Shields. (2008b). Childhood absence epilepsy: behavioral, cognitive, and linguistic comorbidities. *Epilepsia*, 49 1838–1846.

Caplan, R., J. Levitt, P. Siddarth, K.N. Wu, S. Gurbani, R. Sankar, W.D. Shields. (2009a). Frontal and temporal volumes in childhood absence epilepsy. *Epilepsia*, 50(11):2466–2472.

Caplan, R., P. Siddarth, P. Vona, L. Stahl, C.E. Bailey, S. Gurbani, R. Sankar, W.D. Donald. (2009b). Language in pediatric epilepsy. *Epilepsia*, 50(11):2397–2407.

Caplan, R., P. Siddarth, J.G. Levitt, J. Taylor, M. Daley, K.W. Wu, S. Gurbani, R. Sankar, W.D. Donald. (in press). Language and frontotemporal volumes in pediatric epilepsy. *Epilepsy Behav.*

Chetty, S., C. Mirescu, G.E. Bentley, D. Kaufer. (2007). Dose-dependent effects of glucocorticoids on neural precursor cells in the adult dentate gyrus. In *Neuroscience Meeting Planner*, San Diego, CA, Society for Neuroscience, Program No. 298.6.

Daley, M., D. Ott, R. Blanton, P. Siddarth, J. Levitt, E. Mormino, C. Hojatkashani, R. Tenorio, S. Gurbani, W. Shields, R. Sankar, A. Toga, R. Caplan. (2006). Hippocampal volume in childhood complex partial seizures. *Epilepsy Res.*, 72:57–66.

Daley, M., J. Levitt, P. Siddarth, E. Mormino, C. Hojatkashani, S. Gurbani, W.D. Shields, R. Sankar, A. Toga, R. Caplan. (2007). Frontal and temporal volumes in children with complex partial seizures. *Epilepsy Behav.*, 10:470–476.

Daley, M., P. Siddarth, J. Levitt, S. Gurbani, W.D. Shields, R. Sankar, A. Toga, R. Caplan. (2008). Amygdala volume and psychopathology in childhood complex partial seizures. *Epilepsy Behav.*, 13:212–217.

Davies, S., I. Heyman, R. Goodman. (2003). A population survey of mental health problems in children with epilepsy. *Dev. Med. Child Neurol.*, 45:292–295.

Doucette, T., C.L. Ryan, R.A. Tasker. (2007). Gender-based changes in cognition and emotionality in a new rat model of epilepsy. *Amino Acids*, 32:317–322.

Dunn, D., W.G. Kronenberger. (2005). Childhood epilepsy, attention problems, and ADHD: review and practical considerations. *Semin. Pediatr. Neurol.*, 12:222–228.

Dunn, D., J.K. Austin, G. Huster. (1999). Symptoms of depression in adolescents with epilepsy. *J. Am. Acad. Child Adolesc. Psychiatry*, 38:1132–1138.

Dunn, D.W., J.K. Austin, J. Harezlak, W.T. Ambrosius. (2003). ADHD and epilepsy in childhood. *Dev. Med. Child Neurol.*, 45:50–54.

Eroglu, B., S. Kurul, H. Cakmakçi, E. Dirik. (2008). The correlation of seizure characteristics and hippocampal volumetric magnetic resonance imaging findings in children with idiopathic partial epilepsy. *J. Child Neurol.*, 22:348–353.

Ettinger, A., D.M. Weisbrot, E.E. Nolan, K.D. Gadow, S.A. Vitale, M.R. Andriola, N.J. Lenn, G.P. Novak, B. Hermann. (1998). Symptoms of depression and anxiety in pediatric epilepsy patients. *Epilepsia*, 39:595–599.

Fastenau, P., J. Shen, D.W. Dunn, S.M. Perkins, B.P. Hermann, J.K. Austin. (2004). Neuropsychological predictors of academic underachievement in pediatric epilepsy: moderating roles of demographic, seizure, and psychosocial variables. *Epilepsia*, 45:1261–1272.

Fastenau, P., J. Shen, D.W. Dunn, J.K. Austin. (2008). Academic underachievement among children with epilepsy: proportion exceeding psychometric criteria for learning disability and associated risk factors. *J. Learn Disabil.*, 41:195–207.

Gastens, A., C. Brandt, J.P. Bankstahl, W. Loscher. (2008). Predictors of pharmacoresistant epilepsy: pharmacoresistant rats differ from pharmacoresponsive rats in behavioral and cognitive abnormalities associated with experimentally induced epilepsy. *Epilepsia*, 49:1759–1776.

Gélisse, P., P. Genton, J.C. Samuelian, P. Thomas, M. Bureau. (2001). Psychiatric disorders in juvenile myoclonic epilepsy. *Rev. Neurol.*, 157:297–302.

Gloor, P. (1997). *The Temporal Lobe and the Limbic System.* New York: Oxford University Press.

Henkin, Y., L. Kishon-Rabin, H. Pratt, S. Kivity, M. Sadeh, N. Gadoth. (2003). Linguistic processing in idiopathic generalized epilepsy: an auditory event-related potential study. *Epilepsia*, 44:1207–1217.

Hermann, B., S. Whitman, J.R. Hughes, M.M. Melyn, J. Dell. (1988). Multietiological determinants of psychopathology and social competence in children with epilepsy. *Epilepsy Res.*, 2:51–60.

Hermann, B., M. Seidenberg, B. Bell, P. Rutecki, R. Sheth, K. Ruggles, G. Wendt, D. O'Leary, V. Magnotta. (2002). The neurodevelopmental impact of childhood-onset temporal lobe epilepsy on brain structure and function. *Epilepsia*, 43:1062–1071.

Hermann, B., J. Jones, R. Sheth, C. Dow, M. Koehn, M. Seidenberg. (2006). Children with new-onset epilepsy: neuropsychological status and brain structure. *Brain*, 129:2609–2619.

Hermann, B., J. Jones, K. Dabbs, C.A. Allen, R. Sheth, J. Fine, A. McMillan, M. Seidenberg. (2007). The frequency, complications and aetiology of ADHD in new onset paediatric epilepsy. *Brain*, 130:3135–3148.

Hermann, B., J.E. Jones, R. Sheth, M. Koehn, T. Becker, J. Fine, C.A. Allen, M. Seidenberg. (2008). Growing up with epilepsy: a two-year investigation of cognitive development in children with new onset epilepsy. *Epilepsia*, 49:1847–1858.

Hesdorffer, D., P. Ludvigsson, E. Olafsson, G. Gudmundsson, O. Kjartansson, W.A. Hauser. (2004). ADHD as a risk factor for incident unprovoked seizures and epilepsy in children. *Arch. Gen. Psychiatry*, 61:731–736.

Hesdorffer, D., W.A. Hauser, E. Olafsson, P. Ludvigsson, O. Kjartansson. (2006). Depression and suicide attempt as risk factors for incident unprovoked seizures. *Ann. Neurol.*, 59:35–41.

Hesdorffer, D., P. Lúdvígsson, W.A. Hauser, E. Olafsson, O. Kjartansson. (2007). Co-occurrence of major depression or suicide attempt with migraine with aura and risk for unprovoked seizure. *Epilepsy Res.*, 75:220–223.

Hoare, P., S. Kerley. (1991). Psychosocial adjustment of children with chronic epilepsy and their families. *Dev. Med. Child Neurol.*, 33:201–215.

Hoie, B., A. Mykletun, K, Sommerfelt, H. Bjornaes, H. Skeidsvol, P.E. Waaler. (2005). Seizure-related factors and non-verbal intelligence in children with epilepsy: a population-based study from Western Norway. *Seizure*, 14:223–231.

Hoie, B., K. Sommerfelt, P.E. Waaler, F.D. Alsaker, H. Skeidsvoll, A. Mykletun. (2006). Psychosocial problems and seizure-related factors in children with epilepsy. *Dev. Med. Child Neurol.*, 48:213–219.

Holmes, M., M. Brown, D.M. Tucker. (2004). Are "generalized" seizures truly generalized? Evidence of localized mesial frontal and frontopolar discharges in absence. *Epilepsia*, 45:1568–1579.

Jones, J., R. Watson, R. Sheth, R. Caplan, M. Koehn, M. Seidenberg, B. Hermann. (2007). Psychiatric comorbidity in children with new onset epilepsy. *Dev. Med. Child Neurol.*, 49:493–497.

Jones, J., P. Siddarth, S. Gurbani, W.D. Shields, and R. Caplan. (2008a). Follow-up study of comorbidities in pediatric epilepsy. In *Proceedings of AES 2008: American Epilepsy Society 62nd Annual Meeting*, December 5–9, Seattle, WA.

Jones, N., M.R. Salzberg, G. Kumar, A. Couper, M.J. Morris, T.J. O'Brien. (2008b). Elevated anxiety and depressive-like behavior in a rat model of genetic generalized epilepsy suggesting common causation. *Exp. Neurol.*, 209:254–260.

Kavros, P., T. Clarke, L.J. Strug, J.M. Halperin, N.J. Dorta, D.K. Pal. (2008). Attention impairment in Rolandic epilepsy: systematic review. *Epilepsia*, 49:1570–1580.

King, R., M. Schwab-Stone, A.J. Flisher, S. Greenwald, R.A. Krame, S.H. Goodman, B.B. Lahey, D. Shaffer, M.S. Gould. (2001). Psychosocial and risk behavior correlates of youth suicide attempts and suicidal ideation. *J. Am. Acad. Child Adolesc. Psychiatry*, 40:837–846.

Koh, S., R. Magid, H. Chung, S.D. Stine, D.N. Wilson. (2007). Depressive behavior and selective down-regulation of serotonin receptor expression after early-life seizures: reversal by environmental enrichment. *Epilepsy Behav.*, 10:26–31.

Koop, J., P.S. Fastenau, D.W. Dunn, J.K. Austin. (2005). Neuropsychological correlates of electroencephalograms in children with epilepsy. *Epilepsy Res.*, 64:49–62.

Laporte, J., R.F. Ren-Patterson, D.L. Murphy, A.V. Kalueff. (2008). Refining psychiatric genetics: from 'mouse psychiatry' to understanding complex human disorders. *Behav. Pharmacol.*, 19:377–384.

Lawson, J., S. Vogrin, A.F. Bleasel, M.J. Cook, L. Burns, L. McAnally, J. Pereira, A.M. Bye. (2000). Predictors of hippocampal, cerebral, and cerebellar volume reduction in childhood epilepsy. *Epilepsia*, 41:1540–1545.

Lenroot, R.K., J. Giedd. (2006). Brain development in children and adolescents: insights from anatomical magnetic resonance imaging. *Neurosci. Biobehav. Rev.*, 30:718.

Lossius, M., J. Clench-Aas, B. van Roy, P. Mowinckel, L. Gjerstad. (2006). Psychiatric symptoms in adolescents with epilepsy in junior high school in Norway: a population survey. *Epilepsy Behav.*, 9:286–292.

Lundberg, S., J. Weis, O. Eeg-Olofsson, R. Raininko. (2003). Hippocampal region asymmetry assessed by 1H-MRS in Rolandic epilepsy. *Epilepsia*, 44:205–210.

Mazarati, A., P. Siddarth, R.A. Baldwin, D. Shin, R. Caplan, R. Sankar. (2008). Depression after status epilepticus: behavioural and biochemical deficits and effects of fluoxetine. *Brain*, 131:2071–2083.

Mazarati, A., D. Shin, Y.S. Kwon, A. Bragin, E. Pineda, D. Tio, A.N. Taylor, R. Sankar. (2009). Elevated plasma corticosterone level and depressive behavior in experimental temporal lobe epilepsy. *Neurobiol. Dis.*, 34(3):457–461.

McNelis, A., D.W. Dunn, C.S. Johnson, J.K. Austin, S.M. Perkins. (2007). Academic performance in children with new-onset seizures and asthma: a prospective study. *Epilepsy Behav.*, 10:311–318.

Meeren, H., G. van Luijtelaar, F. Lopes da Silva, A. Coenen. (2005). Evolving concepts on the pathophysiology of absence seizures: the cortical focus theory. *Arch. Neurol.*, 62:371–376.

Mesquita, A., H.B. Tavares, R. Silva, N. Sousa. (2006). Febrile convulsions in developing rats induce a hyperanxious phenotype later in life. *Epilepsy Behav.*, 9:401–406.

Mitchell, W., J.M. Chavez, H. Lee, B.L. Guzman. (1991). Academic underachievement in children with epilepsy. *J. Child Neurol.*, 6:65–72.

Nippold, M., L.J. Hesketh, J.K. Duthie, T.C. Mansfield. (2005). Conversational versus expository discourse: a study of syntactic development in children, adolescents, and adults. *J. Speech Lang. Hear Res.*, 48:1048–1064.

Northcott, E., A.M. Connolly, A. Berroya, M. Sabaz, J. McIntyre, J. Christie, A. Taylor, J. Batchelor, A.F. Bleasel, J.A. Lawson, A.M. Bye. (2005). The neuropsychological and language profile of children with benign Rolandic epilepsy. *Epilepsia*, 46:924–930.

Northcott, E., A.M. Connolly, J. McIntyre, J. Christie, A. Berroya, A. Taylor, J. Batchelor, G. Aaron, S. Soe, A.F. Bleasel, J.A. Lawson, A.M. Bye. (2006). Longitudinal assessment of neuropsychologic and language function in children with benign Rolandic epilepsy. *J. Child Neurol.*, 21:518–522.

Northcott, E., A.M. Connolly, A. Berroya, J. McIntyre, J. Christie, A. Taylor, A.F. Bleasel, J.A. Lawson, A.M. Bye. (2007). Memory and phonological awareness in children with benign Rolandic epilepsy compared to a matched control group. *Epilepsy Res.*, 25:57–62.

Oguz, A., S. Kurul, E. Dirik. (2002). Relationship of epilepsy-related factors to anxiety and depression scores in epileptic children. *J. Child Neurol.*, 17:37–40.

Oostrom, K., A. Smeets-Schouten, C.L. Kruitwagen, A.C. Peters, A. Jennekens-Schinkel, Dutch Study Group of Epilepsy in Childhood. (2003). Not only a matter of epilepsy: early problems of cognition and behavior in children with 'epilepsy only'—a prospective, longitudinal, controlled study starting at diagnosis. *Pediatrics*, 112:1338–1344.

Oostrom, K., H. van Teeseling, A. Smeets-Schouten, A. Peters, A. Jennekens-Schinkel, Dutch Study of Epilepsy in Childhood. (2005). Three to four years after diagnosis: cognition and behaviour in children with 'epilepsy only.' A prospective, controlled study. *Brain*, 128:1546–1555.

Pulsipher, D., M. Seidenberg, L. Guidotti, V.N. Tuchscherer, J. Morton, R.D. Sheth, B. Hermann. (2009). Thalamofrontal circuitry and executive dysfunction in recent-onset juvenile myoclonic epilepsy. *Epilepsia*, 50(5):1210–1219.

Ravid, D. (2006). Semantic development in textual contexts during the school years: Noun Scale analyses. *J. Child Lang.*, 33:791–821.

Rodenburg, R., G.J. Stams, A.M. Meijer, A.P. Aldenkamp, M. Dekovic. (2005a). Psychopathology in children with epilepsy: a meta-analysis. *J. Pediatr. Psychol.*, 30:453–468.

Rodenburg, R., A.M. Meijer, M. Dekovic, A.P. Aldenkamp. (2005b). Family factors and psychopathology in children with epilepsy: a literature review. *Epilepsy Behav.*, 6:488–503.

Salzberg, M., G. Kumar, L. Supit, N.C. Jones, M.J. Morris, S. Rees, T.J. O'Brien. (2007). Early postnatal stress confers enduring vulnerability to limbic epileptogenesis. *Epilepsia*, 48:2079–2085.

Sankar, R., J.M. Rho. (2007). Do seizures affect the developing brain? Lessons from the laboratory. *J. Child Neurol.*, 22(5, Suppl.):21S–S29.

Schoenfeld, J., M. Seidenberg, A. Woodard, K. Hecox, C. Inglese, K. Mack, B. Hermann. (1999). Neuropsychological and behavioral status of children with complex partial seizures. *Dev. Med. Child Neurol.*, 41:724–731.

Shaw, P., J. Lerch, D. Greenstein, W. Sharp, L. Clasen, A. Evans, J. Giedd, F.X. Castellanos, J. Rapoport. (2006). Longitudinal mapping of cortical thickness and clinical outcome in children and adolescents with attention-deficit/hyperactivity disorder. *Arch. Gen. Psychiatry*, 65:540–549.

Shore, C., J.K. Austin, G.A. Huster, D.W. Dunn. (2002). Identifying risk factors for maternal depression in families of adolescents with epilepsy. *J. Spec. Pediatr. Nurs.*, 7:71–80.

Siddarth, S., S. Gurbani, S. Koh, R. Jonas, R. Caplan. (2006). Behavior, cognition and thought disorder in siblings of children with complex partial seizures. In *Proceedings of the 60th Annual Meeting of the American Epilepsy Society*, San Diego, CA, December 1–5, 2006.

Stafstrom, C.E. (2007). Neurobiological mechanisms of developmental epilepsy: translating experimental findings into clinical application. *Semin. Pediatr. Neurol.*, 14:164–172.

Steingard, R., P.F. Renshaw, J. Hennen, M. Lennox, C.B. Cintron, A.D. Young, D.F. Connor, T.H. Au, D.A. Yurgelun-Todd. (2002). Smaller frontal lobe white matter volumes in depressed adolescents. *Biol. Psychiatry*, 52:413–417.

Strauss, J., B. Birmaher, J. Bridge, D. Axelson, L. Chiappetta, D. Brent, N. Ryan. (2000). Anxiety disorders in suicidal youth. *Can. J. Psychiatry*, 45:739–745.

Tosun, D., P. Siddarth, M. Seidenberg, A. Toga, R. Caplan, B. Hermann. (2008). The neurodevelopmental relationship between intelligence and cortical morphometry is altered in children with complex partial seizures. In *Proceedings of AES 2008: American Epilepsy Society 62nd Annual Meeting*, December 5–9, Seattle, WA.

Tucker, D., M. Brown, P. Luu, G.L. Holmes. (2007). Discharges in ventromedial frontal cortex during absence spells. *Epilepsy Behav.*, 11:546–547.

van Mil, S., R.P. Reijs, M.H. van Hall, A.P. Aldenkamp. (2008a). Neuropsychological profile of children with cryptogenic localization related epilepsy. *Child Neuropsychol.*, 14:291–301.

van Mil, S., R.P. Reijs, M.H. van Hall, A.P. Aldenkamp. (2008b). The effect of duration of epilepsy on IQ in children with CLRE; a comparison to SLRE and IGE. *Seizure*, 17:308–303.

Wilke, M., I. Krägeloh-Mann, S. Holland. (2007). Global and local development of gray and white matter volume in normal children and adolescents. *Exp. Brain Res.*, 178:296–307.

Williams, J., C. Steel, G.B. Sharp, E. Delos Reyes, T. Phillips, S. Bates, B. Lange, M.L. Griebel. (2003). Anxiety in children with epilepsy. *Epilepsy Behav.*, 4:729–732.

Wirrell, E. (2003). Natural history of absence epilepsy in children. *Can. J. Neurol. Sci.*, 30:184–188.

# Section VI

## The Future of Epilepsy Therapy

# 31 Novel Mechanisms of Drug Delivery

*Vijay Ivaturi and James Cloyd*

## CONTENTS

## INTRODUCTION

A fundamental tenet of antiepileptic drug (AED) therapy is delivery of the right amount of drug at the right time to the site of action. Currently available AED delivery systems used for research or in clinical practice have shortcomings that can hinder laboratory studies or optimizing therapy in patients. This is due, in part, to the fact that most approved and investigational AEDs possess physicochemical and pharmacological properties that affect drug delivery. Among these properties are poor water solubility (e.g., carbamazepine, lamotrigine, oxcarbazepine, phenytoin, rufinamide); slow, incomplete, or irregular oral absorption (e.g., carbamazepine, phenytoin, rufinamide, valproic acid); influx or efflux transport in the gastrointestinal tract or the blood–brain barrier (e.g., gabapentin); and relatively short elimination half-lives necessitating two or more daily doses (e.g., carbamazepine, gabapentin, lacosamide, lamotrigine, levetiracetam, oxcarbazepine, pregabalin, rufinamide, valproic acid). Collectively, these properties complicate the design and interpretation of AED studies in animals and limit the effectiveness and safety therapy in patients due to problems with drug delivery or poor medication adherence. Advances in the pharmaceutical sciences over the last decade have resulted in innovative delivery systems directed at enhancing the effectiveness and safety of drug therapy. These systems are particularly relevant to AED therapy, given the drug delivery challenges this therapeutic class presents. The new delivery platforms can enhance bioavailability, control drug release, overcome transport interactions, enable targeted delivery, reduce dosing frequency, and permit the formulation of safer, more useful enteral and parenteral solutions of poorly soluble AEDs.

This chapter reviews recent advances and developments that are being used, or could be used, to achieve greater control of AED delivery in laboratory studies and improve therapy in patients. The text is divided into different sections, each of which deals with a particular route for drug delivery. This chapter, however, does not consider the currently marketed AED formulations in detail, as there are several excellent reviews to which the reader is referred (Bialer, 2007; Fisher and Chen, 2006; Fisher and Ho, 2002; Pellock et al., 2004; Wheless and Venkataraman, 1999).

## IMPLANTS AND INJECTABLES

*In vivo* preclinical drug studies typically utilize intravenous (IV), intramuscular (IM), subcutaneous (SC), or intraperitoneal (IP) routes of administration. The design and interpretation of such studies may be affected by the limitations inherent in formulations utilizing conventional technologies; for example, organic solvents used to dissolve poorly water soluble drugs may have their own effects on safety and efficacy. The availability of new technologies that either improve solubility or modify drug release would give investigators greater flexibility in study design and greater control over drug delivery. Further, enhanced delivery technologies permit the commercial development of injectable AED formulations for clinical use because they can overcome limitations in drug properties such as poor water solubility or very rapid elimination half-lives (Cleland et al., 2001). Modified-release dosage forms may be used to manipulate systemic or local drug availability, which, in turn, can affect safety and efficacy. In the case where a drug has a short *in vivo* half-life, a modified-release system can be used to extend exposure after a single administration (carbamazepine, levetiracetam, valproic acid). Such systems provide a longer duration of effect, permit fewer injections, and minimize oscillations of AED concentrations, thus reducing the risk of peak-level related toxicities or trough-level loss of seizure control (Okumu and Cleland, 2003). In this section, we discuss novel drug delivery technologies that are either implanted or injected into the body and deliver a controlled release of the drug. Injectables are discussed under the classes of prodrugs, use of solubilization and stability enhancers such as cyclodextrins, liposomes and microspheres, microemulsions, and injectable gels.

### IMPLANTS

Implantable drug delivery systems provide a convenient, patient-friendly alternative to current therapies for long-term treatment of various chronic conditions such as diabetes. Such systems allow automatic, controlled drug delivery without intervention by the patient or caregiver. Precise, controlled delivery can maximize response while limiting toxicity associated with elevated drug concentrations. In addition, implantable drug delivery systems can reduce the cost of care and free the patient from repeated treatments or hospitalization, thus improving quality of life. The patient receiving continuous therapy does not have to worry about taking capsules or tablets, getting injections, or changing transdermal patches. The physician has added assurance that the patient is getting the right dose of medicine and does not have to be concerned with the possibility of a lost prescription, noncompliance, or diversion of an abusable drug substance (Wright et al., 2003).

Implants are most useful when chronic drug administration is required. Devices function for long periods of time once implanted and provide better controlled release than injectable systems. Implants have varying functional lifetimes, which are dictated by the amount of drug contained in the reservoir and the delivery rate. Typically, implants can hold a maximum of 1 g of drug for controlled release (Lane et al., 2008). Direct AED delivery to the central nervous system (CNS) is highly desirable as it can improve the therapeutic index of a drug compared to that found with oral, intravenous, or intramuscular administration by substantially reducing the potential for systemic adverse effects (Fisher and Chen, 2006). Implants provide a good alternative to intrathecal and intraventricular routes for cerebral spinal fluid (CSF) drug delivery. Given the small volume available in implantable devices and the prolonged storage, drugs must be highly soluble and stable in

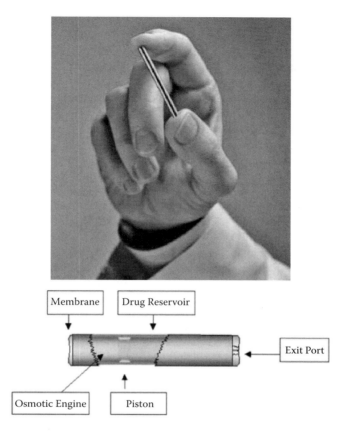

**FIGURE 31.1**   The DUROS® system: (top) the implantable drug-containing device; (bottom) inner cross-section of the device. (From Wright, J.C. et al., *Drug Delivery Tech.*, 3(1), 2003. With permission.)

acceptable solvent systems. Major drawbacks of implantable devices are the risks and costs associated with the onset of serious adverse effects requiring removal of the device.

Several technologies are being employed in implantable drug delivery devices. The DUROS® osmotic pump (DURECT; Cupertino, CA) is a sterile, non-erodible, drug-dedicated system that operates like a miniature syringe implanted under the skin. The device can be as small as 4 mm OD × 44 mm L or smaller (Figure 31.1). It is driven by osmosis, releasing minute quantities of drug for 1 to 12 months or longer (Peery et al., 1999). The DUROS® system provides zero-order delivery capabilities usually found only in syringe pumps that are larger and more complicated. Commercialized and in-development applications of this technology include the Viadur®, CHRONOGESIC®, and DUROS® delivery systems. Viadur® (leuprolide acetate implant) was the first U.S. Food and Drug Administration (FDA)-approved DUROS® product on the market. It delivers leuprolide for one year for palliative treatment of prostate cancer (Fowler, 2001; Fowler et al., 2000; Wright et al., 2001). The CHRONOGESIC® system consists of a small titanium pump, about the size of a matchstick, that is implanted under the skin of a patient in a simple outpatient procedure. The CHRONOGESIC® sufentanil system delivers drug in specified doses for 3 months and has been recently tested in a Phase III clinical trial. DUROS® technology is well suited for the delivery of potent drugs that have narrow therapeutic windows or short half-lives, including biomolecules that require precise control of the rate of delivery. Recent advances in this technology include the development of miniaturized catheter systems for site-directed delivery (i.e., within the CNS) which offer the potential of utilizing smaller drug doses with better selectivity of action and reduction in side effects.

Other implantable systems studied include polymeric wafers that are engineered to release drug over a period of months from a chemical matrix. Implantation of wafers into the brain has been studied as a method of AED drug delivery (Kokaia et al., 1994; Tamargo et al., 2002). Kokaia et al. (1994) implanted γ-aminobutyric acid (GABA)-containing polymers dorsal to the substantia nigra in rats previously kindled through the amygdala. Stimulation of the amygdala produced attenuated seizures compared to an implanted inert polymer wafer. Tamargo et al. (2002) studied the efficacy of intracerebrally administered phenytoin using controlled-release polymer implants in a cobalt-induced rat model of epilepsy. A constant amount of phenytoin was released daily from the polymer matrix for about 4 months. In comparison to animals receiving implants of empty polymer matrices, those given phenytoin improved on a behavioral scale and two measures of electroencephalography (EEG) activity.

The therapeutic potential of a novel silk-protein-based, adenosine-releasing implant was investigated in a dose–response study using a rat model of kindling epileptogenesis (Wilz et al., 2008). The implants were prepared by embedding adenosine containing microspheres into nanofilm-coated silk fibroin scaffolds and were designed to release doses of 0, 40, 200, and 1000 mg/day of adenosine. Four days prior to the onset of kindling, adenosine-releasing polymers were implanted into the infra-hippocampal cleft, and the progressive acquisition of kindled seizures was monitored over a total of 48 stimulations. The investigators reported a dose-dependent retardation of seizure acquisition.

Sierra Neuropharmaceuticals is evaluating another approach to implantable drug delivery for the treatment of various brain disorders, including epilepsy. Their SNP001 system provides direct delivery of highly concentrated, reformulated drugs into the CSF via an implantable pump. As currently configured, the reservoir must be refilled every 3 months. The company has initiated preliminary studies with felbamate, a highly effective drug that has a relatively high risk of hepatic and bone marrow toxicities (http://www.epilepsy.com/pdfs/pipeline_abrams.pdf). Direct delivery into the CSF reduces systemic exposure and thus lowers the potential for serious adverse effects.

Implantable drug-device systems offer new ways to study AED pharmacology in animal models of epilepsy by localizing drug exposure to the CNS and precisely controlling the rate and amount of drug dose. Although implantable devices may not be useful as initial therapy, they may be beneficial in treating patients with severe, refractory epilepsy not amenable to surgery.

## INJECTABLES

Development of modified-release, injectable drug delivery systems suitable for intravenous, intramuscular, and subcutaneous administration requires the use of biocompatible components in matrices because the matrix component of the dosage form remains in the body for hours to days. Many injectable modified-release drug delivery platforms are currently being developed that provide much greater control over the rate and pattern of drug release. These delivery platforms have potential applications in both laboratory research and clinical practice.

The SABER™ Delivery System is an injectable, biodegradable, delivery system that uses a high-viscosity carrier such as sucrose acetate isobutyrate (SAIB) and one or more pharmaceutically acceptable additives. The drug delivered by SABER™ is dissolved or dispersed in a relatively low-viscosity liquid formulation. Upon IM injection, the carrier materials diffuse away, leaving a viscous, but biocompatible, depot from which the drug is released at a controlled rate over periods of a few days to 3 months or more. Both water-soluble and -insoluble drugs of small and large molecules can be formulated, sterilized, and released from SABER™ depot injection systems.

Preclinical studies in rats and rabbits have shown the usefulness of SABER™ as a sustained drug delivery platform for the local and systemic delivery of proteins and peptides (Keskin et al., 2005; Okumu et al., 2002; Pechenov et al., 2004) and for the local delivery of drugs to the CNS (Lee et al., 2006). DURECT (Cupertino, CA) and Nycomed (Denmark) are studying POSIDUR™, a SABER™ bupivacaine formulation, in a Phase II clinical study. The product is designed to treat and control local, postoperative pain by providing continuous 48- to 72-hour delivery of bupivacaine upon instillation into a surgical site. The product has exhibited good tolerability in over 200 patients following exposure to 1.25 to 7.5 mL of POSIDUR™ formulation or placebo vehicle (SAIB/BA) via subcutaneous injections or wound instillations. Dose-linear pharmacokinetics with a rapid onset of delivery and sustained plasma levels for 48 to 72 hours without any drug dumping or drug burst have been observed. DURECT has also conducted additional studies to support the use of SAIB in delivery systems for parenteral and oral applications (e.g., the ORADUR™ oral delivery platform) (Wright et al., 2008).

A drug delivery platform developed by MacroMed, Inc. (recently acquired by Protherics, PLC) is the ReGel® system, which is based on the unique reversible thermal gelation property (sol–gel) of poloxamers. ReGel® significantly increases both the solubility (up to 2000-fold) and stability of hydrophobic drugs, such as paclitaxel and cyclosporine A (Rathi and Fowers, 2008). The ReGel® polymeric drug delivery system has exhibited broad formulation capabilities and has been successfully developed for hydrophobic small molecules, peptides, and proteins. OncoGel™, paclitaxel containing the ReGel® system for local tumor management, is currently in a Phase II clinical trial in the United States and Europe.

DepoFoam® (Pacira Pharmaceuticals; San Diego, CA) is a controlled-release system based on an aqueous suspension of multivesicular liposomes (MVLs) (Kim et al., 1983). DepoFoam® is considered one of the better tolerated, sustained-release injectable formulations (Lambert and Los, 2008). With the exception of intravenous administration, the system can be used for subcutaneous (Katre et al., 1998), intranperitoneal (Bonetti and Kim, 1993; Chatelut et al., 1994), intrathecal, epidural (Viscusi et al., 2005), subconjunctival (Assil and Weinreb, 1987), and intraocular (Howell, 2001) administration. The commercially marketed products using DepoFoam® technologies include DepoCyt®, a sustained-release formulation of cytarabine, and DepoDur™, which provides controlled delivery of morphine for up to 48 hours for epidural pain management following surgeries.

The ALZAMER® Depot™ technology (ALZA Corp.; Mountain View, CA) offers sustained delivery of therapeutic agents in bioerodible dosage forms for durations ranging from days to months with minimal initial drug burst. This technology is primarily used for IM and SQ injections. A particularly important feature of ALZAMER® Depot™ is that the amount of drug in a dose can be as high as 200 to 300 mg/mL. Furthermore, the manufacturing procedure is much simpler than those used in microspheres technologies, which can load only 20 to 60 mg/mL into a delivery system (Chen and Junnarkar, 2008). The system consists of a biodegradable polymer, solvent, and formulated drug particles. In the early phase of absorption, drug release primarily occurs by diffusion followed by slower release due to polymer degradation. ALZAMER® Depot™ technology was initially designed for delivery of therapeutic proteins but is now used to administer peptides and small molecules (Brodbeck et al., 1999; Chen and Junnarkar, 2008; Chen et al., 2003; Jeong et al., 2000). More than 100 different ALZAMER® Depot™ formulations have been tested *in vivo* in animals with no evident adverse events or signs of systemic toxicity. Histopathological evaluations of injection sites have shown simple fibrosis or mild granulomatous inflammation, a response typical to the presence of foreign bodies in tissue (Bannister et al., 2000). The ALZAMER® Depot™ technology, with its flexible drug release profiles and wide array of formulation capabilities, appears to be a particularly promising platform for the sustained delivery of small molecules, peptides, proteins, and other biomolecules.

Cyclodextrins are cyclic oligosaccharides with hydroxyl groups on the outer surface and a lipophilic cavity in the center. This structural feature gives cyclodextrins good water solubility while permitting the formation of inclusion complexes with a wide variety of hydrophobic molecules. The result is a substantial increase in the water solubility of poorly water soluble drugs and, in some cases, improved stability. The commercially available cyclodextrins have no biologic effects at approved doses and are eliminated within a few hours via glomerular filtration. Upon intravenous injection, the drug is very rapidly released (within seconds) from the inclusion complex. The cyclodextrins are largely inert and are eliminated by renal excretion. An example of its use in AED research and therapy is carbamazepine, which is essentially insoluble in water. Cloyd et al. (2007)

created an injectable carbamazepine formulation (10 mg/mL) with 2-hydroxypropyl-β-cyclodextrin (2-HP-β-CD) as the solubilizing agent in order to rigorously investigate the effects of age, sex, and race on disposition of the drug in patients with epilepsy. Building on the research from Cloyd et al., Lundbeck is developing a commercial carbamazepine formulation with a related cyclodextrin, sulfobutylether 7-β-cyclodextrin (Captisol®). Preliminary results were recently reported on 48 patients given the Captisol®-enabled carbamazepine formulation at 70% of their daily dose administered intravenously in equal amounts every 6 hours as either 15- or 30-minute infusions. The formulation maintained carbamazepine exposures (AUC) and trough carbamazepine concentrations ($C_{min}$) that were comparable to steady-state exposure following oral administration (Walzer et al., 2008). No clinically important CNS or systemic adverse effects occurred despite transiently elevated carbamazepine concentrations at the end of the infusion.

The novel, injectable drug delivery technologies described in this section have the potential to enhance and expand the use of parenteral AEDs. The ability to solubilize drugs and control release rates expands the ability of researchers to study the efficacy and toxicity of AEDs in animal models of epilepsy through greater control of drug delivery. At the same time, these technologies have the potential of increasing the number of AEDs with injectable formulations that offer greater safety, better tolerability, and targeted delivery. Thus far, it appears that only the cyclodextrin platform is being used in the development of injectable AEDs.

## ORAL DELIVERY TECHNOLOGIES

Oral delivery is, by far, the most common route of drug administration; however, conventional capsule and tablet formulations possess several shortcomings, such as immediate release, poor or inconsistent bioavailability, undesirable release patterns, and gastrointestinal irritation, that limit realization of the full benefit of the drugs and, in some cases, terminate the development of promising new medications. Innovations in oral delivery systems can overcome many of these problems and enhance the benefit of drug therapy. Commercialization and subsequent clinical use of new technologies can be slowed by regulatory hurdles that require evidence of the safety and efficacy of not only the drug, but also the innovative delivery system containing the drug. Sufficient market demand and interest from regulators in expanding the use of novel drug delivery systems justify industry investments in this field. The published FDA guidelines for the development and evaluation of extended, sustained, and modified-release dosage forms (CDER, 1997a,b), and the proceedings of an American Association of Pharmaceutical Scientists/Food and Drug Administration (AAPS/FDA) workshop discussing the scientific foundations for regulating drug product quality (Lee, 1997) provide an excellent overview in this field, including the need to establish appropriate *in vivo/in vitro* correlations (Walker, 2008).

Approaches to modified oral drug delivery intended to increase solubility, control absorption patterns, and enhance bioavailability rely on the use of such basic concepts as: (1) insoluble, slowly eroding or swelling matrices; (2) polymer-coated tablets, pellets, or granules; (3) osmotically driven systems; (4) systems controlled by ion-exchange mechanisms; and (5) various combinations of these approaches. Modified drug release technologies currently used can be broadly classified into matrix, membrane, and hybrid systems (Lee, 1997). Matrix systems are relatively simple to manufacture and can release the drug in a number of ways, such as by diffusion, dissolution, swelling, or erosion and by changes in geometry of the solid dosage form. Membrane systems release a drug by such mechanisms as osmotic pumping or solution diffusion and include coated pellets or beads, microcapsules or coated capsules, and tablets. Hybrid systems incorporate both membrane and matrix systems and, though technologically challenging to manufacture, provide drug delivery characteristics not possible with either individual system (Walker, 2008).

Carbamazepine, lamotrigine, levetiracetam, phenytoin, and valproic acid are the only AEDs marketed in modified-release dosage forms in addition to immediate-release tablets or capsules. Many of the newer AEDs (lacosamide, levetiracetam, pregabalin, and MHD, the active metabolite of oxcarbazepine) have relatively short half-lives ($t_{1/2}$ < 10 hours) and are absorbed throughout much of the gastrointestinal track (Bialer, 2007). Such characteristics make these drugs good candidates for modified-release delivery systems. Extended-release formulations of oxcarbazepine and topiramate are under clinical development, and that of lamotrigine has been recently approved by the FDA.

## MATRIX SYSTEMS

Matrix systems contain dissolved or dispersed drug within a solid polymer dosage form that swells or slowly erodes when swallowed. The rate of water penetration into the matrix followed by drug diffusion into the surrounding medium, erosion of the matrix, or a combination of the two, governs drug release from these systems. New matrix-based technologies are emerging with innovations in manufacturing and formulation methods. A brief description of the recently patented matrix-based technologies is provided here.

TIMERx® technology incorporates a combination of hydrophilic, xanthan, and locust bean gums together with a saccharide component to form a rate-controlling matrix system. This creates a viscous release-retarding gel upon hydration, capable of delivering drugs with varying physical and chemical characteristics. Also, the TIMERx® platform can accommodate a wide range of doses (4 to 850 mg) and releases drug with first-order, zero-order, and pulsatile release patterns (Staniforth and Baichwal, 1993). One attraction of this technology is the ability to select delivery profiles that coincide with a patient's greatest risk of seizures during a 24-hour period (e.g., seizures upon awakening). Guar gum is utilized in a similar technology, MASR™/COSR™, in which the gum functions as a rate-controlling polymer. The tablets made using this technology can also be film coated, with the net effect of combining the polymer and film coating creating a more precise drug release profile (Altaf and Friend, 2003). Contramid® incorporates a biodegradable, cross-linked, high-amylase starch as the rate-controlling tablet matrix (Lenaerts et al., 1998). As with the other matrix systems, Contramid® is capable of delivering drugs with varying release rates.

Time-dependent changes in the diffusion path length in the gastrointestinal tract and absorptive surface area can alter the zero-order drug release kinetics intended for most oral dosage forms. Systems have been designed to overcome this limitation by modifying the matrix geometry such that the area available for diffusion increases with time to compensate for the increase in diffusional distance (Charman and Charman, 2003). Procise®, developed in 1991 by Glaxo Canada (now GlaxoSmithKline), consists of a compression-coated core whose geometric configuration controls the release profile of drugs. By varying the geometry of the core, the profile of the drug release can be adjusted to follow zero order, first order, or a combination of these orders. The system can also be designed to deliver two drugs at the same time, each having a different release profile (Chopra, 2003; Chopra and Nangia, 1993).

RingCap® (Alkermes; Cambridge, MA) combines matrix tablets and existing capsule banding processes to create solid, controlled-release, oral-dosage forms. The RingCap® system utilizes bands of insoluble polymer on a matrix tablet. The manufacturing process involves compressing the drug into cylindrical matrix tablets that are subsequently film coated. Existing capsule-banding technology is then used to apply two or more polymeric rings around the circumference of the matrix tablet. These bands lower the initial release of the drug by reducing the surface area exposed. As the matrix tablet erodes, new surface areas are created underneath and around the bands, thus increasing the rate of release. The release rate of the drug substance can be designed to remain nearly constant or even increase with time by adjusting the number and width of polymer bands (Figure 31.2) (Verma and Garg, 2001).

A number of novel, oral, controlled-release technologies have been developed by Elan Corporation. These include SODAS®, a multiparticulate delivery system providing tailored, highly flexible release profiles including controlled-release, immediate-release, and pulsatile-release of drug; IPDAS®, a controlled-release technology that reduces gastrointestinal irritation; CODAS®, which provides tailored release of drug for chronotherapeutic delivery; and PRODAS®, a multiparticulate minitab technology system that combines the benefits of basic tablet manufacturing technology within a

**FIGURE 31.2** **(See color insert following page 458.)** Schematic erosion of a RingCap® tablet. (From Rathbone, M.J. et al., Eds., *Modified-Release Drug Delivery Technology*, Marcel Dekker, New York, 2003. With permission.)

capsule (Elan, 2009). The Dual Release Drug Absorption System (DUREDAS™), which uses bilayer tabletting technology, has been specifically developed to provide two different release rates or dual release of a drug from a single dosage form. The system combines an immediate-release granulate (for rapid onset of action) and a controlled-release hydrophilic matrix complex within one tablet. The controlled-release matrix remains intact and slowly absorbs fluid from the gastrointestinal tract, causing the matrix to expand and transforming the hydrophilic polymers into a porous, viscous gel that serves as a barrier between the drug and the surrounding fluid. As the gel continues to expand, fluid penetrates further into the dosage form, dissolving the drug and allowing the resulting solution to diffuse out in a controlled manner (Verma and Garg, 2001).

A recently approved extended-release preparation of levetiracetam utilizes a dual-release mechanism. Its efficacy and tolerability were demonstrated when given once daily ($2 \times 500$ mg) as add-on therapy in patients with partial-onset seizures refractory to one and up to three antiepileptic drugs (Peltola et al., 2009). The formulation is designed to provide a pulsatile, bimodal release from a controlled-release tablet (Figure 31.3). It is comprised of immediate- and modified-release particles, where the modified-release component is an erodible-component or a diffusion-controlled formulation (both of which are matrix-based designs) or an osmotically controlled formulation (Devane et al., 2001, 2006; Jenkins et al., 2007; Rouits et al., 2009).

Several novel, matrix-based, oral technologies have been employed in order to develop a once-daily formulation of oxcarbazepine (OXC). Among these is a matrix formulation containing high-viscosity hydroxypropyl methylcellulose (HPMC). Variations in the proportion of HPMC throughout the dosage form produce different release profiles (Ramakrishnan et al., 2006). Another OXC controlled-release formulation has been developed using a combination of solubilizing agents and release-promoting agents that are pH-dependent polymers. The result is that drug is only released at relatively high pH values. These matrix polymers hydrate and swell in biological fluids, forming a homogeneous matrix structure that maintains its shape throughout the period of time when OXC is being released. The initial polymer hydration releases the drug slowly, and when the polymer matrix is completely hydrated the drug is released at a much faster rate (Bhatt et al., 2007). A third matrix

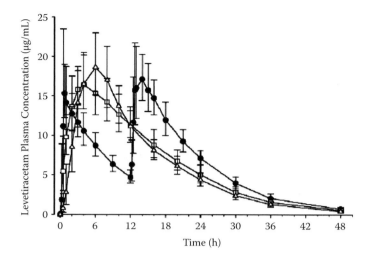

**FIGURE 31.3**   Geometric mean (S.D., $n = 24$) of plasma levetiracetam concentration profiles after single-dose intake of levetiracetam XR, 1000 mg, under fasted (open squares) or fed (open triangles) conditions, and after two doses of levetiracetam IR, 500 mg, 12 hours apart under fasting conditions (filled circles). (From Peltola, J. et al., *Epilepsia*, 50, 406–414, 2009. With permission.)

approach for making an extended-release OXC formulation is one proposed by Desitin Arzneimittel (Hamburg, Germany). This technology uses a homogeneous matrix system containing drug and polymethacrylic acid polymers. A comparative study of this formulation vs. Trileptal™ shows that the extended-release product produces a desirable plasma concentration–time profile for both OXC and MHD, the active metabolite, over longer periods of time than Trileptal™ (Franke and Lennartz, 2003). After completing a proof-of-concept study comparing several modified-release formulations with different dissolution properties, the company is now conducting a clinical trial using the lead extended-release formulation.

Another approach to manufacturing an extended-release product is represented by the DiffCORE® technology developed by GlaxoSmithKline (http://www.gsk.com/infocus/diffcore-tablets.htm). This approach involves creating holes of different number and size into coated tablets. When tablets prepared with this technology are swallowed, gastrointestinal fluid enters the tablet hole in the outer shell and penetrates the core, releasing drug. The rate of release can be modulated by the composition of the internal matrix. GSK is using DiffCORE® technology in products for various indications, including epilepsy and metabolic disorders. Products that use DiffCORE® technology are easily recognizable and look different than conventional tablets (Figure 31.4). An extended-release lamotrigine formulation using DiffCORE® technology has been developed. The product is an eroding matrix formulation that slows the dissolution rate of lamotrigine over a period of approximately 12 to 15 hours, compared to the immediate-release formulation. In addition to controlling the rate of drug release, an enteric coat delays the start of drug release until after the tablet transits through the stomach (Tompson et al., 2008; Werz, 2008).

## MEMBRANE SYSTEMS

Membrane systems contain an encapsulated drug reservoir compartment inside a polymeric membrane that acts as the rate-controlling element. The design of the membrane determines the rate at which drug is released. Multiparticulate dosage delivery systems, for example, contain a subset of membrane-based systems, usually consisting of coated or layered particles—spheres, pellets, granules, or crystals—which are then filled into capsules or compressed into tablets.

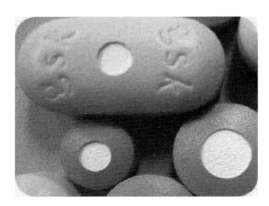

**FIGURE 31.4** **(See color insert following page 458.)** GlaxoSmithKline's DiffCORE® technology. (From http://www.gsk.com/infocus/diffcore-tablets.htm. © 2009 GlaxoSmithKline. Used with permission.)

One example of a membrane-based delivery system is the micropump technology. This approach is useful in controlling delivery of low-molecular-weight drugs that are best absorbed in the small intestine. This is the case for more than 70% of orally administered medications. The Micropump® (Flamel Technologies; Paris, France) technology involves the encapsulation of thousands of microparticles, each measuring 200 to 500 µm in diameter. The microparticles increase the surface area of the drug, thus enhancing the extent of absorption, while the thickness of the particle coating controls drug release (Figure 31.5). The Micropump® platform is well suited for applications geared toward pediatric and geriatric patients because the microparticles can be delivered in suspensions or syrups as well as solid dosage forms. Furthermore, the active ingredients are entirely enclosed in each microparticle, permitting taste masking. This technology also has the capability of delivering more than one drug simultaneously (Autant et al., 1996). Currently, two products use this technology: COREG CR® (carvedilol) and ASACARD® (long-acting aspirin). Each microparticle core contains drug as either crystals or granules and is further covered with a material that controls drug release (www.flamel.com/techAndProd/index.shtml). Two versions of this technology are currently available: Micropump® I for continuous sustained-release of a drug and Micropump® II for delayed release at specific segments of the intestine (Walker, 2008).

Carbatrol®, a carbamazepine extended-release formulation, uses Microtrol® (Shire Pharmaceuticals; Wayne, PA), a technology similar to Micropump®. The delivery system is comprised of immediate-release, extended-release, and enteric-release beads in a fixed ratio (25:40:35) that provide a relatively stable concentration–time profile when drug is taken every 12 hours. The coating on the individual microparticles defines the release pattern of the drug from the dosage form. This extended-release formulation can be administered either orally as an intact capsule or by opening the capsule and sprinkling the contents on food or into a liquid.

## OSMOTIC PRESSURE-BASED TECHNOLOGIES

Osmotic-based delivery systems use osmosis, the movement of water through a membrane, to control drug delivery within the gastrointestinal (GI) tract. This technology consists of a drug-containing core surrounded by a semipermeable membrane with orifices for drug release. In the aqueous environment of the GI tract, water is drawn by osmosis across the semipermeable membrane into the core at a controlled rate dictated by the composition and thickness of the membrane. Drug in solution or suspension is released through the orifices at the same rate that water enters, depending on the dosage form. The OROS® Push-Pull™ (ALZA; Mountain View, CA) system uses a multicompartment core to deliver drugs of any solubility. The basic OROS® Push-Pull™ system resembles a simple tablet in shape and has two layers. The first layer contains the drug substance

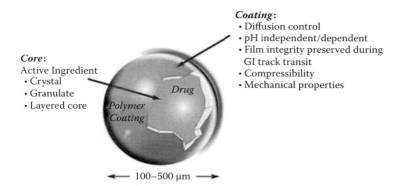

**Core:**
Active Ingredient
• Crystal
• Granulate
• Layered core

*Drug*

*Polymer Coating*

**Coating:**
• Diffusion control
• pH independent/dependent
• Film integrity preserved during GI track transit
• Compressibility
• Mechanical properties

◄── 100–500 µm ──►

**FIGURE 31.5** **(See color insert following page 458.)** Encapsulated microparticles used in micropump technology. (From http://www.flamel.com/techAndProd/micropump.shtml.)

(osmotically active hydrophilic polymers) and the other excipients. The second layer, or push layer, contains a hydrophilic expansion polymer and other osmotically active agents and tablet excipients, as shown in Figure 31.6A. To assist in the transport of drug, the push layer expands, acting like a piston to gently push the drug suspension or solution out through the orifice. Adjustments to the composition and thickness of the system's semipermeable membrane are made to achieve a precise delivery rate that is independent of GI pH and external agitation.

Variations in the Push-Pull™ system include the controlled-onset, extended-release (COER) design and the capsule-shaped tablet, which allows patterned delivery of one or more drugs (examples include ascending and pulsed delivery). The capsule-shaped tablet consists of the multicompartment core, semipermeable membrane, and delivery orifice of the basic Push-Pull™ system; it uses the same osmotic principles, but the capsule shape allows more versatility in patterned delivery. (Figure 31.7B). The Zer-Os tablet technology (Advanced Drug Delivery Technologies AG; Reinach, Switzerland) is an osmotic system developed specifically for the delivery of lipophilic compounds. The tablet consists mainly of a core of poorly water-soluble drug along with gel-forming agents and standard excipients. The gel-forming agent, after coming in contact with water, forms a gel of a defined viscosity, resulting in a suspension that is pushed out of the orifice at a controlled rate. Tegretol-XR® (Novartis Pharmaceuticals; Basel, Switzerland), a product on the U.S. market, is based on this technology (Verma and Garg, 2001).

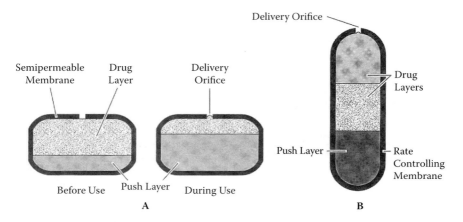

Delivery Orifice

Semipermeable Membrane    Drug Layer    Delivery Orifice

Drug Layers

Push Layer

Rate Controlling Membrane

Before Use    Push Layer    During Use

A

B

**FIGURE 31.6** (A) Cross-section of the OROS® Push-Pull™ bilayer system before and during use, and (B) trilayer capsule-shaped tablet with two drug layers. (From Rathbone, M.J. et al., Eds., *Modified-Release Drug Delivery Technology*, Marcel Dekker, New York, 2003. With permission.)

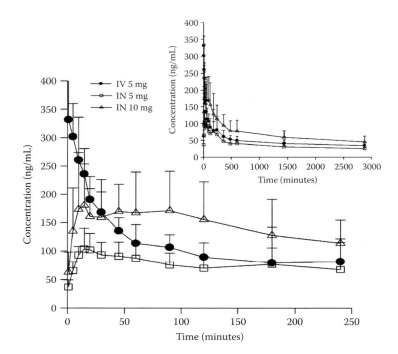

**FIGURE 31.7** Mean plasma diazepam-concentration time profiles after intravenous and intranasal administration in eight subjects (0 to 4 hours). Inset shows the complete profile (0 to 48 hours). (From Rouits, E. et al., *Epilepsy Res.*, 84, 224–231, 2009. With permission.)

## TECHNOLOGIES TO OVERCOME POOR SOLUBILITY

Poor aqueous solubility of drugs poses a particular challenge in the development of new therapies for CNS disorders. Typically, such drugs must be sufficiently lipid soluble to cross the blood–brain barrier. That characteristic usually means that the compound will have poor water solubility. The development of efficient delivery systems for poorly soluble drugs is usually achieved by the use of salts, complexation with cyclodextrin derivatives as solubilizing and stabilizing agents, or techniques such as micronization of drug particles.

The formulation of active drug in an amorphous form is one approach to overcome the dissolution limitation of poorly soluble drugs. The Threeform (Lek Pharmaceuticals; Wilminton, DE) delivery system employs a design in which the amorphous drug is first stabilized by a mixture of surfactant and polymers and is then incorporated into a polymeric matrix that is coated with a rate-controlling film (Kerc, 2003; Kerc et al., 2000).

A reduction in particle size through the preparation of nanoparticles or microparticles is another approach to increase drug solubility. In the DissoCube® technology, high-pressure homogenization is used to form a nanosuspension, which facilitates drug dissolution and increases the saturation solubility. The resulting nanosuspension can then be incorporated into conventional dosage forms to produce tablets or capsules with improved dissolution of active drug. The DissoCube® technology combines the advantages of using particle size reduction with the benefits of the precipitation technique allowing the induction of structural changes resulting in an increased amorphous fraction (Muller et al., 1999, 2003). NanoCrystal® (Elan Corporation; Dublin, Ireland) technology incorporates a proprietary milling technique whereby agglomeration of the resulting particles is prevented by surface adsorption of generally regarded as safe (GRAS) stabilizers (Charman and Charman, 2003). A brief description of NanoCrystal® technology is provided in the section on intranasal drug delivery.

## OTHER NOVEL TECHNOLOGIES

The application of concepts from related fields of technology to the design of modified-release dosage forms for drugs is on the rise. TheriForm™ (Therics, Inc.; Princeton, NJ) is a novel drug delivery device based on three-dimensional printing (3DP) technology that uses a layer-by-layer approach to accurately control the microstructure of the dosage form to provide complex release profiles (Cima, 1993; Kumar et al., 1999). Yet another technology that uses the concept of electrostatic deposition, similar to that used in copy machines for the precise deposition of ink particles, is the Accudep™ technology (Delsys Pharmaceuticals; Monmouth, NJ). The precise and patterned deposition in combination with erosional polymeric film barriers is designed to achieve sustained or pulsatile release profiles of active materials (Chrai et al., 1998). Delsys is applying its proprietary Accudep™ electrostatic deposition technology to overcome the problems of content uniformity usually observed with low-dose, potent drug products. The Accudep™ process involves electrostatically charging dry drug powders followed by deposition onto pharmaceutically acceptable polymeric substrate films. The sites are then sealed with cover films and cut into circular units called Accudep™ cores. This process can deposit as little as a few micrograms of drug powder. The LeQtracoat® and LeQtradose® (Phoqus Pharmaceuticals, Ltd.; West Malling, Kent, U.K.) technologies also utilize dry-powder electrostatic deposition to enable tablet coating and precise drug loading (Charman and Charman, 2003).

A drug delivery system has been developed to address the needs of special populations such as pediatrics, geriatrics, or patients who have difficulty in swallowing or who are unconscious. This is the oral thin-film technology, whose primary advantage is that it does not require any liquids or dose measuring devices to deliver a precise dose. The dosage form is a thin film impregnated with drug that rapidly disintegrates upon contact with saliva in the oral cavity. Another system is the sip system or dose sipping technology (DST), in which granules for suspension are retained in a drinking straw without the need for reconstitution prior to administration. Each individually packed straw is loaded with a single dose of small coated granules containing the prescribed amount of drug in a suitably taste-masked formulation. Drug delivery via DST occurs by swallowing between 50 and 150 mL of fluid either using the straw or directly from the glass following ingestion of the entire contents of the straw. The two commercially available dose sipping technologies are Clarosip®, containing clarithromycin, and Riclasip®, containing amoxicillin/clavulanic acid (Chrai et al., 1998).

Another novel drug delivery technology is the convection-enhanced delivery (CED) system designed to bypass the blood–brain barrier. This technique uses positive hydrostatic pressure to deliver a fluid containing a subtherapeutic substance by bulk flow directly into the interstitial space within a localized region of the brain parenchyma. For the delivery of AEDs, this technology would allow treatment that avoids systemic toxicity of orally administered drugs. A more detailed description of the principles behind this delivery system has been provided by Rogawski (2009).

Collectively, advances in oral drug delivery technologies offer the potential to overcome problems with poor and inconsistent bioavailability and the ability to control the rate and timing of drug delivery to the systemic circulation. These features are especially valuable in AED therapy, where unintended alterations in drug exposure are a common cause of seizure control loss.

## NASAL DRUG DELIVERY

The nasal route offers a good alternative to oral and parenteral drug administration. Advantages of this method include: (1) avoidance of gut wall and hepatic first-pass metabolism, (2) relatively rapid absorption, (3) convenient and easy drug administration, and (4) local, systemic, and, in some cases, direct drug delivery to the CNS. Intranasal administration also presents some significant limitations, such as: (1) most drug molecules diffuse poorly and slowly through the nasal mucosa;

(2) administration volume must be small, usually a maximum of 150 μL per nostril; and (3) nasal mucosa are sensitive to irritants. If a greater volume is administered, the drug solution may drain into the pharynx and be swallowed.

The most obvious indication for intranasal therapy in epilepsy is the treatment of seizure emergencies outside a hospital setting or in a medical center when there is a delay in gaining intravenous access. Benzodiazepines are considered the treatment of choice for seizure emergencies when given IV or rectally. Recently, development of intranasal formulations of midazolam, lorazepam, clonazepam, and diazepam has begun.

One of the biggest challenges to overcome in the formulation of nasal benzodiazepine products is achieving high drug concentrations in small volumes. This is necessitated by the limited capacity of the nasal cavity to hold fluids. The absence of or minimal nasal irritation is another property required of these formulations. Use of an organic solvent, such as glycol, is often necessary in order to achieve the highly concentrated solutions required of nasal products. Alternatives employing methods that cause less irritation are much more desirable. Nonetheless, patients may accept a trade-off between the mild to moderate transient irritation and the benefit of preventing or terminating a seizure (Wermeling, 2009; Wermeling et al., 2009).

Midazolam is, by far, the most extensively studied of all benzodiazepines for intranasal administration. Several open-label studies using injectable midazolam (5 mg/mL) have described rapid onset of seizures cessation; however, bioavailability using the injectable solution intranasally is moderate and highly variable. Also, most investigators report nasal irritation, tearing, and a raw throat sensation immediately after administration, most likely due to a large volume (1 to 2 mL) and low pH (3 to 4) of the administered solution (Bhattacharyya et al., 2006; Fisgin et al., 2002; Harbord et al., 2004; Lahat et al., 2000; Mahmoudian and Zadeh, 2004; Wermeling et al., 2009; Wilson et al., 2004; Wolfe and Macfarlane, 2006). Alternative intranasal midazolam formulations providing better tolerability as well as rapid and improved absorption have been studied (Dale et al., 2006; Gudmundsdottir et al., 2001; Knoester et al., 2002; Loftsson et al., 2001; Wermeling et al., 2006, 2009). The ingredients used in formulating these solutions include cosolvent combinations of polyethylene glycol, propylene glycol, and butylated hydroxytoluene (Knoester et al., 2002; Wermeling et al., 2006, 2009) and cyclodextrin-based solutions (Gudmundsdottir et al., 2001; Loftsson et al., 2001). Midazolam has a relatively fast elimination half-life especially in patients taking enzyme-inducing drugs (0.5 to 2 hours). The rapid elimination of midazolam has the potential to reduce the duration of its effect. A highly concentrated intranasal midazolam formulation utilizing an organic solvent is undergoing a clinical trial (www.clinicaltrials.gov).

Very few studies have investigated different formulation strategies for intranasal diazepam. Some of these include glycol cosolvent systems similar to those used for midazolam (Gizurarson et al., 1999; Lau and Slattery, 1989; Lindhardt et al., 2002). Cloyd et al. (2007) developed a supersaturated solution of diazepam that achieves a 40-mg/mL drug concentration in a 60/40% (v/v) cosolvent mixture of glycofurol and water. This formulation was designed to maximize diazepam concentration, thus allowing administration of small fluid volumes (≤200 μL). The addition of water in the solvent system improves the tolerability while increasing the driving force of permeation across the nasal mucosa (Hou and Siegel, 2006; Ivaturi et al., 2009). The pharmacokinetic profile of this glycofurol-based formulation studied at 5- and 10-mg doses is shown in Figure 31.7 (Ivaturi et al., 2009).

A microemulsion formulation of diazepam has been studied in animals (Choi and Kim, 2004; Li et al., 2002). Microemulsions are clear, stable, isotropic liquid mixtures of oil, water, and surfactant, frequently combined with a cosurfactant. They are employed in drug delivery systems because they are able to substantially increase the solubility of hydrophobic drugs such as benzodiazepines as well as enhance permeation of topical and systemic drug delivery. The physical properties that make microemulsions attractive for drug delivery are their transparent nature, which makes the product aesthetically pleasing and provides easier visualization of contamination. The small droplet size provides a large surface area for rapid drug release and greater bioavailability, enabling a reduction in dose and more consistent drug absorption profiles. Surfactants and cosurfactants in microemulsions can also act as permeation enhancers of the drug across biological membranes which is highly desirable in nasal drug delivery to achieve rapid blood levels (Lawrence and Warisnoicharoen, 2006).

A recently patented intranasal clonazepam formulation uses a binary solvent mixture comprised of diethylene glycol monoethyl ether, triacetin, glycofurol, and propylene glycol in varying proportions (Jamieson et al., 2007). A controlled trial using this formulation in patients with epileptic seizures is underway (www.clinicaltrials.gov).

Amarin is using Elan's proprietary NanoCrystal® technology as the basis for an intranasal lorazepam formulation. NanoCrystals are small particles of drug substance less than 2000 nm in diameter that are prepared by special milling techniques; their stability against agglomeration is maintained by the use of surface adsorption agents. The final product is a dispersion of the drug substance that behaves like a solution but can be processed into a finished dosage form for any route of administration as either a solid or liquid (Elan, 2009).

The nasal cavity has a limited capacity to retain a solution without drainage either anteriorly out the nose or posteriorly to the throat. One of the strategies to overcome this issue is the development of bioadhesive formulations that have a longer residence time in the nasal cavity. Copolymers of polyethylene oxide (PEO), polypropylene oxide (PPO), and polyacrylic acid (PAA) are mucoadhesive and gel at the temperature of human nasal mucosa due to the formation of micellar aggregates acting as cross-links. Formulations using this technology are usually sprayed or injected as a liquid into the nasal cavity; they immediately gel after coming in contact with the nasal mucosa (Bromberg, 2003). Zhou and Donovan (1996) used a rat model to study the putative effects of bioadhesive polymer gels on slowing the nasal mucociliary clearance. Their results showed that all bioadhesive formulations decreased the intranasal mucociliary clearance and resulted in an increased residence time of the formulation in the nasal cavity.

Innovative delivery technologies are being employed to develop new therapies for seizure emergencies. These technologies have overcome problems with drug solubility, nasal irritation, and rate of absorption. As a result, patients with refractory epilepsy will very likely have available better ways of managing their condition that will improve quality of life and lower health care costs.

## CONCLUSION

Technological advances now provide researchers with new ways to administer drugs, control the rate and pattern of absorption, and target drug delivery. These advances will allow researchers to better investigate AED pharmacology in animals as well as facilitate the commercialization of new AED products by improving drug solubility and stability, decreasing the dosing frequency, controlling and defining oscillations in drug concentration, and expanding the routes of drug administration to address unmet clinical needs.

## REFERENCES

Altaf, S.A., D.R. Friend. (2003). MASRx and COSRx sustained-release technology. In *Modified-Release Drug Delivery Technology* (pp. 21–33), M.J. Rathbone, M.S. Roberts, and J. Hadgraft, Eds. New York: Marcel Dekker.

Assil, K.K., R.N. Weinreb. (1987). Multivesicular liposomes: sustained release of the antimetabolite cytarabine in the eye. *Arch. Ophthalmol.*, 105:400–403.

Autant, P., J.P. Selles, G. Soula. (1996). Microcapsules Medicamenteuses et/ou Nutritionnelles pour Administration per os, French Patent No. FP2725623.

Bannister, R., K. Baudouin, E. Maze. (2000). Biological activity, pharmacokinetics, and safety assessment of human growth hormone (hgH) delivered via a subcutaneous depot. *Toxicol. Sci.*, 54(1):407.

Bhatt, P.P., A. Kidane, K. Edwards. (2007). Modified-Release Preparations Containing Oxcarbazepine and Derivatives Thereof, U.S. Patent No. 20070254033.

Bhattacharyya, M., V. Kalra, S. Gulati. (2006). Intranasal midazolam vs. rectal diazepam in acute childhood seizures. *Pediatr. Neurol.*, 34:355–359.

Bialer, M. (2007). Extended-release formulations for the treatment of epilepsy. *CNS Drugs*, 21:765–774.

Bonetti, A., S. Kim. (1993). Pharmacokinetics of an extended-release human interferon alpha-2b formulation. *Cancer Chemother. Pharmacol.*, 33:258–2561.

Brodbeck, K.J., S. Pushpala, A.J. McHugh. (1999). Sustained release of human growth hormone from PLGA solution depots. *Pharm. Res.*, 16:1825–1829.

Bromberg, L. (2003). Poly(ethylene oxide)-*b*-poly(propylene oxide)-*b*-poly(ethylene oxide)-*g*-poly(acrylic acid) copolymers as *in situ* gelling vehicle for nasal drug delivery. In *Modified-Release Drug Delivery Technology* (pp. 749–758), M.J. Rathbone, M.S. Roberts, and J. Hadgraft, Eds. New York: Marcel Dekker.

CDER. (1997a). *Guidance for Industry: Extended Release Oral Dosage Forms: Development, Evaluation, and Application of In Vitro/In Vivo Correlations.* Washington, D.C.: Center for Drug Evaluation and Research, Food and Drug Administration, U.S. Department of Health and Human Services.

CDER. (1997b). *Guidance for Industry: SUPAC-MR: Modified Release Solid Oral Dosage Forms. Scale-Up and Postapproval Changes: Chemistry, Manufacturing, and Controls; In Vitro Dissolution Testing and In Vivo Bioequivalence Documentation.* Washington, D.C.: Center for Drug Evaluation and Research, Food and Drug Administration, U.S. Department of Health and Human Services.

Charman, S.A., W.N. Charman. (2003). Oral modified-release delivery systems. In *Modified-Release Drug Delivery Technology* (pp. 1–10), M.J. Rathbone, M.S. Roberts, and J. Hadgraft, Eds. New York: Marcel Dekker.

Chatelut, E., P. Suh, S. Kim. (1994). Sustained-release methotrexate for intracavity chemotherapy. *J. Pharm. Sci.*, 83:429–432.

Chen, G., G. Junnarkar. (2008). ALZAMER® Depot™ bioerodible polymer technology. In *Modified-Release Drug Delivery Technology*, 2nd ed. (pp. 215–225), M.J. Rathbone, J. Hadgraft, and M.S. Roberts, Eds. New York: Marcel Dekker.

Chen, G.H., L. Kleiner, P. Houston. (2003). Sustained-release of luperolide acetate from ALZAMER® Depot™. In *Proceedings of the 30th Annual Meeting & Exposition of the Controlled Release Society*, July 19–23, Glasgow, Scotland.

Choi, Y.M., K.H. Kim. (2004). Transnasal Microemulsions Containing Diazepam, U.S. Patent No. 2005002987.

Chopra, S.K. (2003). Procise: drug delivery systems based on geometric configuration, modified-release drug delivery technology. In *Modified-Release Drug Delivery Technology* (pp. 35–48), M.J. Rathbone, M.S. Roberts, and J. Hadgraft, Eds. New York: Marcel Dekker.

Chopra, S.K., A.K. Nangia. (1993). Controlled Release Device, European Patent No. 0542364.

Chrai, S.S., B. Singh, M. Kopcha, R. Murarl, S. Sun, N. Kumar, N. Desai, A. Levine, H. Rivenburg, G. Kaganowicz. (1998). Electrostatic dry deposition technology. *Pharm. Technol.*, 4:17–20.

Cima, M.J. (1993). Three-Dimensional Printing Technology, U.S. Patent No. 5204055.

Cleland, J.L., A. Daugherty, R. Mrsny. (2001). Emerging protein delivery methods. *Curr. Opin. Biotechnol.*, 12:212–219.

Cloyd, J.C., S. Marino, A.K. Birnbaum. (2007). Factors affecting antiepileptic drug pharmacokinetics in community-dwelling elderly. *Int. Rev. Neurobiol.*, 81:201–210.

Dale, O., T. Nilsen, T. Loftsson, H. Hjorth Tonnesen, P. Klepstad, S. Kaasa, T. Holand, P.G. Djupesland. (2006). Intranasal midazolam: a comparison of two delivery devices in human volunteers. *J. Pharm. Pharmacol.*, 58:1311–1318.

Devane, J.G., P. Stark, N.M.M. Fanning. (2001). *Multiparticulate Modified-Release Composition.* Dublin, Ireland: Elan Corporation, PLC.

Devane, J.G., P. Stark, N.M.M. Fanning. (2006). Multiparticulate Modified-Release Composition, U.S. Patent No. 20060240105.

Elan. (2009). *Elan: Technology Focus.* Dublin, Ireland: Elan Corporation, PLC.

Fisgin, T., Y. Gurer, T. Tezic, N. Senbil, P. Zorlu, C. Okuyaz, D. Akgun. (2002). Effects of intranasal midazolam and rectal diazepam on acute convulsions in children: prospective randomized study. *J. Child Neurol.*, 17:123–126.

Fisher, R.S., D.K. Chen. (2006). New routes for delivery of anti-epileptic medications. *Acta Neurol. Taiwan*, 15:225–231.

Fisher, R.S., J. Ho. (2002). Potential new methods for antiepileptic drug delivery. *CNS Drugs*, 16:579–593.

Fowler, J.E. (2001). Patient-reported experience with the Viadur 12-month leuprolide implant for prostate cancer. *Urology*, 58:430–434.

Fowler, J.E., M. Flanagan, D.M. Gleason, I.W. Klimberg, J.E. Gottesman, R. Sharifi. (2000). Evaluation of an implant that delivers leuprolide for 1 year for the palliative treatment of prostate cancer. *Urology*, 55:639–642.

Franke, H., P. Lennartz. (2003). Pharmaceutical Composition, Containing Oxcarbazepine with Sustained-Release of an Active Ingredient, U.S. Patent No. 20040142033.

Gizurarson, S., F.K. Gudbrandsson, H. Jonsson, E. Bechgaard. (1999). Intranasal administration of diazepam aiming at the treatment of acute seizures: clinical trials in healthy volunteers. *Biol. Pharm. Bull.*, 22:425–427.

Gudmundsdottir, H., J.F. Sigurjonsdottir, M. Masson, O. Fjalldal, E. Stefansson, T. Loftsson. (2001). Intranasal administration of midazolam in a cyclodextrin based formulation: bioavailability and clinical evaluation in humans. *Pharmazie*, 56(12):963–966.

Harbord, M.G., N.E. Kyrkou, M.R. Kyrkou, D. Kay, K.P. Coulthard. (2004). Use of intranasal midazolam to treat acute seizures in paediatric community settings. *J. Paediatr. Child Health*, 40:556–558.

Hou, H., R.A. Siegel. (2006). Enhanced permeation of diazepam through artificial membranes from supersaturated solutions. *J. Pharm. Sci.*, 95:896–905.

Howell, S.B. (2001). Clinical applications of a novel sustained-release injectable drug delivery system: DepoFoam technology. *Cancer J.*, 7:219–227.

Ivaturi, V.D., J.R. Riss, R.L. Kriel, R.A. Siegel, J.C. Cloyd. (2009). Bioavailability and tolerability of intranasal diazepam in healthy adult volunteers. *Epilepsy Res.*, 84:120–126.

Jamieson, G., M.D. Jardin, C. Allphin. (2007). Pharmaceutical Compositions of Clonazepam and Method of Use Thereof, U.S. Patent No. 20080076761.

Jenkins, S.A., G. Liversidge, Elan Pharma International. (2007). Controlled Release Compositions Comprising Levetiracetam, World Intellectual Property Organization No. WO/2006/088864.

Jeong, B., Y.H. Bae, S.W. Kim. (2000). *In situ* gelation of PEG–PLGA–PEG triblock copolymer aqueous solutions and degradation thereof. *J. Biomed. Mater. Res.*, 50:171–177.

Katre, N.V., J. Asherman, H. Schaefer, M. Hora. (1998). Multivesicular liposome (DepoFoam) technology for the sustained delivery of insulin-like growth factor-I (IGF-I). *J. Pharm. Sci.*, 87:1341–1346.

Kerc, J. (2003). Three-phase pharmaceutical form—Threeform—with controlled release of amorphous active ingredient for once-daily administration. In *Modified-Release Drug Delivery Technology* (pp. 115–123), M.J. Rathbone, M.S. Roberts, and J. Hadgraft, Eds. New York: Marcel Dekker.

Kerc, J., L.B. Rebic, B. Kofler. (2000). Three-Phase Pharmaceutical Form with Constant and Controlled Release of Amorphous Active Ingredient for Single Daily Application. U.S. Patent No. 6042847.

Keskin, D.S., A. Tezcaner, P. Korkusuz, F. Korkusuz, V. Hasirci. (2005). Collagen–chondroitin sulfate-based PLLA-SAIB-coated rhBMP-2 delivery system for bone repair. *Biomaterials*, 26(18):4023–4034.

Kim, S., M.S. Turker, E.Y. Chi, S. Sela, G.M. Martin. (1983). Preparation of multivesicular liposomes. *Biochim. Biophys. Acta*, 728:339–348.

Knoester, P.D., D.M. Jonker, R.T. Van Der Hoeven, T.A. Vermeij, P.M. Edelbroek, G.J. Brekelmans, G.J. de Haan. (2002). Pharmacokinetics and pharmacodynamics of midazolam administered as a concentrated intranasal spray: a study in healthy volunteers. *Br. J. Clin. Pharmacol.*, 53:501–507.

Kokaia, M., P. Aebischer, E. Elmer, J. Bengzon, P. Kalen, Z. Kokaia, O. Lindvall. (1994). Seizure suppression in kindling epilepsy by intracerebral implants of GABA- but not by noradrenaline-releasing polymer matrices. *Exp. Brain Res.*, 100:385–394.

Kumar, S., et al. (1999). *In vivo/in vitro* correlation of ethinyl estradiol release from biodegradable implants fabricated by Theriform process. In *Proceedings of the AAPS Annual Meeting and Exposition*, November 14–18, New Orleans, LA.

Lahat, E., M. Goldman, J. Barr, T. Bistritzer, M. Berkovitch. (2000). Comparison of intranasal midazolam with intravenous diazepam for treating febrile seizures in children: prospective randomised study. *BMJ*, 321:83–86.

Lambert, W.J., K. Los. (2008). DepoFoam® multivesicular liposomes for the sustained release of macromolecules. In *Modified-Release Drug Delivery Technology*, 2nd ed. (pp. 207–214), M.J. Rathbone, J. Hadgraft, and M.S. Roberts, Eds. New York: Marcel Dekker.

Lane, M.E., F.W. Okumu, P. Balasubramanian. (2008). Injections and implants. In *Modified-Release Drug Delivery Technology*, 2nd ed. (pp. 123–131), M.J. Rathbone, J. Hadgraft, and M.S. Roberts, Eds. New York: Marcel Dekker.

Lau, S., J. Slattery. (1989). Absorption of diazepam and lorazepam following intranasal administration. *Int. J. Pharm.*, 54:171–174.

Lawrence, M.J., W. Warisnoicharoen. (2006). Recent advances in microemulsions as drug delivery vehicles. In *Nanoparticulates as Drug Carriers* (pp. 125–171), V.P. Torchilin, Ed. London: Imperial College Press.

Lee, J., G.I. Jallo, M.B. Penno, K.L. Gabrielson, G.D. Young, R.M. Johnson, E.M. Gillis, C. Rampersaud, B.S. Carson, M. Guarnieri. (2006). Intracranial drug-delivery scaffolds: biocompatibility evaluation of sucrose acetate isobutyrate gels. *Toxicol. Appl. Pharmacol.*, 215:64–70.

Lee, P.I. (1997). Oral ER technology: mechanism of release. In *Scientific Foundations for Regulating Drug Product Quality* (pp. 222–231), G.L. Amidon, J.R. Robinson, and R.L. Williams. Eds. Alexandria, VA: AAPS Press.

Lenaerts, V., I. Moussa, Y. Dumoulin, F. Mebsout, F. Chouinard, P. Szabo, M.A. Mateescu, L. Cartilier, R. Marchessault. (1998). Cross-linked high amylose starch for controlled release of drugs: recent advances. *J. Control Release*, 53:225–234.

Li, L., I. Nandi, K.H. Kim. (2002). Development of an ethyl laurate-based microemulsion for rapid-onset intranasal delivery of diazepam. *Int. J. Pharm.*, 237:77–85.

Lindhardt, K., D.R. Olafsson, S. Gizurarson, E. Bechgaard. (2002). Intranasal bioavailability of diazepam in sheep correlated to rabbit and man. *Int. J. Pharm.*, 231:67–72.

Loftsson, T., H. Gudmundsdottir, J.F. Sigurjonsdottir, H.H. Sigurdsson, S.D. Sigfusson, M. Masson, E. Stefansson. (2001). Cyclodextrin solubilization of benzodiazepines: formulation of midazolam nasal spray. *Int. J. Pharm.*, 212:29–40.

Mahmoudian, T., M.M. Zadeh. (2004). Comparison of intranasal midazolam with intravenous diazepam for treating acute seizures in children. *Epilepsy Behav.*, 5:253–255.

Muller, R.H., C. Jacobs, O. Kayser. (1999). Pharmaceutical Nanosuspensions for Medicament Administration as Systems with Increased Saturation Solubility and Rate of Dissolution, U.S. Patent No. 5858410.

Muller, R.H., C. Jacobs, O. Kayser. (2003). DissoCubes: a novel formulation for poorly soluble and poorly bioavailable drugs. In *Modified-Release Drug Delivery Technology* (pp. 135–149), M.J. Rathbone, M.S. Roberts, and J. Hadgraft, Eds. New York: Marcel Dekker.

Okumu, F.W., J. Cleland. (2003). Implants and injectables. In *Modified-Release Drug Delivery Technology* (pp. 633–638), M.J. Rathbone, M.S. Roberts, and J. Hadgraft, Eds. New York: Marcel Dekker.

Okumu, F.W., N. Dao le, P.J. Fielder, N. Dybdal, D. Brooks, S. Sane, J.L. Cleland. (2002). Sustained delivery of human growth hormone from a novel gel system: SABER. *Biomaterials*, 23:4353–4358.

Pechenov, S., B. Shenoy, M.X. Yang, S.K. Basu, A.L. Margolin. (2004). Injectable controlled release formulations incorporating protein crystals. *J. Control. Release*, 96:149–158.

Peery, J.R., K.E. Dionne, J.B. Eckenhoff. (1999). Sustained Delivery of an Active Agent Using an Implantable System, U.S. Patent No. 5985305.

Pellock, J.M., M.C. Smith, J.C. Cloyd, B. Uthman, B.J. Wilder. (2004). Extended-release formulations: simplifying strategies in the management of antiepileptic drug therapy. *Epilepsy Behav.*, 5:301–307.

Peltola, J., C. Coetzee, F. Jimenez, T. Litovchenko, S. Ramaratnam, L. Zaslavaskiy, Z.S. Lu, D. M. Sykes. (2009). Once-daily extended-release levetiracetam as adjunctive treatment of partial-onset seizures in patients with epilepsy: a double-blind, randomized, placebo-controlled trial. *Epilepsia*, 50:406–414.

Ramakrishnan, S. et al. (2006). Sustained-Release Dosage Forms of Oxcarbazepine, U.S. Patent No. 20070059354.

Rathi, R.C., K.D. Fowers. (2008). ReGel depot technology. In *Modified-Release Drug Delivery Technology*, 2nd ed. (pp. 171–181), M.J. Rathbone, J. Hadgraft, and M.S. Roberts, Eds. New York: Marcel Dekker.

Rogawski, M.A. (2009). Convection-enhanced delivery in the treatment of epilepsy. *Neurotherapeutics*, 6:344–351.

Rouits, E., I. Burton, E. Guenole, M.M. Troenaru, A. Stockis, M.L. Sargentini-Maier. (2009). Pharmacokinetics of levetiracetam XR 500 mg tablets. *Epilepsy Res.*, 84:224–231.

Staniforth, J.N., A.R. Baichwal. (1993). Synergistically interacting heterodisperse polysaccharides: function in achieving controllable drug delivery. In *Polymeric Drug Delivery Systems: Properties and Applications* (pp. 327–350), M.A. El-Nokaly, D.M. Piatt, and B.A. Charpentier, Eds. Washington, D.C.: American Chemical Society.

Tamargo, R.J., L.A. Rossell, E.H. Kossoff, B.M. Tyler, M.G. Ewend, J.J. Aryanpur. (2002). The intracerebral administration of phenytoin using controlled-release polymers reduces experimental seizures in rats. *Epilepsy Res.*, 48:145–155.

Tompson, D.J., I. Ali, R. Oliver-Willwong, S. Job, L. Zhu, F. Lemme, A.E. Hammer, A. Vuong, J.A. Messenheimer. (2008). Steady-state pharmacokinetics of lamotrigine when converting from a twice-daily immediate-release to a once-daily extended-release formulation in subjects with epilepsy (the COMPASS Study). *Epilepsia*, 49:410–417.

Verma, R.K., S. Garg. (2001). Current status of drug delivery technologies and future directions. *Pharm. Tech. On-line*, 25(2):1–14.

Viscusi, E.R., G. Martin, C.T. Hartrick, N. Singla, G. Manvelian. (2005). Forty-eight hours of postoperative pain relief after total hip arthroplasty with a novel, extended-release epidural morphine formulation. *Anesthesiology*, 102:1014–1022.

Walker, R.B. (2008). Modified-release delivery systems for oral use. In *Modified-Release Drug Delivery Technology*, 2nd ed. (pp. 131–141), M.J. Rathbone, J. Hadgraft, and M.S. Roberts, Eds. New York: Marcel Dekker.

Walzer, M.A., A. Johnson, D. Baran, C. Cilber, S.P. Wanaski, S. Collins, J.C. Cloyd, D. Tolbert. (2008). Pharmacokinetics of oral and intravenous carbamazepine in adult patients with epilepsy. *Epilepsia*, 49(Suppl. 7):460.

Wermeling, D.P. (2009). Intranasal delivery of antiepileptic medications for treatment of seizures. *Neurotherapeutics*, 6:352–358.

Wermeling, D.P., K.A. Record, T.H. Kelly, S.M. Archer, T. Clinch, A.C. Rudy. (2006). Pharmacokinetics and pharmacodynamics of a new intranasal midazolam formulation in healthy volunteers. *Anesth. Analg.*, 103:344–349.

Wermeling, D.P., K.A. Record, S.M. Archer, A.C. Rudy. (2009). A pharmacokinetic and pharmacodynamic study in healthy volunteers, of a rapidly absorbed intranasal midazolam formulation. *Epilepsy Res.*, 83(2–3):124–132.

Werz, M.A. (2008). Pharmacotherapeutics of epilepsy: use of lamotrigine and expectations for lamotrigine extended release. *Ther. Clin. Risk Manag.*, 4:1035–1046.

Wheless, J.W., V. Venkataraman. (1999). New formulations of drugs in epilepsy. *Expert. Opin. Pharmacother.*, 1:49–60.

Wilson, M.T., S. Macleod, M.E. O'Regan. (2004). Nasal/buccal midazolam use in the community. *Arch. Dis. Child.*, 89:50–51.

Wilz, A., E.M. Pritchard, T. Li, J.Q. Lan, D.L. Kaplan, D. Boison. (2008). Silk polymer-based adenosine release: therapeutic potential for epilepsy. *Biomaterials*, 29:3609–3616.

Wolfe, T.R., T.C. Macfarlane. (2006). Intranasal midazolam therapy for pediatric status epilepticus. *Am. J. Emerg. Med.*, 24:343–346.

Wright, J.C., S. Tao Leonard, C.L. Stevenson, J.C. Beck, G. Chen, R.M. Jao, P.A. Johnson, J. Leonard, R.J. Skowronski. (2001). An *in vivo/in vitro* comparison with a leuprolide osmotic implant for the treatment of prostate cancer. *J. Control. Release*, 75:1–10.

Wright, J.C., R.M. Johnson, S.I. Yum. (2003). Implantable technology. *Drug Delivery Tech.*, 3(1).

Wright, J.C., A.N. Verity, F.W. Okumu. (2008). The SABER™ delivery system for parenteral administration. In *Modified-Release Drug Delivery Technology*, 2nd ed. (pp. 151–158), M.J. Rathbone, J. Hadgraft, and M.S. Roberts, Eds. New York: Marcel Dekker.

Zhou, M., M.D. Donovan. (1996). Intranasal mucociliary clearance of putative bioadhesive polymer gels. *Int. J. Pharm.*, 135(1–2):115–125.

# 32 Early Seizure Detection: Considerations and Applications

*Christophe C. Jouny, Piotr J. Franaszczuk, and Gregory K. Bergey*

## CONTENTS

## POTENTIAL ROLES OF SEIZURE DETECTION

Epileptic seizures are episodic, transient events with clinical manifestations that reflect the regions of seizure onset and subsequent involvement (i.e., propagation with or without secondary generalization). One recent study with intracranial recordings (Afra et al., 2008) found that complex partial seizures of mesial temporal onset had an average duration of only 106 seconds, and those with extratemporal neocortical onset lasted an average of only 78 seconds. This brief duration and the rapidly changing dynamics of the discharge pose various challenges for seizure detection. Even patients with very refractory and frequent seizures spend the vast majority of time in nonseizure (interictal) states. For example, a patient with very refractory complex partial seizures originating from mesial temporal regions who had 20 seizures per month (average duration, 2 minutes) would only be in the ictal state 0.1% of the time, and the other 99.9% would be spent in interictal (and postictal) states. Despite this brief total seizure time, the disruptive and recurrent nature of complex partial seizures (even without secondary generalization) has a major impact on the quality of life of patients with refractory epilepsy. When discussing seizure detection, it is useful to differentiate between early seizure detection, the topic of this chapter, and seizure detection that is not time sensitive.

Early seizure detection can be operationally defined as detection within the first few seconds of electrical seizure onset. This may be before the patient is experiencing any clinical manifestations. Methods of early seizure detection are attracting increased interest with the development of various closed-loop responsive therapies for epilepsy (Bergey, 2009). Any intervention that would hope to benefit the patient would require early detection of either subclinical seizure activity or simple partial seizure activity before it progressed to a disabling seizure (i.e., complex partial symptomatology with altered awareness).

An intervention that prevents the occurrence of a disabling seizure would be quite beneficial for the patient. A therapy that reduced a complex partial seizure duration from 100 seconds to 80 seconds is not likely to provide much practical benefit, as there would still be alteration of consciousness. For a responsive therapy based on early detection to be of benefit it also needs to be very sensitive. Specificity may be of relative importance depending on the therapy employed; for example, specificity may be less important in closed-loop neurostimulation than in focal application of drugs. Repetitive stimulations of the brain appear to be very well tolerated. Because the brain is pain insensitive, the patient is unaware of this therapy, even if administered hundreds of times daily. Repetitive applications of drugs, however, may have potential cumulative side effects.

Ideally, early detection could be extended to seizure prediction—that is, identification of a pre-ictal period minutes or hours before the actual seizure to allow for even earlier intervention (behavioral or therapeutic). In theory, seizure prediction could be quite useful for the patient with epilepsy. The patient could be provided with feedback regarding the likelihood of a seizure in the near future. That patient could then appropriately adjust activities (e.g., refrain from swimming for a few hours) if a seizure was likely. Even a few false positives from a sensitive prediction algorithm would be tolerable if such a detector would minimize the risk of injury. Frequent false positives, however, would potentially interfere with the patient's quality of life, transforming a life of seemingly random seizures into one compromised by repeated anticipation. Conversely, false-negative seizure prediction (i.e., predictor states that seizure is unlikely when in fact it occurs) could result in increased risk for injury if the patient falsely assumes that the risk of a seizure is low and engages in certain activities (e.g., swimming). Despite promising preliminary reports from some laboratories (Sackellares, 2008; Snyder et al., 2008), reliable seizure prediction remains an unfulfilled goal.

When seizure detection is used for diagnostic purposes rather than therapeutic purposes it is much less time sensitive. Seizure detection algorithms have been reported and utilized by various groups (Chan et al., 2008; Gotman, 1999; Osorio et al. 1998), and seizure detection programs are being implemented in commercial monitoring equipment. Detection of seizures in the epilepsy monitoring unit (EMU) or intensive care unit (ICU) can use data from the entire seizure for identification purposes (Friedman et al., 2009). This can be very useful for detection of events where the patient may either be unaware of the seizure (EMU) or obtunded or comatose (ICU). Even using the entire seizure for detection, there are variables that can pose challenges, such as artifacts from scalp recordings. While high sensitivity is always desirable, high specificity may be less critical. For the purposes of presurgical EMU evaluations, capturing sufficient numbers of seizures for presurgical planning fulfills the need, even if a few events are missed. As mentioned above, this chapter will focus on methods specifically directed toward early seizure detection.

## GENERAL CHALLENGES FOR EARLY SEIZURE DETECTION

When does a seizure actually begin? From the perspective of the neurologist reading an electroencephalogram (EEG) it would be the earliest ictal changes noted by visual inspection. This can, however, be influenced by a number of factors independent of the skill of the reviewer. Scalp recordings may not reveal the earliest changes; intracranial EEG (ICEEG) is much more likely to detect activity at seizure onset (Pacia and Ebersole, 1997). Even with intracranial arrays, if the recording contacts are remote from the site of seizure onset the earliest ictal changes may be missed. There are reports from a limited number of patients that high-frequency patterns may be early ictal changes in neocortical onset seizures (Worrell et al., 2004). Such activity, typically in the 80- to 600-Hz range, requires appropriate sampling frequency (at least twice the maximum component of the signal

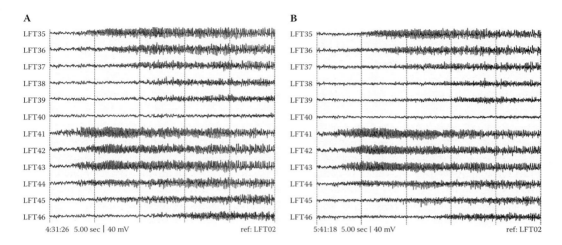

**FIGURE 32.1** Similarity of epileptic seizures. The two panels illustrate two sequential, frontal, neocortical onset, partial seizures recorded from the same patient and separated by ~80 min. Selected channels near the region of seizure onset are shown; extensive subdural grid arrays were employed. Visual inspection indicates similar patterns of onset and evolution. This was further documented by time-frequency decomposition (not shown). The dotted time bars are 1 sec apart.

being studied) and at times special electrodes, but some high-frequency oscillations are normal phenomena (Engel et al., 2009). Even gamma-frequency activity (e.g., 26 to 100 Hz) may be less noticeable to visual inspection because, by definition, it is low-amplitude activity. High-frequency activity (most easily appreciated with ICEEG) may occur in the early ictal broad regional electro-decremental changes that may be characteristic of some neocortical onset seizures.

Any reliable early seizure detection algorithm needs to recognize dynamic early ictal patterns. Often the patterns of seizure initiation are very different from the more organized rhythmic activity (e.g., high theta or alpha activity with mesial temporal onset seizures) seen later in the seizure. Not only is this organized rhythmic activity the most noticeable to visual inspection, but it also facilitates detection because it is regular, rhythmic activity with a predominant (albeit changing) frequency. In contrast, the period of early seizure onset may or may not have a single, high-energy frequency component. Interestingly, both the early seizure onset period and the later organized rhythmic activity are characterized by increased complexity of the signal that can be employed in seizure detection (see below). Another characteristic of human partial seizures is that, despite the considerable variability of patterns of onset and evolution from patient to patient, in any given patient with a single seizure focus the patterns of seizure onset and early evolution are often remarkably stereotyped for all seizures in that patient. Indeed, at times the patterns of propagation may also be similar. This may be apparent from visual inspection, particularly if ICEEG recordings are available (Figure 32.1), but these similarities extend to more detailed time–frequency analyses and composite measures of complexity (discussed below). This can facilitate early seizure detection and also offers the potential benefit of adjusting (tuning) detection algorithms for the seizures of a given patient to enhance early detection as well as specificity and sensitivity.

## THEORETICAL APPROACHES TO SEIZURE DETECTION

To develop a reliable method for detection of any event one needs to have a good definition of the event in terms of some measurable properties. Most of the methods of seizure detection and prediction are based on analysis of measurements of electrical activity of the brain (i.e., scalp EEG or intracranial EEG using subdural or depth electrodes) in patients being evaluated for seizure surgery.

Some attempts have been made to use other measurements to detect seizures, most notably heart-rate changes (Leutmezer et al., 2003), but these methods also proved more reliable when combined with EEG (Greene et al., 2007).

Methods based on analysis of EEG recordings require digitized data and computerized methods of analysis. The detection of seizures using EEG recordings is based on the presence of visible electrographic changes in the signal. Often, in visual analysis, the seizure is identified, perhaps by prominent organized rhythmic activity, and then the recording is scanned in reverse to identify the earlier focal changes. There are many methods of efficient parameterization of these electrographic changes during seizures, and the computed parameters (sometimes called *signal features*) may be subsequently used to classify an epoch of the signal as a seizure. To quantify the efficiency of the detection method we can use standard measures of sensitivity, such as the ratio of detected seizures (called true positives, or TPs) to all seizures in an analyzed record. The detection must be performed on a selected epoch (or window) of a given length. The binary detection method gives only two possible results: detected or not detected. For detection of a seizure as a single event, the window of detection must be large enough to contain the whole seizure or the detector must skip a predefined portion of the signal after detection. The first solution (large window) is preferable because too small of a window may cause the detector to miss the seizure after a false alarm. Unfortunately, this would prevent early detection of seizure because detection will occur after the end of the seizure.

One also needs an independent method (gold standard) to obtain the number of true seizures. Usually, this is obtained by visual screening of EEG by a trained neurologist. By changing the parameters of the detection algorithm, the method can be tuned to obtain sensitivity as close to 100% as possible. Increasing the sensitivity usually leads to detection of some events (e.g., artifacts, normal sleep patterns) that are not seizures. These detections are called false positives (FPs) or false alarms (FAs). Specificity is then defined as the ratio of the number of FPs divided by the sum of FPs plus true negatives. In some instances of early detection with ICEEG, FPs result from actual fragmentary (i.e., not fully evolved) epileptiform activity. In image analysis or in statistical signal processing when statistical properties of the signal and noise can be defined, the performance of the detection can be quantified by constructing the receiver operating characteristics (ROCs) (Kay, 1998). This is a plot of the probability of detection (i.e., sensitivity) vs. the probability of a false alarm (i.e., specificity) for different values of the parameter of detection (threshold) (Figure 32.2). The areas under the ROC curve provide a measure of performance of a detector. This area is equal to 1 for an ideal detector and 1/2 for random detection (i.e., deciding for each epoch randomly if there is a seizure or not).

Figure 32.2 illustrates the balance between sensitivity and specificity in a simple seizure detection paradigm. Whereas true positives are actual seizures, false positives may result from different causes. With scalp EEG recordings, false positives may result from movement or muscle artifact if the detector has a low (e.g., highly sensitive) threshold for detection. With intracranial EEG, there are few muscle or movement artifacts. False positives in ICEEG can result from a highly sensitive detector detecting bursts of epileptiform activity or brief, subclinical, electrical seizures. These false positive detections with ICEEG are false only in that this activity was not truly early ictal activity destined to evolve into a clinical event. Any seizure detector that is designed to detect ictal activity within the first few seconds of a seizure is likely to have a high false positive rate, even with ICEEG.

In practice, seizures make up a very small portion of the monitored signal, so any usable algorithm should be tuned to have a very small proportion of false alarms in relation to the number of epochs used for detection; however, ROC curves have limited value in measuring performance of detection methods over time. Instead, in addition to sensitivity (defined as above), the rate of false positives per hour (FPs/hr) is used as a measure of specificity. Similarly, as with the ROC curve, the designer of the algorithm needs to tune the parameters of detection to reach a compromise between sensitivity and specificity. This decision depends on the purpose of the detection. If the purpose is to assist a neurologist in reviewing the EEG for a preoperative evaluation, missing a seizure, even

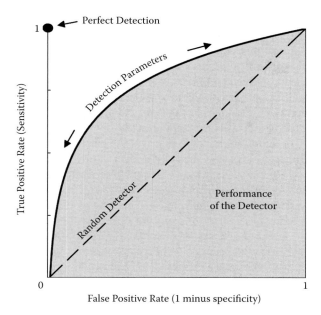

**FIGURE 32.2** Illustration of receiver operating characteristics (ROC) for a hypothetical detector. Variation of threshold changes the sensitivity to specificity ratio along the ROC curve. Dashed line illustrates performance of random detector. The gray area below the ROC curve measures the performance of the detector (larger is better).

a subclinical one, is undesirable, but in a patient with relatively frequent seizures missing a seizure is not critical. False alarms just increase the review time, which in any case will be shorter than reviewing all recorded data. In this case, detection parameters should be tuned to maximize sensitivity even if the false alarm rate is relatively high. On the other hand, when detection is for the purpose of intervention, one wants to detect all disabling clinical events.

## DETECTION VS. PREDICTION

### THE CONCEPT OF A PREICTAL STATE

There is considerable interest in predicting seizures. The concept of seizure prediction or anticipation is a provocative one. Because seizures are manifestations of increased network synchrony, it is a reasonable hypothesis that changes in activity, potentially embedded in the EEG signal, would reflect these synaptic or network changes before the actual seizure occurrence. Interestingly, examination of interictal spikes has not convincingly shown that spike frequency increases prior to ictal initiation (Katz et al., 1991). Other investigators feel, however, that subclinical seizures, bursts, or chirps increase prior to clinical events (Litt et al., 2001).

The concept of these early changes, which conceivably could occur minutes or even hours before a clinical event, is the concept of a preictal state. Any useful discussion of the existence of a preictal state must operationally define it and differentiate it from the interictal period. Appropriate use of the term *preictal state* implies that changes occur in the brain or EEG, not necessarily detectable by simple visual inspection that reliably precede the subsequent ictal event (electrical or clinical). Where the controversy arises is when the preictal state ends and the ictal state begins and when "prediction" is really only early detection. For example, the determination that high- or ultrahigh-frequency activity precedes a focal neocortical seizure would define an earlier ictal onset and not be a preictal state predicting a later seizure.

The above discussion is clearly applicable to detection of a seizure as it is occurring or even retrospectively for diagnostic purposes. Therapeutic applications require different approaches. The goal is to detect a seizure as early as possible to allow intervention to prevent the seizure from developing into a clinical event or to warn a patient in advance. In this case, there is a need for a new definition of what constitutes a seizure and when it starts. Remembering that we are limiting our discussion to analysis of an EEG signal, we need to define the electrographic onset of the seizure. There are different possible definitions. One definition is that any changes in EEG detectable by any current or future method different than the interictal baseline and followed by a clinical seizure are in fact the beginning of a seizure. This definition is an extension of the current practice followed by many neurologists performing visual analysis. They look at the EEG before the seizure for patterns in the signal that differ from those normally occurring during the interictal period. This definition does not allow for a separate definition of the prediction of a seizure; everything is early detection. However, this definition still allows the anticipation of seizure to be defined as the detection of electrographic changes that may indicate increased probability of the seizure in the near future, but not certainty. Another definition of the onset of a seizure is a pattern in the EEG coincident with the clinical onset of the seizure. In this case, if a given method detects some changes in EEG before that, it could be considered prediction if, after this change, a seizure always occurs.

Strictly speaking, a prediction method should also include the time horizon of the prediction. This time horizon can also be associated with the probability of occurrence of the seizure in the given time interval. In this case, the definition of prediction may overlap with the definition of anticipation. Recent papers (Andrzejak et al., 2009; Kreuz et al., 2004) provide guidelines for the evaluation of seizure prediction algorithms. In the literature, there has been a tendency to use the term *prediction*, starting with the pioneering work of Viglione (Viglione and Walsh, 1975) and in many later publications related to nonlinear methods of analysis (Elger and Lehnertz, 1998; Iasemidis et al., 2005; Litt and Echauz, 2002; Salant et al., 1998). Recently, the term *anticipation* has been more frequently used (Le Van Quyen et al., 1999; Navarro et al., 2002; Widman et al., 1999). Prediction is only possible if there is a preictal state. Litt and Lehnertz (2002) suggest that there is enough evidence to support the existence of the preictal state. In this case, the prediction methods can be really treated as detection of a preictal state. Litt et al. (2001) identified prolonged bursts of epileptiform discharges as precursors to seizures. In this case, detection of such precursors will be synonymous with prediction or anticipation. Recently proposed methods of precursor detection include application of a mathematical method called *genetic programming* (Firpi et al., 2007; Smart et al., 2007), a sign periodogram transform (Niederhauser et al., 2003), and phase and lag synchronization measures (Mormann et al., 2003).

It is particularly important when applying methods of seizure prediction that these applications are not influenced by seizure clusters (Jouny et al., 2005). A patient having seizures every 3 hours may have residual changes in the postictal period that might be interpreted as "predicting" the next seizure, as a true baseline interictal period may not be reached during seizure clusters. Any objective test of accurate seizure prediction must be effective in predicting the first seizure of any seizure cluster. The best test of a prediction method is the ability to detect a seizure that occurs after a long (e.g., days) interictal period.

It is tempting to assume that a preictal period, defined above as a period with detectable changes, exists for all seizures and that the difficulty in reliably detecting this period is due to either the nature of the changes or the methods being used. The concept of a preictal period is an attractive hypothesis. An alternative hypothesis, however, is that focal epileptic networks have very slight changes (e.g., synaptic, membrane) in intrinsic network excitability. Most of the time, inputs into the epileptic network do not trigger seizure activity, but on rare occasions this "normal" input can trigger seizures that would not occur in the nonepileptic brain. Neural network models have been constructed to demonstrate that this alternative hypothesis is a plausible one (Anderson et al., unpublished data). In instances where this alternative hypothesis is operational, seizure prediction would be very difficult if not impossible. It is possible that partial seizure occurrence may be governed by different mechanisms in different brains.

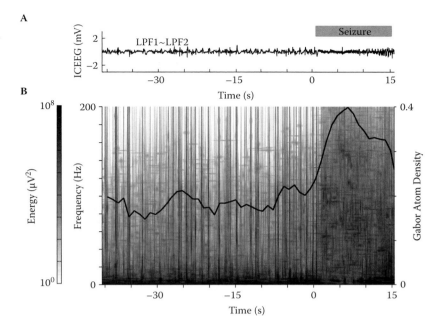

**FIGURE 32.3**  Intracranial EEG, time–frequency representation, and signal complexity at the focus of a neocortical onset seizure. Panel A shows the ICEEG for the electrode closest to the focus ~40 seconds before seizure onset and ~15 seconds into the seizure. Panel B illustrates the time-frequency energy density obtained from the matching pursuit decomposition of this signal (dark gray = high energy, white = low energy). The black trace represents the Gabor atom density (GAD) that quantifies the complexity of the signal. The changes in GAD can be used for early seizure detection.

## SEIZURE DETECTION METHODOLOGY

In our laboratory, we have been studying seizure onset dynamics recorded with ICEEG using the matching pursuit (MP) method (Jouny and Bergey, 2008; Jouny et al., 2003, 2005). MP is a method for time–frequency decomposition of signals that is particularly suited for applications to rapidly changing and transient events such as seizures. This method provides continuous time–frequency decomposition of a signal into components called *atoms*. Both linear and nonlinear signals can be appropriately analyzed using the MP method. Figure 32.3 illustrates MP decomposition of the early period of a neocortical onset seizure recorded near the seizure focus. This example has an active inter-ictal period with frequent spikes. Seizure onset is characterized by increased high-frequency activity.

A composite measure of signal complexity, the Gabor atom density (GAD) has been developed to provide more ready assessment of the signal complexity (Jouny et al., 2003). This composite measure, shown in Figure 32.3 (black trace), indicates that the complexity as measured by GAD increases early in the seizure even though the seizure has what appears to be more organized rhythmic activity. Studies of over 300 seizures in 45 patients reveal that 97% have significant GAD changes associated with seizure onset (Jouny et al., 2010).

Studies of MP time–frequency decompositions of multiple seizures from the same patient (with unifocal partial epilepsy) reveal remarkably similar onset patterns. Figure 32.4 illustrates this using the GAD complexity plots (derived from the time–frequency decompositions) for 36 seizures from a single patient over a 2-day period. Although there is some variability later, the ictal onset period and first 10 seconds have remarkably similar dynamics from seizure to seizure. This similarity of ictal onset makes it possible to adjust various seizure detection algorithms for a given patient to enhance specificity and sensitivity.

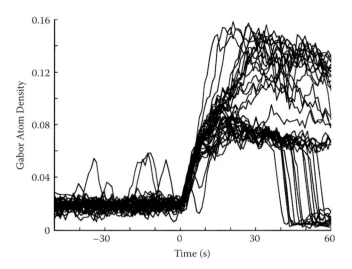

**FIGURE 32.4**  GAD plots for 36 consecutive partial seizures from one patient aligned on the time zero of the electrographical onset of the seizure. All events exhibits a stereotypical evolution early in the seizure and a near 100% GAD increase associated with the early electrographical onset of the seizure.

Trained neurologists are able to detect electrographic signatures of the seizure by visual analysis of EEG recordings. This prompted the development of the expert systems of seizure detection (Liu et al., 2002; Osorio et al., 1998; Pradhan et al., 1996; Si et al., 1998) to reproduce their results. Expert systems attempt to extract from the signal the same features that a trained neurologist uses in visual analysis. These features may include parameters of the shape of the EEG waveform, sometimes as simple as the length of the curve (Esteller et al., 2001), relative amplitude, dominant frequency (Murro et al., 1991), amplitude, slope, curvature, rhythmicity, and frequency components (Webber et al., 1996).

Recently, the detection of multimorphic patterns (alpha, beta, delta, and theta rhythmic activity; amplitude depression; polyspikes) has been implemented for automatic online seizure detection (Meier et al., 2008). Most of these systems use various types of artificial neural network (ANN) algorithms to process the input features (Gabor, 1998; Liu et al., 2002; Petrosian et al., 2001; Pradhan et al., 1996; Webber et al., 1996). Also, the increased speed of computer processors has enabled application of more computationally intensive algorithms for feature extraction; such methods include approximate entropy (Srinivasan et al., 2007), wavelet-chaos methodology (Ghosh-Dastidar et al., 2007), parameters of time–frequency energy distribution (Tzallas et al., 2009; Wilson, 2005), and higher order spectra (Chua et al., 2009). In general, most of these algorithms require a training phase when parameters of the algorithm are tuned to obtain optimal detections in agreement with the analysis of the expert neurologist. These methods can also be described as pattern-matching techniques, where certain patterns defined by specific parameters are to be detected, and the system must be trained to learn how to recognize this pattern. Obviously, these methods can be successful in seizure detection if such patterns of early seizure onset can be defined with sufficient precision.

A different approach to detection has roots in general signal processing detection techniques. The general principle is to use a procedure that will yield a detection function. Values of the detection function in comparison with a certain threshold will indicate occurrence of the event subject to detection. The common characteristic of this approach is the use of the raw EEG signal as the input data and performing numerical computations on this signal to obtain the value of the detection function. It can be performed with every recorded channel of EEG separately or may involve multichannel methods to capture interactions between channels relevant for detection. For seizure detection, the signal processing methods can be classified as nonparametric methods, without any specific assumptions about specific properties of the EEG signal and its generation, or parametric methods, where such assumptions are present. The most commonly applied nonparametric methods of EEG signal analysis include: (1) traditional spectral analysis using fast Fourier transform (FFT) (Chan et al., 2008; Hopfengartner et al., 2007) or autoregressive (AR) spectral analysis (Alkan et al., 2005; Khamis et al., 2009); or (2) time–frequency or time-scale analyses, such as wavelet (Geva and Kerem, 1998; Ghosh-Dastidar et al., 2007; Khan and Gotman, 2003; Senhadji and Wendling, 2002; Shoeb et al., 2004) or matching pursuit (Jouny et al., 2003) decompositions implemented in the commercially available Persyst Reveal© algorithm (Wilson et al., 2004). In detection applications, the multitude of parameters obtained from time–frequency methods can be transformed into one synthetic measure. In the case of wavelet methods, it can be wavelet entropy. The multivariate and multichannel methods use multiple parameters as inputs to multistage classifying procedures (Mitra et al., 2009), and for matching pursuit the Gabor atom density is used to measure the complexity of the signal (Figure 32.3).

Parametric methods of signal processing utilize some assumptions of the EEG signal or its generators. Typical examples of such methods would include single- and multichannel AR models that have the underlying assumption that the background EEG signal is stochastic in nature. These methods allow for computation of spectral properties of the EEG similarly to nonparametric methods, but they also allow for detection of portions of the EEG that are not stochastic. This is believed to be true for most epileptiform activity, so deviation from patterns predicted by AR models can be used as detectors of seizures.

Another parametric approach to EEG analysis successfully applied to early seizure detection or even prediction is based on the assumption that the EEG signal is not stochastic but deterministic at all times; only the underlying process is nonlinear and chaotic. Recently, different nonlinear measures based on this assumption were applied to epileptic EEG signals. Iasemidis et al. (1990) introduced the idea of using nonlinear dynamical parameters, primarily Lyapunov exponents, for characterizing EEG. Elger and Lehnertz (Elger and Lehnertz, 1998; Lehnertz, 1999) used nonlinear measures of EEG complexity such as the correlation dimension to analyze preictal EEG. Le Van Quyen et al. (1998, 1999) used another nonlinear measure: dynamical similarity index. Nonlinear methods based on the dynamics of the leaky integrate-and-fire neuron model were implemented recently in seizure detection and prediction algorithms (Schad et al., 2008). Various measures of synchronization are used to analyze the preictal state (Mormann et al., 2003) and for seizure detection (Slooter et al., 2006; van Putten, 2003). Litt and Lehnertz (2002) reviewed the different nonlinear approaches to seizure prediction, and Jerger et al. (2001) showed that for early seizure detection linear methods based on classical frequency and phase measures can be as effective as nonlinear methods.

In a recent editorial, Lesser and Webber (2008) pointed out the difficulty in designing an automatic detector of seizures due to the lack of a precise definition of a seizure. Even experienced electroencephalographers do not always agree on a seizure definition. In the future, with faster computer processors, seizure detection and prediction methods will probably use combinations of various methods. Linear or nonlinear signal analysis methods will be used for advanced feature extraction with further processing by artificial neural network systems.

## APPLICATIONS OF SEIZURE DETECTION

### EEG Monitoring

With the increased utilization of epilepsy monitoring units it is desirable to have the means for seizure detection by other than human observers. As mentioned above, the most desirable detection algorithms are those that are very sensitive, so as not to miss seizures, and relatively specific to minimize false alarms. Very early detection, although somewhat desirable so staff can respond to the patients, is not essential and certainly whether detection occurs after several or many seconds is not critical. Therefore, detection algorithms for EMU use can utilize relatively long seizure epochs and can be tuned for the organized rhythmic activity of a seizure rather than the more challenging ictal initiation patterns. The similarity of ictal patterns for a given patient allows modification of detection algorithms for the individual patient. The localization of the seizure focus is important for optimal early seizure detection.

Seizure localization, whether with scalp or intracranial electrodes, is an important role of the presurgical evaluation of patients in the EMU. Seizure detection in most monitoring units is done by visual analysis of the multichannel recordings to determine time and location of seizure onset. Because seizures are detected only after regional propagation, if the recording arrays are not optimally placed with electrodes near the seizure focus, early seizure detection cannot be accomplished.

Better methods (compared to mere visual inspection) are needed for localization in seizures that may be difficult to localize even with optimally placed intracranial arrays (e.g., neocortical onset seizures). As discussed above, the Gabor atom density method may permit identification of early changes in seizure complexity, beginning first in the channels of seizure onset. Other methods of seizure detection, particularly if they can be readily applied online, could assist in seizure localization by providing early detection of changes in contacts or channels nearest the seizure focus.

## APPLICATION OF EARLY SEIZURE DETECTION DEVICES TO TREATMENT OF EPILEPSY

As mentioned in the introductory paragraphs, exciting potential applications of early seizure detection include the promise for new methods of seizure therapy utilizing responsive therapy in closed-loop systems. Intracranial recordings offer the earliest detection possible, and it appears that application of these modalities will require chronic intracranial electrodes for these applications to have maximum sensitivity and specificity. With routine utilization of deep brain stimulation for movement disorders, there is a growing body of information indicating that such chronically implanted devices can be safe, effective, and well tolerated (Bergey, 2009).

In contrast to chronic programmed stimulation for epilepsy (e.g., vagus nerve stimulation, anterior thalamic stimulation), responsive neurostimulation is a closed-loop system that links seizure detection to therapeutic intervention. The goal of such therapy is not to prevent seizures *per se*, because the therapy is only administered after seizure onset. Instead, the goal is to provide early seizure termination before a simple partial seizure progresses to a disabling complex partial or secondarily generalized seizure. This requires an automated system with intracranial electrodes near the seizure focus, as seizure detection after regional or bilateral propagation will most likely be after the seizure is already disabling. If future seizure prediction algorithms would allow reliable identification of a preictal state change minutes (or longer) before a seizure, then the window and options for intervention would be considerable greater than with early detection devices.

## RESPONSIVE BRAIN STIMULATION DEVICES

The only FDA-approved neurostimulation device for the treatment of epilepsy is the vagus nerve stimulator (VNS), which stimulates the vagus nerve in the neck (Labar et al., 1999; Morris and Mueller, 1999). A controlled trial of anterior thalamic stimulation (Graves and Fisher, 2005; Kerrigan et al., 2004) was completed and submitted in 2009 for FDA review; the trial results are not yet published. Although very different with regard to the stimulation targets, both the VNS and thalamic stimulation use chronic stimulation paradigms. Theoretically, such chronic stimulation modulates brain activity, potentially affecting background synchronization and reducing seizure frequency. These are open-loop systems, and the stimulating electrode can be at sites remote from the seizure focus. Neither the VNS nor the anterior thalamic stimulator relies on seizure detection;

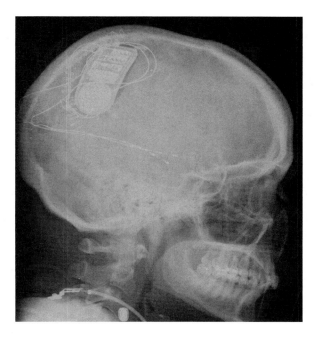

**FIGURE 32.5**   Imaging of a patient implanted with the NeuroPace® device. The device is implanted into the skull in a holder ferrule and connected to bilateral depth electrode arrays that target each hippocampal formation.

stimulus parameters of these devices are largely independent of seizure activity, although the VNS has the option of magnet activation by the patient or observer. No well-controlled studies exist to indicate whether such activation affects seizure duration. Although the VNS has been shown to be well tolerated and has produced over 50% seizure reduction in 40 to 50% of patients in highly refractory populations, utilization of the VNS rarely (<5 to 10%) makes patients seizure free, and it is not known whether anterior thalamic stimulation will improve upon these seizure-free rates.

Responsive therapy for epilepsy employs potential therapy that is triggered by a detected event. The triggered therapy can be stimulation or other modalities; seizure detection is the common component. Computerized detection of ictal onset in rats has triggered application of diazepam to experimental seizure foci (Fisher and Ho, 2002; Stein et al., 2000). Rapid cooling has also been linked with early seizure detection devices (Yang et al., 2002). Animal studies and simulations have employed closed-loop electrical stimulation (Osorio et al., 2001; Peters et al., 2001).

A blinded, controlled trial of responsive neurostimulation (RNS) was completed in approximately 200 patients with medically refractory complex partial epilepsy. This pivotal trial followed initial proof-of-principle studies, external closed-loop stimulation, and then safety and feasibility trials (Bergey et al., 2002, 2006). Figure 32.5 shows the NeuroPace® device, which contains a battery, programmable stimulator, digital processor, and memory chip. This device is implanted into the skull in a holder ferrule and connected to intracranial leads (either depth electrode arrays or subdural strips) that are placed near the seizure focus. The device can store a limited number of events, and the patient can download stored seizures, which can be reviewed remotely. The device utilizes one or more detection algorithms: half-wave, line length, or area under the curve. The detection algorithm is tuned for the specific patient so as to be able to detect events within 2 seconds after seizure onset. The total current delivered (less than 12 mA) and the charge density are modest. Preliminary reports of unblinded trials of the RNS suggesting efficacy have been published (Bergey et al., 2006). The results of a blinded trial demonstrating efficacy in reducing seizures were presented at the American Epilepsy Society in late 2009 but are not published.

The concept behind the application of the responsive neurostimulation is that seizure detection algorithms can be applied to intracranial recordings of seizures. Because of the stereotyped early pattern of seizures from a given focus in a given patient, the detection algorithms can be tuned to optimize sensitivity and specificity. The detected event then triggers a responsive stimulation of the brain with the goal of seizure termination. The idea of delivering excitatory stimuli to disrupt abnormal rhythms as a concept for the treatment of epileptic seizures is relatively novel because one typically thinks of an epileptic seizure as a state of hyperexcitability and treatments to date have focused on reducing this hyperexcitability, either intrinsically (e.g., reduced voltage-dependent sodium conductances) or by increasing inhibition. Model neural networks have shown that external excitatory stimuli can interrupt recurrent bursting if delivered at appropriate times (Franaszczuk et al., 2003; Kudela et al., 2002). Interestingly, network inhibition is not necessary for burst termination in these models. It appears that, when successful, the excitatory stimuli disrupt the seizure dynamics, resulting in premature seizure termination. In contrast to the heart, where a strong depolarizing stimulus is needed to stop ventricular arrhythmias, only small currents in the brain seem to be needed to produce burst termination. Direct application of currents to human afterdischarges produced by cortical stimulation (for presurgical mapping) can terminate some discharges, but these methods have not utilized detection algorithms, relying instead on visual identification of the afterdischarges (Lesser et al., 1999).

The ideal responsive stimulator would accurately detect seizures and be effective in terminating seizures early, preferably a few seconds after the onset, before either subclinical or simple partial seizures become complex partial or secondarily generalized seizures. The question is often asked how one knows that activity prompting a responsive stimulation would have evolved into a seizure? The answer would require extensive blinded random stimulation paradigms. But, if indeed the stimuli are benign, it may not matter whether additional stimuli are delivered. Potentially, such stimuli, if applied to what would be brief subclinical electrical seizures, could still reduce the chance of clinical events. In fact, despite the stereotyped nature of the early ictal pattern, the RNS device typically is triggered many more times than would be expected (e.g., hundreds of detections per day in a patient having five complex partial seizures per month). These are not false positive detections because the detector is detecting epileptiform activity but are due to the need for the detector to detect seizure activity within 2 seconds; therefore, epileptiform bursts or brief runs (e.g., 3 seconds) of rhythmic activity may result in detection. The optimal stimulus parameters for the RNS are still being determined,

The application of responsive devices that utilize preictal events is further from clinical application. A similar implantable device could be employed with intracranial electrodes. At the very least, such a preictal warning could be transmitted to the patient, perhaps minutes (or more) before the seizure, so the patient could move to safety, avoid at-risk behavior, or take or activate medical treatment. It is more difficult to postulate what stimulation parameters would be effective in the preictal period to prevent seizure occurrence.

## CONCLUSION

Seizure detection is attracting considerable attention. Instead of merely applying these methods in the epilepsy monitoring units, there is growing interest in linking detection devices to therapeutic interventions. For responsive therapy (e.g., neurostimulation) to be effective, the detection must be early, within seconds after seizure onset, but, as discussed above, this presents greater challenges. Indeed, the application of such responsive interventions is now technically feasible. The methodology to apply these methods exists; further investigations are necessary to identify the best stimulus parameters, but efficacy has been demonstrated. Responsive brain stimulation holds the promise to be a more specific, targeted, and potentially effective treatment modality than the chronic stimulation treatments currently employed that do not utilize seizure detection.

Seizure prediction depends upon the identification of a true preictal state, not merely the early detection of clear ictal changes. Which sophisticated methods will ultimately prove to be the best at seizure prediction is not known at present. Clearly the ability to predict seizures increases the time for potential therapeutic intervention. What intervention will be best also remains to be determined.

## REFERENCES

Afra, P., C.C. Jouny, G.K. Bergey. (2008). Duration of complex partial seizures: an intracranial EEG study. *Epilepsia*, 49:677–684.

Alkan, A., E. Koklukaya, A. Subasi. (2005). Automatic seizure detection in EEG using logistic regression and artificial neural network. *J. Neurosci. Methods*, 148:167–176.

Andrzejak, R.G., D. Chicharro, C.E. Elger, F. Mormann. (2009). Seizure prediction: any better than chance? *Clin. Neurophysiol.*, 120:1465–1478.

Bergey, G.K. (2009). Brain stimulation. In *The Treatment of Epilepsy* (pp. 1025–1033), S. Shorvon, E. Perucca, and J. Engel, Eds. Oxford, U.K.: Blackwell.

Bergey, G.K., J.W. Britton, G.D. Cascino, H.M. Choi, S.C. Karceski, E. Kossoff. (2002). Implementation of an external responsive neurostimulator system (eRNS) in patients with intractable epilepsy undergoing intracranial seizure monitoring. *Epilepsia*, 43:191.

Bergey, G.K., G. Worrell, D. Chabolla, R. Zimmerman, D. Labar, L. Hirsch, R. Duckrow, A. Murro, M. Smith, D. Vossler, G. Barkley, M. Morrell. (2006). Safety and preliminary efficacy of a responsive neurostimulator (RNS™) for the treatment of intractable epilepsy in adults. *Neurology*, 66:A387.

Chan, A.M., F.T. Sun, E.H. Boto, B.M. Wingeier. (2008). Automated seizure onset detection for accurate onset time determination in intracranial EEG. *Clin. Neurophysiol.*, 119:2687–2696.

Chua, K.C., V. Chandran, U.R. Acharya, C.M. Lim. (2009). Analysis of epileptic EEG signals using higher order spectra. *J. Med. Eng. Technol.*, 33:42–50.

Elger, C.E., K. Lehnertz. (1998). Seizure prediction by non-linear time series analysis of brain electrical activity. *Eur. J. Neurosci.*, 10:786–789.

Engel, J., Jr., A. Bragin, R. Staba, I. Mody. (2009). High-frequency oscillations: what is normal and what is not? *Epilepsia*, 50:598–604.

Esteller, R., J. Echauz, T. Tcheng, B. Litt, B. Pless. (2001). Line length: an efficient feature for seizure onset detection. *IEEE Eng. Med. Biol. Soc.*, 2:1707–1710.

Firpi, H., E.D. Goodman, J. Echauz. (2007). Epileptic seizure detection using genetically programmed artificial features. *IEEE Trans. Biomed. Eng.*, 54:212–224.

Fisher, R.S., J. Ho. (2002). Potential new methods for antiepileptic drug delivery. *CNS Drugs*, 16:579–593.

Franaszczuk, P.J., P. Kudela, G.K. Bergey. (2003). External excitatory stimuli can terminate bursting in neural network models. *Epilepsy Res.*, 53:65–80.

Friedman, D., J. Claassen, L.J. Hirsch. (2009). Continuous electroencephalogram monitoring in the intensive care unit. *Anesth. Analg.*, 109:506–523.

Gabor, A.J. (1998). Seizure detection using a self-organizing neural network: validation and comparison with other detection strategies. *Electroencephalogr. Clin. Neurophysiol.*, 107:27–32.

Geva, A.B., D.H. Kerem. (1998). Forecasting generalized epileptic seizures from the EEG signal by wavelet analysis and dynamic unsupervised fuzzy clustering. *IEEE Trans. Biomed. Eng.*, 45:1205–1216.

Ghosh-Dastidar, S., H. Adeli, N. Dadmehr. (2007). Mixed-band wavelet-chaos-neural network methodology for epilepsy and epileptic seizure detection. *IEEE Trans. Biomed. Eng.*, 54:1545–1551.

Gotman, J. (1999). Automatic detection of seizures and spikes. *J. Clin. Neurophysiol.*, 16:130–140.

Graves, N.M., R.S. Fisher. (2005). Neurostimulation for epilepsy, including a pilot study of anterior thalamic nucleus stimulation. *Clin. Neurosurg.*, 52:127–134.

Greene, B.R., G.B. Boylan, R.B. Reilly, P. de Chazal, S. Connolly. (2007). Combination of EEG and ECG for improved automatic neonatal seizure detection. *Clin. Neurophysiol.*, 118:1348–1359.

Hopfengartner, R., F. Kerling, V. Bauer, H. Stefan. (2007). An efficient, robust and fast method for the offline detection of epileptic seizures in long-term scalp EEG recordings. *Clin. Neurophysiol.*, 118:2332–2343.

Iasemidis, L.D., J.C. Sackellares, H.P. Zaveri, W.J. Williams. (1990). Phase space topography and the Lyapunov exponent of electrocorticograms in partial seizures. *Brain Topogr.*, 2:187–201.

Iasemidis, L.D., D.S. Shiau, P.M. Pardalos, W. Chaovalitwongse, K. Narayanan, A. Prasad, K. Tsakalis, P.R. Carney, J.C. Sackellares. (2005). Long-term prospective on-line real-time seizure prediction. *Clin. Neurophysiol.*, 116:532–544.

Jerger, K.K., T.I. Netoff, J.T. Francis, T. Sauer, L. Pecora, S.L. Weinstein, S.J. Schiff. (2001). Early seizure detection. *J. Clin. Neurophysiol.*, 18:259–268.

Jouny, C.C., G.K. Bergey. (2008). Dynamics of epileptic seizures during evolution and propagation. In *Computational Neuroscience in Epilepsy* (pp. 457–470), I. Soltesz and K. Staley, Eds. London: Academic Press.

Jouny, C.C., P.J. Franaszczuk, G.K. Bergey. (2003). Characterization of epileptic seizure dynamics using Gabor atom density. *Clin. Neurophysiol.*, 114:426–437.

Jouny, C.C., P.J. Franaszczuk, G.K. Bergey. (2005). Signal complexity and synchrony of epileptic seizures: is there an identifiable preictal period? *Clin. Neurophysiol.*, 116:552–558.

Jouny, C.C., G.K. Bergey, P.J. Franaszczuk. (2010). Partial seizures are associated with early increases in signal complexity. *Clin. Neurophysiol.*, 121:7–13.

Katz, A., D.A. Marks, G. McCarthy, S.S. Spencer. (1991). Does interictal spiking change prior to seizures? *Electroencephalogr. Clin. Neurophysiol.*, 79:153–156.

Kay, S.M. (1998). *Fundamentals of Statistical Signal Processing*. Vol. II. *Detection Theory*. Englewood Cliffs, NJ: Prentice Hall.

Kerrigan, J.F., B. Litt, R.S. Fisher, S. Cranstoun, J.A. French, D.E. Blum, M. Dichter, A. Shetter, G. Baltuch, J. Jaggi, S. Krone, M. Brodie, M. Rise, N. Graves. (2004). Electrical stimulation of the anterior nucleus of the thalamus for the treatment of intractable epilepsy. *Epilepsia*, 45:346–354.

Khamis, H., A. Mohamed, S. Simpson. (2009). Seizure state detection of temporal lobe seizures by autoregressive spectral analysis of scalp EEG. *Clin. Neurophysiol.*, 120:1479–1488.

Khan, Y.U., J. Gotman. (2003). Wavelet based automatic seizure detection in intracerebral electroencephalogram. *Clin. Neurophysiol.*, 114:898–908.

Kreuz, T., R.G. Andrzejak, F. Mormann, A. Kraskov, H. Stogbauer, C.E. Elger, K. Lehnertz, P. Grassberger. (2004). Measure profile surrogates: a method to validate the performance of epileptic seizure prediction algorithms. *Phys. Rev. E. Stat. Nonlin. Soft Matter Phys.*, 69:061915.

Kudela, P., P.J. Franaszczuk, G.K. Bergey. (2002). External termination of recurrent bursting in a model of connected local neural sub-networks. *Neurocomputing*, 44:897–905.

Labar, D., J. Murphy, E. Tecoma. (1999). Vagus nerve stimulation for medication-resistant generalized epilepsy. E04 VNS Study Group. *Neurology*, 52:1510–1512.

Le Van Quyen, M., C. Adam, M. Baulac, J. Martinerie, F.J. Varela. (1998). Nonlinear interdependencies of EEG signals in human intracranially recorded temporal lobe seizures. *Brain Res.*, 792:24–40.

Le Van Quyen, M., J. Martinerie, M. Baulac, F. Varela. (1999). Anticipating epileptic seizures in real time by a non-linear analysis of similarity between EEG recordings. *NeuroReport*, 10:2149–2155.

Lehnertz, K. (1999). Non-linear time series analysis of intracranial EEG recordings in patients with epilepsy: an overview. *Int. J. Psychophysiol.*, 34:45–52.

Lesser, R.P., W.R. Webber. (2008). Seizure detection: reaching through the looking glass. *Clin. Neurophysiol.*, 119:2667–2668.

Lesser, R.P., S.H. Kim, L. Beyderman, D.L. Miglioretti, W.R. Webber, M. Bare, B. Cysyk, G. Krauss, B. Gordon. (1999). Brief bursts of pulse stimulation terminate afterdischarges caused by cortical stimulation. *Neurology*, 53:2073–2081.

Leutmezer, F., C. Schernthaner, S. Lurger, K. Potzelberger, C. Baumgartner. (2003). Electrocardiographic changes at the onset of epileptic seizures. *Epilepsia*, 44:348–354.

Litt, B., J. Echauz. (2002). Prediction of epileptic seizures. *Lancet Neurol.*, 1:22–30.

Litt, B., K. Lehnertz. (2002). Seizure prediction and the preseizure period. *Curr. Opin. Neurol.*, 15:173–177.

Litt, B., R. Esteller, J. Echauz, M. D'Alessandro, R. Shor, T. Henry, P. Pennell, C. Epstein, R. Bakay, M. Dichter, G. Vachtsevanos. (2001). Epileptic seizures may begin hours in advance of clinical onset: a report of five patients. *Neuron*, 30:51–64.

Liu, H.S., T. Zhang, F.S. Yang. (2002). A multistage, multimethod approach for automatic detection and classification of epileptiform EEG. *IEEE Trans. Biomed. Eng.*, 49:1557–1566.

Meier, R., H. Dittrich, A. Schulze-Bonhage, A. Aertsen. (2008). Detecting epileptic seizures in long-term human EEG: a new approach to automatic online and real-time detection and classification of polymorphic seizure patterns. *J. Clin. Neurophysiol.*, 25:119–131.

Mitra, J., J.R. Glover, P.Y. Ktonas, A. Thitai Kumar, A. Mukherjee, N.B. Karayiannis, J.D. Frost, Jr., R.A. Hrachovy, E.M. Mizrahi. (2009). A multistage system for the automated detection of epileptic seizures in neonatal electroencephalography. *J. Clin. Neurophysiol.*, 26:218–226.

Mormann, F., R.G. Andrzejak, T. Kreuz, C. Rieke, P. David, C.E. Elger, K. Lehnertz. (2003). Automated detection of a preseizure state based on a decrease in synchronization in intracranial electroencephalogram recordings from epilepsy patients. *Phys. Rev. E. Stat. Nonlin. Soft. Matter Phys.*, 67:021912.

Morris, 3rd, G.L., W.M. Mueller. (1999). Long-term treatment with vagus nerve stimulation in patients with refractory epilepsy. The Vagus Nerve Stimulation Study Group e01–e05. *Neurology*, 53:1731–1735.

Murro, A.M., D.W. King, J.R. Smith, B.B. Gallagher, H.F. Flanigin, K. Meador. (1991). Computerized seizure detection of complex partial seizures. *Electroencephalogr. Clin. Neurophysiol.*, 79:330–333.

Navarro, V., J. Martinerie, M. Le Van Quyen, S. Clemenceau, C. Adam, M. Baulac, F. Varela. (2002). Seizure anticipation in human neocortical partial epilepsy. *Brain*, 125:640–655.

Niederhauser, J.J., R. Esteller, J. Echauz, G. Vachtsevanos, B. Litt. (2003). Detection of seizure precursors from depth-EEG using a sign periodogram transform. *IEEE Trans. Biomed. Eng.*, 50:449–458.

Osorio, I., M.G. Frei, S.B. Wilkinson. (1998). Real-time automated detection and quantitative analysis of seizures and short-term prediction of clinical onset. *Epilepsia*, 39:615–627.

Osorio, I., M.G. Frei, B.F. Manly, S. Sunderam, N.C. Bhavaraju, S.B. Wilkinson. (2001). An introduction to contingent (closed-loop) brain electrical stimulation for seizure blockage, to ultra-short-term clinical trials, and to multidimensional statistical analysis of therapeutic efficacy. *J. Clin. Neurophysiol.*, 18:533–544.

Pacia, S.V., J.S. Ebersole. (1997). Intracranial EEG substrates of scalp ictal patterns from temporal lobe foci. *Epilepsia*, 38:642–654.

Peters, T.E., N.C. Bhavaraju, M.G. Frei, I. Osorio. (2001). Network system for automated seizure detection and contingent delivery of therapy. *J. Clin. Neurophysiol.*, 18:545–549.

Petrosian, A.A., D.V. Prokhorov, W. Lajara-Nanson, R.B. Schiffer. (2001). Recurrent neural network-based approach for early recognition of Alzheimer's disease in EEG. *Clin. Neurophysiol.*, 112:1378–1387.

Pradhan, N., P.K. Sadasivan, G.R. Arunodaya. (1996). Detection of seizure activity in EEG by an artificial neural network: a preliminary study. *Comput. Biomed. Res.*, 29:303–313.

Sackellares, J.C. (2008). Seizure prediction. *Epilepsy Curr.*, 8:55–59.

Salant, Y., I. Gath, O. Henriksen. (1998). Prediction of epileptic seizures from two-channel EEG. *Med. Biol. Eng. Comput.*, 36:549–556.

Schad, A., K. Schindler, B. Schelter, T. Maiwald, A. Brandt, J. Timmer, A. Schulze-Bonhage. (2008). Application of a multivariate seizure detection and prediction method to non-invasive and intracranial long-term EEG recordings. *Clin. Neurophysiol.*, 119:197–211.

Senhadji, L., F. Wendling. (2002). Epileptic transient detection: wavelets and time-frequency approaches. *Neurophysiol. Clin.*, 32:175–192.

Shoeb, A., H. Edwards, J. Connolly, B. Bourgeois, T. Treves, J. Guttag. (2004). Patient-specific seizure onset detection. *Epilepsy Behav.*, 5(4):483–498.

Si, Y., J. Gotman, A. Pasupathy, D. Flanagan, B. Rosenblatt, R. Gottesman. (1998). An expert system for EEG monitoring in the pediatric intensive care unit. *Electroencephalogr. Clin. Neurophysiol.*, 106:488–500.

Slooter, A.J., E.M. Vriens, F.S. Leijten, J.J. Spijkstra, A.R. Girbes, A.C. van Huffelen, C.J. Stam. (2006). Seizure detection in adult ICU patients based on changes in EEG synchronization likelihood. *Neurocrit. Care*, 5:186–192.

Smart, O., H. Firpi, G. Vachtsevanos. (2007). Genetic programming of conventional features to detect seizure precursors. *Eng. Appl. Artif. Intell.*, 20:1070–1085.

Snyder, D.E., J. Echauz, D.B. Grimes, B. Litt. (2008). The statistics of a practical seizure warning system. *J. Neural Eng.*, 5:392–401.

Srinivasan, V., C. Eswaran, N. Sriraam. (2007). Approximate entropy-based epileptic EEG detection using artificial neural networks. *IEEE Trans. Inf. Technol. Biomed.*, 11:288–295.

Stein, A.G., H.G. Eder, D.E. Blum, A. Drachev, R.S. Fisher. (2000). An automated drug delivery system for focal epilepsy. *Epilepsy Res.*, 39:103–114.

Tzallas, A.T., M.G. Tsipouras, D.I. Fotiadis. (2009). Epileptic seizure detection in EEGs using time-frequency analysis. *IEEE Trans. Inf. Technol. Biomed.*, 13:703–710.

van Putten, M.J. (2003). Nearest neighbor phase synchronization as a measure to detect seizure activity from scalp EEG recordings. *J. Clin. Neurophysiol.*, 20:320–325.

Viglione, S.S., G.O. Walsh. (1975). Proceedings: epileptic seizure prediction. *Electroencephalogr. Clin. Neurophysiol.*, 39:435–43å6.

Webber, W.R., R.P. Lesser, R.T. Richardson, K. Wilson. (1996). An approach to seizure detection using an artificial neural network (ANN). *Electroencephalogr. Clin. Neurophysiol.*, 98:250–272.

Widman, G., D. Bingmann, K. Lehnertz, C.E. Elger. (1999). Reduced signal complexity of intracellular recordings: a precursor for epileptiform activity? *Brain Res.*, 836:156–163.

Wilson, S.B. (2005). A neural network method for automatic and incremental learning applied to patient-dependent seizure detection. *Clin. Neurophysiol.*, 116:1785–1795.

Wilson, S.B., M.L. Scheuer, R.G. Emerson, A.J. Gabor. (2004). Seizure detection: evaluation of the reveal algorithm. *Clin. Neurophysiol.*, 115:2280–2291.

Worrell, G.A., L. Parish, S.D. Cranstoun, R. Jonas, G. Baltuch, B. Litt. (2004). High-frequency oscillations and seizure generation in neocortical epilepsy. *Brain*, 127:1496–1506.

Yang, X.F., D.W. Duffy, R.E. Morley, S.M. Rothman. (2002). Neocortical seizure termination by focal cooling: temperature dependence and automated seizure detection. *Epilepsia*, 43:240–245.

# 33 Gene and Stem Cell Therapies for Treating Epilepsy

*Janice R. Naegele and Xu Maisano*

## CONTENTS

## INTRODUCTION

Epilepsy afflicts over 60 million people worldwide, putting it among the most prevalent of neurological disorders. The highest incidence occurs in children under the age of 5 and in the elderly (Theodore et al., 2006). A relatively large proportion of patients with temporal lobe epilepsy (TLE) are resistant to antiepileptic drugs (AEDs) or experience debilitating side effects from long-term treatment such as cognitive impairment, depression, or dementia. Surgery to remove the epileptic tissue may offer an improvement over AEDs, but it is only an option for patients with focal unilateral seizures in brain regions that can be safely removed without causing severe cognitive or sensory deficits. Novel approaches based on gene and stem cell therapy offer the potential for curing epilepsy, rather than treating the symptoms. Coupled with a better understanding of neurological changes caused by seizures, the goals of this emerging area of research are to modify the progression of epilepsy and cure the underlying defects.

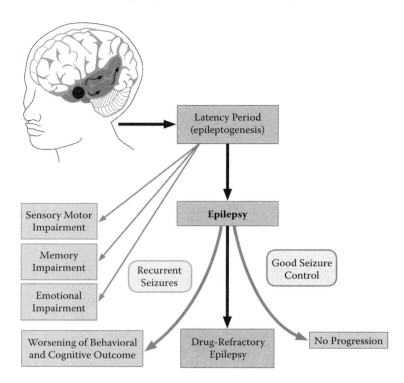

**FIGURE 33.1**    Scenario for patients developing temporal lobe epilepsy (TLE) after seizures, traumatic brain injury, high prolonged fevers, or strokes. Following an initial seizure or seizures, the patient may be seizure free for an indefinite latency period lasting weeks, months, or even years. Neuroplastic changes during this period may make temporal lobe circuits hyperexcitable and lead to the development of spontaneous seizures, a process referred to as *epileptogenesis*. Patients with TLE may be seizure free with antiepileptic drug (AED) medications, but in many individuals the TLE is resistant to AEDs or adverse side effects from these drugs, and seizures can produce cognitive and emotional impairments. (Adapted from Acharya, M.M. et al., *Prog. Neurobiol.*, 84(4), 363–404, 2008.)

This chapter discusses advances in the fields of gene transfer and stem cell therapy that are fueling new treatments for refractory epilepsy. We first discuss gene therapy for treating TLE and inherited forms of epilepsy and then stem-cell-based therapies for TLE. We highlight limitations and important hurdles that must be overcome before patients can be treated. We then describe some of the newest strategies for tailoring embryonic stem (ES) cell-based therapy for TLE, based on recent discoveries about embryonic origins and molecular codes that regulate GABAergic neuron fates.

## BACKGROUND

Developmental disorders of neuronal migration, genetic mutations, and traumatic brain injury are three of the most common causes of epilepsy. Most cases of temporal lobe epilepsy are acquired after an initial episode of status epilepticus (SE) or severe head trauma (Figure 33.1). The risk is particularly high when limbic circuits become hyperexcitable following brain injury that is caused by neurological insults such as prolonged febrile seizures, brain tumors, or spinal meningitis. How an initial precipitating injury alters limbic circuitry to promote epileptogenesis is not well understood. It is now thought that neural plasticity occurs mainly during the latent period (the period between the initial brain insult and the development of spontaneous seizures), which changes the balance between excitation and inhibition, and this imbalance is responsible for epileptogenesis. The latent

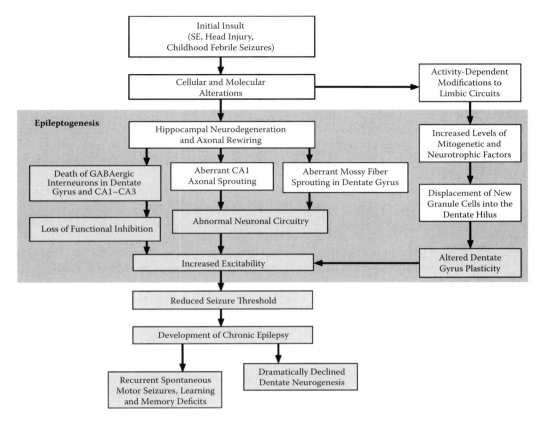

**FIGURE 33.2**   Events in the hippocampus after initial brain injuries that lead to the development of recurrent seizures and temporal lobe epilepsy. Initial damage to the temporal lobes can result from a variety of neurological traumas, including head injury or prolonged seizures such as status epilepticus. Experimental animal models of SE indicate that a large number of cellular and molecular alterations resulting from the initial insult lead to neuroplastic changes and rewiring of hippocampal circuitry. These changes, indicated in the box with gray shading, include changes to inhibitory neurotransmission, axonal sprouting in CA1 and the dentate gyrus, and alterations in adult neurogenesis in the dentate gyrus subgranular zone. Which of these plastic changes critically determine the development of recurrent seizures is not well understood.

period, during which seizures are not evident, can extend for months or years after the initial insult. In childhood-onset TLE, the hippocampus shows progressive damage and sclerosis, whereas adult-onset TLE is not always associated with distinctive neurological signs that are visible with magnetic resonance imaging (Figure 33.2).

Studies of mesial TLE (MTLE) in animal models have extended our understanding of how the limbic lobe is altered by seizure experience, particularly the cellular and molecular changes that take place during epileptogenesis (Pitkanen et al., 2006, 2007). The central structure implicated in MTLE is the hippocampus, an archicortical region of the temporal lobe with three layers and extensive connections to other cortical and subcortical structures. Glutamatergic projections from neocortical pyramidal neurons form the perforant pathway; they become hyperexcitable in MTLE, synapse with granule neurons of the dentate gyrus, and form the first stage in a three-synapse loop. The second connection in the trisynaptic loop is made by mossy fibers from the granule neurons, which form giant synapses on the dendrites of CA3 pyramidal neurons and CA3 interneurons. Schaffer collaterals from the CA3 pyramidal neurons terminating on CA1 pyramidal neurons and interneurons form the third stage of the trisynaptic loop.

The hippocampal granule cells and pyramidal neurons are the principal cells. The pyramidal neurons, located in CA1 to CA3, comprise a relatively homogeneous, excitatory population that releases the neurotransmitter glutamate. Contrasting with the principal neurons in the hippocampus, the nonprincipal cells are morphologically, neurochemically, and electrophysiologically diverse. These GABAergic interneurons are distributed throughout the dentate gyrus and other hippocampal subfields and form short-range connections. The hilus of the dentate gyrus is enriched in interneuron subtypes that appear to be especially important for controlling seizures. Two of the nonprincipal interneuron types in the hilus of the dentate synthesize and co-release somatostatin (SOM) or neuropeptide Y (NPY), generally evoking inhibition of postsynaptic neurons; an additional subset contains the calcium-binding protein parvalbumin. These cells are highly vulnerable to traumatic brain injury and seizures and degenerate in some patients with severe MTLE. As discussed later, GABAergic interneurons play a central role in maintaining inhibitory tone, and their demise following injury is thought to be a critical event leading to MTLE. For this reason, a number of gene and stem cell therapy approaches have focused on the GABAergic system.

In addition to GABAergic interneuron cell death, additional changes include aberrant neurogenesis and displacement of new granule cells in the dentate gyrus, altered expression of neuropeptides and neuromodulators, gliogenesis, axonal injury, mossy fiber sprouting, and immune modulation (Acharya et al., 2008). It is not known which of these changes is responsible for the development of spontaneous seizures; however, recent studies with gene and fetal stem cell therapy in rodent models of epilepsy show that it is possible to inhibit the development of epilepsy or modify epileptogenesis by strategies that increase inhibitory tone (Löscher et al., 2008). Cell transplantation has also demonstrated that genetically engineered cells releasing inhibitory neurotransmitters or neuropeptides can suppress seizures and restore the balance between excitation and inhibition. This work provides "proof of concept" that reducing excitatory neurotransmission in the hippocampus can be beneficial for preventing or controlling epilepsy.

Most studies evaluating gene and stem cell therapies for treating epilepsy have been conducted in a small number of experimental rodent models. These include the electrical kindling model of TLE, in which repetitive electrical stimuli are delivered at subthreshold levels into limbic brain regions until the animals are kindled and show generalized seizures; acute chemoconvulsant seizures induced by systemic or focal brain injections of kainic acid (KA); recurring spontaneous seizures that develop weeks after KA- or pilocarpine-induced status epilepticus; and genetic models of inherited epilepsies. The experimental models that test therapeutic efficacy after SE develops are regarded as being more comparable to human TLE.

Although gene- and cell-based therapies are still in preclinical stages and most require extensive validation in models of chronic epilepsy, safer viral delivery systems for gene therapy have been developed, and clinical trials are now being contemplated. Transplants of ES cell-derived neurons raise the prospect for repairing damaged circuits, eliminating a seizure focus, or correcting abnormal wiring that propagates seizures. As discussed below, however, both safety and ethical issues must be surmounted before ES cell-based therapy becomes possible in patient populations.

## GENE THERAPY FOR TEMPORAL LOBE EPILEPSY

Gene therapy targets endogenous cells to modify gene expression, counteracting the alterations in these cells caused by genetic mutations or seizures. Adeno-associated virus (AAV) is one of the most promising vectors used to convey foreign genes because it can transduce postmitotic neurons (neurotropism), it can promote persistent and long-term expression of transgenes, and it has low toxicity (Figure 33.3) (McCown, 2005; Riban et al., 2008).

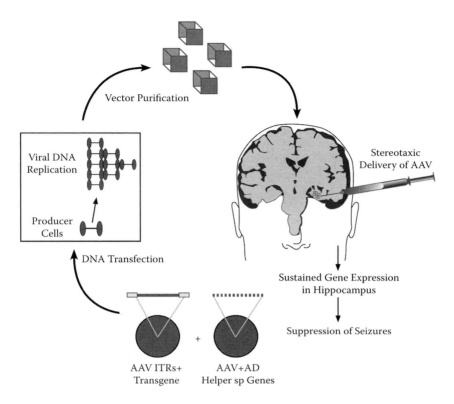

**FIGURE 33.3 (See color insert following page 458.)** General approach for adenovirus-associated gene therapy. Adeno-associated virus (AAV) is a small virus that infects humans and primate tissues. It is not known to cause disease and does not elicit a strong immune response, making it ideal for human gene therapy. Recombinant AAV (rAAV) is coinfected with a helper virus (AD) and a second AAV vector containing inverted terminal repeats (ITRs), which are required for efficient amplification of the AAV genome and for integration of the AAV DNA into the host cell genome. For human gene therapy, ITRs are required *in cis* next to the therapeutic gene. The AAV ITR transgene and AAV+AD genes are cotransfected into a producer cell line for viral DNA replication. The rAAV is purified and delivered by stereotaxic injection into the nervous system to achieve sustained gene expression.

## GABA$_A$ RECEPTORS AND GABA

Experimental analyses of the processes by which recurrent seizures develop in MTLE (epileptogenesis) have not yet explained why the latent period may be decades long after an initial brain injury in human patients. Studies examining the genes altered during the latent period suggest that alterations in the expression of GABA$_A$ receptors are an important determinant in patients and experimental animals (Brooks-Kayal et al., 1999; Hu et al., 2008; Raol et al., 2005, 2006; Roberts et al., 2005). These findings are supported by studies demonstrating that gene therapy for γ-aminobutyric acid type A (GABA$_A$) receptors can suppress seizures in rodents. When the dentate gyrus of rats subjected to pilocarpine-induced status epilepticus was infused with an AAV containing an activity-dependent promoter driving the expression of the GABA$_A$ receptor α$_1$ subunit, the incidence of developing chronic seizures was significantly lower in these animals. Moreover, seizure-free periods were three times longer in rats that did go on to develop chronic epilepsy (Raol et al., 2006).

## Neuropeptide Y

One leading hypothesis to account for hyperexcitability in MTLE is that following the death of mossy cells and hilar GABAergic neurons, granule neurons are disinhibited and become hyperexcitable. Further support of this hypothesis has come from studies showing reduced numbers of GABAergic hippocampal neurons expressing NPY, calbindin, or SOM in mice with experimentally induced epilepsy or genetic mutations associated with a seizure phenotype (Gant et al., 2008; Tuunanen et al., 1997). Building on these findings, gene therapies have been designed to restore the expression of neuropeptides, including NPY, galanin, dynorphin, and SOM, to block epileptogenesis and suppress spontaneous seizures (Wasterlain et al., 2002). NPY is significantly upregulated by seizures (Mazarati et al., 2000) and is an endogenous anticonvulsant (Baraban et al., 1997; El Bahh et al., 2005; Richichi et al., 2004; Vezzani et al., 1999). NPY is released by GABAergic interneurons and arrests the progression of seizures through its inhibitory effect on presynaptic glutamate release by activating NPY-Y2 receptors located on glutamatergic axonal terminals (Vezzani et al., 2000). A strong anticonvulsant effect of a recombinant AAV (rAAV) vector expressing NPY (rAAV–NPY) has been demonstrated in acute models of epilepsy (Haberman et al., 2003; Lin et al., 2003; McCown, 2006; Noe et al., 2007; Richichi et al., 2004). Although these findings were promising, the studies did not test the efficacy of NPY gene therapy in a chronic seizure model in which neuronal death and synaptic reorganization had occurred before gene therapy. To test AAV–NPY gene therapy in neuropathological conditions that resemble human TLE with hippocampal sclerosis, an rAAV–NPY vector expressing the human NPY gene was injected into the brains of rats 3 months after inducing status epilepticus by electrical stimulation of the CA3 subfield of the hippocampus. A majority of these rats developed spontaneous recurrent seizures, and NPY overexpression reduced seizure progression (Noe et al., 2008). Furthermore, compared with rats receiving control rAAV injections, rAAV–NPY injections significantly reduced spontaneous seizure frequency by 40%.

## Galanin

Galanin is another powerful antiepileptic peptide that is produced by multiple neuronal cell types, including noradrenergic neurons in the locus coeruleus and serotonergic neurons in the dorsal raphe nuclei. It is involved in a wide range of behaviors, including feeding, cognition, mood, pain, opiate withdrawal, and seizures. It inhibits neurotransmitter release from neurons expressing norepinephrine, serotonin, or dopamine in the forebrain and inhibits neuronal firing in many classes of neurons (Holmes and Picciotto, 2006; Mazarati et al., 1998, 2000). In the hippocampus, galanin is expressed in cholinergic projections from the septum-basal forebrain and noradrenergic axons originating in the locus coeruleus and is thought to act through subtypes 1 and 2 G-protein-coupled galanin receptors (Lu et al., 2005).

Galanin knockout mice are more susceptible to seizures, and overexpression of galanin results in a seizure-resistant phenotype in mice (Mazarati et al., 2000). In addition, ectopic galanin overexpression in dentate granule cells, as well as hippocampal and cortical pyramidal neurons in transgenic mice, delays seizure development in the kindling model of human temporal lobe epilepsy (Kokaia et al., 2001).

Galanin has powerful anticonvulsant effects after viral gene delivery to the brain (Haberman et al., 2003; Lin et al., 2003; Löscher et al., 2008; McCown, 2006). Hippocampal infusion of galanin-containing AAV leads to a dramatic increase in galanin expression in the dentate molecular layer, hilus region, and mossy fiber tract (Lin et al., 2003). When kainic acid is infused several months later, the experimental mice have fewer seizure events compared with controls, demonstrating that galanin overexpression can have strong anticonvulsant effects.

Overexpression of a secreted form of galanin using an AAV vector with the strong cytomegalovirus promoter and a fibronectin secretion signal suppressed seizures and prevented hilar cell loss (Haberman et al., 2003; Mullen et al., 1992). The neuroprotective effects may be due to the ability of galanin to reduce endogenous excitatory amino acid release in the rat hippocampus (Zini et al., 1993). In a second approach, galanin overexpression was induced in hippocampal mossy fiber–CA3 synapses in an activity-dependent manner with a rAAV vector encoding for galanin under the neuron-specific enolase promoter. This vector delayed the initiation of convulsions but did not alter kindling development (Kanter-Schlifke et al., 2007b). These studies underscore the functional effects of different promoters in driving gene expression levels and the site-specific responses that can be achieved by focal gene delivery.

## GLIAL CELL-DERIVED NEUROTROPHIC FACTOR

Glial cell-derived neurotrophic factor (GDNF), a member of the transforming growth factor-β family of growth factors, promotes neuronal survival by activating the mitogen activated kinase pathway. GDNF has been utilized in gene therapy for temporal lobe epilepsy. In the kindling model of TLE, seizures upregulated GDNF, and delivery of recombinant GDNF suppressed seizures and reduced mossy fiber sprouting but did not alter epileptogenesis (Humpel et al., 1994; Kanter-Schlifke et al., 2007a; Löscher et al., 2008; Martin et al., 1995; Schmidt-Kastner et al., 1994). When rAAV–GDNF was delivered to the hippocampus of adult rats one week before kainic acid-induced SE to increase the levels of GDNF protein, tonic–clonic motor seizures were suppressed and apoptosis inhibitor Bcl-2 was increased. This treatment prevented excitotoxic death of GABAergic neurons in CA3 (Yoo et al., 2006). When rAAV–GDNF was delivered ectopically to an existing seizure focus, it also increased seizure threshold, suggesting that increasing levels of GDNF in the hippocampus may have anticonvulsant and neuroprotective effects (Kanter-Schlifke et al., 2007a).

## GENE THERAPY FOR NEUROPROTECTION

Several strategies employing viral vectors to enhance neuroprotection have been tested in animal models of epilepsy (Lawrence et al., 1995; McLaughlin et al., 2000; Yenari et al., 1998). Sapolsky and his colleagues induced ectopic expression of the glucose transporter-1 or Bcl-2 after seizures with a herpes simplex virus-1 system (McLaughlin et al., 2000). Herpes simplex virus vectors containing the rat brain glucose transporter under the control of strong promoters were shown to increase metabolism and reduce excitotoxic damage in the kainic acid model, even after seizures had occurred.

## PROSPECTS FOR TREATING GENETIC FORMS OF EPILEPSY WITH GENE THERAPY

Although many forms of epilepsy are heterogeneous disorders involving complex changes in the interactions between genes and environmental factors, another category of disorders includes those caused by single gene mutations or well-defined microdeletions that cause epilepsy syndromes. Many of these are channelopathies caused by mutations in genes encoding individual subunits of ion channels that regulate potassium, sodium or calcium flux. Because the expression of ion channels is heterogeneous in different brain regions and different cell types, epilepsy channelopathies may respond to focal gene therapy. Toward this goal, viral-mediated gene transfer has been used to compensate for the loss of the gene *kcna,* encoding the Kv1.1 potassium channel α-subunit in *kcna* knockout mice (Wenzel et al., 2007). This study provided evidence that focal gene therapy can be used for treating specific hippocampal subfields or neuronal populations to reduce seizure activity.

Other categories of mutations associated with inherited epilepsies are loss-of-function mutations in genes controlling GABA levels and transmission. Of a group of 21 genes identified to influence inhibitory synaptic transmission (reviewed in Noebels, 2003), some of these mutations alter interneuron excitability, synaptic levels of GABA, presynaptic release, or the postsynaptic responses of GABA receptors. These include point mutations or truncation mutations in the gamma subunit of GABA receptors that have been linked to childhood tonic–clonic epilepsy with febrile seizures or generalized absence epilepsy and febrile seizures. Other genes were shown to alter the normal balance of excitation and inhibition by interfering with interneuron patterning, fate specification, migration, survival, and distribution. This category includes some of the mutations that are associated with mental retardation, myoclonic epilepsy, lissencephaly, and cortical dysplasia. Studies of gene therapy for treating inherited forms of epilepsy are very limited, but the large number of knockout and transgenic mouse models that have become available will facilitate future work.

From this brief review of the literature, we conclude that gene therapy is a promising avenue for epilepsy treatment in drug-resistant human MTLE. This approach has received considerable validation in experimental animal models, but has not been well studied in monogenic forms of inherited epilepsy. Further work with animal models for human monogenic epilepsies will help to identify which disorders can be treated by focal gene therapy. For those that are associated with early-onset pediatric epilepsy, future research should determine which disorders, if detected during brain development, can be corrected by gene therapy, before widespread damage or compensatory reorganization of neural circuits occurs (Boison, 2007; Conti et al., 2006; Löscher et al., 2008). Further tests for the safety and feasibility of viral vectors are also needed before treatments in human patients. One approach is to test viral gene transfer in cultured human tissue surgically removed to control intractable seizures (Freese et al., 1997). However, uncertainties about the safety of viruses for human gene therapy continue to hamper their development. The potential for viral vectors to transform into malignant types and the impact of foreign gene expression on endogenous gene expression await careful investigation before clinical trials are feasible.

## STEM CELL THERAPY FOR TEMPORAL LOBE EPILEPSY

Cell-based therapies offer several advantages over gene therapy and other more traditional approaches for treating epilepsy, but also pose significant technical challenges. Unlike other tissues of the body, the nervous system has a very limited capacity for self-repair because mature neurons cannot regenerate, and, despite the presence of neural stem cells in the adult brain, their ability to respond to injury is limited. Improving the efficacy of stem cell therapies for replacing neurons or glial cell types destroyed by damage or disease is an extremely active area of investigation. To be successful, grafts of stem cells and their differentiated derivatives in the epileptic brain must not only survive for long periods of time but must also migrate correctly to the appropriate sites, integrate, and establish the correct types of synaptic connections with the host brain. The importance of this last point is underscored by studies showing that seizures induce the genesis of ectopically positioned neurons from endogenous neural stem populations, and these ectopic neurons can contribute to increased excitability and epileptogenesis (Parent, 2007; Scharfman, 2004; Scharfman et al., 2002).

### THE GABA HYPOTHESIS OF TLE

The lateral spread of activity in neocortical circuits is normally restrained by strong surrounding inhibition (Prince and Wilder, 1967). *In vitro* analyses of epileptiform propagation show that seizures can spread through stepwise recruitment of pyramidal neurons (Trevelyan et al., 2006, 2007). Spreading excitation is vetoed by surrounding inhibition; progressive failure of inhibitory networks may account for the slow Jacksonian march of seizures observed in human cortical epilepsy. In the GABA hypothesis, recurrent seizures occur because hippocampal GABAergic interneurons become unable to restrain activity from spreading throughout the limbic circuit. Studies using experimental models of TLE in rodents have supported this hypothesis by showing that a decrease in the functional properties of the GABAergic interneuron networks in the hippocampus is one of the key events underlying temporal lobe epileptogenesis (Cossart et al., 2001).

How can TLE be prevented or reversed by stem cell therapy? The goals of this research are to: (1) modify the fates of embryonic neurons or ES-derived neural stem cells to generate inhibitory neuron precursors; (2) deliver these precursor cells to the appropriate sites of damage in host brain circuits and ensure proper integration; (3) increase their viability in chronic conditions such as TLE; and (4) restore the balance between excitation and inhibition to within the normal range. As discussed below, recent cell transplantation studies in rodent TLE models support this hypothesis by demonstrating that increasing GABAergic neurotransmission or reducing levels of excitation can raise seizure thresholds and reduce both seizure severity and frequency.

## STUDIES OF FETAL NEURONS AS A SOURCE FOR TRANSPLANTATION

Limited repair of the adult hippocampus has been shown with grafts of neural stem cells with more restricted neural or glial fates (Shetty et al., 2008). For example, in the kainic acid model of TLE in rats, comparisons of grafts containing fetal neural precursors obtained from CA1 or CA3 showed that CA3 cell types are more effective for increasing GABAergic function (Shetty et al., 2000). However, detrimental effects of grafts forming abnormal connections with the host brain were found when fetal hippocampal cells were transplanted into the adult hippocampus; these grafts tended to increase epileptic discharges and spontaneous seizures (Buzsaki et al., 1991). By contrast, intrahippocampal grafts of fetal cells from the striatum, when pretreated with growth factors to aid survival of the transplanted cells, successfully reduced seizure frequency and severity in adult seizure models (Buzsaki et al., 1988; Hattiangady et al., 2008).

Fetal stem cell transplants have also included noradrenergic precursors from the locus coeruleus (Bengzon et al., 1993; Buzsaki et al., 1988), the hippocampal CA3 region (Löscher et al., 1998; Shetty and Hattiangady, 2007; Shetty and Turner, 1997), GABAergic precursors from the fetal striatum (Löscher et al., 1998), and GABAergic neural precursors from the ventricular zone of the ganglionic eminence in the embryonic ventral telencephalon (Alvarez-Dolado et al., 2006). When intrahippocampal grafts of striatal neural precursors for GABAergic neurons were transplanted into the hippocampus or substantia nigra in the kindling model of TLE, the grafted cells reduced the incidence of abnormal electrical discharges in response to electrical stimulation of the perforant path and afterdischarges (Buzsaki et al., 1988; Löscher et al., 1998). Moreover, when GABAergic neural precursors from the embryonic medial ganglionic eminence were transplanted into forebrains of young mice, they showed widespread migration across the hippocampus and neocortex. The grafted cells differentiated into distinct types of GABAergic interneurons, functionally integrated, and increased inhibitory tone in the host cortex and hippocampus (Alvarez-Dolado et al., 2006).

These examples generally support the hypothesis that increasing GABAergic neurons in epilepsy models is efficacious for controlling seizures. These studies also underscore the fact that GABAergic precursors and other stem cell types from different regions of the nervous system are not all equally effective and may be detrimental. Moreover, the requirement for harvesting embryonic neurons from aborted human fetal tissue poses a formidable hurdle in the United States. Furthermore, fetal neuronal precursors transplanted into the mature brain do not migrate well, nor do they show long-lasting survival in chronic seizure models. Another hurdle is to stimulate these precursor cells to generate the specific types of neurons or glial cells needed to replace lost or damaged cell types and this has not been possible to date. In addition, immune incompatibility between the donor and host requires that patients receiving fetal stem cells also take immunosuppressive drugs to prevent rejection of fetal cell grafts (Bjorklund and Lindvall, 2000). Autologous stem cell grafts, in which the patient is also the stem cell donor, may help overcome the problem of graft rejection. Skin-derived precursors (SKPs) are one promising type of stem cell located in the hair follicles in adult dermis (Fernandes et al., 2004). SKPs have been derived from rodent and human skin. They can generate immature neurons *in vitro* and *in vivo*, and they survive upon transplantation into normal or kainate-injured hippocampal slices (Fernandes et al., 2006). Work to date, however, indicates that after transplantation they retain immature characteristics and do not fully differentiate into mature neurons.

Another hurdle for stem cell treatments for intractable epilepsy is to discover how to obtain sufficient quantities of particular cell types. Generating neural or glial precursors from pluripotent embryonic stem cells is an alternative to harvesting limited numbers of cells from the embryo. Embryonic stem cell culture methods have the capacity to be adapted for large-scale production of neural precursors *in vitro*. ES cells have the potential to generate any neuronal or glial cell type, because they originate from the inner cell mass of the embryo at the blastocyst stage and give rise to the entire brain and body. It is also relatively straightforward to genetically modify these neural precursors to express recombinant proteins, fluorescent markers, or luciferase to monitor their survival and migration in the living brain. ES cells from mouse, human, and canine lines are available for study, and these cell lines can be genetically engineered to reduce graft rejection.

Embryonic stem cell therapy bypasses some of the ethical roadblocks associated with harvesting stem cells from aborted fetuses, but it has its own ethical problems, including the fact that most of the human ES lines that are being studied were derived from fertilized human embryos. Another major hurdle for ES cell-based therapy is that the risk of tumor formation is high because they are pluripotent and mitotically active. To address this problem, human ES cells have been engineered with suicide genes to allow elimination if the transplanted cells proliferate excessively or form tumors (Schuldiner et al., 2003). As we discuss below, additional strategies for enriching and purifying specific cell types are being developed to help overcome the problem of tumor formation.

## GENETICALLY ENGINEERED CELL LINES AS A SOURCE FOR TRANSPLANTATION

In addition to transplanting acutely dissected fetal cells, transplanted cells have also been used either as a drug-delivery method or to provide local increases in neuromodulators or the neurotransmitter GABA. Most studies have engineered GABA- or adenosine-releasing cells to test for efficacy in controlling seizures (Behrstock et al., 2000; Castillo et al., 2006; Gernert et al., 2002; Li et al., 2007, 2008; Nolte et al., 2008; Ren et al., 2007; Thompson, 2005; Thompson and Suchomelova, 2004). The duration and severity of established seizures have been reduced by transplanting modified GABA-producing cells into the piriform cortex, dentate gyrus, or substantia nigra (Castillo et al., 2008; Gernert et al., 2002; Thompson, 2005). One limitation of these cell lines is that they are not neuronal and do not integrate within the host brain; therefore, the release of neurotransmitters cannot be regulated by neuronal activity. However, genetic modifications, using the Tat-regulatable promoter system (or other comparable regulatable promoters), enable controlled GABA release from the transplanted cells when experimental animals are given doxycycline (Behrstock et al., 2000; Thompson, 2005).

## NEUROMODULATION WITH ADENOSINE

Adenosine is a well-established endogenous neuromodulator that has both neuroprotective and anticonvulsant effects. Adenosine is expressed primarily by astrocytes and modulates neurotransmission. Seizure-induced astrogliosis is thought to increase neuronal excitability by causing adenosine kinase-mediated downregulation of adenosine (Boison, 2006, 2008; Fedele et al., 2005). Adenosine kinase reduces adenosine several hours after SE, and this event is thought to play a decisive role in epileptogenesis (Gouder et al., 2004). Evidence linking loss of adenosine to epileptogenesis suggests that increased expression of adenosine kinase is predictive for epileptogenesis (Li et al., 2008). To test this hypothesis, encapsulated rat fibroblasts or myoblasts, engineered to release adenosine, were transplanted into the lateral ventricles and shown to suppress seizures in kindled rats (Guttinger et al., 2005). These results raise the possibility that focal release of adenosine by intraventricular cell grafts is a feasible option for long-term treatment of TLE. To be used in human patients, this approach would require xenografting or the use of human fibroblasts derived from the patient.

Alternatively, adult stem cells were tested in an autologous grafting protocol that used RNA interference (RNAi) to knock down adenosine kinase. Human mesenchymal stem cells were transduced with a lentivirus containing an anti-adenosine kinase microRNA and grafted into the hippocampal fissure of mice following kainic acid-induced injury to the amygdala. In this paradigm, lentiviral anti-adenosine kinase RNAi reduced adenosine kinase in the stem cells, reduced hippocampal cell loss, and significantly reduced seizure activity (Ren et al., 2007). It has been shown that transplantation of adenosine kinase-deficient neural precursors, derived from modified ES cells, eliminated spontaneous seizures induced by kainic acid (Li et al., 2008).

In summary, engineered cells expressing inhibitory neurotransmitters such as GABA or neuromodulators such as adenosine are providing new vistas for therapies aimed at seizure control and preventing the development of epilepsy. Unless the secretion of GABA or adenosine is activity dependent, however, release will not be compensatory during seizures and may be excessive during interictal periods.

## ES Cell-Derived Neurons for Cell Transplantation Therapies

Studies are now underway investigating the use of ES cell-derived neural stem cells as a source of transplantable neurons for treating epilepsy. When mouse or human ES cell-derived neural precursors were transplanted into the CA3 region of the hippocampus of mice one week following kainic acid-induced seizures, they migrated into the dentate gyrus. Thereafter, they became integrated into the host brain, differentiated into neurons, received synaptic connections, and survived for prolonged periods of up to 2 months (Carpentino et al., 2007). ES cell-derived neural progenitor cells appear to be superior to fetal cells in terms of their ability to adapt to the host environment, as shown by broader migration and more robust effect in controlling seizures (Li et al., 2007). However, only a fraction of the cells survive after they are transplanted into the adult brain. Although very small numbers of surviving cells may be sufficient to produce therapeutic benefits in some neurological conditions, neuroprotective growth factors such as vascular endothelial growth factor or erythropoietin may be required for protecting immature grafted neurons in chronic conditions such as epilepsy (Ferriero, 2005). In addition, gliosis and concomitant inflammation are aspects of chronic epilepsy that may reduce graft survival and impair efforts for repairing the hippocampus.

## Generating Specific Types of Neurons for Transplantation from ES Cells

Before ES cell-based therapies can be used for repairing neural circuits, it must be determined how to generate specific cell types (or their precursors), such as GABAergic neurons or principal neurons that release glutamate and form long-range axonal projections (Figure 33.4 and Figure 33.5). Both of these cell types are damaged or lost in temporal lobe epilepsy. Knowledge gained in the field of developmental neurobiology will greatly aid these efforts. In most studies, transplants were made before the development of recurrent seizures. Validating this approach requires studies of rodents with spontaneous seizures to determine whether the transplants are successful in chronic epilepsy. We now understand that most, if not all, of the functionally distinct subtypes of forebrain GABAergic neurons are derived from microdomains of the ventricular zone in the ventral telencephalon. The different types of interneurons form in the ganglionic eminences, transient bulges in the ventricles of the ventral forebrain, where combinatorial codes of transcription factors regulate their genesis and specify cell fates (Cobos et al., 2005a,b; Gonzalez et al., 2002; Lavdas et al., 1999; Marin et al., 2000; Parnavelas, 1992; Trinh et al., 2006; Wonders and Anderson, 2005a,b, 2006; Wonders et al., 2008).

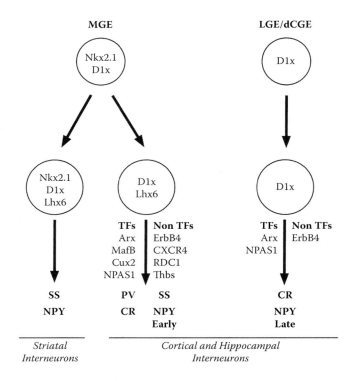

**FIGURE 33.4**  Model of transcriptional regulatory mechanisms controlling the fates and migrations of GABAergic interneurons derived from the ganglionic eminences (see, for example, Zhao et al., 2008).

Interneurons take multiple routes from the ventral telencephalon as they migrate to reach the striatum, cerebral cortex, or hippocampus, and a complex code of adhesive and repulsive cues guides this process (Alifragis et al., 2004; Anderson et al., 2001; Denaxa et al., 2001; Friocourt et al., 2008; Marin et al., 2003). Information about the molecular codes for regulating GABAergic interneuron identity and migration should now make it possible to direct ES cell-derived neural progenitor differentiation into particular fates by stable transfection with expression vectors that drive expression of transcription factors specifying GABAergic interneurons in the embryonic brain (Anderson et al., 1997; Butt et al., 2007; Du et al., 2008; Wonders and Anderson, 2005a; Wonders et al., 2008; Xu et al., 2003, 2004, 2005).

## Role of Epigenetic Factors in Regulating the Production and Differentiation of ES Cell-Derived Neural Precursors for Transplantation

In addition to transcriptional codes that determine cell fates, secreted signaling molecules and epigenetic factors also appear to play a significant role in guiding neural stem cell differentiation. The production of neural stem cells for neuronal replacement therapies may therefore require additional steps that incorporate intercellular communication via cell surface receptors, secreted neurotransmitters, and other soluble factors. For example, the neurotransmitter GABA has long been known to potentiate the maturation of developing neurons (Antonopoulos et al., 1997). The secreted protein Sonic hedgehog (Shh), released from the ventral forebrain, plays a critical role in maintaining the expression of transcription factors in interneuron precursors (Xu et al., 2005). Sequential treatments of ES cells with retinoic acid and Shh can induce caudal and ventral spinal motor neuron

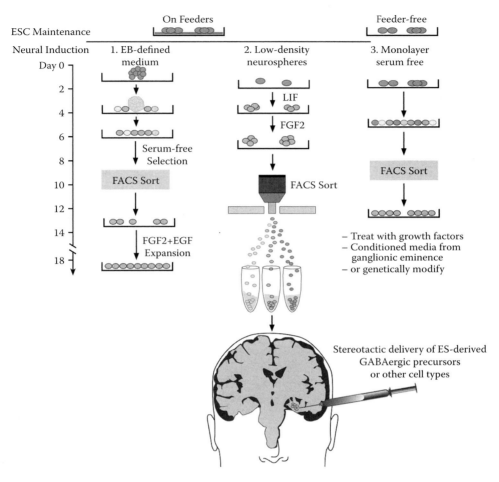

**FIGURE 33.5** **(See color insert following page 458.)** Approaches for producing and purifying embryonic stem (ES) cell-derived neurons for stereotaxic delivery into limbic structures for treating temporal lobe epilepsy, based on studies in many labs. *Abbreviations:* EB, embryoid body; ESC, embryonic stem cell; FACS, fluorescence activated cell sorting; LIF, leukemia inhibitory factor. Figure is based on work in our laboratory and others (see, for example, review by Cai and Grabel, 2007).

phenotypes (Li et al., 2005). Several recent studies described protocols for generating GABAergic and other neuronal cell types by adding a specific sequence of growth factors and diffusible molecules to ES cell cultures (Barberi et al., 2003). Serum treatment and coculturing ES cells with primary fetal cells are two additional approaches employed.

These studies suggest that inductive interactions and the factors secreted by more differentiated cells may influence the choices ES cells make toward specific lineages. In fact, the vast majority of ES cells at the undifferentiated stage are cultured on feeder layer cells such as fibroblast cells or stromal cell lines; therefore, the notion of deriving specific neuronal types from ES cells by coculturing with other cell types is not new. In most cases, however, the elements influencing ES cell differentiation in a coculture system are still undefined. Specific protocols, as well as detailed transcriptomes of interneurons, will soon be available and will ultimately make it possible to evaluate the relative strengths and limitations of different ES cell protocols that employ forced expression of transcriptional factors, defined media, or coculturing for producing specific neuron types.

One powerful tool that is under development is the construction of human and mouse ES cell lines with reporter genes that are driven by lineage specific promoters (Giudice and Trounson, 2008). In this approach, the green fluorescent protein (GFP) gene is turned on only transiently during the neural stem cell stage; the expression is then terminated upon differentiation into postmitotic neurons. Lineage-specific expression is advantageous because neural stem cells can be enriched by methods such as fluorescence-activated cell sorting (FACS), a state-of-the-art technique to pool cells based on their fluorescence expression profiles. FACS purification allows neural stem cells to be separated from undifferentiated and potentially tumorigenic stem cells. For example, transduction of ES cells with lentivirus carrying GFP genes under the phosphoglycerate kinase promoter turns on GFP during differentiation of the ES cells into hematopoietic lineages (Hamaguchi et al., 2000). One of the best characterized genes for this purpose is *Sox1*, the earliest marker currently known for neural stem cells (Ying et al., 2003). When ES cells expressing *Sox1*–GFP were FACS sorted prior to transplanting them, it was found that they produced well-contained grafts without forming tumors (Chung et al., 2006).

## CONCLUSIONS

Only a few years ago, the potential to correct imbalances between excitation and inhibition or to make specific neuronal cell types for transplantation therapies to treat epilepsy seemed out of reach. The rapid progress in technology, including development of safer viral vectors for gene delivery, knowledge about the developmental programs for generating individual classes of neurons from ES cells, and strategies for preventing the death of endogenous neurons, is now opening the door for novel therapies to treat intractable epilepsy.

## ACKNOWLEDGMENTS

Grants to JRN from the National Institute of Neurological Diseases and Stroke (NS42826) and the McKnight Foundation supported this work. XM is supported by a graduate fellowship funded by Connecticut Stem Cell Research funds. The authors thank Marilee Ogren, Paul Lombroso, Laura Grabel, Gloster Aaron, John Kirn, and David Maisano for helpful discussions and comments.

## REFERENCES

Acharya, M.M., B. Hattiangady, A.K. Shetty. (2008). Progress in neuroprotective strategies for preventing epilepsy. *Prog. Neurobiol.*, 84(4):363–404.

Alifragis, P., A. Liapi, J.G. Parnavelas. (2004). Lhx6 regulates the migration of cortical interneurons from the ventral telencephalon but does not specify their GABA phenotype. *J. Neurosci.*, 24(24):5643–5648.

Alvarez-Dolado, M., M.E. Calcagnotto, K.M. Karkar, D.G. Southwell, D.M. Jones-Davis, R.C. Estrada, J.L. Rubenstein, A. Alvarez-Buylla, S.C. Baraban. (2006). Cortical inhibition modified by embryonic neural precursors grafted into the postnatal brain. *J. Neurosci.*, 26(28):7380–7389.

Anderson, S.A., D.D. Eisenstat, L. Shi, J.L. Rubenstein. (1997). Interneuron migration from basal forebrain to neocortex: dependence on *Dlx* genes. *Science*, 278(5337):474–476.

Anderson, S.A., O. Marin, C. Horn, K. Jennings, J.L. Rubenstein. (2001). Distinct cortical migrations from the medial and lateral ganglionic eminences. *Development.*, 128(3):353–363.

Antonopoulos, J., I.S. Pappas, J.G. Parnavelas. (1997). Activation of the GABA$_A$ receptor inhibits the proliferative effects of bFGF in cortical progenitor cells. *Eur. J. Neurosci.*, 9(2):291–298.

Baraban, S.C., G. Hollopeter, J.C. Erickson, P.A. Schwartzkroin, R.D. Palmiter. (1997). Knock-out mice reveal a critical antiepileptic role for neuropeptide Y. *J. Neurosci.*, 17(23):8927–8936.

Barberi, T., P. Klivenyi, N.Y. Calingasan, H. Lee, H. Kawamata, K. Loonam, A.L. Perrier, J. Bruses, M.E. Rubio, N. Topf, V. Tabar, N.L. Harrison, M.F. Beal, M.A. Moore, L. Studer. (2003). Neural subtype specification of fertilization and nuclear transfer embryonic stem cells and application in parkinsonian mice. *Nat. Biotechnol.*, 21(10):1200–1207.

Behrstock, S.P., V. Anantharam, K.W. Thompson, E.S. Schweitzer, A.J. Tobin. (2000). Conditionally-immortalized astrocytic cell line expresses GAD and secretes GABA under tetracycline regulation. *J. Neurosci. Res.*, 60(3):302–310.

Bengzon, J., Z. Kokaia, O. Lindvall. (1993). Specific functions of grafted locus coeruleus neurons in the kindling model of epilepsy. *Exp. Neurol.*, 122(1):143–154.

Bjorklund, A., O. Lindvall. (2000). Cell replacement therapies for central nervous system disorders. *Nat. Neurosci.*, 3(6):537–544.

Boison, D. (2006). Adenosine kinase, epilepsy and stroke: mechanisms and therapies. *Trends Pharmacol. Sci.*, 27(12):652–658.

Boison, D. (2007). Cell and gene therapies for refractory epilepsy. *Curr. Neuropharmacol.*, 5:115–125.

Boison, D. (2008). The adenosine kinase hypothesis of epileptogenesis. *Prog. Neurobiol.*, 84(3):249–262.

Brooks-Kayal, A.R., M.D. Shumate, H. Jin, D.D. Lin, T.Y. Rikhter, K.L. Holloway, D.A. Coulter. (1999). Human neuronal gamma-aminobutyric acid(A) receptors: coordinated subunit mRNA expression and functional correlates in individual dentate granule cells. *J. Neurosci.*, 19(19):8312–8318.

Butt, S.J., I. Cobos, J. Golden, N. Kessaris, V. Pachnis, S. Anderson. (2007). Transcriptional regulation of cortical interneuron development. *J. Neurosci.*, 27(44):11847–11850.

Buzsaki, G., G. Ponomareff, F. Bayardo, T. Shaw, F.H. Gage. (1988). Suppression and induction of epileptic activity by neuronal grafts. *Proc. Natl. Acad. Sci. U.S.A.*, 85(23):9327–9330.

Buzsaki, G., E. Masliah, L.S. Chen, Z. Horvath, R. Terry, F.H. Gage. (1991). Hippocampal grafts into the intact brain induce epileptic patterns. *Brain Res.*, 554(1–2):30–37.

Cai, C., L. Grabel. (2007). Directing the differentiation of embryonic stem cells to neural stem cells. *Dev. Dyn.*, 236(12):3255–3266.

Carpentino, J.E., N.W. Hartman, L.B. Grabel, J.R. Naegele. (2007). Region-specific differentiation of embryonic stem cell-derived neural progenitor transplants into the adult mouse hippocampus following seizures. *J. Neurosci. Res.*, 86(3):512–524.

Castillo, C.G., S. Mendoza, W.J. Freed, M. Giordano. (2006). Intranigral transplants of immortalized GABAergic cells decrease the expression of kainic acid-induced seizures in the rat. *Behav. Brain. Res.*, 171(1):109–115.

Castillo, C.G., S. Mendoza-Trejo, A.B. Aguilar, W.J. Freed, M. Giordano. (2008). Intranigral transplants of a GABAergic cell line produce long-term alleviation of established motor seizures. *Behav. Brain Res.*, 193(1):17–27.

Chung, S., B.S. Shin, E. Hedlund, J. Pruszak, A. Ferree, U.J. Kang, O. Isacson, K.S. Kim. (2006). Genetic selection of sox1GFP-expressing neural precursors removes residual tumorigenic pluripotent stem cells and attenuates tumor formation after transplantation. *J. Neurochem.*, 97(5):1467–1480.

Cobos, I., V. Broccoli, J.L. Rubenstein. (2005a). The vertebrate ortholog of Aristaless is regulated by *Dlx* genes in the developing forebrain. *J. Comp. Neurol.*, 483(3):292–303.

Cobos, I., M.E. Calcagnotto, A.J. Vilaythong, M.T. Thwin, J.L. Noebels, S.C. Baraban, J.L. Rubenstein. (2005b). Mice lacking *Dlx1* show subtype-specific loss of interneurons, reduced inhibition and epilepsy. *Nat. Neurosci.*, 8(8):1059–1068.

Conti, L., E. Reitano, E. Cattaneo. (2006). Neural stem cell systems: diversities and properties after transplantation in animal models of diseases. *Brain Pathol.*, 16(2):143–154.

Cossart, R., C. Dinocourt, J.C. Hirsch, A. Merchan-Perez, J. De Felipe, Y. Ben-Ari, M. Esclapez, C. Bernard. (2001). Dendritic but not somatic GABAergic inhibition is decreased in experimental epilepsy. *Nat. Neurosci.*, 4(1):52–62.

Denaxa, M., C.H. Chan, M. Schachner, J.G. Parnavelas, D. Karagogeos. (2001). The adhesion molecule TAG-1 mediates the migration of cortical interneurons from the ganglionic eminence along the corticofugal fiber system. *Development*, 128(22):4635–4644.

Du, T., Q. Xu, P.J. Ocbina, S.A. Anderson. (2008). NKX2.1 specifies cortical interneuron fate by activating *lhx6*. *Development*, 135(8):1559–1567.

El Bahh, B., S. Balosso, T. Hamilton, H. Herzog, A.G. Beck-Sickinger, G. Sperk, D.R. Gehlert, A. Vezzani, W.F. Colmers. (2005). The anti-epileptic actions of neuropeptide Y in the hippocampus are mediated by Y and not Y receptors. *Eur. J. Neurosci.*, 22(6):1417–1430.

Fedele, D.E., N. Gouder, M. Guttinger, L. Gabernet, L. Scheurer, T. Rulicke, F. Crestani, D. Boison. (2005). Astrogliosis in epilepsy leads to overexpression of adenosine kinase, resulting in seizure aggravation. *Brain*, 128(Pt. 10):2383–2395.

Fernandes, K.J., I.A. McKenzie, P. Mill, K.M. Smith, M. Akhavan, F. Barnabe-Heider, J. Biernaskie, A. Junek, N.R. Kobayashi, J.G. Toma, D.R. Kaplan, P.A. Labosky, V. Rafuse, C.C. Hui, F.D. Miller. (2004). A dermal niche for multipotent adult skin-derived precursor cells. *Nat. Cell Biol.*, 6(11):1082–1093.

Fernandes, K.J., N.R. Kobayashi, C.J. Gallagher, F. Barnabe-Heider, A. Aumont, D.R. Kaplan, F.D. Miller. (2006). Analysis of the neurogenic potential of multipotent skin-derived precursors. *Exp. Neurol.*, 201(1):32–48.

Ferriero, D.M. (2005). Protecting neurons. *Epilepsia*, 46(Suppl. 7):45–51.

Freese, A., M.G. Kaplitt, W.M. O'Connor, M. Abbey, D. Langer, P. Leone, M.J. O'Connor, M.J. During. (1997). Direct gene transfer into human epileptogenic hippocampal tissue with an adeno-associated virus vector: implications for a gene therapy approach to epilepsy. *Epilepsia*, 38(7):759–766.

Friocourt, G., S. Kanatani, H. Tabata, M. Yozu, T. Takahashi, M. Antypa, O. Raguenes, J. Chelly, C. Ferec, K. Nakajima, J.G. Parnavelas. (2008). Cell-autonomous roles of ARX in cell proliferation and neuronal migration during corticogenesis. *J. Neurosci.*, 28(22):5794–5805.

Gant, J.C., O. Thibault, E.M. Blalock, J. Yang, A. Bachstetter, J. Kotick, P.E. Schauwecker, K.F. Hauser, G.M. Smith, R. Mervis, Y. Li, G.N. Barnes. (2008). Decreased number of interneurons and increased seizures in neuropilin 2 deficient mice: implications for autism and epilepsy. *Epilepsia*, 50(4):629–645.

Gernert, M., K.W. Thompson, W. Löscher, A.J. Tobin. (2002). Genetically engineered GABA-producing cells demonstrate anticonvulsant effects and long-term transgene expression when transplanted into the central piriform cortex of rats. *Exp. Neurol.*, 176(1):183–192.

Giudice, A., A. Trounson. (2008). Genetic modification of human embryonic stem cells for derivation of target cells. *Cell Stem Cell*, 2(5):422–433.

Gonzalez, A., J.M. Lopez, C. Sanchez-Camacho, O. Marin. (2002). Regional expression of the homeobox gene *Nkx2-1* defines pallidal and interneuronal populations in the basal ganglia of amphibians. *Neuroscience*, 114(3):567–575.

Gouder, N., L. Scheurer, J.M. Fritschy, D. Boison. (2004). Overexpression of adenosine kinase in epileptic hippocampus contributes to epileptogenesis. *J. Neurosci.*, 24(3):692–701.

Guttinger, M., V. Padrun, W.F. Pralong, D. Boison. (2005). Seizure suppression and lack of adenosine A1 receptor desensitization after focal long-term delivery of adenosine by encapsulated myoblasts. *Exp. Neurol.*, 193(1):53–64.

Haberman, R.P., R.J. Samulski, T.J. McCown. (2003). Attenuation of seizures and neuronal death by adeno-associated virus vector galanin expression and secretion. *Nat. Med.*, 9(8):1076–1080.

Hamaguchi, I., N.B. Woods, I. Panagopoulos, E. Andersson, H. Mikkola, C. Fahlman, R. Zufferey, L. Carlsson, D. Trono, S. Karlsson. (2000). Lentivirus vector gene expression during ES cell-derived hematopoietic development *in vitro. J. Virol.*, 74(22):10778–10784.

Hattiangady, B., M.S. Rao, A.K. Shetty. (2008). Grafting of striatal precursor cells into hippocampus shortly after status epilepticus restrains chronic temporal lobe epilepsy. *Exp. Neurol.*, 212(2):468–481.

Holmes, A., M.R. Picciotto. (2006). Galanin: a novel therapeutic target for depression, anxiety disorders and drug addiction? *CNS Neurol. Disord. Drug Targets*, 5(2):225–232.

Hu, Y., I.V. Lund, M.C. Gravielle, D.H. Farb, A.R. Brooks-Kayal, S.J. Russek. (2008). Surface expression of $GABA_A$ receptors is transcriptionally controlled by the interplay of cAMP-response element-binding protein and its binding partner inducible cAMP early repressor. *J. Biol. Chem.*, 283(14):9328–9340.

Humpel, C., B. Hoffer, I. Stromberg, S. Bektesh, F. Collins, L. Olson. (1994). Neurons of the hippocampal formation express glial cell line-derived neurotrophic factor messenger RNA in response to kainate-induced excitation. *Neuroscience*, 59(4):791–795.

Kanter-Schlifke, I., B. Georgievska, D. Kirik, M. Kokaia. (2007a). Seizure suppression by GDNF gene therapy in animal models of epilepsy. *Mol. Ther.*, 15(6):1106–1113.

Kanter-Schlifke, I., A. Toft Sorensen, M. Ledri, E. Kuteeva, T. Hokfelt, M. Kokaia. (2007b). Galanin gene transfer curtails generalized seizures in kindled rats without altering hippocampal synaptic plasticity. *Neuroscience*, 150(4):984–992.

Kokaia, M., K. Holmberg, A. Nanobashvili, Z.Q. Xu, Z. Kokaia, U. Lendahl, S. Hilke, E. Theodorsson, U. Kahl, T. Bartfai, O. Lindvall, T. Hokfelt. (2001). Suppressed kindling epileptogenesis in mice with ectopic overexpression of galanin. *Proc. Natl. Acad. Sci. U.S.A.*, 98(24):14006–14011.

Lavdas, A.A., M. Grigoriou, V. Pachnis, J.G. Parnavelas. (1999). The medial ganglionic eminence gives rise to a population of early neurons in the developing cerebral cortex. *J. Neurosci.*, 19(18):7881–7888.

Lawrence, M.S., D.Y. Ho, R. Dash, R.M. Sapolsky. (1995). Herpes simplex virus vectors overexpressing the glucose transporter gene protect against seizure-induced neuron loss. *Proc. Natl. Acad. Sci. U.S.A.*, 92(16):7247–7251.

Li, T., J.A. Steinbeck, T. Lusardi, P. Koch, J.Q. Lan, A. Wilz, M. Segschneider, R.P. Simon, O. Brustle, D. Boison. (2007). Suppression of kindling epileptogenesis by adenosine releasing stem cell-derived brain implants. *Brain*, 130(Pt. 5):1276–1288.

Li, T., G. Ren, T. Lusardi, A. Wilz, J.Q. Lan, T. Iwasato, S. Itohara, R.P. Simon, D. Boison. (2008). Adenosine kinase is a target for the prediction and prevention of epileptogenesis in mice. *J. Clin. Invest.*, 118(2):571–582.

Li, X.J., Z.W. Du, E.D. Zarnowska, M. Pankratz, L.O. Hansen, R.A. Pearce, S.C. Zhang. (2005). Specification of motoneurons from human embryonic stem cells. *Nat. Biotechnol.*, 23(2):215–221.

Lin, E.J., C. Richichi, D. Young, K. Baer, A. Vezzani, M.J. During. (2003). Recombinant AAV-mediated expression of galanin in rat hippocampus suppresses seizure development. *Eur. J. Neurosci.*, 18(7):2087–2092.

Löscher, W., U. Ebert, H. Lehmann, C. Rosenthal, G. Nikkhah. (1998). Seizure suppression in kindling epilepsy by grafts of fetal GABAergic neurons in rat substantia nigra. *J. Neurosci. Res.*, 51(2):196–209.

Löscher, W., M. Gernert, U. Heinemann. (2008). Cell and gene therapies in epilepsy: promising avenues or blind alleys? *Trends Neurosci.*, 31(2):62–73.

Lu, X., A. Mazarati, P. Sanna, S. Shinmei, T. Bartfai. (2005). Distribution and differential regulation of galanin receptor subtypes in rat brain: effects of seizure activity. *Neuropeptides*, 39(3):147–152.

Marin, O., S.A. Anderson, J.L. Rubenstein. (2000). Origin and molecular specification of striatal interneurons. *J. Neurosci.*, 20(16):6063–6076.

Marin, O., A.S. Plump, N. Flames, C. Sanchez-Camacho, M. Tessier-Lavigne, J.L. Rubenstein. (2003). Directional guidance of interneuron migration to the cerebral cortex relies on subcortical Slit1/2-independent repulsion and cortical attraction. *Development*, 130(9):1889–1901.

Martin, D., G. Miller, M. Rosendahl, D.A. Russell. (1995). Potent inhibitory effects of glial derived neurotrophic factor against kainic acid mediated seizures in the rat. *Brain Res.*, 683(2):172–178.

Mazarati, A.M., H. Liu, U. Soomets, R. Sankar, D. Shin, H. Katsumori, U. Langel, C.G. Wasterlain. (1998). Galanin modulation of seizures and seizure modulation of hippocampal galanin in animal models of status epilepticus. *J. Neurosci.*, 18(23):10070–10077.

Mazarati, A.M., J.G. Hohmann, A. Bacon, H. Liu, R. Sankar, R.A. Steiner, D. Wynick, C.G. Wasterlain. (2000). Modulation of hippocampal excitability and seizures by galanin. *J. Neurosci.*, 20(16):6276–6281.

McCown, T.J. (2005). Adeno-associated virus (AAV) vectors in the CNS. *Curr. Gene Ther.*, 5(3):333–338.

McCown, T.J. (2006). Adeno-associated virus-mediated expression and constitutive secretion of galanin suppresses limbic seizure activity *in vivo*. *Mol. Ther.*, 14(1):63–68.

McLaughlin, J., B. Roozendaal, T. Dumas, A. Gupta, O. Ajilore, J. Hsieh, D. Ho, M. Lawrence, J.L. McGaugh, R. Sapolsky. (2000). Sparing of neuronal function postseizure with gene therapy. *Proc. Natl. Acad. Sci. U.S.A.*, 97(23):12804–12809.

Mullen, R.J., C.R. Buck, A.M. Smith. (1992). NeuN, a neuronal specific nuclear protein in vertebrates. *Development*, 116(1):201–211.

Noe, F., J. Nissinen, A. Pitkanen, M. Gobbi, G. Sperk, M. During, A. Vezzani. (2007). Gene therapy in epilepsy: the focus on NPY. *Peptides*, 28(2):377–383.

Noe, F., A.H. Pool, J. Nissinen, M. Gobbi, R. Bland, M. Rizzi, C. Balducci, F. Ferraguti, G. Sperk, M.J. During, A. Pitkanen, A. Vezzani. (2008). Neuropeptide Y gene therapy decreases chronic spontaneous seizures in a rat model of temporal lobe epilepsy. *Brain*, 131(Pt. 6):1506–1515.

Noebels, J.L. (2003). The biology of epilepsy genes. *Annu. Rev. Neurosci.*, 26:599–625.

Nolte, M.W., W. Löscher, C. Herden, W.J. Freed, M. Gernert. (2008). Benefits and risks of intranigral transplantation of GABA-producing cells subsequent to the establishment of kindling-induced seizures. *Neurobiol. Dis.*, 31(3):342–354.

Parent, J.M. (2007). Adult neurogenesis in the intact and epileptic dentate gyrus. *Prog. Brain Res.*, 163:529–540.

Parnavelas, J.G. (1992). Development of GABA-containing neurons in the visual cortex. *Prog. Brain Res.*, 90:523–537.

Pitkanen, A., P.A. Schwartzkroin, S.L. Moshe, Eds. (2006). *Models of Seizures and Epilepsy*. New York: Elsevier, 687 pp.

Pitkanen, A., I. Kharatishvili, H. Karhunen, K. Lukasiuk, R. Immonen, J. Nairismagi, O. Grohn, J. Nissinen. (2007). Epileptogenesis in experimental models. *Epilepsia*, 48(Suppl. 2):13–20.

Prince, D.A., B.J. Wilder. (1967). Control mechanisms in cortical epileptogenic foci: "surround" inhibition. *Arch. Neurol.*, 16(2):194–202.

Raol, Y.H., G. Zhang, E.C. Budreck, A.R. Brooks-Kayal. (2005). Long-term effects of diazepam and phenobarbital treatment during development on GABA receptors, transporters and glutamic acid decarboxylase. *Neuroscience*, 132(2):399–407.

Raol, Y.H., I.V. Lund, S. Bandyopadhyay, G. Zhang, D.S. Roberts, J.H. Wolfe, S.J. Russek, A.R. Brooks-Kayal. (2006). Enhancing GABA(A) receptor alpha 1 subunit levels in hippocampal dentate gyrus inhibits epilepsy development in an animal model of temporal lobe epilepsy. *J. Neurosci.*, 26(44):11342–11346.

Ren, G., T. Li, J.Q. Lan, A. Wilz, R.P. Simon, D. Boison. (2007). Lentiviral RNAi-induced downregulation of adenosine kinase in human mesenchymal stem cell grafts: a novel perspective for seizure control. *Exp. Neurol.*, 208(1):26–37.

Riban, V., H.L. Fitzsimons, M.J. During. (2008). Gene therapy in epilepsy. *Epilepsia*, 50(1):24–32.

Richichi, C., E.J. Lin, D. Stefanin, D. Colella, T. Ravizza, G. Grignaschi, P. Veglianese, G. Sperk, M.J. During, A. Vezzani. (2004). Anticonvulsant and antiepileptogenic effects mediated by adeno-associated virus vector neuropeptide Y expression in the rat hippocampus. *J. Neurosci.*, 24(12):3051–3059.

Roberts, D.S., Y.H. Raol, S. Bandyopadhyay, I.V. Lund, E.C. Budreck, M.A. Passini, J.H. Wolfe, A.R. Brooks-Kayal, S.J. Russek. (2005). Egr3 stimulation of GABRA4 promoter activity as a mechanism for seizure-induced up-regulation of GABA(A) receptor alpha4 subunit expression. *Proc. Natl. Acad. Sci. U.S.A.*, 102(33):11894–11899.

Scharfman, H.E. (2004). Functional implications of seizure-induced neurogenesis. *Adv. Exp. Med. Biol.*, 548:192–212.

Scharfman, H.E., J.H. Goodman, A.L. Sollas, S.D. Croll. (2002). Spontaneous limbic seizures after intrahippocampal infusion of brain-derived neurotrophic factor. *Exp. Neurol.*, 174(2):201–214.

Schmidt-Kastner, R., A. Tomac, B. Hoffer, S. Bektesh, B. Rosenzweig, L. Olson. (1994). Glial cell-line derived neurotrophic factor (GDNF) mRNA upregulation in striatum and cortical areas after pilocarpine-induced status epilepticus in rats. *Brain Res. Mol. Brain Res.*, 26(1–2):325–330.

Schuldiner, M., J. Itskovitz-Eldor, N. Benvenisty. (2003). Selective ablation of human embryonic stem cells expressing a "suicide" gene. *Stem Cells*, 21(3):257–265.

Shetty, A.K., B. Hattiangady. (2007). Restoration of calbindin after fetal hippocampal CA3 cell grafting into the injured hippocampus in a rat model of temporal lobe epilepsy. *Hippocampus*, 17(10):943–956.

Shetty, A.K., D.A. Turner. (1997). Fetal hippocampal cells grafted to kainate-lesioned CA3 region of adult hippocampus suppress aberrant supragranular sprouting of host mossy fibers. *Exp. Neurol.*, 143(2):231–245.

Shetty, A.K., V. Zaman, D.A. Turner. (2000). Pattern of long-distance projections from fetal hippocampal field CA3 and CA1 cell grafts in lesioned CA3 of adult hippocampus follows intrinsic character of respective donor cells. *Neuroscience*, 99(2):243–255.

Shetty, A.K., M.S. Rao, B. Hattiangady. (2008). Behavior of hippocampal stem/progenitor cells following grafting into the injured aged hippocampus. *J. Neurosci. Res.*, 86(14):3062–3074.

Theodore, W.H., S.S. Spencer, S. Wiebe, J.T. Langfitt, A. Ali, P.O. Shafer, A.T. Berg, B.G. Vickrey. (2006). Epilepsy in North America: a report prepared under the auspices of the global campaign against epilepsy, the International Bureau for Epilepsy, the International League Against Epilepsy, and the World Health Organization. *Epilepsia*, 47(10):1700–1722.

Thompson, K.W. (2005). Genetically engineered cells with regulatable GABA production can affect after-discharges and behavioral seizures after transplantation into the dentate gyrus. *Neuroscience*, 133(4):1029–1037.

Thompson, K.W., L.M. Suchomelova. (2004). Transplants of cells engineered to produce GABA suppress spontaneous seizures. *Epilepsia*, 45(1):4–12.

Trevelyan, A.J., D. Sussillo, B.O. Watson, R. Yuste. (2006). Modular propagation of epileptiform activity: evidence for an inhibitory veto in neocortex. *J. Neurosci.*, 26(48):12447–12455.

Trevelyan, A.J., D. Sussillo, R. Yuste. (2007). Feedforward inhibition contributes to the control of epileptiform propagation speed. *J. Neurosci.*, 27(13):3383–3387.

Trinh, H.H., J. Reid, E. Shin, A. Liapi, J.G. Parnavelas, B. Nadarajah. (2006). Secreted factors from ventral telencephalon induce the differentiation of GABAergic neurons in cortical cultures. *Eur. J. Neurosci.*, 24(11):2967–2977.

Tuunanen, J., T. Halonen, A. Pitkanen. (1997). Decrease in somatostatin-immunoreactive neurons in the rat amygdaloid complex in a kindling model of temporal lobe epilepsy. *Epilepsy Res.*, 26(2):315–327.

Vezzani, A., G. Sperk, W.F. Colmers. (1999). Neuropeptide Y: emerging evidence for a functional role in seizure modulation. *Trends Neurosci.*, 22(1):25–30.

Vezzani, A., D. Moneta, F. Mule, T. Ravizza, M. Gobbi, J. French-Mullen. (2000). Plastic changes in neuropeptide Y receptor subtypes in experimental models of limbic seizures. *Epilepsia*, 41(Suppl. 6):S115–S121.

Wasterlain, C.G., A.M. Mazarati, D. Naylor, J. Niquet, H. Liu, L. Suchomelova, R. Baldwin, H. Katsumori, Y. Shirasaka, D. Shin, R. Sankar. (2002). Short-term plasticity of hippocampal neuropeptides and neuronal circuitry in experimental status epilepticus. *Epilepsia*, 43(Suppl. 5):20–29.

Wenzel, H.J., H. Vacher, E. Clark, J.S. Trimmer, A.L. Lee, R.M. Sapolsky, B.L. Tempel, P.A. Schwartzkroin. (2007). Structural consequences of *Kcna1* gene deletion and transfer in the mouse hippocampus. *Epilepsia*, 48(11):2023–2046.

Wonders, C., S.A. Anderson. (2005a). Beyond migration: *Dlx1* regulates interneuron differentiation. *Nat. Neurosci.*, 8(8):979–981.

Wonders, C., S.A. Anderson. (2005b). Cortical interneurons and their origins. *Neuroscientist*, 11(3):199–205.

Wonders, C., S.A. Anderson. (2006). The origin and specification of cortical interneurons. *Nat. Rev. Neurosci.*, 7(9):687–696.

Wonders, C.P., L. Taylor, J. Welagen, I.C. Mbata, J.Z. Xiang, S.A. Anderson. (2008). A spatial bias for the origins of interneuron subgroups within the medial ganglionic eminence. *Dev. Biol.*, 314(1):127–136.

Xu, Q., E. de la Cruz, S.A. Anderson. (2003). Cortical interneuron fate determination: diverse sources for distinct subtypes? *Cereb. Cortex*, 13(6):670–676.

Xu, Q., I. Cobos, E. De La Cruz, J.L. Rubenstein, S.A. Anderson. (2004). Origins of cortical interneuron subtypes. *J. Neurosci.*, 24(11):2612–2622.

Xu, Q., C.P. Wonders, S.A. Anderson. (2005). Sonic hedgehog maintains the identity of cortical interneuron progenitors in the ventral telencephalon. *Development*, 132(22):4987–4998.

Yenari, M.A., S.L. Fink, G.H. Sun, L.K. Chang, M.K. Patel, D.M. Kunis, D. Onley, D.Y. Ho, R.M. Sapolsky, G.K. Steinberg. (1998). Gene therapy with *HSP72* is neuroprotective in rat models of stroke and epilepsy. *Ann. Neurol.*, 44(4):584–91.

Ying, Q.L., M. Stavridis, D. Griffiths, M. Li, A. Smith. (2003). Conversion of embryonic stem cells into neuroectodermal precursors in adherent monoculture. *Nat. Biotechnol.*, 21(2):183–186.

Yoo, Y.M., C.J. Lee, U. Lee, Y.J. Kim. (2006). Neuroprotection of adenoviral-vector-mediated GDNF expression against kainic-acid-induced excitotoxicity in the rat hippocampus. *Exp. Neurol.*, 200(2):407–417.

Zhao, Y., P. Flandin, J.E. Long, M.D. Cuesta, H. Westphal, J.L. Rubenstein. (2008). Distinct molecular pathways for development of telencephalic interneuron subtypes revealed through analysis of *Lhx6* mutants. *J. Comp. Neurol.*, 510(1):79–99.

Zini, S., M.P. Roisin, U. Langel, T. Bartfai, Y. Ben-Ari. (1993). Galanin reduces release of endogenous excitatory amino acids in the rat hippocampus. *Eur. J. Pharmacol.*, 245(1):1–7.

# 34 Arresting Epileptogenesis: The Current Challenge

*Philip A. Schwartzkroin*

## CONTENTS

## INTRODUCTION: WHAT IS EPILEPTOGENESIS?

### THE GOAL OF ANTIEPILEPTOGENESIS IS EPILEPSY PREVENTION

Although we have long sought treatments that fall under the rubric of "antiepileptic" (or "anticonvulsant"), attention has now shifted to the concept of *antiepileptogenesis*. Past neglect of this issue has been due largely to a general lack of understanding about the processes of epileptogenesis and thus a practical inability to develop measures that would be effective antiepileptogenic treatments. Our increasing insights into seizure mechanisms and the brain changes that may precede the emergence of an epileptic state have now made it possible to consider seriously the possibility that we may be able to arrest epileptogenesis—that is, actually abort (or at least interfere with) *the processes by which a relatively normal brain becomes capable of generating spontaneous,*

**609**

*repeated seizures.* This more ambitious clinical goal involves focusing on the implementation of treatments that will not only prevent seizure occurrence but also produce a "cure," so the individual being treated no longer has an epileptic condition. As is the case for many medical conditions, the most effective treatment or cure may lie in *prevention*. As epileptologists, we would therefore like to develop procedures that stop the epileptogenic process *before* it becomes a chronic disease state. Antiepileptogenesis is not just a conceptual endpoint but a real and important clinical goal.

## EPILEPTOGENESIS IS A PROCESS, NOT A STATE

Unlike the epileptic state, epileptogenesis is a *process*—or, more likely, a complex set of processes. Epileptogenesis is probably different for different epilepsy endpoints, perhaps different for different individuals, and undoubtedly different depending on the etiological origins of the process. Thus, to think about antiepileptogenesis is to consider hitting not one but several moving targets. Currently, we are significantly worse at dealing with epileptogenesis than we are at treating epilepsy (which, although it comes in multiple forms, is at least a stationary target—or so we'd like to believe). In our attempts to prevent seizures and "cure" epilepsy, we have developed a set of reasonably effective pharmacological and surgical strategies—but these approaches still leave many patients in want of better treatments. Antiepileptogenesis requires a different mindset. For example, the failure of currently effective anticonvulsant medications (antiepileptic drugs, or AEDs) as antiepileptogenic agents (Temkin et al., 1990, 1999) suggests that the mechanisms underlying epileptogenesis are different from those underlying seizure generation in the already epileptic brain. It is worth considering whether we could recognize an antiepileptogenic treatment even if we stumbled over it; for example, is the ketogenic diet—at least on occasion—antiepileptogenic? Chapter 12 in this volume explores some newer pharmacological approaches (based on a preliminary understanding of epileptogenic mechanisms) that are worth further exploration.

Although the task is somewhat intimidating, it is important to gain some understanding of the complexity of epileptogenic processes. Toward such a goal, we could use the same general strategy that has worked reasonably well in developing antiepileptic treatments—empirical trial and error, often arising from a serendipitous finding that serves as a basis for developing potential treatments. However, elucidation of at least some of the processes underlying epileptogenesis would surely provide the basis for a significantly more effective approach to antiepileptogenesis. Indeed, it seems likely that research into underlying epileptogeneic mechanisms would uncover multiple processes, suggesting several points of intervention.

One simple way to view this scenario is to consider the flow diagram in Figure 34.1. Here, the solid arrows linking the boxes represent epileptogenic processes that lead to "seizure/epilepsy." Expanding the details of this diagram leads to some potentially key features of epileptogenesis that can help direct our research efforts (see also Tables 34.1 to 34.3): (1) Epileptogenesis is, by definition, a progressive process (i.e., it takes time); however, the time frame may differ for different conditions. (2) Epileptogenesis is likely to involve multiple contributing and interacting processes. (3) There are likely to be significant feedback loops that facilitate its development. (4) Some of the underlying mechanisms undoubtedly reflect homeostatic efforts to restore normal function—which are ultimately pathological within the context of the insult that initiates the process. (5) The net alterations responsible for seizure activity are quite varied.

The paragraphs below outline at least a few of the candidate processes that should receive attention, but as we consider these possibilities it is important to ask about each of them:

- *Why?* What stimuli trigger these changes, and do these changes actually represent potential epileptogenic precursors?
- *When?* When do these processes come into play, both with respect to the initiating stimuli (e.g., what is the latent period) and with respect to developmental programs (i.e., how do epileptogenic processes interact with the brain's developmental processes)?

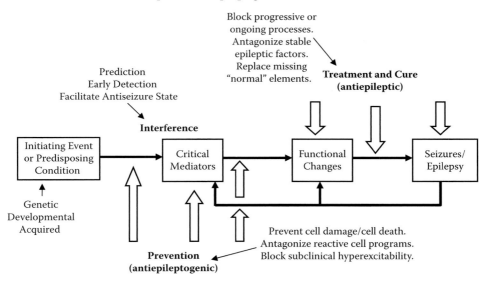

**FIGURE 34.1** Aspects of the epileptogenic process. Solid arrows indicate likely epileptogenic pathways. Open arrows signify opportunities for intervention (i.e., for therapies that might interrupt epileptogenesis). Strategies for prevention, interference, or treatment and cure are expanded in Tables 34.1 to 34.3.

- *Where?* Where in the brain do these changes take place, and what types of cells are affected?
- *How much?* At what level or magnitude do these changes constitute an epileptogenic threat (is there a threshold)?
- *Alone or in concert?* How do these changes interact with each other and with normal elements of the affected brain?
- Are they *necessary or sufficient* to cause epilepsy?

## Ideally, Antiepileptogenesis Involves the Identification of Potential Epilepsy Patients Before They Develop a Chronic Seizure Disorder

A major goal of antiepileptogenesis is to treat at-risk patients *before* a chronic epileptic state is established. This goal is a bit tricky, as epileptologists normally do not see their patients until seizures have become a fact of life—often a chronic disorder that has been in place for years. Two important factors should be considered when dealing with this problem: First, as indicated above, epileptogenesis may well be a progressive process that involves significant feedback; therefore, it is possible that interfering with apparently established epileptic activity may have antiepileptogenic efficacy (especially in the young brain). Second, and perhaps most obvious, we need to be able to identify the at-risk populations in which treatment will decrease the likelihood of future epilepsy. Thus, in approaching antiepileptogenesis, not only are we talking about developing new treatments (directed against "new" mechanisms), but we are also concerned with developing a science of *prediction*. Although a number of epidemiological studies provide a starting point for this effort (Annegers et al., 1996; Hesdorffer and Verity, 1997), this predictive effort is still in its infancy. Intense effort is currently being focused on identifying useful biomarkers that will help reveal those individuals who are at particularly high risk. Such markers include the following:

**TABLE 34.1**
**Prevention**

**Prevent cell damage and cell death via:**

    Glutamate receptor activation

    Rise in intracellular calcium

    Production of reactive oxygen species

    Activation of cell death pathways

    Mitochondrial/DNA degradation

**Antagonize reactive cell programs:**

    Glial reactivity

    Release of cytokines and trophic factors

    Growth of synaptic and aberrant pathways

    Neurogenesis

**Block subclinical hyperexcitability mediated by:**

    Aberrant receptor activation

    Altered ionic balance

    Cell swelling

    pH changes

    Release of modulatory peptides

    Activation of transporters, exchangers, and pumps

- Changes seen in various types of imaging (see Chapter 15, this volume), including hippocampal sclerosis or atrophy (MRI) (Provenzale et al., 2008), changes in metabolic characteristics that are associated with cell loss (NAA in MRS imaging) (Najm et al., 1998), or altered connectivity (as seen in tractography) (Yogarajah and Duncan, 2008)
- Electrophysiological fingerprints, including oscillatory activities in the EEG (see Chapter 21, this volume) such as fast ripples (Engel et al., 2003)
- Gene expression changes for individual genes (Raol et al., 2006) or general patterns of gene expression seen in gene array studies (Aronica and Gorter, 2007; Lukasiuk et al., 2006)

Predictive biomarkers will help us not only to identify those individuals who should be treated but also to develop appropriate treatments and assist in directing the treatments to the key brain regions involved in the epileptogenic process.

## EFFECTIVE ANTIEPILEPTOGENIC TREATMENT AND CLINICAL MANAGEMENT REQUIRE AN EMPHASIS ON EDUCATION FOR FAMILY PRACTITIONERS, PEDIATRICIANS, AND GENERAL NEUROLOGISTS

A large part of any effective antiepileptogenic treatment program will be out of the hands of clinical epileptologists. Certainly, in current practice, epilepsy specialists (often physicians in tertiary care medical centers) do not see their patients until the patient has a chronic (and difficult to control) epileptic condition. The healthcare providers that are central to any efforts at prediction—and thus play a critical role in antiepileptogenesis—are the family practitioners, pediatricians, and general neurologists who first see the at-risk patient. These patients often consult their family doctors in response to the precipitating events that put them at risk, and it is the referral and treatment by these physicians that will determine (and limit) the pathways of treatment for these individuals. Thus, a major aspect of any antiepileptogenic effort is to provide these primary care physicians with information and education—about factors that put their patients at risk, as well as about treatments that will decrease the risk of epilepsy.

---

**TABLE 34.2**
**Interference**

**Early prediction to enable strategies for desynchronization, increasing inhibition, and controlling (clamping) activity levels**

> Implantable stimulating devices
> Focal drug delivery
> Biological pumps
> State-dependent blockers of activity
> Temperature manipulations

**Detection of initiating stimuli and triggering conditions to set a less excitable baseline**

> Hormonal and immunological therapies
> Delivery of peptide modulators
> Dietary changes
> Biofeedback

**Facilitation of the antiseizure state by blocking positive feedback pathways**

> Control transmitter release and uptake
> Maintain blood flow and blood–brain barrier integrity
> Control pH level and ionic balance
> Facilitate normal neuronal and glial metabolism

---

# WHAT FACTORS CONTRIBUTE TO EPILEPTOGENESIS THAT CAN BE USED TO IDENTIFY THE AT-RISK POPULATION?

Although somewhat simplistic in approach, the factors that determine the likelihood of chronic seizures developing from a precipitating event (or baseline condition) can be grouped into three broad categories. It is important to note at the outset that these categories are not clearly separate and that interactions across categories make the search for underlying factors very complicated. Nevertheless, it is useful to consider these general factors, as they are likely to influence virtually every epileptogenic process.

---

**TABLE 34.3**
**Treatments and Cures**

**Blockade of progressive and ongoing epileptogenic processes, such as:**

> Establishment of aberrant circuitry
> Release of trophic factors that support neurogenesis, synaptogenesis, and neurite growth
> Expression of "developmental" genes

**Antagonize stable epileptic processes by:**

> Antisense and knockdown strategies targeting abnormal gene expression
> Controlling energy conditions and blood flow
> Treatment with antibodies against abnormal proteins

**Replace normal missing elements:**

> Neuronal or glial transplantation
> Gene therapy for missing proteins
> Implantation of artificial devices to guide new circuits

---

## GENETICS

It is now very clear that some epilepsies are genetically based. Recent investigations have revealed a number of relatively rare epilepsies that are attributable to single gene mutations, most frequently (but not consistently) involving voltage-gated ion channels or ligand-gated receptor subunits (Heron et al., 2007; Noebels, 2003; Steinlein, 2008). Further, we know that a number of genes responsible for developmental brain abnormalities—such as lissencephaly or tuberous sclerosis—are associated (in a high percentage of cases) with early-onset epilepsies (Guerrini et al., 1999; Schwartzkroin and Walsh, 2000). Finally, we now have rather good evidence that some of the more common epilepsies (e.g., juvenile myoclonic epilepsy) are due to multigenic interactions of a complex nature (Delgado-Escueta et al., 2003). It is important to note, however, that even in those cases in which genes have been clearly implicated, not all individuals with the relevant genetic defect exhibit an epilepsy phenotype.

Increasingly, an individual's overall genetic make-up can be viewed as a predisposing factor for the appearance of epileptic seizures. This concept, that the gene expression pattern of an individual (or family) determines the individual's baseline state with respect to seizure susceptibility, is strongly supported by studies of animal models (Löscher et al., 1998; McKhann et al., 2003; Schauwecker, 2003). For example, experiments involving different mouse strains—none of which is spontaneously epileptic—show that the strain background determines several seizure-related features, including seizure threshold, seizure severity, seizure-induced pathology, and seizure-related mortality.

Genetic determinants of epileptogenesis have not been examined, but it would be truly surprising if genetic make-up did not significantly modulate the epileptogenic process. Indeed, without this genetic influence, it is difficult to explain why a given traumatic insult delivered to a population of individuals (or mice) results in such variable seizure-related endpoints. Exactly what genes might be involved—perhaps genes other than those that we already know influence seizure threshold or susceptibility—remain to be determined. Increased understanding of such genes is a critical step in identifying "at-risk" individuals.

Although there may be a specific lesion or insult that initiates a set of plastic processes, one may argue that some epileptogenic pathways are hardwired. That is, in a given individual these pathways reflect an intrinsic abnormality (i.e., an aberrant gene) that will eventually give rise to a seizure state, irrespective of changes in brain structure or of ongoing brain activity. Should we view such a condition as a "process" or rather as a state (similar to an epileptic state) that requires its own special treatment?

## BRAIN DEVELOPMENTAL PROGRAMS

As indicated above, developmental brain abnormalities are often linked to epileptic syndromes, although we have an only vague idea about why or how such structural abnormalities result in epilepsy. This question is particularly difficult given that apparently similar abnormalities do not necessarily result in the same seizure state. Nevertheless, aberrant brain structure appears to be a definite risk factor, and many current studies now revolve around such difficult questions as: What are the effects of aberrations at different times during brain development? How do abnormalities in different brain regions affect the likelihood of epileptogenesis? Does the size of the structural abnormality determine the level of risk? In dealing with epileptogenesis, such questions demand a rather sophisticated understanding of underlying processes, for the baseline state may well appear relatively normal. In addition, one must factor into this analysis the ongoing *developmental processes*—those dynamic aspects of brain development that, when normal, yield a nonepileptic neuronal system. Slight changes in these dynamic systems could interact with, and support, epileptogenic processes without producing a clear structurally abnormal phenotype.

## ENVIRONMENTAL INFLUENCES

When we think about epileptogenesis, we usually think about the processes associated with such trauma as head injury, stroke, infection, or even seizures (e.g., complex febrile discharges). These events appear to initiate a set of processes that can sometimes result in a chronic seizure state. What we do not know, however, is what features of these insults are truly epileptogenic. Do all head injuries put an individual at risk? Are only some individuals (determined by genetics) or only individuals at some ages (e.g., the young in whom developmental programs are still robust) likely to respond to such trauma with epileptogenic processes? These precipitating events have already received attention (because we can see them and quantify them), but better feature analysis of epileptogenic stimuli is still an important experimental goal. For example, reports have suggested that severe penetrating head injury is much more likely to result in posttraumatic epilepsy than mild or moderate insult (Frey, 2003), but a clear understanding of what constitutes "severe" is still lacking. Is severity defined by the volume of brain involved in the injury? By intraparenchymal bleeding? By the location of the injury? By the types of cells that show particular vulnerability to the insult?

Undoubtedly, as we become more sophisticated in elucidating epileptogenic features of these events, we will begin to appreciate the critical nature of their interactions with baseline genetic and developmental programs, as well as with "normal" factors that alter brain excitability such as arousal and sleep states (see Chapter 29, this volume) and stress levels, such as steroidal hormone concentrations (Frye, 2008; Kotagal and Yardi, 2008). How, for example, does the specific brain region involved by an insult interact with that region's pattern of gene expression (or its stage of maturation) with respect to a possible epileptogenic outcome?

## POTENTIAL MECHANISMS OF EPILEPTOGENESIS: GENERAL THEMES AND ISSUES

### EPILEPTOGENESIS MAY INVOLVE DIFFERENT MECHANISMS FROM THOSE WE HAVE IDENTIFIED AS IMPORTANT IN SEIZURE GENESIS OR INITIATION AND MAINTENANCE OF THE EPILEPTIC STATE

The vast majority of research into the underlying mechanisms of the epilepsies has focused on processes of seizure initiation. Within this context, it has become commonplace to point to the balance between excitatory and inhibitory processes as the key to our understanding (Lothman, 1996). This emphasis, in turn, has led to a considerable preoccupation with mechanisms of inhibitory control and possible compromises in inhibitory status (e.g., loss of inhibitory interneurons, disinhibition, changes in the subunit composition of GABA receptors, altered tonic inhibition) (Bouilleret et al., 2000; Buckmaster and Jongen-Relo, 1999; Coulter, 2001; Zhang et al., 2007), as well as a growing interest in possible changes in glutamate-mediated excitatory events (e.g., changes in the AMPA/NMDA balance, metabotropic receptor-mediated processes, calcium entry through glutamate-gated channels) (DeLorenzo et al., 2005; Karr and Rutecki, 2008; Mody and MacDonald, 1995; Rakhade et al., 2008). These mechanisms undoubtedly have relevance to seizure initiation; for example, there is good evidence that a decrease in inhibition does indeed foreshadow the onset of seizure discharge (Faingold and Anderson, 1991; Min et al., 1999). Such mechanisms may also be important for the maintenance of an underlying epileptic state; for example, fragility of inhibition may characterize some forms of epilepsy (Coulter, 2000; Franck et al., 1995; Liang et al., 2006). However, evidence to support the view that these mechanisms are crucial as epileptogenic processes is still preliminary. Gene expression studies have shown that a precipitating insult may alter the expression of various glutamate or GABA receptor subunits, and that these changes have potential consequences consistent with developing hyperexcitability. What is less certain is whether these seizure-supporting factors are actually *causal*—or whether they reflect the *results* of other changes associated with epileptogenesis. Very few specific epileptogenic mechanisms have been identified to date, and our current understanding of mechanisms underlying seizure activity should not limit our search for those mechanisms. In light of the explosion in molecular biology, we have an amazing assortment of potential mechanisms to explore.

## EPILEPTOGENIC PROCESSES ARE INITIATED BY SOME EVENT OR ABNORMAL CONDITION (E.G., AREAS OF DAMAGE, LESION, STRUCTURAL ABNORMALITY)

We generally think about epileptogenesis having a relatively specific initiation time and site, often associated with a specific brain insult that leads to neuronal cell damage and cell death. Excitotoxicity (e.g., associated with release of excess glutamate) is typically viewed as epileptogenic and is often modeled in experimental studies. What are the changes in brain structure, chemistry, metabolism, and function that reflect the reaction to cell damage, and do they necessarily lead to an epileptic state (i.e., are they epileptogenic)? Among the most obvious of changes associated with brain damage are the reactive glial processes. Although investigators have remarked on reactive gliosis for many years, it is still difficult to identify exactly what constitutes reactive astrocytosis. Are reactive glia simply responding to a change in their environment, or do they actively shape the damaged brain region?

Investigators have recently identified interesting differences between reactive glia in epileptic brain and normal astrocytes; some of these differences (e.g., reduced glutamate transporter expression) could easily be viewed as potentially epileptogenic (Binder and Steinhauser, 2006; Bordey and Sontheimer, 1998). Indeed, the demonstration that astrocytes may play quite an active role in modulating neuronal excitability has shifted a major focus onto glial mechanisms in epileptogenesis. Such changes in glia may be progressive and may involve methods of cell-to-cell communication different from conventional chemical neurotransmission (e.g., electrical coupling) (see Chapter 13, this volume)

Another related aspect of reactivity that has recently received much attention is that of inflammation (see Chapter 4, this volume). We have long recognized that in the injured brain there is infiltration of microglia into traumatized brain regions, presumably to remove damaged neuronal elements. There is now evidence that these cells (along with astrocytes) release powerful factors (e.g., cytokines) that may further influence the viability and excitability of surviving neurons and astrocytes (Choi and Koh, 2008; Vezzani et al., 2000a, 2008). The migration of these cells into regions of injury may be determined, at least in part, by infiltration of foreign substances into the parenchyma as a result of blood–brain barrier (BBB) breakdown (see Chapter 2, this volume). Indeed, the potential importance of BBB integrity in protecting against epileptogenic processes and the potential contribution of BBB disruption to epileptogenesis are only beginning to receive significant attention (Uva et al., 2008).

Glial cells, as well as neurons, may also release a variety of growth factors that influence the patterns of process regeneration and the generation of new synaptic contacts and that support the survival of surrounding cell types (Collin et al., 2001; Patel, 2002; Patel and McNamera, 1995). The receptor for brain-derived neurotrophic factor (BDNF) has been shown convincingly to be necessary for the kindling process (He et al., 2004). Given that kindling has long been viewed as a model of epileptogenesis, it seems likely that BDNF and associated neurotrophic factors are importantly involved in epileptogenesis. Reactive processes may also involve a reversion of cells to immature phenotypes, including the facilitation of neurogenesis (He et al., 2004). Although this latter mechanism may add cells to the affected circuit, it is still unclear (especially in the mature CNS) if or how such newly generated cells (glia and neurons) are, or should be, integrated into existing networks. It appears, for example, that neurogenesis in the dentate gyrus gives rise to a population of heterotopic granule cells that establish aberrant connectivity with the surrounding circuit; such cells are typically seen in the hippocampus resected from temporal lobe epilepsy (TLE) patients (and in animal models of TLE), and their contribution to the epileptogenesis in this region is under critical examination (Parent et al., 1997).

Finally, we know that brain insult—whether it results in cell loss and explicit (morphological) lesion or gives rise to more subtle injury—triggers all kinds of brain changes. What kinds of changes (i.e., plasticity) are appropriately adaptive and what types are epileptogenic? In addition to changes in expression for various ion channel and receptor components, it is clear that the organization (i.e., subunit composition) and localization of channels and receptors may be altered by insult (Jung et al., 2007; Lugo et al., 2008). Such changes can dramatically alter, for example, receptor function (e.g., change the nature of tonic vs. phasic inhibition) (Zhang et al., 2007). Such changes have also been described for voltage-gated ion channels that determine the intrinsic excitability of neurons (e.g., influencing such currents as $I_h$ and $I_A$). A more familiar example of this plasticity is the reorganization of axonal connectivity that we call *sprouting*—as seen in the hippocampus of TLE patients in the form of mossy fiber sprouting (Dudek and Sutula, 2007). Are these changes in circuitry epileptogenic, or are they driven by an already epileptic circuit?

## Are Epileptogenic Processes Progressive? Self-Sustaining? Activity Dependent?

There is considerable interest in the concept that epileptogenesis is a progressive process—that is, it involves a positive feedback mechanism through which a small amount of abnormal activity or function exacerbates the original pathology, leading to more severe abnormal activity or function, and so on (Engel, 1996). Such a progressive change is thought to occur during kindling, which has served as the prototypical model of a progressive epileptogenic process (Cavazos et al., 1994; Sutula, 2001). Whether a kindling-like mechanism underlies human epilepsies is still rather controversial. Reports of the development of secondary, or mirror, foci support this idea of progressive abnormality (Engel, 1996). If some epilepsies are progressive, there are several implications for intervention:

1. It is important to identify and stop or interrupt even the subclinical processes that are initiated by potentially epileptogenic stimuli, as continuing subclinical activity may lead to progressive pathology and, sooner or later, *chronic* seizures. The usual formulation of this view assumes that the progressive process involves some form of electrical discharge—for example, subthreshold discharge activity occurring in the hippocampus and related structures that causes more and more cell death. This version of activity dependency suggests that one way of interfering with the progressive process is to block ongoing electrical discharge. However, the fact that drugs that should reduce electrical excitability (e.g., phenytoin) do not interfere with posttraumatic epileptogenesis suggests that the critical epileptogenic activity may take other forms—perhaps at the molecular level (e.g., gene transcription, protein translation, and related functional changes, which then activate a new cycle of gene transcription). These different forms of activity require rather different interventional approaches.
2. The progressive activity is self-sustaining; that is, it does not require any external boost once the process has been initiated. Thus, once a pathway toward abnormality has been initiated, the pathology that is initially induced is sufficient to generate the ultimate epileptic state. If that is the case, intervention possibilities are significantly more limited than if these processes require continuing interaction with external stimulation. However, if one could identify an (or *the*) underlying process, then its interruption could conceivably constitute a "cure."

## OPPORTUNITIES FOR INTERVENTION

Once we characterize the epileptogenic process, what can we do with this information? The goal for treatment of epileptogenesis is early intervention—to prevent the establishment of the epileptic state before it becomes a stable and difficult to control abnormality.

## EARLY DETECTION

Identification of the epileptogenic process at an early stage would allow us to be more specific in our treatment approaches than to treat all at-risk patients in the same way. As indicated above, development of *markers* of epileptogenesis is therefore crucial. Currently, the most feasible approach to early diagnosis appears to be the detection and prevention of early seizures or abnormal electrical brain activity. Such intervention will depend critically on the further development of techniques to characterize EEG (or other markers) and to detect and predict seizure occurrence (Iasemidis et al., 2005; Lehnertz et al., 2007; Litt et al., 2001). Especially if seizures participate in a progressive process, early seizure detection might allow one to abort oncoming events and thus interfere with the positive feedback process. It is worth noting, however, that non-electrical markers may be more easily obtained or constitute more reliable predictors of epileptogenesis. For example, some markers of cell death may provide precisely the kind of early information that is needed (Sankar et al., 1997).

Disruption of early/initial electrical abnormalities may be effected through stimulation protocols, such as those that have been developed as anticonvulsant treatments (Gluckman et al., 2001). Especially if repeated discharge acts to strengthen abnormal circuits (potentiation, kindling) (Teyler et al., 2001) or establish population synchrony, interfering with such activity could have antiepileptogenic results. Techniques are also being developed to alter current gradients, to deliver pulsed electrical and pharmacological stimuli, and to use electrical stimulation to release endogenous modulatory agents (see Chapter 28, this volume) (Mazarati et al., 2000; Vezzani et al., 2000b). The parameters of such antiepileptogenic stimuli still have to be developed. Further, there is increasing interest in applying such techniques in a regionally restricted manner, so the focus of the epileptogenic process is specifically targeted rather than encompassing the entire brain. The technology for delivering these signals—to disrupt abnormal subclinical activities—is in its infancy.

## PHARMACOLOGICAL APPROACHES

No drugs with an antiepileptogenic mechanism have yet been identified. Indeed, all of the current AEDs are essentially anticonvulsants. Given the possibility—indeed, the likelihood—that epileptogenesis depends on processes different from those involved in seizure generation, there is clearly a need for a new antiepileptogenic pharmacopeia. What mechanisms should new antiepileptogenic drugs target?

If one critical epileptogenic process is ongoing electrical activity, then it may be important to block such activity even if it does not take the form of overt seizures. Our current AEDs are quite good at antagonizing dramatic paroxysmal events, but their use in antagonizing more subtle abnormalities (or for normalizing background discharge) has been poorly explored. The failure of prophylactic phenytoin and valproate to protect against posttraumatic epileptogenesis (Temkin et al., 1990, 1999) has uncertain implications. It suggests that key underlying processes are not electrical, but given the limited protocols that were used in those studies it is still possible that such drugs can be used to control subclinical electrical abnormalities. Further, it may be that once the abnormal electrical activity is established it is too late; that is, we should be targeting earlier posttraumatic changes (e.g., at the level of receptors) to prevent receptor subunit reorganization.

Another major target for new antiepileptogenic therapy is cell *protection*. Many studies have suggested that neuronal cell death can be an epileptogenic stimulus and that the relevant cells die over a long period of time. Thus, pharmacological agents that block this prolonged cell death process (e.g., calcium channel blockers or NMDA-type glutamate receptor antagonists) might reduce the likelihood of epileptogenesis. In considering this strategy, however, it is important to determine whether the rescued cells behave normally or whether they have abnormal properties that might actually contribute to an epileptogenic process. The general issue of neuroprotection is receiving increased attention (see Chapter 11, this volume) and may be a key to early antiepileptogenic efforts.

There are many reports to show that epileptic brain or foci are characterized by significant reorganization of neuronal connections; that is, there is not only cell loss or damage but also robust growth (Cavazos et al., 2004; Prince et al., 1997; Wenzel et al., 2000). Although it is still unclear how (or how much) this sprouting contributes to epileptogenesis, investigators are already targeting antagonists of growth-promoting molecules (e.g., neurotrophic factors) as potentially antiepileptogenic (Binder et al., 1999). Focal interruption of protein synthesis associated with the growth of new axon processes may be antiepileptogenic; experimental efforts to test this hypothesis, however, are highly controversial (Longo et al., 2002).

In addition, as indicated above, studies have suggested that neuronal (and glial) proliferation occurs in the epileptic brain (Parent et al., 1997). Again, the contribution of such proliferative processes to the seizure state has not been clearly determined; however, prevention of neural or glial proliferation may constitute an important antiepileptogenic therapy. Indeed, a somewhat neglected target in epilepsy therapy, and one that should not be overlooked in searching for antiepileptogenic treatments, is the population of brain astrocytes (and microglia). Normal astrocyte function is critical for regulation of the extraneuronal environment, and astrocyte dysfunction has significant effects on neural excitability (Binder and Steinhauser, 2006; Bordey and Songheimer, 1998). Many epileptogenic stimuli alter glial properties—either causing significant glial cell death or producing reactive astrocytes. Preventing such reactivity may be important in antiepileptogenesis. Similarly, preventing the invasion of microglia and the release of microglia-associated agents (e.g., cytokines) may be a significant pharmacological strategy.

Finally, there is accumulating evidence that epileptogenic stimuli cause alterations in cell properties, in part by activating gene transcription programs that are reminiscent of immature or developing neurons (Elliott et al., 2001). Although these changes (reversion to immature states) may have functional homeostatic importance (e.g., increasing the plastic capacity of injured brain), interference with such programs may also have antiepileptogenic consequences. Clearly, it will be important to balance antiepileptogenic treatments with potential enhancement of functional restoration.

## SURGERY AND/OR REPAIR

To date, epilepsy surgery is the only existing antiepileptic treatment we have. Whether surgery can be antiepileptogenic remains to be determined. With the use of appropriate markers to identify tissue that is (or will become) epileptic, there are a number of surgical strategies that may effectively interfere with the epileptogenic process. One approach remains simply to remove "bad brain." If abnormal brain generates activity that can kindle surrounding tissue or if it releases epileptogenic substances (e.g., growth factors) that affect normal neighbors, then removal of this tissue makes good sense (as exemplified by seizure resolution associated with surgical removal of brain tumors). However, the efficacy of such a program for antiepileptogenesis will undoubtedly depend on the integrity of the remaining brain tissue. Currently, the surgical approach is limited to mechanical resections that are relatively gross with respect to the targeted abnormal cells and circuits. New techniques for cell removal or inactivation may provide a more selective approach; for example, non-invasive resections may be made radiologically with impressive accuracy (Regis et al., 2000; Rheims et al., 2008) (see Chapter 18, this volume). Or, as we learn more about what constitutes abnormality in neurons that generate the epileptic state, we may be able to label and kill such cells selectively.

In some cases, the insult that leads to the epileptic state itself removes critical cell types that are essential to the control of brain excitability (e.g., inhibitory interneurons, or glia). In such cases, it may be feasible to introduce appropriate cell types via transplantation techniques (Shetty and Hattiangady, 2007; Zaman and Shetty, 2001) and to direct them to become appropriately integrated within the affected circuitry. New culture, gene expression, and transplantation technologies may allow us to repair injured brain tissue before the epileptogenic process begins.

Unfortunately, for many epilepsies, there is no grossly identifiable cell or tissue abnormality to be removed or replaced; however, recent studies have identified gene abnormalities in many forms of epilepsy, and these genes lead—in a developmental sequence—to epileptic function. In these cases, gene therapy becomes an option worth considering (McCown, 2004; Noe et al., 2008; Wenzel et al., 2007). Indeed, for familial epilepsies in which offspring are at considerable risk, such therapies may in theory be initiated *in utero*, so the epileptogenic process is completely short-circuited.

## OTHER APPROACHES

Given our current knowledge about causes of the epilepsies (i.e., those processes that are involved in generating an epileptic condition), a number of potentially useful therapeutic strategies deserve additional research attention:

1. *Immunological therapies*—Some forms of early-onset epilepsies (e.g., Rasmussen's encephalitis) have been linked to autoimmune processes that can result in altered structure or function (McNamara et al., 1999). For other epilepsies, infection is viewed as the likely initiating insult (Garg et al., 1999; Singh and Prabhakar, 2008; Theodore et al., 2008). In such cases, early treatment with agents that antagonize antibody-mediated effects may have antiepileptic efficacy. Especially since inflammatory reactions involving the immune system appear to be so often associated with epileptogenic processes, it may even be useful to think about epilepsy "vaccinations" that will protect at-risk patients against potential detrimental effects of immune reactions to insult.
2. *Treatment with antioxidants*—Many precipitating injuries (e.g., stroke) are due to excitotoxicity-mediated processes that damage cells in epilepsy-sensitive brain regions (Culmsee et al., 2001). Protection against cell death or injury (e.g., using agents that block the production or action of free radicals) is a rather general therapeutic strategy (Bozzi et al., 2000; Morrison et al., 1996; Patel, 2002) that may have important consequences for the development of a chronic epileptic state.
3. *Development of dietary treatments*—It is already clear that some epilepsies are a result of vitamin (or other dietary) insufficiency and that restoration of the missing element "cures" the seizure disorder (Gospe, 1998). Still unclear is whether more subtle deficiencies (e.g., in critical amino acids) might underlie some epilepsies (Gietzen et al., 1996; Liang et al., 2006). Also still to be explored is whether some dietary regimens, such as the ketogenic diet, might be not only antiepileptic but also antiepileptogenic (Maalouf et al., 2007; Todorova et al., 2000).
4. *Hormonal therapies*—The observations that steroid treatment (e.g., with ACTH) is sometimes effective in the treatment of infantile spasms (Snead, 2001) and that gonadal hormones have powerful epileptic (estrogen) and antiepileptic (progesterone) effects (Reddy and Rogawski, 2001; Rogawski and Reddy, 2002) suggest that some epilepsies may develop as a result of hormone imbalance. Further, an excessive stress hormone level has been linked to cell death (Joels, 2007; Yusim et al., 2000), so that antagonism of this action in critical brain regions (such as hippocampus) could have antiepileptogenic consequences.
5. *Biofeedback (and other behavior-based modification techniques)*—Given that there have been successful demonstrations of biofeedback efficacy in controlling seizure occurrence (Nagai et al., 2004), the usefulness of such behavioral techniques could be assessed within the context of antiepileptogenesis.

## CONCLUSIONS: THE FUTURE FOR RESEARCH IN ANTIEPILEPTOGENESIS

There are a host of challenges—and opportunities—for those hoping to treat epilepsy by preventing its occurrence. What are the immediate issues and achievable goals that challenge us at this early stage of research development?

- Identify at-risk patients. We need not only insightful epidemiological studies but also predictive biological markers of epileptogenic processes.
- Develop treatments that target molecular changes that *precede* seizure appearance. Ideally, we do not want to wait for the appearance of seizures before we begin treatment. At the current time, one useful approach may be to work toward preventing cell death or cell damage and reactivity. We already have a number of pharmacological and molecular tools for enhancing neuronal survival, and a growing number of agents can be used to combat inflammatory reactivity.
- Stop early seizures and try to normalize baseline brain electrical activity. It is unclear whether current AEDs work on this level, but our pharmacological armamentarium is rich, and new techniques are now available for electrical stimulation to interfere with subclinical hyperexcitability and abnormal discharge.
- Evaluate and treat trauma-induced plasticity. For many types of epilepsy, the development of new connections appears to be a salient hallmark. Will blockade of these new or aberrant circuits (even those at the molecular level) disrupt the epileptogenic process? How might such blockade affect normal brain function?
- Finally, there is a need for a reorientation with respect to the goal of treatment. Why is antiepileptogenesis so critical? We think it is important because chronic seizure activity, especially in young brains (in which epileptogenesis is most likely to occur) has significant effects on long-term function and maturation. Thus, the real focus of our work should be on the *consequences* of seizures (e.g., developmental delay)—not the seizures themselves (Sankar et al., 1997). It is prevention of those consequences that must guide our research and our choice of antiepileptogenic treatments.

## REFERENCES

Annegers, J.F., W.A. Rocca, W.A. Hauser. (1996). Causes of epilepsy: contributions of the Rochester epidemiology project. *Mayo Clin. Proc.*, 71:570–575.

Aronica, E., J.A. Gorter. (2007). Gene expression profile in temporal lobe epilepsy. *Neuroscientist*, 13:100–108.

Binder, D.K., AC. Steinhauser. (2006). Functional changes in astroglial cells in epilepsy. *Glia*, 54:358–368.

Binder, D.K., M.J. Routbort, T.E. Ryan, G.D. Yancopoulos, J.O. McNamara. (1999). Selective inhibition of kindling development by intraventricular administration of TrkB receptor body. *J. Neurosci.*, 19:1424–1436.

Bordey, A., H. Sontheimer. (1998). Properties of human glial cells associated with epileptic seizure foci. *Epilepsy Res.*, 32:286–303.

Bouilleret, V., F. Loup, T. Kiener, C. Marescaux, J.M. Fritschy. (2000). Early loss of interneurons and delayed subunit-specific changes in GABA(A)-receptor expression in a mouse model of mesial temporal lobe epilepsy. *Hippocampus*, 10:305–324.

Bozzi, Y., D. Vallone, E. Borrelli. (2000). Neuroprotective role of dopamine against hippocampal cell death. *J. Neurosci.*, 20:8643–8649.

Buckmaster, P.S., A.L. Jongen-Relo. (1999). Highly specific neuron loss preserves lateral inhibitory circuits in the dentate gyrus of kainate-induced epileptic rats. *J. Neurosci.*, 19:9519–9529.

Cavazos, J.E., I. Das, T.P. Sutula. (1994). Neuronal loss induced in limbic pathways by kindling: evidence for induction of hippocampal sclerosis by repeated brief seizures. *J. Neurosci.*, 14:3106–3121.

Cavazos, J.E., S.M. Jones, D.J. Cross. (2004). Sprouting and synaptic reorganization in the subiculum and CA1 region of the hippocampus in acute and chronic models of partial-onset epilepsy. *Neuroscience*, 126:677–688.

Choi, J.,S. Koh. (2008). Role of brain inflammation in epileptogenesis. *Yonsei Med. J.*, 49:1–18.

Collin, C., C. Vicario-Abejon, M.E. Rubio, R.J. Wenthold, R.D. McKay, M. Segal. (2001). Neurotrophins act at presynaptic terminals to activate synapses among cultured hippocampal neurons. *Eur. J. Neurosci.*, 13:1273–1282.

Coulter, D.A. (2000). Mossy fiber zinc and temporal lobe epilepsy: pathological association with altered "epileptic" gamma-aminobutyric acid a receptors in dentate granule cells. *Epilepsia*, 41(Suppl. 6):S96–S99.

Coulter, D.A. (2001). Epilepsy-associated plasticity in gamma-aminobutyric acid receptor expression, function, and inhibitory synaptic properties. *Int. Rev. Neurobiol.*, 45:237–252.

Culmsee, C., S. Bondada, M.P. Mattson. (2001). Hippocampal neurons of mice deficient in DNA-dependent protein kinase exhibit increased vulnerability to DNA damage, oxidative stress and excitotoxicity. *Brain Res. Mol. Brain Res.*, 87:257–262.

Delgado-Escueta, A.V., K.B. Perez-Gosiengfiao, D. Bai, J. Bailey, M.T. Medina, R. Morita, T. Suzuki, S. Ganesh, T. Sugimoto, K. Yamakawa, A. Ochoa, A. Jara-Prado, A. Rasmussen, J. Ramos-Peek, S. Cordova, F. Rubio-Donnadieu, M.E. Alonso. (2003). Recent developments in the quest for myoclonic epilepsy genes. *Epilepsia*, 44(Suppl. 11):13–26.

DeLorenzo, R.J., D.A. Sun, L.S. Deshpande. (2005). Cellular mechanisms underlying acquired epilepsy: the calcium hypothesis of the induction and maintenance of epilepsy. *Pharmacol. Ther.*, 105:229–266.

Dudek, F.E., T.P. Sutula. (2007). Epileptogenesis in the dentate gyrus: a critical perspective. *Prog. Brain Res.*, 163:755–773.

Elliott, R.C., S. Khademi, S.J. Pleasure, J.M. Parent, D.H. Lowenstein. (2001). Differential regulation of basic helix-loop-helix mRNAs in the dentate gyrus following status epilepticus. *Neuroscience*, 106:79–88.

Engel, Jr., J. (1996). Clinical evidence for the progressive nature of epilepsy. *Epilepsy Res. Suppl.*, 12:9–20.

Engel, Jr., J., C. Wilson, A. Bragin. (2003). Advances in understanding the process of epileptogenesis based on patient material: what can the patient tell us? *Epilepsia*, 44(Suppl. 12):60–71.

Faingold, C.L., C.A. Anderson. (1991). Loss of intensity-induced inhibition in inferior colliculus neurons leads to audiogenic seizure susceptibility in behaving genetically epilepsy-prone rats. *Exp. Neurol.*, 113:354–363.

Franck, J.E., J. Pokorny, D.D. Kunkel, P.A. Schwartzkroin. (1995). Physiologic and morphologic characteristics of granule cell circuitry in human epileptic hippocampus. *Epilepsia*, 36:543–558.

Frey, L.C. (2003). Epidemiology of posttraumatic epilepsy: a critical review. *Epilepsia*, 44(Suppl. 10):11–17.

Frye, C.A. (2008). Hormonal influences on seizures: basic neurobiology. *Int. Rev. Neurobiol.*, 83:27–77.

Garg, R.K. (1999). HIV infection and seizures. *Postgrad. Med. J.*, 75:387–390.

Gietzen, D.W., K.D. Dixon, B.G. Truong, A.C. Jones, J.A. Barrett, D.S. Washburn. (1996). Indispensable amino acid deficiency and increased seizure susceptibility in rats. *Am. J. Physiol.*, 271:R18–F24.

Gluckman, B.J., H. Nguyen, S.L. Weinstein, S.J. Schiff. (2001). Adaptive electric field control of epileptic seizures. *J. Neurosci.*, 21:590–600.

Gospe, S.M., Jr. (1998). Current perspectives on pyridoxine-dependent seizures. *J. Pediatr.*, 132:919–923.

Guerrini, R., E. Andermann, M. Avoli, W.B. Dobyns. (1999). Cortical dysplasias, genetics, and epileptogenesis. *Adv. Neurol.*, 79:95–121.

He, X.P., R. Kotloski, S. Nef, B.W. Luikart, L.F. Parada, J.O. McNamara. (2004). Conditional deletion of TrkB but not BDNF prevents epileptogenesis in the kindling model. *Neuron*, 43:31–42.

Heron, S.E., I.E. Scheffer, S.F. Berkovic, L.M. Dibbens, J.C. Mulley. (2007). Channelopathies in idiopathic epilepsy. *Neurotherapeutics*, 4:295–304.

Hesdorffer, D.C., C.M. Verity. (1997). Risk factors. *Epilepsy: A Comprehensive Textbook* (pp. 59–67), J. Engel, Jr., and T.A. Pedley, Eds. Philadelphia, PA: Lippincott-Raven.

Iasemidis, L.D., D.S. Shiau, P.M. Pardalos, W. Chaovalitwongse, K. Narayanan, A. Prasad, K. Tsakalis, P.R. Carney, J.C. Sackellares. (2005). Long-term prospective on-line real-time seizure prediction. *Clin. Neurophysiol.*, 116:532–544.

Joels, M. (2007). Role of corticosteroid hormones in the dentate gyrus. *Prog. Brain Res.*, 163:355–370.

Jung, S., T.D. Jones, J.N. Lugo, Jr., A.H. Sheerin, J.W. Miller, R. D'Ambrosio, A.E. Anderson, N.P. Poolos. (2007). Progressive dendritic HCN channelopathy during epileptogenesis in the rat pilocarpine model of epilepsy. *J. Neurosci.*, 27:13012–13021.

Karr, L., P.A. Rutecki. (2008). Activity-dependent induction and maintenance of epileptiform activity produced by group I metabotropic glutamate receptors in the rat hippocampal slice. *Epilepsy Res.*, 81:14–23.

Kotagal, P., N. Yardi. (2008). The relationship between sleep and epilepsy. *Semin. Pediatr. Neurol.*, 15:42–49.

Lehnertz, K., F. Mormann, H. Osterhage, A. Muller, J. Prusseit, A. Chernihovskyi, M. Staniek, D. Krug, S. Bialonski, C.E. Elger. (2007). State-of-the-art of seizure prediction. *J. Clin. Neurophysiol.*, 24:147–153.

Liang, S.L., G.C. Carlson, D.A. Coulter. (2006). Dynamic regulation of synaptic GABA release by the glutamate–glutamine cycle in hippocampal area CA1. *J. Neurosci.*, 26:8537–8548.

Litt, B., R. Esteller, J. Echauz, M. D'Alessandro, R. Shor, T. Henry, P. Pennell, C. Epstein, R. Bakay, M. Dichter, G. Vachtsevanos. (2001). Epileptic seizures may begin hours in advance of clinical onset: a report of five patients. *Neuron*, 30:51–64.

Longo, B.M., E.R. Sanabria, S. Gabriel, L.E. Mello. (2002). Electrophysiologic abnormalities of the hippocampus in the pilocarpine/cycloheximide model of chronic spontaneous seizures. *Epilepsia*, 43(Suppl. 5):203–208.

Löscher, W., S. Cramer, U. Ebert. (1998). Differences in kindling development in seven outbred and inbred rat strains. *Exp. Neurol.*, 154:551–559.

Lothman, E.W. (1996). Basic mechanisms of seizure expression. *Epilepsy Res. Suppl.*, 11:9–16.

Lugo, J.N., L.F. Barnwell, Y. Ren, W.L. Lee, L.D. Johnston, R. Kim, R.A. Hrachovy, J.D. Sweatt, A.E. Anderson. (2008). Altered phosphorylation and localization of the A-type channel, Kv4.2, in status epilepticus. *J. Neurochem.*, 106:1929–1940.

Lukasiuk, K., M. Dabrowski, A. Adach, A. Pitkanen. (2006). Epileptogenesis-related genes revisited. *Prog. Brain Res.*, 158:223–241.

Maalouf, M., P.G. Sullivan, L. Davis, D.Y. Kim, J.M. Rho. (2007). Ketones inhibit mitochondrial production of reactive oxygen species production following glutamate excitotoxicity by increasing NADH oxidation. *Neuroscience*, 145:256–264.

Mazarati, A.M., J.G. Hohmann, A. Bacon, H. Liu, R. Sankar, R.A. Steiner, D. Wynick, C.G. Wasterlain. (2000). Modulation of hippocampal excitability and seizures by galanin. *J. Neurosci.*, 20:6276–6281.

McCown, T.J. (2004). The clinical potential of antiepileptic gene therapy. *Expert Opin. Biol. Ther.*, 4:1771–1776.

McKhann, 2nd., G.M., H.J. Wenzel, C.A. Robbins, A.A. Sosunov, P.A. Schwartzkroin. (2003). Mouse strain differences in kainic acid sensitivity, seizure behavior, mortality, and hippocampal pathology. *Neuroscience*, 122:551–561.

McNamara, J.O., K.D. Whitney, P.I. Andrews, X.P. He, S. Janumpalli, M.N. Patel. (1999). Evidence for glutamate receptor autoimmunity in the pathogenesis of Rasmussen encephalitis. *Adv. Neurol.*, 79:543–550.

Min, M.Y., Z. Melyan, D.M. Kullmann. (1999). Synaptically released glutamate reduces gamma-aminobutyric acid (GABA)ergic inhibition in the hippocampus via kainate receptors. *Proc. Natl. Acad. Sci. U.S.A.*, 96:9932–9937.

Mody, I., J.F. MacDonald. (1995). NMDA receptor-dependent excitotoxicity: the role of intracellular $Ca^{2+}$ release. *Trends Pharmacol. Sci.*, 16:356–359.

Morrison, R.S., H.J. Wenzel, Y. Kinoshita, C.A. Robbins, L.A. Donehower, P.A. Schwartzkroin. (1996). Loss of the p53 tumor suppressor gene protects neurons from kainate-induced cell death. *J. Neurosci.*, 16:1337–1345.

Nagai, Y., L.H. Goldstein, P.B. Fenwick, M.R. Trimble. (2004). Clinical efficacy of galvanic skin response biofeedback training in reducing seizures in adult epilepsy: a preliminary randomized controlled study. *Epilepsy Behav.*, 5:216–223.

Najm, I.M., Y. Wang, D. Shedid, H.O. Luders, T.C. Ng, Y.G. Comair. (1998). MRS metabolic markers of seizures and seizure-induced neuronal damage. *Epilepsia*, 39:244–250.

Noe, F., A.H. Pool, J. Nissinen, M. Gobbi, R. Bland, M. Rizzi, C. Balducci, F. Ferraguti, G. Sperk, M.J. During, A. Pitkanen, A. Vezzani. (2008). Neuropeptide Y gene therapy decreases chronic spontaneous seizures in a rat model of temporal lobe epilepsy. *Brain*, 131:1506–1515.

Noebels, J.L. (2003). The biology of epilepsy genes. *Annu. Rev. Neurosci.*, 26:599–625.

Parent, J.M., T.W. Yu, R.T. Leibowitz, D.H. Geschwind, R.S. Sloviter, D.H. Lowenstein. (1997). Dentate granule cell neurogenesis is increased by seizures and contributes to aberrant network reorganization in the adult rat hippocampus. *J. Neurosci.*, 17:3727–3738.

Patel, M.N. (2002). Oxidative stress, mitochondrial dysfunction, and epilepsy. *Free Radic. Res.*, 36:1139–1146.

Patel, M.N., J.O. McNamara. (1995). Selective enhancement of axonal branching of cultured dentate gyrus neurons by neurotrophic factors. *Neuroscience*, 69:763–770.

Prince, D.A., P. Salin, G.F. Tseng, S. Hoffman, I. Parada. (1997). Axonal sprouting and epileptogenesis. *Adv. Neurol.*, 72:1–8.

Provenzale, J.M., D.P. Barboriak, K. VanLandingham, J. MacFall, D. Delong, D.V. Lewis. (2008). Hippocampal MRI signal hyperintensity after febrile status epilepticus is predictive of subsequent mesial temporal sclerosis. *AJR Am. J. Roentgenol.*, 190:976–983.

Rakhade, S.N., C. Zhou, P.K. Aujla, R. Fishman, N.J. Sucher, F.E. Jensen. (2008). Early alterations of AMPA receptors mediate synaptic potentiation induced by neonatal seizures. *J. Neurosci.*, 28:7979–7990.

Raol, Y.H., I.V. Lund, S. Bandyopadhyay, G. Zhang, D.S. Roberts, J.H. Wolfe, S.J. Russek, A.R. Brooks-Kayal. (2006). Enhancing GABA(A) receptor alpha 1 subunit levels in hippocampal dentate gyrus inhibits epilepsy development in an animal model of temporal lobe epilepsy. *J. Neurosci.*, 26:11342–11346.

Reddy, D.S., M.A. Rogawski. (2001). Enhanced anticonvulsant activity of neuroactive steroids in a rat model of catamenial epilepsy. *Epilepsia*, 42:337–344.

Regis, J., F. Bartolomei, M. Rey, M. Hayashi, P. Chauvel, J.C. Peragut. (2000). Gamma knife surgery for mesial temporal lobe epilepsy. *J. Neurosurg.*, 93(Suppl. 3):141–146.

Rheims, S., C. Fischer, P. Ryvlin, J. Isnard, M. Guenot, M. Tamura, J. Regis, F. Mauguiere. (2008). Long-term outcome of gamma-knife surgery in temporal lobe epilepsy. *Epilepsy Res.*, 80:23–29.

Rogawski, M.A., D.S. Reddy. (2002). Neurosteroids and infantile spasms: the deoxycorticosterone hypothesis. *Int. Rev. Neurobiol.*, 49:199–219.

Sankar, R., D.H. Shin, C.G. Wasterlain. (1997). Serum neuron-specific enolase is a marker for neuronal damage following status epilepticus in the rat. *Epilepsy Res.*, 28:129–136.

Scharfman, H.E., W.P. Gray. (2007). Relevance of seizure-induced neurogenesis in animal models of epilepsy to the etiology of temporal lobe epilepsy. *Epilepsia*, 48(Suppl. 2):33–41.

Schauwecker, P.E. (2003). Genetic basis of kainate-induced excitotoxicity in mice: phenotypic modulation of seizure-induced cell death. *Epilepsy Res.*, 55:201–210.

Schwartzkroin, P.A., C.A. Walsh. (2000). Cortical malformations and epilepsy. *Ment. Retard. Dev. Disabil. Res. Rev.*, 6:268–280.

Shetty, A.K., B. Hattiangady. (2007). Concise review: prospects of stem cell therapy for temporal lobe epilepsy. *Stem Cells*, 25:2396–2407.

Singh, G., S. Prabhakar. (2008). The association between central nervous system (CNS) infections and epilepsy: epidemiological approaches and microbiological and epileptological perspectives. *Epilepsia*, 49(Suppl. 6):2–7.

Snead, 3rd., O.C. (2001). How does ACTH work against infantile spasms? Bedside to bench. *Ann. Neurol.*, 49:288–289.

Steinlein, O.K. (2008). Genetics and epilepsy. *Dialogues Clin. Neurosci.*, 10:29–38.

Sutula, T.P. (2001). Secondary epileptogenesis, kindling, and intractable epilepsy: a reappraisal from the perspective of neural plasticity. *Int. Rev. Neurobiol.*, 45:355–386.

Temkin, N.R., S.S. Dikmen, A.J. Wilensky, J. Keihm, S. Chabal, H.R. Winn. (1990). A randomized, double-blind study of phenytoin for the prevention of post-traumatic seizures. *N. Engl. J. Med.*, 323:497–502.

Temkin, N.R., S.S. Dikmen, G.D. Anderson, A.J. Wilensky, M.D. Holmes, W. Cohen, D.W. Newell, P. Nelson, A. Awan, H.R. Winn. (1999). Valproate therapy for prevention of posttraumatic seizures: a randomized trial. *J. Neurosurg.*, 91:593–600.

Teyler, T.J., S.L. Morgan, R.N. Russell, B.L. Woodside. (2001). Synaptic plasticity and secondary epileptogenesis. *Int. Rev. Neurobiol.*, 45:253–267.

Theodore, W.H., L. Epstein, W.D. Gaillard, S. Shinnar, M.S. Wainwright, S. Jacobson. (2008). Human herpes virus 6b: a possible role in epilepsy? *Epilepsia*, 49:1828–1837.

Todorova, M.T., P. Tandon, R.A. Madore, C.E. Stafstrom, T.N. Seyfried. (2000). The ketogenic diet inhibits epileptogenesis in El mice: a genetic model for idiopathic epilepsy. *Epilepsia*, 41:933–940.

Uva, L., L. Librizzi, N. Marchi, F. Noe, R. Bongiovanni, A. Vezzani, D. Janigro, M. de Curtis. (2008). Acute induction of epileptiform discharges by pilocarpine in the *in vitro* isolated guinea-pig brain requires enhancement of blood–brain barrier permeability. *Neuroscience*, 151:303–312.

Vezzani, A., D. Moneta, M. Conti, C. Richichi, T. Ravizza, A. De Luigi, M.G. De Simoni, G. Sperk, S. Andell-Jonsson, J. Lundkvist, K. Iverfeldt, T. Bartfai. (2000a). Powerful anticonvulsant action of IL-1 receptor antagonist on intracerebral injection and astrocytic overexpression in mice. *Proc. Natl. Acad. Sci. U.S.A.*, 97:11534–11539.

Vezzani, A., M. Rizzi, M. Conti, R. Samanin. (2000b). Modulatory role of neuropeptides in seizures induced in rats by stimulation of glutamate receptors. *J. Nutr.*, 130:1046S–1048S.

Vezzani, A., T. Ravizza, S. Balosso, E. Aronica. (2008). Glia as a source of cytokines: implications for neuronal excitability and survival. *Epilepsia*, 49(Suppl. 2):24–32.

Wenzel, H.J., C.S. Woolley, C.A. Robbins, P.A. Schwartzkroin. (2000). Kainic acid-induced mossy fiber sprouting and synapse formation in the dentate gyrus of rats. *Hippocampus*, 10:244–260.

Wenzel, H.J., H. Vacher, E. Clark, J.S. Trimmer, A.L. Lee, R.M. Sapolsky, B.L. Tempel, P.A. Schwartzkroin. (2007). Structural consequences of *KCNA1* gene deletion and transfer in the mouse hippocampus. *Epilepsia*, 48:2023–2046.

Yogarajah, M., J.S. Duncan. (2008). Diffusion-based magnetic resonance imaging and tractography in epilepsy. *Epilepsia*, 49:189–200.

Yusim, A., O. Ajilore, T. Bliss, R. Sapolsky. (2000). Glucocorticoids exacerbate insult-induced declines in metabolism in selectively vulnerable hippocampal cell fields. *Brain Res.*, 870:109–117.

Zaman, V., A.K. Shetty. (2001). Fetal hippocampal CA3 cell grafts transplanted to lesioned CA3 region of the adult hippocampus exhibit long-term survival in a rat model of temporal lobe epilepsy. *Neurobiol. Dis.*, 8:942–952.

Zhang, N., W. Wei, I. Mody, C.R. Houser. (2007). Altered localization of GABA(A) receptor subunits on dentate granule cell dendrites influences tonic and phasic inhibition in a mouse model of epilepsy. *J. Neurosci.*, 27:7520–7531.

# Index